Metals

Semimetals

Nonmetals

10 VIII	11 IB	12 II B	13 IIIA	14 IVA	15 VA	16 VIA	17 VIIA	18 VIIIA
								2 **He** 4.00 helium
			5 **B** 10.81 boron	6 **C** 12.01 carbon	7 **N** 14.01 nitrogen	8 **O** 16.00 oxygen	9 **F** 19.00 fluorine	10 **Ne** 20.18 neon
			13 **Al** 26.98 aluminum	14 **Si** 28.09 silicon	15 **P** 30.97 phosphorus	16 **S** 32.07 sulfur	17 **Cl** 35.45 chlorine	18 **Ar** 39.95 argon
28 **Ni** 58.69 nickel	29 **Cu** 63.55 copper	30 **Zn** 65.39 zinc	31 **Ga** 69.72 gallium	32 **Ge** 72.61 germanium	33 **As** 74.92 arsenic	34 **Se** 78.96 selenium	35 **Br** 79.90 bromine	36 **Kr** 83.80 krypton
46 **Pd** 106.42 palladium	47 **Ag** 107.87 silver	48 **Cd** 112.41 cadmium	49 **In** 114.82 indium	50 **Sn** 118.71 tin	51 **Sb** 121.75 antimony	52 **Te** 127.60 tellurium	53 **I** 126.90 iodine	54 **Xe** 131.29 xenon
78 **Pt** 195.08 platinum	79 **Au** 196.97 gold	80 **Hg** 200.59 mercury	81 **Tl** 204.38 thallium	82 **Pb** 207.2 lead	83 **Bi** 208.98 bismuth	84 **Po** (209) polonium	85 **At** (210) astatine	86 **Rn** (222) radon
110 — (269)	111 — (272)	112 — (277)	113	114 — (285)	115	116 — (289)	117	118 — (293)

64 **Gd** 157.25 gadolinium	65 **Tb** 158.93 terbium	66 **Dy** 162.50 dysprosium	67 **Ho** 164.93 holmium	68 **Er** 167.26 erbium	69 **Tm** 168.93 thulium	70 **Yb** 173.04 ytterbium	71 **Lu** 174.97 lutetium
96 **Cm** (247) curium	97 **Bk** (247) berkelium	98 **Cf** (251) californium	99 **Es** (252) einsteinium	100 **Fm** (257) fermium	101 **Md** (258) mendelevium	102 **No** (259) nobelium	103 **Lr** (260) lawrencium

ANNOTATED INSTRUCTOR'S EDITION

Introductory Chemistry

Concepts & Connections

THIRD EDITION

Charles H. Corwin

American River College

PRENTICE HALL

Upper Saddle River, New Jersey 07458

Senior Editor: Kent Porter Hamann
Editorial Director: Paul F. Corey
Development Editor: Joan Kalkut
Editor in Chief, Development: Carol Trueheart
Production Editor: Nicole Bush
Assistant Vice President of Production
and Manufacturing: David W. Riccardi
Executive Managing Editor: Kathleen Schiaparelli
Assistant Managing Editor: Beth Sturla
Marketing Manager: Steve Sartori
Manufacturing Manager: Trudy Pisciotti
Manufacturing Buyer: Michael Bell
Director of Creative Services: Paul Belfanti
Art Director: Joseph Sengotta
Interior Designer: Judith A. Matz-Coniglio
Cover Designer: Stacey Abraham
Cover Molecular Art: © Kenneth Eward/BioGrafx, 2000
Cover Photographs: Vanilla Beans, Kristen Brochmann/Fundamental Photographs;
Vanilla Orchid in Bloom, Karl Wiedmann/Photo Researchers, Inc.
Art Manager: Gus Vibal
Art Editor: Karen Branson
Illustrations: Academy Artworks
Editorial Assistant: Richard Moriarty
Photo Editor: Beth Boyd
Photo Researcher: Mary Teresa Giancoli
Photo Coordinator: Reynold Rieger
Copy Editor: Carol Dean
Composition: Preparé Inc.
Assitant Managing Editor, Science Media: Alison Lorber
Media Editor: Paul Draper
Special Projects Manager: Barbara A. Murray
Associate Editor: Kristen Kaiser

Upper Saddle River, NJ 07458

Printed in the United States of America
10 9 8 7 6 5 4 3 2

ISBN 0-13-030951-6

Prentice-Hall International (UK) Limited, *London*
Prentice-Hall of Australia Pty. Limited, *Sydney*
Prentice-Hall Canada Inc., *Toronto*
Prentice-Hall Hispanoamericana, S.A., *Mexico*
Prentice-Hall of India Private Limited, *New Delhi*
Prentice-Hall of Japan, Inc., *Tokyo*
Pearson Education Asia Pte. Ltd.
Editora Prentice-Hall do Brasil, Ltda., *Rio de Janeiro*

A Guide to the Annotated Instructor's Edition

The *Annotated Instructor's Edition* (AIE) of *Introductory Chemistry: Concepts & Connections*, third edition, consists of the complete student version of the textbook and includes an overlay of commentary written and assembled by Charles H. Corwin, the textbook author, to support your lectures with appropriate supplemental information. Experienced professors of introductory/preparatory chemistry who have established successful teaching strategies may not need this information. But for instructors rotating into introductory/preparatory chemistry once every 3 or 4 years, for those teaching the course for the first time, for graduate assistants, and for those seeking to incorporate new technology into the course, we hope to provide a helpful teaching tool with this AIE.

The following categories of instructor annotations appear in blue type in the margins of the text:

Chapter Overviews and Student Misconceptions: Based on a wealth of classroom experience, this set of remarks summarizes the topical content of each chapter and alerts the instructor to common errors and pitfalls for inexperienced students.

Transparency References: An icon showing an overhead projector provides a quick identifier of the 125 illustrations that are duplicated and enlarged in the accompanying boxed set of full-color acetates. Every transparency is also available on the *Matter 2001 Instructor CD*.

Experiment References: These notations indicate appropriate experiments in the *Prentice Hall Laboratory Manual for Introductory Chemistry*, by Charles H. Corwin.

References to *Introductory Chemistry Media Companion CD-ROM* (For CW: 0-13-032678-X) (For CW+: 0-13-032132-8) Also included in black type are the student edition references to the interactive exercises on the Student CD-ROM that highlight the key concepts in each eChapter and invite students to view animations of processes, video clips of chemistry demonstrations, and simulations. Used to the maximum pedagogical benefit, these activities can be assigned to students. Each eChapter includes several self-assessment exercises. The CD connects seamlessly with the Companion Website and is designed to be complementary to it and to the book.

References to the Introductory Chemistry Companion Website (www.prenhall.com/Corwin):

This interactive resource for students offers additional tutorial and self-assessment opportunities. The site includes:

- Explorer and Master Quizzes—Over 1200 additional problems categorized by level, each containing hints and detailed feedback.
- Web links of interest to introductory chemistry students.
- Online Math Toolkit, available directly, or students can link directly to the appropriate section from each of the problems in the online quizzes.

Brief Contents

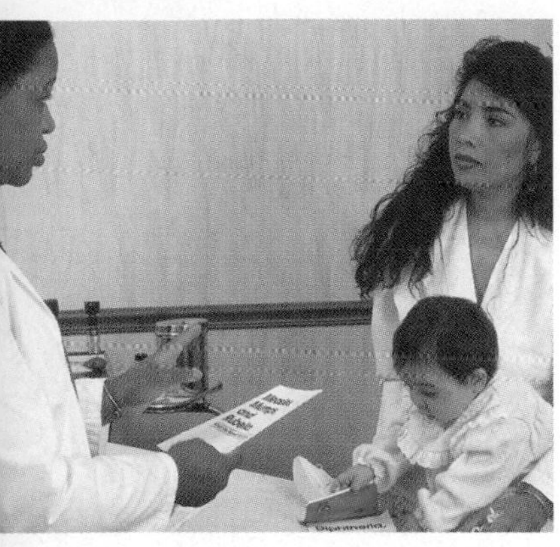

Contents

iMac

1 QUART vs. 1 LITER
1 KILOGRAM vs. 1 POUND
1 YARD vs. 1 METER

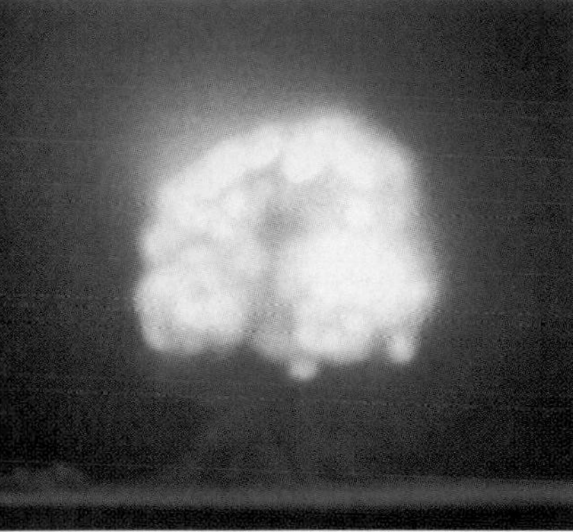

4 Matter and Energy 78

5 Models of the Atom 111

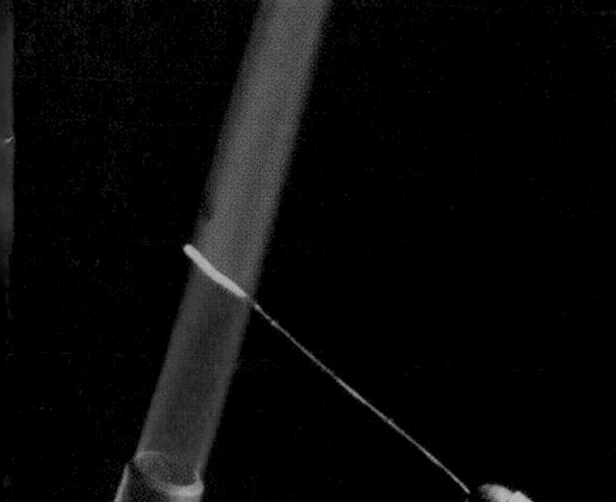

6 The Periodic Table 144

7 Language of Chemistry 175

8 Chemical Reactions 201

Cumulative Review: Chapters 7–8 233

9 The Mole Concept 236

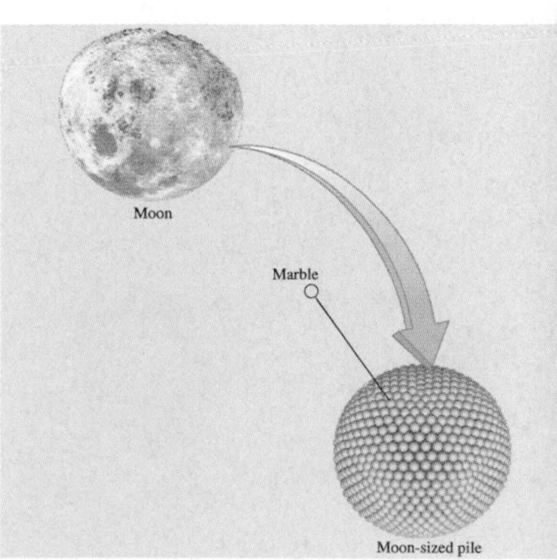

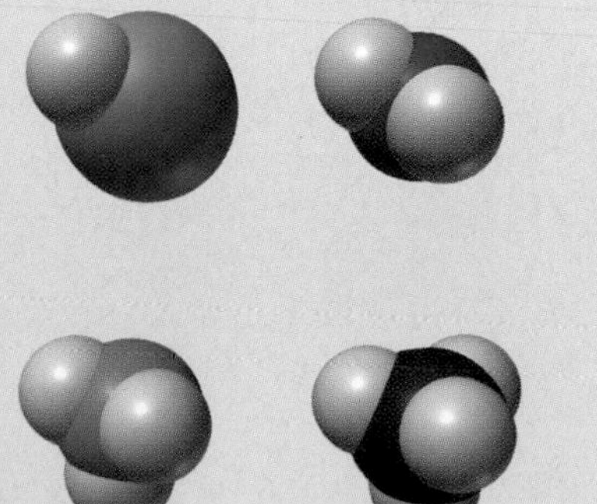

13 Liquids and Solids 356

14 Solutions 385

Cumulative Review: Chapters 12–14 415

15 Acids and Bases 417

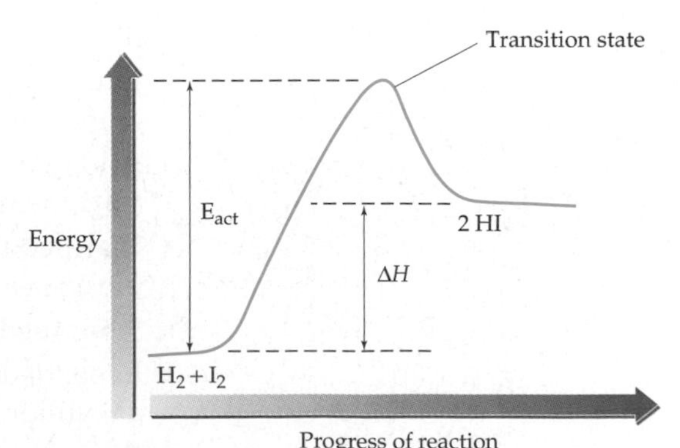

Transition state
Energy
E_{act}
2 HI
ΔH
$H_2 + I_2$
Progress of reaction

Preface

This third edition of *Introductory Chemistry: Concepts & Connections* continues the concepts and connections theme. Chemical educators agree that it is necessary to learn basic *concepts* in order to acquire long-term meaningful understanding, and there is a consensus that motivational teaching requires relevant *connections* to the chemical principles operating in everyday life.

A knowledge of chemistry is an empowering life skill, and this fact demands that textbook explanations be as clear and compelling as possible. Accordingly, the goal of this text is twofold. First, we have utilized numerous strategies to make the material accessible and interesting so that students enjoy their learning experience and acquire a positive initial impression of chemistry. Second, we have redoubled our effort to offer the best ancillaries and supplements available so as to give instructors and students peerless resources.

New Features in the Third Edition

Problem-Solving Enhancements. In this edition, we introduced innovative tools for enabling students to be successful problem solvers. In addition to the systematic application of unit analysis (that is, dimensional analysis), there is global visualization of example problems by frequent incorporation of:

- **Concept-flow-chart diagrams** that help students visualize the overall relationship of each step in the solution to many worked-out example problems.
- **Problem-solving organizers** at the end of the chapters involving calculations and nomenclature that provide a synopsis of the types of problems in the chapter and illustrate the correct procedure for solving each type of question. Every seasoned teacher realizes that beginning chemistry students often have difficulty deciding what is important, and the organizers clearly designate which information is critical for the student to master.

Content Changes. Without compromising the student-friendly style and the original integrity of the presentation, we made the following content changes in response to feedback received from the faculty and students who used the second edition.

- Fine-tuned all chapters based on the feedback from reviewers and users of the second edition.
- Integrated more practical applications of chemistry concepts operating in the daily lives of students.
- Added fresh art to an already acclaimed art program that motivates students by drawing them into the discussions.
- Clarified the stepwise unit analysis problem-solving strategy in each example exercise by single line-statements.
- Simplified the language in the chapter on matter and energy.
- Expanded the chapter on solutions by adding two new sections: "Dilution of a Solution" (Section 14.10) and "Solution Stoichiometry" (Section 14.11).
- Rearranged the sections on acid–base titrations.
- Added new relevant examples to organic chemistry and biochemistry.

Strong Technology Support for Students and Instructors With the advent of the twenty-first century, the role of technology has become increasingly prominent in teaching chemistry. In addition to the traditional format, a chemistry course is likely to involve web-based delivery as well as media support. Thus, this third edition of *Introductory Chemistry: Concepts & Connections* has a strong technology commitment for assisting the learning process, including:

For Students:

- New student CD-ROM (see full description on p. xvii). CD-ROM icons have been added in the margins of the textbook to indicate where students can extend their understanding of a topic by doing the interactive activities and quizzes using this CD-ROM, which is available free to adopters of the text.
- Chemistry SkillBuilder CD-ROM.
- Companion Website (see full description on p. xvii). WWW icons have been added in the margins of the textbook to indicate where students can assess their competency using an explorer quiz or a master quiz at the Companion Website.

For Instructors:

- Matter 2001: A lecture presentation resource for instructors with images for PowerPoint presentations (see full description on p. xvii).
- WebCT/Blackboard Course Management System presentations (see full description on p. xvii).

Please see the following pages for more specific details about how the third edition's integrated learning program will help your students succeed.

Flexible Organization

The text maintains a traditional arrangement of topics for introductory chemistry; however, there is an unusual degree of flexibility owing to its original module format. Initially, this textbook was developed from a series of modules, each of which covered a specific topic. The modules were class-tested by instructors who had different preferences regarding topic sequence and depth of coverage. Through progressive refinement, flexibility was achieved without the sacrifice of continuity. As an example, two of many possible chapter configurations are as follows:

- Emphasis on problem solving in preparation for General Chemistry: Chapters 1, 2, 3, 4, 5, 6, 7, 8, 9, 10, 11, 12, 13, 14, 15, 16, and 17.
- Emphasis on descriptive chemistry leading to Organic Chemistry and Biochemistry: Chapters 1, 2, 3, 4, 5, 6, 7, 8, 9, 10, 11, 12, 14, 15, 18, 19, and 20.

The text offers a balanced approach in the degree of difficulty from chapter to chapter. Similarly, the text is highly consistent in the number of topics per chapter and the number of objectives per section.

The Pedagogically Refined Third Edition

I have instructed nearly 7000 students in lecture and laboratory sections of introductory chemistry. The hours spent in the laboratory have been invaluable because all too often it is in this informal setting where students talk candidly about their difficulties in learning chemistry. The laboratory can also reveal a student's preferred learning style. Any experienced instructor can readily verify that some students who excel in theory do comparatively less well when performing experiments. And the opposite is also true; some students may excel in laboratory even though they are of average accomplishment in lecture.

Visual Problem Solving. In this third edition, we have taken problem solving to yet another level so that each student can piece together the steps necessary to solve a problem. First, the problem is introduced by a "walk-through" example, followed by an example exercise and a self-test. In this edition, we have added concept flowchart diagrams to help students visualize the overall relationship of each step in the solution.

Consistent Problem Solving. All problems are solved systematically in three steps using the unit analysis method. This method of problem solving is introduced in Chapter 2 and reinforced in Chapter 3. In the chapters that follow, unit analysis is applied to mole problems, stoichiometry, and solution calculations. In performing calculations, *the use of algebra is strictly optional*. For gas law problems, solutions are shown by a modified unit analysis approach as well as by an alternate algebraic approach.

Chemical Concepts. The general chemistry community continues to evaluate which topics should be taught to beginning chemistry students. Current thinking is that given an understanding of core chemical concepts, students will be prepared to assimilate the information they encounter in subsequent courses. Accordingly, this text targets the important concepts from which a body of knowledge can be understood. By way of example, the mole concept is explained using verbal and graphic analogies for Avogadro's number.

Length Analogy: If 6.02×10^{23} hydrogen atoms were laid side by side, the total length would be long enough to encircle the Earth about a million times.

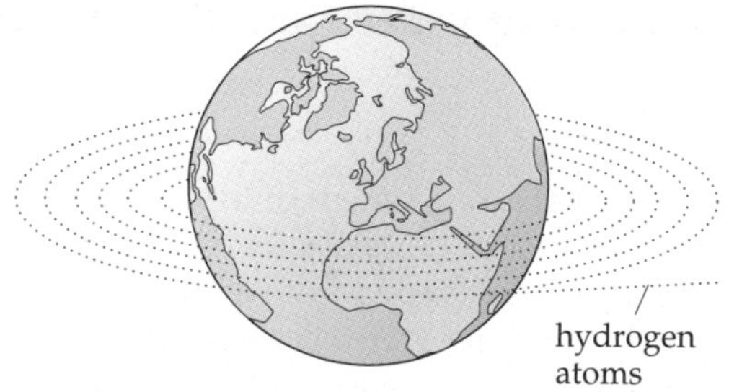

Mass Analogy: The mass of 6.02×10^{23} Olympic shotput balls would be about equal to the mass of the Earth.

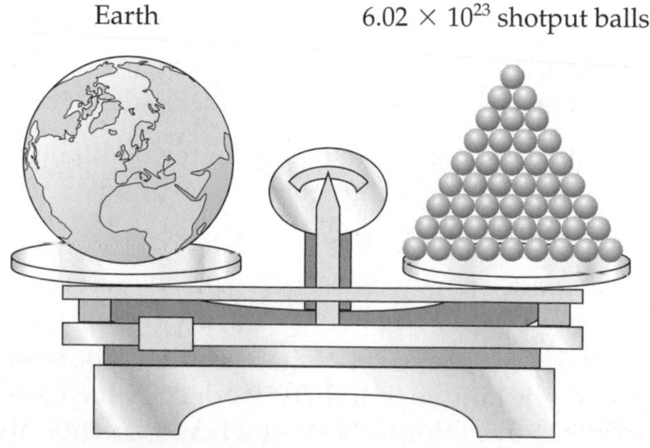

Volume Analogy: The volume occupied by 6.02×10^{23} softballs would be about the size of the Earth.

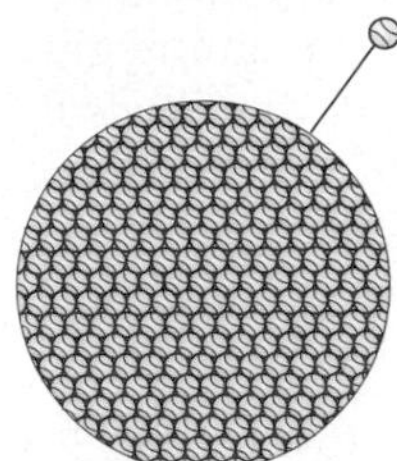

6.02×10^{23} softballs

Chemistry Connections. One of the ways to keep students motivated is by presenting relevant vignettes. There are over 30 "Chemistry Connections" in the text, which range in subject matter from industrial processes to consumer chemistry and from historical profiles to environmental concerns.

Updates. It is important for students and instructors alike that the information in a science textbook be current. In 1999 three new elements were synthesized. Recently, new group designations for the periodic table have been proposed, and there have been changes regarding systematic nomenclature. Special features called "Updates" discuss the latest developments in the field of general chemistry.

Chapter Summaries. Although the chapter summary is intended as a capsule review of each section, some instructors have found it valuable to have students read the summary *before* tackling the chapter at large.

Key Concepts. The third edition has a key concept section following the chapter summary. This feature includes practical questions designed to help a student grasp the fundamental concepts.

Key Terms. An original feature in the text was the matching key term exercises. Students prefer this method of learning terminology to open-ended definitions, which do not furnish a verifiably correct answer. To further promote learning terminology, the *Student Study Guide and Solutions Manual* contains crossword puzzle exercises that offer a unique and fun way to learn the language of chemistry.

Exercises. Students require practice in learning to solve problems, and there are 250 example exercises in the text, each paired with a self-test question. In addition, there are over 1600 end-of-chapter exercises arranged in a matched-pair format. Answers are provided for all the odd-numbered exercises in Appendix J of the text. Complete solutions to the odd-numbered exercises are found in the *Student Study Guide and Solutions Manual*. Complete solutions to even-numbered exercises are found in the *Instructor's Resource Manual*.

Cumulative Reviews. There is a cumulative review of topics every two or three chapters. This feature does not simply consist of additional exercises but rather includes a review of key terms, key concepts, and broad-based questions that stimulate the student to synthesize topics across more than one chapter. Answers to the cumulative review questions are found in the *Instructor's Resource Manual*.

Art Program. The previous edition of the text received high praise for its attractive presentation and nonintimidating tone. In this edition, we again offer copious illustrations as well as the artwork of Michael Goodman, contributing illustrator for *Scientific American*. The use of "molecular art" takes the student from the visible world into the atomic world and portrays chemistry concepts at the molecular level.

Print Supplements

To support the learning process, there are numerous accompanying supplements and ancillaries. To ensure an integrated teaching package, the *Instructor's Resource Manual*, the *Test Item File*, the *Student Study Guide* and *Solutions Manual*, the *Prentice Hall Laboratory Manual*, and the *Annotated Instructor's Edition* have been prepared by the author and extensively class-tested by colleagues.

For the Student

Math Review Toolkit (0-13-031028-X)
This booklet helps students to be successful in chemistry. This fun guide, written in a colloquial tone, offers personal study strategies, an elementary review of numbers, and directions for using a calculator. It can be packaged free with the text.

Full-color Periodic Table (0-13-029099-8)
This handy reference corresponds to the full-color periodic table inside the front cover of the text. It can be packaged free with the text.

Study Guide and Selected Solutions Manual (0-13-031027-1)
This study aid includes diagnostic test questions for each topic covered in the text, crossword puzzles using key terms, and complete solutions to all odd-numbered exercises. Flashcards and study tips are all contained in this essential manual.

Laboratory Manual for Introductory Chemistry (0-13-908922-5)
The *Prentice Hall Laboratory Manual for Introductory Chemistry* contains 25 small-scale experiments that are sensitive to environmental regulation but do not require any special equipment. Each experiment includes safety precautions and pre- and post-laboratory exercises. Written by the author of the text, this lab manual has evolved from experiments performed by over 150,000 students.

For the Instructor

Instructor's Resource Manual with Transparency Masters and Test Item File (0-13-031029-8)
The *Instructor's Resource Manual* lists all learning objectives for course planning, recommended media resources and chemical demonstrations, suggested lecture transparency masters, and the complete solutions to all even-numbered exercises. It also includes the now expanded *Test Item File* with more than 4000 class-tested questions addressing each topic covered in the text. A computerized test generator is available to accompany the printed version.

Prentice Hall Custom Test—WIN (0-13-087481-7)
Prentice Hall Custom Test—MAC (0-13-031031-X)
This powerful testing and grade management software creates exams from an electronic database version of the *Test Item File*. Instructors can generate alternate versions of the same test, add their own material, and edit existing tests effortlessly with this program.

Full-color Transparencies (0-13-031020-4)
Add dimension to your classroom discourse with 125 full-color transparencies from the text, plus 100 added transparency masters from the author's lectures.

Annotated Instructor's Edition of the Laboratory Manual (0-13-908906-3)
This indispensable manual contains a complete listing of chemicals and reagent preparation directions for each experiment. It also provides suggested unknowns and answers to the postlaboratory assignments, and a quiz item file with more than 500 class-tested multiple-choice questions.

Prentice Hall/*New York Times* Themes of the Times Supplement
The New York Times and Prentice Hall bring you "Themes of the Times," a unique newspaper supplement with up-to-the-minute articles relevant to an introductory chemistry course. It is free when obtained through your local Prentice Hall representative.

Technology Supplements

For Students

Concepts & Connections World Wide Website
Student Website http://www.prenhall.com/corwin
This dynamic on-line site allows students to review chapter material and have extra practice with interactive quizzes. For each textbook chapter, there are three explorer quizzes and a slightly harder master quiz. Students can access guiding hints, obtain graded results with detailed feedback on their answers, and e-mail their results to the instructor. A selection of websites appropriate to each chapter is available and updated regularly. There is also a math tutorial for practicing the basic skills necessary for this course. The tutorials are linked from the quizzes when a student has most likely made a math error.

Introductory Chemistry Media Companion CD-ROM: (For CW: 0-13-032678-X) (For CW+: 0-13-031032-8)
This unique CD-ROM invites the student to preview or review the material in the text in an alternate, media-rich way. In a summary format, it highlights the key concepts in each eChapter and invites students to view animations of processes that are difficult to explain with a static image and video clips of chemistry demonstrations and asks them to work with interactive simulations. Each eChapter includes several self-assessment exercises. The CD-ROM connects seamlessly with the Companion Website and is designed to be complementary to it and to the book.

Chemistry SkillBuilder CD-ROM (0-13-660143-X)
Today's visually oriented, computer-literate students will enjoy this interactive environment while mastering introductory chemistry skills. This CD-ROM provides students with an opportunity for extra practice on some of the key concepts critical to their mastery of chemistry, including nomenclature, balancing equations, and stoichiometry.

For the Instructor

Matter 2001 for Corwin: Instructor Presentation CD-ROM (coming in late Fall 2000)
This presentation CD features media assets that can be incorporated into PowerPoint shows or other documents. It includes most of the art from the textbook, as well as animations and video clips selected to be appropriate for each chapter. A prebuilt set of PowerPoint presentations for each chapter is provided, as well as a browsable, searchable catalog of assets on the CD.

Concepts & Connections World Wide Website
Student Website http://www.prenhall.com/corwin
Instructors can use this website as an assessment and communication tool. You can post a calendar-driven syllabus with a simple form interface. You can interact with your students on a bulletin board for virtual office hours. You can assign homework both from the book and from the website with the Syllabus Manager and get a summary report of a student's performance via e-mail. The on-line quizzes serve both as postchapter assignments and for reading comprehension as a way to encourage student preparation.

Acknowledgments

It is my pleasure to recognize several talented people who have contributed to the third edition of this textbook. It is my privilege to have colleagues who are committed to student learning and teaching chemistry effectively, including Norm Allen, Kristin Casale, James Cress, James Fisher, Darren Gottke, Ronald Grider, Tammy Hong, Greg Jorgensen, Afarin Moezzi, Luther Nolen, John Newey, Karen Pesis, Nancy Reitz, Rina Roy, Stephen Ruis, and Linda Zarzana.

Reviewers of the previous editions of *Introductory Chemistry: Concepts & Connections* continue to define that vague line between the simplifications that students require and the explanations that accuracy demands. I have received considerate and reflective comments from the following:

Second- and Third-Edition Reviewers

David Ball
Cleveland State University

Karen Bender
Grossmont College

Henry Brenner
New York University

Nicole K. Carlson
Indiana University

Wesley Fritz
College of Dupage

David R. Gano
Minot State University

Helen Haur
Delaware Technical & Community College

Mary Hickey
Henry Ford Community College

Roy Kennedy
Massachusetts Bay Community College

Anthony Lazzaro
CA University of Pennsylvania

Anne Loeb
College of Lake County

Deborah Miller
Albuquerque TVI

Robert N. Nelson
Georgia Southern University

Jeffrey Rahn
Eastern Washington University

Julianne M. Smist
Springfield College

Gerald C. Swanson
Daytona Beach Community College

Marie Villarba
Albuquerque TVI

Terri Beam
Mount San Antonio College

Ralph Benedetto
Wayne Community College

Joe Brundage
Cuesta College

Robert Fremland
Mesa College

Ana Gaillat
Greenfield Community College

Rebecca K. Hanckel
Charleston Southern University

Leslie Hersh
Delta College

P. D. Hooker
Colby Community College

Richard Kurtik
Moorepark College

Anne Lenhert
Kansas State University

Jerome Gerard May
Southeastern Louisiana University

Richard S. Mitchell
Arkansas State University

Lily Ng
Cleveland State University

Scott M. Savage
Northern State University

Charles E. Sundin
University of Wisconsin–Platteville

Paul E. Tyner
Schenectady County Community College

Justine Walhout
Rockford College

I would also like to recognize several individuals at Prentice Hall for their commitment to the project. Kent Porter Hamann, senior chemistry editor, whose solid reputation precedes her, afforded a heightened level of direction for the third edition and was instrumental in many aspects but especially in the areas of visual problem solving and the application of technology. Joan Kalkut, development editor, demonstrated an unwavering sense of responsibility and motivated me by example. It was most enjoyable working with Nicole Bush, production editor, who exudes a positive presence while firmly keeping the project on schedule. Once again, I would like to thank Joseph Sengotta, art director, for an attractive design that has subliminal nuances in both feeling tone and pedagogical clarity.

Finally, I appreciate the many instructors who have used the textbook and provided invaluable feedback. I have listened carefully to your comments and you may notice your presence in this current edition of the text. In addition, I enjoy comments from students who wage the battle to learn chemistry and share their experience. I invite each of you to contact me at the address below, or send an e-mail to **chcauthor@aol.com.**

Charles H. Corwin
Department of Chemistry
American River College
Sacramento, CA 95841

A Guide to Using this Text

A knowledge of chemistry is an empowering life skill, and the study of chemistry should be as accessible and compelling as possible. The primary objective of this book is to make chemical concepts clear and interesting so you will leave your first course in chemistry with a positive impression, a new set of skills, and a desire to know more.

Problem Solving/Built-In Study Aids

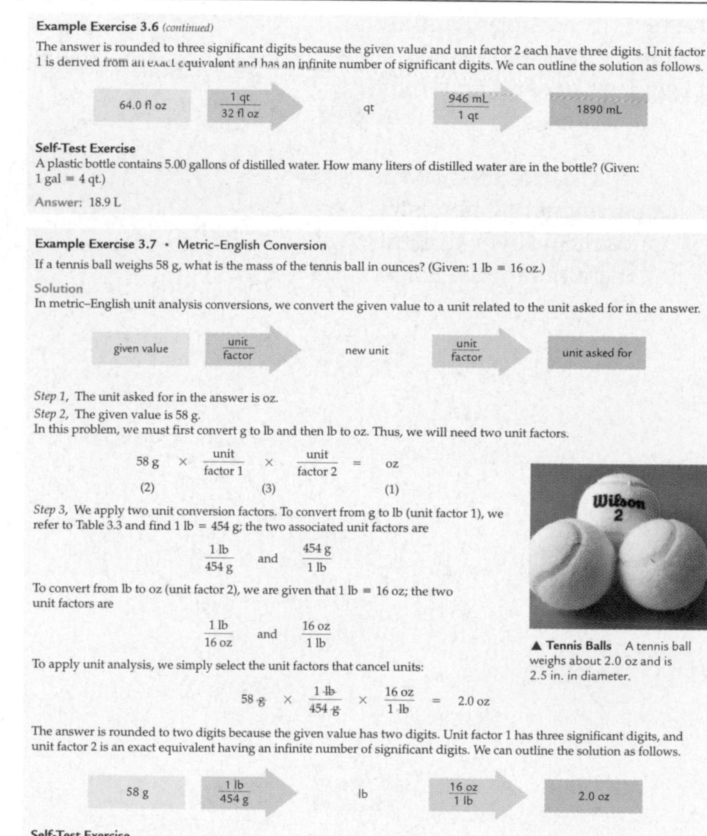

Example Exercise 3.6 *(continued)*

The answer is rounded to three significant digits because the given value and unit factor 2 each have three digits. Unit factor 1 is derived from an exact equivalent and has an infinite number of significant digits. We can outline the solution as follows.

64.0 fl oz → $\frac{1\ \text{qt}}{32\ \text{fl oz}}$ → qt → $\frac{946\ \text{mL}}{1\ \text{qt}}$ → 1890 mL

Self-Test Exercise

A plastic bottle contains 5.00 gallons of distilled water. How many liters of distilled water are in the bottle? (Given: 1 gal = 4 qt.)

Answer: 18.9 L

Example Exercise 3.7 • Metric–English Conversion

If a tennis ball weighs 58 g, what is the mass of the tennis ball in ounces? (Given: 1 lb = 16 oz.)

Solution

In metric–English unit analysis conversions, we convert the given value to a unit related to the unit asked for in the answer.

given value → unit factor → new unit → unit factor → unit asked for

Step 1, The unit asked for in the answer is oz.

Step 2, The given value is 58 g.

In this problem, we must first convert g to lb and then lb to oz. Thus, we will need two unit factors.

$$\underset{(2)}{58\ \text{g}} \times \underset{(3)}{\frac{\text{unit}}{\text{factor 1}}} \times \frac{\text{unit}}{\text{factor 2}} = \underset{(1)}{\text{oz}}$$

Step 3, We apply two unit conversion factors. To convert from g to lb (unit factor 1), we refer to Table 3.3 and find 1 lb = 454 g; the two associated unit factors are

$$\frac{1\ \text{lb}}{454\ \text{g}} \quad \text{and} \quad \frac{454\ \text{g}}{1\ \text{lb}}$$

To convert from lb to oz (unit factor 2), we are given that 1 lb = 16 oz; the two unit factors are

$$\frac{1\ \text{lb}}{16\ \text{oz}} \quad \text{and} \quad \frac{16\ \text{oz}}{1\ \text{lb}}$$

To apply unit analysis, we simply select the unit factors that cancel units:

$$58\ \cancel{\text{g}} \times \frac{1\ \cancel{\text{lb}}}{454\ \cancel{\text{g}}} \times \frac{16\ \text{oz}}{1\ \text{lb}} = 2.0\ \text{oz}$$

▲ **Tennis Balls** A tennis ball weighs about 2.0 oz and is 2.5 in. in diameter.

The answer is rounded to two digits because the given value has two digits. Unit factor 1 has three significant digits, and unit factor 2 is an exact equivalent having an infinite number of significant digits. We can outline the solution as follows.

58 g → $\frac{1\ \text{lb}}{454\ \text{g}}$ → lb → $\frac{16\ \text{oz}}{1\ \text{lb}}$ → 2.0 oz

Self-Test Exercise

If a tennis ball has a diameter of 2.5 inches, what is the diameter in millimeters?

Answer: 64 mm

◀ **Example Exercise** problems are provided throughout the chapters to give you immediate practice on concepts and skills you've just learned. Self-Test Exercises with answers follow each of these. Do these problems as you read to receive instant feedback on your mastery of the concept or skill.

◀ **Conceptual Flow Charts** are often embedded within the worked examples to give you a clear, visual representation of the order of operations and key concepts to use in problem solving. The flow charts consistently reflect the steps in the unit analysis method of problem solving. The first flow chart in a solution helps you identify the value in the problem that is related to the units required in the answer. The second flow chart shows the actual application of the appropriate unit factor to obtain the correct answer. Make sure you understand each flow chart before you proceed.

Summary/Rules Boxes throughout the chapters highlight key rules or problem-solving processes for easy reference. ▶

Applying the Unit Analysis Method

Step 1: Write down the units asked for in the answer.

Step 2: Write down the value given in the problem that is related to the units required in the answer.

Step 3: Apply a unit factor to convert the units in the given value to the units in the answer. Each time we apply unit analysis, we use the following format:

$$\underset{(2)}{\text{given value}} \times \underset{(3)}{\frac{\text{unit}}{\text{factors(s)}}} = \underset{(1)}{\text{units asked for}}$$

Problem-Solving Organizer

Topic	Procedure	Example
Basic Units and Symbols Sec. 3.1	Combine basic metric units and prefixes using symbols.	centimeter, cm (length) kilogram, kg (mass) milliliter, mL (volume)
Metric Conversion Factors Sec. 3.2	(a) Write a unit equation involving basic metric units and prefix units. (b) Write two unit factors for a metric relationship.	1 m = 100 cm $\frac{1\ \text{m}}{100\ \text{cm}}$ and $\frac{100\ \text{cm}}{1\ \text{m}}$
Metric–Metric Conversions Sec. 3.3	1. Write down the unit asked for in the answer. 2. Write down the related given value. 3. Apply a unit factor to convert a given unit to the unit in the answer.	What is the decigram mass of a wheel of cheese that weighs 0.515 kg? $0.515\ \text{kg} \times \frac{1000\ \text{g}}{1\ \text{kg}} \times \frac{10\ \text{dg}}{1\ \text{g}} = 5150\ \text{dg}$
Metric–English Conversions Sec. 3.4	1. Write down the unit asked for in answer. 2. Write down the related given value. 3. Apply a unit factor to convert a given unit to the unit in the answer.	What is the pound mass of a wheel of cheese that weighs 0.515 kg? $0.515\ \text{kg} \times \frac{1000\ \text{g}}{1\ \text{kg}} \times \frac{1\ \text{lb}}{454\ \text{g}} = 113\ \text{lb}$
Volume by Calculation Sec. 3.5	To calculate the volume of a rectangular solid, multiply length by width by thickness: $V = l \times w \times t$.	What is the volume of a rectangular solid measuring 65 mm by 35 mm by 12 mm? $65\ \text{mm} \times 35\ \text{mm} \times 12\ \text{mm} = 27{,}300\ \text{mm}^4$
Volume by Displacement Sec. 3.6	The volume by displacement is the difference between the initial and final readings in a calibrated container.	What is the volume of jade if the water level in a cylinder increases from 25.0 mL to 42.5 mL? 42.5 mL − 25.0 mL = 17.5 mL
The Density	The density of a sample is equal to its mass divided by its volume: density = mass/volume.	What is the density of 10.0 mL of ether if its mass is 7.14

Problem-Solving Organizers, found at the end of computational chapters, provide a synopsis of the types of calculations in the chapter and the correct procedure for solving each type of problem, along with an illustrated example.

Key Concepts *

1. What state-of-the-art instrument is capable of making an exact measurement? What physical quantity can be measured with absolute certainty?
2. Draw a line on a separate sheet of paper and mark the length equal to line *L* shown below. Measure the line using Ruler A and Ruler B as shown in Figure 2.2. Record the length of the line consistent with the uncertainty of each ruler.

L: 

3. Which of the following is the best estimate of the diameter of a 1¢ coin: 0.5 cm, 2 cm, 5 cm, 20 cm, 50 cm? Which of the following is the best estimate of the thickness of a 10¢ coin: 0.001 cm, 0.01 cm, 0.1 cm, 1 cm, 10 cm?

A 1¢ coin   A 10¢ coin   A 25¢ coin

4. Which of the following is the best estimate of the mass of a 1¢ coin: 0.5 g, 1 g, 3 g, 10 g, 20 g? Which of the following is the best estimate of the mass of a 25¢ coin: 1 g, 5 g, 25 g, 50 g, 75 g?
5. Which of the following is the best estimate of the volume corresponding to 10 drops from an eyedropper: 0.5 mL, 1 mL, 5 mL, 10 mL, 20 mL? Which of the following is the best estimate of the volume of milk in a quart (32 fl oz) carton: 1 mL, 10 mL, 100 mL, 1000 mL, 10,000 mL?
6. Is the *mass* of an astronaut more, less, or the same on Uranus as on Mars? The mass of Uranus is 8.66×10^{25} kg, and that of Mars is 6.42×10^{23} kg.
7. Is the *weight* of an astronaut more, less, or the same on Uranus as on Mars? The mass of Uranus is 8.66×10^{25} kg, and that of Mars is 6.42×10^{23} kg.
8. Which of the following relationships is an exact equivalent?
   (a) 1 meter = 1000 millimeters
   (b) 1 meter = 1.09 yards
9. What is the number of significant digits in each of the following relationships?
   (a) 1 meter = 1000 millimeters
   (b) 1 meter = 1.09 yards
10. What are the three steps in the unit analysis method of problem solving?

Key Concept Questions, located in the end-of-chapter problems, test your qualitative mastery of the key chemistry concepts presented in that chapter.

CHAPTERS 1–3
Cumulative Review

▲ **Five-cent Coins** Since a dollar is equivalent to 20 nickels, we can write the unit equation 1 dollar = 20 nickels. The two associated unit factors are: 1 dollar/20 nickels and 20 nickels/1 dollar.

Key Concepts

1. Why is it important to learn chemistry?
2. Why are exact measurements impossible?
3. What is the number of significant digits in each of the following relationships: 1 yd = 36 in., 1 yd = 0.914 m, 1 m = 39.4 in., 1 m = 1000 mm?
4. What is the basic unit of length, mass, and volume in the metric system?
5. What metric unit of liquid volume is equal to $(10\ \text{cm})^3$? to $(1\ \text{cm})^3$?
6. Estimate the diameter of a 5¢ coin (±1 cm).
7. Estimate the mass of a 5¢ coin (±1 g).
8. Estimate the volume of 20 drops of water (±1 mL).
9. In the unit analysis method, what is the first step? the second step?
10. In applying unit analysis, how do you select a unit factor?
11. In a calculation involving a series of multiplication and division operations, is it better to round off after each step or simply to round off the final answer?
12. What is the density of water in metric units?
13. Given that an unknown liquid is chloroform (d = 1.48 g/mL) or ether (d = 0.714 g/mL), how could you quickly identify the liquid?
14. Which of the following temperatures is the coldest: 0°F, 0°C, or 0 K?
15. What is the specific heat of water in metric units?

Key Terms

State the key term that corresponds to each of the following descriptions.

Cumulative Review Sections, found every few chapters, help you integrate, review, and synthesize concepts that apply to multiple chapters.

Applications/Visualization

Chemistry Connection · A Student Success Story

What common metal was more valuable than gold in the nineteenth century?

In 1885 Charles Martin Hall was a 22-year-old student at Oberlin College in Ohio. One day his chemistry teacher told the class that anyone who could discover an inexpensive way to produce aluminum metal would become rich and benefit humanity. At the time, aluminum was a rare and expensive metal. In fact, Napoleon III, a nephew of Napoleon Bonaparte, entertained his most honored guests with utensils made from aluminum while other guests dined with utensils of silver and gold. Although aluminum is the most abundant metal in the Earth's crust, it is not found free in nature; it is usually found combined with oxygen in various minerals.

After graduation, Charles Hall set up a laboratory in a woodshed behind his father's church. Using homemade batteries, he devised a simple method for producing aluminum by passing electricity through a molten mixture of minerals. After only 8 months of experimenting, he invented a successful method for reducing an aluminum mineral to aluminum metal. In February 1886 Charles Hall walked into his former teacher's office with a handful of metallic aluminum globules.

Just as his chemistry teacher had predicted, within a short period of time, Hall became rich and famous. In 1911 he received the Perkin Medal for achievement in chemistry, and in his will, he donated $5 million to Oberlin College. He also helped to establish the Aluminum Company of America (ALCOA), and the process for making aluminum metal gave rise to a huge industry. Aluminum is now second only to steel

covery at the same ti
taining aluminum m

◀ The notebook of Charles Hall along with globules of aluminum.

◀ **Chemistry Connection** vignettes highlight the impact that chemistry has on the consumer, history, the environment, and your everyday life. Each Chemistry Connection box begins with a question; see if you can answer it before you read the box.

CHAPTER 5

Models of the Atom

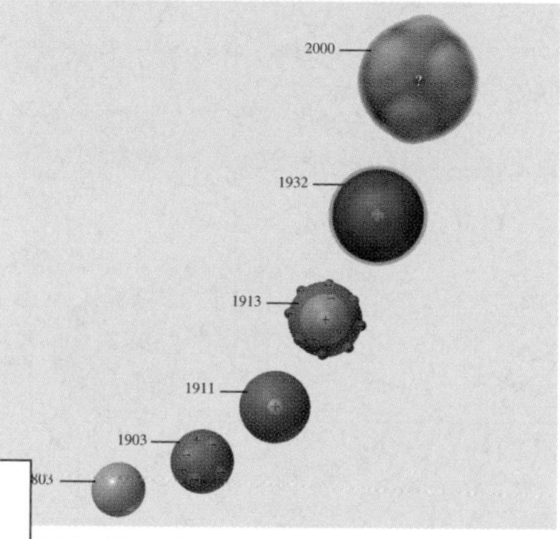

▲ In the evolving model of the atom, what does the question mark (?) represent in the 2000 model?

Molecular Art found throughout the text represents chemical compounds at the atomic scale. Make sure you see the connection between what happens at a molecular level and what you see in the world around you. ▶

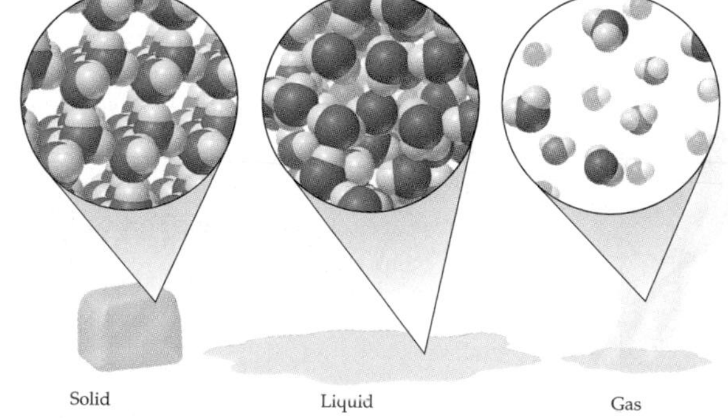

▲ **Figure 4.14 Energy and Physical State** The energy of individual H_2O molecules increases as solid ice is heated. As the attraction between molecules is overcome, solid ice changes to liquid water and then to gaseous steam.

◀ **Micro/Macro Art** helps you connect those items you encounter in everyday life (the macro view) to the depictions of molecules chemists visualize everyday (the micro view). In the text, the two representations are presented together.

Media Resources

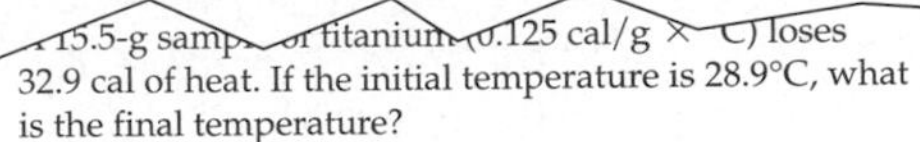

32.9 cal of heat. If the initial temperature is 28.9°C, what is the final temperature?

93. The density of mercury is 13.6 g/mL. Express the density in SI units (kg/m^3).
94. The specific heat of mercury is 0.0331 cal/g × °C. Express the specific heat in SI units (J/kg × K).
95. The radius (r) of the international prototype kilogram cylinder is 1.95 cm. Assuming the density of the kilogram is 21.50 g/cm^3, calculate its height (h). The volume of a cylinder equals $\pi r^2 h$, where π is the constant 3.14.

Explorer Quiz 1
Explorer Quiz 2
Explorer Quiz 3
Master Quiz

Web Activities: Additional exercises are available on the text's Companion Website, indicated by this icon in the text.

5.2 Thomson Model of the Atom

Multiple Proportions Movie

Objectives
- To describe the Thomson plum-pudding model of the atom.
- To state the relative charge and mass of the electron and proton.

About 50 years after Dalton's proposal, there was disturbing evidence that the atom was divisible after all. This evidence came from cathode-ray tubes, which were sealed glass tubes containing a gas at low pressure. When electricity is applied to one end, a cathode-ray tube appears to glow. This phenomenon is referred to as fluorescence, and the glowing ray is a type of light energy. Since the ray emanates from the negative cathode in the tube, the radiation is referred to as a **cathode ray**.

Student CD-ROM Activities: This icon in your text indicates a live, interactive presentation on your *Introductory Chemistry* CD-ROM. Go to this chapter on the CD to see this material in a new, interactive way.

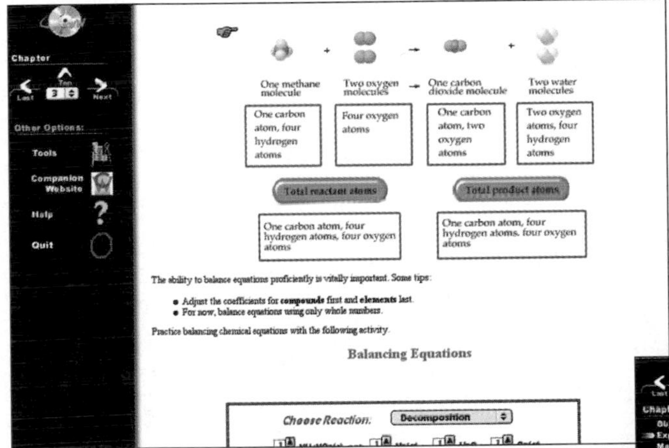

Introductory Chemistry Media Companion: The student CD-ROM follows the book, chapter by chapter, illustrating chemical concepts with interactive simulations, animations, videos, "grabbable" molecules, and practice questions.

Companion Website: Go to www.prenhall.com/corwin for additional practice quizzes/exercises that your instructor may assign with Syllabus Manager. The quizzes/exercises are instantly graded. You also get hints and feedback on your answers.

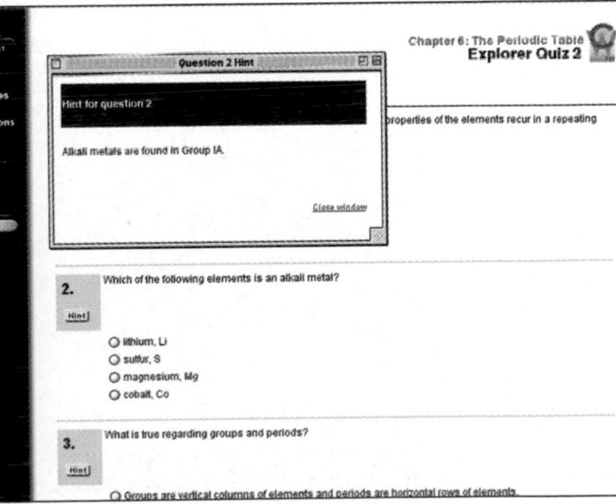

About the Author

Charles H. Corwin

Charles H. Corwin, educated at San Jose State University, has been perfecting his approach to teaching students chemistry for the past two decades. Through special National Science Foundation funding, he studied creative learning systems early in his career, focusing specifically on the introductory chemistry course. It is from these early seeds that *Introductory Chemistry: Concepts & Connections* has grown. At present, his active areas of research are web-based chemical education, conceptual and algorithmic learning models, and collaborative learning and group work.

Currently, Charles Corwin is a professor of chemistry at American River College, where he was recognized with a teacher of the year award for innovative teaching in 1994. He was also the recipient of an alumni teaching award from Purdue University in 1998.

CHAPTER 1

Introduction to Chemistry

▲ Which of the following is installed inside an Apple iMac computer and is not visible in the above photograph: monitor, mouse, or modem?

The initial impression of chemistry that we convey to students sets the tone for the term. Thus, on the first day of class I try to stimulate student interest through a favorite demonstration (see the *Intructor's Resource Manual*) and make positive associations to reassure students that chemistry can be fun and they will be successful if they follow directions and make a reasonable effort to do the assignments. Simultaneously, I provide a syllabus that carefullly organizes both lecture and lab assignments for the entire term. It is convenient to cover one chapter per week as the text is designed to have the same amount of material in each chapter.

In the United States, Canada, and other developed countries, we enjoy a standard of living that could not have been imagined a century ago. Owing to the evolution of science and technology, we have abundant harvests, live in comfortable, climate-controlled buildings, and travel the world via automobiles and airplanes. We also have extended life spans free of many diseases that previously ravaged humanity.

The development of technology has provided machinery and equipment to perform tedious tasks, which gives us time for more interesting activities. The arrival of the computer chip has resulted in electronic appliances that afford ready convenience and dazzling entertainment. We can select from a multitude of audio and video resources that offer crystal-clear sound and brilliant color. Furthermore, many

▲ **Figure 1.1 Compact Disc (CD)** In addition to playing musical sounds on a compact disc, we can view visual images in motion, and store an enormous amount of data.

of us have access to the information superhighway, and moreover, the knowledge in an entire encyclopedia is now available on a single compact disc (Figure 1.1).

Directly or indirectly, our present standard of living requires scientists and technicians with educational training in chemistry. The health sciences—as well as the life sciences, physical sciences, and earth sciences—demand an understanding of chemical principles. In fact, chemistry is sometimes referred to as the central science because it stands at the crossroads of biology, physics, geology, and medicine (Figure 1.2). Just as personal computers are becoming indispensable in our everyday activities, chemistry is assuming an essential role in our daily lives.

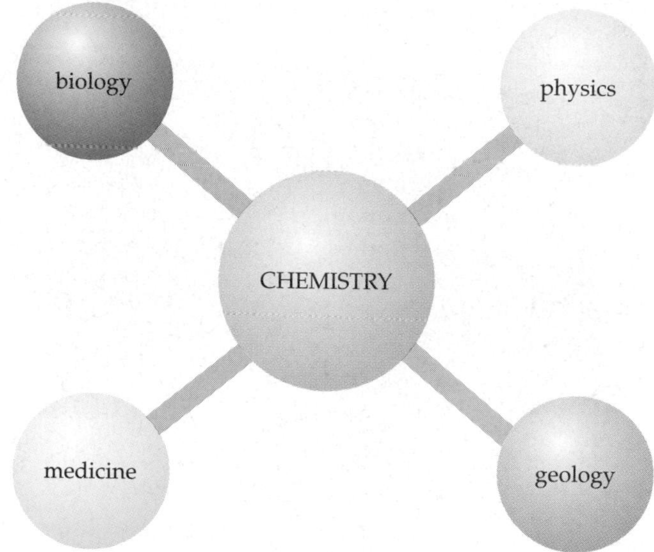

▶ **Figure 1.2 Chemistry—The Central Science** The principles of chemistry are fundamental to our understanding of biology, physics, geology, and medicine, as well as other health sciences.

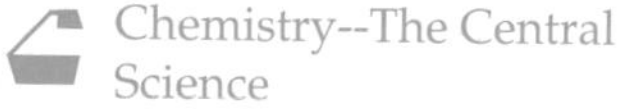

1.1 Evolution of Chemistry

Objectives
- To describe the early practice of chemistry.
- To identify the three steps in the scientific method.

The earliest concept of science began with the ancient Chinese, Egyptian, and Greek civilizations. The Chinese believed that the universe was created from the interaction of two forces. Yin, the feminine force, was manifested in darkness, cold, and wetness. Yang, the masculine force, was manifested in light, heat, and dryness. When the yin and yang forces interacted, they brought the material world into existence and were responsible for everything in nature.

As early as 600 B.C., the Greeks began to speculate that the universe was composed of a single element. Thales, the founder of Greek science, mathematics, and philosophy, suggested that water was the single element. He claimed that the Earth was a dense, flat disk floating in a universe of water. He also believed that air and space were less dense forms of water.

A few years later, another Greek philosopher proposed that air was the basic element. This theory was followed by proposals that fire, and later earth, was the basic element. About 450 B.C., the Greek philosopher Empedocles observed that when wood burned, smoke was released (air), followed by a flame (fire). He also noticed that a cool surface held over a fire collected moisture (water) and that the only remains were ashes (earth). Empedocles interpreted his observations as evidence for air, fire, water, and earth as basic elements. He further speculated that other substances were examples of these four elements combined in varying proportions as illustrated in Figure 1.3.

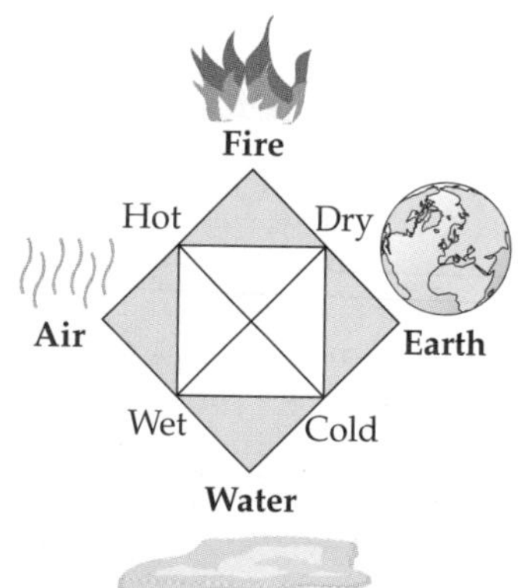

▲ **Figure 1.3 The Four Greek Elements** The four elements—air, earth, fire, water—proposed by the Greeks. Notice the properties—hot, cold, wet, and dry—associated with each element.

The Four Greek Elements

The idea that air, earth, fire, and water were basic elements was later adopted by Aristotle about 350 B.C. In addition, he added a fifth element, ether, that he believed filled all space. Aristotle's influence was so great that his opinions dominated our understanding of nature for nearly 2000 years.

▲ **Robert Boyle** This stamp honors Boyle for his invention of the vacuum pump in 1659. Boyle's classic textbook, *The Sceptical Chymist*, laid the foundation for the scientific method.

The Scientific Method

In 1661 the English scientist Robert Boyle published *The Sceptical Chymist.* In his classic book, Boyle stated that scientific speculation was worthless unless it was supported by experimental evidence. This principle led to development of the scientific method, which marked a turning point and the beginning of modern science.

Science can be defined as the methodical exploration of nature followed by a logical explanation of the observations. The practice of science entails planning an investigation, carefully recording observations, gathering data, and analyzing the results. In an **experiment**, scientists explore nature according to a planned strategy and make observations under controlled conditions.

The **scientific method** is a systematic investigation of nature and requires proposing an explanation for the results of an experiment in the form of a general principle. The initial, tentative proposal of a scientific principle is called a **hypothesis**.

After further experimentation, the initial hypothesis may be rejected, modified, or elevated to the status of a scientific principle. However, for a hypothesis to become a scientific principle, many additional experiments must support and verify the original proposal. Only after there is sufficient evidence does a hypothesis rise to the level of a scientific **theory**. We can summarize the three steps in the scientific method as follows.

Applying The Scientific Method

Step 1: Perform a planned experiment, make observations, and record data.

Step 2: Analyze the data and propose a tentative hypothesis to explain the experimental observations.

Step 3: Conduct additional experiments to test the hypothesis. If the evidence supports the initial proposal, the hypothesis may become a theory.

We should note that scientists exercise caution before accepting a theory. Experience has shown that nature reveals her secrets slowly and only after considerable probing. A scientific theory is not accepted until rigorous testing has established that the hypothesis is a valid interpretation of the evidence. For example, in 1803 John Dalton proposed that all matter was composed of small, indivisible particles called atoms. However, it took nearly 100 years of gathering additional evidence before his proposal was universally accepted and elevated to the status of the atomic theory.

Although the terms "theory" and "law" are related, there is a distinction between a scientific theory and a natural law. A theory is a model that scientifically explains the behavior of nature. To illustrate, the atomic theory states that matter is composed of individual particles, which helps to explain the constant composition of substances as well as the behavior of gases.

A **natural law** does not explain behavior but simply states a measurable relationship under different experimental conditions. To illustrate, Boyle's law states that a decrease in the volume of a gas produces an increase in pressure (if the temperature remains constant). A natural law is often expressed as an equation; for example, Boyle's law can be written as $P_1V_1 = P_2V_2$.

Experiment #1, Prentice Hall Laboratory Manual

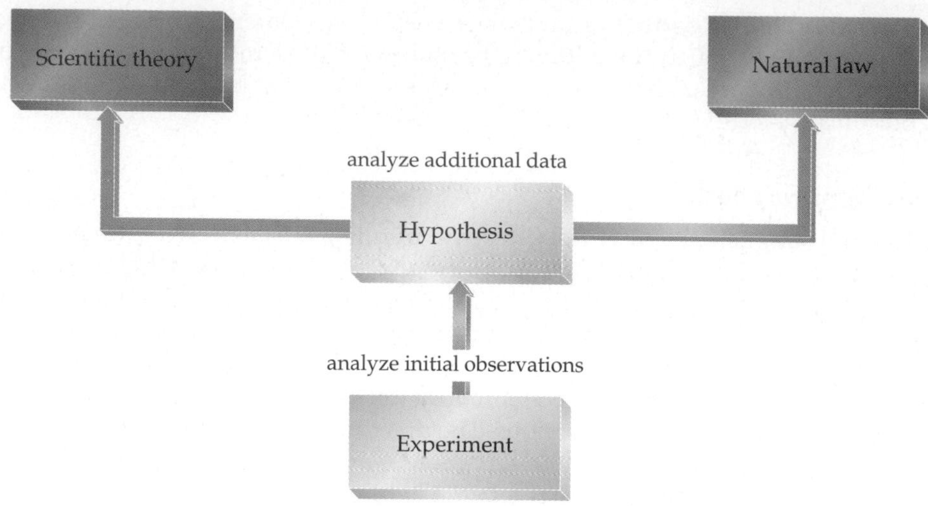

▶ Figure 1.4 The Scientific Method The initial observations from an experiment are analyzed and formulated into a hypothesis. Next, additional data is collected and analyzed from experiments conducted under various conditions. If the additional data supports the initial proposal, the hypothesis may be elevated to a scientific theory or a natural law.

 The Scientific Method

We can distinguish between a theory and a law by simply asking the question Is the proposal measurable? If the answer is yes, the statement is a law; otherwise, the statement is a theory. Figure 1.4 summarizes the relationship between a hypothesis, a scientific theory, and a natural law.

Note A *scientific theory* may be revised in the wake of new evidence. In fact, our understanding of atoms is being modified even as you read this book. On the other hand, a *natural law* cannot be changed because it involves measurable quantities. However, experiments may reveal that a natural law is not valid for a given set of conditions. For instance, the relationship between the pressure and volume of a gas never changes, but this law is invalid for a gas at very low temperatures.

▲ Antoine Lavoisier This stamp honors Lavoisier for his numerous achievements, including the establishment of a magnificent laboratory that attracted scientists from around the world. Lavoisier is generally considered the founder of modern chemistry.

1.2 Modern Chemistry

Objective · To describe the modern practice of chemistry.

In the eighth century A.D., the Arabs introduced the pseudoscience of **alchemy**. Although alchemists conducted experiments, they also believed in a magic potion that had miraculous healing powers and could transmute lead into gold. Although alchemy did not withstand the test of time, its practice did precede the planned, systematic, scientific experiments that are the cornerstone of modern chemical and medical research.

In the late eighteenth century, the French chemist Antoine Lavoisier organized **chemistry** into a comprehensible science and wrote two notable textbooks. Lavoisier also built a magnificent laboratory and invited scientists from around the world to view it; his many visitors included Benjamin Franklin and Thomas Jefferson. Lavoisier was a prolific experimenter and published his work in several languages. For his numerous contributions, he is generally considered to be the founder of modern chemistry.

Today, chemistry is defined as the science that studies the composition of matter and its properties. Chemists have accumulated so much information during the past two centuries that we now divide the subject into several branches or specialties. The branch of chemistry that studies substances containing carbon is called **organic chemistry**. The study of all other substances, those that do not contain carbon, is called **inorganic chemistry**.

The branch of chemistry that studies substances derived from plants and animals is called **biochemistry**. Another branch, analytical chemistry, includes qualitative analysis (which determines what substances are in a sample) and quantitative analysis (which determines how much of each substance is present). Physical

Chemistry Connection · Alchemy

What is the original meaning of the word "elixir," which is a reputed "cure-all" for a variety of ailments?

In the seventh century A.D., the Arabs took control of Persia, northern Africa, and Egypt. In the process, they were exposed to the culture and science of the Greeks who had previously ruled the area. In the eighth century, an Arab known as Geber proposed that the four Greek elements combined to form only two elements—sulfur and mercury. He further suggested that sulfur and mercury could combine to form other substances such as lead, which in turn could be converted to gold. He also believed that changing one substance to another required a mysterious potion the Arabs called *al-iksir*, a term from which we get the word "elixir." Geber's experiments with various substances, as well as his search for a magic potion, eventually became the pseudoscience we call *alchemy*.

There was no evidence that a magic potion ever existed, but alchemists continued their search for over a thousand years. Some alchemists dabbled in the supernatural and believed that a magic potion would have miraculous healing powers and could bestow immortality. Others claimed that a magic potion could convert lead and other metals to gold.

Theophrastus Bombastus von Hohenheim, a powerful, influential alchemist, was born in Europe in 1493. As an adult, he took the name Paracelsus, after the Roman Celsus who had written a complete history of Greek science. Paracelsus was not interested in transmuting lead to gold. Instead, he wanted to discover medicines to cure disease, and he did much to influence other physicians to treat ailments with natural and synthetic drugs. Paracelsus practiced on himself and prescribed opium for pain, and poisonous mercury and antimony substances for other symptoms.

▲ **The Alchemist** This painting by the Dutch artist Cornelius Bega depicts a dark and secretive alchemical laboratory of the seventeenth century.

More than anything, Paracelsus desired immortality and searched ceaselessly for a mystical elixir of life. At one point he even claimed to have found it and proclaimed he would live forever. Although he was mistaken (he died before the age of 50), Paracelsus did leave a medical legacy. He recognized that coal mining was associated with lung disease, he noted that head injuries produced paralysis, and he correctly diagnosed that an abnormal thyroid caused mental and physical retardation.

The word "elixir" is derived from the Arabic word al-iksir, meaning "potion."

---

chemistry is a specialty that proposes theoretical and mathematical explanations for chemical behavior. Recently, environmental chemistry has become an important specialty that focuses on chemical pollution and the safe disposal of chemical waste.

Chemistry plays a meaningful role in medicine, from the training of health care professionals to the dispensing of pharmaceutical prescriptions. Chemists help ensure plentiful agricultural harvests by formulating potent fertilizers and synthesizing pesticides that are not harmful to humans.

Chemical principles are indispensable to many industries including the manufacture of automobiles, electronic components, scientific instruments, aluminum, steel, paper, and plastics. One of the largest industries is the petrochemical industry. Petrochemicals are chemicals derived from petroleum and natural gas. They can be used to manufacture a wide assortment of consumer products including paints, plastics, rubber, textiles, dyes, detergents, aerosols, explosives, and pesticides.

Chemistry encompasses a spectrum of activities from routine laboratory work to ensure quality control to sophisticated investigations seeking to understand the nature of the human body. Figure 1.5 depicts a few of the activities performed by chemists, Figure 1.6 provides a brief glimpse of the role of chemistry in medicine, Figure 1.7 depicts the role of chemistry in agriculture, and Figure 1.8 shows the derivation of consumer products from petroleum.

Chemistry in Action

The Laboratory

Chemists performing precise analyses using a variety of techniques and instruments.

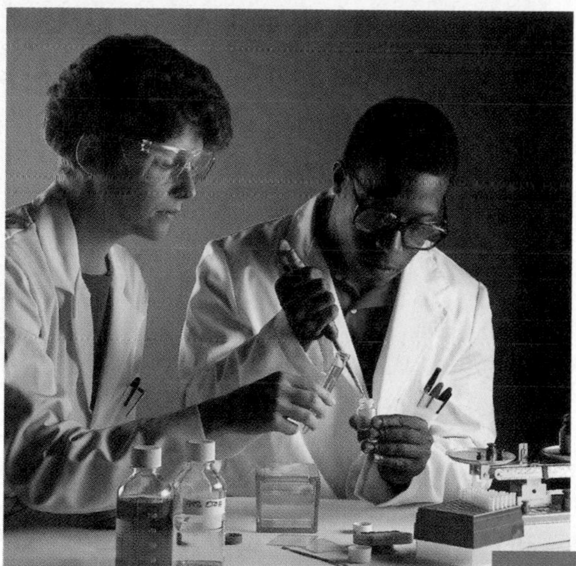

◀ **Figure 1.5 (a)** Chemists perform qualitative analysis for substances in a sample, and quantitative analysis for the amount of each substance.

▼**(b)** Chemists utilize instruments such as an atomic absorption spectrophotometer, which measures the amount of light absorbed by atoms of a sample vaporized in a flame.

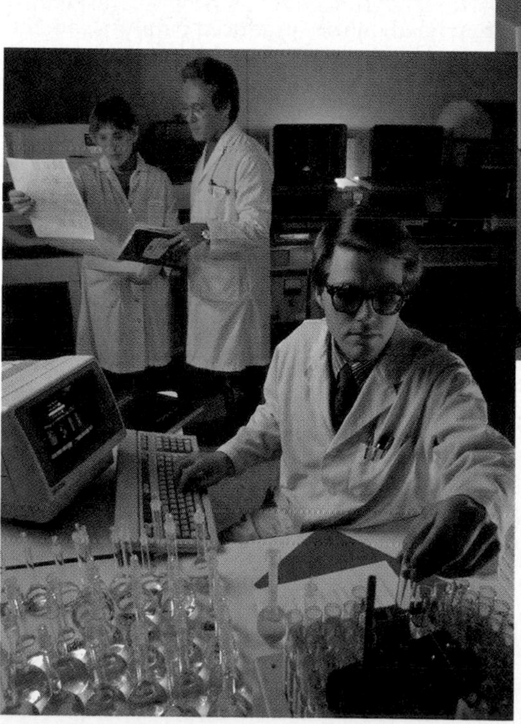

◀**(c)** Chemists often record data on electronic spread sheets and interface computers with instruments that analyze samples.

Chemistry in Medicine

Health Care Practices

Medical treatment begins before birth: diagnosis involves blood analysis; well-being is aided by a variety of drugs.

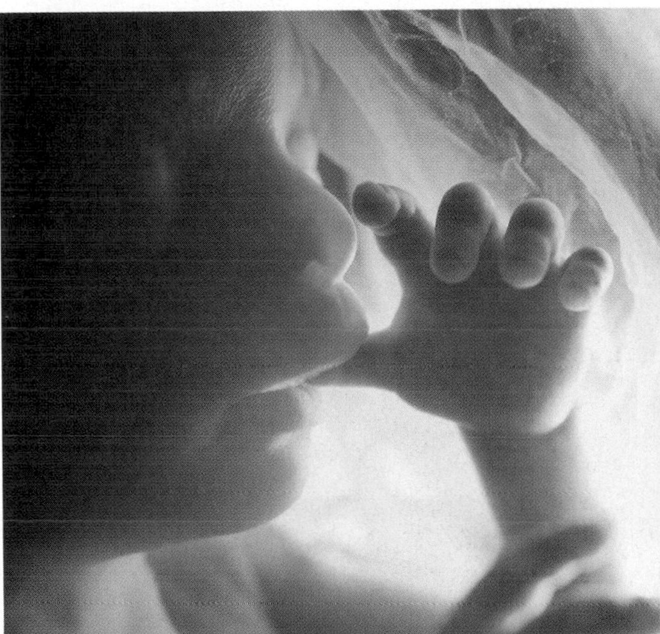

◀ **Figure 1.6 (a)** Chemistry labs regularly examine fetal fluids for abnormalities that may indicate a threat to the health of the fetus.

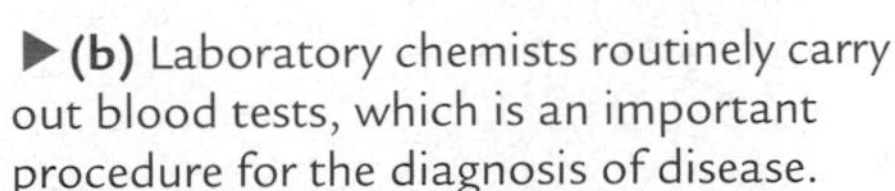

▶ **(b)** Laboratory chemists routinely carry out blood tests, which is an important procedure for the diagnosis of disease.

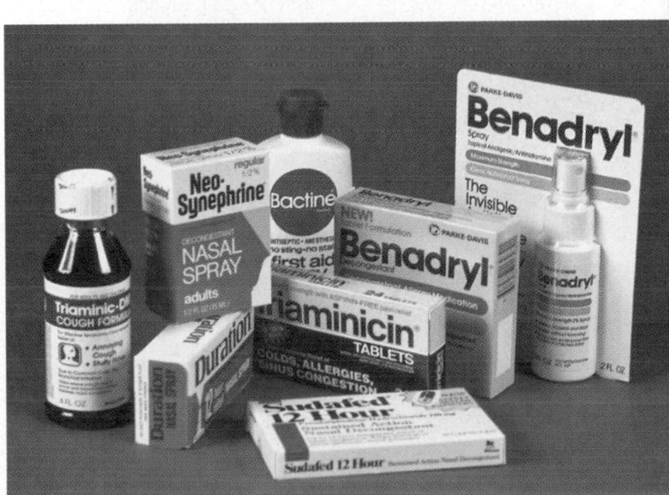

◀ **(c)** Organic chemists and biochemists synthesize prescription and over-the-counter drugs, which have a specific medical purpose.

Chemistry in Agriculture

Agricultural Chemicals

In the early 1900s, it required over 50% of the population to feed the nation. Today, 2% produces a food surplus. Chemicals increase crop yields by replenishing nutrients in the soil and controlling insects and plant diseases.

◀ **Figure 1.7 (a)** Chemists manufacture ammonia, NH_3, which can be used to replenish the nitrogen content of crop soil.

▼**(b)** Lime, CaO, is widely used on acid soil to create a more favorable environment for growing a particular plant.

▼**(c)** Agriculture uses pesticides to control pests and weeds and to eradicate plant diseases, thus increasing harvest yields.

Chemistry in Industry

Petrochemical Industry

An offshore oilwell produces crude oil that is shipped to a refinery. The refinery separates the petroleum into petrochemicals, which in turn, are often used in the manufacture of consumer products.

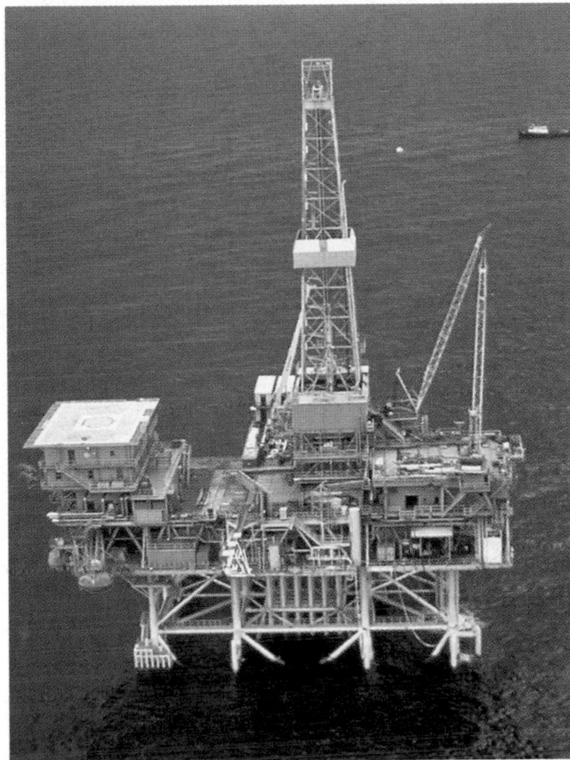

◀ **Figure 1.8 (a)** An offshore platform supports a drilling rig for recovering oil beneath the ocean floor.

▼**(b)** An oil refinery distills petroleum into various fractions such as natural gas, bottled gas, gasoline, kerosene, and jet fuel.

◀**(c)** Petrochemicals are derived from petroleum and in turn can be converted to plastics, paints, and other consumer items.

1.3 Learning Chemistry

Objective · To realize that chemistry is an interesting and relevant subject.

Chemical Reactions in Automotive Air Bags Movie

In a survey published by the American Chemical Society, entering college students were asked to express their attitudes about science courses. The students rated chemistry as the most relevant science course, and 64% rated it as highly relevant to their daily lives. Unfortunately, 83% of the students thought chemistry was a difficult subject. In view of the results of the student survey, perhaps we should take a moment to consider perceptions in general and attitudes about chemistry in particular.

You are probably familiar with the expression that some people see a glass of water as half full, while others see the same glass as half empty. This expression implies that different people can respond to the same experience with optimism or pessimism. Moreover, experimental psychologists have found that they can use abstract visual images to discover underlying attitudes regarding a particular perception. A practical lesson involving two perceptions obtained from the same image is revealed by the following picture.

Dual Perceptions

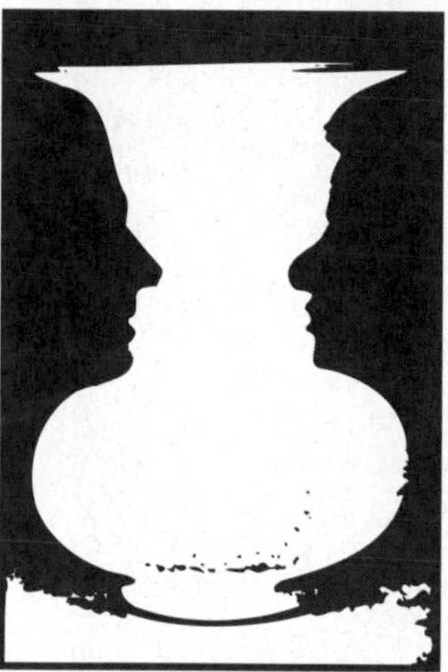

What do you see? Some students see a white vase on a black background; others see two dark profiles facing each other. After a short period of time, one image switches to the other. If you concentrate, can you view only one of the images? Can you choose to switch the images back and forth? This exercise is a classic example of our brains registering dual perceptions of the same image.

Your experience of learning chemistry may be somewhat like the preceding exercise that tests your perspective. That is, sometimes your perception may be that chemistry is challenging, while a short time later your attitude may be that chemistry is easy and fun.

Perception is often affected by unconscious assumptions. Let's consider a type of problem slightly different from the perception of a vase. In the following problem try to connect each of the nine dots using only *four* straight, continuous lines.

We can begin to solve the problem by experimenting. For example, let's start with the upper left dot and draw a line to the upper right dot. We can continue to draw straight lines as follows.

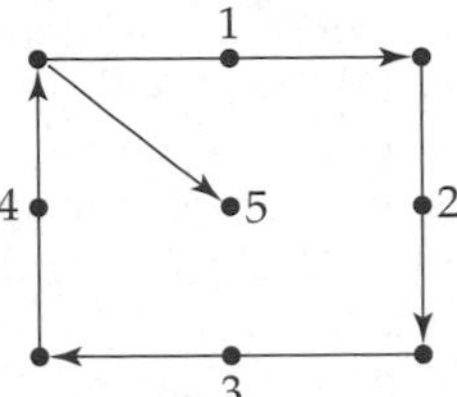

Notice that we connected the nine dots but that it was necessary to use *five* straight, continuous lines. If we start with a different dot, we find that *five* lines are required no matter where we start. Perhaps we are bringing an underlying assumption to the problem. That is, we may be unconsciously confining the nine dots.

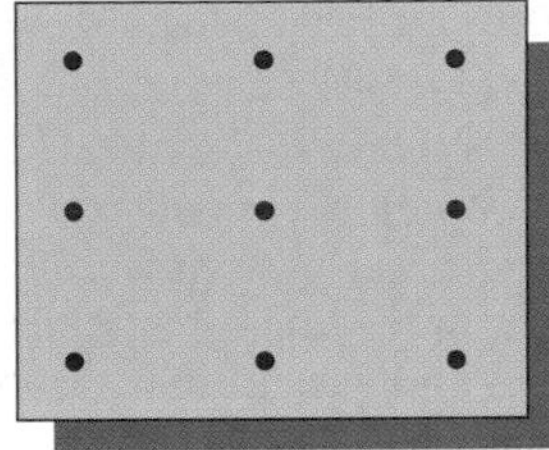

What will happen if we start with the upper left dot and draw a line through the upper right dot? If we continue, we can complete the problem with *four* straight, continuous lines as follows.

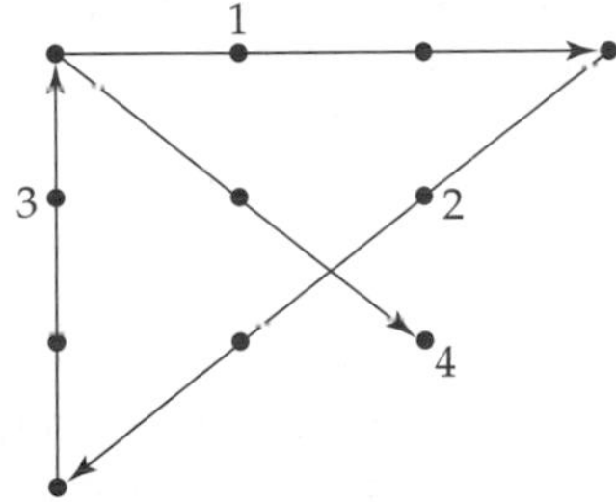

The "secret" to solving this nine-dot problem is to recognize that we may be unconsciously confining the problem and making it impossible to solve. Similarly, we should not confine our concept of chemistry to a preconceived attitude that learning chemistry will be difficult. Or better yet, we should choose positive associations for our concept of chemistry.

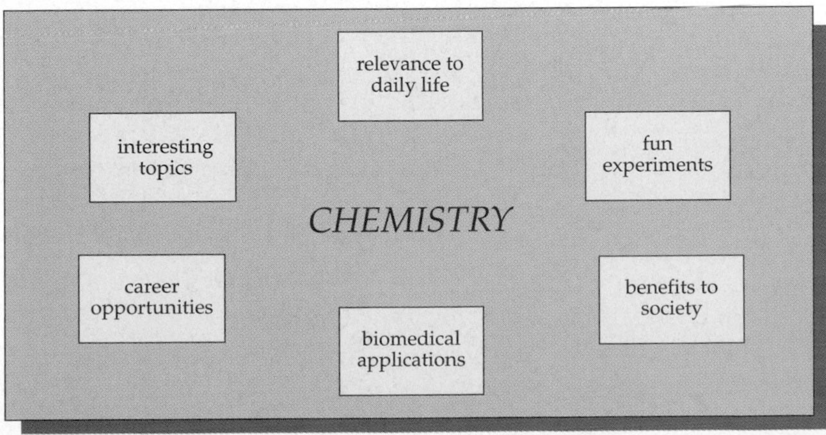

Summary

Section 1.1 This chapter traces the development of chemistry from a historical point of view. Beginning in the period 600–350 B.C., the early Greeks used reason and thoughtful mental exercises to understand the laws of nature. Although they often arrived at conclusions based on speculation, they did unveil some of nature's secrets and had a profound influence on Western civilization that lasted for 20 centuries.

The term **science** implies a rigorous, systematic investigation of nature. Moreover, a scientist must accumulate significant evidence before attempting to explain the results. In the seventeenth century, Robert Boyle founded the **scientific method**, and laboratory experimentation became essential to an investigation. After an **experiment**, scientists use their observations to formulate an initial proposal, which is called a **hypothesis**. However, a hypothesis must be tested repeatedly before it is accepted as valid. After a hypothesis has withstood extensive testing, it becomes either a scientific **theory** or a **natural law**. A scientific theory is an accepted explanation for the behavior of nature, whereas a natural law states a relationship under different experimental conditions and is often expressed as an equation.

Section 1.2 The pseudoscience of **alchemy** introduced the practice of laboratory experimentation and was the forerunner of modern **chemistry**. Today, chemistry is quite diverse and has several branches including **inorganic chemistry**, **organic chemistry**, and **biochemistry**. Currently, the application of chemical principles helps to create a variety of products that provide a standard of living far exceeding that imagined by ancient chemists. The impact of chemistry is felt in medicine and agriculture, as well as in the electronics, pharmaceutical, petrochemical, and other industries.

Section 1.3 In this section we examined some dual perceptions and pointed out that our brains have the ability to respond to the same image in two ways. Before beginning to learn chemistry, most students have already made associations with the subject. It is hoped that you will be able to focus on chemistry as being an interesting and relevant subject, and put aside any preconceived limiting attitudes.

Key Concepts*

1. What is the principal difference between ancient and modern chemistry?
2. What question can we ask to distinguish between a theory and a law?
3. Which of the line segments in the image below, *AB* or *BC*, appears to be longer? If the lines appear to be of equal length, measure each segment with a ruler to verify your perception.

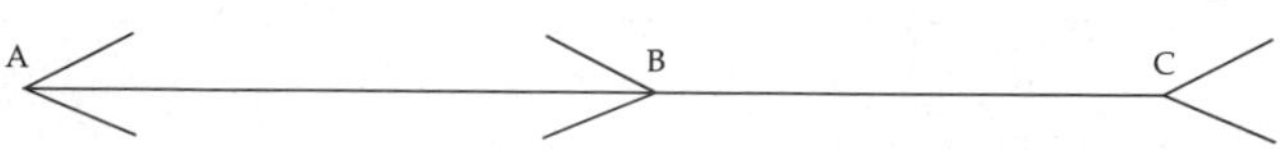

4. Does Box B appear to be identical to Box A (after rotating the image 90 degrees)? If the boxes do not appear to have the same dimensions, measure the length, width, and height with a ruler to verify your perception.

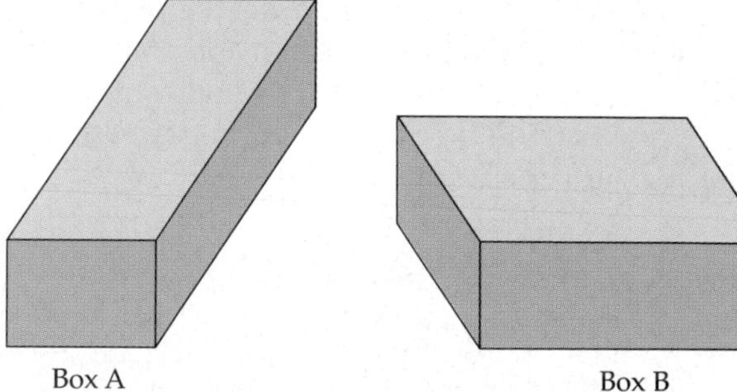

*Answers to Key Concepts are in Appendix H.

Key Terms†

Select the key term below that corresponds to each of the following definitions.

_____ 1. the methodical exploration of nature and logical explanation of the observations

_____ 2. a scientific procedure for gathering data and recording observations under controlled conditions

_____ 3. a systematic investigation that entails performing an experiment, proposing a hypothesis, testing the hypothesis, and stating a theory or law

_____ 4. a tentative proposal of a scientific principle that attempts to explain the meaning of a set of data collected in an experiment

_____ 5. an extensively tested proposal of a scientific principle that explains the behavior of nature

_____ 6. an extensively tested proposal of a scientific principle that states a measurable relationship under different experimental conditions

_____ 7. a pseudoscience that attempted to convert a base metal such as lead to gold; a medieval science that sought to discover a universal cure for disease and a magic potion for immortality

_____ 8. the branch of science that studies the composition and properties of matter

_____ 9. the study of chemical substances that do not contain carbon

_____ 10. the study of chemical substances that contain carbon

_____ 11. the study of chemical substances derived from plants and animals

(a) alchemy (*Sec. 1.2*)
(b) biochemistry (*Sec. 1.2*)
(c) chemistry (*Sec. 1.2*)
(d) experiment (*Sec. 1.1*)
(e) hypothesis (*Sec. 1.1*)
(f) inorganic chemistry (*Sec. 1.2*)
(g) natural law (*Sec. 1.1*)
(h) organic chemistry (*Sec. 1.2*)
(i) science (*Sec. 1.1*)
(j) scientific method (*Sec. 1.1*)
(k) theory (*Sec. 1.1*)

Exercises‡

Evolution of Chemistry (Sec. 1.1)

1. According to the beliefs of the ancient Chinese, what two forces were responsible for bringing the natural world into existence?
2. According to the beliefs of the ancient Greeks, what four elements composed everything in nature?
3. Who is considered the founder of the scientific method?
4. What are the three steps in the scientific method?
5. What is the difference between a hypothesis and a theory?
6. What is the difference between a scientific theory and a natural law?
7. Which of the following statements is a scientific theory?
   (a) Atoms contain protons, neutrons, and electrons.
   (b) A neutron can decay into a proton by emitting a beta particle.
   (c) If the temperature of a gas doubles, the pressure also doubles.
   (d) The volumes of gases in a chemical reaction combine in the ratio of small whole numbers.
8. Which of the following statements is a natural law?
   (a) The nucleus of an atom is composed of protons and neutrons.
   (b) The energy emitted from an atomic nucleus can be found from the equation $E = mc^2$.
   (c) The total mass of substances is the same before and after a chemical change.
   (d) Equal volumes of gases, at the same temperature and pressure, contain equal numbers of molecules.

Modern Chemistry (Sec. 1.2)

9. Who is considered the founder of modern chemistry?
10. Name and describe five branches of chemistry.
11. Name at least five professions in which chemistry plays an important role.
12. Name at least five industries in which chemistry plays an important role.

Learning Chemistry (Sec. 1.3)

13. It is possible to solve the nine-dot problem with only *three* straight, continuous lines. Solve the problem and identify the unconscious assumption.

•   •   •

•   •   •

•   •   •

† Answers to Key Terms are in Appendix I.

‡ Answers to odd-numbered Exercises are in Appendix J.

14. It is possible to solve the nine-dot problem with only *one* straight, continuous line. Solve the problem and identify the unconscious assumption.

15. Stare at the image below and attempt to "flip" the stack of blocks upside down. (*Hint*: If you have difficulty seeing the second perception, stare at the point where the three blocks come together and mentally pull the point toward you.)

Stacks of Blocks

16. Look at the image below and explain how the drawing creates a contradiction in perception.

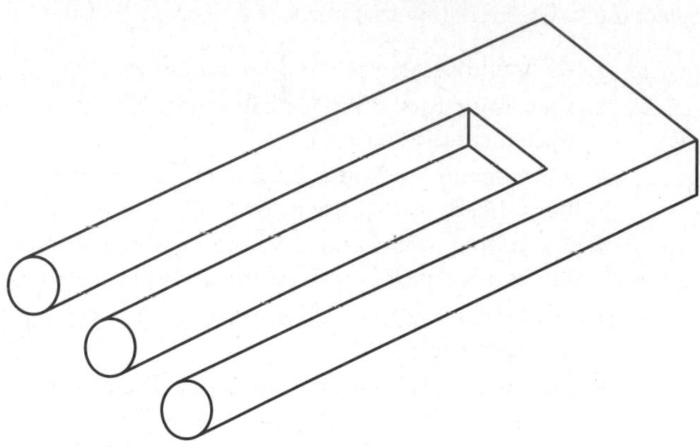

Explorer Quiz 1
Explorer Quiz 2
Explorer Quiz 3
Master Quiz

CHAPTER 2

Scientific Measurements

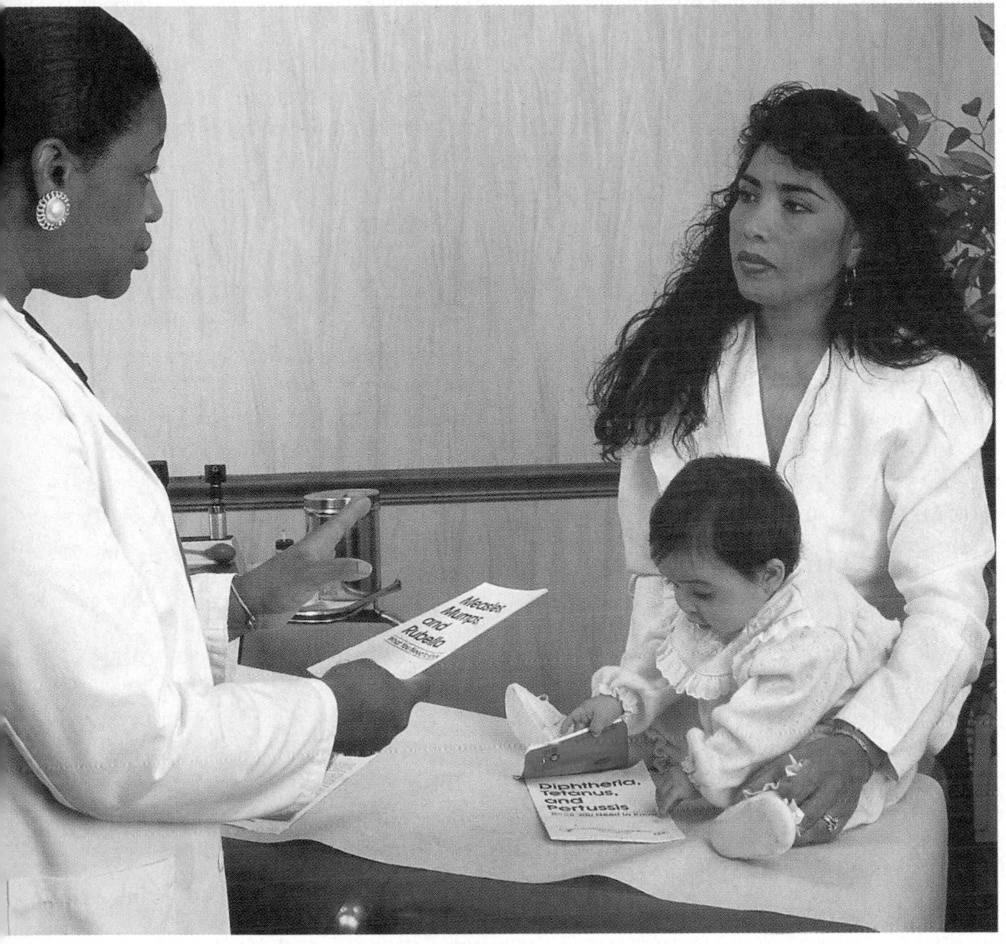

▲ Which measurements are more exact: recording height and weight in a doctor's office, or mass and volume in a chemistry laboratory?

Once again, I try to allay students' fears of chemistry by telling them that we will be solving some problems, but "the method is so powerful that solving problems will be as simple as one, two, three." This statement is a reference to unit analysis, or dimensional analysis, which is used throughout the textbook. In fact, algebra is not required to solve any of the problems in the text.

In this chapter we will establish an important foundation for the chemical concepts and calculations discussed in later chapters. You are aware, of course, that we live in an electronics age where calculators and computers are part of our daily lives. In addition, hundreds of instruments are available that use state-of-the-art technology. In the laboratory, scientists use instruments that provide very sensitive measurements. For instance, chemists routinely use electronic balances that are so sensitive you can weigh your fingerprints!

Our discussion begins with the instruments commonly found in an introductory chemistry laboratory. Later, we will learn to add, subtract, multiply, and divide measurements obtained from these instruments. Last, and perhaps most importantly, we will learn a powerful method for solving problems in three simple steps.

▲ **Figure 2.1 Estimation of Length, Mass, and Volume** The diameter of a 5¢ coin is about 2 cm and its mass is about 5 g. The volume of 20 drops from an eyedropper is about 1 mL.

2.1 Uncertainty in Measurements

Objectives
- To identify typical instruments in a chemistry laboratory.
- To explain why an instrumental measurement is never exact.

We can define a **measurement** as a number with attached units. For example, a chemical sample may weigh 2.005 grams. We measure the mass of a sample with an **instrument** called a balance, and the exactness of the measurement depends on the balance. For instance, electronic balances are common that measure the mass of a sample to 1/1000 of a gram.

Lab technicians routinely inject liquid samples with hypodermic syringes that measure volume to one-millionth of a liter; and electronic stopwatches are available that measure time to a nanosecond, that is, one-billionth of a second. Nevertheless, it is not possible to make exact measurements. An exact measurement is impossible because no instrument measures exactly. That is, an instrument may give a very sensitive reading, but every measurement has **uncertainty**.

A measurement must always include units, such as inches, attached to a numerical value. For example, a measurement of length may be 25 inches. In general, we will avoid English units such as inch (in.), pound (lb), and quart (qt), although they will be used on occasion. Instead we will use metric units such as **centimeter** (symbol **cm**), **gram** (symbol **g**), and **milliliter** (symbol **mL**).

In the metric system, a centimeter is a unit of length, a gram is a unit of mass, and a milliliter is a unit of volume. For reference, it is interesting to note that a 5¢ coin has a diameter of about 2 cm and a mass of about 5 g. Twenty drops from an eyedropper is approximately 1 mL. Figure 2.1 offers some common references for the estimation of length, mass, and volume.

Length Measurements

To help you understand uncertainty, suppose you measure the length of an aluminum rod. You have two metric rulers available that differ as shown in Figure 2.2. Both rulers are satisfactory for the task, however, Ruler B provides a more exact measurement.

Notice that Ruler A has five 1-cm divisions. Since the divisions are large, we can imagine ten subdivisions. Thus, we can estimate to one-tenth of a division, that is, ±0.1 cm. On Ruler A, we see that the aluminum rod measures about 4.2 cm. Since the uncertainty is ±0.1 cm, a reading of 4.1 cm or 4.3 cm is also acceptable.

Notice that Ruler B has five 1-cm divisions and ten 0.1-cm subdivisions. On Ruler B, the subdivisions are quite close together. Thus, we can estimate to only one-half of a division, that is, ±0.05 cm. On Ruler B, we see that the aluminum rod

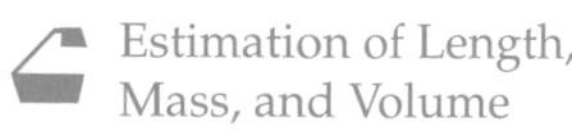
Estimation of Length, Mass, and Volume

Experiment # 2, Prentice Hall Laboratory Manual.

Metric Rulers for Measuring Length

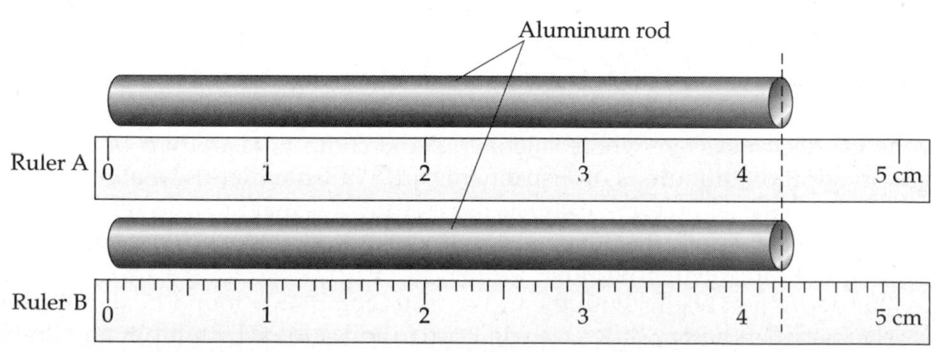

▶ **Figure 2.2 Metric Rulers for Measuring Length** On Ruler A, each division is 1 cm. On Ruler B, a division is 1 cm and each subdivision is 0.1 cm (not to scale).

measures about 4.25 cm. Since the uncertainty is ±0.05 cm, a reading of 4.20 cm or 4.30 cm is also acceptable.

We can compare the length of the aluminum rod measured with Rulers A and B as follows:

Ruler A: 4.2 ± 0.1 cm   Ruler B: 4.25 ± 0.05 cm

In summary, Ruler A has more uncertainty and gives less exact measurements. Conversely, Ruler B has less uncertainty and gives more exact measurements. Example Exercise 2.1 further illustrates the uncertainty in recorded measurements.

Example Exercise 2.1 • Uncertainty in Measurement

Which measurements are consistent with the metric rulers shown in Figure 2.2?

(a) Ruler A: 2 cm, 2.0 cm, 2.05 cm, 2.5 cm, 2.50 cm
(b) Ruler B: 3.0 cm, 3.3 cm, 3.33 cm, 3.35 cm, 3.50 cm

Solution

Ruler A has an uncertainty of ±0.1 cm, and Ruler B has an uncertainty of ±0.05 cm. Thus,

(a) Ruler A can give the measurements 2.0 cm and 2.5 cm.
(b) Ruler B can give the measurements 3.35 cm and 3.50 cm.

Self-Test Exercise

Which measurements are consistent with the metric rulers shown in Figure 2.2?

(a) Ruler A: 1.5 cm, 1.50 cm, 1.55 cm, 1.6 cm, 2.00 cm
(b) Ruler B: 0.5 cm, 0.50 cm, 0.055 cm, 0.75 cm, 0.100 cm

Answers: (a) 1.5 cm, 1.6 cm; (b) 0.50 cm, 0.75 cm

Mass Measurements

The **mass** of an object is a measure of the amount of matter it possesses. Mass is measured with a balance and is not affected by Earth's gravity. You can think of a balance as a teeter-totter with two pans. After an object is placed on one pan, weights are added onto the other pan until the balance is level.

The measurement of mass has uncertainty and varies with the balance. A typical mechanical balance in a laboratory may weigh a sample to 1/100 of a gram. Thus, its mass measurements would have an uncertainty of ±0.01 g. Many laboratories have electronic balances with digital displays. These balances may have uncertainties ranging from ±0.1 g to ±0.0001 g. Figure 2.3 shows three common types of balances.

Although the term "weight" is often used to mean "mass," strictly speaking, the two terms are not interchangeable. **Weight** is the force exerted by gravity on an object. Since the Earth is heavier than the Moon, gravity is greater and objects weigh more on Earth. Similarly, the same object weighs even more on the huge planet Jupiter than on the Earth. On the other hand, the mass of an object obtained using a balance is constant. This is because gravity operates equally on both pans of the balance, thereby canceling its effect. The mass of an object is therefore constant whether it is measured on the Earth, on the Moon, or on any other planet. Figure 2.4 illustrates the distinction between mass and weight.

Volume Measurements

The amount of space occupied by a solid, gas, or liquid is its volume. There are many pieces of laboratory equipment available for measuring the volume of a

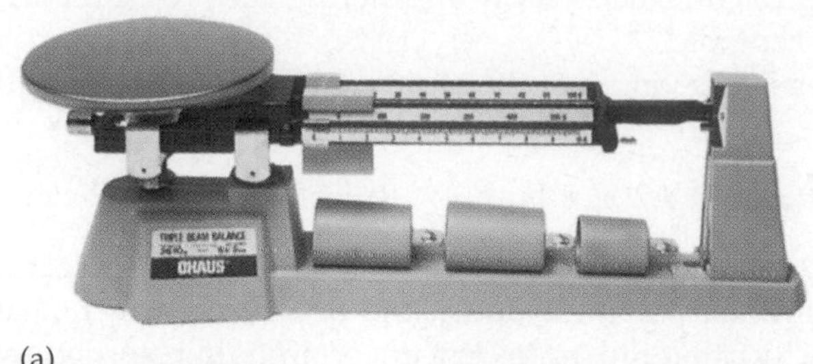

(a)

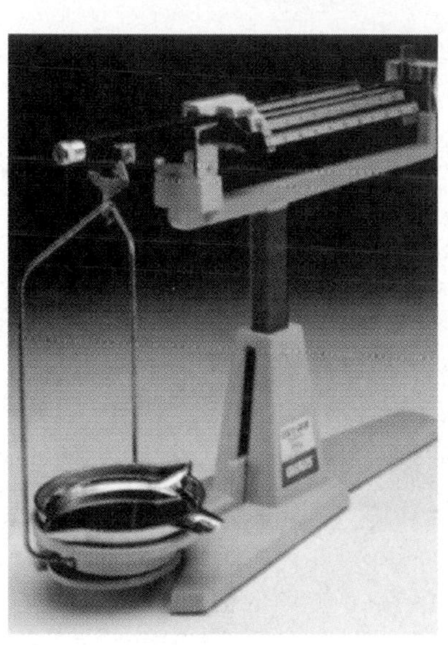

(b)

(c)

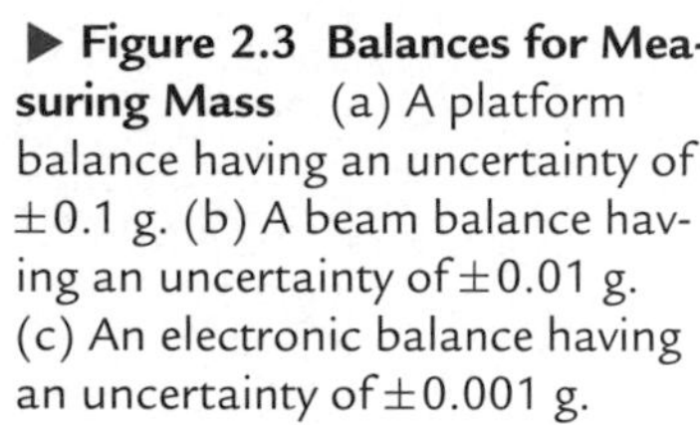

▶ **Figure 2.3 Balances for Measuring Mass** (a) A platform balance having an uncertainty of ±0.1 g. (b) A beam balance having an uncertainty of ±0.01 g. (c) An electronic balance having an uncertainty of ±0.001 g.

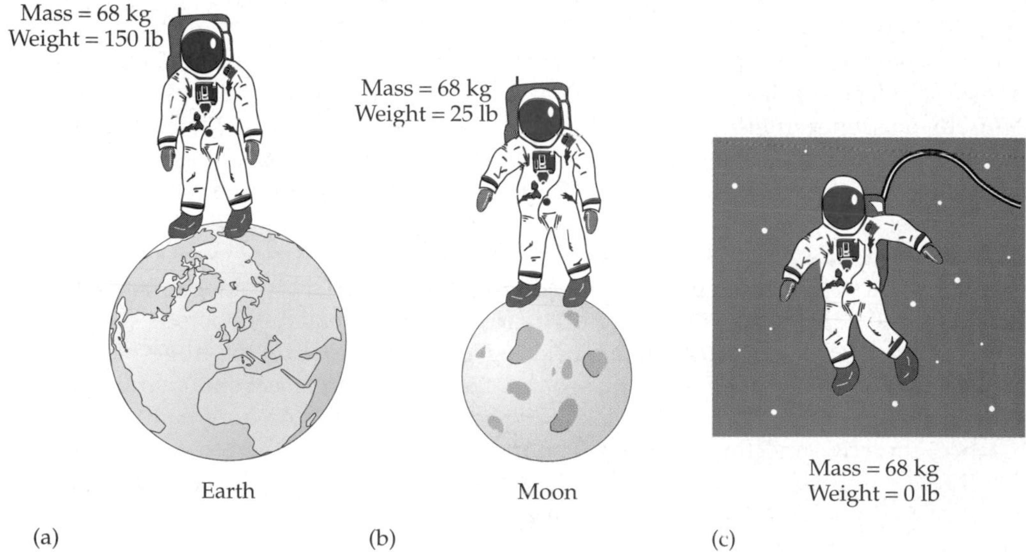

▲ **Figure 2.4 Mass versus Weight** Weight is affected by gravity, whereas mass is not. (a) On the Earth, the astronaut has a mass of 68 kilograms (kg) and a weight of 150 pounds (lb). (b) On the Moon, the mass remains 68 kg, but the weight is only 25 lb. (c) In space, the mass is still 68 kg although the astronaut is weightless.

Instruments for Measuring Volume

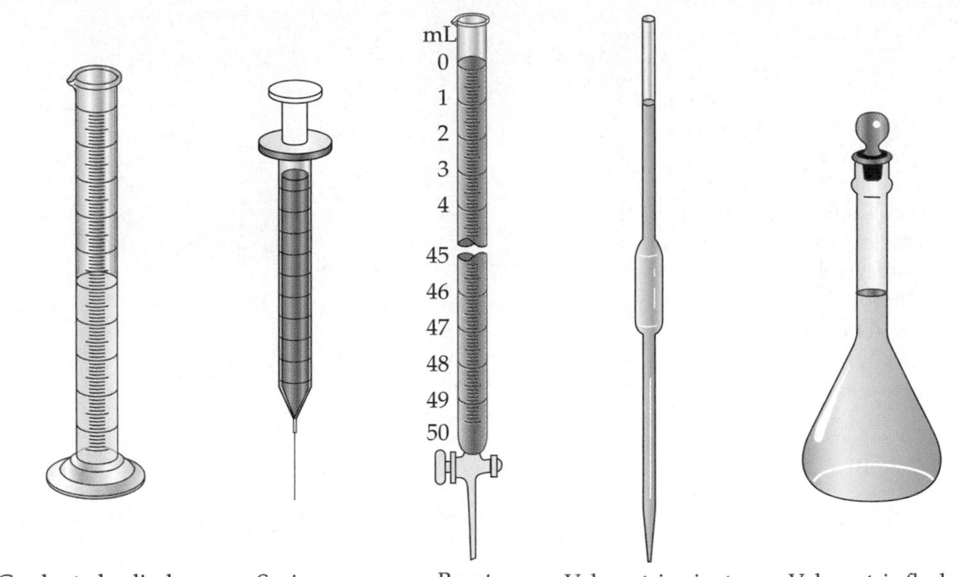

◀ **Figure 2.5 Instruments for Measuring Volume** A graduated cylinder, a syringe, and a buret are calibrated to measure a variable quantity of liquid, whereas a volumetric pipet and a volumetric flask measure only fixed quantities, for example, 10 mL and 250 mL.

liquid. Three of the most common are a graduated cylinder, a pipet, and a buret. Figure 2.5 shows common laboratory equipment for measuring volume.

A graduated cylinder is routinely used to measure a volume of liquid. The most common sizes of graduated cylinders are 10 mL, 50 mL, and 100 mL. The uncertainty of a graduated cylinder measurement varies but usually ranges from one-tenth to one-half of a milliliter (±0.1 mL to ±0.5 mL).

There are many types of pipets. The volumetric pipet shown in Figure 2.5 is used to deliver a fixed volume of liquid. The liquid is drawn up until it reaches a calibration line etched on the pipet. The tip of the pipet is then placed in a container, and the liquid is allowed to drain from the pipet. The volume delivered varies, but the uncertainty usually ranges from 1/10 to 1/100 of a milliliter. For instance, a 10-mL pipet can deliver 10.0 mL (±0.1 mL) or 10.00 mL (±0.01 mL), depending on the uncertainty of the instrument.

A buret is a long, narrow piece of calibrated glass tubing with a valve called a stopcock at the bottom end. The flow of liquid is regulated by opening and closing the stopcock, and the initial and final liquid levels in the buret are observed and recorded. The volume delivered is found by subtracting the initial buret reading from the final buret reading. Burets usually have uncertainties ranging from 1/10 to 1/100 of a milliliter. For instance, the liquid level in a buret can read 22.5 mL (±0.1 mL) or 22.55 mL (±0.01 mL), depending on the uncertainty of the instrument.

2.2 Significant Digits

Objective · To identify the number of significant digits in a given measurement.

In a properly recorded measurement, each number is a **significant digit**, also referred to as a *significant figure*. For instance, suppose we weigh a 5¢ coin on a platform, a beam, and an electronic balance. We may find that the mass of the coin on

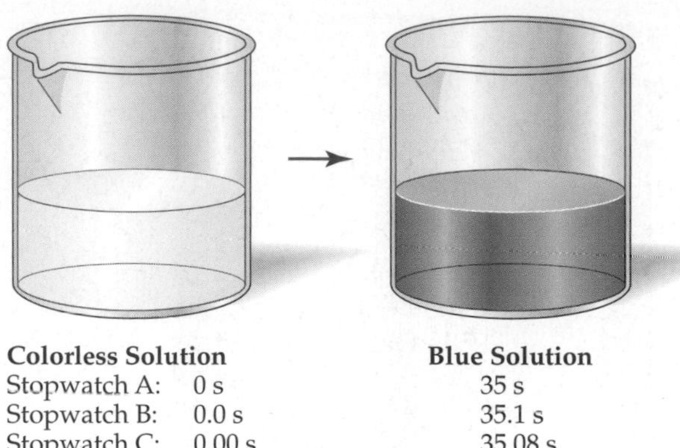

▶ **Figure 2.6 Significant Digits and a Timed Reaction** The data from the timed reaction demonstrates the uncertainty of three different stopwatches. Although each of the measurements is correct, stopwatch C has the least uncertainty.

the different balances is 5.0 g, 5.00 g, and 5.000 g, respectively. Although the uncertainty of the mass varies for the three balances, every digit is significant in each measurement. Removing the last digit from any weighing changes the uncertainty of the measurement. In this example, the measurements of mass have two, three, and four significant digits, respectively.

In every measurement, the significant digits express the uncertainty of the instrument. By way of example, let's examine the chemical reaction shown in Figure 2.6. This is called a "clock reaction," and we note that the solution changes from colorless to blue after about 35 seconds (s). To time the reaction we'll use three different stopwatches. Since a stopwatch can be calibrated in seconds (±1 s), tenths of a second (±0.1 s), or hundredths of a second (±0.01 s), we can use stopwatches having different uncertainties to time the reaction.

Stopwatch A displays 35 s, stopwatch B displays 35.1 s, and stopwatch C displays 35.08 s. Therefore, stopwatch A has more uncertainty than B, and stopwatch B has more uncertainty than C.

To determine the number of significant digits in each measurement, we simply count the number of digits. That is, 35 s has two significant digits, 35.1 s has three significant digits, and 35.08 s has four significant digits. Example Exercise 2.2 further illustrates how to determine the number of significant digits in a measurement.

Example Exercise 2.2 • Significant Digits

State the number of significant digits in the following measurements.

(a) 12,345 cm (b) 0.123 g (c) 0.5 mL (d) 102.0 s

Solution

In each example, we simply count the number of digits. Thus,

(a) 5 (b) 3 (c) 1 (d) 4

Notice that the leading zero in (b) and (c) is not part of the measurement but is inserted to call attention to the decimal point that follows.

Self-Test Exercise

State the number of significant digits in the following measurements.

(a) 2005 cm (b) 25.000 g (c) 25.0 mL (d) 0.25 s

Answers: (a) 4; (b) 5; (c) 3; (d) 2

Significant Digits and Place-Holder Zeros

A measurement may contain place-holder zeros to properly locate the decimal point; for example, 500 cm and 0.005 cm. If the number is less than 1, a place-holder zero is never significant. Thus, 0.5 cm, 0.05 cm, and 0.005 cm each contain only one significant digit.

If the number is greater than 1, a place-holder zero is usually not significant. To avoid confusion, we will assume that place-holder zeros are never significant. Thus, 50 cm, 500 cm, and 5000 cm each contain only one significant digit. Example Exercise 2.3 further illustrates how to determine the number of significant digits.

A demonstration calculator with scientific notation is available for the overhead projector (see the *Instructor's Resource Manual*).

Directions for using a scientific calculator are in Appendix A.

Example Exercise 2.3 • Significant Digits

State the number of significant digits in the following measurements.

(a) 0.025 cm (b) 0.2050 g (c) 25.0 mL (d) 2500 s

Solution

In each example, we count the number of significant digits and disregard place-holder zeros. Thus,

(a) 2 (b) 4 (c) 3 (d) 2

Self-Test Exercise

State the number of significant digits in the following measurements.

(a) 0.050 cm (b) 0.0250 g (c) 50.00 mL (d) 1000 s

Answers: (a) 2; (b) 3; (c) 4; (d) 1

Note If in a special case a place-holder zero is significant, we can express the number using a power of 10. (The power-of-10 concept is reviewed in Section 2.6.) For example, if one zero is significant in 100 cm, we can express the measurement as 1.0×10^2 cm. If both zeros are significant, we can write 1.00×10^2 cm. If neither zero is significant, we can write 1×10^2 cm. A power of 10 has no effect on the number of significant digits; thus, 1.1×10^5 cm has two significant digits, and 1.11×10^{-5} cm has three significant digits.

Significant Digits and Exact Numbers

Since all measurements have uncertainty, they never express exact numbers. This is not true, however, when we simply count items. For instance, a chemistry laboratory may have 30 rulers, 3 balances, and 24 pipets. Since we have simply counted items, 30, 3, and 24 are exact numbers. The rules of significant digits do not apply to exact numbers; they apply only to measurements.

We can summarize the directions for determining the number of significant digits with two simple rules.

Determining the Number of Significant Digits

Rule 1: Count the number of digits in a measurement from left to right:
- (a) Start with the first nonzero digit.
- (b) Do not count place-holder zeros (0.011, 0.00011, and 11,000 each have two significant digits).

Rule 2: The rules for significant digits apply only to measurements and not to exact numbers.

▲ **Scientific Calculator** A calculator often shows nonsignificant digits in the display, which must be rounded off.

2.3 Rounding Off Nonsignificant Digits

Objective · To round off a given value to a stated number of significant digits.

All digits in a correctly recorded measurement are significant. However, we often generate nonsignificant digits when using a calculator. These **nonsignificant digits** should not be reported, but they frequently appear in the calculator display. Since nonsignificant digits are not justified, we must eliminate them. We get rid of nonsignificant digits through a process of **rounding off**. We round off nonsignificant digits by following three simple rules.

Rounding Off Nonsignificant Digits

Rule 1: If the first nonsignificant digit is less than 5, drop all nonsignificant digits.

Rule 2: If the first nonsignificant digit is 5, or greater than 5, increase the last significant digit by 1 and drop all nonsignificant digits.[1]

Rule 3: If a calculation has two or more operations, retain the nonsignificant digits until the final operation. Not only is it much more convenient, but it is also more accurate to round off the final answer.

If a calculator displays 12.846239 and three significant digits are justified, we must round off. Since the first nonsignificant digit is 4 in 12.8**4**6239, we follow rule 1, drop the nonsignificant digits, and round to 12.8. If a calculator displays 12.856239 and three significant digits are justified, we follow rule 2. In this case, since the first nonsignificant digit is 5 in 12.8**5**6239, we round to 12.9.

Rounding Off and Place-Holder Zeros

On occasion, rounding off can create a problem. For example, if we round off 151 to two significant digits, we obtain 15. Since 15 is only a fraction of the original value, we must insert a place-holder zero; thus, rounding off 151 to two significant digits gives 150. Similarly, rounding off 1514 to two significant digits gives 1500. Example Exercise 2.4 further illustrates how to round off numbers.

Example Exercise 2.4 • Rounding Off

Round off the following numbers to three significant digits.

(a) 22.250 (b) 0.34548
(c) 0.072038 (d) 12267

Solution

To locate the first nonsignificant digit, count three digits from left to right. If the first nonsignificant digit is less than 5, drop all nonsignificant digits. If the first nonsignificant digit is 5 or greater, add 1 to the last significant digit.

(a) 22.3 (rule 2) (b) 0.345 (rule 1)
(c) 0.0720 (rule 1) (d) 12,300 (rule 2)

In (d), notice that two place-holder zeros must be added to 123 to obtain the correct decimal place.

Self-Test Exercise

Round off the following numbers to three significant digits.

(a) 12.514748 (b) 0.6015261
(c) 192.49032 (d) 14652.832

Answers: (a) 12.5 (rule 1); (b) 0.602 (rule 2); (c) 192 (rule 1); (d) 14,700 (rule 2)

[1] If the nonsignificant digit is 5, or 5 followed by zeros, an odd-even rule can be applied. That is, if the last significant digit is odd, round up; if it is even, drop the nonsignificant digits.

2.4 Adding and Subtracting Measurements

Objective · To add and subtract measurements and round off the answer to the proper number of significant digits.

When adding or subtracting measurements, *the answer is limited by the value with the most uncertainty*. Let's add the following measurements of mass.

$$\begin{array}{rl} 5 & \text{g} \\ 5.0 & \text{g} \\ +\ 5.00 & \text{g} \\ \hline 15.00 & \text{g} \end{array}$$

The mass of 5 g has the most uncertainty because it measures only ±1 g. Thus, the sum should be limited to the nearest gram. When we round off the answer to the proper significant digit, the correct answer is **15 g**. In addition and subtraction, the unit (cm, g, mL) in the answer is the same as the unit in each piece of data. Example Exercise 2.5 illustrates the addition and subtraction of measurements.

Example Exercise 2.5 • Addition/Subtraction and Rounding Off

Add or subtract the following measurements and round off your answer.

(a) 106.7 g + 0.25 g + 0.195 g (b) 35.45 mL − 30.5 mL

Solution

In addition or subtraction, the answer is limited by the measurement with the most uncertainty.

(a) Let's align the decimal places and perform the addition.

$$\begin{array}{rl} 106.7 & \text{g} \\ 0.25 & \text{g} \\ +\ 0.195 & \text{g} \\ \hline 107.145 & \text{g} \end{array}$$

Since 106.7 g has the most uncertainty (±0.1 g), the answer rounds off to one decimal place. The correct answer is **107.1 g** and is read "one hundred and seven point one grams."

(b) Let's align the decimal places and perform the subtraction.

$$\begin{array}{rl} 35.45 & \text{mL} \\ -\ 30.5 & \text{mL} \\ \hline 4.95 & \text{mL} \end{array}$$

Since 30.5 mL has the most uncertainty (±0.1 mL), the answer rounds off to one decimal place. The correct answer is **5.0 mL** and is read "five point zero milliliters."

Self-Test Exercise

Add or subtract the following measurements and round off your answer.

(a) 8.6 cm + 50.05 cm (b) 34.1 s − 0.55 s

Answers: (a) 58.7 cm; (b) 33.6 s

Note When adding or subtracting measurements, the answer is always limited by the decimal place with the most uncertainty. In other words, the measurement that is least certain in a set of data always limits the answer.

2.5 Multiplying and Dividing Measurements

Objective · To multiply and divide measurements and round off the answer to the proper significant digits.

Significant digits are treated differently in multiplication and division than in addition and subtraction. In multiplication or division, *the answer is limited by the measurement with the least number of significant digits.* Let's multiply the following length measurements.

$$(5.15 \text{ cm})(2.3 \text{ cm}) = 11.845 \text{ cm}^2$$

Significant Digits Activity

The measurement of 5.15 cm has three significant digits, and 2.3 cm has two. Thus, the product should be limited to two digits. When we round off to the proper numbers of significant digits, the correct answer is **12 cm²**. Notice that the units must also be multiplied together, which we have indicated by the superscript 2. Example Exercise 2.6 illustrates the multiplication and division of measurements.

Example Exercise 2.6 • Multiplication/Division and Rounding Off

Multiply or divide the following measurements and round off your answer.

(a) 50.5 cm × 12 cm (b) 103.37 g/20.5 mL

Solution

In multiplication or division, the answer is limited by the measurement with the least number of significant digits.

(a) In this example, 50.5 cm has three significant digits and 12 cm has two.

$$(50.5 \text{ cm})(12 \text{ cm}) = 606 \text{ cm}^2$$

The answer is limited to two significant digits and rounds off to **610 cm²** after inserting a place-holder zero. The answer is read "six hundred ten centimeters squared" or "six hundred ten square centimeters."

(b) In this example, 103.37 g has five significant digits and 20.5 mL has three.

$$\frac{103.37 \text{ g}}{20.5 \text{ mL}} = 5.0424 \text{ g/mL}$$

The answer is limited to three significant digits and rounds off to **5.04 g/mL**. Notice that the unit is a ratio; the answer is read "five point zero four grams per milliliter."

Self-Test Exercise

Multiply or divide the following measurements and round off your answer.

(a) (359 cm)(0.20 cm) (b) 73.950 g/25.5 mL

Answers: (a) 72 cm²; (b) 2.90 g/mL

Note When multiplying or dividing measurements, the answer is always limited by the least number of significant digits. In contrast, recall that when adding or subtracting measurements, the answer is always limited by the decimal place with the most uncertainty.

2.6 Exponential Numbers

Objectives · To explain the concept of exponents and specifically powers of 10.
· To express a value as a power of 10 and as an ordinary number.

When a value is multiplied times itself, the process is indicated by a number written as a superscript. The superscript indicates the number of times the process is repeated. For example, if the number 2 is multiplied two times, the product is expressed as 2^2. Thus, $(2)(2) = 2^2$. If the number 2 is multiplied three times, the product is expressed as 2^3. Thus, $(2)(2)(2) = 2^3$.

A superscript number indicating that a value is multiplied times itself is called an **exponent**. If 2 has the exponent 2, the value 2^2 is read as "2 to the second power" or "2 squared." The value 2^3 is read as "2 to the third power" or "2 cubed."

Powers of 10

A **power of 10** is a number that results when 10 is raised to an exponential power. You know that an exponent raises any number to a higher power, but we are most interested in the base number 10. A power of 10 has the general form

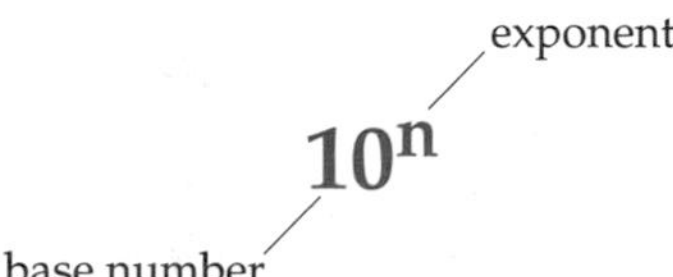

The number 10 raised to the n power is equal to 10 multiplied times itself n times. For instance, 10 to the second power (10^2) is equal to 10 times 10. When we write 10^2 as an ordinary number, we have 100. Notice that the exponent 2 corresponds to the number of zeros in 100. Similarly, 10^3 has three zeros (1000) and 10^6 has six zeros (1,000,000).

The exponent is positive for all numbers greater than 1. Conversely, the exponent is negative for numbers less than 1. For example, 10 to the negative first power (10^{-1}) is equal to 0.1, 10 to the negative second power (10^{-2}) is equal to 0.01, and 10 to the negative third power (10^{-3}) is equal to 0.001. Table 2.1 lists some powers of 10 along with the equivalent ordinary number.

Although you can easily carry out operations with exponents using an inexpensive scientific calculator (Appendix A), you will have greater confidence if you understand exponents. Example Exercises 2.7 and 2.8 further illustrate the relationship between ordinary numbers and exponential numbers.

Table 2.1 Powers of 10

Exponential Number	Ordinary Number
$1 \times 10^6 = 10 \times 10 \times 10 \times 10 \times 10 \times 10$	1,000,000
$1 \times 10^3 = 10 \times 10 \times 10$	1,000
$1 \times 10^2 = 10 \times 10$	100
$1 \times 10^1 = 10$	10
$1 \times 10^0 = 1$	1
$1 \times 10^{-1} = \frac{1}{10}$	0.1
$1 \times 10^{-2} = \frac{1}{10} \times \frac{1}{10}$	0.01
$1 \times 10^{-3} = \frac{1}{10} \times \frac{1}{10} \times \frac{1}{10}$	0.001
$1 \times 10^{-6} = \frac{1}{10} \times \frac{1}{10} \times \frac{1}{10} \times \frac{1}{10} \times \frac{1}{10} \times \frac{1}{10}$	0.000 001

▲ **The Earth** A positive power of 10 can be used to express very large numbers; for example, the diameter of the Earth is about 1×10^7 meters.

 The Earth; Red Blood Cells

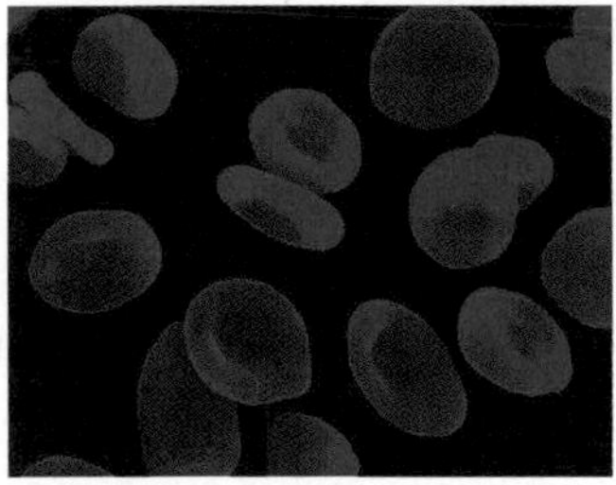

▲ **Red Blood Cells** A negative power of 10 can be used to express very small numbers; for example, the diameter of a red blood cell is about 1×10^{-6} meter.

Example Exercise 2.7 • Converting to Powers of 10

Express each of the following ordinary numbers as a power of 10.

(a) 100,000 (b) 0.000 000 01

Solution

The power of 10 indicates the number of places the decimal point has been moved.

(a) We must move the decimal five places to the left; thus, 1×10^5.
(b) We must move the decimal eight places to the right; thus, 1×10^{-8}.

Self-Test Exercise

Express each of the following ordinary numbers as a power of 10.

(a) 10,000,000 (b) 0.000 000 000 001

Answers: (a) 1×10^7; (b) 1×10^{-12}

Example Exercise 2.8 • Converting to Ordinary Numbers

Express each of the following powers of 10 as an ordinary number.

(a) 1×10^4 (b) 1×10^{-9}

Solution

The power of 10 indicates the number of places the decimal point has been moved.

(a) The exponent in 1×10^4 is positive 4, and so we must move the decimal point four places to the right of 1; thus, 10,000.
(b) The exponent in 1×10^{-9} is negative 9, and so we must move the decimal point nine places to the left of 1; thus, 0.000 000 001.

Self-Test Exercise

Express each of the following powers of 10 as an ordinary number.

(a) 1×10^{10} (b) 1×10^{-5}

Answers: (a) 10,000,000,000; (b) 0.000 01

2.7 Scientific Notation

Objective · To express any number in scientific notation.

Science often deals with numbers that are very large or very small. These numbers may be awkward and incomprehensible because they contain many zeros. The mass of an iron atom, for example, is 0.000 000 000 000 000 000 000 093 g. To overcome this problem, a standard notation has been devised that places the decimal after the first significant digit and sets the size of the number using a power of 10; this method is called **scientific notation**. The scientific notation format is

$$\mathbf{D.DD \times 10^n}$$

(power of 10 → 10^n; significant digits → D.DD)

To use scientific notation, write down all the significant digits in the number. Then, move the decimal point to follow the first nonzero digit. Indicate the number of places the decimal is moved using power-of-10 notation. For instance, you can write 555,000 in scientific notation by first moving the decimal five places to the left to give 5.55 and then adding the appropriate power of 10, which is 10^5. Thus, you can express 555,000 in scientific notation as 5.55×10^5.

You can also express numbers smaller than 1 in scientific notation. For example, you can write 0.000 888 in scientific notation by first moving the decimal four places to the right to give 8.88 and then adding the appropriate power of 10, which is 10^{-4}. Thus, 0.000 888 is expressed in scientific notation as 8.88×10^{-4}.

Regardless of the size of the number, in scientific notation the decimal point is always placed after the first nonzero digit. The size of the number is indicated by a power of 10. A positive exponent indicates a very large number, whereas a negative exponent indicates a very small number. To express a number in scientific notation, follow these two steps.

Applying Scientific Notation

Step 1: Place the decimal point after the first nonzero digit in the number, followed by the remaining significant digits.

Step 2: Indicate how many places the decimal is moved by the power of 10. When the decimal is moved to the left, the power of 10 is positive. When the decimal is moved to the right, the power of 10 is negative.

Example Exercise 2.9 illustrates the conversion between ordinary numbers and exponential numbers.

Example Exercise 2.9 • Scientific Notation

Express each of the following values in scientific notation.

(a) There are 26,800,000,000,000,000,000,000 helium atoms in 1.00 liter (L) of helium gas under standard conditions.

(b) The mass of one helium atom is 0.000 000 000 000 000 000 000 006 65 g.

Solution

We can write each value in scientific notation as follows.

(a) Place the decimal after the 2, followed by the other significant digits (2.68). Next, count the number of places the decimal has moved. The decimal has moved 22 places to the left, and so the exponent is +22. Finally, we have the number of helium atoms in 1.00 L of gas: 2.68×10^{22}.

(b) Place the decimal after the 6, followed by the other significant digits (6.65). Next, count the number of places the decimal has shifted. The decimal has shifted 24 places to the right, and so the exponent is −24. Finally, we have the mass of a helium atom: 6.65×10^{-24} g.

Self-Test Exercise

Express each of the following values as ordinary numbers.

(a) The mass of one mercury atom is 3.33×10^{-22} g.

(b) The number of atoms in 1 mL of liquid mercury is 4.08×10^{22}.

Answers: (a) 0.000 000 000 000 000 000 000 333 g; (b) 40,800,000,000,000,000,000,000 atoms

Calculators and Scientific Notation

A scientific calculator is very helpful in performing the calculations that we will soon encounter. If you need practice or instructions for using your calculator, refer to Appendix A, which explains the operation of a calculator. In addition to general directions, there are specific examples of multiplying and dividing numbers expressed in scientific notation.

▲ **Computer Chip** A personal computer contains an electronic chip that is about the size of a red ant.

Chemistry Connection • A Student Success Story

What common metal was more valuable than gold in the nineteenth century?

In 1885 Charles Martin Hall was a 22-year-old student at Oberlin College in Ohio. One day his chemistry teacher told the class that anyone who could discover an inexpensive way to produce aluminum metal would become rich and benefit humanity. At the time, aluminum was a rare and expensive metal. In fact, Napoleon III, a nephew of Napoleon Bonaparte, entertained his most honored guests with utensils made from aluminum while other guests dined with utensils of silver and gold. Although aluminum is the most abundant metal in the Earth's crust, it is not found free in nature; it is usually found combined with oxygen in various minerals.

After graduation, Charles Hall set up a laboratory in a woodshed behind his father's church. Using homemade batteries, he devised a simple method for producing aluminum by passing electricity through a molten mixture of minerals. After only 8 months of experimenting, he invented a successful method for reducing an aluminum mineral to aluminum metal. In February 1886 Charles Hall walked into his former teacher's office with a handful of metallic aluminum globules.

Just as his chemistry teacher had predicted, within a short period of time, Hall became rich and famous. In 1911 he received the Perkin Medal for achievement in chemistry, and in his will, he donated $5 million to Oberlin College. He also helped to establish the Aluminum Company of America (ALCOA), and the process for making aluminum metal gave rise to a huge industry. Aluminum is now second only to steel as a construction metal.

It is an interesting coincidence that the French chemist Paul Héroult, without knowledge of Hall's work, made a similar discovery at the same time. Thus, the industrial method for obtaining aluminum metal is referred to as the Hall–Héroult process. In 1886, following the discovery of this process, the price of aluminum plummeted from over $100,000 a pound. Today, the price of aluminum is less than $1 a pound.

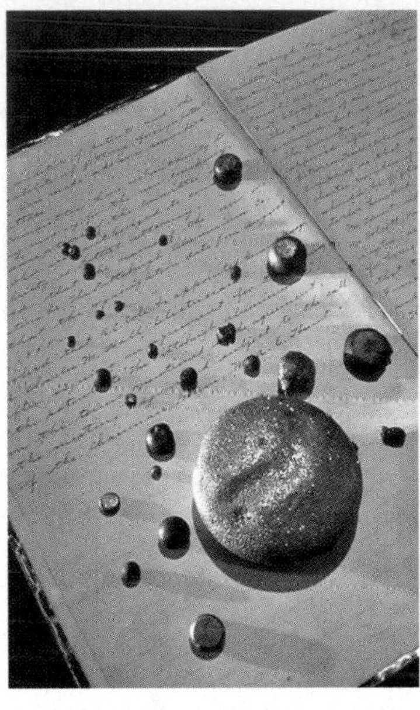

▲ The notebook of Charles Hall along with globules of aluminum.

Before 1885, aluminum was an extremely rare and expensive metal.

2.8 Unit Equations and Unit Factors

Objectives
- To write a unit equation for an equivalent relationship.
- To write two unit conversion factors for a unit equation.

In the next section we will tackle our single most important task: problem solving. We will learn a simple but powerful method called unit analysis, or dimensional analysis. To use the unit analysis method of problem solving, we must understand equivalent relationships. An equivalent relationship provides the unit conversion factors that will enable us to solve problems.

What is an equivalent relationship? An equivalent relationship is a relationship between two quantities that are equal. For instance, 1 dollar equals 10 dimes, and 1 dime equals 10 pennies. Written as an equation, we have

$$\text{dollar} = 10 \text{ dimes}$$
$$1 \text{ dime} = 10 \text{ pennies}$$

These equivalent relationships are called unit equations. A **unit equation** is a simple statement of two equivalent quantities.

◀ **One Hundred Pennies** Since a dollar is equivalent to 100 pennies, we can write the unit equation 1 dollar = 100 pennies. The two associated unit factors are 1 dollar/100 pennies and 100 pennies/1 dollar.

A dime is 1/10 of a dollar, and we can write the unit equation 1 dime = 0.10 dollar. However, this unit equation is not recommended because it is easy to make an error using decimal fractions. Instead, we use whole-number relationships, which are easier to grasp. In this example, we can write the unit equation 1 dollar = 10 dimes.

A unit conversion factor, or **unit factor**, is a ratio of two equivalent quantities, for example, 1 dollar/10 dimes. Since the numerator and denominator are equal, the **reciprocal** ratio, 10 dimes/1 dollar, is an equivalent unit factor. For the unit equation 1 dime = 10 pennies, we can write a unit factor and its reciprocal as follows.

$$\frac{1 \text{ dime}}{10 \text{ pennies}} \quad \text{and} \quad \frac{10 \text{ pennies}}{1 \text{ dime}}$$

Let's write unit factors for the relationship between inches and yards. We know that there are 36 inches in a yard, and so we can write the unit equation 1 yard = 36 inches. The associated unit factor and its reciprocal are

$$\frac{1 \text{ yard}}{36 \text{ inches}} \quad \text{and} \quad \frac{36 \text{ inches}}{1 \text{ yard}}$$

Example Exercise 2.10 further illustrates how to write unit equations and unit factors.

Example Exercise 2.10 • Unit Conversion Factors

Write the unit equation and the two corresponding unit factors for each of the following.

(a) pounds and ounces (b) quarts and gallons

Solution

We first write the unit equation and then the corresponding unit factors.

(a) There are 16 ounces in a pound, and so the unit equation is 1 pound = 16 ounces. The two associated unit factors are

$$\frac{1 \text{ pound}}{16 \text{ ounces}} \quad \text{and} \quad \frac{16 \text{ ounces}}{1 \text{ pound}}$$

(b) There are 4 quarts in a gallon, and so the unit equation is 1 gallon = 4 quarts. The two unit factors are

$$\frac{1 \text{ gallon}}{4 \text{ quarts}} \quad \text{and} \quad \frac{4 \text{ quarts}}{1 \text{ gallon}}$$

Self-Test Exercise

Write the unit equation and the two corresponding unit factors for each of the following.

(a) hours and days (b) hours and minutes

Answers: (a) 1 day = 24 hours; 1 day/24 hours and 24 hours/1 day; (b) 1 hour = 60 minutes; 1 hour/60 minutes and 60 minutes/1 hour

Exactly Equivalent Relationships

By definition, 1 yard is exactly equal to 36 inches. Exact relationships can be shown using a special sign (≡). Thus, we can write 1 yard ≡ 36 inches, which is said to be an exactly equivalent relationship, or simply an **exact equivalent** (symbol ≡).

The rules for significant digits do not apply to unit factors derived from exact equivalents. Therefore, 1 yard/36 inches and 36 inches/1 yard have an infinite number of significant digits. Other examples of exactly equivalent relationships include 1 foot ≡ 12 inches, 1 pound ≡ 16 ounces, and 1 gallon ≡ 4 quarts. Examples of exact metric equivalents include 1 meter ≡ 100 centimeters, 1 kilogram ≡ 1000 grams, and 1 liter ≡ 1000 milliliters.

2.9 Unit Analysis Problem Solving

Objectives
- To state the three steps in the unit analysis method.
- To apply the unit analysis method of problem solving.

Science has the reputation for tackling difficult problems. Fortunately, you are about to learn a powerful technique for solving problems. It is called the **unit analysis method**, or sometimes the *dimensional analysis method* of problem solving. Unit analysis cannot solve every problem, but it is very effective for the problems we encounter in introductory chemistry.

Solving problems by unit analysis is a simple three-step process. Step 1: Read the problem and determine the units required in the answer. Step 2: Analyze the problem and determine which given value is related to the answer. Step 3: Write one or more unit factors to convert the given value to the units in the answer. You should follow these three steps consistently and systematically when using the unit analysis method of problem solving.

Applying the Unit Analysis Method

Step 1: Write down the units asked for in the answer.

Step 2: Write down the value given in the problem that is related to the units required in the answer.

Step 3: Apply a unit factor to convert the units in the given value to the units in the answer. Each time we apply unit analysis, we use the following format:

$$\underset{(2)}{\text{given value}} \times \underset{(3)}{\frac{\text{unit}}{\text{factors(s)}}} = \underset{(1)}{\text{units asked for}}$$

Suppose we wish to find the number of days in 2.5 years. *First*, we read the problem and determine the unit asked for (days). *Second*, we analyze the given information and select the value related to days (2.5 years). *Third*, we apply a unit factor using the following format.

$$\underset{(2)}{2.5\text{ years}} \times \underset{(3)}{\frac{\text{unit}}{\text{factor}}} = \underset{(1)}{\text{days}}$$

The unit equation that relates steps (1) and (2) is 1 year = 365 days. Therefore, the two associated unit factors are

$$\frac{1\text{ year}}{365\text{ days}} \quad \text{and} \quad \frac{365\text{ days}}{1\text{ year}}$$

We choose the unit factor whose denominator has the same unit as the unit of the given value (years). After canceling (~~years~~), we have the unit asked for in this problem (days). Thus, we choose the unit factor 365 days/1 year.

$$2.5 \cancel{\text{years}} \times \frac{365 \text{ days}}{1 \cancel{\text{year}}} = 912.5 \text{ days}$$

After rounding off to two significant digits, we find that the answer is 910 days. Example Exercises 2.11 and 2.12 provide practical examples of the unit analysis method of problem solving.

Example Exercise 2.11 • Unit Analysis Problem Solving

A can of Coca-Cola contains 12 fluid ounces (fl oz). What is the volume in quarts? (Given: 1 qt ≡ 32 fl oz.)

Solution

Let's show the format for the unit analysis method of problem solving.

given value → $\frac{\text{unit}}{\text{factor}}$ → unit asked for

Step 1, The unit asked for in the answer is **qt**.

Step 2, The relevant given value is **12 fl oz**.

Applying the standard format for solving problems by unit analysis, we have

$$\underset{(2)}{12 \text{ fl oz}} \times \underset{(3)}{\frac{\text{unit}}{\text{factor}}} = \underset{(1)}{\text{qt}}$$

Step 3, We apply a unit conversion factor. Since the unit equation is 1 qt ≡ 32 fl oz, the two possible unit factors are

$$\frac{1 \text{ qt}}{32 \text{ fl oz}} \quad \text{and} \quad \frac{32 \text{ fl oz}}{1 \text{ qt}}$$

To apply the proper unit factor, we choose 1 qt/32 fl oz because the unit we wish to cancel (fl oz) appears in the denominator. Thus,

$$12 \cancel{\text{fl oz}} \times \frac{1 \text{ qt}}{32 \cancel{\text{fl oz}}} = 0.38 \text{ qt}$$

The given value, 12 fl oz, limits the answer to two significant digits. The unit factor 1 qt/32 fl oz is an exact equivalent and therefore does not affect the number of significant digits. We can summarize the solution as follows.

12 fl oz → $\frac{1 \text{ qt}}{32 \text{ fl oz}}$ → 0.38 qt

Self-Test Exercise

A can of Pepsi-Cola contains 355 mL. What is the volume in liters? (Given: 1 L ≡ 1000 mL.)

Answer: 0.355 L

▲ **Coca-Cola** A 12-fl-oz can of Coke contains 355 mL.

▲ **Pepsi-Cola** A 12-fl-oz can of Pepsi contains 355 mL.

Example Exercise 2.12 • Unit Analysis Problem Solving

A marathon covers a distance of 26.2 miles (mi). If 1 mi is equivalent to exactly 1760 yd, what is the distance of the race in yards?

Solution

Let's show the format for the unit analysis method of problem solving.

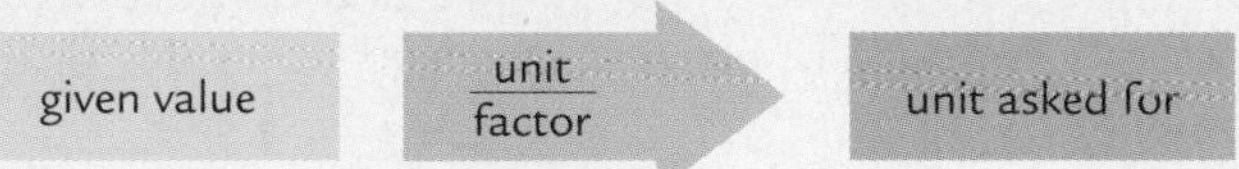

Step 1, The unit asked for in the answer is **yd**.

Step 2, The relevant given value is **26.2 mi**.

Applying the standard format for unit analysis, we have

$$26.2 \text{ mi} \times \frac{\text{unit}}{\text{factor}} = \text{yd}$$

(2) (3) (1)

Step 3, We apply a unit conversion factor.

From the given information, we can write the unit equation: 1 mile ≡ 1760 yards. It follows that the two unit factors are

$$\frac{1 \text{ mi}}{1760 \text{ yd}} \quad \text{and} \quad \frac{1760 \text{ yd}}{1 \text{ mi}}$$

We will choose 1760 yards/1 mile, because the unit we wish to cancel (~~miles~~) appears in the denominator.

$$26.2 \cancel{\text{mi}} \times \frac{1760 \text{ yd}}{1 \cancel{\text{mi}}} = 46{,}100 \text{ yd}$$

Since the given value, 26.2 mi, has three significant digits, we round off the answer to three digits. We can summarize the solution as follows.

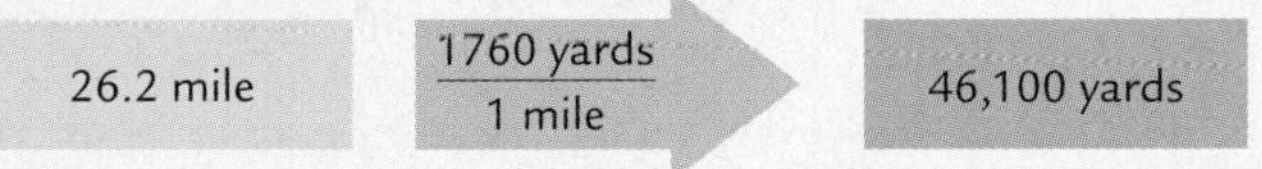

Self-Test Exercise

Given that a marathon is 26.2 mi, what is the distance in kilometers? (Given: 1 km = 0.62 mi.)

Answer: 42 km

Note Example Exercise 2.12 provides an interesting lesson on significant digits. Since 1 mile is exactly equivalent to 1760 yards, we can write 1 mile ≡ 1760 yards. The corresponding unit factor, 1760 yards/1 mile, has an infinite number of significant digits and has no effect on the number of significant digits in the answer.

In Self-Test Exercise 2.12 we are given that 1 kilometer = 0.62 mile. This is not an exact equivalent because it relates a metric unit (kilometer) to an English unit (mile). Since the metric and English measuring systems have different reference standards, the relationship between metric and English units is always *approximate*.

2.10 The Percent Concept

Objectives
- To explain the concept of percent.
- To apply percent as a unit factor.

Percent expresses the amount of a single quantity compared to an entire sample. The ratio of one quantity to a total sample, all multiplied by 100, equals a **percent** (symbol %).[2] For example, a dime is 10% of a dollar and a quarter is 25% of a dollar.

A percent can be referred to as parts per 100 parts. To calculate a percent, divide one quantity by the total sample and multiply that ratio by 100.

$$\frac{\text{one quantity}}{\text{total sample}} \times 100 = \%$$

As an example, consider a student who has a large coin collection. If the collection contains 1195 pennies, 403 nickels, 215 dimes, and 187 quarters, what percent of the collection is pennies? We can calculate the percent of pennies in the coin collection as follows.

$$\frac{1195 \text{ pennies}}{(1195 + 403 + 215 + 187) \text{ coins}} \times 100 = 59.75\%$$

Example Exercise 2.13 • The Percent Concept

Sterling silver contains silver and copper metals. If a sterling silver chain contains 18.5 g of silver and 1.5 g of copper, what is the percent of silver?

Solution

We compare the mass of silver to the total mass of the sterling silver chain. Thus,

$$\frac{18.5 \text{ g}}{18.5 \text{ g} + 1.5 \text{ g}} \times 100 = 92.5\%$$

Genuine sterling silver is cast from 92.5% silver and 7.5% copper. In fact, if you carefully examine a piece of sterling silver, you may see the notation .925, which indicates that the item is genuine sterling silver.

Self-Test Exercise

A 14-karat gold ring contains 7.45 g of gold, 2.66 g of silver, and 2.66 g of copper. Calculate the percent of gold in the ring.

Answer: 58.3%

▲ **Silver** A bar of pure silver and coins containing the metal.

Percent Unit Factors

A percent can be expressed as parts per 100 parts. For instance, 10% and 25% correspond to 10 and 25 parts per 100 parts, respectively. Thus, we can write a percent as a ratio fraction; that is, we can write 10% as 10/100, and 25% as 25/100.

Let's try a type of problem in which the percent is given. *Apollo 11* traveled 394,000 km from the Earth to land the first astronaut on the Moon. How many

[2] The percent symbol (%) is derived from combining the ratio sign (/) and the two zeros in 100.

kilometers did the astronauts travel before they had gone 15.5% of the distance? By definition, 15.5% means 15.5 parts per 100 parts. In this example, 15.5% is 15.5 km per 100 km. Thus, we can write the ratio as a unit factor and apply the unit analysis format.

$$394{,}000\ \cancel{\text{km}} \times \frac{15.5\ \text{km}}{100\ \cancel{\text{km}}} = 61{,}100\ \text{km}$$

In this problem, notice that we used percent as a unit factor. Example Exercise 2.14 further illustrates the application of percent as a unit factor.

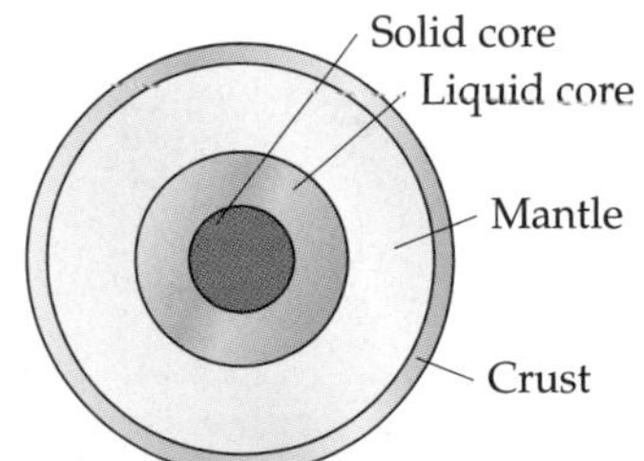

▲ **Earth's Crust** The radius of the Earth is about 4000 mi; the outer 5- to 25-mi layer of oceans and continents is referred to as the crust.

Example Exercise 2.14 • Percent as a Unit Factor

The Earth's crust is analyzed and found to contain 7.50% aluminum. What is the mass of the Earth's crust if the amount of aluminum is 1.90×10^{24} g?

Solution

We know that 7.50% means 7.50 parts per 100 parts. In this case, 7.50% is 7.50 g of aluminum per 100 g of the Earth's crust. Thus, we can write the ratio as a unit factor (7.50 g aluminum/100 g Earth's crust). If we apply the reciprocal unit factor to cancel units, we can solve the problem using the unit analysis format.

$$1.90 \times 10^{24}\ \cancel{\text{g aluminum}} \times \frac{100\ \text{g Earth's crust}}{7.50\ \cancel{\text{g aluminum}}} = 2.53 \times 10^{25}\ \text{g Earth's crust}$$

Self-Test Exercise

A lunar sample is analyzed and found to contain 7.50% aluminum. What is the mass of the sample if 0.232 kg of aluminum is found?

Answer: 3.09 kg

Chemistry Connection • The Coinage Metals

What common metal makes up most of the mass of a penny minted before 1982? After 1982?

The Earth's core is thought to be composed of only two metals, iron and nickel. On the other hand, the Earth's crust contains dozens of metals including aluminum, iron, calcium, sodium, potassium, magnesium, and titanium. These metals, however, do not occur naturally in the free state. They are found combined with other elements.

Copper, silver, and gold are found free in nature. Since ancient times, these metals have been used to make jewelry and precious objects. Not only do they have an attractive appearance, but they are relatively soft and can be easily altered into various shapes. Thousands of years ago in India and Egypt, silver and gold tokens were used for trade. Later, the Romans used silver and gold to make coins for monetary exchange. Today, copper, silver, and gold are often referred to as *coinage metals*.

The term "coinage metals" is somewhat misleading. In 1934 the United States took gold coins out of circulation. In 1971 the United States stopped using silver except for 50¢ coins and special collector issues. A 5¢ coin is minted from 75% copper and 25% nickel. Both 10¢ and 25¢ coins are minted as a "sandwich" composed of an outer layer of copper–nickel and an inner core

◀ **Coinage Metals** Copper, silver, and gold are found in coins minted in the United States and Canada.

of pure copper. When you examine a 10¢ or a 25¢ coin closely, the reddish-brown color of copper is clearly visible along the serrated edge.

The first 1¢ coin was minted in 1793 from pure copper and was the size of a half-dollar. The copper penny continued to be minted until 1856 when it was replaced by a penny made from a mixture of 88% copper and 12% nickel. The addition of nickel metal gave a white appearance to the coin. Five years later, the penny was changed to a mixture of copper, tin, and zinc metals. From 1962 to 1982, the penny was cast from a mixture of 95% copper and 5% zinc.

In 1982 the high price of copper forced the U.S. Mint to alter the composition once again. In lieu of copper, zinc is the main component of the new 1¢ coin. These pennies are minted from a zinc core plated with an ultrathin coating of copper metal. In 1995 the United States minted 13.5 billion 1¢ coins, but most pennies do not stay in active circulation. Currently, the U.S. Treasury is considering elimination of the penny from our monetary system.

Pre-1982 pennies are mostly copper; post-1982 pennies are almost entirely zinc.

Summary 

Section 2.1 This chapter introduced the basic laboratory equipment for obtaining a **measurement**. We often record measurements in units of **centimeter (cm)**, **gram (g)**, **milliliter (mL)**, or second (s). No measurement is exact because every **instrument** has **uncertainty**. We use a balance to measure the **mass** of an object. Although we commonly say that we weigh an object, strictly speaking, we determine the mass of an object. **Weight** is affected by the Earth's gravity, and mass is not.

Section 2.2 The **significant digits** in a measurement are those known with certainty from the instrument, plus one digit that is estimated.

Section 2.3 **Nonsignificant digits** exceed the certainty of the instrument and result from computations. The process of eliminating nonsignificant digits is called **rounding off**.

Section 2.4 When adding or subtracting measurements, the answer is limited by the measurement having the most uncertainty. If the uncertainty of three mass measurements is ± 0.1 g, ± 0.01 g, and ± 0.001 g, respectively, the measurement with the most uncertainty, ± 0.1 g, limits the answer.

Section 2.5 When multiplying or dividing measurements, the answer is limited by the least number of significant digits in the data. Every calculated answer must include the correct number of significant digits as well as the proper units.

Section 2.6 A number written as a superscript indicating that a value is multiplied times itself is called an **exponent**. If the value is 10, the exponent is a **power of 10**. A power of 10 can be positive or negative. If the exponent is positive (10^3), the number is greater than 1; if the exponent is negative (10^{-3}), it is less than 1.

Section 2.7 In science we use **scientific notation** to express numbers that are very large or very small. Scientific notation uses the following format to express a number with three significant digits: $D.DD \times 10^n$. The three significant digits $(D.DD)$ are multiplied by a power of 10 (10^n) to set the decimal point.

Section 2.8 We introduced our single most important task, namely, problem solving. We write a **unit equation** to generate a **unit factor**. For example, the unit equation 1 yard = 3 feet gives two unit factors: 1 yard/3 feet and its **reciprocal**, 3 feet/1 yard. An **exact equivalent** is an exactly equal relationship such as 1 yard ≡ 36 inches.

Section 2.9 We systematically used the **unit analysis method** to solve problems. Step 1: Write down the units asked for in the answer. Step 2: Write down the given value related to the answer. Step 3: Apply one or more unit factors to cancel the units in the given value and obtain the units in the answer.

Section 2.10 The ratio of a given quantity to the whole sample—all multiplied by 100—equals a **percent (%)**. Since percent is a ratio, we can apply percent unit factors to perform calculations using the unit analysis method.

Problem-Solving Organizer

Topic	Procedure	Example
Uncertainty in Measurements Sec. 2.1	State the measurement and indicate the uncertainty.	11.50 cm $\pm$ 0.05 cm 0.545 g $\pm$ 0.001 g 10.0 mL $\pm$ 0.5 mL
Significant Digits Sec. 2.2	Count the significant digits in a measurement from left to right: (a) start with the first nonzero digit, (b) do not count place-holder zeros.	0.05 g (1 significant digit) 50.0 g (3 significant digits) 505 g (3 significant digits) 50,000 g (1 significant digit)
Rounding Off Sec. 2.3	1. If the first nonsignificant digit is less than 5, drop all nonsignificant digits. 2. If the first nonsignificant digit is 5 or greater, add 1 to the last significant digit.	505.05 g rounds to 505 g (3 significant digits) 505.50 g rounds to 506 g (3 significant digits)
Adding and Subtracting Measurements Sec. 2.4	The answer is limited by the least certain measurement in a set of data.	5.05 g + 5.005 g = 10.06 g 5.05 mL − 2.0 mL = 3.0 mL
Multiplying and Dividing Measurements Sec. 2.5	The answer is limited by the least number of significant digits in the given data.	5.05 cm $\times$ 5.005 cm = 25.3 cm^2 5.0 g/10.0 mL = 0.50 g/mL
Exponential Numbers Sec. 2.6	(a) Numbers greater than 1 have a positive exponent. (b) Numbers less than 1 have a negative exponent.	1,000,000 = 1×10^6 0.000 001 = 1×10^{-6}
Scientific Notation Sec. 2.7	Place the decimal point after the first nonzero digit. Indicate how many places the decimal is moved using an exponent.	55,500 g = 5.55×10^4 g 0.000 555 g = 5.55×10^{-4} g
Unit Equations and Unit Factors Sec. 2.8	(a) Write a unit equation that shows an equivalent relationship. (b) Write two unit factors corresponding to the unit equation.	1 quart = 2 pints $\frac{1 \text{ quart}}{2 \text{ pints}}$ and $\frac{2 \text{ pints}}{1 \text{ quart}}$
Unit Analysis Problem Solving Sec. 2.9	1. Write down the units asked for in the answer. 2. Write down the related given value. 3. Apply a units factor to convert a given value to the units in the answer.	What is the number of feet in 1.75 yd? Given: 1 yard = 3 feet $1.75 \cancel{\text{yd}} \times \frac{3 \text{ ft}}{1 \cancel{\text{yd}}} = 5.25 \text{ ft}$
The Percent Concept Sec. 2.10	Compare the amount of a single quantity to the total sample and multiply by 100.	If a 14-karat ring contains 20.0 g of gold and 14.3 g of silver, what is the percent of gold? $\frac{20.0 \text{ g}}{20.0 \text{ g} + 14.3 \text{ g}} \times 100 = 58.3\%$

Key Concepts *

1. What state-of-the-art instrument is capable of making an exact measurement? What physical quantity can be measured with absolute certainty?
2. Draw a line on a separate sheet of paper and mark the length equal to line L shown below. Measure the line using Ruler A and Ruler B as shown in Figure 2.2. Record the length of the line consistent with the uncertainty of each ruler.

   L: ______________________

3. Which of the following is the best estimate of the diameter of a 1¢ coin: 0.5 cm, 2 cm, 5 cm, 20 cm, 50 cm? Which of the following is the best estimate of the thickness of a 10¢ coin: 0.001 cm, 0.01 cm, 0.1 cm, 1 cm, 10 cm?

A 1¢ coin   A 10¢ coin   A 25¢ coin

4. Which of the following is the best estimate of the mass of a 1¢ coin: 0.5 g, 1 g, 3 g, 10 g, 20 g? Which of the following is the best estimate of the mass of a 25¢ coin: 1 g, 5 g, 25 g, 50 g, 75 g?
5. Which of the following is the best estimate of the volume corresponding to 10 drops from an eyedropper: 0.5 mL, 1 mL, 5 mL, 10 mL, 20 mL? Which of the following is the best estimate of the volume of milk in a quart (32 fl oz) carton: 1 mL, 10 mL, 100 mL, 1000 mL, 10,000 mL?
6. Is the *mass* of an astronaut more, less, or the same on Uranus as on Mars? The mass of Uranus is 8.66×10^{25} kg, and that of Mars is 6.42×10^{23} kg.
7. Is the *weight* of an astronaut more, less, or the same on Uranus as on Mars? The mass of Uranus is 8.66×10^{25} kg, and that of Mars is 6.42×10^{23} kg.
8. Which of the following relationships is an exact equivalent?
   (a) 1 meter = 1000 millimeters
   (b) 1 meter = 1.09 yards
9. What is the number of significant digits in each of the following relationships?
   (a) 1 meter = 1000 millimeters
   (b) 1 meter = 1.09 yards
10. What are the three steps in the unit analysis method of problem solving?

Key Terms †

Select the key term below that corresponds to each of the following definitions.

_____ 1. a numerical value with units that expresses a physical quantity such as length, mass, or volume
_____ 2. a common metric unit of length
_____ 3. a common metric unit of mass
_____ 4. a common metric unit of volume
_____ 5. a device for recording a measurement such as length, mass, or volume
_____ 6. the degree of inexactness in an instrumental measurement
_____ 7. the quantity of matter in an object
_____ 8. the force exerted by gravity on an object
_____ 9. the certain digits in a measurement plus one estimated digit
_____ 10. the digits in a measurement that exceed the certainty of the instrument
_____ 11. the process of eliminating digits that are not significant
_____ 12. a number written as a superscript indicating that a value is multiplied times itself, for example, $10^4 = 10 \times 10 \times 10 \times 10$, and $cm^3 = cm \times cm \times cm$
_____ 13. a positive or negative exponent of 10
_____ 14. a method for expressing numbers by moving the decimal place after the first significant digit and indicating the number of places the decimal moves by a power of 10

(a) centimeter (cm) (*Sec. 2.1*)
(b) exact equivalent (≡) (*Sec. 2.8*)
(c) exponent (*Sec. 2.6*)
(d) gram (g) (*Sec. 2.1*)
(e) instrument (*Sec. 2.1*)
(f) mass (*Sec. 2.1*)
(g) measurement (*Sec. 2.1*)
(h) milliliter (mL) (*Sec. 2.1*)
(i) nonsignificant digits (*Sec. 2.3*)
(j) percent (%) (*Sec. 2.10*)
(k) power of 10 (*Sec. 2.6*)
(l) reciprocal (*Sec. 2.8*)
(m) rounding off (*Sec. 2.3*)
(n) scientific notation (*Sec. 2.7*)
(o) significant digits (*Sec. 2.2*)
(p) uncertainty (*Sec. 2.1*)

* Answers to Key Concepts are in Appendix H.

† Answers to Key Terms are in Appendix I.

_____ 15. a statement of two equivalent values, for example, 1 foot = 12 inches
_____ 16. a ratio of two quantities that are equivalent and can be applied to convert from one unit to another, for example, 1 foot/12 inches
_____ 17. the relationship of a fraction and its inverse, for example, 1 yard/3 feet and 3 feet/1 yard
_____ 18. a statement of two exactly equivalent values, for example, 1 yard ≡ 36 inches
_____ 19. a systematic procedure for solving problems that proceeds from a given value to a related answer by the conversion of units
_____ 20. the ratio of a single quantity compared to an entire sample, all multiplied by 100

(q) unit analysis method (*Sec. 2.9*)
(r) unit equation (*Sec. 2.8*)
(s) unit factor (*Sec. 2.8*)
(t) weight (*Sec. 2.1*)

Exercises‡

Uncertainty in Measurements (Sec. 2.1)

1. What quantity (length, mass, volume, time) is expressed by the following instruments?
(a) metric ruler **(b)** buret
(c) balance **(d)** pipet
(e) stopwatch **(f)** graduated cylinder

2. What quantity (length, mass, volume, time) is expressed by the following units?
(a) gram (g) **(b)** milliliter (mL)
(c) centimeter (cm) **(d)** second (s)

3. The uncertainty is given for each of the following measurements. State the maximum and minimum values that are acceptable.
(a) 6.5 ± 0.1 cm **(b)** 0.51 ± 0.01 g
(c) 10.0 ± 0.1 mL **(d)** 35.5 ± 0.1 s

4. The uncertainty is given for each of the following measurements. State the maximum and minimum values that are acceptable.
(a) 6.35 ± 0.05 cm **(b)** 1.556 ± 0.001 g
(c) 30.05 ± 0.05 mL **(d)** 60.01 ± 0.01 s

Significant Digits (Sec. 2.2)

5. State the number of significant digits in each of the following.
(a) 0.05 cm **(b)** 0.707 g
(c) 83.0 mL **(d)** 34.60 s

6. State the number of significant digits in each of the following.
(a) 2.50 cm **(b)** 3.060 g
(c) 0.2 mL **(d)** 10,025 s

7. State the number of significant digits in each of the following.
(a) 280 cm **(b)** 1200 g
(c) 100 mL **(d)** 3000 s

8. State the number of significant digits in each of the following.
(a) 570 cm **(b)** 100 g
(c) 2050 mL **(d)** 101,500 s

9. State the number of significant digits in each of the following.
(a) 3.71×10^3 cm **(b)** 9.5×10^{-1} g
(c) 2.000×10^4 mL **(d)** 1×10^{-9} s

10. State the number of significant digits in each of the following.
(a) 9.61×10^3 cm **(b)** 4.500×10^{-1} g
(c) 8×10^4 mL **(d)** 5.0×10^{-6} s

11. State the number of significant digits in each of the following.
(a) 25 metric rulers **(b)** 2 metric balances
(c) 12 beakers **(d)** 10 stopwatches

12. State the number of significant digits in each of the following.
(a) 50 pennies **(b)** 35 nickels
(c) 40 dimes **(d)** 75 quarters

Rounding Off Nonsignificant Digits (Sec. 2.3)

13. Round off the following values to three significant digits.
(a) 31.505 **(b)** 213,600
(c) 5.155 **(d)** 77.504

14. Round off the following values to three significant digits.
(a) 0.01842 **(b)** 0.000 000 484 500
(c) 2.571×10^5 **(d)** 5.6954×10^{-2}

15. Round off the following values to three significant digits.
(a) 61.15 **(b)** 362.01
(c) 2155 **(d)** 0.3665

16. Round off the following values to three significant digits.
(a) 12.59 **(b)** 35.55
(c) 1.598×10^9 **(d)** 2.6514×10^{-4}

Adding and Subtracting Measurements (Sec. 2.4)

17. Add or subtract the following measurements and round off the answer.
(a) 31.15 cm + 41.000 cm
(b) 50.2 cm − 0.500 cm

‡Answers to odd-numbered Exercises are in Appendix J.

18. Add or subtract the following measurements and round off the answer.
(a) 0.35 g + 0.250 g
(b) 24.90 g − 0.550 g

19. Add the following measurements and round off the answer.
(a) 0.4 g, 0.44 g, + 0.444 g
(b) 15.5 g, 7.50 g, + 0.050 g

20. Add the following measurements and round off the answer.
(a) 0.55 cm, 36.15 cm, + 17.3 cm
(b) 4 cm, 16.3 cm, + 0.95 cm

21. Subtract the following measurements and round off the answer.
(a) 242.167 g − 175 g
(b) 27.55 g − 14.545 g

22. Subtract the following measurements and round off the answer.
(a) 22.10 mL − 10.5 mL
(b) 10.0 mL − 0.15 mL

Multiplying and Dividing Measurements (Sec. 2.5)

23. Multiply the following measurements and round off the answer.
(a) 3.65 cm × 2.10 cm (b) 8.75 cm × 1.15 cm
(c) 16.5 cm × 1.7 cm (d) 21.1 cm × 20 cm

24. Multiply the following measurements and round off the answer.
(a) 5.1 cm × 1.25 cm × 0.5 cm
(b) 5.15 cm × 2.55 cm × 1.1 cm
(c) 12.0 cm^2 × 1.00 cm
(d) 22.1 cm^2 × 0.75 cm

25. Divide the following measurements and round off the answer.
(a) 66.3 g/7.5 mL (b) 12.5 g/4.1 mL
(c) 42.620 g/10.0 mL (d) 91.235 g/10.00 mL

26. Divide the following measurements and round off the answer.
(a) 26.0 cm^2/10.1 cm (b) 9.95 cm^2/0.15 cm
(c) 131.78 cm^3/19.25 cm (d) 131.78 cm^3/19.26 cm

Exponential Numbers (Sec. 2.6)

27. Express each of the following products in exponential form.
(a) 2 × 2 × 2 × 2 × 2 × 2
(b) 1/2 × 1/2 × 1/2

28. Express each of the following products in exponential form.
(a) 3 × 3 × 3 × 3
(b) 1/3 × 1/3 × 1/3 × 1/3 × 1/3

29. Express each of the following products as a power of 10.
(a) 10 × 10 × 10 × 10 × 10
(b) 1/10 × 1/10 × 1/10

30. Express each of the following products as a power of 10.
(a) 10 × 10 × 10 × 10 × 10 × 10 × 10
(b) 1/10 × 1/10 × 1/10 × 1/10

31. Express each of the following powers of 10 as a product.
(a) 1×10^3 (b) 1×10^{-7}

32. Express each of the following powers of 10 as a product.
(a) 1×10^{12} (b) 1×10^{-22}

33. Express each of the following ordinary numbers as a power of 10.
(a) 1,000,000,000 (b) 0.000 000 01

34. Express each of the following ordinary numbers as a power of 10.
(a) 100,000,000,000,000,000 (b) 0.000 000 000 000 001

35. Express each of the following powers of 10 as an ordinary number.
(a) 1×10^1 (b) 1×10^{-1}

36. Express each of the following powers of 10 as an ordinary number.
(a) 1×10^0 (b) 1×10^{-10}

Scientific Notation (Sec. 2.7)

37. Express the following ordinary numbers in scientific notation.
(a) 80,916,000 (b) 0.000 000 015
(c) 335,600,000,000,000 (d) 0.000 000 000 000 927

38. Express the following ordinary numbers in scientific notation.
(a) 1,010,000,000,000,000
(b) 0.000 000 000 000 456
(c) 94,500,000,000,000,000
(d) 0.000 000 000 000 000 019 50

39. There are 26,900,000,000,000,000,000,000 neon atoms in 1 L of neon gas under standard conditions. Express this number of atoms in scientific notation.

40. The mass of a neon atom is 0.000 000 000 000 000 000 000 0335 g. Express the mass in scientific notation.

Unit Equations and Unit Factors (Sec. 2.8)

41. Write a unit equation for each of the following.
(a) nickels and dimes (b) nickels and pennies

42. Write a unit equation for each of the following.
(a) quarters and pennies (b) quarters and nickels

43. Write a unit equation for each of the following.
(a) miles and feet (b) tons and pounds

44. Write a unit equation for each of the following.
(a) quarts and pints (b) years and centuries

45. Write the two unit factors associated with each of the following.
(a) nickels and dimes (b) nickels and pennies

46. Write the two unit factors associated with each of the following.
(a) quarters and pennies (b) quarters and nickels

47. Write the two unit factors associated with each of the following.
(a) miles and feet (b) tons and pounds
48. Write the two unit factors associated with each of the following.
(a) quarts and pints (b) years and centuries
49. Which of the following relationships is an exact equivalent?
(a) 1 mile = 5280 feet
(b) 1 kilogram = 2.20 pounds
(c) 1 liter = 1.06 quarts
(d) 1 week = 7 days
50. Which of the following relationships is an exact equivalent?
(a) 1 mile = 1.61 kilometers
(b) 1 ton = 2000 pounds
(c) 1 gallon = 16 cups
(d) 1 month = 30 days
51. What is the number of significant digits in each of the following relationships?
(a) 1 meter = 39.4 inches
(b) 1 carat ≡ 200 milligrams
(c) 1 bushel ≡ 32 quarts
(d) 1 year = 52 weeks
52. What is the number of significant digits in each of the following relationships?
(a) 1 rod ≡ 16.5 feet (b) 1 gram = 15.4 grains
(c) 1 foot^3 = 7.479 gallons (d) 1 minute ≡ 60 seconds

Unit Analysis Problem Solving (Sec. 2.9)

53. Using the unit analysis method, find the number of furlongs in 15 miles. (Given: 1 mile ≡ 8 furlongs.)
54. Using the unit analysis method, find the number of furlongs in 750 rods. (Given: 1 furlong ≡ 40 rods.)
55. Using the unit analysis method, find the mass in grams of a diamond that is 3/4 carat (0.750 carat). (Given: 1 carat ≡ 0.200 gram.)
56. Using the unit analysis method, find the mass in grains of a 325-milligram aspirin tablet. (Given: 1000 milligrams = 15.4 grains.)
57. Using the unit analysis method, find the number of cups of cream in 2.5 pints. (Given: 1 pint ≡ 2 cups.)
58. Using the unit analysis method, find the number of tablespoons of salad oil in 15 teaspoons. (Given: 1 tablespoon ≡ 3 teaspoons.)
59. Using the unit analysis method, calculate the number of months in 3.25 years.
60. Using the unit analysis method, calculate the number of days in 12.0 weeks.
61. If the United States has 263,000,000,000 troy ounces of gold on reserve, what is the mass of gold in troy pounds? (Given: 1 troy pound ≡ 12 troy ounces.)
62. If a bar of silver weighs 50.0 troy ounces, what is the mass of silver in grams? (Given: 1 troy ounce = 31.1 grams.)

▲ **Gold** A gold nugget and bars of the pure metal.

63. How many miles does light travel in 1 year? (Given: The velocity of light is 186,000 mi/s.)
64. How many meters does light travel in 1 year? (Given: The velocity of light is 3.00×10^8 m/s.)

The Percent Concept (Sec. 2.10)

65. In a freshman class of 5846 students, 101 select chemistry as a major. What is the percent of chemistry majors in the freshman class?
66. Blood bank records show blood type and RH factor for 55,368 patients as follows: 21,594 O+, 18,825 A+, 4706 B+, 1938 AB+, 3876 O−, 3322 A−, 831 B−, 276 AB−. What is the percent of each blood type group?
67. A 5.750-g sample of bauxite, an aluminum-containing ore, was found by analysis to contain 4.10% aluminum. Find the mass of aluminum.
68. Air is 20.9% oxygen by volume. Find the volume of air that contains 225 mL of oxygen.
69. If a cool-burning solution contains 255 mL of ethanol and 375 mL of water, what is the percent of ethanol in the solution?
70. If 12.5 gallons (gal) of gasohol contains 1.50 gal of ethyl alcohol, what is the percent of alcohol in the gasohol?
71. Water is composed of 11.2% hydrogen and 88.8% oxygen. What mass of water contains 15.0 g of oxygen?
72. Ordinary table salt, sodium chloride, is composed of 39.3% sodium and 60.7% chlorine. Calculate the mass of sodium in 0.375 g of salt.
73. The Earth's crust has a mass of 2.37×10^{25} g and contains 49.2% oxygen, 25.7% silicon, and 7.50% aluminum. Calculate the mass of each element in the crust.
74. Uranus has five moons with a combined mass of 3.87×10^{21} kg. What is the mass of the largest moon, Titania, if it is 54.3% of the total mass?

75. Before 1982 the U.S. Mint cast penny coins from a copper and zinc mixture. If a 1980 penny weighs 3.051 g and contains 2.898 g copper, what are the percentages of copper and zinc in the coin?

76. In 1982 the U.S. Mint stopped making copper pennies because of the price of copper and began phasing in pennies made of zinc plated with a thin layer of copper. If a 1990 penny weighs 2.554 g and contains 2.490 g zinc, what are the percentages of copper and zinc in the coin?

General Exercises

77. If a 10-mL pipet has an uncertainty of 1/10 mL ($\pm$0.1 mL), indicate the volume of the pipet in ordinary numbers.

78. If a 1000-mL flask has an uncertainty of 1 mL ($\pm$1 mL), express the volume of the flask in scientific notation.

79. Round off the mass of a sodium atom, 22.989768 atomic mass units (amu), to three significant digits.

80. Round off the velocity of light, 2.997925×10^{10} cm/s, to three significant digits.

81. Find the total mass of two brass cylinders that weigh 126.457 g and 131.6 g.

82. Find the length of magnesium ribbon that remains after two 25.0-cm strips are cut from a 255-cm length of the metal.

83. Convert the following exponential numbers to scientific notation.
(a) 352×10^4 (b) 0.191×10^{-5}

84. Convert the following exponential numbers to scientific notation.
(a) 0.170×10^2 (b) 0.00350×10^{-1}

85. The mass of an electron is 9.10953×10^{-28} g, and a proton is 1.67265×10^{-24} g. What is the total mass of an electron and a proton?

86. The mass of a neutron is 1.67495×10^{-24} g, and a proton is 1.67265×10^{-24} g. What is the difference in mass between a neutron and a proton?

87. A metric ton is defined as 1000 kg or 2.20×10^3 lb. An English ton is 2000 lb. What is the difference in mass between a metric ton and an English ton expressed in pounds?

88. The distance from the Earth to the Moon is 2.39×10^5 mi; from the Moon to Mars it is 4.84×10^7 mi. What is the total distance a space probe travels from the Earth to the Moon to Mars?

▲ **Earthrise** This photo taken on the Moon shows the Earth rising over the lunar horizon.

89. The oldest rock found on the Earth was discovered in South Africa and is estimated to be 4.0 billion years old. Express the age of the rock in hours using scientific notation.

90. *Apollo 15* brought back a lunar rock estimated to be 4.1 billion years old. Express the age of the rock in minutes using scientific notation.

91. If an ounce equals 28.4 g, what is the mass in grams of a pound of feathers? (Given: 1 pound = 16 ounces.)

92. If a troy ounce equals 31.1 g, what is the mass in grams of a pound of gold? (Given: 1 troy pound = 12 troy ounces.)

93. Refer to Exercises 91 and 92 and answer the timeless question "Which weighs more, a pound of feathers, or a pound of gold?"

Explorer Quiz 1
Explorer Quiz 2
Explorer Quiz 3
Master Quiz

CHAPTER 3

The Metric System

▲ Which unit of measurement is greater: 1 kg or 1 lb of jelly beans, 1 yd or 1 m, 1 qt of milk or 1 L of orange drink?

This chapter formally introduces the metric system along with the concepts of density, temperature, and specific heat. More importantly, this chapter is designed to reinforce the unit analysis method of problem solving. Once again, I address students' fears of chemistry by repeating that solving problems is "as simple as one, two, three." I use visual aids to support the metric system and perform demonstrations of density to make the concept more interesting (refer to the *Instructor's Resource Manual*).

Since the beginning of civilization there has been a need to convey measurement. When exchanging goods, traders had to agree on standards for length, weight, and volume. Historically, length was often defined in terms of hands and feet because it was convenient. For instance, a digit was defined as the thickness of the index finger, a span was the width of four fingers, and a cubit was the distance from the elbow to the end of the middle finger.

The Greeks proposed the foot as a unit of length and divided it into 12 parts. The Romans in turn adopted the unit and referred to 1/12 of a foot as an *unicae*. The Anglo-Saxons changed the word from *unicae* to "inch," which was later adopted by the English and eventually passed on to the United States.

The English system of measurement originated in 1215 with the signing of the Magna Carta. The Magna Carta attempted to bring uniform measurements to world trade and decreed: "Throughout the kingdom there shall be standard measures of wine, ale, and corn." To ensure uniform measurements, the English used reference standards. An inch was defined as the length of three barleycorns. A bushel of barley had a mass of 50 pounds and a volume of 8 gallons.

3.1 Basic Units and Symbols

Objectives
- To state the basic units and symbols of the metric system.
- To state the prefixes for multiples and fractions of basic units.

Discontent over the lack of worldwide standards in measurement reached a peak in the late 1700s. Although the **English system** was common, it was used primarily within the British empire. In 1790 the French government appointed a committee of scientists to investigate the possibility of a universal measuring system.

The committee spent nearly 10 years devising and agreeing on a new system. The resulting **metric system** offers simplicity and basic units. The basic unit of length is the **meter** (symbol **m**), the basic unit of mass is the **gram** (symbol **g**), the basic unit of volume is the **liter** (symbol **L**), and the basic unit of time is the **second** (symbol **s**). The basic units and symbols of the metric system are shown in Table 3.1.

Antoine Lavoisier, the founder of modern chemistry, praised the metric system as follows: "Never has anything more grand and more simple, more coherent in all of its parts, issued from the hand of man."

Table 3.1 The Metric System

Physical Quantity	Basic Unit	Symbol
length	meter*	m
mass	gram	g
volume	liter*	L
time	second	s

*The U.S. Metric Association recommends the spellings "meter" and "liter." All other English-speaking nations, however, use the spellings "metre" and "litre."

The metric system is simple and coherent for two reasons. First, it uses a single basic unit for each quantity measured. Second, it is a decimal system that uses prefixes to enlarge or reduce a basic unit. For instance, a kilometer is a 1000 m, and a centimeter is 1/100 m.

The metric committee defined a meter, kilogram, and liter as follows. A meter was equal to one ten-millionth of the distance from the North Pole to the equator. A kilogram was equal to the mass of a cube of water one-tenth of a meter on a side. A liter was equal to the volume occupied by a kilogram of water (at 4°C).

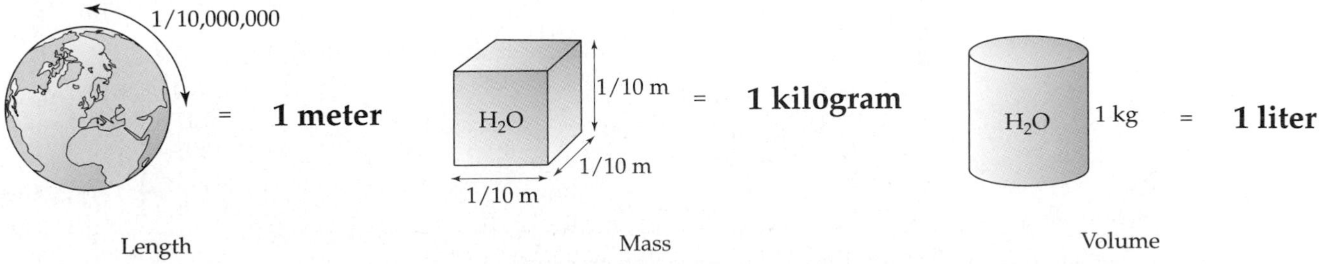

▲ **The Original Metric References** A meter was originally based on one ten-millionth of the distance from the North Pole to the equator. A kilogram is equal to the mass of a cube of water one-tenth of a meter on each side. A liter is equal to the volume of one kilogram of water at 4°C.

The Original Metric References

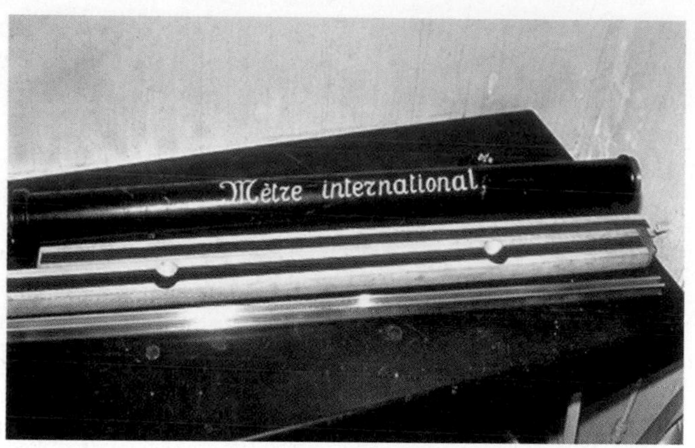

▶ **Figure 3.1 The Original Standard Meter** The prototype reference meter is an X-shaped bar cast from platinum and iridium metals (shown in the foreground). The original meter was cast in 1790 and still remains at the International Bureau of Weights and Measures outside Paris, France.

In addition, the metric committee made reference standards for the meter, kilogram, and liter. They cast a platinum bar as the reference standard for one meter. They cast a solid platinum cylinder as the reference for one kilogram, and a platinum container as the standard for one liter. These platinum castings became the reference standards for world measurement.[1] In 1875 all the major nations of the world, including the United States, signed the Treaty of the Meter. Duplicates of the original castings were made and distributed to all member nations. A duplicate casting of the standard reference meter is shown in Figure 3.1.

Metric Prefixes

The United States uses the dollar ($1.00) as the basic unit of currency. In our monetary system we move the decimal to express a multiple or fraction of a dollar ($10.00 and $0.10). Similarly, the metric system is a decimal system but uses a prefix to express a multiple or a fraction of a basic unit. A metric prefix increases or decreases a basic unit by a factor of 10. For instance, the prefix *kilo-* increases a basic unit by 1000, whereas the prefix *centi-* decreases a basic unit by a 100. Thus, a kilometer is 1000 m, and a centimeter is 0.01 m. Table 3.2 lists some common prefixes for basic metric units.

Table 3.2 Metric Prefixes

Prefix	Symbol	Multiple/Fraction
giga-	G	$1{,}000{,}000{,}000 = 1 \times 10^{9}$
mega-	M	$1{,}000{,}000 = 1 \times 10^{6}$
kilo-	k	$1{,}000 = 1 \times 10^{3}$
	Basic unit: meter, gram, liter, second	
deci-	d	$0.1 = 1 \times 10^{-1}$
centi-	c	$0.01 = 1 \times 10^{-2}$
milli-	m	$0.001 = 1 \times 10^{-3}$
micro-	μ*	$0.000\ 001 = 1 \times 10^{-6}$
nano-	n	$0.000\ 000\ 001 = 1 \times 10^{-9}$

*The Greek letter mu (μ) is the symbol for *micro*.

[1] More precisely, the metric reference standards were cast from a mixture of 90% platinum and 10% iridium metals.

The names of metric units can be abbreviated using symbols. We can illustrate this by writing symbols for the following metric units: kilometer (km), milligram (mg), microliter (μL), and nanosecond (ns). Example Exercise 3.1 further illustrates how to write names of metric units and symbols.

Example Exercise 3.1 • Metric Basic Units and Prefixes

Give the symbol for each of the following metric units and state the quantity measured by each unit.

(a) gigameter (b) kilogram
(c) centiliter (d) microsecond

Solution

We compose the symbol for each unit by combining the prefix symbol and the basic unit symbol. If we refer to Tables 3.1 and 3.2, we have

(a) Gm, length (b) kg, mass
(c) cL, volume (d) μs, time

Self-Test Exercise

Give the symbol for each of the following metric units and state the quantity measured by each unit.

(a) nanosecond (b) milliliter
(c) decigram (d) megameter

Answers: (a) ns, time; (b) mL, volume; (c) dg, mass; (d) Mm, length

The metric system is sometimes referred to as the International System of measurement. Although the **International System** (symbol **SI**) is similar to the metric system, SI is more comprehensive and sophisticated. In this text, we will use the simpler metric system and utilize SI units only in special instances.

3.2 Metric Conversion Factors

Objectives
- To write the unit equation for a basic metric unit and a prefix unit.
- To write the two unit factors derived from a metric unit equation.

Recall from Section 2.9 that a **unit equation** relates two quantities that are equal. As an example, let's find two equal quantities involving kilometers and meters. Since the metric prefix *kilo-* means 1000 basic units, 1 kilometer is 1000 meters. We can write the **exact equivalent** as 1 kilometer ≡ 1000 meters, or simply 1 km ≡ 1000 m.

A centimeter is 1/100 of a meter, and we could write the relationship as 1 cm = 0.01 m. However, working with fractions is more confusing than working with multiples of basic units. Fortunately, we can avoid decimal fractions by simply restating the relationship. Since there are 100 cm in a meter, we can write the exact equivalent as 1 m ≡ 100 cm. Example Exercise 3.2 further illustrates how to write unit equations for metric relationships.

Example Exercise 3.2 • Metric Unit Equations

Complete the unit equation for each of the following exact metric equivalents.

(a) 1 Mm = ? m (b) 1 kg = ? g
(c) 1 L = ? dL (d) 1 s = ? ns

Solution

We can refer to Table 3.2 as necessary.

(a) The prefix *mega-* means 1,000,000 basic units; thus, 1 Mm ≡ 1,000,000 m.
(b) The prefix *kilo-* means 1000 basic units; thus, 1 kg ≡ 1000 g.

(continued)

Example Exercise 3.2 *(continued)*

(c) The prefix *deci-* means 0.1 of a basic unit; thus, 1 L ≡ 10 dL.
(d) The prefix *nano-* means 0.000 000 001 of a basic unit; thus, 1 s ≡ 1,000,000,000 ns.

Alternatively, we can express the unit equation using exponential numbers.

(a) 1 Mm ≡ 1×10^6 m (b) 1 kg ≡ 1×10^3 g
(c) 1 L ≡ 1×10^1 dL (d) 1 s ≡ 1×10^9 ns

Self-Test Exercise

Complete the unit equation for each of the following exact metric equivalents.

(a) 1 Gm = ? m (b) 1 g = ? cg
(c) 1 L = ? μL (d) 1 s = ? ms

Answers: (a) 1 Gm ≡ 1×10^9 m; (b) 1 g ≡ 1×10^2 cg; (c) 1 L ≡ 1×10^6 μL; (d) 1 s ≡ 1×10^3 ms

Update • The International System (SI)

What is the only metric system reference standard that is also utilized by SI?

In 1875 representatives of several nations, including the United States, signed a pact creating the International Bureau of Weights and Measures. This organization was given authority to establish worldwide standards of measurement, and in 1960 it approved a resolution to establish an international system of units.

The International System (called SI for *Système Internationale*) is an extension of the metric system. SI, however, has seven base units and numerous derived units. The base unit of length remains the meter, but the base unit of mass is the kilogram—not the gram. The five other SI quantities are time, temperature, electric current, light intensity, and amount of substance.

The International System of Units (SI)

Quantity	Base Unit	Symbol
length	meter	m
mass	kilogram	kg
time	second	s
temperature	Kelvin	K
electric current	ampere	A
light intensity	candela	cd
amount of substance	mole	mol

The Official Standard Kilogram

The official reference kilogram was cast from platinum and iridium metals in 1790. The original casting is protected under glass vacuum jars at the International Bureau of Weights and Measures outside Paris, France.

It is interesting to note that SI units have been defined in terms of natural phenomena so that measurements can be reproduced anywhere in the world without the need for physical reference standards. Mass is the sole exception and still relies on the original international prototype kilogram in France.

In 1960 SI defined a meter as 1,650,763.73 wavelengths of the orange-red light emitted from excited atoms of krypton gas. In 1983 SI redefined a meter as the distance light travels in a vacuum in 1/299,792,458 s. Other SI base units have similarly technical definitions.

Currently, the U.S. Bureau of Weights and Measures is encouraging the use of SI units. In practice, however, scientists do not always follow SI convention. It requires considerable effort to apply the system correctly, and many instruments are not calibrated in SI units. Moreover, SI disallows use of the familiar term "liter" and discourages use of the common prefixes *deci-* and *centi-*.

▲ **The Official Standard Kilogram** The official reference kilogram was cast from platinum and iridium metals in 1790. The original casting is protected under glass vacuum jars at the International Bureau of Weights and Measures outside Paris, France.

The international prototype kilogram is the reference standard for mass in both the metric system and SI.

Writing Metric Unit Factors

Once we are able to write unit equations, we can easily generate two unit factors. A **unit factor** is the ratio of the two equivalent quantities. That is, the quantity in the numerator is equal to the quantity in the denominator.

Since the numerator and the denominator are equivalent, we can generate a second unit factor by inverting the ratio and writing the **reciprocal**. As an example, consider the unit equation 1 m ≡ 100 cm.[2] The two associated unit factors are

$$\frac{1 \text{ m}}{100 \text{ cm}} \quad \text{and} \quad \frac{100 \text{ cm}}{1 \text{ m}}$$

Example Exercise 3.3 further illustrates how to write unit factors corresponding to a given metric relationship.

Example Exercise 3.3 • Metric Unit Factors

Write two unit factors for each of the following metric relationships.

(a) kilometers and meters (b) grams and decigrams

Solution

We start by writing the unit equation to generate the two unit factors.

(a) The prefix *kilo-* means 1000 basic units; thus, 1 km ≡ 1000 m. The two unit factors are

$$\frac{1 \text{ km}}{1000 \text{ m}} \quad \text{and} \quad \frac{1000 \text{ m}}{1 \text{ km}}$$

(b) The prefix *deci-* means 0.1 basic unit; thus, 1 g ≡ 10 dg. The two unit factors are

$$\frac{1 \text{ g}}{10 \text{ dg}} \quad \text{and} \quad \frac{10 \text{ dg}}{1 \text{ g}}$$

Self-Test Exercise

Write two unit factors for each of the following metric relationships.

(a) liters and milliliters (b) megaseconds and seconds

Answers: (a) 1 L/1000 mL and 1000 mL/1 L; (b) 1 Ms/1,000,000 s and 1,000,000 s/1 Ms

3.3 Metric–Metric Conversions

Objective · To express a given metric measurement with a different metric prefix.

In Section 2.9, we learned a powerful method for solving problems. It is called the **unit analysis method**. In Chapter 2, we solved problems requiring only one unit factor. In this chapter we solve problems requiring two or more unit factors. We will be successful, however, if we apply the following three steps consistently.

Applying the Unit Analysis Method

Step 1: Write down the units asked for in the answer.

Step 2: Write down the value given in the problem that is related to the units required in the answer.

Step 3: Apply one or more unit factors to cancel the units in the given value and convert to the units in the answer.

[2] It is helpful to consistently write the larger metric unit first and the smaller metric unit second in a unit equation; for example, 1 m ≡ 100 cm, 1 kg ≡ 1000 g, and 1 L ≡ 1000 mL.

Let's begin with a simple problem and find the mass in grams of a 325-mg aspirin tablet. *First*, we write down the unit asked for in the answer (g). *Second*, we write down the related given value (325 mg). *Third*, we apply a unit factor using the following format.

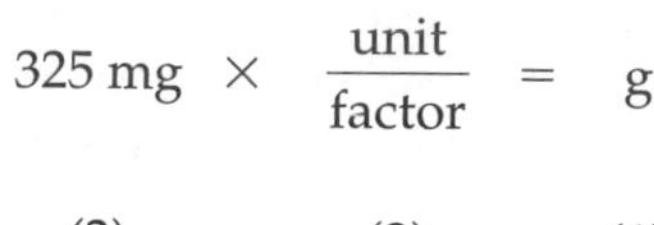

$$325 \text{ mg} \times \frac{\text{unit}}{\text{factor}} = \text{g}$$

(2) (3) (1)

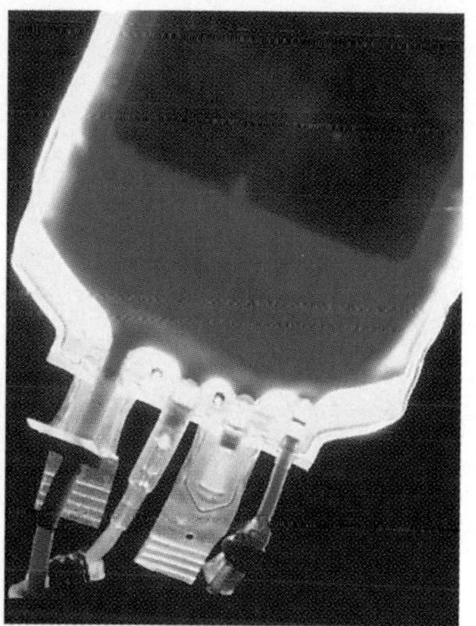

▲ **Blood Plasma** A blood plasma bag has a volume of 1 deciliter (dL), that is, 0.1 L.

The unit equation that relates steps 1 and 2 is 1 g ≡ 1000 mg. Therefore, the two associated unit factors are

$$\frac{1 \text{ g}}{1000 \text{ mg}} \quad \text{and} \quad \frac{1000 \text{ mg}}{1 \text{ g}}$$

To cancel the unit in the given value (325 mg), we select the unit factor 1 g/1000 mg. After substituting, we have

$$325 \cancel{\text{mg}} \times \frac{1 \text{ g}}{1000 \cancel{\text{mg}}} = 0.325 \text{ g}$$

The given value has three significant digits, and the unit factor is an exact equivalent. Thus, the answer is rounded off to three significant digits.

Example Exercises 3.4 and 3.5 further illustrate the unit analysis method of problem solving for metric–English conversions.

Example Exercise 3.4 • Metric–Metric Conversion

A hospital has 125 deciliter bags of blood plasma. What is the volume of plasma expressed in milliliters?

Solution

In metric–metric unit analysis conversions, we relate the given value to a basic unit (m, g, L, s) and then convert to the unit asked for in the answer.

Step 1, The unit asked for in the answer is **mL**.
Step 2, The given value is **125 dL**.
We must convert dL to the basic unit L and then convert L to mL. The unit analysis format is

$$125 \text{ dL} \times \frac{\text{unit}}{\text{factor 1}} \times \frac{\text{unit}}{\text{factor 2}} = \text{mL}$$

(2) (3) (1)

Step 3, We apply unit conversion factors. To convert from dL to L (unit factor 1), we use the relationship 1 L ≡ 10 dL. The two possible unit factors are

$$\frac{1 \text{ L}}{10 \text{ dL}} \quad \text{and} \quad \frac{10 \text{ dL}}{1 \text{ L}}$$

To convert from L to mL (unit factor 2), we use the relationship 1 L ≡ 1000 mL. The two unit factors are

$$\frac{1 \text{ L}}{1000 \text{ mL}} \quad \text{and} \quad \frac{1000 \text{ mL}}{1 \text{ L}}$$

To apply unit analysis, we simply select the unit factors that cancel units.

$$125\ \cancel{\text{dL}} \times \frac{1\ \cancel{\text{L}}}{10\ \cancel{\text{dL}}} \times \frac{1000\ \text{mL}}{1\ \cancel{\text{L}}} = 12{,}500\ \text{mL}$$

The given value has three significant digits, and so the answer is limited to three digits. The unit factors are exact equivalents, and so they do not affect significant digits. We can show the solution visually as follows:

125 dL → $\frac{1\ \text{L}}{10\ \text{dL}}$ → L → $\frac{1000\ \text{mL}}{1\ \text{L}}$ → 12,500 mL

Self-Test Exercise

A dermatology patient is treated with ultraviolet light having a wavelength of 375 nm. What is the wavelength expressed in centimeters?

Answer: 0.0000375 cm $(3.75 \times 10^{-5}\ \text{cm})$

Example Exercise 3.5 • Metric–Metric Conversion

The mass of the Earth is 5.98×10^{24} kg. What is the mass expressed in megagrams?

▶ **The Earth** The mass of the Earth is about 6,000,000,000,000,000,000,000,000 kg.

Earth

Solution

In metric–metric unit analysis conversions, we relate the given value to a basic unit (m, g, L, s) and then convert to the unit asked for in the answer.

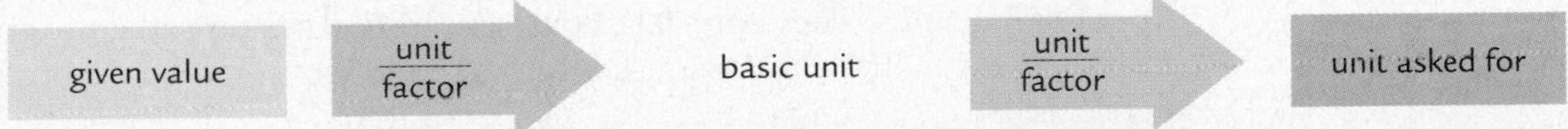

Step 1, The unit asked for in the answer is **Mg**.

Step 2, The given value is **5.98×10^{24} kg**.

In this problem, we must first convert kg to g and then convert g to Mg. Thus, we will need two unit factors.

$$5.98 \times 10^{24}\ \text{kg} \times \frac{\text{unit}}{\text{factor 1}} \times \frac{\text{unit}}{\text{factor 2}} = \text{Mg}$$

Step 3, We apply unit conversion factors. To convert from kg to g (unit factor 1), we need the unit equation 1 kg ≡ 1000 g; the two associated unit factors are

$$\frac{1\ \text{kg}}{1000\ \text{g}} \quad \text{and} \quad \frac{1000\ \text{g}}{1\ \text{kg}}$$

To convert from g to Mg (unit factor 2), we use the relationship 1 Mg ≡ 1,000,000 g; the two corresponding unit factors are

$$\frac{1\ \text{Mg}}{1{,}000{,}000\ \text{g}} \quad \text{and} \quad \frac{1{,}000{,}000\ \text{g}}{1\ \text{Mg}}$$

To correctly apply unit analysis, we simply choose unit factors that cancel units.

$$5.98 \times 10^{24}\ \cancel{\text{kg}} \times \frac{1000\ \cancel{\text{g}}}{1\ \cancel{\text{kg}}} \times \frac{1\ \text{Mg}}{1{,}000{,}000\ \cancel{\text{g}}} = 5.98 \times 10^{21}\ \text{Mg}$$

(continued)

Example Exercise 3.5 *(continued)*

The given value allows three significant digits in the answer because the unit factors have an infinite number of significant digits. We can outline the solution as follows:

5.98×10^{24} kg → $\frac{1000 \text{ g}}{1 \text{ kg}}$ → g → $\frac{1 \text{ Mg}}{1{,}000{,}000 \text{ g}}$ → 5.98×10^{21} Mg

Self-Test Exercise

Light travels through the universe at a velocity of 3.00×10^{10} cm/s. How many gigameters does light travel in 1 s?

Answer: 0.300 Gm (3.00×10^{-1} Gm)

3.4 Metric–English Conversions

Objective · To express a given measurement in metric units or English units.

Despite the superiority of the metric system, its adoption has been resisted from its inception. When the French asked the Americans and the British for their cooperation in creating a worldwide system of measurement, both countries declined. The initial objection, interestingly, was political rather than scientific. Since the meter was defined along a longitude that passed through Dunkirk in France, the French had control of the metric standards.

In 1975 President Gerald Ford signed an official metric act. Thus, the United States became the last major nation in the world to formally adopt the metric system. Although the metric system is now taught in public schools, the United States is making slow progress in achieving full compliance. Economically, it will cost billions of dollars to retool machinery and convert to metric measurements.

In science, we regularly record measurements in metric units. In everyday activities, however, you are probably more familiar with inches, pounds, quarts, and other English units. Therefore, let's compare the two systems to gain a practical appreciation for metric dimensions. Table 3.3 contrasts some common English and metric units.

Table 3.3 Metric–English Equivalents

Physical Quantity	English Unit	Metric Equivalent*
length	1 inch (in.)	1 in. = 2.54 cm
mass	1 pound (lb)	1 lb = 454 g
volume	1 quart (qt)	1 qt = 946 mL
time	1 second (sec)	1 sec = 1.00 s

*When applying these metric equivalents, assume three significant digits. Since the metric and English systems have different reference standards, these are not exact equivalents; however, the U.S. Bureau of Weights and Measures has redefined 1 in. as exactly equal to 2.54 cm.

Which foot race do you think is run faster: 100 meters or 100 yards? It is obvious that the faster race has the shorter distance. We can calculate the distance in yards for a 100.0-m race given that 1 yd = 0.914 m.

First, we write down the unit asked for in the answer (yd). *Second,* we write down the related given value (100.0 m). *Third,* we apply a unit factor using the following format.

$$\underset{(2)}{100.0 \text{ m}} \times \underset{(3)}{\frac{\text{unit}}{\text{factor}}} = \underset{(1)}{\text{yd}}$$

The unit equation that relates steps 1 and 2 is 1 yd = 0.914 m. Therefore, the two unit factors are

$$\frac{1 \text{ yd}}{0.914 \text{ m}} \quad \text{and} \quad \frac{0.914 \text{ m}}{1 \text{ yd}}$$

To cancel the unit in the given value (100.0 m), we select the unit factor (1 yd/0.914 m). After substituting, we have

$$100.0 \cancel{\text{m}} \times \frac{1 \text{ yd}}{0.914 \cancel{\text{m}}} = 109 \text{ yd}$$

Although the given value (100.0 m) has four significant digits, the unit factor has only three because the unit equation (1 yd = 0.914 m) is not an exact equivalent. Thus, the answer is rounded to three significant digits. Since 100 yd is a shorter distance than 100 meters, the 100-yard race is faster. In fact, the world record for 100 yards is less than 9 seconds!

Example Exercises 3.6 and 3.7 further illustrate the unit analysis method of problem solving for metric–English conversions.

Example Exercise 3.6 • Metric–English Conversion

A half-gallon carton contains 64.0 fl oz of milk. How many milliliters of milk are in the carton? (Given: 1 qt ≡ 32 fl oz.)

Solution

In metric–English unit analysis conversions, we convert the given value to a unit related to the unit asked for in the answer.

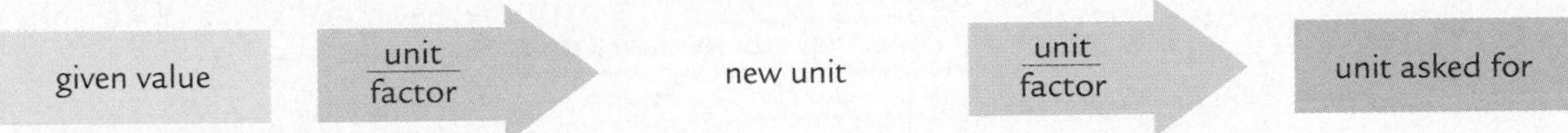

Step 1, The unit asked for in the answer is **mL**.
Step 2, The given value is **64.0 fl oz**.
In this problem, we must first convert fl oz to qt and then qt to mL. Thus, we will need two unit factors.

$$\underset{(2)}{64.0 \text{ fl oz}} \times \frac{\text{unit}}{\text{factor 1}} \underset{(3)}{\times} \frac{\text{unit}}{\text{factor 2}} = \underset{(1)}{\text{mL}}$$

Step 3, We apply two unit conversion factors.
To convert from fl oz to qt (unit factor 1), we need the unit equation 1 qt ≡ 32 fl oz; the two associated unit factors are

$$\frac{1 \text{ qt}}{32 \text{ fl oz}} \quad \text{and} \quad \frac{32 \text{ fl oz}}{1 \text{ qt}}$$

To convert from qt to mL (unit factor 2), we refer to Table 3.3 and find the relationship 1 qt = 946 mL; the two corresponding unit factors are

$$\frac{1 \text{ qt}}{946 \text{ mL}} \quad \text{and} \quad \frac{946 \text{ mL}}{1 \text{ qt}}$$

To apply unit analysis, we simply select the unit factors that cancel units.

$$64.0 \cancel{\text{fl oz}} \times \frac{1 \cancel{\text{qt}}}{32 \cancel{\text{fl oz}}} \times \frac{946 \text{ mL}}{1 \cancel{\text{qt}}} = 1890 \text{ mL}$$

▲ **Milk Carton** A half-gallon carton of milk has a volume of 64.0 fl oz.

(continued)

Example Exercise 3.6 *(continued)*

The answer is rounded to three significant digits because the given value and unit factor 2 each have three digits. Unit factor 1 is derived from an exact equivalent and has an infinite number of significant digits. We can outline the solution as follows.

64.0 fl oz → $\frac{1\text{ qt}}{32\text{ fl oz}}$ → qt → $\frac{946\text{ mL}}{1\text{ qt}}$ → 1890 mL

Self-Test Exercise

A plastic bottle contains 5.00 gallons of distilled water. How many liters of distilled water are in the bottle? (Given: 1 gal ≡ 4 qt.)

Answer: 18.9 L

Example Exercise 3.7 • Metric–English Conversion

If a tennis ball weighs 58 g, what is the mass of the tennis ball in ounces? (Given: 1 lb ≡ 16 oz.)

Solution

In metric–English unit analysis conversions, we convert the given value to a unit related to the unit asked for in the answer.

Step 1, The unit asked for in the answer is **oz**.

Step 2, The given value is **58 g**.

In this problem, we must first convert g to lb and then lb to oz. Thus, we will need two unit factors.

$$\underset{(2)}{58\text{ g}} \times \frac{\text{unit}}{\text{factor 1}} \underset{(3)}{\times} \frac{\text{unit}}{\text{factor 2}} = \underset{(1)}{\text{oz}}$$

Step 3, We apply two unit conversion factors. To convert from g to lb (unit factor 1), we refer to Table 3.3 and find 1 lb = 454 g; the two associated unit factors are

$$\frac{1\text{ lb}}{454\text{ g}} \quad \text{and} \quad \frac{454\text{ g}}{1\text{ lb}}$$

To convert from lb to oz (unit factor 2), we are given that 1 lb ≡ 16 oz; the two unit factors are

$$\frac{1\text{ lb}}{16\text{ oz}} \quad \text{and} \quad \frac{16\text{ oz}}{1\text{ lb}}$$

To apply unit analysis, we simply select the unit factors that cancel units:

$$58\,\cancel{\text{g}} \times \frac{1\,\cancel{\text{lb}}}{454\,\cancel{\text{g}}} \times \frac{16\text{ oz}}{1\,\cancel{\text{lb}}} = 2.0\text{ oz}$$

▲ **Tennis Balls** A tennis ball weighs about 2.0 oz and is 2.5 in. in diameter.

The answer is rounded to two digits because the given value has two digits. Unit factor 1 has three significant digits, and unit factor 2 is an exact equivalent having an infinite number of significant digits. We can outline the solution as follows.

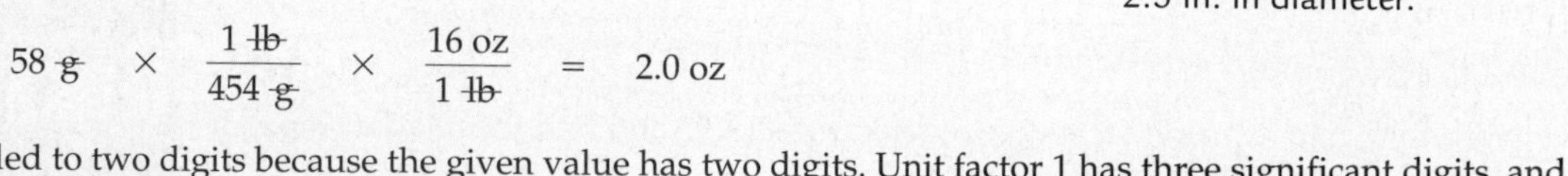

Self-Test Exercise

If a tennis ball has a diameter of 2.5 inches, what is the diameter in millimeters?

Answer: 64 mm

Answers with Compound Units

Now, let's try a problem in which the answer involves a ratio of two units. For example, if a highway speed limit is posted at 55 miles per hour (mi/h), do you know the metric speed limit in kilometers per hour? (Given: 1 mi = 1.61 km.)

First, we write down the ratio of units asked for in the answer (km/h). *Second*, we write down the related given value (55 mi/h). *Third*, we apply a unit factor using the standard format.

$$\underset{(2)}{\frac{55\text{ mi}}{1\text{ h}}} \times \underset{(3)}{\frac{\text{unit}}{\text{factor}}} = \underset{(1)}{\frac{\text{km}}{\text{h}}}$$

We can use the given unit equation 1 mi = 1.61 km to write the following unit factors: 1 mi/1.61 km and 1.61 km/1 mi. To cancel miles in the given value, we apply the second unit factor. Thus, we have

$$\frac{55\text{ }\cancel{\text{mi}}}{1\text{ h}} \times \frac{1.61\text{ km}}{1\text{ }\cancel{\text{mi}}} = 89\text{ km/h}$$

Where the metric speed limit is posted, the value is usually 90 km/h. Example Exercise 3.8 illustrates problems requiring compound units in the answer.

Example Exercise 3.8 • Conversion Involving Compound Units

If an automobile is traveling at 65 mi/h, what is the velocity in feet per second? (Given: 1 mi ≡ 5280 ft and 1 h ≡ 3600 s.)

Solution

In conversions involving compound units, we must convert the given value to the units asked for in the answer.

Step 1, The unit asked for in the answer is **ft/s**.
Step 2, The related given value is **65 mi/h**.
In this problem, we must convert mi to ft and h to s. Thus, we will need two unit factors.

$$\underset{(2)}{\frac{65\text{ mi}}{1\text{ h}}} \times \frac{\text{unit}}{\text{factor 1}} \underset{(3)}{\times} \frac{\text{unit}}{\text{factor 2}} = \underset{(1)}{\frac{\text{ft}}{\text{s}}}$$

Step 3, We apply unit conversion factors.
To convert from mi to ft, we need the equivalent 1 mi ≡ 5280 ft. To cancel miles in the given value, we apply the unit factor as follows.

$$\frac{65\text{ }\cancel{\text{mi}}}{1\text{ h}} \times \frac{5280\text{ ft}}{1\text{ }\cancel{\text{mi}}} \times \frac{\text{unit}}{\text{factor 2}} = \frac{\text{ft}}{\text{s}}$$

To convert from h to s, we use the equivalent 1 h ≡ 3600 s. To cancel hours in the given value, we apply the unit factor as follows.

$$\frac{65\text{ }\cancel{\text{mi}}}{1\text{ }\cancel{\text{h}}} \times \frac{5280\text{ ft}}{1\text{ }\cancel{\text{mi}}} \times \frac{1\text{ }\cancel{\text{h}}}{3600\text{ s}} = 95\text{ ft/s}$$

The answer is rounded off to two digits because the given value has two significant digits. Unit factors 1 and 2 each have an infinite number of significant digits. We can outline the solution as follows.

(continued)

Example Exercise 3.8 *(continued)*

$$\frac{65\text{ mi}}{1\text{ h}} \times \frac{5280\text{ ft}}{1\text{ mi}} \rightarrow \frac{\text{ft}}{\text{h}} \times \frac{1\text{ h}}{3600\text{ s}} \rightarrow 95\frac{\text{ft}}{\text{s}}$$

Self-Test Exercise

If a runner completes a 10K race in 32.50 minutes (min), what is the 10.0-km pace in miles per hour? (Given: 1 mi = 1.61 km.)

Answer: 11.5 mi/h

Chemistry Connection · The Olympics

Which Olympic race is nearly equal in length to a quarter-mile, that is, 440 yards?

The Olympics began in 776 B.C. with a single event, a 200-yd dash called a *stadion*. In time, 2-stadia and 24-stadia events were added, and wrestling appeared in 708 B.C. Eventually, interest in the games declined, and they were discontinued in 394 A.D.

In the summer of 1896, the modern Olympic games were initiated and held in Greece. The games attracted about 500 athletes from 13 nations. Since then, the games have been held every 4 years in various cities around the world.

The games have steadily increased in number of participants, as well as in number of events. In 1896, 311 men from 13 nations competed, and the United States won gold medals in 9 of the 12 events. In 1996 over 10,000 men and women athletes from 197 nations competed.

In the past, U.S. track and field competitions were conducted using the English system of measurement. That is, American runners competed at distances of 100 yards, 440 yards, and 1 mile. In Olympic competitions, events are conducted using the metric system owing to its international acceptance. Olympic runners compete at comparable distances such as 100, 400, and 1500 meters. The marathon, however, is an exception. It corresponds to the distance run by the Greek messenger who carried news of the Athenian victory on the plains of Marathon in 490 B.C. That legendary distance was 26 miles, 385 yards.

The Olympic winter games were introduced at Chamonix, France, and have been held three times in the United States, twice at Lake Placid and once at Squaw Valley. The winter games feature skating, skiing, bobsledding, luge, tobogganing, and ice hockey. In Nordic cross-country skiing, the races are 10, 20, 30, and 50 kilometers; in ski jumping, the ramps are 70 and 90 meters. Thus, Olympic competitions—winter and summer—employ the metric system of measurement.

◀ **World Record** Maurice Greene of the United States shatters the world record for the 100-m dash by recording a time of 9.79 s.

The 400-meter Olympic race is nearly identical to 440 yards, and the world records for the two races are within tenths of a second.

3.5 Volume by Calculation

Objectives
- To state the relationship of length, width, and thickness to the volume of a rectangular solid.
- To express a given volume in units of milliliters, cubic centimeters, or cubic inches.

The amount of space occupied by a solid, liquid, or gas is referred to as volume. The volume of a regular solid can be determined by calculation. The volume of a rectangular solid is found by multiplying length (l) times width (w) times thickness (t). We can express this relationship by the following equation.

$$l \times w \times t = \text{volume}$$

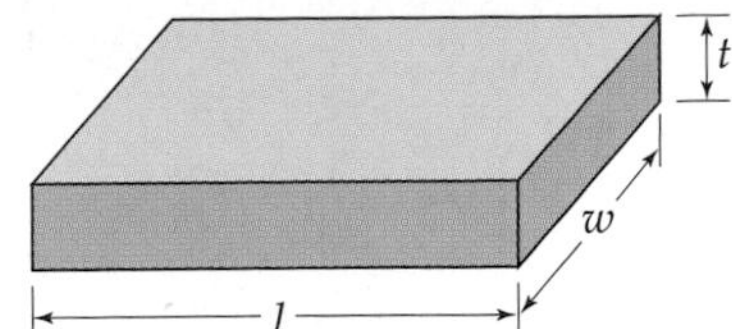

▲ **Volume by Calculation** The volume of a rectangular solid is equal to its length (l) times its width (w) times its thickness (t) or height.

Before we calculate a volume, we must express each dimension in the same units. Suppose the length and width are in centimeters, but the thickness is in millimeters. Before calculating the volume, we must express all three dimensions in either centimeters or millimeters. When we express length, width, and thickness in centimeters, the volume of the solid is expressed in **cubic centimeters** (symbol **cm^3**). For example, if a rectangular solid measures 3 cm by 2 cm by 1 cm, we multiply and find the volume is 6 cm^3. Example Exercise 3.9 further illustrates how to determine the volume of a rectangular solid by calculation.

Example Exercise 3.9 • Volume by Calculation

If a rectangular stainless-steel solid measures 5.55 cm long, 3.75 cm wide, and 2.25 cm thick, what is the volume?

Solution

Since all three dimensions are given in cm, we will calculate the volume in cm^3. Substituting into the formula for volume, we have

$$\begin{array}{ccccccc} l & \times & w & \times & t & = & \text{volume} \\ (5.55 \text{ cm}) & & (3.75 \text{ cm}) & & (2.25 \text{ cm}) & = & 46.8 \text{ cm}^3 \end{array}$$

Self-Test Exercise

If a rectangular brass solid measures 52.0 mm by 25.0 mm by 15.0 mm, what is the volume in cubic millimeters?

Answer: 19,500 mm^3 (1.95×10^4 mm^3)

Let's consider a variation of the preceding problem. For example, a rectangular solid has a length of 4 cm, a width of 3 cm, and a volume of 24 cm^3. If we are given the length, width, and volume of a rectangular solid, we can calculate its thickness. We obtain the thickness of the solid by dividing its volume by its length and its width; that is, we divide 24 cm^3 by 4 cm and 3 cm. In this example, the thickness of the solid is 2 cm. Example Exercise 3.10 further illustrates how to find the thickness of a rectangular solid.

Example Exercise 3.10 • Calculation of Thickness

A sheet of tin foil measures 25.0 cm by 10.0 cm, and the volume is 3.12 cm^3. What is the thickness of the foil in millimeters?

Solution

The volume (V) of the tin foil is equal to the product of the three dimensions. To obtain the thickness of the foil, we divide the volume by the length and the width. Thus,

$$\frac{V}{l \times w} = \text{thickness}$$

$$\frac{3.12 \text{ cm}^3}{25.0 \text{ cm} \times 10.0 \text{ cm}} = 0.0125 \text{ cm}$$

Since the problem asked for the thickness in mm, we convert as follows.

$$0.0125 \cancel{\text{cm}} \times \frac{1 \cancel{\text{m}}}{100 \cancel{\text{cm}}} \times \frac{1000 \text{ mm}}{1 \cancel{\text{m}}} = 0.125 \text{ mm}$$

(continued)

Example Exercise 3.10 *(continued)*

Self-Test Exercise

A sheet of aluminum foil measures 35.0 cm by 25.0 cm, and the volume is 1.36 cm^3. What is the thickness of the foil in centimeters?

Answer: 0.00155 cm $(1.55 \times 10^{-3}$ cm)

Cubic Volume and Liquid Volume

In the metric system, the basic unit of liquid volume is the liter. A liter is equivalent to the volume occupied by a cube exactly 10 cm on a side.

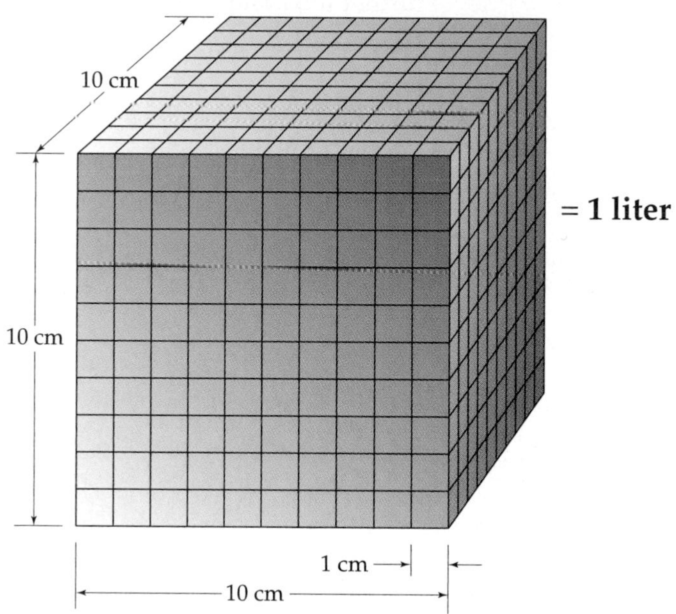

One liter cube

▶ **One Liter** A liter equals the volume of a cube 10 cm on a side, that is, 1000 cm^3.

A cube is a regular solid with its length, width, and thickness being equal. We calculate the volume of the cube by multiplying length by width by thickness.

$$(10 \text{ cm})(10 \text{ cm})(10 \text{ cm}) = 1000 \text{ cm}^3$$

The calculated volume of 1 L is 1000 cm^3. Moreover, we recall the exact equivalent: 1 L ≡ 1000 mL. We can combine the two equivalents as follows.

$$1000 \text{ cm}^3 \equiv 1 \text{ L} \equiv 1000 \text{ mL}$$

Simplifying, we have

$$1 \text{ cm}^3 \equiv 1 \text{ mL}$$

We see that 1 cubic centimeter (cm^3) is exactly equivalent to 1 mL. Example Exercise 3.11 further illustrates the calculation of volume.

▲ **The Beetle** The original Volkswagen Beetle was the most widely sold automobile in the world, and the newest model is also proving popular.

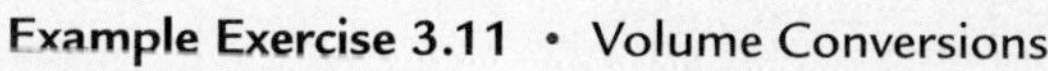

Example Exercise 3.11 • Volume Conversions

A Volkswagen Beetle has a 1.8-L engine. What is the engine volume in (a) milliliters, (b) cubic centimeters, and (c) cubic inches?

Solution

Since 1 L ≡ 1000 mL, we can easily convert from L to mL.

$$1.8 \not{\text{L}} \times \frac{1000 \text{ mL}}{1 \not{\text{L}}} = 1800 \text{ mL}$$

Since 1 mL $\equiv$ 1 cm^3, we can directly convert from mL to cm^3.

$$1800 \cancel{mL} \times \frac{1\ cm^3}{1\ \cancel{mL}} = 1800\ cm^3$$

The conversion of cm^3 to $in.^3$ is more challenging. We recall from Table 3.3 that 1 in. = 2.54 cm. Therefore, we can begin the calculation as follows.

$$1800\ cm^3 \times \frac{1\ in.}{2.54\ cm} \neq in.^3$$

Notice that cm units do not cancel cm^3 and that $in.^3$ is not produced. To obtain cubic units, we must use the unit factor three times. That is,

$$1800\ \cancel{cm}^3 \times \frac{1\ in.}{2.54\ \cancel{cm}} \times \frac{1\ in.}{2.54\ \cancel{cm}} \times \frac{1\ in.}{2.54\ \cancel{cm}} = 110\ in.^3$$

Self-Test Exercise

A sport utility vehicle has a 245-$in.^3$ engine. What is the engine volume in (a) cubic centimeters, (b) milliliters, and (c) liters?

Answers: (a) 4010 cm^3; (b) 4010 mL; (c) 4.01 L

Note A cubic centimeter is sometimes abbreviated cc, as in a 1 cc medical injection. Since it is not obvious that cc stands for cubic centimeters, use of this abbreviation is discouraged in scientific measurements.

3.6 Volume by Displacement

Objective · To describe the technique of determining a volume by displacement.

The volumes of liquids and solids can be determined by several methods. The volumes of liquids can be determined directly with calibrated glassware, such as graduated cylinders, pipets, and burets. The volumes of regular solids (cubes, cylinders, spheres) can be calculated from measurements.

The volume of an irregular solid cannot be determined directly by multiplying its measurements. However, the volume can be determined *indirectly* by measuring the amount of water the solid displaces. This technique is called determination of **volume by displacement**.

Suppose we wish to determine the volume of a piece of jade. Since it is an irregular solid object, we cannot measure its volume directly. We have to use the technique of determining volume by displacement. We first fill a graduated cylinder halfway with water (Figure 3.2) and record the water level. Next, we carefully slip the piece of jade into the graduated cylinder and record the new water level. The difference between the initial and final water levels represents the volume of the piece of jade.

We can also use the volume by displacement method to determine the volume of a gas. When a gas is produced from a chemical reaction, its volume is equal to the amount of water it displaces. This technique provides good results for gases that are not soluble, or only slightly soluble, in water. For instance, heating a solid sample may release oxygen gas, which displaces water from a closed container. Since oxygen is not very soluble in water, the volume of water displaced is equal to the volume of gas released. This technique is called *collecting a gas over water* and is illustrated in Figure 3.3.

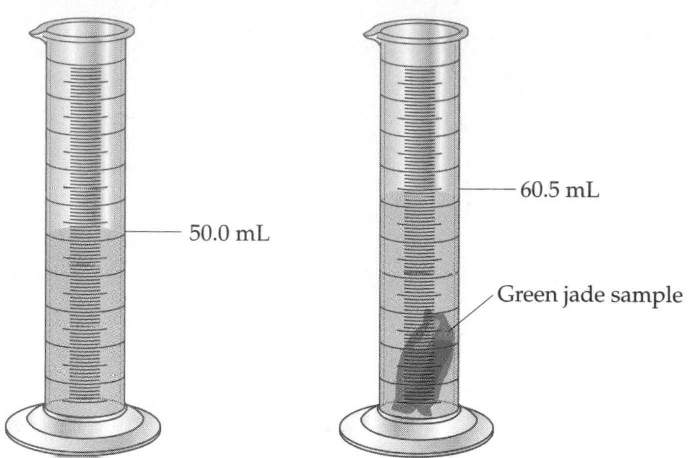

▶ **Figure 3.2 Volume of a Solid by Displacement** The difference between the initial and final water levels (10.5 mL) in the graduated cylinder is equal to the volume of the solid piece of jade.

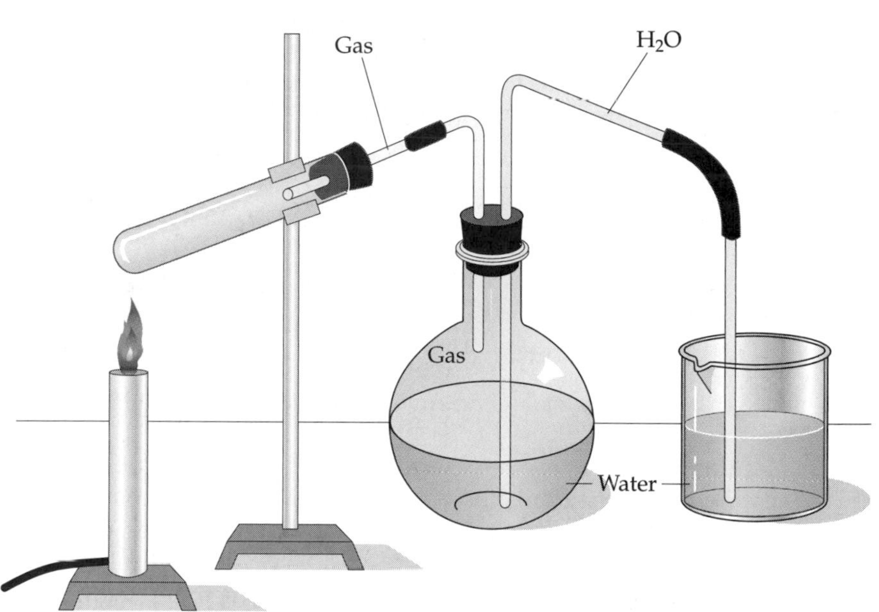

▶ **Figure 3.3 Volume of a Gas by Displacement** A flask is filled with water, and a sample is placed in the test tube. After heating, the sample decomposes and produces oxygen gas, which in turn displaces water from the flask. The volume of water displaced into the beaker equals the volume of oxygen gas produced by the sample.

Example Exercise 3.12 • Determining Volume by Displacement

A quartz stone weighing 30.475 g is dropped into a graduated cylinder. If the water level increases from 25.0 mL to 36.5 mL, what is the volume of the quartz?

Solution

We find the volume of the quartz by displacement. The volume is equal to the difference between the two water levels. Thus,

$$36.5 \text{ mL} - 25.0 \text{ mL} = 11.5 \text{ mL}$$

Self-Test Exercise

Hydrogen peroxide decomposes to give oxygen gas, which displaces an equivalent volume of water into a beaker. If the water level in the beaker increases from 50.0 mL to 105.5 mL, what is the volume of oxygen gas?

Answer: 55.5 mL

3.7 The Density Concept

Experiment #3, Prentice Hall Laboratory Manual

Objectives
- To explain the concept of density.
- To state the value for the density of water: 1.00 g/mL.
- To perform calculations that relate density to mass and volume.

We use the term **density** (symbol ***d***) to express the concentration of mass. The more concentrated the mass is in a sample, the greater its density. We sometimes say an object is heavy when we are actually trying to convey the concept of density. A block of lead and a block of ice, each weighing 1 kg, have the same mass. The lead may seem heavier, however, because its density is 11.3 times greater.

Density is formally defined as the amount of mass per unit volume. The term "per" indicates a ratio, as in "miles per hour." Thus, mass divided by volume equals density. We can write this relationship mathematically as

$$\frac{\text{mass}}{\text{volume}} = \text{density}$$

Density is expressed in different units. For solids and liquids, it is usually expressed in grams per cubic centimeter (g/cm^3) or grams per milliliter (g/mL). For gases, density is usually expressed in grams per liter (g/L). Figure 3.4 illustrates the concept of density.

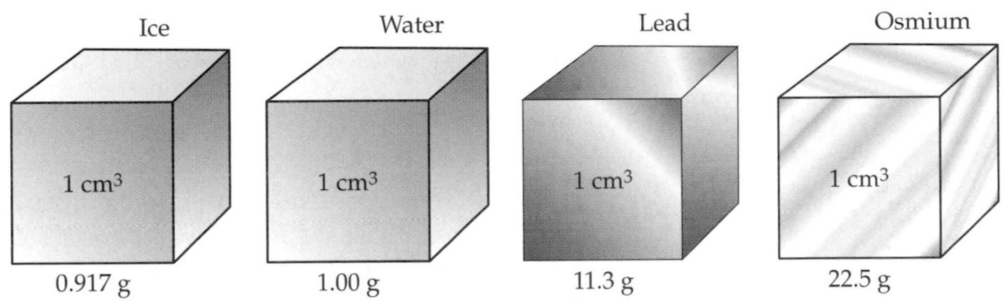

Figure 3.4 Illustration of Density Each cube represents a volume of 1.00 cm^3. Notice that the mass of each cube increases as the density becomes greater. Osmium has the highest density of any metal, 22.5 g/cm^3.

Recall that the original metric standards for mass and volume were based on water. A mass of 1 kg was set equal to a 1-L volume of water. Therefore, the density of water is 1 kg/L. A density of 1 kg/L is exactly equivalent to 1 g/mL.

Although the density of solids and liquids varies slightly with temperature, we will ignore this small effect. For our purposes, we will consider the density of water to be 1.00 g/mL. The density of gases, however, is an exception. The density of a gas varies greatly with changes in temperature and pressure. Table 3.4 lists the density for selected solids, liquids, and gases.

We can drop an object into water and estimate its density. An object that sinks in water has a density greater than 1.00 g/mL and an object that floats on water has a density less than 1.00 g/mL.

Consider the following experiment. Water, chloroform, and ethyl ether are poured into a tall glass cylinder. The liquids form three layers as shown in Figure 3.5. Identify liquids L_1 and L_2 by referring to the density of the liquids listed in Table 3.4. Next, ice, rubber, and aluminum cubes are dropped into the cylinder. Identify solids S_1, S_2, and S_3 by referring to the density of the solids listed in Table 3.4.

We know that liquid L_1 must be less dense than water (d = 1.00 g/mL) because it floats; liquid L_2 must be more dense than water because it sinks. Thus, liquid L_1 is ethyl ether (d = 0.714 g/mL), which is less dense than water, and liquid L_2 is chloroform (d = 1.48 g/mL), which is more dense than water.

We see in Figure 3.5 that S_1 floats on water, S_2 floats on chloroform, and S_3 sinks in chloroform. Table 3.4 indicates that the densities of ice, rubber, and aluminum are

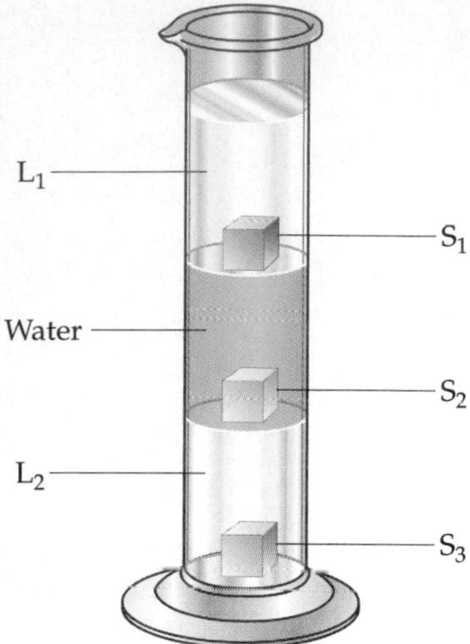

▲ **Figure 3.5 The Concept of Density for Liquids and Solids** In a tall glass cylinder, liquid L_1 is shown floating on water which is floating on liquid L_2. Solid S_1 is floating on water, solid S_2 is floating on liquid L_2, and solid S_3 has sunk to the bottom of the cylinder.

Table 3.4 Density of Selected Solids, Liquids, and Gases

Example	Density (*d*)
Solids	
ice	0.917 g/cm^3 or g/mL
rubber	1.19
magnesium	1.74
aluminum	2.70
iron	7.87
lead	11.3
gold	18.9
Liquids	
ethyl ether	0.714 g/cm^3 or g/mL
ethyl alcohol	0.789
water	1.00
chloroform	1.48
mercury	13.6
Gases*	
hydrogen	0.090 g/L
helium	0.179
ammonia	0.760
air	1.29
oxygen	1.43

*The listed density value of each gas is given at 0°C and normal atmospheric pressure.

0.917, 1.19, and 2.70 g/cm^3, respectively. Thus, S_1 must be ice because it sinks in ether and floats on water, S_2 must be rubber because it sinks in water and floats on chloroform, and S_3 must be aluminum because it sinks in chloroform.

Now that we understand the concept of density, let's try some calculations. We can find the density of a sample by simply dividing its mass by its volume. Example Exercise 3.13 illustrates the calculation of density.

Example Exercise 3.13 • Density Calculation

The density of platinum metal is about twice the density of lead. If a platinum bar measures 5.00 cm by 2.00 cm by 1.00 cm and has a mass of 224.50 g, what is the density of the metal?

Solution

Density is defined as the ratio of mass to volume.

$$\frac{\text{mass}}{\text{volume}} = \text{density}$$

The mass is given, but we must calculate the volume. That is,

$$l \times w \times t = \text{volume}$$

$$(5.00\ \text{cm})\quad(2.00\ \text{cm})\quad(1.00\ \text{cm}) = 10.0\ \text{cm}^3$$

Substituting into the equation for density and solving, we have

$$\frac{224.50\ \text{g}}{10.0\ \text{cm}^3} = 22.5\ \text{g/cm}^3$$

It is interesting to note that platinum metal is more dense than gold ($d = 18.9\ g/cm^3$); it is also slightly more valuable.

Self-Test Exercise

Carbon tetrachloride is a solvent used for degreasing electronic parts. If 25.0 mL of carbon tetrachloride has a mass of 39.75 g, what is the density of the liquid?

Answer: 1.59 g/mL

Applying Density as a Unit Factor

We can also solve problems that ask for the mass or volume of a sample by applying the unit analysis method of problem solving. The secret to solving these problems is to realize that density can be used as a unit factor because it relates mass and volume. For example, the density of mercury is 13.6 g/mL, and so we can write

$$\frac{13.6\ g}{1\ mL} \quad \text{and} \quad \frac{1\ mL}{13.6\ g}$$

Let's calculate the milliliter volume of 75.5 g of liquid mercury. *First*, we write down the unit asked for in the answer (mL). *Second*, we write down the related given value (75.5 g). *Third*, we apply a unit factor using the following format.

$$\underset{(2)}{75.5\ g} \times \underset{(3)}{\frac{\text{unit}}{\text{factor}}} = \underset{(1)}{mL}$$

To cancel the units in the given value (75.5 g), we select the unit factor 1 mL/13.6 g. After substituting, we have

$$75.5\ \cancel{g} \times \frac{1\ mL}{13.6\ \cancel{g}} = 5.55\ mL$$

The given value and the unit factor each have three significant digits; thus, the answer is rounded to three digits. Example Exercises 3.14 and 3.15 further illustrate the use of density as a unit factor.

Example Exercise 3.14 • Density as a Unit Factor

An automobile battery contains 1275 mL of sulfuric acid. If the density of battery acid is 1.84 g/mL, how many grams of acid are in the battery?

Solution

The unit asked for in the answer is g, and the given value is 1275 mL; therefore,

$$1275\ mL \times \frac{\text{unit}}{\text{factor}} = g$$

Since the density is 1.84 g/mL, the two unit factors are

$$\frac{1.84\ g}{1\ mL} \quad \text{and} \quad \frac{1\ mL}{1.84\ g}$$

We select the first unit factor to cancel mL. Thus,

$$1275\ \cancel{mL} \times \frac{1.84\ g}{1\ \cancel{mL}} = 2350\ g$$

(continued)

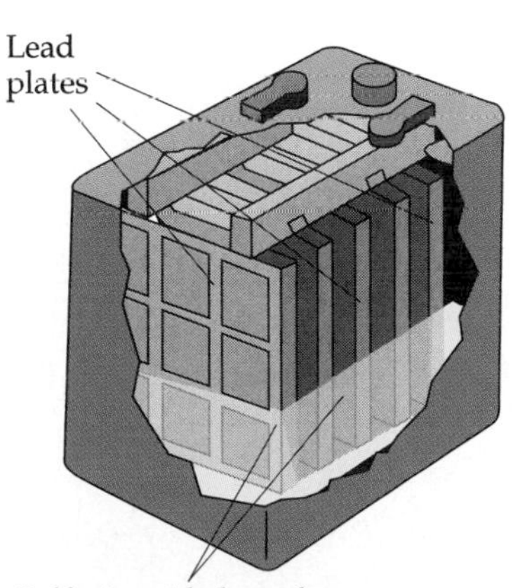

▲ **Automobile Battery** An automobile battery contains lead plates and is filled with sulfuric acid.

Example Exercise 3.14 *(continued)*

Self-Test Exercise

The most abundant gases in our atmosphere are nitrogen, oxygen, and argon. What is the volume of 1.00 kg of air? (Assume that the density of air is 1.29 g/L.)

Answer: 775 L

Example Exercise 3.15 • Density as a Unit Factor

A 1.00-in. cube of copper measures 2.54 cm on a side. What is the mass of the copper cube? (Given: *d* of copper = 8.96 g/cm^3.)

Solution

We are asked for mass, and so we choose gram as the unknown unit. The given value is a cube 2.54 cm on a side. We must first calculate the volume by multiplying side times side times side. Thus,

$$(2.54 \text{ cm})\,(2.54 \text{ cm})\,(2.54 \text{ cm}) = 16.4 \text{ cm}^3$$

The unit analysis format is

$$16.4 \text{ cm}^3 \times \frac{\text{unit}}{\text{factor}} = \text{g}$$

We are given that the density of copper is 8.96 g/cm^3. We can apply a density unit factor to cancel units. Thus,

$$16.4 \cancel{\text{cm}^3} \times \frac{8.96 \text{ g}}{1 \cancel{\text{cm}^3}} = 147 \text{ g}$$

Self-Test Exercise

A cube of silver is 5.00 cm on a side and has a mass of 1312.5 g. What is the density of silver?

Answer: 10.5 g/cm^3

Specific Gravity

The ratio of the density of a liquid to the density of water at 4°C is called **specific gravity** (symbol **sp gr**). Since specific gravity is a ratio of two densities, the units cancel. Thus, specific gravity is a unitless quantity. The density of water is 1.00 g/mL, and the specific gravity of water is 1.00.

Diagnostic medical testing often includes measurement of the specific gravity of body fluids. For instance, the specific gravity of urine may be 1.02, and the specific gravity of blood may be 1.06. Both values are considered to be in the normal range.

3.8 Temperature

Objectives

- To state the values for the freezing point and boiling point of water on the Fahrenheit, Celsius, and Kelvin scales.
- To express a given temperature in degrees Fahrenheit (°F), degrees Celsius (°C), or Kelvin units (K).

The "hotness" or "coolness" of the atmosphere is determined by how fast air molecules are moving. If the temperature is warmer, molecules move faster and have more energy. If the temperature is cooler, molecules move slower and have less energy. **Temperature** is a measure of the average energy of individual particles in a system. The system may be air molecules in the atmosphere, water molecules in the ocean, or iron atoms in molten metal. We measure temperature using a thermometer.

In 1724 Daniel Gabriel Fahrenheit (1686–1736), a German physicist, invented the mercury thermometer. In attempting to produce as cold a temperature as possible, Fahrenheit prepared an ice bath to which he added salt to lower the temperature further. He then assigned a value of zero to that temperature and marked his Fahrenheit scale accordingly.

Fahrenheit obtained a second reference point by recording his body temperature; he assigned that temperature a value of 96. The distance between the two reference points was divided into 96 equal units, and each division was termed a **Fahrenheit degree** (symbol °F). Later, the freezing and boiling points of water were selected as the standard reference points. The freezing point of water was assigned a value of 32°F, and the boiling point a value of 212°F. The Fahrenheit degree eventually became a basic unit in the English system of measurement.

In 1742 Anders Celsius (1701–1744), a Swedish astronomer, proposed a scale similar in principle to the Fahrenheit scale. On the Celsius scale, the freezing point of water was assigned a value of 0° and the boiling point of water a value of 100°. The scale was then divided into 100 equal divisions, each division representing a **Celsius degree** (symbol °C). The Celsius degree is also referred to as a centigrade degree and is a basic unit in the metric system.

In 1848 William Thomson (1824–1907), an English physicist knighted Lord Kelvin, proposed a scale based on the lowest possible temperature. The unit of temperature was the **Kelvin** (symbol **K**), which is a basic unit in the SI system. On the Kelvin scale, the coldest temperature was assigned a value of 0, and each division on the scale is equal to one Celsius degree. The lowest temperature is referred to as absolute zero and corresponds to −273.15°C. Although there is no highest temperature on the Kelvin scale, the interior of the Sun reaches about 10,000,000 K.

A unit on the absolute temperature scale is abbreviated K, not °K

Fahrenheit and Celsius Temperature Conversions

One division on the Kelvin scale is equivalent to one degree on the Celsius scale. Since 0 K is equivalent to −273°C, the freezing point of water is 273 K and the boiling point of water is 373 K. Figure 3.6 illustrates how the three temperature scales are related. Notice that 180 Fahrenheit units are equivalent to 100 Celsius units. Therefore, to convert from degrees Fahrenheit to degrees Celsius, we first subtract 32 (the difference between the freezing point of water on the two scales) and then multiply by 100°C/180°F. That is,

$$(°\text{F} - 32°\text{F}) \times \frac{100°\text{C}}{180°\text{F}} = °\text{C}$$

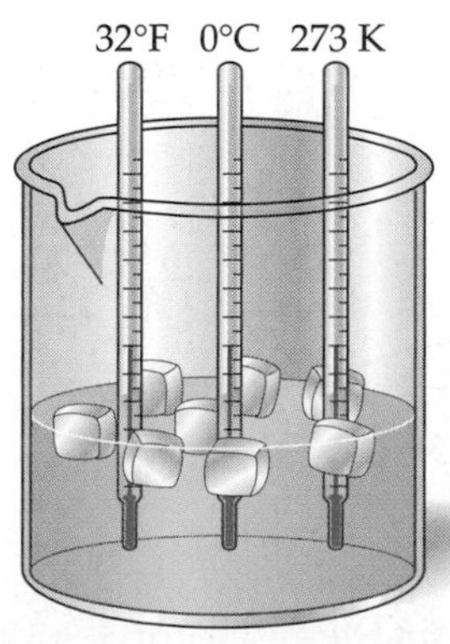

(a)

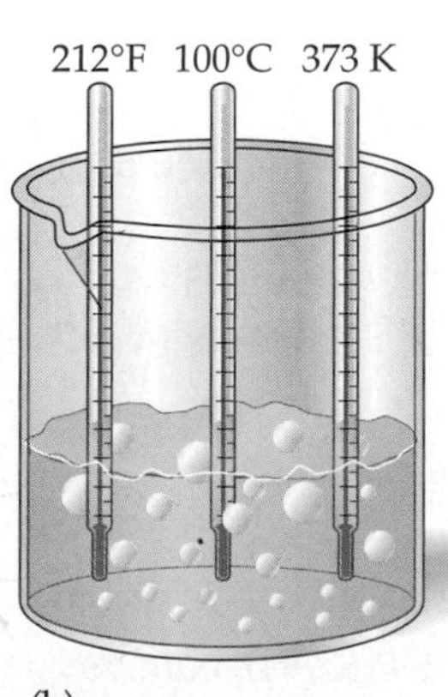

(b)

Fahrenheit, Celsius, and Kelvin Scales

◀ **Figure 3.6 Fahrenheit, Celsius, and Kelvin Temperature Scales** A Fahrenheit, a Celsius, and a Kelvin thermometer are placed in (a) ice water and (b) boiling water. Notice the freezing point and boiling point on each scale. The number of divisions is 180 units on the Fahrenheit scale, 100 units on the Celsius scale, and 100 units on the Kelvin scale.

▲ **Australian Stamp** The caricature illustrates that 38°C is approximately equal to 100°F.

Example Exercise 3.16 • °F–°C Temperature Conversion

Normal human body temperature is 98.6°F. What is normal body temperature in degrees Celsius?

Solution

To calculate °C, we refer to Figure 3.6 and compare the Celsius and Fahrenheit temperature scales. The conversion from °F to °C is as follows.

$$(98.6°\text{F} - 32°\text{F}) \times \frac{100°\text{C}}{180°\text{F}} = °\text{C}$$

Simplifying and canceling units gives

$$66.6°\cancel{\text{F}} \times \frac{100°\text{C}}{180°\cancel{\text{F}}} = 37.0°\text{C}$$

Since 32°F and 100°C/180°F are derived from definitions, each value has an infinite number of significant digits. The given value, 98.6°F, has three significant digits, and so the answer is rounded to three digits.

Self-Test Exercise

A temperature of 5°F combined with a 10-mi/h wind, chills the temperature of the air to −15°F. What is a −15°F wind chill temperature on the Celsius scale?

Answer: −26°C

To convert from °C to °F, we reverse the foregoing procedure of converting °F to °C. That is, we first multiply the Celsius temperature by the ratio 180°F/100°C, and then add 32°F. That is,

$$\left(°\text{C} \times \frac{180°\text{F}}{100°\text{C}}\right) + 32°\text{F} = °\text{F}$$

Example Exercise 3.17 • °C–°F Temperature Conversion

The average surface temperature of Mars is −55°C. What is the equivalent Fahrenheit temperature?

Solution

To calculate °F, we examine Figure 3.5 and compare the Celsius and Fahrenheit temperature scales. The conversion formula is as follows.

$$\left(-55°\cancel{\text{C}} \times \frac{180°\text{F}}{100°\cancel{\text{C}}}\right) + 32°\text{F} = °\text{F}$$

Simplifying and canceling units, we have

$$-99°\text{F} + 32°\text{F} = -67°\text{F}$$

Self-Test Exercise

The average surface temperature of Venus is 457°C. What is the equivalent Fahrenheit temperature?

Answer: 855°F

Celsius and Kelvin Temperature Conversions

Let's reexamine the temperature scales in Figure 3.6. Notice that the Kelvin scale is 273 units above the Celsius scale. Therefore, to convert from degrees Celsius to Kelvin, we must add 273 units to the Celsius temperature.

$$°\text{C} + 273 = \text{K}$$

Conversely, to convert from a Kelvin to a Celsius temperature, subtract 273 units from the Kelvin temperature. It is helpful to remember that negative Kelvin temperatures are nonexistent. By definition, the lowest possible temperature is assigned a value of 0 K. Example Exercise 3.18 illustrates the conversion of Celsius and Kelvin temperatures.

Example Exercise 3.18 • °C–K Temperature Conversion

Liquid nitrogen is used in medicine to freeze skin tissue. If liquid nitrogen is −196°C, what is the temperature on the Kelvin scale?

Solution

Given the Celsius boiling point temperature, we add 273 units to find the corresponding Kelvin temperature. Thus, we have

$$-196°C + 273 = 77 \text{ K}$$

Self-Test Exercise

The secret to "fire-walking" is to first walk barefoot through damp grass and then step lively on the red-hot coals. If the bed of coals is at 1475 K, what is the temperature on the Celsius scale?

Answer: 1202°C

▲ **Thermos of Liquid Nitrogen** Although nitrogen is ordinarily a gas, it liquefies at −196°C. This temperature is cold enough to freeze the moisture in air and form a white mist.

Note Ideally, you should be able to convert Fahrenheit and Celsius temperatures by understanding the relationship of one temperature scale to another (Figure 3.6). In practice, temperatures are routinely converted with the use of a reference table or a wall chart. Moreover, it is possible to convert temperatures with a single keystroke using an inexpensive scientific calculator.

3.9 Heat and Specific Heat

Objectives
- To explain the concepts of heat and specific heat.
- To state the value for the specific heat of water: 1.00 cal/g × °C.
- To perform calculations that relate heat to the mass, specific heat, and temperature change of a substance.

Heat and temperature both measure the energy in a solid, liquid, or gas. The distinction is that **heat** measures the *total energy*, whereas **temperature** measures the *average energy*. To grasp the difference, let's consider a cup of coffee and a teaspoon of coffee, each of which contains a hot liquid at 100°C. What would be more dangerous: drinking the cup of coffee or drinking the teaspoon of coffee? Obviously, an entire cup of coffee is more harmful than a teaspoon. The scientific reason is that the cup of coffee has much more heat than the teaspoon of coffee, even though each contains liquid at the same temperature. Figure 3.7 further illustrates the concepts of heat and temperature.

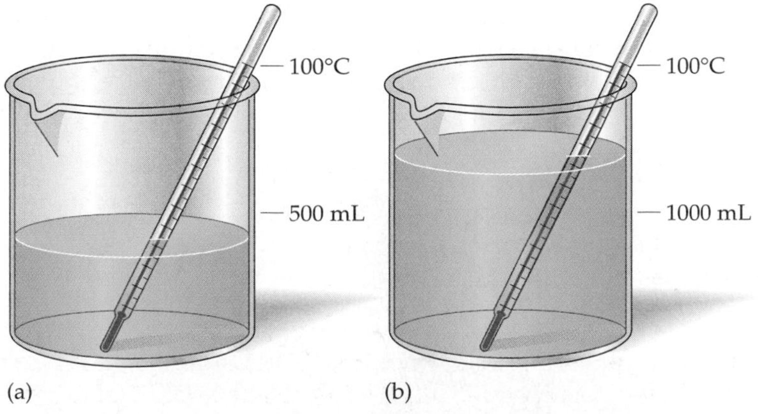

◀ **Figure 3.7 Heat versus Temperature** In (a) 500 mL of water is heated to 100°C, and in (b) 1000 mL is heated to 100°C. Although the temperatures are the same, the second beaker has twice the amount of heat.

Heat energy is often expressed in units of calories or kilocalories. A **calorie** (symbol **cal**) is the amount of heat necessary to raise 1 gram of water 1 degree on the Celsius scale. A kilocalorie (kcal) is the amount of heat necessary to raise 1000 grams of water 1 degree on the Celsius scale. A food Calorie (Cal) is spelled with a capital letter to distinguish it from a metric calorie. One food Calorie is equal to 1 kilocalorie, that is, 1000 calories.

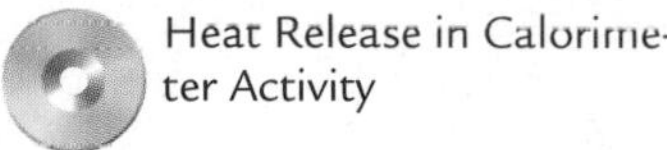

A SI unit of energy is the **joule** (symbol **J**), where 1 cal = 4.184 J. The heat produced by chemical reactions is often expressed in kilocalories, as well as in kilojoules (kJ), where 1 kcal = 4.184 kJ. Example Exercise 3.19 illustrates the conversion of calories, kilocalories, and joules.

Example Exercise 3.19 • Energy Unit Conversion

Burning one gram of natural gas produces 13,200 cal of heat energy. Express the heat of combustion in (a) kilocalories and (b) kilojoules.

Solution

We can convert cal to kcal and kJ as follows.

(a) The unit asked for is kcal, and the given value is 13,200 cal. Since 1 kcal ≡ 1000 cal, we have

$$13{,}200\ \cancel{\text{cal}} \times \frac{1\ \text{kcal}}{1000\ \cancel{\text{cal}}} = 12.2\ \text{kcal}$$

(b) The unit asked for is kJ, and the given value is 13,200 cal. Since 1 kcal = 4.184 kJ, we have

$$13{,}200\ \cancel{\text{cal}} \times \frac{1\ \cancel{\text{kcal}}}{1000\ \cancel{\text{cal}}} \times \frac{4.184\ \text{kJ}}{1\ \cancel{\text{kcal}}} = 55.2\ \text{kJ}$$

Self-Test Exercise

Burning one gram of gasoline produces 47.9 kJ of heat energy. Express the heat of combustion in (a) kilocalories and (b) calories.

Answers: (a) 11.4 kcal; (b) 11,400 cal

Specific Heat

We can define **specific heat** as the amount of heat required to bring about a given change in temperature. It is observed that the amount of heat necessary is unique for each substance. The specific heat of water is relatively high, and the change in temperature is minimal as water gains or loses heat. The surface of the Earth is covered with water, and fortunately its high specific heat helps to regulate climates and maintain moderate temperatures. Figure 3.8 illustrates the increase in temperature for four substances each receiving 1 cal of heat.

We can define specific heat as the amount of heat required to raise the temperature of one gram of substance one degree Celsius; the units of specific heat are often given in calories per gram per degree Celsius. For reference, the specific heat

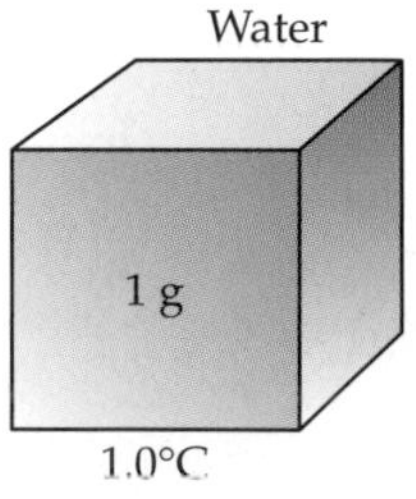

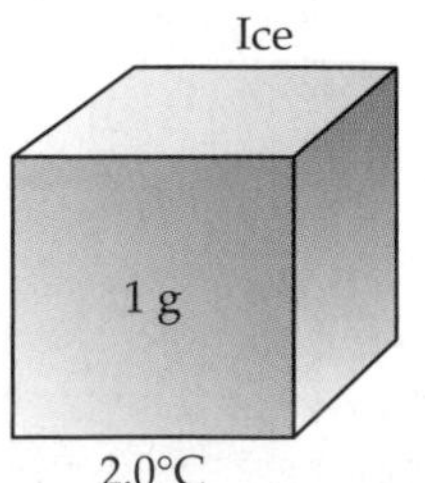

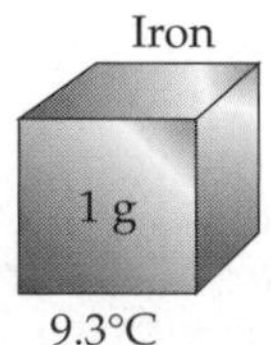

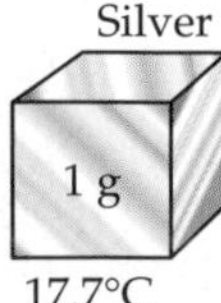

▶ **Figure 3.8 Illustration of Specific Heat** Each cube represents 1 g of substance receiving 1 cal of heat. The temperature change, which varies with the substance, increases the number of degrees shown.

of water is relatively high and has a value of 1.00 cal/g × °C. The specific heats of ice and steam are much lower. Table 3.5 lists the specific heat for selected solids, liquids, and gases.

Table 3.5 Specific Heat for Selected Solids, Liquids, and Gases

Example	Specific Heat
Solids	
ice	0.492 cal/g × °C
aluminum	0.215
carbon (graphite)	0.170
carbon (diamond)	0.124
iron	0.108
copper	0.0920
silver	0.0566
gold	0.0305
Liquids	
water	1.00 cal/g × °C
ethyl alcohol (ethanol)	0.587
methyl alcohol (methanol)	0.424
freon (a refrigerant)	0.232
mercury	0.0331
Gases	
steam	0.485 cal/g × °C
nitrogen	0.249
oxygen	0.219
argon	0.124
radon	0.0224

The units of specific heat can be expressed as calories per gram per degree Celsius. Thus, a gain or loss of heat divided by mass and temperature change (Δt) equals specific heat. We can write this relationship mathematically as

$$\frac{\text{heat (cal)}}{\text{mass (g)} \times \text{temperature change } (\Delta t)} = \text{specific heat} \left(\frac{\text{cal}}{\text{g} \times {}^\circ\text{C}} \right)$$

Let's calculate the specific heat for zinc given that a 65.5-g sample of the metal loses 480 cal of heat when cooled from 100.0° to 21.0°C. We divide the heat loss of the zinc metal by its mass and temperature change. Thus,

$$\frac{480 \text{ cal}}{65.5 \text{ g} \times (100.0 - 21.0)^\circ\text{C}} = \frac{\text{cal}}{\text{g} \times {}^\circ\text{C}}$$

We must first find the temperature change and then calculate the specific heat of the metal. Thus,

$$\frac{480 \text{ cal}}{65.5 \text{ g} \times 79.0^\circ\text{C}} = 0.093 \text{ cal/g} \times {}^\circ\text{C}$$

Applying Specific Heat as a Unit Factor

We can also solve problems that ask for a heat change or the mass of sample, given the specific heat. Once again, we will apply the unit analysis method of problem solving. Like density, specific heat can be used as a unit factor. For example, the specific heat of water is 1.00 cal/g × °C, and so the two unit factors are

$$\frac{1.00 \text{ cal}}{1 \text{ g} \times 1°\text{C}} \quad \text{and} \quad \frac{1 \text{ g} \times 1°\text{C}}{1.00 \text{ cal}}$$

Let's calculate the calories required to heat 125 g of water from 19.5° to 75.0°C. *First*, we write down the unit asked for in the answer (cal). *Second*, we write down the related given value (125 g). *Third*, we apply a unit factor using the following format.

$$\underset{(2)}{125 \text{ g}} \quad \times \quad \underset{(3)}{\frac{\text{unit}}{\text{factor}}} \quad = \quad \underset{(1)}{\text{cal}}$$

In order to cancel the units in the given value (125 g), we select the unit factor 1.00 cal/g × °C. After substituting, we have

$$125 \cancel{\text{g}} \quad \times \quad \frac{1.00 \text{ cal}}{1 \cancel{\text{g}} \times 1°\text{C}} \quad = \quad \text{cal}$$

To cancel the units of °C, we multiply by the temperature change. Thus,

$$125 \cancel{\text{g}} \times \frac{1.00 \text{ cal}}{1 \cancel{\text{g}} \times 1°\cancel{\text{C}}} \times (75.0 - 19.5)°\cancel{\text{C}} = 6940 \text{ cal}$$

Since all the data contains three significant digits, the answer is rounded to three digits. Example Exercise 3.20 further illustrates the use of specific heat as a unit factor.

Example Exercise 3.20 • Specific Heat Calculation

In an energy-efficient home, a solar collector is used to provide heat energy. If the solar system releases 42,000 cal of heat when it cools from 50.0° to 48.0°C, what is the mass of water in the solar collector (specific heat = 1.00 cal/g × °C)?

Solution

The problem asks for mass, and so we write down units of g. We are given the specific heat of water (1.00 cal/g × °C), the heat loss (42,000 cal), and the temperature change (from 50.0° to 48.0°C). Applying the problem-solving format,

$$42{,}000 \text{ cal} \times \frac{\text{unit}}{\text{factors}} = \text{g}$$

Substituting 1.00 cal/g × °C as a unit factor to cancel cal, we have

$$42{,}000 \cancel{\text{cal}} \times \frac{1 \text{ g} \times 1°\text{C}}{1.00 \cancel{\text{cal}}} \times \frac{\text{unit}}{\text{factor}} = \text{g}$$

Multiplying by the reciprocal of the temperature change to cancel °C, we have

$$42{,}000\ \cancel{\text{cal}} \times \frac{1\text{ g} \times 1\cancel{^\circ\text{C}}}{1.00\ \cancel{\text{cal}}} \times \frac{1}{(50.0 - 48.0)\cancel{^\circ\text{C}}} = 21{,}000\text{ g (21 kg)}$$

Notice that 21 kg of water must cool only 2.0°C to supply 42,000 cal of heat. The mass of water, 21 kg, corresponds to 21 L of water in the solar energy collector.

Self-Test Exercise
A 725-g steel horseshoe is heated to 425°C and dropped into a bucket of cold water. If the horseshoe cools to 20°C and the specific heat of steel is 0.11 cal/g × °C, how much heat is released?

Answer: 32,000 cal (32 kcal)

Summary 

Section 3.1 The **English system** of measurement has many unrelated units. On the other hand, the **metric system** is a decimal system of measurement with basic units: **meter (m)**, **gram (g)**, **liter (L)**, and **second (s)**. Metric prefixes provide multiples and fractions of basic units. The common prefixes include *giga-* (G), *mega-* (M), *kilo-* (k), *deci-* (d), *centi-* (c), *milli-* (m), *micro-* (μ), and *nano-* (n). The **International System (SI)** is based on the metric system but is more comprehensive.

Section 3.2 Metric conversion problems are solved by systematically writing a **unit equation** or an **exact equivalent** (1 m ≡ 100 cm). An equal or equivalent relationship generates a **unit factor** and its **reciprocal** (1 m/100 cm and 100 cm/1 m).

Section 3.3 Metric problems are solved systematically by applying the **unit analysis method**. Step 1: Write down the units of the unknown. Step 2: Write down a given value related to the unknown. Step 3: Apply one or more unit factors to convert the units of the given value to the units asked for in the answer.

Section 3.4 With a few exceptions, every nation in the world uses the metric system; however, the English system is still common in the United States. You should memorize the following relationships: 2.54 cm = 1 in., 454 g = 1 lb, and 946 mL = 1 qt. These equivalents can be used for metric–English conversions.

Section 3.5 The volume of a rectangular solid is equal to length times width times thickness. The calculated volume is reported in cubic units, such as **cubic centimeters** ($\mathbf{cm^3}$). You should memorize the exact equivalent: 1 mL ≡ 1 cm^3.

Section 3.6 For irregularly shaped objects, as well as gases, we must determine volume indirectly. We can find the volume of an object or a gas from the amount of water it displaces; this technique involves **volume by displacement**.

Section 3.7 The **density** of water is 1.00 g/mL and serves as a reference. A liquid or solid that floats on water is less dense than water; if it sinks, it is more dense. Density can be used as a unit factor, for example, 1.00 g/1 mL, or 1 mL/1.00 g. **Specific gravity (sp gr)** is unitless and expresses the ratio of the density of a liquid to the density of water.

Section 3.8 **Temperature** measures the average energy of particles in a system. The **Fahrenheit, Celsius,** and **Kelvin** scales have related reference points. The freezing point of water is 32°F, 0°C, or 273 K, and the boiling point of water is 212°F, 100°C, or 373 K.

Section 3.9 **Heat** is a measure of the total energy of particles in a system. Heat changes are expressed in the metric units **calories (cal)** or kilocalories (kcal). Heat changes can also be expressed in the SI unit **joule (J)**, where 1 cal = 4.184 J. **Specific heat** is the amount of heat necessary to raise one gram of substance 1°C; for water, the value is 1.00 cal/g × °C.

Problem-Solving Organizer

Topic	Procedure	Example
Basic Units and Symbols Sec. 3.1	Combine basic metric units and prefixes using symbols.	centimeter, cm (length) kilogram, kg (mass) milliliter, mL (volume)
Metric Conversion Factors Sec. 3.2	(a) Write a unit equation involving basic metric units and prefix units. (b) Write two unit factors for a metric relationship.	1 m = 100 cm $\frac{1\ \text{m}}{100\ \text{cm}}$ and $\frac{100\ \text{cm}}{1\ \text{m}}$
Metric–Metric Conversions Sec. 3.3	1. Write down the units asked for in the answer. 2. Write down the related given value. 3. Apply a unit factor to convert a given value to the units in the answer.	What is the decigram mass of a wheel of cheese that weighs 0.515 kg? $0.515\ \cancel{\text{kg}} \times \frac{1000\ \cancel{\text{g}}}{1\ \cancel{\text{kg}}} \times \frac{10\ \text{dg}}{1\ \cancel{\text{g}}} = 5150\ \text{dg}$
Metric–English Conversions Sec. 3.4	1. Write down the units asked for in the answer. 2. Write down the related given value. 3. Apply a unit factor to convert a given value to the units in the answer.	What is the pound mass of a wheel of cheese that weighs 0.515 kg? $0.515\ \cancel{\text{kg}} \times \frac{1000\ \cancel{\text{g}}}{1\ \cancel{\text{kg}}} \times \frac{1\ \text{lb}}{454\ \cancel{\text{g}}} = 1.13\ \text{lb}$
Volume by Calculation Sec. 3.5	To calculate the volume of a rectangular solid, multiply length by width by thickness: $V = l \times w \times t$.	What is the volume of a rectangular solid measuring 65 mm by 35 mm by 12 mm? $65\ \text{mm} \times 35\ \text{mm} \times 12\ \text{mm} = 27{,}300\ \text{mm}^4$
Volume by Displacement Sec. 3.6	The volume by displacement is the difference between the initial and final readings in a calibrated container.	What is the volume of jade if the water level in a cylinder increases from 25.0 mL to 42.5 mL? 42.5 mL − 25.0 mL = 17.5 mL
The Density Concept Sec. 3.7	The density of a sample is equal to its mass divided by its volume: density = mass/volume.	What is the density of 10.0 mL of ether if its mass is 7.142 g? $\frac{7.142\ \text{g}}{10.0\ \text{mL}} = 0.714\ \text{g/mL}$
Temperature Conversion Sec. 3.8	(a) To convert from Celsius to Kelvin, add 273 units. (b) To convert from Kelvin to Celsius, subtract 273 units.	−88°C + 273 = 185 K 1265 K − 273 = 992°C
Heat and Specific Heat Sec. 3.9	1. Write down the units asked for in the answer. 2. Write down the related given value. 3. Apply specific heat as a unit factor to convert the units.	What is the calorie heat loss when 275 g of water cools from 75.0° to 20.0°C? Specific heat: 1.00 cal/g × °C $275\ \cancel{\text{g}} \times \frac{1.00\ \text{cal}}{1\ \cancel{\text{g}} \times 1\cancel{°\text{C}}} \times (75.0 - 20.0°\text{C})$ $= 15{,}100\ \text{cal}$

Key Concepts*

1. Which of the following statements are true regarding the metric system?
   (a) It is an official system of measurement in the United States.
   (b) It uses one basic unit for length, mass, and volume.
   (c) It is a decimal system.
   (d) It is used by scientists.
   (e) It is used for international trade.
2. Which of the following is a basic unit of length in the English system: inch, foot, mile, rod, yard?
3. Which of the following is a basic unit of length in the metric system: meter, centimeter, millimeter, micrometer, nanometer?
4. How many significant digits are justified by the unit factor 100 cm/1 m?
5. How many significant digits are justified by the unit factor 39.4 in./1 m?
6. Which of the following is equivalent to the volume of a cube 1 cm on a side: 1 L, 1 cL, 1 dL, 1 kL, 1 mL?
7. Which of the following is equivalent to the volume of a cube 10 cm on a side: 1 L, 1 cL, 1 dL, 1 kL, 1 mL?
8. A glass cylinder contains three liquids as shown in the illustration: water, ether (d = 0.714 g/mL), and mercury (d = 13.6 g/mL). If a cork (d = 0.25 g/mL) is dropped into the cylinder, where does it come to rest? Where does beeswax (d = 0.97 g/mL) come to rest? Where does a silver coin (d = 10.5 g/mL) come to rest? Where does a gold coin (d = 18.9 g/mL) come to rest?

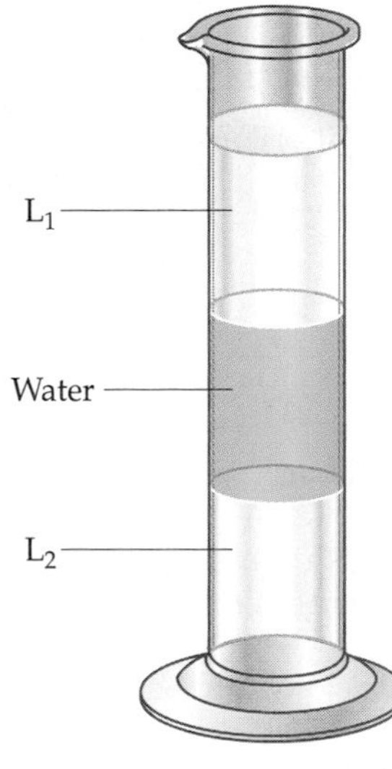

◀ **The Density Concept** A tall glass cylinder is shown containing liquid L_1, water, and liquid L_2.

9. Which of the following temperatures does not exist: −100°F, −100°C, −100 K?
10. If the crust of an apple pie cooks faster than the filling, which has the higher specific heat?

Key Terms†

Select the key term below that corresponds to each of the following definitions.

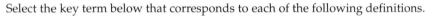

___ 1. a nondecimal system of measurement without basic units
___ 2. a decimal system of measurement with basic units
___ 3. the basic unit of length in the metric system
___ 4. the basic unit of mass in the metric system
___ 5. the basic unit of volume in the metric system
___ 6. the basic unit of time in the metric system
___ 7. a comprehensive system of measurement with seven base units
___ 8. a statement of two equivalent values, for example, 1 in. = 2.54 cm
___ 9. a statement of two exactly equal values, for example, 1 m ≡ 100 cm
___ 10. a ratio of two quantities that are equivalent, for example, 1 lb/454 g
___ 11. the relationship between a fraction and its inverse, for example, 1 qt/946 mL and 946 mL/1 qt
___ 12. a procedure for solving problems that proceeds from a given value to a related answer by the conversion of units
___ 13. the volume occupied by a cube 1 cm on a side
___ 14. a technique for determining volume from the amount of water displaced
___ 15. the amount of mass in one unit of volume
___ 16. the ratio of the density of a liquid compared to the density of water at 4°C
___ 17. a measure of the total energy of particles in a system

(a) calorie (cal) (*Sec. 3.9*)
(b) Celsius degree (°C) (*Sec. 3.8*)
(c) cubic centimeter (cm^3) (*Sec. 3.5*)
(d) density (d) (*Sec. 3.7*)
(e) English system (*Sec. 3.1*)
(f) exact equivalent (*Sec. 3.2*)
(g) Fahrenheit degree (°F) (*Sec. 3.8*)
(h) gram (g) (*Sec. 3.1*)
(i) heat (*Sec. 3.9*)
(j) International System (SI) (*Sec. 3.1*)
(k) joule (J) (*Sec. 3.9*)
(l) Kelvin (K) (*Sec. 3.8*)
(m) liter (L) (*Sec. 3.1*)
(n) meter (m) (*Sec. 3.1*)
(o) metric system (*Sec. 3.1*)
(p) reciprocal (*Sec. 3.2*)
(q) second (s) (*Sec. 3.1*)

*Answers to Key Concepts are in Appendix H.
†Answers to Key Terms are in Appendix I.

_____ **18.** a measure of the average energy of particles in a system
_____ **19.** the basic unit of temperature in the English system
_____ **20.** the basic unit of temperature in the metric system
_____ **21.** the basic unit of temperature in the SI system
_____ **22.** the amount of heat required to raise 1 g of substance 1°C
_____ **23.** the amount of heat required to raise 1 g of water 1°C
_____ **24.** a unit of energy in the SI system

(r) specific gravity (sp gr) (*Sec. 3.7*)
(s) specific heat (*Sec. 3.9*)
(t) temperature (*Sec. 3.8*)
(u) unit analysis method (*Sec. 3.3*)
(v) unit equation (*Sec. 3.2*)
(w) unit factor (*Sec. 3.2*)
(x) volume by displacement (*Sec. 3.6*)

Exercises‡

Basic Units and Symbols (Sec. 3.1)

1. Which of the following statements concerning the metric system are true?
   (a) It uses a basic unit for length, mass, and volume.
   (b) It is a decimal system that uses prefixes related by a power of 10.
   (c) It is used exclusively throughout the scientific community.
   (d) It is an official system of measurement in the United States.
2. Which of the following statements concerning the English system are true?
   (a) It uses a basic unit for length, mass, and volume.
   (b) It is a decimal system that uses prefixes related by a power of 10.
   (c) It is used exclusively throughout the scientific community.
   (d) It is an official system of measurement in the United States.
3. State the basic unit and symbol for the following metric quantities.
   (a) length **(b)** mass
   (c) volume **(d)** time
4. State the physical quantity expressed by the following metric symbols.
   (a) dm **(b)** ns
   (c) kg **(d)** mL
5. State the name and symbol for the following metric prefixes.
   (a) 1000 **(b)** 1,000,000
   (c) 0.1 **(d)** 0.000 001
6. State the name and symbol for the following metric prefixes.
   (a) 1×10^{9} **(b)** 1×10^{3}
   (c) 1×10^{-3} **(d)** 1×10^{-9}
7. Write the symbol for the following metric units.
   (a) kilometer **(b)** gigagram
   (c) microliter **(d)** millisecond
8. Write the symbol for the following metric units
   (a) megameter **(b)** centigram
   (c) deciliter **(d)** nanosecond
9. Write the name of the unit indicated by the following symbols.
   (a) mm **(b)** kg
   (c) mL **(d)** μs
10. Write the name of the unit indicated by the following symbols.
   (a) μm **(b)** dg
   (c) nL **(d)** Ms

Metric Conversion Factors (Sec. 3.2)

11. Write a unit equation for each of the following exact metric equivalents.
   (a) m and mm **(b)** g and μg
   (c) L and cL **(d)** Gs and s
12. Write a unit equation for each of the following exact metric equivalents.
   (a) m and nm **(b)** Mg and g
   (c) kL and L **(d)** s and ds
13. Write two unit factors for each of the following metric relationships.
   (a) Gm and m **(b)** g and mg
   (c) L and μL **(d)** s and cs
14. Write two unit factors for each of the following metric relationships.
   (a) m and dm **(b)** g and ng
   (c) ML and L **(d)** ks and s

Metric–Metric Conversions (Sec. 3.3)

15. Perform the following metric–metric conversions.
   (a) 1.55 km to m **(b)** 0.486 g to cg
   (c) 125 mL to L **(d)** 100 ns to s
16. Perform the following metric–metric conversions.
   (a) 0.388 Mm to m **(b)** 10.6 μg to g
   (c) 1.885 L to dL **(d)** 0.000 125 Gs to s

‡Answers to odd-numbered Exercises are in Appendix J.

17. Perform the following metric–metric conversions.
    (a) 125 Gm to Mm **(b)** 255 mg to dg
    (c) 14.5 μL to cL **(d)** 1.56×10^{-3} ks to ns
18. Perform the following metric–metric conversions.
    (a) 0.555 km to Mm **(b)** 0.327 kg to cg
    (c) 1.85 mL to μL **(d)** 8.15×10^{4} Gs to ns
19. An automobile airbag inflates to 50.0 L in 35 ms. What is the time of inflation in microseconds?
20. An automobile antilock brake system (ABS) operates the brakes at 30 pulses per second. How many times do the brakes pulse in 1.00 ds?

Metric–English Conversions (Sec. 3.4)

21. State the following metric—English equivalents.
    (a) ? cm = 1 in. **(b)** ? g = 1 lb
    (c) ? mL = 1 qt **(d)** ? s = 1 sec
22. Calculate the following metric—English relationships.
    (a) ? cm = 12.0 in. **(b)** ? g = 2.20 lb
    (c) ? mL = 1.06 qt **(d)** ? s = 60 sec
23. Perform the following metric—English conversions.
    (a) 66 in. to cm **(b)** 1.01 lb to g
    (c) 0.500 qt to mL **(d)** 8.00×10^{2} sec to s
24. Perform the following metric—English conversions.
    (a) 86 cm to in. **(b)** 36 g to lb
    (c) 750 mL to qt **(d)** 5.52×10^{-3} s to sec
25. Perform the following metric—English conversions.
    (a) 72 in. to m **(b)** 175 lb to kg
    (c) 1250 mL to gallons **(d)** 1.52×10^{3} ds to min
26. Perform the following metric—English conversions.
    (a) 800.0 m to yd **(b)** 0.375 kg to lb
    (c) 0.500 qt to L **(d)** 1.05×10^{-4} min to ks
27. The EPA highway mileage for a subcompact car is 52 miles per gallon. What is the mileage in kilometers per liter? (Given: 1 mi = 1.61 km and 1 gal = 3.784 L.)
28. A .357 Magnum bullet has a muzzle velocity of 1200 ft/sec. Express the velocity in meters per minute.

Volume by Calculation (Sec. 3.5)

29. A rectangular piece of solid mahogany measures 5.08 cm by 10.2 cm by 3.05 m. What is the volume in cubic centimeters?
30. A quartz rock was cut into a rectangular solid that measures 5.00 cm by 5.00 cm by 25.0 mm. What is the volume in cubic millimeters?
31. A rectangular solid piece of brass measures 4.95 cm by 2.45 cm and has a volume of 15.3 cm^3. What is the thickness of the brass?
32. A sheet of aluminum foil measures 30.5 cm by 75.0 cm and has a mass of 9.94 g. What is the thickness of the foil if the volume is 3.68 cm^3?
33. Complete the following volume equivalents.
    (a) 1 L = ? mL **(b)** 1 L = ? cm^3
34. Complete the following volume equivalents.
    (a) 1 mL = ? cm^3 **(b)** 1 in.3 = ? cm^3
35. A midsize vehicle has a 415-in.3 engine. Express the engine volume in liters.
36. A subcompact automobile has a 1.20-L engine. Express the engine volume in cubic inches.

Volume by Displacement (Sec. 3.6)

37. The initial water level in a 10-mL graduated cylinder reads 4.5 mL. After a ruby is dropped into the cylinder, the water level reads 5.0 mL. What is the volume of the gemstone?
38. The initial water level in a 100-mL graduated cylinder is 44.5 mL. After a large opal is added into the cylinder, the water level is 55.0 mL. What is the volume of the gemstone?
39. Magnesium metal reacts with hydrochloric acid to produce hydrogen gas. The gas displaces water into a cylinder and the level increases from 125 mL to 255 mL. What volume of hydrogen is produced by the reaction?
40. Calcium metal reacts with water to produce hydrogen gas. If the gas displaces 212 mL of water into a beaker, how many cubic centimeters of hydrogen are produced by the reaction?

The Density Concept (Sec. 3.7)

41. State whether the following will sink or float when dropped into water.
    (a) ebony (d = 1.2 g/mL)
    (b) bamboo (d = 0.40 g/mL)
42. State whether the following will sink or float when dropped into water.
    (a) paraffin wax (d = 0.90 g/cm^3)
    (b) limestone (d = 2.8 g/cm^3)
43. State whether a balloon filled with the following gas will rise in the air or fall to the ground. (Assume the mass of the balloon is negligible and the density of air is 1.29 g/L.)
    (a) helium (d = 0.178 g/L)
    (b) laughing gas (d = 1.96 g/L)
44. State whether a balloon filled with the following gas will rise in the air or fall to the ground. (Assume the mass of the balloon is negligible and the density of air is 1.29 g/L.)
    (a) argon (d = 1.78 g/L)
    (b) ammonia (d = 0.759 g/L)
45. Calculate the mass in grams for each of the following.
    (a) 250 mL of gasoline (d = 0.69 g/mL)
    (b) 0.75 cm^3 of rock salt (d = 2.18 g/cm^3)
46. Calculate the mass in grams for each of the following.
    (a) 36.5 mL of methanol (d = 0.791 g/mL)
    (b) 455 cm^3 of borax (d = 1.715 g/cm^3)
47. Calculate the volume in milliliters for each of the following.
    (a) 0.500 g of bromine (d = 3.12 g/mL)
    (b) 10.0 g of nickel (d = 8.90 g/cm^3)

48. Calculate the volume in milliliters for each of the following.
(a) 0.899 kg of acetone (d = 0.792 g/mL)
(b) 1.00 kg of cork (d = 0.25 g/cm^3)

49. Calculate the density in grams per milliliter for each of the following.
(a) 25.0 mL of ethyl alcohol having a mass of 19.7 g
(b) 11.6-g marble whose volume is found by displacement to be 4.1 mL

50. Calculate the density in grams per milliliter for each of the following.
(a) 10.0 g of ether having a volume of 14.0 mL
(b) 131.5-g bronze rectangular solid measuring 3.55 cm × 2.50 cm × 1.75 cm

Temperature (Sec. 3.8)

51. State the freezing point of water on the following temperature scales.
(a) Fahrenheit (b) Kelvin
(c) Celsius

52. State the boiling point of water on the following temperature scales.
(a) Fahrenheit (b) Kelvin
(c) Celsius

53. Express the following Fahrenheit temperatures in degrees Celsius.
(a) 100°F (b) −215°F

54. Express the following Celsius temperatures in degrees Fahrenheit.
(a) 19°C (b) −175°C

55. Express the following Celsius temperatures in Kelvin units.
(a) 495°C (b) −185°C

56. Express the following Kelvin temperatures in degrees Celsius.
(a) 273 K (b) 100 K

Heat and Specific Heat (Sec. 3.9)

57. Distinguish between the terms "temperature" and "heat".

58. State the metric value for the specific heat of water.

59. A beaker containing 250 g of water is heated from 23° to 100°C. What is the heat gain in calories?

60. A pool containing 2.5 × 10^8 g of water cools from 25.0° to 17.0°C. What is the heat loss in kilocalories?

61. Calculate the heat required to raise 25.0 g of iron (0.108 cal/g × °C) from 25.0° to 50.0°C.

62. Calculate the heat released as 35.5 g of copper (0.0920 cal/g × °C) cools from 50.0° to 45.0°C.

63. Find the specific heat of gold if 25.0 cal is required to heat 30.0 g of gold from 27.7° to 54.9°C.

64. Find the specific heat of platinum if 35.7 cal is lost as 75.0 g of the metal cools from 43.9° to 28.9°C.

65. Calculate the mass of titanium (0.125 cal/g × °C) that requires 75.6 cal to heat the metal from 20.7° to 31.4°C.

66. Calculate the mass of lead (0.0308 cal/g × °C) that releases 52.5 cal as the metal cools from 35.7° to 25.1°C.

67. What is the temperature change when 10.5 g of silver (0.0566 cal/g × °C) absorbs 35.2 cal of heat?

68. What is the temperature change when 8.92 g of sodium (0.293 cal/g × °C) loses 750 cal of heat energy?

General Exercises

69. A double-density floppy disk can store 720 kilobytes (kB). If a computer hard disk has a capacity of 270 megabytes (MB), how many floppy disks of information can be loaded onto the hard disk?

70. A high-density floppy disk can store 1.40 MB. If a computer hard disk has a capacity of 1.2 gigabytes (GB), how many floppy disks of information can be loaded onto the hard disk?

▲ **Computer Floppy Disks** A floppy disk contains a circular disk inside a square plastic holder. Although 3.5-in. floppy disks are rigid, the original 5.25-in. disks were flexible, thus, the term "floppy disk."

71. How many significant digits are in the following unit factors?
(a) 1 m/10 dm (b) 1 lb/454 g
(c) 1 L/1000 mL (d) 1 qt/946 mL

72. Which of the following are exact equivalent relationships?
(a) 1 m = 10 dm (b) 1 lb = 454 g
(c) 1 L = 1000 mL (d) 1 qt = 946 mL

73. A light-year is the distance light travels in 1.00 year. Given the velocity of light, 186,000 mi/s, how many miles does light travel in a light-year?

74. A parsec is the distance light travels in 3.26 years. Given the velocity of light, 3.00 × 10^8 m/s, how many kilometers does light travel in a parsec?

75. Olympic athletes compete in a 1500-m event but not in a mile event. Which race is faster: 1500 meters or 1 mile?

76. An oxygen molecule travels 975 mi/h at room temperature. What is the velocity in meters per second?

77. How many 5-grain tablets can be produced from 2.50 kg of powdered aspirin. (Given: 1 grain = 64.8 mg.)

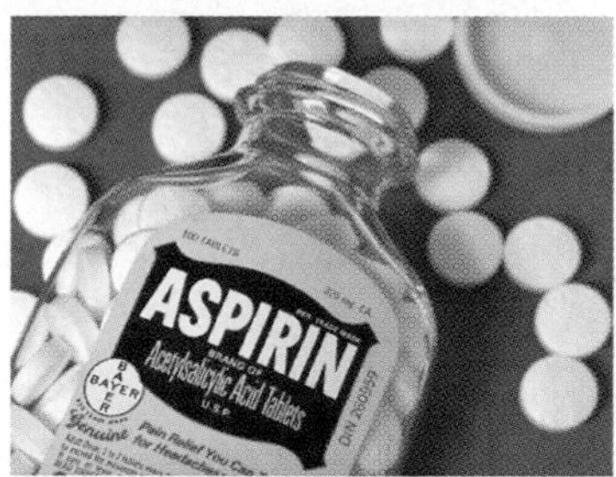

▲ **Aspirin Tablets** A 5-grain tablet contains 325 mg of aspirin.

78. How many molecules of water are in one drop if there are 3.34×10^{22} molecules in 1.00 g? (Given: 1 mL = 20 drops.)
79. A football field measures 100.0 yards by 160.0 feet. Calculate the playing area in square yards.
80. A basketball court measures 94.0 feet by 50.0 feet. Calculate the playing area in square meters. (Given: 1 yd = 0.914 m.)
81. What is the density of fool's gold if a 37.51-g sample added into a graduated cylinder increased the liquid level from 50.0 mL to 57.5 mL?
82. What is the mass of 275 L of seawater if the density is 1.025 g/cm^3?
83. What is the specific gravity of gasohol if the density is 0.801 g/mL?
84. What is the density of jet fuel if the specific gravity is 0.775?
85. Express the density of water in the English units of pounds per cubic foot.
86. Express the density of water in the English units of pounds per gallon.
87. The space shuttle uses liquefied hydrogen at a temperature of −422°F. What is the equivalent Kelvin temperature?
88. The temperature in the Mojave Desert in California has reached 324 K. What is the equivalent Fahrenheit temperature?
89. A red-hot fireplace poker is plunged into 3500 g of water at 22°C. If the water temperature rises to 51°C, how many kilocalories of heat are lost by the poker?
90. A 10.0-g sample of metal at 19.0°C is dropped into 75.0 g of water at 80.0°C. If the resulting temperature is 78.3°C, what is the metal? (*Hint:* Refer to Table 3.5.)
91. A 10.0-g sample of copper (0.0920 cal/g × °C) gains 27.8 cal of heat. If the initial temperature is 22.7°C, what is the final temperature?
92. A 15.5-g sample of titanium (0.125 cal/g × °C) loses 32.9 cal of heat. If the initial temperature is 28.9°C, what is the final temperature?
93. The density of mercury is 13.6 g/mL. Express the density in SI units (kg/m^3).
94. The specific heat of mercury is 0.0331 cal/g × °C. Express the specific heat in SI units (J/kg × K).
95. The radius (r) of the international prototype kilogram cylinder is 1.95 cm. Assuming the density of the kilogram is 21.50 g/cm^3, calculate its height (h). The volume of a cylinder equals $\pi r^2 h$, where π is the constant 3.14.

Explorer Quiz 1
Explorer Quiz 2
Explorer Quiz 3
Master Quiz

CHAPTERS 1–3
Cumulative Review

▲ **Five-cent Coins** Since a dollar is equivalent to 20 nickels, we can write the unit equation 1 dollar = 20 nickels. The two associated unit factors are: 1 dollar/20 nickels and 20 nickels/1 dollar.

Key Concepts

1. Why is it important to learn chemistry?
2. Why are exact measurements impossible?
3. What is the number of significant digits in each of the following relationships: 1 yd ≡ 36 in., 1 yd = 0.914 m, 1 m = 39.4 in., 1 m ≡ 1000 mm?
4. What is the basic unit of length, mass, and volume in the metric system?
5. What metric unit of liquid volume is equal to $(10\text{ cm})^3$? to $(1\text{ cm})^3$?
6. Estimate the diameter of a 5¢ coin (±1 cm).
7. Estimate the mass of a 5¢ coin (±1 g).
8. Estimate the volume of 20 drops of water (±1 mL).
9. In the unit analysis method, what is the first step? the second step?
10. In applying unit analysis, how do you select a unit factor?
11. In a calculation involving a series of multiplication and division operations, is it better to round off after each step or simply to round off the final answer?
12. What is the density of water in metric units?
13. Given that an unknown liquid is chloroform (d = 1.48 g/mL) or ether (d = 0.714 g/mL), how could you quickly identify the liquid?
14. Which of the following temperatures is the coldest: 0°F, 0°C, or 0 K?
15. What is the specific heat of water in metric units?

Key Terms

State the key term that corresponds to each of the following descriptions.

______ 1. the study of nature that is practical and logical
______ 2. the study of the composition of matter and its properties
______ 3. a systematic procedure that collects data and observes changes
______ 4. a numerical value with units
______ 5. the degree of inexactness in an instrumental measurement
______ 6. the certain digits in a measurement plus one estimated digit
______ 7. a method for expressing numbers using significant digits and powers of 10
______ 8. the ratio of a single quantity compared to an entire sample, all multiplied by 100
______ 9. a relationship between two equivalent values
______ 10. a relationship between two exactly equal values
______ 11. a ratio between two quantities that are equivalent
______ 12. the relationship between a fraction and its inverse
______ 13. a systematic procedure for solving problems by the conversion of units
______ 14. a decimal system of measurement with basic units
______ 15. the volume occupied by a cube 1 cm on a side
______ 16. the volume occupied by a cube 10 cm on a side
______ 17. the amount of mass per unit volume
______ 18. a measure of the total energy of particles in a system
______ 19. a measure of the average energy of particles in a system
______ 20. the amount of heat required to raise 1 g of water 1°C

Review Exercises

▲ **Fool's Gold** Iron pyrite, FeS_2, is commonly referred to as fool's gold because of its yellow metallic luster.

1. State the number of significant digits in each of the following.
   (a) 680.0 cm **(b)** 0.300 g
2. Round off the following values to three significant digits.
   (a) 90.050 **(b)** 125,499
3. The mass of a proton is 1.67265×10^{-24} g and an electron is 9.10953×10^{-28} g. What is the difference in mass between a proton and an electron?
4. There are 135,000,000,000,000,000,000,000 atoms in a 5.00-L tank of helium gas at 25°C. Express the number of helium atoms in scientific notation.
5. Table sugar is composed of 42.1% carbon, 6.5% hydrogen, and 51.4% oxygen. State the mass of carbon, hydrogen, and oxygen in 1.00 g of sugar.
6. Write the unit equation that relates each of the following.
   (a) m and dm **(b)** Mg and g
7. Write the two unit factors that correspond to each of the following.
   (a) 1 L ≡ 100 cL **(b)** 1 s ≡ 1×10^6 μs
8. Using unit analysis, calculate the millimeter length of a pencil that is 19.05 cm.
9. Using unit analysis, calculate the kilogram mass of the 842 lb of lunar samples that have been brought back from the Moon. (Given: 1 lb = 454 g.)
10. Using unit analysis, calculate the number of milliliters of water in 1.00 pint. (Given: 1 qt ≡ 2 pt and 1 qt = 946 mL.)
11. What is the average speed in feet per second of a world-class swimmer who swims a 400.0-m freestyle race in 3 min, 49.05 s? (Given: 1 m = 3.281 ft.)
12. A piece of gold foil measures 4.95 cm by 2.45 cm and has a volume of 15.3 cm^3. What is the thickness of the gold foil?
13. If 100 lb of water occupies 1.60 ft^3, what is the volume of the water in liters?
14. What is the density of iron pyrite (fool's gold) if a 37.51-g sample added into a graduated cylinder displaced the water level from 50.0 mL to 57.5 mL?
15. What is the mass in grams of 20.0 L of gasoline (d = 0.695 g/mL)?
16. What is the volume in milliliters of 1.00 kg of mercury (d = 13.6 g/mL)?
17. The space shuttle uses liquefied hydrogen at a temperature of −422°F. What is the equivalent Kelvin temperature?
18. A beaker containing 1.00 kg of water is heated from 10.0° to 20.0°C. What is the heat gain in kilocalories?
19. Calculate the heat required to raise 25.0 g of iron (0.106 cal/g × °C) from 25.0° to 50.0°C.
20. Calculate the volume of the Earth assuming it is spherical and has a radius (r) of 6370 km. The volume of a sphere equals $4\pi r^3/3$, where π is the constant 3.14.

CHAPTER 4

Matter and Energy

▲ Given Einstein's famous equation, $E = mc^2$, is the mass of the bomb detonated in an atomic explosion greater than, equal to, or less than the mass of its fragments?

For many students, this chapter is a welcome change of pace as there are no calculations and all the material is qualitative. I ask students to memorize the names and symbols of the elements in Table 4.3, which is an easy task that helps to build confidence. I also hand out periodic tables with the lectures; it not only facilitates note taking, but it acts as an educational security blanket. I perform numerous demonstrations to make the concepts more interesting (refer to the *Instructor's Resource Manual*).

In the seventeenth century, the British scientist Robert Boyle established the importance of experiments in the study of science. Boyle realized the value of laboratory research and described his work thoroughly so that other scientists could repeat his procedures and confirm his observations. In 1661 he published *The Sceptical Chymist*, which marked a turning point in science by suggesting that theories were no better than the experimental methods that supported them.

Boyle rejected the Greek notion that air, earth, fire, and water were basic elements. Instead, he proposed that an element was a substance that could not be broken down further. Thus, an element must always gain weight, and never lose weight, when undergoing a chemical change. This was a practical proposal that could be tested in the laboratory.

Following the lead of Boyle, others established scientific principles by performing experiments that could be verified by fellow scientists. Although scientists continued to propose theories about the behavior of matter, their theories were now based on laboratory evidence.

4.1 Physical States of Matter

Objective · To describe the three physical states of matter in terms of the motion of particles.

Matter is any substance that has mass and occupies volume. It exists in one of three **physical states**: solid, liquid, or gas. In the solid state, matter has a fixed shape and a definite volume; it cannot be compressed because its particles are tightly packed. In the liquid state, particles of matter are free to move past one another. The particles are loosely packed but can be slightly compressed. The shape of a liquid may vary, but its volume is fixed.

In the gaseous state, tiny particles of matter are widely spaced and uniformly distributed throughout the container. If the volume increases, the gas expands and the particles move farther away from each other. If the volume decreases, the gas compresses and the particles move closer to one another. Table 4.1 summarizes the properties of the three states of matter.

▲ **Vaporization** Reddish-brown liquid bromine vaporizes to a gas.

Table 4.1 Physical States of Matter

Property	Solid	Liquid	Gas
shape	definite	indefinite	indefinite
volume	fixed	fixed	variable
compressibility	negligible	negligible	significant

Even though we may describe a substance as being a solid, every substance can exist as a solid, liquid, or gas. We can change the physical state of a substance by changing the temperature. Water, for example, can be changed from a liquid to solid ice at 0°C, and from a liquid to gaseous steam at 100°C. Other substances behave similarly at different temperatures. Iron, for example, can be changed to a molten liquid at 1535°C and to a gas at 2750°C.

We can describe changes of physical state as follows. As temperature increases, a solid *melts* to a liquid, and then the liquid *vaporizes* to a gas. A direct change of state from a solid to a gas is called **sublimation**. Mothballs, for example, undergo sublimation and disappear as they change directly from a solid to a gas.

Conversely, as temperature decreases, a gas *condenses* to a liquid, and the liquid then *freezes* to a solid. A direct change of state from a gas to a solid is called **deposition**. The freezer compartment in a refrigerator, for example, may demonstrate deposition by collecting ice. Opening the refrigerator door allows moist air to enter which deposits frost without a trace of liquid water. Figure 4.1 shows the relationship between temperature and physical state.

▲ **Sublimation and Deposition** Solid iodine crystals in the bottom of the beaker are heated until they sublime to a purple gas. The purple gas deposits on the cool flask at the top of the beaker.

Changes of State Movie

Example Exercise 4.1 • Change of Physical State

State the term that applies to each of the following changes of physical state.

(a) Snow changes from a solid to a liquid.

(b) Gasoline changes from a liquid to a gas.

(c) Dry ice changes from a solid to a gas.

(continued)

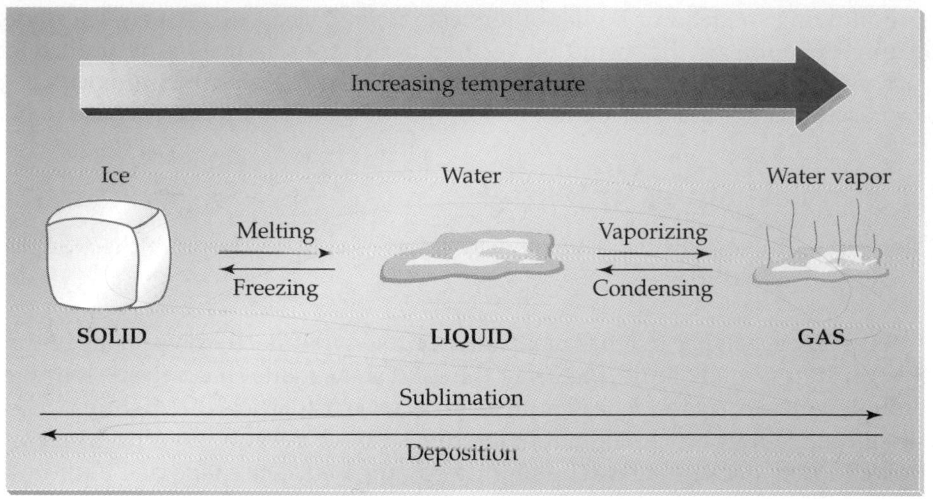

Changes in Physical State.

▶ **Figure 4.1 Changes in Physical State** As temperature increases, a solid melts to a liquid and then vaporizes to a gas. As temperature decreases, a gas condenses to a liquid and then freezes to a solid.

Experiment # 4, Prentice Hall Laboratory Manual.

Vaporization, Sublimation, and Deposition Movie

Solution

Refer to Figure 4.1 for the changes of physical state.

(a) The change from solid to liquid is called *melting*.
(b) The change from liquid to gas is called *vaporizing*.
(c) The change from solid to gas is called *sublimation*.

Self-Test Exercise

State the term that applies to each of the following changes of physical state.

(a) Freon refrigerant changes from a gas to a liquid.
(b) Milk changes from a liquid to a solid.
(c) Iodine vapor changes from a gas to a solid.

Answers: (a) condensing; (b) freezing; (c) deposition

Note A fourth state of matter exists at very high temperatures. Under extreme conditions, matter can separate into positive and negative subatomic particles and is referred to as *plasma*. At the surface of the Sun, at temperatures of about 6000°C, hydrogen and helium exist as plasma. Solar flares are streams of plasma shooting out from the surface of the Sun.

4.2 Elements, Compounds, and Mixtures

Objective · To classify a sample of matter as an element, compound, or mixture.

A sample of matter may have properties that are consistent throughout, or they may vary. One way to tell if the properties of matter are consistent is to melt the sample. A sample of pure gold always melts at 1064°C and does not vary. It makes no difference whether it is a large gold nugget or a small gold flake, the yellow metal consistently melts at 1064°C.

Now consider a sample of quartz rock that contains a vein of pure gold. Although the quartz sample is observed to melt over a broad range from 1000° to 1600°C, pure gold always melts precisely at 1064°C. Scientists classify the quartz rock as heterogeneous and gold as homogeneous. A sample of matter is said to be *heterogeneous* if its properties are indefinite and vary. A sample of pure gold is said to be *homogeneous* because its properties are definite and constant.

A **heterogeneous mixture** can be separated into pure substances by physical methods. For example, panning for gold uses the property of density to separate pure gold from sand and rock. Gold is quite dense (18.9 g/cm^3) and remains at the bottom of the pan, while the less dense sand and rock are swirled away with water.

Air is not a heterogeneous mixture but rather a homogeneous mixture of nitrogen, oxygen, and other gases. Similarly, saline water is a homogeneous mixture of salt and water. Unlike those of a heterogeneous mixture, the properties of a **homogeneous mixture** are constant for a given sample. However, a homogeneous mixture may have properties that vary from sample to sample. For example, samples of seawater from the Pacific Ocean and from the Dead Sea have different properties. Seawater from the Dead Sea has a higher density and a greater concentration of dissolved minerals.

An **alloy** is a homogeneous mixture of two or more metals. Examples of alloys include 10K, 14K, and 18K gold jewelry. Although 10K, 14K, and 18K jewelry may contain only gold, silver, and copper metals, the amount of gold varies from 42% to 75%. Gold alloy is a homogeneous mixture; therefore the properties from different samples can vary. For instance, 10K gold is a harder alloy and is more scratch-resistant than 18K gold.

A **substance** is matter that has definite composition and constant properties. A substance is either a compound or an element. A **compound** has predictable properties but can be broken down into elements by an ordinary chemical reaction. Table sugar is an example of a compound that can be broken down into carbon, hydrogen, and oxygen.

An **element** is a substance that cannot be broken down further by a chemical reaction. For example, table salt is a compound composed of the elements sodium and chlorine, which cannot be broken down any further. Figure 4.2 shows the overall classification of matter.

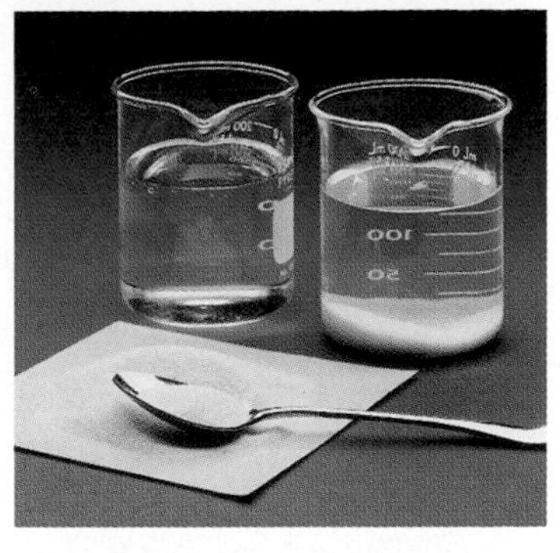

▲ **Heterogeneous Mixture**
Sand and water form a heterogeneous mixture, and the properties vary throughout the sample.

Mixtures and Compounds Movie

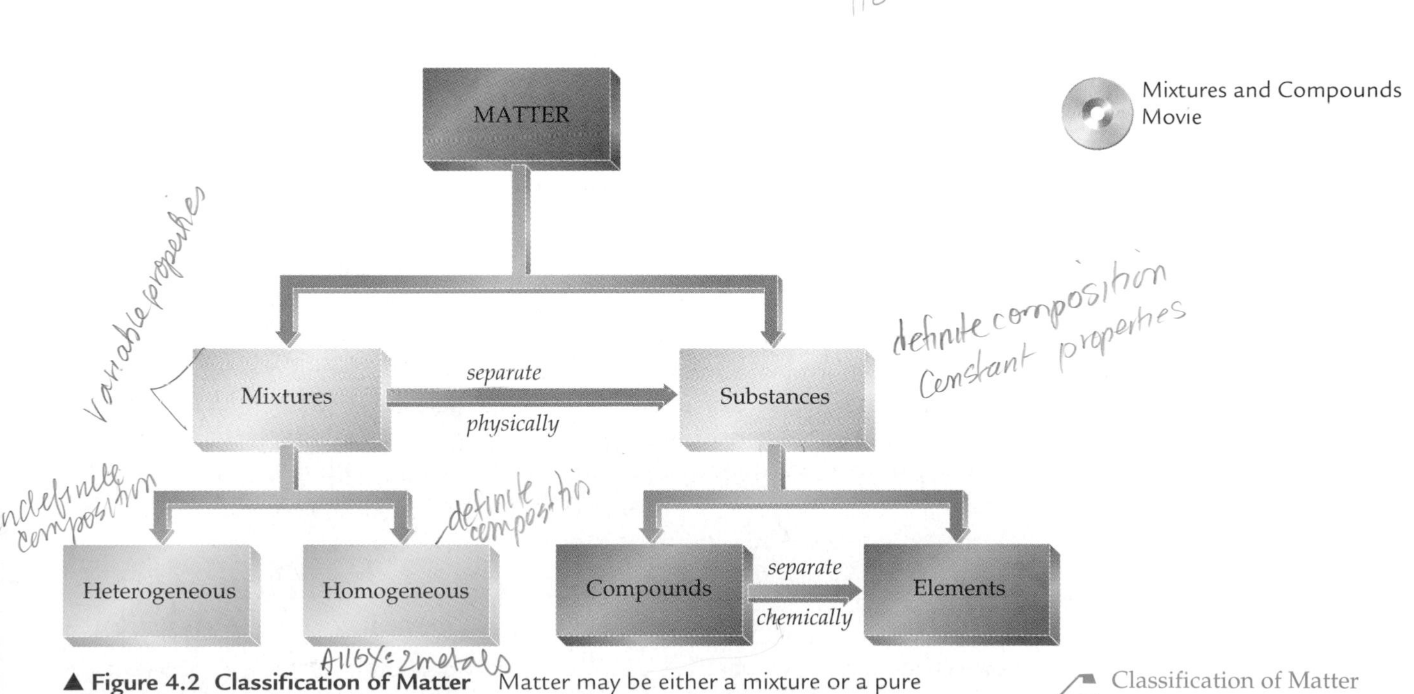

▲ **Figure 4.2 Classification of Matter** Matter may be either a mixture or a pure substance. The properties of a heterogeneous mixture vary within the sample. The properties of a homogeneous mixture are constant but can vary from sample to sample. A substance may be either a compound or an element. The properties of a substance are predictable and constant.

Classification of Matter

Example Exercise 4.2 • Element, Compound, or Mixture

Consider the following properties of the element copper.

(a) Copper metal cannot be broken down by a chemical change.
(b) Copper reacts with oxygen in air to give copper oxide.
(c) Copper, in the form of malachite ore, is found in South America.
(d) Copper and tin compose bronze alloy.

Classify each of the following copper samples as an element, a compound, a homogeneous mixture, or a heterogeneous mixture.

(a) copper wire (b) copper oxide
(c) malachite ore (d) bronze alloy

Solution

Refer to Figure 4.2 to classify each sample.

(a) Copper wire is a metallic *element*.
(b) Copper oxide is a *compound* of the elements copper and oxygen.
(c) Malachite ore is a *heterogeneous mixture* of copper and other substances.
(d) Bronze alloy is a *homogeneous mixture* of copper and tin.

Self-Test Exercise

Consider the following properties of the element mercury.

(a) Mercury liquid cannot be broken down by a chemical change.
(b) Mercury oxide can be heated to give mercury and oxygen gas.
(c) Mercury, in the form of cinnabar ore, is found in Spain and Italy.
(d) Mercury and silver compose amalgam alloy used for dental fillings.

Classify each of the following mercury samples as an element, a compound, a homogeneous mixture, or heterogeneous mixture.

(a) mercury liquid (b) mercury oxide
(c) cinnabar ore (d) amalgam alloy

Answers: (a) element; (b) compound; (c) heterogeneous mixture; (d) homogeneous mixture

▲ Compound, Element, Mixture The orange compound mercury oxide decomposes to give the element mercury which is a silver liquid. Adding the orange powder and the metallic liquid together produces a heterogeneous mixture.

4.3 Names and Symbols of the Elements

Objective · To state the names and symbols of selected elements.

There are 81 stable elements that occur in nature. In addition, there are a few other naturally occurring elements, such as uranium, which are unstable and decay radioactively. In total, there are over 100 elements, but only 10 account for 95% of the mass of the Earth's crust, water, and atmosphere.

Oxygen is the most abundant element in nature. It is found combined with hydrogen in water and with silicon in sand and rocks. However, oxygen is found uncombined in air and constitutes about 21% of our atmosphere. In various forms, the mass of oxygen is about equal to the total mass of all other elements in the Earth's crust, water, and atmosphere. Table 4.2 lists the ten most abundant elements.

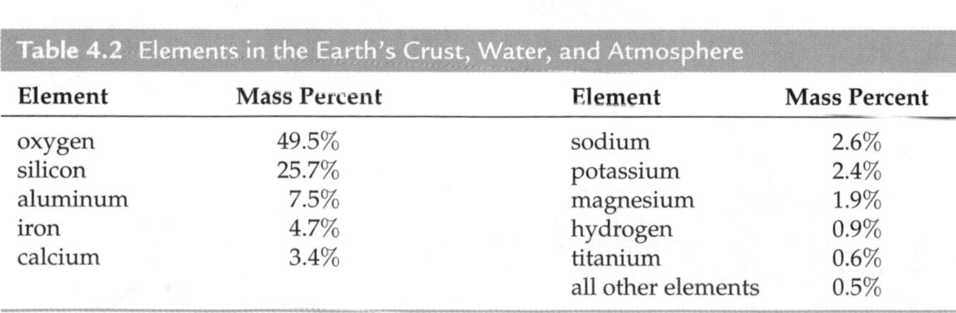

Table 4.2 Elements in the Earth's Crust, Water, and Atmosphere

Element	Mass Percent	Element	Mass Percent
oxygen	49.5%	sodium	2.6%
silicon	25.7%	potassium	2.4%
aluminum	7.5%	magnesium	1.9%
iron	4.7%	hydrogen	0.9%
calcium	3.4%	titanium	0.6%
		all other elements	0.5%

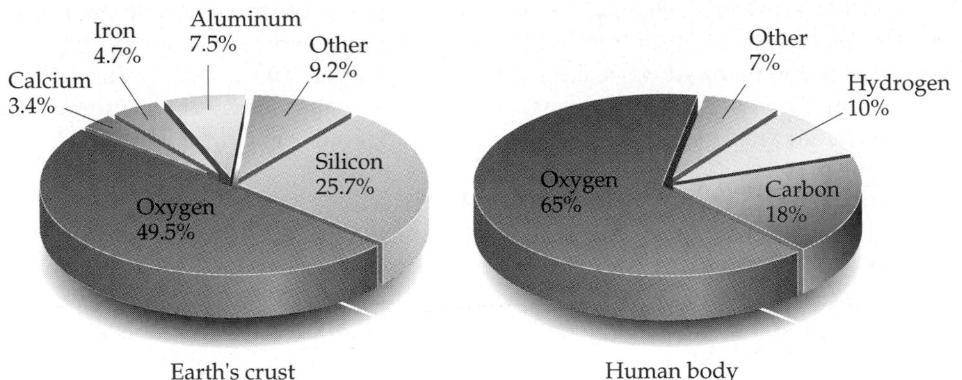

◀ **Figure 4.3 Abundance of Elements** Notice that oxygen is the most abundant element in both the Earth's crust and the human body. Although aluminum is abundant in the Earth, there is evidence that this element is toxic to the human body.

Oxygen, silicon, and aluminum are the three most abundant elements in the Earth's crust, water, and atmosphere. The elements oxygen, carbon, hydrogen, nitrogen, calcium, and phosphorus account for over 99% of the mass of the human body. The remaining 1% consists of trace elements, many of which are essential to human life. For example, a trace of iron is necessary to bind oxygen to hemoglobin in blood. Figure 4.3 compares the distribution of elements in the Earth's crust and in the human body.

The names of elements are derived from various sources. For example, hydrogen is derived from the Greek *hydro*, meaning "water former." Carbon is derived from the Latin *carbo*, meaning "coal." Calcium is derived from the Latin *calcis*, which translates as "lime," a mineral source of calcium. Some elements are named for their region of discovery. For example, germanium comes from Germany, and scandium from Scandinavia. Several elements are named for a famous scientist, such as curium (Marie Curie) and nobelium (Alfred Nobel).

The name of each element is abbreviated using a **chemical symbol**. In 1803 the English chemist John Dalton proposed that elements are composed of indivisible spherical particles. Dalton referred to each of these individual particles as an **atom**, from the Greek *atomos*, meaning "indivisible." He suggested the use of circles with enclosed markings as symbols for the elements. Figure 4.4 shows selected symbols chosen by Dalton to represent elements.

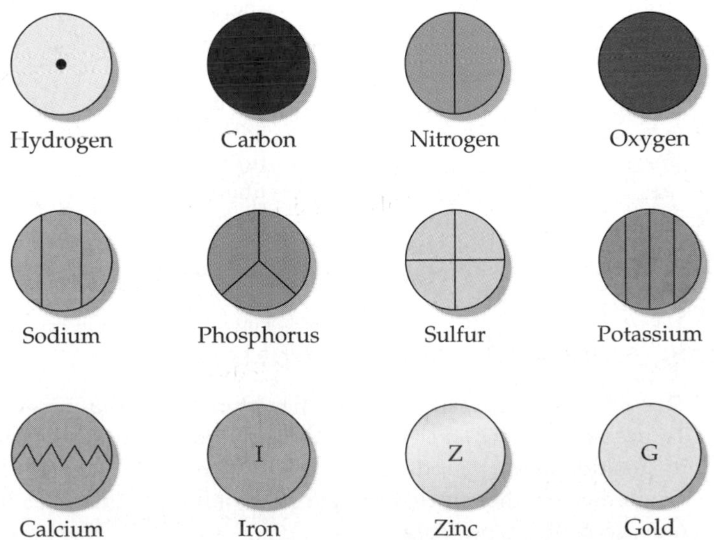

◀ **Figure 4.4 Dalton's Symbols for Selected Elements** Dalton symbolized each element using a circle with an inscribed marking. Dalton's system was not practical, but it did convey the idea that elements were composed of atomic spheres.

In 1813 the Swedish chemist J. J. Berzelius proposed our current system of symbols for the elements. He suggested that a symbol correspond to the first letter of the name of the element, for example, H for hydrogen, O for oxygen, and C for carbon. Furthermore, when elements start with the same letter, he recommended using two letters in the name, for example, Ca for calcium, Cd for cadmium, and Cl for chlorine.

In some instances, the chemical symbol is derived from the original Latin name of the element. For example, the symbol Pb for lead is derived from the Latin *plumbum*. Interestingly, the Romans used lead pipe to transport water, which explains the derivation of our word "plumbing." The symbol Au for gold is derived from the Latin *aurum*, meaning "golden dawn." Similarly, the symbols for silver (Ag), copper (Cu), iron (Fe), mercury (Hg), potassium (K), sodium (Na), antimony (Sb), and tin (Sn) are derived from the original Latin names.

When writing symbols for the names of elements, it is important to follow convention to avoid confusion. The first letter is always capitalized, and the second is lowercase. Thus, the symbol for the metallic element cobalt is Co. In contrast, notice that the formula for the deadly gas carbon monoxide is CO. Table 4.3 lists the names and symbols of selected elements.

Table 4.3 Names and Symbols of Selected Elements

Name of Element	Symbol	Name of Element	Symbol
aluminum	Al	lead	Pb
antimony	Sb	lithium	Li
argon	Ar	magnesium	Mg
arsenic	As	manganese	Mn
barium	Ba	mercury	Hg
beryllium	Be	neon	Ne
bismuth	Bi	nickel	Ni
boron	B	nitrogen	N
bromine	Br	oxygen	O
cadmium	Cd	phosphorus	P
calcium	Ca	platinum	Pt
carbon	C	potassium	K
chlorine	Cl	radium	Ra
chromium	Cr	selenium	Se
cobalt	Co	silicon	Si
copper	Cu	silver	Ag
fluorine	F	sodium	Na
germanium	Ge	strontium	Sr
gold	Au	sulfur	S
helium	He	tellurium	Te
hydrogen	H	tin	Sn
iodine	I	titanium	Ti
iron	Fe	xenon	Xe
krypton	Kr	zinc	Zn

4.4 Metals, Nonmetals, and Semimetals

Objectives
- To distinguish between the properties of metals and nonmetals.
- To predict whether an element is a metal, nonmetal, or semimetal given its position in the periodic table.
- To predict whether an element is a solid, liquid, or gas at 25°C and normal atmospheric pressure.

An element is a substance that cannot be broken down further and still maintain its unique properties. A **metal** is an element that typically is a solid, has a bright metallic luster, a high density, and a high melting point, and is a good conductor of heat and electricity. A metal can usually be hammered into a thin sheet of foil and is said to be **malleable**. If it can be drawn into a fine wire, it is said to be **ductile**. Aluminum, copper, and silver are familiar examples of metals.

A **nonmetal** is an element that usually has a low density, a low melting point, and is a poor conductor of heat and electricity. Many nonmetals occur naturally in the solid state and have a dull appearance, for example, carbon and sulfur. These solid nonmetals are neither malleable nor ductile and crush to a powder if hammered. Eleven nonmetals occur naturally in the gaseous state; for example, hydrogen and oxygen are familiar examples that are colorless, gaseous nonmetals. Table 4.4 summarizes the properties of metals and nonmetals.

Table 4.4 General Characteristics of Metals and Nonmetals*

Property	Metals	Nonmetals
physical state	solid	solid, gas
appearance	metallic luster	dull
pliability	malleable, ductile	brittle
conductivity	heat, electricity	nonconductor
density	usually high	usually low
melting point	usually high	usually low
chemical reactivity	react with nonmetals	react with metals and nonmetals

*There are numerous exceptions to these general characteristics. For example, mercury and bromine are liquids, magnesium metal has a low density, and gallium metal has a melting point below 30°C.

A **semimetal** (or metalloid) is an element that typically has properties midway between those of metals and nonmetals. Silicon is a familiar semimetal used in the semiconductor industry for making transistors and integrated circuits. Example Exercise 4.3 further illustrates the identification of properties of metals and semimetals.

Example Exercise 4.3 • Properties of Metals

Which of the following properties is *not* characteristic of a metal?

(a) good conductor of heat
(b) malleable
(c) high melting point
(d) reacts with other metals

Solution

Refer to Table 4.4 to classify each of the following properties.

(a) Metals are good conductors of heat.
(b) Metals are malleable.
(c) Metals usually have high melting points.
(d) Metals do not react with other metals; they form alloys.

Self-Test Exercise

Which of the following properties is *not* characteristic of a nonmetal?

(a) insulator of electricity
(b) ductile
(c) low density
(d) reacts with nonmetals

Answer: (b) Nonmetals crush to a powder and are neither malleable nor ductile.

Interactive Periodic Table

Periodic Table of the Elements

In Chapter 6 we will devote our entire discussion to the chemical elements. For now, realize that each element is assigned a characteristic whole number. The number that identifies a particular element is called the **atomic number**. For example, the atomic number of hydrogen is 1, helium is 2, lithium is 3, and so on, and the atomic number of uranium is 92.

All the elements have been arranged by atomic number and placed in a special chart. This chart is called the periodic table of the elements, or simply the **periodic table**. Metals are placed on the left side of the table, and nonmetals are placed on the right side. Hydrogen is an exception; although it is a nonmetal, hydrogen has unusual properties and has been placed by itself at the top center of the periodic tables in this text.

Metals and nonmetals are separated by semimetals that include boron, silicon, germanium, arsenic, antimony, and tellurium. The radioactive elements polonium and astatine are also considered semimetals. Figure 4.5 shows the overall arrangement of metals, nonmetals, and semimetals in the periodic table of the elements.

Physical States of the Elements

The periodic table can help you master a large amount of information. With little effort, you can correctly predict the physical state of most elements. Excluding mercury, all the metals are in the solid state at normal conditions of 25°C and normal atmospheric pressure. All semimetals are also in the solid state at normal temperature and pressure.

Nonmetals, on the other hand, show great diversity in physical state. At 25°C and normal pressure, 5 nonmetals are solids, 1 is a liquid, and 11 are gases. The solids are carbon, phosphorus, sulfur, selenium, and iodine. The only nonmetal liquid is reddish-brown bromine. The 11 gases are hydrogen, nitrogen, oxygen, fluo-

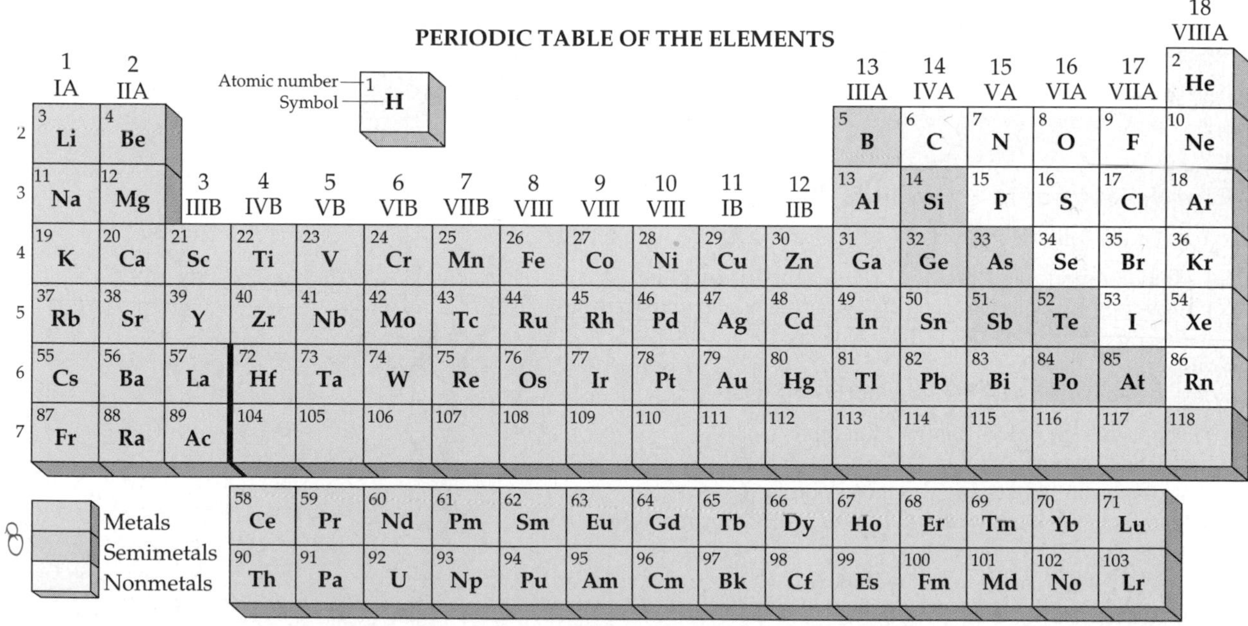

Metals, Nonmetals, and Semimetals

▲ **Figure 4.5 Metals, Nonmetals, and Semimetals** The symbols of elements having metallic properties are on the left side of the periodic table, nonmetallic are on the right side, and semimetallic are midway between. Notice the special placement of hydrogen, a nonmetallic element.

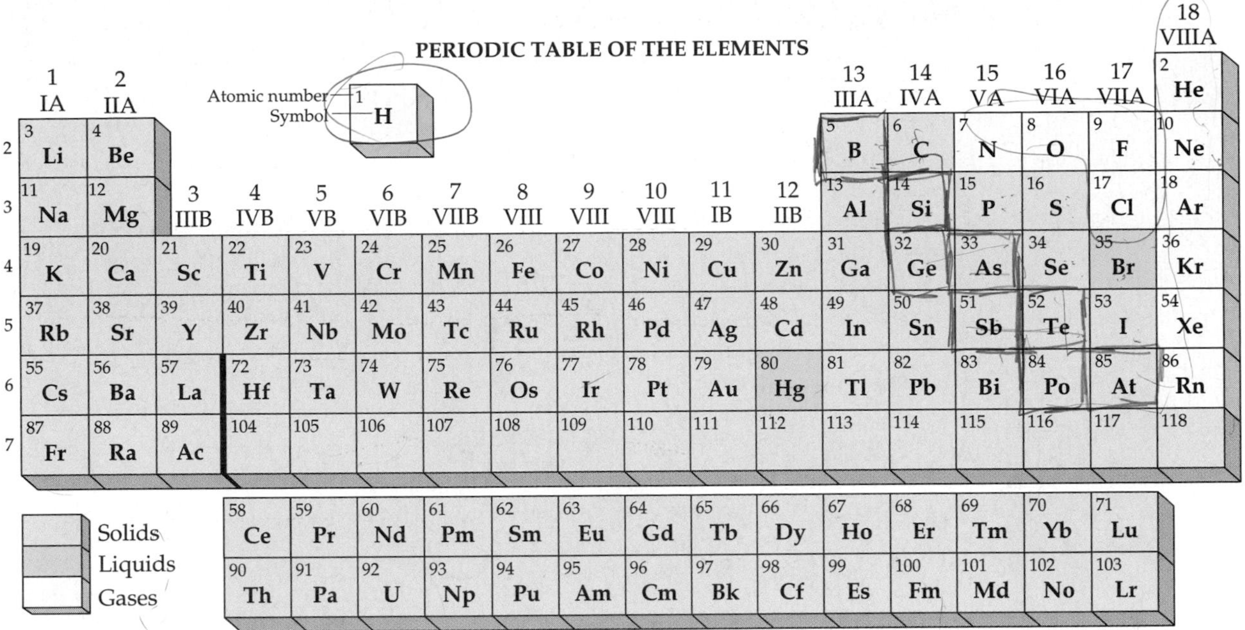

▲ **Figure 4.6 Physical States of the Elements** At 25°C and normal atmospheric pressure, all metals are in the solid state except Hg. Most nonmetals are gases except C, P, S, Se, and I, which are solids. The only elements in the liquid state are Hg and Br.

rine, chlorine, helium, neon, argon, krypton, xenon, and radon. All these gases are colorless except fluorine and chlorine, which are greenish-yellow. Figure 4.6 illustrates the normal physical state of each element.

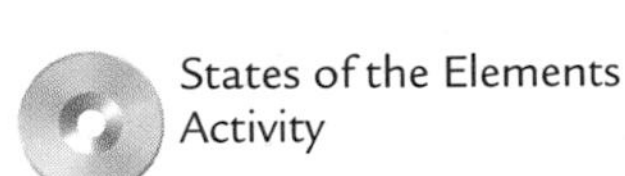
States of the Elements Activity

Example Exercise 4.4 • Physical States of the Elements

Indicate the physical state for each of the following elements at 25°C and normal pressure; classify each element as a metal, nonmetal, or semimetal.

(a) barium (b) boron
(c) bismuth (d) bromine

Solution

Referring to Figures 4.5 and 4.6, we observe the following.

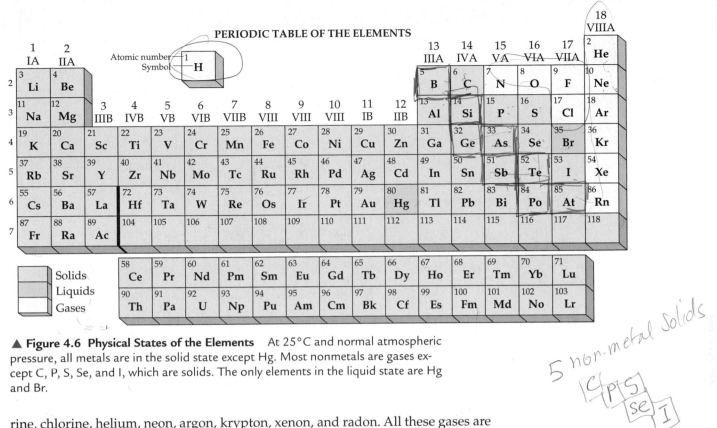

(a) Barium (Ba) is on the left side of the periodic table; it is a *solid metal* under normal conditions.
(b) Boron (B) is in the middle of the periodic table; it is a *solid semimetal*.
(c) Bismuth (Bi) is to the right but below the semimetals in the periodic table; it is a *solid metal*.
(d) Bromine (Br) is on the right side of the periodic table; it is a *liquid nonmetal* at normal conditions.

Self-Test Exercise

Indicate the physical state for the following elements at 25°C and normal pressure; classify each element as a metal, nonmetal, or semimetal.

(a) aluminum (b) hydrogen
(c) helium (d) radium

Answers: (a) solid metal; (b) gaseous nonmetal; (c) gaseous nonmetal; (d) solid metal

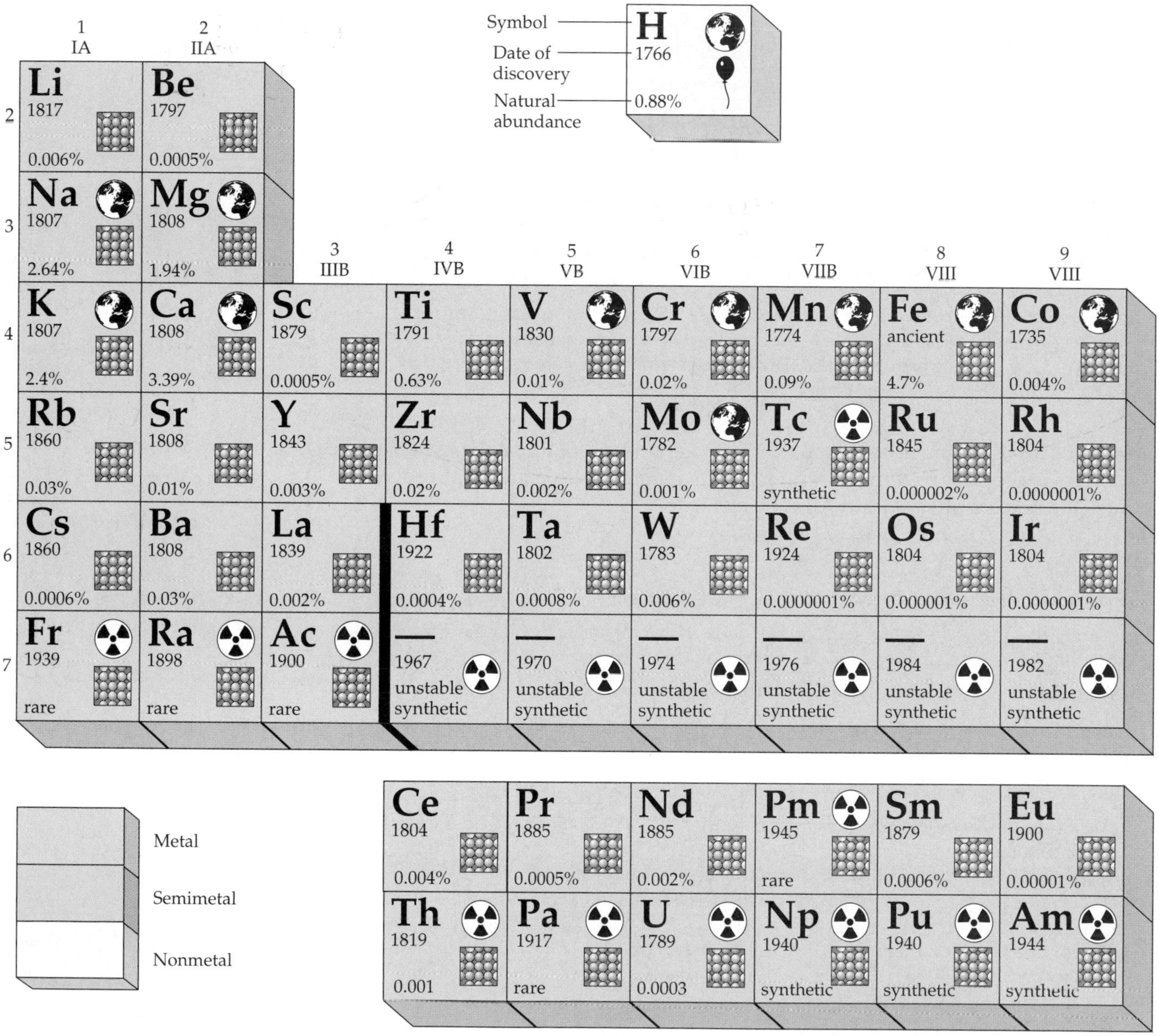

Symbol	H
Date of discovery	1766
Natural abundance	0.88%

	1 IA	2 IIA	3 IIIB	4 IVB	5 VB	6 VIB	7 VIIB	8 VIII	9 VIII
2	Li 1817 0.006%	Be 1797 0.0005%							
3	Na 1807 2.64%	Mg 1808 1.94%							
4	K 1807 2.4%	Ca 1808 3.39%	Sc 1879 0.0005%	Ti 1791 0.63%	V 1830 0.01%	Cr 1797 0.02%	Mn 1774 0.09%	Fe ancient 4.7%	Co 1735 0.004%
5	Rb 1860 0.03%	Sr 1808 0.01%	Y 1843 0.003%	Zr 1824 0.02%	Nb 1801 0.002%	Mo 1782 0.001%	Tc 1937 synthetic	Ru 1845 0.000002%	Rh 1804 0.0000001%
6	Cs 1860 0.0006%	Ba 1808 0.03%	La 1839 0.002%	Hf 1922 0.0004%	Ta 1802 0.0008%	W 1783 0.006%	Re 1924 0.0000001%	Os 1804 0.000001%	Ir 1804 0.0000001%
7	Fr 1939 rare	Ra 1898 rare	Ac 1900 rare	— 1967 unstable synthetic	— 1970 unstable synthetic	— 1974 unstable synthetic	— 1976 unstable synthetic	— 1984 unstable synthetic	— 1982 unstable synthetic

Metal

Semimetal

Nonmetal

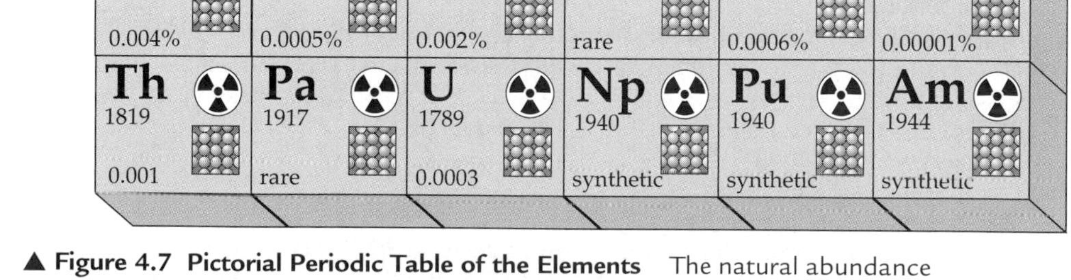

Ce 1804 0.004%	Pr 1885 0.0005%	Nd 1885 0.002%	Pm 1945 rare	Sm 1879 0.0006%	Eu 1900 0.00001%
Th 1819 0.001	Pa 1917 rare	U 1789 0.0003	Np 1940 synthetic	Pu 1940 synthetic	Am 1944 synthetic

Pictorial Periodic Table of the Elements

▲ **Figure 4.7 Pictorial Periodic Table of the Elements** The natural abundance is the percent by mass of an element in the Earth's crust, oceans, and atmosphere. The natural abundance of an element listed as rare is less than 1 mg per metric ton (1000 kg). An element listed as synthetic is made artificially and does not occur naturally. An element listed as unstable disintegrates in a fraction of a second.

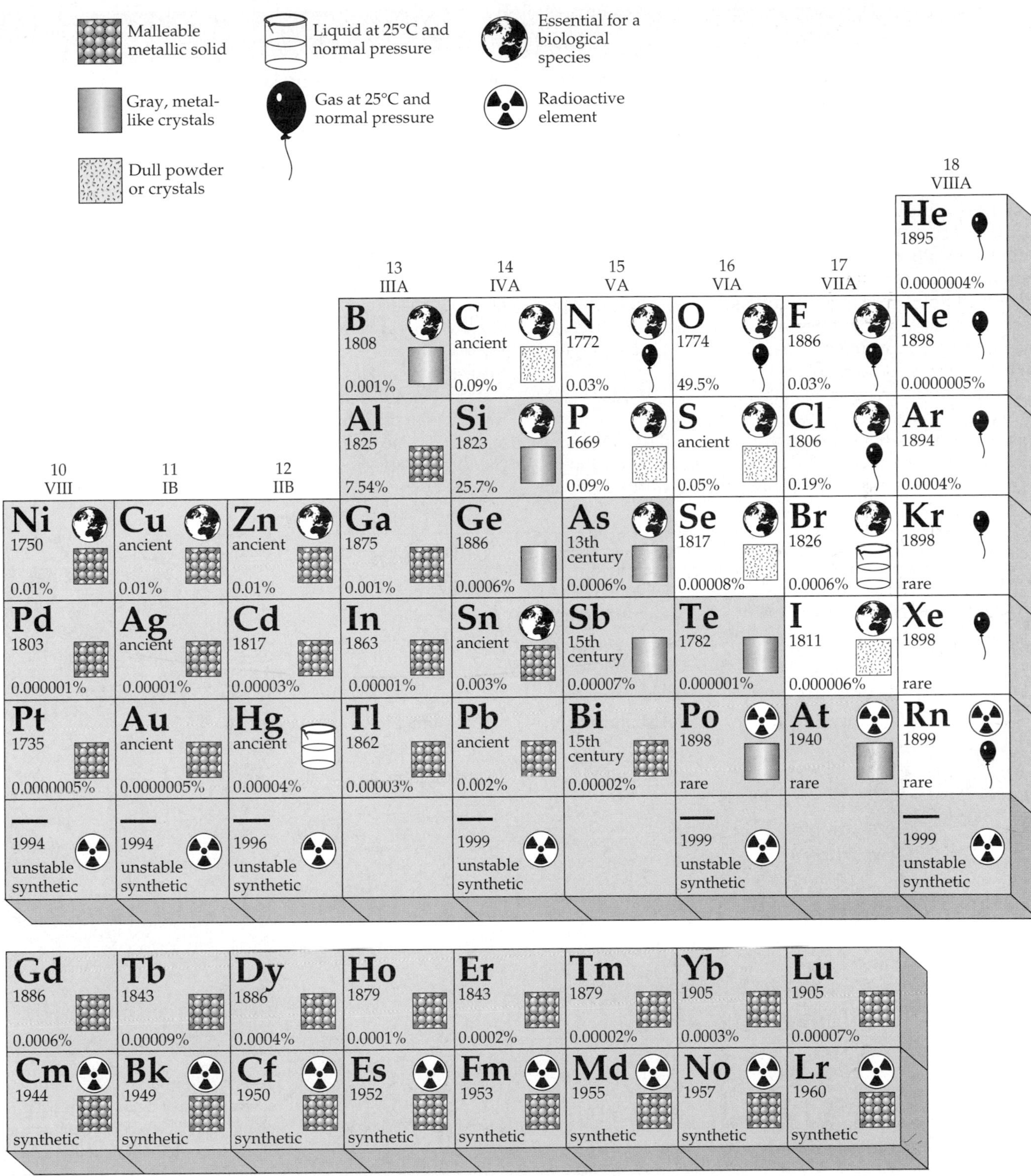
Malleable metallic solid
Gray, metal-like crystals
Dull powder or crystals
Liquid at 25°C and normal pressure
Gas at 25°C and normal pressure
Essential for a biological species
Radioactive element
18 VIIIA
He 1895 0.0000004%
13 IIIA
14 IVA
15 VA
16 VIA
17 VIIA
B 1808 0.001%
C ancient 0.09%
N 1772 0.03%
O 1774 49.5%
F 1886 0.03%
Ne 1898 0.0000005%
Al 1825 7.54%
Si 1823 25.7%
P 1669 0.09%
S ancient 0.05%
Cl 1806 0.19%
Ar 1894 0.0004%
10 VIII
11 IB
12 IIB
Ni 1750 0.01%
Cu ancient 0.01%
Zn ancient 0.01%
Ga 1875 0.001%
Ge 1886 0.0006%
As 13th century 0.0006%
Se 1817 0.00008%
Br 1826 0.0006%
Kr 1898 rare
Pd 1803 0.000001%
Ag ancient 0.00001%
Cd 1817 0.00003%
In 1863 0.00001%
Sn ancient 0.003%
Sb 15th century 0.00007%
Te 1782 0.000001%
I 1811 0.000006%
Xe 1898 rare
Pt 1735 0.0000005%
Au ancient 0.0000005%
Hg ancient 0.00004%
Tl 1862 0.00003%
Pb ancient 0.002%
Bi 15th century 0.00002%
Po 1898 rare
At 1940 rare
Rn 1899 rare
— 1994 unstable synthetic
— 1994 unstable synthetic
— 1996 unstable synthetic
— 1999 unstable synthetic
— 1999 unstable synthetic
— 1999 unstable synthetic
Gd 1886 0.0006%
Tb 1843 0.00009%
Dy 1886 0.0004%
Ho 1879 0.0001%
Er 1843 0.0002%
Tm 1879 0.00002%
Yb 1905 0.0003%
Lu 1905 0.00007%
Cm 1944 synthetic
Bk 1949 synthetic
Cf 1950 synthetic
Es 1952 synthetic
Fm 1953 synthetic
Md 1955 synthetic
No 1957 synthetic
Lr 1960 synthetic

Update • Elements 104 and Beyond

What is the atomic number of the element that was first named kurchatovium by Russian scientists and rutherfordium by Americans?

In 1964 a team of Russian scientists reported the first synthesis of element 104. The Russians proposed the name "kurchatovium" for Igor Kurchatov, a Soviet physicist. American scientists could not confirm the results, however, and denied the Russian claim of discovery. In 1969 a heavy-element team at the University of California, Berkeley, synthesized element 104 and proposed the name "rutherfordium" after the English physicist Ernest Rutherford. A few years later, a similar disagreement arose over element 105. Russian and American scientists each claimed they discovered element 105, and each team proposed different names and symbols.

In 1985 the International Union of Pure and Applied Chemistry (IUPAC) attempted to resolve the controversy by recommending systematic names. According to IUPAC, the names for elements 104 and beyond were to be formed from Latin prefixes plus the suffix -ium. The name for element 104, for example, was to be unnilquadium. Interestingly, there was a strong objection to the IUPAC recommendation because it did not honor scientists who had made significant contributions to our understanding of heavy elements.

In 1997 IUPAC issued a new set of official names. Element 104 is named rutherfordium for Ernest Rutherford; element 105 is named dubnium for the Russian nuclear research facility; element 106 is named seaborgium for the American physicist, Glenn Seaborg; element 107 is named bohrium in honor of the Danish physicist Niels Bohr; element 108 is named hassium for the province of Hesse in Germany where new elements have been made; and element 109 is named meitnerium in honor of the Austrian physicist Lise Meitner. In the summer of 1999, elements 114, 116, and 118 were synthesized at the University of California, Berkeley.

Element	Discovery	Proposed Name	IUPAC Name
104	Russia, 1964	kurchatovium, Ku	rutherfordium, Rf
	United States, 1969	rutherfordium, Rf	
105	Russia, 1967	nielsbohrium, Ns	dubnium, Db
	United States, 1970	hahnium, Ha	
106	United States, 1974	seaborgium, Sg	seaborgium, Sg
107	Germany, 1974	nielsbohrium, Ns	bohrium, Bh
108	Germany, 1984	hassium, Hs	hassium, Hs
109	Germany, 1982	meitnerium, Mt	meitnerium, Mt
110	Germany, 1994	—	—
111	Germany, 1994	—	—
112	Germany, 1996	—	—
113	not discovered	—	—
114	United States, 1999	—	—
115	not discovered	—	—
116	United States, 1999	—	—
117	not discovered	—	—
118	United States, 1999	—	—

▲ **New Elements** Several new elements have been synthesized using a particle accelerator at the University of California, Berkeley.

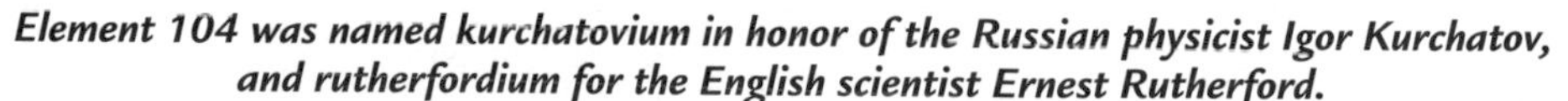

Element 104 was named kurchatovium in honor of the Russian physicist Igor Kurchatov, and rutherfordium for the English scientist Ernest Rutherford.

4.5 Compounds and Chemical Formulas

Objectives
- To explain the law of definite composition for a compound.
- To state the number of atoms of each element in a compound given the chemical formula.

In the late 1700s, the French chemist Joseph Louis Proust (1754–1826) painstakingly analyzed the compound copper carbonate. No matter how he prepared the compound, he found that the elements copper, carbon, and oxygen were always present in the same proportion by mass. Proust studied many other compounds and obtained similar constant proportions for the elements. In 1799 Proust stated that "Compounds always contain the same elements in a constant proportion by mass." This statement is now called the **law of definite composition** or the law of constant proportion.

Ordinary table salt is the compound sodium chloride, NaCl. According to the law of definite composition, salt contains sodium and chlorine in a constant proportion by mass. For NaCl, the ratio is 39.3% sodium and 60.7% chlorine. Moreover, the proportion of sodium and chlorine is the same whether we have a tiny crystal of salt, a block of salt, or a mountain of salt.

Similarly, the compound water contains 11.2% hydrogen and 88.8% oxygen by mass. Regardless of the amount, water always contains hydrogen and oxygen in a constant proportion by mass. Figure 4.8 illustrates the law of definite composition with water as an example.

Chemical Formulas

Most elements occur naturally as a collection of individual atoms. A few nonmetal elements, such as hydrogen and oxygen, occur naturally as particles composed of two or more atoms. A single particle composed of two or more nonmetal atoms is called a **molecule**. Hydrogen (H_2) and oxygen (O_2) occur naturally as molecules containing two atoms.

We have learned to use chemical symbols for the names of elements. In a similar fashion, we use chemical formulas for the names of compounds. A **chemical formula** expresses the number of atoms of each element in a compound. The number of atoms is indicated with a subscript, unless the number is 1, in which case the subscript is omitted. A molecule of water contains two atoms of hydrogen and one atom of oxygen. Therefore, its chemical formula is H_2O. A molecule of ammonia has one atom of nitrogen and three atoms of hydrogen; its chemical formula is NH_3. The chemical formula of sulfuric acid, H_2SO_4, is interpreted in Figure 4.9.

Example Exercise 4.5 • Composition of Chemical Formulas

State the atomic composition for a molecule of niacin, vitamin B_3, $C_6H_6N_2O$.

Solution

The chemical formula for niacin indicates 6 carbon atoms, 6 hydrogen atoms, 2 nitrogen atoms, and 1 oxygen atom. Thus, $C_6H_6N_2O$ has a total of 15 atoms.

Self-Test Exercise

Write the chemical formula for vitamin B_6 if a molecule is composed of 8 carbon atoms, 11 hydrogen atoms, 1 nitrogen atom, and 3 oxygen atoms.

Answer: $C_8H_{11}NO_3$ (total of 23 atoms)

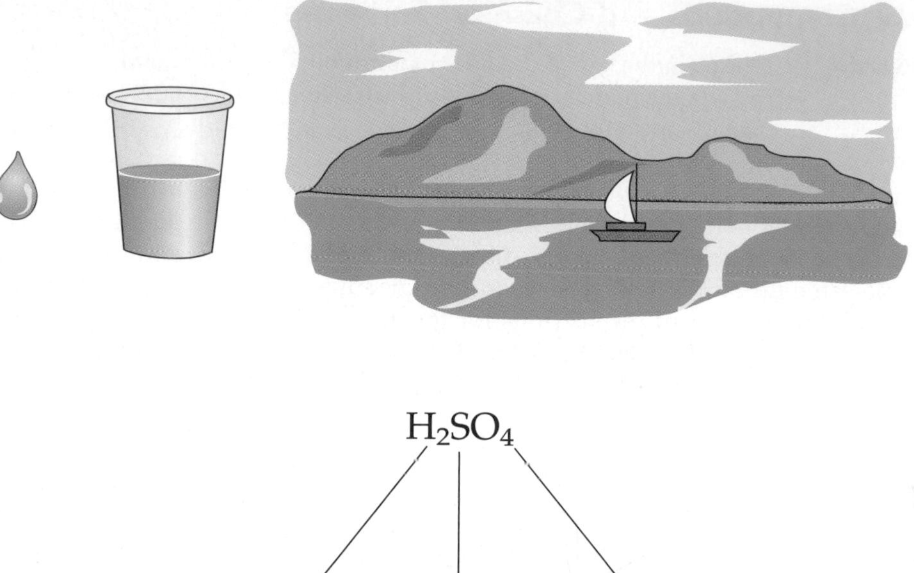

▶ **Figure 4.8 Law of Definite Composition** A drop of water, a glass of water, and a lake of water all contain hydrogen and oxygen in the same percent by mass, that is, 11.2% hydrogen and 88.8% oxygen.

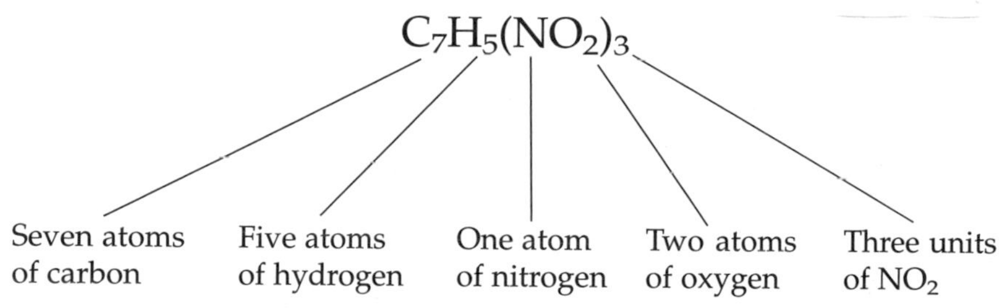

▶ **Figure 4.9 Interpretation of a Chemical Formula** The chemical formula for sulfuric acid, H_2SO_4, indicates 2 atoms of hydrogen (H), 1 atom of sulfur (S), and 4 atoms of oxygen (O). Thus, the formula of sulfuric acid has a total of 7 atoms. Sulfuric acid is the corrosive electrolyte found in an automobile battery.

Some chemical formulas use parentheses to clarify the atomic composition. For example, antifreeze has the chemical formula $C_2H_4(OH)_2$. A molecule is composed of 2 atoms of carbon, 4 atoms of hydrogen, and 2 units of OH. The parentheses emphasize that OH is a fundamental part of the formula and is so written in the chemical formula. Figure 4.10 illustrates the chemical formula for the explosive trinitrotoluene (TNT), which has the formula $C_7H_5(NO_2)_3$.

$C_7H_5(NO_2)_3$

Seven atoms of carbon | Five atoms of hydrogen | One atom of nitrogen | Two atoms of oxygen | Three units of NO_2

▶ **Figure 4.10 Interpretation of a Chemical Formula** The chemical formula of TNT, $C_7H_5(NO_2)_3$, indicates 7 carbon atoms, 5 hydrogen atoms, and 3 NO_2 units. The nitrogen atom and 2 oxygen atoms are a single unit that appears three times in a molecule of trinitrotoluene.

Example Exercise 4.6 • Composition of Chemical Formulas

State the atomic composition for a molecule of glycerin, $C_3H_5(OH)_3$.

Solution

The chemical formula for glycerin indicates 3 carbon atoms, 5 hydrogen atoms, and 3 OH units. Thus, $C_3H_5(OH)_3$ has a total of 14 atoms.

Self-Test Exercise

Write the chemical formula for nitroglycerin if a molecule is composed of 3 carbon atoms, 5 hydrogen atoms, 3 oxygen atoms, and 3 NO_2 units.

Answers: $C_3H_5O_3(NO_2)_3$ (20 atoms total)

Note In Chapter 7 we will learn how to systematically name compounds. It is helpful to begin referring to compounds by pronouncing their chemical formulas. Even professional chemists pronounce chemical formulas of compounds; for instance, water is H_2O and is referred to as "H two O." Sulfuric acid is H_2SO_4 and is verbalized "H two, S, O four." Glucose, $C_6H_{12}O_6$, is verbalized "C six, H twelve, O six. Milk of magnesia, $Mg(OH)_2$, is pronounced "Mg, OH taken twice."

4.6 Physical and Chemical Properties

Objective · To classify a property of a substance as physical or chemical.

Previously we classified matter as either a mixture or a substance. Recall that the physical and chemical properties of a substance are consistent throughout a sample. Moreover, no two substances have the same physical and chemical properties.

Furthermore, we said that a pure substance is either a compound or an element. However, the physical and chemical properties of a compound are never the same as the properties of its constituent elements. For example, sodium is a soft, silver metal, and chlorine is a poisonous, yellow gas. In a reaction, these two elements combine to form white crystals of table salt, NaCl (Figure 4.11).

A **physical property** refers to those characteristics of a substance that we can observe without changing the composition of the substance. The list of physical properties is extensive, but those considered important are appearance, melting and boiling points, density, heat and electrical conductivity, solubility, and physical state under normal conditions. Table 4.5 lists some of the physical properties of sodium, chlorine, and sodium chloride at 25°C and normal atmospheric pressure.

Table 4.5 Physical Properties of Sodium, Chlorine, and Sodium Chloride at Normal Conditions (25°C and 1.00 atm)

Property	Sodium	Chlorine	Sodium Chloride
appearance	silver metal	yellowish gas	colorless crystals
melting point	98°C	−101°C	801°C
boiling point	883°C	−35°C	1413°C
density	0.97 g/cm^3	2.90 g/L	2.17 g/cm^3
solubility in 100 g water	reacts with water	0.51 g at 30°C	35.7 g at 0°C

A **chemical property** of a substance describes its chemical reactions with other substances. The chemical properties of sodium, for example, include its reaction with oxygen to form sodium oxide and its reaction with water to produce hydrogen gas. There are many reactions for the elements, more than we can possibly consider. We can, however, use the periodic table of elements to predict chemical behavior.

The periodic table is arranged according to families of elements that in general give similar reactions. For instance, the metals in Group IA/1 on the far left side of the periodic table react with oxygen gas to give products with similar chemical formulas, namely, Li_2O, Na_2O, K_2O, Rb_2O, and Cs_2O. Notice that in each case, the number of metal to oxygen atoms is in the ratio of 2 to 1.

In Chapter 6 we will study families of elements. For now, review Table 4.6, which illustrates that elements in the same family have similar chemical properties and give products with similar chemical formulas.

Table 4.6 Chemical Properties for Families of Elements

Family and Element	Reaction with Oxygen	Reaction with Water	Reaction with Hydrochloric Acid
Group IA/1			
lithium	Li_2O	$H_2 + LiOH$	$H_2 + LiCl$
sodium	Na_2O	$H_2 + NaOH$	$H_2 + NaCl$
potassium	K_2O	$H_2 + KOH$	$H_2 + KCl$
Group IIA/2			
calcium	CaO	$H_2 + Ca(OH)_2$	$H_2 + CaCl_2$
strontium	SrO	$H_2 + Sr(OH)_2$	$H_2 + SrCl_2$
barium	BaO	$H_2 + Ba(OH)_2$	$H_2 + BaCl_2$
Group IVA/14			
carbon	CO_2	NR*	NR
silicon	SiO_2	NR	NR
germanium	GeO_2	NR	NR
Group VIIIA/18			
helium	NR	NR	NR
neon	NR	NR	NR
argon	NR	NR	NR

*NR, no reaction.

Classifying Physical and Chemical Properties

We can distinguish between a physical and a chemical property as follows. Physical properties are observed without altering the composition of the substance, whereas chemical properties always involve a chemical change. When a chemical change occurs, the composition of a substance changes and another substance is formed. Example Exercise 4.7 illustrates the distinction between physical and chemical properties.

Example Exercise 4.7 • Physical and Chemical Properties

Classify each of the following properties as physical or chemical.

(a) Water appears colorless and odorless at 20°C.
(b) Water dissolves sucrose crystals.
(c) Water produces a gas with calcium metal.
(d) Water exists as ice at −10°C.

Solution

If a reaction occurs, there is a change in composition and the property is chemical. Otherwise, the property is physical.

(a) Color and odor are *physical* properties.
(b) Solubility is a *physical* property.
(c) A chemical reaction is a *chemical* property.
(d) A physical state is a *physical* property.

Self-Test Exercise

Classify each of the following properties as physical or chemical.

(a) Water appears hard and crystalline at 0°C.
(b) Water is insoluble in gasoline.
(c) Water is a very weak conductor of electricity.
(d) Water produces a gas with calcium metal.

Answers: (a) physical; (b) physical; (c) physical; (d) chemical

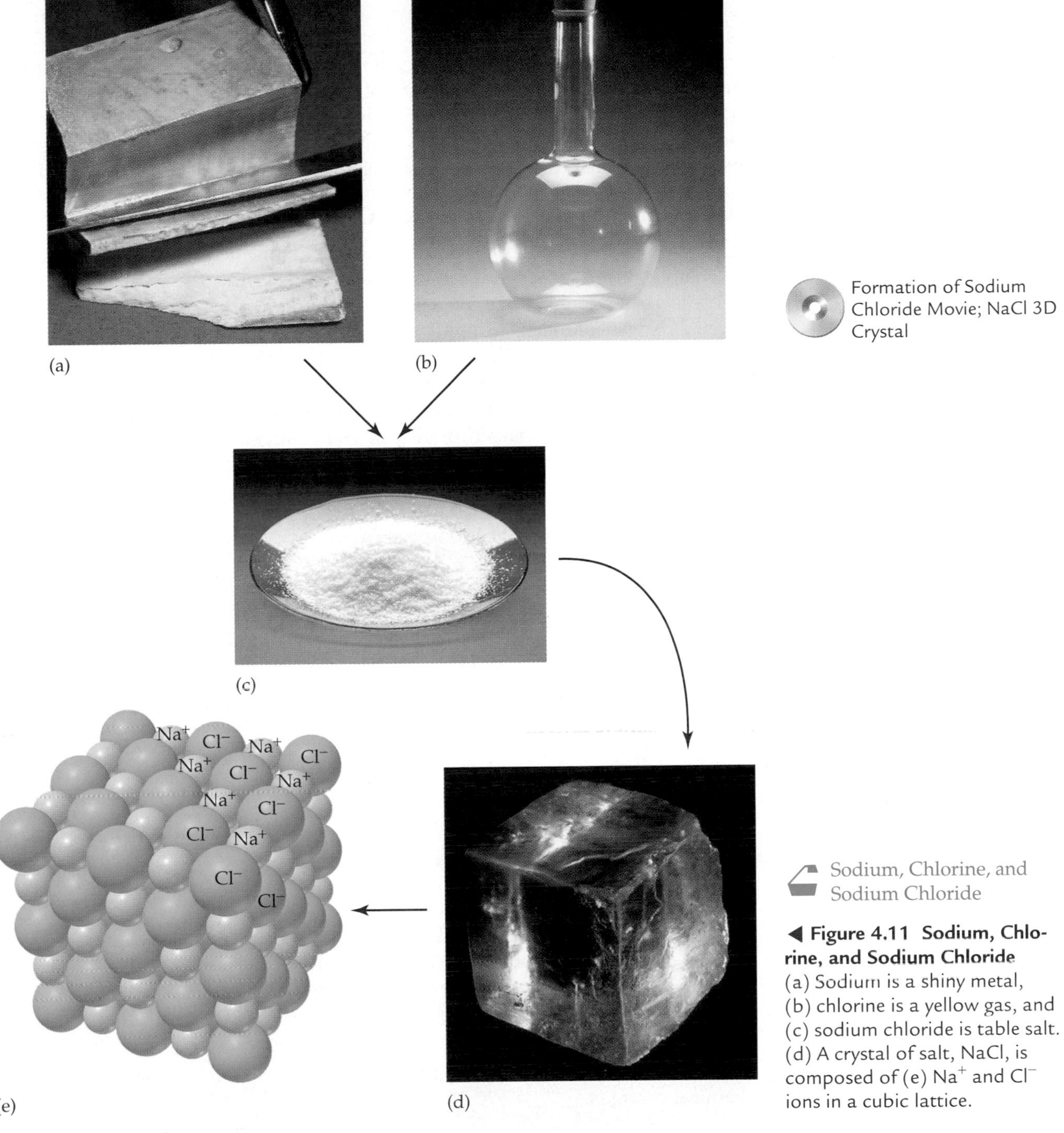

Formation of Sodium Chloride Movie; NaCl 3D Crystal

Sodium, Chlorine, and Sodium Chloride

◀ **Figure 4.11 Sodium, Chlorine, and Sodium Chloride**
(a) Sodium is a shiny metal, (b) chlorine is a yellow gas, and (c) sodium chloride is table salt. (d) A crystal of salt, NaCl, is composed of (e) Na^+ and Cl^- ions in a cubic lattice.

4.7 Physical and Chemical Changes

Objective · To classify a change in a substance as physical or chemical.

Experiment # 5, Prentice Hall Laboratory Manual

When a substance is altered, the alteration is classified as a physical or a chemical change. In a **physical change**, the chemical composition of the sample does not change. If we melt ice to water, heat alcohol to a vapor, or recycle aluminum cans into aluminum foil, we are performing a physical change. The chemical composition of the ice, alcohol, and aluminum is the same before and after the change. In other words, the formula of each substance (H_2O, C_2H_5OH, and Al) remains constant. Thus, altering the shape or physical state of a substance is an indication of a physical change.

In a **chemical change**, there is a chemical reaction. The composition of the sample changes, and we observe a new set of properties. If a banana ripens from green to yellow, an antacid tablet in water evolves gas bubbles, or a fireworks displays a shower of colorful lights, we are observing a chemical change. The composition of the banana, antacid tablet, and fireworks is different before and after the change.

Classifying Physical and Chemical Changes

We can distinguish between a physical and a chemical change as follows: a physical change can be observed without altering the composition of the substance, whereas a chemical change always involves the formation of a new substance. When a new substance is formed, we usually observe one of the following: a permanent change in color, an odor or bubbles from the release of a gas, or light or heat from the release of energy.

Precipitation Reactions Movie

In the laboratory, we may observe that two solutions added together produce an insoluble substance. The formation of a solid substance forming in solution is a practical observation that indicates a chemical change. Figure 4.12 illustrates three examples of chemical changes.

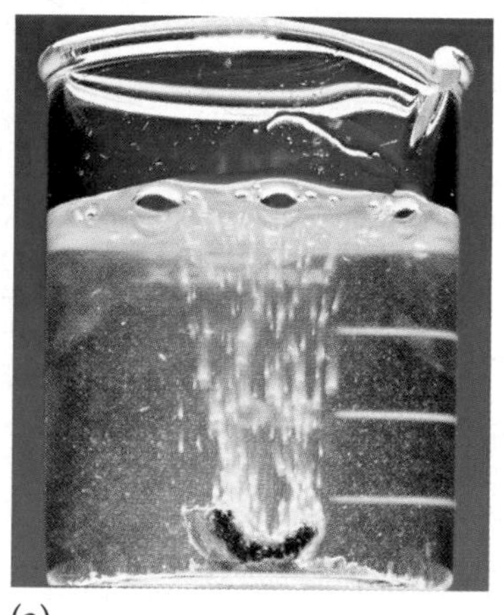

(a)

(b)

(c)

▲ **Figure 4.12 Evidence of Chemical Change** The evidence for a chemical change includes (a) gas bubbles from the reaction of calcium metal in water, (b) heat and light energy from the reaction of potassium metal in water, and (c) a colorful, insoluble substance from the reaction of two solutions.

Example Exercise 4.8 further illustrates the distinction between physical and chemical changes.

Example Exercise 4.8 • Physical and Chemical Changes

Classify each of the following observations as a physical or a chemical change.

(a) Touching a match to hydrogen soap bubbles results in an explosion.
(b) Heating water in a flask produces moisture on the glass.
(c) Adding together two colorless solutions gives a yellow solid.
(d) Pouring vinegar on baking soda produces gas bubbles.

Solution

The observations that indicate a physical change include changing shape, volume, or physical state. The observations that suggest a chemical change include burning, fizzing, changing color, or forming an insoluble substance in solution.

(a) Hydrogen explodes; thus, it is a *chemical* change.
(b) Water is boiled; thus, it is a *physical* change.
(c) Two solutions give an insoluble substance; thus, it is a *chemical* change.
(d) Baking soda fizzes; thus, it is a *chemical* change.

Self-Test Exercise

Classify each of the following observations as a physical or a chemical change.

(a) Freezing water in a refrigerator makes cubes of ice.
(b) Adding silver nitrate to tap water gives a cloudy solution.
(c) Burning sulfur gives a light-blue flame.
(d) Grinding aspirin tablets produces a powder.

Answers: (a) physical; (b) chemical; (c) chemical; (d) physical

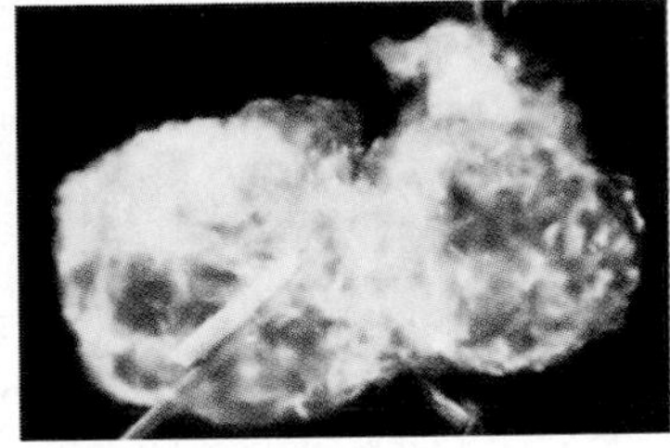

▲ **Hydrogen Explosion** Soap bubbles filled with hydrogen gas undergo a chemical change after being ignited by a candle flame.

4.8 Conservation of Mass

Objective · To apply the conservation of mass law to chemical changes.

In the late 1800s, the French chemist Antoine Lavoisier performed exacting laboratory experiments and found that the mass of substances before a chemical change was always equal to the mass of substances after the change. He concluded that matter was neither created nor destroyed during a chemical reaction. In 1789 he announced the conservation of mass principle. This principle has become known as the **law of conservation of mass**.

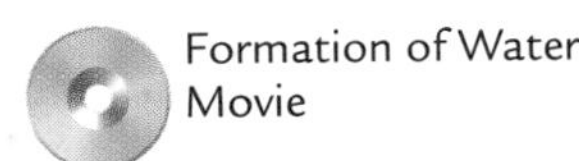

Formation of Water Movie

As an example of conservation of mass, consider the reaction of hydrogen and oxygen. Hydrogen and oxygen always combine in the same ratio by mass to form water. That is, if 1 g of hydrogen combines with 8 g of oxygen, then 2 g of hydrogen combines with 16 g of oxygen, 3 g of hydrogen reacts with 24 g of oxygen, and so on. We can predict the mass of water produced by each of these reactions. From the conservation of mass law, we know that the mass of hydrogen plus the mass of oxygen must equal the mass of water produced. That is,

$$1.0 \text{ g hydrogen} + 8.0 \text{ g oxygen} = 9.0 \text{ g water}$$

Conversely, we can predict the masses of hydrogen and oxygen produced by the decomposition of water. Passing an electric current through water produces hydrogen gas and oxygen gas. This process is called electrolysis. If the electrolysis of 45.0 g of water produces 5.0 g of hydrogen, how many grams of oxygen are evolved?

The conservation of mass law states that the mass of the hydrogen and oxygen is equal to the mass of the water, 45.0 g. Therefore, the mass of oxygen must equal the mass of the water minus the mass of the hydrogen. That is,

$$45.0 \text{ g water} - 5.0 \text{ g hydrogen} = 40.0 \text{ g oxygen}$$

The conservation of mass law is one of the most important principles in chemistry. Historically, it was used to determine the formulas of compounds and to develop the first periodic table of elements. As we will discuss in Chapter 10, it is the cornerstone of all quantitative relationships in chemistry. Example Exercise 4.9 further illustrates the conservation of mass law.

Example Exercise 4.9 • Conservation of Mass Law

In an experiment, 2.430 g of magnesium metal was ignited and burned with oxygen in the air. If 4.030 g of white magnesium oxide powder, MgO, was collected, what was the mass of oxygen gas that reacted?

Solution

Applying the conservation of mass law, we find that the mass of the magnesium metal plus the mass of the oxygen gas equals the mass of the magnesium oxide powder. That is,

$$2.430 \text{ g mg} + \text{mass of oxygen} = 4.030 \text{ g MgO}$$
$$\text{mass of oxygen} = 4.030 \text{ g MgO} - 2.430 \text{ g Mg}$$
$$\text{mass of oxygen} = 1.600 \text{ g}$$

Self-Test Exercise

If 0.654 g of zinc metal reacts with 0.321 g of yellow powdered sulfur, what is the mass of the zinc sulfide produced?

Answer: 0.975 g

4.9 Potential and Kinetic Energy

Objectives
- To distinguish between potential and kinetic energy.
- To relate kinetic energy, temperature, and physical state.

Energy can be defined as the ability to do work. There are two types of energy: potential and kinetic. **Potential energy** is stored energy that matter possesses as a result of its position or composition. **Kinetic energy** is the energy matter has as a result of its motion. A boulder perched at the top of a mountain has potential energy (PE). If the boulder rolls down the mountain, it loses potential energy but gains kinetic energy (KE). In the process of rolling down the mountain, the potential energy of the boulder is converted to kinetic energy. Figure 4.13 illustrates the conversion from potential to kinetic energy.

In our study of chemistry, we are usually interested in the potential energy associated with the chemical composition of a substance. For example, gasoline has potential energy that is converted to kinetic energy when it undergoes combustion. As gasoline burns, the potential energy of the cool liquid is converted to the kinetic energy of the hot, expanding gas mixture.

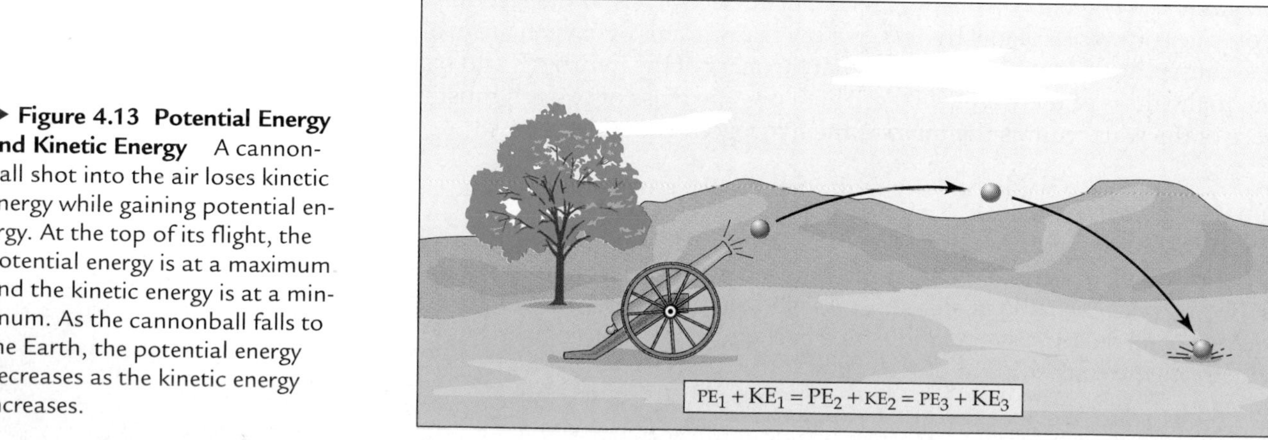

$PE_1 + KE_1 = PE_2 + KE_2 = PE_3 + KE_3$

▶ **Figure 4.13 Potential Energy and Kinetic Energy** A cannonball shot into the air loses kinetic energy while gaining potential energy. At the top of its flight, the potential energy is at a maximum and the kinetic energy is at a minimum. As the cannonball falls to the Earth, the potential energy decreases as the kinetic energy increases.

Kinetic Energy, Temperature, and Physical State

We can view a sample of matter as a large collection of very small particles. The attraction between particles is one of the criteria that influences the physical state. In solids, this force of attraction holds particles in fixed positions and allows little or no movement.

When a substance in the *solid state* is heated, the temperature increases and particles begin to randomly vibrate in fixed positions. As a solid acquires more heat energy, particles vibrate so intensely that the force of attraction is overcome and the particles break free of each other. This is a description of the *liquid state*.

A liquid is therefore a state of matter in which particles possess sufficient energy to overcome their mutual attraction and have limited movement inside a container. When a liquid has enough energy to completely overcome the force of attraction, the particles are free to fly about within the walls of the container. At this point, the liquid has changed to the *gaseous state*. Table 4.7 summarizes the relationship between the properties of particles and the physical state.

Table 4.7 Properties and Physical States

Property of Particles	Solid	Liquid	Gas
kinetic energy	very low	high	very high
movement	none	restricted	unrestricted

We can review the relationship between kinetic energy and physical state as follows. If we heat a solid, it acquires enough energy to overcome the attraction between particles and melts to a liquid. As the temperature increases, particles gradually gain more energy and demonstrate increasing motion. If we continue to heat the liquid, it acquires enough energy to vaporize to a gas. Figure 4.14 illustrates the relationship between increasing energy and physical state.

Changes of State Movie

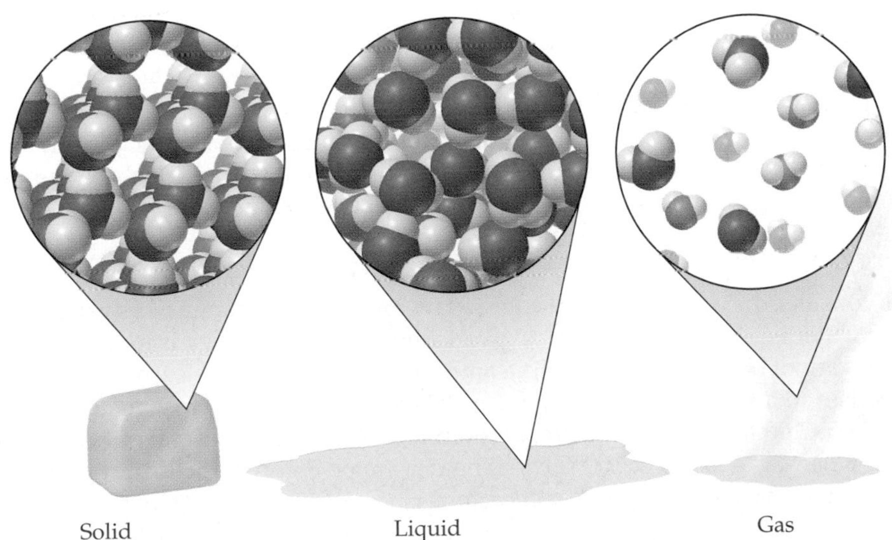

▲ **Figure 4.14 Energy and Physical State** The energy of individual H_2O molecules increases as solid ice is heated. As the attraction between molecules is overcome, solid ice changes to liquid water and then to gaseous steam.

The kinetic energy of a gas is directly related to its temperature. If the temperature increases, the kinetic energy increases; if the temperature decreases, the kinetic energy decreases. Moreover, if the temperature increases, the motion of particles increases; if the temperature decreases, the motion decreases.

Example Exercise 4.10 further illustrates how temperature, kinetic energy, and the motion of gas particles are related.

Example Exercise 4.10 • Kinetic Energy and Molecular Motion

A balloon filled with helium gas is cooled from 25° to −25°C. State the change in each of the following.

(a) kinetic energy of the gas (b) motion of helium atoms

Solution

Temperature, kinetic energy, and velocity are related as follows.

(a) As the temperature cools from 25° to −25°C, the kinetic energy of helium atoms *decreases*.

(b) Since a drop in temperature produces an decrease in kinetic energy, the motion of helium atoms *decreases*.

Self-Test Exercise

A steel cylinder containing air is heated from 25° to 50°C. State the change in each of the following.

(a) kinetic energy of the gas (b) motion of air molecules

Answers: (a) increases; (b) increases

4.10 Conservation of Energy

Objectives
- To apply the conservation of energy law to physical and chemical changes.
- To identify the following forms of energy: chemical, electrical, mechanical, nuclear, heat, and light.

In the middle of the nineteenth century, several English and German scientists proposed that the total energy in the universe is constant. This proposal was worded in different ways, but it essentially stated that energy cannot be created nor destroyed. Energy can, however, be converted from one form to another. This principle is known as the **law of conservation of energy**. This principle applies experimentally, and we can make energy predictions for physical and chemical changes.

Energy and Physical Changes

Let's consider the conservation of energy law applied to a change of physical state. When water is converted to steam, a given amount of heat energy is required. Since energy cannot be destroyed, the same amount of heat must be released if the steam condenses to a liquid. In practice, it requires 540 cal of heat energy to convert 1 g of water at 100°C to steam at 100°C. It therefore follows that 540 cal of heat are released when 1 g of steam at 100°C condenses to water. Figure 4.15 illustrates the conservation of energy for a physical change.

Energy and Chemical Changes

The conservation of energy law applied to a chemical change is analogous to the conservation of mass law. That is, the total energy before a reaction is the same as the total energy after the reaction. It is important to understand, however, that some chemical changes release energy while others absorb energy. Yet, the total energy before and after a reaction must always be constant.

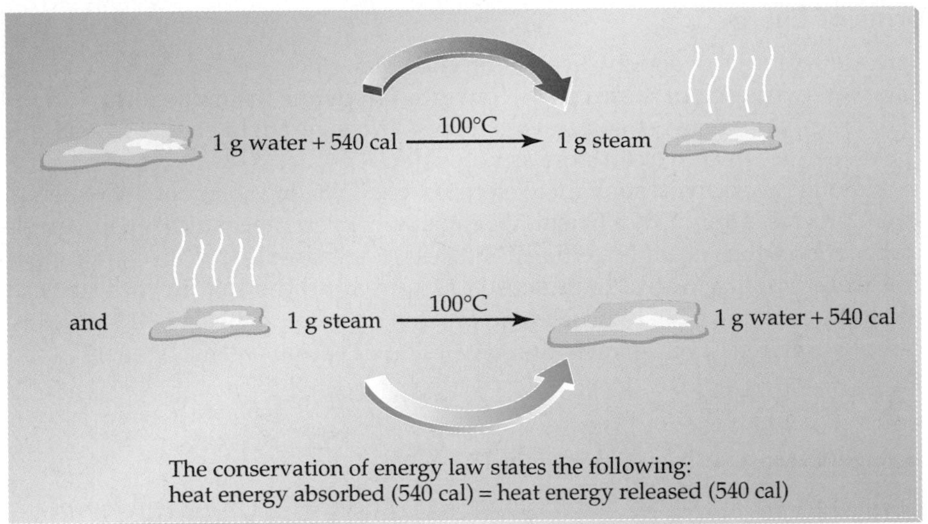

◀ **Figure 4.15 Conservation of Energy for a Physical Change** The physical change of state from water to steam, and the reverse process, involve the same amount of heat energy.

How can the energy be constant for a reaction that involves a change in heat? The explanation is that a reaction involves the two types of energy discussed previously, *potential* chemical energy and *kinetic* heat energy. In a reaction, the initial reactants may lose potential energy, but the resulting products gain kinetic energy. In reactions that produce heat, the initial reactants have more potential energy than the products. Conversely, in reactions that absorb heat, the initial reactants have less potential energy than the products.

If we consider the conservation of energy law, we can say that the energy released from a given change is equal to the energy required for the reverse change. For example, we know from experiment that 3200 cal of heat is released when hydrogen gas and oxygen gas react to produce 1 g of water. Thus, 3200 cal of heat is required for the reverse process of decomposing 1 g of water into hydrogen gas and oxygen gas. If we decompose water into the two gases using an electric current, the electrical energy required is equivalent to 3200 cal. Figure 4.16 illustrates the conservation of energy for a chemical change.

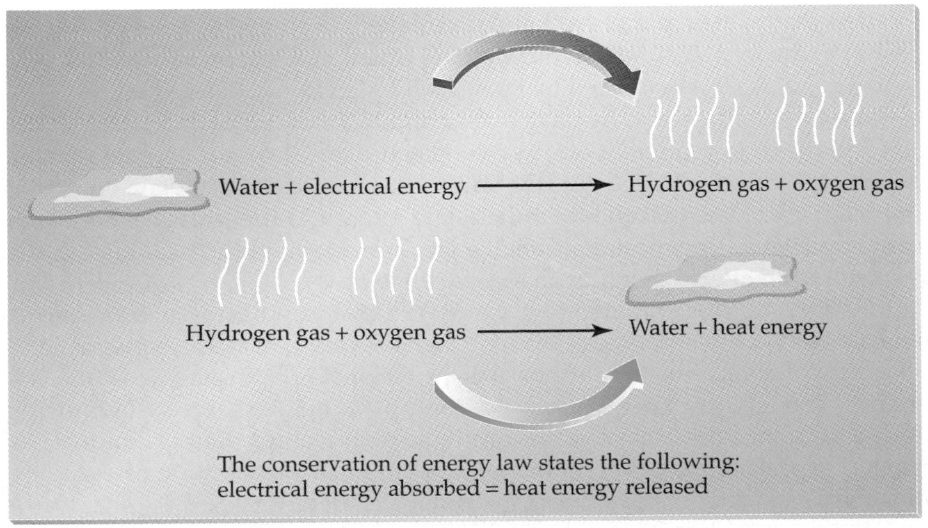

◀ **Figure 4.16 Conservation of Energy for a Chemical Change** The chemical change from water to hydrogen and oxygen gases, and the reverse process, involve the same amount of energy.

Forms of Energy

There are six forms of energy: heat, light, chemical, electrical, mechanical, and nuclear. Moreover, each of these can be converted from one form of energy to another. For example, we can use solar energy panels to convert light energy from the Sun to heat energy, which we can store as hot water.

Although we convert sunlight to electrical energy using solar energy panels, the current process is only 15% efficient. We can also convert mechanical energy to electrical energy, which is a more efficient process. For example, the mechanical energy of water spilling over a dam is used to spin a turbine that in turn drives an electrical generator. This hydroelectric process operates at nearly 90% efficiency. Example Exercise 4.11 illustrates the conversion of various forms of energy.

Example Exercise 4.11 • Forms of Energy

Identify two forms of energy that are involved in each of the following conversions.

(a) radioactive uranium vaporizes water to steam
(b) steam drives a turbine
(c) a turbine spins and drives an electrical generator

Solution

We can refer to the six forms of energy listed above. It follows that

(a) *nuclear* energy is converted to *heat* energy.
(b) *heat* energy is converted to *mechanical* energy.
(c) *mechanical* energy is converted to *electrical* energy.

Self-Test Exercise

Identify two forms of energy that are involved in each of the following devices.

(a) flashlight (b) solar calculator
(c) lead–acid battery

Answers: (a) chemical and light; (b) light and electrical; (c) chemical and electrical

▲ Albert Einstein Stamp The relationship between mass and energy is expressed by Einstein's equation: $E = mc^2$.

Conservation of Mass and Energy

In 1905 Albert Einstein presented a paper on the special theory of relativity that proposed that matter and energy are interrelated. Moreover, he stated this relationship in the form $E = mc^2$. In this famous equation, Einstein showed that energy (E) and mass (m) are related by the velocity of light (c) squared.

Since mass and energy are related, it naturally follows that the conservation of mass and the conservation of energy are related. Thus, a more accurate statement combines these two principles into the **law of conservation of mass and energy**. The combined law states that the total mass and energy in the universe is constant. When applied to a chemical reaction, the law indicates that the total mass and energy before and after a chemical change is constant.

By applying the combined law, we realize that any decrease in mass corresponds to an increase in energy. Even a slight decrease in mass, however, corresponds to an enormous amount of energy. For instance, a mass loss of 0.001 g produces enough heat energy to raise the temperature of a large swimming pool about 30°C. Since the mass loss for ordinary chemical reactions is undetectable from a practical laboratory point of view, the original conservation of mass law is still valid.

Chemistry Connection · Recycling Aluminum

How many aluminum cans must be recycled to effect an energy savings equal to the energy produced by burning a gallon of gasoline?

Aluminum is a very abundant metal; in fact, it is the most common metal in the Earth's crust. Why, then, is there so much concern about recycling aluminum? One reason is that it takes a great deal more energy to extract aluminum from the Earth than to recycle it. To produce 1 ton of aluminum from its raw ore requires the combustion of about 8 tons of coal. However, to produce 1 ton of aluminum from recycled scrap requires only 0.4 ton of coal. This is a 95% savings in energy. By comparison, recycling iron provides a 75% savings in energy, and recycling paper provides about a 70% savings.

In a recent publication, the Environmental Protection Agency (EPA) estimated that more than 1 million tons of aluminum cans are produced annually in the United States. Although we are using more aluminum each year, we are also recycling more. From 1970 to 1988, we increased our use of aluminum from 0.10 to 1.25 million tons for beverage cans alone. But during that same period of time, we increased the amount of aluminum recycled from essentially zero to 0.8 million tons. The bar graph to the left shows the rising percentage of recycled aluminum.

▲ **Aluminum Scrap Metal** In addition to aluminum cans, aluminum scrap metal is also recycled.

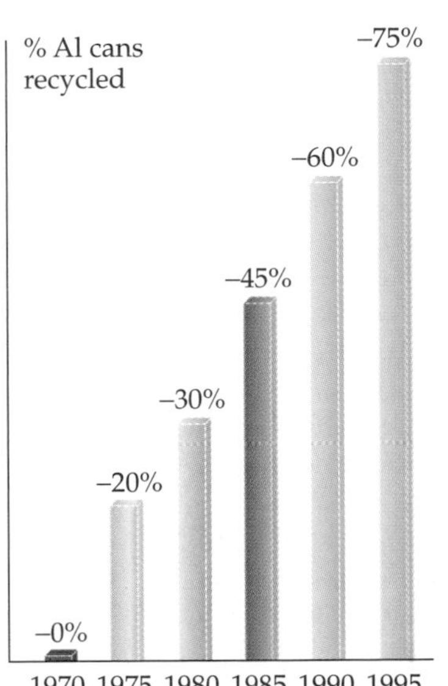

▲ **Aluminum Can Recycling** Notice the dramatic increase in the percentage of recycled aluminum cans.

Aluminum is often found combined with other scrap metals, creating a serious problem for the recycling process. For example, aluminum scrap may contain iron, copper, zinc, and lead. Aluminum can be separated from these other metals in a process called the froth flotation method. This method takes advantage of the fact that each metal has a unique density. During froth flotation, a mixture of scrap metals is first melted and then systematically separated as the least dense metal floats to the surface. Since aluminum has a low density, it can be removed before other metals, that are more dense.

By the year 2001, it is estimated that we will be recycling more than 75% of the aluminum used for disposable containers in the United States. Thus, about 1.2 million tons of aluminum metal will be recycled rather than being processed from aluminum ore. Recycling 24 aluminum cans provides an energy savings approximately equivalent to the energy provided by a gallon of gasoline.

The energy saved by recycling 24 aluminum cans, rather than using aluminum ore, is about equal to the energy produced by burning a gallon of gasoline.

Summary 

Section 4.1 Matter exists in three **physical states**: solid, liquid, and gas. A *solid* is characterized by a definite shape and a fixed volume. A *liquid* has an indefinite shape, but a fixed volume. A *gas* has an indefinite shape and a variable volume.

As temperature increases, a solid *melts* to a liquid and then *vaporizes* to a gas. A direct change of state from a solid to a gas is called **sublimation**. Conversely, as temperature decreases, a gas *condenses* to a liquid and then *freezes* to a solid. A direct change of state from a gas to a solid is called **deposition**.

Section 4.2 We can classify matter according to its properties. A **heterogeneous mixture** has an indefinite composition and variable properties; it can be separated into two or more substances. A **homogeneous mixture** has a definite composition but variable properties; examples include **alloys**, solutions, and mixtures of gases. A **substance** can be either an **element** or a **compound**. A substance has a definite composition and constant properties.

Section 4.3 In 1803 the English chemist John Dalton proposed that elements are composed of indivisible spherical particles called **atoms**. In 1813 the Swedish chemist J. J. Berzelius suggested that the name of an element be abbreviated with a **chemical symbol** corresponding to the first letter of the name of the element, for instance, N for nitrogen. If two or more names start with the same letter, he recommended using two letters in the name, for example, Ni for nickel.

Section 4.4 Most elements can be classified as **metals** or **nonmetals**. If an element can be hammered into a thin sheet, it is said to **malleable**. If it can be drawn into a fine wire, it is said to be **ductile**. A few elements have properties between those of metals and nonmetals and are termed **semimetals**, or metalloids. All elements can be ordered by **atomic number** and arranged in a **periodic table**. The periodic table shows a general trend in properties, with metals on the left side and nonmetals on the right side.

Section 4.5 The **law of definite composition** states that a compound always contains the same elements in the same proportion by mass. The composition of a compound is given by its **chemical formula**. Vitamin C has the formula $C_6H_8O_6$; and one **molecule** contains 6 carbon atoms, 8 hydrogen atoms, and 6 oxygen atoms. Thus, the total number of atoms in the chemical formula of vitamin C is 20.

Section 4.6 The properties of substances are classified as physical or chemical. A **physical property** refers to color, density, melting point, conductivity, solubility, and physical state. A **chemical property** refers only to the chemical reactions of a substance with other substances.

Section 4.7 The changes in substances are classified as physical or chemical. A **physical change** involves a change in mass, volume, or physical state. A **chemical change** involves a change in chemical composition. Evidence for a chemical change includes burning, producing a gas, forming a solid substance in solution, and releasing heat or light energy.

Section 4.8 The **law of conservation of mass** states that matter is neither created nor destroyed. This fundamental principle has practical application in the chemistry laboratory; that is, the total mass of substances produced from a reaction can be calculated because the mass of the products must equal the mass of the reactants.

Section 4.9 The **potential energy** of a substance is related to its chemical composition, whereas the **kinetic energy** of a substance is related to the motion of its particles. Kinetic energy is directly related to temperature. That is, when we heat a solid substance and increase its temperature, we also increase its kinetic energy. As the temperature increases, the particles gain more energy and begin to overcome their attraction for each other. If we continue heating, we will eventually produce a change of state from a solid to a liquid, and from a liquid to a gas.

Section 4.10 The six forms of energy are heat, light, chemical, electrical, mechanical, and nuclear. The **law of conservation of energy** states that energy is neither created nor destroyed. Energy may be converted, however, from one form to another. In practice, chemical energy is used to provide heat and electrical energy. The laws of conservation of mass and conservation of energy can be treated as a single principle—the **law of conservation of mass and energy**. For ordinary chemical reactions, energy changes are sufficiently small that we can ignore this principle and apply conservation of mass and conservation of energy individually.

Key Concepts*

1. Identify the physical state (solid, liquid, or gas) that corresponds to each of the following pictorial representations.

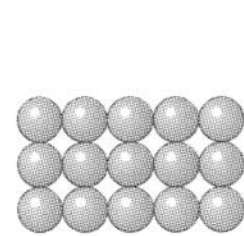
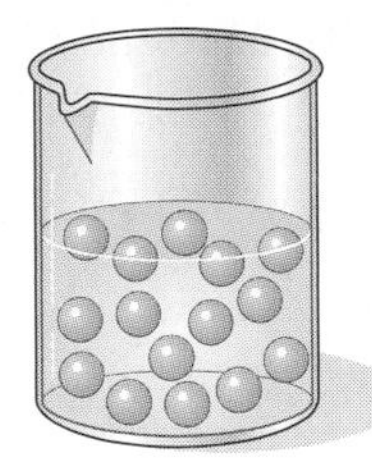
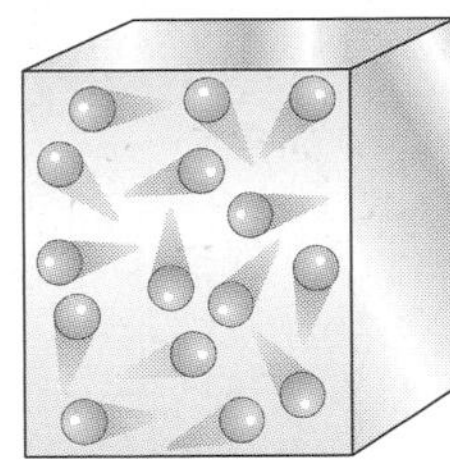

2. Classify each of the following as an element, compound, homogeneous mixture, or heterogeneous mixture.
   (a) water vapor **(b)** oxygen gas
   (c) alcohol dissolved in water
   (d) oil floating on water
3. Classify each of the following as an element, compound, homogeneous mixture, or heterogeneous mixture according to the pictorial representation.

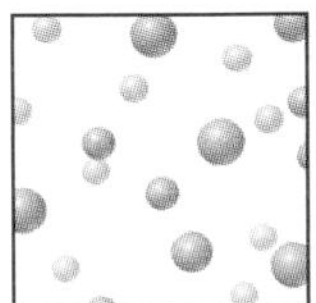
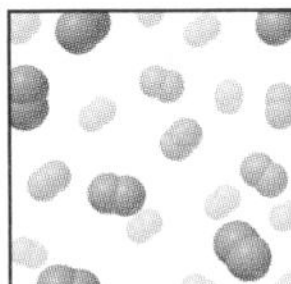
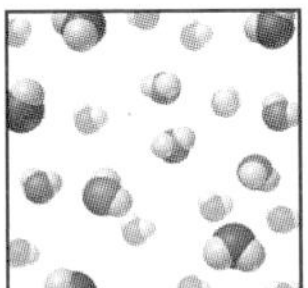
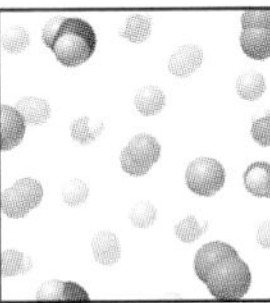

4. Which of the following is a solid metal under normal conditions: lithium, sulfur, bromine, krypton, mercury? (Refer to the periodic table.)
5. Which of the following is a gaseous nonmetal under normal conditions: cobalt, phosphorus, arsenic, chlorine, iodine? (Refer to the periodic table.)
6. Which of the following compounds has physical and chemical properties similar to those of metallic silver: AgCl, AgI, $AgNO_3$, Ag_2CO_3, AgCN?
7. An aspirin tablet dissolves in water without producing bubbles. Is this an example of a physical change or a chemical change?
8. An Alka-Seltzer tablet dissolves in water and produces bubbles. Is this an example of a physical change or a chemical change?

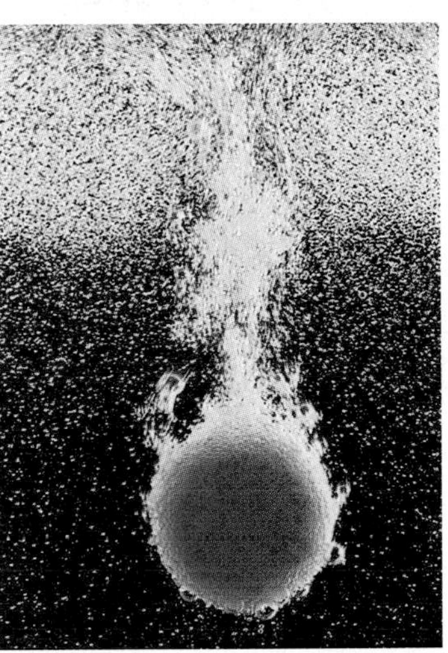

▲ **Alka-Seltzer** Alka-Seltzer fizzes in water when baking soda in the tablet releases carbon dioxide gas.

9. Igniting a 1.000-g magnesium flare produces 1.658 g of magnesium oxide smoke. What is the mass of oxygen in the air that reacted with the magnesium metal?
10. What happens to the kinetic energy and velocity of gaseous molecules when the temperature increases?

Key Terms†

Select the key term below that corresponds to each of the following definitions.

____ **1.** the condition of matter existing as a solid, liquid, or gas
____ **2.** a direct change of state from a solid to a gas
____ **3.** a direct change of state from a gas to a solid
____ **4.** matter having an indefinite composition and variable properties
____ **5.** matter having a definite composition, but variable properties
____ **6.** matter having a definite composition and constant properties
____ **7.** a homogeneous mixture of two or more metals
____ **8.** a substance that can be broken down into simpler substances
____ **9.** a substance that cannot be broken down by chemical reaction
____ **10.** the smallest particle that represents an element

(a) alloy (*Sec. 4.2*)
(b) atom (*Sec. 4.3*)
(c) atomic number (*Sec. 4.4*)
(d) chemical change (*Sec. 4.7*)
(e) chemical formula (*Sec. 4.5*)
(f) chemical property (*Sec. 4.6*)
(g) chemical symbol (*Sec. 4.3*)
(h) compound (*Sec. 4.2*)
(i) conservation of energy (*Sec. 4.10*)

*Answers to Key Concepts are in Appendix H.
† Answers to Key Terms are in Appendix I.

_____ 11. an abbreviation for the name of an element

_____ 12. a high-density element that is a good conductor of heat and electricity

_____ 13. the property of a metal that allows it to be machined into a foil

_____ 14. the property of a metal that allows it to be drawn into a wire

_____ 15. an element that has a low melting point and is not a good conductor

_____ 16. an element that has properties midway between those of a metal and a nonmetal

_____ 17. a number that identifies a particular element

_____ 18. a chart that arranges elements according to their properties

_____ 19. the principle that a compound always contains the same mass ratio of elements

_____ 20. a single particle composed of two or more nonmetal atoms

_____ 21. an abbreviation for the name of a compound

_____ 22. a property that can be observed without changing the formula of a substance

_____ 23. a property that cannot be observed without changing the formula of a substance

_____ 24. a modification of a substance that alters the chemical composition

_____ 25. a modification of a substance that does not alter the chemical composition

_____ 26. the stored energy that matter possesses owing to its position or composition

_____ 27. the energy associated with the mass and velocity of a particle

_____ 28. the law that states mass cannot be created or destroyed

_____ 29. the law that states energy cannot be created or destroyed

_____ 30. the law that states the total mass and energy is constant

(j) conservation of mass (*Sec. 4.8*)
(k) conservation of mass and energy (*Sec. 4.10*)
(l) deposition (*Sec. 4.1*)
(m) ductile (*Sec. 4.4*)
(n) element (*Sec. 4.2*)
(o) heterogeneous mixture (*Sec. 4.2*)
(p) homogeneous mixture (*Sec. 4.2*)
(q) kinetic energy (*Sec. 4.9*)
(r) law of definite composition (*Sec. 4.5*)
(s) malleable (*Sec. 4.4*)
(t) metal (*Sec. 4.4*)
(u) molecule (*Sec. 4.5*)
(v) nonmetal (*Sec. 4.4*)
(w) periodic table (*Sec. 4.4*)
(x) physical change (*Sec. 4.7*)
(y) physical property (*Sec. 4.6*)
(z) physical state (*Sec. 4.1*)
(aa) potential energy (*Sec. 4.9*)
(bb) semimetal (*Sec. 4.4*)
(cc) sublimation (*Sec. 4.1*)
(dd) substance (*Sec. 4.2*)

Exercises‡

Physical States of Matter (Sec. 4.1)

1. Indicate whether the shape is definite or indefinite for each of the following.
   (a) solids **(b)** liquids
   (c) gases
2. Indicate whether the volume is fixed or variable for each of the following.
   (a) solids **(b)** liquids
   (c) gases
3. Indicate whether the compressibility is negligible or significant for each of the following states of matter.
   (a) solids **(b)** liquids
   (c) gases
4. Describe the movement of individual particles in each of the following states of matter.
   (a) solids **(b)** liquids
   (c) gases
5. Supply the term that describes each of the following changes of physical state.
   (a) solid to liquid **(b)** liquid to gas
   (c) gas to solid
6. Supply the term that describes each of the following changes of physical state.
   (a) gas to liquid **(b)** liquid to solid
   (c) solid to gas
7. Indicate whether heat energy is absorbed or released for the following changes of physical state.
   (a) solid to liquid **(b)** liquid to gas
   (c) gas to solid
8. Indicate whether heat energy is absorbed or released for the following changes of physical state.
   (a) gas to liquid **(b)** liquid to solid
   (c) solid to gas

Elements, Compounds, and Mixtures (Sec. 4.2)

9. Distinguish between a pure substance and a mixture.
10. Distinguish between a homogeneous mixture and a heterogeneous mixture.
11. Classify each of the following as a pure substance or a mixture.
    (a) chlorine gas **(b)** oxyacetylene gas
    (c) salt water **(d)** table salt

‡Answers to odd-numbered Exercises are in Appendix J.

12. Classify each of the following as a homogeneous mixture or a heterogeneous mixture.
(a) Earth's crust (b) Earth's atmosphere
(c) Earth's oceans (d) Earth's rivers

13. Classify each of the following as an element, compound, homogeneous mixture, or heterogeneous mixture.
(a) iron metal (b) iron ore
(c) iron oxide (d) steel alloy

14. Classify each of the following as an element, compound, homogeneous mixture, or heterogeneous mixture.
(a) oxygen gas (b) brass alloy
(c) diesel emissions (d) aluminum oxide

Names and Symbols of the Elements (Sec. 4.3)

15. Name the three most abundant elements in Earth's crust.

16. Name the three most abundant elements in the human body.

17. Write the chemical symbol for each of the following elements.
(a) lithium (b) argon
(c) magnesium (d) manganese
(e) fluorine (f) sodium
(g) copper (h) nickel

▲ **Copper Wire** Copper is ductile and can be drawn into thin metal wire.

18. Write the chemical symbol for each of the following elements.
(a) strontium (b) xenon
(c) oxygen (d) zinc
(e) beryllium (f) silicon
(g) mercury (h) titanium

19. Write the name of the element for each of the following chemical symbols.
(a) Cl (b) Ne
(c) Cd (d) Ge
(e) Co (f) Ra
(g) Cr (h) Te

20. Write the name of the element for each of the following chemical symbols.
(a) P (b) S
(c) Kr (d) Fe
(e) Sb (f) N
(g) Pt (h) Ag

21. Refer to the periodic table and find the atomic number for each of the following.
(a) hydrogen (b) barium
(c) gold (d) iodine
(e) bromine (f) aluminum
(g) potassium (h) tin

22. Refer to the periodic table and find the atomic number for each of the following.
(a) helium (b) arsenic
(c) boron (d) selenium
(e) carbon (f) bismuth
(g) lead (h) calcium

Metals, Nonmetals, and Semimetals (Sec. 4.4)

23. State whether each of the following properties is more typical of a metal or of a nonmetal.
(a) shiny luster (b) low melting point
(c) malleability (d) reacts with metals and nonmetals

24. State whether each of the following properties is more typical of a metal or of a nonmetal.
(a) dull powder (b) gaseous state
(c) high density (d) reacts with nonmetals only

25. State whether each the following properties is more typical of a metal or of a nonmetal.
(a) shiny solid (b) brittle solid
(c) ductility (d) reacts with calcium

26. State whether each of the following properties is more typical of a metal or of a nonmetal.
(a) shiny liquid at 50°C (b) reddish liquid
(c) low density (d) forms alloys

27. Refer to the periodic table and classify each of the following elements as a metal, nonmetal, or semimetal.
(a) Na (b) Br
(c) Ar (d) Ge

28. Refer to the periodic table and classify each of the following elements as a metal, nonmetal, or semimetal.
(a) Sb (b) S
(c) Hg (d) Kr

29. Refer to the periodic table and classify each of the following elements as a metal, nonmetal, or semimetal.
(a) beryllium (b) germanium
(c) phosphorus (d) manganese

30. Refer to the periodic table and classify each of the following elements as a metal, nonmetal, or semimetal.
(a) radium (b) fluorine
(c) selenium (d) xenon

31. Refer to the periodic table and indicate the physical state for each of the following elements at 25°C and normal pressure.
(a) Na (b) N
(c) Ar (d) Ge

32. Refer to the periodic table and indicate the physical state for each of the following elements at 25°C and normal pressure.
(a) Sb (b) S
(c) Hg (d) Kr

33. Refer to the periodic table and indicate the physical state for each of the following elements at 25°C and normal pressure.
(a) cobalt (b) bromine
(c) helium (d) titanium

34. Refer to the periodic table and indicate the physical state for each of the following elements at 25°C and normal pressure.
(a) hydrogen (b) neon
(c) barium (d) arsenic

Compounds and Chemical Formulas (Sec. 4.5)

35. Joseph Proust prepared copper carbonate in the laboratory and determined the Cu:C:O mass ratio to be 5:1:4. Predict the composition of copper carbonate in a naturally occurring mineral.

36. The mass ratio of C:H:O in natural vitamin C is 9:1:12. Predict the composition of synthetic ascorbic acid (vitamin C) produced in a laboratory.

37. State the atomic composition for a molecule of each of the following compounds.
(a) aspirin, $C_9H_8O_4$ (b) riboflavin, $C_{17}H_{20}N_4O_6$

38. State the atomic composition for a molecule of each of the following compounds.
(a) cholesterol, $C_{27}H_{45}OH$ (b) quinine, $C_{20}H_{24}N_2O_2$

39. Write the chemical formula for each of the following compounds.
(a) retinol (vitamin A): 20 carbon, 30 hydrogen, and 1 oxygen atom
(b) thymine (vitamin B_1): 12 carbon, 18 hydrogen, 2 chlorine, 4 nitrogen, 1 oxygen, and 1 sulfur atom

40. Write the chemical formula for the each of following compounds.
(a) lysine (an amino acid): 3 carbon, 8 hydrogen, 2 oxygen, and 2 nitrogen atoms
(b) methionine (an amino acid): 5 carbon, 11 hydrogen, 2 oxygen, 1 nitrogen, and 1 sulfur atom

41. What is the total number of atoms in a molecule of each of the following compounds?
(a) citric acid (in citrus fruit), $C_3H_4OH(COOH)_3$
(b) lactic acid (in sour milk), $C_2H_4(OH)COOH$

42. What is the total number of atoms in a molecule of each of the following compounds?
(a) stearic acid (a fatty acid), $CH_3(CH_2)_{16}COOH$
(b) tristearin (a triglyceride), $C_3H_5(C_{17}H_{35}COO)_3$

Physical and Chemical Properties (Sec. 4.6)

43. State whether each of the following is a physical or a chemical property of copper.
(a) reddish-brown color
(b) high melting point
(c) conducts heat and electricity
(d) oxidizes in air

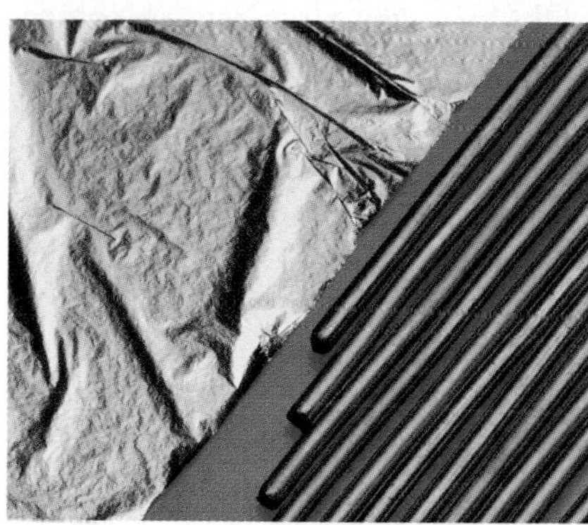

▲ **Gold Foil and Copper Rods** Gold is malleable and can be beaten into a foil; copper is ductile and can be drawn into thin rods.

44. State whether each of the following is a physical or a chemical property of cobalt.
(a) shiny metallic luster
(b) high density
(c) malleability and ductility
(d) reacts with acid

45. State whether each of the following is a physical or a chemical property.
(a) Titanium pulls into a fine wire.
(b) Zinc generates an odorless gas in acid.
(c) Yellow phosphorus ignites in air.
(d) Chromium conducts heat.

46. State whether each of the following is a physical or a chemical property.
(a) Magnesium is a lightweight metal.
(b) Silver shows no reaction in acid.
(c) Lead oxidizes slowly in air.
(d) Carbon crystallizes as diamond.

47. State whether each of the following is a physical or a chemical property.
(a) Ether is flammable.
(b) Bromine is a reddish-orange liquid.
(c) Ethanol vaporizes at 78°C.
(d) Steam condenses to water at 100°C.

48. State whether each of the following is a physical or a chemical property.
(a) Electricity decomposes water.
(b) Ice floats on water.
(c) Potassium metal fizzes in water.
(d) Acetone dissolves in water.

49. State whether each of the following is a physical or a chemical property.
(a) Dry ice pellets disappear.
(b) Iron oxidizes to rust.
(c) Brandy has a fragrant bouquet.
(d) Sugar ferments to alcohol.

50. State whether each of the following is a physical or a chemical property.
(a) Methanol dissolves in gasoline.
(b) Potassium metal turns gray in air.
(c) Uranium decays by radioactive emission.
(d) Antifreeze crystallizes at −55°C.

Physical and Chemical Changes (Sec. 4.7)

51. State whether each of the following is a physical or a chemical change.
(a) dissolving
(b) burning
(c) rusting
(d) freezing

52. State whether each of the following is a physical or a chemical change.
(a) oxidation
(b) vaporization
(c) sublimation
(d) combustion

53. State whether each of the following is a physical or a chemical change.
(a) slicing an orange into wedges
(b) fermenting apples into apple cider
(c) grinding sugar crystals into powder
(d) digesting carbohydrates for energy

54. State whether each of the following is a physical or a chemical change.
(a) cutting aluminum foil
(b) caramelizing sugar
(c) adding air to a tire
(d) igniting a sparkler

55. State whether each of the following is a physical or a chemical change.
(a) nickel metal alloys with copper in a 5¢ coin
(b) tin metal produces a colorless gas in acid
(c) beryllium metal melts at 1278°C
(d) lithium metal reacts with chlorine gas

56. State whether each of the following is a physical or a chemical change.
(a) hydrogen and oxygen explode
(b) bromine vaporizes to a gas
(c) methanol freezes to a solid
(d) calcium oxidizes to a white powder

57. State whether each of the following is a physical or a chemical change.
(a) baking soda fizzes in vinegar
(b) vinegar and oil separate into two layers
(c) a flaming fireplace releases heat
(d) a helium balloon decreases in size

58. State whether each of the following is a physical or a chemical change.
(a) mercury cools to a silvery solid
(b) natural gas burns with a blue flame
(c) ammonia and copper solution turns deep-blue
(d) sugar crystals dissolve in water

Conservation of Mass (Sec. 4.8)

59. If 2.50 g of iron powder reacts with 1.44 g of yellow sulfur powder, what is the mass of the iron sulfide product?

60. If 10.11 g of limestone decomposes by heat to 8.51 g of calcium oxide and carbon dioxide gas, what is the mass of carbon dioxide produced?

61. If 0.750 g of orange mercury oxide decomposes to 0.695 g of liquid mercury and oxygen gas, what is the mass of oxygen produced?

62. If ammonia gas and 0.365 g of hydrogen chloride gas react to give 0.535 g of solid ammonium chloride, what is the mass of ammonia that reacted?

Potential and Kinetic Energy (Sec. 4.9)

63. Using a roller coaster as an example, when is the potential energy maximum and the kinetic energy minimum?

64. Using a swing as an example, when is the potential energy minimum and the kinetic energy maximum?

65. Which physical state has particles with the highest kinetic energy?

66. Which physical state has particles with the most restricted movement?

67. If the temperature increases, does the kinetic energy increase or decrease?

68. If the temperature increases, does the motion of particles increase or decrease?

69. If a steel cylinder of argon gas is heated from 50° to 100°C, does the kinetic energy and motion of argon atoms increase or decrease?

70. If a steel cylinder of radon gas is cooled from 50° to 0°C, does the kinetic energy and motion of radon atoms increase or decrease?

Conservation of Energy (Sec. 4.10)

71. Steam at 100°C passing through a radiator condenses to water and releases heat. If 1500 kcal of heat are released, how much energy is required to vaporize the water, at 100°C, to steam?

72. An ice cube at 0°C melts to water by absorbing heat. If 10.5 kc of heat are required to melt the ice, how much energy must be lost to freeze the water, at 0°C, to ice?

73. A 10.0-g sample of mercury absorbs 110 cal as it is heated from 25°C to its boiling point at 356°C. It then requires an additional 697 cal to vaporize. How much energy is released as the mercury vapor cools from 356° to 25°C?

74. A 25.0-g sample of molten iron releases 1230 cal as it cools from 2000°C to its freezing point, 1535°C. It then releases an additional 1590 cal as it solidifies. How much energy is required to heat the solid iron from 1535° to 2000°C?

75. The reaction of hydrogen and oxygen gases produces

15.0 g of water and releases 48.0 kcal of heat. How much energy is required to decompose 15.0 g of water into hydrogen and oxygen gases?

76. The electrolysis of 25.0 g of water requires 335 kJ of energy to produce hydrogen and oxygen gases. How much energy is released when hydrogen and oxygen gases react to produce 25.0 g of water?

77. List the six different forms of energy.

78. What form of energy is solar radiation?

79. Identify two forms of energy that are involved in each of the following energy conversions relating to a nuclear power plant.
   (a) uranium converts water to steam
   (b) steam drives a turbine
   (c) a turbine drives a generator
   (d) a generator spins to make electricity

80. Identify two forms of energy that are involved in each of the following energy conversions relating to a fossil fuel power plant.
   (a) burning coal converts water to steam
   (b) steam drives a turbine
   (c) a turbine drives a generator
   (d) a generator spins to make electricity

81. Identify two forms of energy that are involved in each of the following energy conversions relating to an automobile.
   (a) lead–acid battery discharging electricity
   (b) electricity powering a starter motor
   (c) starter motor turning a flywheel
   (d) gasoline exploding into hot gases

82. Identify two forms of energy that are involved in each of the following energy conversions relating to an automobile.
   (a) hot gases expanding to move pistons
   (b) pistons rotating the crankshaft
   (c) crankshaft belt turning the generator
   (d) generator charging the lead–acid battery

83. Define each symbol in Einstein's equation: $E = mc^2$.

84. State the law of conservation of mass and energy as it applies to a chemical reaction.

General Exercises

85. Gasoline contains a complex mixture of hydrocarbons. Is gasoline an example of a heterogeneous or a homogeneous mixture?

86. A dry cleaning solvent contains chlorofluorocarbons (CFCs). Is this dry cleaning solvent an example of a homogeneous mixture or a pure substance?

87. The electronics industry manufactures semiconductor chips from silicon. Refer to the periodic table and predict an element that can be substituted for silicon.

88. The electronics industry manufactures transistors using arsenic diffusion. Refer to the periodic table and predict an element that can be substituted for arsenic.

89. Write the chemical symbol for each of the following elements.
   (a) ferrum **(b)** plumbum
   (c) stannum **(d)** aurum

90. State the atomic number for each of the following elements.
   (a) rutherfordium **(b)** dubnium
   (c) seaborgium **(d)** bohrium

91. State whether the following describes a physical or a chemical change: changing physical state but not chemical formula.

92. State whether the following describes a physical or a chemical change: changing chemical formula but not physical state.

93. The reaction of hydrogen and nitrogen gases produces 5.00 g of ammonia and releases 3250 cal of heat. How much energy is required to decompose 5.00 g of ammonia into hydrogen and nitrogen gases?

94. The decomposition of 10.0 g of ammonia requires 27,200 J of energy to give hydrogen and nitrogen gases. How much energy is released when hydrogen and nitrogen gases react to produce 10.0 g of ammonia?

95. Iron and sulfur react to produce iron sulfide and heat energy. An experiment shows that the mass of iron and sulfur is equal to the mass of the iron sulfide. In *theory*, should the products weigh slightly more or slightly less than the reactants?

96. Hydrogen and iodine react to give hydrogen iodide while absorbing heat energy. An experiment shows that the total mass of the reactants is equal to the product. In *theory*, should the product weigh slightly more or slightly less than the reactants?

Explorer Quiz 1
Explorer Quiz 2
Explorer Quiz 3
Master Quiz

CHAPTER 5

Models of the Atom

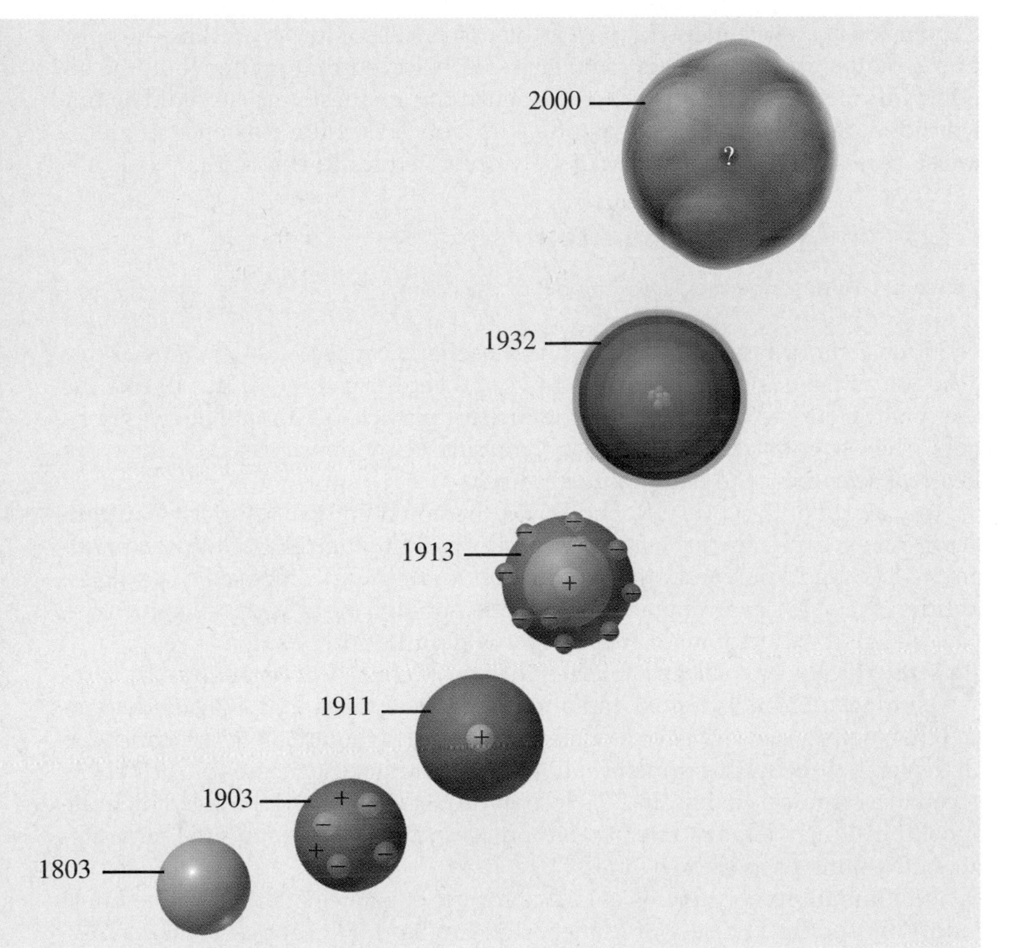

▲ In the evolving model of the atom, what does the question mark (?) represent in the 2000 model?

This chapter is carefully designed to gradually increase in sophistication of treatment for atomic theory. That is, starting with the simplistic Dalton model, the explanations evolve through the Thomson model, Rutherford model, and Bohr model. The treatment of the quantum mechanical atom, which can be considered optional, is reserved for the end of the chapter. I often choose not to cover the concepts of electron waves and orbitals in introductory chemistry as the theory may confuse many students who are trying to grasp the concepts of energy levels and sublevels. I perform demonstrations with gas discharge tubes and atomic models to help students visualize the concepts (refer to the *Instructor's Resource Manual*).

The concept of an atom was born in Greece about 450 B.C. One of the great issues of the day was whether matter and motion were continuous or discontinuous. Zeno, a Greek philosopher, pointed out that to travel any distance, you first must cover half the distance, then half the remaining distance, and so on. This paradox suggests that you will never arrive at your destination if motion is continuous (Figure 5.1). Zeno therefore reasoned that motion is *discontinuous* and occurs by a series of tiny leaps.

Democritus, another Greek philosopher, argued that matter is *discontinuous* and could not be infinitely divided. He believed that at some point a fundamental indivisible particle would emerge. He called this ultimate particle an atom, from the Greek word *atomos*, which means "indivisible."

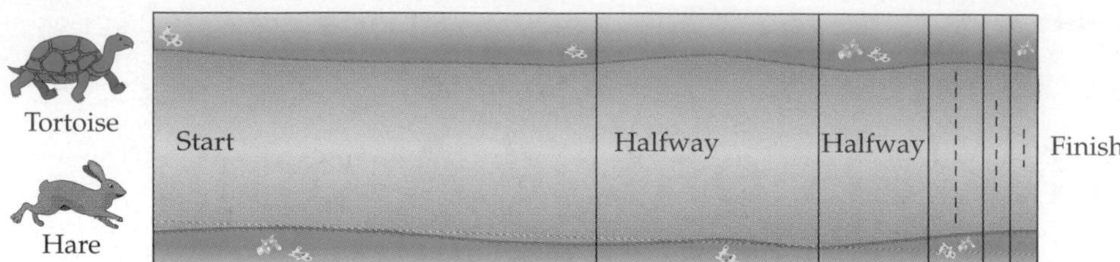

▶ **Figure 5.1 Zeno's Paradox** Motion must occur by discontinuous jumps; otherwise, the race can never be completed. Neither the tortoise nor the hare will complete the race if they continue to move half the distance to the finish.

About 350 B.C., Aristotle—the most influential philosopher of his time—argued that both matter and motion are *continuous*. He believed that matter could be divided an infinite number of times and that even the smallest particle could be further divided. So powerful was Aristotle's influence that the existence of atoms became a closed issue and his point of view prevailed for 21 centuries.

5.1 Dalton Model of the Atom

Objective · To describe the Dalton model of the atom.

In 1803 John Dalton, a modest English schoolteacher, proposed that matter was discontinuous as Democritus had argued over 20 centuries before. But unlike the Greeks, Dalton offered experimental evidence for particles. Although he was not a distinguished scientist, Dalton presented convincing evidence based on the work of the great scientists who had preceded him.

In the seventeenth century, Robert Boyle, the noted English physicist, had concluded from experiments that a gas was made up of tiny particles. Dalton expanded on Boyle's conclusion and proposed that *all* matter was composed of particles. In addition to Boyle's experiments, Dalton relied on two other scientific principles, the law of conservation of mass and the law of definite composition.

In 1789 Antoine Lavoisier established the law of conservation of mass. By carefully weighing substances before and after a chemical reaction, Lavoisier demonstrated that matter was neither created nor destroyed. Although atoms are much too small to weigh directly, this principle allows us to predict the masses of substances involved in a chemical change. In 1799 Joseph Proust established the law of definite composition, which demonstrated that a compound always contains the same elements in the same proportion by mass.

Dalton initially presented his evidence for the existence of atoms to the Literary and Philosophical Society of Manchester, England. He proposed that an element is composed of tiny, indivisible, indestructible particles. Furthermore, he argued that compounds are simply combinations of two or more atoms of different elements. In 1808 Dalton published the atomic theory in his classic textbook *A New System of Chemical Philosophy*.

In a surprisingly short period of time, the atomic theory was generally accepted by the scientific community. The theory can be summarized as follows.

1. An element is composed of tiny, indivisible, indestructible particles called atoms.
2. All atoms of an element are identical and have the same properties.
3. Atoms of different elements combine to form compounds.
4. Compounds contain atoms in small whole-number ratios.
5. Atoms can combine in more than one ratio to form different compounds.

As we will learn, Dalton's first two proposals were incorrect, but the atomic theory was nonetheless, an important step toward understanding the nature of matter.

Chemistry Connection · John Dalton

How was Dalton able to accept an honorary degree from Oxford University dressed in a scarlet robe when wearing scarlet was forbidden by his Quaker faith?

John Dalton was born the son of a weaver into a devoutly religious family. At the age of 12, he began teaching at a Quaker school and developed an interest in science. He was most interested in meteorology and kept a lifelong daily journal of atmospheric conditions for his hometown of Manchester, England.

◀John Dalton (1766–1844)

By all accounts, Dalton was not an inspiring lecturer, and he was hindered as a researcher by being color-blind. He had only a minimal education and limited finances, but he compensated for this with persistence and meticulous work habits. Dalton's daily study of the weather led him to conclude, like Robert Boyle and Isaac Newton before him, that the air was made up of gas particles. Over time, he began to construct his atomic theory as follows.

According to the law of definite composition, carbon and oxygen always react in the same mass ratio to produce carbon dioxide. Dalton proposed that one atom of carbon combines with two atoms of oxygen to produce a molecule of CO_2.

Carbon dioxide, CO_2

Similarly, he proposed that two atoms of hydrogen combine with one atom of oxygen to give a molecule of H_2O.

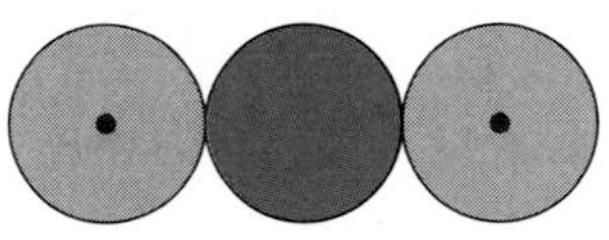

Water, H_2O

Finally, Dalton reasoned that two atoms of hydrogen in water could substitute for each of the oxygen atoms in carbon dioxide. This would result in a molecule with one carbon atom and four hydrogen atoms, that is, CH_4. At the time, this compound was known as marsh gas, but we now refer to it as methane.

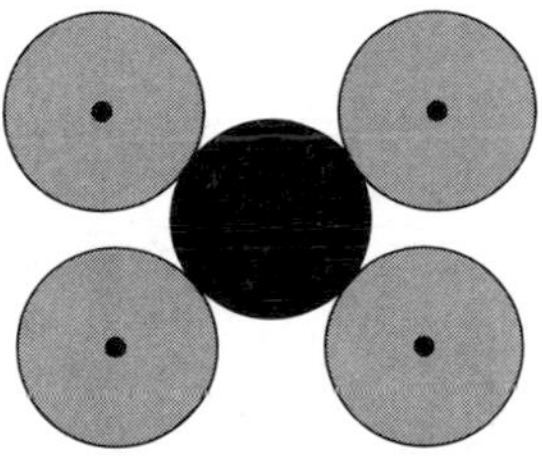

Methane, CH_4

Through experimentation, Dalton indeed found that the combining ratio of carbon to hydrogen in methane was 1 to 4; this agreed perfectly with his prediction. Thus, Dalton had laboratory evidence to support the atomic theory.

Dalton rationalized that since he was color-blind and saw the ceremonial robe as gray, he was not violating Quaker doctrine.

5.2 Thomson Model of the Atom

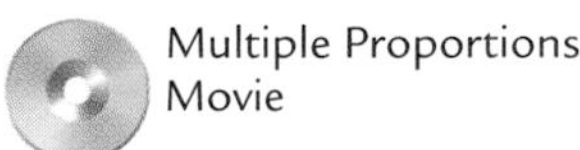

Multiple Proportions Movie

Objectives
- To describe the Thomson plum-pudding model of the atom.
- To state the relative charge and mass of the electron and proton.

About 50 years after Dalton's proposal, there was disturbing evidence that the atom was divisible after all. This evidence came from cathode-ray tubes, which were sealed glass tubes containing a gas at low pressure. When electricity is applied to one end, a cathode-ray tube appears to glow. This phenomenon is referred to as fluorescence, and the glowing ray is a type of light energy. Since the ray emanates from the negative cathode in the tube, the radiation is referred to as a **cathode ray**.

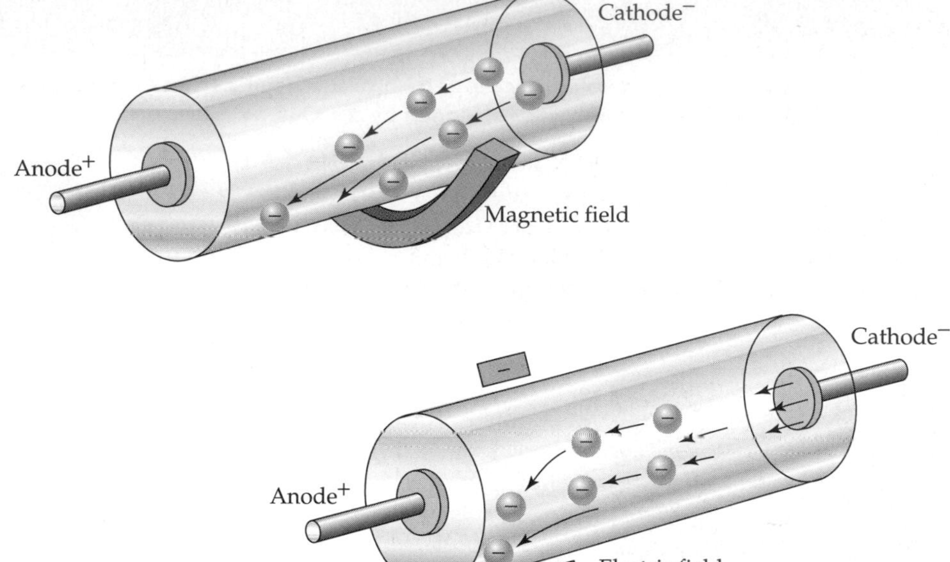

▶ **Figure 5.2 Cathode-Ray Tubes** Note the influence on cathode rays by a magnetic field and an electric field. Several English and German physicists share the credit for these experiments.

In the late 1870s, physicists observed that cathode rays were attracted by a magnetic field (Figure 5.2). This observation suggested that cathode rays are actually particles, not radiation. When different gases are put in the tubes, the results do not change. This led to the theory that cathode rays are composed of tiny, negatively charged particles. One of these subatomic particles was called an **electron** (symbol $\mathbf{e}^-$).

In 1886 a German physicist experimented with cathode-ray tubes that had small holes, or channels, in the cathode. The results showed that positive rays, as well as negative rays, were produced at the cathode. The positive rays, however, moved in the direction opposite the negative cathode rays. The positive rays were referred to as channel rays, which through translation, became known as canal rays (from the German *kanal*, meaning "channel").

Further experiments revealed that canal rays were composed of small positively charged particles (Figure 5.3). The smallest particle was discovered with hydrogen gas in the tube. This particle had a charge equal, but opposite in sign, to that of an electron. One of these subatomic particles in hydrogen gas was called a **proton** (symbol $\mathbf{p}^+$).

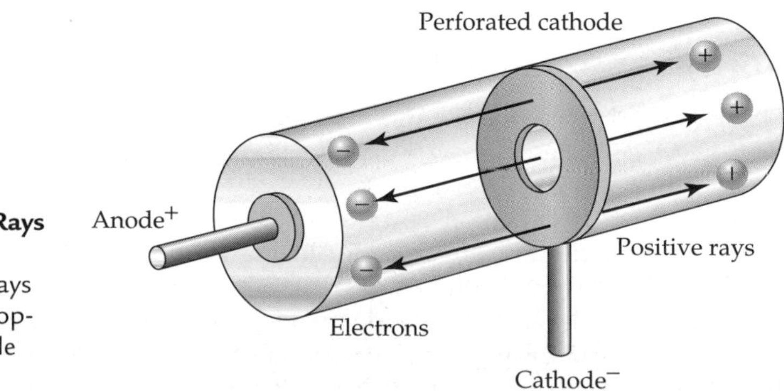

▶ **Figure 5.3 Canal Rays** This cathode-ray tube shows positive canal rays move in the direction opposite negative cathode rays.

In 1897 the English physicist J. J. Thomson demonstrated that cathode rays are deflected by an electric field as well as by a magnetic field (Figure 5.2). This was the final piece of evidence confirming the notion that electrons are particles. Although evidence had accumulated for 20 years, Thomson is usually given credit for the discovery of the electron.

Since the electron is so tiny, Thomson could not measure its actual charge or mass. He was able, however, to determine its charge-to-mass ratio. Thomson continued his experiments and in a short time obtained a value for the charge-to-mass ratio for the proton as well as the electron. Table 5.1 lists the relative charge and mass of the electron and the proton.

Table 5.1 Relative Charge and Mass of the Electron and Proton

Subatomic Particle	Symbol	Relative Charge	Relative Mass
electron	e^-	1−	1/1836
proton	p^+	1+	1

In 1903 J. J. Thomson proposed a subatomic model of the atom. The model pictured a positively charged atom containing negatively charged electrons. Thomson visualized electrons in homogeneous spheres of positive charge in a way that was analogous to raisins in English plum pudding (Figure 5.4). Thus, the Thomson proposal became popularly known as the *plum-pudding model* or *raisin-pudding model* of the atom.

Originally, Thomson was only able to determine the relative charge-to-mass ratio of the electron and the proton. However in 1911, after 5 years of tedious observations, the American physicist Robert Millikan determined the actual charge on the electron. This allowed Thomson to calculate the actual mass of the electron and the proton. Thomson calculated that the mass of the electron is 9.11×10^{-28} g and that the mass of the proton is 1.67×10^{-24} g.

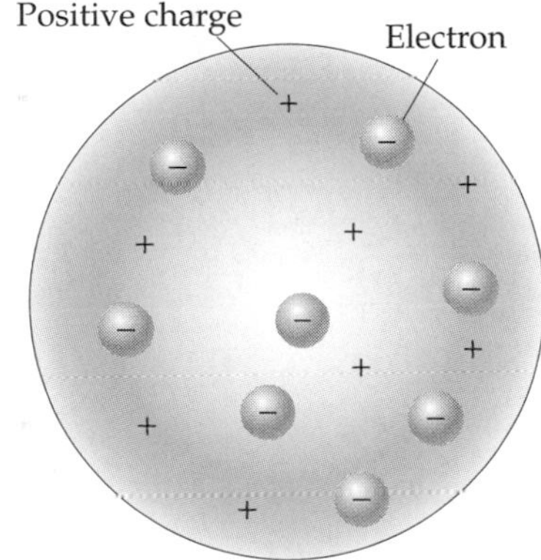

▲ **Figure 5.4 Thomson Model of the Atom** Atoms are pictured as spheres of positive charge. The small negative particles in the sphere represent electrons.

Note J. J. Thomson spent his academic career at Cambridge University and, at the age of 27, became director of the prestigious Cavendish Laboratory. In 1906 he won the Nobel prize in physics for his work on the electron, and 2 years later he was knighted. Thomson's contribution was far-reaching—seven of his former students and assistants went on to receive Nobel prizes.

5.3 Rutherford Model of the Atom

Objectives
- To describe the Rutherford nuclear model of the atom.
- To state the relative charge and approximate mass of the electron, proton, and neutron.

Ernest Rutherford (1871–1937) was digging potatoes on his father's farm in New Zealand when he received the news that he had won a scholarship to Cambridge University. He had actually come in second, but the winner declined the scholarship to get married. The 24-year-old Rutherford postponed his own wedding plans and immediately set out for England.

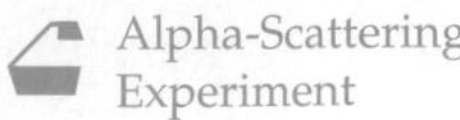

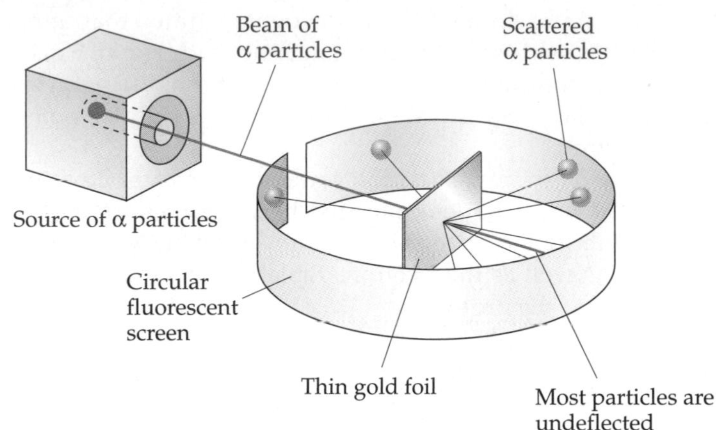

▶ **Figure 5.5 Alpha-Scattering Experiment** The diagram shows the deflection of alpha particles by a thin gold foil. Although the foil was only 0.5 μm thick, a few alpha particles actually rebounded backward.

At Cambridge, Ernest Rutherford studied subatomic particles and earned his doctorate working for J. J. Thomson. Upon graduation, Rutherford went to McGill University in Canada and began to work in the field of radioactivity. After a year he went back to New Zealand, married, and returned to Manchester University in England. There he continued to study radioactivity and coined the terms "alpha ray" and "beta ray" for two types of radiation. He discovered a third type of radiation that was not affected by a magnetic field and gave it the name "gamma ray." In 1908 Rutherford received the Nobel prize in chemistry for his work on radioactivity.

In 1906 Rutherford found that alpha rays contained particles identical to those of helium atoms stripped of electrons. He experimented with alpha rays by firing them at thin gold foils. As expected, the particles passed straight through the foil or, on occasion, deflected slightly (Figure 5.5). This observation seemed reasonable since the plum-pudding model of the atom pictured homogeneous spheres.

A few years later the true picture of the atom was unveiled. The person who did the experiment was Hans Geiger, Rutherford's assistant and the inventor of the Geiger counter. Here is a description of the experiment in Rutherford's own words:

> One day Geiger came to me and said, "Don't you think that young Marsden, whom I am training in radioactive methods, ought to begin a small research?" Now I had thought that too, so I said, "Why not let him see if any alpha particles can be scattered through a large angle?" I may tell you in confidence that I did not believe that they would be, since we knew that the alpha particle was a very fast massive particle, with a great deal of energy. Then I remember two or three days later Geiger coming to me in great excitement and saying, "We have been able to get some of the alpha particles coming backwards." It was quite the most incredible event that has ever happened to me in my life. It was almost as incredible as if you fired a 15-inch shell at a piece of tissue paper and it came back and hit you.

Rutherford interpreted the alpha-scattering results as follows. He believed that most of the alpha particles passed directly through the foil because an atom is largely empty space with electrons moving about. But in the center of the atom is the **atomic nucleus** containing protons. Rutherford reasoned that, compared to the atom, the nucleus is tiny and has a very high density. The alpha particles that

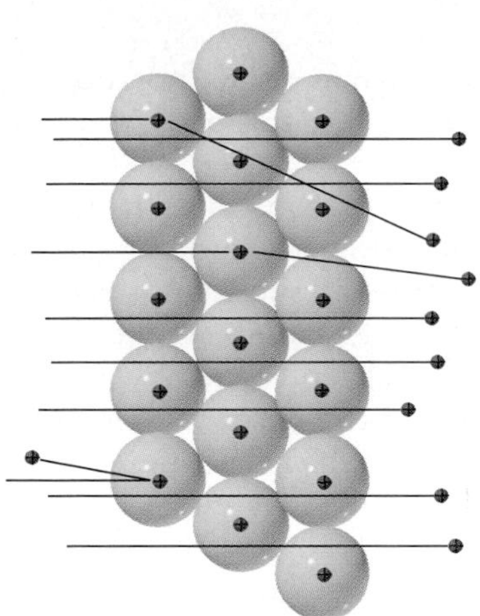

▲ **Figure 5.6 Explanation of the Alpha-Scattering Results** Atoms are represented by circles and nuclei by small dots. An alpha particle is positively charged and is deflected by a heavy, positive, gold nucleus.

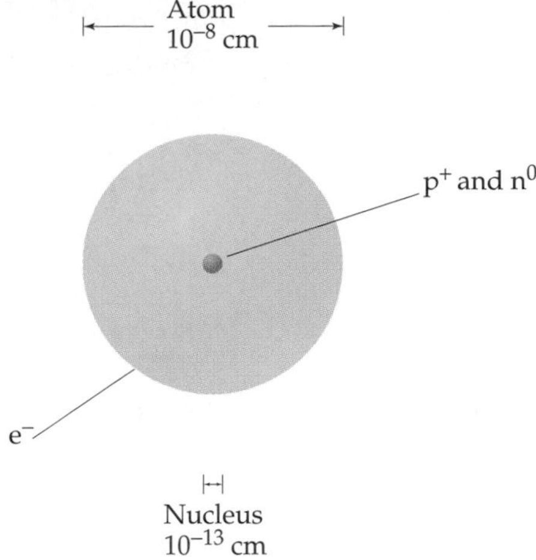

▲ **Figure 5.7 Rutherford Model of the Atom** Most of the mass of the atom is found in the nucleus. Note that the atoms and nuclei are not to scale, as the size of an atom is about 100,000 times larger than the size of the nucleus.

bounced backward recoiled after striking the dense nucleus. The scattering of alpha particles by the atomic nuclei in the gold foil is illustrated in Figure 5.6.

Explanation of Alpha-Scattering Results

In 1911 Rutherford proposed a new model of the atom. He suggested that negatively charged electrons were distributed about a positively charged nucleus. Rutherford was able to estimate the size of the atom and its nucleus. He calculated that an atom has a diameter of about 1×10^{-8} cm and that the nucleus has a diameter of about 1×10^{-13} cm. The Rutherford model of the atom is illustrated in Figure 5.7.

Rutherford Model of the Atom

Since the diameters of atoms and nuclei are extremely small and difficult to comprehend, consider the following analogy to visualize an atom. The size of an atom compared to the size of its nucleus is similar to a kilometer compared to a centimeter. Moreover, if the atom were as large as the Astrodome, the nucleus would be the size of a small marble. This analogy clarifies why so few alpha particles were scattered by gold nuclei in the thin foil.

Owing to the heaviness of the nucleus, Rutherford predicted that it contained neutral particles in addition to protons. Twenty years later, one of Rutherford's former students found the elusive neutral particle. In 1932 James Chadwick discovered the **neutron** (symbol $\mathbf{n^0}$) and 3 years later was awarded the Nobel prize in physics. We can summarize the data for the electron, proton, and neutron as shown in Table 5.2.

Table 5.2 Subatomic Particles

Subatomic Particle	Symbol	Location	Relative Charge	Relative Mass
electron	e^-	outside nucleus	1−	1/1836
proton	p^+	inside nucleus	1+	1
neutron	n^0	inside nucleus	0	1

5.4 Atomic Notation

Objectives
- To draw a diagram of an atom given its atomic notation.
- To explain and illustrate the concept of isotopes.

Each element has a characteristic number of protons in its atomic nucleus. This value is called the **atomic number**. The total number of protons and neutrons in the nucleus of an atom is called the **mass number**. A shorthand method for keeping track of protons and neutrons in the nucleus of an atom is called **atomic notation**. By convention, the symbol of the element (Sy) is preceded by a subscript and superscript. The subscript, designated the *Z* value, represents the atomic number. The superscript, designated the *A* value, represents the mass number. Thus,

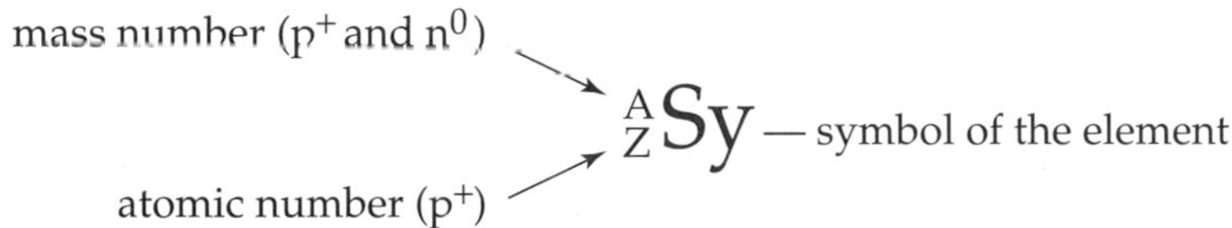

As an example, an atom of sodium can be written: $^{23}_{11}Na$ Here the atomic number (protons) is 11 and the mass number (protons + neutrons) is 23. From this information, we can determine the number of neutrons. The nucleus of this sodium atom contains 12 neutrons ($A - Z = 23 - 11 = 12$). Since atoms are neutral, the number of negative electrons must equal the number of positive protons. Thus, there are 11 electrons surrounding the nucleus. A simple diagram of the sodium atom is as follows.

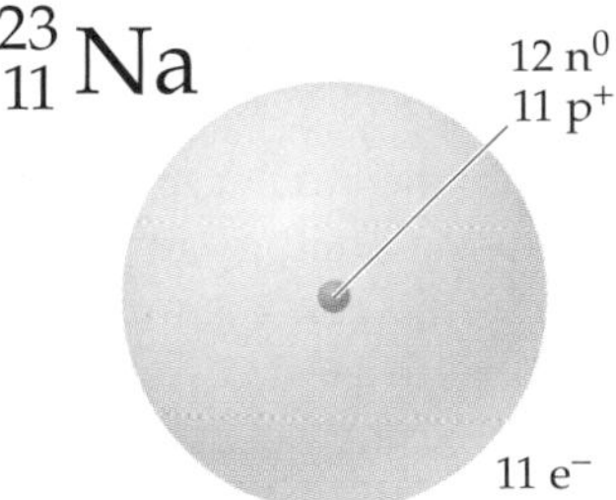

Example Exercise 5.1 further illustrates the interpretation of atomic notation to state the composition of an atom.

Example Exercise 5.1 • Atomic Notation

Given the atomic notation for the following atoms, draw a diagram showing the arrangement of protons, neutrons, and electrons.

(a) $^{19}_{9}F$ (b) $^{109}_{47}Ag$

Solution

We can draw a diagram of an atom by showing protons and neutrons in the nucleus surrounded by electrons.

(a) Since the atomic number is 9 and the mass number is 19, the number of neutrons is 10 (19 − 9). If there are 9 protons, there must be 9 electrons.

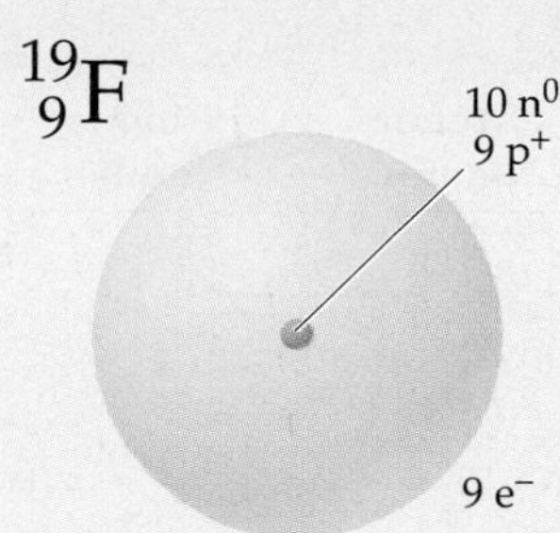

(b) Since the atomic number is 47 and the mass number is 109, the number of neutrons is 62 (109 − 47). If there are 47 protons, there must be 47 electrons.

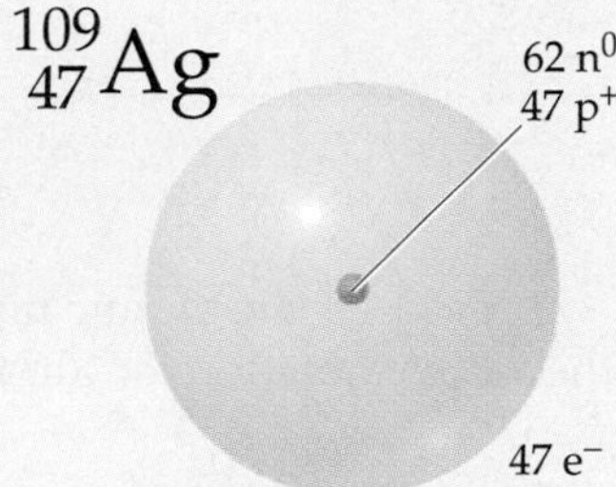

Self-Test Exercise

Given the following diagram, indicate the nucleus using atomic notation.

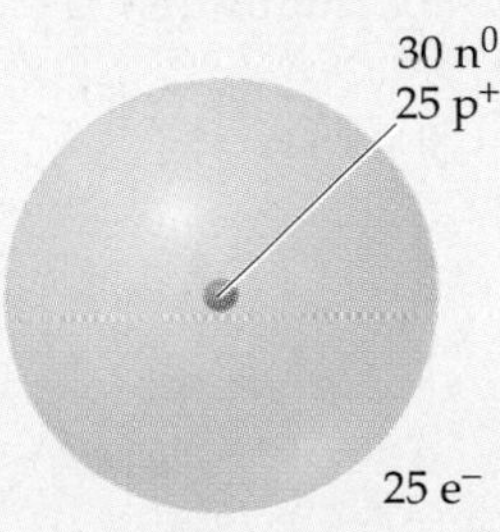

Answer: $^{55}_{25}Mn$

Isotopes

Approximately 20 elements occur naturally with a fixed number of neutrons. For most elements, the number of neutrons in the nucleus varies. Atoms of the same element that have a different number of neutrons in the nucleus are called **isotopes**.

In isotopes, as the number of neutrons varies, so does the mass number. For example, hydrogen occurs naturally as two stable isotopes, protium ($^{1}_{1}H$) and deuterium ($^{2}_{1}H$). Protium has only one proton in its nucleus, whereas deuterium has a proton and a neutron. A third isotope of hydrogen, tritium ($^{3}_{1}H$), is unstable and radioactive. Tritium has one proton and two neutrons in its nucleus. Table 5.3 lists the naturally occurring stable isotopes of the first ten elements.

We often refer to an isotope by stating the name of the element followed by its mass number, for example, carbon-14 and cobalt-60. To write the atomic notation, we must also know the atomic number. In this textbook, the atomic number is found above the symbol of the element in the periodic table. If we refer to the inside

Table 5.3 Naturally Occurring Isotopes of the First 10 Elements

Atomic Number	Mass Number	Atomic Notation	Atomic Number	Mass Number	Atomic Notation
1	1	$^{1}_{1}\text{H}$	6	13	$^{13}_{6}\text{C}$
1	2	$^{2}_{1}\text{H}$	7	14	$^{14}_{7}\text{N}$
2	3	$^{3}_{2}\text{He}$	7	15	$^{15}_{7}\text{N}$
2	4	$^{4}_{2}\text{He}$	8	16	$^{16}_{8}\text{O}$
3	6	$^{6}_{3}\text{Li}$	8	17	$^{17}_{8}\text{O}$
3	7	$^{7}_{3}\text{Li}$	8	18	$^{18}_{8}\text{O}$
4	9	$^{9}_{4}\text{Be}$	9	19	$^{19}_{9}\text{F}$
5	10	$^{10}_{5}\text{B}$	10	20	$^{20}_{10}\text{Ne}$
5	11	$^{11}_{5}\text{B}$	10	21	$^{21}_{10}\text{Ne}$
6	12	$^{12}_{6}\text{C}$	10	22	$^{22}_{10}\text{Ne}$

front cover of this textbook, we find that the atomic number of C is 6 and that of Co is 27. Thus, we can write the atomic notation for carbon-14 as $^{14}_{6}\text{C}$ and for cobalt-60 as $^{60}_{27}\text{Co}$.

Example Exercise 5.2 further illustrates the relationship between atomic notation and the composition of an atomic nucleus.

Example Exercise 5.2 • Nuclear Composition of Isotopes

State the number of protons and the number of neutrons in an atom of each of the following isotopes.

(a) $^{37}_{17}\text{Cl}$ (b) mercury-202

Solution

The subscript value refers to the atomic number (p^+), and the superscript value refers to the mass number (p^+ and n^0).

(a) Thus, $^{37}_{17}\text{Cl}$ has 17 p^+ and 20 n^0($37 - 17 = 20$).
(b) In the periodic table, we find that the atomic number of mercury is 80. Thus, the atomic notation, $^{202}_{80}\text{Hg}$, indicates 80 p^+ and 122 n^0($202 - 80 = 122$).

Self-Test Exercise

State the number of protons and the number of neutrons in an atom of each of the following isotopes.

(a) $^{120}_{50}\text{Sn}$ (b) uranium-238

Answers: (a) 50 p^+ and 70 n^0; (b) 92 p^+ and 146 n^0

Note In general, the properties of isotopes of an element are similar. Consider a carbon-12 atom, which has six protons and six neutrons in the nucleus. If the mass number increases by one neutron, we have a carbon-13 atom. Other than a slight change in mass, the properties of carbon-12 and carbon-13 are nearly identical.

On the other hand, when the atomic number of carbon-12 increases by one proton, we have a nitrogen-13 atom. The additional proton changes the properties radically. Carbon occurs naturally as coal, diamond, and graphite, whereas nitrogen is a colorless, odorless gas in the atmosphere.

5.5 Atomic Mass

Objectives
- To explain the concept of relative atomic mass.
- To calculate the atomic mass for an element given the mass and abundance of the naturally occurring isotopes.

We realize, of course, that atoms are much too small to weigh directly on a balance. A carbon atom, for instance, has a mass of only 1.99×10^{-23} g. Instead of weighing atoms, scientists determine the mass of atoms relative to each other. With a special magnetic-field instrument, the mass of an atom can be compared to the mass of a carbon-12 atom. The carbon-12 isotope has been chosen as a reference standard and is assigned a value of 12 atomic mass units. Stated differently, an **atomic mass unit** (symbol **amu**) is equivalent to 1/12 the mass of a carbon-12 atom.

Simple and Weighted Averages

Before proceeding any further, let's explain averages using an interesting analogy. Let's suppose a manufacturer produces both 12-lb and 16-lb metal shotput balls. Of the total shotput balls manufactured, 25% are 12-lb and 75% are 16-lb. What is the average mass of a shotput ball? If we add the two masses together and divide by 2, we will find the simple average.

$$\frac{12 \text{ lb} + 16 \text{ lb}}{2} = 14 \text{ lb}$$

The simple average mass of a shotput ball is 14 lb, but the true average mass must be weighted in favor of the higher percentage of shotput balls. Since the percentages are 25% and 75%, we can write the decimal fractions 0.25 and 0.75. To calculate the weighted average mass, we must consider the mass and percentage of the 12-lb and 16-lb balls. We can proceed as follows.

$$\begin{aligned} 12 \text{ lb: } 12 \text{ lb} \times 0.25 &= 3 \text{ lb} \\ 16 \text{ lb: } 16 \text{ lb} \times 0.75 &= \underline{12 \text{ lb}} \\ & 15 \text{ lb} \end{aligned}$$

▲ **Olympic Shotput** In the Olympic shotput event, athletes toss a steel ball.

Although the weighted average mass is 15 lb, we should note that no shotput ball weighs 15 lb. An actual shotput ball weighs either 12 lb or 16 lb. A mass of 15 lb represents the theoretical mass of an average shotput ball.

Atomic Mass of an Element

The **atomic mass** of an element is the weighted average mass of all naturally occurring isotopes. To calculate the atomic mass, we will use a method similar to the one used in the shotput example. That is, to calculate the weighted average mass of an atom, we must consider the mass as well as the percent abundance of each isotope.

Carbon has two naturally occurring stable isotopes: carbon-12 and carbon-13. We can calculate the atomic mass of carbon given the mass and natural abundance of each isotope.

Isotope	Mass	Abundance
^{12}C	12.000 amu	98.89%
^{13}C	13.003 amu	1.11%

The natural abundance of each isotope, expressed as a decimal, is 0.9889 and 0.0111. To calculate the true weighted average mass of a carbon atom, we must consider the mass and percentage of each isotope. We proceed as follows.

$$
\begin{aligned}
{}^{12}\text{C}:&\ 12.000\ \text{amu} \times 0.9889 = 11.87\ \text{amu}\\
{}^{13}\text{C}:&\ 13.003\ \text{amu} \times 0.0111 = \underline{\ 0.144\ \text{amu}}\\
&\qquad\qquad\qquad\qquad\quad 12.01\ \text{amu}
\end{aligned}
$$

Even though the atomic mass of carbon is 12.01 amu, we should note that no carbon atom weighs 12.01 amu. An actual carbon atom weighs either 12.000 amu or 13.003 amu. A mass of 12.01 amu represents the mass of a hypothetical average carbon atom. Example Exercise 5.3 further illustrates the calculation of atomic mass from isotopic mass and abundance data.

Example Exercise 5.3 • Calculation of Atomic Mass

Silicon is the second most abundant element in the Earth's crust. Calculate the atomic mass of silicon given the following data for its three natural isotopes.

Isotope	Mass	Abundance
^{28}Si	27.977 amu	92.21%
^{29}Si	28.976 amu	4.70%
^{30}Si	29.974 amu	3.09%

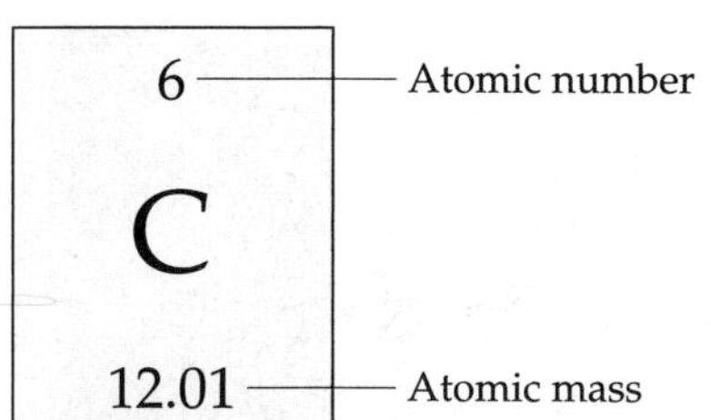

▲ Figure 5.8 Carbon from the Periodic Table of Elements
The atomic number designates the number of protons (6). The atomic mass designates the weighted average mass of the naturally occurring stable isotopes of the element.

Solution
We can find the atomic mass of silicon as follows.

$$
\begin{aligned}
{}^{28}\text{Si}:&\ 27.977\ \text{amu} \times 0.9221 = 25.80\ \text{amu}\\
{}^{29}\text{Si}:&\ 28.976\ \text{amu} \times 0.0470 = 1.36\ \text{amu}\\
{}^{30}\text{Si}:&\ 29.974\ \text{amu} \times 0.0309 = \underline{0.926\ \text{amu}}\\
&\qquad\qquad\qquad\qquad\quad 28.09\ \text{amu}
\end{aligned}
$$

The average mass of a silicon atom is 28.09 amu, although we should note that there are no silicon atoms with a mass of 28.09 amu.

Self-Test Exercise
Calculate the atomic mass of copper given the following data.

Isotope	Mass	Abundance
^{63}Cu	62.930 amu	69.09%
^{65}Cu	64.928 amu	30.91%

Answer: 63.55 amu

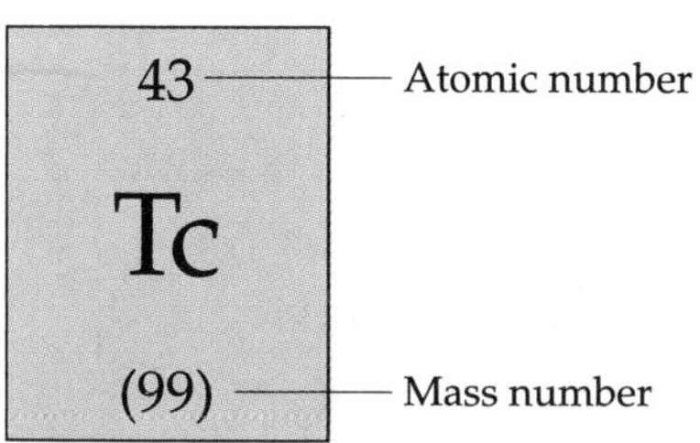

▲ Figure 5.9 Technetium from the Periodic Table of Elements
The atomic number designates the number of protons (43). The mass number for this radioactive isotope is 99. Technetium-99 has a total of 99 protons and neutrons in its nucleus.

The Periodic Table

We often need to refer to the atomic number and atomic mass of an element. For convenience, this information is listed in the periodic table. In this textbook the atomic number is indicated above the symbol, and the atomic mass is given below the symbol. Figure 5.8 illustrates carbon as it appears in the periodic table.

When we refer to the periodic table, it is not possible to determine the number of naturally occurring isotopes for a given element. Of the first 83 elements, however, 81 elements have one or more stable isotopes. Only technetium (element 43) and promethium (element 61) are radioactive and unstable. In the periodic table on the inside front cover of this textbook, the mass number of unstable elements is given in parentheses as shown in Figure 5.9.

The value in parentheses below the symbol of the element indicates the mass number of the most stable or best known radioactive isotope. The mass number refers to the total number of protons and neutrons in the unstable nucleus. From the periodic table you can easily distinguish the elements that are stable from those that are unstable and radioactive. For unstable elements, there is a whole number in parentheses below the symbol of the element. Example Exercise 5.4 further illustrates how to use the periodic table as a reference.

Example Exercise 5.4 • Nuclear Composition from the Periodic Table

Refer to the periodic table on the inside cover of this textbook and determine the atomic number and atomic mass for iron.

Solution

In the periodic table we observe

The atomic number of iron is 26, and the atomic mass is 55.85 amu. From the periodic table information, we should note that it is not possible to determine the number of isotopes for iron or their mass numbers.

Self-Test Exercise

Refer to the periodic table on the inside cover of this text and determine the atomic number and mass number for the given radioactive isotope of radon gas.

Answer: 86 and (222)

5.6 The Wave Nature of Light

Objectives
- To explain the wave nature of light.
- To state the relationship of wavelength, frequency, and energy of light.

We can use our imagination to visualize light traveling through space in a fashion similar to an ocean wave. **Wavelength** refers to the distance the light wave travels to complete one cycle. **Frequency** refers to the number of wave cycles completed in 1 s. The velocity of light is constant, as all wavelengths and frequencies travel at 3.00×10^8 m/s. Figure 5.10 illustrates the wave nature of light.

As the wavelength of light decreases, the frequency of a light wave increases. Conversely, as the wavelength increases, the frequency decreases. As an example, the wavelength decreases from red to violet as the frequency increases. We can visualize a high-frequency light wave moving rapidly up and down while completing many cycles per second. Consequently, high-frequency light is more energetic than low-frequency light, and short-wavelength light is more energetic than long-wavelength light.

Light—A Continuous Spectrum

A rainbow is created when sunlight passes through raindrops. Each raindrop acts as a miniature prism and separates sunlight into various bands of color. What we

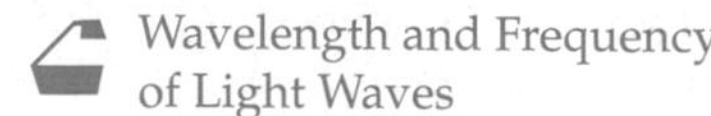

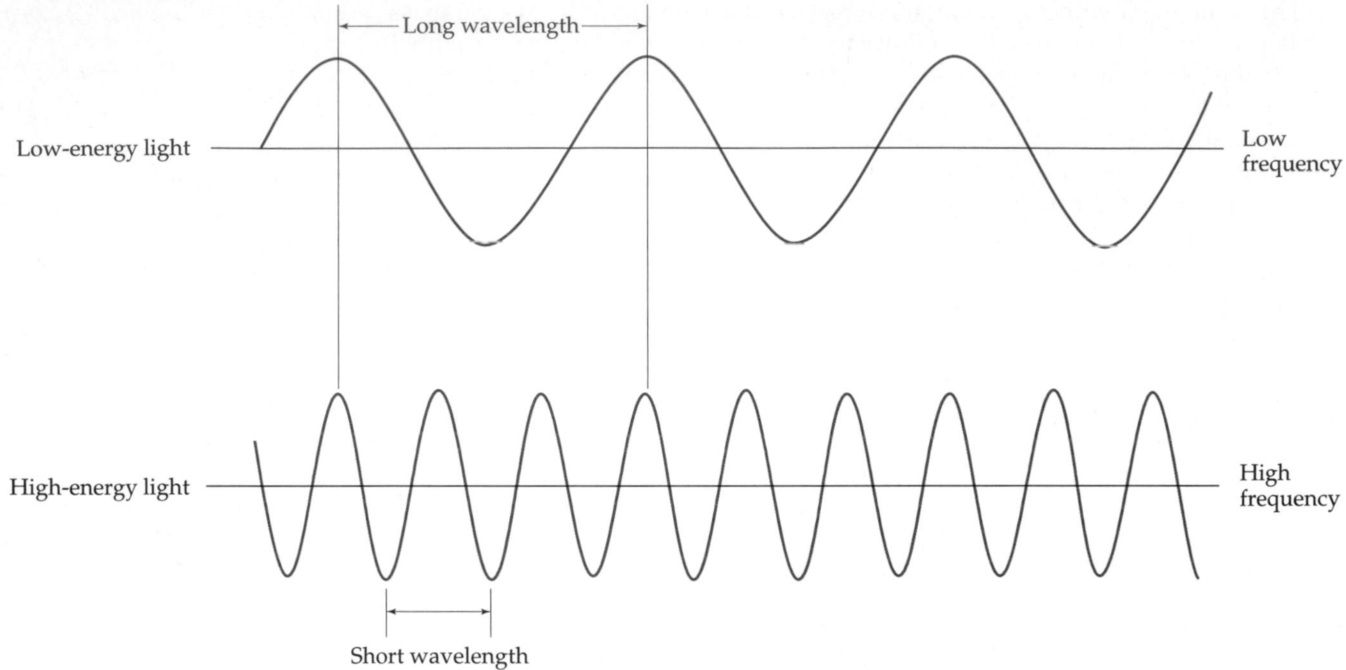

▲ **Figure 5.10 Wavelength and Frequency of Light Waves** Notice that the wavelength is longer for low-energy light than for high-energy light. Also notice that the frequency is greater for high-energy light than for low-energy light.

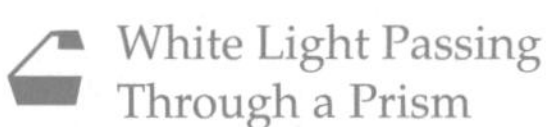

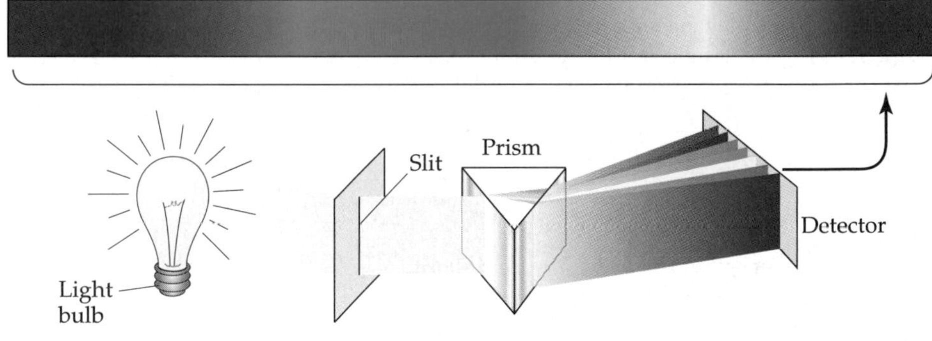

▶ **Figure 5.11 White Light Passing Through a Prism** Notice that white light separates into a rainbow of colors when it passes through a glass prism. Similarly, sunlight produces a rainbow when it passes through raindrops.

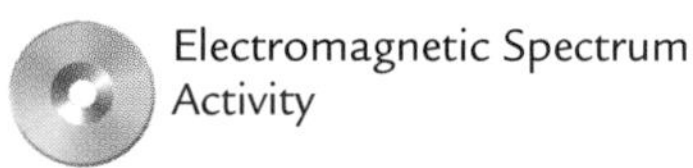

ordinarily observe as white light is actually several colors of light mixed together. When white light passes through a glass prism, it separates into all the colors of the rainbow, that is, red, orange, yellow, green, blue, and violet (Figure 5.11).

The term **light** usually refers to radiant energy that is visible. Our eyes can see light in the **visible spectrum** (400–700 nm) but not in the ultraviolet and infrared regions. The wavelength of ultraviolet radiation is too short to be visible (below 400 nm), and infrared radiation is too long (above 700 nm).

We also, on occasion, use the term "light" when referring to radiant energy that is not visible. The complete **radiant energy spectrum** is an uninterrupted band, or **continuous spectrum**, of visible and invisible light that ranges from short through long wavelengths. The radiant energy spectrum includes invisible gamma rays, X-rays, and microwaves, as well as visible light (Figure 5.12).

Example Exercise 5.5 illustrates how the wavelength, frequency, and energy of light are related.

Wavelength increases

Cosmic rays | Gamma rays | X rays | Ultra-violet | Visible | Infrared | Microwaves | TV | Radio

400 nm | 500 nm | 600 nm | 700 nm

Visible spectrum

◀ **Figure 5.12 The Radiant Energy Spectrum** The complete radiant energy spectrum includes short-wavelength gamma rays through long-wavelength microwaves. Notice that the visible spectrum is only a narrow window in a broad band of radiant energy.

The Radiant Energy Spectrum

Example Exercise 5.5 • Properties of Light

Considering blue light and yellow light, which has the

(a) longer wavelength? (b) higher frequency? (c) higher energy?

Solution

Referring to Figure 5.12, we notice that the wavelength of yellow light is about 600 nm and that of blue light, about 500 nm. Thus,

(a) *yellow light* has a longer wavelength than blue light.
(b) *blue light* has a higher frequency because it has a shorter wavelength.
(c) *blue light* has a higher energy because it has a higher frequency.

Self-Test Exercise

Considering infrared light and ultraviolet light, which has the

(a) longer wavelength? (b) higher frequency? (c) higher energy?

Answers: (a) infrared; (b) ultraviolet; (c) ultraviolet

5.7 The Quantum Concept

Objective · To explain the quantum concept.

In 1900 Max Planck (1858–1947), a German physicist, introduced a revolutionary idea. Planck proposed that the energy radiated by a heated object is not continuous, but rather that the radiation is emitted in small bundles. The idea that energy is released in discrete units is referred to as the *quantum concept*.

When an object radiates light, it releases a unit of radiant energy called a **photon**. According to the quantum theory, a beam of light actually consists of a stream of individual particles. By way of example, an ordinary light bulb radiates energy in the form of tiny individual photons.

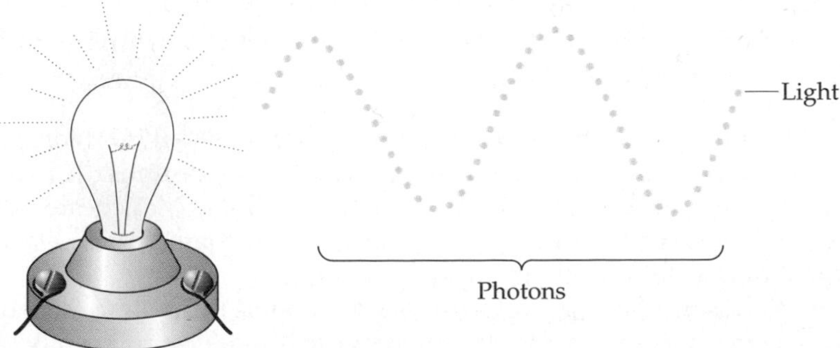

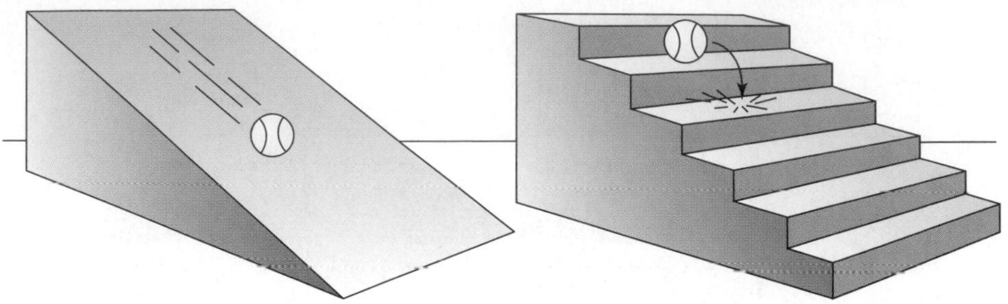

▶ **Figure 5.13 Stair Analogy for the Quantum Principle** A ball rolling down a ramp loses potential energy continuously. Conversely, a ball rolling down a flight of stairs loses potential energy in quantized amounts each time it drops from one step to another.

Stair Analogy for the Quantum Principle

We can illustrate the quantum concept with the following analogy. A ball rolling down a ramp loses potential energy continuously. Conversely, a ball rolling down a flight of stairs loses potential energy in discrete units each time it drops from one step to another. In this example, the loss of potential energy is continuous as the ball rolls down the ramp, and the loss is quantized as the ball rolls down the stairs. Figure 5.13 contrasts a continuous change in energy versus a quantized energy change.

The following Example Exercise 5.6 further illustrates practical applications of the quantum theory.

Example Exercise 5.6 • Quantum Concept

State whether each of the following scientific instruments gives a continuous or a quantized measurement of mass.

(a) triple-beam balance (b) digital electronic balance

Solution

Refer to Figure 2.3 if you have not used these balances in the laboratory.

(a) On a triple-beam balance a small metal rider is moved along a beam. Since the metal rider can be moved to any position on the beam, a triple-beam balance gives a *continuous* mass measurement.

(b) On a digital electronic balance the display indicates the mass of an object to a particular decimal place, for example, 5.015 g. Since the last digit in the display must be a whole number, a digital balance gives a *quantized* mass measurement.

Self-Test Exercise

State whether each of the following musical instruments produces continuous or quantized musical notes.

(a) acoustic guitar (b) electronic keyboard

Answers: (a) continuous; (b) quantized

5.8 Bohr Model of the Atom

Objectives
- To describe the Bohr planetary model of the atom.
- To explain the relationship between energy levels in an atom and lines in an emission spectrum.

▲ **Niels Bohr Stamp** The planetary model of the atom is illustrated for the model proposed by Niels Bohr.

In 1911 a brilliant young Danish physicist, Niels Bohr (1885–1962), completed his doctorate and left for England to begin postdoctoral work under J. J. Thomson. After a few months, he left Cambridge to join Rutherford at Manchester where the atomic nucleus had been discovered. In took only 2 years for the young Dane to raise our understanding of the atom to yet another level.

In 1913 Bohr speculated that electrons orbit around the atomic nucleus just as planets circle around the sun. He further suggested that electron orbits were at a fixed distance from the nucleus and had a definite energy. The electron was said to

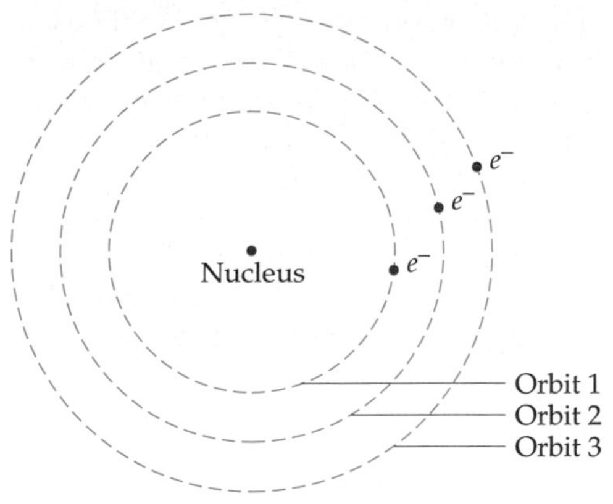

◀ **Figure 5.14 The Bohr Model of the Hydrogen Atom** The electron is a specific distance from the nucleus and occupies an orbit of discrete energy. According to the Bohr model, the electron Is found only in a given energy level.

The Bohr Model of the Atom

travel in a fixed-energy orbit referred to as an **energy level**. Moreover, electrons could be found only in specific energy levels and nowhere else. Figure 5.14 illustrates the model of the hydrogen atom proposed by Niels Bohr. This model is referred to as the Bohr planetary model of the atom, or simply, the **Bohr atom**.

Evidence for Energy Levels

The Bohr model was a beautiful mental picture of electrons in atoms. However, no one knew whether the model was right or wrong because there was no experimental evidence to support the theory. Coincidentally, Bohr received a paper on the emission of light from hydrogen gas. The paper showed that excited hydrogen gas emits separate emission lines of light rather than a continuous band of color.

An emission spectrum is produced when hydrogen gas is excited by an electrical voltage. To do so, hydrogen gas is sealed in a gas discharge tube and energized by electricity. The discharge tube then emits light, which separates into a series of narrow lines when passed through a prism. This collection of narrow bands of light energy is referred to as an **emission line spectrum**, and the individual bands of light are called spectral lines. The emission spectrum of hydrogen gas is shown in Figure 5.15.

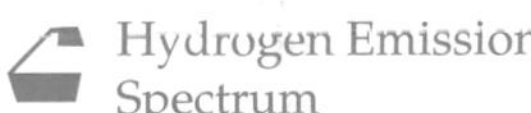

Hydrogen Emission Spectrum

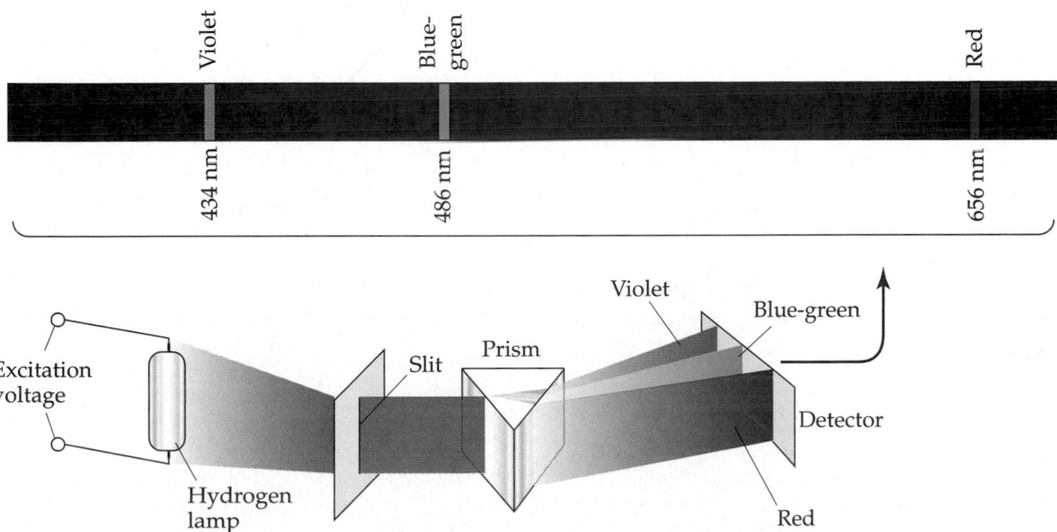

▲ **Figure 5.15 Hydrogen Emission Spectrum** An emission line spectrum is produced when hydrogen gas is excited by an electrical voltage. After the emitted light is passed through a prism, three discrete vivid lines are observed.

After examining the emission spectrum of hydrogen, Bohr realized that he had experimental evidence to support his model of the atom. The concept of electron energy levels was supported by the line spectrum of hydrogen. In a gas discharge tube, excited atoms of hydrogen have electrons in a high-energy orbit; for example, the electron may temporarily occupy the second orbit. Since this state is unstable, the electron quickly drops from the higher level back to a lower level, that is, from level 2 to level 1. In the process, the electron loses a discrete amount of energy. This discrete energy loss corresponds to a photon of light energy. The energy of the photon of light equals the amount of energy lost by the electron as it drops from the higher to the lower energy level.

In the hydrogen spectrum, there are three bright lines: red, blue-green, and violet. The red line has the longest wavelength of the three and the lowest energy. Bohr found that the red line corresponds to an excited electron dropping from energy level 3 to 2. The blue-green line is more energetic than the red line and corresponds to an excited electron dropping from energy level 4 to 2. The violet line is the most energetic of the three and is produced when an electron drops from energy level 5 to 2. Figure 5.16 illustrates the correlation between Bohr's electron energy levels and the observed lines in the hydrogen spectrum.

Since a single photon is emitted each time an electron drops to a lower level, it follows that several photons are emitted when several electrons change levels. For example, if the electron drops from energy level 5 to 2 in ten hydrogen atoms, ten photons would be emitted. Each of the photons would have the same energy, and collectively they would be observed as a violet line in the emission spectrum. Example Exercise 5.7 further illustrates the relationship between energy levels and emission lines.

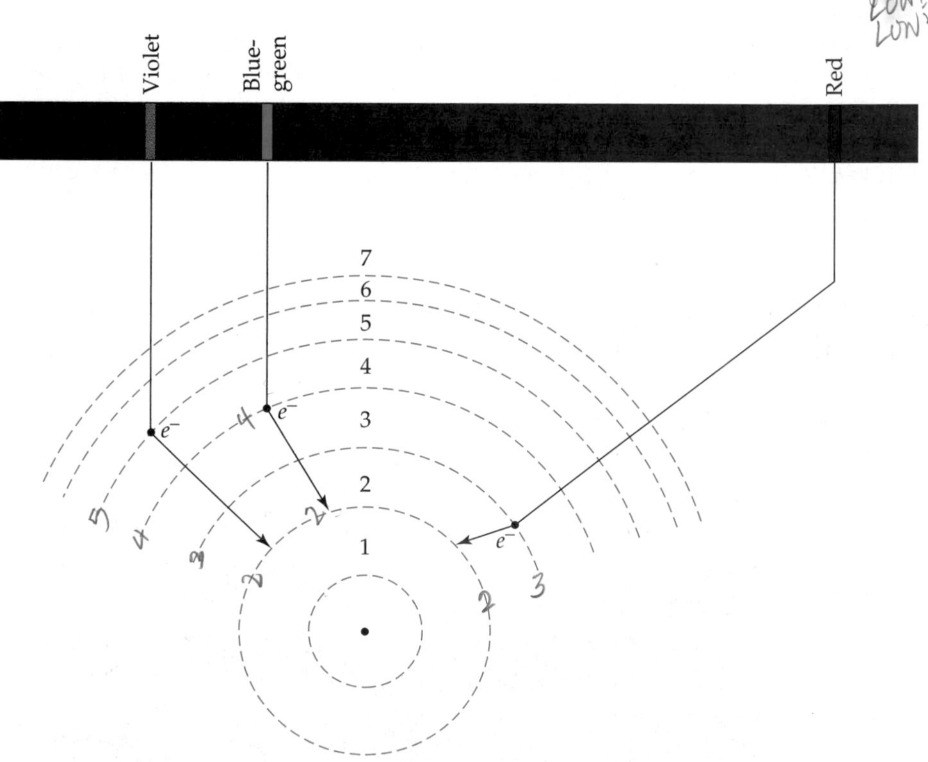

Spectral Lines and Energy Levels in Hydrogen

▲ Figure 5.16 Spectral Lines and Energy Levels in Hydrogen When electrons drop from energy level 5 to 2, we see a violet line. When electrons drop from level 4 to 2, we see a blue-green line; from level 3 to 2, we observe a red line.

Example Exercise 5.7 • Emission Spectrum and Energy Levels

Explain the relationship between an observed emission line in a spectrum and electron energy levels.

Solution

When an electron drops from a higher to a lower energy level, light is emitted. For each electron that drops, a single photon of light energy is emitted. The energy lost by the electron that drops equals the energy of the photon that is emitted. Several photons of light having the same energy are observed as a spectral line.

Self-Test Exercise

Indicate the number and color of the photons emitted for each of the following electron transitions in hydrogen atoms.

(a) 1 electron dropping from energy level 3 to 2
(b) 10 electrons dropping from energy level 3 to 2
(c) 100 electrons dropping from energy level 4 to 2
(d) 500 electrons dropping from energy level 5 to 2

Answers: (a) 1 red photon; (b) 10 red photons; (c) 100 blue-green photons; (d) 500 violet photons

"Atomic Fingerprints"

Further study of emission spectra revealed that each element produced a unique set of spectral lines. This observation indicated that the energy levels must be unique for atoms of each element. Therefore, a line spectrum is sometimes referred to as an "atomic fingerprint."

Atomic fingerprints are useful in the identification of elements. For instance, in 1868 the atomic fingerprint of a new element was observed in the spectrum from the Sun. This element was named helium, after *helios*, the Greek word for "sun." In 1895 an element was discovered in uranium ore with an atomic fingerprint identical to that observed for helium in the Sun's spectrum. Thus, helium was discovered on Earth 27 years after it had first been observed in the solar spectrum. Figure 5.17 compares a continuous spectrum to the emission line spectra of four elements.

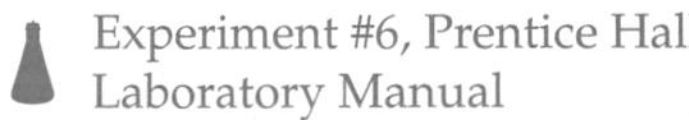
Experiment #6, Prentice Hall Laboratory Manual

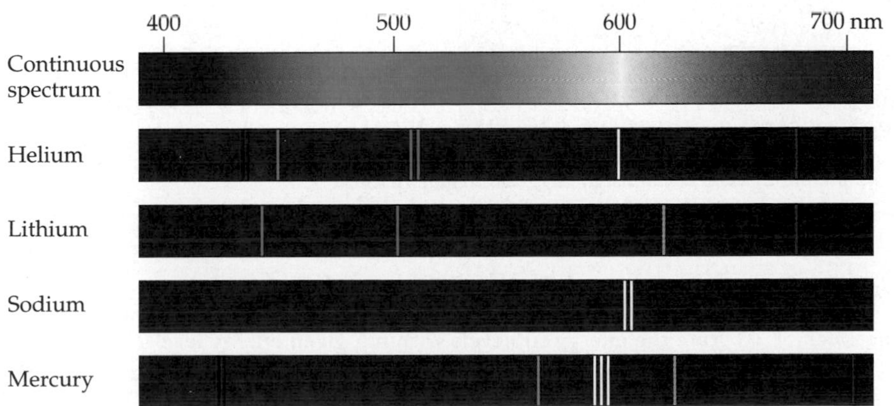

◀ Figure 5.17 A Continuous Spectrum versus Line Spectra A continuous spectrum is produced from an ordinary light bulb. The emission line spectra are produced by excited atoms of elements In the gaseous state.

Chemistry Connection • Neon Lights

What element in the "neon" sign is emitting the purple light?

At the beginning of the twentieth century, J. J. Thomson discovered the electron using a cathode-ray tube. Thomson constructed his cathode-ray tube out of thin glass and placed a metal electrode in each end. After evacuating air from the glass tube, he introduced a small amount of gas. When the metal cathode and anode electrodes were electrically excited, he noticed that the tube glowed. He identified the glowing rays from the cathode as a stream of small negative particles. These particles were named electrons, and Thomson is given credit for their discovery. The glass tubes used by Thomson were forerunners of cathode-ray tubes, which are used today in television sets and computer monitor screens.

In 1913 Niels Bohr "explained" that exciting gases with electricity caused electrons to be temporarily promoted to higher energy states within the atom. The excited electrons, however, quickly lose energy by emitting light and returning to their original state. The light emitted by different gases varies because the energy levels within atoms vary for each element. For example, a gas discharge tube containing mercury vapor gives off a blue glow, whereas nitrogen gas gives off a yellowish-orange glow.

In 1898 the Scottish chemist William Ramsay discovered the noble gas neon. Unlike argon, which comprises about 1% of air, neon is much more rare. It is about a thousand times less concentrated in the atmosphere. When neon gas is placed in a narrow glass tube and electrically excited, it produces a reddish-orange light that is very arresting to the eye. The fact that gas discharge tubes produce an attractive array of colors led naturally to their use as advertising lights. Light from excited neon gas is very intense, and the term "neon light" has become a generic term for all lighted advertising displays.

Obviously, not all lights used in advertising are the same color. That is, "neon lights" can be red, green, blue, and so on. To produce a given color, a gas discharge tube must be filled with a specific gas. For a purple light, argon gas can be used, and for a pink light, helium gas is used. Only when we wish to produce a reddish-orange light is an advertising sign actually filled with neon gas.

◀ **Neon Light** The reddish-orange glow from neon gas is illuminating the word NEON.

The purple light in a neon sign suggests argon gas.

5.9 Energy Levels and Sublevels

Objectives
- To state the energy sublevels within a given energy level.
- To state the maximum number of electrons that can occupy a given energy level and sublevel.

As we learned in the previous section, in 1913 Niels Bohr proposed a model for the atom that pictured electrons circling around the nucleus in fixed-energy levels. His proposal was supported experimentally by the lines in the emission spectrum of

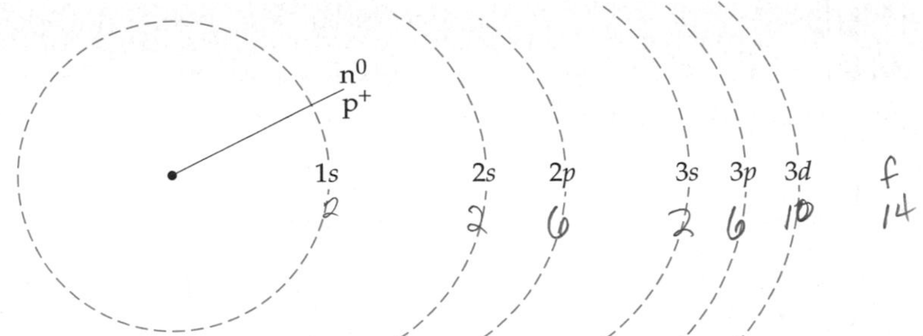

◀ Figure 5.18 A Cross Section of an Atom The first energy level has only one sublevel (1*s*). The second energy level has two sublevels (2*s* and 2*p*). The third energy level has three sublevels (3*s*, 3*p*, and 3*d*). Although the diagram suggests that electrons travel in circular orbits, this is a simplification and is not actually the case.

 Cross-Section of an Atom

hydrogen. The emission spectra of other elements, however, had far too many lines to interpret. Although Bohr could not explain the spectra of other elements, he did suggest the idea of sublevels within a main energy level. The model that eventually emerged had electrons occupying an **energy sublevel** within a main energy level. These energy sublevels were designated *s*, *p*, *d*, and *f*—in reference to the *sharp*, *principal*, *diffuse*, and *fine* lines, respectively, in the emission spectra of the elements.

The number of sublevels in each level corresponds to the number of the main energy level. That is, the first energy level (1) has one sublevel and is designated 1*s*. The second energy level (2) has two sublevels designated 2*s* and 2*p*. The third energy level (3) has three sublevels designated 3*s*, 3*p*, and 3*d*. The fourth energy level (4) is composed of 4*s*, 4*p*, 4*d*, and 4*f* sublevels (Figure 5.18).

The maximum number of electrons in each of the energy sublevels depends on the type of sublevel. That is, an *s* sublevel can hold a maximum of 2 electrons. A *p* sublevel can have a maximum of 6 electrons. A *d* sublevel can have 10 electrons, and an *f* sublevel can hold a maximum of 14 electrons.

To find the maximum number of electrons in a main energy level, we add up the electrons in each sublevel. The first energy level has one *s* sublevel; it can contain only $2e^-$. The second major energy level has two sublevels, 2*s* and 2*p*. The 2*s* can hold $2e^-$, and the 2*p* can hold $6e^-$. Thus, the second energy level can hold a maximum of $8e^-$. Example Exercise 5.8 further illustrates how energy levels, sublevels, and number of electrons are related.

Example Exercise 5.8 • Energy Levels, Sublevels, and Electrons

How many sublevels are in the third energy level. What is the maximum number of electrons that can occupy the third energy level?

Solution

The number of sublevels in an energy level corresponds to the number of the energy level. The third energy level is split into three sublevels: 3*s*, 3*p*, and 3*d*. The maximum number of electrons that can occupy a given sublevel is as follows.

$$s \text{ sublevel} = 2e^-$$
$$p \text{ sublevel} = 6e^-$$
$$d \text{ sublevel} = 10e^-$$

The maximum number of electrons in the third energy level is found by adding the three sublevels.

$$3s + 3p + 3d = \text{total electrons}$$
$$2e^- + 6e^- + 10e^- = 18e^-$$

The third energy level can hold a maximum of 18 electrons. Of course, in elements where the third energy level of an atom is not filled, there are fewer than 18 electrons.

Self-Test Exercise

How many sublevels are in the fourth energy level? What is the maximum number of electrons that can occupy the fourth energy level?

Answers: 4*s*, 4*p*, 4*d*, 4*f*; $32e^-$ $(2e^- + 6e^- + 10e^- + 14e^-)$

Table 5.4 Distribution of Electrons by Energy Level

Energy Level	Energy Sublevel	Maximum e^- in Sublevel	Maximum e^- in Energy Level
1	1*s*	$2e^-$	$2e^-$
2	2*s*	$2e^-$	
	2*p*	$6e^-$	$8e^-$
3	3*s*	$2e^-$	
	3*p*	$6e^-$	
	3*d*	$10e^-$	$18e^-$
4	4*s*	$2e^-$	
	4*p*	$6e^-$	
	4*d*	$10e^-$	
	4*f*	$14e^-$	$32e^-$

In summary, electrons are arranged around the nucleus in sublevels that have a specific fixed energy. Electrons farther from the nucleus occupy higher energy levels than those closer to the nucleus. Table 5.4 summarizes how main energy levels, sublevels, and number of electrons are related.

5.10 Electron Configuration

Objectives
- To list the order of sublevels according to increasing energy.
- To write the predicted electron configurations for selected elements.

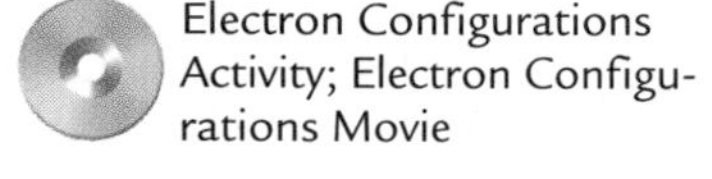
Electron Configurations Activity; Electron Configurations Movie

Electrons are arranged about the nucleus in a regular manner. The first electrons fill the energy sublevel closest to the nucleus. Additional electrons fill energy sublevels further and further from the nucleus. In other words, each energy level is filled sublevel by sublevel. The *s* sublevel is filled before a *p* sublevel, a *p* sublevel is filled before a *d* sublevel, and a *d* sublevel is filled before an *f* sublevel.

In general, sublevels are higher in energy as the energy level increases. Therefore, we would expect the order of sublevel filling to be 1*s*, 2*s*, 2*p*, 3*s*, 3*p*, 3*d*, 4*s*, and so on. This is not quite accurate because there are exceptions. For instance, the 4*s* sublevel is lower in energy than the 3*d* sublevel, and the 5*s* is lower than the 4*d* sublevel. A partial list of sublevels in order of increasing energy is $1s < 2s < 2p < 3s < 3p < 4s < 3d < 4p < 5s < 4d < 5p < 6s$.

In Chapter 6 we will learn how to predict the order in which sublevels are filled from the position of the element in the periodic table. In fact, the unusual shape of the periodic table reflects the order of sublevels according to increasing energy. For now, you should memorize the order of sublevel filling or refer to Figure 5.19.

Filling Diagram for Energy Sublevels

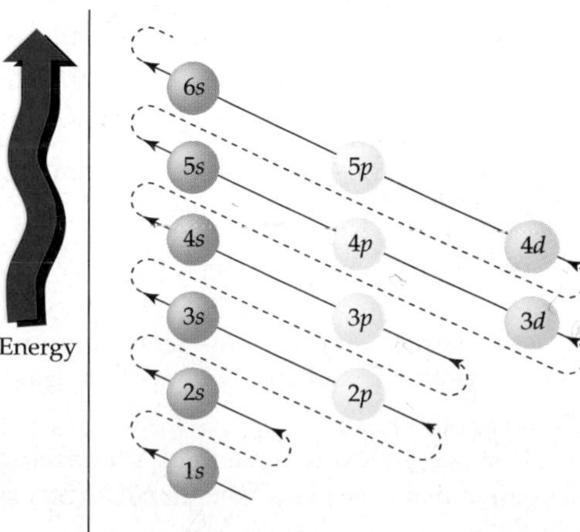

▶ **Figure 5.19 Filling Diagram for Energy Sublevels** The order of sublevel filling is arranged according to increasing energy. Electrons first fill the 1*s* sublevel followed by the 2*s*, 2*p*, 3*s*, 3*p*, 4*s*, 3*d*, 4*p*, 5*s*, 4*d*, 5*p*, and 6*s* sublevels.

Example Exercise 5.9 • Order of Sublevels

According to increasing energy, what is the next energy sublevel after each of the following sublevels?

(a) $3p$ (b) $4d$

Solution

If you have not memorized the order of sublevels, refer to the filling diagram in Figure 5.19.

(a) Although the third energy level has $3s$, $3p$, and $3d$ sublevels, the $3d$ sublevel does not immediately follow the $3p$. Instead, the $4s$ sublevel follows the $3p$ and precedes the $3d$. Thus,

$$3s, 3p, 4s$$

(b) Although the fourth energy level has $4s$, $4p$, $4d$, and $4f$ sublevels, the $4f$ sublevel does not immediately follow the $4d$. Instead, the $5p$ sublevel begins accepting electrons after the $4d$ is filled. Thus,

$$4p, 5s, 4d, 5p$$

Self-Test Exercise

Which sublevel gains electrons after each of the following sublevels is filled.

(a) $2s$ (b) $5p$

Answers: (a) $2p$; (b) $6s$

Electron Configuration

The **electron configuration** of an atom is a shorthand statement describing the location of electrons by sublevel. First, the sublevel is written, followed by a superscript indicating the number of electrons. For example, if the $2p$ sublevel contains two electrons, the standard notation is $2p^2$. Thus,

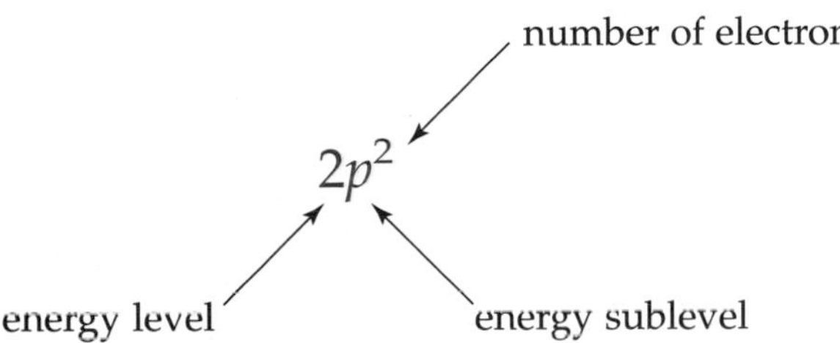

Writing the electron configuration for an atom is a straightforward procedure. First, find the atomic number of the element in the periodic table; this corresponds to the number of electrons in a neutral atom. Then write the sublevels according to increasing energy. Each sublevel is filled with electrons in sequence until the total of all the superscripts equals the atomic number of the element.

As an example, let's write the electron configuration for iron. If we refer to the periodic table, we find that the atomic number of iron is 26. Given the atomic number, we know that an iron nucleus must have 26 protons and is surrounded by 26 electrons. The order in which the electron sublevels are arranged according to increasing energy is as follows.

$$1s\ 2s\ 2p\ 3s\ 3p\ 4s\ 3d \ldots$$

The energy sublevels in iron are filled beginning with the $1s$ sublevel and ending when there is a total of 26 electrons. The electron configuration for iron is as follows.

$$\text{Fe:}\quad 1s^2\ 2s^2\ 2p^6\ 3s^2\ 3p^6\ 4s^2\ 3d^6$$

Notice that the sum of the superscripts equals the atomic number of iron (26). Example Exercise 5.10 illustrates the electron configuration for other elements.

Example Exercise 5.10 • Electron Configuration

Write the electron configuration for each of the following elements given the atomic number.

(a) Ne (b) Sr

Solution

We refer to the periodic table to find the atomic number of an element.

(a) The atomic number of neon is 10; therefore, the number of electrons is 10. We can fill sublevels until 10 electrons are present as follows.

$$\text{Ne:}\quad 1s^2\,2s^2\,2p^6$$

(b) From the periodic table, we find that the atomic number for strontium is 38. The number of electrons in a neutral atom of strontium is 38. Thus,

$$\text{Sr:}\quad 1s^2\,2s^2\,2p^6\,3s^2\,3p^6\,4s^2\,3d^{10}\,4p^6\,5s^2$$

To check your answer, find the total number of electrons by adding up the superscripts. The total is $38e^-$; this agrees with the atomic number for Sr.

Self-Test Exercise

Write the electron configuration for each of the following elements. Use standard notation, grouping electrons together according to sublevel.

(a) argon (b) cadmium

Answers: (a) $1s^2\,2s^2\,2p^6\,3s^2\,3p^6$; (b) $1s^2\,2s^2\,2p^6\,3s^2\,3p^6\,4s^2\,3d^{10}\,4p^6\,5s^2\,4d^{10}$

5.11 Quantum Mechanical Model of the Atom

Objectives
- To describe the quantum mechanical model of the atom.
- To describe the relative sizes and shapes of *s* and *p* orbitals.

▲ **Albert Einstein and Niels Bohr** The two famous scientists provided important insights to further our understanding of the atom.

In the mid-1920s, a new model of the atom began to emerge. A more powerful theory evolved because the behavior of electrons could not be fully explained by the Bohr model of the atom. The German physicist Werner Heisenberg concluded that it was not possible to accurately determine both the position and energy of an electron. In his **uncertainty principle**, Heisenberg stated that it is impossible to precisely measure both the location and energy of a small particle simultaneously. In fact, the more accurately the position of an electron in an atom is known, the less precisely its energy can be determined.

In 1932 Heisenberg won the Nobel prize in physics for his uncertainty principle. Not everyone, however, subscribed to the principle of uncertainty. Some physicists found it unsettling to consider that they might live in a universe ruled by chance. Albert Einstein was sufficiently troubled by the uncertainty principle that he offered the famous quote: "It seems hard to look into God's cards but I cannot for a moment believe He plays dice as the current quantum theory alleges He does." Although the uncertainty principle was initially controversial, it was an essential contribution to the new view of the atom.

Gradually, the deeper nature of the atom came into focus. The new model retained the idea of quantized energy levels but incorporated the concept of uncertainty. The new model that emerged became known as the **quantum mechanical atom**. Recall that in the Bohr model the energy of an electron is defined in terms of a fixed-energy orbit about the nucleus. In the quantum mechanical model the energy of an electron can be described in terms of the probability of it being within a spatial volume surrounding the nucleus. This region of high probability (~95%) for finding an electron of given energy is called an **orbital**.

Sizes and Shapes of *s* and *p* Orbitals

In the quantum mechanical atom, orbitals are arranged about the nucleus according to their size and shape. In general, electrons having higher energy are found in larger orbitals. Similar to the energy levels in the Bohr atom, the energy of orbitals is quantized and assigned a whole-number value such as 1, 2, 3, 4, ... As the number increases, the energy and size of an orbital also increases.

We can describe the shapes of orbitals by the letters *s*, *p*, *d*, and *f*. For example, the shape of an *s* orbital is that of a sphere, and the shape of a *p* orbital is that of a dumbbell. The shapes of *d* and *f* orbitals are too complex for our discussion. We can designate the size and shape of an orbital by combining the number that indicates its energy, and the letter that indicates its shape. For example, the designations 1*s*, 2*p*, and 3*d* indicate three orbitals that differ in size, energy, and shape. All *s* orbitals are spherical, but they are not all the same size. A 3*s* orbital is a larger sphere than a 2*s*, and a 2*s* orbital is larger than a 1*s*. That is, the size and energy of the orbital increase as the energy level increases. Figure 5.20 illustrates the relationship between *s* orbitals about the nucleus.

All *p* orbitals have the shape of a dumbbell, but they are not all equal in size or energy. A 3*p* orbital is larger than a 2*p* orbital and is at a higher energy level. A *p* orbital is said to resemble a dumbbell because it has two lobes. Electrons in a *p* orbital can occupy either of the lobes.

There are three different 2*p* orbitals. Although these three orbitals are identical in size and shape, they differ in their orientation to each other. That is, the three 2*p* orbitals intersect at the nucleus, but they are oriented at right angles to each other. Figure 5.21 illustrates the relationships between the $2p_x$, $2p_y$, and $2p_z$ orbitals. The

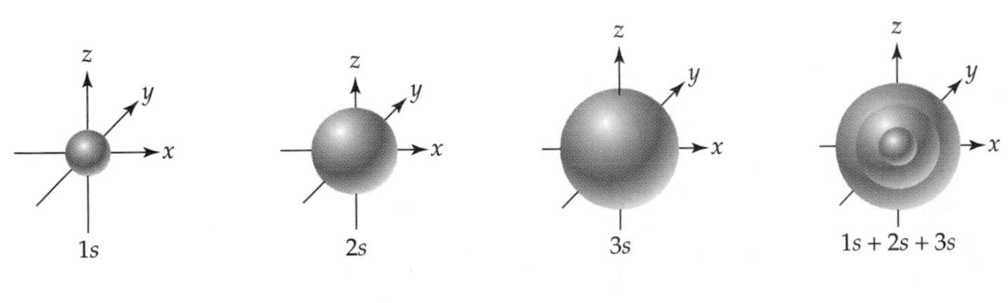

◀ **Figure 5.20 Relative Sizes of *s* Orbitals** The relative sizes of 1*s*, 2*s*, and 3*s* orbitals are shown. As the main energy level increases, the size and energy of the orbital increases. The nucleus of the atom is located in the center where the three axes intersect. The sketch on the far right illustrates the relationship of the 1*s*, 2*s*, and 3*s* orbitals.

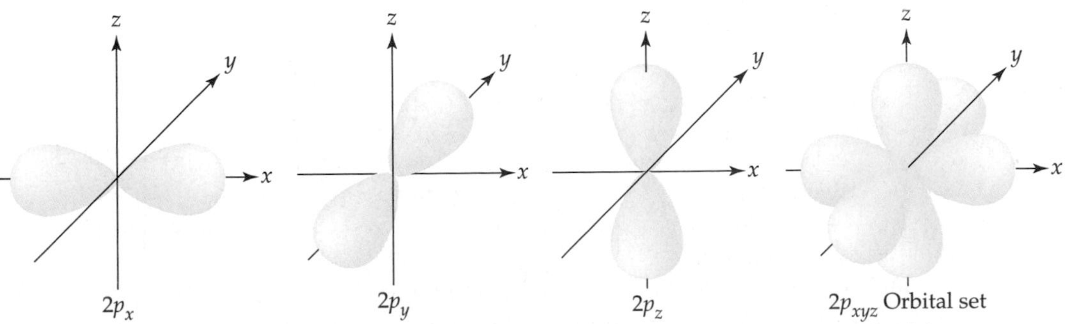

▲ **Figure 5.21 Orientation of 2*p* Orbitals** The size and shape of the three 2*p* orbitals are identical. The 2*p* orbitals do not have a fixed orientation, but rather are perpendicular to each other. An electron in a $2p_x$ orbital has the same probability of occupying the $2p_y$ or $2p_z$ orbital.

p_x orbital is oriented along the x-axis of a three-dimensional axes system, and the p_y and p_z orbitals are oriented along the y-axis and z-axis, respectively.

Example Exercise 5.11 provides further practice in describing the relative sizes and shapes of orbitals.

Example Exercise 5.11 • Atomic Orbitals

Describe the relative size, energy, and shape for each of the following orbitals.

(a) 4*s* versus 3*s* and 5*s* (b) 4*p* versus 3*p* and 5*p*

Solution

The size and energy of an orbital is indicated by the number; the shape of the orbital is designated by the letter.

(a) Size and energy are greater for a 4*s* orbital than for a 3*s* orbital, but less than for a 5*s* orbital. The shape of a 4*s* orbital—and all *s* orbitals—is similar to the shape of a sphere.

(b) Size and energy are greater for a 4*p* orbital than for a 3*p* orbital but less than for a 5*p* orbital. The shape of a 4*p* orbital—and all *p* orbitals—is similar to the shape of a dumbbell.

Self-Test Exercise

Select the orbital in each of the following pairs that fits the description.

(a) the higher energy orbital: 3*p* or 4*p*

(b) the larger-size orbital: 4*d* or 5*d*

Answers: (a) 4*p*; (b) 5*d*

As an analogy, try to visualize a flying insect trapped inside two bottles with the open ends held together. The insect is free to fly about the entire inner volume of the two bottles. In this analogy, the insect represents an electron, and the two bottles represent the two lobes of a *p* orbital. Thus, there is a high probability of finding the electron anywhere within the volume of the entire *p* orbital. Figure 5.22 illustrates this analogy.

Sizes and Shapes of d Orbitals

So far we have shown the shapes of only *s* and *p* orbitals. However, there are many other orbitals; for example, in the third energy level there are different orbitals in the 3*s*, 3*p*, and 3*d* sublevels (Figure 5.19).

The 3*s* sublevel contains a single spherical 3*s* orbital. The 3*p* sublevel contains three dumbbell-shaped orbitals: $3p_x$, $3p_y$, and $3p_z$. The 3*p* orbitals are similar in shape, but 3*p* orbitals are larger than 2*p* orbitals. The 3*d* sublevel contains five 3*d* orbitals. Although it is not necessary to know the shapes of *d* orbitals, the five different 3*d* orbitals are shown in Figure 5.23.

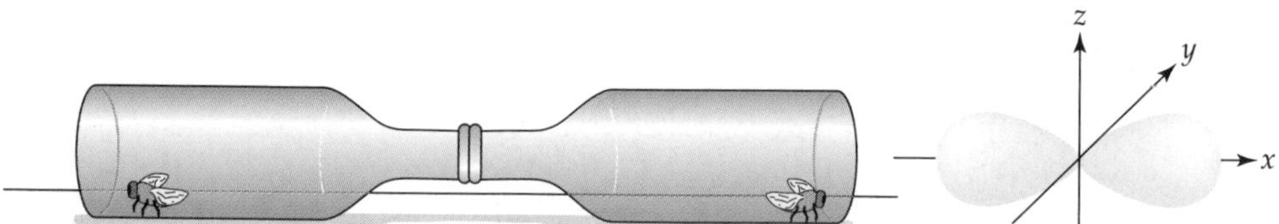

▲ **Figure 5.22 Analogy for a *p* Orbital** (a) Notice the two insects trapped within the bottles held end to end. The two insects can both be in the left bottle, the right bottle, or one insect can be in each bottle. (b) Similarly, two electrons have a probability of being found anywhere within the two lobes of the *p* orbital.

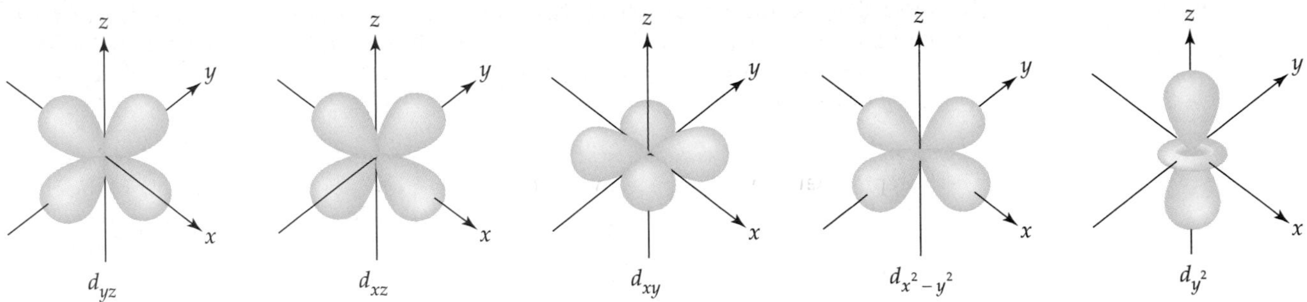

▲ **Figure 5.23 Shapes of 3*d* Orbitals** The shapes of the five 3*d* orbitals are not the same. Although three of the 3*d* orbitals are similar in shape, they have a different orientation to each other.

Summary 

Section 5.1 In 1803 John Dalton proposed that matter consisted of atoms and supported the atomic theory with experiments on the behavior of gases and the laws of definite composition and conservation of mass.

Section 5.2 Toward the end of the 1800s, there was evidence that the atom was divisible. When electricity was applied to a sealed glass tube containing a gas at low pressure, negative and positive rays were observed. Scientists discovered that a **cathode ray** was composed of tiny negatively charged particles they called **electrons**. When the tube was filled with hydrogen gas, they found that a positive canal ray was composed of the smallest positively charged particles. These particles were named **protons**.

Section 5.3 In 1911 Rutherford performed a classic experiment in which alpha particles were fired at a thin sheet of gold foil. Much to his astonishment, some of the alpha particles bounced backward. He interpreted the results as evidence for a tiny, dense **atomic nucleus** at the center of the atom. The nucleus contains positively charged protons surrounded by negatively charged electrons. Twenty years later, the nucleus was also found to contain a neutral particle called a **neutron**.

Section 5.4 Chemists use a symbolic shorthand called **atomic notation** to designate the composition of a nucleus. Atoms of an element always contain the same number of protons, but the number of neutrons in the nucleus can vary. The number of protons is called the atomic number (Z), and the sum of the protons and neutrons is called the mass number (A). Atoms with the same **atomic number** but a different **mass number** are called **isotopes**.

Section 5.5 Although the mass of an atom is much too small to measure directly, we can determine its relative mass. Carbon-12 is used as the reference isotope and is assigned a mass of exactly 12 **atomic mass units (amu)**. The masses of all other atoms are related to the mass of carbon-12. To find a representative value for the atomic mass of an element, we average the mass of each isotope. The weighted average mass of all isotopes is termed the **atomic mass** of the element.

Section 5.6 **Light** travels through space as a wave of radiant energy. The crest-to-crest distance between waves is the **wavelength**, and the number of cycles completed in a second is the **frequency**. As the wavelength decreases, the frequency and energy of light increase. The **visible spectrum** extends from 400–700 nm, but the entire **radiant energy spectrum** also includes gamma rays, X rays, and microwaves. Thus, radiant energy is a **continuous spectrum** of visible and invisible light.

Section 5.7 In 1900 Max Planck introduced the quantum concept. Planck stated that the energy radiated by a object is not continuous, but rather that the radiation is emitted in small bundles. When an object radiates light, it releases a unit of radiant energy called a **photon**.

Section 5.8 In 1913 Niels Bohr suggested that electrons travel in circular orbits about the nucleus. The electron possesses a specific energy and is said to occupy an **energy level**. If an electron changes orbits in the **Bohr atom**, there is a quantum energy change. Bohr argued that an **emission line spectrum** results from electrons dropping from higher energy levels to lower levels. Each time an electron drops, a photon of light is released whose energy corresponds to the difference in energy between the two levels.

Section 5.9 A closer examination of emission line spectra from gases revealed **energy sublevels** within main levels. The number of sublevels corresponds to the number of the energy level. For example, the fourth energy level has four sublevels (s, p, d, f). An s sublevel can hold 2 electrons, a p sublevel 6 electrons, a d sublevel 10 electrons, and an f sublevel a maximum of 14 electrons.

Section 5.10 Electrons fill sublevels in order of increasing energy as follows: $1s < 2s < 2p < 3s < 3p < 4s < 3d < 4p < 5s < 4d < 5p < 6s$. Notice that the $4s$ sublevel fills before the $3d$ and the $5s$ sublevel before the $4d$. A description of sublevel filling for an element is given by the **electron configuration**. A superscript following each sublevel indicates the number of electrons in a sublevel; for instance, $1s^2\ 2s^2\ 2p^6\ 3s^1$ is the electron configuration for sodium (atomic number 11).

Section 5.11 In the 1920s our understanding of electrons in atoms became very sophisticated. In 1925 Werner Heisenberg suggested the **uncertainty principle**; that is, it is impossible to simultaneously know both the precise location and the energy of an electron. Instead, the energy of an electron can be known only in terms of its probability of being located somewhere within the atom. This description gave rise to the **quantum mechanical atom**. A location within the atom where there is a high probability of finding an electron having a certain energy is called an **orbital**. An orbital is a region about the nucleus having a given energy, size, and shape. The shape of an s orbital is spherical, and a p orbital resembles the shape of a dumbbell.

Key Concepts *

1. If an atom is magnified to the size of a golf ball, and a golf ball is equally magnified, what is the approximate size of the enlarged golf ball? (tennis ball, basketball, the Earth, the universe)
2. An atomic nucleus has been described by the analogy: "like a marble in the Astrodome." If a marble represents the atomic nucleus, what does the Astrodome represent?
3. The scattering of alpha particles by a thin gold foil has been described by the analogy: "like missiles shot through the solar system." If a missile represents an alpha particle, what do the planets represent?
4. Can atoms of different elements have the same atomic number? Can atoms of different elements have the same mass number?
5. Complete the following analogy. An ocean wave is to a drop of water as a light wave is to a ________.

◀ **Astrodome** Imagine how small a marble (nucleus) is compared to the Astrodome (atom).

* Answers to Key Concepts are in Appendix H.

6. Which of the following statements is false according to the Bohr model of the atom?
   (a) Electrons are attracted to protons in the nucleus.
   (b) Electrons circle the nucleus in the same way that planets circle the Sun.
   (c) Electrons lose energy as they circle the nucleus.
   (d) Electrons lose energy as they drop to an orbit closer to the nucleus.
7. Which of the following statements is false according to the quantum mechanical model of the atom?
   (a) Orbitals represent quantum energy levels.
   (b) Orbitals represent probability boundaries.
   (c) Orbitals can contain from 1 to 14 electrons.
   (d) Orbitals can have different shapes.
8. Briefly describe the characteristics of (a) the 1803 Dalton model of the atom, (b) the 1903 Thomson model, (c) the 1911 Rutherford model, (d) the 1913 Bohr model, (e) the 1926 Heisenberg model, and (f) the current model of the atom.

Key Terms†

Select the key term below that corresponds to each of the following definitions.

_____ 1. a stream of negative particles produced in a cathode-ray tube
_____ 2. a negatively charged subatomic particle having a negligible mass
_____ 3. a positively charged subatomic particle having an approximate mass of 1 amu
_____ 4. a neutral subatomic particle having an approximate mass of 1 amu
_____ 5. a region in the center of an atom containing protons and neutrons
_____ 6. a value indicating the number of protons in the nucleus of an atom
_____ 7. a value indicating the number of protons and neutrons in the nucleus of an atom
_____ 8. a symbolic method for expressing the composition of an atomic nucleus
_____ 9. atoms of the same element that have a different number of neutrons
_____ 10. a unit of mass exactly equal to 1/12 the mass of a carbon-12 atom
_____ 11. the average mass of all the naturally occurring isotopes of an element
_____ 12. the distance a light wave travels to complete one cycle
_____ 13. the number of times a light wave completes a cycle in 1 second
_____ 14. a general term that can refer to either visible or invisible radiant energy
_____ 15. a range of light energy from violet through red, that is, 400–700 nm
_____ 16. a range of light energy extending from gamma rays through microwaves
_____ 17. a band of light energy that is uninterrupted
_____ 18. a particle of radiant energy
_____ 19. a model of the atom that describes electrons circling the nucleus in orbits
_____ 20. a fixed-energy orbit that electrons occupy as they circle the nucleus
_____ 21. a collection of narrow bands of light produced by atoms of a given element releasing energy
_____ 22. an electron energy level resulting from splitting a main energy level
_____ 23. a shorthand description of the arrangement of electrons by sublevels according to increasing energy
_____ 24. the statement that it is impossible to precisely measure the location and energy of a particle at the same time
_____ 25. a sophisticated model of the atom that describes the energy of an electron in terms of its probability of being found in a particular location about the nucleus
_____ 26. a region about the nucleus in which there is a high probability of finding an electron with a given energy

(a) atomic mass (*Sec. 5.5*)
(b) atomic mass unit (amu) (*Sec. 5.5*)
(c) atomic notation (*Sec. 5.4*)
(d) atomic nucleus (*Sec. 5.3*)
(e) atomic number (Z) (*Sec. 5.4*)
(f) Bohr atom (*Sec. 5.8*)
(g) cathode ray (*Sec. 5.2*)
(h) continuous spectrum (*Sec. 5.6*)
(i) electron (e^-) (*Sec. 5.2*)
(j) electron configuration (*Sec. 5.10*)
(k) emission line spectrum (*Sec. 5.8*)
(l) energy level (*Sec. 5.8*)
(m) energy sublevel (*Sec. 5.9*)
(n) frequency (*Sec. 5.6*)
(o) isotopes (*Sec. 5.4*)
(p) light (*Sec. 5.6*)
(q) mass number (A) (*Sec. 5.4*)
(r) neutron (n^0) (*Sec. 5.3*)
(s) orbital (*Sec. 5.11*)
(t) photon (*Sec. 5.7*)
(u) proton (p^+) (*Sec. 5.2*)
(v) quantum mechanical atom (*Sec. 5.11*)
(w) radiant energy spectrum (*Sec. 5.6*)
(x) uncertainty principle (*Sec. 5.11*)
(y) visible spectrum (*Sec. 5.6*)
(z) wavelength (*Sec. 5.6*)

† Answers to Key Terms are in Appendix I.

Exercises‡

Dalton Model of the Atom (Sec. 5.1)

1. State Dalton's five proposals regarding the atomic theory.
2. State the two experimental laws Dalton used to support the atomic theory.
3. Which two of Dalton's proposals were later shown to be invalid?
4. Are atoms indestructible? Explain.

Thomson Model of the Atom (Sec. 5.2)

5. What was the simplest particle observed in cathode rays?
6. What was the simplest particle observed in canal rays?
7. What are the relative charges of the electron and the proton?
8. What are the relative masses of the electron and the proton?
9. What do the raisins represent in the plum-pudding analogy of the atom?
10. Where is the mass of an atom found according to the plum-pudding model?

Rutherford Model of the Atom (Sec. 5.3)

11. What did Rutherford conclude about the atom when alpha particles recoiled backward after striking a thin gold foil?
12. Describe an atom according to the Rutherford model.
13. State the location of electrons, protons, and neutrons in the Rutherford model of the atom.
14. State the approximate size of an atom and its nucleus in centimeters.
15. State the relative charges of the electron, proton, and neutron.
16. State the relative masses of the electron, proton, and neutron.

Atomic Notation (Sec. 5.4)

17. State the number of neutrons in an atom of each of the following isotopes.
    (a) $^{4}_{2}He$ (b) $^{32}_{16}S$
    (c) $^{10}_{5}B$ (d) $^{44}_{20}Ca$
18. State the number of neutrons in an atom of each of the following isotopes.
    (a) ^{15}N (b) ^{52}Cr
    (c) ^{26}Mg (d) ^{58}Ni
19. State the number of neutrons in an atom of each of the following isotopes.
    (a) lithium-7 (b) potassium-40
    (c) strontium-88 (d) platinum-195
20. State the number of neutrons in an atom of each of the following isotopes.
    (a) hydrogen-3 (b) cobalt-60
    (c) silicon-28 (d) iodine-131

21. Complete the following table and provide the missing information.

Atomic Notation	Atomic Number	Mass Number	Number of Protons	Number of Neutrons	Number of Electrons
$^{4}_{2}He$					
$^{21}_{10}Ne$					
$^{50}_{22}Ti$					
$^{197}_{79}Au$					

22. Complete the following table and provide the missing information.

Atomic Notation	Atomic Number	Mass Number	Number of Protons	Number of Neutrons	Number of Electrons
$^{A}_{Z}Se$		78			
$^{A}_{Z}X$	38			50	
$^{A}_{Z}Sn$		120			
$^{A}_{Z}X$	54			77	

23. Draw a diagram of the arrangement of protons, neutrons, and electrons in an atom of each of the following isotopes.
    (a) $^{7}_{3}Li$ (b) $^{13}_{6}C$
    (c) $^{16}_{8}O$ (d) $^{20}_{10}Ne$
24. Draw a diagram of the arrangement of protons, neutrons, and electrons in an atom of each of the following isotopes.
    (a) ^{31}P (b) ^{35}Cl
    (c) ^{40}Ar (f) ^{131}I

Atomic Mass (Sec. 5.5)

25. What is the reference isotope for the atomic mass scale?
26. What is the assigned mass for the reference isotope?
27. Why are atomic masses expressed on a *relative* atomic mass scale?
28. Distinguish between isotopic mass and atomic mass.
29. Given that the only naturally occurring isotope of aluminum is ^{27}Al, determine its mass from the periodic table.
30. Given that the only naturally occurring isotope of phosphorus is ^{31}P, determine its mass from the periodic table.
31. A marble collection has 100 large marbles with a mass of 5.0 g each and 200 small marbles with a mass of 2.0 g each. Calculate (a) the simple average mass, and (b) the weighted average mass of the marble collection.
32. The final grade in a chemistry class is based on homework (10%), quizzes (10%), tests (40%), experiments (20%), and a final exam (20%). What is the weighted av-

‡Answers to odd-numbered Exercises are in Appendix J.

erage score of a student with the following grades: homework, 95; quizzes, 83; tests, 75; experiments, 92; and final exam, 68?

33. Calculate the atomic mass for lithium given the following data for its natural isotopes.

^{6}Li:	6.015 amu	7.42%
^{7}Li:	7.016 amu	92.58%

34. Calculate the atomic mass for magnesium given the following data for its natural isotopes.

^{24}Mg:	23.985 amu	78.70%
^{25}Mg:	24.986 amu	10.13%
^{26}Mg:	25.983 amu	11.17%

35. Calculate the atomic mass for iron given the following data for its natural isotopes.

^{54}Fe:	53.940 amu	5.82%
^{56}Fe:	55.935 amu	91.66%
^{57}Fe:	56.935 amu	2.19%
^{58}Fe:	57.933 amu	0.33%

36. Calculate the atomic mass for zinc given the following data for its natural isotopes.

^{64}Zn:	63.929 amu	48.89%
^{66}Zn:	65.926 amu	27.81%
^{67}Zn:	66.927 amu	4.11%
^{68}Zn:	67.925 amu	18.57%
^{70}Zn:	69.925 amu	0.62%

37. Chlorine has two naturally occurring isotopes: ^{35}Cl and ^{37}Cl. Which isotope is more abundant if the atomic mass of chlorine is 35.45 amu?

38. Bromine has two natural isotopes that occur in approximately equal abundance. If ^{79}Br is one of the isotopes, what is the other isotope if the atomic mass of bromine is 79.90 amu?

The Wave Nature of Light (Sec. 5.6)

39. Which of the following is most energetic: violet, green, or orange light?

40. Which of the following is least energetic: blue, yellow, or red light?

41. Which of the following has the longest wavelength: violet, green, or orange light?

42. Which of the following has the shortest wavelength: blue, yellow, or red light?

43. Which of the following wavelengths of light is most energetic: 650 nm, 550 nm, or 450 nm?

44. Which of the following wavelengths of light is least energetic: 425 nm, 525 nm, or 625 nm?

45. Which of the following wavelengths of light has the highest frequency: 650 nm, 550 nm, or 450 nm?

46. Which of the following wavelengths of light has the lowest frequency: 425 nm, 525 nm, or 625 nm?

The Quantum Concept (Sec. 5.7)

47. What is the quantum particle in light energy?

48. What is the quantum particle in electrical energy?

49. State whether each of the following is continuous or quantized.
(a) a rainbow **(b)** a line spectrum

50. State whether each of the following is continuous or quantized.
(a) a spiral staircase **(b)** an elevated ramp

51. State whether each of the following instruments gives a continuous or a quantized measurement of length.
(a) a metric ruler **(b)** a digital laser

52. State whether each of the following instruments gives a continuous or a quantized measurement of volume.
(a) 10-mL volumetric pipet
(b) 10-mL graduated cylinder

Bohr Model of the Atom (Sec. 5.8)

53. Draw the Bohr model of the atom.

54. What is the experimental evidence for electron energy levels in an atom?

55. Which of the following energy level changes for an electron is most energetic: $5 \rightarrow 2$, $4 \rightarrow 2$, or $3 \rightarrow 2$?

56. Which of the following energy level changes for an electron is least energetic: $4 \rightarrow 1$, $3 \rightarrow 1$, or $\rightarrow 1$?

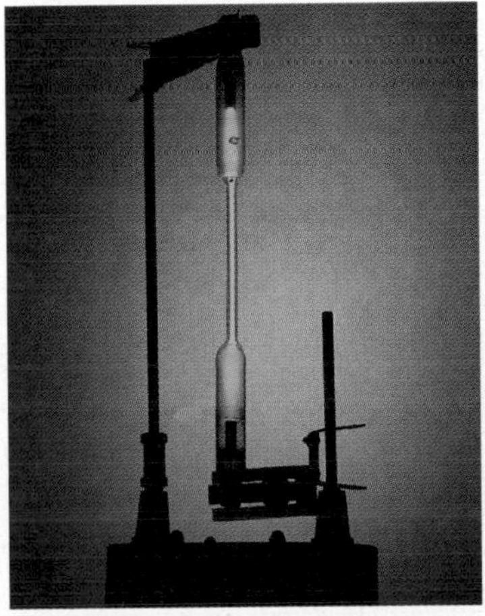

▲ **Mercury Light** The blue glow from mercury vapor illuminates the gas discharge tube.

57. In a hydrogen atom, what color is the emission line observed when the electron drops from the fourth to the second level?
58. In a hydrogen atom, what color is the emission line observed when the electron drops from the fifth to the second level?
59. An electron in a hydrogen atom drops from the fifth energy level to which lower level to emit an ultraviolet photon?
60. An electron in a hydrogen atom drops from the fifth energy level to which lower level to emit an infrared photon?
61. Which of the following lines in the emission spectrum of hydrogen is most energetic: red, blue-green, or violet?
62. Which of the following lines in the emission spectrum of hydrogen has the longest wavelength: red, blue-green, or violet?
63. How many photons of light are emitted for each of the following?
    (a) $1e^-$ drops from energy level 3 to 1
    (b) $1e^-$ drops from energy level 3 to 2
64. How many photons of light are emitted for each of the following?
    (a) $100e^-$ drop from energy level 3 to 2
    (b) $100e^-$ drop from energy level 4 to 2
65. What is the color of the spectral line emitted for each of the following electron energy changes in excited hydrogen gas?
    (a) Electrons drop from energy level 2 to 1.
    (b) Electrons drop from energy level 3 to 2.
    (c) Electrons drop from energy level 4 to 3.
66. What is the color of the spectral line emitted for each of the following electron energy changes in excited hydrogen gas?
    (a) Electrons drop from energy level 5 to 1.
    (b) Electrons drop from energy level 5 to 2.
    (c) Electrons drop from energy level 5 to 3.

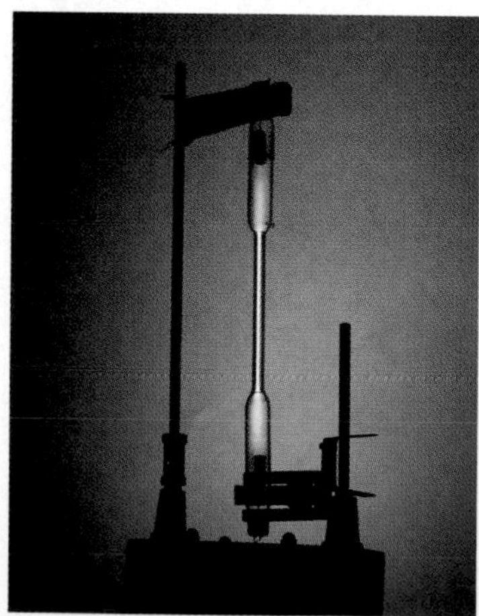

▲ **Nitrogen Light** The yellowish-orange glow from nitrogen gas illuminates the gas discharge tube.

Energy Levels and Sublevels (Sec. 5.9)

67. What experimental evidence suggests the concept of electrons in energy levels?
68. What experimental evidence suggests main energy levels split into sublevels?
69. Designate all the sublevels within each of the following energy levels.
    (a) 1st (b) 2nd
    (c) 3rd (d) 4th
70. State the number of sublevels in each of the following energy levels.
    (a) 1st (b) 3rd
    (c) 5th (d) 6th
71. What is the maximum number of electrons in each of the following sublevels?
    (a) $2s$ (b) $4p$
    (c) $3d$ (d) $5f$
72. What is the maximum number of electrons in each of the following?
    (a) an *s* sublevel (b) a *p* sublevel
    (c) a *d* sublevel (d) an *f* sublevel
73. What is the maximum number of electrons in the second energy level?
74. What is the maximum number of electrons in the fourth energy level?

Electron Configuration (Sec. 5.10)

75. List the order of sublevels from $1s$ through $5p$ according to increasing energy. (*Hint:* Draw a filling diagram.)
76. Draw a filling diagram and predict the sublevel that follows $5p$.
77. Write the predicted electron configuration for each of the following elements.
    (a) He (b) Be
    (c) Co (d) Cd
78. Write the predicted electron configuration for the following elements.
    (a) boron (b) argon
    (c) manganese (d) nickel
79. Which element corresponds to each of the following electron configurations?
    (a) $1s^2\ 2s^1$
    (b) $1s^2\ 2s^2\ 2p^6\ 3s^2\ 3p^2$
    (c) $1s^2\ 2s^2\ 2p^6\ 3s^2\ 3p^6\ 4s^2\ 3d^2$
    (d) $1s^2\ 2s^2\ 2p^6\ 3s^2\ 3p^6\ 4s^2\ 3d^{10}\ 4p^6\ 5s^2$
80. Which element corresponds to each of the following electron configurations?
    (a) $1s^2\ 2s^2\ 2p^5$
    (b) $1s^2\ 2s^2\ 2p^6\ 3s^2\ 3p^6$
    (c) $1s^2\ 2s^2\ 2p^6\ 3s^2\ 3p^6\ 4s^2\ 3d^{10}\ 4p^6\ 5s^2\ 4d^5$
    (d) $1s^2\ 2s^2\ 2p^6\ 3s^2\ 3p^6\ 4s^2\ 3d^{10}\ 4p^6\ 5s^2\ 4d^{10}\ 5p^5$

Quantum Mechanical Model of the Atom (Sec. 5.11)

81. What is the distinction between an orbit and an orbital?
82. What are two significant differences between the Bohr model of the atom and the quantum mechanical model?
83. Sketch a three-dimensional representation for each of the following orbitals. Label the x-, y-, and z-axes.
    (a) $1s$ (b) $2p_x$
    (c) $3p_y$ (d) $4p_z$
84. Sketch a three-dimensional representation for each of the following orbital sets. Label the x-, y-, and z-axes.
    (a) $1s, 2s, 2p_x$ (b) $3p_x, 3p_y, 3p_z$
85. Which orbital in each of the following pairs has the higher energy?
    (a) $2s$ or $3s$ (b) $2p_x$ or $3p_x$
    (c) $2p_x$ or $2p_y$ (d) $4p_y$ or $4p_z$
86. Which orbital in each of the following pairs has the larger size?
    (a) $2s$ or $3s$ (b) $2p_x$ or $3p_x$
    (c) $2p_x$ or $2p_y$ (d) $4p_y$ or $4p_z$
87. Designate the orbital that fits each of the following descriptions.
    (a) a spherical orbital in the fifth energy level
    (b) a dumbbell-shaped orbital in the fourth energy level
88. Designate the orbital that fits each of the following descriptions.
    (a) a spherical orbital in the sixth energy level
    (b) a dumbbell-shaped orbital in the third energy level
89. State the maximum number of electrons that can occupy each of the following orbitals.
    (a) $1s$ (b) $2p$
    (c) $3d$ (d) $4f$
90. State the maximum number of electrons that can occupy each of the following sublevels.
    (a) $1s$ (b) $2p$
    (c) $3d$ (d) $4f$

General Exercises

91. If the electron charge-to-mass ratio is 1.76×10^8 coulomb/g and the absolute charge is 1.60×10^{-19} coulomb, what is the mass of an electron in grams?
92. If the electron charge-to-mass ratio is 9.57×10^4 coulomb/g and the absolute charge is 1.60×10^{-19} coulomb, what is the mass of a proton in grams?
93. Gallium occurs naturally as ^{69}Ga and ^{71}Ga. Given the mass and abundance of ^{69}Ga (68.926 amu and 60.10%), what is the mass of ^{71}Ga?
94. Boron (atomic mass 10.811 amu) occurs naturally as ^{10}B and ^{11}B. Given the mass of the two isotopes (10.013 amu and 11.009 amu), what is the percentage abundance of each isotope?
95. Element 61 was named for the mythological Prometheus who stole fire from heaven. Refer to the periodic table and state whether Pm has any stable isotopes.
96. Element 84 was discovered by Marie Curie and named polonium for her native Poland. Refer to the periodic table and state whether Po is stable or radioactive.
97. Indicate the region of the spectrum (infrared, visible, or ultraviolet) for each of the following wavelengths of light.
    (a) 200 nm (b) 500 nm
    (c) 1200 nm
98. Which of the following light frequencies has the higher energy, 5×10^{10} cycles/s or 5×10^{11} cycles/s?
99. Explain why the electron configuration for copper is $1s^2\,2s^2\,2p^6\,3s^2\,3p^6\,4s^1\,3d^{10}$ rather than the predicted $1s^2\,2s^2\,2p^6\,3s^2\,3p^6\,4s^2\,3d^9$.
100. Explain why the electron configuration for silver is $1s^2\,2s^2\,2p^6\,3s^2\,3p^6\,4s^2\,3d^{10}\,4p^6\,5s^1\,4d^{10}$ rather than the predicted $1s^2\,2s^2\,2p^6\,3s^2\,3p^6\,4s^2\,3d^{10}\,4p^6\,5s^2\,4d^9$.

Explorer Quiz 1
Explorer Quiz 2
Explorer Quiz 3
Master Quiz

CHAPTER 6

The Periodic Table

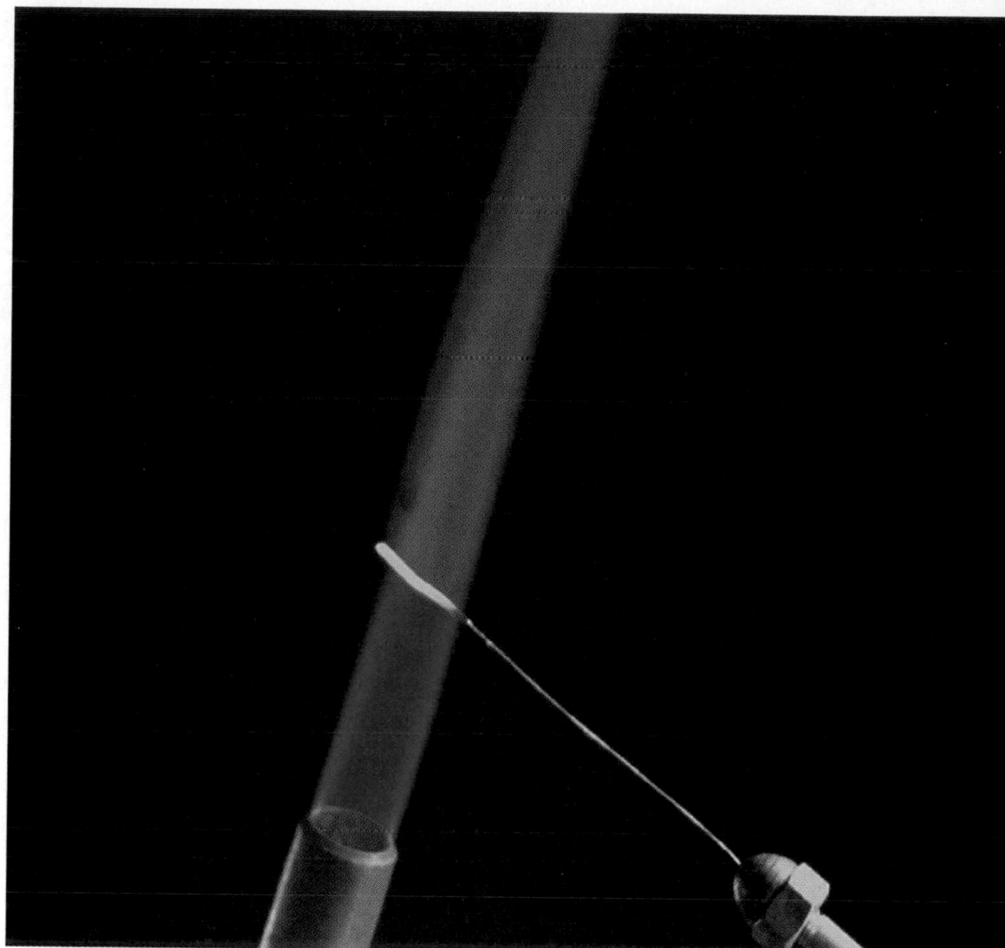

▲ How do electrons produce the color of the flame test given by the element on the wire placed in the flame?

Students enjoy this chapter as the periodic table is the single most identifiable symbol of chemical education. In addition, it reinforces and synthesizes much of the information in previous chapters with a minimum effort. For example, electron configuration is introduced in Chapter 5 after memorizing the filling order of sublevels; in Chapter 6 electron configuration is explained using the positions of sublevels in the periodic table. I hand out a blank periodic table in each lecture and have students take notes and participate in class by filling out trends and asking them to make predictions based on trends in properties.

Flame Tests for Metals Movie

Copper, silver, and gold have been used since prehistoric time. However, not until the 1700s were these metals classified as elements. By 1800, 30 elements had been isolated and identified, and by 1870, there were more than 60 known elements.

In 1829 the German chemist J. W. Döbereiner observed that several elements could be classified into groups of three, or *triads*. All three elements in a triad showed remarkable similarity in their chemical reactions. They also revealed an orderly trend in physical properties such as density, melting point, and especially atomic mass.

In 1865 the English chemist J. A. R. Newlands suggested that the 62 known elements be arranged into groups of seven according to increasing atomic mass. He proposed that an eighth element would then repeat the properties of the first element in the previous group. Interestingly, his theory, called the *law of octaves*, was received with ridicule and was not accepted for publication. Although Newlands' insight into the periodic relationships of the elements was essentially correct, it took 20 years for him to receive professional recognition. In 1887 Newlands graciously accepted the Davy Medal from the Royal Society of Great Britain.

6.1 Classification of Elements

Objective · To state the original periodic law proposed by Mendeleev.

In the 1860s the Russian chemist Dmitri Mendeleev proposed that properties of elements repeat at regular intervals when they are arranged in order of *increasing atomic mass*. Today, Mendeleev is regarded as the architect of the modern periodic table, which depicts a systematic arrangement of the elements.

Mendeleev's achievement was the result of his patient and systematic study of the physical and chemical properties of the elements. Furthermore, he had the insight and courage to predict the existence and properties of three elements before their discovery. In 1869 Mendeleev predicted the existence of an element he called ekasilicon. In 1886 ekasilicon was discovered in Germany and given the name germanium. The properties of ekasilicon predicted by Mendeleev and the observed properties of germanium are compared in Table 6.1.

Table 6.1 Mendeleev's Predictions of the Properties for Ekasilicon (Ek)

Property	Ekasilicon Predicted (1869)	Germanium Discovered (1886)
color	gray	gray
atomic mass	72 amu	72.6 amu
density	5.5 g/mL	5.32 g/mL
melting point	very high	937°C
formula of oxide	EkO_2	GeO_2
density of oxide	4.7 g/mL	4.70 g/mL
formula of chloride	$EkCl_4$	$GeCl_4$
boiling point of chloride	100°C	86°C

In 1869 Mendeleev published his original periodic table with the elements arranged in vertical columns. Two years later, he created an updated version showing the elements in horizontal rows. He arranged the elements by increasing atomic mass and began a new row with an element that repeated the properties of a previous element. Figure 6.1 illustrates a portion of Mendeleev's 1871 periodic table of elements.

The Noble Gases

The arrangement of the elements expanded significantly with the discovery of the group of similar elements that appear on the far right side of the periodic table. Argon was isolated from air in 1894, and helium, neon, krypton, xenon, and radon were discovered in the ensuing 5 years. Originally, this collection of gaseous elements was referred to as the *inert gases* because they showed no chemical reactivity.

In 1962 a compound containing xenon gas was synthesized at the University of British Columbia. After the first compound was made, $PtXeF_6$, several other compounds containing inert gases quickly followed. Although compounds of xenon and krypton have been prepared, the elements argon, neon, and helium have yet to be combined. More recently, the term "noble gas" has been substituted for "inert gas"

Chemistry Connection · Dmitri Mendeleev

Which element in the periodic table is named in honor of Dmitri Mendeleev?

Dmitri Ivanovich Mendeleev (1834–1907) was born in Siberia, the youngest of 14–17 children (records vary). Mendeleev's father was a high school principal, but blindness ended his career when he was quite young. Mendeleev's mother, a woman of remarkable energy and determination, subsequently started a glass factory to support the family. About the time Mendeleev graduated from high school, his father died and his mother's factory burned down. Although destitute, his mother used her influence to get him into college just a few months before her own death.

Mendeleev enrolled at the University of St. Petersburg where he graduated first in his class. Following graduate work in Europe, he returned to St. Petersburg as a professor of chemistry. Mendeleev's consuming interest was finding a common thread that linked the rapidly growing number of elements. In 1869 he published a table that related elements according to increasing atomic mass. He even predicted the existence and properties of three undiscovered elements. In 1874 the predicted element *gallium* was discovered in France (Gallia); in 1876 *scandium* was found in Scandinavia; and in 1886 *germanium* was discovered in Germany.

Gradually, scientists recognized the importance of the periodic table, and Mendeleev became the most famous chemist in the world. His professional reputation thrived, and he was invited to major universities throughout Europe and gave lectures in the United States and Canada.

His personal reputation suffered when he divorced his wife and married a young art student. In the eyes of the Russian Orthodox Church he was a bigamist because he did not wait the required 7 years before he remarried. When Czar Alexander II was questioned about Mendeleev's bigamy, he replied: "Yes, Mendeleev has two wives but I have only one Mendeleev."

In the late nineteenth century, Russia was engulfed in political turmoil. Mendeleev was an outspoken liberal and voiced his concern about the rights of students in general and women in particular. This led to his forced resignation from the university in 1890. Fortunately, Mendeleev still had influential friends, and he was named director of the Bureau of Weights and Measures.

In 1906 he missed winning the Nobel prize by a single vote. Historians speculate that Mendeleev was more deserving of the recognition, but owing to his controversial personality, the prize was awarded instead to Henri Moissan, the French chemist who discovered the element fluorine.

Element 101 is named mendelevium (symbol Md) in honor of Mendeleev.

▶ Figure 6.1 Mendeleev's Classification of the Elements (1871) The formula for the oxide of an element is indicated by the notation R_2O, RO, and so on. For example, Group I oxides include Li_2O, Na_2O, and K_2O; and Group II oxides include BeO, Mgo, and CaO. In Groups III and IV, Mendeleev indicated the existence of three undiscovered elements, which he named ekaboron, ekaaluminum, and ekasilicon.

Group	I	II	III	IV	V	VI	VII	VIII
Formula of Oxide	R_2O	RO	R_2O_3	RO_2	R_2O_5	RO_3	R_2O_7	RO_4
	H							
	Li	Be	B	C	N	O	F	
	Na	Mg	Al	Si	P	S	Cl	
	K	Ca	**eka-**	Ti	V	Cr	Mn	Fe, Co
	Cu	Zn	**eka-**	**eka-**	As	Se	Br	& Ni
	Rb	Sr	Yt	Zr	Nb	Mo	—	Ru, Rh
	Ag	Cd	In	Sn	Sb	Te	I	& Pd
	Cs	Ba	Di	Ce	—	—	—	
	—	—	—	—	—	—	—	
	—	—	Er	La	Ta	W	—	Os, Ir
	Au	Hg	Tl	Pb	Bi	—	—	& Pt
	—	—	—	Th	—	U	—	

to convey the unreactive nature of the gases in this group. Similarly, copper, silver, and gold are referred to as noble metals because of their resistance to chemical reaction.

6.2 The Periodic Law Concept

Objective · To explain the modern periodic law concept proposed by Moseley.

In 1869 Mendeleev proposed that elements showed recurring properties according to increasing atomic mass. In 1913 H. G. J. Moseley (1887–1915), a postdoctoral student, bombarded atomic nuclei with high-energy radiation. By studying the X-rays that were subsequently emitted, Moseley discovered that the nuclear charge increased by 1 for each element in the periodic table. Thus,

H = 1							He = 2
Li = 3	Be = 4	B = 5	C = 6	N = 7	O = 8	F = 9	Ne = 10
Na = 11	Mg = 12	Al = 13	Si = 14	P = 15	S = 16	Cl = 17	Ar = 18

Moseley correctly concluded that arranging elements according to increasing nuclear charge, rather than atomic mass, more clearly explained their repeating properties. That is, arranging elements according to atomic number better explained the trends found in the periodic table. Elements, therefore, should be arranged according to the number of protons in their nucleus and not their atomic mass. The **periodic law** states that the properties of elements recur in a repeating pattern when arranged according to *increasing atomic number*. As it so happens, with only a few exceptions, these trends are identical.

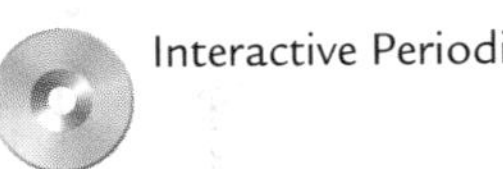
Interactive Periodic Table

In the 1920s, Niels Bohr introduced the concept of electron energy levels, and the periodic table took on a new shape. The new shape resembled the familiar arrangement used today, as shown in Figure 6.2.

Modern Periodic Table of the Elements

Key: Atomic number — 1; Symbol — H; Atomic mass — 1.01

	1 IA	2 IIA	3 IIIB	4 IVB	5 VB	6 VIB	7 VIIB	8 VIII	9 VIII	10 VIII	11 IB	12 IIB	13 IIIA	14 IVA	15 VA	16 VIA	17 VIIA	18 VIIIA
1																		2 He 4.00
2	3 Li 6.94	4 Be 9.01											5 B 10.81	6 C 12.01	7 N 14.01	8 O 16.00	9 F 19.00	10 Ne 20.18
3	11 Na 22.99	12 Mg 24.31											13 Al 26.98	14 Si 28.09	15 P 30.97	16 S 32.07	17 Cl 35.45	18 Ar 39.95
4	19 K 39.10	20 Ca 40.08	21 Sc 44.96	22 Ti 47.88	23 V 50.94	24 Cr 52.00	25 Mn 54.94	26 Fe 55.85	27 Co 58.93	28 Ni 58.69	29 Cu 63.55	30 Zn 65.39	31 Ga 69.72	32 Ge 72.61	33 As 74.92	34 Se 78.96	35 Br 79.90	36 Kr 83.80
5	37 Rb 85.47	38 Sr 87.62	39 Y 88.91	40 Zr 91.22	41 Nb 92.91	42 Mo 95.94	43 Tc (99)	44 Ru 101.07	45 Rh 102.91	46 Pd 106.42	47 Ag 107.87	48 Cd 112.41	49 In 114.82	50 Sn 118.71	51 Sb 121.75	52 Te 127.60	53 I 126.90	54 Xe 131.29
6	55 Cs 132.91	56 Ba 137.33	57 La 138.91	72 Hf 178.49	73 Ta 180.95	74 W 183.85	75 Re 186.21	76 Os 190.2	77 Ir 192.22	78 Pt 195.08	79 Au 196.97	80 Hg 200.59	81 Tl 204.38	82 Pb 207.2	83 Bi 208.98	84 Po (209)	85 At (210)	86 Rn (222)
7	87 Fr (223)	88 Ra (226)	89 Ac (227)	104 — (261)	105 — (262)	106 — (263)	107 — (262)	108 — (265)	109 — (266)	110 — (269)	111 — (272)	112 — (277)		114 — (285)		116 — (289)		118 — (293)

58 Ce 140.12	59 Pr 140.91	60 Nd 144.24	61 Pm (147)	62 Sm 150.36	63 Eu 151.97	64 Gd 157.25	65 Tb 158.93	66 Dy 162.50	67 Ho 164.93	68 Er 167.26	69 Tm 168.93	70 Yb 173.04	71 Lu 174.97
90 Th 232.04	91 Pa (231)	92 U 238.03	93 Np (237)	94 Pu (244)	95 Am (243)	96 Cm (247)	97 Bk (247)	98 Cf (251)	99 Es (252)	100 Fm (257)	101 Md (258)	102 No (259)	103 Lr (260)

▲ **Figure 6.2 The Modern Periodic Table of Elements** Atomic numbers increase stepwise throughout the periodic table. Note that atomic masses, with few exceptions, also increase. However, the nuclear charge (that is, the atomic number), more clearly explains the recurring properties of elements.

If you closely examine the fourth row in the periodic table, you will notice that the sequence of atomic masses does not increase from Co to Ni, even though Co precedes Ni in the periodic table. This is because the atomic number of Co (27) is less than the atomic number of Ni (28). It is also true that the properties of cobalt resemble those of the elements in Group 9 and that the properties of Ni are similar to those of the elements in Group 10.

Example Exercise 6.1 • Periodic Law

Find the two elements in the fifth row of the periodic table that violate the original periodic law proposed by Mendeleev.

Solution

Mendeleev proposed that elements be arranged according to increasing atomic mass. Beginning with Rb, each of the elements in the fifth row increases in atomic mass until iodine. Although the atomic numbers of Te (52) and I (53) increase, the atomic masses of Te (127.60) and I (126.90) do not. Experimentally, it is I, and not Te, whose properties are similar to those of F, Cl, and Br.

Self-Test Exercise

Find a pair of elements in the periodic table with atomic numbers less than 20 that are an exception to the original periodic law.

Answers: Ar and K

6.3 Groups and Periods of Elements

Objectives

- To apply the following terms to the periodic table of elements:
  - (a) groups (families) and periods (series)
  - (b) representative elements and transition elements
  - (c) metals, semimetals, and nonmetals
  - (d) alkali metals, alkaline earth metals, halogens, and noble gases
  - (e) lanthanide series and actinide series
  - (f) rare earth elements and transuranium elements
- To designate a group of elements in the periodic table using both the American convention (IA–VIIIA) and the IUPAC convention (1–18).

A vertical column in the periodic table is called a **group** or family of elements, and a horizontal row is called a **period** or series. When we examine the periodic table on the inside front cover of this textbook, we notice there are seven horizontal rows of elements. These periods of elements are numbered 1–7. The first period has only 2 elements, H and He. The second and third periods each have 8 elements, Li through Ne, and Na through Ar. The fourth and fifth periods have 18 elements, K through Kr and Rb through Xe.

The element hydrogen occupies a special position in the periodic table. We know that hydrogen is a gas and has properties similar to those of other nonmetals. But hydrogen can react by losing its one electron—and giving up electrons is a property of metals. In most references H appears on the far left of the periodic table. Alternately, some textbooks place H on both sides of the periodic table. In this book, we will recognize the ambiguous behavior of hydrogen and place H in the middle of the periodic table.

There are 18 vertical columns in the periodic table. In the past, American chemists have used a Roman numeral followed by the letter A or B to designate a group of elements, for example, IA, IIA, IIB. In the periodic table, you will notice that Group IA contains the elements Li to Fr, Group IIA has the elements Be through Ra, and Group IIB has the elements Zn, Cd, Hg.

Update • Official IUPAC Group Numbers

In which two groups in the periodic table is hydrogen often placed?

Recently, the designation of groups of elements in the periodic table has been a topic of interest. In the 1970s a lingering controversy surfaced that was addressed by IUPAC. Previously, in the United States, elements on the left side and right side of the periodic table had been designated A groups. The elements in the middle of the periodic table were labeled B groups. This so-called American convention is shown below.

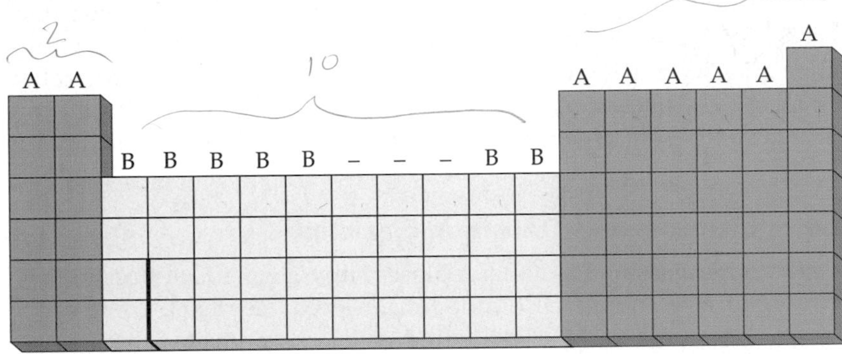

◀ **Periodic Table of Elements—American Convention** Representative elements are placed in A groups and transition elements in B groups.

In England and in the rest of Europe, elements on the left side of the periodic table had been previously designated A groups, and those on the right side had been labeled B groups. The so-called European convention is shown in the following illustration.

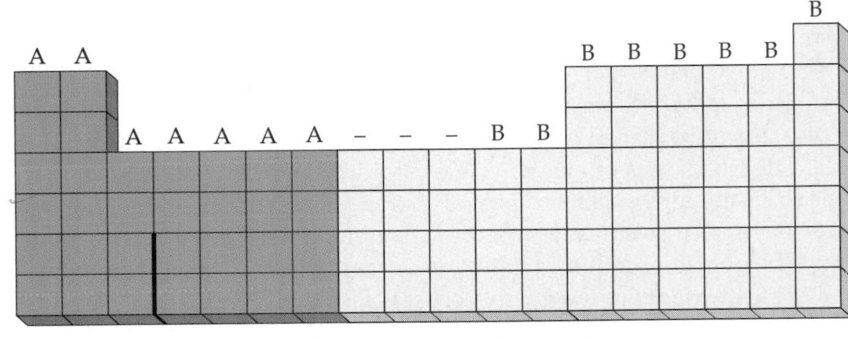

◀ **Periodic Table of Elements—European Convention** The elements on the left are A groups, and the elements on the right are B groups.

After much discussion, in 1985 IUPAC proposed that groups of elements be designated by the numerals 1 through 18. Currently, IUPAC recommends that groups of elements be numbered as shown in the table below.

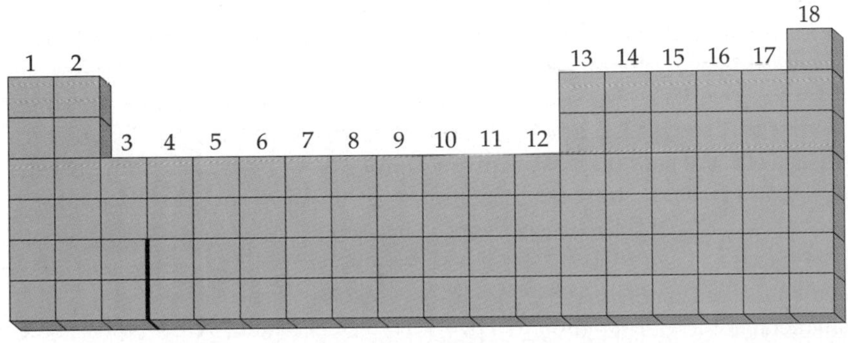

◀ **Periodic Table of Elements—IUPAC Convention** Groups of elements are indicated by the sequential numerals 1 through 18.

One of the reasons for IUPAC's recommendation is that the last digit in the group number corresponds to the Roman numeral in both the American and European conventions. For example, in group 13, the number 3 corresponds to the Roman numeral III in both the American (IIIA) and European (IIIB) versions.

Hydrogen is often placed in Groups IA/1 and VIIA/17; however, hydrogen is neither an alkali metal nor a halogen.

Since the 1920s, a controversy has existed over the numbering of the groups in the periodic table. Recently, the International Union of Pure and Applied Chemistry (IUPAC) resolved the dispute by proposing to number the groups using the numerals 1 to 18. For example, IUPAC recommends that Group IA be designated Group 1, Group IIB be designated Group 12, and Group VIIIA be designated Group 18. Currently, chemists are in the process of adopting the IUPAC convention. Hence, the periodic tables in this textbook list both conventions, for example, IA/1, IIB/12, and VIIIA/18.

In the periodic table, groups of elements can also be referred to by their family names. The family name for Group IA/1 is **alkali metals**. The elements in Group IIA/2 are called **alkaline earth metals**. The elements in Group VIIA/17 are called **halogens**. Group VIIIA/18 elements on the far right side of the periodic table are all gases that are usually unreactive under normal conditions; they are called **noble gases**.

Representative Elements and Transition Elements

Groups of elements are classified as either representative elements or transition elements. **Representative elements** (also called main-group elements) are found in A groups, on the left and right sides of the periodic table. As a rule, the chemical behavior of representative elements is predictable. For example, magnesium always reacts with oxygen to produce MgO.

Transition elements are found in the B groups in the middle of the periodic table. The chemical behavior of transition elements is not as predictable as that of representative elements. For instance, in the presence of limited oxygen, iron reacts to form FeO; if excess oxygen gas is available, the product is Fe_2O_3.

Inner transition elements are found beneath the main portion of the periodic table. These two series are placed below the main portion to avoid an unduly wide periodic table. The first series of elements, Ce through Lu, lie between La and Hf in the periodic table and are considered part of Period 6. This series is called the **lanthanide series** because it follows lanthanum. The elements Ce–Lu have similar properties, and they occur together in nature along with Sc, Y, and La. The natural abundance of these elements in the Earth's crust is less than 0.005%, and they are collectively referred to as **rare earth elements**.

The second series of inner transition elements, Th through Lr, is called the **actinide series** because it follows element 89, actinium. This series, Th–Lr, is considered a part of Period 7. All the elements in this series are radioactive; in fact, except for trace amounts, none of the elements past uranium is naturally occurring. Element 93 and beyond are the result of high-energy synthesis in particle accelerators called "atom smashers." The isotopes of these elements have very short lifetimes, often less than a millisecond. The elements following uranium, Np through Lr through element 118, are called **transuranium elements**. The periodic table shown in Figure 6.3 summarizes the names of the important groups and periods.

Example Exercise 6.2 • Groups and Periods of Elements

Select the symbol of the element that fits each of the following descriptions.

(a) the alkali metal in the fourth period
(b) the halogen in the third period
(c) the rare earth with the lowest atomic mass
(d) the metal in Group VIIB/7 and Period 4

Solution

Referring to the periodic table in Figure 6.3, we have

(a) K (b) Cl (c) Sc (d) Mn

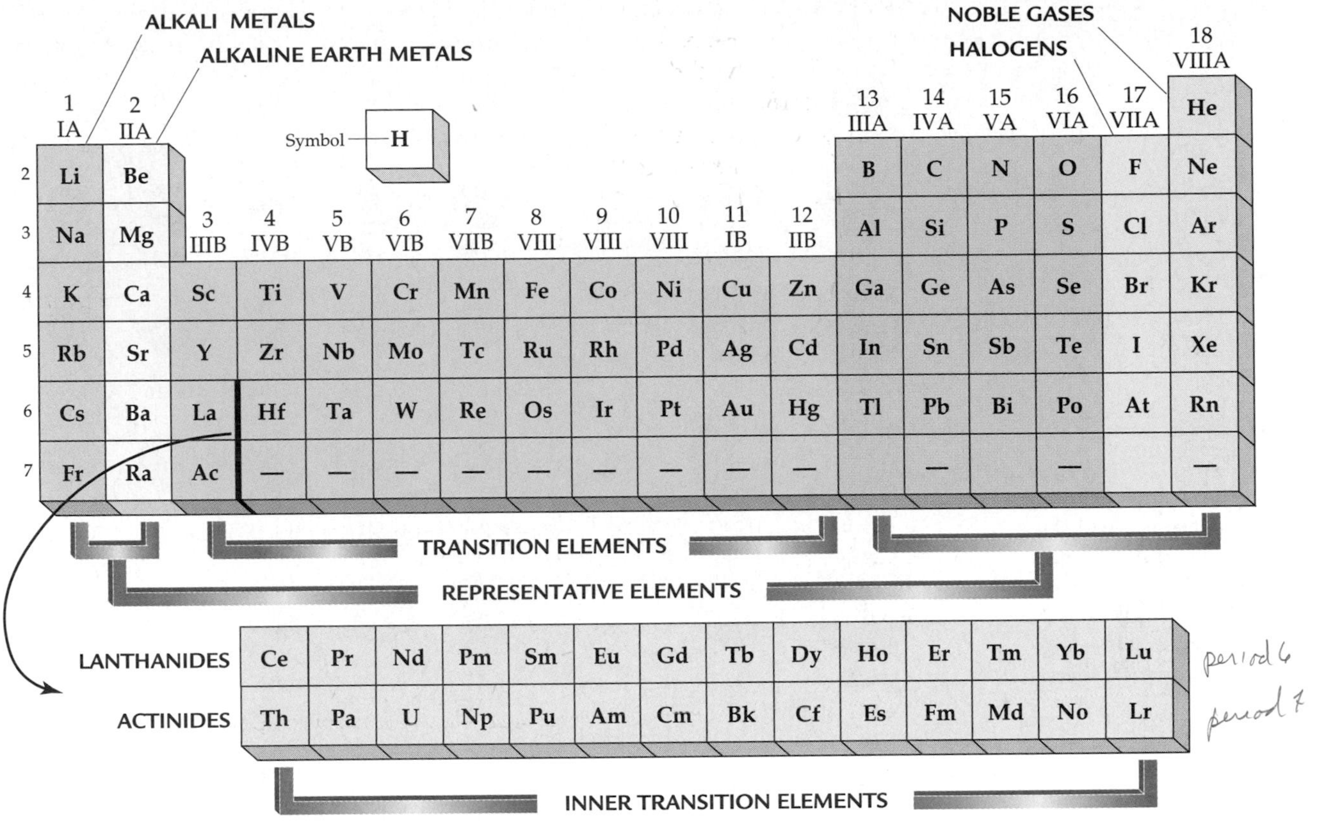

ALKALI METALS
ALKALINE EARTH METALS
NOBLE GASES
HALOGENS

Symbol — H

	1 IA	2 IIA	3 IIIB	4 IVB	5 VB	6 VIB	7 VIIB	8 VIII	9 VIII	10 VIII	11 IB	12 IIB	13 IIIA	14 IVA	15 VA	16 VIA	17 VIIA	18 VIIIA
																		He
2	Li	Be											B	C	N	O	F	Ne
3	Na	Mg											Al	Si	P	S	Cl	Ar
4	K	Ca	Sc	Ti	V	Cr	Mn	Fe	Co	Ni	Cu	Zn	Ga	Ge	As	Se	Br	Kr
5	Rb	Sr	Y	Zr	Nb	Mo	Tc	Ru	Rh	Pd	Ag	Cd	In	Sn	Sb	Te	I	Xe
6	Cs	Ba	La	Hf	Ta	W	Re	Os	Ir	Pt	Au	Hg	Tl	Pb	Bi	Po	At	Rn
7	Fr	Ra	Ac	—	—	—	—	—	—	—	—	—		—		—		—

TRANSITION ELEMENTS

REPRESENTATIVE ELEMENTS

LANTHANIDES	Ce	Pr	Nd	Pm	Sm	Eu	Gd	Tb	Dy	Ho	Er	Tm	Yb	Lu
ACTINIDES	Th	Pa	U	Np	Pu	Am	Cm	Bk	Cf	Es	Fm	Md	No	Lr

INNER TRANSITION ELEMENTS

▲ **Figure 6.3 Names of Groups and Periods** The common names of groups and periods are shown for selected families and series.

Names of Groups and Periods

Self-Test Exercise

Select the symbol of the element that fits each of the following descriptions.

(a) the alkaline earth metal in the sixth period
(b) the noble gas in the third period
(c) the actinide with the highest atomic mass
(d) the semimetal in Group IIIA/13

Answers: (a) Ba; (b) Ar; (c) Lr; (d) B

6.4 Periodic Trends

Objectives
- To state the trend in atomic size within a group or period of elements.
- To state the trend in metallic character within a group or period.

We can visualize atoms as spheres and express their size in terms of atomic radius. The atomic radius is the distance from the nucleus to the outermost electrons. As this distance is very small, we will express the atomic radius in units of nanometers.

Periodic Properties Movie; Periodic Trends: Atomic Radii Movie

In the periodic table there are two general trends in atomic size. First, as we move up a group of elements from bottom to top, the radii of the atoms decrease. We can explain the decrease in radius because there are fewer energy levels of electrons surrounding the nucleus. As the number of energy levels decreases, the distance from the nucleus to the outermost electrons decreases. Therefore, the *atomic radius decreases up a group*.

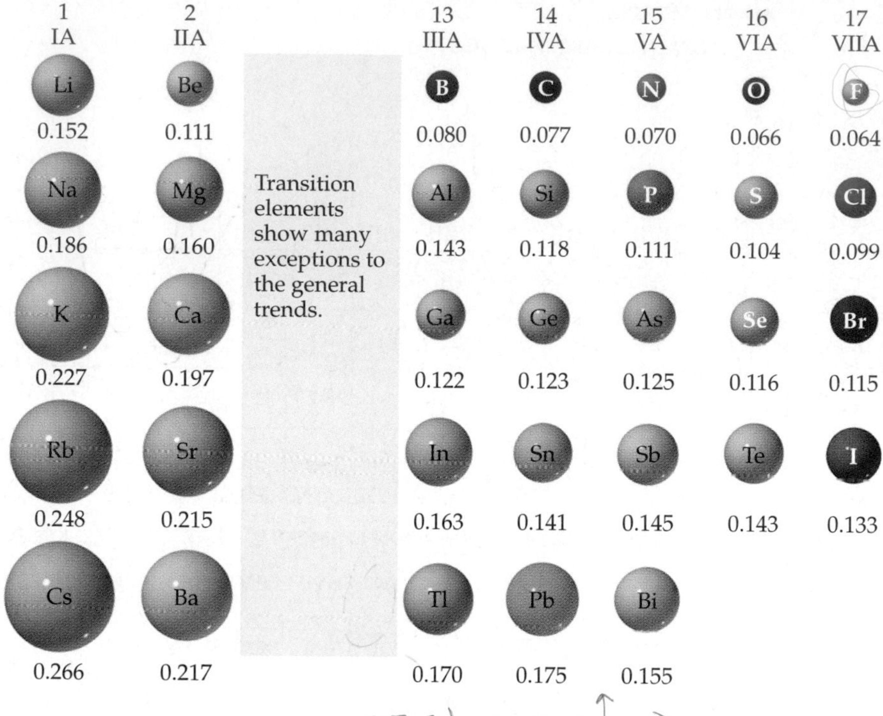

Atomic Radii of Representative Elements

▶ **Figure 6.4 Atomic Radii of Selected Representative Elements** Atomic radii decrease up a group and across a period. The values for atomic radii are given in nanometer; there are 1,000,000,000 nm in a meter.

Second, as we move left to right within a period of elements, the radii of the atoms decrease. The explanation for this decrease is as follows. The atomic number increases from left to right, and therefore the number of protons increases. As the number of protons in the nucleus increases across a period, the nuclear charge of the elements increases. This has the effect of pulling the electrons closer to the nucleus and reducing the size of the atom. Thus, the *atomic radius decreases across a period* from left to right.

Figure 6.4 illustrates trends in the size of atomic radii for a portion of the periodic table. Transition elements and inner transition elements have been omitted as there are numerous exceptions to the general trends. Moreover, the atomic radii of transition elements in a given sublevel are reasonably similar in size.

The periodic table also shows two general trends in the degree of metallic character of the elements. First, recall that metals are on the left side of the periodic table and nonmetals are on the right side. Thus, as we move from left to right, *metallic character decreases across a period.*

Second, a metal characteristically reacts by losing one or more of its outermost electrons to a nonmetal. As we move up a group, the outermost electrons of an atom are closer to the nucleus. When the electrons and the nucleus are closer together, it is more difficult for an atom to lose an electron. That is, as the distance between the negatively charged electrons and the positively charged nucleus becomes less, the tendency for the metal to lose an electron is less. Therefore, *metallic character decreases up a group.* Figure 6.5 illustrates the general trends in the periodic table for atomic radius and metallic character.

Example Exercise 6.3 • Periodic Table Predictions

Based on the general trends in the periodic table, predict which element in each of the following pairs has the smaller atomic radius.

(a) Na or K (b) P or N (c) Ca or Ni (d) Si or S

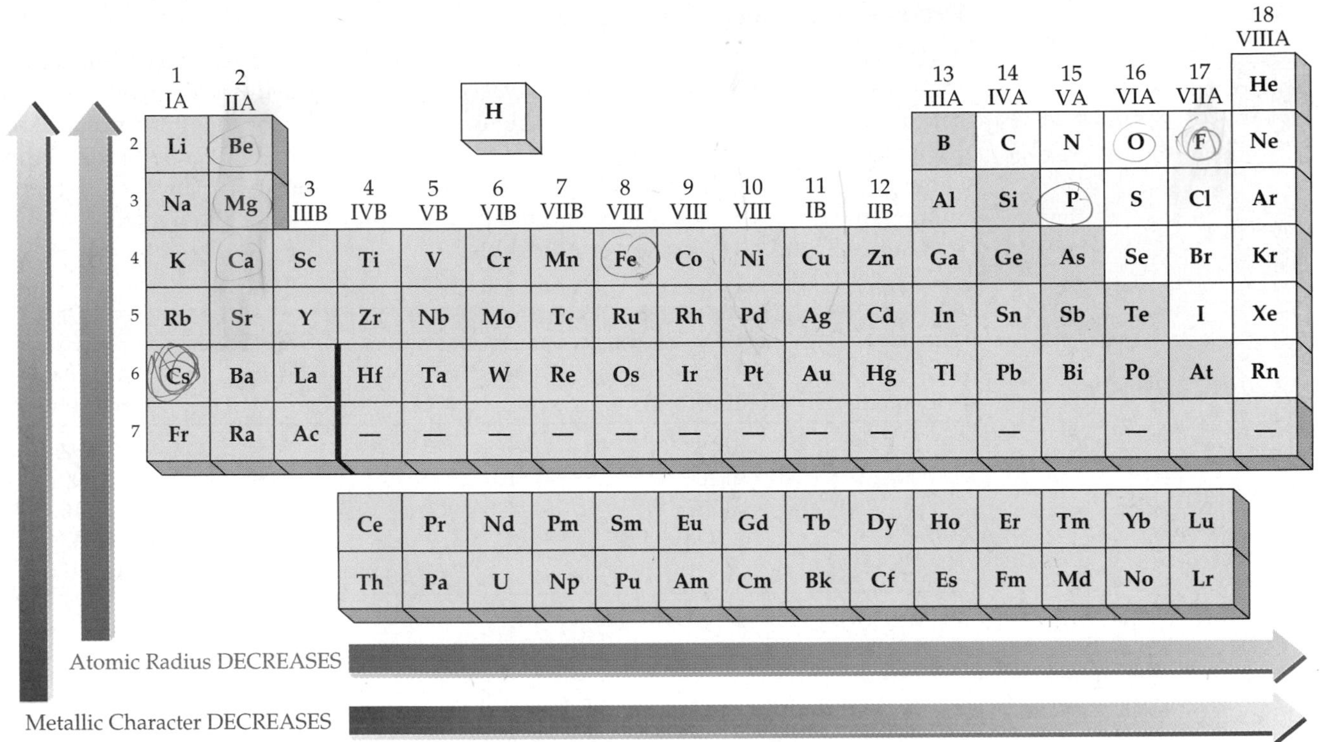

▲ **Figure 6.5 General Trends in Atomic Radii and Metallic Character** The general trend for the atomic radius is to decrease both up a group and across a period from left to right. The metallic character trend in the periodic table is similar.

Solution

In general, atomic radius decreases up a group and across a period from left to right. Referring to the periodic table,

(a) Na is above K in Group IA/1; the atomic radius of Na is smaller.
(b) N is above P in Group VA/5; the atomic radius of N is smaller.
(c) Ni is to the right of Ca in Period 4; the atomic radius of Ni is smaller.
(d) S is to the right of Si in Period 3; the atomic radius of s is smaller.

Self-Test Exercise

Based on the general trends in the periodic table, predict which element in each of the following pairs has the most metallic character.

(a) Sn or Pb (b) Ag or Sr (c) Al or B (d) Br or As

Answers: (a) Pb; (b) Sr; (c) Al; (d) As

6.5 Properties of Elements

Objectives
- To predict a physical property for an element given the values of other elements in the same group.
- To predict a chemical formula for a compound given the formulas of other compounds containing an element in the same group.

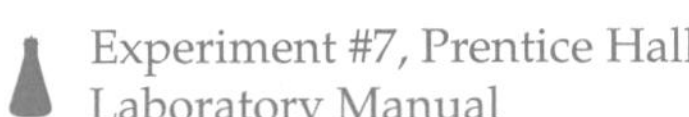
Experiment #7, Prentice Hall Laboratory Manual

Mendeleev designed the periodic table so that elements with similar properties were placed in the same group. He was then able to predict yet undiscovered elements from the gaps in his periodic table. He successfully predicted the properties of ekasilicon, for example, after noticing a trend in properties such as color, density, melting point, and atomic mass.

Predicting Physical Properties

Sodium and Potassium in Water Movie

Today, chemists have access to information regarding properties of elements cataloged in reference books such as the *Handbook of Chemistry and Physics*. However, we can use the same procedure that Mendeleev used to estimate the numerical value for a physical property. For example, if we know the density of two elements in the same group, we can make a reasonable estimate of the density for another element in the same group. Table 6.2 lists selected physical properties of alkali metals, which appear in Group IA/1 in the periodic table.

Table 6.2 Physical Properties of Alkali Metals

Element	Atomic Radius (nm)	Density (g/mL at 20°C)	Melting Point (°C)	Atomic Mass (amu)
Li	0.152	0.53	180.5	6.94
Na	0.186	0.97	97.8	22.99
K	0.227	0.86*	63.3	39.10
Rb	0.248	1.53	38.9	85.47
Cs	0.266	1.87	28.4	132.91

*Notice that the density of K is less than that of Na. Small irregularities in group trends are not unusual.

Notice that the properties of the radioactive element francium, Fr, are not included with those of the other alkali metals in Table 6.2. We can, however, make some reasonable predictions about its physical properties based on the trends shown by the other elements. Since Fr is below Cs in Group IA/1, we can predict that its atomic radius is greater than 0.266 nm, its density is greater than 1.87 g/mL, its melting point is less than 28.4°C, and its atomic mass is greater than 132.91 amu.

Example Exercise 6.4 • Predicting Physical Properties

Predict the missing value (?) for each physical property listed below. The (a) atomic radius, (b) density, and (c) melting point are given for two of three alkaline earth metals in Group IIA/2.

Element	Atomic Radius	Density at 20°C	Melting Point
Ca	0.197 nm	1.54 g/mL	(?)°C
Sr	0.215 nm	(?) g/mL	769°C
Ba	(?) nm	3.65 g/mL	725°C

Solution

We can estimate a value for the physical property of an element by observing the trend in values for other elements within the same group.

(a) To determine the atomic radius value for Ba, we first find the increase from Ca to Sr; that is, 0.215 nm − 0.197 nm = 0.018 nm. We then add this difference (0.018 nm) to the atomic radius of Sr and obtain 0.215 nm + 0.018 nm = 0.233 nm. Note that we assumed that the atomic radius increases the same amount from Sr to Ba as it does from Ca to Ba. (*The literature value is 0.217 nm.*)

(b) Notice that Sr lies between Ca and Ba in Group IIA/2. We can thus estimate that the density of Sr lies midway between that of Ca and Ba. To find the density of Sr, we calculate the average value for Ca and Ba; that is, (1.54 + 3.65)/2 = 2.60 g/mL. (*The literature value is 2.63 g/mL.*)

(c) From the general trend, we can predict that the melting point of Ca is greater than that of Sr. To determine the value, let's find the increase in melting point from Ba to Sr. It is 769°C − 725°C − 44°C. Then we add 44°C to the value of Sr:

$769°C + 44°C = 813°C$. We therefore predict the melting point of Ca as 813°C. (*The literature value is 839°C.*)

Self-Test Exercise

Predict the missing value (?) for each physical property listed below. The (a) atomic radius, (b) density, and (c) melting point are given for two of the metals in Group VIII/10.

Element	Atomic Radius	Density at 20°C	Melting Point
Ni	0.125 nm	8.91 g/cm^3	(?)°C
Pd	0.138 nm	(?) g/cm^3	1554°C
Pt	(?) nm	21.5 g/cm^3	1772°C

Answers: (a) 0.151 nm; (b) 15.2 g/cm^3; (c) 1336°C

Predicting Chemical Properties

In Chapter 8 we will study chemical reactions systematically. However, it is possible to predict the products of chemical reactions by understanding the organization of elements in the periodic table. For example, if we know that magnesium and oxygen react to give magnesium oxide (MgO), we can predict that the other Group IIA/2 elements react in a similar fashion. That is, calcium, strontium, and barium should react with oxygen to give similar oxides (CaO, SrO, and BaO), and they do. There are exceptions to the general rule, but the principle of using the periodic table is helpful to our understanding.

Example Exercise 6.5 • Predicting Chemical Properties

Metallic sodium reacts with chlorine gas to give sodium chloride, NaCl. Predict the products formed when (a) lithium and (b) potassium react with chlorine gas.

Solution

Since Li and K are in the same group as Na (Group IA/1), we can predict that the products are similar to NaCl. Thus,

(a) Lithium metal should react with chlorine gas to give LiCl.
(b) Potassium metal should react with chlorine gas to give KCl.

Self-Test Exercise

The chemical formulas for the oxides of potassium, calcium, gallium, and germanium are, respectively: K_2O, CaO, Ga_2O_3, and GeO_2. Refer to the periodic table and predict the chemical formula for each of the following compounds.

(a) rubidium oxide (b) strontium oxide
(c) indium oxide (d) tin oxide

Answers: (a) Rb_2O; (b) SrO; (c) In_2O_3; (d) SnO_2

6.6 Blocks of Elements

Objectives
- To predict the highest energy sublevel for an element given its position in the periodic table.
- To predict the electron configuration for an element given its position in the periodic table.

In Section 5.9 we learned that the order of electron energy sublevels according to increasing energy is $1s < 2s < 2p < 3s < 3p < 4s < 3d < 4p < 5s < 4d < 5p$. In this section, we will see that the unusual shape of the periodic table is the result of the ordering of energy sublevels. That is, the order of energy sublevels follows the systematic arrangement of elements by groups.

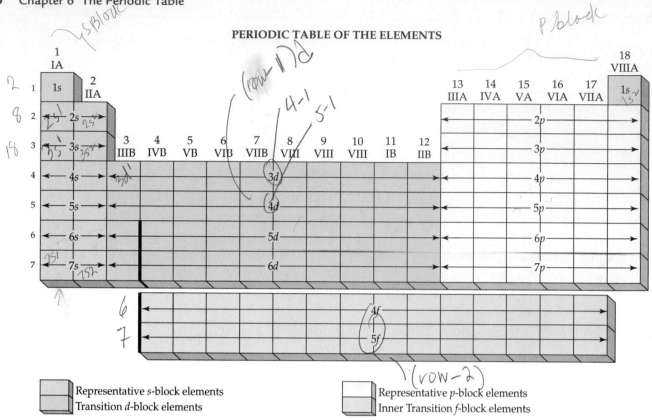

PERIODIC TABLE OF THE ELEMENTS

Representative *s*-block elements
Transition *d*-block elements
Representative *p*-block elements
Inner Transition *f*-block elements

Blocks of Elements in the Periodic Table

▲ **Figure 6.6 Blocks of Elements In the Periodic Table** The relationship between energy sublevels and the *s*, *p*, *d*, and *f* blocks of elements in the periodic table.

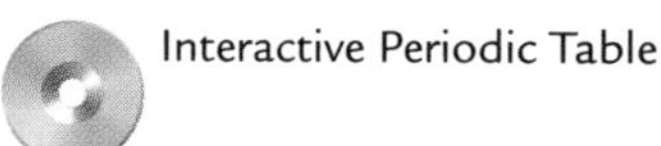
Interactive Periodic Table

Elements in Groups IA/1 and IIA/2 fill *s* sublevels. The *s* sublevels are to the left in the periodic table and are collectively called the *s block* of elements. Elements in Groups IIIA/13 through VIIIA/18 fill *p* sublevels, and are referred to as the *p block* of elements. Transition elements in Groups IIIB/3 through IIB/12 fill *d* sublevels and likewise make up the *d block* of elements. Finally, the inner transition elements (Ce–Lu and Th–Lr) fill *f* sublevels and comprise the *f block* of elements. Figure 6.6 illustrates the *s*-, *p*-, *d*-, and *f*-block elements in the periodic table.

Previously, we had to memorize that the 3*d* energy sublevel comes after the 4*s* energy sublevel. Now, we can simply refer to the periodic table and see that *s*-block elements precede each *d* sublevel. Also, recall that the number of sublevels corresponds to number of the main energy level. For example, in the third main level, there are three sublevels (3*s*, 3*p*, and 3*d*). In the fourth main level, there are four sublevels (4*s*, 4*p*, 4*d*, and 4*f*).

We can see in Figure 6.6 that members of the lanthanide and actinide series are *f*-block elements. The lanthanides are filling the 4*f* energy sublevel, and the actinides are filling the 5*f* sublevel.

Example Exercise 6.6 • Energy Sublevels and the Periodic Table

State the highest energy sublevel in each of the following elements.

(a) H (b) S (c) Ni (d) U

Solution

Refer to the periodic table and determine the energy sublevel based on the period and the block of elements.

(a) Hydrogen has only one electron; thus, H is filling a 1*s* sublevel.
(b) Sulfur is in third period and is a *p*-block element; S is filling a 3*p* sublevel.
(c) Nickel is in the first series of *d*-block elements; Ni is filling a 3*d* sublevel.
(d) Uranium is in the second series of *f*-block elements; U is filling a 5*f* sublevel.

Self-Test Exercise

State the energy sublevel being filled in each of the following series of elements.

(a) Cs–Ba (b) Y–Cd (c) In–Xe (d) Ce–Lu

Answers: (a) 6*s*; (b) 4*d*; (c) 5*p*; (d) 4*f*

In Section 5.10, we learned to write the **electron configuration** for an element after memorizing the order of sublevel filling. Now that we understand the relationship of sublevels in the periodic table, we can write electron configurations using the periodic table.

The electron configuration for Na (atomic number 11) is $1s^2\ 2s^2\ 2p^6\ 3s^1$. As a shorthand method, we can abbreviate the electron configuration by indicating the innermost electrons with the symbol of the preceding noble gas. The preceding noble gas with an atomic number less than 11 is neon (atomic number 10). The symbol for neon is placed in brackets, [Ne], followed by the outermost electrons. That is, the electron configuration for Na can be written as [Ne] $3s^1$.

Example Exercise 6.7 • Electron Configuration and the Periodic Table

Refer to a periodic table and write the predicted electron configuration for each of the following elements.

(a) P (b) Co

Solution

Now that you understand blocks of elements in the periodic table, you can predict the order of sublevels according to increasing energy.

(a) Phosphorus is the third element in the 3*p* sublevel. The electron configuration for P is $1s^2\ 2s^2\ 2p^6\ 3s^2\ 3p^3$, or [Ne] $3s^2\ 3p^3$.
(b) Cobalt is the seventh element in the 3*d* sublevel. The electron configuration for Co is $1s^2\ 2s^2\ 2p^6\ 3s^2\ 3p^6\ 4s^2\ 3d^7$ or [Ar] $4s^2\ 3d^7$.

Self-Test Exercise

Refer to a periodic table and write the predicted electron configuration for each of the following elements.

(a) Zn (b) I

Answers:

(a) $1s^2\ 2s^2\ 2p^6\ 3s^2\ 3p^6\ 4s^2\ 3d^{10}$ or [Ar] $4s^2\ 3d^{10}$
(b) $1s^2\ 2s^2\ 2p^6\ 3s^2\ 3p^6\ 4s^2\ 3d^{10}\ 4p^6\ 5s^2\ 4d^{10}\ 5p^5$ or [Kr] $5s^2\ 4d^{10}\ 5p^5$

Note There are a few exceptions in the predicted electron configurations for some of the elements. That is, some elements have actual electron configurations that vary from their predicted arrangements. On occasion, electrons in a lower energy sublevel jump to an unfilled sublevel of slightly higher energy. Some of these exceptions are easy to explain; others are not. For our purposes, we will disregard these exceptions and write electron configurations based on the general predictions for the order of sublevel filling.

6.7 Valence Electrons

Objective · To predict the number of valence electrons for any representative element.

When an atom undergoes a chemical reaction, only the outermost electrons are involved. These electrons are highest in energy, farthest from the nucleus, and more available to react with another atom. These outermost electrons are called **valence electrons**. Valence electrons form chemical bonds between atoms and are responsible for the chemical behavior of the element. For our purposes, we will ignore transition elements and consider only representative elements. In these cases, the number of valence electrons is equal to the total number of electrons in the outermost *s* and *p* sublevels.

We can predict the number of valence electrons by noting the group number in the periodic table. Using the Roman numeral convention, we see that the group number is identical to the total number of valence electrons. That is, elements in Group IA have 1 valence electron, and elements in Group IIA have 2 valence electrons. Group IIIA elements have 2 electrons from an *s* sublevel plus 1 electron from a *p* sublevel, for a total of 3 valence electrons. Group IVA elements have 4 valence electrons; Group VA elements have 5 valence electrons; and so on.

With the IUPAC designations for group numbers, the last digit (for example, the 3 in 13) indicates the number of valence electrons. Group 1 elements have 1 valence electron; Group 2 elements have 2 valence electrons; Group 13 elements have 3 valence electrons; Group 14 elements have 4 valence electrons; and Group 18 elements have 8 valence electrons. Note that the *s* and *p* sublevels can have a maximum of 2 and 6 electrons, respectively. Thus, the maximum number of valence electrons is $8(2 + 6)$.

Lithium is in Group IA/1, and so a lithium atom has only 1 valence electron. Oxygen is in Group VIA/16 and has 6 valence electrons. The following example further illustrates the relationship between group number and number of valence electrons.

Example Exercise 6.8 • Valence Electrons and the Periodic Table

Refer to the periodic table and predict the number of valence electrons for an atom of each of the following representative elements.

(a) Na (b) Al (c) S (d) Xe

Solution

Find the element in the periodic table, note the group number, and indicate the number of valence electrons.

(a) Since sodium is in Group IA/1, Na has 1 valence electron.
(b) Aluminum is in Group IIIA/13, and so Al has 3 valence electrons.
(c) Sulfur is in Group VIA/16, and so S has 6 valence electrons.
(d) Xenon is in Group VIIIA/18, and so Xe has 8 valence electrons.

Self-Test Exercise

Refer to the periodic table and state the number of valence electrons for any element in each of the following groups.

(a) Group IIA (b) Group VA (c) Group 14 (d) Group 17

Answers: (a) 2; (b) 5; (c) 4; (d) 7

Note We can quickly predict the number of valence electrons for any main-group element by referring to its group number in the periodic table. However, the transition elements are filling *d* sublevels, not an *s* or *p* sublevel, which complicates the discussion. Therefore, we will not predict the number of valence electrons for a transition element.

6.8 Electron Dot Formulas

Objective · To draw the electron dot formula for any representative element.

We said previously that only valence electrons are involved when an element undergoes a chemical reaction. That's because valence electrons are farthest from the nucleus and are the most accessible. To keep track of valence electrons, chemists have devised a notation called the electron dot formula. Electron dot formulas are also referred to as Lewis structures in honor of G. N. Lewis, a famous American chemist who conceived the notion of outer electrons.

An **electron dot formula** shows the symbol of an element surrounded by its valence electrons. The symbol of the element represents the **core** (or kernel) of the atom. That is, it represents the nucleus and the inner electrons. Dots are placed around the symbol to represent the valence electrons. Figure 6.7 illustrates the general symbol for an electron dot formula.

In practice, we will use the following guidelines for drawing electron dot formulas.

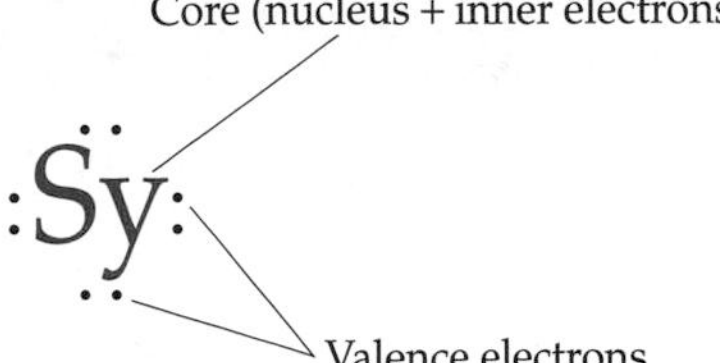

▲ **Figure 6.7 Representation of Electron Dot Formulas** The electron dot formula for an atom of an element uses the symbol for the element to represent the core of the atom. Each valence electron is represented by a dot arranged around the symbol.

Guidelines for Drawing Electron Dot Formulas of Atoms

1. Write the symbol of the element to represent the core of the atom.
2. By convention, assume that we can draw a maximum of two dots on each side of the symbol. A dot represents 1 valence electron, and the maximum number is 8 electrons (s sublevel + p sublevel = $2e^- + 6e^- = 8e^-$).
3. Determine the number of valence electrons from the group number of the element in the periodic table. Draw one dot around the symbol for each valence electron.

Although there is no absolute rule for placing dots, the first dot is usually placed to the right of the symbol, the second dot beneath the symbol, the third dot to the left of the symbol, and the fourth dot above the symbol. The fifth dot pairs with the first dot. The sixth, seventh, and eighth dots pair with the second, third, and fourth dots, respectively, moving clockwise around the symbol.

As an example, consider the element phosphorus. Phosphorus is in Group VA/15 and therefore has 5 valence electrons. We can draw the electron dot formula for phosphorus as follows. First, write the symbol for phosphorus; then add five dots, one at a time, as follows.

$$\underset{\text{core}}{\text{P}} \; > \; \underset{+\,1\,e^-}{\text{P}\cdot} \; > \; \underset{+\,2\,e^-}{\underset{\bullet}{\text{P}}\cdot} \; > \; \underset{+\,3\,e^-}{\cdot\underset{\bullet}{\text{P}}\cdot} \; > \; \underset{+\,4\,e^-}{\cdot\overset{\bullet}{\underset{\bullet}{\text{P}}}\cdot} \; > \; \underset{+\,5\,e^-}{\cdot\overset{\bullet}{\underset{\bullet}{\text{P}}}:}$$

Keep in mind that the placement of dots is not intended to show the actual positions of the electrons about the core of the atom. Actually, electrons are in constant motion, moving about the nucleus. Electron dot formulas are intended only to help us keep track of valence electrons. The following example exercise further illustrates electron dot formulas.

Example Exercise 6.9 • Electron Dot Formulas

Draw the electron dot formula for each of the following elements.

(a) Si (b) Xe

Solution

Let's find the group number of the element in the periodic table and note the number of valence electrons. We write the symbol of the element and place the same number of

(continued)

Example Exercise 6.9 *(continued)*

dots around the symbol as there are valence electrons. In these examples, Si has 4 valence electrons and Xe has 8. The electron dot formulas are as follows.

(a) ·$\dot{\underset{\cdot}{\text{Si}}}$· (b) :$\ddot{\underset{\cdot\cdot}{\text{Xe}}}$:

Self-Test Exercise

Draw the electron dot formula for each of the following elements.

(a) K (b) I

Answers: (a) K· (b) :$\dot{\underset{\cdot\cdot}{\text{I}}}$:

Note Chemists do not always draw electron dot formulas by adding one dot at a time. Consider the element Mg, which has two valence electrons in an *s* sublevel. We can draw the two valence electrons together on one side of the symbol, indicating that the element has two valence electrons in the same sublevel. Thus, the Lewis structure of Mg can be drawn as

Mg: which may be preferable to ·Mg·

Either method is acceptable; however, remember that valence electrons are not stationary particles on four sides of the nucleus. Electrons are in constant motion, and the electron dot formula is only a tool for keeping track of valence electrons. This is an important issue when we consider the formation of chemical bonds by the interaction between valence electrons of atoms during a chemical reaction.

6.9 Ionization Energy

Objectives
- **To state the general trends of ionization energy in the periodic table.**
- **To state the group with the highest and the lowest ionization energy.**

Although electrons can be removed from all elements, metals lose electrons more easily than nonmetals. In fact, metals undergo chemical reactions by losing one or more valence electrons. Since electrons are negatively charged, metals become positively charged after losing an electron. Any atom bearing a charge is called an **ion**.

Ionization Energy Movie; Periodic Trends: Ionization Energies Movie

Energy is always required to remove an electron from an atom. By definition, the amount of energy necessary to remove an electron from a neutral atom in the gaseous state is called the **ionization energy**, or ionization potential. We can illustrate ionization of a sodium atom as follows.

$$\text{Na} \xrightarrow{\text{ionization energy}} \text{Na}^+ + e^-$$

Figure 6.8 shows the relative energy required to remove a single electron from an atom of elements 1 through 86.

Notice in Figure 6.8 that the elements having the highest ionization energy belong to the same group—noble gases. One reason this group shows little tendency to undergo reaction is the difficulty with which an electron is removed. We can also reason that noble gases have high ionization energies because their valence shells are completely filled, and so they do not need to gain or lose electrons in order to become stable.

On the other hand, notice that the elements having the lowest ionization energy are alkali metals. Since they are in the same group, all alkali metals have a similar electron configuration. That is, they all have one electron in an *s* sublevel. If we remove an electron from an alkali metal, the resulting ion will have the same number of electrons as the preceding noble gas. For example, a lithium ion has 2 electrons (the same as He), a sodium ion has 10 electrons (the same as Ne), and a

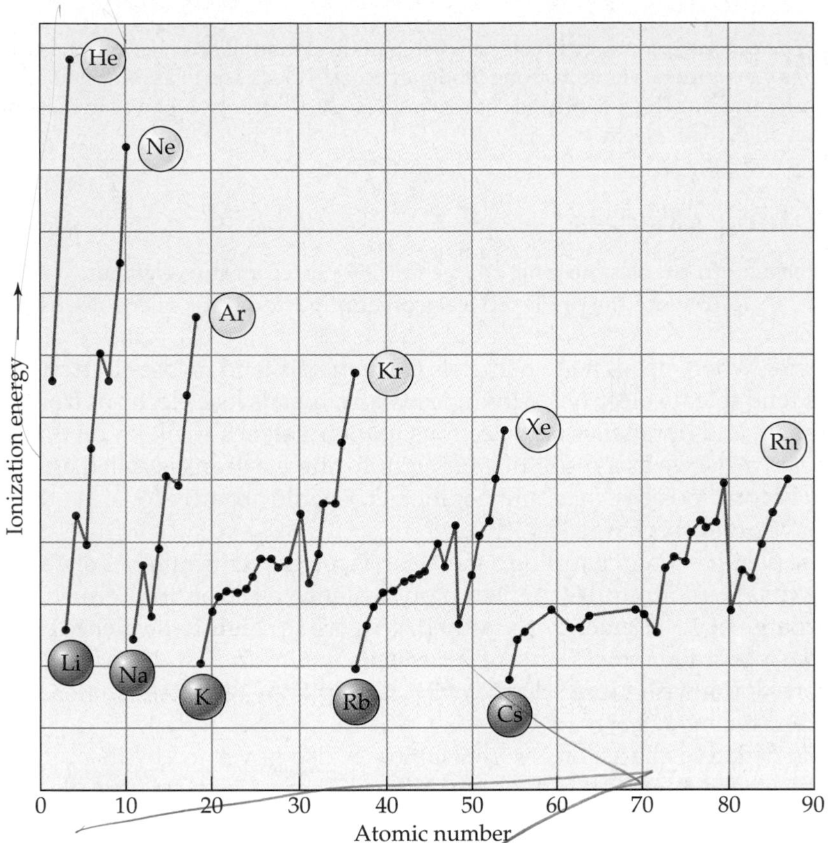

Trends in Ionization Energy

◀ **Figure 6.8 Ionization Energy** The diagram shows the energy required to remove a single electron from a neutral atom of an element in the gaseous state.

potassium ion has 18 electrons (the same as Ar). When an alkali metal loses an electron, it assumes a noble gas electron configuration which is exceedingly stable.

There are general trends in the periodic table regarding ionization energy. Within a group of elements, the energy required to remove an electron increases as the atomic radius decreases. As we proceed up a group, the atomic radius decreases since the valence electrons are closer to the nucleus and are more tightly held. Thus, the ionization energy *increases up a group of elements*. As we move from left to right in the periodic table, the nuclear charge becomes greater. The energy required to remove an electron, therefore, increases. In general, the ionization energy *increases from left to right across a period of elements*.

Example Exercise 6.10 • Ionization Energy and the Periodic Table

Based on the general trends in the periodic table, predict which element in each of the following pairs has the higher ionization energy.

(a) Li or Na (b) N or O

Solution

Let's refer to the periodic table and apply the general trends in ionization energy, which increases up a group and across a period.

(a) Li is above Na in Group IA/1, and so Li has the higher ionization energy.
(b) O is right of N in Period 2, and so O has the higher ionization energy.

Self-Test Exercise

Based on the general trends in the periodic table, predict which element in each of the following pairs has the higher ionization energy.

(a) Na or Mg (b) O or S

Answers: (a) Mg; (b) O

Note The amount of energy involved when a neutral atom gains an electron is called *electron affinity*. Nonmetals have a strong tendency to gain electrons to assume a noble gas electron configuration. Thus, nonmetals have a high electron affinity, whereas metals show little tendency to gain an electron.

6.10 Ionic Charges

Objectives
- To predict the ionic charge for any representative element.
- To write the predicted electron configuration for selected ions.

In general, when metals and nonmetals react, metals tend to lose electrons and nonmetals tend to gain electrons. More specifically, metals lose electrons from their valence shell, and nonmetals add electrons to their valence shell. Recall that an atom that bears a charge as a result of gaining or losing electrons is called an ion. Thus, metals become positive ions, and nonmetals become negative ions; in other words, they have an **ionic charge**.

The positive ionic charge on a metal ion is related to its number of valence electrons. Metals in Group IA/1 give up one valence electron to produce a positive ionic charge of 1+. Elements in Group IIA/2 give up their two valence electrons to produce a positive ionic charge of 2+. Metals in Group IIIA/13 usually lose three electrons and have an ionic charge of 3+. Group IVA/14 metals can lose four electrons, producing a charge of 4+.

The negative charge on a nonmetal ion is also governed by its number of valence electrons. Nonmetals in Group VIIA/17 have seven valence electrons and tend to add one electron to assume a noble gas configuration with eight valence electrons. After gaining one electron, the nonmetal has a negative ionic charge of 1−. Elements in Group VIA/16 gain two valence electrons, which produces an ionic charge of 2−. Nonmetals in Group VA/15 add three valence electrons, which give an ionic charge of 3−.

Figure 6.9 shows several common elements and their ionic charge as predicted by the position of the element in the periodic table.

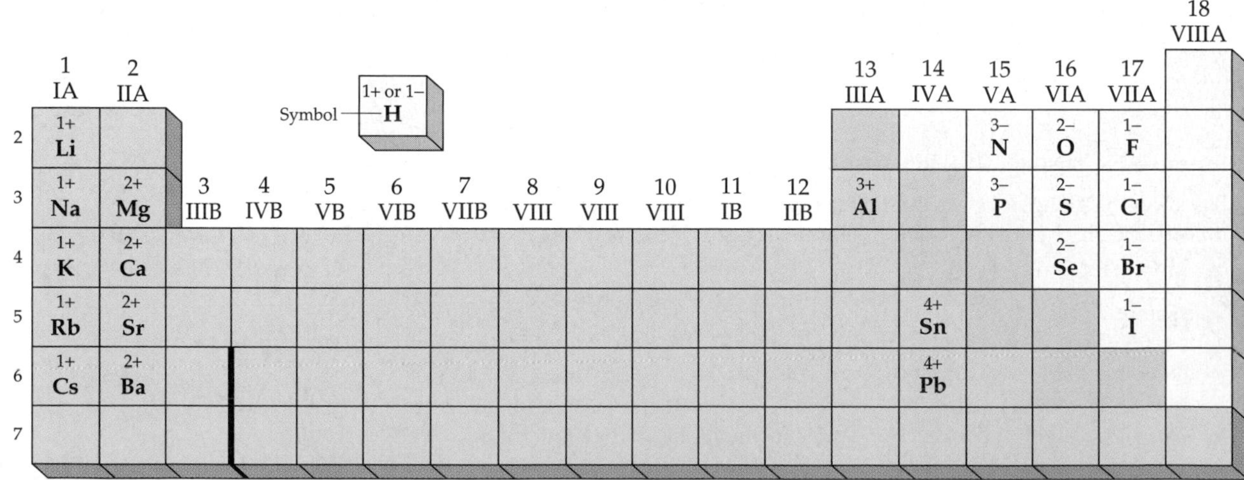

▲ **Figure 6.9 Periodic Table of Selected Ions** Notice that metals have a positive ionic charge equal to their group number. Nonmetals have a negative ionic charge equal to 8 minus their group number in the periodic table. Both metals and nonmetals attempt to assume a stable noble gas electron configuration.

Lithium is in Group IA/1, and so a lithium ion has an ionic charge of 1+. The lithium ion is written Li^+, with the 1 (in 1+) understood. Oxygen is in Group VIA/16, has an ionic charge of 2−, and is shown as O^{2-}. The following examples further illustrate the relationship between group number and ionic charge.

Example Exercise 6.11 • Ionic Charge and the Periodic Table

Predict the ionic charge for each of the following ions based on the group number of the element in the periodic table.

(a) Al ion (b) S ion

Solution

We refer to the periodic table to find the group number of the element.

(a) Aluminum is in Group IIIA/13. The metal atom loses 3 e^-, and so the ionic charge is 3+, that is, Al^{3+}.
(b) Sulfur is in Group VIA/16. The nonmetal atom gains 2 e^-, and so the ionic charge is 2−, that is, S^{2-}.

Self-Test Exercise

Predict the ionic charge for each of the following ions based on the group number of the element in the periodic table.

(a) Mg ion (b) Br ion

Answers: (a) Mg^{2+}; (b) Br^-

In Section 6.9 we learned that Group IA/1 elements form ions by losing one electron. We showed that each Group IA/1 ion has an electron configuration identical to that of the previous noble gas element. Similarly, Group VIIA/17 elements gain one electron to assume an electron configuration identical to that of the next noble gas element.

For example, a sodium ion (Na^+) and a fluoride ion (F^-) each have ten electrons, as does a neutral neon atom. Although their properties are not related, their electron configurations are identical. By definition, two or more ions having the same number of electrons are said to be **isoelectronic**. The sodium ion, the fluoride ion, and the neon atom are said to be members of an isoelectronic series.

Example Exercise 6.12 • Predicting Isoelectronic Ions

Refer to the periodic table and predict which of the following ions are isoelectronic with the noble gas argon.

(a) K^+ (b) Br^- (c) Ca^{2+} (d) O^{2-}

Solution

We refer to the periodic table to find the atomic number of the element.

(a) K^+ has 18 electrons (19 − 1); it is isoelectronic with argon (18 e^-).
(b) Br^- has 36 electrons (35 + 1); it is isoelectronic with krypton (36 e^-).
(c) Ca^{2+} has 18 electrons (20 − 2); it is isoelectronic with argon (18 e^-).
(d) O^{2-} has 10 electrons (8 + 2); it is isoelectronic with neon (10 e^-).

Self-Test Exercise

Refer to the periodic table and predict which of the following ions are isoelectronic with the noble gas xenon.

(a) Cs^+ (b) Cl^- (c) La^{3+} (d) Se^{2-}

Answers: (a) Cs^+ and (c) La^{3+} are isoelectronic with Xe; (b) Cl^- is isoelectronic with argon; (d) Se^{2-} is isoelectronic with krypton.

Electron Configuration of Ions

In Section 6.6 we learned how to predict the electron configuration of an element from the blocks of elements in the periodic table. Now we will learn how to write the electron configuration of an ion. First we find the element in the periodic table, and then we write out the electron configuration as we did in Section 6.6.

When writing the electron configuration for positive ions, we remove the number of electrons that corresponds to its positive ionic charge. For example, the electron configuration for a sodium atom is $1s^2\ 2s^2\ 2p^6\ 3s^1$. Since the sodium ion is Na^+, it loses one electron. That is,

$$\begin{array}{ccc} \text{Na atom} & \xrightarrow{\text{loses 1 e}^-} & \text{Na}^+ \\ 1s^2\ 2s^2\ 2p^6\ 3s^1 & & 1s^2\ 2s^2\ 2p^6 \end{array}$$

We should note that transition metals lose two electrons from the highest *s* sublevel first before losing electrons from their outer *d* sublevel. In the fourth-period elements, for example, 4*s* electrons are lost before 3*d* electrons. The electron configuration for a manganese ion is $1s^2\ 2s^2\ 2p^6\ 3s^2\ 3p^6\ 3d^5$. Note that manganese forms an Mn^{2+} ion by losing two electrons from the 4*s* sublevel rather than the 3*d* sublevel. That is,

$$\begin{array}{ccc} \text{Mn atom} & \xrightarrow{\text{loses 2 e}^-} & \text{Mn}^{2+} \\ 1s^2\ 2s^2\ 2p^6\ 3s^2\ 3p^6\ 4s^2\ 3d^5 & & 1s^2\ 2s^2\ 2p^6\ 3s^2\ 3p^6\ 3d^5 \end{array}$$

Electron configurations can be simplified by using a noble gas symbol to represent the nucleus and inner electrons. This method of showing the core of an atom is called **core notation**. In the preceding example with an Mn atom, we can write the electron configuration as [Ar] $4s^2\ 3d^5$ using core notation. For Mn^{2+}, we can write the electron configuration as [Ar] $3d^5$.

When writing the electron configuration for negative nonmetal ions, we must add the number of electrons that corresponds to its negative ionic charge. For example, the electron configuration for chlorine is $1s^2\ 2s^2\ 2p^6\ 3s^2\ 3p^5$. Since chloride ion is Cl^-, it gains one electron. That is,

$$\begin{array}{ccc} \text{Cl atom} & \xrightarrow{\text{gains 1 e}^-} & \text{Cl}^- \\ 1s^2\ 2s^2\ 2p^6\ 3s^2\ 3p^5 & & 1s^2\ 2s^2\ 2p^6\ 3s^2\ 3p^6 \end{array}$$

In core notation, the electron configuration of Cl^- is [Ar]. The following example exercise further illustrates how to write the electron configuration of ions.

Example Exercise 6.13 • Electron Configuration of Ions

Refer to the periodic table and write the predicted electron configuration for each of the following ions using core notation.

(a) Fe^{3+} (b) Se^{2-}

Solution

We refer to the periodic table to recall the blocks of elements so that we can write the electron configuration for the element.

(a) Fe is $1s^2\ 2s^2\ 2p^6\ 3s^2\ 3p^6\ 4s^2\ 3d^6$ or [Ar] $4s^2\ 3d^6$. The electron configuration for the Fe^{3+} ion is [Ar] $3d^5$.

(b) Se is [Ar] $4s^2\ 3d^{10}\ 4p^4$. For Se^{2-}, the electron configuration can be written as [Ar] $4s^2\ 3d^{10}\ 4p^6$, or simply [Kr].

Self-Test Exercise

Refer to the periodic table and write the predicted electron configuration for each of the following ions using core notation.

(a) Cd^{2+} (b) P^{3-}

Answers: (a) [Kr] $4d^{10}$; (b) [Ne] $3s^2\ 3p^6$ or [Ar].

Summary 

Section 6.1 In 1869 Dmitri Mendeleev explained the recurring properties of more than 60 elements by arranging them according to increasing atomic mass. Mendeleev published his table of elements, which even included undiscovered elements and their predicted properties.

Section 6.2 In 1913 H. G. J. Moseley showed that the positively charged nucleus of each element in the periodic table increases progressively. As a result, the **periodic law** was rewritten, and it is now stated that physical and chemical properties repeat periodically when the elements are arranged in order of increasing atomic number.

Section 6.3 The periodic table is organized by vertical columns called **groups** and horizontal rows called **periods**. The main-group elements in Groups 1, 2, and 13–18 are referred to as **representative elements**, and the elements in Groups 3–12 are called **transition elements**. The elements in Groups 1, 2, 17, and 18 are referred to as **alkali metals, alkaline earth metals, halogens**, and **noble gases**, respectively. The elements following lanthanum (Ce–Lu) are referred to as **lanthanides**, and the elements following actinium (Th–Lr) are called **actinides**. Collectively, lanthanides and actinides are known as **inner transition elements**. The elements Sc, Y, La, and Ce–Lu are referred to as **rare earth elements**. The elements beyond uranium are called **transuranium elements**.

Section 6.4 The periodic table is divided into metals on the left side and nonmetals on the right. Semimetals have intermediate properties and are found between the metals and the nonmetals. The atomic radius decreases up a group and from left to right. The metallic character also decreases up a group and from left to right.

Section 6.5 The trends in the periodic table enable us to predict physical and chemical properties. If we are given the atomic radius and oxide formula of two elements in a group, we can make a reasonable prediction for the atomic radius and oxide formula of another element in the same group.

Section 6.6 In this chapter we learned that elements in the periodic table are arranged by sublevels of increasing energy. We can easily predict the energy sublevel containing the outermost electrons by referring to the periodic table. We can write the **electron configuration** by noting the sublevels in the *s*, *p*, *d*, and *f* blocks of elements.

Section 6.7 We can use the periodic table to quickly predict the number of electrons in the outermost *s* and *p* sublevels of an element; that is, the number of **valence electrons**. The number of valence electrons corresponds to the group number in the periodic table. For example, a Group IA/1 element has 1 valence electron, and a Group VIIA/17 has 7 valence electrons.

Section 6.8 We can diagram an atom by letting the symbol of the element represent the **core** of the atom, and drawing dots to represent the valence electrons. This diagram is referred to as the **electron dot formula** of the element.

Section 6.9 When an element loses or gains electrons, the resulting positively or negatively charged atoms are called **ions**. The amount of energy required to remove an electron from an atom is called the **ionization energy**. Since each noble gas has a filled valence level, their ionization energy is extremely high. Conversely, each alkali metal has a low ionization energy because it has to lose only 1 valence electron to acquire a noble gas electron structure. Ionization energy increases up a group and from left to right in the periodic table.

Section 6.10 We can predict the **ionic charge** for a representative element from its group number. Metals lose valence electrons and nonmetals gain electrons in order to obtain a noble gas structure. For example, potassium is in Group IA/1 and loses 1 valence electron to form K^+. Sulfur is in Group VIA/16, has 6 valence electrons, and gains 2 electrons to form S^{2-}. The resulting K^+ and S^{2-} ions each have 18 electrons and are said to be **isoelectronic**. To write the electron configuration for an ion, we add or subtract electrons corresponding to the ionic charge. For the magnesium ion, Mg^{2+}, we subtract 2 electrons, and for the fluoride ion, F^-, we add 1 electron. Since Mg^{2+} and F^- each have 10 electrons, they are isoelectronic and we can write their electron configurations as $1s^2\ 2s^2\ 2p^6$, or simply in **core notation** as [Ne].

Key Concepts*

1. The modern periodic law states that the properties of elements repeat when the periodic table is arranged according to what trend?
2. Which element is the lowest atomic mass alkali metal? alkaline earth metal? halogen? noble gas? transuranium element?
3. Which element is the lowest atomic mass semimetal? lanthanide? actinide? rare earth metal? inner transition element?
4. Which alkali metal has the most metallic character? the largest atomic radius? Which halogen has the least metallic character? the smallest atomic radius?
5. Given the atomic radius of Se (0.116 nm) and Te (0.143 nm), what is the predicted atomic radius for radioactive Po?
6. Given the boiling points of Kr (−152°C) and Xe (−107°C), what is the predicted boiling point for radioactive Rn?
7. Given the chemical formulas MgO, Al_2O_3, and SiO_2, what is the predicted formula for boron oxide?
8. What is the predicted number of valence electrons for an atom of boron?
9. Which of the following ions is isoelectronic with the noble gas krypton: As^{3+}, I^-, K^+, La^{3+}, or Se^{2-}?
10. What is the predicted electron configuration for the cobalt ion, Co^{3+}?
11. What is the predicted electron configuration for a bromide ion, Br^-?
12. The following photograph shows a watchglass containing antimony, sulfur, and antimony sulfide. Based on the characteristics of semimetals and nonmetals, identify the (a) orange, (b) yellow, and (c) gray substances.

▲ Antimony, Sulfur, and Antimony Sulfide

Key Terms†

Select the key term below that corresponds to each of the following definitions.

_____ 1. the properties of elements recur in a repeating pattern when arranged by increasing atomic number
_____ 2. a vertical column of elements in the periodic table
_____ 3. a horizontal row of elements in the periodic table
_____ 4. the Group IA/1 elements, excluding hydrogen
_____ 5. the Group IIA/2 elements
_____ 6. the Group VIIA/17 elements
_____ 7. the relatively unreactive Group VIIIA/18 elements
_____ 8. the Group A (1, 2, and 13–18) elements in the periodic table
_____ 9. the Group B (3–12) elements in the periodic table

(a) actinide series (*Sec. 6.3*)
(b) alkali metals (*Sec. 6.3*)
(c) alkaline earth metals (*Sec. 6.3*)
(d) core (kernel) (*Sec. 6.8*)
(e) core notation (*Sec. 6.10*)
(f) electron configuration (*Sec. 6.6*)
(g) electron dot formula (*Sec. 6.8*)

* Answers to Key Concepts are in Appendix H.

† Answers to Key Terms are in Appendix I.

_____ 10. the elements with atomic numbers 58–71

_____ 11. the elements with atomic numbers 90–103

_____ 12. the elements in the lanthanide and actinide series

_____ 13. the elements with atomic numbers 21, 39, 57, and 58–71

_____ 14. the elements beyond atomic number 92

_____ 15. a shorthand description of the arrangement of electrons by sublevels according to increasing energy

_____ 16. the portion of the atom that includes the nucleus and inner electrons

_____ 17. the electrons that occupy the outermost *s* and *p* sublevels of an atom

_____ 18. a symbolic diagram for an element and its valence electrons; the chemical symbol is surrounded by a dot for each valence electron

_____ 19. an atom that bears a charge as a result of gaining or losing valence electrons

_____ 20. the amount of energy necessary to remove an electron from an atom

_____ 21. refers to the positive charge on a metal atom that has lost electrons or to the negative charge on a nonmetal atom that has gained electrons

_____ 22. refers to ions having the same electron configuration; for example, Mg^{2+} and O^{2-} each have ten electrons

_____ 23. a method of writing an electron configuration in which core electrons are represented by a noble gas symbol in brackets followed by the valence electrons, for example, [Ne] $3s^2$

(h) group (*Sec. 6.3*)
(i) halogens (*Sec. 6.3*)
(j) inner transition elements (*Sec. 6.3*)
(k) ion (*Sec. 6.9*)
(l) ionic charge (*Sec. 6.10*)
(m) ionization energy (*Sec. 6.9*)
(n) isoelectronic (*Sec. 6.10*)
(o) lanthanide series (*Sec. 6.3*)
(p) noble gases (*Sec. 6.3*)
(q) period (*Sec. 6.3*)
(r) periodic law (*Sec. 6.2*)
(s) rare earth elements (*Sec. 6.3*)
(t) representative elements (*Sec. 6.3*)
(u) transition elements (*Sec. 6.3*)
(v) transuranium elements (*Sec. 6.3*)
(w) valence electrons (*Sec. 6.7*)

Exercises‡

Classification of Elements (Sec. 6.1)

1. According to Mendeleev's periodic table of 1871, which element would begin a new period following the potassium series?
2. According to Mendeleev's periodic table of 1871, which element would begin a new period following the rubidium series?
3. Why did Mendeleev not include germanium in his periodic table of 1871?
4. Why did Mendeleev not include noble gases in his periodic table of 1871?

The Periodic Law Concept (Sec. 6.2)

5. Mendeleev suggested that physical and chemical properties tend to repeat periodically when elements are arranged according to what trend?
6. Consider only the third and fourth rows in the periodic table and find two pairs of elements that obey the modern periodic law but violate the original periodic law as stated by Mendeleev.
7. Following Moseley's discovery in 1913, the periodic law states that physical and chemical properties tend to recur periodically when elements are arranged according to what trend?
8. By studying the X-ray emission from excited nuclei, Moseley discovered that elements in the periodic table have a stepwise increase in what property?

Groups and Periods of Elements (Sec. 6.3)

9. Vertical columns in the periodic table are referred to by what two terms?
10. Horizontal rows in the periodic table are referred to by what two terms?
11. What is the collective term for the main-group elements that appear in Groups IA–VIIIA (that is, Groups 1, 2, and 13–18)?
12. What is the collective term for the elements that belong to Groups IIIB–IIB (that is, Groups 3–12)?
13. What is the collective term for the two series of elements that include Ce–Lu and Th–Lr?
14. What is the collective term for the elements on the left side of the periodic table?
15. What is the collective term for the elements on the right side of the periodic table?
16. What is the collective term for the elements having properties that lie between those of metals and nonmetals?
17. Identify the group number corresponding to each of the following families of elements?
    (a) alkali metals **(b)** alkaline earth metals
    (c) halogens **(d)** noble gases
18. Identify the group number corresponding to each of the following families of elements?
    (a) boron group **(b)** oxygen group
    (c) nickel group **(d)** copper group

‡Answers to odd-numbered Exercises are in Appendix J.

19. What is the collective term for the elements in the series that follows element 57?
20. What is the collective term for the elements in the series that follows element 89?
21. What is the collective term for Sc, Y, La, and Ce through Lu?
22. What is the collective term for all the synthetic elements beyond uranium?
23. According to IUPAC, what is the designation for each of the following groups indicated by the American convention?
    (a) Group IA (b) Group IB
    (c) Group IIIA (d) Group IIIB
    (e) Group VA (f) Group VB
    (g) Group VIIA (h) Group VIIB
24. According to the American convention using Roman numerals, what is the group number designation for each of the following?
    (a) Group 2 (b) Group 5
    (c) Group 6 (d) Group 11
    (e) Group 12 (f) Group 14
    (g) Group 16 (h) Group 18
25. Refer to the periodic table and select the symbol of the element that fits the each of following descriptions.
    (a) the Group IVA/14 semimetal in the fourth period
    (b) the third-period alkali metal
    (c) the nonradioactive halogen that normally exists as a solid
    (d) the lanthanide that is not naturally occurring
26. Refer to the periodic table and select the symbol of the element that fits each of the following descriptions.
    (a) the rare earth element whose atomic number is greatest
    (b) the sixth-period representative element with properties similar to Be
    (c) the fifth-period transition element with properties similar to Ti
    (d) the lighter-than-air noble gas used in blimps and balloons
27. Refer to the periodic table and select the symbol of the element that fits the each of following descriptions.
    (a) the Group VA/15 semimetal in the fifth period
    (b) the third-period alkaline earth metal
    (c) the halogen that exists as a reddish-brown liquid at normal conditions
    (d) the actinide with properties similar to Ce
28. Refer to the periodic table and select the symbol of the element that fits each of the following descriptions.
    (a) the rare earth element whose atomic mass is lowest
    (b) the fourth-period representative element with properties similar to O
    (c) the sixth-period transition element with properties similar to Ni
    (d) the radioactive noble gas

Periodic Trends (Sec. 6.4)

29. According to the general trend, the atomic radius (increases/decreases) proceeding down a group of elements in the periodic table.
30. According to the general trend, the atomic radius for a period of elements (increases/decreases) proceeding from left to right in the periodic table.
31. According to the general trend, metallic character (increases/decreases) proceeding down a group of elements in the periodic table.
32. According to the general trend, metallic character for a period of elements (increases/decreases) proceeding from left to right in the periodic table.
33. According to general trends in the periodic table, predict which element in each of the following pairs has the larger atomic radius.
    (a) Li or Na (b) N or P
    (c) Mg or Ca (d) Ar or Kr
34. According to general trends in the periodic table, predict which element in each of the following pairs has the larger atomic radius.
    (a) Rb or Sr (b) As or Se
    (c) Pb or Bi (d) I or Xe
35. According to general trends in the periodic table, predict which element in each of the following pairs has greater metallic character.
    (a) B or Al (b) Na or K
    (c) Mg or Ba (d) H or Fe
36. According to general trends in the periodic table, predict which element in each of the following pairs has greater metallic character.
    (a) K or Ca (b) Mg or Al
    (c) Fe or Cu (d) S or Ar

Properties of Elements (Sec. 6.5)

37. Predict the missing value (?) for each property listed below. The atomic radius, density, and melting point are given for elements in Group IA/1.

Element	Atomic Radius	Density at 20°C	Melting Point
K	(?) nm	0.86 g/mL	63.3°C
Rb	0.248 nm	(?) g/mL	38.9°C
Cs	0.266 nm	1.90 g/mL	(?)°C

38. Predict the missing value (?) for each property listed below. The atomic radius, density, and melting point are given for elements in Group VIB/6.

Element	Atomic Radius	Density at 20°C	Melting Point
Cr	0.125 nm	7.14 g/mL	(?)°C
Mo	(?) nm	10.28 g/mL	2617°C
W	0.137 nm	(?) g/mL	3410°C

39. Predict the missing value (?) for each property listed below. The atomic radius, density, and boiling point are given for elements in Group VIIA/17.

Element	Atomic Radius	Density at Bp	Boiling Point
Cl	(?) nm	1.56 g/mL	−34.6°C
Br	0.115 nm	(?) g/mL	58.8°C
I	0.133 nm	4.97 g/mL	(?)°C

40. Predict the missing value (?) for each property listed below. The atomic radius, density, and boiling point are given for elements in Group VIIIA/18.

Element	Atomic Radius	Density at STP	Boiling Point
Ar	0.180 nm	1.78 g/L	(?)°C
Kr	(?) nm	3.74 g/L	−152°C
Xe	0.210 nm	(?) g/L	−107°C

41. The formulas for the oxides of sodium, magnesium, aluminum, and silicon are, respectively, Na_2O, MgO, Al_2O_3, and SiO_2. Using the periodic table, predict the chemical formulas for each of the following similar compounds.
   (a) lithium oxide (b) calcium oxide
   (c) gallium oxide (d) tin oxide

42. The formulas for the chlorides of potassium, calcium, boron, and germanium are, respectively, KCl, $CaCl_2$, BCl_3, and $GeCl_4$. Using the periodic table, predict the chemical formula for each of the following similar compounds.
   (a) potassium fluoride (b) calcium fluoride
   (c) boron bromide (d) germanium iodide

43. The chemical formula for zinc oxide is ZnO. Predict the formula for each of the following similar compounds.
   (a) cadmium oxide (b) zinc sulfide
   (c) mercury sulfide (d) cadmium selenide

44. The chemical formula for barium chloride is $BaCl_2$. Predict the formula for each of the following similar compounds.
   (a) strontium chloride (b) strontium bromide
   (c) magnesium iodide (d) calcium fluoride

45. Selenium reacts with oxygen to produce SeO_3. Predict the chemical formula for each of the following similar compounds.
   (a) sulfur oxide (b) tellurium oxide
   (c) selenium sulfide (d) tellurium sulfide

46. Phosphorus reacts with oxygen to produce both P_2O_3 and P_2O_5. Predict two formulas for each of the following similar compounds.
   (a) nitrogen oxide (b) arsenic oxide
   (c) phosphorus sulfide (d) antimony sulfide

Blocks of Elements (Sec. 6.6)

47. What type of energy sublevel is being filled by the elements in Groups IA/1 and IIA/2?

48. What type of energy sublevel is being filled by the elements in Groups IIIA/13 through VIIIA/18?

49. What type of energy sublevel is being filled by the elements in Groups IIIB/3 through IIB/12?

50. What type of energy sublevel is being filled by the inner transition elements?

51. Which energy sublevel is being filled by the lanthanide series?

52. Which energy sublevel is being filled by the actinide series?

53. Refer to the periodic table and state the highest energy sublevel for each of the following elements.
   (a) H (b) Na
   (c) Sm (d) Br
   (e) Sr (f) C
   (g) Sn (h) Cs

54. Refer to the periodic table and state the highest energy sublevel for each of the following elements.
   (a) He (b) K
   (c) U (d) Pd
   (e) Be (f) Co
   (g) Si (h) Pt

55. Refer to the periodic table and write the predicted electron configuration for each of the following elements.
   (a) Li (b) F
   (c) Mg (d) P
   (e) Ca (f) Mn
   (g) Ga (h) Rb

56. Refer to the periodic table and write the predicted electron configuration for each of the following elements.
   (a) B (b) Ti
   (c) Na (d) O
   (e) Ge (f) Ba
   (g) Pd (h) Kr

Valence Electrons (Sec. 6.7)

57. State the number of valence electrons in each of the following groups as predicted from the periodic table.
(a) Group IA/1 (b) Group IIIA/3
(c) Group VA/15 (d) Group VIIA/17

58. State the number of valence electrons in each of the following groups as predicted from the periodic table.
(a) Group IIA/2 (b) Group IVA/4
(c) Group VIA/16 (d) Group VIIIA/18

59. State the number of valence electrons for each of the following elements.
(a) H (b) B
(c) N (d) F
(e) Ca (f) Si
(g) O (h) Ar

60. State the number of valence electrons for each of the following elements.
(a) He (b) Pb
(c) Se (d) Ne
(e) Cs (f) Ga
(g) Sb (h) Br

Electron Dot Formulas (Sec. 6.8)

61. Draw the electron dot formula for each of the following elements.
(a) H (b) B
(c) N (d) F
(e) Ca (f) Si
(g) O (h) Ar

62. Draw the electron dot formula for each of the following elements.
(a) He (b) Pb
(c) Se (d) Ne
(e) Cs (f) Ga
(g) Sb (h) Br

Ionization Energy (Sec. 6.9)

63. According to the general trend, the ionization energy for a group of elements (increases/decreases) proceeding down a group in the periodic table.

64. According to the general trend, the ionization energy for a period of elements (increases/decreases) proceeding from left to right in the periodic table.

65. Which group of elements has the highest ionization energy?

66. Which group of elements has the lowest ionization energy?

67. Refer to the periodic table and predict which element in each of the following pairs has the higher ionization energy.
(a) Mg or Ca (b) S or Se
(c) Sn or Pb (d) N or P

68. Refer to the periodic table and predict which element in each of the following pairs has the higher ionization energy.
(a) Ga or Ge (b) Si or P
(c) Br or Cl (d) As or Sb

69. Refer to the periodic table and predict which element in each of the following pairs has the lower ionization energy.
(a) Rb or Cs (b) He or Ar
(c) B or Al (d) F or I

70. Refer to the periodic table and predict which element in each of the following pairs has the lower ionization energy.
(a) Mg or Si (b) Pb or Bi
(c) Ca or Ga (d) P or Cl

Ionic Charges (Sec. 6.10)

71. State the predicted ionic charge of metal ions in each of the following groups of elements.
(a) Group IA/1 (b) Group IIA/2
(c) Group IIIA/13 (d) Group IVA/14

72. State the predicted ionic charge of nonmetal ions in each of the following groups of elements.
(a) Group IVA/14 (b) Group VA/15
(c) Group VIA/16 (d) Group VIIA/17

73. Write the ionic charge for each of the following ions as predicted from the group number in the periodic table.
(a) Cs ion (b) Ga ion
(c) O ion (d) I ion

74. Write the ionic charge for each of the following ions as predicted from the group number in the periodic table.
(a) Be ion (b) Sn ion
(c) P ion (d) S ion

75. Refer to the periodic table and predict which of the following ions are isoelectronic with the noble gas argon.
(a) Al^{3+} (b) Ca^{2+}
(c) S^{2-} (d) N^{3-}

76. Refer to the periodic table and predict which of the following ions are isoelectronic with the noble gas krypton.
(a) K^{+} (b) Sr^{2+}
(c) Cl^{-} (d) Se^{2-}

77. Refer to the periodic table and write the predicted electron configuration for each of the following positive ions.
(a) Mg^{2+} (b) K^{+}
(c) Fe^{2+} (d) Cs^{+}

78. Refer to the periodic table and write the predicted electron configuration for each of the following positive ions.
(a) Sr^{2+} (b) Y^{3+}
(c) Zn^{2+} (d) Ti^{4+}

79. Refer to the periodic table and write the predicted electron configuration for each of the following negative ions.

(a) F^- (b) S^{2-}

(c) N^{3-} (d) I^-

80. Refer to the periodic table and write the predicted electron configuration for each of the following negative ions.

(a) Br^- (b) Te^{2-}

(c) As^{3-} (d) O^{2-}

General Exercises

81. Examine Figure 6.1 and determine the name of the element that Mendeleev predicted in 1871 and called ekaboron.

82. Examine Figure 6.1 and determine the name of the element that Mendeleev predicted in 1871 and called ekaaluminum.

83. Use the American convention to designate the group number corresponding to each of the following groups listed by the European convention.

(a) Group IA (b) Group IB

(c) Group IIIA (d) Group IIIB

84. Use the IUPAC convention to designate the group number corresponding to each of the following groups listed by the European convention?

(a) Group IIA (b) Group IIB

(c) Group IVA (d) Group IVB

85. Predict the atomic radius, density, and melting point for radioactive francium, Fr.

Element	Atomic Radius	Density at 20°C	Melting Point
Rb	0.248 nm	1.53 g/mL	38.9°C
Cs	0.266 nm	1.87 g/mL	28.4°C
Fr	(?) nm	(?) g/mL	(?)°C

86. Predict the atomic radius, density, and melting point for radioactive radium, Ra.

Element	Atomic Radius	Density at 20°C	Melting Point
Sr	0.215 nm	2.63 g/mL	769°C
Ba	0.217 nm	3.65 g/mL	725°C
Ra	(?) nm	(?) g/mL	(?)°C

87. Refer to the periodic table and write the predicted electron configuration for each of the following elements using core notation.

(a) Sr (b) Ru

(c) Sb (d) Cs

88. Refer to the periodic table and write the predicted electron configuration for each of the following elements using core notation.

(a) W (b) Bi

(c) Ra (d) Ac

89. Explain why the ionization energy for the alkaline earth metals is higher than the ionization energy for the alkali metals.

90. Explain why the ionization energy for aluminum, contrary to the general trend, is less than the ionization energy for magnesium.

91. Explain why the ionization energy for hydrogen is much higher than that of other Group IA/1 elements.

92. Predict two ionic charges for hydrogen. Write the formulas of the two ions and explain the ionic charges.

Explorer Quiz 1
Explorer Quiz 2
Explorer Quiz 3
Master Quiz

CHAPTERS 4–6
Cumulative Review

Key Concepts

1. Identify the physical state (solid, liquid, or gas) that corresponds to each of the following pictorial representations.

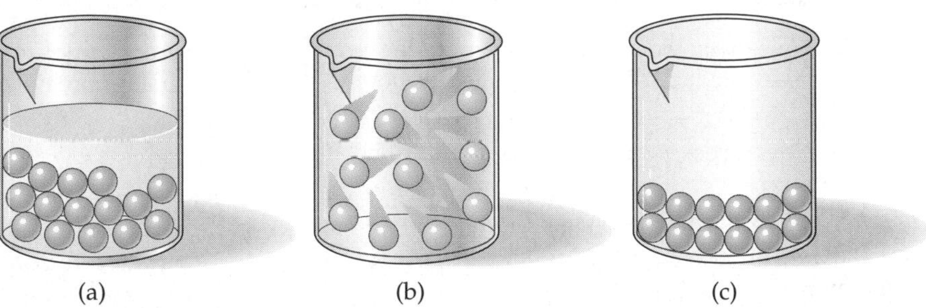

2. Classify each of the following as an element, compound, homogeneous mixture, or heterogeneous mixture.
   (a) silver metal **(b)** silver ore
   (c) silver oxide **(d)** silver alloy
3. Classify each of the following as an element, compound, homogeneous mixture, or heterogeneous mixture according to the pictorial representations.

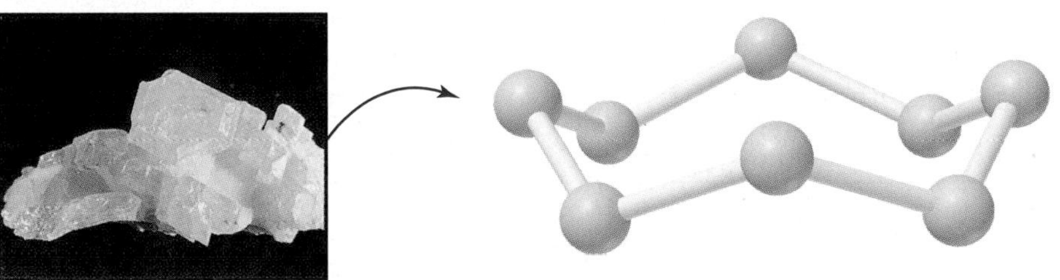

4. State whether each of the following is a physical or a chemical change.
   (a) Silver metal alloys with gold to give 18K jewelry.
   (b) Silver metal produces a brown gas in nitric acid.
   (c) Silver metal melts at 962°C.
   (d) Silver metal forms a dark tarnish in air.
5. Which has greater mass: a gallon of gasoline or the gaseous products from the combustion of a gallon of gasoline?
6. If a nucleus is described by the analogy "a tiny lead pellet in a huge indoor sports stadium," what does the stadium represent if the lead pellet represents an atomic nucleus?
7. Can ions of different elements have the same number of protons? Can ions of different elements have the same number of electrons?
8. Complete the following analogy: electrons are to electricity as photons are to _________.
9. Which of the following statements is true of the Bohr model of the atom?
   (a) Electrons are repelled by protons in the nucleus.
   (b) Electrons move randomly about the atom.
   (c) Electrons gain energy as they jump to an orbit farther from the nucleus.

10. Which of the following statements is true of the quantum mechanical model of the atom?
    (a) Electrons are attracted to neutrons in the nucleus.
    (b) Electrons can be found anywhere within an orbital.
    (c) Electrons lose energy when they drop from a 2*s* to a 1*s* orbital.
11. Which trend in the periodic table explains the recurring properties of elements?
12. What are the general trends in the periodic table for decreasing metallic character of the elements?
13. What are the general trends in the periodic table for increasing ionization energy of the elements?
14. Given the chemical formulas Na_2O, MgO, and Al_2O_3, what is the predicted chemical formula for radioactive radium oxide?
15. Which of the following ions is not isoelectronic with the noble gas xenon: Ba^{2+}, La^{3+}, I^-, or Se^{2-}?

Key Terms

State the key term that corresponds to each of the following descriptions.

_____ 1. matter having a definite composition and constant properties
_____ 2. a substance that cannot be broken down by chemical reaction
_____ 3. a homogeneous mixture of two or more metals
_____ 4. a property that can be observed without changing the formula of a substance
_____ 5. a modification of a substance that alters the chemical composition
_____ 6. the stored energy that matter possesses owing to its composition
_____ 7. the energy associated with the mass and velocity of a particle
_____ 8. the law that states mass cannot be created nor destroyed
_____ 9. a value that indicates the number of protons in the nucleus of an atom
_____ 10. atoms having the same atomic number but a different mass number
_____ 11. a unit of mass exactly equal to $\frac{1}{12}$ the mass of a carbon-12 atom
_____ 12. the average mass of all the naturally occurring isotopes of an element
_____ 13. a range of light energy extending from violet through red (~400–700 nm)
_____ 14. a particle of radiant energy
_____ 15. a model of the atom that describes electrons circling the nucleus in orbits
_____ 16. a fixed-energy orbit that electrons occupy as they circle the nucleus
_____ 17. an electron energy level that results from splitting a main-energy level
_____ 18. a shorthand description of the arrangement of electrons in an atom by sublevels according to increasing energy
_____ 19. a model of the atom that describes the energy of an electron in terms of its probability of being found in a particular location around the nucleus
_____ 20. a region around the nucleus in which there is a high probability of finding an electron with a given energy
_____ 21. the properties of elements recur in a repeating pattern when arranged by increasing atomic number
_____ 22. the Group A (1, 2, and 13–18) elements in the periodic table
_____ 23. the Group B (3–12) elements in the periodic table
_____ 24. the electrons that occupy the outermost *s* and *p* sublevels of an atom
_____ 25. an atom that bears a charge as a result of gaining or losing valence electrons

Review Exercises

1. Which physical state has a random arrangement of particles that can be greatly compressed and reduced in volume?
2. What is the chemical symbol for titanium? for manganese? for xenon?

3. What law is illustrated by the statement that ethyl alcohol is always composed of 52% carbon, 13% hydrogen, and 35% oxygen by mass?
4. Alum, $Al_2(SO_4)_3$, is used in styptic pencils to stop minor bleeding. What is the total number of atoms in one formula unit of alum?
5. Is a carbonated beverage classified as a heterogeneous mixture, a homogeneous mixture, or a pure substance?
6. In a nuclear power plant, uranium fuel converts water to steam, the steam drives a turbine, and the turbine spins a generator that produces electricity. What are the four forms of energy involved in generating electricity from uranium?
7. Using atomic notation, indicate the isotope having 25 p^+ and 30 n^0.
8. Element Z has two natural isotopes: Z-79 (78.918 amu) and Z-81 (80.916 amu). Calculate the atomic mass of element Z given that the abundance of Z-79 is 50.69%.
9. How many photons are emitted when an electron drops from energy level 6 to energy level 2 in a hydrogen atom?
10. What is the maximum number of electrons that can occupy the fourth principal energy level?
11. Which sixth period representative element has the highest atomic number?
12. Which sixth-period transition element has the highest atomic number?
13. Predict the melting point of tungsten, W, given the melting points of chromium, Cr (1857°C), and molybdenum, Mo (2617°C).
14. Which energy sublevel is being filled by the elements Y through Cd?
15. Predict the number of valence electrons for a Group VA/15 element.
16. What is the electron dot formula for an atom of bromine?
17. Which noble gas has the largest atomic radius? the highest ionization energy?
18. Which group has a predictable ionic charge of 3+? of 3−?
19. In core notation, what is the predicted electron configuration for a strontium atom? for a selenium atom?
20. In core notation, what is the predicted electron configuration for a strontium ion, Sr^{2+}? for the selenide ion, Se^{2-}?

CHAPTER 7

The Language of Chemistry

▲ The "language of chemistry" refers to the names and formulas of chemicals. What common supermarket item has the systematic name and formula of sodium hydrogen carbonate, $NaHCO_3$?

Every experienced instructor has encountered difficulty in motivating students to learn nomenclature. Typical inducements include flashcards, "ion arrays," and frequent quizzing. This chapter offers a unique "divide and conquer" strategy. That is, before naming is attempted, Section 7.1 affords students practice in determining whether a formula represents a binary ionic, ternary ionic, binary molecular, binary acid, or ternary oxyacid. Then, each of these types of classifications is presented systematically. Flashcards of ions are found in the student *Study Guide and Solutions Manual*.

During the Middle Ages, alchemists identified many substances, which they named in a haphazard fashion. Since there were only a few dozen substances, this did not pose a great problem. Toward the end of the 1700s, however, they had identified more than 10,000 substances; and the number of compounds was growing rapidly. Chemists then faced the staggering task of providing names for all these substances. The problem of assigning names was eventually solved by using a set of systematic rules.

The French chemist Antoine Lavoisier developed the first systematic method of naming substances. He proposed that chemical names refer to the composition of the compound and be derived from Latin or Greek. In 1787 Lavoisier, with the aid of others, published *Methods of Chemical Nomenclature*. The naming system he proposed was so clear and logical that it was universally accepted in a short time. In fact, it was so thoughtful that it became the basis for our current system of naming.

7.1 Classification of Compounds

Objectives
- To classify a compound as a binary ionic, a ternary ionic, or a binary molecular compound.
- To classify an acid as a binary acid or a ternary oxyacid.
- To classify an ion as a monoatomic cation, a monoatomic anion, a polyatomic cation, or a polyatomic anion.

In 1921 the International Union of Pure and Applied Chemistry (IUPAC) formed the Commission on the Nomenclature of Inorganic Chemistry. In 1938 the IUPAC Committee on the Reform of Inorganic Nomenclature met in Berlin. Two years later it released a comprehensive set of rules. Although the rules have been expanded and revised, the 1940 Rules remain as the official international system for naming chemical compounds. These rules are referred to as **IUPAC nomenclature**.

Classification of Inorganic Compounds

With a few exceptions, an **inorganic compound** does not contain the element carbon. According to the 1940 IUPAC rules, inorganic compounds can be placed in one of several categories. Five common classes are binary ionic, ternary ionic, binary molecular, binary acid, and ternary oxyacid.

A **binary ionic compound** contains two elements, a metal and a nonmetal. Examples of binary ionic compounds are KCl and $AlCl_3$. A **ternary ionic compound** contains three elements, with at least one metal and one nonmetal. Examples of ternary ionic compounds are KNO_3 and $Al(NO_3)_3$. A **binary molecular compound** contains two elements that are both nonmetals. Water, H_2O, is a common example, as is ammonia, NH_3.

An **aqueous solution** is produced when a compound dissolves in water. Aqueous solutions are indicated by the symbol (aq). A **binary acid** is an aqueous solution of a compound containing hydrogen and one other nonmetal. Formulas of acids begin with H, and examples of binary acids are HCl(aq) and H_2S(aq).

A **ternary oxyacid** is an aqueous solution of a compound containing hydrogen, a nonmetal, and oxygen. Examples of ternary oxyacids are HNO_3(aq) and H_2SO_4(aq). Figure 7.1 illustrates the relationships among the five types of compounds we will classify according to IUPAC rules.

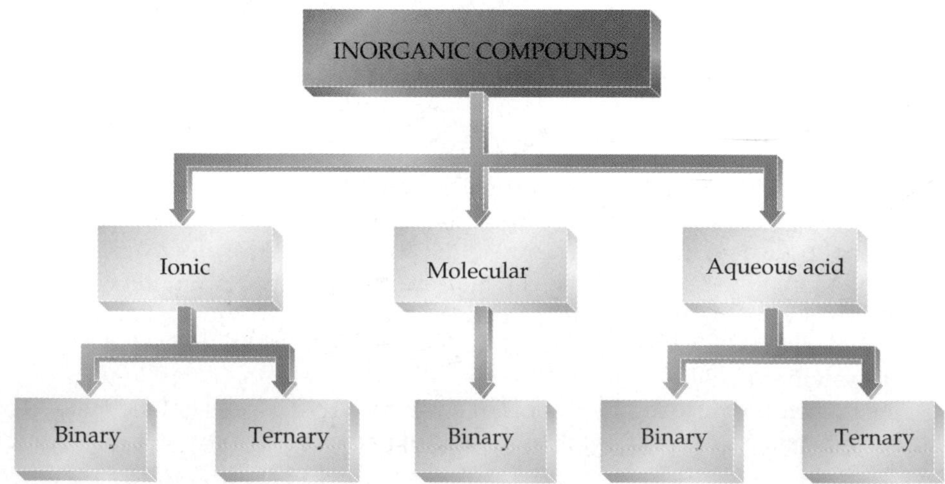

Classification of Inorganic Compounds

▶ **Figure 7.1 Classification of Inorganic Compounds** According to IUPAC nomenclature, inorganic compounds are divided into five categories: binary ionic, ternary ionic, binary molecular, binary acid, and ternary oxyacid.

The following example exercise illustrates the classification of different types of compounds and acids.

Example Exercise 7.1 • Classifying Compounds

Classify each of the following as a binary ionic compound, ternary ionic compound, binary molecular compound, binary acid, or ternary oxyacid.

(a) titanium oxide, TiO_2
(b) sulfur dioxide, SO_2
(c) silver chromate, Ag_2CrO_4
(d) hydrofluoric acid, HF(aq)
(e) carbonic acid, H_2CO_3(aq)

Solution

We can refer to the periodic table and classify each compound or solution as follows.

(a) TiO_2 contains two elements, a metal and nonmetal. Thus, TiO_2 is a *binary ionic compound.*
(b) SO_2 contains two elements, both nonmetals. Thus, SO_2 is a *binary molecular compound.*
(c) Ag_2CrO_4 contains three elements, two metals and a nonmetal. Thus, Ag_2CrO_4 is a *ternary ionic compound.*
(d) HF(aq) is a compound of hydrogen and a nonmetal dissolved in water. Thus, HF(aq) is a *binary acid.*
(e) H_2CO_3(aq) is a compound containing three elements, including hydrogen and oxygen, dissolved in water. Thus, H_2CO_3(aq) is a *ternary oxyacid.*

Self-Test Exercise

Classify each of the following as a binary ionic compound, ternary ionic compound, binary molecular compound, binary acid, or ternary oxyacid.

(a) carbon disulfide, CS_2
(b) lithium dichromate, $Li_2Cr_2O_7$
(c) magnesium iodide, MgI_2
(d) nitric acid, HNO_3(aq)
(e) hydrochloric acid, HCl(aq)

Answers: (a) binary molecular compound; (b) ternary ionic compound; (c) binary ionic compound; (d) ternary oxyacid; (e) binary acid

▲ **Rutile, TiO_2** Rutile is a natural mineral that contains titanium.

Classification of Ions

According to IUPAC nomenclature, ions are named systematically, depending on the category in which they are placed. A positive ion is referred to as a **cation**, and a negative ion is referred to as an **anion**. A single atom bearing a positive or negative charge is called a **monoatomic ion**. A particle containing two or more atoms having a positive or negative charge is called a **polyatomic ion**. Figure 7.2 illustrates the classification of and relationship of these ions.

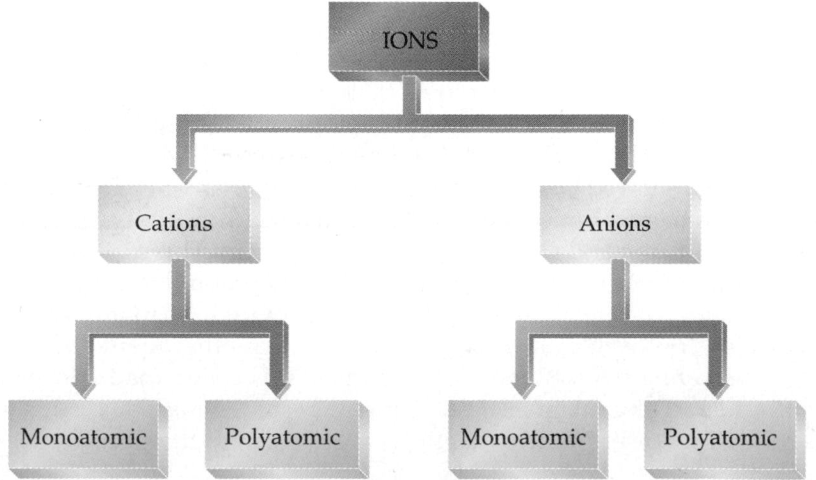

◀ **Figure 7.2 Classification of Ions** According to the IUPAC nomenclature, ions are divided into four categories. They are monoatomic cations, polyatomic cations, monoatomic anions, and polyatomic anions.

The following example exercise illustrates the classification of different types of ions.

Example Exercise 7.2 • Classifying Cations and Anions

Classify each of the following ions as a monoatomic cation, monoatomic anion, polyatomic cation, or polyatomic anion.

(a) chromium ion, Cr^{3+}
(b) chloride ion, Cl^-
(c) mercurous ion, Hg_2^{2+}
(d) sulfate ion, SO_4^{2-}

Solution

We can classify each ion as follows.

(a) Cr^{3+} is a single atom with a positive three charge. Thus, Cr^{3+} is a *monoatomic cation.*
(b) Cl^- is a single atom with a negative charge. Thus, Cl^- is a *monoatomic anion.*
(c) Hg_2^{2+} contains two atoms and has a positive charge. Thus, Hg_2^{2+} is a *polyatomic cation.*
(d) SO_4^{2-} contains five atoms and has a negative two charge. Thus, SO_4^{2-} is a *polyatomic anion.*

Self-Test Exercise

Classify each of the following ions as a monoatomic cation, monoatomic anion, polyatomic cation, or polyatomic anion.

(a) ammonium ion, NH_4^+
(b) sulfide ion, S^{2-}
(c) permanganate ion, MnO_4^-
(d) lithium ion, Li^+

Answers: (a) polyatomic cation; (b) monoatomic anion; (c) polyatomic anion; (d) monoatomic cation

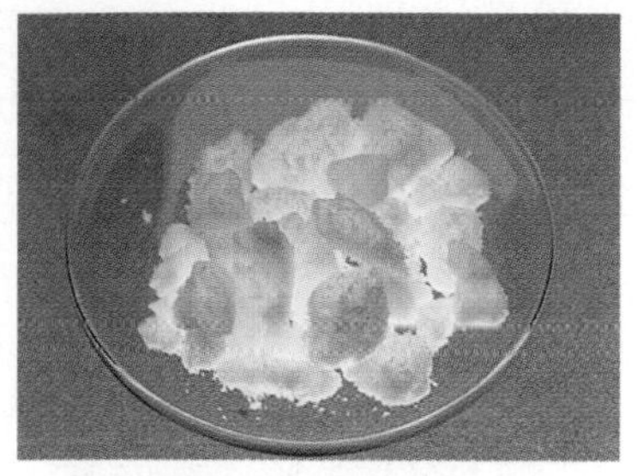

▲ **Ammonium Carbonate, $(NH_4)_2CO_3$** Ammonium carbonate decomposes in air to give ammonia, NH_3, and carbon dioxide, CO_2; hence, it is used in smelling salts and baking powders.

7.2 Monoatomic Ions

Objectives
· To write systematic names and formulas for common monoatomic ions.
· To predict the ionic charge for ions of representative elements.

As we now know, metal atoms lose valence electrons and become positively charged ions. Another name for positively charged ions is cathode ions. This name comes from the fact that in a battery containing an aqueous solution, positive ions are attracted to the negative electrode, or cathode. Rather than referring to these positive ions as cathode ions, they are simply called cations.

Naming Monoatomic Cations

According to IUPAC nomenclature rules, cations are named for the parent metal followed by the word "ion." For example, Na^+ is named sodium ion, Mg^{2+} is named magnesium ion, and Al^{3+} is named aluminum ion.

Main-group metals usually form one cation; tin and lead are exceptions. Transition metals, however, often form more than one cation. Iron, for example, can form Fe^{2+} and Fe^{3+}. To name a metal cation having more than one ionic charge, it is necessary to specify the charge. IUPAC recommends that the cation be named for the parent metal followed by its charge specified by Roman numerals in parentheses. Thus, Fe^{2+} is named iron(II) ion, and Fe^{3+} is named iron(III) ion. Similarly, Cu^+ is named copper(I) ion and Cu^{2+} is named copper(II) ion. This method of naming transition metal cations is called the **Stock system**. We should note that the names of Ag^+, Zn^{2+}, and Cd^{2+} are exceptions and do not require Roman numerals.

In addition to the Stock system, IUPAC allows another method for naming metal cations having two common ionic charges. It is called the **Latin system** or the suffix system. This system takes the Latin name of the metal and adds the suffix *-ous* or *-ic*. The lower of the two ionic charges receives the suffix *-ous*, and the

Chemistry Connection • Antoine Lavoisier

Who provided invaluable laboratory assistance and helped spread the enormous body of work published by Lavoisier?

Antoine Laurent Lavoisier (1743–1794) was born into an affluent French family and received an excellent education. He initially became a lawyer like his father, but later turned to science. Lavoisier was a gifted experimenter and is generally considered to be the founder of modern chemistry. He established a magnificent laboratory that attracted scientists from all over the world; even Benjamin Franklin and Thomas Jefferson were among its visitors.

In 1787 Lavoisier published a book entitled *Methods of Chemical Nomenclature.* The book designated systematic principles for naming chemical substances and was so logical it became the basis for our present rules of nomenclature. In 1789 he published the landmark textbook *Elementary Treatise on Chemistry,* which offered the first modern view of chemistry.

◀ Antoine Lavoisier and his wife, Marie

Before the time of Lavoisier, it was commonly believed that for a substance to burn, it must contain the element *phlogiston.* Lavoisier suggested that it was not phlogiston but an element in the air, that was responsible for combustion. Lavoisier named the element oxygen, but it was the English chemist Joseph Priestley who first discovered it.

In 1771, when he was 28, Lavoisier married Marie-Anne Pierrette, who at the time was only 14 years old. Despite her youth, she proved invaluable to his work and joined alongside him in the laboratory. Unfortunately, she was also the daughter of an important official in a firm that collected taxes. Although Lavoisier did not collect taxes, he received royalties from the same firm. When the French Revolution broke out, revolutionaries threw tax collectors into jail. Lavoisier was first barred from his laboratory and then arrested. After a farce trial, he was sentenced to the guillotine along with his father-in-law and other tax collectors.

On May 8, 1794, Lavoisier was executed and buried in an unmarked grave. The French court declared that France had no need for scientists, but the famous mathematician Lagrange lamented: "A moment was all that was necessary to strike off his head, and probably a hundred years will not be sufficient to produce another like it." Within 2 years, the French people regretted their mistake and began unveiling statues of Lavoisier.

Lavoisier's wife, Marie-Anne, is credited with keeping laboratory records and translating scientific papers into other languages.

higher charge receives the suffix *-ic*. For example, the Latin name for iron is *ferrum*. To name an iron ion, we take the stem *ferr-* and add the suffix *-ous* or *-ic*. Thus, Fe^{2+} is named ferrous ion, and Fe^{3+} is named ferric ion. The Latin name for copper is *cuprum*. Similarly, we add the suffix *-ous* or *-ic* to the stem *cupr-*. Thus, we have Cu^{+} for cuprous ion, and Cu^{2+} for cupric ion.

The formula for the mercury(I) ion is an exception. When mercury loses an electron, the resulting ion, Hg^{+}, becomes more stable by combining with another Hg^{+} ion. Thus, the mercury(I) ion is written Hg_2^{2+}. The Latin name for mercury, *hydrargyrum*, is difficult to pronounce, and so IUPAC recommends adding the suffix to the English name. Thus, Hg_2^{2+} is named mercurous ion, and Hg^{2+} is named mercuric ion. Table 7.1 lists the systematic names for common cations.

Naming Monoatomic Anions

As we now know, nonmetal atoms can gain valence electrons and become negatively charged ions called anions. According to IUPAC rules, nonmetal ions are named by using the *nonmetal stem* plus the suffix *-ide*. Examples of nonmetal ions are Cl^{-}, chloride ion; S^{2-}, sulfide ion; and P^{3-}, phosphide ion. Table 7.2 lists the systematic names for common anions.

Table 7.1 Common Monoatomic Cations

Cation	Stock System	Latin System	Cation	Stock System	Latin System
Al^{3+}	aluminum ion		Li^{+}	lithium ion	
Ba^{2+}	barium ion		Mg^{2+}	magnesium ion	
Cd^{2+}	cadmium ion		Mn^{2+}	manganese(II) ion	
Ca^{2+}	calcium ion		Hg_2^{2+}	mercury(I) ion*	mercurous ion
Co^{2+}	cobalt(II) ion		Hg^{2+}	mercury(II) ion	mercuric ion
Co^{3+}	cobalt(III) ion		Ni^{2+}	nickel(II) ion	
Cu^{+}	copper(I) ion	cuprous ion	K^{+}	potassium ion	
Cu^{2+}	copper(II) ion	cupric ion	Ag^{+}	silver ion	
Cr^{3+}	chromium(III) ion		Na^{+}	sodium ion	
H^{+}	hydrogen ion		Sr^{2+}	strontium ion	
Fe^{2+}	iron(II) ion	ferrous ion	Sn^{2+}	tin(II) ion	stannous ion
Fe^{3+}	iron(III) ion	ferric ion	Sn^{4+}	tin(IV) ion	stannic ion
Pb^{2+}	lead(II) ion	plumbous ion	Zn^{2+}	zinc ion	
Pb^{4+}	lead(IV) ion	plumbic ion			

*Note that the mercury(I) ion is diatomic and is written Hg_2^{2+}.

Table 7.2 Common Monoatomic Anions

Anion	IUPAC Name	Anion	IUPAC Name
Br^{-}	bromide ion	N^{3-}	nitride ion
Cl^{-}	chloride ion	O^{2-}	oxide ion
F^{-}	fluoride ion	P^{3-}	phosphide ion
I^{-}	iodide ion	S^{2-}	sulfide ion

Experiment #8, Prentice Hall Laboratory Manual

Predicting Formulas of Monoatomic Cations

We can use the periodic table to help learn the names and formulas of the ions and predict their charge based on the group number of the element. Recall that Group IA/1 metals always form 1+ ions. Thus, we have Li^{+}, Na^{+}, and K^{+}. Also recall that the Group IIA/2 elements always form 2+ ions. Thus, we have Mg^{2+}, Ca^{2+}, Sr^{2+}, and Ba^{2+}. Similarly, we can predict that aluminum in Group IIIA/13 forms a 3+ ion, that is, Al^{3+}.

Not all metal ions are predictable from the periodic table. Tin and lead in Group IVA/14 can each form two ions: Sn^{2+} or Sn^{4+} and Pb^{2+} or Pb^{4+}. Therefore, you will have to memorize the formulas for these ions.

With few exceptions all transition elements have two *s* electrons as well as a variable number of *d* electrons. In general, a transition metal loses its two *s* electrons to form an ion with a 2+ charge, for example, Ni^{2+}. Many transition metals form additional ions having charges that are unpredictable. You will have to memorize the names of these ions. Figure 7.3 shows the relationship of ionic charge to the position of the element in the periodic table.

Although we should memorize the names and formulas of ions, the periodic table is a valuable resource for verifying ionic charges. The following example exercise illustrates the names and formulas of monoatomic cations.

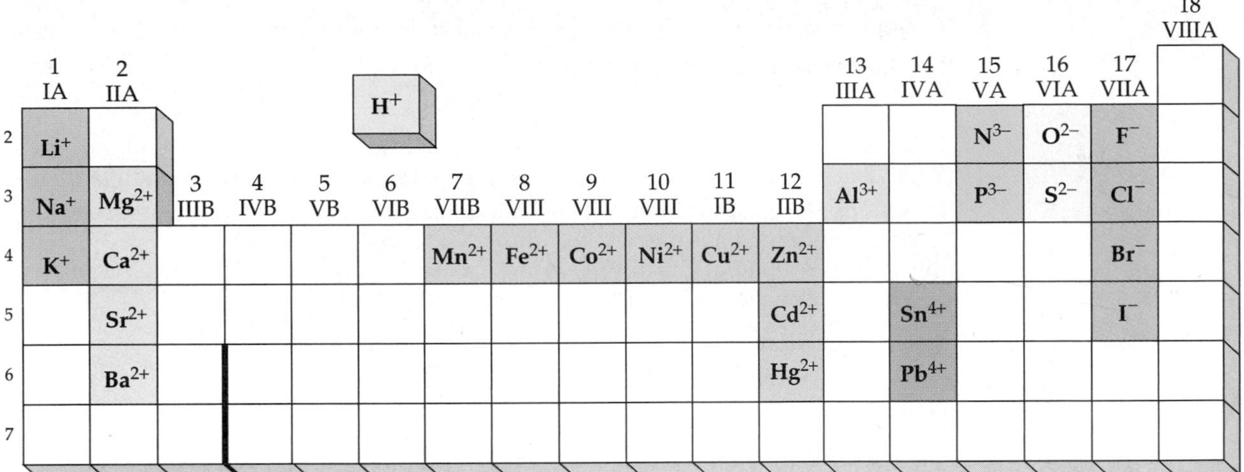

▲ **Figure 7.3 Periodic Table of Selected Ions** Note the correlation of ionic charge and group number. Although the transition elements exhibit more than one ionic charge, nearly all these metals exhibit a charge of 2+.

Periodic Table of Selected Ions

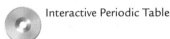

Interactive Periodic Table

Example Exercise 7.3 • Names and Formulas of Monoatomic Cations

Provide the formula for the following monoatomic cations.

(a) barium ion (b) cobalt(II) ion

Solution

We can use the periodic table to predict the charge on a cation.

(a) Barium is found in Group IIA/2 and can lose two valence electrons. We predict the formula of the ion is Ba^{2+}.

(b) Cobalt is a transition metal and can lose two or more valence electrons. The Roman number (II) indicates the loss of two electrons, and so the formula of the ion is Co^{2+}.

Self-Test Exercise

Supply a systematic name for the following monoatomic cations.

(a) Al^{3+} (b) Hg^{2+}

Answers: (a) aluminum ion; (b) mercury(II) ion or mercuric ion

Predicting Formulas of Monoatomic Anions

Nonmetals gain electrons to form negative ions in a predictable fashion. The halogens in Group VIIA/17 need only one electron to become isoelectronic with a noble gas. Thus, the halogens have a 1− charge and the formulas F^-, Cl^-, Br^-, and I^-. The elements in Group VIA/16 gain two electrons to become isoelectronic with a noble gas. Thus, these elements have a 2− charge and the formulas O^{2-}, S^{2-}, and Se^{2-}. Nitrogen and phosphorus are in Group VA/15 and gain three electrons. These elements have a 3− charge and the formulas N^{3-} and P^{3-}.

Example Exercise 7.4 • Names and Formulas of Monoatomic Anions

Provide the formula for each of the following monoatomic anions.

(a) fluoride ion (b) oxide ion

Solution

Recall that nonmetals gain electrons so as to acquire a noble gas electron configuration.

(a) Fluorine is found in Group VIIA/17 and can gain one electron to become isoelectronic with neon. We predict the formula for the fluoride ion is F^-.

(b) Oxygen is found in Group VIA/16 and can gain two electrons to become isoelectronic with neon. We predict the formula for the oxide ion is O^{2-}.

Self-Test Exercise

Supply a systematic name for each of the following monoatomic anions.

(a) Br^- (b) N^{3-}

Answers: (a) bromide ion; (b) nitride ion

Experiment #9, Prentice Hall Laboratory Manual

7.3 Polyatomic Ions

Objective · To write systematic names and formulas for common polyatomic ions.

Polyatomic anions generally contain one or more elements combined with oxygen. Such an anion is referred to as an **oxyanion**, a term that reflects its composition. Most oxyanions have names ending in the suffix *-ate*. Examples include the nitrate ion, NO_3^-, and the sulfate ion, SO_4^{2-}. A few oxyanions have names ending in the suffix *-ite*. Examples include the nitrite ion, NO_2^-, and the sulfite ion, SO_3^{2-}. Notice that in each case the formula for ions with the suffix *-ite* has one less oxygen than that for ions with the suffix *-ate*.

This suffix pattern can provide clues when naming some polyatomic ions. Given the formula for the chlorate ion ClO_3^-, we can predict the formula for the chlorite ion. Since the suffix *-ate* has changed to the suffix *-ite*, the formula has one less oxygen. Therefore, the formula for the chlorite ion is ClO_2^-. The general relationship between the suffixes *-ate* and *-ite* allows us to simplify our task of memorizing formulas for ions.

There are two important polyatomic ions that have the suffix *-ide*. The cyanide ion, CN^-, and the hydroxide ion, OH^-, are exceptions, and their names should be memorized. In addition to these two polyatomic anions, there is the ammonium cation, NH_4^+, whose formula is derived from that for ammonia gas, NH_3. Table 7.3 lists the common polyatomic ions whose formulas should be learned.

The following example exercise illustrates how general principles can be used for assistance in memorizing the names and formulas of polyatomic ions.

Table 7.3 Common Polyatomic Ions

Cation	IUPAC Name
NH_4^+	ammonium ion
Anion	**IUPAC Name**
$C_2H_3O_2^-$	acetate ion
CO_3^{2-}	carbonate ion
ClO_3^-	chlorate ion
ClO_2^-	chlorite ion
CrO_4^{2-}	chromate ion
CN^-	cyanide ion*
$Cr_2O_7^{2-}$	dichromate ion
HCO_3^-	hydrogen carbonate ion
HSO_4^-	hydrogen sulfate ion
OH^-	hydroxide ion*
ClO^-	hypochlorite ion
NO_3^-	nitrate ion
NO_2^-	nitrite ion
ClO_4^-	perchlorate ion
MnO_4^-	permanganate ion
PO_4^{3-}	phosphate ion
SO_4^{2-}	sulfate ion
SO_3^{2-}	sulfite ion

*Note that the suffix *-ide* is an exception to the general *-ate* and *-ite* rule.

Example Exercise 7.5 • Names and Formulas of Polyatomic Ions

Provide a systematic name for each of the following polyatomic oxyanions.

(a) CO_3^{2-} (b) CrO_4^{2-} (c) ClO_2^- (d) HSO_4^-

Solution

We can make reasonable predictions for the names of many polyatomic ions. This makes the task of memorization much easier.

(a) CO_3^{2-} contains carbon, and we predict the name has the suffix *-ate*. Thus, we predict CO_3^{2-} is named the *carbonate ion*.

(b) CrO_4^{2-} contains chromium, and we predict the name has the suffix *-ate*. Thus, we predict CrO_4^{2-} is named the *chromate ion*.

(c) ClO_2^- is related to ClO_3^-, which is named the chlorate ion. Since ClO_2^- has one less oxygen atom, the suffix changes to *-ite*. Thus, we predict ClO_2^- is named the *chlorite ion*.
(d) HSO_4^- is related to the sulfate ion, SO_4^{2-}. With the addition of hydrogen, the name becomes the *hydrogen sulfate ion*.

Self-Test Exercise

Provide the formula for each of the following polyatomic oxyanions.

(a) acetate ion (b) dichromate ion
(c) perchlorate ion (d) hydrogen carbonate ion

Answers: (a) $C_2H_3O_2^-$; (b) $Cr_2O_7^{2-}$; (c) ClO_4^-; (d) HCO_3^-

---

Note You may find it helpful to make flashcards to help you memorize ions. Write the name of the ion on one side of a card and the formula of the ion on the other side. Your task will be easier, however, if you recall the following.

1. There is one common polyatomic cation, NH_4^+.
2. Most oxyanions have the suffix *-ate*.
3. The suffix *-ate* changes to the suffix *-ite* for an oxyanion with one less oxygen atom; for example, nitrate, NO_3^-, changes to nitrite, NO_2^-.
4. There are two common polyatomic anions that are exceptions: OH^- and CN^- each have the suffix *-ide*.

7.4 Writing Chemical Formulas

Objective · To write formula units for compounds composed of monoatomic and polyatomic ions.

An ionic compound is composed of positive and negative ions. A **formula unit** is the simplest representative particle in an ionic compound. Since a formula unit is neutral, the total positive charge must equal the total negative charge in both the formula unit and the ionic compound. That is, the total positive charge from the metal ions must be the same as the total negative charge from the nonmetal ions.

A formula unit of ordinary salt contains Na^+ and Cl^-. Since the positive and negative ions have equal but opposite charges, the formula for the compound is NaCl. In a formula unit containing Ca^{2+} and Cl^- the charges are not the same. It is necessary to have two Cl^- ions for each Ca^{2+} ion to balance the charges and have a neutral compound; therefore, the formula is $CaCl_2$. In a formula unit containing Al^{3+} and Cl^-, it is necessary to have three Cl^- ions for each Al^{3+} ion, and the formula for the compound is $AlCl_3$.

The following example exercise provides additional illustrations of how to write formulas for ionic compounds.

▲ **Salt, NaCl** NaCl is so essential to health that, historically, laborers were paid in salt. In fact, the term *salary* translates as "salt money."

Example Exercise 7.6 • Writing Formulas of Binary Ionic Compounds

Write the chemical formula for the following binary compounds given their constituent ions.

(a) copper(I) oxide, Cu^+ and O^{2-} (b) cadmium oxide, Cd^{2+} and O^{2-}
(c) cobalt(III) oxide, Co^{3+} and O^{2-}

Solution

(a) The copper(I) ion has a charge of 1+, and the oxide ion has a charge of 2−. Thus, two positive ions are required for each negative ion in a neutral formula unit. The formula of copper(I) oxide is written Cu_2O.

(continued)

Example Exercise 7.6 *(continued)*

(b) Since the cadmium ion and oxide ion each have a charge of 2, the ratio is 1:1, that is, Cd_1O_1. It is not necessary to write the subscript 1, and so the formula of cadmium oxide is simply CdO.

(c) This example is more difficult. The cobalt(III) ion has a charge of 3+, and the oxide ion has a charge of 2−. Since the lowest common multiple is 6, two 3+ ions are required to cancel the charge of three 2− ions. The ratio is 2:3, and the formula of cobalt(III) oxide is written Co_2O_3.

Self-Test Exercise

Write the chemical formula for the following binary compounds given their constituent ions.

(a) iron(II) sulfide, Fe^{2+} and S^{2-}
(b) mercury(I) fluoride, $Hg_2{}^{2+}$ and F^-
(c) lead(IV) oxide, Pb^{4+} and O^{2-}

Answers: (a) FeS; (b) Hg_2F_2; (c) PbO_2

Note You can quickly verify that the chemical formula is written correctly by crossing-over the charge on each ion. Consider aluminum oxide, which contains Al^{3+} and O^{2-}. The 3+ charge on the aluminum ion becomes the subscript for the oxygen, and the 2− charge on the oxide ion becomes the subscript for the aluminum ion. That is,

$$Al^{(3+)}O^{(2-)} = Al_2O_3$$

Formula Units Containing Polyatomic Ions

As mentioned earlier, formula units are the simplest particles representing an ionic compound, and the total positive charge is equal to the total negative charge. Previously, we learned to write neutral formula units for binary ionic compounds. Similarly, in a formula unit containing K^+ and $SO_4{}^{2-}$, two K^+ ions are required to balance the charge for each $SO_4{}^{2-}$. Thus, a neutral formula unit is written K_2SO_4.

Magnesium sulfate is found in Epsom salts and contains Mg^{2+} and $SO_4{}^{2-}$. Since the magnitude of the charge is the same on each ion, the ratio of positive ion to negative ion is 1:1. The formula unit is written $MgSO_4$.

Ammonium sulfate is a nitrogen-supplying component in fertilizer and contains $NH_4{}^+$ and $SO_4{}^{2-}$ ions. Since the negative charge is greater for $SO_4{}^{2-}$, two $NH_4{}^+$ ions are necessary to give a neutral formula unit. To avoid misunderstanding, parentheses are placed around the $NH_4{}^+$ ion. The correct formula is written $(NH_4)_2SO_4$. The following example exercise provides additional illustrations of how to write formulas for ionic compounds.

▲ **Chalk, $CaCO_3$** Ordinary classroom chalk is composed of calcium carbonate, whose chemical formula is $CaCO_3$.

Example Exercise 7.7 • Writing Formulas of Ternary Ionic Compounds

Write the chemical formula for each of the following ternary compounds given their constituent ions.

(a) calcium carbonate, Ca^{2+} and $CO_3{}^{2-}$
(b) calcium hydroxide, Ca^{2+} and OH^-
(c) calcium phosphate, Ca^{2+} and $PO_4{}^{3-}$

Solution

(a) Since the positive and negative ions each have a charge of 2, one positive ion and one negative ion are required to produce a neutral formula unit, and the formula is $CaCO_3$. Calcium carbonate occurs naturally as ordinary chalk.

(b) The positive ion has a charge of 2+, and the negative ion has a charge of 1−. Therefore, one positive ion and two negative ions are required to produce a neutral formula unit. Since OH^- is a polyatomic ion, parentheses are required, and the formula is written $Ca(OH)_2$. Calcium hydroxide is known as slaked lime and is sometimes used to mark the boundaries of an athletic field.

(c) The positive ion has a charge of 2+, and the negative ion has a charge of 3−. The lowest common multiple of the charges is 6. Three positive ions are required for every two negative ions to produce a neutral formula unit. A calcium phosphate formula unit is written $Ca_3(PO_4)_2$. Calcium phosphate is found in tooth enamel.

Self-Test Exercise

Write the chemical formula for each of the following ternary compounds given their constituent ions.

(a) copper(II) permanganate, Cu^{2+} and MnO_4^-
(b) iron(III) carbonate, Fe^{3+} and CO_3^{2-}
(c) ammonium dichromate, NH_4^+ and $Cr_2O_7^{2-}$

Answers: (a) $Cu(MnO_4)_2$; (b) $Fe_2(CO_3)_3$; (c) $(NH_4)_2Cr_2O_7$

Note As before, we can verify that the formula is correct by simply crossing-over the charge on each ion. Consider calcium phosphate, which contains Ca^{2+} and PO_4^{3-}. The 2+ charge on the calcium ion becomes the subscript for the phosphate ion. Since phosphate is a polyatomic ion, we must use parentheses, that is, (PO_4). Conversely, the 3− charge on the phosphate ion becomes the subscript for the calcium ion.

$$Ca^{(2+)}\ PO_4^{(3-)} = Ca_3(PO_4)_2$$

Update • IUPAC Nomenclature

What is the systematic name for stomach acid?

Alchemists, the forerunners of modern chemists, are usually associated with attempts to convert lead to gold. They did, however, produce a rich vocabulary describing chemical substances. Unfortunately, their names for substances did not offer a clue to the chemical composition.

Antoine Lavoisier, the founder of modern chemistry, provided the first published naming system that indicated the chemical composition of substances. Lavoisier, along with several collaborators, published *Méthode de Nomenclature Chimique* in 1787. The great Swedish chemist J. J. Berzelius adopted Lavoisier's ideas and extended the system into the Germanic languages, which in turn, were translated into English.

In 1892 a conference was held in Geneva, Switzerland, that laid the foundation for an internationally accepted system of nomenclature for organic compounds. Subsequently, the International Union of Pure and Applied Chemistry (IUPAC) was formed.

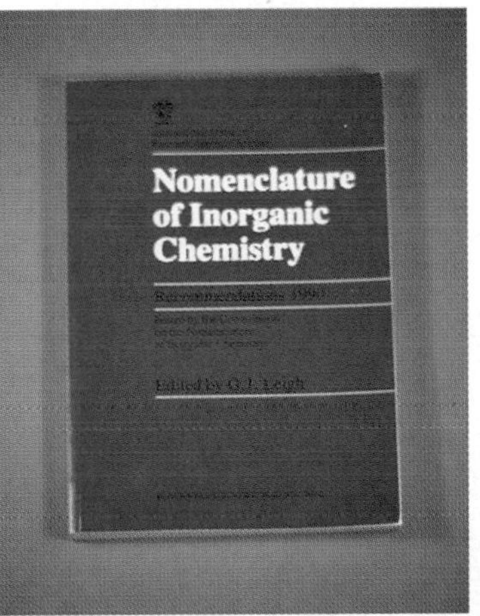

◀ The Official IUPAC Red Book

In 1921 IUPAC addressed the issue of systematic nomenclature for *inorganic compounds*. Almost 20 years elapsed before IUPAC issued a systematic set of rules. Interestingly, these so-called 1940 IUPAC rules were consistent with the methods proposed in Lavoisier's text 150 years earlier. The official recommendations suggest that there is more than one acceptable system of nomenclature. For example, $CuSO_4$ can be named copper(II) sulfate according to the preferred Stock system of nomenclature or cupric sulfate according to the Latin system. Although common names still persist, such as baking soda for $NaHCO_3$, IUPAC disapproves of the use of common names and discourages the use of names that do not indicate the chemical composition.

In truth, the chemistry community is not in complete agreement on subtle aspects of nomenclature. Currently, the most authoritative guide to systematic naming is the 1990 IUPAC recommendations contained in *Nomenclature of Inorganic Chemistry*. This official publication is referred to as the *Red Book*.

Stomach acid is aqueous HCl, and its systematic name is hydrochloric acid.

7.5 Binary Ionic Compounds

Objectives
- To determine the ionic charge on a cation in a binary ionic compound.
- To write systematic names and formulas for binary ionic compounds.

Positively charged metal ions and negatively charged nonmetal ions compose neutral binary ionic compounds. The oppositely charged cation and anion are attracted to each other and create compounds in which the overall charge is neutral. That is, an ionic compound has an overall charge of zero even though it contains charged ions.

Binary Compounds Containing a Transition Metal

Some binary ionic compounds contain transition metal ions, which have a variable ionic charge, for instance, Fe^{2+} and Fe^{3+}. To name such a compound correctly, we first have to determine the ionic charge of the metal cation. The following example exercise illustrates the determination of ionic charge.

▲ **Hematite, Fe_2O_3** Hematite is a natural mineral that contains iron.

Example Exercise 7.8 • Determining Ionic Charge in a Compound

Determine the ionic charge for iron in the mineral hematite, Fe_2O_3.

Solution

The charge on an oxide ion is 2−, and there are three oxide ions. The total negative charge must be equal to six negative:

$$O^{2-} + O^{2-} + O^{2-} = 6 \text{ negative}$$

Since all compounds are electrically neutral, the total positive charge must equal the total negative charge: 6 negative = 6 positive. Thus, the two iron ions have a charge of six positive.

$$Fe^{x+} + Fe^{x+} = 6 \text{ positive}$$

$$Fe^{x+} = 3 \text{ positive}$$

The iron ion is therefore Fe^{3+}. The name of Fe_2O_3 is iron(III) oxide according to the Stock system. It is named ferric oxide according to the Latin system.

Self-Test Exercise

Determine the ionic charge for the transition metal in each of the following compounds.

(a) Cu_3P (b) CoN

Answers: (a) Cu+; (b) Co^{3+}

▲ **Cinnabar, Hgs** Cinnabar is a natural mineral that contains mercury.

Naming Binary Ionic Compounds

When writing a chemical formula, the positive cation is always written before the negative anion. Therefore, in naming binary ionic compounds, the metal cation is named first followed by the nonmetal anion. We learned in Section 7.2 that monoatomic anions end with the suffix *-ide*. For example, MgO is composed of the magnesium ion and the oxide ion and is named magnesium oxide.

The mineral cinnabar, HgS, is composed of the mercury ion and the sulfide ion. Mercury has two possible ions, Hg_2^{2+} and Hg^{2+}. Since the ionic charge on a sulfide ion is 2−, the mercury ion is Hg^{2+}. Thus, we can name HgS mercury(II) sulfide according to the Stock system. Alternatively, we can name HgS mercuric sulfide according to the Latin system.

Example Exercise 7.9 • Names of Binary Ionic Compounds

Supply a systematic name for each of the following binary ionic compounds.

(a) ZnO (b) SnF_2

Solution

We can name an ionic compound by designating the two ions.

(a) ZnO contains the zinc ion and the oxide ion; ZnO is named zinc oxide.

(b) SnF_2 contains the tin(II) ion and the fluoride ion. Thus, SnF_2 is named tin(II) fluoride. The Latin system name is stannous fluoride, which is an active ingredient in a popular toothpaste.

Self-Test Exercise

Supply a systematic name for each of the following binary ionic compounds.

(a) Mn_3P_2 (b) Fe_2S_3

Answers: (a) manganese(II) phosphide; (b) iron(III) sulfide or ferric sulfide

Example Exercise 7.10 • Formulas of Binary Ionic Compounds

Provide the formula for each of the following binary ionic compounds.

(a) lithium fluoride (b) plumbous sulfide

Solution

We can write the formula by combining the cation and the anion into a neutral formula unit. Refer to Section 7.4 to review the writing of formula units.

(a) Lithium fluoride is composed of Li^+ and F^-; thus, the formula of the compound is written LiF.

(b) Plumbous sulfide is composed of Pb^{2+} and S^{2-}; thus, the formula of the compound is written PbS.

Self-Test Exercise

Provide the formula for each of the following binary ionic compounds.

(a) copper(II) iodide (b) mercuric oxide

Answers: (a) CuI_2; (b) HgO

▲ **Galena, PbS** Galena is a natural mineral that contains lead.

Predicting Formulas of Binary Ionic Compounds

We have learned the value of the periodic table in mastering the formulas of ions. Now we are going to use the periodic table to predict the chemical formulas of binary ionic compounds.

We can predict the formulas of compounds based on the formula of a similar compound. Let's begin with alkali metal chlorides. Sodium is in Group IA/1. The formula of sodium chloride is $NaCl$. All alkali metal chlorides, in fact, have a similar chemical formula. For the alkali metal family, the chloride compounds have the following formulas: $LiCl$, $NaCl$, and KCl.

We can predict the formulas for alkaline earth metal chlorides in a similar manner. Barium is in Group IIA/2, and the formula for barium chloride is $BaCl_2$. Therefore, the formulas for the other alkaline earth metal chlorides are $MgCl_2$, $CaCl_2$, and $SrCl_2$.

Example Exercise 7.11 • Predicting Formulas of Binary Ionic Compounds

Predict the chemical formula for each of the following binary compounds, given the formula of aluminum oxide, Al_2O_3.

(a) gallium oxide (b) aluminum sulfide

Solution

To predict the chemical formula, we compare the elements that are different in the similar compounds.

(a) The elements Ga and Al are both in Group IIIA/3, and so the formula is Ga_2O_3.
(b) The elements S and O are both in Group VIA/16, and so the formula is Al_2S_3.

Self-Test Exercise

Predict the chemical formula for each of the following binary compounds, given the formula of magnesium chloride, $MgCl_2$.

(a) radium chloride (b) magnesium fluoride

Answers: (a) $RaCl_2$; (b) MgF_2

7.6 Ternary Ionic Compounds

Objectives
- To determine the ionic charge on a cation in a ternary ionic compound.
- To write systematic names and formulas for ternary ionic compounds.

Compounds containing a metal and two other elements are classified as ternary ionic compounds. Typically, ternary ionic compounds have a monoatomic metal cation and a polyatomic oxyanion, for example, $NaNO_3$.

Like all ionic compounds, the cation is written first in the formula. The names of ternary ionic compounds usually have the suffix *-ate* or *-ite*. For example, $KClO_3$ is potassium chlorate, and $KClO_2$ is potassium chlorite.

Ternary Compounds Containing a Transition Metal

Some ternary compounds contain transition metal ions that have a variable ionic charge, for instance, Fe^{2+} and Fe^{3+}. To name such a compound correctly, we first have to determine the ionic charge of the transition metal cation.

Example Exercise 7.12 • Determining Ionic Charge in a Compound

Determine the ionic charge for iron in iron phosphate, $Fe_3(PO_4)_2$.

Solution

The charge on a phosphate ion is 3−, and there are two phosphate ions. Therefore, the total negative charge must be equal to six negative:

$$PO_4^{3-} + PO_4^{3-} = 6 \text{ negative}$$

Since all compounds are electrically neutral, the total positive charge must equal the total negative charge: 6 negative = 6 positive. Thus, the three iron ions have a charge of six positive.

$$Fe^{x+} + Fe^{x+} + Fe^{x+} = 6 \text{ positive}$$

$$Fe^{x+} = 2 \text{ positive}$$

The iron ion is therefore Fe^{2+}. The name of $Fe_3(PO_4)_2$ is iron(II) phosphate according to the Stock system. It is named ferrous phosphate according to the Latin system.

Self-Test Exercise

Determine the ionic charge for the metal cation in each of the following compounds.

(a) $Hg(OH)_2$ (b) $Co(ClO_3)_3$

Answers: (a) Hg^{2+}; (b) Co^{3+}

Naming Ternary Ionic Compounds

Like all ionic compounds, ternary ionic compounds are named by designating the cation followed by the anion. The mineral marble, $CaCO_3$, is composed of the calcium ion and the carbonate ion and is named calcium carbonate. $Ca(NO_3)_2$ is composed of the calcium ion and two nitrate ions and is named calcium nitrate.

The compound $CuSO_4$ is composed of a copper ion and a sulfate ion. Copper, however, has two possible ions, Cu^+ and Cu^{2+}. Since the sulfate charge is 2−, the charge on copper must be 2+ because compounds are neutral. Since the copper ion is Cu^{2+}, the name of the compound is copper(II) sulfate, or cupric sulfate.

▲ **Marble, $CaCO_3$** The Taj Mahal in India is made of the natural mineral marble.

Example Exercise 7.13 • Names of Ternary Ionic Compounds

Supply a systematic name for the following ternary ionic compounds.

(a) $KMnO_4$ (b) $Hg(NO_3)_2$

Solution

We can name an ionic compound by designating the two ions.

(a) $KMnO_4$ is composed of the potassium ion and the permanganate ion. Thus, the compound is named potassium permanganate.

(b) $Hg(NO_3)_2$ contains the mercury(II) ion and the nitrate ion. Therefore, it is named mercury(II) nitrate or mercuric nitrate.

Self-Test Exercise

Supply a systematic name for each of the following ternary ionic compounds.

(a) $BaCrO_4$ (b) $Cu(NO_2)_2$

Answers: (a) barium chromate; (b) copper(II) nitrite or cupric nitrite

Example Exercise 7.14 • Formulas of Ternary Ionic Compounds

Provide the formula for each of the following ternary ionic compounds.

(a) nickel(II) acetate (b) ferric sulfate

Solution

We can write the formula by combining the cation and the polyatomic anion into a neutral formula unit.

(a) Nickel(II) acetate is composed of Ni^{2+} and $C_2H_3O_2^-$. The formula of the compound is written $Ni(C_2H_3O_2)_2$.

(b) Ferric sulfate contains Fe^{3+} and SO_4^{2-}; the formula is written $Fe_2(SO_4)_3$.

Self-Test Exercise

Provide the formula for each of the following ternary ionic compounds.

(a) mercury(I) nitrite (b) tin(IV) permanganate

Answers: (a) $Hg_2(NO_2)_2$; (b) $Sn(MnO_4)_4$

Predicting Formulas of Ternary Ionic Compounds

The periodic table can be used to predict the formulas of ternary ionic compounds. For example, if we know that the formula of potassium nitrate is KNO_3, we can predict the formulas of sodium nitrate and lithium nitrate. Since Li, Na, and K are all in Group IA/1, we can predict that the compounds have similar chemical formulas, that is, $LiNO_3$, $NaNO_3$, and KNO_3. The following example exercise illustrates how to predict formulas of ternary ionic compounds.

Example Exercise 7.15 • Predicting Formulas of Ternary Ionic Compounds

Predict the chemical formula for each of the following ternary ionic compounds given the formula of calcium carbonate, $CaCO_3$.

(a) radium carbonate (b) calcium silicate

Solution

To predict the formula, we compare the elements that are different in the similar compounds.

(a) The elements Ra and Ca are both in Group IIA/2. Thus, the formula for radium carbonate is $RaCO_3$.

(b) The elements Si and C are both in Group IVA/14. Therefore, the formula for calcium silicate is $CaSiO_3$.

Self-Test Exercise

Predict the chemical formula for each of the following ternary compounds given the formula of potassium chlorate, $KClO_3$.

(a) lithium chlorate (b) potassium bromate

Answers: (a) $LiClO_3$; (b) $KBrO_3$

7.7 Binary Molecular Compounds

Objective · To write systematic names and formulas for binary molecular compounds.

Recall that a binary molecular compound is composed of two nonmetal elements. The simplest representative particle in a binary molecular compound is a **molecule**. In general, the more nonmetallic element is written second in the chemical formula. IUPAC prescribes the following order for writing the elements in a compound: C, P, N, H, S, I, Br, Cl, O, F. Notice that hydrogen is in the middle of the series. Thus, the binary compounds of hydrogen are written as follows: CH_4, PH_3, NH_3, H_2S, HI, HBr, HCl, H_2O, and HF.

Naming Binary Molecular Compounds

In the naming of binary molecular compounds, IUPAC specifies that the first element in the compound be named first and that the second element have the suffix *-ide*. Thus, binary ionic and binary molecular compounds both end in the suffix *-ide*. It also specifies that the number of atoms of each element be indicated by Greek prefixes. The Greek prefixes in Table 7.4 indicate the number of atoms of each element that correspond to the subscripts in the chemical formula.

According to the official 1990 *Red Book* recommendations, the prefix *mono-* is always omitted unless its presence is necessary to avoid confusion. The only common exceptions you are responsible for are CO and NO. The name of CO includes the prefix *mono-* in front of oxygen and is written carbon monoxide. Similarly, NO is named nitrogen monoxide.

Table 7.4 Greek Prefixes for Binary Molecular Compounds

Atoms	Prefix	Atoms	Prefix
1	*mono-*	6	*hexa-*
2	*di-*	7	*hepta-*
3	*tri-*	8	*octa-*
4	*tetra-*	9	*nona-**
5	*penta-*	10	*deca-*

* Although the Latin prefix *nona-* is commonly used, IUPAC prefers the Greek prefix *ennea-*.

Let's consider the binary molecular compound P_4S_3. This compound is found on match tips and ignites in air when struck on a rough surface (Figure 7.4). Since the ratio is four phosphorus atoms to three sulfur atoms, the Greek prefixes are *tetra-* and *tri-*, respectively. Thus, the name of P_4S_3 is tetraphosphorus trisulfide. A different compound, P_4S_7, has the same elements but is named tetraphosphorus heptasulfide.

▲ **Figure 7.4 Chemistry of Matches** The substances P_4S_3 and $KClO_3$ are present on the tip of a match. When the match is struck on a rough surface, the two chemicals ignite and produce a flame. The products from the reaction are P_2O_5, KCl, and SO_2, the last of which is responsible for the characteristic sulfur smell.

Example Exercise 7.16 • Names of Binary Molecular Compounds

Give the IUPAC systematic name for each of the following binary molecular compounds.

(a) IF_6 (b) Br_3O_8

Solution

We name binary molecular compounds by attaching the suffix *-ide* to the second nonmetal and indicate the atomic ratios by Greek prefixes.

(a) IF_6 is first named iodine fluoride. After supplying the Greek prefixes for the atomic ratios, we have *iodine hexafluoride*.

(b) Br_3O_8 is first named bromine oxide. After supplying the Greek prefixes for the atomic ratios, we have *tribromine octaoxide*.

Self-Test Exercise

Give the IUPAC systematic name for each of the following binary molecular compounds.

(a) Cl_2O_5 (b) P_4S_{10}

Answers: (a) dichlorine pentaoxide; (b) tetraphosphorus decasulfide

Example Exercise 7.17 • Formulas of Binary Molecular Compounds

Provide the formula for each of the following binary molecular compounds.

(a) diphosphorus pentasulfide (b) tetraiodine nonaoxide

Solution

To write the formula, we give the symbol for each element followed by a subscript indicating the number of atoms.

(a) Diphosphorus pentasulfide is composed of two phosphorus atoms and five sulfur atoms. The formula of the compound is written P_2S_5.

(b) Tetraiodine nonaoxide is composed of four iodine atoms and nine oxygen atoms. The formula of the compound is written I_4O_9.

Self-Test Exercise

Provide the formula for each of the following binary molecular compounds.

(a) diphosphorus tetraiodide (b) sulfur hexafluoride

Answers: (a) P_2I_4; (b) SF_6

Note In the past, for ease of pronunciation, double vowels were avoided in naming compounds with Greek prefixes. For example, if the Greek prefix ended in an "a" or an "o" and the nonmetal was oxygen, the first vowel was dropped. Thus, tetroxide, not tetraoxide, was preferred. According to the 1990 *Red Book*, however, vowels are not to be dropped with one exception. If *mono-* is used as a prefix before oxygen, then monoxide, not monooxide, is recommended.

7.8 Binary Acids

Objective · To write systematic names and formulas for binary acids.

A binary acid is an aqueous solution of a compound containing hydrogen and a nonmetal. The chemical formulas of acids always begin with H, for example, HF(aq). A binary acid is produced by dissolving a binary molecular compound, such as HF, in water. The resulting aqueous solution, HF(aq), is a binary acid.

Binary acids are systematically named by using the prefix *hydro-* before the nonmetal stem and adding the suffix *-ic acid*. As an example, consider the acid used to treat swimming pools and sold in supermarkets under the common name of muriatic acid. Muriatic acid is aqueous hydrogen chloride, that is, HCl(aq). The IUPAC name for muriatic acid is formed as follows: *hydro- chlor(ine)* + *-ic* acid; the systematic name for HCl(aq) is hydrochloric acid.

▲ **Hydrofluoric Acid, HF** An aqueous solution of HF is the acid used to etch silicon in the manufacture of computer chips.

Example Exercise 7.18 • Names of Binary Acids

Give the IUPAC systematic name for HF(aq), a binary acid.

Solution

We name a binary acid as *hydro-* plus *nonmetal stem* plus *-ic acid*. HF(aq) contains the nonmetal fluorine, and we construct its name as follows: hydro + fluor + ic acid. HF(aq) is named *hydrofluoric acid*.

Self-Test Exercise

Give the IUPAC systematic name for H_2S(aq).

Answers: hydrosulfuric acid

If a particular class is responding slowly to nomenclature, the naming of binary acids and ternary oxyacids can be postponed to Chapter 15, *Acids and Bases*.

Note Be careful not to confuse the names of a binary acid and a binary molecular compound. A binary acid is a compound of hydrogen and a nonmetal dissolved in water. For example, HCl(aq) is a binary acid and is named hydrochloric acid. On the other hand, HCl(g) is a gaseous binary molecular compound and is named hydrogen chloride.

7.9 Ternary Oxyacids

Objective · To write systematic names and formulas for ternary oxyacids.

A ternary oxyacid is an aqueous solution of a compound containing hydrogen and an oxyanion. Most ternary oxyacids are named by attaching the suffix *-ic acid* to the nonmetal stem of the oxyanion. For example, the name of HNO_3(aq) is systematically formed as follows: *nonmetal stem* plus *-ic acid*. The name of HNO_3(aq) is nitr + ic acid, or *nitric acid*.

Some ternary oxyacids are named by attaching the suffix *-ous acid* to the nonmetal stem of the oxyanion. For example, the name of $HNO_2(aq)$ is systematically formed as follows: nonmetal stem + *-ous acid*. The name of $HNO_2(aq)$ is nitr + ous acid, or *nitrous acid*.

A ternary oxyacid with the suffix *-ic acid* contains an oxyanion with the suffix *-ate*. Consider automobile battery acid, or sulfuric acid, $H_2SO_4(aq)$, which contains the sulfate oxyanion, SO_4^{2-}. In an analogous way, a ternary oxyacid with the suffix *-ous acid* contains an oxyanion ending in the suffix *-ite*. Consider sulfurous acid, $H_2SO_3(aq)$, which contains the sulfite oxyanion, SO_3^{2-}. In summary,

Ternary Oxyacid	Polyatomic Oxyanion
sulfur**ic acid**, H_2SO_4	sulf**ate**, SO_4^{2-}
sulfur**ous acid**, H_2SO_3	sulf**ite**, SO_3^{2-}

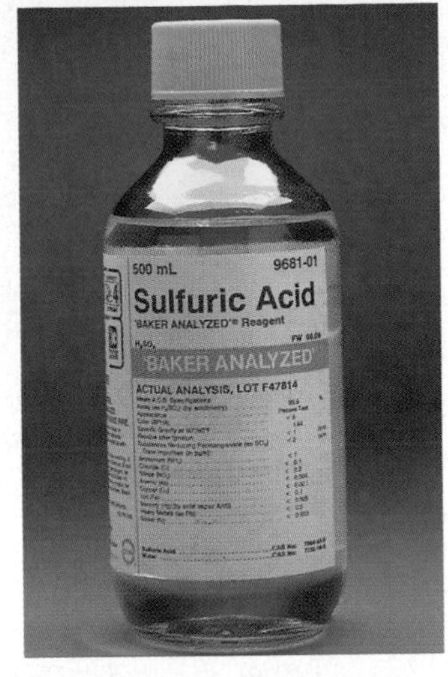

▲ **Sulfuric Acid, H_2SO_4** Aqueous H_2SO_4 is found in the chemistry laboratory and has the systematic name of sulfuric acid. Since H_2SO_4 is used in automobile batteries, it is commonly referred to as battery acid.

Now that we have provided an overview of ternary oxyacids, let's try the following example exercise which illustrates systematic naming.

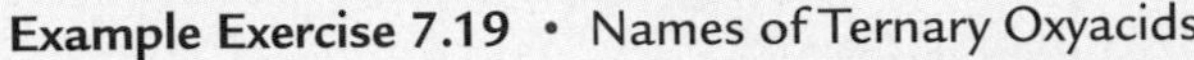

Example Exercise 7.19 • Names of Ternary Oxyacids

Give the IUPAC systematic name for $HClO_3(aq)$, a ternary oxyacid.

Solution

Ternary oxyacids are named as *-ic* acids or *-ous* acids. Since $HClO_3(aq)$ contains the chlorate oxyanion, ClO_3^-, it is an *-ic* acid. We can construct the name as follows: chlor + ic acid; $HClO_3(aq)$ is named *chloric acid*.

Self-Test Exercise

Give the IUPAC systematic name for $HClO_2(aq)$, a ternary oxyacid.

Answer: chlorous acid

Summary

Section 7.1 **Inorganic compounds** do not contain carbon and are named according to a systematic set of rules referred to as **IUPAC nomenclature**. The IUPAC rules classify substances as belonging to one of five categories: **binary ionic**, **ternary ionic**, **binary molecular**, **binary acid**, or **ternary oxyacid**. A substance dissolved in water is referred to as an **aqueous solution**. Ionic compounds are composed of positive **cations** and negative **anions**. A **monoatomic ion** is a single atom having a charge, whereas a **polyatomic ion** is a group of two or more atoms bearing a charge.

Section 7.2 The names of most monoatomic cations are derived from the parent metal; for example, the sodium ion, Na^+. Cations having two possible ionic charges require further identification. The ionic charge may be indicated using either the Stock system or the Latin system. The **Stock system** indicates the charge on the metal with Roman numerals in parentheses. The **Latin system** attaches the suffix *-ous* or *-ic* to the Latin name of the element. Thus, Cu^{2+} is named copper(II) ion according to the Stock system, and cupric ion using the Latin system.

The names of monoatomic anions are derived from the parent nonmetal. A monoatomic anion is named using the nonmetal stem and attaching the suffix *-ide*. Examples of nonmetal ions include the chloride ion, Cl^-, and the oxide ion, O^{2-}.

Section 7.3 A polyatomic anion that contains one or more elements combined with oxygen is called an **oxyanion**. The names of most polyatomic oxyanions have the suffix *-ate*, for example, the nitrate ion, NO_3^-. A few oxyanions have the suffix *-ite*, for example, the nitrite ion, NO_2^-. The nitrite ion has one less oxygen atom than the nitrate ion. Oxyanions having one less oxygen atom end with the suffix *-ite,* for instance, the sulfate ion, SO_4^{2-} and the sulfite ion, SO_3^{2-}.

Section 7.4 The simplest representative particle in an ionic compound is called a **formula unit**. The net charge for a formula unit is zero. That is, the total positive charge for the cations must be the same as the total negative charge for the anions.

Section 7.5 Binary ionic compounds have names ending with the suffix *-ide*. For example, NaCl is named sodium chloride, $CaBr_2$ is named calcium bromide, and AlF_3 is named aluminum fluoride.

Section 7.6 Ternary ionic compounds have names ending with the suffix *-ate* or *-ite*. For example, $CaSO_4$ is named calcium sulfate, and $CaSO_3$ is named calcium sulfite. A ternary compound containing CN^- or OH^- is an exception, and each ends with the suffix *-ide*. For example, NaOH is named sodium hydroxide and KCN is named potassium cyanide.

Section 7.7 Binary molecular compounds contain two nonmetals, and the simplest representative particle is called a **molecule**. Binary molecular compounds are named using a Greek prefix and the suffix *-ide*; for example, SO_2 and SO_3 are named sulfur dioxide and sulfur trioxide, respectively.

Section 7.8 Binary acids are named *hydro- + nonmetal stem + -ic acid*. A binary acid is an aqueous solution of a hydrogen-containing compound. For example, hydrogen chloride dissolved in water, HCl(aq), is named hydrochloric acid.

Section 7.9 Ternary oxyacids are usually named *nonmetal stem + -ic acid*. For example, an aqueous solution of HNO_3(aq) is named nitric acid. A few ternary acids are named *nonmetal stem + -ous acid*. An aqueous solution of HNO_2(aq) is named nitrous acid.

Nomenclature Organizer

Topic	Procedure	Example
Classification of Compounds Sec. 7.1	When a formula has two elements, a metal and a nonmetal, it is a **binary ionic compound**.	binary ionic compound, KCl
	When a formula has three elements and at least one metal, it is a **ternary ionic compound**.	ternary ionic compound, $KClO_3$
	When a formula has two nonmetal elements, it is a **binary molecular compound**.	binary molecular compound, HCl
	When a formula has H and a nonmetal in aqueous solution, it is a **binary acid**.	binary acid, HCl(aq)
	When a formula has H, a nonmetal, and oxygen in aqueous solution, it is a **ternary oxyacid**.	ternary oxyacid, $HClO_3$(aq)
Monoatomic Ions Sec. 7.2	Representative group metal ions are named for the element. Transition group metal ions are named using (1) the **Stock system**, or (2) the **Latin system**. Nonmetals are named for the element plus the **suffix *-ide***.	K^+, potassium ion Fe^{2+}, iron(II) ion Fe^{2+}, ferrous ion Cl^-, chloride ion
Polyatomic Ions Sec. 7.3	Polyatomic oxyanions are named with (1) the **suffix *-ate***, or (2) the **suffix *-ite*** when there is one less oxygen atom in the oxyanion.	ClO_3^-, chlorate ion ClO_2^-, chlorite ion
	There are two common polyatomic anion exceptions, which have the **suffix *-ide***.	OH^-, hydroxide ion CN^-, cyanide ion
Writing Chemical Formulas Sec. 7.4	In writing formulas of ionic compounds, the ratio of cations and anions must provide a total positive charge equal to the total negative charge so that the compound is electrically neutral.	Fe^{2+} and O^{2-} = FeO Fe^{3+} and O^{2-} = Fe_2O_3 Fe^{2+} and ClO_3^- = $Fe(ClO_3)_2$ Fe^{3+} and ClO_3^- = $Fe(ClO_3)_3$

Nomenclature Organizer *(continued)*		
Topic	**Procedure**	**Example**
Binary Ionic Compounds Sec. 7.5	Binary ionic compounds end in the **suffix** ***-ide***. The metal cation is named first followed by the name of the anion.	KCl, potassium chloride $FeCl_2$, iron(II) chloride or ferrous chloride
Ternary Ionic Compounds Sec. 7.6	Ternary ionic compounds end in the **suffix** ***-ate***, or the **suffix** ***-ite***. The metal cation is named first followed by the name of the oxyanion.	$KClO_3$, potassium chlorate $KClO_2$, potassium chlorite
Binary Molecular Compounds Sec. 7.7	Binary molecular compounds end in the **suffix** ***-ide***. The number of nonmetal atoms is indicated by a Greek prefix.	ClO_2, chlorine dioxide Cl_2O, dichlorine monoxide
Binary Acids Sec. 7.8	Binary acids are named ***hydro-*** plus ***nonmetal*** **stem** plus ***-ic acid***.	HCl(aq), hydrochloric acid
Ternary Oxyacids Sec. 7.9	Most ternary oxyacids are named ***nonmetal*** **stem** plus ***-ic acid***.	$HClO_3$(aq), chloric acid

Key Concepts*

1. Classify each of the following as binary ionic, ternary ionic, binary molecular, binary acid, or ternary oxyacid: NaCl, HCl, HCl(aq), NaClO, HClO(aq).
2. Classify each of the following as a monoatomic cation, monoatomic anion, polyatomic cation, or polyatomic anion: Na^+, Cl^-, ClO^-, ClO_3^-.
3. What is the formula for the ionic compound composed of the mercuric ion, Hg^{2+}, and the chloride ion, Cl^-?
4. What is the formula for the ionic compound composed of the mercuric ion, Hg^{2+}, and the chlorate ion, ClO_3^-?
5. Which of the following compounds is named using the suffix *-ide*: NaCl, HCl, NaClO, $NaClO_2$, $NaClO_3$, $NaClO_4$?
6. Which of the following compounds is named using the suffix *-ate*: NaCl, HCl, NaClO, $NaClO_2$, $NaClO_3$, $NaClO_4$?
7. Which of the following compounds is named using the suffix *-ite*: NaCl, HCl, NaClO, $NaClO_2$, $NaClO_3$, $NaClO_4$?
8. Which of the following acids is named *hydro-* plus *nonmetal stem* plus *-ic acid*: HCl(aq), HClO(aq), $HClO_2$(aq), $HClO_3$(aq), $HClO_4$(aq)?
9. Which of the following acids is named *nonmetal stem* plus *-ic acid*: HCl(aq), HClO(aq), $HClO_2$(aq), $HClO_3$(aq), $HClO_4$(aq)?
10. Which of the following acids is named *nonmetal stem* plus *-ous acid*: HCl(aq), HClO(aq), $HClO_2$(aq), $HClO_3$(aq), $HClO_4$(aq)?

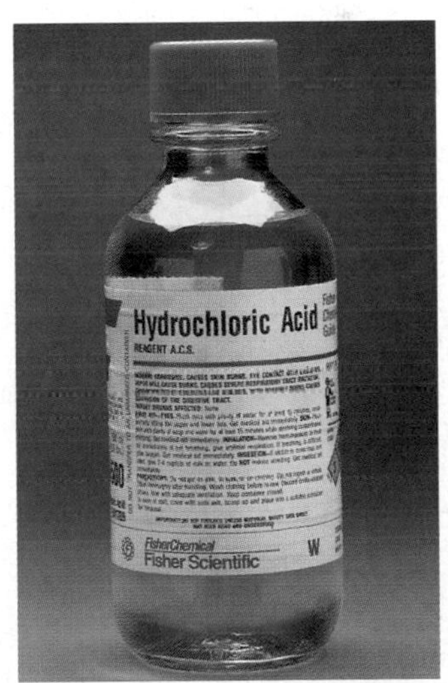

▲ **Hydrochloric Acid, HCl** Aqueous HCl is found in the chemistry laboratory and has the systematic name of hydrochloric acid. Since HCl(aq) is produced in the stomach and aids in digestion, hydrochloric acid is commonly referred to as stomach acid.

*Answers to Key Concepts are in Appendix H.

Key Terms†

Select the key term below that corresponds to each of the following definitions.

_____ 1. the international system of rules for naming chemical compounds
_____ 2. a compound that does not contain the element carbon
_____ 3. a compound that contains one metal and one nonmetal
_____ 4. a compound that contains three elements including at least one metal
_____ 5. a compound that contains two nonmetals
_____ 6. a substance dissolved in water
_____ 7. an aqueous compound that contains hydrogen and a nonmetal
_____ 8. an aqueous compound that contains hydrogen, a nonmetal, and oxygen
_____ 9. any positively charged ion
_____ 10. any negatively charged ion
_____ 11. a single atom that has a positive or negative charge
_____ 12. a group of atoms that has a positive or a negative charge
_____ 13. a system that designates the charge on a cation with Roman numerals
_____ 14. a system that designates the charge on a cation with the suffix *-ic* or *-ous*.
_____ 15. a polyatomic anion that contains one or more elements combined with oxygen; for example, CO_3^{2-}
_____ 16. the simplest representative particle in a compound composed of ions
_____ 17. the simplest representative particle in a compound composed of nonmetals

(a) anion (*Sec. 7.1*)
(b) aqueous solution (*Sec. 7.1*)
(c) binary acid (*Sec. 7.1*)
(d) binary ionic compound (*Sec. 7.1*)
(e) binary molecular compound (*Sec. 7.1*)
(f) cation (*Sec. 7.1*)
(g) formula unit (*Sec. 7.4*)
(h) IUPAC nomenclature (*Sec. 7.1*)
(i) inorganic compound (*Sec. 7.1*)
(j) Latin system (*Sec. 7.2*)
(k) molecule (*Sec. 7.7*)
(l) monoatomic ion (*Sec. 7.1*)
(m) oxyanion (*Sec. 7.3*)
(n) polyatomic ion (*Sec. 7.1*)
(o) Stock system (*Sec. 7.2*)
(p) ternary ionic compound (*Sec. 7.1*)
(q) ternary oxyacid (*Sec. 7.1*)

Exercises‡

Classification of Compounds (Sec. 7.1)

1. Classify each of the following as a binary ionic compound, ternary ionic compound, binary molecular compound, binary acid, or ternary oxyacid.
   (a) lead(II) sulfide, PbS (b) ammonia, NH_3
   (c) silver nitrate, $AgNO_3$ (d) sulfurous acid, $H_2SO_3(aq)$
   (e) hydrobromic acid, HBr(aq)
2. Classify each of the following as a binary ionic compound, ternary ionic compound, binary molecular compound, binary acid, or ternary oxyacid.
   (a) carbon dioxide, CO_2
   (b) hydrofluoric acid, HF(aq)
   (c) phosphoric acid, $H_3PO_4(aq)$
   (d) cobalt(III) oxide, Co_2O_3
   (e) zinc phosphate, $Zn_3(PO_4)_2$
3. Classify each of the following as a monoatomic cation, monoatomic anion, polyatomic cation, or polyatomic anion.
   (a) bromide ion, Br^- (b) ammonium ion, NH_4^+
   (c) acetate ion, $C_2H_3O_2^-$ (d) mercuric ion, Hg^{2+}
4. Classify each of the following as a monoatomic cation, monoatomic anion, polyatomic cation, or polyatomic anion.
   (a) cobalt ion, Co^{3+} (b) cyanide ion, CN^-
   (c) hydronium ion, H_3O^+ (d) nitride ion, N^{3-}

Monoatomic Ions (Sec. 7.2)

5. Supply a systematic name for each of the following monoatomic cations.
   (a) K^+ (b) Ba^{2+}
   (c) Ag^+ (d) Cd^{2+}
6. Provide the formula for each of the following monoatomic cations.
   (a) lithium ion (b) aluminum ion
   (c) zinc ion (d) strontium ion
7. Supply the Stock system name for each of the following monoatomic cations.
   (a) Hg^{2+} (b) Cu^{2+}
   (c) Fe^{2+} (d) Co^{3+}

† Answers to Key Terms are in Appendix I.

‡ Answers to odd-numbered Exercises are in Appendix J.

8. Provide the formula for each of the following monoatomic cations.
(a) lead(II) ion (b) nickel(II) ion
(c) tin(IV) ion (d) manganese(II) ion

9. Supply the Latin system name for each of the following monoatomic cations.
(a) Cu^+ (b) Fe^{3+}
(c) Sn^{2+} (d) Pb^{4+}

10. Provide the formula for each of the following monoatomic cations.
(a) cupric ion (b) mercuric ion
(c) ferrous ion (d) plumbic ion

11. Supply a systematic name for each of the following monoatomic anions.
(a) F^- (b) I^-
(c) O^{2-} (d) P^{3-}

12. Provide the formula for each of the following monoatomic anions.
(a) chloride ion (b) bromide ion
(c) sulfide ion (d) nitride ion

Polyatomic Ions (Sec. 7.3)

13. Supply a systematic name for each of the following oxyanions.
(a) ClO^- (b) SO_3^{2-}
(c) $C_2H_3O_2^-$ (d) CO_3^{2-}

14. Supply a systematic name for each of the following oxyanions.
(a) SO_4^{2-} (b) NO_3^-
(c) CrO_4^{2-} (d) HSO_4^-

15. Provide the formula for each of the following polyatomic ions.
(a) hydroxide ion (b) nitrite ion
(c) dichromate ion (d) hydrogen carbonate ion

▲ **Caustic Soda, NaOH** Caustic soda is an alkaline chemical used as drain cleaner.

16. Provide the formula for each of the following polyatomic ions.
(a) ammonium ion (b) chlorite ion
(c) cyanide ion (d) perchlorate ion

Writing Chemical Formulas (Sec. 7.4)

17. Write the chemical formula for each of the following binary compounds given their constituent ions.
(a) lithium chloride, Li^+ and Cl^-
(b) silver oxide, Ag^+ and O^{2-}
(c) chromium oxide, Cr^{3+} and O^{2-}
(d) tin(IV) iodide, Sn^{4+} and I^-

18. Write the chemical formula for each of the following binary compounds given their constituent ions.
(a) aluminum bromide, Al^{3+} and Br^-
(b) cadmium sulfide, Cd^{2+} and S^{2-}
(c) magnesium phosphide, Mg^{2+} and P^{3-}
(d) lead(IV) sulfide, Pb^{4+} and S^{2-}

19. Write the chemical formula for each of the following ternary compounds given their constituent ions.
(a) potassium nitrate, K^+ and NO_3^-
(b) ammonium dichromate, NH_4^+ and $Cr_2O_7^{2-}$
(c) aluminum sulfite, Al^{3+} and SO_3^{2-}
(d) chromium(III) hypochlorite, Cr^{3+} and ClO^-

20. Write the chemical formula for each of the following ternary compounds given their constituent ions.
(a) iron(III) hydrogen carbonate, Fe^{3+} and HCO_3^-
(b) cuprous carbonate, Cu^+ and CO_3^{2-}
(c) mercury(II) cyanide, Hg^{2+} and CN^-
(d) stannic acetate, Sn^{4+} and $C_2H_3O_2^-$

21. Write the chemical formula for each of the following ternary compounds given their constituent ions.
(a) strontium nitrite, Sr^{2+} and NO_2^-
(b) zinc permanganate, Zn^{2+} and MnO_4^-
(c) calcium chromate, Ca^{2+} and CrO_4^{2-}
(d) chromium(III) perchlorate, Cr^{3+} and ClO_4^-

22. Write the chemical formula for each of the following ternary compounds given their constituent ions.
(a) lead(IV) sulfate, Pb^{4+} and SO_4^{2-}
(b) stannous chlorite, Sn^{2+} and ClO_2^-
(c) cobalt(II) hydroxide, Co^{2+} and OH^-
(d) mercurous phosphate, Hg_2^{2+} and PO_4^{3-}

Binary Ionic Compounds (Sec. 7.5)

23. Supply a systematic name for each of the following binary ionic compounds.
(a) MgO (b) $AgBr$
(c) $CdCl_2$ (d) Al_2S_3

24. Provide the formula for each of the following binary ionic compounds.
(a) lithium oxide (b) strontium oxide
(c) zinc chloride (d) nickel(II) fluoride

25. Supply a Stock system name for each of the following binary ionic compounds.
(a) CuO (b) FeO
(c) HgO (d) SnO

26. Provide the formula for each of the following binary ionic compounds.
(a) manganese(II) nitride (b) iron(II) bromide
(c) tin(IV) fluoride (d) cobalt(II) sulfide

27. Supply a Latin system name for each of the following binary ionic compounds.
(a) Cu_2O (b) Fe_2O_3
(c) Hg_2O (d) SnO_2

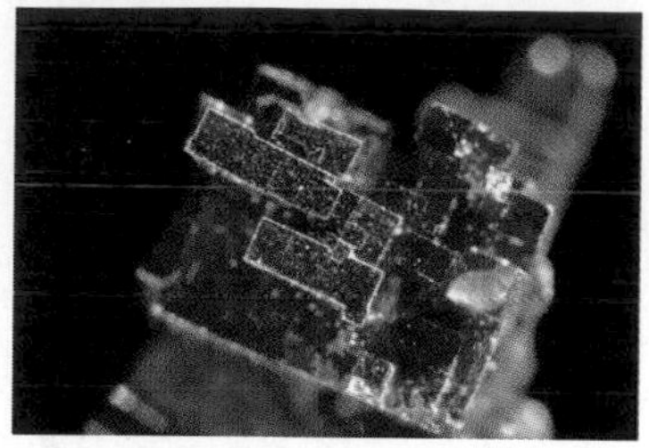

▲ **Cuprite, Cu_2O** Cuprite is a natural mineral that contains copper.

28. Provide the formula for each of the following binary ionic compounds.
(a) plumbous oxide (b) ferric phosphide
(c) mercuric iodide (d) copper(I) chloride

29. Predict the chemical formula for each of the following binary ionic compounds given the formula of sodium chloride, NaCl.
(a) rubidium chloride (b) sodium bromide

30. Predict the chemical formula for each of the following binary ionic compounds given the formula of calcium oxide, CaO.
(a) beryllium oxide (b) calcium selenide

31. Predict the chemical formula for each of the following binary ionic compounds given the formula of aluminum nitride, AlN.
(a) gallium nitride (b) aluminum arsenide

32. Predict the chemical formula for each of the following binary ionic compounds given the formula of titanium oxide, TiO_2.
(a) zirconium oxide (b) titanium sulfide

Ternary Ionic Compounds (Sec. 7.6)

33. Supply a systematic name for each of the following ternary ionic compounds.
(a) $LiMnO_4$ (b) $Sr(ClO_4)_2$
(c) $CaCrO_4$ (d) $Cd(CN)_2$

34. Provide the formula for each of the following ternary ionic compounds.
(a) potassium nitrate (b) magnesium perchlorate
(c) silver sulfate (d) aluminum dichromate

35. Supply a Stock system name for each of the following ternary ionic compounds.
(a) $CuSO_4$ (b) $FeCrO_4$
(c) $Hg(NO_2)_2$ (d) $Pb(C_2H_3O_2)_2$

36. Provide the formula for each of the following ternary ionic compounds.
(a) manganese(II) acetate (b) copper(II) chlorite
(c) tin(II) phosphate (d) iron(III) hypochlorite

37. Supply a Latin system name for each of the following ternary ionic compounds.
(a) Cu_2SO_4 (b) $Fe_2(CrO_4)_3$
(c) $Hg_2(NO_2)_2$ (d) $Pb(C_2H_3O_2)_4$

38. Provide the formula for each of the following ternary ionic compounds.
(a) cuprous chlorite (b) plumbic sulfite
(c) mercuric chlorate (d) ferrous chromate

39. Predict the chemical formula for each of the following ternary ionic compounds given the formula of sodium sulfate, Na_2SO_4.
(a) francium sulfate (b) sodium sulfite

40. Predict the chemical formula for each of the following ternary ionic compounds given the formula of scandium nitrate, $Sc(NO_3)_3$.
(a) lanthanum nitrate (b) scandium nitrite

41. Predict the chemical formula for each of the following ternary ionic compounds given the formula of barium chlorate, $Ba(ClO_3)_2$.
(a) radium chlorate (b) barium bromate

42. Predict the chemical formula for each of the following ternary ionic compounds given the formula of iron(III) sulfate, $Fe_2(SO_4)_3$.
(a) cobalt(III) sulfate (b) iron(III) selenate

Binary Molecular Compounds (Sec. 7.7)

43. Give a systematic name for each of the following binary molecular compounds.
(a) SO_3 (b) P_2O_3
(c) N_2O (d) C_3O_2

44. Give a systematic name for each of the following binary molecular compounds.
(a) HCl (b) BrF_3
(c) I_2O_4 (d) Cl_2O_3

45. Provide the formula for each of the following binary molecular compounds.
(a) dinitrogen pentaoxide (b) carbon tetrachloride
(c) iodine monobromide (d) dihydrogen sulfide

46. Provide the formula for each of the following binary molecular compounds.
(a) chlorine dioxide (b) sulfur tetrafluoride
(c) iodine monochloride (d) nitrogen monoxide

Binary Acids (Sec. 7.8)

47. Give the IUPAC systematic name for each of the following binary acids.
(a) HBr(aq) (b) HI(aq)

48. Provide the formula for each of the following binary acids.
(a) hydrosulfuric acid (b) hydroselenic acid

Ternary Oxyacids (Sec. 7.9)

49. Give the IUPAC systematic name for each of the following ternary oxyacids.
(a) $HClO_2(aq)$ (b) $H_3PO_4(aq)$

50. Give the IUPAC systematic name for each of the following ternary oxyacids.
(a) $HClO_4(aq)$ (b) $H_2SO_3(aq)$

51. Provide the formula for each of the following ternary oxyacids.
(a) acetic acid (b) phosphorous acid

52. Provide the formula for each of the following ternary oxyacids.
(a) carbonic acid (b) nitrous acid

53. Predict the chemical formula for each of the following ternary oxyacids given the formula of hypochlorous acid, $HClO(aq)$.
(a) chlorous acid (b) hypobromous acid

54. Predict the chemical formula for each of the following ternary oxyacids given the formula of phosphoric acid, $H_3PO_4(aq)$.
(a) arsenic acid (b) phosphorus acid

General Exercises

55. State the ionic charge for each of the following substances.
(a) iron metal atoms (b) ferrous ions
(c) iron(III) ions (d) iron compounds

56. State the ionic charge for each of the following substances.
(a) chlorine gas molecules (b) chloride ions
(c) hypochlorite ion (d) chlorine compounds

57. Predict which of the following polyatomic anions has an ionic charge of 2−. (*Hint*: The total number of valence electrons must be an even number.)
(a) periodate ion, $IO_4^{?-}$ (b) silicate ion, $SiO_3^{?-}$

58. Predict which of the following polyatomic anions has an ionic charge of 1−. (*Hint*: The total number of valence electrons must be an even number.)
(a) thiocyanate ion, $CNS^{?-}$
(b) thiosulfate ion, $S_2O_3^{?-}$

59. Complete the table below by combining cations and anions into chemical formulas. Give a systematic name for each of the compounds.

Ions	F^-	O^{2-}	N^{3-}
Ag^+	AgF silver fluoride		
Hg_2^{2+}			
Al^{3+}			

60. Complete the table below by writing the formulas of the cations and anions. Combine the cations with each of the anions to give a correct chemical formula.

Ions	chloride ion	sulfide ion	phosphide ion
lithium ion	Li^+ Cl^- LiCl		
copper(II) ion			
iron(III) ion			

61. Complete the table below by combining cations and oxyanions into chemical formulas. Give a systematic name for each of the compounds.

Ions	MnO_4^-	SO_3^{2-}	PO_4^{3-}
Cu^+	$CuMnO_4$ copper(I) permanganate		
Cd^{2+}			
Cr^{3+}			

62. Complete the table below by writing the formula of each cation and oxyanion. Combine cations and oxyanions to give correct chemical formulas.

Ions	chlorite ion	sulfate ion	phosphate ion
sodium ion	Na^+ ClO_2^- $NaClO_2$		
iron(II) ion			
cobalt(III) ion			

63. State the suffix in the name for each of the following.
(a) Na_2S (b) $H_2S(aq)$

64. State the suffix in the name for each of the following.
(a) NaI (b) HI(aq)

65. State the suffix in the name for each of the following.
(a) Na_2SO_3 (b) $H_2SO_3(aq)$

66. State the suffix in the name for each of the following.
(a) $NaIO_2$ (b) $HIO_2(aq)$

67. State the suffix in the name for each of the following.
(a) Na_2SO_4 (b) $H_2SO_4(aq)$

68. State the suffix in the name for each of the following.
(a) $NaIO_3$ (b) $HIO_3(aq)$

69. Write the chemical formula for each of the following household chemicals.
(a) dihydrogen oxide (common liquid)
(b) sodium hypochlorite (bleach)
(c) sodium hydroxide (caustic soda)
(d) sodium bicarbonate (baking soda)

70. Write the chemical formula for each of the following household chemicals.
   (a) acetic acid (vinegar solution)
   (b) aqueous nitrogen trihydide (ammonia solution)
   (c) aqueous magnesium hydroxide (milk of magnesia)
   (d) aqueous sodium bisulfate (bowl cleaner)

71. Give the name for each of the following compounds containing a semimetal. (*Hint*: Name the formula as a binary molecular compound.)
   (a) BF_3 (b) $SiCl_4$
   (c) As_2O_5 (d) Sb_2O_3

72. Write the formula for each of the following compounds containing a semimetal.
   (a) boron tribromide (b) trisilicon tetranitride
   (c) diarsenic trioxide (d) diantimony pentaoxide

73. "Canned heat" is a flammable gel containing calcium acetate dissolved in a 50% alcohol solution. What is the formula of calcium acetate?

74. Oxalic acid, $H_2C_2O_4$, is used to clean jewelry. Sodium hydrogen oxalate is found in spinach and produces an abrasive sensation on teeth. What is the formula of sodium hydrogen oxalate?

75. The transuranium element lawrencium is unstable and has a lifetime of only a few minutes. The chemical formula of lawrencium chloride was determined at the University of California. Predict the formula for lawrencium chloride given the formula of lutetium chloride, $LuCl_3$.

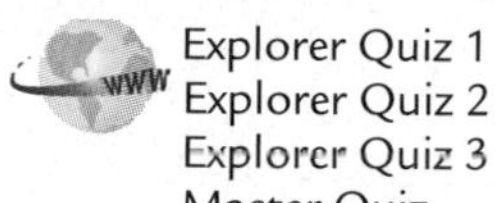

Explorer Quiz 1
Explorer Quiz 2
Explorer Quiz 3
Master Quiz

CHAPTER 8

Chemical Reactions

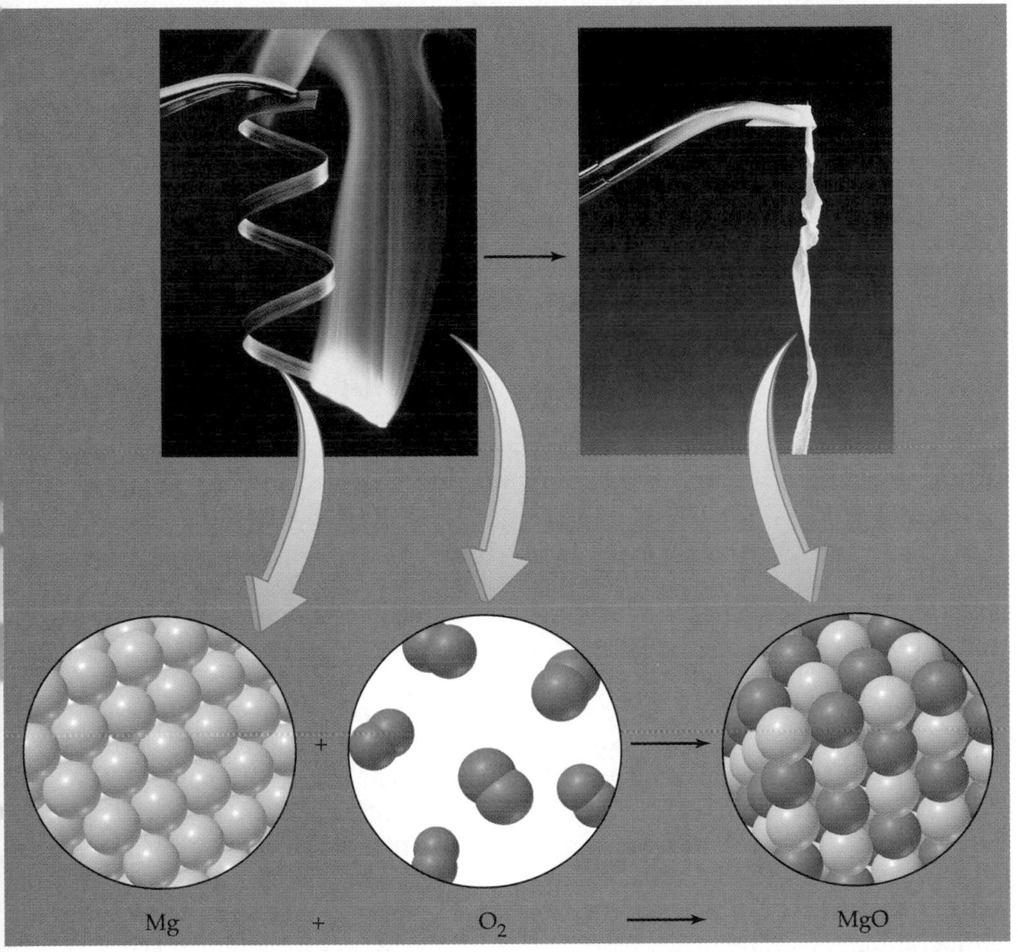

▲ In a chemical reaction, two substances can react to form a new compound. In this example, the coil of metal is Mg and the white ribbon is MgO. What is the source of O_2?

Pedagogically, this chapter naturally follows Chapter 7 as it reinforces nomenclature and the "language of chemistry." Section 8.2 introduces the process of transforming a chemical reaction expressed in words into a chemical equation. All the examples in Section 8.2 represent a balanced equation with the coefficients having the default value of one. Section 8.3 illustrates the task of transforming chemical equations into balanced equations using the method of balancing by inspection.

Everyday substances are constantly being altered in some way. These alterations can be classified as a physical or a chemical change. In a *physical change*, the chemical composition of the substance remains constant. For instance, when we melt ice or boil water, we cause a physical change. The chemical composition of the ice and the water, H_2O, is the same before and after the change .

In a *chemical change*, the chemical composition of the substance does not remain constant. For example, when we drop an antacid tablet into a glass of water, we produce a chemical change. The chemical composition of the antacid tablet is not the same before and after the change. The evidence for the formation of a new substance is the gas bubbles that are released in the glass of water.

We may observe a *chemical reaction* when a substance undergoes a chemical change and forms a new substance. When we form a new substance, we usually observe one of the following: a change in color, an odor or bubbles indicating the liberation of a gas, or light or heat from the release of energy. For instance, we can mix flour, yeast, and water and put the mixture into an oven, and in a short time, smell the odor of bread baking. A chemical reaction has occurred in which the flour, yeast, and water have combined to produce a new substance, evidenced by a change in color and smell.

In the laboratory, we may observe that two colorless solutions added together produce a solid substance that is not soluble. The formation of an insoluble solid in solution is a practical observation that indicates a chemical reaction has occurred.

8.1 Evidence for Chemical Reactions

Objective · To state four observations that are evidence for a chemical reaction.

If we combine substances, the resulting change can be physical or chemical. We can distinguish between a physical and a chemical change by observing the change. Each of the following indicates a chemical change; that is, a **chemical reaction** has taken place.

Reaction of Tin and Zinc Movie

Precipitation Reactions Movie

Sodium and Potassium in Water Movie

1. *A gas is released.* We can observe a gas in a number of ways; for example, we may see bubbles in solution. When we add an Alka–Seltzer tablet to water, it begins to fizz. The fizzing is evidence for a chemical reaction, and the release of carbon dioxide gas, which is produced by the reaction of citric acid and baking soda in the tablet.
2. *An insoluble solid is produced.* A substance dissolves in water to give an **aqueous solution**. If we add two aqueous solutions together, we may observe solid particles in solution. This insoluble solid is called a **precipitate**. The formation of a precipitate is evidence for a reaction.
3. *A permanent color change is observed.* Many chemical reactions involve a permanent change in color. For example, adding aqueous ammonia to a copper(II) sulfate solution changes the color from light blue to deep blue. To observe an acid reacting with a base, we use an indicator that changes color. The indicator enables us to indirectly follow a reaction that would otherwise not be visible.
4. *A heat energy change is noted.* In chemical reactions there is often a change in temperature. A reaction that releases heat is said to be an **exothermic reaction**; conversely, a reaction that absorbs heat is said to be an **endothermic reaction**. For example, when an emergency flare is ignited, heat and light are observed. Heat and light are two forms of energy indicating that an exothermic chemical reaction has taken place.

In summary, each of the above four criteria—the production of a gas, the formation of a precipitate, a change in color, or a change in energy—is an indication that a chemical reaction has occurred. This is shown in Figure 8.1. A chemical reaction, however, may have occurred even though no change is obvious. In some reactions, for example, acid–base reactions, the energy change may be too subtle to notice and an external indicator may be necessary.

Example Exercise 8.1 • Evidence for a Reaction

Which of the following is experimental evidence for a chemical reaction?

(a) Pouring vinegar on baking soda causes foamy bubbles.

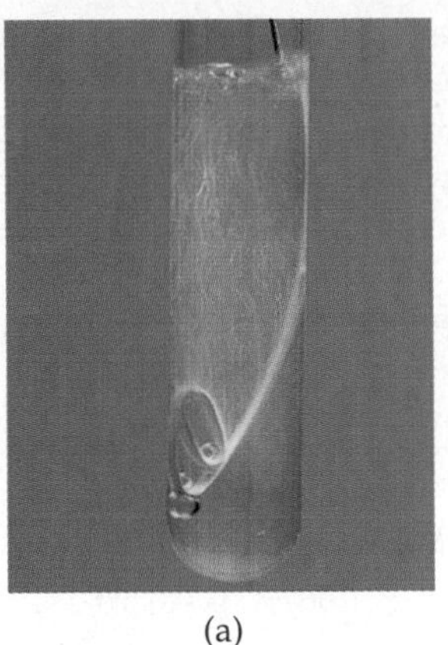
(a)

(b)

(c)

◀ **Figure 8.1 Evidence for a Chemical Reaction** Each of the following is evidence for a chemical reaction: (a) the gas bubbles released from magnesium metal in acid; (b) the heat and light released from sodium metal in water; and (c) the color change and the precipitate produced by two aqueous solutions.

(b) Mixing two solutions produces insoluble particles.
(c) Mixing two colorless solutions gives a pink solution.
(d) Mixing two solutions produces a temperature increase.

Solution

We can analyze each of these observations based on the criteria for a chemical reaction.

(a) The bubbles produced indicate a chemical reaction is occurring.
(b) The insoluble particles formed indicate a chemical reaction.
(c) The pink color produced indicates a chemical reaction.
(d) The temperature increase indicates heat energy is being released and, thus, an exothermic chemical reaction.

Self-Test Exercise

What are four indications that a chemical reaction has occurred?

Answers: (a) a gas is released; (b) a precipitate is produced; (c) a permanent color change is observed; (d) an energy change is noted

8.2 Writing Chemical Equations

Objectives
- To identify seven elements that occur naturally as diatomic molecules: H_2, N_2, O_2, F_2, Cl_2, Br_2, I_2.
- To write a chemical equation from the description of a chemical reaction.

In this chapter we will use formulas and symbols to describe a chemical reaction; that is, we will write a **chemical equation**. Consider the following description of a chemical reaction: substance A and substance B undergo a chemical change that produces substances C and D. We can state this description in symbols as follows.

$$A + B \longrightarrow C + D$$

In this general chemical equation, A and B are called the **reactants**, and C and D are called the **products**.

We can provide more information about the reaction if we specify the physical state of each substance, that is, solid, liquid, or gas. The physical state is indicated by the abbreviation (s), (l), or (g). By convention, we indicate an aqueous solution by the symbol (aq) and a precipitate by the symbol (s). For example, an aqueous solution of A reacts with a gaseous substance B to yield a precipitate of C and an aqueous solution D. This statement can be written as follows.

$$A(aq) + B(g) \longrightarrow C(s) + D(aq)$$

A **catalyst** is a substance that speeds up a reaction without being consumed or permanently altered. For example, the catalytic converter in an automobile contains a metallic catalyst that speeds up the conversion of unburned fuel to carbon dioxide and water. The presence of a catalyst is indicated by placing its formula above the arrow. Table 8.1 lists the symbols that are used to describe chemical reactions.

Table 8.1 Chemical Equation Symbols

Symbol	Interpretation of Chemical Equation Symbol
$\longrightarrow$	produces, yields, gives (points from reactants to products)
$+$	reacts with, added to, plus (separates two or more reactants or two or more products)
$\xrightarrow{\Delta}$	heat catalyst
$\xrightarrow{Fe}$	metallic iron catalyst
NR	no reaction
(s)	solid substance or precipitate
(l)	liquid substance
(g)	gaseous substance
(aq)	aqueous solution

Let's interpret the symbols in the following chemical equation for the reaction of acetic acid and baking soda.

$$HC_2H_3O_2(aq) + NaHCO_3(s) \longrightarrow NaC_2H_3O_2(aq) + H_2O(l) + CO_2(g)$$

The formulas and symbols are read as follows: aqueous acetic acid is added to solid sodium hydrogen carbonate, which produces aqueous sodium acetate, water, and carbon dioxide gas.

Chemical reactions often involve hydrogen, nitrogen, or oxygen gases. These nonmetals occur naturally as **diatomic molecules**, that is, H_2, N_2, and O_2. Moreover, the halogens also occur naturally as diatomic molecules and are therefore written as F_2, Cl_2, Br_2, and I_2. If one of these elements appears in a chemical equation, it is written as a diatomic molecule.

List of Naturally Occurring Diatomic Molecules

H_2, N_2, O_2, F_2, Cl_2, Br_2, and I_2

The following example exercise further illustrates how to write a chemical equation from the description of a chemical reaction.

Example Exercise 8.2 • Writing Chemical Equations

Write a chemical equation for each of the following chemical reactions.

(a) Iron metal is heated with sulfur powder to produce solid iron(II) sulfide.
(b) Zinc metal reacts with sulfuric acid to give aqueous zinc sulfate and hydrogen gas.

Solution

To write the chemical equation, we must provide formulas and symbols for each substance. We can describe each of the above chemical reactions as follows.

(a) $Fe(s) + S(s) \longrightarrow FeS(s)$
(b) $Zn(s) + H_2SO_4(aq) \longrightarrow ZnSO_4(aq) + H_2(g)$

Self-Test Exercise
Write a chemical equation for each of the following chemical reactions.
(a) Aqueous solutions of sodium iodide and silver nitrate yield silver iodide precipitate and aqueous sodium nitrate.
(b) Acetic acid reacts with aqueous potassium hydroxide to give aqueous potassium acetate plus water.

Answers: (a) $NaI(aq) + AgNO_3(aq) \longrightarrow AgI(s) + NaNO_3(aq)$; (b) $HC_2H_3O_2(aq) + KOH(aq) \longrightarrow KC_2H_3O_2(aq) + H_2O(l)$

8.3 Balancing Chemical Equations

Objective · To write balanced chemical equations by inspection.

In the previous section we translated the description of a chemical reaction into a chemical equation. In each case, the number of atoms of each element was equal in the reactants and products. Because the number of atoms of each element is the same on each side of the arrow, the chemical reaction is said to be a balanced chemical equation.

More often, the number of atoms of each element in the reactants and products is not the same. Therefore, it is necessary to balance the number of atoms of each element on each side of the chemical reaction. We balance a chemical reaction by placing a whole-number **coefficient** in front of each substance. It is important to note that a coefficient multiplies all subscripts in the chemical formula that follows. That is, 3 H_2O indicates six hydrogen atoms and three oxygen atoms.

Let's consider the reaction of hydrogen gas with chlorine gas to give hydrogen chloride, a gas with a sharp odor. We can write the chemical reaction as follows.

$$H_2(g) + Cl_2(g) \longrightarrow HCl(g)$$

Notice that the **subscript** for H_2 and Cl_2 is 2, but that only one H atom and one Cl atom appear in the product. Thus, the equation is not balanced. To balance the number of H atoms on both sides of the equation, we place the coefficient 2 in front of the HCl. This gives us

$$H_2(g) + Cl_2(g) \longrightarrow 2\ HCl(g)$$

After adding the reaction coefficient 2, we have two molecules of HCl. That is, we have two H atoms and two Cl atoms on each side of the equation. The equation is now balanced.

Let's try a more difficult example. Aluminum metal is heated with oxygen gas to give solid aluminum oxide. The formula for aluminum oxide is Al_2O_3. The chemical equation is

$$Al(s) + O_2(g) \longrightarrow Al_2O_3(s)$$

Notice that two O atoms appear in the reactants but that three O atoms appear in the product. This reaction is not balanced. To balance the numbers of O atoms, we will use the lowest common multiple of 2 and 3; the number is 6. We place the coefficient 3 in front of the O_2 to give six oxygen atoms in the reactants and the coefficient 2 in front of Al_2O_3 to give six oxygen atoms in the products. That gives us

$$Al(s) + 3\ O_2(g) \longrightarrow 2\ Al_2O_3(s)$$

The numbers of Al atoms, however, are not balanced. On the reactant side we have one Al atom. On the product side, we have two units of Al_2O_3, for a total of four Al atoms. We can place the coefficient 4 in front of the reacting Al metal. This gives a balanced chemical equation.

$$4\,Al(s) + 3\,O_2(g) \longrightarrow 2\,Al_2O_3(s)$$

This method of placing coefficients by systematically analyzing each side of an equation is called *balancing by inspection*. Although there is no formal prescription for balancing an equation by inspection, we can list some general guidelines.

It is very helpful to students to list these guidelines for balancing a chemical equation. That is, (1) verify the chemical formulas in the equation; (2a) begin balancing with the most complex formula; (2b) balance polyatomic ions as a single unit; and (2c) discuss the use of a fractional coefficient.

General Guidelines for Balancing a Chemical Equation

1. Before placing a coefficient in an equation, check the formula of each substance to make sure that the subscripts are correct.*
2. Balance each element in the equation by placing a coefficient in front of each substance. Coefficients of 1 are assumed and do not appear in the balanced chemical equation.
   (a) Begin balancing the equation starting with the most complex formula.
   (b) Balance polyatomic ions as a single unit. However, if a polyatomic ion decomposes, you must balance each atom separately.
   (c) The coefficients in an equation must be whole numbers. Occasionally, it is helpful to use a fractional coefficient to balance an element in a diatomic molecule; for example,

   $$H_2(g) + \tfrac{1}{2}O_2(g) \longrightarrow H_2O(l)$$

   If a fraction is used, such as $\frac{1}{2}$, we must then multiply the equation by 2 to obtain whole-number coefficients. Thus, the balanced equation is

   $$2\,H_2(g) + O_2(g) \longrightarrow 2\,H_2O(l)$$

3. After balancing the equation, check (✓) each symbol of every element (or polyatomic ion) to verify that the coefficients are correct. Proceed back and forth between reactants and products. The procedure for verification is to multiply the coefficient times the subscript of each element; the totals should be the same on both sides of the equation.

   Finally, check the coefficients to make sure they represent the smallest whole-number ratio. It may be possible to obtain a smaller set of coefficients. Although the following equation is balanced,

   $$6\,H_2(g) + 2\,N_2(g) \longrightarrow 4\,NH_3(g)$$

   you must reduce the coefficients by a factor of 2 to obtain the correct balanced equation.

   $$3\,H_2(g) + N_2(g) \longrightarrow 2\,NH_3(g)$$

*Subscripts of correct formulas are never changed to balance an equation. In fact, if an equation is difficult to balance, it is most often because of an incorrect chemical formula.

We can write an equation for the formation of the yellow precipitate shown in Figure 8.1c. In the reaction, aqueous solutions of lead(II) nitrate and potassium iodide produce a yellow precipitate of lead(II) iodide and an aqueous solution of potassium nitrate. We can write a chemical equation for the reaction as follows.

$$Pb(NO_3)_2(aq) + KI(aq) \longrightarrow PbI_2(s) + KNO_3(aq)$$

Let's begin balancing with $Pb(NO_3)_2$ because it is the most complex formula. Notice that one Pb atom appears on each side of the equation. However, there are two NO_3^- ions on the reactant side and only one NO_3^- ion on the product side. We can balance the NO_3 polyatomic ion by placing a 2 in front of KNO_3. Thus,

$$Pb(NO_3)_2(aq) + KI(aq) \longrightarrow PbI_2(s) + 2\ KNO_3(aq)$$

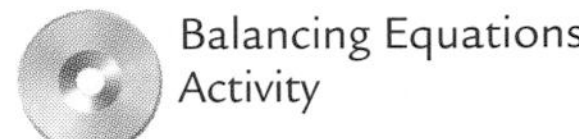

Balancing Equations Activity

We now have two K on the right side of the equation and only one K on the left. Thus, we need the coefficient 2 in front of KI.

$$Pb(NO_3)_2(aq) + 2\ KI(aq) \longrightarrow PbI_2(s) + 2\ KNO_3(aq)$$

Notice that the coefficient 2 in front of KI gives two I, which is equal to the two I in PbI_2. As a final inspection, let's check off each element and polyatomic ion to verify that we indeed have a balanced equation.

✓ ✓ ✓✓ ✓ ✓ ✓✓

$$Pb(NO_3)_2(aq) + 2\ KI(aq) \longrightarrow PbI_2(s) + 2\ KNO_3(aq)$$

The following example exercises further illustrate the general guidelines for balancing a chemical equation.

Example Exercise 8.3 • Balancing Chemical Equations

Aqueous solutions of calcium acetate and potassium phosphate react to give a white precipitate of calcium phosphate and aqueous potassium acetate. Write a balanced chemical equation given

$$Ca(C_2H_3O_2)_2(aq) + K_3PO_4(aq) \longrightarrow Ca_3(PO_4)_2(s) + KC_2H_3O_2(aq)$$

Solution

We see that $Ca_3(PO_4)_2$ is the most complicated formula because each ion in the formula is followed by a subscript. There are three Ca on the right side of the equation and only one Ca on the left side. Placing the coefficient 3, we have

$$3\ Ca(C_2H_3O_2)_2(aq) + K_3PO_4(aq) \longrightarrow Ca_3(PO_4)_2(s) + KC_2H_3O_2(aq)$$

We now have six $C_2H_3O_2$ on the left side of the equation and only one on the right. Thus, we need the coefficient 6 in front of $KC_2H_3O_2$.

$$3\ Ca(C_2H_3O_2)_2(aq) + K_3PO_4(aq) \longrightarrow Ca_3(PO_4)_2(s) + 6\ KC_2H_3O_2(aq)$$

The coefficient 6 in front of $KC_2H_3O_2$ generates six K. We now need six K on the left side. We can place a 2 in front of K_3PO_4. The coefficient 2 gives two PO_4 on the left side, which equals the number in $Ca_3(PO_4)_2$. Finally, we can check off each element and polyatomic ion to verify that the chemical equation is balanced.

✓ ✓ ✓ ✓ ✓ ✓ ✓ ✓

$$3\ Ca(C_2H_3O_2)_2(aq) + 2\ K_3PO_4(aq) \longrightarrow Ca_3(PO_4)_2(s) + 6\ KC_2H_3O_2(aq)$$

Self-Test Exercise

Aqueous solutions of aluminum sulfate and barium nitrate react to yield a white precipitate of barium sulfate and aqueous aluminum nitrate. Write a balanced chemical equation given

$$Al_2(SO_4)_3(aq) + Ba(NO_3)_2(aq) \longrightarrow BaSO_4(s) + Al(NO_3)_3(aq)$$

✓ ✓ ✓ ✓ ✓ ✓ ✓ ✓

Answer: $Al_2(SO_4)_3(aq) + 3\ Ba(NO_3)_2(aq) \longrightarrow 3\ BaSO_4(s) + 2\ Al(NO_3)_3(aq)$

Example Exercise 8.4 • Balancing Chemical Equations

Sulfuric acid reacts with aqueous sodium hydroxide to give aqueous sodium sulfate and water. Write a balanced chemical equation given

$$H_2SO_4(aq) + NaOH(aq) \longrightarrow Na_2SO_4(aq) + H_2O(l)$$

Solution

Let's start with Na_2SO_4, which contains the same number of atoms as H_2SO_4. There is one SO_4 on each side of the equation, and so SO_4 is balanced. However, there are two Na on the right side of the equation and one Na on the left. We must place the coefficient 2 in front of NaOH.

$$H_2SO_4(aq) + 2\,NaOH(aq) \longrightarrow Na_2SO_4(aq) + H_2O(l)$$

To balance this reaction more easily, we can write the formula for water as HOH. Notice that we now have two OH on the left side of the equation and HOH on the right side. By placing a 2 in front of the HOH, we can balance the OH. The two H on the right side are balanced by the two H in H_2SO_4. Finally, we check off each atom and polyatomic ion to verify that we have a balanced chemical equation.

$$\overset{\checkmark\checkmark}{H_2SO_4(aq)} + \overset{\checkmark\checkmark}{2\,NaOH(aq)} \longrightarrow \overset{\checkmark\checkmark}{Na_2SO_4(aq)} + \overset{\checkmark\checkmark}{2\,HOH(l)}$$

Self-Test Exercise

Carbonic acid reacts with aqueous ammonium hydroxide to give aqueous ammonium carbonate and water. Write a balanced chemical equation given

$$H_2CO_3(aq) + NH_4OH(aq) \longrightarrow (NH_4)_2CO_3(aq) + HOH(l)$$

Answer: $\overset{\checkmark\checkmark}{H_2CO_3(aq)} + \overset{\checkmark\checkmark}{2\,NH_4OH(aq)} \longrightarrow \overset{\checkmark\checkmark}{(NH_4)_2CO_3(aq)} + \overset{\checkmark\checkmark}{2\,HOH(l)}$

Note Balancing an equation is a straightforward task. Start with the most complex formula and balance systematically, back and forth, between reactants and products. If you encounter difficulty in the final step, it is usually because the original equation had an incorrect subscript in a chemical formula.

8.4 Classifying Chemical Reactions

The mastery of chemical reactions is greatly aided by an initial overview classification scheme for reactions as: (a) combination, (b) decomposition, (c) single replacement, (d) double replacement, or (e) neutralization.

Objective • To classify a chemical reaction as one of the following types: combination, decomposition, single replacement, double replacement, or neutralization.

There are millions of chemical reactions. How can we attempt to master such a large number of reactions? The answer lies in the same way that we learned previously to master chemical formulas. We *classify chemical reactions* and put them in categories. There are many ways to classify reactions, and later we will study redox reactions (Chapter 17) and combustion reactions (Chapter 19).

In this chapter, we begin our study with five simple types of reactions: combination, decomposition, single replacement, double replacement, and neutralization. We can briefly describe these five types as follows.

1. A **combination reaction** involves simpler substances being combined into a more complex compound. It is also referred to as a *synthesis reaction*. In the general case, substance A combines with substance Z to produce compound AZ. The chemical equation is

$$A + Z \longrightarrow AZ$$

2. In a **decomposition reaction** a single compound is broken down into two or more simpler substances. In this case, heat or light is usually applied to decompose the compound. In the general case, compound AZ decomposes into substances A and Z. The chemical equation is

$$AZ \longrightarrow A + Z$$

3. In a **single-replacement reaction** one metal displaces another metal, or hydrogen, from a compound or aqueous solution. The substance that is displaced shows less tendency to undergo reaction; that is, it is less active. In the general case, metal A replaces metal B in BZ to give AZ and B. The chemical equation is

$$A + BZ \longrightarrow AZ + B$$

4. In a **double-replacement reaction** two compounds exchange anions. Compound AX reacts with compound BZ to yield the products AZ and BX. The general form of the reaction is

$$AX + BZ \longrightarrow AZ + BX$$

5. An acid and a base react with each other in a **neutralization reaction**. Acid HX reacts with base BOH to give ionic compound BX and water. We can write the general form of the reaction as

$$HX + BOH \longrightarrow BX + HOH$$

When we examine a neutralization reaction, we see that it is actually a special type of double-replacement reaction. An acid and a base simply switch anion partners. Notice that we wrote water as HOH rather than H_2O. Use of the formula HOH shows the double-replacement nature of the neutralization reaction and makes it easier to balance the equation.

Experiment #10, Prentice Hall Laboratory Manual

Example Exercise 8.5 • Classifying Chemical Reactions

Classify each of the following reactions as combination, decomposition, single replacement, double replacement, or neutralization.

(a) Copper metal heated with oxygen gas produces solid copper(II) oxide.

$$2\,Cu(s) + O_2(g) \longrightarrow 2\,CuO(s)$$

(b) Heating powdered iron(III) carbonate produces solid iron(III) oxide and carbon dioxide gas.

$$Fe_2(CO_3)_3(s) \longrightarrow Fe_2O_3(s) + 3\,CO_2(g)$$

(c) Aluminum metal reacts with aqueous manganese(II) sulfate to give aqueous aluminum sulfate and manganese metal.

$$2\,Al(s) + 3\,MnSO_4(aq) \longrightarrow Al_2(SO_4)_3(aq) + 3\,Mn(s)$$

(d) Aqueous sodium chromate reacts with aqueous barium chloride to give insoluble barium chromate and aqueous sodium chloride.

$$Na_2CrO_4(aq) + BaCl_2(aq) \longrightarrow BaCrO_4(s) + 2\,NaCl(aq)$$

(e) Nitric acid reacts with aqueous potassium hydroxide to give aqueous potassium nitrate and water.

$$HNO_3(aq) + KOH(aq) \longrightarrow KNO_3(aq) + H_2O(l)$$

(continued)

Chemistry Connection • Household Chemicals

What two common household chemicals, when mixed, can produce a deadly gas?

Salt and sugar are two familiar chemicals on your kitchen table. Vinegar, another kitchen chemical, is a solution of acetic acid. Citrus fruit contains citric acid and is responsible for the sour taste of lemons and limes. Aspirin contains acetylsalicylic acid, which can irritate the lining of the stomach if taken in excess.

Hydrochloric acid, used to acidify swimming pools, is sold in the supermarket as muriatic acid. Sulfuric acid, found in lead storage batteries, is a dangerous chemical that should be handled with great caution. Sulfuric acid is strongly corrosive and can cause skin ulcers. Another potential danger of sulfuric acid is associated with "jump-starting" an automobile. When an electric current passes through battery acid, hydrogen gas is evolved. If the jumper cables spark, hydrogen can react explosively with oxygen in the air.

Perhaps the most dangerous chemical in the home is caustic soda, NaOH, sold under various trade names as a drain cleaner. If it contacts your skin, it gives a slippery feeling as a small amount of skin tissue sloughs off. If taken internally by a child, caustic soda could be lethal. Household ammonia is potentially dangerous and should not be used in conjunction with bleach. Together, these two chemicals produce a poisonous gas.

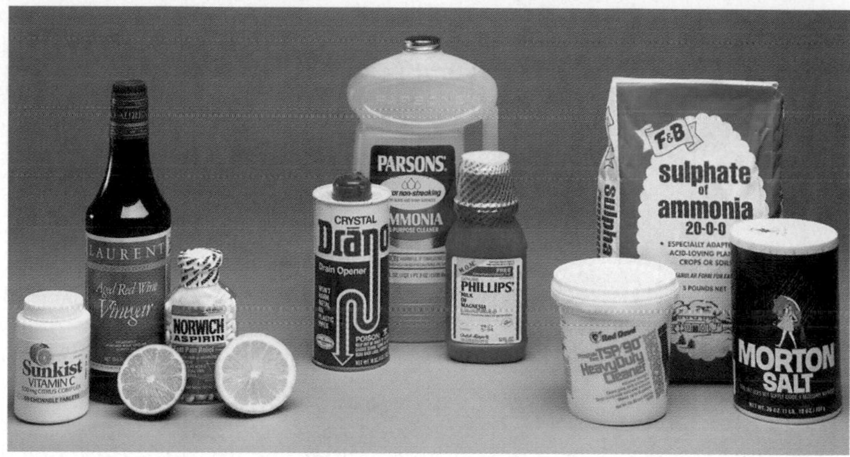

▲ **Common Household Chemicals**

Now that you have been alerted to some dangers of common chemicals, you should also remember that each time you sip sweet, tart lemonade, it is household chemicals that produce the pleasant taste.

Common Household Chemicals

Name	Formula	Product/Use	Safety
Acids			
acetic acid	$HC_2H_3O_2$	vinegar	
carbonic acid	H_2CO_3	carbonated drinks	
hydrochloric acid	HCl	swimming pools	avoid contact*
sulfuric acid	H_2SO_4	battery acid	avoid contact*†
Bases			
ammonia	NH_3	cleaning solutions	
magnesium hydroxide	$Mg(OH)_2$	milk of magnesia	
sodium bicarbonate	$NaHCO_3$	antacid, fire extinguisher	
sodium hydroxide	$NaOH$	drain and oven cleaner	avoid contact*†
Miscellaneous			
aluminum hydroxide	$Al(OH)_3$	antacid tablets	
carbon dioxide (solid)	CO_2	dry ice	avoid contact
epsom salts	$MgSO_4 \cdot 7H_2O$	cathartic, laxative	
sodium hypochlorite	$NaClO$	bleach	avoid contact*
Organic			
ethylene glycol	$C_2H_4(OH)_2$	antifreeze	avoid ingestion†
methanol	CH_3OH	solvent, antifreeze	avoid ingestion†
naphthalene	$C_{10}H_8$	mothballs	avoid ingestion†
trichloroethane	$C_2H_3Cl_3$	spot remover	avoid ingestion†

*In the event of contact, flush with water. †Seek medical attention immediately.

Solutions of household ammonia and ordinary bleach can react to produce a lethal gas.

Example Exercise 8.5 *(continued)*

Solution

We can classify the type of each reaction as follows.

(a) The two elements Cu and O_2 synthesize a single compound; this is an example of a *combination* reaction.

(b) The compound $Fe_2(CO_3)_3$ is heated and breaks down into a simpler compound and a gas; this is an example of a *decomposition* reaction.

(c) The metal Al displaces the metal Mn from aqueous $MnSO_4$; this is an example of a *single-replacement* reaction.

(d) The two compounds Na_2CrO_4 and $BaCl_2$ exchange anions; this is an example of a *double-replacement* reaction.

(e) The acid HNO_3 reacts with the base KOH to form KNO_3 and water; this is an example of a *neutralization* reaction.

Self-Test Exercise

Classify the following types of reactions as combination, decomposition, single replacement, double replacement, or neutralization.

(a) $Zn(s) + CuSO_4(aq) \longrightarrow ZnSO_4(aq) + Cu(s)$

(b) $2\,Sr(s) + O_2(g) \longrightarrow 2\,SrO(s)$

(c) $Cd(HCO_3)_2(s) \longrightarrow CdCO_3(s) + H_2O(g) + CO_2(g)$

(d) $HC_2H_3O_2(aq) + NaOH(aq) \longrightarrow NaC_2H_3O_2(aq) + H_2O(l)$

(e) $AgNO_3(aq) + KCl(aq) \longrightarrow AgCl(s) + KNO_3(aq)$

Answers: (a) single replacement; (b) combination; (c) decomposition; (d) neutralization; (e) double replacement

8.5 Combination Reactions

Objectives
- To write a balanced chemical equation for the reaction of a metal and oxygen gas.
- To write a balanced chemical equation for the reaction of a nonmetal and oxygen gas.
- To write a balanced chemical equation for the reaction of a metal and a nonmetal.

Reactions with Oxygen Movie

There are many examples of combination reactions. For instance, heating two substances may cause them to combine into a single compound. In this section, we will study three important kinds of combination reactions.

Metal and Oxygen Gas

In one kind of combination reaction, a metal and oxygen gas react to give a metal oxide. In the following example, magnesium metal and oxygen gas combine to give magnesium oxide.

$$\textbf{metal} + \textbf{oxygen gas} \longrightarrow \textbf{metal oxide}$$
$$2\,Mg(s) + O_2(g) \longrightarrow 2\,MgO(s)$$

A metal and oxygen react to give a metal oxide compound. We can usually predict the formulas of metal oxides containing a main-group metal. On the other hand, we cannot predict the formulas of most metal oxides containing a transition metal. For transition metal oxides, the ionic charge on the metal must be given in order to write the chemical formula of the metal oxide.

The following example exercise further illustrates how to write balanced equations for a combination reaction of a metal and oxygen gas.

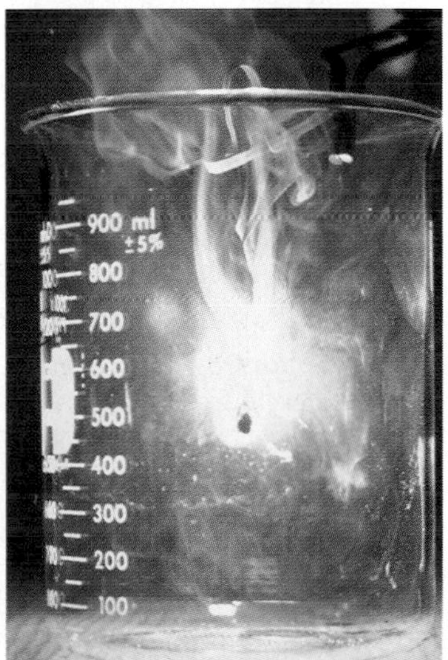

▲ **Igniting Magnesium Ribbon** A strip of magnesium metal reacts with oxygen in air to give white smoke, MgO, and a bright white light.

Example Exercise 8.6 • Combination Reactions

Write a balanced chemical equation for each of the following combination reactions.

(a) Zinc metal is heated with oxygen gas in air to yield solid zinc oxide.

(b) Chromium metal is heated with oxygen gas to give chromium(III) oxide.

Solution

A metal and oxygen react to produce a metal oxide.

(a) Zinc is a metal with a predictable charge, that is, Zn^{2+}. The formula of zinc oxide is ZnO. The balanced equation for the reaction is

$$2\,Zn(s) + O_2(g) \longrightarrow 2\,ZnO(s)$$

(b) Chromium is a metal with a variable charge. From the name chromium(III) oxide, we know the ion is Cr^{3+}. The formula of the oxide is, therefore, Cr_2O_3, and the balanced equation for the reaction is

$$4\,Cr(s) + 3\,O_2(g) \longrightarrow 2\,Cr_2O_3(s)$$

Self-Test Exercise

Write a balanced chemical equation for each of the following combination reactions.

(a) Lead metal is heated with oxygen in air to yield solid lead(IV) oxide.

(b) Cobalt metal is heated with oxygen gas to give solid cobalt(III) oxide.

Answers: (a) $Pb(s) + O_2(g) \longrightarrow PbO_2(s)$; (b) $4\,Co(s) + 3\,O_2(g) \longrightarrow 2\,Co_2O_3(s)$

Nonmetal and Oxygen Gas

In this kind of combination reaction, a nonmetal and oxygen gas react to give a nonmetal oxide. In the following example, yellow sulfur powder and oxygen gas combine to give sulfur dioxide gas.

$$\textbf{nonmetal} + \textbf{oxygen gas} \longrightarrow \textbf{nonmetal oxide}$$

$$S(s) + O_2(g) \longrightarrow SO_2(g)$$

Igniting Sulfur Powder A small portion of yellow sulfur burns with the oxygen in air to give an intense blue flame and sulfur dioxide gas, SO_2, which has a sharp odor.

Nonmetal oxides demonstrate multiple combining capacities. In general, the formula of a nonmetal oxide product is unpredictable and varies with temperature and pressure. For example, nitrogen and oxygen gases can combine to give all of the following: NO, NO_2, N_2O, N_2O_3, N_2O_4, and N_2O_5. Therefore, to complete and balance an equation of a nonmetal and oxygen gas, we must be given the formula of the nonmetal oxide.

The following example exercise further illustrates how to write balanced equations for a combination reaction of a nonmetal and oxygen gas.

Example Exercise 8.7 • Combination Reactions

Write a balanced chemical equation for each of the following combination reactions.

(a) Carbon is heated with oxygen gas to produce carbon dioxide gas.

(b) Phosphorus and oxygen gas react to give solid diphosphorus pentaoxide.

Solution

A nonmetal and oxygen combine to produce a nonmetal oxide.

(a) The formula of the nonmetal oxide is unpredictable. We are given that the product is carbon dioxide and not carbon monoxide. The balanced equation for the reaction is

$$C(s) + O_2(g) \longrightarrow CO_2(g)$$

(b) The formula for the oxide of phosphorus is not predictable, but we have the name of the nonmetal oxide product. The formula for diphosphorus pentaoxide is P_2O_5. The balanced equation is

$$4\ P(s) + 5\ O_2(g) \longrightarrow 2\ P_2O_5(s)$$

Self-Test Exercise

Write a balanced chemical equation for each of the following combination reactions.

(a) Nitrogen gas is heated with oxygen to give dinitrogen trioxide gas.
(b) Chlorine gas is heated with oxygen to give dichlorine monoxide gas.

Answers: (a) $2\ N_2(g) + 3\ O_2(g) \longrightarrow 2\ N_2O_3(g)$; (b) $2\ Cl_2(g) + O_2(g) \longrightarrow 2\ Cl_2O(g)$

Metal and Nonmetal

In this kind of combination reaction, a metal and a nonmetal react to give an ionic compound. In the following example, silvery sodium metal and yellow chlorine gas combine to give solid sodium chloride.

$$\textbf{metal} + \textbf{nonmetal} \longrightarrow \textbf{ionic compound}$$

$$2\ Na(s) + Cl_2(g) \longrightarrow 2\ NaCl(s)$$

When a metal and a nonmetal react, they produce a binary ionic compound. When the compound contains a main-group metal, the formula is usually predictable. The formulas for compounds containing a transition metal are often not predictable.

The following example exercise further illustrates how to write balanced equations for a combination reaction of a metal and a nonmetal.

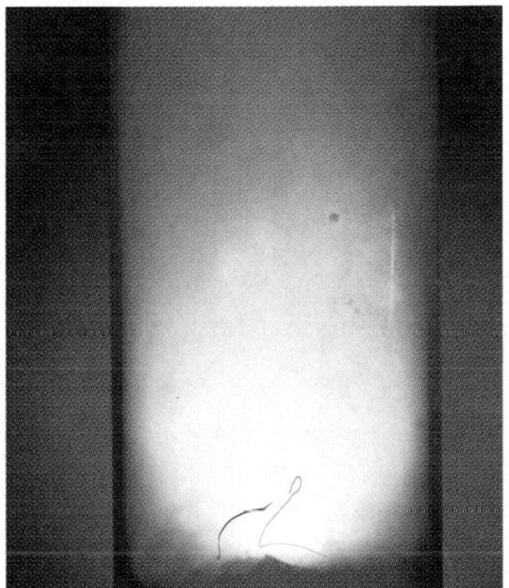

Igniting Sodium and Chlorine Sodium metal reacts strongly with chlorine gas to give a bright yellow light and sodium chloride, NaCl, powder.

Formation of Sodium Chloride Movie

Example Exercise 8.8 • Combination Reactions

Write a balanced chemical equation for each of the following combination reactions.

(a) Aluminum metal is heated with sulfur and gives a solid product.
(b) Chromium metal is heated with iodine and produces powdered chromium(III) iodide.

Solution

A metal and a nonmetal react to produce an ionic compound.

(a) The formula of the product is predictable. Aluminum combines with sulfur to give aluminum sulfide, Al_2S_3. The balanced equation for the reaction is

$$2\ Al(s) + 3\ S(s) \longrightarrow Al_2S_3(s)$$

(b) Chromium is a transition metal, and so we cannot predict the formula for the product. We are given the name of the compound, chromium(III) iodide, and so the formula is CrI_3. The equation is

$$2\ Cr(s) + 3\ I_2(s) \longrightarrow 2\ CrI_3(s)$$

Self-Test Exercise

Write a balanced chemical equation for each of the following combination reactions.

(a) Calcium metal is heated with fluorine gas to yield solid calcium fluoride.
(b) Manganese metal reacts with bromine vapor to give crystalline manganese(IV) bromide.

Answers: (a) $Ca(s) + F_2(g) \longrightarrow CaF_2(s)$; (b) $Mn(s) + 2\ Br_2(g) \longrightarrow MnBr_4(s)$

8.6 Decomposition Reactions

Objectives
- To write a balanced equation for the decomposition of a metal hydrogen carbonate.
- To write a balanced equation for the decomposition of a metal carbonate.
- To write a balanced equation for the decomposition of a compound that releases oxygen gas.

There are many examples of decomposition reactions. For instance, heating a compound may cause it to decompose into two or more simpler substances. In this section, we will study three important kinds of decomposition reactions.

Metal Hydrogen Carbonates

A metal hydrogen carbonate undergoes a decomposition reaction when it is heated. For example, you may be aware that baking soda is a natural fire extinguisher. That is, heat from a fire decomposes baking soda (sodium hydrogen carbonate) and releases carbon dioxide gas. Since carbon dioxide is more dense than air, it can smother a fire by excluding oxygen from the flame.

The decomposition of a metal hydrogen carbonate yields a metal carbonate, water, and carbon dioxide. In the following example, sodium hydrogen carbonate is decomposed by heating to give sodium carbonate, water, and carbon dioxide.

metal hydrogen carbonate $\longrightarrow$ **metal carbonate + water + carbon dioxide**

$$2\ NaHCO_3(s) \xrightarrow{\Delta} Na_2CO_3(s) + H_2O(g) + CO_2(g)$$

During the decomposition reaction of a metal hydrogen carbonate, the ionic charge of the metal does not change. Therefore, the formula of the metal carbonate is predictable. Even transition metal hydrogen carbonates are predictable. If nickel(II) hydrogen carbonate decomposes, one of the products is nickel(II) carbonate. That is,

$$Ni(HCO_3)_2(s) \xrightarrow{\Delta} NiCO_3(s) + H_2O(g) + CO_2(g)$$

The following example exercise further illustrates how to write balanced equations for a decomposition reaction of a metal hydrogen carbonate.

Example Exercise 8.9 • Decomposition Reactions

Write a balanced chemical equation for each of the following decomposition reactions.

(a) Lithium hydrogen carbonate decomposes on heating.
(b) Lead(II) hydrogen carbonate decomposes on heating.

Solution

A metal hydrogen carbonate decomposes with heat to give a metal carbonate, water, and carbon dioxide gas.

(a) All the formulas are predictable including that of the product, Li_2CO_3. The balanced equation for the reaction is

$$2\ LiHCO_3(s) \xrightarrow{\Delta} Li_2CO_3(s) + H_2O(g) + CO_2(g)$$

(b) Although the ionic charge on lead is variable, we are given that lead(II) hydrogen carbonate is the reactant. Therefore, the product is lead(II) carbonate, $PbCO_3$. The balanced equation for the reaction is

$$Pb(HCO_3)_2(s) \xrightarrow{\Delta} PbCO_3(s) + H_2O(g) + CO_2(g)$$

Self-Test Exercise

Write a balanced chemical equation for each of the following decomposition reactions.

(a) Barium hydrogen carbonate is decomposed by heating the compound.

(b) Copper(I) hydrogen carbonate is decomposed with heat.

Answers: (a) $Ba(HCO_3)_2(s) \xrightarrow{\Delta} BaCO_3(s) + H_2O(g) + CO_2(g)$;

(b) $2\ CuHCO_3(s) \xrightarrow{\Delta} Cu_2CO_3(s) + H_2O(g) + CO_2(g)$

Metal Carbonates

After a metal hydrogen carbonate decomposes into a metal carbonate, the metal carbonate can further decompose with prolonged heating. The carbonate can decompose into a metal oxide while releasing carbon dioxide gas.

In the following example, calcium carbonate is decomposed by heating to give calcium oxide and carbon dioxide.

$$\textbf{metal carbonate} \longrightarrow \textbf{metal oxide} + \textbf{carbon dioxide}$$

$$CaCO_3(s) \xrightarrow{\Delta} CaO(s) + CO_2(g)$$

The ionic charge on the metal does not change during the decomposition reaction of a metal carbonate. Therefore, we can predict the formula for the metal oxide product from the decomposition of a metal carbonate. If nickel(II) carbonate is decomposed, one of the products is nickel(II) oxide. That is,

$$NiCO_3(s) \xrightarrow{\Delta} NiO(s) + CO_2(g)$$

Example Exercise 8.10 • Decomposition Reactions

Write a balanced chemical equation for each of the following decomposition reactions.

(a) Magnesium carbonate decomposes on heating.

(b) Copper(I) carbonate decomposes on heating.

Solution

A metal carbonate decomposes with heat to a metal oxide and carbon dioxide gas.

(a) All the formulas are predictable including that of the metal oxide, MgO. The balanced equation for the reaction is

$$MgCO_3(s) \xrightarrow{\Delta} MgO(s) + CO_2(g)$$

(b) The ionic charge on copper can be either 1+ or 2+. Since the reactant is copper(I) carbonate, the product is copper(I) oxide, that is, Cu_2O. Thus, the balanced equation for the reaction is

$$Cu_2CO_3(s) \xrightarrow{\Delta} Cu_2O(s) + CO_2(g)$$

Self-Test Exercise

Write a balanced chemical equation for each of the following decomposition reactions.

(a) Aluminum carbonate is decomposed by heating the compound.

(b) Iron(II) carbonate is decomposed by heating.

Answers: (a) $Al_2(CO_3)_3(s) \xrightarrow{\Delta} Al_2O_3(s) + 3\ CO_2(g)$; (b) $FeCO_3(s) \xrightarrow{\Delta} FeO(s) + CO_2(g)$

▲ **Decomposing Mercury Oxide** A small portion of orange HgO powder decomposes with heat to give oxygen gas, O_2, and beads of silvery, metallic mercury.

Decomposition of Mercury Oxide Movie

Miscellaneous Oxygen-Containing Compounds

We can often decompose an oxygen-containing ionic compound, which releases oxygen gas. However, we cannot always predict the products. In this kind of decomposition reaction, we must be given the formulas of the products.

In the following example, mercury(II) oxide is decomposed by heating to give liquid mercury metal and oxygen gas.

$$\text{oxygen-containing compound} \longrightarrow \text{substance} + \text{oxygen gas}$$

$$2\,HgO(s) \xrightarrow{\Delta} 2\,Hg(l) + O_2(g)$$

Example Exercise 8.11 • Decomposition Reactions

Write a balanced chemical equation for each of the following decomposition reactions.

(a) Sodium nitrate is a white crystalline substance that decomposes on heating to give solid sodium nitrite and oxygen gas.

(b) Manganese(II) sulfate is a pink powder that decomposes on heating to give solid manganese(II) oxide and sulfur trioxide gas.

Solution

There is no general format for these reactions; however, we are given the names of the products.

(a) We can write the formula for each substance and then balance the chemical equation. The balanced equation for the reaction is

$$2\,NaNO_3(s) \xrightarrow{\Delta} 2\,NaNO_2(s) + O_2(g)$$

(b) The decomposition of manganese(II) sulfate gives MnO and SO_3. Thus, the balanced equation for the reaction is

$$MnSO_4(s) \xrightarrow{\Delta} MnO(s) + SO_3(g)$$

Self-Test Exercise

Write a balanced chemical equation for each of the following decomposition reactions.

(a) White crystals of potassium chlorate decompose on heating to give solid potassium chloride and oxygen gas.

(b) A colorless solution of hydrogen peroxide, H_2O_2, decomposes on heating to give water and oxygen gas.

Answers: (a) $2\,KClO_3(s) \xrightarrow{\Delta} 2\,KCl(s) + 3\,O_2(g)$; (b) $2\,H_2O_2(aq) \xrightarrow{\Delta} 2\,H_2O(l) + O_2(g)$

8.7 The Activity Series Concept

Objectives
- To explain the concept of an activity series for metals.
- To predict whether a single-replacement reaction occurs by referring to the activity series.

When a metal undergoes a replacement reaction, it displaces another metal from a compound or aqueous solution. The metal that displaces the other metal does so because it has a greater tendency to undergo a reaction; in other words, it is more active. If metal A is more active than metal B, metal A will displace metal B from compound BZ. We can show the reaction as

$$A + BZ \longrightarrow AZ + B$$

Activity Series for Metals

Li > K > Ba > Sr > Ca > Na > Mg > Al > Mn > Zn > Fe > Cd > Co > Ni > Sn > Pb > (H) > Cu > Ag > Hg > Au

The activity of a metal is a measure of its ability to compete in a replacement reaction. In an **activity series** (or electromotive series), a sequence of metals is arranged according to their ability to undergo reaction. Metals that are most reactive appear first in the activity series. Metals that are less reactive appear last in the series. The relative activity of several metals is listed in the margin.

We can see that lithium precedes potassium in the series, which in turn precedes barium, and so on. Although hydrogen (H) is not a metal, it is included in the series as a reference. Metals that precede (H) in the series react with an aqueous acid; metals that follow (H) in the series do not react with acids. Notice that the metals Cu, Ag, Hg, and Au do not react with acids.

Consider the reaction of iron metal in an aqueous solution of copper(II) sulfate.

$$Fe(s) + CuSO_4(aq) \longrightarrow FeSO_4(aq) + Cu(s)$$

Iron precedes copper in the series, and so Fe displaces Cu from aqueous solution. If we perform the reaction in the laboratory, a reddish-brown copper deposit appears on the iron metal. It is this type of experiment that establishes the order of metals in the activity series; that is, Fe $> > >$ Cu.

Conversely, there is no reaction if we place copper metal in an aqueous solution of iron(II) sulfate. That is, Cu cannot displace Fe from solution because copper follows iron in the activity series. Thus, if we put copper wire into a solution of iron(II) sulfate, we observe

$$Cu(s) + FeSO_4(aq) \longrightarrow NR$$

Now let's consider the reaction of iron wire with sulfuric acid. The reaction releases gas bubbles in solution, and the equation is

$$Fe(s) + H_2SO_4(aq) \longrightarrow FeSO_4(aq) + H_2(g)$$

Iron precedes (H) in the activity series, and so Fe displaces H_2 gas from a sulfuric acid solution. When we observe the reaction, we see tiny bubbles of hydrogen gas on the iron wire. Now consider the reaction of copper metal and sulfuric acid.

$$Cu(s) + H_2SO_4(aq) \longrightarrow NR$$

Copper is after (H) in the activity series, and so Cu cannot displace H_2 gas from sulfuric acid. When we observe the copper wire in sulfuric acid, there is no evidence for reaction.

▲ **Iron in a Copper Solution**
An iron nail reacts in an aqueous blue $CuSO_4$ solution to give a reddish-brown deposit of Cu metal on the nail.

Active Metals

There are a few metals that are so reactive that they react directly with water at room temperature. These metals are called **active metals** and include most of the metals in Groups IA/1 and IIA/2. Specifically, Li, Na, K, Rb, Cs, Ca, Sr, and Ba react with water. The most common active metals are listed in the margin.

Consider an experiment in which sodium metal is dropped into water. Sodium is an active metal and reacts with water according to the following equation.

$$2\,Na(s) + 2\,H_2O(l) \longrightarrow 2\,NaOH(aq) + H_2(g)$$

The following example exercise further illustrates how to predict the reactivity of metals with aqueous solutions, acids, and water.

List of Active Metals

Li $>$ K $>$ Ba $>$ Sr $>$ Ca $>$ Na

As a teaching aid, point out the position of the active metals in the periodic table.

Example Exercise 8.12 • **Predictions Based on the Activity Series**

Predict whether or not a reaction occurs for each of the following.

(a) Aluminum foil is added to an iron(II) sulfate solution.
(b) Iron wire is added to an aluminum sulfate solution.
(c) Manganese metal is added to acetic acid.
(d) Magnesium metal is added to water.

(continued)

Example Exercise 8.12 *(continued)*

Solution

We refer to the activity series for the reactivity of each of the metals.

(a) Aluminum precedes iron in the activity series: Al > Fe. Thus, a reaction occurs, and Al displaces Fe from solution.

(b) Conversely, iron follows aluminum in the activity series: Al > Fe. Thus, there is *no reaction*.

(c) Manganese precedes hydrogen in the series: Mn > (H). Thus, a reaction occurs and Mn produces H_2 gas that bubbles from solution.

(d) Magnesium is not an active metal. Therefore, magnesium does not react with water, and there is *no reaction*.

Self-Test Exercise

Predict whether or not a reaction occurs for each of the following.

(a) A gold ring is dropped into sulfuric acid.

(b) A zinc granule is dropped into hydrochloric acid.

(c) A cadmium foil is put into a lead(II) nitrate solution.

(d) A chromium strip is put into water.

Answers: (a) There is *no reaction* because Au follows (H) in the activity series.
(b) There is a reaction because Zn precedes (H) in the activity series.
(c) There is a reaction because Cd precedes Pb in the activity series.
(d) There is *no reaction* because Cr is not an active metal.

▲ **Gold in Acid** A gold ring gives no reaction in an aqueous sulfuric acid, H_2SO_4, solution.

Note In this introduction to the concept of an activity series, we have considered only the activity of metals. In addition, the ability of nonmetals to undergo a reaction also gives rise to an activity series. For example, the halogens demonstrate the following activity series: F > Cl > Br > I. Consider the reaction

$$Cl_2(g) + 2\,NaBr(aq) \longrightarrow 2\,NaCl(aq) + Br_2(l)$$

A reaction occurs because Cl precedes Br in the activity series; thus, the more active Cl displaces the less active Br from solution. Conversely, Cl follows F in the series. Therefore, Cl is less reactive than F and cannot displace it from solution. Thus,

$$Cl_2(g) + NaF(aq) \longrightarrow NR$$

8.8 Single-Replacement Reactions

Objectives
- To write a balanced chemical equation for the reaction of a metal in an aqueous solution of an ionic compound.
- To write a balanced chemical equation for the reaction of a metal in an acid.
- To write a balanced chemical equation for the reaction of an active metal in water.

There are many examples of single-replacement reactions in which a more active metal displaces a less active metal from solution. In general, the activity of a metal follows the metallic character trend in the periodic table. To determine the relative reactivity of a specific metal, we refer to the activity series.

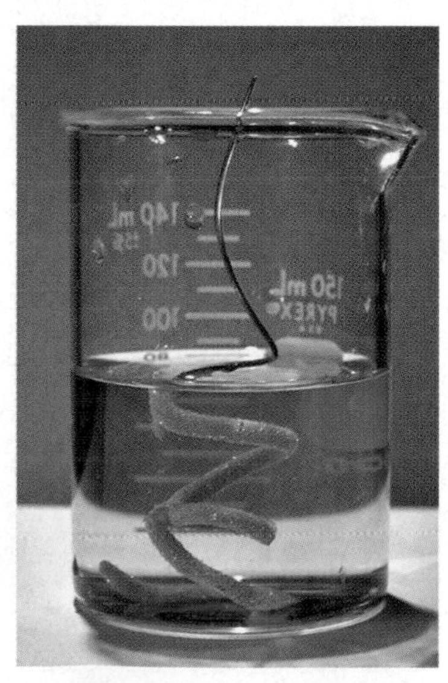

▲ **Copper in a Silver Solution** A copper wire reacts in a colorless $AgNO_3$ solution to give white crystals of Ag and a light blue aqueous solution.

Metal and Aqueous Solution

One kind of single-replacement reaction involves a metal reacting with an aqueous solution of an ionic compound. The general equation for the reaction is

$$\textbf{metal}_1 + \textbf{aqueous solution}_1 \longrightarrow \textbf{metal}_2 + \textbf{aqueous solution}_2$$

Consider the reaction of metallic copper wire in an aqueous solution of silver nitrate. The balanced equation for the reaction is

$$Cu(s) + 2\,AgNO_3(aq) \longrightarrow 2\,Ag(s) + Cu(NO_3)_2(aq)$$

If we refer to the activity series, we note that copper is the more active metal. That is, copper lies above silver in the series. Therefore, Cu metal can displace Ag metal from aqueous solution.

Now let's consider the reaction of metallic silver wire in an aqueous solution of copper nitrate. Since Ag metal is less active than Cu metal, we can predict that there is no reaction. That is,

$$Ag(s) + Cu(NO_3)_2(aq) \longrightarrow NR$$

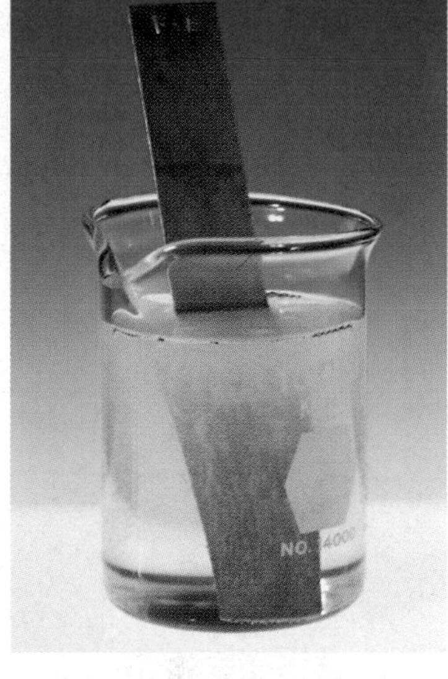

▲ **Iron in Acid** A strip of iron metal reacts in an aqueous sulfuric acid, H_2SO_4, solution to give bubbles of hydrogen gas.

Metal and Aqueous Acid Solution

Another kind of single-replacement reaction is that of a metal with an aqueous acid. A metal and an aqueous acid react to produce hydrogen gas and an aqueous solution of an ionic compound. The general equation for the reaction is

metal + aqueous acid ⟶ aqueous solution + hydrogen gas

Consider the reaction of iron metal and sulfuric acid. First, refer to the activity series to check the position of iron. Iron is above hydrogen in the series. Therefore, Fe displaces hydrogen gas from the acid. If we assume that the reaction produces iron(II) sulfate, we can write the balanced chemical equation.

$$Fe(s) + H_2SO_4(aq) \longrightarrow FeSO_4(aq) + H_2(g)$$

Active Metal and Water

In this kind of single-replacement reaction, an active metal and water react to produce a metal hydroxide and hydrogen gas. The general equation for the reaction is

metal + water ⟶ metal hydroxide + hydrogen gas

Consider the reaction of calcium metal in water to give calcium hydroxide and hydrogen gas. The balanced equation for the reaction is

$$Ca(s) + 2\,H_2O(l) \longrightarrow Ca(OH)_2(aq) + H_2(g)$$

The following example exercise further illustrates how to write balanced chemical equations for the three types of single-replacement reactions.

▲ **Calcium in Water** A piece of calcium metal in water reacts to give bubbles of hydrogen gas.

Formation of Silver Crystals Movie

Example Exercise 8.13 • Single-Replacement Reactions

Write a balanced chemical equation for each of the following single-replacement reactions.

(a) Nickel metal is placed in a tin(II) sulfate solution.
(b) Cobalt metal is put in a cadmium(II) nitrate solution.
(c) Manganese chips are added to sulfuric acid.
(d) A small, soft chunk of potassium is dropped into water.

Solution

First, we will refer to the activity series for the relative positions of each of the metals. Then we will write the equations.

(a) Nickel is above tin in the activity series: Ni > Sn. Therefore, a reaction occurs and Ni displaces Sn from solution.

$$Ni(s) + SnSO_4(aq) \longrightarrow NiSO_4(aq) + Sn(s)$$

(continued)

Example Exercise 8.13 *(continued)*

(b) Cobalt is below cadmium in the series: Cd > Co. Therefore, there is no reaction.

$$Co(s) + Cd(NO_3)_2(aq) \longrightarrow NR$$

(c) Manganese is above hydrogen in the series: Mn > (H). Therefore, a reaction occurs, and Mn releases H_2 gas bubbles from solution.

$$Mn(s) + H_2SO_4(aq) \longrightarrow MnSO_4(aq) + H_2(g)$$

(d) Potassium is a Group IA/1 active metal. Therefore, a reaction occurs, and K evolves H_2 gas from solution.

$$2\,K(s) + 2\,H_2O(l) \longrightarrow 2\,KOH(aq) + H_2(g)$$

Self-Test Exercise

Write a balanced chemical equation for each of the following single-replacement reactions.

(a) Nickel metal is placed in a silver nitrate solution.
(b) Gold metal is placed in a silver nitrate solution.
(c) A chunk of cadmium metal is dropped into hydrochloric acid.
(d) A small piece of strontium metal is dropped into water.

Answers: (a) $Ni(s) + 2\,AgNO_3(aq) \longrightarrow 2\,Ag(s) + Ni(NO_3)_2(aq)$; (b) $Au(s) + AgNO_3(aq) \longrightarrow NR$; (c) $Cd(s) + 2\,HCl(aq) \longrightarrow CdCl_2(aq) + H_2(g)$; (d) $Sr(s) + 2\,H_2O(l) \longrightarrow Sr(OH)_2(aq) + H_2(g)$

Note In a single-replacement reaction, it may not be possible to predict the ionic charge on the resulting metal cation. For example, when copper metal reacts with silver nitrate solution, the resulting compound may contain either Cu^+ or Cu^{2+}.

8.9 Solubility Rules

Objective · To predict whether an ionic compound dissolves in water given the general rules for solubility.

In a double-replacement reaction, which we will discuss next in Section 8.10, two aqueous solutions may react to form a compound that is not soluble, that is, a precipitate. When we observe that a precipitate is produced in solution, we have evidence for a chemical reaction.

It is important to identify an insoluble compound in solution. In Table 8.2, there are general rules for predicting the solubility of ionic compounds in water. However, there are exceptions to the general rules, and a few of these exceptions are indicated in the table; for example, Na_2CO_3 is soluble in water (rule 6).

Now let's apply the solubility rules to selected compounds and determine whether or not the compound is soluble in water.

Example Exercise 8.14 • **Applying Solubility Rules**

State whether each of the following compounds is soluble or insoluble in water.

(a) sodium sulfate, Na_2SO_4
(b) aluminum nitrate, $Al(NO_3)_3$
(c) barium sulfate, $BaSO_4$
(d) potassium chromate, K_2CrO_4
(e) ammonium sulfide, $(NH_4)_2S$

Solution

We refer to the solubility rules for ionic compounds.

(a) Sodium sulfate contains the alkali metal ion Na^+. According to rule 1, Na_2SO_4 is *soluble*.

Table 8.2 Solubility Rules for Ionic Compounds

Compounds containing the following ions are generally *soluble* in water:

1. alkali metal ions and the ammonium ion, Li^+, Na^+, K^+, NH_4^+
2. acetate ion, $C_2H_3O_2^-$
3. nitrate ion, NO_3^-
4. halide ions (X), Cl^-, Br^-, I^- (AgX, Hg_2X_2, and PbX_2 are insoluble exceptions.)
5. sulfate ion, SO_4^{2-} ($SrSO_4$, $BaSO_4$, and $PbSO_4$ are insoluble exceptions.)

Compounds containing the following ions are generally *insoluble* in water:*

6. carbonate ion, CO_3^{2-} (See rule 1 exceptions that are soluble.)
7. chromate ion, CrO_4^{2-} (See rule 1 exceptions that are soluble.)
8. phosphate ion, PO_4^{3-} (See rule 1 exceptions that are soluble.)
9. sulfide ion, S^{2-} (CaS, SrS, BaS, and rule 1 exceptions that are soluble)
10. hydroxide ion, OH^- [$Ca(OH)_2$, $Sr(OH)_2$, $Ba(OH)_2$, and rule 1 exceptions that are soluble]

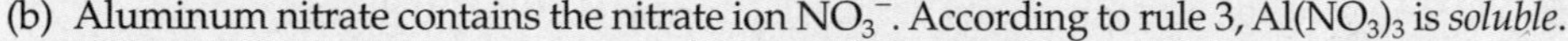

*These compounds are actually slightly soluble, or very slightly soluble, in water.

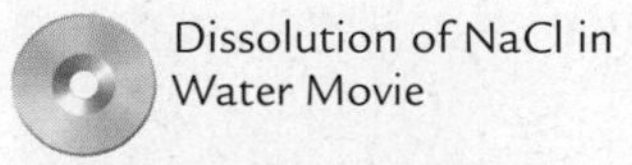

Dissolution of NaCl in Water Movie

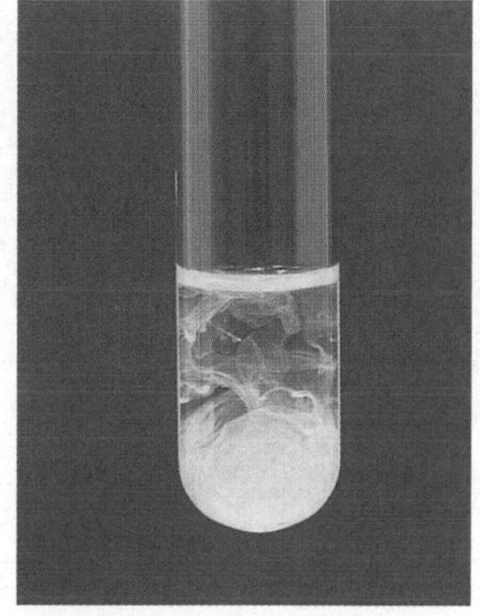

▲ **Insoluble Silver Chloride** Aqueous solutions of silver nitrate and sodium chloride react to give white, insoluble AgCl.

(b) Aluminum nitrate contains the nitrate ion NO_3^-. According to rule 3, $Al(NO_3)_3$ is *soluble*.
(c) Barium sulfate contains the sulfate ion, SO_4^{2-}. According to the rule 5 exception, $BaSO_4$ is *insoluble*.
(d) Potassium chromate contains the chromate ion CrO_4^{2-}. According to the rule 7 exception, K_2CrO_4 is *soluble*.
(e) Ammonium sulfide contains the ammonium ion NH_4^+. According to either rule 1 or rule 9 exception, $(NH_4)_2S$ is *soluble*.

Self-Test Exercise

State whether each the following compounds is soluble or insoluble in water.

(a) lead(II) acetate, $Pb(C_2H_3O_2)_2$
(b) mercury(II) bromide, $HgBr_2$
(c) magnesium carbonate, $MgCO_3$
(d) zinc phosphate, $Zn_3(PO_4)_2$
(e) calcium hydroxide, $Ca(OH)_2$

Answers: (a) soluble (rule 2); (b) soluble (rule 4); (c) insoluble (rule 6); (d) insoluble (rule 8); (e) soluble (rule 10 exception)

8.10 Double-Replacement Reactions

Objective · To write a balanced chemical equation for the reaction of two aqueous solutions of ionic compounds.

In a double-replacement reaction, which is sometimes referred to as a precipitation reaction, two ionic compounds in aqueous solution switch anions and produce two new compounds. The general form of the reaction is

$$AX + BZ \longrightarrow AZ + BX$$

In the reaction of AX and BZ, we can assume that either AZ or BX is an insoluble precipitate. If both AZ and BX are soluble, there is no precipitate, and hence there is no chemical reaction. Consider the double-replacement reaction of two aqueous solutions of ionic compounds.

$$\textbf{aqueous solution}_1 + \textbf{aqueous solution}_2 \longrightarrow \textbf{precipitate} + \textbf{aqueous solution}_3$$

As an example, when we add an aqueous solution of silver nitrate to an aqueous solution of sodium carbonate, we observe a white precipitate. In this reaction,

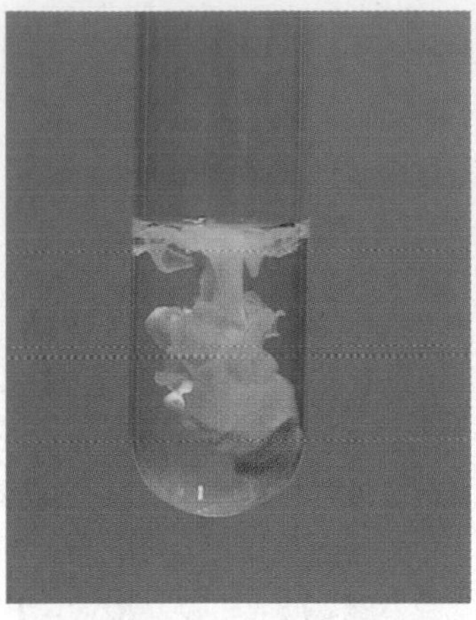

▲ **Insoluble Silver Carbonate** Aqueous solutions of silver nitrate and sodium carbonate react to give an insoluble precipitate of Ag_2CO_3.

the two ionic compounds switch anions to give insoluble silver carbonate and sodium nitrate. We can write the balanced chemical equation as follows.

$$2\,AgNO_3(aq) + Na_2CO_3(aq) \longrightarrow Ag_2CO_3(s) + 2\,NaNO_3(aq)$$

By referring to solubility rule 6, we find that most compounds containing the carbonate ion, CO_3^{2-}, are insoluble. Since silver carbonate is insoluble, it is written as $Ag_2CO_3(s)$ in the above equation. The (s) indicates that Ag_2CO_3 is an insoluble precipitate. The following example exercise further illustrates how to write balanced equations for double-replacement reactions.

Example Exercise 8.15 • Double-Replacement Reactions

Write a balanced chemical equation for each of the following double-replacement reactions.

(a) Aqueous barium chloride is added to a potassium chromate solution.

(b) Aqueous strontium acetate is added to a lithium hydroxide solution.

Solution

For double-replacement reactions, we switch anions for the two compounds and check the solubility rules for an insoluble compound.

(a) Barium chloride and potassium chromate give barium chromate and potassium chloride. According to solubility rule 7, we find that barium chromate is insoluble. The balanced equation is

$$BaCl_2(aq) + K_2CrO_4(aq) \longrightarrow BaCrO_4(s) + 2\,KCl(aq)$$

(b) Strontium acetate and lithium hydroxide react as follows.

$$Sr(C_2H_3O_2)_2(aq) + LiOH(aq) \longrightarrow Sr(OH)_2(aq) + LiC_2H_3O_2(aq)$$

However, the solubility rules indicate that $Sr(OH)_2$ and $LiC_2H_3O_2$ are soluble. Therefore, the equation is written

$$Sr(C_2H_3O_2)_2(aq) + LiOH(aq) \longrightarrow NR$$

Self-Test Exercise

Write a balanced chemical equation for each of the following double-replacement reactions.

(a) Aqueous zinc sulfate is added to a sodium carbonate solution.

(b) Aqueous manganese(II) nitrate is added to a potassium hydroxide solution.

Answers: (a) $ZnSO_4(aq) + Na_2CO_3(aq) \longrightarrow ZnCO_3(s) + Na_2SO_4(aq)$; (b) $Mn(NO_3)_2(aq) + 2\,KOH(aq) \longrightarrow Mn(OH)_2(s) + 2\,KNO_3(aq)$

Note In a double-replacement reaction, the ionic charge of a metal cation does not change. For example, if a reactant contains Fe^{2+}, a product will contain Fe^{2+}. Although transition metals have cations with variable charge, the ionic charge does not change during a double-replacement reaction.

8.11 Neutralization Reactions

Objective · To write a balanced chemical equation for the reaction of an acid and a base.

In a neutralization reaction, an acid and a base react to produce an aqueous ionic compound and water. An **acid** is a substance that releases hydrogen ions, H^+, in solution; a **base** is a substance that releases hydroxide ions, OH^-. The resulting ionic compound is called a **salt** and is composed of the cation from the base and the anion from the acid. The general form of the reaction is

$$\text{H}\textbf{X} + \textbf{B}\text{OH} \longrightarrow \textbf{BX} + \text{HOH}$$

A neutralization reaction is a special case of a double-replacement reaction. Notice that we have written water as HOH to emphasize the switching of ions. You will also discover that writing water as HOH is helpful in balancing equations for neutralization reactions. In a neutralization reaction, an acid and a base react to produce a salt and water.

aqueous acid + aqueous base ⟶ aqueous salt + water

As an example, when we add hydrochloric acid to sodium hydroxide solution, we obtain sodium chloride and water. We can write the balanced chemical equation as follows.

$$HCl(aq) + NaOH(aq) \longrightarrow NaCl(aq) + HOH(l)$$

Other than a slight warming, there is no evidence of reaction. To provide evidence for this neutralization reaction, we can add a drop of the acid–base indicator phenolphthalein to the HCl. Phenolphthalein is colorless in acid and pink in base. Initially, the HCl acid solution is colorless; however, as we add NaOH base, we see flashes of color and eventually the solution turns pink. With an acid–base indicator we can observe a color change, which is evidence for a chemical reaction.

The following example exercise provides additional practice in writing balanced equations for neutralization reactions.

▲ **Phenolphthalein Acid–Base Indicator** In an acid–base neutralization reaction, a drop of phenolphthalein indicator signals a change. If we have an acid in a flask and we add a base, the indicator turns from colorless to pink.

Example Exercise 8.16 • Neutralization Reactions

Write a balanced chemical equation for each of the following neutralization reactions.

(a) Nitric acid neutralizes an ammonium hydroxide solution.
(b) Sulfuric acid neutralizes a potassium hydroxide solution.

Solution

A neutralization reaction produces a salt and water.

(a) Nitric acid and ammonium hydroxide produce ammonium nitrate and water. The balanced equation is

$$HNO_3(aq) + NH_4OH(aq) \longrightarrow NH_4NO_3(aq) + HOH(l)$$

(b) Sulfuric acid and potassium hydroxide produce potassium sulfate and water. The balanced equation is

$$H_2SO_4(aq) + 2\ KOH(aq) \longrightarrow K_2SO_4(aq) + 2\ HOH(l)$$

Self-Test Exercise

Write a balanced chemical equation for each of the following neutralization reactions.

(a) Chloric acid neutralizes a strontium hydroxide solution.
(b) Phosphoric acid neutralizes a sodium hydroxide solution.

Answers: (a) $2\ HClO_3(aq) + Sr(OH)_2(aq) \longrightarrow Sr(ClO_3)_2(aq) + 2\ HOH(l)$; (b) $H_3PO_4(aq) + 3\ NaOH(aq) \longrightarrow Na_3PO_4(aq) + 3\ HOH(l)$

Chemistry Connection • Fireworks

What element is responsible for the orange color in a fireworks display?

Have you ever watched a Fourth of July fireworks display and wondered how the colors were produced? The color of fireworks is produced by the chemicals in a rocket shell, which is fitted with a fuse and fired into the air. When the fuse burns down, gunpowder ignites and sets off an explosion that shoots chemicals through the sky. If the rocket shell is packed with a sodium compound, a yellow color is observed. If the shell is packed with a barium compound, a green color is observed.

◀ **Fourth of July Celebration**

Not all chemicals produce a colored display. You may recall having seen fireworks that simply produce a shower of white sparks. Powdered magnesium metal can be used to produce this effect. The following table lists the chemicals used to produce various colors in fireworks.

Chemical	Fireworks Color
Na compounds	yellow
Ba compounds	green
Ca compounds	orange
Sr compounds	red
Li compounds	crimson
Cu compounds	blue
Al or Mg metals	white sparks

Interestingly, the colors of fireworks and the colors of neon signs are based on the same principle. In a fireworks display an element is energized by heat, whereas in a "neon" light, an element is energized by electricity. In either case, electrons are first excited and then immediately lose energy by emitting light. The observed color corresponds to the wavelength of the light emitted. For example, if the emitted light has a wavelength near 650 nm, it is observed as red. If the light has a wavelength near 450 nm, it appears blue.

Chemists can often identify an element by the color of its flame test. For example, sodium gives a yellow flame test, and barium a green flame test. Flame-test colors are identical to fireworks colors containing the same element. In the laboratory, a flame test is performed by placing a small amount of chemical on the tip of a wire and holding the wire in a hot flame. The flame test for calcium is illustrated below.

Flame Tests for Metals Movie

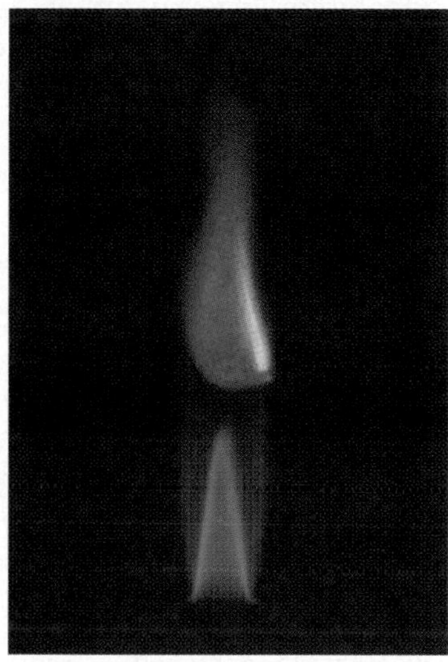

▲ **Calcium Flame Test** The flame test for calcium is described as orange.

The orange color suggests that the element is calcium.

Summary 

Section 8.1 In this chapter we learned that the evidence for a **chemical reaction** is any of the following: a gas is detected, a **precipitate** is formed in **aqueous solution**, or a color change or energy change is observed. An **exothermic reaction** releases heat energy, and an **endothermic reaction** absorbs heat energy.

Section 8.2 A **reactant** substance undergoes a chemical change to become a **product**. A **catalyst** is a substance that speeds up a reaction. A **chemical equation** describes a reaction

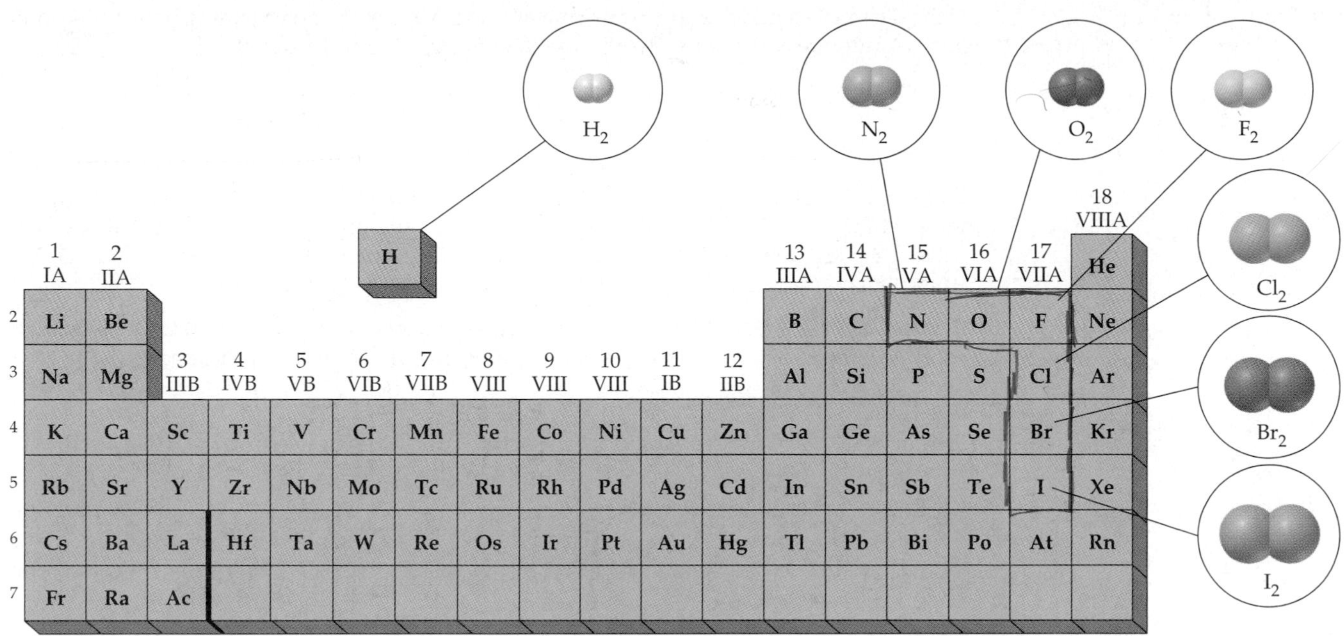

▲ **Diatomic Molecules** Seven elements occur naturally as diatomic molecules: H_2, N_2, O_2, F_2, and Cl_2 are gases. Bromine, Br_2, occurs as a liquid and iodine, I_2, as solid crystals at normal conditions.

using formulas and symbols. Seven nonmetal elements occur naturally as **diatomic molecules**, and their formulas are written H_2, N_2, O_2, F_2, Cl_2, Br_2, and I_2.

Section 8.3 To balance a chemical equation, **coefficients** are systematically placed in front of each substance until there is the same number of atoms of each element on both sides of the equation. If the reaction coefficient is 1, it is omitted. Similarly, if a formula **subscript** is 1, as in the case of oxygen in H_2O, it is omitted.

Section 8.4 Although there are thousands of chemical reactions, most can be classified as one of the following: combination, decomposition, single replacement, double replacement, or neutralization. Table 8.3 lists these five basic types.

Table 8.3 Basic Types of Chemical Reactions

Reaction Type	General Format		Reactions
combination reaction	A + Z	⟶	AZ
decomposition reaction	AZ	⟶	A + Z
single-replacement reaction	A + BZ	⟶	AZ + B
double-replacement reaction	AX + BZ	⟶	AZ + BX
neutralization reaction	HX + BOH	⟶	BX + HOH

Section 8.5 There are three kinds of combination reactions: a metal and oxygen; a nonmetal and oxygen; a metal and a nonmetal. Often, we can predict the products from the reaction. Nonmetals, however, combine with oxygen to give a variety of oxides, and so the product is not predictable. In summary,

1. **metal + oxygen gas ⟶ metal oxide**
2. **nonmetal + oxygen gas ⟶ nonmetal oxide**
3. **metal + nonmetal ⟶ ionic compound**

Section 8.6 There are three kinds of decomposition reactions. In the first type, a metal hydrogen carbonate is decomposed by heating to give a metal carbonate product. In the second, a metal carbonate is decomposed to a metal oxide. Most oxygen-containing polyatomic ions, such as chlorate and nitrate, decompose to give oxygen gas. In summary,

1. **metal hydrogen carbonate** $\longrightarrow$ **metal carbonate + water + carbon dioxide**
2. **metal carbonate** $\longrightarrow$ **metal oxide + carbon dioxide**
3. **oxygen compounds** $\longrightarrow$ **oxygen gas**

Section 8.7 The **activity series** lists the relative ability of a metal to undergo reaction. A metal higher in the series always displaces a metal lower in the series. Metals that are more active than (H) displace hydrogen gas from an acid solution. **Active metals** include Li, Na, K, Ba, Sr, and Ca; each of these metals displaces hydrogen gas from water.

Section 8.8 There are three kinds of single-replacement reactions: a metal and aqueous solution, a metal and a dilute acid, and an active metal and water. In summary,

1. **metal_1 + $\text{aqueous solution}_1$** $\longrightarrow$ **metal_2 + $\text{aqueous solution}_2$**
2. **metal + aqueous acid** $\longrightarrow$ **aqueous solution + hydrogen gas**
3. **metal + water** $\longrightarrow$ **metal hydroxide + hydrogen gas**

Section 8.9 An insoluble substance does not dissolve extensively in water. Ionic compounds do, however, dissolve slightly or very slightly in aqueous solution. To determine whether a substance is soluble or insoluble, we refer to the solubility rules for ionic compounds in Table 8.2.

Section 8.10 In a double-replacement reaction, two ionic compounds react in aqueous solution. The two compounds react by switching cation and anion partners and typically form a precipitate. In general,

$\text{aqueous solution}_1$ + $\text{aqueous solution}_2$ $\longrightarrow$ precipitate + $\text{aqueous solution}_3$

Section 8.11 In a neutralization reaction an **acid** and a **base** react to yield a **salt** and water. A neutralization reaction is a special type of double-replacement reaction. It is helpful to write the formula for water as HOH in balancing neutralization reactions.

aqueous acid + aqueous base $\longrightarrow$ aqueous salt + water

Key Concepts*

1. Which of the following are evidence for a chemical reaction producing a gas?
   (a) A glowing wooden splint bursts into flames.
   (b) A heated solid emits an ammonia smell.
   (c) A white solid fizzes in solution.
2. Which of the following are evidence for a chemical reaction when two aqueous solutions are added together?
   (a) The solution releases bubbles.
   (b) The solution forms insoluble particles.
   (c) The solution turns from colorless to pink.
3. Which of the following is *not* evidence for an exothermic chemical reaction: heat, light, explosion, color change?
4. Which of the following nonmetals does *not* occur naturally as a diatomic molecule: H, C, N, O, F, Cl, Br, I?
5. Ethanol, C_2H_5OH, is blended with gasoline to give *gasohol.* Balance the following equation for the combustion of ethanol.

$$C_2H_5OH(g) + O_2(g) \xrightarrow{\text{spark}} CO_2(g) + H_2O(g)$$

6. Which of the following metals reacts with aqueous $CuSO_4$: Ag, Ca, Mg, Zn? (Refer to the activity series in Section 8.7.)
7. Which of the following metals reacts with an aqueous acid: Ag, Ca, Mg, Zn? (Refer to the activity series in Section 8.7.)
8. Which of the following metals reacts with water at room temperature: Ag, Ca, Mg, Zn?
9. Which of the following compounds is insoluble in water: $AgC_2H_3O_2$, $AgNO_3$, AgCl, Ag_2CO_3? (Refer to the solubility rules in Table 8.2.)
10. In the following double-replacement reaction, what are the two products?

$$KI(aq) + Pb(NO_3)_2(aq) \longrightarrow$$

11. In the following neutralization reaction, what are the two products?

$$HCl(aq) + NaOH(aq) \longrightarrow$$

*Answers to Key Concepts are in Appendix H.

Key Terms†

Select the key term below that corresponds to each of the following definitions.

_____ 1. the process of undergoing a chemical change
_____ 2. a chemical reaction that evolves heat energy
_____ 3. a chemical reaction that absorbs heat energy
_____ 4. a substance dissolved in water
_____ 5. an insoluble solid substance produced from a reaction in aqueous solution
_____ 6. a representation using formulas and symbols to describe a chemical change
_____ 7. a substance undergoing a chemical reaction
_____ 8. a substance resulting from a chemical reaction
_____ 9. a substance that increases the rate of reaction but is usually recovered without being permanently changed
_____ 10. a particle composed of two nonmetal atoms
_____ 11. a digit placed in front of a formula to balance a chemical equation
_____ 12. a digit in a chemical formula that represents the number of atoms or ions appearing in the substance
_____ 13. a type of reaction in which two substances produce a single compound
_____ 14. a type of reaction in which a single compound produces two or more substances
_____ 15. a type of reaction in which a more active metal displaces a less active metal from a solution or compound
_____ 16. a type of reaction in which two cations in different compounds exchange anions
_____ 17. a type of reaction in which an acid and a base produce a salt and water
_____ 18. a relative order of metals arranged by their ability to undergo reaction
_____ 19. a metal that reacts with water at room temperature
_____ 20. a substance that releases hydrogen ions in aqueous solution
_____ 21. a substance that releases hydroxide ions in aqueous solution
_____ 22. an ionic compound produced by an acid–base reaction

(a) acid (*Sec. 8.11*)
(b) active metal (*Sec. 8.7*)
(c) activity series (*Sec. 8.7*)
(d) aqueous solution (aq) (*Sec. 8.1*)
(e) base (*Sec. 8.11*)
(f) catalyst (*Sec. 8.2*)
(g) chemical equation (*Sec. 8.2*)
(h) chemical reaction (*Sec. 8.1*)
(i) coefficient (*Sec. 8.3*)
(j) combination (*Sec. 8.4*)
(k) decomposition (*Sec. 8.4*)
(l) diatomic molecule (*Sec. 8.2*)
(m) double replacement (*Sec. 8.4*)
(n) endothermic reaction (*Sec. 8.1*)
(o) exothermic reaction (*Sec. 8.1*)
(p) neutralization (*Sec. 8.4*)
(q) precipitate (*Sec. 8.1*)
(r) product (*Sec. 8.2*)
(s) reactant (*Sec. 8.2*)
(t) salt (*Sec. 8.11*)
(u) single replacement (*Sec. 8.4*)
(v) subscript (*Sec. 8.3*)

Exercises‡

Evidence for Chemical Reactions (Sec. 8.1)

1. Which of the following are evidence for a chemical reaction?
   (a) Mixing two aqueous solutions produces gas bubbles.
   (b) Mixing two aqueous solutions produces a precipitate.
   (c) Mixing two colorless solutions gives a yellow solution.
   (d) Mixing two 10.0-mL solutions gives a volume of 19.5 mL.
2. Which of the following are evidence for a chemical reaction?
   (a) Heating a compound gives a gas with a sharp odor.
   (b) Heating a compound gives a gas that extinguishes a fire.
   (c) Heating magnesium in air produces white smoke.
   (d) Heating magnesium in a vacuum produces a metallic film.
3. Which of the following are evidence for a chemical reaction?
   (a) light **(b)** heat
   (c) burning **(d)** explosion
4. Which of the following are evidence for a chemical reaction?
   (a) a gas condenses **(b)** a liquid evaporates
   (c) a solid melts **(d)** a solid sublimates

Writing Chemical Equations (Sec. 8.2)

5. Which seven nonmetallic elements occur naturally as diatomic molecules?

†Answers to Key Terms are in Appendix I.

‡Answers to odd-numbered Exercises are in Appendix J.

6. Write the formulas for the halogens that occur naturally as diatomic molecules.
7. Write a chemical equation for iron metal combining with chlorine gas to give a yellow iron(III) chloride solid.

▲ **Iron in Chlorine Gas** Steel wool, Fe, reacts in Cl_2 gas to give sparks and a yellow-brown $FeCl_3$ residue.

8. Write a chemical equation for tin metal combining with oxygen gas to give a white tin(IV) oxide solid.
9. Write a chemical equation for solid zinc carbonate decomposing to yield zinc oxide powder and carbon dioxide gas.
10. Write a chemical equation for solid iron(II) hydrogen carbonate decomposing to yield iron(II) carbonate powder, water, and carbon dioxide gas.
11. Write a chemical equation for the reaction of magnesium metal with aqueous cobalt(II) nitrate to produce aqueous magnesium nitrate and cobalt metal.
12. Write a chemical equation for the reaction of manganese metal with sulfuric acid to produce aqueous manganese(II) sulfate and hydrogen gas.
13. Write a chemical equation for the reaction of aqueous solutions of lithium bromide and silver nitrate to give a silver bromide precipitate and aqueous lithium nitrate.
14. Write a chemical equation for the reaction of aqueous solutions of sodium chromate with calcium sulfide to give a calcium chromate precipitate and aqueous sodium sulfide.
15. Write a chemical equation for the neutralization of aqueous potassium hydroxide by acetic acid to give aqueous potassium acetate plus water.
16. Write a chemical equation for the neutralization of aqueous ammonium hydroxide by nitric acid to give aqueous ammonium nitrate plus water.

Balancing Chemical Equations (Sec. 8.3)

17. Which of the following is *not* a general guideline for balancing a chemical equation by inspection.
    (a) Use coefficients to balance the elements in each substance.
    (b) Begin balancing the equation by starting with the most complex formula.
    (c) Balance polyatomic ions as a unit rather than by the elements separately.
    (d) After balancing the equation, verify the coefficients.
    (e) If an equation coefficient is a fraction, change a subscript in the chemical formula to give whole-number coefficients.
18. If a chemical equation is difficult to balance, what is most likely the problem?
19. Balance each of the following chemical equations by inspection.
    (a) $Co(s) + O_2(g) \longrightarrow Co_2O_3(s)$
    (b) $LiClO_3(s) \longrightarrow LiCl(s) + O_2(g)$
    (c) $Cu(s) + AgC_2H_3O_2(aq) \longrightarrow Cu(C_2H_3O_2)_2(aq) + Ag(s)$
    (d) $Pb(NO_3)_2(aq) + LiCl(aq) \longrightarrow PbCl_2(s) + LiNO_3(aq)$
    (e) $H_2SO_4(aq) + Al(OH)_3(aq) \longrightarrow Al_2(SO_4)_3(aq) + H_2O(l)$
20. Balance each of the following chemical equations by inspection.
    (a) $H_2(g) + N_2(g) \longrightarrow NH_3(g)$
    (b) $Al(HCO_3)_3(s) \longrightarrow Al_2(CO_3)_3(s) + CO_2(g) + H_2O(g)$
    (c) $Sr(s) + H_2O(l) \longrightarrow Sr(OH)_2(aq) + H_2(g)$
    (d) $K_2SO_4(aq) + Ba(OH)_2(aq) \longrightarrow BaSO_4(s) + KOH(aq)$
    (e) $H_3PO_4(aq) + Mn(OH)_2(s) \longrightarrow Mn_3(PO_4)_2(s) + H_2O(l)$
21. Balance each of the following chemical equations by inspection.
    (a) $H_2CO_3(aq) + NH_4OH(aq) \longrightarrow (NH_4)_2CO_3(aq) + H_2O(l)$
    (b) $Hg_2(NO_3)_2(aq) + NaBr(aq) \longrightarrow Hg_2Br_2(s) + NaNO_3(aq)$
    (c) $Mg(s) + HC_2H_3O_2(aq) \longrightarrow Mg(C_2H_3O_2)_2(aq) + H_2(g)$
    (d) $LiNO_3(s) \longrightarrow LiNO_2(s) + O_2(g)$
    (e) $Pb(s) + O_2(g) \longrightarrow PbO(s)$
22. Balance each of the following chemical equations by inspection.
    (a) $HClO_4(aq) + Ba(OH)_2(s) \longrightarrow Ba(ClO_4)_2(s) + H_2O$
    (b) $Co(NO_3)_2(aq) + H_2S(g) \longrightarrow CoS(s) + HNO_3(aq)$
    (c) $Fe(s) + Cd(NO_3)_2(aq) \longrightarrow Fe(NO_3)_3(aq) + Cd(s)$
    (d) $Fe_2(CO_3)_3(s) \longrightarrow Fe_2O_3(s) + CO_2(g)$
    (e) $Sn(s) + P(s) \longrightarrow Sn_3P_2(s)$

Classifying Chemical Reactions (Sec. 8.4)

23. Classify each reaction in Exercise 19 as one of the following: combination, decomposition, single replacement, double replacement, or neutralization.

24. Classify each reaction in Exercise 20 as one of the following: combination, decomposition, single replacement, double replacement, or neutralization.

25. Classify each reaction in Exercise 21 as one of the following: combination, decomposition, single replacement, double replacement, or neutralization.

26. Classify each reaction in Exercise 22 as one of the following: combination, decomposition, single replacement, double replacement, or neutralization.

Combination Reactions (Sec. 8.5)

27. Write a balanced equation for each of the following combination reactions.

▲ **Iron in Oxygen Gas** Steel wool, Fe, reacts in O_2 gas to give a bright white light and a gray Fe_2O_3 residue.

(a) Iron metal is heated with oxygen gas to yield iron(III) oxide.
(b) Tin metal is heated with oxygen gas to produce tin(II) oxide.

28. Write a balanced equation for each of the following combination reactions.
(a) Copper metal is heated with oxygen gas to produce copper(I) oxide.
(b) Titanium metal is heated with oxygen gas to yield titanium(IV) oxide.

29. Write a balanced equation for each of the following combination reactions.
(a) Carbon is heated with oxygen to give carbon monoxide gas.
(b) Phosphorus is heated with oxygen to give diphosphorus pentaoxide.

30. Write a balanced equation for each of the following combination reactions.
(a) Nitrogen is heated with oxygen to form dinitrogen pentaoxide gas.
(b) Chlorine is heated with oxygen to form dichlorine trioxide gas.

31. Write a balanced equation for each of the following combination reactions.
(a) Copper is heated with chlorine gas to produce solid copper(I) chloride.
(b) Cobalt is heated with sulfur powder to produce cobalt(II) sulfide.

32. Write a balanced equation for each of the following combination reactions.
(a) Iron is heated with fluorine gas to yield iron(III) fluoride.
(b) Lead is heated with phosphorus powder to yield lead(IV) phosphide.

33. Write a balanced equation for each of the following combination reactions.
(a) Chromium is heated with oxygen to give chromium(III) oxide.
(b) Chromium is heated with nitrogen to give chromium(III) nitride.

34. Write a balanced equation for each of the following combination reactions.
(a) Sulfur is heated with oxygen to form sulfur dioxide gas.
(b) Sulfur is heated with oxygen and Pt catalyst to form sulfur trioxide gas.

35. Complete and balance each of the following combination reactions.
(a) $Li + O_2 \longrightarrow$ (b) $Ca + O_2 \longrightarrow$

36. Complete and balance each of the following combination reactions.
(a) $Sr + O_2 \longrightarrow$ (b) $Al + O_2 \longrightarrow$

37. Complete and balance each of the following combination reactions.
(a) $Na + I_2 \longrightarrow$ (b) $Ba + N_2 \longrightarrow$

38. Complete and balance each of the following combination reactions.
(a) $Zn + P \longrightarrow$ (b) $Al + S \longrightarrow$

Decomposition Reactions (Sec. 8.6)

39. Write a balanced equation for each of the following decomposition reactions.
(a) Silver hydrogen carbonate decomposes on heating to give solid silver carbonate, water, and carbon dioxide gas.
(b) Barium hydrogen carbonate decomposes on heating to give solid barium carbonate, water, and carbon dioxide gas.

40. Write a balanced equation for each of the following decomposition reactions.
(a) Cobalt(III) hydrogen carbonate decomposes by heating to give solid cobalt(III) carbonate, water, and carbon dioxide gas.
(b) Tin(IV) hydrogen carbonate decomposes by heating to give solid tin(IV) carbonate, water, and carbon dioxide gas.

41. Write a balanced equation for each of the following decomposition reactions.
(a) Potassium carbonate decomposes on heating to give potassium oxide and carbon dioxide gas.
(b) Manganese(II) carbonate decomposes on heating to give manganese(II) oxide and carbon dioxide gas.

42. Write a balanced equation for each of the following decomposition reactions.
(a) Chromium(III) carbonate decomposes on heating to give chromium(III) oxide and carbon dioxide gas.
(b) Lead(IV) carbonate decomposes on heating to give lead(IV) oxide and carbon dioxide gas.

43. Write a balanced equation for each of the following decomposition reactions.
(a) Calcium nitrate decomposes on heating to give calcium nitrite and oxygen gas.
(b) Silver sulfate decomposes on heating to give silver sulfite and oxygen gas.

44. Write a balanced equation for each of the following decomposition reactions.
(a) Stannous chlorate decomposes on heating to give stannous chloride and oxygen gas.
(b) Plumbic oxide decomposes on heating to give plumbous oxide and oxygen gas.

45. Complete and balance each of the following decomposition reactions.
(a) $KHCO_3 \longrightarrow$ (b) $Zn(HCO_3)_2 \longrightarrow$

46. Complete and balance each of the following decomposition reactions.
(a) $Li_2CO_3 \longrightarrow$ (b) $CdCO_3 \longrightarrow$

47. Balance each of the following decomposition reactions.
(a) $NaClO_3 \longrightarrow NaCl + O_2$
(b) $Ca(NO_3)_2 \longrightarrow Ca(NO_2)_2 + O_2$

48. Balance each of the following decomposition reactions.
(a) $AlPO_4 \longrightarrow AlPO_3 + O_2$
(b) $SnSO_4 \longrightarrow SnSO_3 + O_2$

The Activity Series Concept (Sec. 8.7)

(Refer to the activity series to answer the following questions.)

49. Predict which of the following metals reacts with aqueous iron(II) nitrate.
(a) Hg (b) Zn
(c) Cd (d) Mg

50. Predict which of the following metals reacts with aqueous nickel(II) nitrate.
(a) Ag (b) Sn
(c) Co (d) Mn

51. Predict which of the following metals reacts with dilute hydrochloric acid.
(a) Ni (b) Zn
(c) Cu (d) Al

52. Predict which of the following metals reacts with dilute sulfuric acid.
(a) Fe (b) Mn
(c) Cd (d) Au

53. Predict which of the following metals reacts with water at room temperature.
(a) Li (b) Mg
(c) Ca (d) Al

54. Predict which of the following metals reacts with water at room temperature.
(a) Ba (b) Mn
(c) Sn (d) K

Single-Replacement Reactions (Sec. 8.8)

55. Write a balanced equation for each of the following single-replacement reactions.
(a) Copper wire is placed in an aluminum nitrate solution.
(b) Aluminum wire is placed in a copper(II) nitrate solution.

56. Write a balanced equation for each of the following single-replacement reactions.
(a) Cadmium metal is put into an iron(II) sulfate solution.
(b) Iron metal is put into a cadmium sulfate solution.

57. Write a balanced equation for each of the following single-replacement reactions.
(a) Nickel metal is put into a lead(II) acetate solution.
(b) Lead metal is put into a nickel(II) acetate solution.

58. Write a balanced equation for each of the following single-replacement reactions.
(a) Iron filings are added to a mercury(II) sulfate solution.
(b) Drops of liquid mercury are added to a ferrous sulfate solution.

59. Write a balanced equation for each of the following single-replacement reactions.
(a) Magnesium ribbon is added to hydrochloric acid.
(b) Manganese chips are added to nitric acid.

60. Write a balanced equation for each of the following single-replacement reactions.
(a) Zinc granules are placed in acetic acid.
(b) Cadmium metal is added to carbonic acid.

61. Write a balanced equation for each of the following single-replacement reactions.
(a) A soft, gray chunk of lithium is added to water.
(b) A small, shiny chunk of barium is added to water.

62. Write a balanced equation for each of the following single-replacement reactions.
(a) A gray chunk of cesium is added to water.
(b) A piece of radioactive radium is added to water.

63. Complete and balance each of the following single-replacement reactions.
(a) $Zn(s) + Pb(NO_3)_2(aq) \longrightarrow$
(b) $Cd(s) + Fe(NO_3)_2(aq) \longrightarrow$

▲ **Zinc in a Lead Solution** A zinc strip reacts in an aqueous $Pb(NO_3)_2$ solution to give a gray deposit of spongy Pb metal.

64. Complete and balance each of the following single-replacement reactions.
(a) $Mg(s) + NiSO_4(aq) \longrightarrow$
(b) $Al(s) + SnSO_4(aq) \longrightarrow$

65. Complete and balance each of the following single-replacement reactions.
(a) $Zn(s) + HNO_3(aq) \longrightarrow$
(b) $Cd(s) + HNO_3(aq) \longrightarrow$

66. Complete and balance each of the following single-replacement reactions.
(a) $Mg(s) + H_2SO_4(aq) \longrightarrow$
(b) $Al(s) + H_2SO_4(aq) \longrightarrow$

67. Complete and balance each of the following single-replacement reactions.
(a) $K(s) + H_2O(l) \longrightarrow$
(b) $Ba(s) + H_2O(l) \longrightarrow$

68. Complete and balance each of the following single-replacement reactions.
(a) $Mg(s) + H_2O(l) \longrightarrow$
(b) $Ca(s) + H_2O(l) \longrightarrow$

Solubility Rules (Sec. 8.9)

(Refer to the solubility rules in Table 8.2 to answer the following questions.)

69. Predict which of the following compounds are soluble in water.
(a) cobalt(II) hydroxide, $Co(OH)_2$
(b) iron(II) sulfate, $FeSO_4$
(c) tin(II) chromate, $SnCrO_4$
(d) lead(II) acetate, $Pb(C_2H_3O_2)_2$

70. Predict which of the following compounds are soluble in water.
(a) aluminum nitrate, $Al(NO_3)_3$
(b) lead(II) sulfate, $PbSO_4$
(c) ammonium sulfide, $(NH_4)_2S$
(d) iron(III) phosphate, $FePO_4$

71. Predict which of the following compounds are insoluble in water.
(a) mercury(I) chloride, Hg_2Cl_2
(b) mercury(II) chloride, $HgCl_2$
(c) silver bromide, AgBr
(d) lead(II) iodide, PbI_2

72. Predict which of the following compounds are insoluble in water.
(a) strontium carbonate, $SrCO_3$
(b) calcium hydroxide, $Ca(OH)_2$
(c) nickel(II) sulfide, NiS
(d) mercury(II) bromide, $HgBr_2$

Double-Replacement Reactions (Sec. 8.10)

73. Write a balanced equation for each of the following double-replacement reactions.
(a) Aqueous solutions of zinc chloride and ammonium hydroxide react to give aqueous ammonium chloride and a zinc hydroxide precipitate.
(b) Aqueous solutions of nickel(II) sulfate and mercury(I) nitrate react to give aqueous nickel(II) nitrate and a mercury(I) sulfate precipitate.

74. Write a balanced equation for each of the following double-replacement reactions.
(a) Aqueous solutions of tin(II) chloride and sodium sulfide react to give aqueous sodium chloride and a tin(II) sulfide precipitate.
(b) Aqueous solutions of cobalt(II) nitrate and potassium chromate react to give aqueous potassium nitrate and a cobalt(II) chromate precipitate.

75. Complete and balance each of the following double-replacement reactions.
(a) $MgSO_4(aq) + BaCl_2(aq) \longrightarrow$
(b) $AlBr_3(aq) + Na_2CO_3(aq) \longrightarrow$

76. Complete and balance each of the following double-replacement reactions.
(a) $AgC_2H_3O_2(aq) + SrI_2(aq) \longrightarrow$
(b) $FeSO_4(aq) + Ca(OH)_2(aq) \longrightarrow$

Neutralization Reactions (Sec. 8.11)

77. Write a balanced equation for each of the following neutralization reactions.
(a) Sodium hydroxide solution is added to nitric acid.
(b) Barium hydroxide solution is added to phosphoric acid.

78. Write a balanced equation for each of the following ne tralization reactions.
(a) Potassium hydroxide solution is added to car acid.
(b) Strontium hydroxide solution is added t

79. Complete and balance each of the following neutralization reactions.
(a) $HF(aq) + Ca(OH)_2(aq) \longrightarrow$
(b) $H_2SO_4(aq) + LiOH(aq) \longrightarrow$

80. Complete and balance each of the following neutralization reactions.
(a) $HNO_3(aq) + Sr(OH)_2(aq) \longrightarrow$
(b) $H_3PO_4(aq) + Ba(OH)_2(aq) \longrightarrow$

General Exercises

81. Balance each of the following chemical equations by inspection.
(a) $Fe(s) + H_2O(g) \longrightarrow Fe_3O_4(s) + H_2(g)$
(b) $FeS(s) + O_2(g) \longrightarrow Fe_2O_3(s) + SO_2(g)$

82. Balance each of the following chemical equations by inspection.
(a) $FeO(l) + Al(l) \longrightarrow Al_2O_3(l) + Fe(l)$
(b) $MnO_2(l) + Al(l) \longrightarrow Al_2O_3(l) + Mn(l)$

83. Balance each of the following chemical equations by inspection.
(a) $F_2(g) + NaBr(aq) \longrightarrow Br_2(l) + NaF(aq)$
(b) $Sb_2S_3(s) + HCl(aq) \longrightarrow SbCl_3(aq) + H_2S(g)$

84. Balance each of the following chemical equations by inspection.
(a) $PCl_5(s) + H_2O(l) \longrightarrow H_3PO_4(aq) + HCl(aq)$
(b) $TiCl_4(s) + H_2O(g) \longrightarrow TiO_2(s) + HCl(g)$

85. Balance each of the following combustion reactions by inspection.
(a) $CH_4(g) + O_2(g) \longrightarrow CO_2(g) + H_2O(g)$
(b) $C_3H_8(g) + O_2(g) \longrightarrow CO_2(g) + H_2O(g)$

86. Balance each of the following combustion reactions by inspection.
(a) $CH_4O(l) + O_2(g) \longrightarrow CO_2(g) + H_2O(g)$
(b) $C_3H_8O(l) + O_2(g) \longrightarrow CO_2(g) + H_2O(g)$

87. Chlorine is prepared industrially by heating hydrogen chloride gas with oxygen gas. Assuming the only products are water and chlorine gas, write a balanced chemical equation for the manufacture of chlorine.

88. Iron is prepared industrially by passing carbon monoxide gas through molten iron ore, Fe_2O_3, in a blast furnace at 1500°C. Assuming the only products are molten iron and carbon dioxide gas, write a balanced chemical equation for the manufacture of iron.

89. The Contact process is the industrial method for manufacturing sulfuric acid. In the first step, sulfur is burned with oxygen gas to give sulfur dioxide. In the second step, sulfur dioxide and oxygen gas react in the presence of a platinum catalyst to produce sulfur trioxide. In the third step, sulfur trioxide gas is passed through water to yield sulfuric acid. Write three balanced equations for the Contact process.

90. The Ostwald process is an important industrial method for making nitric acid. In the first step, ammonia and oxygen are heated to give nitrogen monoxide gas and water. In the second step, nitrogen monoxide and oxygen react to produce nitrogen dioxide. In the third step, nitrogen dioxide gas is passed through water to yield nitric acid and nitrogen monoxide gas. Write three balanced equations for the Ostwald process.

WWW
Explorer Quiz 1
Explorer Quiz 2
Explorer Quiz 3
Master Quiz

CHAPTERS 7–8
Cumulative Review

Key Concepts

1. Classify each of the following as binary ionic, ternary ionic, binary molecular, binary acid, or ternary oxyacid: K_2S, H_2S, $H_2S(aq)$, K_2SO_4, $H_2SO_4(aq)$.
2. Which of the following compounds is named using the suffix *-ide*: K_2S, H_2S, K_2SO_3, K_2SO_4?
3. Which of the following compounds is named using the suffix *-ate*: K_2S, H_2S, K_2SO_3, K_2SO_4?
4. Which of the following compounds is named using the suffix *-ite*: K_2S, H_2S, K_2SO_3, K_2SO_4?
5. Which of the following acids is named *hydro* + *nonmetal stem* plus *-ic acid*: $H_2S(g)$, $H_2S(aq)$, $H_2SO_3(aq)$, $H_2SO_4(aq)$?
6. Which of the following acids is named *nonmetal stem* plus *-ic acid*: $H_2S(g)$, $H_2S(aq)$, $H_2SO_3(aq)$, $H_2SO_4(aq)$?
7. Which of the following acids is named *nonmetal stem* plus *-ous acid*: $H_2S(g)$, $H_2S(aq)$, $H_2SO_3(aq)$, $H_2SO_4(aq)$?
8. Which of the following metals does **not** react with $AgNO_3(aq)$: Al, Hg, Mn, Zn? (Refer to the activity series.)
9. Which of the following metals does **not** react with $HNO_3(aq)$: Ag, Fe, Pb, Sn? (Refer to the activity series.)
10. Explain the following illustrated reaction that shows hydrochloric acid dripping into a test tube containing white baking soda powder.

11. Complete and balance the following combination reaction.

$$K(s) + O_2(g) \longrightarrow$$

12. Complete and balance the following decomposition reaction.

$$KHCO_3(s) \xrightarrow{\Delta}$$

13. Complete and balance the following single-replacement reaction.

$$Al(s) + HNO_3(aq) \longrightarrow$$

14. Complete and balance the following double-replacement reaction.

$$Al(NO_3)_3(aq) + K_3PO_4(aq) \longrightarrow$$

15. Complete and balance the following neutralization reaction.

$$HNO_3(aq) + KOH(aq) \longrightarrow$$

Key Terms

State the key term that corresponds to each of the following descriptions.

_____ 1. the international system of rules for naming chemical compounds
_____ 2. a single atom that has a positive charge
_____ 3. a group of atoms that has a negative charge
_____ 4. a system that designates the charge on a cation with Roman numerals
_____ 5. a system that designates the charge on a cation by the suffix *-ic* or *-ous*
_____ 6. the simplest representative particle in a compound composed of ions
_____ 7. the simplest representative particle in a compound composed of nonmetals
_____ 8. an insoluble solid substance produced by an aqueous solution reaction
_____ 9. a substance that increases the rate of a chemical reaction
_____ 10. a digit placed in front of a formula to balance a chemical equation
_____ 11. a type of reaction in which two substances produce a single compound
_____ 12. a type of reaction in which a single compound produces two or more substances
_____ 13. a type of reaction in which a more active metal displaces a less active metal from a solution or compound
_____ 14. a type of reaction in which two cations in different compounds exchange anions
_____ 15. a type of reaction in which an acid and a base produce a salt and water
_____ 16. a relative order of metals arranged by their ability to undergo reaction
_____ 17. a metal that reacts with water at room temperature
_____ 18. a substance that releases hydrogen ions in aqueous solution
_____ 19. a substance that releases hydroxide ions in aqueous solution
_____ 20. an ionic compound produced from an acid–base reaction

Review Exercises

1. Is potassium permanganate, $KMnO_4$, classified as a binary or a ternary compound? an ionic or a molecular compound?
2. What is the formula for the ionic compound composed of the ferrous ion, Fe^{2+}, and the phosphide ion, P^{3-}? the ferrous ion and the phosphate ion, PO_4^{3-}?
3. What is the name of Fe_2S_3 according to the Stock system of nomenclature?
4. What is the name of FeS according to the Latin system of nomenclature?
5. What is the name of $Fe_2(SO_4)_3$ according to the Stock system of nomenclature?
6. What is the name of $FeSO_4$ according to the Latin system of nomenclature?
7. What is the name of P_4S_3 according to IUPAC nomenclature?
8. What is the name of aqueous H_2S according to IUPAC nomenclature?
9. What is the name of aqueous H_2SO_3 according to IUPAC nomenclature?
10. What is the name of aqueous H_2SO_4 according to IUPAC nomenclature?
11. Write a balanced equation for the combination of cobalt metal with oxygen gas to give solid cobalt(III) oxide.
12. Write a balanced equation for the decomposition of solid cobalt(III) hydrogen carbonate on heating to give cobalt(III) carbonate powder, steam, and carbon dioxide gas.

13. Write a balanced equation for the reaction of cobalt metal with aqueous nickel(II) nitrate to produce aqueous cobalt(II) nitrate and nickel metal.
14. Write a balanced equation for the reaction between aqueous solutions of lead(II) nitrate and lithium iodide as shown in the following illustration.

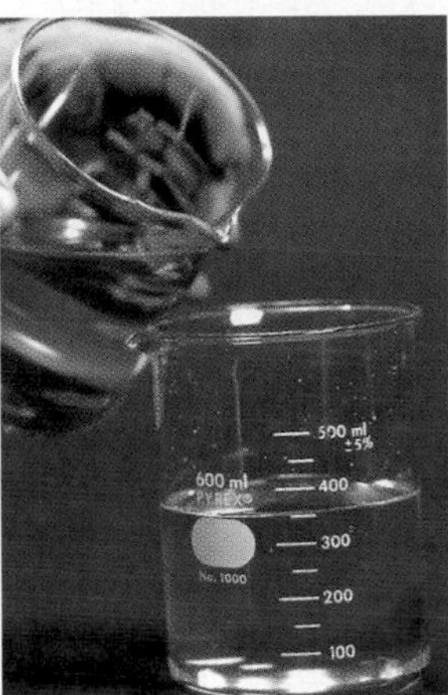

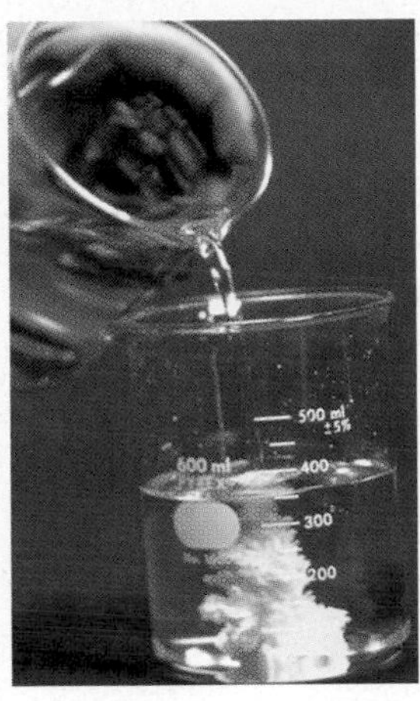

15. Write a balanced equation for the neutralization of aqueous ammonium hydroxide by carbonic acid to give aqueous ammonium carbonate plus water.

CHAPTER 9

The Mole Concept

After providing a nomenclature and reaction foundation, this chapter begins chemical calculations using the mole concept. Once again, I address students fears of mathematical operations by reminding them that solving problems is "as simple as one, two, three," a reference to unit analysis. Since the material in Chapters 4–8 is more qualitative than quantitative, it is helpful to move methodically into the calculations. In addition to an algorithmic explanation of problem solving, I simultaneously use concept maps to relate the mole concept to Avogadro's number, molar mass, and molar volume.

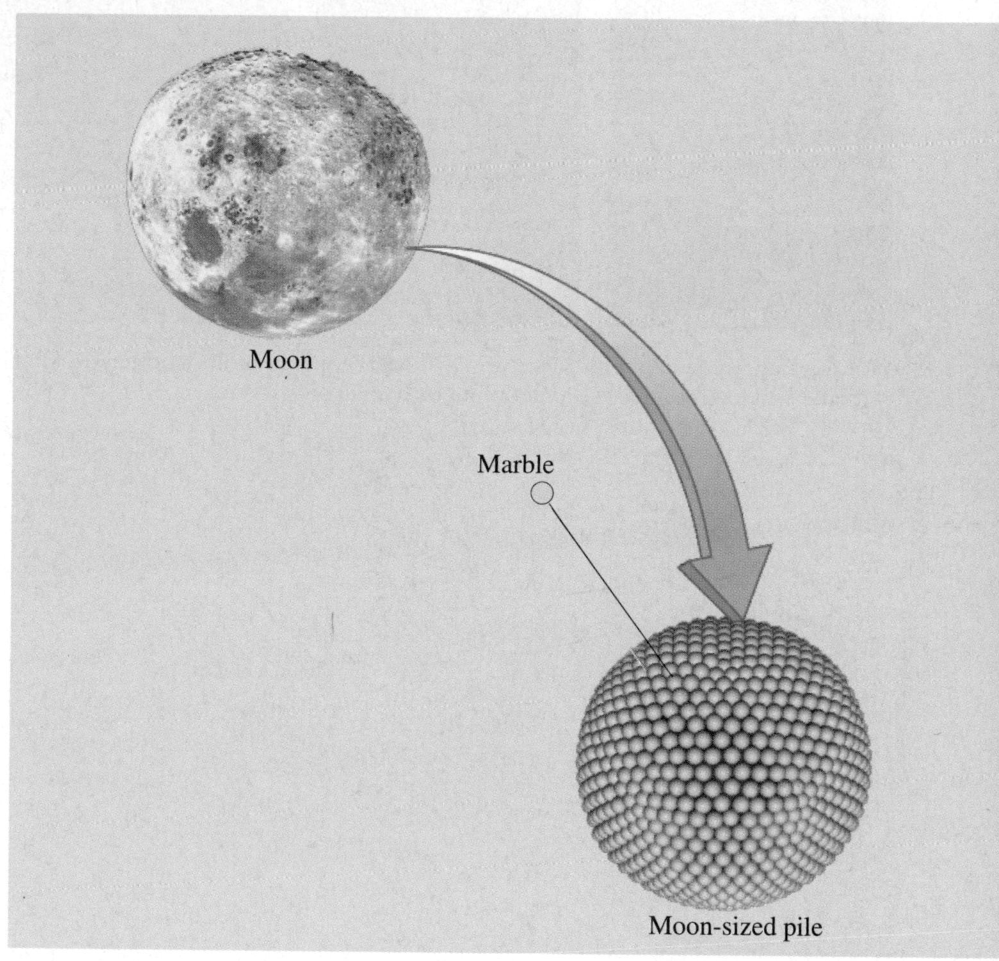

▲ If you had a pile of marbles that was as big as the Moon, how many marbles would be in the pile? (Assume each marble is about an inch in diameter.)

Is it possible to keep track of atoms by counting them? The answer is yes, although we cannot count them directly. Atoms are much too small to count individually, and so we count them in groups. That is, we count atoms in the same way we count eggs by the dozen (12), pencils by the gross (144), and sheets of paper by the ream (500). Since atoms are so tiny, we must consider a very large group of atoms.

When considering a very large number, it may be difficult to comprehend its magnitude. In this chapter, we will encounter an extremely large number—602,000,000,000,000,000,000,000. This number refers to a group of atoms and is called Avogadro's number. To appreciate the enormity of this number, consider the following analogy. If you had Avogadro's number of marbles, the size of the group of marbles would be the size of the Moon.

9.1 Avogadro's Number

Experiment #11, Prentice Hall Laboratory Manual

Objectives
- To state the value of Avogadro's number: 6.02×10^{23}.
- To state the mass of Avogadro's number of atoms for any element by referring to the periodic table.

Before explaining how to count atoms, let's explain how to count eggs by weighing. Suppose we weigh several cartons of eggs and the total mass is 15,000 g. Moreover, each carton of eggs has an average mass of 750 g and contains 12 eggs. Using unit analysis, we can easily find the number of eggs.

$$15{,}000 \,\cancel{\text{g}} \times \frac{1 \,\cancel{\text{carton}}}{750 \,\cancel{\text{g}}} \times \frac{12 \text{ eggs}}{1 \,\cancel{\text{carton}}} = 240 \text{ eggs}$$

Similarly, we can count atoms if we weigh a large group of atoms and know the average mass and number of atoms in the group. Let's weigh enough atoms so that the mass in grams is the same numerical value as the atomic mass expressed in atomic mass units. For carbon, let's weigh enough atoms to equal a mass of 12.01 g (atomic mass = 12.01 amu).

But how many atoms are in 12.01 g of carbon? Since atoms are so tiny, the number of carbon atoms in 12.01 g must be extremely large. Experiments have shown that there are 6.02×10^{23} carbon atoms in 12.01 g. This very large number is referred to as **Avogadro's number** (symbol *N*) in honor of the Italian scientist Amedeo Avogadro (1776–1856). Avogadro's number is sometimes referred to as the chemist's dozen.

Thus, to count Avogadro's number of atoms, we only have to know the atomic mass of the element. That is, 6.02×10^{23} carbon atoms have a mass of 12.01 g, 6.02×10^{23} neon atoms have a mass of 20.18 g (20.18 amu), and 6.02×10^{23} chromium atoms have a mass of 52.00 g (52.00 amu). The following example exercise further illustrates how to count Avogadro's number of atoms.

Example Exercise 9.1 • Atomic Mass and Avogadro's Number

Refer to the atomic masses in the periodic table on the inside front cover of this textbook. State the mass of Avogadro's number of atoms for each of the following elements.

(a) copper (b) mercury
(c) sulfur (d) helium

Solution

The atomic mass of each element is listed below the symbol of the element in the periodic table: Cu = 63.55 amu, Hg = 200.59 amu, S = 32.07 amu, and He = 4.00 amu. The mass of Avogadro's number of atoms is the atomic mass expressed in grams. Therefore, 6.02×10^{23} atoms of

(a) Cu = 63.55 g (b) Hg = 200.59 g
(c) S = 32.07 g (d) He = 4.00 g

Self-Test Exercise

Refer to the periodic table and state the mass for each of the following.

(a) 1 atom of Au
(b) 6.02×10^{23} atoms of Au

Answers: (a) 196.97 amu; (b) 196.97 g

▲ **Figure 9.1 Avogadro's Number** The amount of copper (63.55 g), mercury (200.59 g), and sulfur (32.07 g) that contains 6.02×10^{23} atoms of each element. The balloon contains 6.02×10^{23} atoms of helium gas.

In Example Exercise 9.1, we found the mass of Avogadro's number of copper, mercury, sulfur, and helium atoms. To further appreciate the concept of this huge number, Figure 9.1 shows the amount of each element that contains Avogadro's number of atoms.

Chemistry Connection · Analogies for Avogadro's Number

If 6.02×10^{23} dollars earns 5% interest, how many dollars does the account earn every nanosecond?

Length Analogy: If 6.02×10^{23} hydrogen atoms were laid side by side, the total length would be long enough to encircle the Earth about a million times.

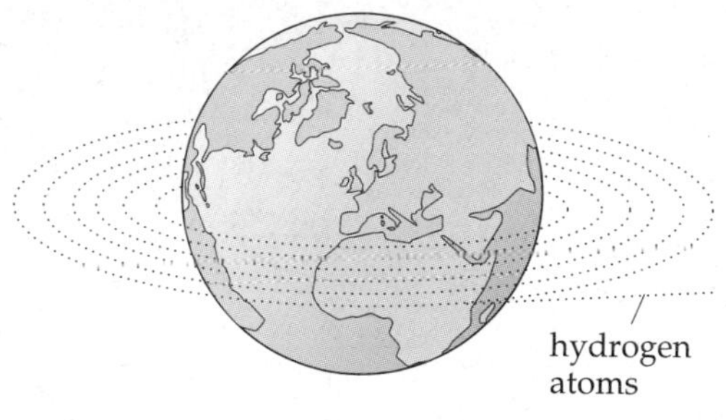

Mass Analogy: The mass of 6.02×10^{23} Olympic shotput balls would be about equal to the mass of the Earth.

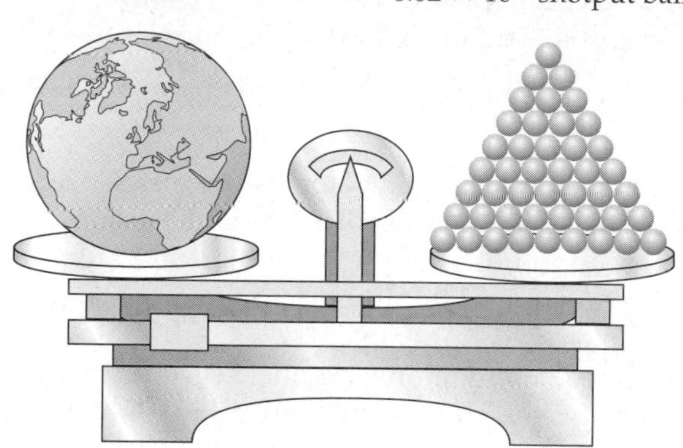

Volume Analogy: The volume occupied by 6.02×10^{23} softballs would be about the size of the Earth.

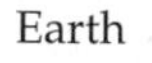

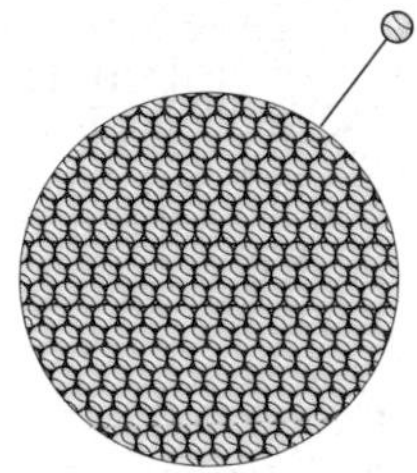

The account earns $1,000,000 every nanosecond; that is, it earns a million dollars every billionth of a second.

9.2 Mole Calculations I

Objective · To relate the moles of a substance to the number of particles.

The **mole** (symbol **mol**) is a unit of measure for an amount of a chemical substance. We define a mole as the amount of substance that contains Avogadro's number of particles, that is, 6.02×10^{23} particles. The individual particles may be atoms, molecules, formula units, or any other particles. Thus

$$1 \text{ mole} = \text{Avogadro's number } (N) = 6.02 \times 10^{23} \text{ particles}$$

The mole relationship allows us to convert between the number of particles and the mass of an amount of substance. Now we are ready to try some mole calculations. Remember to apply the three steps in the unit analysis method of problem solving.

Applying the Unit Analysis Method

Step 1: Write down the units asked for in the answer.
Step 2: Write down the given value in the problem that is related to the units in the answer.
Step 3: Apply one or more unit factors to cancel the units in the given value and convert to the units in the answer.

We can perform calculations that relate number of moles and number of particles. For instance, we can determine how many molecules of chlorine are in a 0.250 mol of the yellow gas. By applying the unit analysis method of problem solving, we have

$$0.250\ \cancel{\text{mol Cl}_2} \times \frac{6.02 \times 10^{23}\ \text{molecules Cl}_2}{1\ \cancel{\text{mol Cl}_2}} = 1.51 \times 10^{23}\ \text{molecules Cl}_2$$

The following example exercises will help to reinforce the unit analysis method of problem solving for calculations involving the mole concept.

▲ The Mole Concept—Avogadro's Number A mole of helium, a mole of water, and a mole of table salt. The balloon contains 6.02×10^{23} atoms of He gas; the graduated cylinder holds 6.02×10^{23} molecules of H_2O, and the watchglass has 6.02×10^{23} formula units of NaCl.

The Mole Concept—Avogadro's Number

Example Exercise 9.2 • Moles and Avogadro's Number

Calculate the number of sodium atoms in 0.120 mol Na.

Solution

We can relate moles of substance to the number of particles using Avogadro's number. Let's recall the format for the unit analysis method of problem solving.

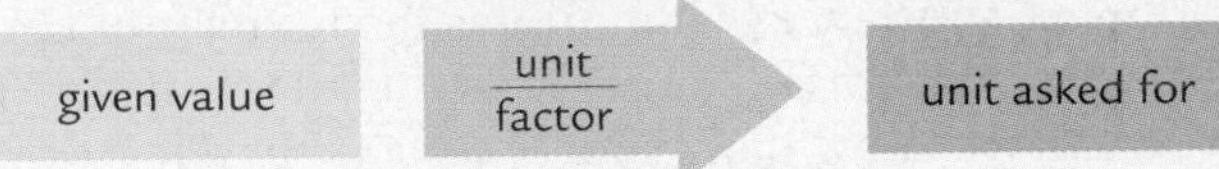

Step 1, The unit asked for in the answer is number of **atoms Na**.
Step 2, The given value is **0.120 mol Na**.
Step 3, We apply a conversion factor to cancel units. We know that 1 mol of substance contains Avogadro's number of atoms, molecules, or formula units. In this example, the unit factor is derived from 1 mol = 6.02×10^{23} atoms.

$$0.120\ \cancel{\text{mol Na}} \times \frac{6.02 \times 10^{23}\ \text{atoms Na}}{1\ \cancel{\text{mol Na}}} = 7.22 \times 10^{22}\ \text{atoms Na}$$

We can summarize the solution as follows.

0.120 mol → $\dfrac{6.02 \times 10^{23}\ \text{atoms}}{1\ \text{mol}}$ → 7.22×10^{22} atoms

Self-Test Exercise

Calculate the number of formula units in 0.0763 mol of sodium chloride, NaCl.

Answer: 4.59×10^{22} formula units NaCl

We can also perform the reverse procedure and determine the amount of substance given the number of particles. For instance, we can calculate the number of moles of violet iodine crystals corresponding to 2.50×10^{23} molecules of I_2. Applying the unit analysis method of problem solving, we have

$$2.50 \times 10^{23}\ \cancel{\text{molecules I}_2} \times \frac{1\ \text{mol I}_2}{6.02 \times 10^{23}\ \cancel{\text{molecules I}_2}} = 0.415\ \text{mol I}_2$$

Example Exercise 9.3 • Avogadro's Number and Moles

Calculate the number of moles of potassium in 1.25×10^{21} atoms of K.

Solution

Step 1, The unit asked for in the answer is **mol K**.
Step 2, The given value is **1.25×10^{21} atoms K**.
Step 3, We apply a conversion factor. In this example, 1 mol = 6.02×10^{23} atoms. After applying a unit factor to cancel units, we have

$$1.25 \times 10^{21} \cancel{\text{atoms K}} \times \frac{1 \text{ mol K}}{6.02 \times 10^{23} \cancel{\text{atoms K}}} = 2.08 \times 10^{-3} \text{ mol K}$$

We can summarize the solution as follows.

$$1.25 \times 10^{21} \text{ atoms} \rightarrow \frac{1 \text{ mol}}{6.02 \times 10^{23} \text{ atoms}} \rightarrow 2.08 \times 10^{-3} \text{ mol}$$

Self-Test Exercise

Calculate the number of moles of potassium iodide in 5.34×10^{25} formula units of KI.

Answer: 88.7 mol KI

9.3 Molar Mass

Objective · To calculate the molar mass of a substance given its chemical formula.

A mole is an amount of substance that indicates the number of particles in a sample. That is, 1 mol of an element contains 6.02×10^{23} atoms. Since the atomic mass of carbon is 12.01 amu, we know that 1 mol of carbon has a mass of 12.01 g. In fact, the atomic mass of any substance expressed in grams corresponds to 1 mol of the substance. In our discussion, we will refer to the atomic mass of a substance expressed in grams as the **molar mass** (symbol **MM**).

The molar mass of an element is equal to its atomic mass. By referring to the periodic table, we find that the atomic mass of iron is 55.85 g. Thus, the molar mass of iron is 55.85 g/mol. Since naturally occurring oxygen is O_2, the molar mass of oxygen gas is equal to twice 16.00 g, or 32.00 g/mol.

The Mole Concept—Molar Mass

We can calculate the molar mass of a compound by adding the molar masses of each element. For example, we can find the molar mass of iron(III) oxide, Fe_2O_3, by summing the atomic masses. Thus,

$$Fe_2O_3\text{: } 2(55.85 \text{ g Fe}) + 3(16.00 \text{ g O}) = 159.70 \text{ g}$$

The calculated molar mass of Fe_2O_3 is 159.70 g/mol. Notice that we used 16.00 g as the molar mass of oxygen. Even though the element occurs naturally as molecules of oxygen, it is atoms of oxygen that are combined in compounds. The following examples further illustrate the calculation of molar mass.

▲ **The Mole Concept—Molar Mass** A mole of water, sulfur, table sugar, mercury, and copper (clockwise). The molar masses are as follows: 18.02 g of H_2O, 32.07 g of S, 342.34 g of $C_{12}H_{22}O_{11}$, 200.59 g of Hg, and 63.55 g of Cu.

Example Exercise 9.4 • Molar Mass

Calculate the molar mass for each of the following substances.

(a) silver metal, Ag
(b) ammonia gas, NH_3
(c) magnesium nitrate, $Mg(NO_3)_2$

Solution

We begin by finding the atomic mass in the periodic table. The atomic mass value expressed in grams is the mass of 1 mol; that is, the molar mass.

(a) The atomic mass of Ag is 107.87 amu, and the molar mass is 107.87 g/mol.
(b) The atomic mass of nitrogen is 14.01 amu and that of hydrogen is 1.01 amu. The sum of the atomic masses for NH_3 is $14.01 + 1.01 + 1.01 + 1.01 = 17.04$ amu. Hence, the molar mass is 17.04 g/mol.
(c) The sum of the atomic masses for $Mg(NO_3)_2$ is $24.31 + 2(14.01 + 16.00 + 16.00 + 16.00) = 24.31+2(62.01) = 148.33$ amu. Therefore, the molar mass is 148.33 g/mol.

Self-Test Exercise
Calculate the molar mass for each of the following substances.

(a) manganese metal, Mn (b) sulfur hexafluoride, SF_6
(c) strontium acetate, $Sr(C_2H_3O_2)_2$

Answers: (a) 54.94 g/mol; (b) 146.07 g/mol; (c) 205.72 g/mol

9.4 Mole Calculations II

Calculating with Moles Activity

Objective · To relate the mass of a substance to the number of particles.

The mole is the central unit in chemistry. It is an amount of a substance and relates the number of particles to the mass of the substance; that is,

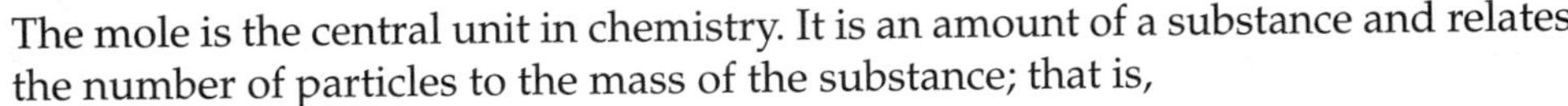

$$6.02 \times 10^{23} \text{ particles} = 1 \text{ mole} = \text{molar mass of substance}$$

We can perform calculations that relate mass to the number of particles. For instance, we can find the mass of 2.55×10^{23} atoms of lead. By applying the unit analysis method of problem solving, we first find the number of moles.

$$2.55 \times 10^{23} \cancel{\text{atoms Pb}} \times \frac{1 \text{ mol Pb}}{6.02 \times 10^{23} \cancel{\text{atoms Pb}}} = 0.424 \text{ mol Pb}$$

To calculate the mass of lead in grams, we multiply the mol Pb by its molar mass. From the periodic table we find that the molar mass of lead is 207.2 g/mol.

$$0.424 \cancel{\text{mol Pb}} \times \frac{207.2 \text{ g Pb}}{1 \cancel{\text{mol Pb}}} = 87.8 \text{ g Pb}$$

The following example exercises help to reinforce mole calculations involving Avogadro's number and molar mass.

Example Exercise 9.5 • Number of Particles and Mass

What is the mass in grams of 2.01×10^{22} atoms of sulfur?

Solution
In mole calculations, we relate the given value to moles and then convert to the unit asked for in the answer.

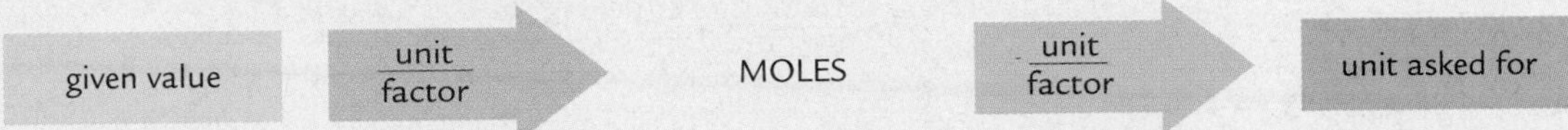

Step 1, The unit asked for is **g S**.
Step 2, The given value is **2.01×10^{22} atoms S**.
Step 3, We apply unit conversion factors to convert the number of atoms to mass. We can use the relationship 1 mol = 6.02×10^{23} atoms, and the molar mass of sulfur is 32.07 g/mol from the periodic table. Applying unit factors to cancel units, we have

(continued)

Example Exercise 9.5 *(continued)*

$$2.01 \times 10^{22} \cancel{\text{atoms S}} \times \frac{1 \text{ mol S}}{6.02 \times 10^{23} \cancel{\text{atoms S}}} \times \frac{32.07 \text{ g S}}{1 \cancel{\text{mol S}}} = 1.07 \text{ g S}$$

We can summarize the steps in the solution as follows.

Self-Test Exercise

What is the mass of 7.75×10^{22} formula units of lead(II) sulfide, PbS?

Answer: 30.8 g PbS

Example Exercise 9.6 • Mass and Number of Particles

How many O_2 molecules are present in 0.470 g of oxygen gas?

Solution

In mole calculations, we relate the given value to moles and then convert to the unit asked for in the answer.
Step 1, The unit asked for is **molecules O_2**.
Step 2, The given value is **0.470 g O_2**.
Step 3, We apply unit factors to convert the mass of O_2 to the number of molecules. We find that the molar mass of O_2 in the periodic table is 32.00 g/mol, and 1 mol O_2 = 6.02×10^{23} molecules of O_2. Applying unit factors to cancel units, we have

$$0.470 \cancel{\text{g } O_2} \times \frac{1 \cancel{\text{mol } O_2}}{32.00 \cancel{\text{g } O_2}} \times \frac{6.02 \times 10^{23} \text{ molecules } O_2}{1 \cancel{\text{mol } O_2}} = 8.84 \times 10^{21} \text{ molecules } O_2$$

We can summarize the steps in the solution as follows.

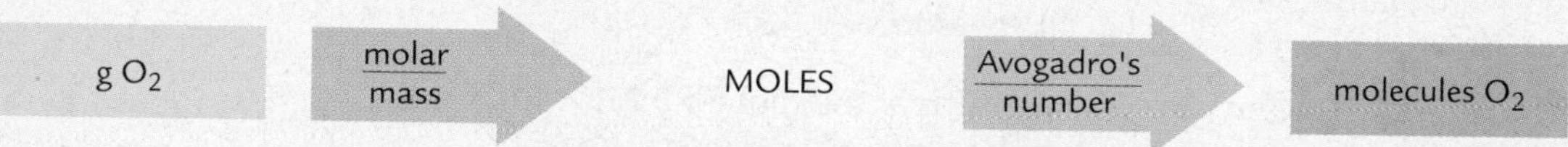

Self-Test Exercise

How many formula units of lithium fluoride are found in 0.175 g LiF?

Answer: 4.06×10^{21} formula units LiF

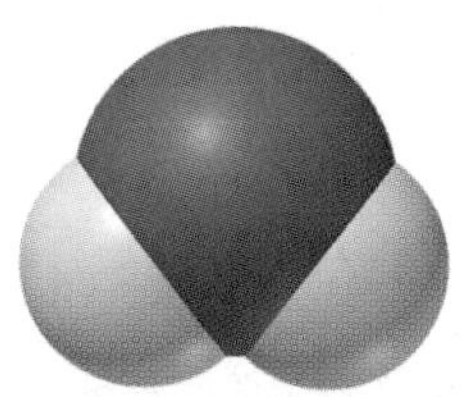

▲ Water molecule—H_2O

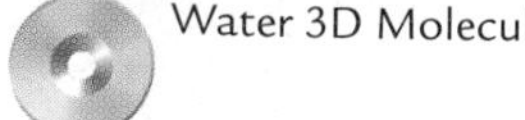

Mass of an Atom or Molecule

Now, let's try a different type of problem. If the molar mass of water is 18.02 g/mol, what is the mass of a single molecule of water? This problem requires compound units in the answer; that is, g H_2O/molecule. To arrive at an answer with compound units, we start with a ratio of two units, that is,

$$\frac{18.02 \text{ g } H_2O}{1 \text{ mol } H_2O} \times \frac{\text{unit}}{\text{factor}} = \frac{\text{g } H_2O}{\text{molecule } H_2O}$$

The unit factor is provided by Avogadro's number: 1 mol = 6.02×10^{23} molecules.

$$\frac{18.02 \text{ g } H_2O}{1 \cancel{\text{mol } H_2O}} \times \frac{1 \cancel{\text{mol } H_2O}}{6.02 \times 10^{23} \text{ molecules}} = 2.99 \times 10^{-23} \text{ g } H_2O/\text{molecule}$$

The following example exercise further illustrates how to apply the mole concept to determine the mass of an individual atom or molecule.

Example Exercise 9.7 • Mass of a Molecule

Calculate the mass in grams for a single molecule of carbon dioxide, CO_2.

Solution

The atomic mass of carbon is 12.01 g and that of oxygen is 16.00 g; therefore, the molar mass of CO_2 is 44.01 g/mol. We can set up the problem as follows.

$$\frac{44.01 \text{ g } CO_2}{1 \text{ mol } CO_2} \times \frac{\text{unit}}{\text{factor}} = \frac{\text{g } CO_2}{\text{molecule}}$$

The unit factor is provided by the relationship 1 mol = 6.02×10^{23} molecules.

$$\frac{44.01 \text{ g } CO_2}{1 \cancel{\text{mol } CO_2}} \times \frac{1 \cancel{\text{mol } CO_2}}{6.02 \times 10^{23} \text{ molecules}} = 7.31 \times 10^{-23} \text{ g } CO_2/\text{molecule}$$

We can summarize the solution as follows.

$$\frac{44.01 \text{ g } CO_2}{1 \text{ mol}} \xrightarrow{\text{Avogadro's number}} \frac{7.31 \times 10^{-23} \text{ g } CO_2}{\text{molecule}}$$

Self-Test Exercise

Calculate the mass in grams for a single atom of iron, Fe.

Answer: 9.28×10^{-23} g Fe/atom

▲ Carbon dioxide molecule—CO_2

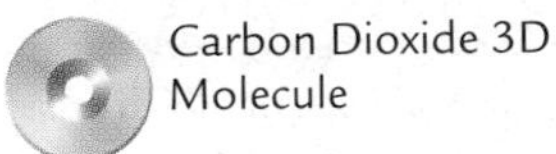

Carbon Dioxide 3D Molecule

9.5 Molar Volume

Objectives
- To state the value for the molar volume of any gas at STP: 22.4 L/mol.
- To relate the density of a gas at STP to its molar mass and volume.

Recall that 1 mol of any gas contains 6.02×10^{23} molecules. The gas can be hydrogen, oxygen, or any other gas except an inert gas, which is composed of atoms. In 1811 Avogadro proposed that two gases containing equal numbers of molecules occupy equal volumes under similar conditions. This statement is known as **Avogadro's theory**. It naturally follows, therefore, that 1 mol of hydrogen gas and 1 mol of oxygen gas occupy the same volume. Stated differently, 6.02×10^{23} molecules of H_2 occupy the same volume as 6.02×10^{23} molecules of O_2. In fact, 6.02×10^{23} molecules of any gas occupy the same volume as 6.02×10^{23} molecules of hydrogen or oxygen.

What volume of gas contains Avogadro's number of molecules? At **standard temperature and pressure** (symbol **STP**) the volume is 22.4 L. Standard temperature has been chosen to be 0°C. Standard pressure has been chosen to be 1 atm, which is the atmospheric pressure exerted by air at sea level. The volume occupied by 1 mol of any gas at STP is called the **molar volume** (Figure 9.2).

Table 9.1 compares 1 mole of five different gases. Notice that the number of molecules and the volume are constant, whereas the molar mass varies.

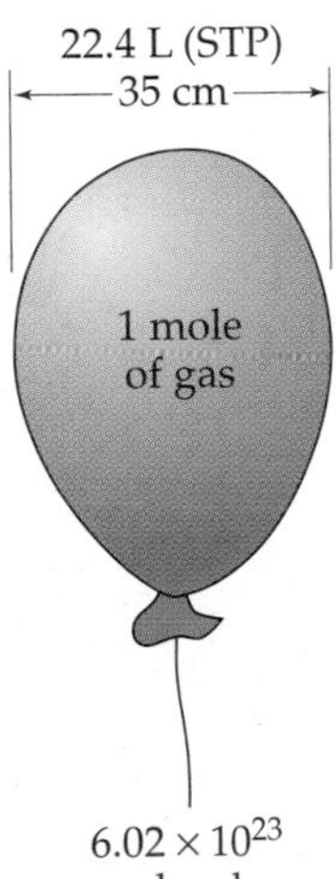

▲ **Figure 9.2 Molar Volume of a Gas** A balloon containing 1 mol of gas has a diameter of about 35 cm. A mole of any gas contains Avogadro's number of molecules and occupies a volume of 22.4 L at STP.

Gas Density

As you may recall, the density of a gas is much less than the density of a liquid or a solid. In Section 3.7 we found that at 3.98°C the density of water is 1.00 g/mL. However, the density of air at 3.98°C is only 1.29 g/L. Therefore, the density of water is almost a 1000 times greater than that of air.

We also learned in Section 3.7 that the definition of density is mass divided by volume. Likewise, the density of a gas, such as hydrogen or oxygen, is mass divided by volume. We can easily calculate the density of a gas because 1 mol of any

▲ **The Mole Concept—Molar Volume** A 22.4-L box. Thus, a basketball, a football, and a soccer ball each contain somewhat less than 1 mol of gas at STP.

Table 9.1 Mole Relationships for Selected Gases

Gas	No. of Moles	No. of Molecules	Molar Mass	Molar Volume at STP
hydrogen, H_2	1.00	6.02×10^{23}	2.02 g/mol	22.4 L/mol
oxygen, O_2	1.00	6.02×10^{23}	32.00 g/mol	22.4 L/mol
carbon dioxide, CO_2	1.00	6.02×10^{23}	44.01 g/mol	22.4 L/mol
ammonia, NH_3	1.00	6.02×10^{23}	17.04 g/mol	22.4 L/mol
argon, Ar*	1.00	6.02×10^{23}	39.95 g/mol	22.4 L/mol

*Argon gas is composed of atoms rather than molecules.

gas has a mass equal to its molar mass and a volume equal to its molar volume at STP. The formula for **gas density** is

$$\frac{\text{molar mass in grams}}{\text{molar volume in liters}} = \text{density, g/L (at STP)}$$

The following example exercises illustrate how the density, molar mass, and molar volume of a gas are related.

Example Exercise 9.8 • Density of a Gas at STP

Calculate the density of ammonia gas, NH_3, at STP.

Solution

To find the density of a gas at STP, we divide the molar mass by the molar volume.

$$\frac{\text{molar mass } NH_3}{\text{molar volume } NH_3} = \text{density, g/L}$$

From the periodic table, we calculate the mass of ammonia. The molar mass of NH_3 is 17.04 g. The molar volume is 22.4 L at STP.

$$\frac{17.04 \text{ g}}{22.4 \text{ L}} = 0.761 \text{ g/L}$$

If we filled a balloon with ammonia gas (0.761 g/L), it would float in the air (1.29 g/L). This is similar to the observation that all fluids that are less dense float on fluids that are more dense; for example, gasoline floats on water.

Self-Test Exercise

Calculate the density of ozone, O_3, at STP.

Answer: 2.14 g/L

Example Exercise 9.9 • Molar Mass and Gas Density at STP

A fire extinguisher releases 1.96 g of an unknown gas that occupies 1.00 L at STP. What is the molar mass of the unknown gas?

Solution

Let's first write down the units of molar mass, that is, g/mol. In this problem we are given the density of the unknown gas: 1.96 g per 1.00 L. We can outline the calculation as follows.

$$\frac{1.96 \text{ g}}{1.00 \text{ L}} \times \frac{\text{unit}}{\text{factor}} = \frac{\text{g}}{\text{mol}}$$

To convert the units of L to mol, we apply the unit factor 22.4 L/1 mol.

$$\frac{1.96\ g}{1.00\ \cancel{L}} \times \frac{22.4\ \cancel{L}}{1\ mol} = 43.9\ g/mol$$

Because the unknown gas is from a fire extinguisher, we might suspect that it is carbon dioxide. By adding up the molar mass of CO_2 (44.01 g/mol), we help confirm that the unknown gas is carbon dioxide.

We can summarize the solution as follows.

$\frac{1.96\ g}{1.00\ L}$ → molar volume → $\frac{43.9\ g}{mol}$

Self-Test Exercise

Boron trifluoride gas is used to manufacture computer chips. Given that 1.51 g of the gas occupies 500.0 mL at STP, what is the molar mass of boron trifluoride?

Answer: 67.6 g/mol

9.6 Mole Calculations III

Objective · To relate the volume of a gas at STP to its mass and number of particles.

As mentioned previously, the mole is a central unit in chemical calculations. A mole has three interpretations.

1. A mole is Avogadro's number of particles.
2. A mole of a substance has a mass equal to its atomic mass in grams.
3. A mole of any gas at STP occupies a volume of 22.4 L.

We can state the three interpretations in the form of equations as follows.

$$1\ \text{mole} = 6.02 \times 10^{23}\ \text{particles}$$

$$1\ \text{mole} = \text{molar mass}$$

$$1\ \text{mole} = 22.4\ \text{L at STP}$$

From the preceding mole equations, we can relate the number of a particles of a substance to its mass or to its gaseous volume at STP. The process of interchanging these three quantities involves mole calculations.

The following example exercises illustrate calculations involving the number of moles of a substance. In these calculations, we will use the concepts of Avogadro's number, molar mass, and molar volume.

Example Exercise 9.10 • Mass and Volume of a Gas at STP

What is the mass of 3.36 L of ozone gas, O_3, at STP?

Solution

In mole calculations, we relate the given value to moles and then convert moles to the unit asked for in the answer.

Step 1, The unit asked for is **g O_3**.

Step 2, The given value is **3.36 L O_3**.

Step 3, We apply unit factors to convert the volume of ozone to the mass of O_3. The molar volume gives the relationship 1 mol = 22.4L; and the molar mass of O_3 from the periodic table is 1mol = 48.00g. Applying unit factors to cancel units, we have

(continued)

Example Exercise 9.10 *(continued)*

$$3.36 \cancel{L\ O_3} \times \frac{1\ mol\ O_3}{22.4 \cancel{L\ O_3}} \times \frac{48.00\ g\ O_3}{1 \cancel{mol\ O_3}} = 7.20\ g\ O_3$$

We can summarize the solution as follows.

Self-Test Exercise

What volume is occupied by 0.125 g of hydrogen sulfide gas, H_2S, at STP.

Answer: 0.0821 L H_2S (82.1 mL)

Example Exercise 9.11 • **Molecules and Gas Volume at STP**

How many molecules of hydrogen gas, H_2, occupy 0.500 L at STP?

Solution

In mole calculations, we relate the given value to moles and then convert moles to the unit asked for in the answer.
Step 1, The unit asked for is **molecules H_2**.
Step 2, The given value is **0.500 L H_2**.
Step 3, We apply unit factors to convert the volume of hydrogen to molecules of H_2. The molar volume gives the relationship 1 mol = 22.4 L; and Avogadro's number gives the relationship 1 mol = 6.02×10^{23} molecules. Applying unit factors to cancel units, we have

$$0.500 \cancel{L}\ H_2 \times \frac{1 \cancel{mol}}{22.4 \cancel{L}} \times \frac{6.02 \times 10^{23}\ molecules}{1 \cancel{mol}} = 1.34 \times 10^{22}\ molecules\ H_2$$

We can summarize the steps in the solution as follows.

Self-Test Exercise

What volume is occupied by 3.33×10^{21} atoms of helium gas, He, at STP?

Answer: 0.124 L He (124 mL)

9.7 Percent Composition

Objective · To calculate the percent composition of a compound given its chemical formula.

Previously, in Section 2.10 we defined percent as parts per 100 parts. Another way to define percent is to say that it is the ratio of a given quantity to the entire sample, all multiplied by 100. Now, let's apply this concept to the composition of a compound.

The **percent composition** is a list of the mass percent of each element in a compound. The percent composition of water, H_2O, for example, is 11% hydrogen and 89% oxygen. According to the law of definite composition discussed in Section 4.5, the elements in a compound are always present in the same proportion by mass. Therefore, water always contains 11% hydrogen and 89% oxygen regardless of the amount. A drop of water, a milliliter of water, and a pool of water all contain 11% hydrogen and 89% oxygen.

The percent of each element in water is given, but how are the values obtained? We can calculate the percent composition of water as follows. Let's begin by assuming that we have 1 mol of H_2O. A mole of H_2O contains 2 mol of hydrogen and 1 mol of oxygen. Thus,

$$2(\text{mol H}) + 1 \text{ mol O} = 1 \text{ mol } H_2O$$
$$2(\text{molar mass H}) + 1 \text{ molar mass O} = 1 \text{ molar mass } H_2O$$
$$2(1.01 \text{ g H}) + 16.00 \text{ g O} = \text{g } H_2O$$
$$2.02 \text{ g H} + 16.00 \text{ g O} = 18.02 \text{ g } H_2O$$

▲ **Percent Composition** The percent composition of H_2O is the same for a drop of water and a glass of water, that is, 11% hydrogen and 89% oxygen.

Next, we find the percent composition of water by comparing the molar masses of hydrogen and oxygen to the molar mass of the compound.

$$\frac{2.02 \text{ g H}}{18.02 \text{ g } H_2O} \times 100 = 11.2\% \text{ H}$$
$$\frac{16.00 \text{ g O}}{18.02 \text{ g } H_2O} \times 100 = 88.79\% \text{ O}$$

The following example exercise further illustrates calculation of the percent composition of a compound.

Example Exercise 9.12 • Percent Composition of a Substance

Trinitrotoluene (TNT) is a white, crystalline substance that explodes at 240°C. Calculate the percent composition of TNT, $C_7H_5(NO_2)_3$.

Solution

Let's calculate the percent composition assuming there is 1 mol of TNT. For compounds with parentheses, it is necessary to count the number of atoms of each element carefully. That is, 1 mol of $C_7H_5(NO_2)_3$ contains 7 mol of C atoms, 5 mol of H atoms, 3 mol of N atoms, and 6 mol of O atoms.

We begin the calculation by finding the molar mass of $C_7H_5(NO_2)_3$ as follows.

$$7(12.01 \text{ g C}) + 5(1.01 \text{ g H}) + 3(14.01 \text{ g N} + 32.00 \text{ g O}) = \text{g } C_7H_5(NO_2)_3$$
$$84.07 \text{ g C} + 5.05 \text{ g H} + 42.03 \text{ g N} + 96.00 \text{ g O} = 227.15 \text{ g } C_7H_5(NO_2)_3$$

Now, let's compare the mass of each element to the total molar mass of the compound, that is, 227.15 g.

$$\frac{84.07 \text{ g C}}{227.15 \text{ g } C_7H_5(NO_2)_3} \times 100 = 37.01\% \text{ C}$$
$$\frac{5.05 \text{ g H}}{227.15 \text{ g } C_7H_5(NO_2)_3} \times 100 = 2.22\% \text{ H}$$
$$\frac{42.03 \text{ g N}}{227.15 \text{ g } C_7H_5(NO_2)_3} \times 100 = 18.50\% \text{ N}$$
$$\frac{96.00 \text{ g O}}{227.15 \text{ g } C_7H_5(NO_2)_3} \times 100 = 42.26\% \text{ O}$$

The percent composition of TNT reveals that the explosive is mostly oxygen by mass. When we sum the individual percents (37.01 + 2.22 + 18.50 + 42.26 = 99.99%), we verify our calculation and find that the total is approximately 100%.

(continued)

Example Exercise 9.12 *(continued)*

Self-Test Exercise
Ethylenediaminetetraacetic acid (EDTA) is used as a food preservative and in the treatment of heavy-metal poisoning. Calculate the percent composition of EDTA, $C_{10}H_{16}N_2O_8$.

Answer: 41.09% C, 5.53% H, 9.59% N, and 43.79% O

Experiment #12, Prentice Hall Laboratory Manual

9.8 Empirical Formula

Objectives
- To calculate the empirical formula of a compound given experimental data for its synthesis.
- To calculate the empirical formula of a compound given its percent composition.

During the late 1700s, chemists experimented to see how elements reacted to form compounds. In particular, they were interested in the reactions of elements with oxygen to form oxides. By measuring the mass of an element and the mass of the oxide after reaction, chemists could determine the formula of the compound.

The **empirical formula** of a compound corresponds to the simplest whole–number ratio of ions in a formula unit or to the simplest ratio of atoms of each element in a molecule. Compounds with similar empirical formulas contain elements in the same group in the periodic table. Magnesium, calcium, and barium are in Group IIA/2 because their compounds have similar empirical formulas. For example, the empirical formulas of their oxides all have a metal-to-oxygen ratio of 1:1, that is, MgO, CaO, and BaO. In general, members of a family of elements in the periodic table all have oxides with similar empirical formulas.

The following example illustrates the determination of an empirical formula. A 1.640-g sample of radioactive radium was heated with oxygen to produce 1.755 g of radium oxide. By subtracting the mass of radium from the mass of the oxide (1.755 g–1.640 g), we find that the mass gain was 0.115 g. Thus, the radium reacted with 0.115 g of oxygen gas.

What is the empirical formula for radium oxide, Ra_xO_y? The empirical formula is the simplest whole-number ratio of radium and oxide ions in a formula unit of the compound. We can determine this ratio from the moles of each reactant. If the molar mass of radium is given as 226.03 g/mol, we can proceed as follows.

$$1.640 \cancel{\text{g Ra}} \times \frac{1 \text{ mol Ra}}{226.03 \cancel{\text{g Ra}}} = 0.00726 \text{ mol Ra}$$

We can calculate the moles of oxygen (16.00 g/mol) as follows.

$$0.115 \cancel{\text{g O}} \times \frac{1 \text{ mol O}}{16.00 \cancel{\text{g O}}} = 0.00719 \text{ mol O}$$

The mole ratio of the elements in radium oxide is: $Ra_{0.00726}O_{0.00719}$. We can simplify the mole ratio by dividing by the smaller number.

$$Ra_{\frac{0.00726}{0.00719}}O_{\frac{0.00719}{0.00719}} = Ra_{1.01}O_{1.00}$$

Since the empirical formula must be a small whole-number ratio, we round $Ra_{1.01}O_{1.00}$ to RaO. The slight discrepancy from whole numbers is due to experimental error. Notice that radium has an empirical formula similar to that of other Group IIA/2 elements.

Example Exercise 9.13 • Empirical Formula from Mass

In a laboratory experiment, 0.500 g of scandium was heated and allowed to react with oxygen from the air. The resulting product oxide had a mass of 0.767 g. What is the empirical formula for scandium oxide, Sc_xO_y?

Solution

The empirical formula is the whole-number ratio of scandium and oxygen in the compound scandium oxide. This ratio is experimentally determined from the moles of each reactant. The moles of scandium are calculated as follows.

$$0.500\ \cancel{g\ Sc} \times \frac{1\ mol\ Sc}{44.96\ \cancel{g\ Sc}} = 0.0111\ mol\ Sc$$

The moles of oxygen are calculated after first subtracting the mass of Sc from the product, Sc_xO_y.

$$0.767\ g\ Sc_xO_y - 0.500\ g\ Sc = 0.267\ g\ O$$

The moles of oxygen are calculated from the mass of oxygen that reacted.

$$0.267\ \cancel{g\ O} \times \frac{1\ mol\ O}{16.00\ \cancel{g\ O}} = 0.0167\ mol\ O$$

The mole ratio in scandium oxide is $Sc_{0.0111}O_{0.0167}$. To simplify the ratio and obtain small whole numbers, we divide by the smaller number.

$$Sc_{\frac{0.0111}{0.0111}}O_{\frac{0.0167}{0.0111}} = Sc_{1.00}O_{1.50}$$

We cannot round off the experimental ratio, $Sc_{1.00}O_{1.50}$, but we can double the ratio to obtain $Sc_{2.00}O_{3.00}$. Thus, the empirical formula is Sc_2O_3.

Self-Test Exercise

Iron can react with chlorine gas to give two different compounds, $FeCl_2$ and $FeCl_3$. Under given conditions, 0.558 g of metallic iron reacts with chlorine gas to yield 1.621 g of iron chloride. Which iron compound is produced in the experiment?

Answer: $FeCl_3$

Empirical Formulas from Percent Composition

Experiment #13, Prentice Hall Laboratory Manual

Benzene was a common liquid solvent until the Environmental Protection Agency (EPA) discovered that it was a carcinogen. Let's calculate the formula for benzene if its percent composition is 92.2% carbon and 7.83% hydrogen.

Empirical Formula Activity

The empirical formula expresses the simplest whole-number ratio of carbon to hydrogen atoms in a molecule of benzene. To calculate the moles of each element, we will assume we have 100 g of sample. Thus, in 100 g of benzene there are 92.2 g of carbon and 7.83 g of hydrogen. That is, the percentage of each element corresponds to its mass in 100 g of the compound. We find the moles of C and H as we did in the previous examples.

$$92.2\ \cancel{g\ C} \times \frac{1\ mol\ C}{12.01\ \cancel{g\ C}} = 7.68\ mol\ C$$

$$7.83\ \cancel{g\ H} \times \frac{1\ mol\ H}{1.01\ \cancel{g\ H}} = 7.75\ mol\ H$$

The mole ratio of the elements in benzene is $C_{7.68}H_{7.75}$. We can simplify the mole ratio by dividing both values by the smaller number, 7.68.

$$C_{\frac{7.68}{7.68}}H_{\frac{7.75}{7.68}} = C_{1.00}H_{1.01}$$

We can round off the ratio $C_{1.00}H_{1.01}$ to CH in order to obtain the empirical formula for benzene. The following example exercise further illustrates calculation of an empirical formula from percent composition data.

Example Exercise 9.14 • Empirical Formula from Percent Composition

Glycine is an amino acid found in protein. An analysis of glycine gave the following data: 32.0% carbon, 6.7% hydrogen, 18.7% nitrogen, and 42.6% oxygen. Calculate the empirical formula of the amino acid.

Solution

If we assume a 100-g sample, the percentage of each element equals its mass in 100 g of glycine, that is, 32.0 g C, 6.7 g H, 18.7 g N, and 42.6 g O. We can determine the empirical formula as follows.

$$32.0 \cancel{\text{g C}} \times \frac{1 \text{ mol C}}{12.01 \cancel{\text{g C}}} = 2.66 \text{ mol C}$$

$$6.7 \cancel{\text{g H}} \times \frac{1 \text{ mol H}}{1.01 \cancel{\text{g H}}} = 6.6 \text{ mol H}$$

$$18.7 \cancel{\text{g N}} \times \frac{1 \text{ mol N}}{14.01 \cancel{\text{g N}}} = 1.33 \text{ mol N}$$

$$42.6 \cancel{\text{g O}} \times \frac{1 \text{ mol O}}{16.00 \cancel{\text{g O}}} = 2.66 \text{ mol O}$$

The mole ratio of the elements in the amino acid is $C_{2.66}H_{6.6}N_{1.33}O_{2.66}$. We can find a small whole-number ratio by dividing by the smallest number.

$$C_{\frac{2.66}{1.33}}H_{\frac{6.6}{1.33}}N_{\frac{1.33}{1.33}}O_{\frac{2.66}{1.33}} = C_{2.00}H_{5.0}N_{1.00}O_{2.00}$$

Simplifying, we find that the empirical formula for the amino acid glycine is $C_2H_5NO_2$.

Self-Test Exercise

Calculate the empirical formula for caffeine given the following percent composition: 49.5% C, 5.15% H, 28.9% N, and 16.5% O.

Answer: $C_4H_5N_2O$

9.9 Molecular Formula

Objective • To calculate the molecular formula for a compound given its empirical formula and molar mass.

Molecular compounds are represented by individual molecules composed of nonmetal atoms. To see what this means in terms of their empirical formulas, recall that in the previous section we found that the empirical formula for benzene is CH. The empirical formula for benzene—1 C atom and 1 H atom—does not represent a stable molecule. Instead, it only illustrates the relative number of carbon and hydrogen atoms in a molecule. The actual **molecular formula** for benzene must therefore be some multiple of the empirical formula. We can represent the formula as $(CH)_n$, where n indicates some multiple of CH that is 2 or more.

Now consider acetylene, used in oxyacetylene welding, which also has the empirical formula CH. Even though acetylene and benzene have totally unrelated properties, each shares the same empirical formula, CH. In addition, the compound styrene, used in manufacturing Styrofoam cups, also shares the same empirical formula, CH.

If benzene, acetylene, and styrene are different compounds, their molecular formulas must be different. Experiments have provided the molar mass for each of these compounds. For benzene, the molar mass is 78 g/mol; for acetylene, 26 g/mol; and for styrene, 104 g/mol. We can indicate the number of multiples of the empirical formula for each compound as follows.

$$\text{Benzene:} \quad (CH)_n = 78\ \text{g/mol}$$
$$\text{Acetylene:} \quad (CH)_n = 26\ \text{g/mol}$$
$$\text{Styrene:} \quad (CH)_n = 104\ \text{g/mol}$$

The mass of the empirical formula CH is found by adding the molar masses of C and H; that is, 12 g + 1 g = 13 g/mol. Now we can determine how many multiples of the empirical formula are in benzene. We have

$$\text{Benzene:} \quad \frac{(CH)_n}{CH} = \frac{78\ \text{g/mol}}{13\ \text{g/mol}}$$
$$n = 6$$

Therefore, the molecular formula of benzene is $(CH)_6$, which we can write as C_6H_6. Similarly, we can find the molecular formula for acetylene.

$$\text{Acetylene:} \quad \frac{(CH)_n}{CH} = \frac{26\ \text{g/mol}}{13\ \text{g/mol}}$$
$$n = 2$$

The molecular formula of acetylene is $(CH)_2$ or C_2H_2. In a similar fashion, we can find the number of multiples of the empirical formula in styrene.

$$\text{Styrene:} \quad \frac{(CH)_n}{CH} = \frac{104\ \text{g/mol}}{13\ \text{g/mol}}$$
$$n = 8$$

Thus, the molecular formula of styrene is $(CH)_8$ or C_8H_8.

Even though benzene, acetylene, and styrene have the same empirical formula, their molecular formulas are different. Thus, benzene, acetylene, and styrene are different compounds. The following example exercise further illustrates how to determine a molecular formula from an empirical formula.

Example Exercise 9.15 • Determing the Molecular Formula from the Empirical Formula

The empirical formula for fructose, or fruit sugar, is CH_2O. If the molar mass of fructose is 180 g/mol, find the actual molecular formula for the sugar.

Solution

We can indicate the molecular formula of fructose as $(CH_2O)_n$. The molar mass of the empirical formula CH_2O is 12 g C + 2(1 g H) + 16 g O = 30 g/mol. Thus, the number of multiples of the empirical formula is

(continued)

Example Exercise 9.15 *(continued)*

$$\text{Fructose: } \frac{(CH)_n}{CH} = \frac{180\text{ g/mol}}{30\text{ g/mol}}$$

$$n = 6$$

Thus, the molecular formula of fructose is $(CH_2O)_6$ or $C_6H_{12}O_6$.

Self-Test Exercise

Ethylene dibromide was used as a grain pesticide until it was banned. Calculate (a) the empirical formula, and (b) the molecular formula for ethylene dibromide given its approximate molar mass of 190 g/mol and its percent composition: 12.7% C, 2.1% H, and 85.1% Br.

Answers: (a) CH_2Br; (b) $C_2H_4Br_2$

Update • Avogadro's Number

What mass of pure silicon contains Avogadro's number of atoms?

In 1911 Ernest Rutherford determined a value for Avogadro's number based on the principle of radioactivity. Rutherford was studying the emission of alpha particles from radioactive radium and was able to count these particles using a Geiger counter, which had been invented by his assistant Hans Geiger. After the alpha particles decayed into helium, he measured the volume of the helium gas. From the data, Rutherford calculated a value of 6.11×10^{23} for Avogadro's number.

Currently, the most accurate value for Avogadro's number is 6.0221367×10^{23}. This experimental value was found with an X-ray method. X-rays were used to determine the spacing of atoms in an ultrapure crystal of silicon (see figure). After striking the nuclei of silicon atoms, the X rays were diffracted; that is, the path of the X-ray beam was altered. The volume occupied by a single atom can be calculated from the spacing between two adjacent X rays. The value for Avogadro's number is obtained by dividing the volume of an individual silicon atom into the total volume of a crystal with a mass equal to the atomic mass of silicon.

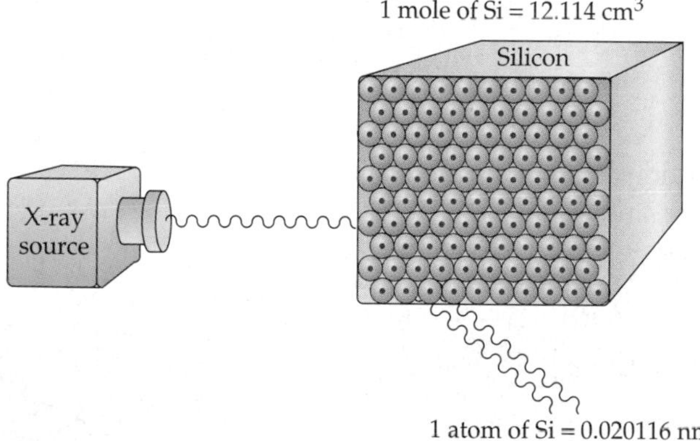

▲ **X-ray Diffraction of a Silicon Crystal** We can calculate the volume of a Si atom after using an X-ray beam to find the spacing of atoms in a silicon crystal.

We can illustrate how to calculate a value for Avogadro's number as follows. In the adjacent figure, notice that the volume occupied by each silicon atom is 0.020116 nm^3 and that the volume of the crystal is 12.114 cm^3. To find Avogadro's number we simply divide the volume of one atom of silicon into the volume of the silicon crystal. However, we must use the same units for the volumes of the atom and the crystal. For example, we can convert the volume of the crystal from cm^3 to nm^3.

$$12.114\text{ cm}^3 \times \left(\frac{1\text{ m}}{10^2\text{ cm}}\right)^3 \times \left(\frac{10^9\text{ nm}}{1\text{ m}}\right)^3 = \text{nm}^3$$

$$12.114\ \cancel{\text{cm}^3} \times \frac{1\ \cancel{\text{m}^3}}{10^6\ \cancel{\text{cm}^3}} \times \frac{10^{27}\text{ nm}^3}{1\ \cancel{\text{m}^3}} = 1.2114 \times 10^{22}\text{ nm}^3$$

When we divide the volume of the silicon atom into the volume of the crystal, we find the number of silicon atoms.

$$1.2114 \times 10^{22}\ \cancel{\text{nm}^3} \times \frac{1\text{ atom Si}}{0.020116\ \cancel{\text{nm}^3}} = 6.0221 \times 10^{23}\text{ atoms Si}$$

In this example, we obtained an experimental value of 6.0221×10^{23} atoms for Avogadro's number.

The mass of silicon containing Avogadro's number of atoms corresponds to the molar mass, that is, 28.09 g.

Note For molecular compounds, the empirical and molecular formulas are usually different. For ionic compounds, the actual formulas are almost always identical to the empirical formulas. For example, the actual formula for sodium chloride is NaCl, not Na_2Cl_2 or some other multiple. There is one common exception, however, mercury(I) compounds. Notice that the actual formula of mercury(I) chloride, Hg_2Cl_2, is twice the empirical formula.

Summary 

Section 9.1 **Avogadro's number (N)** is the value that corresponds to the number of atoms in 12.01 g of carbon. Moreover, the number of atoms in a gram atomic mass of any element is 6.02×10^{23}.

Section 9.2 A **mole (mol)** is the amount of substance that contains Avogadro's number of particles, that is, 6.02×10^{23} particles. A mole is the central unit in performing chemical formula calculations. These computations, based on moles of substance, are known as mole calculations.

Section 9.3 The mass of 1 mol of any substance is called the **molar mass**. The molar mass of an element corresponds to the atomic mass of the element. For instance, the molar mass of carbon is 12.01 g/mol, and the molar mass of oxygen is 16.00 g/mol. The molar mass of a compound corresponds to the sum of the gram atomic masses for all the atoms of each element in the compound. For example, the molar mass of carbon monoxide, CO, equals 28.01 g/mol.

Section 9.4 The mole is perhaps the most important unit in chemistry. A mole relates the number of particles (atoms, molecules, or formula units) to the amount of a substance. Given the moles of a substance, we can calculate the mass of the substance as well as the number of particles.

Section 9.5 **Avogadro's theory** states that equal volumes of gases, at the same conditions, contain equal numbers of molecules. The volume of 1 mol of a gaseous substance at standard conditions is called the **molar volume**. The molar volume any gas at **standard temperature and pressure (STP)** is 22.4 L/mol. Standard conditions are 0°C and 1 atm pressure. **Gas density** is found by dividing the molar mass of a gas by the molar volume; that is, we divide the molar mass by 22.4 L/mol.

Section 9.6 The mole concept relates the number of particles, molar mass, and molar volume of a gas. Given the numbers of liters of a gaseous substance, we can apply the mole concept to find the mass of the substance as well as the number of molecules.

◀ **The Mole Concept** A mole of sugar, water, mercury, sulfur, salt, copper, lead (clockwise), and orange potassium dichromate (center). Each sample contains 6.02×10^{23} particles; however, the molar masses differ as follows: 342.34 g of $C_{12}H_{22}O_{11}$, 18.02 g of H_2O, 200.59 g of Hg, 32.07 g of S, 58.44 g of NaCl, 63.55 g of Cu, 207.2 g of Pb, and 294.20 g of $K_2Cr_2O_7$.

Section 9.7 The **percent composition** of a compound is found by comparing the mass contributed by each element to the molar mass of the substance. A list of the resulting percentages of each element is the percent composition of the compound.

Section 9.8 The **empirical formula** is the simplest whole-number ratio of the atoms of the elements in a compound. The empirical formula can be calculated from (1) experimental synthesis data, or (2) the percent composition of the compound. After calculating the moles of each element in a compound, the mole ratio is simplified to small whole numbers.

Section 9.9 The **molecular formula** is the actual ratio of the atoms in a molecular compound. For example, lactic acid and glucose each have the empirical formula CH_2O. The molecular formula of lactic acid, $C_3H_6O_3$, is three times the empirical formula. The actual formula of glucose, $C_6H_{12}O_6$, is six times the empirical formula.

Problem-Solving Organizer

Topic	Procedure	Example
Avogadro's Number Sec. 9.1	Express the atomic mass of an element in grams to find the mass of Avogadro's number of atoms.	Avogadro's number of Fe atoms has a mass of 55.85 g.
Mole Calculations I Sec. 9.2	**1.** Write down the unit asked for in the answer. **2.** Write down the related given value. **3.** Apply a unit factor to convert the given units to the units in the answer.	How many iron atoms are in 0.500 mol of Fe? $0.500 \cancel{\text{mol}} \times \frac{6.02 \times 10^{23} \text{ atoms}}{1 \cancel{\text{mol}}} = 3.01 \times 10^{23} \text{ atoms}$
Molar Mass Sec. 9.3	Sum the atomic masses of each atom of each element in a substance to calculate the molar mass.	What is the molar mass of iron(III) oxide, Fe_2O_3? $55.85\text{ g} + 55.85\text{ g} + 16.00\text{ g} + 16.00\text{ g} + 16.00\text{ g} = 159.70\text{ g}$
Mole Calculations II Sec. 9.4	**1.** Write down the unit asked for in the answer. **2.** Write down the related given value. **3.** Apply a unit factor to convert the given units to the units in the answer.	What is the mass of 5.00×10^{23} formula units of Fe_2O_3? $5.00 \times 10^{23} \cancel{\text{formula units}} \times \frac{1 \cancel{\text{mol}}}{6.02 \times 10^{23} \cancel{\text{formula units}}}$ $\times \frac{159.70 \text{ g } Fe_2O_3}{1 \cancel{\text{mol}}} = 132 \text{ g } Fe_2O_3$
Molar Volume Sec. 9.5	The volume occupied by 1 mol of any gas at STP is 22.4 L.	The volume of 1.00 mol of O_2 at STP is 22.4 L.
Mole Calculations III Sec. 9.6	**1.** Write down the unit asked for in the answer. **2.** Write down the related given value. **3.** Apply a unit factor to convert the given units to the units in the answer.	What is the mass of 2.50 L of O_2 gas at STP? $2.50 \cancel{\text{L}} \times \frac{1 \cancel{\text{mol}}}{22.4 \cancel{\text{L}}} \times \frac{32.00 \text{ g } O_2}{1 \cancel{\text{mol}}} = 3.57 \text{ g } O_2$
Percent Composition Sec. 9.7	Find the mass of each element compared to the total molar mass of the compound, all multiplied by 100.	What is the percent composition of hydrogen peroxide, H_2O_2 (MM = 34.02 g/mol)? $\frac{2(1.01) \text{ g}}{34.02 \text{ g}} \times 100 = 5.94\%\text{ H}$ $\frac{2(16.00) \text{ g}}{34.02 \text{ g}} \times 100 = 94.06\%\text{ O}$

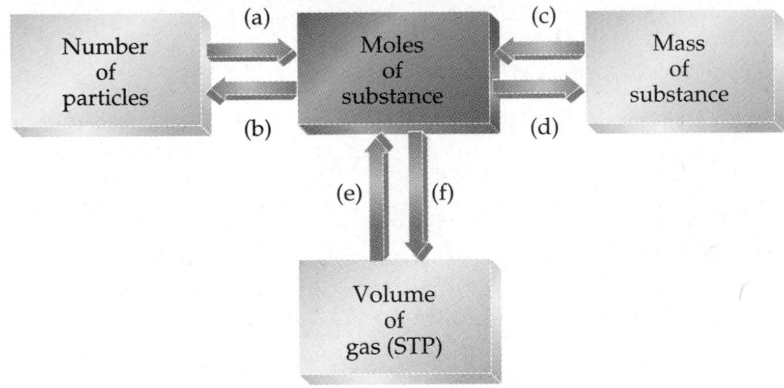

(a) Use N as a unit factor: multiply by $1\ \text{mol}/6.02 \times 10^{23}$
(b) Use N as a unit factor: multiply by $6.02 \times 10^{23}/1\ \text{mol}$
(c) Use molar mass as a unit factor: multiply by 1 mol/g
(d) Use molar mass as a unit factor: multiply by g/1 mol
(e) Use molar volume as a unit factor: multiply by 1 mol/22.4 L
(f) Use molar volume as a unit factor: multiply by 22.4 L/1 mol

◀ **Mole Calculations** The six possible steps (a), (b), (c), (d), (e), and (f) outline the relationships between mole quantities. The mole is the central unit for these conversions.

Key Concepts*

1. If a computer can count at the rate of one number per nanosecond, can it count from 1 to 6.02×10^{23} in a hundred years?
2. If an extremely sensitive electronic balance can weigh samples having a mass of 1 μg or greater, can it weigh a billion iron atoms?
3. What is the mass of an average iron atom expressed in atomic mass units?
4. What is the mass of Avogadro's number of iron atoms expressed in grams?
5. What is the number of particles in (a) 1.00 mol of Fe, (b) 1.00 mol of O_2 gas, and (c) 1.00 mol of FeO?
6. What is the mass of 6.02×10^{23} molecules of glucose, $C_6H_{12}O_6$?
7. Which of the following 1.00-mol samples of gas has a volume of 22.4 L at STP: He, H_2, CH_4?
8. What is the volume of 6.02×10^{23} molecules of carbon dioxide gas, CO_2, at STP?
9. Assume the two balloons have equal volumes and explain why the N_2 balloon sinks while the He balloon floats.

▲ Gas Balloons

10. If 0.500 mol of yellow sulfur powder reacts with 0.500 mol of oxygen gas, what is the empirical formula of the product?
11. Galactose is a sugar found in milk. What is the molecular formula of galactose if the empirical formula is $C_1H_2O_1$ and the approximate molar mass is 180 g/mol?

*Answers to Key Concepts are in Appendix H.

Key Terms†

Select the key term below that corresponds to each of the following definitions.

______ 1. the value that corresponds to the number of atoms in 12.01 g of carbon, that is, 6.02×10^{23} individual particles

______ 2. the amount of substance that contains 6.02×10^{23} particles

______ 3. the mass of 1 mol of pure substance expressed in grams; the mass of Avogadro's number of atoms, molecules, or formula units

______ 4. the volume occupied by 1 mol of any gas at standard conditions

______ 5. a temperature of 0°C and 1 atm pressure that has been chosen as standard conditions

______ 6. equal volumes of gases, under the same conditions of temperature and pressure, contain equal numbers of molecules

______ 7. the ratio of mass per unit volume for a gas, expressed in grams per liter

______ 8. a list of the elements in a compound and the mass percent of each

______ 9. the chemical formula of a compound that expresses the simplest ratio of the atoms in a molecule, or ions in a formula unit

______ 10. the chemical formula of a compound that expresses the actual number of atoms present in a molecule

(a) Avogadro's number (*N*) (*Sec. 9.1*)
(b) Avogadro's theory (*Sec. 9.5*)
(c) empirical formula (*Sec. 9.8*)
(d) gas density (*Sec. 9.5*)
(e) molar mass (MM) (*Sec. 9.3*)
(f) molar volume (*Sec. 9.5*)
(g) mole (mol) (*Sec. 9.2*)
(h) molecular formula (*Sec. 9.9*)
(i) percent composition (*Sec. 9.7*)
(j) standard temperature and pressure (STP) (*Sec. 9.5*)

Exercises‡

Avogadro's Number (Sec. 9.1)

1. Refer to the periodic table and state the average mass (in amu) of one atom for each of the following elements.
   (a) H **(b)** Li
   (c) C **(d)** P
2. Refer to the periodic table and state the average mass (in amu) of one atom for each of the following elements.
   (a) Ca **(b)** Zn
   (c) As **(d)** Bi
3. Refer to the periodic table and state the average mass of 6.02×10^{23} atoms of each of the following elements.
   (a) H **(b)** Li
   (c) C **(d)** P
4. Refer to the periodic table and state the average mass of 6.02×10^{23} atoms of each of the following elements.
   (a) Ca **(b)** Zn
   (c) As **(d)** Bi

Mole Calculations I (Sec. 9.2)

5. State the number of particles in 1 mol of each of the following.
   (a) 1 mol of manganese atoms, Mn
   (b) 1 mol of manganese nitrate formula units, $Mn(NO_3)_2$
6. State the number of particles in 1 mol of each of the following.
   (a) 1 mol of nitrogen molecules, N_2
   (b) 1 mol of nitrogen dioxide molecules, NO_2
7. State the number of moles represented by each of the following.
   (a) 6.02×10^{23} atoms of copper, Cu
   (b) 6.02×10^{23} formula units of copper(II) sulfate, $CuSO_4$
8. State the number of moles represented by each of the following.
   (a) 6.02×10^{23} molecules of fluorine, F_2
   (b) 6.02×10^{23} molecules of sulfur hexafluoride, SF_6
9. Calculate the number of particles in each of the following.
   (a) 0.335 mol titanium atoms, Ti
   (b) 0.112 mol carbon dioxide molecules, CO_2
   (c) 1.935 mol zinc chloride formula units, $ZnCl_2$
10. Calculate the number of particles in each of the following.
    (a) 2.12 mol argon atoms, Ar
    (b) 7.10 mol nitrogen trifluoride molecules, NF_3
    (c) 0.552 mol silver sulfate formula units, Ag_2SO_4
11. Calculate the number of moles containing each of the following.

† Answers to Key Terms are in Appendix I.

‡ Answers to odd-numbered Exercises are in Appendix J.

(a) 4.15×10^{22} atoms of iron, Fe
(b) 3.31×10^{21} molecules of bromine, Br_2
(c) 4.19×10^{20} formula units of cadmium nitrate, $Cd(NO_3)_2$

12. Calculate the number of moles containing each of the following.
(a) 7.88×10^{24} atoms of selenium, Se
(b) 5.55×10^{25} molecules of hydrogen sulfide, H_2S
(c) 2.25×10^{22} formula units of strontium carbonate, $SrCO_3$

Molar Mass (Sec. 9.3)

13. State the molar mass for each of the following elements.
(a) mercury, Hg 200.59 (b) silicon, Si
(c) bromine, Br_2 160 (d) phosphorus, P_4

14. State the molar mass for each of the following elements.
(a) copper, Cu (b) selenium, Se
(c) iodine, I_2 (d) sulfur, S_8

15. Calculate the molar mass for each of the following ionic compounds.
(a) barium fluoride, BaF_2
(b) potassium sulfide, K_2S
(c) iron(III) acetate, $Fe(C_2H_3O_2)_3$
(d) strontium phosphate, $Sr_3(PO_4)_2$

16. Calculate the molar mass for each of the following molecular compounds.
(a) methane, CH_4
(b) phosphorus triiodide, PI_3
(c) diarsenic pentaoxide, As_2O_5
(d) glycerin, $C_3H_5(OH)_3$

Mole Calculations II (Sec. 9.4)

17. Calculate the mass in grams for each of the following.
(a) 2.95×10^{23} atoms of mercury, Hg
(b) 1.16×10^{22} molecules of nitrogen, N_2
(c) 5.05×10^{21} formula units of barium chloride, $BaCl_2$

18. Calculate the mass in grams for each of the following.
(a) 1.21×10^{24} atoms krypton, Kr
(b) 6.33×10^{22} molecules of dinitrogen oxide, N_2O
(c) 4.17×10^{21} formula units of magnesium perchlorate, $Mg(ClO_4)_2$

19. Calculate the number of particles in each of the following.
(a) 1.50 g potassium, K
(b) 0.470 g oxygen, O_2
(c) 0.555 g silver chlorate, $AgClO_3$

20. Calculate the number of particles in each of the following.
(a) 7.57 g platinum, Pt
(b) 3.88 g ethane, C_2H_6
(c) 0.152 g aluminum chloride, $AlCl_3$

21. Calculate the mass in grams for a single atom of the following elements.
(a) beryllium, Be (b) sodium, Na
(c) cobalt, Co (d) arsenic, As

22. Calculate the mass in grams for a single molecule of the following compounds.
(a) methane, CH_4 (b) ammonia, NH_3
(c) sulfur trioxide, SO_3 (d) dinitrogen pentaoxide, N_2O_5

Molar Volume (Sec. 9.5)

23. State standard conditions for a gas in degrees Celsius and atmospheres.

24. State standard conditions for a gas in Kelvin units and millimeters Hg. (*Hint:* 1 atm = 76 cm Hg)

25. Calculate the density for each of the following gases at STP.
(a) neon, Ne (b) chlorine, Cl_2
(c) nitrogen dioxide, NO_2 (d) hydrogen iodide, HI

26. Calculate the density for each of the following gases at STP.
(a) xenon, Xe (b) fluorine, F_2
(c) propane, C_3H_8 (d) sulfur trioxide, SO_3

27. Calculate the molar mass for each of the following gases given the STP density.
(a) ethane, 1.34 g/L (b) diborane, 1.23 g/L
(c) Freon-12, 5.40 g/L (d) nitrous oxide, 2.05 g/L

28. Calculate the molar mass for each of the following gases given the STP density.
(a) isobutane, 2.59 g/L (b) silane, 1.43 g/L
(c) Freon-22, 3.86 g/L (d) nitric oxide, 1.34 g/L

29. Given 1 mol of each gas listed, complete the following table.

Gas	Molecules	Mass	Volume at STP
fluorine, F_2			
hydrogen fluoride, HF			
silicon tetrafluoride, SiF_4			
oxygen difluoride, OF_2			

30. Given 1 mol of each gas listed, complete the following table.

Gas	Molecules	Mass	Volume at STP
bromine, Br_2			
hydrogen sulfide, H_2S			
phosphine, PH_3			
butane, C_4H_{10}			

Mole Calculations III (Sec. 9.6)

31. Calculate the volume in liters for each of the following gases at STP.

(a) 0.250 g of helium, He
(b) 5.05 g of nitrogen, N_2

32. Calculate the volume in liters for each of the following gases at STP.
(a) 2.22×10^{22} molecules of methane, CH_4
(b) 4.18×10^{24} molecules of ethane, C_2H_6

33. Calculate the mass in grams for each of the following gases at STP.
(a) 1.05 L of hydrogen sulfide, H_2S
(b) 5.33 L of dinitrogen trioxide, N_2O_3

34. Calculate the mass in grams for each of the following gases at STP.
(a) 5.42×10^{22} molecules of propane, C_3H_8
(b) 1.82×10^{23} molecules of butane, C_4H_{10}

35. Calculate the number of molecules in each of the following gases at STP.
(a) 100.0 mL of hydrogen, H_2
(b) 70.5 mL of ammonia, NH_3

36. Calculate the number of molecules in each of the following gases at STP.
(a) 0.150 g of carbon monoxide, CO
(b) 2.75 g of nitrogen monoxide, NO

37. Use the given quantity for each gas listed to complete the following table.

Gas	Molecules	Atoms	Mass	Volume at STP
N_2	1.35×10^{23}			
NO_2		4.06×10^{23}		
NO			6.75 g	
N_2O_4				5.02 L

38. Use the given quantity for each gas listed to complete the following table.

Gas	Molecules	Atoms	Mass	Volume at STP
HCl	1.15×10^{22}			
Cl_2		4.27×10^{24}		
Cl_2O			10.0 g	
ClO_2				0.282 L

Percent Composition (Sec. 9.7)

39. Benzoyl peroxide is the active ingredient in a popular acne cream. Calculate the percent composition of benzoyl peroxide, $C_7H_6O_3$.

40. Dynamite is nitroglycerin in a porous material such as cellulose. Calculate the percent composition of nitroglycerin, $C_3H_5O_3(NO_2)_3$.

41. The illegal drug cocaine has the chemical formula $C_{17}H_{21}NO_4$. Calculate the percent composition of the compound.

42. The amino acid methionine has the chemical formula $C_5H_{11}NSO_2$. Calculate the percent composition of the compound.

43. Mustard gas has been used as a weapon in chemical warfare. Find the percent composition of the active ingredient, $C_4H_8SCl_2$, in mustard gas.

44. Chlorophyll is a dark-green plant pigment. Calculate the percent composition of chlorophyll, $C_{55}H_{70}MgN_4O_6$.

45. Monosodium glutamate (MSG) is added to food to enhance the flavor. Find the percent composition of MSG given its formula, $NaC_5H_8NO_4$.

46. Mercurochrome has the chemical formula $HgNa_2C_{20}H_8Br_2O_6$. Calculate the percent composition for the compound.

Empirical Formula (Sec. 9.8)

47. In an experiment 0.500 g of tin reacted with nitric acid to give tin oxide. If the oxide had a mass of 0.635 g, what is the empirical formula of tin oxide?

48. In an experiment 0.500 g of nickel reacted with air to give 0.704 g of nickel oxide. What is the empirical formula of the oxide?

49. In an experiment 1.550 g of mercury oxide decomposed to give oxygen gas and 1.435 g of liquid mercury. What is the empirical formula of mercury oxide?

50. In an experiment 2.410 g of copper oxide produced 1.925 g of copper metal after heating with hydrogen gas. What is the empirical formula of copper oxide?

51. A 1.115-g sample of cobalt was heated with sulfur to give 2.025 g of cobalt sulfide. What is the empirical formula of cobalt sulfide?

52. A 0.715-g sample of titanium was heated with chlorine gas to give 2.836 g of titanium chloride. What is the empirical formula of titanium chloride?

53. Calculate the empirical formula for the following ionic compounds given their percent composition.
(a) manganese fluoride, 59.1% Mn and 40.9% F
(b) copper chloride, 64.1% Cu and 35.9% Cl
(c) tin bromide, 42.6% Sn and 57.4% Br

54. Calculate the empirical formula for the following ionic compounds given their percent composition.
(a) potassium superoxide, 55.0% K and 45.0% O
(b) vanadium oxide, 68.0% V and 32.0% O
(c) bismuth oxide, 89.7% Bi and 10.3% O

55. Trichloroethylene (TCE) is a common solvent used to degrease machine parts. Calculate the empirical formula for TCE if the percent composition is 18.25% C, 0.77% H, and 80.99% Cl.

56. Dimethyl sulfoxide (DMSO) is a liniment for horses and has been used in the treatment of arthritis for humans. Calculate the empirical formula for DMSO, given that the percent composition is 30.7% C, 7.74% H, 20.5% O, and 41.0% S.

Molecular Formula (Sec. 9.9)

57. Aspirin has a molar mass of 180 g/mol. If the empirical formula is $C_9H_8O_4$, what is the molecular formula of aspirin?

58. Quinine is used to treat malaria and has a molar mass of 325 g/mol. If the empirical formula is $C_{10}H_{12}NO$, what is the molecular formula of quinine?

59. Adipic acid is used to manufacture nylon. If the molar mass is 147 g/mol and the empirical formula is $C_3H_5O_2$, what is the molecular formula of adipic acid?

60. Hexamethylenediamine is used to manufacture nylon-6. If the molar mass is approximately 115 g/mol and the empirical formula is C_3H_8N, what is the molecular formula of the compound?

61. Ethylene glycol is used as permanent antifreeze. If the molar mass is 62 g/mol and the percent composition is 38.7% C, 9.74% H, and 51.6% O, what is the molecular formula of ethylene glycol?

62. Dioxane is a common solvent for plastics. If the molar mass is 88 g/mol and the percent composition is 54.5% C, 9.15% H, and 36.3% O, what is the molecular formula of dioxane?

63. Lindane is an insecticide. If the molar mass is 290 g/mol and the percent composition of 24.8% C, 2.08% H, and 73.1% Cl, what is the molecular formula of lindane?

64. Mercurous chloride is a fungicide. If the molar mass is 470 g/mol and the percent composition is 85.0% Hg and 15.0% Cl, what is the molecular formula of mercurous chloride?

65. Nicotine is found in tobacco and has a molar mass of 160 g/mol. If the percent composition is 74.0% C, 8.70% H, and 17.3% N, what is the molecular formula of nicotine?

66. Galactose is referred to as cerebrose or brain sugar and has a molar mass of 180 g/mol. If the percent composition is 40.0% C, 6.72% H, and 53.3% O, what is the molecular formula of galactose?

General Exercises

67. A mole of electrons is referred to as a Faraday after the English scientist Michael Faraday. How many electrons are in a Faraday?

68. A mole of photons is referred to as an Einstein after the German physicist Albert Einstein. How many photons are in an Einstein?

69. A quadrillion is the approximate number of red blood cells in 50,000 people. Which is greater: a quadrillion, 1×10^{15}, red blood cells or the number of nickel atoms in a 5-g nickel coin?

70. By rubbing the Lincoln profile on a penny coin, a student removed just enough mass to be detected by an analytical balance, 0.0001 g. How many copper atoms were rubbed off the coin?

71. Which weighs more: a mole of furry moles or the Earth? Assume the average rodent mole has a mass of 100 g. The mass of the Earth is 6×10^{24} kg.

72. Which is longer: a mole of furry moles (head to tail) or 10 light-years? Assume the average rodent mole has a length of 17 cm. A light-year is the distance light travels in a year, 9.5×10^{12} km.

73. In 1871 Mendeleev predicted the undiscovered element *ekaaluminum*. In 1875 the element was discovered in Gaul (France) and given the name gallium. If 0.500 g of gallium reacts with oxygen gas to give 0.672 g of gallium oxide, what is its empirical formula?

74. In 1871 Mendeleev predicted the undiscovered element *ekasilicon*. In 1886 the element was discovered in Germany and given the name germanium. If 0.500 g of germanium reacts with chlorine gas to give 1.456 g of germanium chloride, what is its empirical formula?

75. Calculate the cubic centimeter volume occupied by one molecule of water in a mole of water. Recall that the density of water is 1.00 g/cm^3.

76. Calculate the milliliter volume occupied by one molecule of ethyl alcohol, C_2H_5OH, in a beaker of alcohol. The density of ethyl alcohol is 0.789 g/mL.

77. Calculate the number of carbon atoms in 1.00 g of table sugar, $C_{12}H_{22}O_{11}$.

78. Calculate the number of carbon atoms in 1.00 g of blood sugar, $C_6H_{12}O_6$.

79. Vitamin K has a molar mass of 173 g/mol and is 76.3% carbon by mass. How many carbon atoms are in one molecule of vitamin K?

80. What is the mass of rust, Fe_2O_3, that contains 10.0 g of iron?

81. The volume occupied by each copper atom in a 1-mol crystal is 0.0118 nm^3. If the density of the copper crystal is 8.92 g/cm^3, what is the experimental value of Avogadro's number?

82. Each atom in a crystal of aluminum metal occupies a theoretical cube that is 0.255 nm on a side. If the density of the aluminum crystal is 2.70 g/cm^3, what is the experimental value of Avogadro's number?

Explorer Quiz 1
Explorer Quiz 2
Explorer Quiz 3
Master Quiz

CHAPTER 10

Stoichiometry

▲ In the volcano reaction, a small amount of orange ammonium dichromate decomposes into an overflow amount of green chromium oxide, while releasing nitrogen gas and water vapor. Which weighs more, the orange crystals or the green powder?

After providing a mole concept foundation in Chapter 9, this chapter begins chemical equation calculations. After writing a balanced chemical equation for a reaction, stoichiometry calculations can be easily solved using unit analysis problem solving. It may be helpful to point out that stoichiometry problems may be direct (reactant to product) or the reverse (product to reactant). Note that all the mass–volume problems in this chapter involve gases at STP. The following Chapter 11 deals exclusively with the gas laws and introduces gases at non-STP conditions.

The manufacturing of chemicals is one of the most important industries in the United States. Each year billions of pounds of chemicals are produced for use both in the United States and around the world. These chemicals are used to manufacture medicines, computer chips and electronic instruments, fertilizers and pesticides, glass, paper, plastics, synthetic fibers, and other products.

Chemists and chemical engineers must routinely perform calculations based on balanced chemical equations to determine the cost of producing chemicals. These calculations are critical to the manufacture of products for the agricultural, electronic, medical, pharmaceutical, plastics, textile, and other industries. Each year the amounts of chemicals used industrially are published, and sulfuric acid consistently leads this list. In fact, the amount of sulfuric acid produced indicates the general activity of manufacturing and, for that reason is referred to as the barometer of the chemical industry.

10.1 Interpreting a Chemical Equation

Objective
- To relate the coefficients in a balanced chemical equation to:
  (a) moles of reactants and products
  (b) liters of gaseous reactants and products

Let's examine the information we can obtain from a balanced chemical equation. Consider nitrogen monoxide gas, which is present in automobile emissions. In the atmosphere, ultraviolet light catalyzes the reaction of nitrogen monoxide and oxygen to produce reddish-brown nitrogen dioxide smog. The balanced chemical equation is

$$2\,NO(g) + O_2(g) \xrightarrow{UV} 2\,NO_2(g)$$

Interpreting a Chemical Equation, $NO+O_2$

The Mole Interpretation of Equation Coefficients

The coefficients in the balanced equation indicate that 2 molecules of NO react with 1 molecule of O_2 to produce 2 molecules of NO_2. These coefficients indicate relative numbers of reactant and product molecules. It therefore follows that multiples of these coefficients are in the same ratios. For example, 2000 molecules of NO react with 1000 molecules of O_2 to give 2000 molecules of NO_2. Let's consider an even larger number than 2000; let's consider Avogadro's number (symbol *N*) of molecules, that is, 6.02×10^{23} molecules. When we substitute this very large number of molecules for the coefficients, the equation remains balanced.

$$2\,N\,NO + 1\,N\,O_2 \longrightarrow 2\,N\,NO_2$$

We read the above equation as "two times Avogadro's number of nitrogen monoxide molecules reacts with Avogadro's number of oxygen molecules to produce two times Avogadro's number of nitrogen dioxide molecules." Since Avogadro's number is the number of molecules in 1 mol, we can write the equation in terms of moles of substance.

$$2\text{ mol NO} + 1\text{ mol }O_2 \longrightarrow 2\text{ mol }NO_2$$

Similarly, this equation reads "two moles of nitrogen monoxide react with one mole of oxygen to give two moles of nitrogen dioxide." Furthermore, the coefficients indicate the ratio of moles, or the **mole ratio**, of reactants and products in every balanced chemical equation.

▲ **Amedeo Avogadro** This stamp commemorates Avogadro and states his theory in Italian.

The Volume Interpretation of Equation Coefficients

According to **Avogadro's theory**, there are equal numbers of molecules in equal volumes of gas at the same temperature and pressure. It therefore follows that twice the number of molecules occupies twice the volume of gas. In the preceding equation, 2 molecules of NO react with 1 molecule of O_2 to give 2 molecules of NO_2.

Accordingly, 2 volumes of NO react with 1 volume of O_2 gas to give 2 volumes of NO_2 gas. If we choose the liter as our unit of volume, we can write the ratio of volumes of gas for the reaction.

$$2\,\text{L NO} + 1\,\text{L O}_2 \longrightarrow 2\,\text{L NO}_2$$

Regardless of the units, the numbers in the ratio of volumes correspond to the coefficients of the balanced chemical equation, that is, 2:1:2. For example, 2 mL of NO reacts with 1 mL of O_2 to give 2 mL of NO_2. Similarly, 20 mL of NO reacts with 10 mL of O_2 to give 20 mL of NO_2. Although the ratio of gaseous volumes is 20:10:20, the smallest whole-number ratio is 2:1:2. Table 10.1 summarizes the information that can be obtained from the coefficients of a balanced chemical equation.

Table 10.1 Interpretation of Chemical Equation Coefficients

For the General Equation	2A	+	3B	⟶	C	+	2D
The ratio of molecules is	2	:	3	:	1	:	2
The ratio of moles is	2	:	3	:	1	:	2
The ratio of volumes of gas is	2	:	3	:	1	:	2

The following example exercise illustrates the information that can be obtained by interpreting the coefficients in a balanced chemical equation.

Example Exercise 10.1 • Interpreting Chemical Equation Coefficients

Given the chemical equation for the combustion of methane, CH_4, balance the equation and interpret the coefficients in terms of (a) moles and (b) liters.

$$CH_4(g) + O_2(g) \xrightarrow{\text{spark}} CO_2(g) + H_2O(g)$$

Solution

The balanced chemical equation is

$$CH_4(g) + 2\,O_2(g) \xrightarrow{\text{spark}} CO_2(g) + 2\,H_2O(g)$$

(a) The coefficients in the equation (1:2:1:2) indicate the ratio of moles as well as molecules.

$$1\text{ mol } CH_4 + 2\text{ mol } O_2 \longrightarrow 1\text{ mol } CO_2 + 2\text{ mol } H_2O$$

(b) The coefficients in the equation (1:2:1:2) indicate the ratio of volumes of gases. If we express the volume in liters, we have

$$1\text{ L } CH_4 + 2\text{ L } O_2 \longrightarrow 1\text{ L } CO_2 + 2\text{ L } H_2O$$

Self-Test Exercise

Given the chemical equation for the combustion of propane, C_3H_8, balance the equation and interpret the coefficients in terms of (a) moles and (b) milliliters.

$$C_3H_8(g) + O_2(g) \xrightarrow{\text{spark}} CO_2(g) + H_2O(g)$$

Answers: (a) $1\text{ mol } C_3H_8 + 5\text{ mol } O_2 \longrightarrow 3\text{ mol } CO_2 + 4\text{ mol } H_2O$;
(b) $1\text{ mL } C_3H_8 + 5\text{ mL } O_2 \longrightarrow 3\text{ mL } CO_2 + 4\text{ mL } H_2O$

Verifying the Conservation of Mass Law

In 1787, after conducting numerous experiments, Antoine Lavoisier formulated the **law of conservation of mass**. He stated that mass is neither created nor destroyed during a chemical reaction. As a result, the combined masses of the reactants must equal the combined masses of the products.

We can verify the conservation of mass law for any balanced equation. Let's consider the previous equation for the formation of nitrogen dioxide.

$$2\ NO(g) + O_2(g) \longrightarrow 2\ NO_2(g)$$

According to the coefficients in the balanced equation, the ratio of the substances is 2:1:2. Therefore, the mole ratio of each substance is 2:1:2.

$$2\ \text{mol NO} + 1\ \text{mol } O_2 \longrightarrow 2\ \text{mol } NO_2$$

By summing the atomic masses for each substance, we find that the **molar mass** for NO is 30.01 g/mol; for O_2, 32.00 g/mol; and for NO_2, 46.01 g/mol. Substituting the molar mass for a mole of each substance gives

$$2\ (30.01\ \text{g}) + 1\ (32.00\ \text{g}) \longrightarrow 2\ (46.01\ \text{g})$$

Simplifying,

$$60.02\ \text{g} + 32.00\ \text{g} \longrightarrow 92.02\ \text{g}$$

and

$$92.02\ \text{g} \longrightarrow 92.02\ \text{g}$$

In this example, the total mass of the reactants is 92.02 g, which is equal to the total mass of the products. Thus, from our calculation, we have verified the conservation of mass law. The following example exercise further illustrates the conservation of mass law.

Example Exercise 10.2 • Verifying the Conservation of Mass Law

Verify the conservation of mass law using the molar masses of the reactants and products in the following chemical equation.

$$NaNO_3(s) \xrightarrow{\Delta} NaNO_2(s) + O_2(g)$$

Solution

Before we can verify the conservation of mass law, we must balance the chemical equation for the reaction

$$2\ NaNO_3(s) \longrightarrow 2\ NaNO_2(s) + O_2(g)$$

Next, we substitute the molar mass of each substance into the equation. The molar mass for $NaNO_3$ is 85.00 g/mol; for $NaNO_2$, 69.00 g/mol; and for O_2, 32.00 g/mol.

$$2\ (85.00\ \text{g}) \longrightarrow 2\ (69.00\ \text{g}) + 1\ (32.00\ \text{g})$$
$$170.00\ \text{g} \longrightarrow 138.00\ \text{g} + 32.00\ \text{g}$$
$$170.00\ \text{g} \longrightarrow 170.00\ \text{g}$$

We have verified the conservation of mass law because the mass of the reactant (170.00 g) is equal to the total mass of the products (170.00 g).

(continued)

Example Exercise 10.2 *(continued)*

Self-Test Exercise
Verify the conservation of mass law using the molar masses of the reactants and products in the following chemical equation.

$$KClO_3(s) \xrightarrow{\Delta} KCl(s) + O_2(g)$$

Answer: $2(122.55\text{ g}) \longrightarrow 2(74.55\text{ g}) + 3(32.00\text{ g}); 245.10\text{ g} \longrightarrow 245.10\text{ g}$

10.2 Mole–Mole Relationships

Objective · To relate the number of moles of two substances in a balanced chemical equation.

As we demonstrated in Section 10.1, the coefficients in a chemical equation indicate the mole ratio of the reactants and products. Consider the combination reaction of nitrogen and oxygen to give nitrogen monoxide. The equation is

$$N_2(g) + O_2(g) \xrightarrow{\Delta} 2\,NO(g)$$

From the coefficients in the balanced equation, we see that 1 mol of nitrogen reacts with 1 mol of oxygen to produce 2 mol of nitrogen monoxide. We can write several mole ratios for this equation.

Since 1 mol of N_2 reacts with 1 mol of O_2, the corresponding mole ratios are

$$\frac{1\text{ mol }N_2}{1\text{ mol }O_2} \quad \text{and} \quad \frac{1\text{ mol }O_2}{1\text{ mol }N_2}$$

Since 1 mol of N_2 produces 2 mol of NO, the corresponding mole ratios are

$$\frac{1\text{ mol }N_2}{2\text{ mol NO}} \quad \text{and} \quad \frac{2\text{ mol NO}}{1\text{ mol }N_2}$$

Since 1 mol of O_2 produces 2 mol of NO, the corresponding mole ratios are

$$\frac{1\text{ mol }O_2}{2\text{ mol NO}} \quad \text{and} \quad \frac{2\text{ mol NO}}{1\text{ mol }O_2}$$

Suppose we want to find out how many moles of oxygen react with 2.25 mol of nitrogen. As usual, we apply the unit analysis method of problem solving.

$$2.25\text{ mol }N_2 \times \frac{\text{unit}}{\text{factor}} = \text{mol }O_2$$

In order to cancel units, we select the mole ratio 1 mol O_2/1 mol N_2 as our unit factor.

$$2.25\ \cancel{\text{mol }N_2} \times \frac{1\text{ mol }O_2}{1\ \cancel{\text{mol }N_2}} = 2.25\text{ mol }O_2$$

We can also calculate the moles of nitrogen monoxide produced by the reaction. In order to cancel units, we select the mole ratio 2 mol NO/1 mol N_2 as our unit factor.

$$2.25\ \cancel{\text{mol }N_2} \times \frac{2\text{ mol NO}}{1\ \cancel{\text{mol }N_2}} = 4.50\text{ mol NO}$$

Whenever we have a balanced chemical equation, we can always convert from moles of one substance to moles of another substance using a mole ratio as a unit factor. Example Exercise 10.3 further illustrates the relationship between moles of reactants and moles of products.

Example Exercise 10.3 • Mole–Mole Relationship

Carbon monoxide is produced in an industrial blast furnace by passing oxygen gas over hot coal. The balanced equation is

$$2\,C(s) + O_2(g) \xrightarrow{\Delta} 2\,CO(g)$$

(a) How many moles of oxygen react with 2.50 mol of carbon?
(b) How many moles of CO are produced from 2.50 mol of carbon?

Solution

The mole ratio for reactants and products corresponds to the coefficients in the balanced chemical equation.

(a) From the balanced equation, 2 mol C = 1 mol O_2.

$$2.50\ \cancel{\text{mol C}} \times \frac{1\ \text{mol}\ O_2}{2\ \cancel{\text{mol C}}} = 1.25\ \text{mol}\ O_2$$

(b) From the balanced equation, 2 mol C = 2 mol CO.

$$2.50\ \cancel{\text{mol C}} \times \frac{2\ \text{mol CO}}{2\ \cancel{\text{mol C}}} = 2.50\ \text{mol CO}$$

We can summarize the solution visually as follows

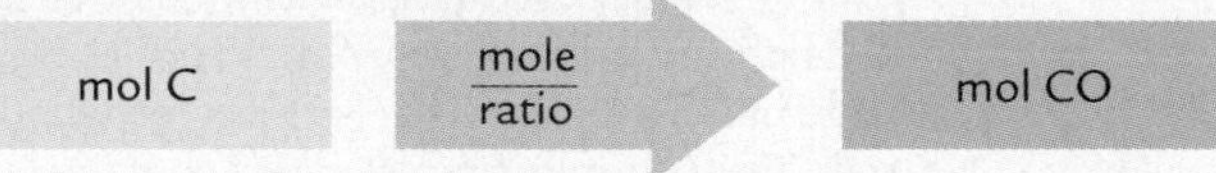

Self-Test Exercise

Iron is produced in an industrial blast furnace by passing carbon monoxide gas through molten iron(III) oxide. The balanced equation is

$$Fe_2O_3(l) + 3\,CO(g) \xrightarrow{\Delta} 2\,Fe(l) + 3\,CO_2(g)$$

(a) How many moles of carbon monoxide react with 2.50 mol of Fe_2O_3?
(b) How many moles of iron are produced from 2.50 mol Fe_2O_3?

Answers: (a) 7.50 mol CO; (b) 5.00 mol Fe

10.3 Types of Stoichiometry Problems

Objectives
- To classify the three basic types of stoichiometry problems: mass–mass, mass–volume, and volume–volume.
- To state the procedure for solving a stoichiometry problem given the balanced equation.

In Chapter 9, we performed mole calculations based on chemical formulas. In this chapter, we will perform mole calculations based on chemical equations. That is, we will apply mole ratios in order to relate quantities of reactants and products. The term **stoichiometry** is used to refer to the relationship between quantities in a chemical reaction according to the balance chemical equation. The term is derived from the Greek words *stoicheion*, meaning "element," and *metron*, meaning "measure."

Chemistry Connection · Manufacturing Iron

What is the difference between iron and steel?

The Iron Age began about 1000 B.C. in western Asia and Europe. The Iron Age followed the Bronze Age after it was discovered that iron could be made by heating iron ore with coal. Interestingly, iron is still obtained by a similar method.

Today, an industrial blast furnace converts iron ore to iron. In the process, iron ore, coke (a form of coal), and limestone, $CaCO_3$, are added into the top of a heated furnace while a blast of hot air is blown in at the bottom.

▲ **Blast Furnace** Molten pig iron reacts with a "blast" of oxygen gas which converts carbon and sulfur impurities into CO_2 and SO_2 gases.

Many reactions take place in the process, but first, oxygen in the hot air reacts with carbon in coke to produce carbon monoxide.

$$2\,C(s) + O_2(g) \longrightarrow 2\,CO(g)$$

Iron ore is an impure mixture of hematite, Fe_2O_3, and magnetite, Fe_3O_4. In the blast furnace hematite and magnetite are converted to iron by carbon monoxide gas. The reaction of hematite is as follows.

$$Fe_2O_3(l) + 3\,CO(g) \longrightarrow 2\,Fe(l) + 3\,CO_2(g)$$

The iron ore also contains silicon impurities which must be removed. This is done by reacting the impurities with CaO, which is produced from limestone, that is, $CaCO_3$. Silicon compounds such as $CaSiO_3$ form and are referred to as *slag*. This less-dense slag is easily removed, as it floats on the molten iron at the bottom of the furnace.

The molten iron obtained from the blast furnace is still not pure and is referred to as *pig iron*. Pig iron contains impurities such as carbon and sulfur. In a second stage of the process, these impurities are removed by blowing oxygen gas through the molten pig iron. The impurities are converted to oxides, and some are released as gases such as CO_2 and SO_2.

The iron obtained after the oxygen process contains a small amount of carbon. Manganese is then often added for strength and flexibility, and the resulting product is referred to as *steel*. A large number of alloy steels can be produced by adding other metals. For example, stainless steel contains traces of chromium and nickel.

Iron is a pure element,whereas steel is an alloy of iron with traces of carbon and other metals such as manganese.

Some stoichiometric calculations relate a given amount of reactant to an unknown amount of product. Conversely, other stoichiometric calculations relate a given amount of product to an unknown amount of reactant. Whether we relate a reactant to a product, or vice versa, the calculations use the same basic method of problem solving.

We can classify three types of stoichiometry problems based on the given quantity and the quantity to be calculated. Consider the general equation for any reaction, where a, b, c, and d are the respective coefficients for the substances A, B, C, and D.

$$a\,\mathrm{A} + b\,\mathrm{B} \longrightarrow c\,\mathrm{C} + d\,\mathrm{D}$$

If we are given grams of A and asked to find how many grams of C are produced, we classify this as a **mass–mass problem**. If we are given grams of A and asked to calculate the milliliters of a gas D, we classify this as a **mass–volume problem**. If we are given liters of a gas A and asked to find the liters of a gas D, we classify this as a **volume–volume problem**. In summary, stoichiometry problems can be classified as one of these basic types: mass–mass, mass–volume, or volume–volume.

Once we have classified a problem, we apply the unit analysis method of problem solving to its solution. Each type of problem has a solution that is characteristic of the unknown and given quantities. The following example exercise offers practice in classifying types of stoichiometry problems.

Example Exercise 10.4 • Classifying Stoichiometry Problems

Classify the type of stoichiometry problem for each of the following.

(a) How many grams of Zn metal react with hydrochloric acid to give 0.500 g of zinc chloride?

(b) How many liters of H_2 gas react with chlorine gas to yield 50.0 cm^3 of hydrogen chloride gas?

(c) How many kilograms of Fe react with sulfuric acid to produce 50.0 mL of hydrogen gas?

Solution

After analyzing a problem for the unknown quantity and the relevant given value, we classify the type of problem.

(a) The problem asks for grams of Zn (mass) that react to give 0.500 g of $ZnCl_2$ (mass). This is a *mass–mass* type of problem.

(b) The problem asks for liters of H_2 gas (volume) that react to yield 50.0 cm^3 of HCl gas (volume). This is a *volume–volume* type of problem.

(c) The problem asks for kilograms of Fe (mass) that react to produce 50.0 mL of H_2 gas (volume). This is a *mass–volume* type of problem.

Self-Test Exercise

Classify the type of stoichiometry problem for each of the following.

(a) How many grams of HgO decompose to give 0.500 L of oxygen gas at STP?

(b) How many grams of AgCl are produced from the reaction of 0.500 g of solid sodium chloride with silver nitrate solution?

(c) How many milliliters of H_2 gas react with nitrogen gas to yield 1.00 L of ammonia gas?

Answers: (a) mass–volume; (b) mass–mass; (c) volume–volume

▲ **Reaction of Zn Metal** Zinc metal reacts with hydrochloric acid to give $ZnCl_2$ and H_2 bubbles.

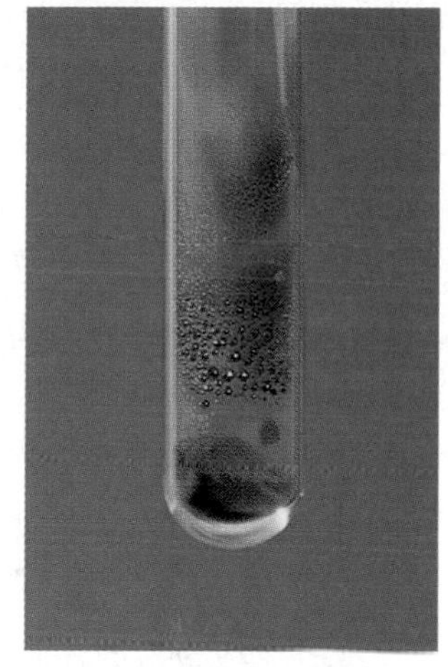

▲ **Decomposition of HgO** Orange mercury(II) oxide decomposes to give silver Hg liquid and O_2 gas.

10.4 Mass–Mass Problems

Objective · To perform mass–mass stoichiometry calculations.

A mass–mass stoichiometry problem is so named because an *unknown mass* of substance is calculated from a *given mass* of reactant or product in a chemical equation. After balancing the chemical equation, we proceed as follows.

(a) Convert the given mass of substance to moles using the molar mass of the substance as a unit factor.

(b) Convert the moles of the given to moles of the unknown using the coefficients in the balanced equation.

(c) Convert the moles of the unknown to grams using the molar mass of the substance as a unit factor.

To see how to solve a mass–mass problem, consider the high-temperature conversion of 14.4 g of iron(II) oxide to elemental iron with aluminum metal. The balanced equation for the reaction is

Experiment #14, Prentice Hall Laboratory Manual

$$3\,FeO(l) + 2\,Al(l) \xrightarrow{\Delta} 3\,Fe(l) + Al_2O_3(l)$$

To calculate the mass of aluminum necessary for the reaction, we must first find the moles of iron(II) oxide. Given that the molar mass of FeO is 71.85 g/mol, the unit factor conversion is as follows.

Experiment #15, Prentice Hall Laboratory Manual

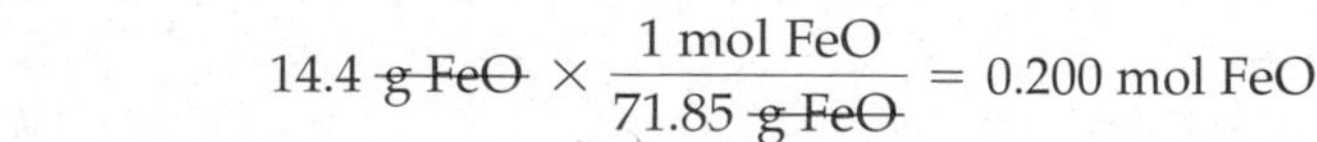

(a) $$14.4\ \cancel{g\ FeO} \times \frac{1\ mol\ FeO}{71.85\ \cancel{g\ FeO}} = 0.200\ mol\ FeO$$

Next, we convert moles of FeO to moles of Al by applying a mole ratio using the coefficients in the balanced equation. Since 3 mol FeO = 2 mol Al, the mole ratio unit factor is

(b) $$0.200 \cancel{\text{mol FeO}} \times \frac{2 \text{ mol Al}}{3 \cancel{\text{mol FeO}}} = 0.134 \text{ mol Al}$$

Last, we use the molar mass of aluminum (26.98 g/mol) as a unit factor to obtain the mass of Al reacting with 14.4 g of FeO.

(c) $$0.134 \cancel{\text{mol Al}} \times \frac{26.98 \text{ g Al}}{1 \cancel{\text{mol Al}}} = 3.60 \text{ g Al}$$

After you gain confidence in solving stoichiometry problems, it will be more convenient to perform one continuous calculation. For example, the solution to the above problem can be shown as

$$14.4 \cancel{\text{g FeO}} \times \underset{(a)}{\frac{1 \cancel{\text{mol FeO}}}{71.85 \cancel{\text{g FeO}}}} \times \underset{(b)}{\frac{2 \cancel{\text{mol Al}}}{3 \cancel{\text{mol FeO}}}} \times \underset{(c)}{\frac{26.98 \text{ g Al}}{1 \cancel{\text{mol Al}}}} = 3.60 \text{ g Al}$$

The following example exercises give additional illustrations of mass–mass stoichiometry calculations using unit factors.

Example Exercise 10.5 • Mass–Mass Stoichiometry

Calculate the mass of tin(IV) chloride produced by the reaction of 1.25 g of metallic tin with yellow chlorine gas.

$$Sn(s) + 2\,Cl_2(g) \longrightarrow SnCl_4(s)$$

Solution

After verifying that the equation is balanced, we first calculate the moles of tin. The molar mass of Sn, as shown in the periodic table, is 118.71 g/mol.

(a) $$1.25 \cancel{\text{g Sn}} \times \frac{1 \text{ mol Sn}}{118.71 \cancel{\text{g Sn}}} = 0.0105 \text{ mol Sn}$$

Next, we find the moles of $SnCl_4$ using the coefficients in the balanced equation. In this example, 1 mol Sn = 1 mol $SnCl_4$. Applying the mole ratio,

(b) $$0.0105 \cancel{\text{mol Sn}} \times \frac{1 \text{ mol SnCl}_4}{1 \cancel{\text{mol Sn}}} = 0.0105 \text{ mol SnCl}_4$$

Last, we calculate the mass of the product. The molar mass of $SnCl_4$ is 260.51 g/mol (118.71 + 35.45 + 35.45 + 35.45 + 35.45 = 260.51 g/mol).

(c) $$0.0105 \cancel{\text{mol SnCl}_4} \times \frac{260.51 \text{ g SnCl}_4}{1 \cancel{\text{mol SnCl}_4}} = 2.74 \text{ g SnCl}_4$$

Alternatively, we can solve this problem by one continuous calculation.

$$1.25 \cancel{\text{g Sn}} \times \underset{(a)}{\frac{1 \cancel{\text{mol Sn}}}{118.71 \cancel{\text{g Sn}}}} \times \underset{(b)}{\frac{1 \cancel{\text{mol SnCl}_4}}{1 \cancel{\text{mol Sn}}}} \times \underset{(c)}{\frac{260.51 \text{ g SnCl}_4}{1 \cancel{\text{mol SnCl}_4}}} = 2.74 \text{ g SnCl}_4$$

We can summarize the solution visually as follows.

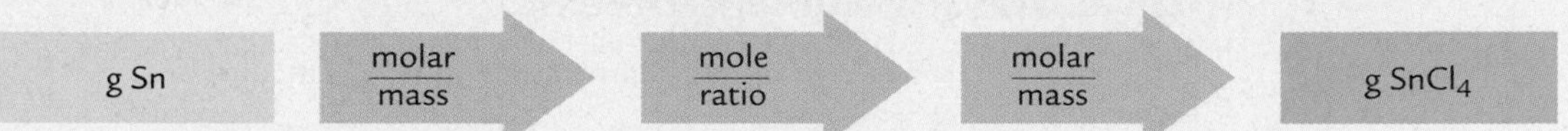

Self-Test Exercise

Calculate the mass of carbon dioxide released from the decomposition of 10.0 g of cobalt(III) carbonate given the *unbalanced* equation for the reaction.

$$Co_2(CO_3)_3(s) \xrightarrow{\Delta} Co_2O_3(aq) + CO_2(g)$$

Answer: 4.43 g CO_2

Example Exercise 10.6 • Mass–Mass Stoichiometry

Calculate the mass of potassium iodide (166.00 g/mol) required to yield 1.78 g of mercury(II) iodide precipitate (454.39 g/mol).

$$2\,KI(aq) + Hg(NO_3)_2(aq) \longrightarrow HgI_2(s) + 2\,KNO_3(aq)$$

Solution

After verifying that the equation is balanced, we first calculate the moles of mercury(II) iodide. The molar mass of HgI_2 is given as 454.39 g/mol, and so

(a) $$1.78\ \cancel{\text{g HgI}_2} \times \frac{1\ \text{mol HgI}_2}{454.39\ \cancel{\text{g HgI}_2}} = 0.00392\ \text{mol HgI}_2$$

Next, we calculate the moles of potassium iodide required for the reaction. From the coefficients in the balanced equation we see that 2 mol KI = 1 mol HgI_2. Therefore, the mole ratio unit factor is

(b) $$0.00392\ \cancel{\text{mol HgI}_2} \times \frac{2\ \text{mol KI}}{1\ \cancel{\text{mol HgI}_2}} = 0.00783\ \text{mol KI}$$

Last, we calculate the mass of potassium iodide. The molar mass of KI is given as 166.00 g/mol, and we have

(c) $$0.00783\ \cancel{\text{mol KI}} \times \frac{166.00\ \text{g KI}}{1\ \cancel{\text{mol KI}}} = 1.30\ \text{g KI}$$

Alternatively, we can solve this problem in one continuous calculation.

$$1.78\ \cancel{\text{g HgI}_2} \times \underbrace{\frac{1\ \cancel{\text{mol HgI}_2}}{454.39\ \cancel{\text{g HgI}_2}}}_{(a)} \times \underbrace{\frac{2\ \cancel{\text{mol KI}}}{1\ \cancel{\text{mol HgI}_2}}}_{(b)} \times \underbrace{\frac{166.00\ \text{g KI}}{1\ \cancel{\text{mol KI}}}}_{(c)} = 1.30\ \text{g KI}$$

We can summarize the solution visually as follows.

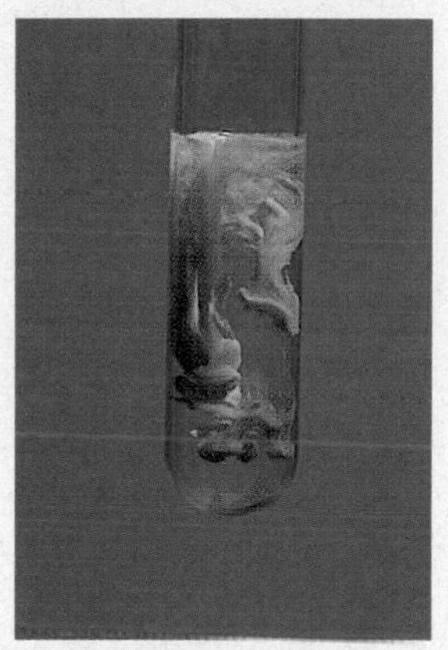

▲ **Precipitation of HgI_2** Potassium iodide and mercury nitrate react to give an orange precipitate of HgI_2.

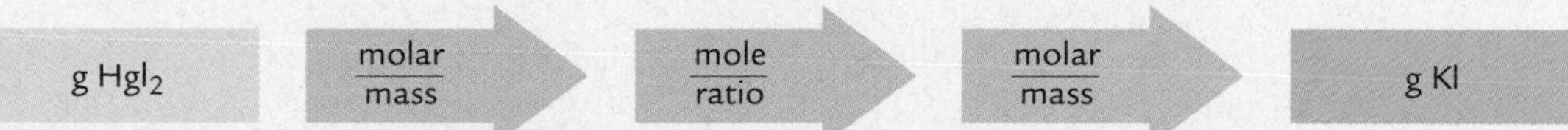

Self-Test Exercise

Calculate the mass of iron filings required to produce 0.455 g of silver metal given the *unbalanced* equation for the reaction.

$$Fe(s) + AgNO_3(aq) \longrightarrow Fe(NO_3)_3(aq) + Ag(s)$$

Answer: 0.0785 g Fe

10.5 Mass–Volume Problems

Objective · To perform mass–volume stoichiometry calculations.

A mass–volume stoichiometry problem is so named because an *unknown volume* of substance is calculated from a *given mass* of reactant or product in a chemical equation. After balancing the chemical equation, we proceed as follows.

(a) Convert the given mass of substance to moles using the molar mass of the substance as a unit factor.

(b) Convert the moles of the given to moles of the unknown using the coefficients in the balanced equation.

(c) Convert the moles of the unknown to liters using the **molar volume** of a gas as a unit factor. At **standard temperature and pressure (STP)**, the molar volume of a gas is 22.4 L/mol.

Of course, it is also possible to perform the reverse calculation; that is, we can find the mass of an unknown substance from a given volume of gas. For simplicity, this reverse process is also referred to as a mass–volume stoichiometry problem.

To understand how to solve a mass–volume stoichiometry problem, consider the reaction of 0.165 g of aluminum metal with dilute hydrochloric acid. The balanced equation for the reaction is

$$2\ Al(s) + 6\ HCl(aq) \longrightarrow 2\ AlCl_3(aq) + 3\ H_2(g)$$

Let's calculate the volume of hydrogen gas produced by the reaction at STP. This is a mass–volume stoichiometry problem. First, we calculate the moles of aluminum whose molar mass is 26.98 g/mol.

(a)
$$0.165\ \cancel{g\ Al} \times \frac{1\ mol\ Al}{26.98\ \cancel{g\ Al}} = 0.00611\ mol\ Al$$

Next, we use the coefficients in the balanced equation to find the moles of hydrogen gas. According to the balanced equation, 2 mol Al = 3 mol H_2. Applying the mole ratio unit factor we have

(b)
$$0.00611\ \cancel{mol\ Al} \times \frac{3\ mol\ H_2}{2\ \cancel{mol\ Al}} = 0.00917\ mol\ H_2$$

Last, we multiply by the molar volume, 22.4 L/mol, to obtain the volume of H_2 gas produced at STP.

(c)
$$0.00917\ \cancel{mol\ H_2} \times \frac{22.4\ L\ H_2}{1\ \cancel{mol\ H_2}} = 0.205\ L\ H_2$$

The solution to the above problem can also be shown as one continuous calculation.

$$0.165\ \cancel{g\ Al} \times \underset{(a)}{\frac{1\ \cancel{mol\ Al}}{26.98\ \cancel{g\ Al}}} \times \underset{(b)}{\frac{3\ \cancel{mol\ H_2}}{2\ \cancel{mol\ Al}}} \times \underset{(c)}{\frac{22.4\ L\ H_2}{1\ \cancel{mol\ H_2}}} = 0.205\ L\ H_2$$

The following example exercises provides additional practice in solving mass–volume stoichiometry problems.

Example Exercise 10.7 • Mass–Volume Stoichiometry

Inflatable air bags are a required safety feature in new automobiles. In the event of a collision, the unstable compound sodium azide, NaN_3, decomposes explosively and fills the air bag with nitrogen gas in about 30 ms. If an air bag contains 100.0 g of NaN_3 (65.02 g/mol), what is the volume of nitrogen gas produced at STP?

$$2\ NaN_3(s) \xrightarrow{\text{spark}} 2\ Na(s) + 3\ N_2(g)$$

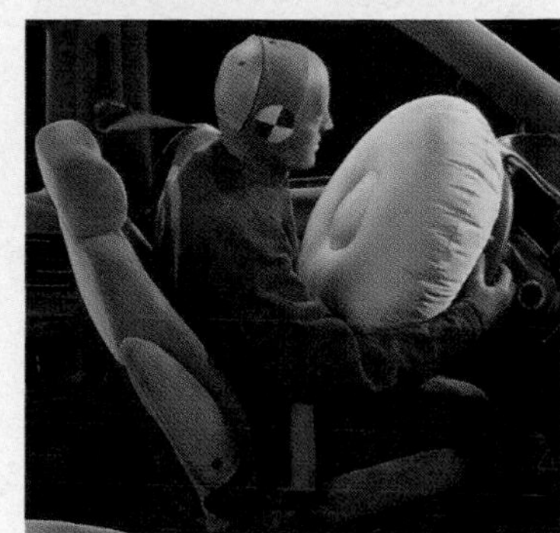

▲ **Automobile Air Bag** The decomposition of sodium azide releases N_2 gas, which inflates the air bag.

Solution

After verifying the coefficients in the equation, we calculate the moles of NaN_3 using the given molar mass, 65.02 g/mol.

(a) $$100.0\ \cancel{g\ NaN_3} \times \frac{1\ mol\ NaN_3}{65.02\ \cancel{g\ NaN_3}} = 1.54\ mol\ NaN_3$$

Next, we calculate the moles of nitrogen gas produced. According to the balanced equation, we see that 2 mol NaN_3 = 3 mol N_2. Applying the mole ratio, we have

(b) $$1.54\ \cancel{mol\ NaN_3} \times \frac{3\ mol\ N_2}{2\ \cancel{mol\ NaN_3}} = 2.31\ mol\ N_2$$

Last, we find the volume of nitrogen gas produced. Recall that the molar volume of a gas at STP is 22.4 L/mol.

(c) $$2.31\ \cancel{mol\ N_2} \times \frac{22.4\ L\ N_2}{1\ \cancel{mol\ N_2}} = 51.7\ L\ N_2$$

Alternatively, we can solve this problem in one continuous calculation.

$$100.0\ \cancel{g\ NaN_3} \times \underbrace{\frac{1\ \cancel{mol\ NaN_3}}{65.02\ \cancel{g\ NaN_3}}}_{(a)} \times \underbrace{\frac{3\ \cancel{mol\ N_2}}{2\ \cancel{mol\ NaN_3}}}_{(b)} \times \underbrace{\frac{22.4\ L\ N_2}{1\ \cancel{mol\ N_2}}}_{(c)} = 51.7\ L\ N_2$$

We can summarize the solution visually as follows.

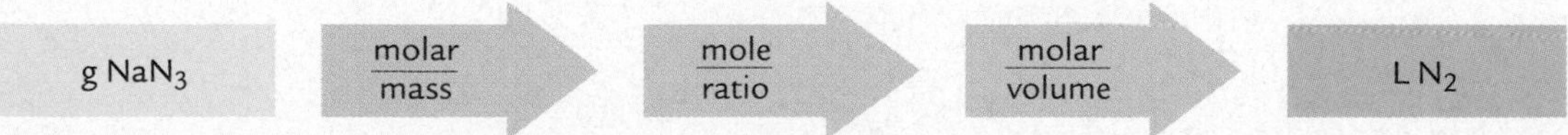

Self-Test Exercise

Calculate the volume of hydrogen gas produced at STP from 1.55 g of sodium metal in water given the *unbalanced* equation for the reaction.

$$Na(s) + H_2O(l) \longrightarrow NaOH(aq) + H_2(g)$$

Answer: 0.755 L H_2

Example Exercise 10.8 • Mass–Volume Stoichiometry

Baking soda can be used as a fire extinguisher. When heated, it decomposes to carbon dioxide gas which can smother a fire. If a sample of $NaHCO_3$ (84.01 g/mol) produces 0.500 L of carbon dioxide gas at STP, what is the mass of the sample?

$$2\ NaHCO_3(s) \xrightarrow{\Delta} Na_2CO_3(s) + H_2O(g) + CO_2(g)$$

Solution

After verifying that the equation is balanced, we will find the moles of CO_2. The volume occupied by 1 mol of CO_2 at STP is 22.4 L.

(continued)

Example Exercise 10.8 *(continued)*

▲ **Baking Soda, $NaHCO_3$** Ordinary baking soda can be used to extinguish a fire. When heated, it decomposes and releases carbon dioxide gas, which is more dense than air and displaces oxygen from the flame.

(a) $$0.500 \cancel{\text{L CO}_2} \times \frac{1 \text{ mol CO}_2}{22.4 \cancel{\text{L CO}_2}} = 0.0223 \text{ mol CO}_2$$

Next, we find the moles of sodium bicarbonate used. According to the balanced equation, 2 mol of $NaHCO_3$ produces 1 mol of CO_2. Applying the mole ratio unit factor, we have

(b) $$0.0223 \cancel{\text{mol CO}_2} \times \frac{2 \text{ mol NaHCO}_3}{1 \cancel{\text{mol CO}_2}} = 0.0446 \text{ mol NaHCO}_3$$

Last, we find the mass of sodium hydrogen carbonate. The molar mass is given, 84.01 g/mol.

(c) $$0.0446 \cancel{\text{mol NaHCO}_3} \times \frac{84.01 \text{ g NaHCO}_3}{1 \cancel{\text{mol NaHCO}_3}} = 3.75 \text{ g NaHCO}_3$$

Alternatively, we can solve this problem in one continuous calculation.

$$0.500 \cancel{\text{L CO}_2} \times \underbrace{\frac{1 \cancel{\text{mol CO}_2}}{22.4 \cancel{\text{L CO}_2}}}_{(a)} \times \underbrace{\frac{2 \cancel{\text{mol NaHCO}_3}}{1 \cancel{\text{mol CO}_2}}}_{(b)} \times \underbrace{\frac{84.01 \text{ g NaHCO}_3}{1 \cancel{\text{mol NaHCO}_3}}}_{(c)} = 3.75 \text{ g NaHCO}_3$$

We can summarize the solution visually as follows.

L CO_2 → molar volume → mole ratio → molar mass → g $NaHCO_3$

Self-Test Exercise

Calculate the mass of aluminum metal required to release 2160 mL of hydrogen gas at STP from sulfuric acid given the *unbalanced* equation for the reaction.

$$\text{Al(s)} + \text{H}_2\text{SO}_4\text{(aq)} \longrightarrow \text{Al}_2(\text{SO}_4)_3\text{(aq)} + \text{H}_2\text{(g)}$$

Answer: 1.73 g Al

10.6 Volume–Volume Problems

Objective · To perform volume–volume stoichiometry calculations.

In 1804 the French chemist Joseph Gay-Lussac ascended to 23,000 feet in a hydrogen-filled balloon. During his a record-setting ascent, he took samples of the atmosphere and later studied their composition.

In 1808 Gay-Lussac proposed the **law of combining volumes**. This law states that volumes of gases, under similar conditions, combine in small whole-number ratios. For example, 10.0 mL of hydrogen gas combines with 10.0 mL of chlorine gas to give 20.0 mL of hydrogen chloride gas.

$$\underset{10.0 \text{ mL}}{\text{H}_2\text{(g)}} + \underset{10.0 \text{ mL}}{\text{Cl}_2\text{(g)}} \longrightarrow \underset{20.0 \text{ mL}}{2 \text{ HCl(g)}}$$

Notice that the volumes of gas (10.0 mL : 10.0 mL : 20.0 mL) are in the ratio of small whole numbers (1 : 1 : 2). Moreover, the ratio of the volumes is identical to the mole ratio in the balanced chemical equation (1 : 1 : 2).

Law of Combining Volumes, H_2 + Cl_2

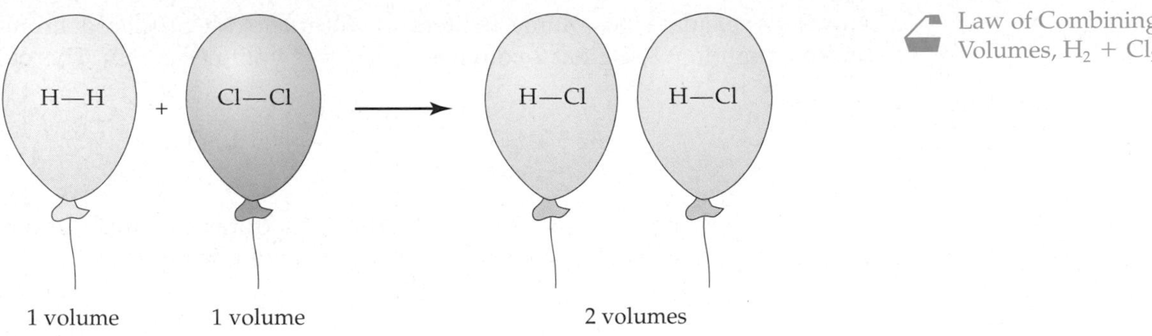

Similarly, 100.0 mL of hydrogen gas reacts with 50.0 mL of oxygen gas to produce 100.0 mL of water vapor.

$$2\ H_2(g) + O_2(g) \longrightarrow 2\ H_2O(g)$$
$$100.0\ mL\quad 50.0\ mL\qquad 100.0\ mL$$

In this example, the volumes of gas (100.0 mL : 50.0 mL : 100.0 mL) are also in the ratio of small whole numbers (2 : 1 : 2). Once again, the ratio of the volumes is identical to the mole ratio in the balanced equation (2 : 1 : 2).

Law of Combining Volumes, H_2 + O_2

Electrolysis of Water Movie

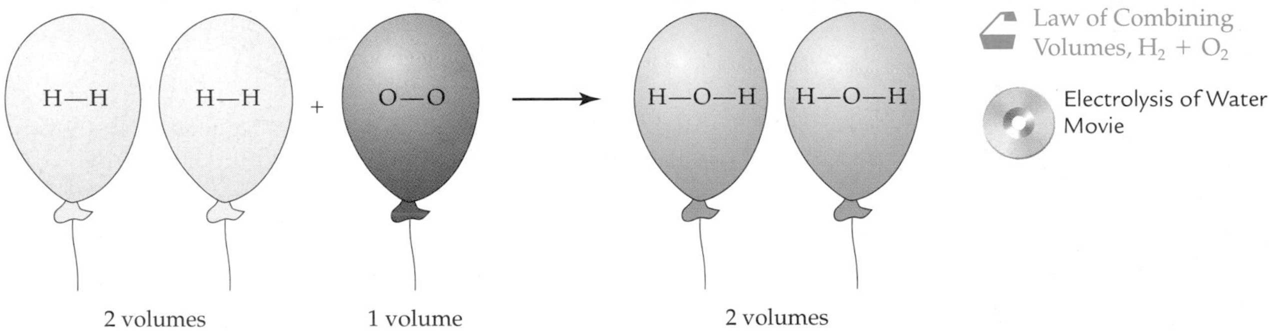

According to the law of combining volumes, gases combine in small whole-number ratios. Since the numbers in this volume ratio are identical to the coefficients in the balanced equation, we can convert from a given volume to an unknown volume of gas in a single conversion using the mole ratio. A volume–volume stoichiometry problem is so named because an *unknown volume* of gas is calculated from a *given volume* of gaseous reactant or product in a chemical equation.

We can illustrate a volume–volume problem using the industrial method for preparing sulfuric acid. As mentioned earlier, H_2SO_4 production is the barometer of the chemical industry. Sulfuric acid is manufactured by the Contact process, which begins with the conversion of sulfur dioxide to sulfur trioxide using heat and a platinum catalyst. The sulfur trioxide gas is then passed through water to produce sulfuric acid. The balanced equation for the conversion of SO_2 to SO_3 is

$$2\ SO_2(g) + O_2(g) \xrightarrow{Pt/\Delta} 2\ SO_3(g)$$

Let's calculate the liters of oxygen gas that react with 37.5 L of sulfur dioxide. From the balanced equation, we see that 2 volumes of SO_2 = 1 volume of O_2. When we express the volume in liters, we have 2 L SO_2 = 1 L O_2. Using the volume ratio as a unit factor, we have

$$37.5\ \cancel{L\ SO_2} \times \frac{1\ L\ O_2}{2\ \cancel{L\ SO_2}} = 18.8\ L\ O_2$$

Now let's calculate the volume, in liters, of sulfur trioxide produced. From the balanced equation we see that 2 volumes of SO_2 = 2 volumes of SO_3. Therefore,

$$37.5 \, \cancel{\text{L SO}_2} \times \frac{2 \text{ L SO}_3}{2 \, \cancel{\text{L SO}_2}} = 37.5 \text{ L SO}_3$$

The following example exercise further illustrates the application of stoichiometry principles to a problem of the volume–volume type.

Example Exercise 10.9 • Volume–Volume Stoichiometry

In the Haber process, nitrogen and hydrogen gases combine to give ammonia gas. If 5.55 L of nitrogen gas is available, calculate the volume of hydrogen gas that reacts and the volume of ammonia that is produced. Assume that all volumes of gas are measured under constant conditions of 500°C and 300 atm pressure.

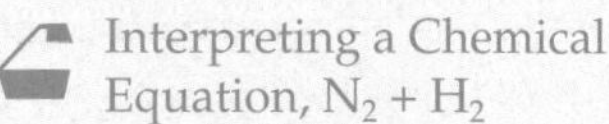

$$N_2(g) + 3\,H_2(g) \xrightarrow{\text{Fe/Al}_2\text{O}_3} 2\,NH_3(g)$$

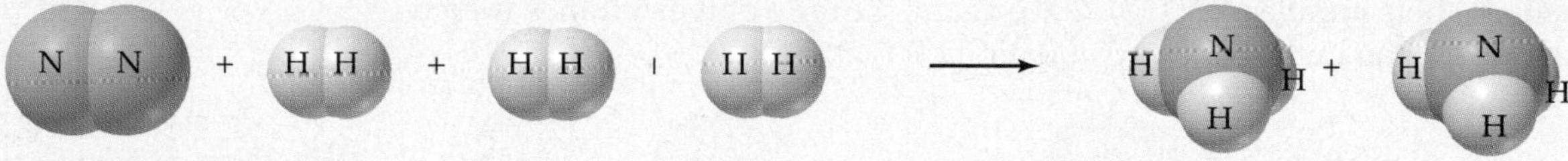

Solution

The coefficients of the balanced chemical equation indicate that 1 volume of N_2 = 3 volumes of H_2. We can calculate the volume of H_2 as follows.

$$5.55 \, \cancel{\text{L N}_2} \times \frac{3 \text{ L H}_2}{1 \, \cancel{\text{L N}_2}} = 16.7 \text{ L H}_2$$

The equation also indicates that 1 volume of N_2 = 2 volumes of NH_3. Therefore,

$$5.55 \, \cancel{\text{L N}_2} \times \frac{2 \text{ L NH}_3}{1 \, \cancel{\text{L N}_2}} = 11.1 \text{ L NH}_3$$

▲ **Fritz Haber** This stamp commemorates Haber's 1918 Nobel prize.

We can summarize the solution visually as follows.

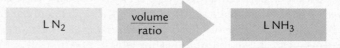

L N_2 → volume ratio → L NH_3

Self-Test Exercise

Calculate the volumes of (a) hydrogen chloride gas and (b) oxygen gas that react to yield 50.00 mL of chlorine gas given the following unbalanced equation. Assume all gases are at the same temperature and pressure.

$$HCl(g) + O_2(g) \longrightarrow Cl_2(g) + H_2O(g)$$

Answers: (a) 100.0 mL HCl; (b) 25.00 mL O_2

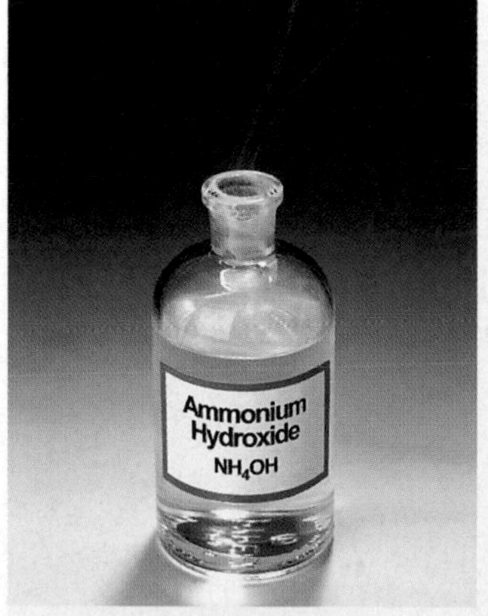

◀ **Ammonia, NH_3** Aqueous ammonia, NH_3, is found in laboratory reagent bottles labeled ammonium hydroxide, NH_4OH. The white fumes escaping from the bottle are ammonia gas released from the aqueous solution.

Chemistry Connection • Manufacturing Ammonia

What are the main uses for ammonia, which is one of the ten most important industrial chemicals?

Over 15 million tons of ammonia are produced annually in the United States. Ammonia is primarily used in agriculture to fertilize the soil and replenish the nitrogen depleted by crops. Although our atmosphere is a vast potential source of nitrogen, the gaseous element is unreactive and does not readily combine with other elements. Nitrogen is essentially a colorless, odorless, inert gas.

▲ **Soil Nitrogen** After liquid ammonia is injected, it vaporizes to a gas and acts as a nitrogen fertilizer.

In 1905 the German chemist Fritz Haber (1868–1934) successfully prepared ammonia in the laboratory for the first time. He discovered that nitrogen and hydrogen gases combine directly at high temperatures and pressures in the presence of metal oxide catalysts. The Haber process thus provides a method for converting unreactive atmospheric nitrogen to the versatile compound ammonia, NH_3. Ammonia in turn is easily converted to important ammonium compounds and nitric acid, HNO_3. Today, the Haber process is the main source of manufacturing nitrogen compounds throughout the world.

Until the beginning of the 1900s, the bulk of the world's nitrate was supplied by the rich saltpeter, KNO_3, deposits in South America. Although nitrates were used principally as fertilizer, they were also used to make explosives. Indirectly, the Haber process had a pronounced effect on World War I. Soon after the outbreak of the war, the British blockade halted the supply of Chilean saltpeter to Germany. Without a source of nitrate, Germany would have soon run out of ammunition. But by late 1914, Germany had built factories that applied the Haber process for the manufacture of ammonia and nitrate compounds.

It is often noted that Haber was sympathetic to Germany's war effort. He actively contributed to chemical warfare by helping develop poisonous chlorine gas and the even more lethal mustard gas. In 1918, the same year World War I ended, Haber was awarded the Nobel prize in chemistry. Following the announcement of the Nobel prize, he was denounced by American, English, and French scientists for his involvement in the war.

It is ironic that the patriotic Haber was later condemned by his own country. When the Nazis rose to power in 1933, Haber had to resign as director of the Kaiser Wilhelm Institute because he was Jewish. In distress, he fled to Cambridge University in England where he spent the last few months of his life.

The chief use of ammonia, NH_3, is in agriculture, but it is also used in explosives, ammunition, and household products such as window cleaner.

10.7 The Limiting Reactant Concept

Objectives
- To explain the concept of a limiting reactant.
- To identify the limiting reactant in a chemical reaction given the number of moles of each reactant.

In world-class cycling, competitors carry extra bicycle frames and wheels. If a cyclist participating in the Tour de France has 5 frames and 20 wheels, how many complete bicycles can be assembled? Obviously, only 5 complete bicycles can be assembled, and there are 10 spare wheels. In this example, the frames limit the number of bicycles that can be assembled. Analogously, we will learn that one of the reactants in a chemical reaction can limit the amount of product.

We have learned how to perform calculations using information from a balanced chemical equation. For example, we can calculate the amount of iron(II) sulfide produced by the reaction of metallic iron and powdered sulfur.

$$Fe(s) + S(s) \xrightarrow{\Delta} FeS(s)$$

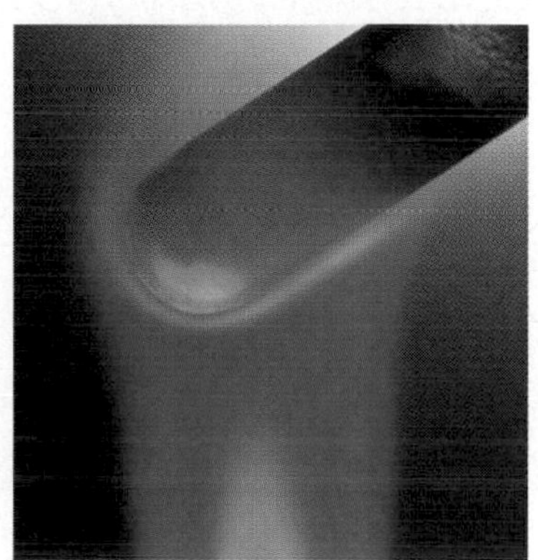

▲ **Producing FeS** Heating a mixture of iron powder and sulfur powder produces iron(II) sulfide; either the iron or the sulfur can be the limiting reactant.

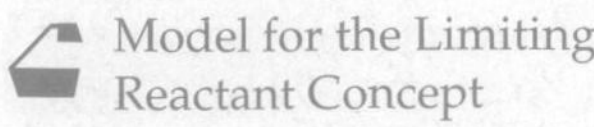
Model for the Limiting Reactant Concept

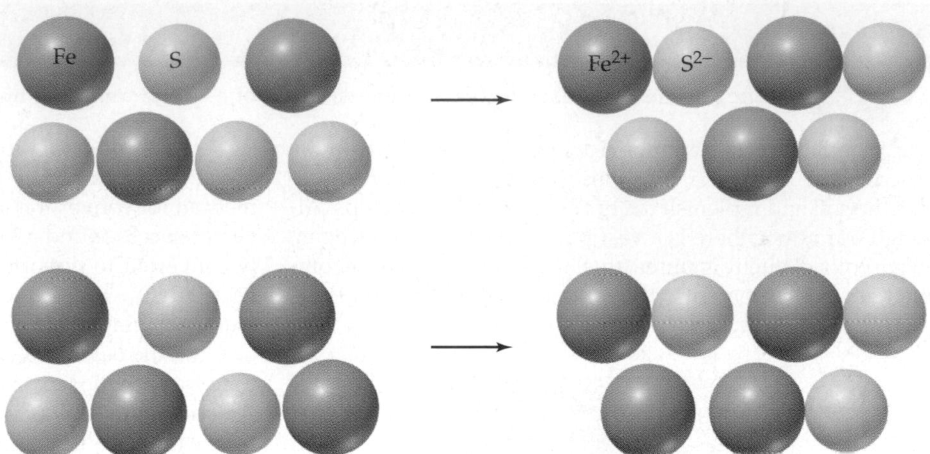

▶ **Figure 10.1 Model for the Limiting Reactant Concept**
(a) As Fe and S react to give FeS, there is excess S; thus, Fe is the limiting reactant. Notice the unreacted S that remains. (b) In another reaction, there is an excess of Fe; thus, S is the limiting reactant. Notice the unreacted Fe that remains.

Previously, we have assumed that there is sufficient iron and sulfur to react completely with each other. However, this assumption may not be true. What if there is not enough sulfur to react with the iron? If that were the case, the sulfur would *limit* the amount of FeS produced. In a stoichiometry problem, the reactant that controls the amount of product is called the **limiting reactant** and must be identified in order to correctly calculate the amount of product. Figure 10.1 is an example model to help you visualize the concept of a limiting reactant.

Determining the Limiting Reactant

Limiting Reactant Movie; Limiting Reactants Activity

Now that we have a visual model for the limiting reactant concept, let's examine data from an experiment and solve a problem. Suppose we heat 2.50 mol of iron with 3.00 mol of sulfur. How many moles of FeS are formed? We will analyze the problem by considering the amount of each substance before and after the reaction.

Before the reaction begins, there is 2.50 mol of Fe, 3.00 mol of S, and 0.00 mol FeS. According to the balanced equation, 1 mol of Fe reacts with 1 mol of S to give 1 mol of FeS. Therefore, 2.50 mol of Fe reacts with 2.50 mol of S to give 2.50 mol of FeS. At the outset we have 3.00 mol of S. Therefore, iron is the limiting reactant and sulfur is the excess reactant. After the reaction, the excess sulfur is 0.50 mol (3.00 mol − 2.50 mol). Table 10.2 summarizes the amounts of each substance before and after the reaction.

Table 10.2 Summary of Experimental Synthesis of FeS

Experiment	mol Fe	mol S	mol FeS
before reaction	2.50	3.00	0.00
after reaction	0.00	0.50	2.50

The following example exercise further illustrates the concept of a limiting reactant.

Example Exercise 10.10 • Limiting Reactant Concept

A 1.00-mol sample of iron(II) oxide is heated with 1.00 mol of aluminum metal and converted to molten iron. Identify the limiting reactant and calculate the moles of iron produced given the equation for the reaction.

$$FeO(l) + Al(l) \xrightarrow{\Delta} Fe(l) + Al_2O_3(s)$$

Solution
Since all stoichiometry problems require a balanced equation, we must write the correct coefficients for the reactants and products.

$$3\ FeO(l) + 2\ Al(l) \xrightarrow{\Delta} 3\ Fe(l) + Al_2O_3(s)$$

First, let's calculate the amount of Fe produced from 1.00 mol of FeO.

$$1.00\ \cancel{\text{mol FeO}} \times \frac{3\ \text{mol Fe}}{3\ \cancel{\text{mol FeO}}} = 1.00\ \text{mol Fe}$$

Second, we must calculate the amount of Fe produced from 1.00 mol of Al.

$$1.00\ \cancel{\text{mol Al}} \times \frac{3\ \text{mol Fe}}{2\ \cancel{\text{mol Al}}} = 1.50\ \text{mol Fe}$$

Notice that FeO produces only 1.00 mol of Fe, whereas Al produces 1.50 mol of Fe. Thus, FeO is the *limiting reactant* and the amount of product is 1.00 mol Fe.

Self-Test Exercise
A 5.00-mol sample of iron(III) oxide is heated with 5.00 mol of aluminum metal and converted to molten iron. Identify the limiting reactant and calculate the moles of iron produced given the unbalanced equation for the reaction.

$$Fe_2O_3(l) + Al(l) \xrightarrow{\Delta} Fe(l) + Al_2O_3(s)$$

Answer: The limiting reactant is Al, which produces 5.00 mol of Fe.

10.8 Limiting Reactant Problems

Objectives
- To perform mass–mass stoichiometry calculations involving a limiting reactant.
- To perform volume–volume stoichiometry calculations involving a gaseous limiting reactant.

Suppose we wish to find how much molten iron is produced from the reaction of 25.0 g of FeO with 25.0 g of Al. The balanced equation for the reaction is

$$3\ FeO(l) + 2\ Al(l) \xrightarrow{\Delta} 3\ Fe(l) + Al_2O_3(s)$$

Since the mass of each reactant is given, it is not obvious whether FeO or Al limits the amount of iron produced. In problems where we are given the amounts of two reactants, we determine the limiting reactant as follows

I. *Calculate the mass of product that can be produced from the first reactant.*
   (a) Calculate the moles of reactant.
   (b) Calculate the moles of product.
   (c) Calculate the mass of product.

II. *Calculate the mass of product that can be produced from the second reactant.*
   (a) Calculate the moles of reactant.
   (b) Calculate the moles of product.
   (c) Calculate the mass of product.

III. *State the limiting reactant and the corresponding mass of product formed.*
The limiting reactant gives the least amount of product. The actual mass of product obtained from the chemical reaction is the lesser of the two masses calculated in steps I and II.

▲ **Reaction of FeO and Al** The reaction of iron oxide and aluminum metal produces molten iron and a shower of sparks.

Let's apply this process to the above problem. Step I: We calculate the amount of iron that can be produced from the ferrous oxide. Using the balanced equation for the reaction, we can find the mass of Fe produced from 25.0 g of FeO. Let's outline the solution to the problem.

$$\text{g FeO} \xrightarrow[\text{I (a)}]{} \text{mol FeO} \xrightarrow[\text{I (b)}]{} \text{mol Fe} \xrightarrow[\text{I (c)}]{} \text{g Fe}$$

From the periodic table, we find that for Fe, the molar mass is 55.85 g/mol, and for FeO, 71.85 g/mol. The unit analysis solution to the problem is

$$25.0 \cancel{\text{g FeO}} \times \frac{1 \cancel{\text{mol FeO}}}{71.85 \cancel{\text{g FeO}}} \times \frac{3 \cancel{\text{mol Fe}}}{3 \cancel{\text{mol FeO}}} \times \frac{55.85 \text{ g Fe}}{1 \cancel{\text{mol Fe}}} = 19.4 \text{ g Fe}$$

Step II: We calculate the amount of iron that can be produced by the aluminum. Using the balanced equation, we can find the mass of Fe produced from 25.0 g of Al. We can outline the solution to the problem as follows.

$$\text{g Al} \xrightarrow[\text{II (a)}]{} \text{mol Al} \xrightarrow[\text{II (b)}]{} \text{mol Fe} \xrightarrow[\text{II (c)}]{} \text{g Fe}$$

From the periodic table, we find that the molar mass of Al is 26.98 g/mol. The unit analysis solution to the problem is

$$25.0 \cancel{\text{g Al}} \times \frac{1 \cancel{\text{mol Al}}}{26.98 \cancel{\text{g Al}}} \times \frac{3 \cancel{\text{mol Fe}}}{2 \cancel{\text{mol Al}}} \times \frac{55.85 \text{ g Fe}}{1 \cancel{\text{mol Fe}}} = 77.6 \text{ g Fe}$$

Step III: We compare the mass of iron produced from each of the two reactants.

$$25.0 \text{ g FeO} \longrightarrow 19.4 \text{ g Fe}$$
$$25.0 \text{ g Al} \longrightarrow 77.6 \text{ g Fe}$$

We see that the FeO yields less mass of product; thus, FeO is the *limiting reactant*. Conversely, aluminum metal is the excess reactant, which means that not all the Al is used in the reaction. Therefore, the maximum yield of metallic iron from the reaction is limited to 19.4 g of Fe.

Example Exercise 10.11 provides additional practice in solving a limiting reactant problem.

Example Exercise 10.11 • Mass–Mass Limiting Reactant

In a reaction, 50.0 g of manganese(IV) oxide reacts with 25.0 g of aluminum. Identify the limiting reactant and calculate the mass of manganese metal produced. The equation for the reaction is

$$3\ MnO_2(l) + 4\ Al(l) \xrightarrow{\Delta} 3\ Mn(l) + 2\ Al_2O_3(s)$$

Solution

We begin by verifying that the equation is balanced. Step I: We calculate the mass of Mn produced from 50.0 g of MnO_2. We can outline the solution as follows.

$$\text{g } MnO_2 \longrightarrow \text{mol } MnO_2 \longrightarrow \text{mol Mn} \longrightarrow \text{g Mn}$$

From the periodic table, we find that for MnO_2 the molar mass is 86.94 g/mol, and for Mn, 54.94 g/mol. The unit analysis solution to the problem is

$$50.0 \cancel{\text{g } MnO_2} \times \frac{1 \cancel{\text{mol } MnO_2}}{86.94 \cancel{\text{g } MnO_2}} \times \frac{3 \cancel{\text{mol Mn}}}{3 \cancel{\text{mol } MnO_2}} \times \frac{54.94 \text{ g Mn}}{1 \cancel{\text{mol Mn}}} = 31.6 \text{ g Mn}$$

Step II: We calculate the mass of Mn produced from 25.0 g of Al. We outline the solution as follows.

$$\text{g Al} \longrightarrow \text{mol Al} \longrightarrow \text{mol Mn} \longrightarrow \text{g Mn}$$

From the periodic table, we find that the molar mass of Al is 26.98 g/mol. Starting with 50.0 g of Al, the unit analysis solution is

$$25.0 \cancel{\text{g Al}} \times \frac{1 \cancel{\text{mol Al}}}{26.98 \cancel{\text{g Al}}} \times \frac{3 \cancel{\text{mol Mn}}}{4 \cancel{\text{mol Al}}} \times \frac{54.94 \text{ g Mn}}{1 \cancel{\text{mol Mn}}} = 38.1 \text{ g Mn}$$

Step III: We compare the mass of product obtained from each of the reactants.

$$50.0 \text{ g } MnO_2 \longrightarrow 31.6 \text{ g Mn}$$

$$25.0 \text{ g Al} \longrightarrow 38.1 \text{ g Mn}$$

In this example, MnO_2 is the *limiting reactant* because it yields less product. Al is the excess reactant and is not completely consumed in the reaction. Thus, the maximum yield from the reaction is 31.6 g of Mn.

Self-Test Exercise

In a reaction, 75.0 g of manganese(IV) oxide reacts with 30.0 g of aluminum. Identify the limiting reactant and calculate the mass of aluminum oxide produced.

$$3\ MnO_2(l) + 4\ Al(l) \xrightarrow{\Delta} 3\ Mn(l) + 2\ Al_2O_3(s)$$

Answer: The limiting reactant is Al, which gives 56.7 g of Al_2O_3.

Limiting Reactants Involving Volumes of Gas

Now let's try a limiting reactant problem that involves reactants and products in the gaseous state. For example, consider the reaction of nitrogen monoxide and oxygen to give nitrogen dioxide. The balanced equation for the reaction is

$$2\ NO(g) + O_2(g) \xrightarrow{\text{UV}} 2\ NO_2(g)$$

Suppose 5.00 L of NO gas reacts with 5.00 L of O_2 gas. Since the volumes of both reactants are given, it is not obvious which gas limits the amount of product. We assume that the temperature and pressure remain constant and use the following approach to determine the limiting reactant.

I. *Calculate the volume of product that can be produced from the first reactant.*
Recall that the coefficients in the balanced equation are in the same mole ratio as the volumes of gas.

II. *Calculate the volume of product that can be produced from the second reactant.*
Again, the coefficients in the balanced equation are in the same mole ratio as the volumes of gas.

III. *State the limiting reactant and the corresponding volume of product.*
The limiting reactant is the gas that gives the least amount of product. The actual volume of product obtained from the chemical reaction is the lesser of the two volumes calculated in steps I and II.

Step I: We calculate the volume of nitrogen dioxide produced from 5.00 L of nitrogen monoxide gas. From the balanced equation, we see that 2 mol of NO produces 2 mol of NO_2. Accordingly, 2 L of NO produces 2 L of NO_2. Hence,

$$5.00 \cancel{\text{L NO}} \times \frac{2 \text{ L } NO_2}{2 \cancel{\text{L NO}}} = 5.00 \text{ L } NO_2$$

Step II: We calculate the volume of nitrogen dioxide produced from 5.00 L of oxygen gas. From the balanced equation, we see that 1 mol of O_2 produces 2 mol of NO_2. Accordingly, 1 L of O_2 produces 2 L of NO_2. Thus,

$$5.00\ \cancel{L\ O_2} \times \frac{2\ L\ NO_2}{1\ \cancel{L\ O_2}} = 10.0\ L\ NO_2$$

Step III: We compare the volume of gaseous NO_2 product from each of the gaseous reactants.

$$5.00\ L\ NO \longrightarrow 5.00\ L\ NO_2$$

$$5.00\ L\ O_2 \longrightarrow 10.0\ L\ NO_2$$

We see that the NO yields less volume of product; thus, NO is the *limiting reactant*. Conversely, O_2 is the excess reactant, and not all the O_2 gas is used in the reaction. Therefore, the maximum yield of product from the reaction is 5.00 L of NO_2. The following example exercise provides additional practice in solving a limiting reactant problem involving gases.

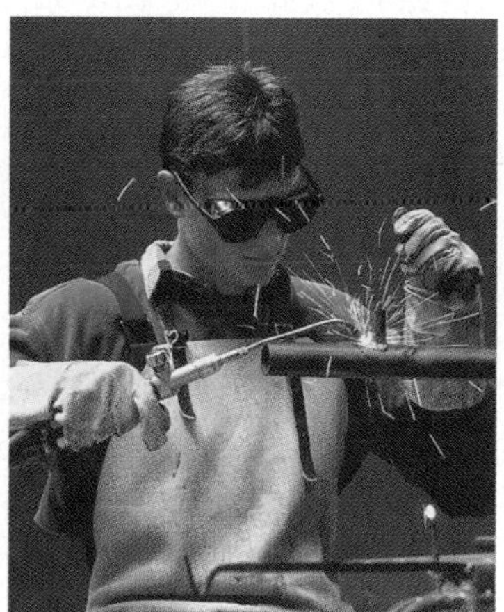

▲ **Oxyacetylene Welding** The reaction of oxygen and acetylene gases releases enough heat to melt iron.

Example Exercise 10.12 • Volume–Volume Limiting Reactant

In oxyacetylene welding, acetylene reacts with oxygen to give carbon dioxide and water. If 25.0 mL of C_2H_2 reacts with 75.0 mL of O_2, what is the limiting reactant? Assuming constant conditions, what is the volume of CO_2 produced?

$$2\ C_2H_2(g) + 5\ O_2(g) \xrightarrow{\text{spark}} 4\ CO_2(g) + 2\ H_2O(g)$$

Solution

After checking the coefficients, we see that the equation is balanced. Step I: We calculate the volume of CO_2 gas produced from 25.0 mL of C_2H_2. From the equation we see that 2 mL of C_2H_2 gives 4 mL of CO_2. Therefore,

$$25.0\ \cancel{mL\ C_2H_2} \times \frac{4\ mL\ CO_2}{2\ \cancel{mL\ C_2H_2}} = 50.0\ mL\ CO_2$$

Step II: We calculate the volume of CO_2 gas produced from 75.0 mL of O_2. From the equation we see that 5 mL of O_2 gives 4 mL of CO_2. Therefore,

$$75.0\ \cancel{mL\ O_2} \times \frac{4\ mL\ CO_2}{5\ \cancel{mL\ O_2}} = 60.0\ mL\ CO_2$$

Step III: We compare the volume of gaseous CO_2 product from each of the reactants.

$$25.0\ mL\ C_2H_2 \longrightarrow 50.0\ mL\ CO_2$$

$$75.0\ mL\ O_2 \longrightarrow 60.0\ mL\ CO_2$$

The limiting reactant is C_2H_2 because it yields less volume of product. Thus, the maximum volume of CO_2 produced from the reaction is 50.0 mL of CO_2.

Self-Test Exercise

Ethane undergoes combustion to give carbon dioxide and water. If 10.0 L of C_2H_6 reacts with 25.0 L of O_2, what is the limiting reactant? Assuming constant conditions, what is the volume of CO_2 produced? The equation is

$$2\ C_2H_6(g) + 7\ O_2(g) \xrightarrow{\text{spark}} 4\ CO_2(g) + 6\ H_2O(g)$$

Answer: The limiting reactant is O_2, which gives 14.3 L of CO_2.

10.9 Percent Yield

Objective · To calculate the percent yield for a reaction given the actual yield and theoretical yield.

▲ **Percent Yield of CdS** The aqueous solution reaction of $CdCl_2$ and Na_2S yields a yellow precipitate of CdS.

To understand the concept of yield, consider the following laboratory experiment. First, a student weighs a 1.000-g sample of cadmium chloride and dissolves it in water. Then, the student adds aqueous sodium sulfide solution to obtain a precipitate of cadmium sulfide, CdS. The balanced equation for the reaction is

$$CdCl_2(aq) + Na_2S(aq) \longrightarrow CdS(s) + 2\,NaCl(aq)$$

The student collects the insoluble CdS precipitate in filter paper and determines its mass on a balance. The experimental mass of the precipitate is called the **actual yield**.

The mass of cadmium sulfide obtained, starting with the 1.000-g sample of cadmium chloride, can also be calculated as a mass–mass problem. The calculated amount of precipitate is called the **theoretical yield**. In practice, there is always inherent experimental error. Thus, we compare the actual and theoretical yields and express the results in terms of percent yield. The **percent yield** is the actual yield divided by the theoretical yield expressed as a percent.

$$\frac{\text{actual yield}}{\text{theoretical yield}} \times 100 = \text{percent yield}$$

To find the percent yield for this laboratory experiment, we start with the 1.000-g sample of cadmium chloride. Applying stoichiometry to the balanced equation, we calculate that the amount of cadmium sulfide precipitate is 0.788 g. Suppose the student weighs the CdS precipitate and finds that it has an actual mass of 0.775 g. To find the percent yield, we compare the actual yield (0.775 g) to the theoretical yield (0.788 g) as follows:

$$\frac{0.775\text{ g}}{0.788\text{ g}} \times 100 = 98.4\%$$

Suppose the student performs a second trial for the same experiment, again using a 1.000-g sample of cadmium chloride. In this second trial, the precipitate has a mass of 0.805 g. Since the student started with 1.000 g of cadmium chloride, the theoretical yield for the two trials is the same, 0.788 g. As before, to find the percent yield we compare the actual and theoretical yields.

$$\frac{0.805\text{ g}}{0.788\text{ g}} \times 100 = 102.2\%$$

Notice that the percent yield for this second trial is greater than 100%. Although surprising, this result is possible. In an experiment, some errors lead to high results and other errors lead to low results. In the two trials, the student made different types of experimental errors. For example, high results are obtained when impurities are trapped or when the precipitate is not completely dry.

The following example exercise provides additional practice in calculating percent yield.

▲ **Percent Yield of $CuCO_3$** The aqueous solution reaction of $Cu(NO_3)_2$ and Na_2CO_3 yields a blue precipitate of $CuCO_3$.

Example Exercise 10.13 • Percent Yield

A student dissolves 1.500-g of copper(II) nitrate in water. After adding aqueous sodium carbonate solution, the student obtains 0.875 g of $CuCO_3$ precipitate. If the theoretical yield is 0.988 g, what is the percent yield?

$$Cu(NO_3)_2(aq) + Na_2CO_3(aq) \longrightarrow CuCO_3(s) + 2\,NaNO_3(aq)$$

Solution

The percent yield is the ratio of the actual yield compared to the theoretical yield. In this experiment, the actual yield is 0.875 g and the theoretical yield is 0.988 g. The percent yield therefore is

$$\frac{\text{actual yield}}{\text{theoretical yield}} \times 100 = \text{percent yield}$$

$$\frac{0.875\text{ g}}{0.988\text{ g}} \times 100 = 88.6\%$$

The percent yield obtained by the student is 88.6%.

Self-Test Exercise

Ammonium nitrate is used in explosives and is produced by the reaction of ammonia, NH_3, and nitric acid. The equation for the reaction is

$$NH_3(g) + HNO_3(aq) \longrightarrow NH_4NO_3(s)$$

If 15.0 kg of ammonia gives an actual yield of 65.3 kg of ammonium nitrate, what is the percent yield? The calculated yield of ammonium nitrate for the experiment is 70.5 kg.

Answer: 92.6%

Summary 

Section 10.1 In this chapter we learned to interpret the coefficients in a balanced equation as a **mole ratio**. We also showed that by applying **Avogadro's theory** we can express the mole ratio in terms of volumes of gases. Furthermore, we verified the **law of conservation of mass** by substituting the **molar mass** for the moles of each reactant and product in the chemical equation.

Section 10.2 Before studying the relationship between masses and volumes of substances according to a balanced equation, we first learned to relate the moles of reactants to the moles of products. Given the balanced equation, we can use the coefficients to relate the moles of each substance in a chemical reaction.

Section 10.3 **Stoichiometry** relates the quantities of substances involved in a chemical reaction. There are three basic types of stoichiometry problems. In a **mass–mass problem**, the mass of one substance is related to the mass of another substance according to a balanced chemical equation. In a **mass–volume problem**, the mass of a substance is related to the volume of a gas. In a **volume–volume problem**, the volumes of two gaseous substances are related.

Section 10.4 In a mass–mass problem, we first convert the given mass of a substance to the number of moles using the molar mass. If the molar mass is not shown in the problem, we refer to the periodic table and add up the gram atomic masses for each element in the substance. Next, we apply the mole ratio to convert the moles of the given substance to the moles of the unknown substance, using the coefficients from the balanced equation. Finally, we determine the mass of the unknown by multiplying by the molar mass of the unknown.

Section 10.5 In a mass–volume problem, we first convert the given mass of substance to the number of moles using the molar mass. Next, we apply the mole ratio to convert the moles of the given substance to the moles of the unknown substance, using the coefficients from the balanced equation. Finally, we determine the volume of the unknown gas at STP by multiplying by the **molar volume**. The volume occupied by 1 mol of gas is 22.4 L at standard conditions, that is, 0°C and 1 atm pressure.

Section 10.6 In a volume–volume problem, we can take advantage of Gay-Lussac's **law of combining volumes**. Since the volumes of gases combine in the same whole-number ratio as the coefficients in the balanced equation, we can solve this type of problem in one conversion step. We simply multiply the given volume by the ratio of the coefficients in the equation.

Section 10.7 In more difficult stoichiometry problems, we cannot assume that there is a sufficient quantity of each reactant available. The amount of one reactant may not be sufficient to completely consume another reactant that is in excess. The reactant that controls the reaction and limits the amount of product formed is referred to as the **limiting reactant**.

Section 10.8 If we are given the amounts of two reactants, we can determine the limiting reactant by finding the amount of product yield from each of the reactants. The reactant that produces the least amount of product controls the reaction and is the limiting reactant. A limiting reactant problem may involve the masses of two reacting substances as well as the volumes of two gases.

Section 10.9 The percent yield provides an evaluation of the laboratory method and the skill of the analyst. The **percent yield** from an experiment is the ratio between the **actual yield** and the **theoretical yield**, all multiplied by 100.

Problem-Solving Organizer

Topic	Procedure	Example
Interpreting a Chemical Equation Sec. 10.1	Express the mole ratio, and volume ratio of gaseous substances, in a balanced equation using the coefficients.	*Given*: $N_2(g) + O_2(g) \longrightarrow 2\,NO(g)$ Interpret the mole ratio and the volume ratio in liters. $1\text{ mol } N_2 + 1\text{ mol } O_2 \longrightarrow 2\text{ mol NO}$ $1\text{ L } N_2 + 1\text{ L } O_2 \longrightarrow 2\text{ L NO}$
Mole-Mole Relationships Sec. 10.2	Convert moles of a given substance to moles of an unknown substance using the mole ratio.	How many moles of NO are produced from 0.500 mol N_2? $0.500\ \cancel{\text{mol } N_2} \times \frac{2\text{ mol NO}}{1\ \cancel{\text{mol } N_2}} = 1.00\text{ mol NO}$
Types of Stoichiometry Problems Sec. 10.3	Classify a stoichiometry problem as a mass–mass, mass–volume, or volume–volume type of problem.	How many grams of NO are produced from 11.2 L N_2 is what type of problem? (mass–volume problem)
Mass-Mass Problems Sec. 10.4	(a) Convert mass of given to moles of given. (b) Convert moles of given to moles of unknown. (c) Convert moles of unknown to mass of unknown.	How many grams of NO are produced from 10.0 g N_2? $10.0\ \cancel{\text{g } N_2} \times \frac{1\ \cancel{\text{mol } N_2}}{28.02\ \cancel{\text{g } N_2}} \times \frac{2\ \cancel{\text{mol NO}}}{1\ \cancel{\text{mol } N_2}} \times \frac{30.01\text{ g NO}}{1\ \cancel{\text{mol NO}}} = 21.4\text{ g NO}$

Moles of given
(b)
Moles of unknown
(a)
(c)
Mass of given
Mass of unknown

◀ **Mass–Mass Stoichiometry Format** We can see from the diagram that the mass of given substance is related to the mass of the unknown. In performing mass-mass stoichiometry calculations, the unit factor method can be used for each step of the calculation.

(continued)

Problem-Solving Organizer *(continued)*

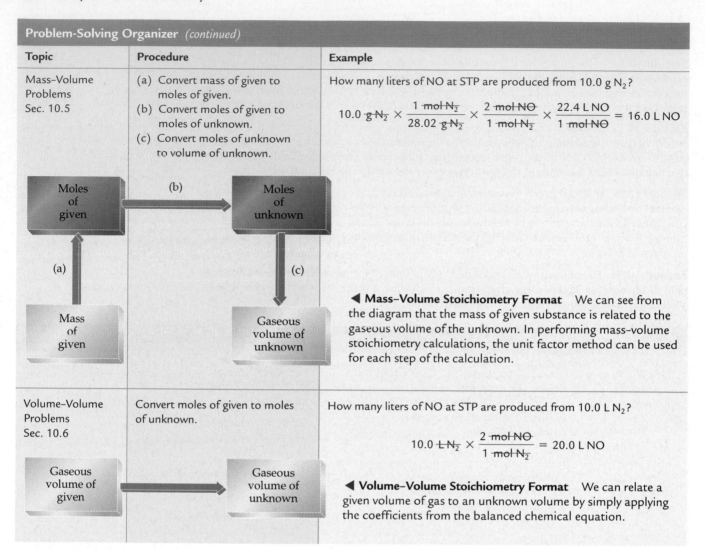

Topic	Procedure	Example
Mass–Volume Problems Sec. 10.5	(a) Convert mass of given to moles of given. (b) Convert moles of given to moles of unknown. (c) Convert moles of unknown to volume of unknown.	How many liters of NO at STP are produced from 10.0 g N_2? $10.0\ \cancel{g\ N_2} \times \frac{1\ \cancel{mol\ N_2}}{28.02\ \cancel{g\ N_2}} \times \frac{2\ \cancel{mol\ NO}}{1\ \cancel{mol\ N_2}} \times \frac{22.4\ L\ NO}{1\ \cancel{mol\ NO}} = 16.0\ L\ NO$
Moles of given →(b)→ Moles of unknown; Mass of given →(a)→ Moles of given; Moles of unknown →(c)→ Gaseous volume of unknown		◀ **Mass–Volume Stoichiometry Format** We can see from the diagram that the mass of given substance is related to the gaseous volume of the unknown. In performing mass-volume stoichiometry calculations, the unit factor method can be used for each step of the calculation.
Volume–Volume Problems Sec. 10.6	Convert moles of given to moles of unknown.	How many liters of NO at STP are produced from 10.0 L N_2? $10.0\ \cancel{L\ N_2} \times \frac{2\ \cancel{mol\ NO}}{1\ \cancel{mol\ N_2}} = 20.0\ L\ NO$
Gaseous volume of given → Gaseous volume of unknown		◀ **Volume–Volume Stoichiometry Format** We can relate a given volume of gas to an unknown volume by simply applying the coefficients from the balanced chemical equation.

Key Concepts*

1. Which of the following quantities are in the same ratio as the coefficients in a balanced chemical equation?
   (a) moles of reactants and products
   (b) masses of reactants and products
   (c) volumes of gaseous reactants and products

2. In the first step of the Ostwald process for manufacturing nitric acid, ammonia and oxygen gases are heated with a platinum catalyst to yield nitrogen monoxide and water. The balanced equation for the chemical reaction is

$$4\ NH_3(g) + 5\ O_2(g) \xrightarrow{Pt/825°C} 4\ NO(g) + 6\ H_2O(g)$$

   (a) How many moles of oxygen react with 4.00 mol of ammonia?
   (b) How many moles of NO are produced from 4.00 mol of NH_3?
   (c) How many grams of oxygen react with 2.00 g of NH_3?
   (d) How many liters of NO are produced from 2.00 g of NH_3?
   (e) How many liters of oxygen react with 5.00 L of NH_3?
   (f) How many liters of NO are produced from 5.00 L of NH_3?

3. In the second step of the Ostwald process, nitrogen monoxide and oxygen gases are heated to give nitrogen dioxide gas. The balanced equation is

$$2\ NO(g) + O_2(g) \longrightarrow 2\ NO_2(g)$$

   (a) How many moles of nitrogen dioxide are produced from 2.00 mol of NO and 2.00 mol of O_2?
   (b) How many grams of nitrogen dioxide are produced from 2.00 g of NO and 2.00 g of O_2?
   (c) How many liters of nitrogen dioxide are produced from 2.00 L of NO and 2.00 L of O_2?

*Answers to Key Concepts are in Appendix H.

4. In the third step of the Ostwald process, nitrogen dioxide dissolves in water to give nitric acid and nitrogen monoxide gas. The balanced equation is

$$3\,NO_2(g) + H_2O(l) \longrightarrow 2\,HNO_3(aq) + NO(g)$$

If the experimental yield of nitric acid is 35.0 kg and the calculated yield of nitric acid is 37.5 kg what is the percent yield?

5. Zinc metal reacts with sulfur to give ZnS. The balanced equation is

$$Zn(s) + S(s) \longrightarrow ZnS(s)$$

Complete the following illustrations by writing formula units of product and atoms of excess reactant. Indicate which substance is the limiting reactant.

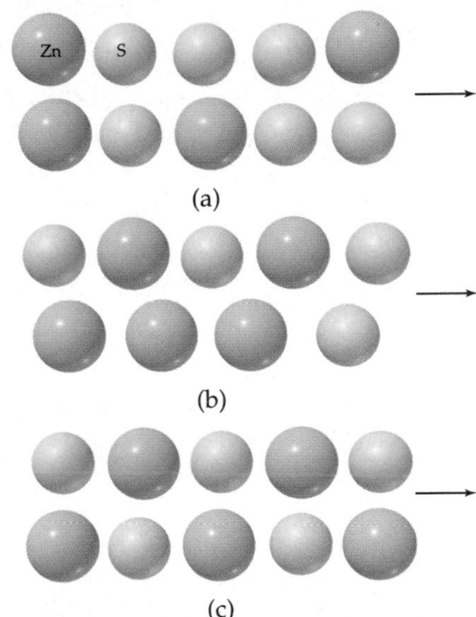

(a) (b) (c)

6. Copper metal can react with sulfur to give copper(II) sulfide. Complete the following table by stating the moles of Cu, S, and CuS after the reaction is complete. The balanced equation for the chemical reaction is

$$Cu(s) + S(s) \longrightarrow CuS(s)$$

Experiment	mol Cu	mol S	mol CuS
before reaction	3.00	2.00	0.00
after reaction			

7. Copper metal can also react with sulfur to yield copper(I) sulfide. Complete the following table by stating the number of moles of Cu, S, and Cu_2S after the reaction is complete. The balanced equation for the chemical reaction is

$$2\,Cu(s) + S(s) \longrightarrow Cu_2S(s)$$

Experiment	mol Cu	mol S	mol Cu_2S
before reaction	3.00	2.00	0.00
after reaction			

Key Terms†

Select the key term below that corresponds to each of the following definitions.

_____ 1. the ratio of moles of reactants and products according to the coefficients in the balanced chemical equation

_____ 2. the principle that equal volumes of gases contain equal numbers of molecules, at the same temperature and pressure

_____ 3. the law that states that the total mass of reactants in a chemical reaction is equal to the total mass of products

_____ 4. the mass in grams of 1 mol of substance

_____ 5. the relationship between amounts of substance (mass of a substance or volume of a gas) in a chemical reaction according to the balanced equation

_____ 6. a type of stoichiometry calculation that relates the masses of two substances

_____ 7. a type of stoichiometry calculation that relates the mass of a substance to the volume of a gas

_____ 8. a type of stoichiometry calculation that relates the volumes of two gases at the same temperature and pressure

_____ 9. the volume of 1 mol of gas, that is, 22.4 L/mol

(a) actual yield (*Sec. 10.9*)
(b) Avogadro's theory (*Sec. 10.1*)
(c) combining volumes (*Sec. 10.6*)
(d) conservation of mass (*Sec. 10.1*)
(e) limiting reactant (*Sec. 10.7*)
(f) mass–mass problem (*Sec. 10.3*)
(g) mass–volume problem (*Sec. 10.3*)
(h) molar mass (*Sec. 10.1*)
(i) mole ratio (*Sec. 10.1*)
(j) molar volume (*Sec. 10.5*)

† Answers to Key Terms are in Appendix I.

______ 10. a temperature of 0°C and a pressure of 1 atm

______ 11. the law that states that the volumes of gases that combine in a chemical reaction are in the ratio of small whole numbers

______ 12. the substance in a chemical reaction that controls the maximum amount of product formed

______ 13. the amount of product experimentally obtained from a given amount of reactant

______ 14. the amount of product calculated to be obtained from a given amount of reactant

______ 15. the ratio of actual yield to theoretical yield, all multiplied by 100

(k) percent yield *(Sec. 10.9)*
(l) standard conditions *(Sec. 10.5)*
(m) stoichiometry *(Sec. 10.3)*
(n) theoretical yield *(Sec. 10.9)*
(o) volume–volume problem *(Sec. 10.3)*

Exercises‡

Interpreting a Chemical Equation (Sec. 10.1)

1. Consider the general chemical equation:
$2\,A + 3\,B \longrightarrow C + 2\,D$
(a) How many moles of C are produced from 2 mol of A?
(b) How many liters of gas D are produced from 2 L of gas A?

2. Consider the general chemical equation:
$A + 2\,B \longrightarrow C + 3\,D$
(a) How many moles of A must react to give 3 mol of D?
(b) How many liters of gas B must react to give 3 L of gas D?

3. Consider the general chemical equation:
$A + 3\,B \longrightarrow 2\,C$
(a) If 10.0 g of A reacts with 15.0 g of B, what is the mass of C?
(b) If 50.0 g of A reacts to produce 75.0 g of C, what is the mass of B?

4. Consider the general chemical equation:
$3\,A + B \longrightarrow 2\,C$
(a) If 5.00 g of A reacts with 1.50 g of B, what is the mass of C?
(b) If 2.50 g of A reacts to produce 6.50 g of C, what is the mass of B?

5. Verify the conservation of mass law using the molar masses of reactants and products for each substance in the following balanced equations.
(a) $2\,KNO_3(s) \longrightarrow 2\,KNO_2(s) + O_2(g)$
(b) $2\,Al(s) + 3\,Br_2(s) \longrightarrow 2\,AlBr_3(s)$

6. Verify the conservation of mass law using the molar masses of reactants and products for each substance in the following balanced equations.
(a) $CH_4(g) + 2\,O_2(g) \xrightarrow{\text{spark}} CO_2(g) + 2\,H_2O(g)$
(b) $2\,C_2H_6(g) + 7\,O_2(g) \xrightarrow{\text{spark}} 4\,CO_2(g) + 6\,H_2O(g)$

Mole–Mole Relationships (Sec. 10.2)

7. Given the balanced equation, calculate the moles of oxygen gas that react with 0.500 mol of hydrogen gas. How many moles of water are produced?
$$2\,H_2(g) + O_2(g) \xrightarrow{\text{spark}} 2\,H_2O(g)$$

8. Given the balanced equation, calculate the moles of oxygen gas that react with 2.50 mol of phosphorus? How many moles of diphosphorus pentaoxide are produced?
$$4\,P(s) + 5\,O_2(g) \longrightarrow 2\,P_2O_5(s)$$

9. How many moles of chlorine gas react with 0.333 mol of metallic iron? How many moles of iron(III) chloride are produced?
$$Fe(s) + Cl_2(g) \longrightarrow FeCl_3(s)$$

10. How many moles of barium metal react to produce 0.333 mol of barium nitride? How many moles of nitrogen gas react?
$$Ba(s) + N_2(g) \longrightarrow Ba_3N_2(s)$$

11. How many moles of propane gas, C_3H_8, react with 1.75 mol of oxygen gas? How many moles of carbon dioxide are produced?
$$C_3H_8(g) + O_2(g) \longrightarrow CO_2(g) + H_2O(g)$$

▲ **Propane Torch** The reaction of propane gas, C_3H_8, and air releases heat energy.

12. How many moles of butane gas, C_4H_{10}, react to produce 0.125 mol of water? How many moles of oxygen gas react?
$$C_4H_{10}(g) + O_2(g) \longrightarrow CO_2(g) + H_2O(g)$$

‡Answers to odd-numbered Exercises are in Appendix J.

Types of Stoichiometry Problems (Sec. 10.3)

13. Classify the type of stoichiometry problem: How many kilograms of iron oxide are produced from the reaction of steam and 256.4 g of iron?
14. Classify the type of stoichiometry problem: How many milliliters of nitrogen dioxide gas are produced from the reaction of oxygen gas and 1.00 L of nitrogen monoxide gas?
15. Classify the type of stoichiometry problem: How many milliliters of hydrogen are produced from the reaction of sulfuric acid and 1.414 g of cadmium metal?
16. Classify the type of stoichiometry problem: How many liters of hydrogen gas react with bromine liquid to yield 25.0 cm^3 of HBr gas?
17. Classify the type of stoichiometry problem: How many grams of aluminum metal react with nitric acid to give 0.403 g of $Al(NO_3)_3$?
18. Classify the type of stoichiometry problem: How many grams of potassium chlorate are decomposed by heating to yield 45.5 cm^3 of O_2 gas?

Mass–Mass Problems (Sec. 10.4)

19. Given the balanced equation, calculate the mass of product that can be prepared from 2.36 g of zinc metal.

$$2\,Zn(s) + O_2(g) \longrightarrow 2\,ZnO(s)$$

20. How many grams of oxygen gas must react to give 1.28 g of ZnO?
21. Given the balanced equation, calculate the mass of product that can be prepared from 3.45 g of bismuth metal.

$$2\,Bi(s) + 3\,Cl_2(g) \longrightarrow 2\,BiCl_3(s)$$

22. How many grams of chlorine gas must react to give 3.52 g of $BiCl_3$?
23. What is the mass of silver that can be prepared from 0.615 g of copper metal?

$$Cu(s) + AgNO_3(aq) \longrightarrow Cu(NO_3)_2(aq) + Ag(s)$$

24. How many grams of silver nitrate must react to give 1.00 g Ag?
25. What is the mass of mercury that can be prepared from 1.25 g of cobalt metal?

$$Co(s) + HgCl_2(aq) \longrightarrow CoCl_3(aq) + Hg(l)$$

26. How many grams of mercuric chloride must react to give 5.11 g Hg?
27. What is the mass of calcium phosphate that can be prepared from 1.78 g of Na_3PO_4?

$$Na_3PO_4(aq) + Ca(OH)_2(aq) \longrightarrow Ca_3(PO_4)_2(s) + NaOH(aq)$$

28. How many grams of calcium hydroxide must react to give 2.39 g of $Ca_3(PO_4)_2$?

Mass–Volume Problems (Sec. 10.5)

29. Given the balanced equation, how many milliliters of carbon dioxide gas at STP are produced from the decomposition of 1.59 g of ferric carbonate?

$$Fe_2(CO_3)_3(s) \longrightarrow Fe_2O_3(s) + 3\,CO_2(g)$$

30. Given the balanced equation, how many milliliters of oxygen gas at STP are released from the decomposition of 2.57 g of calcium chlorate?

$$Ca(ClO_3)_2(s) \longrightarrow CaCl_2(s) + 3\,O_2(g)$$

31. How many milliliters of carbon dioxide gas at STP are liberated from the decomposition of 1.59 g of lithium hydrogen carbonate?

$$LiHCO_3(s) \longrightarrow Li_2CO_3(s) + H_2O(l) + CO_2(g)$$

32. How many milliliters of oxygen gas at STP are released from the decomposition of 2.50 g of mercuric oxide?

$$HgO(s) \longrightarrow Hg(l) + O_2(g)$$

33. What mass of magnesium metal reacts with sulfuric acid to evolve 225 mL of hydrogen gas at STP?

$$Mg(s) + H_2SO_4(aq) \longrightarrow MgSO_4(aq) + H_2(g)$$

34. What mass of sodium metal reacts with water to give 75.0 mL of hydrogen gas at STP?

$$Na(s) + H_2O(l) \longrightarrow NaOH(aq) + H_2(g)$$

35. How many grams of hydrogen peroxide must decompose to give 55.0 mL of oxygen gas at STP?

$$H_2O_2(l) \longrightarrow H_2O(l) + O_2(g)$$

▲ **Hydrogen Peroxide** Hydrogen peroxide is unstable and breaks down to give water and oxygen gas. When H_2O_2 is poured on a wound, the fizzing bubbles are O_2 gas.

287

grams of manganese(II) chloride must react uric acid to release 49.5 mL of hydrogen chlo- s at STP?

$$..Cl_2(s) + H_2SO_4(aq) \longrightarrow MnSO_4(aq) + 2\ HCl(g)$$

Volume–Volume Problems (Sec. 10.6)

37. Assuming all gas volumes are measured at the same temperature and pressure, how many liters of oxygen gas react with 2.00 L of carbon monoxide?

$$CO(g) + O_2(g) \longrightarrow CO_2(g)$$

38. How many liters of carbon dioxide are produced from 2.00 L of CO?

39. Assuming all gas volumes are measured at the same temperature and pressure, how many milliliters of iodine vapor react with 125 mL of hydrogen gas?

$$H_2(g) + I_2(g) \longrightarrow HI(g)$$

40. How many milliliters of hydrogen iodide are produced from 125 mL of H_2?

41. Assuming all gas volumes are measured at the same temperature and pressure, how many milliliters of nitrogen gas react to give 45.0 mL of ammonia gas?

$$H_2(g) + N_2(g) \longrightarrow NH_3(g)$$

42. How many milliliters of hydrogen gas must react to give 45.0 mL of NH_3?

43. Assuming all gas volumes are measured at the same temperature and pressure, how many milliliters of chlorine gas react to yield 1.75 L of dichlorine trioxide?

$$Cl_2(g) + O_2(g) \longrightarrow Cl_2O_3(g)$$

44. How many milliliters of oxygen gas must react to give 1.75 L of Cl_2O_3?

45. What volume of sulfur trioxide gas is produced from 25.0 L of oxygen gas? The volume of each gas is measured at 875°C and 1.00 atm pressure.

$$SO_2(g) + O_2(g) \longrightarrow SO_3(g)$$

46. What volume of sulfur dioxide gas reacts with 25.0 L of O_2?

47. What volume of nitrogen gas reacts to yield 500.0 cm^3 of dinitrogen pentaoxide? The volume of each gas is measured at 350°C and 2.50 atm pressure.

$$N_2(g) + O_2(g) \longrightarrow N_2O_5(g)$$

48. What volume of oxygen gas reacts to yield 500.0 cm^3 of N_2O_5?

The Limiting Reactant Concept (Sec. 10.7)

49. If 1.00 mol of nitrogen gas and 1.50 mol of oxygen gas react, what is the limiting reactant and how many moles of NO are produced from the reaction?

$$N_2(g) + O_2(g) \longrightarrow 2\ NO(g)$$

50. If 1.00 mol of nitrogen gas and 0.500 mol of oxygen gas react, what is the limiting reactant and how many moles of NO are produced from the reaction?

51. If 1.00 mol of nitrogen monoxide gas and 1.00 mol of oxygen gas react, what is the limiting reactant and how many moles of NO_2 are produced from the reaction?

$$2\ NO(g) + O_2(g) \longrightarrow 2\ NO_2(g)$$

52. If 1.00 mol of nitrogen monoxide gas and 0.250 mol of oxygen gas react, what is the limiting reactant and how many moles of NO_2 are produced from the reaction?

53. If 5.00 mol of hydrogen gas and 5.00 mol of oxygen gas react, what is the limiting reactant and how many moles of water are produced from the reaction?

$$2\ H_2(g) + O_2(g) \longrightarrow 2\ H_2O(l)$$

54. If 5.00 mol of hydrogen gas and 1.50 mol of oxygen gas react, what is the limiting reactant and how many moles of water are produced from the reaction?

55. If 1.00 mol of ethane gas and 5.00 mol of oxygen gas react, what is the limiting reactant and how many moles of water are produced from the reaction?

$$2\ C_2H_6(g) + 7\ O_2(g) \longrightarrow 4\ CO_2(g) + 6\ H_2O(g)$$

56. If 1.00 mol of ethane gas and 1.50 mol of oxygen gas react, what is the limiting reactant and how many moles of water are produced from the reaction?

57. Cobalt metal can react with sulfur to give cobalt(II) sulfide. Complete the following table by stating the moles of Co, S, and CoS after the reaction is complete. The balanced equation for the chemical reaction is

$$Co(s) + S(s) \longrightarrow CoS(s)$$

Experiment	mol Co	mol S	mol CoS
1. before reaction	1.50	2.00	0.00
after reaction			
2. before reaction	3.00	2.00	0.00
after reaction			

58. Cobalt metal can also react with sulfur to give cobalt(III) sulfide. Complete the following table by stating the moles of Co, S, and Co_2S_3 after the reaction is complete. The balanced equation for the chemical reaction is

$$2\ Co(s) + 3\ S(s) \longrightarrow Co_2S_3(s)$$

Experiment	mol Co	mol S	mol Co_2S_3
1. before reaction	1.50	2.00	0.00
after reaction			
2. before reaction	3.00	2.00	0.00
after reaction			

Limiting Reactant Problems (Sec. 10.8)

59. If 40.0 g of molten iron(II) oxide reacts with 10.0 g of magnesium according to the following equation, what is the mass of iron produced?

$$FeO(l) + Mg(l) \longrightarrow Fe(l) + MgO(s)$$

60. If 10.0 g of molten iron(II) oxide reacts with of 40.0 g magnesium according to the above equation, what is the mass of iron produced?

61. If 175 g of molten iron(III) oxide reacts with 37.5 g of aluminum according to the following equation, what is the mass of iron produced?

$$Fe_2O_3(l) + Al(l) \longrightarrow Fe(l) + Al_2O_3(s)$$

62. If 37.5 g of molten iron(III) oxide reacts with 175 g of aluminum according to the above equation, what is the mass of iron produced?

63. If 1.00 g of magnesium hydroxide reacts with 0.605 g of sulfuric acid according to the following equation, what is the mass of magnesium sulfate produced?

$$Mg(OH)_2(s) + H_2SO_4(l) \longrightarrow MgSO_4(s) + H_2O(l)$$

64. If 0.605 g of magnesium hydroxide reacts with 1.00 g of sulfuric acid according to the above equation, what is the mass of magnesium sulfate produced?

65. If 1.00 g of aluminum hydroxide reacts with 3.00 g of sulfuric acid according to the following equation, what is the mass of water produced?

$$Al(OH)_3(s) + H_2SO_4(l) \longrightarrow Al_2(SO_4)_3(aq) + H_2O(l)$$

66. If 3.00 g of aluminum hydroxide reacts with 1.00 g of sulfuric acid according to the above equation, what is the mass of water produced?

67. If 45.0 mL of nitrogen gas reacts with 95.0 mL of oxygen gas according to the following equation, what is the volume of NO_2 produced? Assume all gases are at the same conditions.

$$N_2(g) + O_2(g) \longrightarrow NO_2(g)$$

68. If 95.0 mL of nitrogen gas reacts with 45.0 mL of oxygen gas according to the above equation, what is the volume of NO_2 produced?

69. If 70.0 mL of nitrogen gas reacts with 45.0 mL of oxygen gas according to the following equation, what is the volume of N_2O_3 produced? Assume all gases are at the same conditions.

$$N_2(g) + O_2(g) \longrightarrow N_2O_3(g)$$

70. If 45.0 mL of nitrogen gas reacts with 70.0 mL of oxygen gas according to the above equation, what is the volume of N_2O_3 produced? Assume all gases are at the same conditions.

71. If 3.00 L of sulfur dioxide gas reacts with 1.25 L of oxygen gas, what is the volume of sulfur trioxide gas produced? Assume all gases are at the same conditions.

$$SO_2(g) + O_2(g) \longrightarrow SO_3(g)$$

72. If 1.25 L of sulfur dioxide gas reacts with 3.00 L of oxygen gas according to the above equation, what is the volume of sulfur trioxide gas produced?

73. If 50.0 L of hydrogen chloride gas reacts with 10.0 L of oxygen gas, what is the volume of chlorine gas produced? Assume all gases are at the same conditions.

$$HCl(g) + O_2(g) \longrightarrow Cl_2(g) + H_2O(g)$$

74. If 10.0 L of hydrogen chloride gas reacts with 50.0 L of oxygen gas according to the above equation, what is the volume of chlorine gas produced?

Percent Yield (Sec. 10.9)

75. A chemistry student prepares acetone by decomposing 31.6 g of calcium acetate. If the student collected 10.4 g of acetone and the theoretical yield is 11.6 g, what is the percent yield?

76. A chemistry student prepares lead(II) iodide from 1.55 g of lead(II) nitrate and excess aqueous potassium iodide. If the student collected 2.01 g of PbI_2 and the theoretical yield is 11.6 g, what is the percent yield?

77. A 1.50-g sample of sodium nitrate is decomposed by heating. If the resulting sodium nitrite has a mass of 1.29 g and the calculated yield is 1.22 g, what is the percent yield?

78. A 1.000-g sample of potassium bicarbonate is decomposed by heating. If the resulting potassium carbonate weighs 0.725 g and the calculated yield is 0.716 g, what is the percent yield?

General Exercises

79. What units are associated with molar mass?

80. What units are associated with molar volume?

81. The "volcano reaction" produces chromium(III) oxide, water, and nitrogen gas from the decomposition of ammonium dichromate (chapter opening photo). Calculate the mass of green chromium(III) oxide produced by the decomposition of 1.54 g of orange ammonium dichromate.

82. What volume of nitrogen gas at STP is produced in the "volcano reaction" from the decomposition of 1.54 g of ammonium dichromate according to the reaction described in Exercise 81?

83. Manganese metal and aluminum oxide are produced by the reaction of manganese(IV) oxide and aluminum metal. Calculate the mass of aluminum necessary to yield 1.00 kg of manganese metal.

nass of aluminum oxide must react to produce kg of manganese metal according to the reaction escribed in Exercise 83?

. The mineral stibnite, antimony(III) sulfide, is treated with hydrochloric acid to give antimony(III) chloride and hydrogen sulfide gas. What STP volume of H_2S is evolved from a 3.00-g sample of stibnite?

86. What mass of $SbCl_3$ is produced from a 3.00-g sample of stibnite according to the reaction described in Exercise 85?

87. When an electric current is passed through water, hydrogen and oxygen gases are liberated. What STP volume of hydrogen gas is obtained from the electrolysis of 100.0 mL of water? (*Hint:* The density of water is 1.00 g/mL.)

88. What STP volume of oxygen gas is obtained from the electrolysis of 100.0 mL of water according to the reaction described in Exercise 87?

89. Recreational vehicles use propane for cooking and heating. The combustion of propane and oxygen produces carbon dioxide and water. What mass of water is produced from the combustion of 10.0 g of propane, C_3H_8?

90. What STP volume of carbon dioxide gas is produced from the combustion of 10.0 g of propane according to the reaction described in Exercise 89?

91. A chemical reaction stops when one of the reactants is consumed. For example, the combustion reaction in an automobile stops when the gasoline "runs out." In this case, what is the limiting reactant? What is the excess reactant?

WWW
Explorer Quiz 1
Explorer Quiz 2
Explorer Quiz 3
Master Quiz

CHAPTER 11

The Gaseous State

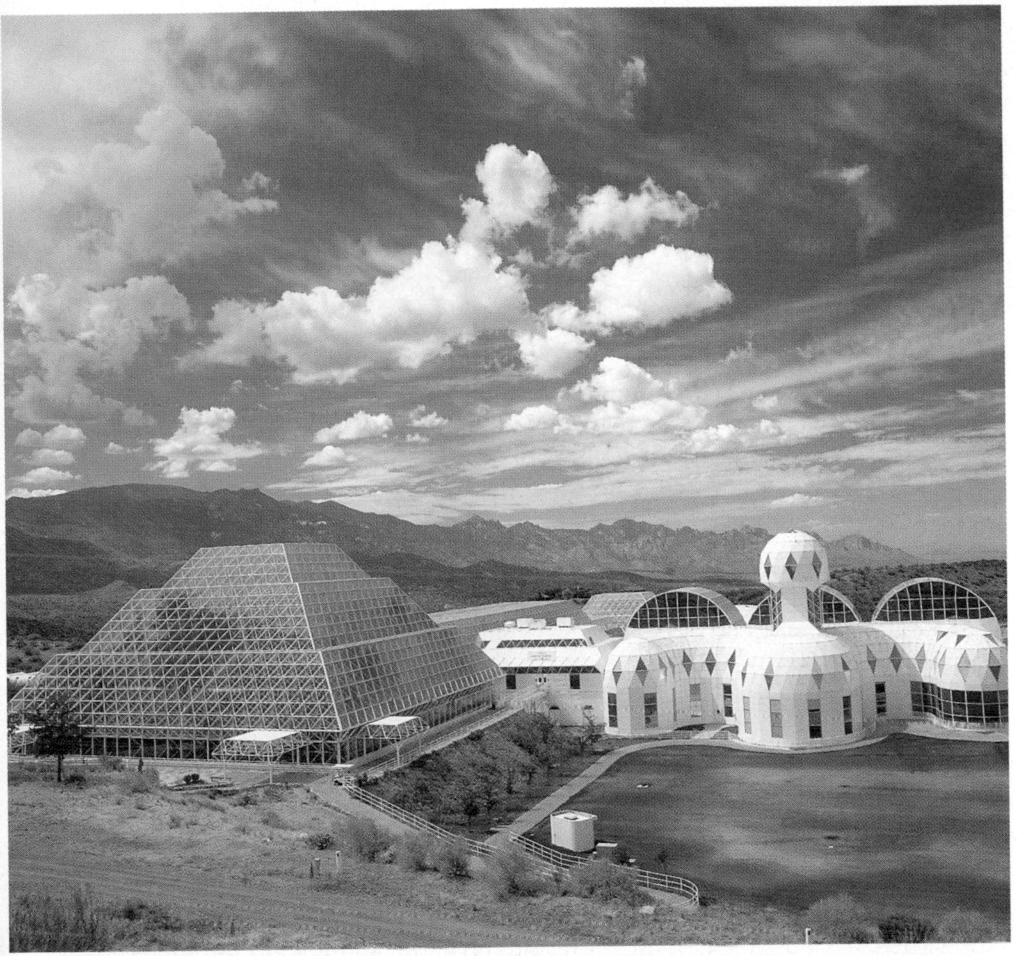

▲ The Biosphere 2 complex in Oracle, Arizona, near Tucson, was operated by Columbia University in a two-year study to test the feasibility of a self-sustaining space colony. What gas is produced by the plants in the Biosphere greenhouse that is necessary to sustain human life?

The calculations in this chapter involving gas laws are presented using the unit analysis method of problem solving. However, if preferred, the gas law calculations can be performed using algebra. In fact, each of the example exercise calculations shows both a unit analysis solution and an alternate algebraic solution.

In early Greek civilization, logic and reason dominated science. The Greek philosophers valued thoughtful mental exercises, but they showed little interest in practical experiments. The Greeks believed that all matter was composed of four basic elements—air, earth, fire, and water. Much later, European scientists discovered that matter is composed of many basic elements including sulfur, arsenic, zinc, tin, and lead. In the 1600s, the English chemist Robert Boyle proposed that these and other elements could be identified by experiments performed in the laboratory.

Robert Boyle is best known for his work with gases. He studied the atmosphere and found that a volume of air could be compressed. He envisioned air as being

▲ **Hot-Air Balloon** The hot air inside the balloon is less dense than the air surrounding the balloon. As the less dense air rises, the balloon ascends.

composed of invisible particles that can be squeezed closer together. Furthermore, he found that other gases could also be compressed and reasoned that all gases were composed of discrete particles. His experiments were remarkable because they offered the first evidence of the particle nature of matter. In 1803 John Dalton proposed the atomic theory, which was supported in part by Boyle's experiments with gases.

11.1 Properties of Gases

Objective · To list five observed properties of a gas.

In the late 1700s, high-altitude ballooning became popular in France. Even French scientists filled balloons with hydrogen gas and ascended miles into the sky. Ballooning in turn helped to create a great deal of interest in the study of gases. After numerous experiments, scientists concluded that all gases have common characteristics. Even mixtures of gases, such as air, have similar properties. We can summarize these properties as follows.

1. *Gases have an indefinite shape.* A gas takes on the shape of its container and fills it uniformly. If the shape of the container changes, so does the shape of the gas.
2. *Gases can expand.* A gas continuously expands and distributes itself throughout a closed container. This means that the volume of gas in a closed cylinder increases when we enlarge the volume of the container.
3. *Gases can compress.* The volume of gas in a closed cylinder decreases when we reduce the volume. If we reduce the volume sufficiently, the gas will eventually liquefy.
4. *Gases have low densities.* The density of air is about 1.3 g/L. The density of water, on the other hand, is 1.0 g/mL. Thus, air is almost 1000 times less dense than water.
5. *Gases diffuse uniformly throughout their containers to form homogeneous mixtures.* Air is a common example of a gaseous mixture. During photosynthesis, green plants release oxygen gas which mixes homogeneously with other gases in the atmosphere. Similarly, automobiles emit oxides of nitrogen which diffuse throughout the atmosphere.

Although we cannot see most gases, we can observe their properties. For example, when we change the temperature of oxygen gas, we observe a change in pressure. Thus, it is possible to gather information about an invisible gaseous substance by studying its properties.

By the mid-1800s, scientists began to formulate a model for the observable behavior of a gas based on the behavior of *individual* gas molecules. In Section 11.10, we will study this model which we call the kinetic theory of gases. For now, we will consider only the practical, observable properties of gases.

11.2 Atmospheric Pressure

Objectives · To state standard atmospheric pressure in the following units: atm, mm Hg, torr, cm Hg, in. Hg, psi, and kPa.
· To convert a given gas pressure to a different unit of measurement.

Gas pressure is the result of constantly moving molecules striking the inside surface of its container. The pressure that a gas exerts depends on how often and how hard these molecules strike the walls of the container.

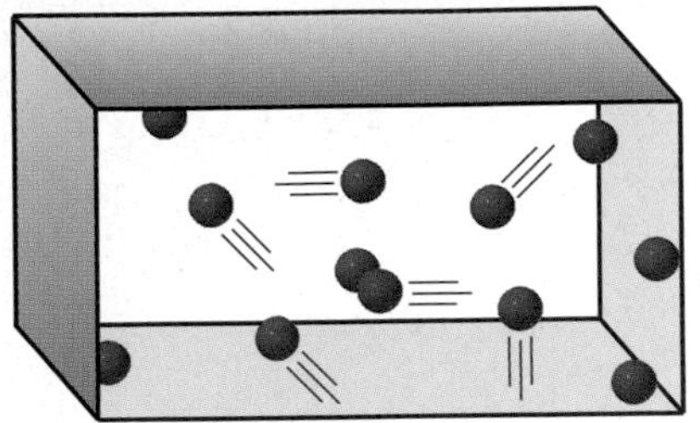
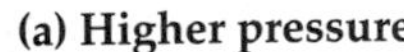
(a) Higher pressure

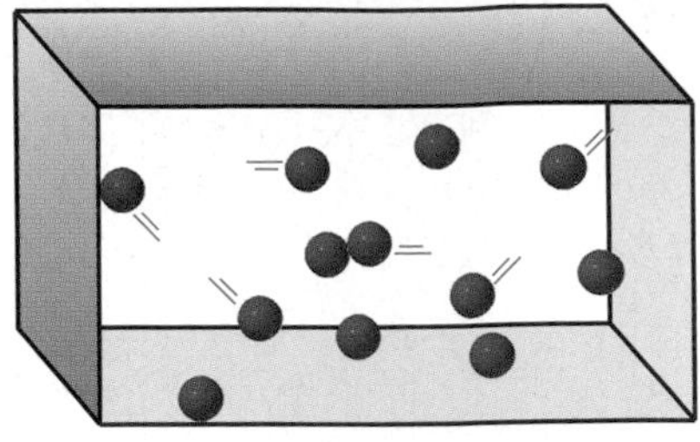
(b) Lower pressure

◀ **Figure 11.1 Pressure Depends on the Frequency and the Energy of Collisions** (a) The molecules are moving faster and colliding more frequently and more energetically with the walls of the container. (b) The molecules are moving slower and colliding less frequently and less energetically with the walls of the container.

The Concept of Gas Pressure

1. If the molecules collide more often, the gas pressure *increases*.
2. If the molecules collide with more energy, the pressure *increases*.

As the temperature increases, the molecules move faster and collide more frequently and with more energy. Thus, the gas pressure increases (Figure 11.1).

Atmospheric Pressure

The ancient Greeks observed that a wine barrel with a single hole emptied slowly. To explain this observation, Aristotle suggested that a vacuum is created as wine empties from a barrel. He further stated that "nature abhors a vacuum" because a vacuum violates natural principles.

The Italian scientist Evangelista Torricelli (1608–1647) published the first scientific explanation of a vacuum. He proposed that a "sea of air" surrounds the Earth. He argued that this sea of air exerts a pressure on everything it contacts. Therefore, air pressure is responsible for slowing the flow of wine from a barrel. Torricelli reasoned that a second hole in the top of the barrel would allow air to rush in and the liquid to quickly rush out. He opened a second hole in the top of the barrel, and sure enough his prediction was correct. We now understand that there are no gas molecules in a **vacuum** and that the gas pressure is therefore zero.

By the end of the 1600s, the concept of atmospheric pressure was well established. Today, we understand that **atmospheric pressure** is the result of air molecules striking various surfaces in the environment. The pressure of the atmosphere is considerable, about 15 pounds on every square inch of surface. In fact, the atmosphere exerts a total weight of nearly 20 tons on an average-size human body! More effects of atmospheric pressure are illustrated in Figure 11.2.

In 1643 Torricelli invented the **barometer** to measure atmospheric pressure (Figure 11.3). He took a four-foot glass tube, sealed one end, and filled it with mercury. He put a stopper in the open end of the tube, turned it upside down, and inserted the stoppered end into a dish of mercury. When Torricelli removed the stopper, he found that the height of mercury inside the tube fell to about 30 inches. Each day he read the height of mercury and found that it varied slightly. He even took his barometer on hikes in the Italian Alps and recorded the height of mercury at different elevations.

Torricelli claimed that atmospheric pressure is responsible for the fluctuations in the height of mercury in the column. At sea level, the height of mercury measures 29.9 inches which we define as standard atmospheric pressure, that is, 1 atmosphere (symbol atm). Above sea level, the height of mercury is less than 29.9 inches, indicating that atmospheric pressure is less at higher elevations.We can express standard pressure in many other units besides atmospheres and inches of mercury. For example, 29.9 in. of Hg is equivalent to 760 mm Hg. In honor of Torricelli,

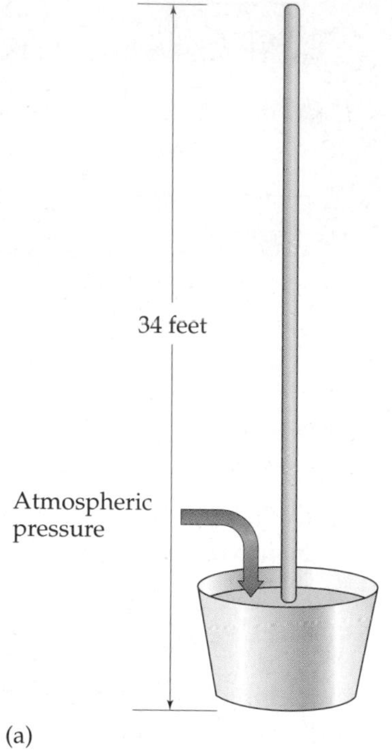

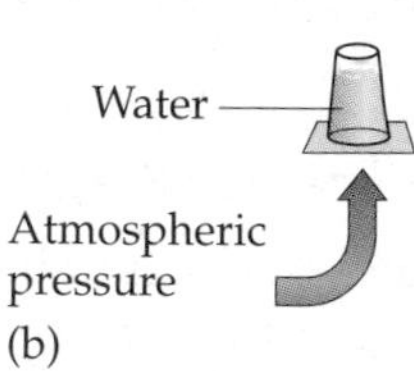

Figure 11.2 Atmospheric Pressure (a) The pressure of air molecules in the atmosphere supports a 34-foot column of water. (b) Since atmospheric pressure operates in all directions, air molecules strike the bottom of the card and support the water in the inverted glass.

standard pressure is 760 torr, where 1 **torr** equals 1 mm of Hg. Table 11.1 lists units that are frequently used to express pressure.

Since each of the values in Table 11.1 is an expression for the same standard pressure, we can relate one unit to another. For example, if the pressure of propane gas in a steel tank is 2550 torr, we can relate the pressure to atmospheres. We can perform the conversion by applying the unit analysis method. The unknown unit is atm, and the given value is 2550 torr. Thus,

$$2550 \text{ torr} \times \frac{\text{unit}}{\text{factor}} = \text{atm}$$

To convert from one unit of pressure to another, we derive a unit factor from the relationship among units of standard pressure. In this example, 1 atm = 760 torr, so the two related unit factors are 1 atm/760 torr and 760 torr/1 atm. We select the first unit factor in order to cancel units.

▲ **Gauge Pressure** When measuring tire pressure, we obtain a so-called gauge pressure. Gauge pressure is equal to the pressure inside the tire plus the pressure of the atmosphere on the outside of the tire. Thus, the actual pressure inside the tire equals the gauge pressure minus about 14.7 psi.

Table 11.1 Units of Gas Pressure

Unit	Standard Pressure
atmosphere	1 atm (exactly)
inches of mercury	29.9 in. Hg
centimeters of mercury	76 cm Hg (exactly)
millimeters of mercury	760 mm Hg (exactly)
torr*	760 torr (exactly)
pounds per square inch	14.7 psi
kilopascal†	101 kPa

* A millimeter of mercury (mm Hg) is defined as exactly equal to 1 torr pressure in honor of Torricelli who invented the barometer. Thus, standard pressure can be given as 760 torr.

† The kilopascal (kPa) is the standard SI unit of pressure.

$$2550 \text{ \sout{torr}} \times \frac{1 \text{ atm}}{760 \text{ \sout{torr}}} = 3.36 \text{ atm}$$

The following example exercise further illustrates the conversion of gas pressures.

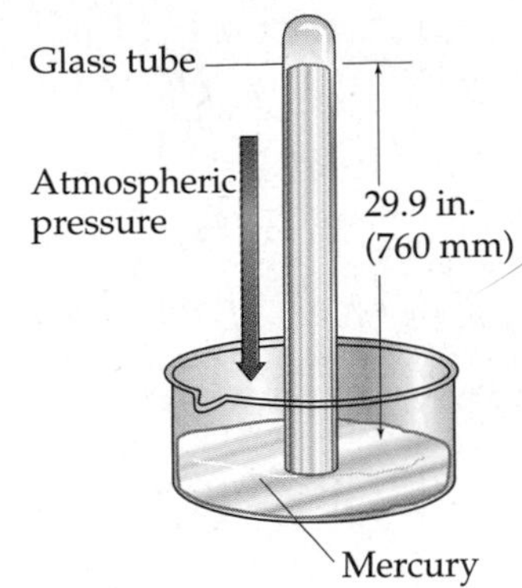

▲ **Figure 11.3 Torricelli's Mercury Barometer** A sealed glass tube is filled with liquid mercury, inverted, and placed into a dish of mercury. There are no gas molecules above the column of mercury inside the glass tube. The pressure of the atmosphere at sea level supports a column of mercury 29.9 in. (760 mm) high.

Example Exercise 11.1 • Gas Pressure Conversion

Meteorologists say that a "falling" barometer indicates that a storm is moving into a region. Given a barometric pressure of 27.5 in. Hg, express the pressure in each of the following units.

(a) atm
(b) mm Hg
(c) psi
(d) kPa

Solution

For each conversion, we apply a unit conversion factor related to units of standard pressure.

(a) To express the pressure in atmospheres, we derive a unit factor related to the equivalent relationship 29.9 in. Hg = 1 atm.

$$27.5 \text{ \sout{in. Hg}} \times \frac{1 \text{ atm}}{29.9 \text{ \sout{in. Hg}}} = 0.920 \text{ atm}$$

(b) To convert to millimeters of mercury, we derive a unit factor related to the equivalent relationship 29.9 in. Hg = 760 mm Hg.

$$27.5 \text{ \sout{in. Hg}} \times \frac{760 \text{ mm Hg}}{29.9 \text{ \sout{in. Hg}}} = 699 \text{ mm Hg}$$

(c) To calculate the pressure in pounds per square inch, we derive a unit factor related to the equivalent relationship 29.9 in. Hg = 14.7 psi.

$$27.5 \text{ \sout{in. Hg}} \times \frac{14.7 \text{ psi}}{29.9 \text{ \sout{in. Hg}}} = 13.5 \text{ psi}$$

(d) To find the pressure in kilopascals, we derive a unit factor related to the equivalent relationship 29.9 in. Hg = 101 kPa.

$$27.5 \text{ \sout{in. Hg}} \times \frac{101 \text{ kPa}}{29.9 \text{ \sout{in. Hg}}} = 92.9 \text{ kPa}$$

Since 1 atm and 760 mm Hg are exact values, the answers have been rounded to three significant digits, which is consistent with the given value, 27.5 in. Hg.

Self-Test Exercise

Given that the gauge pressure of an automobile tire is 34.0 psi, express the pressure in each of the following units.

(a) atm
(b) cm Hg
(c) torr
(d) kPa

Answers: (a) 2.31 atm; (b) 176 cm Hg; (c) 1760 torr; (d) 234 kPa

11.3 Variables Affecting Gas Pressure

Objectives
- To identify the three variables that affect the pressure of a gas.
- To state whether gas pressure increases or decreases for a given change in the volume, the temperature, or the number of moles of gas.

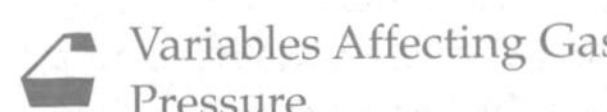

In Section 11.2, we learned that gas pressure is related to the frequency and energy of molecular collisions. Experimentally, we cannot directly change the frequency or energy of collisions. Therefore, we must indirectly affect collisions to change the pressure. In any gaseous system, there are only three ways to change the pressure.

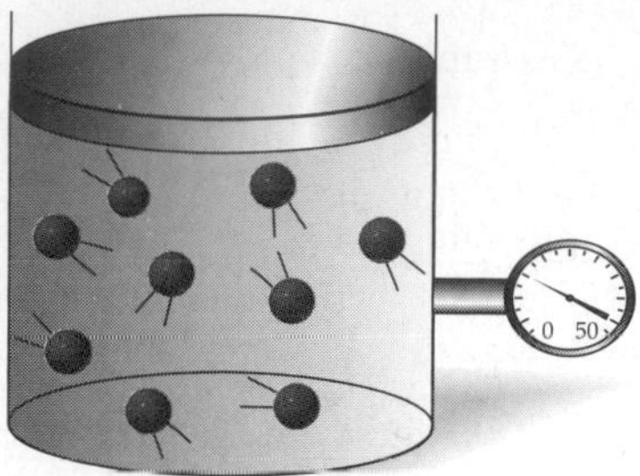

Initial 1.0 L of gas

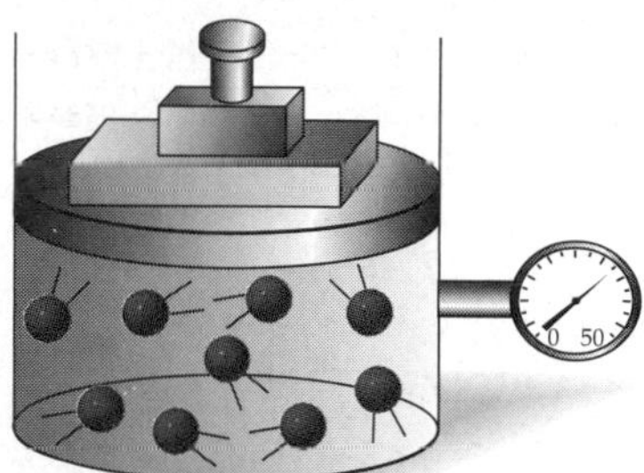

(a) 0.5 L of gas

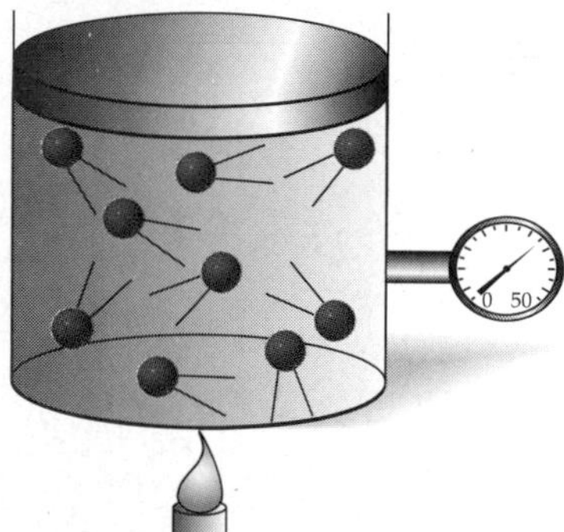

(b) 1.0 L of gas

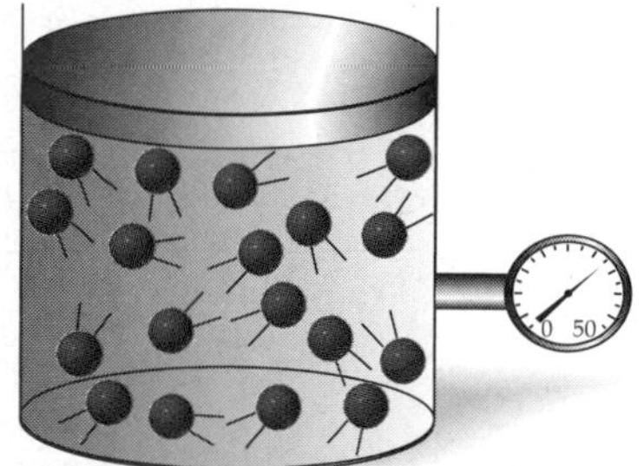

(c) 1.0 L of gas

▲ **Figure 11.4 Variables Affecting Gas Pressure** A gas is contained in a cylinder with a moving piston. The pressure of the gas is affected by changing (a) the volume, (b) the temperature, and (c) the number of gas molecules in the cylinder. (Note the pressure gauge in each case.)

1. *Increase or decrease the volume of the container.* When we increase the volume, gas molecules are further apart, collide less frequently, and the pressure decreases. When we decrease the volume, gas molecules are closer together, collide more frequently, and the pressure increases. The pressure is said to be *inversely* related to the volume. That is, when the volume increases, the pressure decreases. When the volume decreases, the pressure increases.
2. *Increase or decrease the temperature of the gas.* When we increase the temperature, gas molecules move faster and collide with a greater frequency and energy. When we decrease the temperature, gas molecules move slower and collide less frequently. The pressure is said to be *directly* related to the temperature. When the temperature increases, the pressure increases. When the temperature decreases, the pressure decreases.
3. *Increase or decrease the number of molecules in the container.* When we increase the number of gas molecules, there are more collisions and the pressure increases. When we decrease the number of gas molecules, there are fewer collisions and the pressure decreases. The pressure is said to be *directly* related to the number of moles of gas, or number of gas molecules. When the number of gas molecules increases, the pressure increases. When the number of molecules decreases, the pressure decreases.

The effects of these three variables, volume, temperature, and number of molecules, on pressure are illustrated in Figure 11.4.

Example Exercise 11.2 • Gas Pressure Changes

State whether the pressure of a gas in a closed system increases or decreases with each of the following changes.

(a) The volume changes from 250 mL to 500 mL.
(b) The temperature changes from 20° to −80°C.
(c) The moles of gas change from 1.00 mol to 1.50 mol.

Solution

For each of the changes we must consider whether the number of molecular collisions increases or decreases.

(a) The volume increases, and so the number of collisions decreases; thus, the *pressure decreases.*
(b) The temperature decreases, and so the number of collisions decreases; thus, the *pressure decreases.*
(c) When the moles of gas increase, the number of molecules increases. With more molecules, there are more collisions and the *pressure increases.*

Self-Test Exercise

Indicate whether gas pressure increases or decreases with each of the following changes in a closed gaseous system.

(a) increasing the temperature
(b) increasing the volume
(c) increasing the number of gas molecules

Answers: (a) pressure increases; (b) pressure decreases; (c) pressure increases

11.4 Boyle's Law

Objectives
- To sketch a graph of the pressure–volume relationship for a gas.
- To calculate the pressure or volume of a gas after a change in conditions.

In Chapter 1, we mentioned that Robert Boyle was the founder of the scientific method. In one of his experiments, Boyle trapped air in a J-shaped tube with liquid

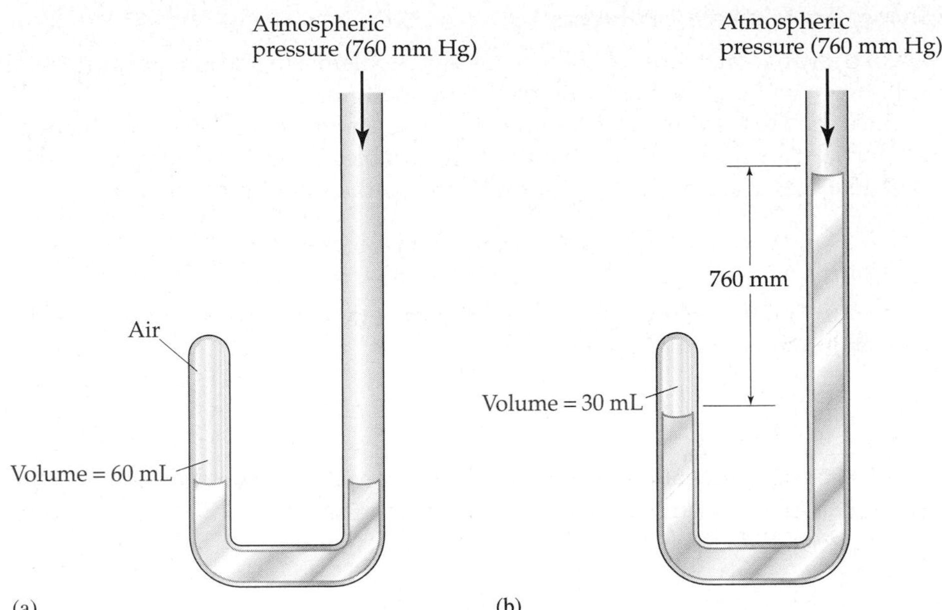

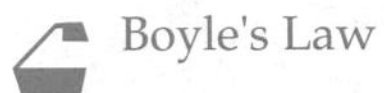

Figure 11.5 Boyle's Law (a) The initial volume of gas at atmospheric pressure is 60 mL. (b) After adding mercury, the volume of gas is 30 mL. Note that the volume is halved when the pressure is doubled. Initially, the pressure is 760 mm Hg; after adding mercury, the pressure is 1520 mm Hg (760 mm + 760 mm).

mercury. As he added mercury into the tube, he found that the volume of trapped air decreased (Figure 11.5). If he doubled the pressure, the volume was reduced to a half. When he tripled the pressure, the volume was reduced to a third. Thus, Boyle established the relationship between the pressure and the volume of a gas.

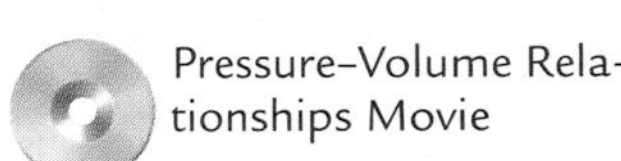

The results of Boyle's experiments are known as **Boyle's law**, which states that the volume of a gas is inversely proportional to the pressure when the temperature remains constant. We can express the relationship as

$$V \propto \frac{1}{P} \qquad (T \text{ constant})$$

That is, the volume (V) is inversely proportional to (α) the pressure ($1/P$) when the temperature (T) remains constant. **Inversely proportional** means that two variables have a reciprocal relationship. As one variable gets larger, the other variable gets smaller. Using experimental data, we can plot pressure versus volume and obtain the curve shown in Figure 11.6.

The relationship between pressure and volume can be written as an equation by using a proportionality constant (k).

$$V = k \times \frac{1}{P}$$

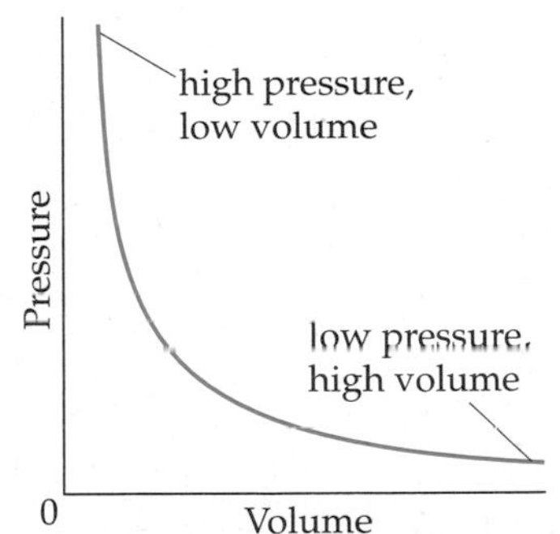

Figure 11.6 Gas Pressure versus Volume As the volume of a gas increases, the pressure decreases. Conversely, as the volume decreases, the pressure increases. The volume and the pressure of a gas are inversely proportional.

Multiplying both sides of the equation by P, we see that the product of pressure and volume is equal to the constant (k). The equation becomes

$$PV = k$$

We can consider a sample of a gas under different conditions. Let the initial conditions of pressure and volume be P_1 and V_1, respectively. After a change in conditions, we can indicate the final pressure and volume as P_2 and V_2. Since the product of pressure and volume is a constant, we write

$$P_1V_1 = k = P_2V_2$$

Solving Boyle's Law Problems

We will show two methods for solving gas law problems. In addition to an algebraic solution, we will use the following reasoning method.

Since the pressure and the volume of a gas are inversely related, we can apply a proportionality factor to find an unknown variable. To find the gas pressure after a change in volume, we apply a volume factor to the initial pressure (P_1).

$$P_1 \times V_{\text{factor}} = P_2$$

Similarly, to find the volume after a change in gas pressure, we apply a pressure factor to the initial volume (V_1).

$$V_1 \times P_{\text{factor}} = V_2$$

Let's consider the following Boyle's law gas problem. In an experiment, 5.00 L of propane gas is compressed, and the pressure increases from 1.00 atm to 1.50 atm. To calculate the final volume (V_2), we write

$$5.00 \text{ L} \times P_{\text{factor}} = V_2$$

Since the pressure of the propane gas increases from 1.00 atm to 1.50 atm, the final volume must decrease. Therefore, the P_{factor} must be less than 1. That is, the smaller pressure value must appear in the numerator.

$$5.00 \text{ L} \times \frac{1.00 \cancel{\text{atm}}}{1.50 \cancel{\text{atm}}} = 3.33 \text{ L}$$

Notice that the cancellation of units takes place in the unit factor (P_{factor}). Although the numerator and the denominator are not equal, this method uses a format very much like that of the unit analysis method of problem solving. Now, let's reinforce our understanding of Boyle's law with an example exercise.

Example Exercise 11.3 • Boyle's Law Problem

A 1.50-L sample of methane gas exerts a pressure of 1650 mm Hg. Calculate the final pressure if the volume changes to 7.00 L. Assume the temperature remains constant.

Solution

We can find the final pressure (P_2) by applying Boyle's law and using the relationship

$$P_1 \times V_{\text{factor}} = P_2$$

The volume increases from 1.50 L to 7.00 L. Thus, the pressure decreases. The V_{factor} must be less than 1. Hence,

$$1650 \text{ mm Hg} \times \frac{1.50 \cancel{\text{L}}}{7.00 \cancel{\text{L}}} = 354 \text{ mm Hg}$$

We can visually summarize the Boyle's law solution as follows.

1650 mm Hg → volume factor → 354 mm Hg

Algebraic Solution

Alternatively, we can solve this problem using the equation

$$P_1V_1 = P_2V_2$$

Solving for P_2 gives

$$\frac{P_1 V_1}{V_2} = P_2$$

Substituting for each variable and simplifying, we obtain

$$\frac{1650 \text{ mm Hg} \times 1.50 \cancel{\text{L}}}{7.00 \cancel{\text{L}}} = 354 \text{ mm Hg}$$

Self-Test Exercise

A sample of ethane gas has a volume of 125 mL at 20°C and 725 torr. What is the volume of the gas at 20°C when the pressure decreases to 475 torr?

Answer: 191 mL

Note The reasoning method for solving gas law problems is valuable because it reinforces our conceptual understanding. We tend to become more involved in how changes in pressure, volume, or temperature actually affect a gas.

Chemistry Connection · Robert Boyle

What instrument did Robert Boyle invent to demonstrate that a feather and a lump of lead are affected identically by gravity?

Robert Boyle was a child prodigy who was speaking Latin and Greek by the age of 8. He was the son of wealthy English aristocrats and traveled throughout Europe with a private tutor to gain a broad education. When he was 18 years old, his father died and left him with a lifetime income. At 27, he enrolled in Oxford University to continue his studies. There he became interested in experimentation, although laboratory work was considered to be of minor importance at that time. Most influential scientists believed, like the Greeks, that reason was far superior to experimentation.

In 1657 Boyle designed a vacuum pump that was an improvement over the first one developed a few years earlier. In one experiment, he evacuated most of the air from a sealed chamber and showed that a ticking clock could not be heard in a vacuum. He then correctly concluded that sound does not exist in the absence of air. Boyle was fascinated by the behavior of gases and formulated the law stating that the pressure and volume of a gas are inversely related.

In 1661 Boyle published *The Sceptical Chymist* and argued that theories were no better than the experiments on which they were based. Gradually, this point of view was accepted, and his text marked a turning point for recognition of the importance of experimentation. Because of his numerous contributions to chemistry and physics, Boyle is generally regarded as the founder of the modern scientific method.

Boyle felt strongly about keeping meticulous notes and reporting experimental results. He thought that everyone should be able to learn from published accounts of laboratory inquiry. Boyle felt that experiments should be held up to scientific scrutiny and that others should have the opportunity to confirm or disprove the results. This practice has become a cornerstone of science and, except for information involving military and industrial secrets, research experiments are published and made available to the scientific community as well as to the public at large.

▲ Robert Boyle (1627–1691)

Historians are quick to point out that Boyle was devoutly religious, studied the Bible, wrote essays on religion, and personally financed Christian missionary work. After his death, funds from his will supported the Boyle Lectures, which publicly defended Christianity.

Boyle invented a vacuum pump to remove air from a cylinder. When he released a feather and a lump of lead from the same height, he found that they landed simultaneously at the bottom of the cylinder.

11.5 Charles' Law

Objectives
- To sketch a graph of the volume–temperature relationship for a gas.
- To calculate the volume or temperature of a gas after a change in conditions.

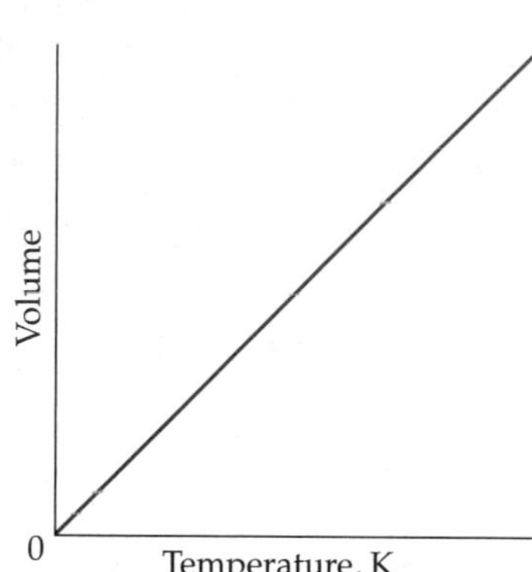

▲ **Figure 11.7 Gas Volume versus Temperature** As the Kelvin temperature decreases, the volume of a gas decreases. The Kelvin temperature and the volume of a gas are directly proportional.

In 1783 the French scientist Jacques Charles (1746–1823) filled a balloon with hydrogen gas and ascended to 10,000 feet—nearly two miles! Although hot-air ballooning had been pioneered by the Montgolfier brothers, Jacques Charles was the first to make an ascent in a hydrogen-filled balloon.

In 1787 Charles discovered the effect of temperature on the volume of a gas. From his experiments, he correctly deduced that the volume of a gas is **directly proportional** to its absolute temperature. That is, when we double the Kelvin temperature of a gas, we double its volume. When we halve the Kelvin temperature of a gas, we halve its volume. We can state **Charles' law** as follows: the volume of a gas is directly proportional to the absolute temperature if the pressure remains constant. Using experimental data, we can plot volume versus Kelvin temperature and obtain the straight-line linear relationship shown in Figure 11.7.

We can express the straight-line linear relationship between volume and Kelvin temperature as

$$V \propto T \qquad (P \text{ constant})$$

The relationship between volume and Kelvin temperature can also be written as an equation using a proportionality constant (k).

$$V = kT$$

Dividing both sides of the equation by T, we see that the ratio of volume to temperature equals the constant k. The equation becomes

$$\frac{V}{T} = k$$

Let's consider a sample of gas under different conditions. We indicate the initial conditions of volume and temperature as V_1 and T_1, respectively. After a change in conditions, we indicate the final volume and temperature as V_2 and T_2. Since the ratio of volume to temperature equals the constant (k), we write

$$\frac{V_1}{T_1} = k = \frac{V_2}{T_2}$$

Solving Charles' Law Problems

Since the pressure and volume of a gas are directly related, we can apply a proportionality factor to find an unknown variable. To find the volume after a change in temperature, we apply a temperature factor to the initial volume (V_1).

$$V_1 \times T_{factor} = V_2$$

Similarly, to find the temperature after a change in volume, we apply a volume factor to the initial temperature (T_1).

$$T_1 \times V_{factor} = T_2$$

(a)

(b)

◀ **Illustration of Charles' Law** (a) A balloon contains air at room temperature. (b) As liquid nitrogen at −196°C is poured on the balloon, the air inside the balloon cools, causing it to decrease in volume.

Let's consider the following Charles' law gas problem. In an experiment, a sample of argon gas at 225 K is heated and the volume increases from 3.50 L to 12.5 L. To calculate the final temperature (T_2), we write

$$225 \text{ K} \times V_{\text{factor}} = T_2$$

Since the volume of argon gas increases from 3.50 to 12.5 L, the final temperature must increase. Therefore, V_{factor} must be greater than 1. That is, the larger value must appear in the numerator.

$$225 \text{ K} \times \frac{12.5 \not{\text{L}}}{3.50 \not{\text{L}}} = 804 \text{ K}$$

The final temperature is 804 K, or 531°C. Now, let's reinforce our understanding of Charles' law with an example exercise.

Example Exercise 11.4 • Charles' Law

A 275-L helium balloon is heated from 20° to 40°C. Calculate the final volume assuming the pressure remains constant.

Solution

We first convert the Celsius temperatures to Kelvin by adding 273 units.

$$20°\text{C} + 273 = 293 \text{ K}$$
$$40°\text{C} + 273 = 313 \text{ K}$$

We can find the final volume (V_2) by applying Charles' law and using the relationship

$$V_1 \times T_{\text{factor}} = V_2$$

The temperature increases from 293 K to 313 K. It follows that the volume increases and the T_{factor} must be greater than 1.

$$275 \text{ L} \times \frac{313 \not{\text{K}}}{293 \not{\text{K}}} = 294 \text{ L}$$

We can visually summarize the Charles' law solution as follows.

275 L → temperature factor → 294 L

(continued)

Example Exercise 11.4 *(continued)*

Algebraic Solution

Alternatively, we can solve this problem using the equation

$$\frac{V_1}{T_1} = \frac{V_2}{T_2}$$

Solving for V_2 gives

$$\frac{V_1 T_2}{T_1} = V_2$$

Substituting for each variable and simplifying, we obtain

$$\frac{275 \text{ L} \times 313 \cancel{\text{K}}}{293 \cancel{\text{K}}} - 294 \text{ L}$$

Self-Test Exercise

A krypton balloon has a volume of 555 mL at 21°C. If the balloon is cooled and the volume decreases to 475 mL, what is the final temperature? Assume the pressure remains constant.

Answer: −21°C

Note In Charles' law problems we can state the volume in any units. For example, we can express the volume in liters or cubic centimeters. The temperature, however, must *always* be in Kelvin units.

11.6 Gay-Lussac's Law

Objectives
- To sketch a graph of the pressure–temperature relationship for a gas.
- To calculate the pressure or temperature of a gas after a change in conditions.

Like Jacques Charles, the French scientist Joseph Gay-Lussac (1778–1850) was also interested in ballooning and made a record ascent to 23,000 feet in a hydrogen-filled balloon. During his ascent, he collected samples of the atmosphere for study in his laboratory.

In 1802 Gay-Lussac established the relationship between pressure and temperature. He correctly interpreted his experiments to mean that the pressure of a gas is directly proportional to its absolute temperature. That is, when we double the Kelvin temperature of a gas, we double its pressure. When we halve the Kelvin temperature of a gas, we halve its pressure. We can state **Gay-Lussac's law** as follows: the pressure of a gas is directly proportional to its absolute temperature if the volume remains constant. Using experimental data, we can plot pressure versus Kelvin temperature and obtain the straight-line linear relationship shown in Figure 11.8.

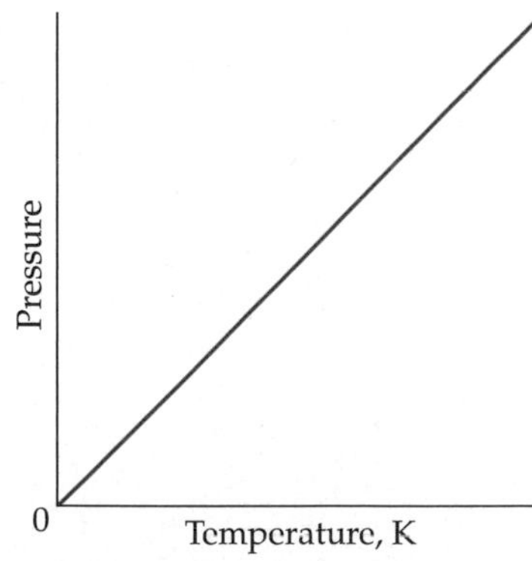

▲ **Figure 11.8 Gas Pressure versus Temperature** As the Kelvin temperature decreases, the pressure of a gas decreases. The Kelvin temperature and the pressure exerted by a gas are directly proportional.

We can express the relationship between pressure and Kelvin temperature as

$$P \propto T \qquad (V \text{ constant})$$

The relationship between pressure and temperature can also be written as an equation using the proportionality constant (k).

$$P = kT$$

Dividing both sides of the equation by T, we see that the ratio of pressure to temperature equals the constant k, The equation becomes

$$\frac{P}{T} = k$$

Let's consider a sample of a gas under different conditions. We indicate the initial conditions of pressure and temperature as P_1 and T_1, respectively. After a change in conditions, we indicate the final pressure and temperature as P_2 and T_2. Since the ratio of pressure to temperature equals the constant (k), we can write

$$\frac{P_1}{T_1} = k = \frac{P_2}{T_2}$$

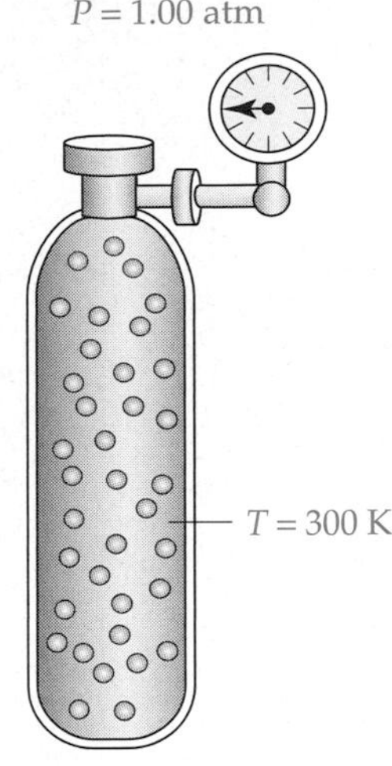

(a)

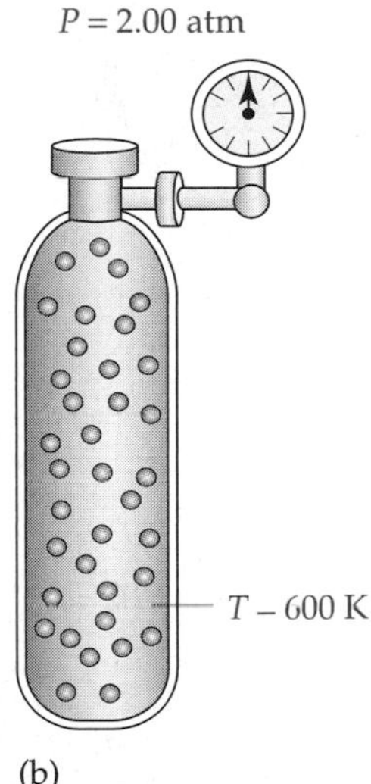

(b)

▲ Illustration of Gay-Lussac's Law (a) A cylinder contains helium gas at 300 K and a pressure of 1.00 atm. (b) After the temperature is doubled to 600 K, the pressure doubles to 2.00 atm.

Solving Gay-Lussac's Law Problems

Since the pressure and temperature of a gas are directly related, we can apply a proportionality factor to find an unknown variable. To find the pressure after a change in temperature, we apply a temperature factor to the initial pressure (P_1).

$$P_1 \times T_{\text{factor}} = P_2$$

Similarly, to find the temperature after a change in pressure, we apply a pressure factor to the initial temperature (T_1).

$$T_1 \times P_{\text{factor}} = T_2$$

Let's consider the following problem involving Gay-Lussac's law. An automobile tire is inflated to 28.0 psi at a temperature of 20°C. After it has traveled at high speed, the tire pressure is 36.0 psi. To calculate the final temperature (T_2) we first convert the initial temperature to Kelvin (20°C + 273 = 293 K) and then write

$$293 \text{ K} \times P_{\text{factor}} = T_2$$

Since the pressure of air increases from 28.0 psi to 36.0 psi, the final temperature must increase. Therefore, the P_{factor} must be greater than 1. That is, the larger pressure value must appear in the numerator.

$$293 \text{ K} \times \frac{36.0 \cancel{\text{psi}}}{28.0 \cancel{\text{psi}}} = 377 \text{ K}$$

The final temperature is 377 K, or 104°C. Now, let's reinforce our understanding of Gay-Lussac's law with an example exercise.

Example Exercise 11.5 • Gay-Lussac's Law

A steel cylinder filled with nitrous oxide at 15.0 atm is cooled from 25° to −40°C. Calculate the final pressure assuming the volume remains constant.

Solution

We must first convert the Celsius temperatures to Kelvin by adding 273.

$$25°\text{C} + 273 = 298 \text{ K}$$

$$-40°\text{C} + 273 = 233 \text{ K}$$

(continued)

Example Exercise 11.5 *(continued)*

We can find the final pressure (P_2) by applying Gay-Lussac's law and using the relationship

$$P_1 \times T_{\text{factor}} = P_2$$

The temperature decreases from 298 K to 233 K. Therefore, the pressure decreases. The T_{factor} must be less than 1. Hence,

$$15.0 \text{ atm} \times \frac{233 \cancel{\text{K}}}{298 \cancel{\text{K}}} = 11.7 \text{ atm}$$

We can visually summarize the Gay-Lussac's law solution as follows.

15.0 atm → $\frac{\text{temperature}}{\text{factor}}$ → 11.7 atm

Algebraic Solution

Alternatively, we can solve this problem using the equation

$$\frac{P_1}{T_1} = \frac{P_2}{T_2}$$

Solving for P_2 gives

$$\frac{P_1 T_2}{T_1} = P_2$$

Substituting for each variable and simplifying, we have

$$\frac{15.0 \text{ atm} \times 233 \cancel{\text{K}}}{298 \cancel{\text{K}}} = 11.7 \text{ atm}$$

Self-Test Exercise

A copper sphere has a volume of 555 mL and is filled with air at 25°C. The sphere is immersed in dry ice, and the pressure of the gas drops from 761 torr to 495 torr. What is the final temperature of the air in the copper sphere?

Answer: 194 K (−79°C)

11.7 Combined Gas Law

Objective · **To calculate a pressure, volume, or temperature for a gas after a change in conditions.**

Experiment #16, Prentice Hall Laboratory Manual

We began our discussion of gas laws with a simplifying assumption. In discussing Boyle's, Charles', and Gay-Lussac's laws, we assumed that we could limit our treatment of gases to two variables. Experimentally, all three variables (pressure, volume, and temperature) usually change simultaneously. Now, we will bring all three variables together in a single expression. The resulting expression is called the **combined gas law**, and the equation is

$$\frac{P_1 V_1}{T_1} = \frac{P_2 V_2}{T_2}$$

Since the variables P, V, and T are proportional, we can solve combined gas law problems using the unit analysis approach. To calculate a final pressure (P_2) we apply a volume factor and a temperature factor to the initial pressure (P_1).

$$P_1 \times V_{factor} \times T_{factor} = P_2$$

To calculate a final volume (V_2), we ,apply a pressure factor and a temperature factor to the initial volume (V_1).

$$V_1 \times P_{factor} \times T_{factor} = V_2$$

To calculate a final temperature (T_2), we apply a pressure factor and a volume factor to the initial temperature (T_1).

$$T_1 \times P_{factor} \times V_{factor} = T_2$$

In a combined gas law problem, we have three variables at initial and final conditions. That gives us six pieces of data, and so it is helpful to use a table. We place the data in a table and organize the information as follows.

Conditions	***P***	***V***	***T***
initial	P_1	V_1	T_1
final	P_2	V_2	T_2

Let's apply the combined gas law to 10.0 L of carbon dioxide gas at 300 K and 1.00 atm. When both the volume and the Kelvin temperature double, what is the final pressure in atmospheres? The conditions are 300 K and 1.00 atm. Doubling the temperature gives 600 K; doubling the volume gives 20.0 L. Now let's enter the data in a table.

Conditions	***P***	***V***	***T***
initial	1.00 atm	10.0 L	300 K
final	P_2	20.0 L	600 K

We calculate the final pressure (P_2) by applying a V_{factor} and a T_{factor} to the initial pressure, 1.00 atm.

$$1.00 \text{ atm} \times V_{factor} \times T_{factor} = P_2$$

Since the volume of the gas increases from 10.0 L to 20.0 L, P_2 must decrease. The V_{factor} is less than 1, and the smaller value is placed in the numerator.

$$1.00 \text{ atm} \times \frac{10.0 \not{\text{L}}}{20.0 \not{\text{L}}} \times T_{factor} = P_2$$

The absolute temperature of the gas increases from 300 K to 600 K; therefore, P_2 increases. The T_{factor} is greater than 1, and the larger value is placed in the numerator.

$$1.00 \text{ atm} \times \frac{10.0 \not{\text{L}}}{20.0 \not{\text{L}}} \times \frac{600 \not{\text{K}}}{300 \not{\text{K}}} = 1.00 \text{ atm}$$

Notice that the pressure did not change even though the volume and the temperature each doubled. The pressure remained unchanged because pressure and volume are inversely proportional; pressure and temperature are directly related.

Standard Conditions of Temperature and Pressure

The **standard temperature and pressure** (symbol **STP**) for a gas are 0°C and 1 atm. Alternatively, we can express standard temperature as 273 K and standard pressure in other units such as 760 mm Hg, 760 torr, and 76 cm Hg. The following example exercise illustrates the combined gas law at STP conditions.

Example Exercise 11.6 • Combined Gas Law

A nitrogen gas sample occupies 50.5 mL at −80°C and 1250 torr. What is the volume at STP?

Solution

Although the final conditions are not given, we know that STP conditions are 273 K and 760 torr. We can summarize the information as follows.

Conditions	P	V	T
initial	1250 torr	50.5 mL	−80 + 273 = 193 K
final	760 torr	V_2	273 K

We can calculate the final volume by applying a P_{factor} and a T_{factor} to the initial volume.

$$V_1 \times P_{factor} \times T_{factor} = V_2$$

The pressure decreases, and so the volume increases; thus, the P_{factor} is greater than 1. The temperature increases, and so the volume increases; thus, the T_{factor} is also greater than 1.

$$50.5 \text{ mL} \times \frac{1250 \cancel{\text{torr}}}{760 \cancel{\text{torr}}} \times \frac{273 \cancel{\text{K}}}{193 \cancel{\text{K}}} = 117 \text{ mL}$$

We can visually summarize the combined gas law solution as follows.

50.5 mL → pressure factor → temperature factor → 117 mL

Algebraic Solution

Alternatively, we can solve this problem using the equation

$$\frac{P_1V_1}{T_1} = \frac{P_2V_2}{T_2}$$

Rearranging variables and solving for V_2,

$$\frac{P_1V_1T_2}{T_1P_2} = V_2$$

Substituting for each variable and simplifying, we obtain

$$\frac{1250 \cancel{\text{torr}} \times 50.5 \text{ mL} \times 273 \cancel{\text{K}}}{193 \cancel{\text{K}} \times 760 \cancel{\text{torr}}} = 117 \text{ mL}$$

Self-Test Exercise

An oxygen gas sample occupies 50.0 mL at 27°C and 765 mm Hg. What is the final temperature if the gas is cooled to 35.5 mL and a pressure of 455 mm Hg?

Answer: 127 K (−146°C)

11.8 The Vapor Pressure Concept

Objectives
- To explain the concept of vapor pressure.
- To state the relationship between vapor pressure and temperature.

Everyone has observed that water in an open container evaporates. Vaporization occurs when molecules have enough energy to escape from the liquid. When the container is enclosed, the vapor molecules are trapped above the liquid. We should note that molecules continuously escape from the liquid into the vapor. Simultaneously, other molecules return from the vapor to the liquid. The **vapor pressure** is the pressure exerted by molecules in the vapor above a liquid when the rate of evaporation and condensation are equal (Figure 11.9).

Measuring the Vapor Pressure of a Liquid

The vapor pressure of a liquid can be determined with a mercury barometer (Figure 11.10). A drop of liquid is introduced at the bottom of the barometer. The less dense liquid droplet floats to the top of the mercury and vaporizes. The vaporized liquid exerts a gas pressure, thus driving the column of mercury down. *The decrease in the mercury level corresponds to the vapor pressure of the liquid.* If the mercury level drops from 760 mm to 740 mm, the vapor pressure of the liquid is recorded as 20 mm Hg.

Temperature Dependence of Vapor Pressure Movie

We can observe that vapor pressure increases as the temperature increases. This is because liquid molecules evaporate faster and vapor molecules have more kinetic energy. At 25°C the vapor pressure of water is 23.8 mm Hg. At 50°C the vapor pressure increases to 92.5 mm Hg. Table 11.2 lists selected values for the vapor pressure of water.

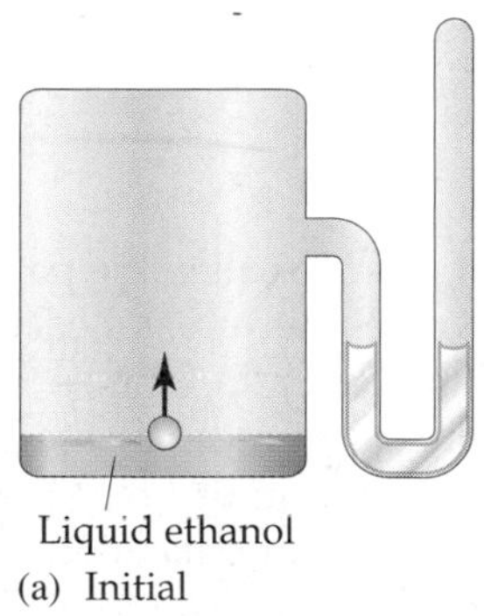

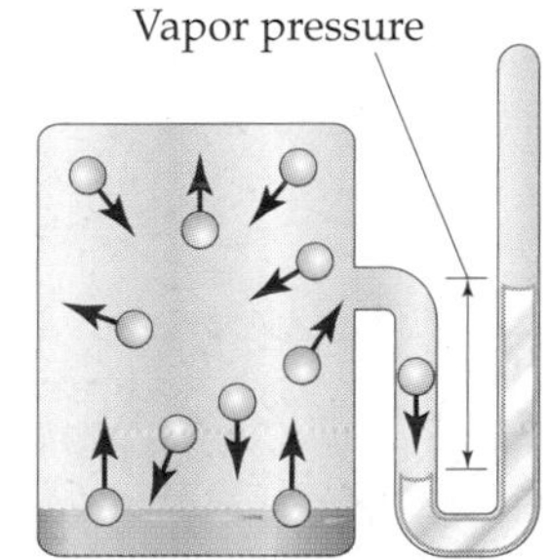

Figure 11.9 Vapor Pressure of a Liquid (a) Initially, ethanol is placed in a closed container and molecules begin to escape from the liquid. After awhile, some of the ethanol molecules in the vapor return to the liquid. (b) Eventually, the ethanol molecules are evaporating and condensing at the same rate. The vapor pressure is measured by the difference in the height of mercury in the side tube.

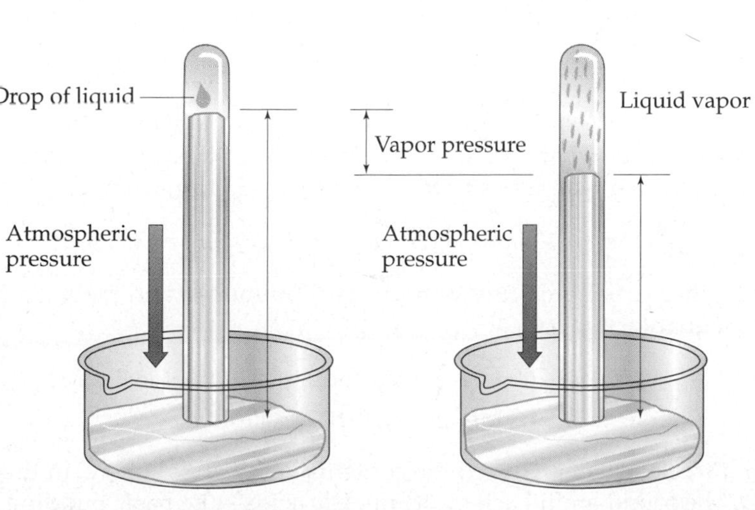

Figure 11.10 Measuring the Vapor Pressure of a Liquid (a) Initially, a single drop of liquid is introduced into the tube of mercury. As molecules of the liquid evaporate, the resulting gaseous vapor molecules begin to exert a pressure on the column of mercury. (b) Eventually, molecules are evaporating and condensing at the same rate. The difference in the height of the mercury before and after a drop of liquid is introduced is the vapor pressure of the liquid at the given temperature.

Table 11.2 Vapor Pressure of Water

Temperature (°C)	Pressure (mm Hg)	Temperature (°C)	Pressure (mm Hg)
5	6.5	55	118.0
10	9.2	60	149.4
15	12.8	65	187.5
20	17.5	70	233.7
25	23.8	75	289.1
30	31.8	80	355.1
35	41.2	85	433.6
40	55.3	90	525.8
45	71.9	95	633.9
50	92.5	100	760.0

11.9 Dalton's Law

Objective · To apply Dalton's law of partial pressures to a mixture of gases.

When John Dalton proposed the atomic theory, his evidence was largely based on the behavior of gases. His knowledge of gases was related to his interest in meteorology, and in fact he kept daily records of the weather throughout his life. In 1801 Dalton proposed that a gas in a mixture exerted the same pressure that it would if it were the only gas present. He then formulated **Dalton's law of partial pressures,** which states that the total pressure of a gaseous mixture is equal to the sum of the individual pressures of each gas. We write this relationship as follows.

$$P_1 + P_2 + P_3 + \cdots = P_{total}$$

The pressure exerted by each gas in a mixture of gases is called the **partial pressure.** In the equation, P_1, P_2, and P_3 represent the partial pressures of each of the gases in the mixture; P_{total} symbolizes the total of all the partial pressures.

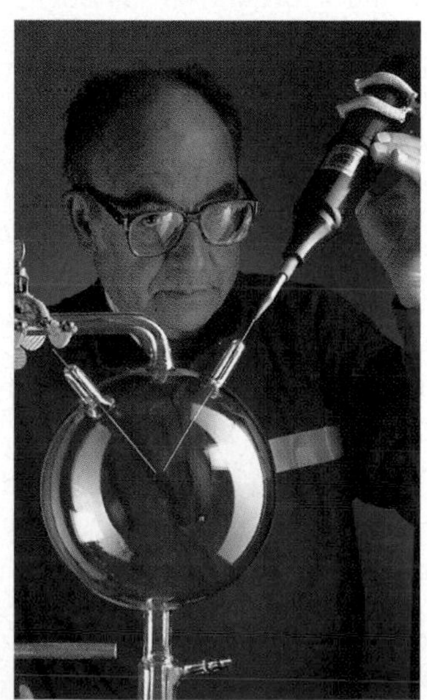

▲ **Ammonia, NH_3** Stanley Miller placed H_2, NH_3, CH_4, and H_2O vapor in his apparatus and energized the gaseous mixture with ultraviolet light and electric sparks.

In 1952 Stanley Miller conducted a historic experiment at the University of Chicago. In an attempt to prove the chemical evolution of life, Miller simulated Earth's primordial atmosphere. He placed hydrogen, ammonia, methane, and water vapor in a glass sphere and energized the mixture with ultraviolet light and electric sparks for 7 days. Applying Dalton's law, we write

$$P_{hydrogen} + P_{ammonia} + P_{methane} + P_{water\ vapor} = P_{total}$$

If the partial pressures of hydrogen, ammonia, and methane were 275, 125, and 340 torr, respectively, what was the partial pressure of the water vapor? Assuming the pressure in the sphere was standard pressure, 760 torr, we have

$$275\text{ torr} + 125\text{ torr} + 340\text{ torr} + P_{water\ vapor} = 760\text{ torr}$$

Subtracting the partial pressures of hydrogen, ammonia, and methane from each side of the equation, we have

$$P_{water\ vapor} = 760\text{ torr} - (275\text{ torr} + 125\text{ torr} + 340\text{ torr})$$

$$P_{water\ vapor} = 760\text{ torr} - 740\text{ torr} = 20\text{ torr}$$

After a week, Miller analyzed the contents of the glass sphere. In the "primordial soup," he found amino acids and nucleic acids—the basic building blocks of life! Biologist frequently cite this famous experiment as evidence for the chemical evolution of life.

Example Exercise 11.7 • Dalton's Law

An atmospheric sample contains nitrogen, oxygen, argon, and traces of other gases. If the partial pressure of nitrogen is 587 mm Hg, oxygen is 158 mm Hg, and argon is 7 mm Hg, what is the observed pressure as read on the barometer?

Solution

The sum of the individual partial pressures equals the total atmospheric pressure; therefore,

$$P_{\text{nitrogen}} + P_{\text{oxygen}} + P_{\text{argon}} = P_{\text{total}}$$

Substituting the values for the partial gas pressures, we have

$$587 \text{ mm Hg} + 158 \text{ mm Hg} + 7 \text{ mm Hg} = 752 \text{ mm Hg}$$

Thus, the atmospheric pressure as read on the barometer is 752 mm Hg.

Self-Test Exercise

The regulator on a steel scuba tank containing compressed air indicates that the pressure is 2250 psi. If the partial pressure of nitrogen is 1755 psi and that of argon is 22 psi, what is the partial pressure of oxygen in the tank?

Answer: 473 psi

Collecting a Gas Over Water

We cannot measure the volume of a gas directly, but we can determine the volume indirectly. That is, we can determine the volume of a gas from the amount of water it displaces. This method is called determining **volume by displacement**, and the volume of gas equals the volume of water displaced.

Let's consider a laboratory experiment that produces hydrogen gas from the reaction of zinc metal and sulfuric acid.

$$Zn(s) + H_2SO_4(aq) \longrightarrow ZnSO_4(aq) + H_2(g)$$

To collect the hydrogen gas evolved, we invert a graduated cylinder filled with water over the zinc metal. Figure 11.11 shows the experimental apparatus.

When the reaction is complete, the pressure exerted by the hydrogen gas inside the graduated cylinder is equal to the atmospheric pressure. That is, the hydrogen gas pressure on the water inside the graduated cylinder is equal to the atmospheric pressure on the surface of the solution in the beaker.

Experiments #17 and #18, Prentice Hall Laboratory Manual

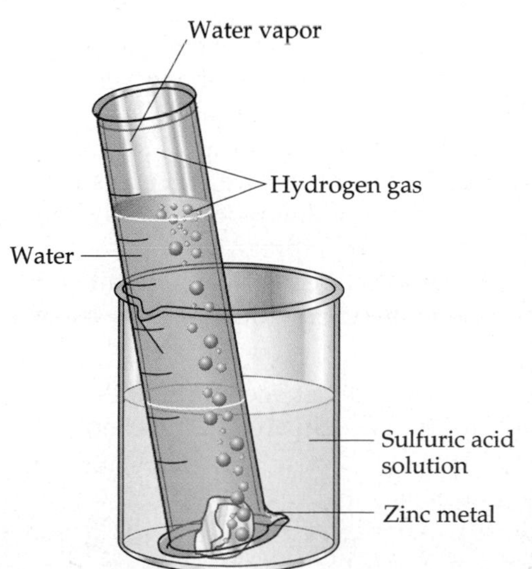

◀ Figure 11.11 Collecting a Gas over Water Zinc metal reacts with sulfuric acid to give bubbles of hydrogen gas. A graduated cylinder full of water is placed over the metal to collect the gas bubbles. The volume of water displaced from the graduated cylinder equals the volume of hydrogen gas liberated from the acid.

However, there are actually two gases inside the graduated cylinder. One of the gases is hydrogen produced by the reaction, and the other gas is water vapor. Since the hydrogen gas is collected over an aqueous solution, water vapor is present. We use Dalton's law of partial pressures to determine the pressure of the hydrogen.

$$P_{\text{hydrogen}} + P_{\text{water vapor}} = P_{\text{atmosphere}}$$

If the reaction takes place at 25°C and the barometer reads 767 mm Hg, what is the partial pressure of the hydrogen gas? We refer to Table 11.2 and find that the vapor pressure of water is 23.8 mm Hg at 25°C. Then we calculate the pressure of the hydrogen gas.

$$P_{\text{hydrogen}} + 23.8 \text{ mm Hg} = 767 \text{ mm Hg}$$

After rearranging, we have

$$P_{\text{hydrogen}} = 767 \text{ mm Hg} - 23.8 \text{ mm Hg} = 743 \text{ mm Hg}$$

Note When a gas is collected over water, it is often called a "wet" gas. That is, the collected gas contains water vapor. A gas that does not contain water vapor is sometimes referred to as a "dry" gas.

11.10 Ideal Gas Behavior

Objectives
- To list five characteristics of an ideal gas according to the kinetic theory of gases.
- To determine the value of absolute zero from a graph of volume or pressure versus temperature.

By the early 1800s, experiments had provided much information about the behavior of gases. There was, however, no clear way to interpret this information. About 1850 a theory began to emerge. The British physicist James Joule (1818–1889) suggested that the temperature of a gas is related to the energy of molecules. In other words, Joule proposed that heat and the motion of molecules in a gas are related. At higher temperatures, gas molecules move faster, and at lower temperatures, gas molecules move slower.

Kinetic Theory of Gases

In the period from 1850 to 1870, scientists attempted to construct a model that would explain the behavior of individual gas molecules. They reasoned that given a theoretical model, they would be able to explain the behavior of real gases. This model would also allow predictions of ideal gas behavior. An **ideal gas** is a gas that always behaves in a consistent and predictable manner. Experimentally, a **real gas**, such as hydrogen or oxygen, does not behave ideally under all conditions. Specifically, a real gas does not behave ideally at low temperatures and high pressures.

In due time, the behavior of an ideal gas was described by a model called the **kinetic theory** of gases. According to the kinetic theory, an ideal gas has the following characteristics:

1. *Gases are made up of very tiny molecules.* The distance between molecules is quite large. Therefore gases are mostly empty space. In an ideal gas, molecules occupy a negligible volume.

Chemistry Connection • The Greenhouse Effect

Why has planting millions of trees been suggested as a means of reducing global warming?

When you read about the greenhouse effect, you may think that it is life-threatening to our planet. However, the greenhouse effect is actually beneficial and helps to maintain the mild climatic temperature of the Earth. The current fear is that our atmosphere has an abundance of so-called greenhouse gases that will cause drastic global warming. Although carbon dioxide is singled out as the chief offender, other gases such as chlorofluorocarbons, methane, CH_4, and the oxides of nitrogen, NO_x, also contribute to global warming.

Scientists explain the greenhouse effect as follows. The Earth's atmosphere is transparent to short-wavelength radiation from the Sun. After striking the Earth's surface, the radiation from the Sun is emitted back into space as long-wavelength heat energy. Long-wavelength heat energy can be absorbed by carbon dioxide and other trace gases in the atmosphere. Subsequently, these gas molecules release heat energy in all directions, some of which is radiated back toward the Earth. The net effect is that carbon dioxide and other gases in the atmosphere act as a giant canopy that traps heat energy.

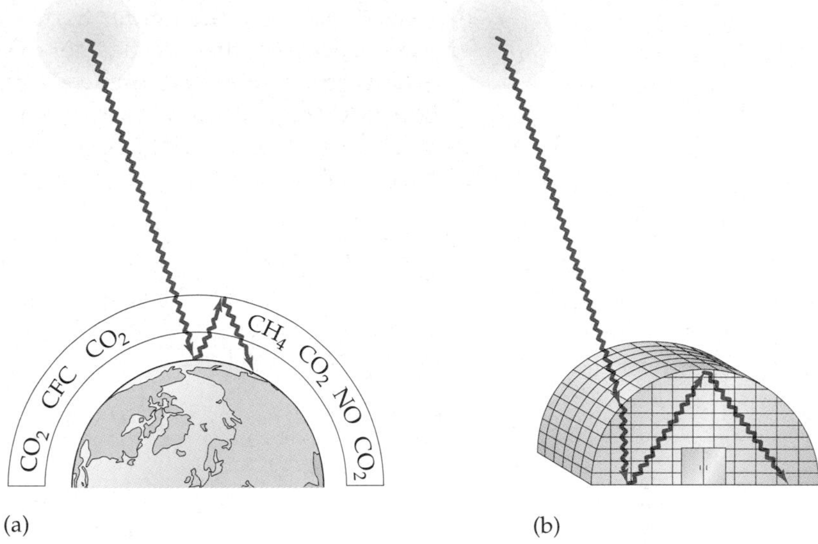

▲ Greenhouse gases can trap heat energy from the Sun in much the same way as a plant greenhouse.

Our present concern is that trace gases in the atmosphere are increasing the temperature of our planet at an alarming rate. Global warming can result from high concentrations of carbon dioxide in the atmosphere. Carbon dioxide is produced by burning fossil fuels (coal, oil, and gas) in automobiles, homes, and power plants. The problem is further accelerated by massive deforestation, especially in South America. The loss of forests is especially harmful because trees convert carbon dioxide in the atmosphere to oxygen.

While some scientists do not think global warming is as severe as the dire predictions, others believe we must do something immediately. One futuristic suggestion is to add chemicals to the oceans that will stimulate the growth of plankton. Plankton remove carbon dioxide from the air through the process of photosynthesis. Currently, our efforts are directed at planting millions of trees and reducing the emission of greenhouse gases into the atmosphere.

The greenhouse gas that contributes most to global warming is carbon dioxide. The planting of millions of trees will allow the process of photosynthesis to reduce the amount of carbon dioxide and return oxygen to the atmosphere.

---

2. *Gas molecules demonstrate rapid motion, move in straight lines, and travel in random directions.*
3. *Gas molecules have no attraction for one another.* After colliding with each other, molecules simply bounce off in different directions.
4. *Gas molecules undergo* ***elastic collisions***. That is, they do not lose kinetic energy after colliding. If a high-energy molecule strikes a less energetic molecule, part of the energy can be transferred. The total energy of both molecules, before and after the collision, does not change.
5. *The average kinetic energy of gas molecules is proportional to the Kelvin temperature; that is,* $KE \propto T$. At the same temperature, all gas molecules have equal kinetic energy.

Even for different gases, the average kinetic energy is equal at the same temperature. For example, the kinetic energy of hydrogen and oxygen gases is equal at 25°C. Although the kinetic energy is equal, hydrogen molecules move faster than oxygen molecules because they are lighter. Given the equation for kinetic energy, $KE = \frac{1}{2}mv^2$, we can explain this mathematically. If the kinetic energy of the two gases is equal, then the lighter hydrogen molecules ($m = 2$ amu) must have a greater velocity (v) than the heavier oxygen molecules ($m = 32$ amu).

At higher temperatures, molecules have more kinetic energy, move faster, and collide more frequently. At lower temperatures, molecules have less kinetic energy, move slower, and collide less frequently. The following example exercise further illustrates the characteristics of an ideal gas according to the kinetic theory.

150
125
100
75
50
25
Volume (mL)
−273 −173 −73 27 127 227 °C
0 100 200 300 400 500 K
Temperature

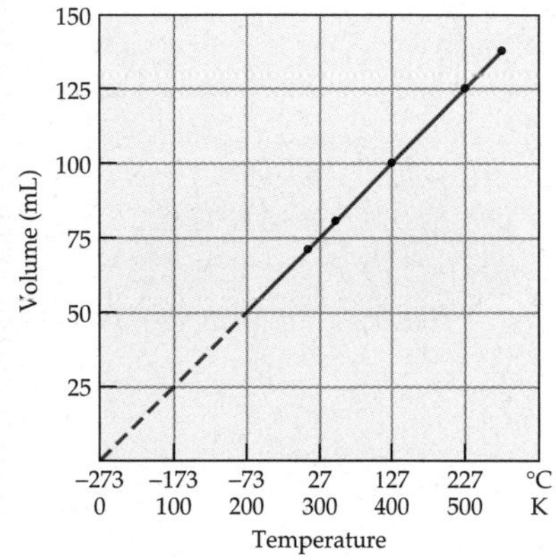

▲ Figure 11.12 Absolute Zero at Zero Volume As the temperature of the gas decreases, the volume decreases. As an ideal gas approaches zero volume, the temperature approaches absolute zero.

Example Exercise 11.8 • Ideal Gas Behavior

Suppose we have two 5.00-L samples of gas at 25°C. One sample is ammonia, NH_3, and the other nitrogen dioxide, NO_2. Which gas has the greater kinetic energy? Which gas has the faster molecules?

Solution

Since the temperature of each gas is 25°C, we know that the kinetic energy is the same for NH_3 and NO_2.

At the same temperature, we know that lighter molecules move faster than heavier molecules. The molecular mass of NH_3 is 17 amu, and NO_2 is 46 amu. Since NH_3 is lighter than NO_2, the ammonia molecules have a higher velocity than the nitrogen dioxide molecules.

Self-Test Exercise

Which of the following statements is *not* true according to the kinetic theory of gases?

(a) Molecules occupy a negligible volume.
(b) Molecules move in straight-line paths.
(c) Molecules are attracted to each other.
(d) Molecules undergo elastic collisions.
(e) Molecules of different gases at the same temperature have the same average kinetic energy.

Answer: All these statements are true except (c). Molecules of an ideal gas are not attracted to each other and behave as independent particles.

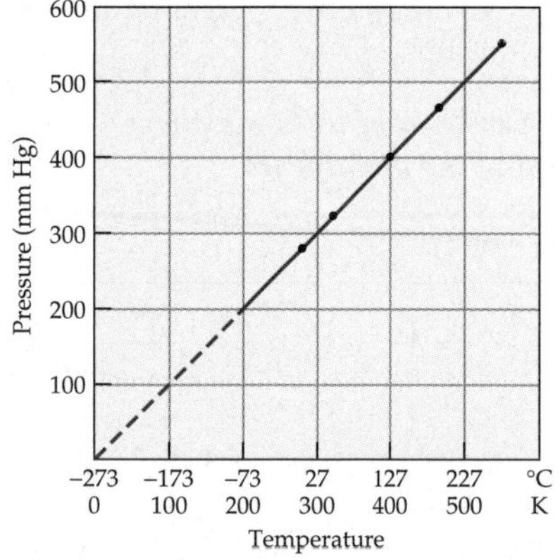

▲ Figure 11.13 Absolute Zero at Zero Pressure As the temperature of the gas decreases, the pressure decreases. As an Ideal gas approaches zero pressure, the temperature approaches absolute zero.

Absolute Zero

The temperature at which the pressure and volume of a gas theoretically reach zero is referred to as **absolute zero**. Absolute zero is the coldest possible temperature and corresponds to −273°C, or 0 K. An ideal gas at absolute zero has no kinetic energy, and therefore no molecular motion.

How do we determine the value of absolute zero if −273°C is impossible to attain experimentally? We determine absolute zero from the data for the volume of a gas over a range of temperatures above 0 K. When we plot the volume and temperature data points and extrapolate the graph to zero volume, we have the temperature corresponding to absolute zero. Figure 11.12 illustrates an experimental value for absolute zero.

We can also determine absolute zero by obtaining data for the pressure and temperature of a gas. We can plot the pressure and temperature data and extrapolate the graph to absolute zero. Figure 11.13 illustrates the determination of absolute zero.

11.11 Ideal Gas Law

Objectives
- To calculate the pressure, volume, temperature, or moles of gas from the ideal gas equation.
- To calculate the molar mass of a gas from the ideal gas equation.

In Section 11.3 we learned that the pressure (P) of a gas is inversely proportional to the volume (V). We also learned that pressure is directly proportional to the number of molecules or moles of gas (n), and directly proportional to the Kelvin temperature (T). We can therefore write a relationship that pressure is proportional to the moles of gas, multiplied by the temperature, divided by the volume.

$$P \propto \frac{nT}{V}$$

By introducing the proportionality constant (R), we can write this relationship as an equation.

$$P = \frac{RnT}{V}$$

After rearranging,

$$PV = nRT$$

This equation is called the **ideal gas law**. The constant R is the **ideal gas constant** and can be expressed as 0.0821 atm · L/mol · K. To use this value of R in a gas law calculation, the pressure must be expressed in atmospheres, the volume in liters, and the temperature in Kelvin. The ideal gas law is a powerful equation, which summarizes the behavior of gases. That is, any change in the pressure, volume, temperature, or moles of a gas is related through this single equation.

Let's try a calculation using the ideal gas law. If a cylinder contains 1.10 mol of nitrogen gas at 25°C and 3.75 atm, what is the volume in liters? We begin by rearranging the ideal gas equation to solve for volume (V). If $PV = nRT$, then

$$V = \frac{nRT}{P}$$

To convert the temperature to Kelvin, we add 273 to the Celsius temperature, which gives 298 K. Then we can substitute for each variable in the ideal gas equation.

$$V = \frac{1.10\ \cancel{\text{mol}} \times \dfrac{0.0821\ \cancel{\text{atm}} \cdot \text{L}}{1\ \cancel{\text{mol}} \cdot \cancel{\text{K}}} \times 298\ \cancel{\text{K}}}{3.75\ \cancel{\text{atm}}} = 7.18\ \text{L}$$

The following example exercise further illustrates the ideal gas law.

Example Exercise 11.9 • Ideal Gas Law

How many moles of hydrogen gas occupy a volume of 0.500 L at STP?

Solution

We begin by rearranging the ideal gas equation, $PV = nRT$, and solving for n.

$$n = \frac{PV}{RT}$$

(continued)

Example Exercise 11.9 *(continued)*

The temperature at STP is 273 K, and the pressure is 1 atm. Then we substitute for each variable in the ideal gas equation.

$$n = \frac{1\ \cancel{\text{atm}} \times 0.500\ \cancel{\text{L}}}{\dfrac{0.0821\ \cancel{\text{atm}} \cdot \cancel{\text{L}}}{1\ \text{mol} \cdot \cancel{\text{K}}} \times 273\ \cancel{\text{K}}} = 0.0223\ \text{mol}$$

Alternatively, we can display the value for the ideal gas constant separately to more clearly show the cancellation of units.

$$n = \frac{1\ \cancel{\text{atm}} \times 0.500\ \cancel{\text{L}}}{273\ \cancel{\text{K}}} \times \frac{1\ \text{mol} \cdot \cancel{\text{K}}}{0.0821\ \cancel{\text{atm}} \cdot \cancel{\text{L}}} = 0.0223\ \text{mol}$$

Self-Test Exercise
What is the temperature of 0.250 mol of chlorine gas at 655 torr if the volume is 3.50 L?

Answer: 147 K (−126°C)

Molar Mass of a Gas

We can use the ideal gas law to compute the molar mass (symbol MM) of a gas. Since the moles of a gas equal its mass in grams divided by its molar mass (g/MM), we can substitute for moles (n) in the ideal gas equation, $PV = nRT$.

$$PV = \frac{\text{g}}{\text{MM}} \times RT$$

When we are given the values for P, V, and T, we can calculate the molar mass of a gas from the ideal gas law. The following example exercise illustrates.

Example Exercise 11.10 • Ideal Gas Law

An unknown gas having a mass of 2.041 g occupies a volume of 1.15 L at 0.974 atm and 20°C. Calculate the molar mass of the unknown gas.

Solution
We begin by rearranging the equation $PV = \text{g}\ RT/\text{MM}$ and solve for MM.

$$\text{MM} = \frac{\text{g}}{PV} \times RT$$

We must convert the temperature to Kelvin; that is, 20°C + 273 = 293 K. Substituting for each variable and simplifying the equation, we have

$$\text{MM} = \frac{2.041\ \text{g}}{0.974\ \cancel{\text{atm}} \times 1.15\ \cancel{\text{L}}} \times \frac{0.0821\ \cancel{\text{atm}} \cdot \cancel{\text{L}}}{1\ \text{mol} \cdot \cancel{\text{K}}} \times 293\ \cancel{\text{K}} = 43.8\ \text{g/mol}$$

Self-Test Exercise
Find the molar mass of an unknown gas given that 0.320 g has a volume of 275 mL at 35°C and 745 mm Hg.

Answer: 30.0 g/mol

Note We can often identify a substance by determining its properties. For example, let's suppose the fuel from a recreational vehicle has a molar mass of 43.8 g/mol. Recreational vehicles use propane, C_3H_8 (44 g/mol), or butane, C_4H_{10} (58 g/mol) as fuel. Since the fuel has a molar mass of 43.8 g/mol, we predict that the recreational vehicle fuel is propane.

Summary

Section 11.1 The shape and volume of a gas are variable. A gas can expand and compress, but the gaseous state has a density much less than that of the liquid state. A gas diffuses in its container and mixes uniformly with other gases to form a homogeneous mixture.

Section 11.2 **Gas pressure** results from molecules striking the inside surface of its container. The pressure exerted by a gas depends on how often and how hard the molecules strike the container. In a **vacuum**, the gas pressure is zero. **Atmospheric pressure** results from molecules in the air and is measured with a **barometer**. Standard pressure can be expressed as 1 atm, 760 mm Hg, 760 **torr**, 76 cm Hg, 29.9 in. Hg, 14.7 psi, or 101 kPa.

Section 11.3 We can increase gas pressure by changing one of three variables. First, if we *decrease the volume*, molecules will collide more frequently and the pressure will increase. Second, if we *increase the temperature*, molecules will collide more frequently and the will pressure increase. Third, if we *increase the number of molecules*, there will be more collisions and the pressure will increase.

Section 11.4 The relationship between the pressure and the volume of a gas is **inversely proportional**. In solving problems involving **Boyle's law**, the pressure increases when the volume decreases, while the temperature remains constant.

Section 11.5 The relationship between the volume and the temperature of a gas is **directly proportional**. In solving problems involving **Charles' law**, the volume increases when the temperature increases, while the pressure remains constant.

Section 11.6 The relationship between the pressure and the temperature of a gas is directly proportional. In solving problems involving **Gay-Lussac's law**, the pressure increases when the temperature increases, while the volume remains constant. We can summarize all the relationships for the gas laws as shown in Table 11.3.

Table 11.3 Summary of Gas Law Variables

Gas Law	Pressure	Volume	Temperature
Boyle's	increases	decreases	constant
	decreases	increases	constant
Charles'	constant	increases	increases
	constant	decreases	decreases
Gay-Lussac's	increases	constant	increases
	decreases	constant	decreases

Section 11.7 **Standard temperature and pressure** (**STP**) is 273 K and exactly 1 atm. In solving problem involving the **combined gas law**, simultaneous changes occur in pressure, volume, and temperature.

Section 11.8 **Vapor pressure** is the pressure exerted by vapor molecules above a liquid when the rates of evaporation and condensation are equal. A familiar example is the pressure of H_2O molecules in the vapor above water in a sealed bottle. When the temperature increases, the vapor pressure increases. Conversely, when the temperature decreases, the vapor pressure decreases.

Section 11.9 **Dalton's law of partial pressures** states that the total pressure in a gaseous system is equal to the sum of the **partial pressures** of the gases in the mixture. We can collect a gas over water to determine the **volume by displacement**. A "wet" gas collected over water contains water vapor. The partial pressure of the gas is found by subtracting the vapor pressure of water from the total pressure.

Section 11.10 According to the **kinetic theory**, a gas is made up of individual molecules distributed in empty space. Molecules in an **ideal gas** move about rapidly and randomly in straight-line paths. Gas molecules show no attraction for one another and undergo **elastic collisions**. The kinetic energy is directly proportional to the Kelvin temperature. At **absolute zero** (0 K), the pressure of an ideal gas is zero. The velocity of gas molecules increases with the temperature and decreases with the molecular mass. The behavior of a **real gas** can deviate substantially at low temperatures and high pressures from that predicted by the kinetic theory.

Section 11.11 A more mathematical method for solving gas problems involves the **ideal gas law**. In the equation $PV = nRT$, the **ideal gas constant** (symbol $\mathbf{R}$) has a value of 0.0821 atm · L/mol · K. By substituting g/MM in the equation for n, we can calculate the molar mass (MM) of a gas.

Problem-Solving Organizer

Topic	Procedure	Example
Properties of Gases Sec. 11.1	Predict the shape and volume of a gas.	The shape and volume of argon gas are variable.
Atmospheric Pressure Sec. 11.2	Express a given gas pressure in units of atm, mm Hg, torr, cm Hg, in. Hg, psi, or kPa.	If argon gas is at 655 mm Hg, what is the pressure in atm? $655\ \cancel{\text{mm Hg}} \times \frac{1\ \text{atm}}{760\ \cancel{\text{mm Hg}}} = 0.862\ \text{atm}$
Variables Affecting Gas Pressure Sec. 11.3	Predict an increase or decrease in gas pressure for a change in volume, temperature, or number of molecules.	An increase in which variable causes a decrease in pressure? Answer: volume
Boyle's Law Sec. 11.4	Calculate a final pressure after a change in volume (temperature constant). Pressure and volume are inversely proportional.	If a gas is at 1.00 atm, what is the pressure when the volume decreases from 5.00 L to 2.00 L? $1.00\ \text{atm} \times \frac{5.00\ \cancel{\text{L}}}{2.00\ \cancel{\text{L}}} = 2.50\ \text{atm}$
Charles' Law Sec. 11.5	Calculate a final volume after a change in temperature (pressure constant). Volume and temperature are directly proportional.	If a gas occupies 50.0 mL, what is the volume when the gas is heated from 300 K to 450 K? $50.0\ \text{mL} \times \frac{450\ \cancel{\text{K}}}{300\ \cancel{\text{K}}} = 75.0\ \text{mL}$
Gay-Lussac's Law Sec. 11.6	Calculate a final pressure after a change in temperature (volume constant). Pressure and temperature are directly proportional.	If a gas is at 1.00 atm, what is the pressure when the gas is heated from 300 K to 450 K? $1.00\ \text{atm} \times \frac{450\ \cancel{\text{K}}}{300\ \cancel{\text{K}}} = 1.50\ \text{atm}$
Combined Gas Law Sec. 11.7	Calculate a final volume after a change in pressure and temperature. Volume and pressure are inversely proportional; volume and temperature are directly proportional.	If 50.0 mL of gas is at 1.50 atm and 450 K, what is the volume at 1.00 atm and 273 K? $50.0\ \text{mL} \times \frac{1.50\ \cancel{\text{atm}}}{1.00\ \cancel{\text{atm}}} \times \frac{273\ \cancel{\text{K}}}{450\ \cancel{\text{K}}} = 45.5\ \text{mL}$

Key Concepts*

1. A few drops of water in a metal can are heated to steam, and the can is promptly sealed. As the steam cools from 100° to 20°C, the can is crushed as shown in the diagram. Explain the observation.

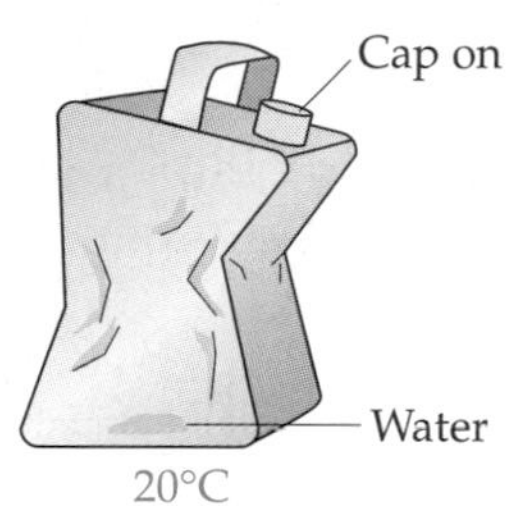

2. Which of the following is responsible for the liquid rising when you drink through a straw: atmospheric pressure, suction, vacuum, vapor pressure?
3. Which of the following variables remains constant in a closed gaseous system: pressure, volume, temperature, or number of molecules?
4. When helium in a steel cylinder expands from 1.00 L to 5.00 L and the temperature remains constant, the pressure inside the cylinder (increases/decreases).
5. When helium in a balloon warms from 25° to 100°C and the gas pressure remains constant, the volume of the balloon (increases/decreases).
6. When helium in a steel bulb changes from 15.0 psi to 25.0 psi and the volume remains constant, the temperature inside the bulb (increases/decreases).
7. Ammonia gas, NH_3, makes you cry, and nitrous oxide, N_2O, makes you laugh. Would you laugh first or cry first if both gases were released at the same time from the same distance?

Key Terms†

Select the key term below that corresponds to each of the following definitions.

_____ 1. a measure of the frequency and energy of molecules colliding against the walls of a container
_____ 2. a gaseous volume that does not contain molecules
_____ 3. the pressure exerted by the molecules in air
_____ 4. an instrument for measuring atmospheric pressure
_____ 5. a unit of pressure equal to 1 mm Hg
_____ 6. an association between two variables such that when one doubles, the other doubles
_____ 7. an association between two variables such that when one doubles, the other halves
_____ 8. the statement that the pressure and volume of a gas are inversely proportional at constant temperature
_____ 9. the statement that the volume and the Kelvin temperature of a gas are directly proportional at constant pressure
_____ 10. the statement that the pressure and the Kelvin temperature of a gas are directly proportional at constant volume
_____ 11. the statement that the pressure exerted by a gas is inversely proportional to its volume and directly proportional to its Kelvin temperature
_____ 12. the conditions for a gas at 273 K and 760 mm Hg
_____ 13. the pressure exerted by vapor molecules above a liquid in a closed container when the rates of evaporation and condensation are equal

(a) absolute zero (*Sec. 11.10*)
(b) atmospheric pressure (*Sec. 11.2*)
(c) barometer (*Sec. 11.2*)
(d) Boyle's law (*Sec. 11.4*)
(e) Charles' law (*Sec. 11.5*)
(f) combined gas law (*Sec. 11.7*)
(g) Dalton's law of partial pressures (*Sec. 11.9*)
(h) directly proportional (*Sec. 11.5*)
(i) elastic collision (*Sec. 11.10*)
(j) gas pressure (*Sec. 11.2*)
(k) Gay-Lussac's law (*Sec. 11.6*)
(l) ideal gas (*Sec. 11.10*)
(m) ideal gas constant (*Sec. 11.11*)
(n) ideal gas law (*Sec. 11.11*)
(o) inversely proportional (*Sec. 11.4*)

*Answers to Key Concepts are in Appendix H.
†Answers to Key Terms are in Appendix I.

_____ 14. the statement that the pressure exerted by a mixture of gases is equal to the sum of the individual pressures exerted by each gas

_____ 15. the pressure exerted by an individual gas in a mixture of two or more gases

_____ 16. a technique for determining the amount of gas from the amount of water it displaces

_____ 17. a theoretical description of gas molecules demonstrating ideal behavior

_____ 18. a theoretical gas that obeys the kinetic theory under all conditions

_____ 19. an actual gas that deviates from ideal behavior under certain conditions

_____ 20. an impact between gas molecules with no change in total energy

_____ 21. the theoretical temperature at which the kinetic energy of a gas is zero

_____ 22. the principle stated by the relationship $PV = nRT$

_____ 23. the proportionality constant R in the equation $PV = nRT$

(p) kinetic theory (*Sec. 11.10*)
(q) partial pressure (*Sec. 11.9*)
(r) real gas (*Sec. 11.10*)
(s) standard temperature and pressure (*Sec. 11.7*)
(t) torr (*Sec. 11.2*)
(u) vacuum (*Sec. 11.2*)
(v) vapor pressure (*Sec. 11.8*)
(w) volume by displacement (*Sec. 11.9*)

Exercises‡

Properties of Gases (Sec. 11.1)

1. What are the five observed properties of gases?
2. Approximately how many times more dense is water than air?

▲ **The Atmosphere** The atmospheric pressure progressively decreases at higher altitudes. By way of example, the atmospheric pressure on Mt. Everest is about one-third the atmospheric pressure at sea level.

Atmospheric Pressure (Sec. 11.2)

3. Give the value for standard atmospheric pressure in each of the following units.
   (a) atmospheres **(b)** millimeters of mercury
   (c) torr **(d)** centimeters of mercury
4. Give the value for standard atmospheric pressure in each of the following units.
   (a) inches of mercury **(b)** pounds per square inch
   (c) kilopascals
5. If oxygen gas in a steel cylinder is at a pressure of 5.25 atm, what is the pressure expressed in each of the following units?
   (a) mm Hg **(b)** torr
   (c) cm Hg **(d)** in. Hg
6. If an automobile piston compresses a fuel–air mixture to a pressure of 7555 torr, what is the pressure expressed in each of the following units?
   (a) cm Hg **(b)** in. Hg
   (c) psi **(d)** kPa
7. An American newscast states that the barometer reads 28.8 in. Hg. Express the atmospheric pressure in each of the following units.
   (a) atm **(b)** mm Hg
   (c) cm Hg **(d)** torr
8. A Canadian newscast states that the barometer reads 99.9 kPa. Express the atmospheric pressure in each of the following units.
   (a) atm **(b)** mm Hg
   (c) psi **(d)** in. Hg

Variables Affecting Gas Pressure (Sec. 11.3)

9. State the three variables that can directly affect the pressure of a gas.
10. Explain how increasing the temperature of a gas increases its pressure.
11. Indicate what happens to the pressure of a gas with the following changes.
    (a) The volume increases.
    (b) The temperature increases.
    (c) The moles of gas increase.
12. Indicate what happens to the pressure of a gas with the following changes.
    (a) The volume decreases.
    (b) The temperature decreases.
    (c) The moles of gas decrease.

‡Answers to odd-numbered Exercises are in Appendix J.

13. State whether the pressure of a gas in a closed system increases or decreases with the following changes.
   (a) The volume changes from 2.50 L to 5.00 L.
   (b) The temperature changes from 20° to 100°C.
   (c) The moles of gas changes from 0.500 mol to 0.250 mol.
14. State whether the pressure of a gas in a closed system increases or decreases with the following changes.
   (a) The volume changes from 75.0 mL to 50.0 mL.
   (b) The temperature changes from 0° to −195°C.
   (c) The moles of gas change from 1.00 mol to 5.00 mol.

▲ **Tire Pressure** The pressure inside the bicycle tire increases as the number of air molecules in the tire increases.

Boyle's Law (Sec. 11.4)

15. Sketch a graph of pressure versus volume, assuming temperature is constant. Label the vertical axis P, and the horizontal axis V.
16. Sketch a graph of pressure versus inverse volume, assuming temperature is constant. Label the vertical axis P, and the horizontal axis $1/V$.
17. A sample of air at 0.750 atm is expanded from 250.0 mL to 655.0 mL. If the temperature remains constant, what is the final pressure in atm?
18. What is the final volume of argon gas if 2.50 L at 705 torr is compressed to a pressure of 1550 torr? Assume the temperature remains constant.
19. A 50.0-mL sample of carbon monoxide gas at 25°C has a pressure of 15.0 psi. If the final volume is 44.0 mL at 25°C, what is the final pressure in psi?
20. Calculate the volume of chlorine gas at 20°C and 75.0 cm Hg if the volume of the gas is 1.10 L at 20°C and 95.5 cm Hg.

Charles' Law (Sec. 11.5)

21. Sketch a graph of volume versus Kelvin temperature, assuming pressure is constant. Label the vertical axis V, and the horizontal axis T (K).
22. Sketch a graph of volume versus Celsius temperature, assuming pressure is constant. Label the vertical axis V, and the horizontal axis t, (°C). Assume the Celsius temperature approaches zero at the origin.
23. A 335-mL sample of oxygen at 25°C is heated to 50°C. If the pressure remains constant, what is the final volume in milliliters?
24. What is the final Celsius temperature if 4.50 L of nitric oxide gas at 35°C is cooled until the volume reaches 1.00 L? Assume the pressure remains constant.
25. A 80.0-cm^3 sample of fluorine gas at 0°C has a pressure of 761 torr. If the gas is heated to 100°C at 761 torr, what is the final volume in cubic centimeters?
26. Calculate the final Celsius temperature of hydrogen chloride gas if 0.500 L at 35°C and 0.950 atm is heated until the volume reaches 1.26 L at 0.950 atm.

Gay-Lussac's Law (Sec. 11.6)

27. Sketch a graph of pressure versus Kelvin temperature, assuming volume is constant. Label the vertical axis P, and the horizontal axis T (K).
28. Sketch a graph of pressure versus Celsius temperature, assuming volume is constant. Label the vertical axis P, and the horizontal axis t (°C). Assume the Celsius temperature approaches zero at the origin.
29. A sample of ammonia gas at 760 torr is heated from 20° to 200°C. If the volume remains constant, what is the final pressure in torr?
30. A sample of xenon gas at 20°C and 0.570 atm is cooled to a pressure of 0.100 atm. If the volume remains constant, what is the final Celsius temperature?
31. A 1.00-L sample of neon gas at 0°C has a pressure of 76.0 cm Hg. If the gas is heated to 100°C, what is the final pressure in cm Hg if the volume remains constant?
32. Calculate the final Celsius temperature of sulfur dioxide if 0.500 L of the gas at 35°C and 650 mm Hg is heated until the pressure reaches 745 mm Hg. Assume the volume remains 0.500 L.

Combined Gas Law (Sec. 11.7)

33. A 100.0-mL sample of hydrogen gas is collected at 772 mm Hg and 21°C. Calculate the volume of hydrogen at STP.
34. A 5.00-L sample of nitrogen dioxide gas is collected at 5.00 atm and 500°C. What is the volume of nitrogen dioxide under standard conditions?

2.00 L

35. If a sample of air occupies 2.00 L at STP, what is the volume at 75°C and 365 torr?

36. If a sample of gas occupies 25.0 mL at −25°C and 650 mm Hg, what is the volume at 25°C and 350 mm Hg?

37. A sample of hydrogen fluoride gas has a volume of 1250 mL at STP. What is the pressure in torr if the volume is 255 mL at 300°C?

38. A sample of air occupies 0.750 L at standard conditions. What is the pressure in atm if the volume is 100.0 mL at 25°C?

39. A sample of krypton gas has a volume of 500.0 mL at 225 mm Hg and −125°C. Calculate the pressure in mm Hg if the gas occupies 220.0 mL at 100°C.

40. A sample of gas has a volume of 1.00 L at STP. What is the temperature in °C if the volume is 10.0 L at 2.00 atm?

41. A sample of air occupies 50.0 mL at standard conditions. What is the Celsius temperature if the volume is 350.0 mL at 350 torr?

42. A sample of oxygen gas occupies 500.0 mL at 75.0 cm Hg and −185°C. Calculate the temperature in °C if the gas has a volume of 225.0 mL at a 55.0 cm Hg.

The Vapor Pressure Concept (Sec. 11.8)

43. What is the general relationship between the temperature of a liquid and its vapor pressure?

44. Explain how a glass of water evaporates using the concept of vapor pressure.

45. Refer to Table 11.2 and state the vapor pressure for water at each of the following temperatures in mm Hg.
(a) 25°C (b) 50°C

46. Refer to Table 11.2 and state the vapor pressure for water at each of the following temperatures in atm.
(a) 75°C (b) 100°C

Dalton's Law (Sec. 11.9)

47. Air is composed of nitrogen, oxygen, and argon. If the partial pressure of nitrogen is 587 mm Hg, oxygen is 158 mm Hg, and argon is 7 mm Hg, what is the atmospheric pressure?

48. Air is composed of nitrogen, oxygen, argon, and trace gases. If the pressure of nitrogen is 592 torr, oxygen is 160 torr, and argon is 7 torr, what is the partial pressure of the trace gases? Assume the atmospheric pressure is 760 torr.

49. An alloy cylinder contains sulfur dioxide, sulfur trioxide, and oxygen gases at 825°C and 1.00 atm. If the partial pressure of sulfur dioxide is 150 mm Hg and sulfur trioxide is 475 mm Hg, what is the partial pressure of oxygen in mm Hg?

50. An alloy cylinder contains nitrogen, hydrogen, and ammonia gases at 500 K and 5.00 atm. If the partial pressure of nitrogen is 1850 torr and hydrogen is 1150 torr, what is the partial pressure of ammonia in torr?

51. Define the expression "collecting a gas over water."

52. Distinguish between a "wet" gas and a "dry" gas.

53. If oxygen is collected over water at 20°C and 766 torr, what is the partial pressure of the oxygen? Refer to Table 11.2 for the vapor pressure of water.

54. If xenon is collected over water at 30°C and 755 torr, what is the partial pressure of the xenon? Refer to Table 11.2 for the vapor pressure of water.

Ideal Gas Behavior (Sec. 11.10)

55. State the five characteristics of an ideal gas according to the kinetic theory.

56. Distinguish between a real gas and an ideal gas.

57. Under what conditions of temperature and pressure does a real gas behave most like an ideal gas?

58. At what Celsius temperature does a gas possess zero kinetic energy?

59. A stainless-steel cylinder contains the noble gases He, Ne, and Ar. Which of these gases fits each of the following descriptions?
(a) highest kinetic energy (b) lowest kinetic energy
(c) highest atomic velocity (d) lowest atomic velocity

60. A stainless-steel cylinder contains the gases H_2, N_2, and O_2 gases. Which of these gases fits each of the following descriptions?
(a) highest kinetic energy (b) lowest kinetic energy
(c) highest molecular velocity (d) lowest molecular velocity

61. What pressure is exerted by an ideal gas at absolute zero?

62. What volume is occupied by an ideal gas at absolute zero?

Ideal Gas Law (Sec. 11.11)

63. If 0.500 mol of hydrogen gas occupies 50.0 mL at 25°C, what is the pressure in atmospheres?

64. If 1.25 mol of oxygen gas exerts a pressure of 1200 mm Hg at 25°C, what is the volume in liters?

65. If 10.0 L of nitrous oxide exerts a pressure of 125 psi at 373 K, what is the number of moles of gas?

66. If 0.100 mol of argon gas occupies 2.15 L at 725 torr, what is the temperature in degrees Celsius?

67. If the density of ozone is 2.14 g/L at STP, what is the molar mass of ozone?

68. If the density of Freon is 5.40 g/L at STP, what is the molar mass of Freon?

69. A sample of unknown gas weighs 1.95 g and occupies 3.00 L at 1.25 atm and 20°C. What is the molar mass of the unknown gas?

70. A sample of unknown gas weighs 2.85 g and occupies 750 mL at 760 mm Hg and 100°C. What is the molar mass of the unknown gas?

71. A sample of chlorine gas occupies 1550 mL at 0.945 atm and 50°C. What is the mass of the sample?

72. A sample of fluorine gas occupies 855 mL at 710 mm Hg and 155°C. What is the mass of the sample?

General Exercises

73. If the surface area of a human body is 2500 square inches, what is the total weight of the atmosphere on a human expressed in pounds?

74. If the atmospheric pressure is 760 torr, what is the height in feet of a barometer filled with water? (*Given*: Hg is 13.6 times more dense than water.)

75. An automobile tire contains nitrogen, oxygen, and argon. If the partial pressures of the gases are 0.79 atm, 3.07 psi, and 7.55 torr, respectively, what is the total pressure (in psi) inside the tire?

76. A steel cylinder contains hydrogen, chlorine, and hydrogen chloride gases. If the partial pressures of the three gases are 3.15 atm, 50.0 psi, and 2500 torr, respectively, what is the total pressure (in atm) inside the cylinder?

77. The decomposition of potassium chlorate produces oxygen gas. If 42.5 mL of "wet" oxygen gas is collected over water at 22°C and 764 mm Hg, what is the volume of "dry" oxygen gas at STP conditions? (The vapor pressure of water at 22°C is 19.8 mm Hg.)

78. Zinc metal reacts with hydrochloric acid to produce hydrogen gas. If the volume of hydrogen gas collected over water is 79.9 mL at 16°C and 758 mm Hg, what is the volume of gas at STP conditions? (The vapor pressure of water at 16°C is 13.6 mm Hg.)

79. A 5.00-L sample of krypton gas contains 1.51×10^{24} atoms at 25°C. What is the pressure of the gas in atmospheres?

80. How many molecules of carbon monoxide, CO, are present in 1.00 cm^3 of gas at STP?

81. What volume is occupied by 3.38×10^{22} molecules of nitrogen monoxide, NO, at 100°C and 255 torr?

82. Given samples of propane gas, C_3H_8, and butane gas, C_4H_{10}, each at 100°C, which gas has the greater kinetic energy? Which gas has the greater molecular velocity?

83. Express the ideal gas constant in the following units: torr · L/mol · K.

84. Express the ideal gas constant in the following SI units: J/mol · K. (Given: 1 atm · L = 101.27 J)

85. Deep-sea divers breathe a helium–oxygen mixture, rather than compressed air, at depths below 200 feet. Explain why divers have high, squeaky voices while breathing the mixture. (*Hint*: As molecules move faster across the vocal cords, the voice pitch rises.)

▲ **Deep-Sea Divers** These divers are breathing a mixture of helium and oxygen gases.

Explorer Quiz 1
Explorer Quiz 2
Explorer Quiz 3
Master Quiz

CHAPTERS 9–11
Cumulative Review

Key Concepts

1. Which of the following is the best estimate for the volume of 1 mol of marbles: 1000 mL, the New Orleans Superdome, the Grand Canyon, or the Moon?
2. If a 1-carat diamond is 0.200 g of crystalline carbon, does it contain more or less than a trillion (1×10^{12}) carbon atoms?
3. The only natural isotope of sodium is sodium-23. What are the masses of one Na atom and of Avogadro's number of Na atoms, respectively.
4. Which of the following gases contains more molecules: 1.00 mol of CO or 44.0 g of CO_2?
5. Which of the following gases contains more molecules: 1.00 mol of oxygen, O_2, or 22.4 L of ozone, O_3, at STP?
6. Which of the following are not in the same ratio as the coefficients in a balanced chemical equation: moles, grams, or liters of gas?
7. In the Contact process for manufacturing sulfuric acid, sulfur dioxide and oxygen gases are heated with a platinum catalyst to give sulfur trioxide.

$$2\ SO_2(g) + O_2(g) \xrightarrow{\text{Pt/825°C}} 2\ SO_3(g)$$

   (a) How many moles of O_2 react with 5.00 mol of SO_2?
   (b) How many moles of SO_3 are produced from 5.00 mol of SO_2?

8. In the Contact process for manufacturing sulfuric acid:
   (a) How many liters of O_2 react with 5.00 L of SO_2?
   (b) How many liters of SO_3 are produced from 5.00 L of SO_2?
9. Silver reacts with sulfur to give Ag_2S. The balanced equation is

$$2\ Ag(s) + S(s) \xrightarrow{\Delta} Ag_2S\ (s)$$

   Complete the following illustrations by writing formula units of product and atoms of excess reactant. Indicate which substance is the limiting reactant.

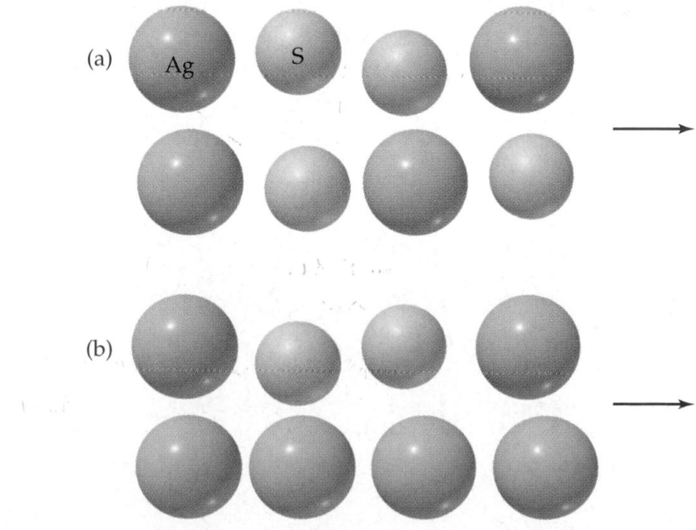

10. What must be done first before applying unit analysis to solve a stoichiometry problem?

11. In addition to modified unit analysis, what alternate method can be applied to solve a gas law problem?
12. If you release a balloon and it rises into the atmosphere, is the volume greater at sea level or 5000 feet?
13. If you use a gauge to check tire pressure, is the pressure greater when the tire is warm or cold?
14. If you heat the air in a hot-air balloon, the pressure inside the balloon is the same as that of the surrounding air. Why does the hot-air balloon rise?
15. Methyl acetate vapor, $C_3H_6O_2$, is fragrant, and hydrogen sulfide gas, H_2S, smells like rotten eggs. Would you be attracted first or repulsed first if both gases were released at the same time from the same distance?

Key Terms

State the key term that corresponds to each of the following descriptions.

______ 1. the value corresponding to 6.02×10^{23} atoms, molecules, or formula units
______ 2. the amount of substance that contains 6.02×10^{23} particles
______ 3. the mass of 1 mol of pure substance expressed in grams
______ 4. the volume occupied by 1 mol of any gas at standard conditions
______ 5. the chemical formula of a compound that expresses the simplest ratio of atoms in a molecule, or ions in a formula unit
______ 6. the relationship of amounts of substance according to a balanced equation
______ 7. the law stating that the mass of reactants is equal to the mass of products
______ 8. the substance in a reaction that controls the maximum amount of product
______ 9. the amount of product experimentally obtained from a reactant
______ 10. the amount of product calculated to be obtained from a reactant
______ 11. the conditions for a gas at 273 K and 760 mm Hg
______ 12. the frequency and energy of gas molecules colliding with the walls of its container
______ 13. the pressure exerted by vapor molecules above a liquid in a closed container when the rates of evaporation and condensation are equal
______ 14. the pressure exerted by an individual gas in a mixture of two or more gases
______ 15. a technique for determining an amount of gas by measuring the water it displaces
______ 16. a theoretical description of gas molecules that demonstrate ideal behavior
______ 17. a theoretical gas that obeys the kinetic theory under all conditions
______ 18. an actual gas that deviates from ideal behavior under certain conditions
______ 19. an impact between gas molecules that results in no change in total energy
______ 20. the theoretical temperature at which the kinetic energy of a gas is zero

Review Exercises

1. What is the number of molecules in 0.375 mol of nitrogen gas, N_2, at STP?
2. What is the mass of 0.375 mol of nitrogen gas, N_2, at STP?
3. What is the volume of 0.375 mol of nitrogen gas, N_2, at STP?
4. How many water molecules are in a 1.00-cm^3 cube of ice ($d = 0.917\ g/cm^3$)?
5. If 0.250 mol of red phosphorus reacts with 0.625 mol of yellow sulfur, what is the empirical formula of the phosphorus sulfide product?
6. What mass of methane gas, CH_4, reacts with 1.55 g of oxygen at STP?

$$CH_4(g) + O_2(g) \longrightarrow CO_2(g) + H_2O(l)$$

7. What mass of methane gas, CH_4, reacts with 1.55 L of oxygen at STP?
8. What volume of methane gas, CH_4, reacts with 1.55 L of oxygen at STP?

9. What mass of iron metal is produced by the reaction of 75.0 g of iron(II) oxide and 25.0 g of magnesium metal according to the following equation?

$$FeO(s) + Mg(s) \longrightarrow Fe(l) + MgO(s)$$

10. If 3.00 L of sulfur dioxide gas reacts with 4.50 L of oxygen gas, what volume of sulfur trioxide gas is produced (assume constant conditions)?

$$SO_2(g) + O_2(g) \longrightarrow SO_3(g)$$

11. If a sample of helium at 1.00 atm in a cylinder expands from 1.00 L to 5.00 L and the temperature remains constant, what is the new pressure inside the cylinder?
12. If a 1.00-L sample of helium in a balloon warms from 25° to 100°C and the pressure remains constant, what is the new volume of the balloon?
13. If the pressure of a sample of helium at 50°C in a steel bulb changes from 15.0 psi to 25.0 psi and the volume remains constant, what is the new temperature inside the bulb?
14. Draw the graph of kinetic energy versus Kelvin temperature for an ideal gas.
15. An unknown gas occupies a volume of 1.50 L at 21°C and 0.950 atm. If the mass is 2.01 g, what is the molar mass of the gas?

CHAPTER 12

Chemical Bonding

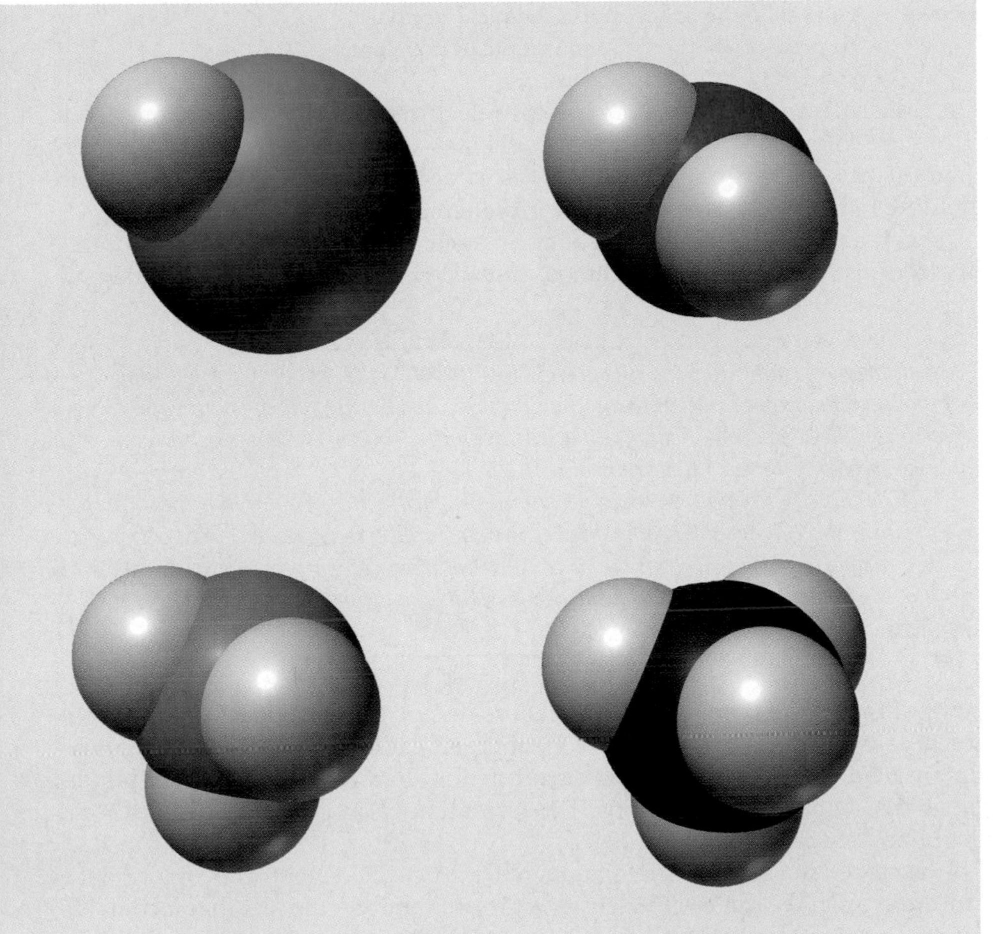

▲ These space-filling models represent the shapes of four simple molecules. Can you identify the ammonia, hydrogen chloride, methane, and water molecule?

Since understanding chemical bonding is a conceptual task, this topic has been placed later in the text, after the treatment of gases. The discussion of polar and nonpolar bonds provides the foundation for the concepts of polar and nonpolar solvents in liquids and solutions (Chapters 13 and 14). However, the flexibility of the textbook allows for the treatment of chemical bonding immediately following atomic theory and periodicity (Chapters 5 and 6).

At the beginning of the twentieth century, scientists were rapidly unraveling the mystery of the atom. In 1897 J. J. Thomson discovered the electron and, shortly thereafter, the proton. In 1911 Ernest Rutherford unveiled the atomic nucleus. And in 1913 Niels Bohr established that electrons circled the nucleus in planetary orbits. During this same period of time, scientists began to speculate that electrons were responsible for holding atoms together. They proposed that electrons provided the "glue" that bonded one atom to another.

In 1916 the American chemist G. N. Lewis formulated one of the first theories of chemical bonding. He noted that noble gases were unusually stable and that, with the exception of helium, all these gases had eight electrons in their outer shell.

Lewis theorized that atoms bonded together to attain a stable noble gas electron structure. This principle, referred to as the octet rule, states that atoms bond in such a way that each atom attains eight electrons in its outer shell. Lewis initially envisioned bonding electrons surrounding the nucleus and located at the corners of an imaginary cube.

12.1 The Chemical Bond Concept

Objectives
- To explain the concept of a chemical bond.
- To predict whether a bond is ionic or covalent.

We previously learned that an atom has core electrons and valence electrons. Core electrons are found close to the nucleus, whereas **valence electrons** are found in the highest *s* and *p* energy subshells. It is the valence electrons that are responsible for holding two or more atoms together in a **chemical bond**.

The **octet rule** states that atoms bond in such a way that each atom acquires eight electrons in its outer shell. In the words of G. N. Lewis, the chemical bond concept can be described as follows.

> *Two atoms may conform to the rule of eight, or the octet rule, not only by the transfer of electrons from one atom to another, but also by sharing one or more pairs of electrons. These electrons which are held in common by two atoms may be considered to belong to the outer shells of both atoms.*
>
> *The simplest explanation of the predominant occurrence of an even number of electrons in the valence shells of molecules is that the electrons are definitely paired with one another. Two electrons thus coupled together, when lying between two atomic centers, and held jointly in the shells of the two atoms, I have considered to be the chemical bond.*

In an **ionic bond**, a metal atom transfers one or more valence electrons to a nonmetal atom. The resulting positively charged cation and negatively charged anion are held together by electrostatic attraction. In a **covalent bond**, two nonmetal atoms share valence electrons. The formation of an ionic bond and of a covalent bond are shown in Figure 12.1.

As discussed in Section 7.4, a fundamental particle held together by ionic bonds is a **formula unit**. We can easily identify a formula unit by the fact that it contains metal and nonmetal ions. NaCl, MgO, and AlN all contain a metal cation and a nonmetal anion and are examples of formula units held together by ionic bonds.

A fundamental particle held together by covalent bonds is a **molecule**. We can easily identify a molecule by the fact that it contains nonmetal atoms. H_2O, NH_3,

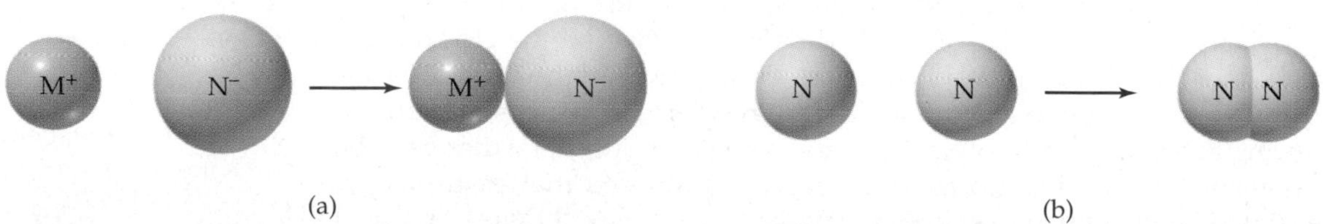

▲ **Figure 12.1 Chemical Bonds** (a) An ionic bond forms when a metal cation (M^+) is attracted to a nonmetal anion (N^-). (b) A covalent bond forms when valence electrons are shared by two nonmetal atoms (N).

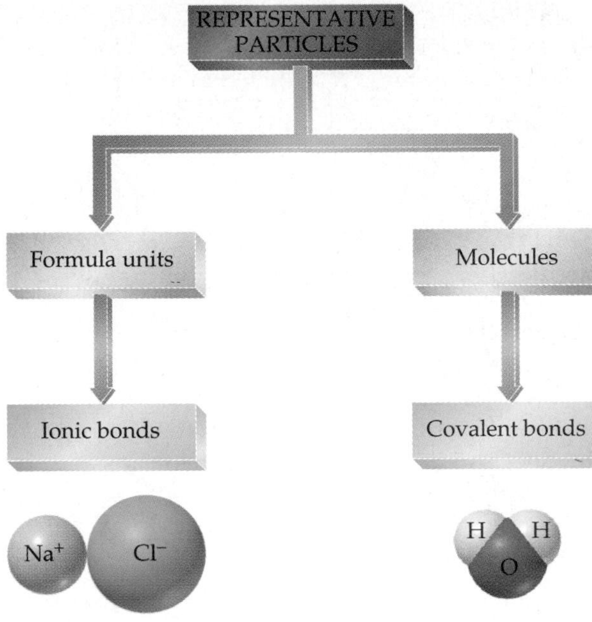

◀ **Figure 12.2 Classification of Representative Particles** Particles can be formula units (for example, NaCl) held together by ionic bonds or molecules (for example, H_2O) held together by covalent bonds.

and CH_4 all contain nonmetal atoms and are examples of molecules held together by covalent bonds. A substance can be composed of either formula units or molecules. The distinction between the types of representative particles found in a substance is shown in Figure 12.2.

The following example exercise further illustrates the relationship between bonding and the type of fundamental particles.

Example Exercise 12.1 • Bond Predictions

Predict whether each of the following is held together by ionic or by covalent bonds.

(a) ammonia, NH_3 (b) magnesium nitride, Mg_3N_2

Solution

A metal ion and a nonmetal ion are attracted in ionic bonds; two or more nonmetal atoms are attracted in covalent bonds.

(a) Ammonia contains the nonmetals nitrogen and hydrogen. It follows that NH_3 has *covalent* bonds.

(b) Magnesium nitride contains a metal (Mg) and a nonmetal (N). It follows that Mg_3N_2 has *ionic* bonds.

Self-Test Exercise

Predict whether each of the following is held together by ionic or by covalent bonds.

(a) aluminum oxide, Al_2O_3 (b) sulfur dioxide, SO_2

Answers: (a) ionic; (b) covalent

12.2 Ionic Bonds

Objective · To describe the formation of an ionic bond between a metal atom and a nonmetal atom.

An ionic bond results from the attraction between a positively charged cation and a negatively charged anion (Figure 12.1). In the bonding process, energy is released. This phenomenon is termed electrostatic attraction and is similar to the attraction between opposite ends of two magnets. In a crystal of ordinary table salt,

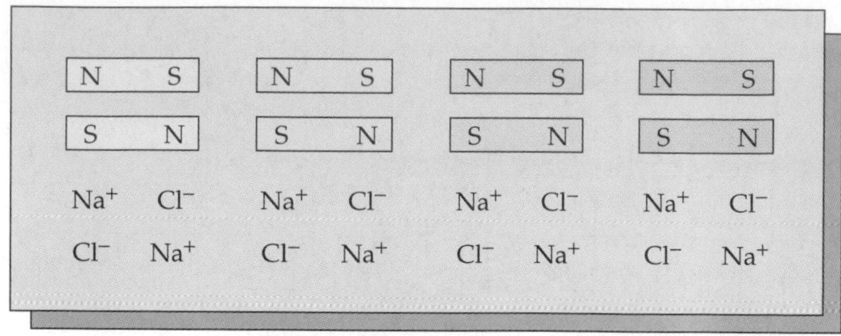

▶ **Figure 12.3 Electrostatic Attraction** Opposite ends of ordinary magnets are attracted to each other in much the same way that oppositely charged cations and anions are attracted. Notice that attraction can take place vertically as well as horizontally.

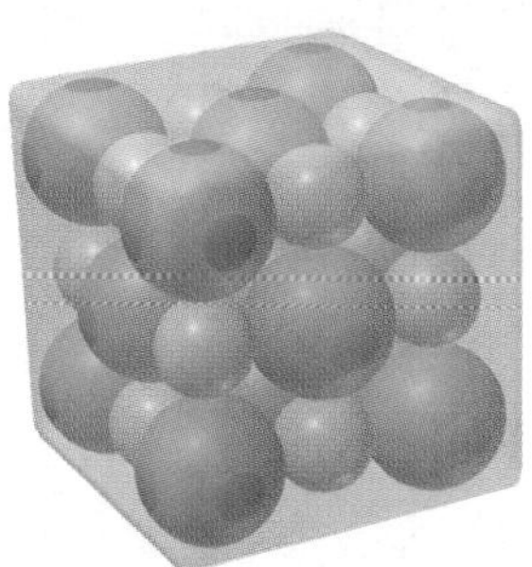

▲ **Figure 12.4 Crystalline Structure of Sodium Chloride** A crystal of sodium chloride is composed of sodium and chloride ions which repeat regularly. Notice that each sodium ion is attracted to several nearby chloride ions. The alternating pattern of cations and anions gives rise to a three-dimensional crystalline structure.

NaCl, there is an attraction between the Na^+ ions and a Cl^- ions. Figure 12.3 illustrates electrostatic attraction between ions, which is analogous to the attraction between magnets.

An ionic bond results from the combination of cations and anions. In ordinary table salt, NaCl, the ionic bonds between sodium ion and chloride ion are so strong that they create a rigid crystalline structure (Figure 12.4).

Formation of Cations

When a metal atom loses valence electrons, it becomes positively charged. Recall that the number of valence electrons corresponds to the group number of the element in the periodic table. When a sodium atom (Group IA/1) loses its 1 valence electron, it becomes Na^+. Referring to the periodic table, we notice that Na^+ has 10 electrons $(11 - 1 = 10\ e^-)$. This is the same number of electrons found in the preceding noble gas neon $(10\ e^-)$.

Although there are exceptions, metals usually achieve a noble gas electron configuration after losing their valence electrons. Stated differently, metal ions are generally *isoelectronic* with a noble gas. When a magnesium atom (Group IIA/12) loses its 2 valence electrons, it becomes Mg^{2+}. A magnesium ion has 10 electrons $(12 - 2 = 10\ e^-)$ and is also isoelectronic with neon.

In Section 6.8 we learned how to write electron dot formulas for atoms of elements. We can now use electron dot diagrams to clarify the formation of ions. Electron dot diagrams allow us to focus our attention on valence electrons involved in this process of forming ions. Figure 12.5 illustrates the formation of ions by third-period metals.

Note Notice that the electron dot diagrams of the ions in Figure 12.5 do not have any dots. Each of these ions has the electron configuration of neon. And each of these ions, like neon, has a stable electron configuration consisting of 8 electrons. By convention, chemists do not usually draw the underlying octet of electrons because they are considered core electrons.

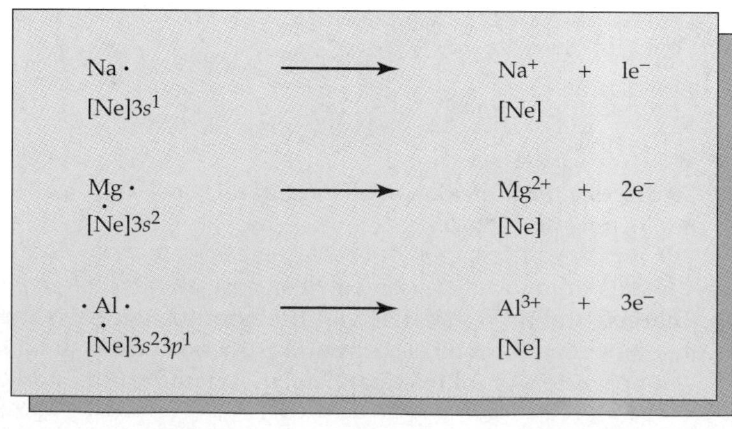

▶ **Figure 12.5 Formation of Metal Ions** The atoms of the metals in Period 3 form cations by losing 1, 2, and 3 electrons, respectively. In each case the metal atom becomes isoelectronic with the preceding noble gas neon, [Ne].

:Cl: $[Ne]3s^23p^5$	+ $1e^-$	$\longrightarrow$	$[:Cl:]^-$ $[Ne]3s^23p^6 = [Ar]$
·S: $[Ne]3s^23p^4$	+ $2e^-$	$\longrightarrow$	$[:S:]^{2-}$ $[Ne]3s^23p^6 = [Ar]$
·P: $[Ne]3s^23p^3$	+ $3e^-$	$\longrightarrow$	$[:P:]^{3-}$ $[Ne]3s^23p^6 = [Ar]$

◀ **Figure 12.6 Formation of Nonmetal Ions** The nonmetals in Period 3 gain 1, 2, and 3 electrons, respectively. In each case the nonmetal ion becomes isoelectronic with the noble gas argon, [Ar].

Formation of Anions

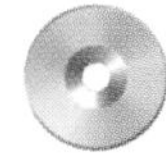

Dissolution of NaCl in Water Movie

When a nonmetal atom gains valence electrons, it becomes negatively charged. When a chlorine atom (Group VIIA/17) gains 1 valence electron, it becomes Cl^-. Referring to the periodic table, we notice that Cl^- has 18 electrons $(17 + 1 = 18\ e^-)$. This is the same number of electrons found in the next noble gas argon $(18\ e^-)$. When a sulfur atom (Group VIA/16) gains two valence electrons, it becomes S^{2-}. A sulfide ion has 18 electrons $(16 + 2 = 18\ e^-)$ and is also isoelectronic with argon. Figure 12.6 illustrates the formation of ions by third-period nonmetals.

The following example exercise further illustrates the formation of cations and anions.

Example Exercise 12.2 • Electron Configuration of Ions

Which noble gas has an electron configuration identical to that of each of the following ions?

(a) lithium ion (b) oxide ion (c) calcium ion (d) bromide ion

Solution

Refer to the group number in the periodic table for the number of valence electrons.

(a) Lithium (Group IA/1) forms an ion by losing 1 valence electron; Li^+ has only 2 electrons remaining and is isoelectronic with *He*.
(b) Oxygen (Group VIA/16) forms an ion by gaining 2 electrons; O^{2-} has 10 electrons and is isoelectronic with *Ne*.
(c) Calcium (Group IIA/2) forms an ion by losing 2 electrons; Ca^{2+} has 18 electrons remaining and is isoelectronic with *Ar*.
(d) Bromine (Group VIIA/17) forms an ion by gaining 1 electron; Br^- has 36 electrons and is isoelectronic with *Kr*.

Self-Test Exercise

Which noble gas element has an electron configuration identical to that of each of the following ions?

(a) potassium ion (b) nitride ion (c) strontium ion (d) iodide ion

Answers: (a) K^+ is isoelectronic with Ar; (b) N^{3-} is isoelectronic with Ne; (c) Sr^{2+} is isoelectronic with Kr; (d) I^- is isoelectronic with Xe

Ionic Radii

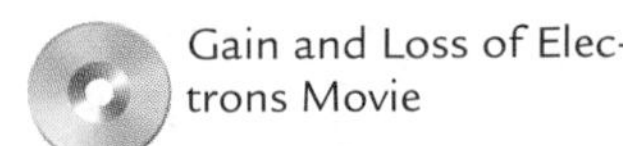

Gain and Loss of Electrons Movie

To visualize an ionic bond more clearly, we can picture a sodium atom becoming smaller after losing its valence electron. The radius of a sodium atom is 0.186 nm, whereas the radius of the sodium ion is 0.095 nm. The reason for this decrease in radius is that the sodium atom has lost its 3*s* energy sublevel. Moreover, in the sodium ion the positive nuclear charge (11+) is greater than the negative electron charge (10−) and pulls the electrons closer to the nucleus; thus, the radius decreases.

Conversely, we can picture a chlorine atom becoming larger after gaining a valence electron. The radius of a chlorine atom is 0.099 nm, whereas the radius of the chloride ion is 0.181 nm. The reason for this increase in radius is that the additional electron repels the electrons already present. In fact, the radius of the charged chloride ion is almost twice as large as that of the neutral chlorine atom.

The behavior of metallic sodium and nonmetallic chlorine shows the same general trend observed for all metals and nonmetals with respect to atomic radius. The radius of a *cation is smaller* than that of the corresponding metal atom; the radius of an *anion is larger* than that of the corresponding nonmetal atom. The specific changes in the radius of a sodium atom and a chlorine atom are depicted in Figure 12.7.

The following example exercise further illustrates the characteristics associated with the formation of an ionic bond.

Example Exercise 12.3 • Ionic Bonds

Which of the following statements is correct regarding the formation of an ionic bond from iron and chlorine atoms?

(a) An iron atom loses electrons, and a chlorine atom gains electrons.
(b) The iron atom is larger in radius than the iron ion.
(c) The chlorine atom is smaller in radius than the chloride ion.
(d) The iron and chloride ions form a bond by electrostatic attraction.

Solution

All these statements are correct regarding an ionic bond.

(a) The metal atom loses valence electrons, and the nonmetal atom gains valence electrons.
(b) The radius of a metal atom is greater than its ionic radius.
(c) The radius of a nonmetal atom is less than its ionic radius.
(d) The formation of an ionic bond is due to an attraction between ions.

Self-Test Exercise

Which of the following statements is correct regarding the formation of an ionic bond by zinc atoms and sulfur atoms?

(a) A zinc atom gains electrons, and a sulfur atom loses electrons.
(b) The zinc atom is smaller in radius than the zinc ion.
(c) The sulfur atom is larger in radius than the sulfide ion.
(d) The zinc and sulfide ions form a bond by electrostatic attraction.

Answer: Only statement (d) is correct.

12.3 Covalent Bonds

Objective · To describe the formation of a covalent bond between two nonmetal atoms.

A covalent bond results from the sharing of electrons by two nonmetal atoms. To create an accurate picture, we can visualize the electrons roaming about each atom. The electrons belong to both nonmetal atoms, and each atom uses these bonding

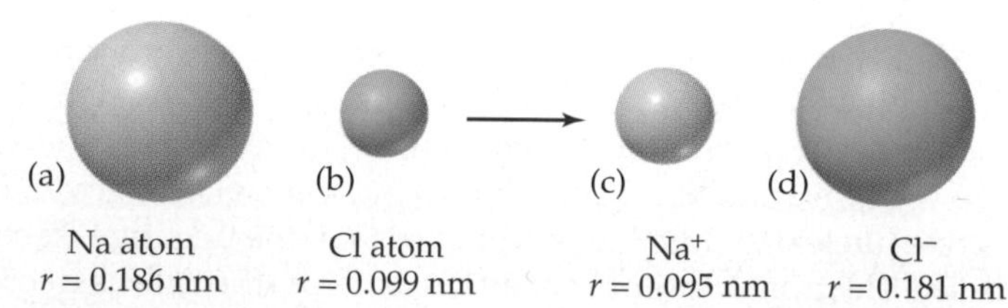

▶ Figure 12.7 Formation of Sodium and Chloride Ions The radius of the sodium atom decreases (a and c) as a result of losing an electron, whereas the radius of the chlorine atom increases (b and d) after gaining an electron.

electrons to complete an octet. That is, each nonmetal atom uses shared electrons to complete its valence shells. Since a filled valence shell is very stable, the resulting bond is also extremely stable.

Now, let's see what happens during the formation of a covalent bond. As an example, consider the formation of hydrogen chloride, HCl, from a hydrogen atom and a chlorine atom. During bond formation, the hydrogen atom shares its one valence electron with the chlorine atom. This additional electron gives chlorine eight electrons in its valence shell. Chlorine thus becomes isoelectronic with the next noble gas, argon, and is stable.

In the same process, the chlorine atom shares one of its valence electrons with the hydrogen atom. The additional electron gives the hydrogen atom two electrons in its valence shell. Hydrogen becomes isoelectronic with the next noble gas, helium, and is therefore stable. Both the hydrogen atom and the chlorine atom become stable in the process of sharing an electron pair and forming a covalent bond. The bonding electrons are distributed over each atom. In fact, the electrons are free to move about the entire molecule; thus, the electrons are said to be *delocalized*.

Bond Length

To understand bond formation more clearly, picture the valence shells of the two atoms overlapping each other. For instance, in the example above, overlapping occurs because the 1*s* energy sublevel of the hydrogen atom mixes with the 3*p* sublevel of the chlorine atom. This mixing of sublevels draws the two nuclei closer together. The radius of the hydrogen atom is 0.037 nm, and that of the chlorine atom is 0.099 nm. If the two atoms were simply next to each other, the distance from the hydrogen nucleus to the chlorine nucleus would be (0.037 + 0.099) nm or 0.136 nm. But experiments reveal that the distance between the two nuclei is actually 0.127 nm. Thus, the shells overlap as shown in Figure 12.8. The distance between the two nuclei is called the **bond length**.

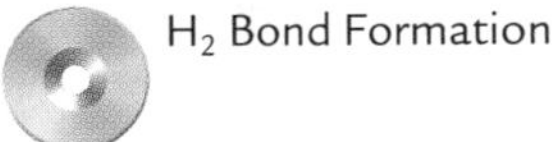

Bond Energy

Energy is released when two ions are attracted to one another and form an ionic bond. Correspondingly, energy is released when two atoms form a covalent bond. When hydrogen and chlorine atoms combine to form an HCl molecule, heat energy is released.

$$\text{H(g)} + \text{Cl(g)} \longrightarrow \text{HCl(g)} + \text{heat}$$

Conversely, energy is necessary to break an H—Cl bond. The amount of energy required to break the bond is called the bond dissociation energy, or simply the **bond energy**. The amount of energy necessary to break the H—Cl bond is identical to the amount of heat released when H and Cl form a hydrogen chloride molecule.

$$\text{HCl(g)} + \text{heat} \longrightarrow \text{H(g)} + \text{Cl(g)}$$

The following example exercise summarizes the formation of a covalent bond.

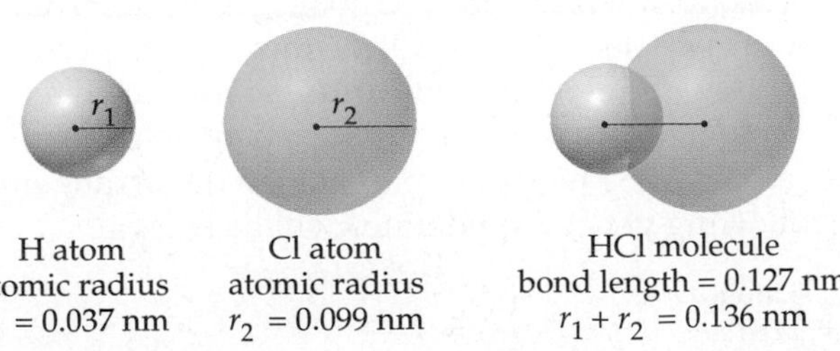

Figure 12.8 Formation of a Covalent Bond The sharing of an electron pair by a hydrogen atom and a chlorine atom produces a covalent bond. The two atoms are held together in a molecule of HCl. The actual bond length (0.127 nm) is less than the sum of the two atomic radii (0.037 + 0.099 = 0.136 nm).

Example Exercise 12.4 • Covalent Bonds

Which of the following statements is correct regarding the formation of a covalent bond between hydrogen and oxygen atoms?

(a) Valence electrons are shared by a hydrogen atom and an oxygen atom.
(b) Bonding electrons are distributed around the hydrogen and oxygen atoms.
(c) The H—O bond length is less than the sum of the atomic radii of a hydrogen atom and an oxygen atom.
(d) Energy is required to break the H—O covalent bond.

Solution

From the discussion in this section, we can assert that all of the above statements are correct.

(a) A covalent bond is formed by the sharing of valence electrons by the two nonmetal atoms.
(b) The bonding electrons are spread out over both atoms in the covalent bond and are delocalized throughout the entire molecule.
(c) The bond length in a covalent bond is less than the sum of the two atomic radii.
(d) The energy required to break a covalent bond is equal to the bond energy.

Self-Test Exercise

Which of the following statements is *not* correct regarding the formation of a covalent bond between carbon and sulfur?

(a) Valence electrons are shared by carbon and sulfur atoms.
(b) Bonding electrons are distributed over both carbon and sulfur atoms.
(c) The bond length between carbon and sulfur atoms is equal to the sum of the two atomic radii.
(d) The formation of covalent bonds between carbon and sulfur atoms releases energy.

Answer: (c) The bond length is less than the sum of the atomic radii.

Note It is always true that energy is released when two *atoms* form a covalent bond. On occasion, however, energy is required to form a covalent bond. For example, the formation of a H—I bond from *molecules* of hydrogen, H_2, and iodine, I_2, requires energy. We should explain that energy is first required to break H—H and I—I bonds before H—I bonds are formed. The formation of hydrogen iodide is endothermic because the energy required to break H—H and I—I bonds is greater than the energy released in the formation of H—I bonds.

Experiment #19, Prentice Hall Laboratory Manual

12.4 Electron Dot Formulas of Molecules

Objectives
- To draw the electron dot formula for a molecule.
- To draw the structural formula for a molecule.

In Section 6.8, we drew the electron dot formula for an atom of a representative element. We found the element in the periodic table, observed its group number, and recorded the number of valence electrons. For example, in the second period, we found that Li has 1 valence electron (Group IA/1), Be has 2 (Group IIA/2), B has 3 (Group IIIA/13), C has 4 (Group IVA/14), N has 5 (Group VA/15), O has 6 (Group VIA/16), F has 7 (Group VIIA/17), and the noble gas Ne has 8 (Group VIIIA/18). We represented each valence electron by a dot surrounding the symbol for the element.

Now, we will draw the **electron dot formula** (also called a Lewis structure) for a molecule. An electron dot formula shows the valence electrons, which indicate the bonds between atoms. This helps us to visualize the arrangement of atoms in a molecule. The following general guidelines will be helpful.

Guidelines for Drawing Electron Dot Formulas of Molecules

1. Calculate the total number of valence electrons by adding all the valence electrons for each atom in the molecule. The total should be an even number in keeping with the octet rule. If the total is an odd number, check your calculations.
2. Divide the total number of valence electrons by 2 to find the number of electron pairs in the molecule.
3. Surround the central atom with four electron pairs. Use the remaining electron pairs to complete an octet around each of the other atoms. (Hydrogen is the sole exception, as it requires only one electron pair.) The electron pairs shared by atoms are called **bonding electrons**. The other electron pairs simply complete the octet and are called **nonbonding electrons**, or lone pairs.
   In a molecule, usually two or more atoms are bonded to a central atom. Since it may be difficult for you to recognize the central atom, this textbook indicates the central atom in bold type.
4. If there are not enough electron pairs to provide an octet for each atom, move a nonbonding electron pair between two atoms that already share an electron pair.

Electron Dot Formula for Water

To learn how to draw electron dot formulas for molecules, we will draw the electron dot formula for water, H_2O. The total number of valence electrons in the molecule is $2(1\ e^-) + 6\ e^- = 8\ e^-$. The number of electron pairs is 4 $(8/2 = 4)$.

In the formula H_2**O**, oxygen is the central atom and is indicated in bold. We place the 4 electron pairs around the oxygen to provide the necessary octet.

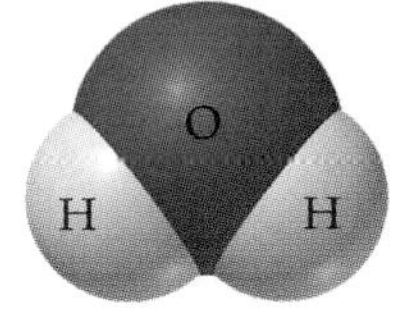

▲ Water, H_2O

We then place the two hydrogen atoms in any of the four electron pair positions.

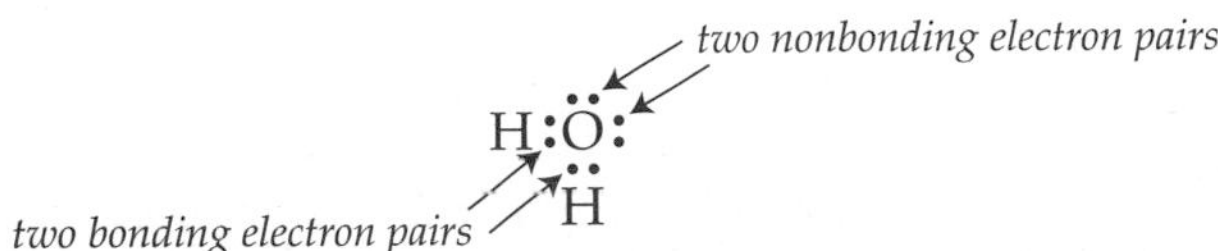

Notice that there are two bonding and two nonbonding electron pairs. All the electrons, however, are identical, and the 4 electron pairs (8 valence electrons) move around the entire molecule. This is why bonding and nonbonding valence electrons in a molecule are said to be delocalized.

To simplify, we represent each pair of bonding electrons by a single dash called a **single bond**. The resulting structure is referred to as the **structural formula** of the molecule.

H—O
   |
   H

Electron Dot Formula for Sulfur Trioxide

Now we'll try a more difficult example. Let's draw the electron dot formula for sulfur trioxide, $\mathbf{S}O_3$. Sulfur is the central atom in the molecule and is indicated in bold. The total number of valence electrons in the molecule is $6\,e^- + 3(6\,e^-) = 24\,e^-$. The number of electron pairs is 12 $(24/2 = 12)$.

Since sulfur is the central atom, we can begin by placing 4 electron pairs around the sulfur and then attaching the three oxygen atoms. That gives us

```
   ..
 O:S:O
   ..
   O
```

We started with 12 electron pairs, and so we have 8 pairs remaining. We place the remaining pairs around the oxygen atoms to complete each octet.

```
  ..  ..  ..
 :O : S : O:
  ..  ..  ..
     :O
      ..
```

Notice that the bottom oxygen atom does not have an octet; it has only 3 electron pairs. According to the guidelines, we can move a nonbonding electron pair from the S atom so as to provide 2 bonding electron pairs between the sulfur and the oxygen atoms.

```
  ..      ..
 :O : S : O:
  ..  ..  ..
      ..
     :O
      ..
```

The two shared electron pairs constitute a **double bond**. All 4 electrons in the double bond are shared by the S and the O atoms. Therefore, the sulfur atom still has an octet and the bottom oxygen atom has gained 2 electrons to complete its octet. The structural formula for SO_3 has one double bond and two single bonds. We represent the double bond by a double dash, and the single bond by a single dash.

```
O—S—O
  ‖
  O
```

Since we can rotate the structural formula, the double bond can be written on the left side as well as on the right side. Thus, the structural formula for a SO_3 molecule can be shown by any one of the following. This phenomenon is known as *resonance*, and we can draw three resonance structures for an SO_3 molecule.

```
O═S—O     O—S—O     O—S═O
  |         ‖         |
  O         O         O
```

Although these structures may appear to be different, all three formulas are identical. If we were to construct the molecule using models, we could easily verify this statement. Each molecular model can be rotated and shown to be identical to the other two.

Electron Dot Formula for Hydrogen Cyanide

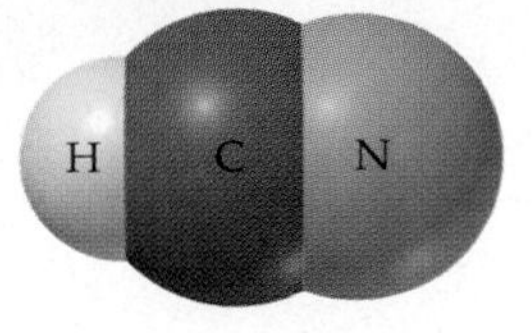

▲ Hydrogen Cyanide, HCN

Let's draw the electron dot formula for hydrogen cyanide, HCN. Carbon is the central atom in the molecule and is indicated in bold. The total number of valence electrons in the molecule is 1 e^- + 4 e^- + 5 e^- = 10 e^-. The number of electron pairs is 5 (10/2 = 5). Since carbon is the central atom, we begin by placing 4 electron pairs around the carbon and attaching the hydrogen and nitrogen atoms.

$$\mathrm{H}:\overset{\cdot\cdot}{\underset{\cdot\cdot}{\mathrm{C}}}:\mathrm{N}$$

We started with 5 electron pairs, and there is only 1 pair remaining. Since the hydrogen atom has its two required electrons, we place the remaining pair next to the nitrogen atom.

$$\mathrm{H}:\overset{\cdot\cdot}{\underset{\cdot\cdot}{\mathrm{C}}}:\mathrm{N}:$$

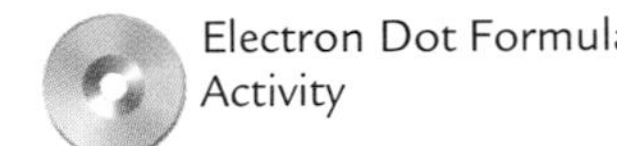

Electron Dot Formula Activity

The nitrogen atom now has 4 electrons, which is 4 e^- less than an octet. According to the guidelines, we can move nonbonding electron pairs between the C and the N atoms.

$$\mathrm{H}:\underset{\cdot\cdot}{\mathrm{C}}::\mathrm{N}:$$

The nitrogen atom has only 6 electrons. If we move the other nonbonding electron pair onto the carbon atom, we have

$$\mathrm{H}:\mathrm{C}:::\mathrm{N}:$$

The 3 electron pairs produce a **triple bond**. All 6 electrons in the triple bond are shared by the C and the N atoms. Therefore, both the carbon atom and the nitrogen atom have obtained an octet. A triple bond is indicated by a triple dash. The structural formula for HCN is therefore written as

$$\mathrm{H{-}C{\equiv}N}$$

The following example exercises further illustrate how to draw electron dot and structural formulas of molecules.

Example Exercise 12.5 • Electron Dot Formulas of Molecules

Draw the electron dot formula and the structural formula for a chloroform molecule, $CHCl_3$.

Solution

Carbon is the central atom in the chloroform molecule and is indicated in bold. The total number of valence electrons is 4 e^- + 1 e^- + 3(7 e^-) = 26 e^-. The number of electron pairs is 13 (26/2 = 13). We begin by placing 4 electron pairs around the carbon and adding the one H and three Cl atoms.

$$\begin{array}{c} \mathrm{H} \\ \mathrm{Cl}:\overset{\cdot\cdot}{\underset{\cdot\cdot}{\mathrm{C}}}:\mathrm{Cl} \\ \mathrm{Cl} \end{array}$$

We have 13 electron pairs minus the 4 pairs we used for the carbon. Since 9 electron pairs remain, let's place 3 pairs around each chlorine.

(continued)

Example Exercise 12.5 *(continued)*

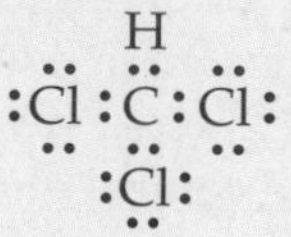

```
        H
   ..   ..   ..
  :Cl : C : Cl:
   ..   ..   ..
       :Cl:
        ..
```

Each atom is surrounded by an octet of electrons, except hydrogen which has 2. This is the correct electron dot formula. In the corresponding structural formula we replace each bonding electron pair by a single dash. The structural formula is

```
      H
      |
  Cl—C—Cl
      |
      Cl
```

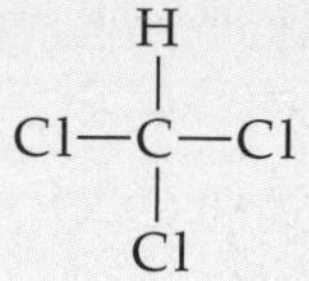

Note that the structural formula is free to rotate. We could have also written the hydrogen atom below or at the side of the carbon atom.

Self-Test Exercise

Draw the electron dot formula and the structural formula for a molecule of **SiHClBrI**.

Answer:

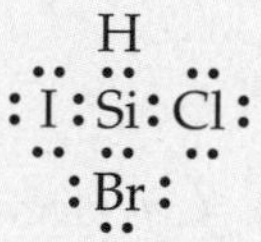

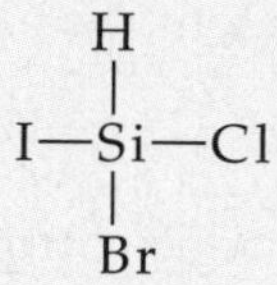

```
       H                   H
  ..   ..  ..              |
 : I : Si : Cl:         I—Si—Cl
  ..   ..  ..              |
     : Br :                Br
       ..
```

▲ Carbon Dioxide, CO_2

Example Exercise 12.6 • Electron Dot Formulas of Molecules

Draw the electron dot formula and the structural formula for a carbon dioxide molecule, CO_2.

Solution

Carbon is the central atom in the carbon dioxide molecule as indicated in bold. The total number of valence electrons is $4\ e^- + 2(6\ e^-) = 16\ e^-$. The number of electron pairs is 8 ($16/2 = 8$). We begin by placing 4 electron pairs around the carbon and adding the two O atoms.

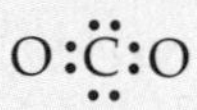

```
   ..
 O:C:O
   ..
```

The number of remaining electron pairs is $8 - 4 = 4$ pairs. We can add 2 pairs to each oxygen.

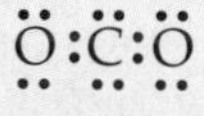

```
 .. .. ..
 O:C:O
 .. .. ..
```

Each oxygen atom shares 6 electrons, 2 less than an octet. We can use the nonbonding electron pairs around the carbon. We move one pair to the oxygen on the left, and the other pair to the oxygen on the right. Each carbon–oxygen bond shares 2 electron pairs.

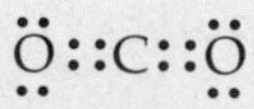

```
 ..      ..
 O::C::O
 ..      ..
```

Now the octet rule is satisfied for each atom. This is the correct electron dot formula. There are 2 electron pairs between each oxygen and carbon. Thus, there are two double bonds in the carbon dioxide molecule. In the structural formula each double bond is indicated by a double dash.

$$O{=}C{=}O$$

Self-Test Exercise
Draw the electron dot formula and the structural formula for a molecule of SiO_2.

Answer:

$$:\ddot{O}::Si::\ddot{O}: \qquad O{=}Si{=}O$$

Note We should mention that writing electron dot formulas is a trial-and-error procedure that involves moving electron pairs within a molecule to satisfy the octet rule. However, there are exceptions to the rule besides hydrogen. For example, in $BeCl_2$ there are only 4 electrons around the Be atom; and in BF_3 there are only 6 electrons around the B atom. Moreover, there are exceptions that violate the octet rule by exceeding 8 electrons, For example, in PCl_5 there are 10 electrons around the P atom; and in SF_6 there are 12 electrons around the S atom.

12.5 Electron Dot Formulas of Polyatomic Ions

Objectives
- To draw the electron dot formula for a polyatomic ion.
- To draw the structural formula for a polyatomic ion.

So far in our discussion of ions, we have considered only single atoms that have gained or lost electrons. This type of simple ion having a positive or negative charge is called a **monoatomic ion**. An ion can also contain two or more atoms held together by covalent bonds; this type is called a **polyatomic ion**. Some common polyatomic ions are the ammonium ion, NH_4^+, found in fertilizer; the hydroxide ion, OH^-, found in caustic lye; and the hydrogen carbonate ion, HCO_3^-, found in baking soda.

In Section 12.4 we learned to draw the electron dot formula for molecules. Now, we can draw the electron dot formula for polyatomic ions. The following directions for ions are similar to the previous guidelines for molecules.

Guidelines for Drawing Electron Dot Formulas of Polyatomic Ions

1. Calculate the valence electrons by adding the valence electrons for each atom in the polyatomic ion. If the ion is negatively charged, add electrons—equal to the charge—to the number of valence electrons. If the ion is positively charged, subtract electrons—equal to the charge—from the number of valence electrons. In each case, the result will be the total number of valence electrons.
2. Divide the total number of valence electrons and charges by 2 to find the number of electron pairs in the polyatomic ion.
3. Surround the central atom with four electron pairs. Use the remaining electron pairs to complete an octet around each of the other atoms. (Hydrogen is the sole exception, as H requires only one electron pair.)
4. If there are not enough electron pairs to provide an octet for each atom, move a nonbonding electron pair between two atoms that already share an electron pair.

We can draw the electron dot formula for the ammonium ion, NH_4^+, as follows. The total number of valence electrons in this positive polyatomic ion is 5 e^- + $4(1\ e^-) - 1\ e^- = 8\ e^-$. The number of electron pairs is 4 ($8/2 = 4$). In the formula NH_4^+, nitrogen is the central atom. We place the 4 electron pairs around the nitrogen atom to provide the necessary octet.

$$\left[\begin{array}{c}\mathrm{H}\\ \mathrm{H}:\underset{\cdot\cdot}{\overset{\cdot\cdot}{\mathrm{N}}}:\mathrm{H}\\ \mathrm{H}\end{array}\right]^+$$

Notice that the electron dot formula is written correctly for the ammonium ion. The nitrogen atom has an octet, and each hydrogen atom shares 2 electrons.

Now, let's simplify the electron dot formula and draw the structural formula of this polyatomic ion. We represent each bonding pair of electrons by a single dash and draw the structural formula for the ammonium ion as follows.

$$\left[\begin{array}{c}\mathrm{H}\\ |\\ \mathrm{H}-\mathrm{N}-\mathrm{H}\\ |\\ \mathrm{H}\end{array}\right]^+$$

The usual way to draw formulas of polyatomic ions is to enclose them in brackets. The overall charge on the polyatomic ion is indicated outside the brackets. Placing the charge outside the brackets emphasizes that the positive charge is distributed over the entire ion.

Now, let's draw the electron dot formula for the chlorate ion, ClO_3^-. The total number of valence electrons for this negative ion is $7\ e^- + 3(6\ e^-) + 1\ e^- = 26\ e^-$. The number of electron pairs is 13 ($26/2 = 13$). Since chlorine is the central atom, we begin by placing 4 electron pairs around the Cl atom and then add the three O atoms.

$$\begin{array}{c}\mathrm{O}:\underset{\cdot\cdot}{\overset{\cdot\cdot}{\mathrm{Cl}}}:\mathrm{O}\\ \mathrm{O}\end{array}$$

Since we started with 13 electron pairs, we place the remaining 9 electron pairs around the oxygen atoms to complete each octet.

$$\left[\begin{array}{c}:\underset{\cdot\cdot}{\overset{\cdot\cdot}{\mathrm{O}}}:\underset{\cdot\cdot}{\overset{\cdot\cdot}{\mathrm{Cl}}}:\underset{\cdot\cdot}{\overset{\cdot\cdot}{\mathrm{O}}}:\\ :\underset{\cdot\cdot}{\mathrm{O}}:\end{array}\right]^-$$

In the structural formula, each bonding electron pair is represented by a single dash, and nonbonding electrons are usually omitted. Thus, the structural formula for ClO_3^- is as follows.

$$\left[\begin{array}{c}\mathrm{O}-\mathrm{Cl}-\mathrm{O}\\ |\\ \mathrm{O}\end{array}\right]^-$$

Let's try a more difficult example and draw the electron dot formula for the carbonate ion, CO_3^{2-}. The total number of valence electrons in this polyatomic ion is $4\ e^- + 3(6\ e^-) + 2\ e^- = 24\ e^-$. The total number of electron pairs is 12 ($24/2 = 12$). Since carbon is the central atom, we begin by placing 4 electron pairs around the C atom and then add three O atoms. We have

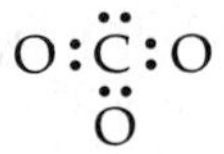

```
  ••
O:C:O
  ••
  O
```

We started with 12 electron pairs, and so we have 8 pairs remaining. We place the remaining pairs around the oxygen atoms to give octets as follows.

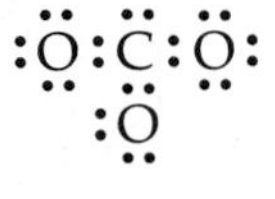

```
 ••  ••  ••
:O:C:O:
 ••  ••  ••
    :O
     ••
```

Notice that the bottom oxygen atom does not have an octet; it has only 3 electron pairs. According to our guidelines, we can move a nonbonding electron pair. Let's shift the nonbonding pair between the C and the O atoms.

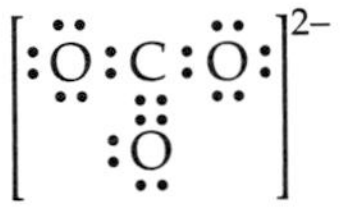

```
[ ••      ••  ]2−
[:O:C:O:]
[ ••  ••  ••  ]
[     ::      ]
[    :O       ]
[     ••      ]
```

The 2 electron pairs constitute a double bond. All 4 electrons in the double bond are shared by the C and the O atoms. Therefore, the central carbon atom has an octet and the oxygen atom has gained 2 electrons to complete its octet.

We use brackets to convey the idea that valence electrons are free to move over the entire polyatomic ion. Consequently, there appear to be three different structural formulas for the carbonate ion.

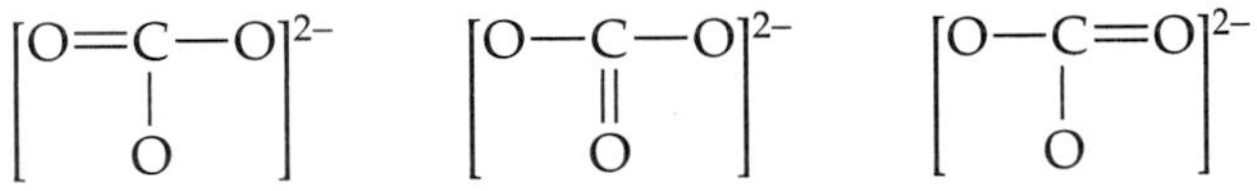

```
[O=C—O]2−     [O—C—O]2−     [O—C=O]2−
[  |  ]       [  ||  ]      [  |  ]
[  O  ]       [  O   ]      [  O  ]
```

Although the above structures for the carbonate ion appear to be different, they are all equivalent. No matter how we represent CO_3^{2-}, the structural formula has one double bond and two single bonds. This phenomenon is known as *resonance*, and we can draw three resonance structures for the carbonate ion. The following example exercise further illustrates how to draw electron dot and structural formulas for polyatomic ions.

Example Exercise 12.7 • Electron Dot Formulas of Polyatomic Ions

Draw the electron dot formula and the structural formula for the sulfate ion, SO_4^{2-}.

Solution

The total number of valence electrons is the sum of each atom plus 2 for the negative charge: $6\,e^- + 4(6\,e^-) + 2\,e^- = 32\,e^-$. The number of electron pairs is 16 ($32/2 = 16$). We begin by placing 4 electron pairs around the central sulfur atom and attaching the four oxygen atoms as follows.

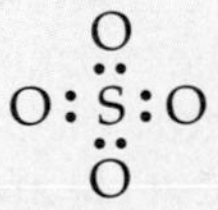

```
  O
  ••
O:S:O
  ••
  O
```

We have 12 remaining electron pairs, and so we place 3 electron pairs around each oxygen atom.

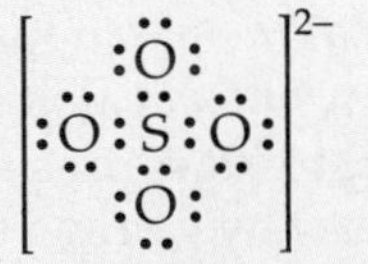

```
[     ••      ]2−
[    :O:      ]
[ ••  ••  ••  ]
[:O:S:O:]
[ ••  ••  ••  ]
[    :O:      ]
[     ••      ]
```

(continued)

Example Exercise 12.7 *(continued)*

Notice that each atom is surrounded by an octet of electrons. This is the correct electron dot formula. In the structural formula each bonding electron pair is replaced by a single dash. The structural formula is

$$\left[\begin{array}{ccc} & O & \\ & | & \\ O- & S & -O \\ & | & \\ & O & \end{array}\right]^{2-}$$

Self-Test Exercise
Draw the electron dot formula and the structural formula for the nitrite ion, NO_2^-.

Answer:

$[:\ddot{\underset{..}{O}}:\ddot{N}::\ddot{O}:]^-$ $[O—N{=}O]^-$

12.6 Polar Covalent Bonds

Objectives
- To describe and identify a polar covalent bond.
- To state the electronegativity trends in the periodic table.
- To apply delta notation (δ^+ and δ^-) to a polar bond.

Covalent bonds result from the sharing of valence electrons. Up to this point in our discussion we have assumed that electrons are shared equally in a covalent bond. What about a covalent bond in which the two atoms do not share electrons equally? In many instances one of the two atoms holds the electron pair more tightly. When the electrons are drawn more closely to one of the atoms, the bond is said to be polarized. This type of bond is called a polar covalent bond, or simply a **polar bond**.

Electronegativity Trends

Each element has an innate ability to attract valence electrons. This ability is related to the nearness of the valence shell to the nucleus. It is also related to the magnitude of the positive charge in the nucleus. The ability of an atom to attract electrons in a chemical bond is referred to as its **electronegativity**. Atoms of elements that strongly attract bonding electrons are said to be highly electronegative.

Linus Pauling, an American chemist, devised a method for measuring the electronegativity values for each of the elements. He assigned carbon a value of 2.5 and then determined the ability of other elements to attract bonding electrons relative to carbon. He found that fluorine is the most electronegative element. It has a value of 4.0 compared to carbon. Other highly electronegative elements are oxygen, 3.5; nitrogen, 3.0; and chlorine, 3.0. The most electronegative elements are the nonmetals on the far right of the periodic table. Figure 12.9 shows selected elements and their Pauling electronegativity value.

Note the electronegativity trends in Figure 12.9. First, elements generally become more electronegative from *left to right* in the periodic table. Second, elements become more electronegative from *bottom to top* within a group.

Periodic Trends
Electronegativity Movie

We can see that the horizontal trend is consistent with the chemical properties of elements. That is, nonmetals react by gaining electrons, and metals react by los-

PERIODIC TABLE OF THE ELEMENTS

	1 IA	2 IIA	3 IIIB	4 IVB	5 VB	6 VIB	7 VIIB	8 VIII	9 VIII	10 VIII	11 IB	12 IIB	13 IIIA	14 IVA	15 VA	16 VIA	17 VIIA	18 VIIIA
1																		He —
2	Li 1.0	Be 1.5											B 2.0	C 2.5	N 3.0	O 3.5	F 4.0	Ne —
3	Na 0.9	Mg 1.2											Al 1.5	Si 1.8	P 2.1	S 2.5	Cl 3.0	Ar —
4	K 0.8	Ca 1.0	Sc 1.3	Ti 1.5	V 1.6	Cr 1.6	Mn 1.8	Fe 1.8	Co 1.8	Ni 1.8	Cu 1.9	Zn 1.6	Ga 1.6	Ge 1.8	As 2.0	Se 2.4	Br 2.8	Kr —
5	Rb 0.8	Sr 1.0	Y 1.2	Zr 1.4	Nb 1.6	Mo 1.8	Tc 1.9	Ru 2.2	Rh 2.2	Pd 2.2	Ag 1.9	Cd 1.7	In 1.7	Sn 1.8	Sb 1.9	Te 2.1	I 2.5	Xe —
6	Cs 0.7	Ba 0.9	La 1.1	Hf 1.3	Ta 1.5	W 1.7	Re 1.9	Os 2.2	Ir 2.2	Pt 2.2	Au 2.4	Hg 1.9	Tl 1.8	Pb 1.8	Bi 1.9	Po 2.0	At 2.2	Rn —
7	Fr	Ra	Ac															

Electronegativity increases

▲ **Figure 12.9 Electronegativity Values for the Elements** The Pauling electronegativity value for an element is shown below the symbol. In general, electronegativity increases across a period and up a group. Since noble gases do not form bonds that are stable, electronegativity values are not given for Group 18.

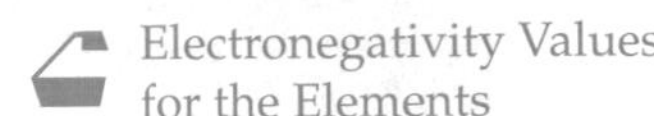

ing electrons. The vertical trend follows the trend in nonmetallic character. As an example, examine Group VA/15. We see that Bi is a metal, Sb and As are semimetals, and P and N are nonmetals. Therefore, the trend in electronegativity increases while corresponding to the trend in nonmetallic character. That is, nonmetals are more electronegative than semimetals, which in turn are more electronegative than metals.

The following example exercise illustrates the prediction of electronegativity values from the general trends in the periodic table.

Example Exercise 12.8 • Electronegativity Predictions

Predict which element in each of the following pairs is more electronegative according to the general electronegativity trends in the periodic table.

(a) N or O (b) Br or Se (c) F or Cl (d) Si or C

Solution

According to the trends, in the periodic table, elements that lie to the right in a series or at the top of a group are more electronegative.

(a) O is more electronegative than N.

(b) Br is more electronegative than Se.

(c) F is more electronegative than Cl.

(d) C is more electronegative than Si.

Self-Test Exercise

Predict which element in each of the following pairs is more electronegative according to the general trends in the periodic table.

(a) H or Cl (b) Br or I (c) P or S (d) As or Sb

Answers: (a) Cl; (b) Br; (c) S; (d) As

Delta Notation for Polar Bonds

In Section 12.1 we introduced the covalent bond. In Figure 12.9, we see that the electronegativity value of H is 2.1 and Cl is 3.0. Since there is a difference in electronegativity between the two elements ($3.0 - 2.1 = 0.9$), the bond in a H—Cl molecule is polar. Moreover, since Cl is the more electronegative element, the bonding electron pair is attracted toward the Cl atom and away from the H atom. The Cl atom thus becomes slightly negatively charged, whereas the H atom becomes slightly positively charged.

We can identify a polar bond with a special symbol. We indicate the atom having a partially negative charge by the symbol δ^-. Similarly, we indicate the atom having a partially positive atom by the symbol δ^+. These symbols, δ^- and δ^+, use the Greek letter delta (δ) and are referred to as **delta notation**. We use delta notation to illustrate the polar bond in a HCl molecule as follows.

$$\delta^+ \text{ H—Cl } \delta^-$$

Since the chlorine atom is the more electronegative atom, it draws the electron pair closer. As a result, the hydrogen atom becomes slightly positively charged. The following example exercise further illustrates the application of delta notation.

H—F

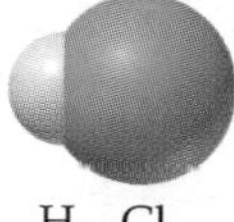

H—Cl

H—Br

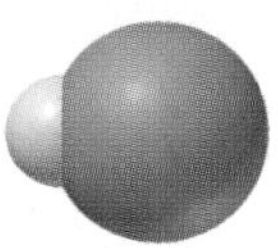

H—I

▲ **Hydrogen Halides** The two atoms in each hydrogen halide molecule are joined by a polar covalent bond.

Hydrogen Halides

Example Exercise 12.9 • Delta Notation

Calculate the electronegativity difference and apply delta notation to the bond between carbon and oxygen, C—O.

Solution

From Figure 12.9, we find that the electronegativity value of C is 2.5 and of O is 3.5. By convention, we always subtract the lesser value from the greater. The difference between the two elements is $3.5 - 2.5 = 1.0$. The result shows that the C—O bond is slightly polarized.

Since O is the more electronegative element, the bonding electron pair is drawn away from the C atom and toward the O atom. The O atom thus becomes slightly negatively charged, whereas the C atom becomes slightly positively charged. Applying the delta convention, we have

$$\delta^+ \text{ C—O } \delta^-$$

Self-Test Exercise

Using delta notation (δ^+ and δ^-), label each atom in the following polar covalent bonds.

(a) N—O (b) H—F

Answers: (a) δ^+ N—O δ^-; (b) δ^+ H—F δ^-

Note Polar molecules are different from ionic formula units, and the delta convention simply indicates partially negative and partially positive atoms in a polar covalent bond. On the other hand, ionic charges indicate negative and positive ions in an ionic bond. An electronegativity difference greater than 1.8 is usually considered ionic; for example, we know that NaCl is ionic and that the electronegativity difference between the two elements is greater than 1.8; that is, $3.0 - 0.9 = 2.1$.

12.7 Nonpolar Covalent Bonds

Objectives
- To describe and identify a nonpolar covalent bond.
- To identify seven elements that occur naturally as diatomic molecules: H_2, N_2, O_2, F_2, Cl_2, Br_2, I_2.

Chemistry Connection · Linus Pauling

What famous chemist won a Nobel peace prize as well as a Nobel prize for chemistry?

Linus Pauling (1901–1994) is regarded by many as the most influential chemist of the twentieth century. Pauling was born in Portland, Oregon, to a family of modest means. His father died when Linus was quite young, and so he had to take on a series of menial jobs to help support the family. At age 16, he enrolled at Oregon State University where he received a bachelor's degree in chemical engineering. There he met his wife and acknowledged inspiration, Ava Helen Miller, while teaching a chemistry class to home economics students.

▲ Linus Pauling (1901–1994)

Pauling decided on the California Institute of Technology for graduate school. In 1925 he received a Ph.D, and following graduation, was awarded a Guggenheim fellowship to study in Europe. In 1927 he returned to Caltech as a professor of chemistry and remained there for most of his career. Pauling received immediate recognition and was described by his department chairman in the following words: "Were all the rest of the chemistry department wiped away except Pauling, it would still be one of the most important departments of chemistry in the world."

Among his many accomplishments, Linus Pauling invented the concept of electronegativity and provided an ingenious method for calculating the electronegativity values of elements. One of his most brilliant achievements was an explanation of the unusual stability of the benzene molecule, C_6H_6. His most celebrated work was investigating the shapes of proteins. He found that their structures resembled a helix and further established the relationship between abnormal molecular structure and genetic disease. His work in this area earned him the 1954 Nobel prize in chemistry.

In the 1960s Pauling focused his energies on banning aboveground testing of nuclear weapons and pointed out the potential dangers of radiation. He spoke passionately against war and encouraged global disarmament. For his efforts, Pauling was awarded the 1962 Nobel peace prize. In the 1970s he stirred controversy when he publicly advocated massive doses of vitamin C to combat the common cold. In addition to two Nobel prizes, Linus Pauling received numerous medals and awards, accepted honorary memberships in several scientific societies, published more than 400 research papers, and wrote the classic textbook *The Nature of the Chemical Bond.*

Linus Pauling won a Nobel prize in chemistry and a Nobel peace prize for his efforts to ban aboveground testing of nuclear weapons.

In Section 12.6, we learned that polar bonds result from the unequal sharing of bonding electrons. In a polar bond, one of the two atoms has a greater tendency to attract electrons. The atom having the greater electronegativity attracts the bonded electron pair more strongly and has a partial negative charge.

But what about a covalent bond in which each atom has the same or similar electronegativity? To answer this question, let's begin by reexamining the Pauling electronegativity values in Figure 12.9. We see that electronegativity increases across a period and up a group. Notice, however, that several elements have the same Pauling electronegativity value. For example, N and Cl have a value of 3.0, and C, S, and I all have a value of 2.5.

How then would we describe a S—I bond? Since the electronegativity of each atom is the same, the bond is not polarized. A covalent bond between two atoms with the same electronegativity is referred to as a nonpolar covalent bond, or simply a **nonpolar bond**. The following example exercise illustrates the classification of polar and nonpolar bonds.

Example Exercise 12.10 • Polar versus Nonpolar Bonds

Classify each of the following as a polar or a nonpolar bond.

(a) Cl—Cl (b) Cl—N (c) Cl—Br (d) Cl—H

Solution

From Figure 12.9 we find the following electronegativity values: Cl = 3.0, N = 3.0, Br = 2.8, and H = 2.1.

(a) The Cl—Cl bond (3.0 − 3.0 = 0) is *nonpolar.*
(b) The Cl—N bond (3.0 − 3.0 = 0) is *nonpolar.*
(c) The Cl—Br bond (3.0 − 2.8 = 0.2) is *slightly polar.*
(d) The Cl—H bond (3.0 − 2.1 = 0.9) is *polar.*

Self-Test Exercise

Classify each of the following as a polar or a nonpolar bonds.

(a) C—C (b) C—O (c) C—S (d) C—H

Answers: (a) nonpolar; (b) polar; (c) nonpolar; (d) polar

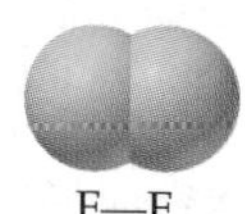

F—F

Cl—Cl

Br—Br

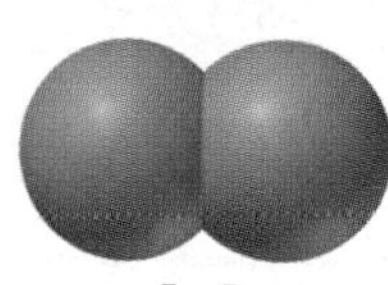

I—I

▲ **Halogens** The two atoms in each halogen molecule are joined by a nonpolar covalent bond.

The Halogens

Diatomic Nonpolar Molecules

A **diatomic molecule** consists of two nonmetal atoms joined by a covalent bond. The simplest example of a nonpolar bond is found between two identical atoms. For example, oxygen occurs naturally as a gas and contains O_2 molecules. Even though oxygen is quite electronegative, the bond in an O_2 molecule is nonpolar. This is because each oxygen atom has the same tendency to attract shared electrons.

In Section 6.3, we briefly studied the elements in Group VIIA/17. Recall that, collectively, these elements are called halogens. The halogens (F, Cl, Br, and I) are a family and have similar properties. All four elements occur naturally as diatomic molecules: F_2, Cl_2, Br_2, and I_2. Elemental fluorine is a yellowish-green gas, chlorine is a greenish-yellow gas, bromine is a reddish-brown liquid, and iodine is a dark-violet crystalline solid.

There are seven elements that exist naturally as diatomic molecules: H_2, N_2, O_2, F_2, Cl_2, Br_2, and I_2. Although each molecule is distinctly different, each exhibits a nonpolar bond that is perfectly covalent. For our discussion, we classify a bond as nonpolar when there is no electronegativity difference between the two bonded atoms. In practice, a bond between two atoms having an electronegativity difference of 0.5 or less is usually considered a nonpolar bond.

12.8 Coordinate Covalent Bonds

Objective · To describe and identify a coordinate covalent bond.

We have described a covalent bond as an electron pair that is shared by two nonmetal atoms. Furthermore, each nonmetal atom is surrounded by nonbonding electron pairs to complete its octet.

Now, let's consider a special case of sharing electrons. An atom such as oxygen (six valence electrons) can attain an octet by sharing a nonbonding electron pair on another atom. A covalent bond resulting from one atom donating an electron pair to another atom is called a **coordinate covalent bond**. A good example of a coordinate covalent bond is found in a molecule of ozone, O_3. Here, an oxygen molecule donates a nonbonding electron pair to an oxygen atom to produce ozone.

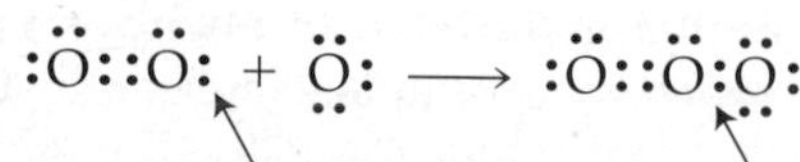

We can identify a coordinate covalent bond as follows. First, we draw the electron dot formula for the given molecule. If we can remove an atom and still have an octet around the remaining atoms, we know that the atom was joined by a coordinate covalent bond. In the above illustration for ozone, O_3, we remove an O atom and are left with a stable electron dot formula for O_2. Thus, O_3 has a coordinate covalent bond. The following example exercise further illustrates coordinate covalent bonds.

Example Exercise 12.11 • Coordinate Covalent Bonds

Burning yellow sulfur powder produces sulfur dioxide, SO_2. Sulfur dioxide is a colorless gas with a suffocating odor. It is used to kill insect larvae and produces the odor released when a match is ignited. Draw the electron dot formula for SO_2. Then show the formation of a coordinate covalent bond in SO_3.

Solution

The total number of valence electrons in one molecule of SO_2 is $6\ e^- + 2(6\ e^-) = 18\ e^-$. After experimenting, we find that the electron dot formula for SO_2 has a double bond. Since the sulfur atom has a nonbonding electron pair, it can bond to an additional oxygen atom. A diagram of the formation of the coordinate covalent bond is as follows.

:Ö:S: + Ö: ⟶ :Ö:S:Ö:
:Ö   :Ö

nonbonding electron pair *coordinate covalent bond*

Self-Test Exercise

A nitrogen molecule can form a coordinate covalent bond with an oxygen atom to give nitrous oxide, N_2O, that is, laughing gas. Draw the electron dot formula for a nitrogen molecule and attach an oxygen atom.

Answer:

(a) :N:::N: (b) :N:::N:Ö:

12.9 Shapes of Molecules

Objectives
- To determine the shape of a molecule by applying VSEPR theory.
- To explain how a molecule with polar bonds can be nonpolar.

In the 1950s a simple theory was proposed to explain the shapes of molecules. This theory is an extension of electron dot formulas in which pairs of electrons surround a central atom. In essence, the theory states that the electron pairs surrounding an atom tend to repel each other and that the shape of the molecule is the result of this electron pair repulsion. This model is referred to as the **valence shell electron pair repulsion (VSEPR**, pronounced "vesper") theory. The VSEPR model portrays bonding and nonbonding electron pairs as occupying specific positions around the central atom in a molecule.

The VSEPR theory uses the term **electron pair geometry** to indicate the arrangement of bonding and nonbonding electron pairs around the central atom. VSEPR theory uses the term "molecular geometry," or **molecular shape**, to indicate the arrangement of atoms around the central atom as a result of electron pair repulsion. The angle formed by any two atoms bonded to the central atom is referred to as the **bond angle**.

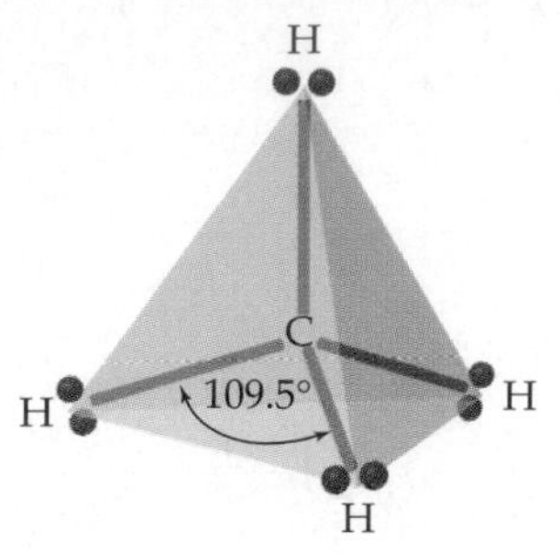

▲ The electron pair geometry in methane, CH_4, is *tetrahedral*.

Methane, CH_4

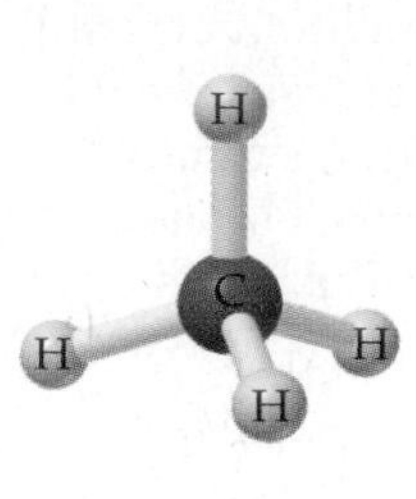

▲ The molecular shape of CH_4 is *tetrahedral*; the bond angle is 109.5°.

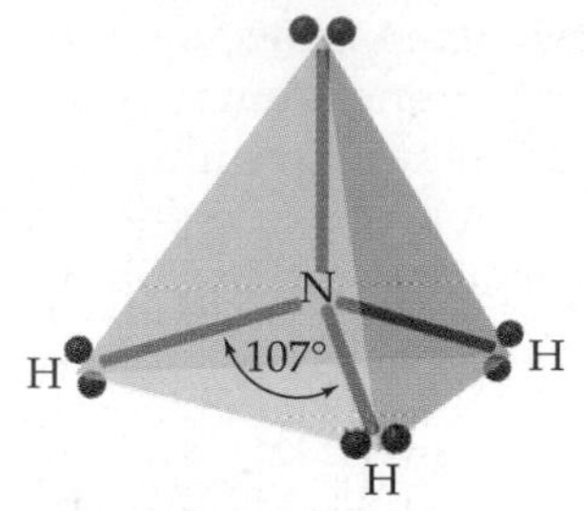

▲ The electron pair geometry in ammonia, NH_3, is *tetrahedral*.

Ammonia, NH_3

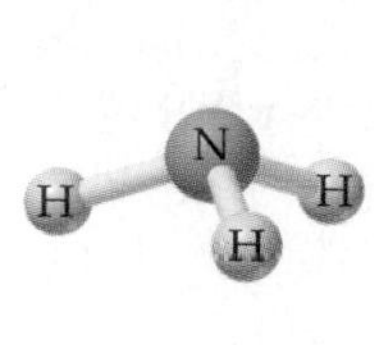

▲ The molecular shape of NH_3 is *trigonal pyramidal*; the bond angle is 107°.

Let's apply VSEPR theory to a methane molecule, CH_4, which has four pairs of bonding electrons. In a molecule of CH_4, the central C atom is surrounded by four electron pairs located at the corners of a three-dimensional figure called a tetrahedron. Each electron pair is bonded to a hydrogen atom, and so the shape of the molecule is said to be *tetrahedral*. Moreover, according to VSEPR theory, any molecule with four electron pairs around the central atom has tetrahedral electron pair geometry.

In a molecule of ammonia, NH_3, the central N atom is surrounded by three bonding electron pairs and one nonbonding electron pair. The electron pair geometry is once again tetrahedral, and ammonia should have bond angles of 109.5°. Experimentally, the bond angle is found to be 107°. VSEPR theory explains the smaller bond angle by suggesting that the nonbonding electron pair exerts a stronger repelling force than the bonding pairs. Hence, the hydrogen atoms are pushed closer together, that is, from 109.5° to 107°. The molecular shape is said to be *trigonal pyramidal*.

In a molecule of water, H_2O, the central O atom is surrounded by two bonding electron pairs and two nonbonding electron pairs. Since the electron pair geometry is tetrahedral, the predicted bond angle between the two hydrogen atoms is 109.5°. The experimentally measured bond angle is only 104.5°. As in the preceding example, it is proposed that the nonbonding electrons in water exert a greater repelling force than the bonding pairs. The resulting bond angle is smaller, allowing more space for the unshared electron pairs. The molecular shape is said to be *bent* or V-shaped.

Water, H_2O

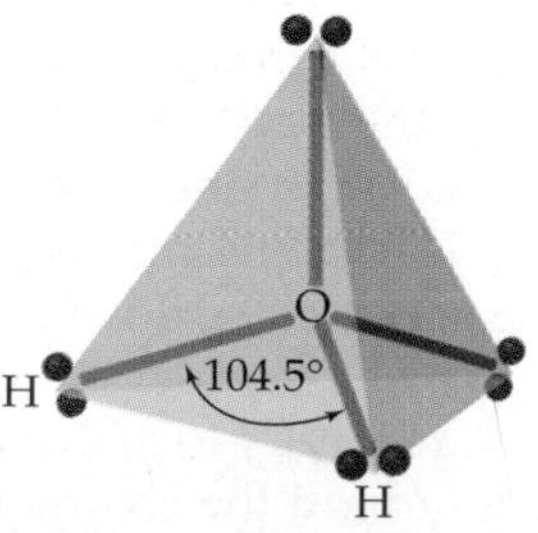

▲ The electron pair geometry in water, H_2O, is *tetrahedral*.

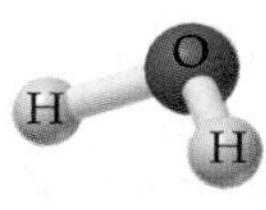

▲ The molecular shape of H_2O is *bent*; the bond angle is 104.5°.

We can summarize the information obtained from VSEPR theory about three types of molecules as shown in Table 12.1.

Table 12.1 Summary of VSEPR Theory

Bonding/Nonbonding Electron Pairs	Electron Pair Geometry	Molecular Shape	Bond Angle	Example Molecule
4/0	tetrahedral	tetrahedral	109.5°	CH_4
3/1	tetrahedral	trigonal pyramidal	107°	NH_3
2/2	tetrahedral	bent	104.5°	H_2O

Nonpolar Molecules with Polar Bonds

Why is it that carbon tetrachloride, CCl_4, is a nonpolar molecule even though it has polar bonds? Recall that a chlorine atom is more electronegative than a carbon atom; hence, the C—Cl bond is polar. Furthermore, there are four C—Cl polar bonds in a CCl_4 molecule.

To explain why a CCl_4 molecule is nonpolar, we can apply VSEPR theory. There are four bonding electron pairs in CCl_4; thus, the electron pair geometry and the molecular shape are both tetrahedral. We illustrate the CCl_4 molecule as having four chlorine atoms at the corners of a tetrahedron.

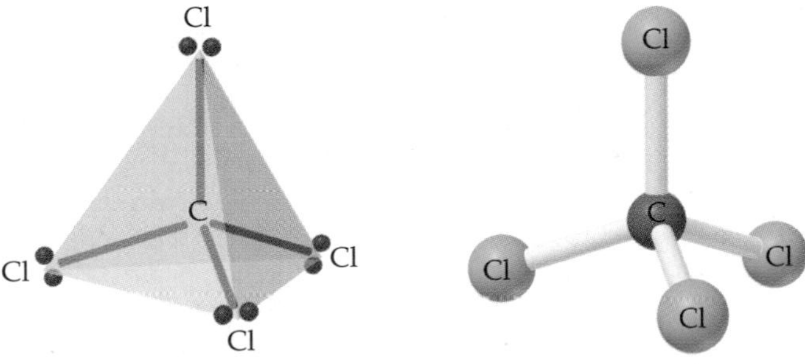

Note that each chlorine atom exerts equal attractive forces in four opposing directions. Thus, the attractions exerted by the four polar bonds cancel each other. The net effect of the polar bonds is zero, and overall, the molecule is considered nonpolar. We can illustrate that the CCl_4 molecule is nonpolar as follows.

Carbon tetrachloride, CCl_4

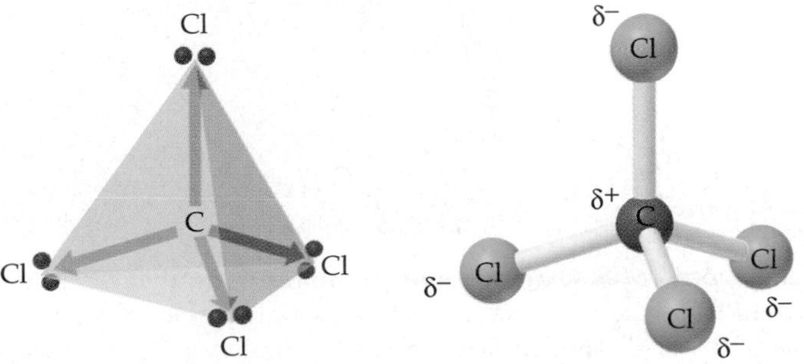

Consequently, a molecule may contain polar bonds and yet be nonpolar. In carbon dioxide, CO_2, there are two polar bonds. An O=C=O molecule is nonpolar because the more electronegative oxygen atoms pull equally in opposite directions. The molecular shape of CO_2 is described as *linear*.

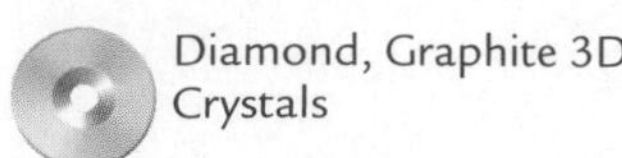

On the other hand, sulfur dioxide, SO_2, is a polar molecule. The reason that SO_2 is polar is that the O—S=O molecule is not linear but rather has a bent shape. The two oxygen atoms do not pull in opposite directions, and the polar bonds do not cancel each other. Thus, molecular shape can affect the polarity of the molecule.

Chemistry Connection · Diamonds

What is the difference between a precious diamond and the graphite in an ordinary pencil?

Diamonds have been prized for centuries owing to their clarity and fiery sparkle. Diamonds originate deep within the Earth where they are formed under conditions of extreme heat and pressure. A diamond is the hardest known substance; its melting point is above 3500°C, and it is inert to corrosive chemicals.

The hardness and crystalline nature of a diamond enables a rough stone to be cut so as to give numerous tiny facets. On average, an uncut stone loses 50% of its original weight in becoming a finished diamond. Moreover, a misdirected blow by a diamond cutter can shatter the stone into small shards.

Diamond is a crystalline solid containing carbon atoms arranged in a three-dimensional network. Carbon atoms are bonded in a repeating tetrahedral structure as shown in the illustration. The more extensive the repeating network, the larger the actual diamond.

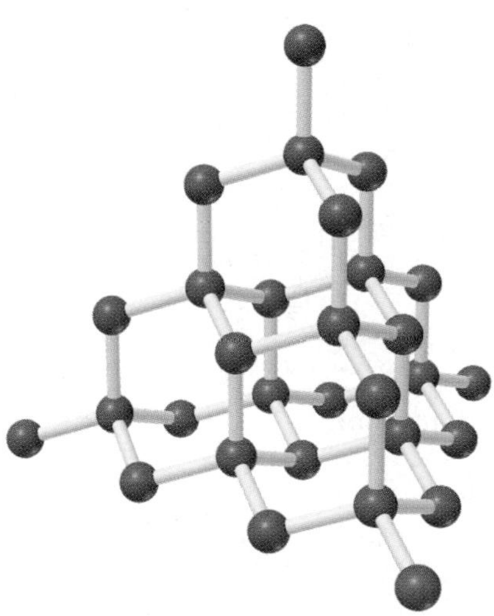

▲ **The Structure of Diamond**

Like diamond, graphite is also a crystalline solid composed exclusively of carbon atoms. Unlike diamond, graphite is a soft substance which we know as the dark gray "lead" in an ordinary pencil. The structure of graphite consists of fused hexagon rings, each containing six carbon atoms. Unlike diamond, graphite is arranged in a flat, two-dimensional, crystalline network. The lubricating property of graphite is due to "sheets" of carbons rings sliding past one another. Two sheets of graphite are shown in the illustration.

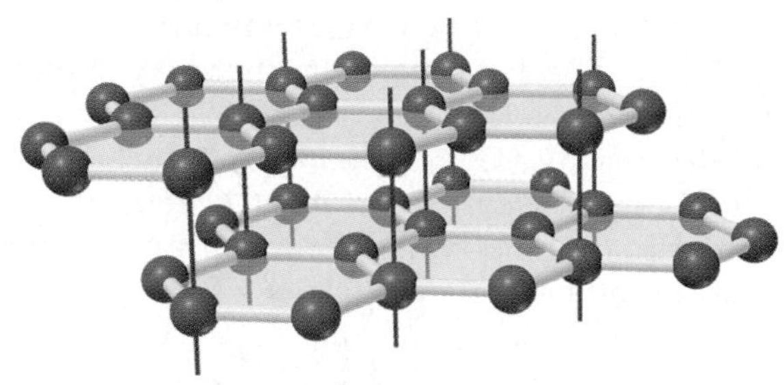

▲ **The Structure of Graphite**

Diamonds can be synthesized from graphite by mimicking the conditions under which natural diamonds are formed. A sample of graphite is placed in a hydraulic press and converted to diamond in a few hours at high temperature and pressure. Typically, synthetic diamonds are only a few millimeters in size and are too flawed to be used as gemstones. However, synthetic diamonds are valuable industrially for use in coating drill bits and cutting tools.

▲ **Diamond and Graphite** A synthetic diamond made from powdered graphite shown alongside the tip of a pencil. Diamond and graphite are both examples of crystalline carbon.

Although diamond and graphite are both composed of carbon, the structure of diamond is a repeating tetrahedron, whereas that of graphite is a repeating hexagon.

Summary 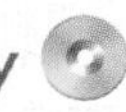

Section 12.1 In 1916 G. N. Lewis described the **chemical bond** when he proposed that **valence electrons** in different atoms interact in such a way that each atom completes its valence shell. Since eight electrons are found in the valence shells of noble gas atoms, the Lewis theory became known as the rule of eight or, the **octet rule**. The only exception to the octet rule that we will encounter is the hydrogen atom, which requires two electrons.
When a metal atom loses valence electrons as a nonmetal atom gains electrons, an **ionic bond** results. When valence electrons are shared by two nonmetal atoms, a **covalent bond** results. The simplest representative particle in a substance held together by ionic bonds is a **formula unit**. The simplest representative particle held together by covalent bonds is a **molecule**.

Section 12.2 In the formation of ionic bonds, metal atoms lose valence electrons and are often left with a noble gas structure. Simultaneously, nonmetal atoms gain electrons to complete their valence shells. In the process of forming ions, the radii of metal atoms become smaller while the radii of nonmetal atoms become larger. The formation of ionic bonds from metal and nonmetal atoms always releases heat energy.

Section 12.3 In the formation of covalent bonds, the outer valence shells of two nonmetal atoms overlap and share electrons. Each of the atoms in the bond shares the number of electrons necessary to complete their individual octets. The **bond length** between the two atoms is always less than the sum of their radii. When atoms of different elements react, energy is always released. The amount of energy released during the formation of a bond is exactly equal to the amount of energy required to break that bond. The amount of energy required to separate two bonded atoms is called the **bond energy**.

Section 12.4 To write an **electron dot formula**, we first calculate the total number of valence electrons from all the atoms in the molecule. We then divide the valence electron total by 2 to find the number of electron pairs. The electron pairs are placed around the central atom and then the remaining atoms in the molecule so as to provide octets. One pair of electrons shared by two atoms is a **single bond**. A molecule can also contain two or three electron pairs between two atoms. These bonds are referred to as a **double bond** and a **triple bond**, respectively.
Double and triple bonds result from an insufficient number of valence electrons around each atom in a molecule. To provide an octet, it may be necessary to move **nonbonding electrons** between two atoms, these electrons then become **bonding electrons**. In the **structural formula** of a molecule, a single bond is shown as a dash, a double bond as two dashes, and a triple bond as three dashes.

Section 12.5 An ionic bond can be formed by the attraction between a simple **monoatomic ion** and a complex **polyatomic ion**. Ammonium chloride, NH_4Cl, and copper sulfate, $CuSO_4$, each contain a polyatomic ion. To draw the electron dot formula for a polyatomic ion, we must find the total number of valence electrons for each atom and then add or subtract the number of electrons equal to its ionic charge. For the ammonium cation, NH_4^+, the total number of valence electrons is 8 $(9\ e^-$ minus $1\ e^-)$. For the sulfate ion, SO_4^{2-}, the total number of valence electrons is 32 e^- $(30\ e^-$ plus $2\ e^-)$.

Section 12.6 If one atom in a bond attracts electrons more strongly than the other, it is said to have a higher **electronegativity**. A **polar bond** results when one of the bonded atoms has a greater attraction for the electron pair. We can indicate a polar bond using **delta notation**: The more electronegative atom is labeled δ^-, and the less electronegative atom is labeled δ^+.

Section 12.7 If each atom in a bond has an equal attraction for bonding electrons, the bond is said to be a **nonpolar bond**. The simplest example of a nonpolar bond occurs in a **diatomic molecule** having two identical atoms: The bond in H_2, N_2, O_2, F_2, Cl_2, Br_2, and I_2 is the best example of a nonpolar bond.

Section 12.8 A **coordinate covalent bond** is a special type of covalent bond and is formed when one atom donates an electron pair to another atom. A simple illustration of a coordinate covalent bond occurs when the chlorine atom in HCl donates an electron pair to an oxygen atom to give HClO.

Section 12.9 After writing the electron dot formula for a molecule, we can draw its shape. That is, we can predict the shape of a molecule by applying valence shell electron pair repulsion (VSEPR, pronounced "vesper") theory. According to **VSEPR theory**, bonding and nonbonding electron pairs occupy positions around the central atom. If there are four electron pairs, they are located at the corners of a three-dimensional figure called a tetrahedron. Examples of molecules having four electron pairs around the central atom include methane, CH_4, ammonia, NH_3, and water, H_2O. We refer to the arrangement of electron pairs around the central atom as the **electron pair geometry**. We refer to the arrangement of atoms around the central atom as the **molecular shape**. We refer to the angle formed by two atoms bonded to the central atom as the **bond angle**.

Key Concepts*

1. What type of chemical bond results from the attraction between a metal cation and a nonmetal anion? between two nonmetal atoms?
2. Which noble gas is isoelectronic with a magnesium ion? with a bromide ion?
3. Which of the following is held together by ionic bonds: LiCl, MgO, CuS, HBr, AlP?
4. Which of the following is held together by covalent bonds: AgBr, HCl, NO, CO, IF?
5. Classify each of the following diagrams as illustrating a substance composed of atoms, formula units, or molecules.
6. Draw the electron dot formula for SO_2. How many pairs of nonbonding electrons are in one molecule of sulfur dioxide?
7. Draw the electron dot formula for SO_3^{2-}. How many pairs of nonbonding electrons are in one sulfite ion?
8. According to the general trends in the periodic table, predict which of the following molecules is nonpolar: H_2O, NH_3, CH_4, N_2O, O_3.
9. Which of the following molecules contains a coordinate covalent bond: HCl, HClO, $HClO_2$, $HClO_3$, $HClO_4$?
10. What is the electron pair geometry for a hydrogen sulfide, H_2S, molecule? What is the molecular shape of a H_2S molecule?

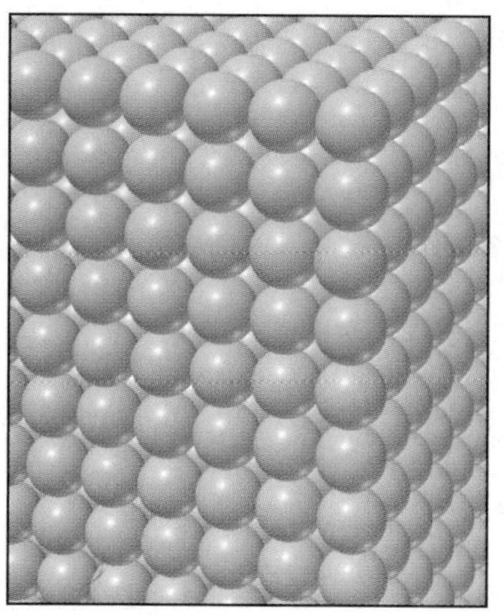
(a)

(b)

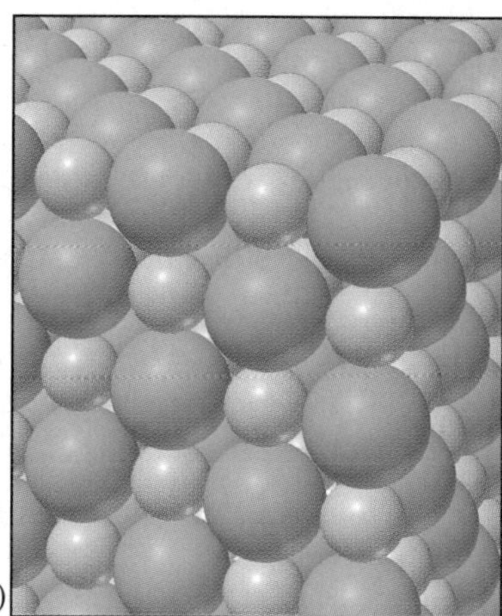
(c)

Key Terms† 

Select the key term below that corresponds to each of the following definitions.

_____ 1. the general term for the attraction between two ions or two atoms
_____ 2. the electrons in the outermost energy level that are available for bonding
_____ 3. the statement that an atom must be surrounded by eight valence electrons to be stable
_____ 4. a chemical bond characterized by the attraction between a cation and an anion
_____ 5. a chemical bond characterized by the sharing of one or more pairs of valence electrons

(a) bond angle (*Sec. 12.9*)
(b) bond energy (*Sec. 12.3*)
(c) bonding electrons (*Sec. 12.4*)
(d) bond length (*Sec. 12.3*)
(e) chemical bond (*Sec. 12.1*)
(f) coordinate covalent bond (*Sec. 12.8*)

*Answers to Key Concepts are in Appendix H.
†Answers to Key Terms are in Appendix I.

_____	6.	the simplest representative particle in a substance held together by ionic bonds	(g) covalent bond (*Sec. 12.1*)
_____	7.	the simplest representative particle in a substance held together by covalent bonds	(h) delta (δ) notation (*Sec. 12.6*)
_____	8.	the distance between the nuclei of two atoms joined by a covalent bond	(i) diatomic molecule (*Sec. 12.7*)
_____	9.	the amount of energy required to break a given bond in a gaseous substance	(j) double bond (*Sec. 12.4*)
_____	10.	the valence electrons in a molecule that are shared	(k) electron dot formula (*Sec. 12.4*)
_____	11.	the valence electrons in a molecule that are not shared	(l) electron pair geometry (*Sec. 12.9*)
_____	12.	a diagram of a molecule in which each atom is surrounded by two dots for each pair of bonding or nonbonding electrons	(m) electronegativity (*Sec. 12.6*)
_____	13.	a diagram of a molecule in which each atom is represented by its chemical symbol and a dash for each pair of bonding electrons	(n) formula unit (*Sec. 12.1*)
_____	14.	a bond composed of one electron pair shared by two atoms	(o) ionic bond (*Sec. 12.1*)
_____	15.	a bond composed of two electron pairs shared by two atoms	(p) molecular shape (*Sec. 12.9*)
_____	16.	a bond composed of three electron pairs shared by two atoms	(q) molecule (*Sec. 12.1*)
_____	17.	a single atom that bears a charge as the result of gaining or losing electrons	(r) monoatomic ion (*Sec. 12.5*)
_____	18.	a group of atoms held together by covalent bonds that has an overall positive or negative charge	(s) nonbonding electrons (*Sec. 12.4*)
_____	19.	the ability of an atom to attract a shared pair of electrons	(t) nonpolar bond (*Sec. 12.7*)
_____	20.	a method of indicating a partial positive and a partial negative charge in a chemical bond	(u) octet rule (*Sec. 12.1*)
_____	21.	a bond in which a pair of electrons is shared unequally	(v) polar bond (*Sec. 12.6*)
_____	22.	a bond in which a pair of electrons is shared equally	(w) polyatomic ion (*Sec. 12.5*)
_____	23.	a molecule composed of two nonmetal atoms held together by a covalent bond	(x) single bond (*Sec. 12.4*)
_____	24.	a bond in which an electron pair is shared but both electrons have been donated by a single atom	(y) structural formula (*Sec. 12.4*)
_____	25.	a model that explains the shapes of molecules as a result of electron pairs around the central atom repelling each other	(z) triple bond (*Sec. 12.4*)
_____	26.	the geometric shape formed by bonding and nonbonding electron pairs around the central atom in a molecule	(aa) valence electrons (*Sec. 12.1*)
_____	27.	the geometric shape formed by atoms bonded to the central atom in a molecule	(bb) VSEPR theory (*Sec. 12.9*)
_____	28.	the angle formed by two atoms attached to the central atom in a molecule	

Exercises‡

The Chemical Bond Concept (Sec. 12.1)

1. Describe the formation of an ionic bond in terms of valence electrons.
2. Describe the formation of a covalent bond in terms of valence electrons.
3. State the number of valence electrons for a magnesium atom and a sulfur atom. State the number of valence electrons for a magnesium ion and a sulfide ion in an ionic bond.
4. State the number of valence electrons for a hydrogen atom and an iodine atom. State the number of valence electrons for a hydrogen atom and an iodine atom in a covalent bond.
5. Predict whether each of the following is held together by ionic or by covalent bonds.
   (a) water, H_2O **(b)** sodium chloride, NaCl
   (c) methane, CH_4 **(d)** zinc oxide, ZnO
6. Predict whether each of the following is held together by ionic or by covalent bonds.
   (a) nitrogen dioxide, NO_2 **(b)** lithium chloride, LiCl
   (c) ferrous sulfate, $FeSO_4$ **(d)** iodine heptafluoride, IF_7
7. State whether the representative particle in each of the following substances is a a formula unit or a molecule.
   (a) ethyl alcohol, C_2H_5OH **(b)** strontium iodide, SrI_2
   (c) carbon monoxide, CO **(d)** calcium carbonate, $CaCO_3$
8. State whether the representative particle in each of the following substances is a formula unit or a molecule.
   (a) acetone, C_3H_6O **(b)** silver sulfate, Ag_2SO_4
   (c) potassium bromide, KBr **(d)** titanium oxide, TiO_2
9. State whether the representative particle in each of the following substances is an atom, a formula unit, or a molecule.

‡Answers to odd-numbered Exercises are in Appendix J.

(a) neon, Ne (b) fluorine, F_2
(c) freon, CF_2Cl_2 (d) uranium fluoride, UF_6

10. State whether the representative particle in each of the following substances is an atom, a formula unit, or a molecule.
(a) ethane, C_2H_6 (b) copper, Cu
(c) magnetite, Fe_3O_4 (d) sulfur, S_8

Ionic Bonds (Sec. 12.2)

11. Use the periodic table to predict the ionic charge for each of the following metal ions.
(a) Na ion (b) Mg ion
(c) Sn ion (d) Al ion

12. Use the periodic table to predict the ionic charge for each of the following metal ions.
(a) Be ion (b) Cs ion
(c) Ga ion (d) Pb ion

13. Use the periodic table to predict the ionic charge for each of the following nonmetal ions.
(a) F ion (b) Br ion
(c) S ion (d) N ion

14. Use the periodic table to predict the ionic charge for each of the following nonmetal ions.
(a) I ion (b) S ion
(c) Se ion (d) P ion

15. Write out the electron configuration for each of following metal ions.
(a) Li^+ (b) Al^{3+}
(c) Ca^{2+} (d) Mg^{2+}

16. Write out the electron configuration for each of following metal ions.
(a) Sc^{3+} (b) K^+
(c) Ti^{4+} (d) Ba^{2+}

17. Write out the electron configuration for each of following nonmetal ions.
(a) Cl^- (b) I^-
(c) S^{2-} (d) P^{3-}

18. Write out the electron configuration for each of following nonmetal ions.
(a) Br^- (b) O^{2-}
(c) Se^{2-} (d) N^{3-}

19. Which noble gas has an electron configuration identical to that of each of the following metal ions?
(a) Li^+ (b) K^+
(c) Ca^{2+} (d) Ra^{2+}

20. Which noble gas has an electron configuration identical to that of each of the following nonmetal ions?
(a) Cl^- (b) I^-
(c) O^{2-} (d) P^{3-}

21. Which noble gas is isoelectronic with each of the following metal ions?
(a) Li^+ (b) Al^{3+}
(c) Ca^{2+} (d) Mg^{2+}

22. Which noble gas is isoelectronic with each of the following metal ions?
(a) Sc^{3+} (b) K^+
(c) Ti^{4+} (d) Ba^{2+}

23. Which noble gas is isoelectronic with each of the following nonmetal ions?
(a) Cl^- (b) I^-
(c) S^{2-} (d) P^{3-}

24. Which noble gas is isoelectronic with each of the following nonmetal ions?
(a) Br^- (b) O^{2-}
(c) Se^{2-} (d) N^{3-}

25. In each of the following pairs, which has the larger radius?
(a) Li atom or Li ion (b) Mg atom or Mg ion
(c) F atom or F ion (d) O atom or O ion

26. In each of the following pairs, which has a larger radius?
(a) Al atom or Al ion (b) Pb atom or Pb ion
(c) Se atom or Se ion (d) N atom or N ion

27. Which of the following statements is true regarding the formation of an ionic bond between a metal and a nonmetal?
(a) An ionic bond is formed by the attraction of metal and nonmetal ions.
(b) The ionic radius of a metal atom is greater than its atomic radius.
(c) The ionic radius of a nonmetal atom is less than its atomic radius.
(d) The simplest representative particle is a molecule.

28. Which of the following statements is true regarding the formation of an ionic bond between cobalt and bromine?
(a) Cobalt atoms lose electrons, and bromine atoms gain electrons.
(b) The cobalt atom is larger in radius than the cobalt ion.
(c) The bromine atom is smaller in radius than the bromide ion.
(d) Cobalt and bromide ions bond as a result of electrostatic attraction.

Covalent Bonds (Sec. 12.3)

29. Which of the following is greater?
(a) the sum of the H and I atomic radii or the bond length in H—I
(b) the sum of the N and O atomic radii or the bond length in N—O

30. Which of the following is greater?
(a) the sum of the C and Cl atomic radii or the bond length in C—Cl
(b) the sum of the S and F atomic radii or the bond length in S—F

31. Which of the following statements is true regarding the formation of a covalent bond between two nonmetal atoms?
    (a) Valence electrons are transferred to the more electronegative atom.
    (b) Bonding electrons are found only between the bonded atoms.
    (c) The bond length is equal to the sum of the two atomic radii.
    (d) Energy is released when a covalent bond is broken.
32. Which of the following statements is true regarding the formation of a covalent bond between nitrogen and oxygen to give nitric oxide, NO?
    (a) Valence electrons are shared between nitrogen and oxygen atoms.
    (b) Bonding electrons are distributed over the entire NO molecule.
    (c) The bond length is greater than the sum of the two atomic radii.
    (d) The formation of a N—O bond releases energy.

Electron Dot Formulas of Molecules (Sec. 12.4)

33. Write the electron dot formula and draw the structural formula for each of the following molecules. *(The central atom is indicated in bold.)*
    (a) H_2 (b) F_2
    (c) H**Br** (d) **N**H_3
34. Write the electron dot formula and draw the structural formula for each of the following molecules.
    (a) Cl_2 (b) O_2
    (c) HI (d) PH_3
35. Write the electron dot formula and draw the structural formula for each of the following molecules.
    (a) HONO (b) SO_2
    (c) C_2H_4 (d) C_2H_2
36. Write the electron dot formula and draw the structural formula for each of the following molecules.
    (a) N_2 (b) PI_3
    (c) CS_2 (d) HOCl
37. Write the electron dot formula and draw the structural formula for each of the following molecules.
    (a) CH_4 (b) OF_2
    (c) H_2O_2 (d) NF_3
38. Write the electron dot formula and draw the structural formula for each of the following molecules.
    (a) CCl_4 (b) $HONO_2$
    (c) CH_3OH (d) HOCN

Electron Dot Formulas of Polyatomic Ions (Sec. 12.5)

39. Write the electron dot formula and draw the structural formula for each of the following polyatomic ions. *(Central atoms are indicated in bold.)*
    (a) BrO^- (b) BrO_2^-
    (c) BrO_3^- (d) BrO_4^-
40. Write the electron dot formula and draw the structural formula for each of the following polyatomic ions.
    (a) IO^- (b) IO_2^-
    (c) IO_3^- (d) IO_4^-
41. Write the electron dot formula and draw the structural formula for each of the following polyatomic ions.
    (a) SO_4^{2-} (b) HSO_4^-
    (c) SO_3^{2-} (d) HSO_3^-
42. Write the electron dot formula and draw the structural formula for each of the following polyatomic ions.
    (a) PO_4^{3-} (b) HPO_4^{2-}
    (c) PO_3^{3-} (d) HPO_3^{2-}
43. Write the electron dot formula and draw the structural formula for each of the following polyatomic ions.
    (a) H_3O^+ (b) OH^-
    (c) HS^- (d) CN^-
44. Write the electron dot formula and draw the structural formula for each of the following polyatomic ions.
    (a) PH_4^+ (b) SeO_3^{2-}
    (c) CO_3^{2-} (d) BO_3^{3-}

Polar Covalent Bonds (Sec. 12.6)

45. What is the general trend in electronegativity down a group in the periodic table?
46. What is the general trend in electronegativity across a series in the periodic table?
47. Which elements are more electronegative, metals or nonmetals?
48. Which elements are more electronegative, semimetals or nonmetals?
49. Predict which element in each of the following pairs is more electronegative according to the general electronegativity trends in the periodic table.
    (a) Br or Cl (b) O or S
    (c) Se or As (d) N or F
50. Predict which element in each of the following pairs is more electronegative according to the general electronegativity trends in the periodic table.
    (a) Se or Br (b) C or B
    (c) Te or S (d) Ba or Be
51. Refer to the electronegativity values in Figure 12.9 and calculate the polarity for each of the following bonds.
    (a) Br—Cl (b) Br—F
    (c) I—Cl (d) I—Br
52. Refer to the electronegativity values in Figure 12.9 and calculate the polarity for each of the following bonds.
    (a) H—Cl (b) H—Br
    (c) N—O (d) C—O
53. Refer to Figure 12.9 and label each atom in the following polar covalent bonds using delta notation (δ^+ and δ^-).

(a) H—S (b) O—S
(c) N—F (d) S—Cl

54. Refer to Figure 12.9 and label each atom in the following polar covalent bonds using delta notation.
(a) C—H (b) Se—O
(c) P—I (d) H—Br

Nonpolar Covalent Bonds (Sec. 12.7)

55. Refer to Figure 12.9 and indicate which of the following are nonpolar covalent bonds.
(a) Cl—Cl (b) Cl—N
(c) N—H (d) H—P

56. Refer to Figure 12.9 and indicate which of the following are nonpolar covalent bonds.
(a) I—C (b) C—S
(c) S—H (d) H—Br

57. Which of the following elements occurs naturally as diatomic molecules: H, He, N, P, O, S?

58. Write the chemical formula for the seven nonmetals that occur naturally as diatomic molecules.

Coordinate Covalent Bonds (Sec. 12.8)

59. An oxygen atom can bond to a hydrogen bromide molecule to give HBrO. Draw the electron dot formula for HBrO and label a coordinate covalent bond.

60. An oxygen atom can bond to a hydrogen iodide molecule to give HIO. Draw the electron dot formula for HIO and label a coordinate covalent bond.

61. An oxygen atom can bond to a HBrO molecule to give $HBrO_2$. Draw the electron dot formula for $HBrO_2$ and label two coordinate covalent bonds.

62. An oxygen atom can bond to a HIO molecule to give HIO_2. Draw the electron dot formula for HIO_2 and label two coordinate covalent bonds.

63. A hydrogen ion can bond to an ammonia molecule, NH_3, forming NH_4^+. Draw the electron dot formula for NH_4^+ and label a coordinate covalent bond.

64. A hydrogen ion can bond to a phosphine molecule, PH_3, forming PH_4^+. Draw the electron dot formula for PH_4^+ and label a coordinate covalent bond.

65. A nitrite ion, NO_2^-, can bond to an oxygen atom to form the nitrate ion. Draw the electron dot formula for NO_3^- and label a coordinate covalent bond.

66. A phosphite ion, PO_3^{3-}, can bond to an oxygen atom to form the phosphate ion. Draw the electron dot formula for PO_4^{3-} and label a coordinate covalent bond.

Shapes of Molecules (Sec. 12.9)

67. Predict the electron pair geometry, the molecular shape, and the bond angle for a silane molecule, SiH_4, using VSEPR theory.

68. Predict the electron pair geometry, the molecular shape, and the bond angle for a carbon tetrabromide molecule, CBr_4, using VSEPR theory.

69. Predict the electron pair geometry, the molecular shape, and the bond angle for a phosphine molecule, PH_3, using VSEPR theory.

70. Predict the electron pair geometry, the molecular shape, and the bond angle for a nitrogen triiodide molecule, NI_3, using VSEPR theory.

71. Predict the electron pair geometry, the molecular shape, and the bond angle for a hydrogen sulfide molecule, H_2S, using VSEPR theory.

72. Predict the electron pair geometry, the molecular shape, and the bond angle for a dichlorine monoxide molecule, Cl_2O, using VSEPR theory.

73. Apply VSEPR theory to explain why CF_4 is a nonpolar molecule even though it has four polar bonds.

74. Apply VSEPR theory to explain why CO_2 is a nonpolar molecule even though it has two polar bonds.

General Exercises

75. State whether the representative particle in each of the following substances is an atom, a molecule, or a formula unit.
(a) chromium, Cr (b) phosphorus, P_4
(c) chromium phosphide, CrP

76. State whether the representative particle in each of the following substances is an atom, a molecule, or a formula unit.
(a) plutonium, Pu (b) ozone, O_3
(c) plutonium oxide, Pu_2O_3

77. Write formula units by combining the cations and anions in each of the following pairs.
(a) Sr^{2+} and As^{3-} (b) Ra^{2+} and O^{2-}
(c) Al^{3+} and CO_3^{2-} (d) Cd^{2+} and OH^-

78. Write formula units by combining the cations and anions in each of the following pairs.
(a) Sc^{3+} and N^{3-} (b) Ti^{4+} and O^{2-}
(c) NH_4^+ and CO_3^{2-} (d) Hg_2^{2+} and PO_4^{3-}

79. Explain why the radius of a sodium ion (0.095 nm) is about half that of a sodium atom (0.186 nm).

80. Explain why the radius of a chloride ion (0.181 nm) is about twice that of a chlorine atom (0.099 nm).

81. Refer to the electronegativity values in Figure 12.9 and calculate the polarity of a B—Cl bond.

82. Refer to the electronegativity values in Figure 12.9 and calculate the polarity of a Sb—Cl bond.

83. Refer to the electronegativity values in Figure 12.9 and calculate the polarity of a H—P bond.

84. Refer to the electronegativity values in Figure 12.9 and calculate the polarity of a S—I bond.

85. Label the polar Ge—Cl bond using delta notation (δ^+ and δ^-).

86. Label the polar As—Cl bond using delta notation (δ^+ and δ^-).

87. Write the electron dot formula and draw the structural formula for silane, SiH_4, whose central atom is a semimetal.

88. Write the electron dot formula and draw the structural formula for stibine, SbH_3, whose central atom is a semimetal.

89. Write the electron dot formula and draw the structural formula for the arsenate ion, AsO_3^{3-}, whose central atom is a semimetal.

90. Write the electron dot formula and draw the structural formula for the silicate ion, SiO_3^{2-}, whose central atom is a semimetal.

91. Explain the concept of delocalized valence electrons in a molecule.

92. Explain the concept of delocalized ionic charge in a polyatomic ion.

93. There are two resonance structural formulas for the sulfur dioxide molecule. Write the two structural formulas for SO_2.

94. There are two resonance structural formulas for the nitrite ion. Write the two structural formulas for NO_2^-.

95. Boron trifluoride is a stable molecule that violates the octet rule (the B atom shares only three electron pairs). Write the electron dot formula and draw the structural formula for BF_3.

▲ Boron Trifluoride, BF_3

96. Sulfur hexafluoride is a stable molecule that violates the octet rule (the S atom shares six electron pairs). Write the electron dot formula and draw the structural formula for SF_6.

▲ Sulfur Hexafluoride, SF_6

97. Xenon dioxide is a slightly stable molecule that contains a noble gas. Write the electron dot formula and draw the structural formula for XeO_2.

98. Xenon trioxide is a slightly stable molecule that contains a noble gas. Write the electron dot formula and draw the structural formula for XeO_3.

99. Using VSEPR theory, contrast the molecular shape of a water molecule, H_2O, with that of a hydronium ion, H_3O^+.

100. Using VSEPR theory, contrast the molecular shape of an ammonia molecule, NH_3, with that of an ammonium ion, NH_4^+.

Explorer Quiz 1
Explorer Quiz 2
Explorer Quiz 3
Master Quiz

CHAPTER 13

Liquids and Solids

▲ Notice that the iceberg is less dense and floats on water. Why is H_2O less dense in the solid state than the liquid state? (*Hint*: Examine the spacing of molecules in the LIQUID and SOLID states.)

Liquids and Solids provides an excellent overview and reinforcement of several topics including: states of matter, physical and chemical properties, specific heat calculations, and the composition of hydrates. The discussion of intermolecular bonding provides the foundation for understanding the concepts of solubility and miscibility of polar and nonpolar solutes and solvents (Chapter 14).

In Chapter 11 we discussed the gaseous state of matter, where individual molecules are relatively distant from one another. In this chapter, we will discuss the liquid and solid states of matter in which particles contact one another. We will use water in much of our discussion to reinforce the concepts of specific heat, nomenclature, mole calculations, and bonding that we discussed in earlier chapters.

Water is the most important liquid on Earth and covers about three-fourths of the surface of our planet. It is necessary for all the chemical reactions that support plant and animal life. Although the percentage of water in animals varies, about two-thirds of the mass of the human body is water. With the exception of oxygen, water is our most critical substance. We can survive quite some time without food, but only a few days without water.

The physical properties of water are unusual in several ways. One unusual property is its density. The density of a substance is generally greater in the solid state than in the liquid state. The two exceptions to this generalization are water and ammonia. Since the density of solid ice is less than that of liquid water, ice floats on water. If ice were more dense than water, marine life would not survive. Rivers and lakes would freeze into solid chunks of ice. Initially, ice would form on the surface and then sink, and eventually the entire body of water would change to solid ice.

▲ **Liquid Glass** Over the past 500 years, these stained glass windows have become measurably thicker at the bottom.

13.1 Properties of Liquids

Objective · To list five observed properties of a liquid.

Unlike gases, liquids do not respond dramatically to temperature and pressure changes. Also, the mathematical relationships that apply to gases, such as the combined gas law, do not apply to liquids. Rather, when we study the liquid state, we observe the following general properties.

1. *Liquids have an indefinite shape but a fixed volume.* The shape of a liquid conforms to the shape of its container.
2. *Liquids usually flow readily.* Liquids flow at different rates; for example, petroleum flows more slowly than water. Glass is actually a liquid that flows very slowly. Some panes of glass in the Sistine Chapel (built in 1473) are now thicker at the bottom than at the top.
3. *Liquids do not compress or expand to any degree.* The volume of a liquid changes very little with changes in temperature or pressure.
4. *Liquids have a high density compared to gases.* Gases and liquids are both fluids; that is, the shape is not definite, and individual particles are free to move throughout the container. However, the liquid state is about 1000 times more dense than the gaseous state because particles are closer together. For example, the density of water is 1.00 g/mL, whereas the density of air is about 0.001 g/mL.
5. *Liquids that are soluble mix uniformly.* Liquids diffuse more slowly than gases. Soluble liquids, however, eventually form a homogeneous mixture. When alcohol is added to water, for example, the liquids slowly diffuse and mix uniformly.

13.2 The Intermolecular Bond Concept

Objectives · To explain the concept of an intermolecular bond.
· To describe three types of attraction between molecules in a liquid.

Various properties of liquids, such as vapor pressure, viscosity, and surface tension, are determined by the attraction between molecules. Although this attraction is considered an intermolecular bond, it is much weaker than an ordinary covalent bond. The three types of intermolecular attraction are based on temporary dipoles, permanent dipoles, and hydrogen bonds.

In Section 12.9, we learned that a polar molecule results from one or more polar bonds. In a polar molecule, positive and negative charges are concentrated in different regions. These two charged regions are created by an uneven distribution of electrons around the molecule. A molecule with these two regions, one positive and one negative, is called a *dipole*.

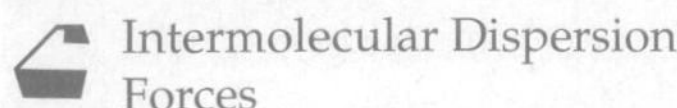

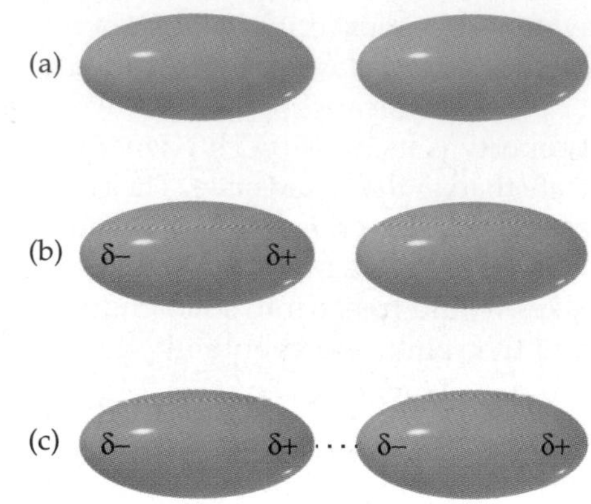

▶ **Figure 13.1 Intermolecular Dispersion Forces** (a) Two nonpolar molecules. (b) The nonpolar molecule on the left forms a temporary dipole, which can induce a dipole in the molecule on the right. (c) The two nonpolar molecules are temporarily attracted to each other. In turn, these molecules can induce temporary dipole attractions in surrounding molecules of the liquid.

Let's first consider temporary dipole attraction. Temporary dipole attraction is found between nonpolar molecules. Even though atoms in a nonpolar molecule share electrons equally, the electrons are constantly shifting about. This shifting about of electrons produces regions in the molecule that are temporarily "electronrich" and slightly negative. Simultaneously, another region of the molecule is temporarily "electronpoor" and slightly positive. A negative region in one molecule has a weak attraction for a positive region in another molecule. This temporary attraction between molecules is referred to as a **dispersion force**, or a London dispersion force.

Although dispersion forces last for only brief periods of time, they occur frequently. At any one time, there are so many temporary dipoles that there is significant attraction between the molecules in a liquid. Figure 13.1 illustrates the nature of temporary dispersion forces.

Because of one or more polar bonds, a polar molecule has regions of positive and negative charges. Similarly to miniature magnets, these molecules are attracted to each other. In contrast to temporary dipole attraction, permanent dipole attraction operates continuously. A permanent **dipole force** operates between polar molecules as shown in Figure 13.2.

The dipole attraction between certain molecules can be especially strong. For instance, when a hydrogen atom is bonded to an oxygen atom or a nitrogen atom, a highly polar bond results. This highly polar bond is capable of producing a strong dipole force between molecules. This special type of dipole attraction is called a **hydrogen bond** (Figure 13.3). Although a hydrogen bond between two molecules is strong, it is typically less than 10% of the strength of a normal covalent bond.

The following example exercise illustrates the characteristics of intermolecular bonds.

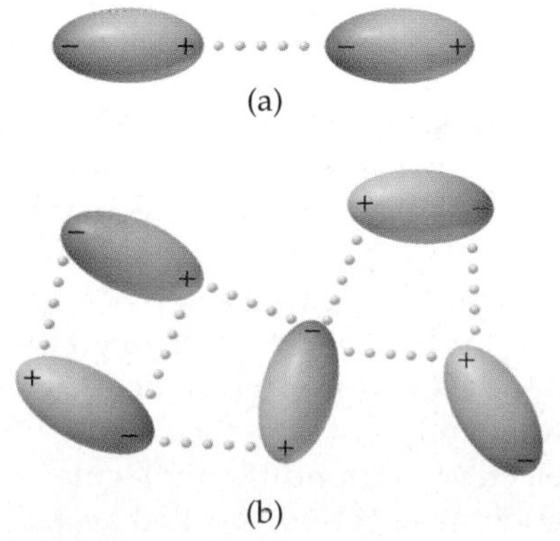

▶ **Figure 13.2 Intermolecular Dipole Forces** (a) A dipole force results from a dipole–dipole attraction between two molecules. (b) Dipole forces between several molecules in a liquid are shown.

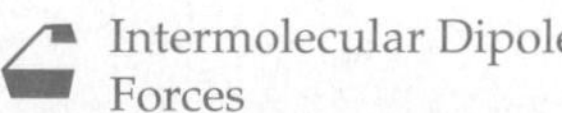

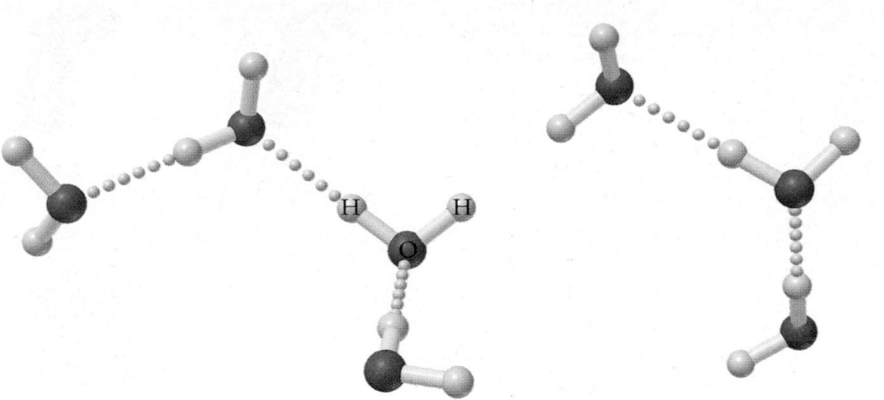

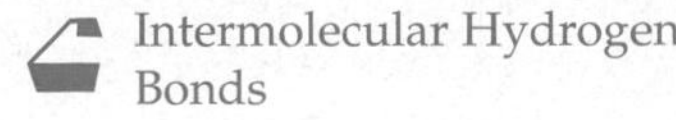

◀ **Figure 13.3 Intermolecular Hydrogen Bonds** Polar water molecules form intermolecular hydrogen bonds. Notice that the hydrogen atom is attracted to the nonbonding electrons on the highly electronegative oxygen atom.

Example Exercise 13.1 • Intermolecular Attraction

In a liquid with polar molecules, what is the strongest type of intermolecular attraction?

Solution

In a liquid composed of polar molecules, the intermolecular attraction is the result of both permanent and temporary dipoles. Molecules that contain either H—O or H—N bonds have the strongest type of intermolecular attraction, that is, hydrogen bonds. Water, H_2O, and ammonia, NH_3, are examples of liquids with hydrogen bonds.

Self-Test Exercise

In a liquid with nonpolar molecules, what is the strongest type of intermolecular attraction?

Answer: The only type of attraction that can exist between nonpolar molecules is a dispersion force resulting from temporary dipoles.

13.3 Vapor Pressure, Boiling Point, Viscosity, Surface Tension

Objective · To relate the vapor pressure, boiling point, viscosity, and surface tension of a liquid to the strength of attraction between molecules.

In Chapter 11, we introduced the kinetic theory of gases to explain the behavior of gases. Now, we will extend this theory to explain the properties of liquids.

According to the kinetic theory the attraction between gas molecules is negligible. This is not the case for liquids. Molecules in the liquid state are in contact, and their attraction for each other restricts their movement. Individual molecules do have enough energy to move about one another; that is, a liquid is free to flow. The liquid state and the gaseous state are analogous to honeybees swarming in a hive (liquid state) and individual bees gathering pollen (gaseous state). Now let's consider some properties of liquids.

Vapor Pressure

The kinetic theory relates the average energy of molecules to temperature. The higher the temperature, the greater the kinetic molecular energy. However, not all molecules at the same temperature have identical energies. Some are more energetic than others. At the surface of a liquid, for example, some molecules have enough energy to completely escape the attraction of neighboring molecules. The molecules that escape are in the vapor state. This process is called *evaporation*. In the reverse process, some molecules in the vapor state return to the liquid. This process is called *condensation*.

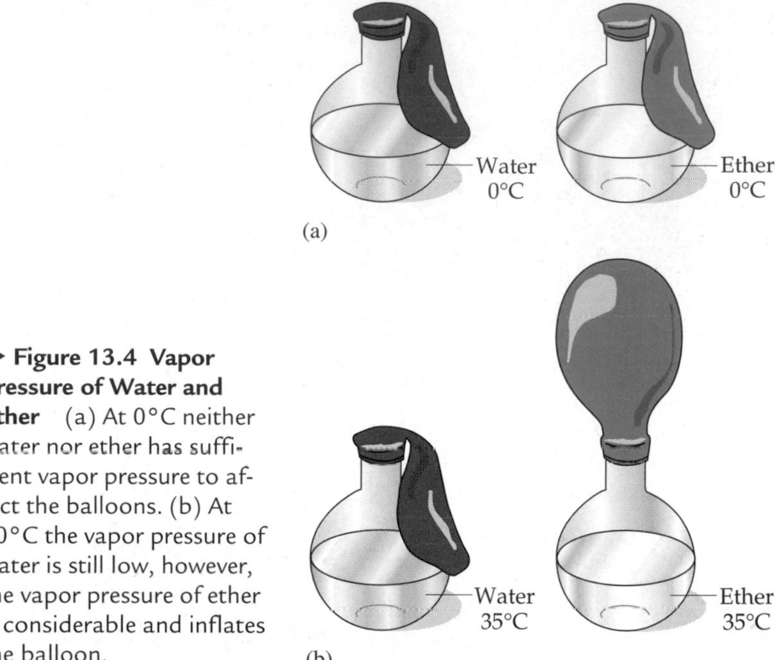

▶ **Figure 13.4 Vapor Pressure of Water and Ether** (a) At 0°C neither water nor ether has sufficient vapor pressure to affect the balloons. (b) At 20°C the vapor pressure of water is still low, however, the vapor pressure of ether is considerable and inflates the balloon.

When the rates of evaporation and condensation are equal, the pressure exerted by the gas molecules above a liquid is called the **vapor pressure**. However, vapor pressure depends on the attraction between molecules in the liquid. Water molecules have a strong attraction for each other and ethyl ether molecules have a relatively weak attraction. Because of the lesser attraction between molecules, ether molecules escape from the liquid state more readily than do water molecules. Therefore, the vapor pressure of ether is greater than that of water at the same temperature (Figure 13.4).

Table 13.1 compares the attraction between molecules, that is, *intermolecular attraction*, and the vapor pressure for a few selected liquids.

Table 13.1 Vapor Pressure of Selected Liquids

Liquid	Approximate Molar Mass	Intermolecular Attraction	Vapor Pressure at 20°C
water	18 g/mol	strong	18 mm Hg
propionic acid, C_2H_5COOH	74 g/mol	strong	5 mm Hg
butyl alcohol, C_4H_9OH	74 g/mol	strong	6 mm Hg
propyl chloride, C_3H_7Cl	79 g/mol	weak	300 mm Hg
ethyl ether, $C_2H_5OC_2H_5$	74 g/mol	weak	450 mm Hg

In general, as the attraction between molecules increases, the vapor pressure decreases. In Table 13.1 we see that propionic acid and butyl alcohol each has a strong intermolecular attraction. Thus, their vapor pressures at 20°C are quite low. Conversely, propyl chloride and ethyl ether each has a weak intermolecular attraction. Thus, their vapor pressures are considerably higher. When comparing vapor pressure values, it is necessary to choose two liquids that have approximately the same molar mass. Note that in these comparisons, each liquid has a molar mass of ~ 74 g/mol.

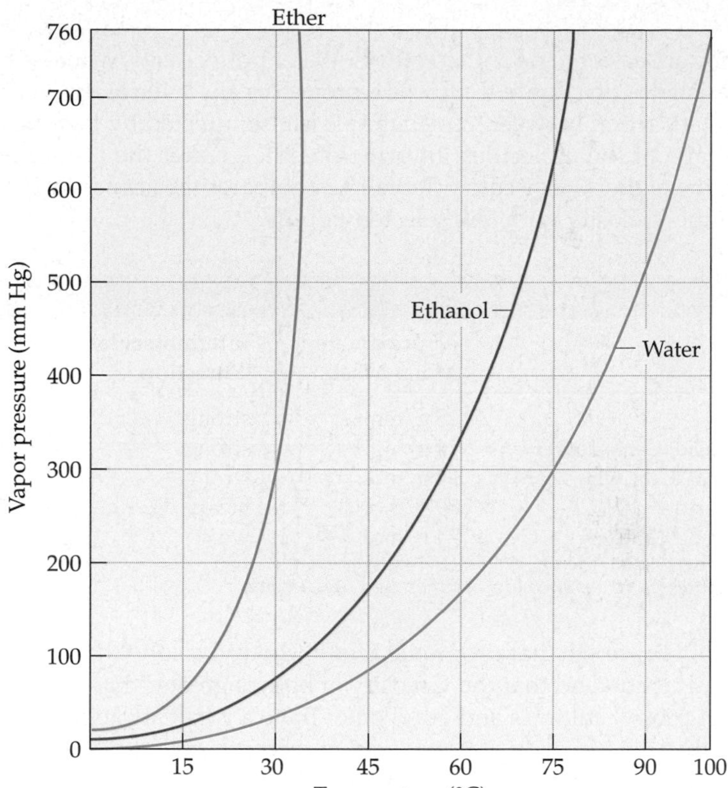

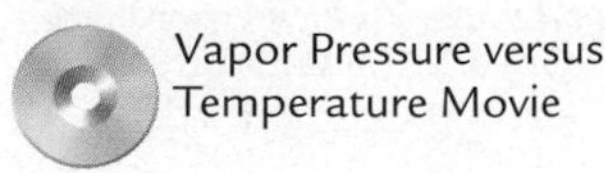

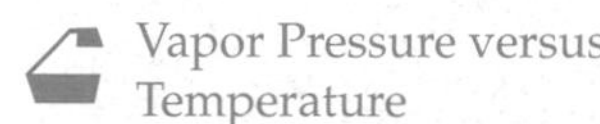

◀ **Figure 13.5 Vapor Pressure versus Temperature** The vapor pressure of a liquid increases as the temperature increases. The vapor pressure of water equals 760 mm Hg at 100°C. Thus, at 100°C the vapor pressure of water equals the atmospheric pressure (760 mm Hg) and water begins to boil.

Boiling Point

As the temperature of a liquid increases, its vapor pressure increases. At 30°C the vapor pressure of water is about 30 mm Hg, and at 60°C it is 150 mm Hg. Figure 13.5 shows the relationship between vapor pressure and temperature for ether, ethanol, and water.

A liquid begins to boil when the pressure of the vapor above the liquid equals the pressure of the atmosphere. Thus, we define the **boiling point** as the temperature at which the vapor pressure equals atmospheric pressure. Accordingly, liquids that have *high* boiling points must have *low* vapor pressures. The following example exercise illustrates the determination of boiling points from vapor pressure data.

Example Exercise 13.2 • Boiling Point Predictions

Refer to Figure 13.5 and determine the approximate boiling point of ether.

Solution

The normal boiling point of a liquid is the temperature at which the vapor pressure equals standard atmospheric pressure, that is, 760 mm Hg. From the graph, we find that the vapor pressure of ether is 760 mm Hg at about 35°C. Thus, the normal boiling point of ether is about 35°C. The actual observed value for the boiling point of ether is 36°C.

Self-Test Exercise

Refer to Figure 13.5 and determine the approximate boiling point of ethanol.

Answer: ~ 80°C (The actual boiling point of ethanol is 78°C.)

▲ **Viscosity** The molasses on the left is more viscous and flows more slowly than the water on the right.

Viscosity

Some liquids are easier to pour than others. Water pours easily, whereas honey does not. The resistance of a liquid to flow is a property called **viscosity**. Viscosity is the result of an attraction between molecules. It is also affected by factors such as the size and shape of the molecules. In principle, the greater the attraction between molecules, the higher the viscosity. Table 13.2 compares the attraction between molecules and the viscosity for a few selected liquids.

Table 13.2 Viscosity of Selected Liquids

Liquid	Approximate Molar Mass	Intermolecular Attraction	Viscosity* at 20°C
water	18 g/mol	strong	1.00
propionic acid, C_2H_5COOH	74 g/mol	strong	1.10
butyl alcohol, C_4H_9OH	74 g/mol	strong	2.95
propyl chloride, C_3H_7Cl	79 g/mol	weak	0.35
ethyl ether, $C_2H_5OC_2H_5$	74 g/mol	weak	0.23

*Values are expressed in centipoise, a common unit of viscosity.

In Table 13.2 we see that propionic acid and butyl alcohol have a strong intermolecular attraction and that the viscosity of both is greater than that of water. We also see that propyl chloride and ethyl ether have a weak attraction between molecules and that the viscosity of both is much lower than that of water. Honey and molasses are examples of viscous liquids. We can therefore predict that there is a strong attraction between molecules in each of these liquids.

Surface Tension

At some time you have probably noticed a small insect or some other object floating on water. For an insect or any object to sink in a liquid, it has to break through the surface. But the molecules on the surface of a liquid resist being pushed apart. The attraction between the surface molecules in a liquid is called **surface tension**. There are other factors to consider, but in general the greater the intermolecular attraction, the higher the surface tension. Table 13.3 compares the intermolecular attraction and surface tension for a few selected liquids.

We all know that rain forms drops as it falls. In fact, when we spray any liquid, it forms drops and each drop has the shape of a small sphere. Drops of liquid are spherical because surface tension causes them to have the smallest possible surface

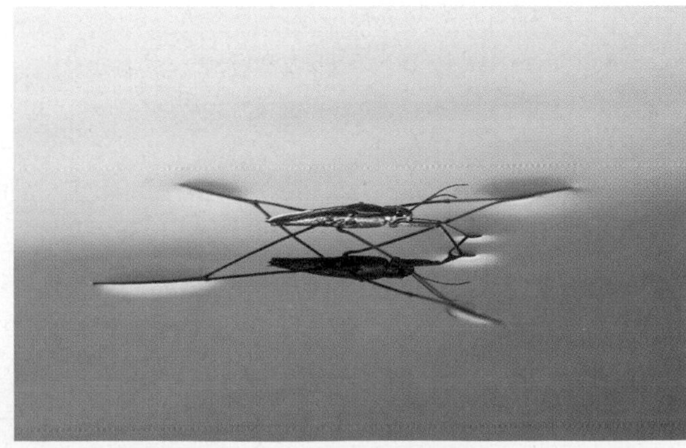

▶ **Surface Tension** Water has a high surface tension, and this allows the insect to "walk" on water.

Table 13.3 Surface Tension of Selected Liquids

Liquid	Approximate Molar Mass	Intermolecular Attraction	Surface Tension* at 20°C
water	18 g/mol	strong	70
propionic acid, C_2H_5COOH	74 g/mol	strong	27
butyl alcohol, C_4H_9OH	74 g/mol	strong	25
propyl chloride, C_3H_7Cl	79 g/mol	weak	18
ethyl ether, $C_2H_5OC_2H_5$	74 g/mol	weak	17

*Values are expressed in dynes per square centimeter, a common unit of surface tension.

area, and that area corresponds to a spherical droplet. Table 13.3 shows that water has an unusually high surface tension.

The following example exercise illustrates the relationships between the properties of liquids as they relate to intermolecular attraction.

Example Exercise 13.3 • Physical Property Predictions

Consider the following properties of liquids. State whether the value for each property is high or low for a liquid with a strong intermolecular attraction.

(a) vapor pressure (b) boiling point (c) viscosity (d) surface tension

Solution

For a liquid having a strong attraction between molecules, properties (b), (c), and (d) are generally high; property (a) is low.

(a) Molecular attraction slows evaporation. Therefore, vapor pressure is *low* for liquids with a strong intermolecular attraction.
(b) Attraction between molecules inhibits boiling. Thus, the boiling point is *high* for liquids with a strong intermolecular attraction.
(c) Molecular attraction increases the resistance of a liquid to flow. Viscosity is *high* for liquids with a strong intermolecular attraction.
(d) Attraction between molecules draws a drop of liquid into a sphere. Surface tension is *high* for liquids with a strong intermolecular attraction.

Self-Test Exercise

In pentane, C_5H_{12}, the intermolecular attraction is less than that in isopropyl alcohol, C_3H_7OH. Predict which liquid has the higher value for each of the following.

(a) vapor pressure (b) boiling point (c) viscosity (d) surface tension

Answers: (a) C_5H_{12}; (b) C_3H_7OH; (c) C_3H_7OH; (d) C_3H_7OH

13.4 Properties of Solids

Objective · To list five observed properties of a solid.

Unlike liquids, solids have a fixed shape. The reason for this is that the individual particles of the solid are not free to move. Unlike gases, the volume of a solid shows very little response to changes in temperature or pressure. The observed properties of the solid state are as follows.

1. *Solids have a definite shape and a fixed volume.* Unlike liquids, solids are rigid and their shape is fixed.
2. *Solids are either crystalline or noncrystalline.* A **crystalline solid** contains particles arranged in a regular repeating pattern. Each particle occupies a fixed position in the crystal. The high degree of order of the molecules can produce

a beautiful clear crystal, for example, a diamond or a ruby. An ordinary stone is a noncrystalline solid and is nontransparent.

3. *Solids do not compress or expand to any degree.* Assuming no change in physical state, temperature and pressure have a negligible effect on the volume of a solid.
4. *Solids have a slightly higher density than their corresponding liquids.* For example, solid chunks of iron sink in a high-temperature furnace containing molten iron. One important exception to this rule is water; ice is less dense than liquid water. As a result, ice floats on water.
5. *Solids do not mix by diffusion.* In a solid heterogeneous mixture, the particles are not free to diffuse and they cannot mix uniformly. In an alloy, which is a homogeneous mixture of metals, the particles mix uniformly in the molten liquid state before cooling to a solid.

13.5 Crystalline Solids

Objective · To describe three types of crystalline solids: ionic, molecular, and metallic.

We know that the particles of a crystalline solid are arranged in a regular geometric pattern. However, the particles can be of different types. They can be ionic, molecular, or metallic (Figure 13.6). In ionic solids the crystals are composed of regular patterns of ions. In molecular solids the molecules form repeating patterns. In metallic solids metal atoms are arranged geometrically.

Ionic Solids

A crystalline **ionic solid** is an ionic compound composed of positive and negative ions. Salt, for example, is a crystalline solid of NaCl. Here, sodium ions, Na^+, and chloride ions, Cl^-, are arranged in a regular three-dimensional structure referred to as a crystal lattice. Notice that each crystal of table salt shown in Figure 13.7 has the same characteristic shape. Other ionic crystalline compounds, such as NaF, CaF_2, and $CaCO_3$, have different geometric shapes.

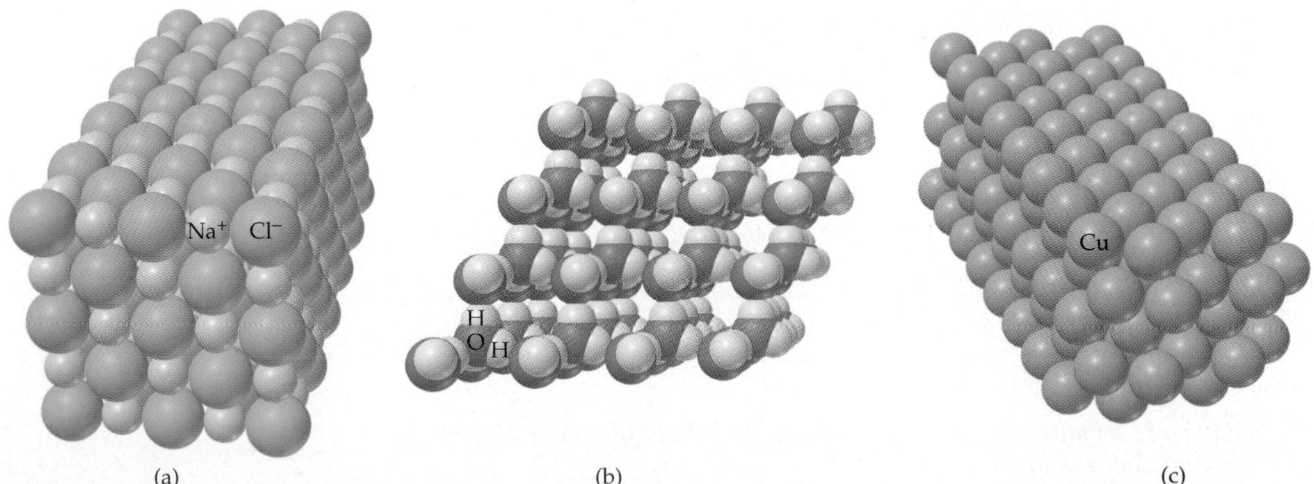

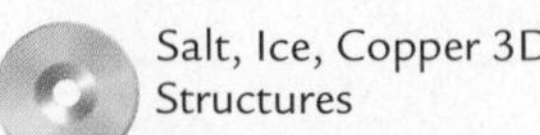

▲ **Figure 13.6 Ionic, Molecular, and Metallic Solids** (a) The arrangement of ions in salt, NaCl. (b) The pattern of water molecules in ice. (c) The geometry of copper atoms in a metallic crystal.

▲ **Figure 13.7 A Crystalline Ionic Solid** Sodium chloride, NaCl, crystals are an example of a crystalline ionic solid.

▲ **Figure 13.8 A Crystalline Molecular Solid** Sulfur, S_8, crystals are an example of a crystalline molecular solid.

Molecular Solids

Sulfur 3D Molecule

A crystalline **molecular solid** has molecules arranged in a particular configuration. Crystalline sucrose (table sugar), for example, is composed of $C_{12}H_{22}O_{11}$ molecules. The sucrose molecules are arranged in a regular order that allows light to pass through the crystal. Therefore, a large crystal of sucrose appears transparent. Other molecular solids are sulfur and phosphorus. Sulfur crystals (Figure 13.8) are made from S_8 molecules held together by covalent bonds. Phosphorus powder is made from P_4 molecules.

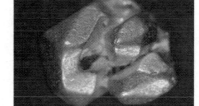

Metallic Solids

A crystalline **metallic solid** has atoms of metals arranged in a definite pattern. That is, a metallic crystal is made up of positive metal ions surrounded by valence electrons. This pattern is referred to as the "electron sea" model. The flow of electricity is associated with the movement of electrons through a metal. Metals are good conductors of electricity because the valence electrons are free to move about the crystal. Figure 13.9 shows a crystal of gold.

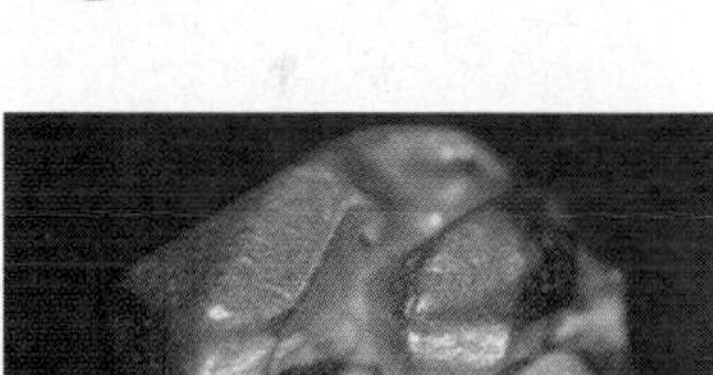

▲ **Figure 13.9 A Crystalline Metallic Solid** A gold crystal is an example of a crystalline metallic solid.

Classifying Crystalline Solids

It is helpful to classify crystalline solids in order to predict their properties. In general, the properties of ionic solids such as melting point, hardness, conductivity, and solubility are similar. The properties of molecular solids are usually similar as well. Metals show a range of physical properties, but all are malleable and ductile and are good conductors of electricity. Table 13.4 lists the general properties for each of the three types of crystalline solids.

Table 13.4 General Properties of Crystalline Solids

Type of Solid	General Properties	Examples
ionic	high melting point, hard, brittle, at least slightly soluble in water, conductor of electricity when melted or in solution	NaCl, $CaCO_3$, $MgSO_4$
molecular	low melting point, generally insoluble in water, nonconductor of electricity	S_8, $C_{10}H_8$, $C_6H_{12}O_6$
metallic	low to high melting point, malleable, ductile, conductor of electricity insoluble in most solvents	Fe, Ag, Au

▲ **Network Solid** Diamond is a network crystalline solid of carbon.

Diamond is composed of carbon and is a special type of crystalline solid that has covalent bonds between a vast number of C atoms. This type of crystalline solid is referred to as a *network solid*. Diamond is a hard, brittle, crystalline network solid with unusual properties, including a very high melting point (above 3500°C).

The following example exercise further illustrates the classification of ionic, molecular, and metallic types of crystalline solids.

Example Exercise 13.4 • Classifying Crystalline Solids

Classify each of the following crystalline solids as ionic, molecular, or metallic.

(a) nickel, Ni (b) nickel oxide, NiO

Solution

The type of crystalline solid is dictated by the type of particle in the solid.

(a) Nickel is a metal composed of atoms; thus, Ni is a *metallic solid*.

(b) Nickel oxide contains ions and is therefore an *ionic solid*.

Self-Test Exercise

Classify each of the following crystalline solids as ionic, molecular, or metallic.

(a) iodine, I_2 (b) silver iodide, AgI

Answers: (a) molecular; (b) ionic

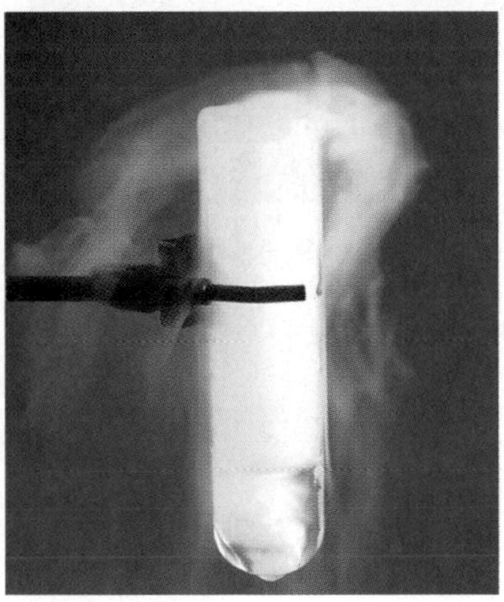

▲ **Dry Ice** Solid CO_2 is undergoing sublimation, a direct change in state from a solid to a vapor.

Note Dry ice is a crystalline solid composed of CO_2 molecules having covalent bonds. In turn, each CO_2 molecule is attracted to other CO_2 molecules by dipole forces. Dry ice sublimes at −80°C because dipole forces are relatively weak compared to covalent bonds.

13.6 Changes of Physical State

Objective · **To calculate heat changes that involve the heat of fusion, specific heat, and heat of vaporization for a given substance.**

Heat is necessary to change the physical state of a substance. Specific heat is the amount of heat required to raise 1.00 g of a substance 1°C. Every substance has a unique value for its specific heat. Water is considered a reference, and its specific heat is 1.00 calorie per gram per degree Celsius (1.00 cal/g × °C). The specific heats for ice and steam are approximately half that of water.

Next, let's consider a substance changing state from a solid to a liquid at its melting point. The amount of heat required to melt 1.00 g of substance is called the **heat of fusion** (H_{fusion}). For water, the heat of fusion is 80.0 cal/g. Water releases the same amount of heat energy, 80 cal/g, when it changes from the liquid to the solid state. This heat change is called the *heat of solidification* (H_{solid}). The heats of fusion and solidification are equal for all substances.

A substance rapidly changes state from a liquid to a vapor at its boiling point. The amount of heat required to vaporize 1.00 g of a substance is called the **heat of vaporization** (H_{vapor}). For water, it is 540 cal/g. Conversely, water releases the same amount of heat energy, 540 cal/g, when it condenses from a gas to a liquid. This heat change is called the *heat of condensation* (H_{cond}). The heats of vaporization and condensation are equal for all substances. Table 13.5 lists the heat values for water, ice, and steam.

Changes of State Movie

Table 13.5 Heat Values for Water

Substance	Specific heat (cal/g × °C)	H_{fusion} (cal/g)	H_{solid} (cal/g)	H_{vapor} (cal/g)	H_{cond} (cal/g)
ice, $H_2O(s)$	0.50	80.0	—	—	—
water, $H_2O(l)$	1.00	—	80.0	540	—
steam, $H_2O(g)$	0.48	—	—	—	540

To see the change in temperature with a constant application of heat, we draw a temperature–energy graph, sometimes called a heating curve. The heating curve for water is shown in Figure 13.10.

Let's combine the concept of a heating curve with the heat values in Table 13.5. For example, let's find the amount of heat energy necessary to convert 25.0 g of ice at −5.0°C to steam at 100°C. This problem requires four steps: (1) heat the ice from −5.0° to 0.0°C; (2) convert the ice to water at 0.0°C; (3) heat the water from 0.0° to 100.0°C; and (4) convert the water to steam at 100.0°C.

1. To calculate the amount of energy required to heat the ice we use its mass (25.0 g), the temperature change (−5.0° to 0.0°C), and the specific heat of ice (0.50 cal/g × °C). Using 0.50 cal/g × °C as a unit factor, we have

$$25.0\ \cancel{g} \times [0.0 - (-5.0)]\ \cancel{^\circ C} \times \frac{0.50\ cal}{1\ \cancel{g} \times \cancel{^\circ C}} = 63\ cal$$

2. The heat of fusion for ice, 80.0 cal/g, is found in Table 13.5. The energy required to melt 25.0 g of ice is

$$25.0\ \cancel{g} \times \frac{80.0\ cal}{1\ \cancel{g}} = 2000\ cal$$

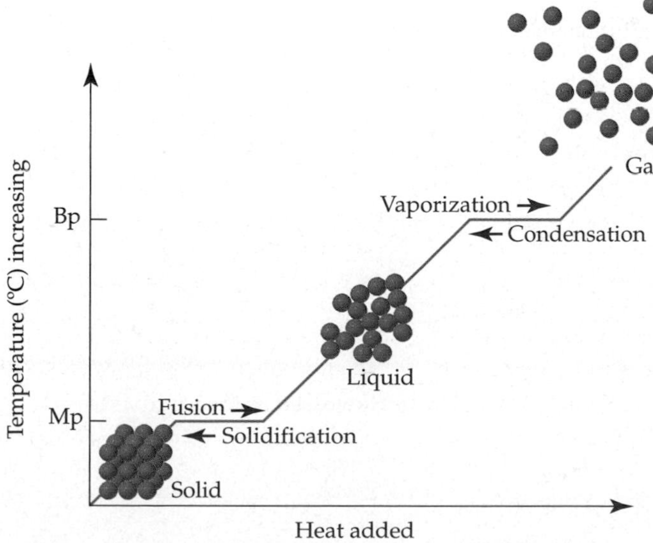

◀ **Figure 13.10 Temperature–Energy Graph** As heat is continuously added to a substance, the substance eventually changes its physical state. Notice that the temperature remains constant during the change of state from solid to liquid and from liquid to gas.

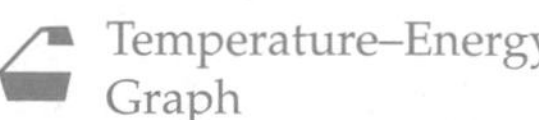
Temperature–Energy Graph

3. To calculate the amount of energy required to heat the water we must know its mass, the temperature change (0.0° to 100.0°C), and the specific heat of water; therefore, we have

$$25.0\ \cancel{g} \times (100.0 - 0.0)\ \cancel{^\circ C} \times \frac{1.00\ \text{cal}}{1\ \cancel{g} \times \cancel{^\circ C}} = 2500\ \text{cal}$$

4. The heat of vaporization, 540 cal/g, is found in Table 13.5. The energy required to vaporize the water to steam is

$$25.0\ \cancel{g} \times \frac{540\ \text{cal}}{1\ \cancel{g}} = 13{,}500\ \text{cal}$$

The total heat energy required to heat and vaporize the water is equal to the sum of the values obtained in steps 1 through 4.

$$63\ \text{cal} + 2000\ \text{cal} + 2500\ \text{cal} + 13{,}500\ \text{cal} = 18{,}100\ \text{cal}$$

The heat required to raise the temperature of the ice at −5.0°C to steam at 100.0°C is 18,100 cal, or 18.1 kcal. The following example exercise illustrates the heat changes associated with the cooling of water and its solidification to ice.

Example Exercise 13.5 • Change in Heat

Calculate the amount of heat released when 15.5 g of water at 22.5°C cools to ice at −10.0°C.

Solution

In this problem we have to consider (1) the specific heat of water, (2) the heat of solidification, and (3) the specific heat of ice.

1. To calculate the amount of heat released when cooling the water, consider the mass, the temperature change (22.5° to 0.0°C), and the specific heat of water, 1.00 cal/g × °C.

$$15.5\ \cancel{g} \times (22.5 - 0.0)\ \cancel{^\circ C} \times \frac{1.00\ \text{cal}}{1\ \cancel{g} \times \cancel{^\circ C}} = 349\ \text{cal}$$

We can summarize the solution visually as follows.

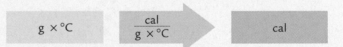

$$\text{g} \times {}^\circ\text{C} \quad \xrightarrow{\ \frac{\text{cal}}{\text{g} \times {}^\circ\text{C}}\ } \quad \text{cal}$$

2. The heat of solidification, found in Table 13.5, is 80.0 cal/g. The heat released when water solidifies to ice is

$$15.5\ \cancel{g} \times \frac{80.0\ \text{cal}}{1\ \cancel{g}} = 1240\ \text{cal}$$

3. The specific heat of ice is 0.50 cal/g × °C. The heat released as the ice cools to −10.0°C is found as follows.

$$15.5\ \cancel{g} \times [0.0 - (-10.0)]\ \cancel{^\circ C} \times \frac{0.50\ \text{cal}}{1\ \cancel{g} \times \cancel{^\circ C}} = 78\ \text{cal}$$

The total heat energy released when the water cools to ice at −10.0°C equals the sum of the values obtained in steps 1 through 3.

$$349 \text{ cal} + 1240 \text{ cal} + 78 \text{ cal} = 1670 \text{ cal}$$

Thus, the heat released when the water cools is 1670 cal, or 1.67 kcal.

Self-Test Exercise
Calculate the amount of heat required to convert 50.0 g of steam at 100.0°C to ice at 0.0°C.

Answer: 36,000 cal (36.0 kcal)

13.7 Structure of Water

Objective · To illustrate the bond angle and net dipole in a water molecule.

We learned in Section 13.3 that the boiling point and surface tension of water are unusually high. We also saw that water has a strong intermolecular attraction because of hydrogen bonding. To understand these concepts more completely, let's review the covalent bonding in a water molecule. Specifically, let's consider the electron dot formula, the structural formula, the observed bond angle, polar covalent bonds, and the net dipole of water.

Electron Dot Formula

We begin by writing the electron dot formula for water, H_2O. The total number of valence electrons in one molecule is $2(1\ e^-) + 6\ e^- = 8\ e^-$. Thus, there are four electron pairs. We can place the four pairs of electrons around the oxygen to provide the necessary octet.

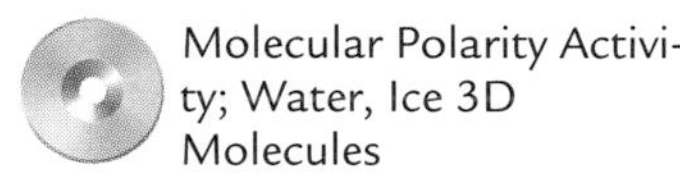

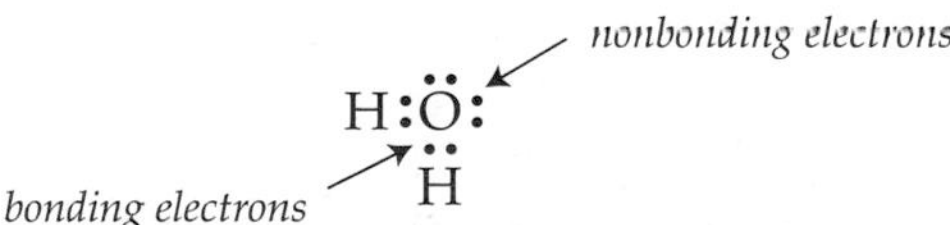

Notice that there are two bonding and two nonbonding electron pairs. We learned in Section 12.9 that the molecular shape of a water molecule is bent, with a bond angle of 104.5°. Figure 13.11 shows the arrangement of bonding and nonbonding electron pairs.

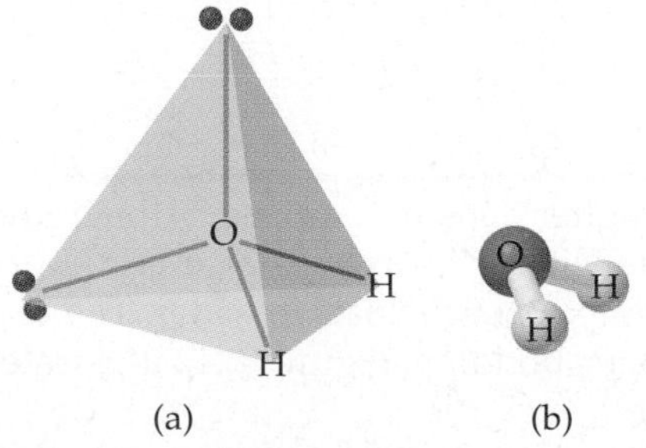

◀ **Figure 13.11 Bonding in a Water Molecule** (a) The arrangement of the electron pairs in a water molecule forms the four corners of a tetrahedron. (b) A water molecule can be described as bent or V-shaped. The two hydrogen atoms are repelled slightly toward each other by the two nonbonding electron pairs.

Structural Formula and Bond Angle

The structural formula for a molecule uses dashes to represent covalent bonds. With this notation, the unusual properties of water are best explained if the two hydrogen atoms are at an angle to each other. Experimental evidence shows that the angle between the two hydrogen atoms is 104.5°. The angle formed by the H—O—H bonds is referred to as the **bond angle**.

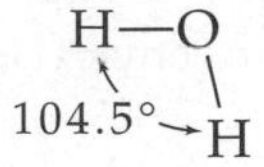

H—O
104.5° H

Polar Covalent Bonds

In a water molecule the two covalent bonds are polarized. That is, the more electronegative oxygen atom draws the bonding electrons closer to its nucleus. In turn, the hydrogen atoms become slightly positive. In Chapter 12 we used delta notation (δ^+ and δ^-) to indicate bond polarity. Since the covalent bond between H and O has a partially positive and a partially negative aspect, there is a dipole.

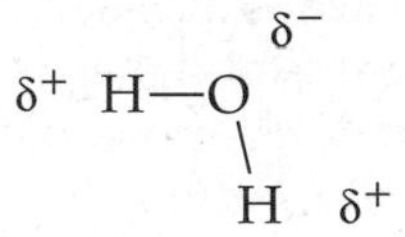

δ^-
δ^+ H—O
H δ^+

Net Dipole

Notice that a water molecule has two dipoles, each pulling an electron pair toward the central atom. The effect of the two dipoles results in a single dipole that passes through the center of the molecule. The single, overall dipole for a molecule having two or more dipoles is called the **net dipole**.

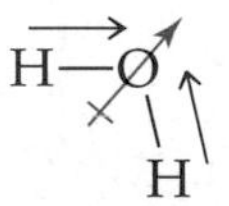

H—O
H

The net dipole produces a negative and a positive end in the water molecule. The negative end of the molecule is indicated by the arrow. The positive end is indicated by the plus sign on the opposite end of the net dipole.

13.8 Physical Properties of Water

Objective · To explain the unusual properties of water.

Water is a colorless, odorless, tasteless liquid and a powerful solvent. At room temperature water has the highest specific heat, heat of fusion (except for ammonia), and heat of vaporization of any liquid.

Density

Generally, a substance in the solid state has a higher density than one in the liquid state. Therefore, we would predict that the density of ice is greater than that of water. But it's obvious that this is not correct. We know ice floats in water. The reason ice is less dense than water relates to hydrogen bonding. Figure 13.12 illustrates three-dimensional hydrogen bonding in water.

When water freezes to solid ice, the hydrogen bonds produce a three-dimensional crystal. Figure 13.13 illustrates the structure of an ice crystal. Because of the

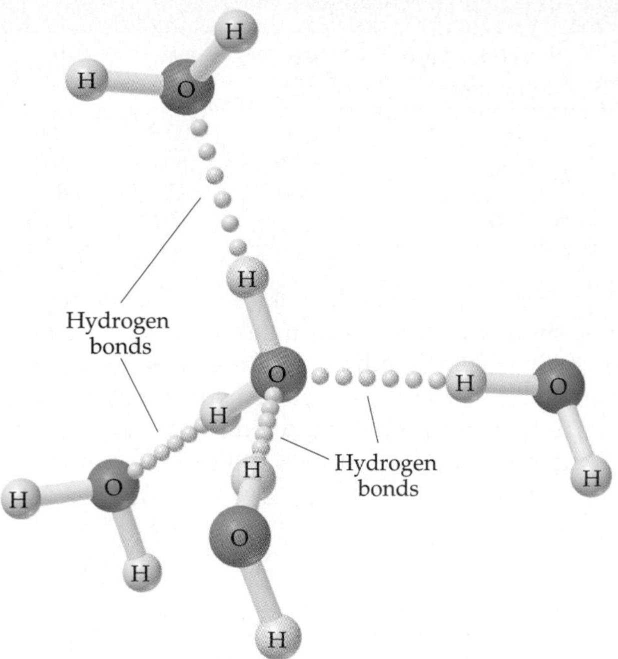

◀ **Figure 13.12 Hydrogen Bonding** Each water molecule is attached to four other molecules. The intermolecular hydrogen bond is about 50% longer than an ordinary covalent bond. Since a hydrogen bond is longer, it is weaker and requires much less energy to break.

◀ **Figure 13.13 Structure of Ice Crystals** Water molecules hydrogen-bond to form six-member rings. The rings in turn hydrogen-bond to other rings, producing large, three-dimensional crystalline structures.

arrangement of water molecules, however, the crystal has holes. These holes create a volume for ice that is greater than that for an equal mass of liquid water. Furthermore, since the volume of ice is greater than that of water, its density is less than that of water. At 0°C, the density of ice is 0.917 g/mL. The density of water is 1.00 g/mL.

Melting and Boiling Points

Water has an unusually high melting point and boiling point for a small molecule. To see how unusual these properties are, compare them with those of some hydrogen compounds of Group VIA/16 as outlined in Table 13.6.

If we ignore water, we see a clear trend in the melting and boiling points of Group VIA/16 hydrogen compounds. First, notice the increase in molar mass for

Table 13.6 Group VIA/16 Hydrogen Compounds

Compound	Molar Mass (g/mol)	Mp (°C)	Bp (°C)	H_{fusion} (cal/mol)	H_{vapor} (cal/mol)
H_2O	18	0.0	100.0	1440	9720
H_2S	34	−85.5	−60.7	568	4450
H_2Se	81	−60.4	−41.5	899	4620
H_2Te	130	−48.9	−2.2	1670	5570

H_2S through H_2Te. Next, notice that the values for the melting point and the boiling point increase simultaneously. As with density, the properties of water are unusual because of hydrogen bonding. Hydrogen bonding produces a strong intermolecular attraction that resists the movement of molecules. Therefore, a higher temperature is needed to melt ice and to boil water. Figure 13.14 illustrates the melting and boiling points of Group VIA/16 hydrogen compounds.

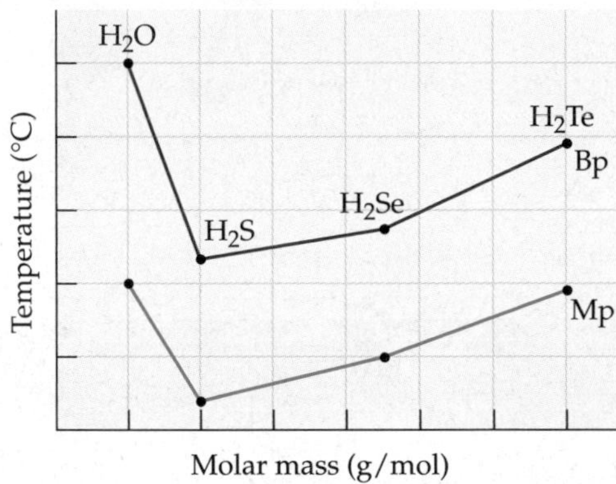

▶ Figure 13.14 Mp and Bp of Group VIA/16 Hydrogen Compounds Notice the systematic trend in melting point and boiling point as the molar mass increases. Water is a striking exception because of strong intermolecular hydrogen bonds.

Heats of Fusion and Vaporization

Water also has surprisingly high values for heat of fusion and heat of vaporization. This is illustrated in Table 13.6. When we ignore water, we notice that as the molar mass of H_2S through H_2Te increases, the values for the heat of fusion and heat of vaporization increase. The explanation for this trend is that, as the molecular size increases, the attractive forces increase slightly. Therefore, more energy is required to melt a solid or vaporize a liquid. The unusually high values for water are due to hydrogen bonding.

13.9 Chemical Properties of Water

Objective · To write balanced chemical equations for reactions that involve water.

Recall that in Chapter 8 we studied five basic types of chemical reactions. Water is usually only a solvent for these chemical reactions, although it does react under selected conditions. One such condition is the **electrolysis** of water. For example, passing an electric current through water decomposes H_2O into hydrogen and oxygen gases. From the balanced chemical equation, we notice that two volumes of hydrogen are produced for every volume of oxygen.

$$2\,H_2O(l) \xrightarrow{\text{electricity}} 2\,H_2(g) + O_2(g)$$

One of the five basic types of reactions is a replacement reaction. In this reaction an active metal (Li, Na, K, Ca, Sr, or Ba) reacts directly with water to give a metal hydroxide and hydrogen gas. These reactions occur rapidly at room temperature. At 25°C, potassium metal reacts violently with water as follows.

$$2\ K(s) + 2\ H_2O(l) \longrightarrow 2\ KOH(aq) + H_2(g)$$

▲ **Potassium in Water** The reaction of potassium metal in water produces hydrogen gas, which is flammable and gives a violet flame test for potassium.

The oxides of many metals can react with water to yield a metal hydroxide. Hydroxide compounds are said to be basic or alkaline. Since a **metal oxide** reacts with water to yield a basic solution, a metal oxide is referred to as a basic oxide. For example, calcium oxide reacts with water as follows.

$$CaO(s) + H_2O(l) \longrightarrow Ca(OH)_2(aq) \qquad \text{(a base)}$$

The oxides of most nonmetals react with water to yield an acidic solution. Since a **nonmetal oxide** reacts with water to yield an acid, a nonmetal oxide is referred to as an acidic oxide. For example, carbon dioxide reacts with water as follows.

$$CO_2(g) + H_2O(l) \longrightarrow H_2CO_3(aq) \qquad \text{(an acid)}$$

Reactions that Produce Water

Water is produced by several types of reactions. The simplest reaction is the formation of water directly from hydrogen and oxygen. In this reaction, hydrogen and oxygen gases react to give H_2O. The reaction takes place very slowly at room temperature, but explosively if exposed to a flame. From the balanced chemical equation, we note that 2 volumes of hydrogen react with 1 volume of oxygen.

$$2\ H_2(g) + O_2(g) \xrightarrow{\text{spark}} 2\ H_2O(l)$$

Another reaction that produces water is the combustion of hydrocarbons. Hydrocarbons are organic compounds containing hydrogen and carbon. They burn in oxygen to give carbon dioxide and water. For example, propane, C_3H_8, undergoes combustion as follows.

$$C_3H_8(g) + 5\ O_2(g) \xrightarrow{\text{spark}} 3\ CO_2(g) + 4\ H_2O(g)$$

Hydrocarbons containing oxygen also undergo combustion to give carbon dioxide and water. Ethanol, C_2H_6O, for example, is currently blended with gasoline to produce gasohol. It undergoes combustion to give carbon dioxide and water as follows.

$$C_2H_6O(g) + 3\ O_2(g) \xrightarrow{\text{spark}} 2\ CO_2(g) + 3\ H_2O(g)$$

Recall that neutralization reactions also produce water. An acid neutralizes a base to produce an aqueous salt and water. For example, battery acid, H_2SO_4, reacts with aqueous lye, NaOH, to produce sodium sulfate and water.

$$H_2SO_4(aq) + 2\ NaOH(aq) \longrightarrow Na_2SO_4(aq) + 2\ H_2O(l)$$

The decomposition of a hydrate compound also produces water. A **hydrate** is a crystalline compound that contains a specific number of water molecules attached to an ionic formula unit. Gypsum is a hydrate of calcium sulfate, $CaSO_4$. The formula $CaSO_4 \cdot 2H_2O$ indicates that two water molecules are attached to each formula unit. Heating a hydrate releases water from the compound. For example, heat decomposes gypsum to give $CaSO_4$ and two molecules of water.

$$CaSO_4 \cdot 2H_2O(s) \xrightarrow{\Delta} CaSO_4(s) + 2\ H_2O(g)$$

Chemistry Connection · Heavy Water

What are the name and formula for heavy water?

Most hydrogen atoms have one proton as their nuclei. However, about 1 in 6000 hydrogen atoms has both a proton and a neutron in its nucleus. This isotope of hydrogen is called deuterium (symbol, D). The mass of deuterium is about twice that of hydrogen. When deuterium atoms replace the hydrogen atoms in water, the resulting compound is called **heavy water**. The systematic name of heavy water is deuterium oxide, and its formula is D_2O.

Heavy water was discovered by the American chemist Harold Urey, who won the Nobel prize in 1934 for his discovery. Heavy water is colorless, odorless, and tasteless, but animals find it toxic. The physical and chemical properties of heavy water are similar to those of ordinary light water. The following table lists the properties of light water and heavy water.

Owing to its greater density, heavy water is used as a moderator and coolant in nuclear fission reactors. When an atomic nucleus splits apart, that is, undergoes fission, it releases neutrons that must be slowed down to cause further nuclear fission in a radioactive substance such as uranium-235. Heavy water is ideal for slowing down neutrons released during the fission process.

Properties of Water versus Heavy Water

Property	Light Water, H_2O	Heavy Water, D_2O
appearance	colorless, odorless liquid	colorless, odorless liquid
molar mass	18.02 g/mol	20.03 g/mol
density	1.000 g/mL at 4°C	1.105 g/mL at 4°C
freezing point	0.00°C	3.82°C
boiling point	100.00°C	101.42°C
heat of fusion	1436 cal/mol	1516 cal/mol

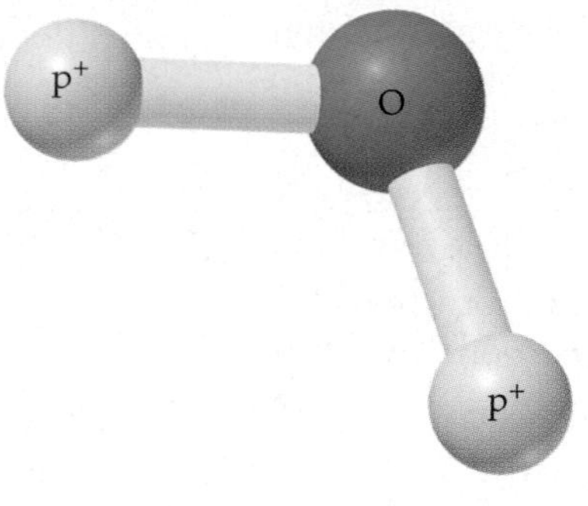

▲ Water, H_2O

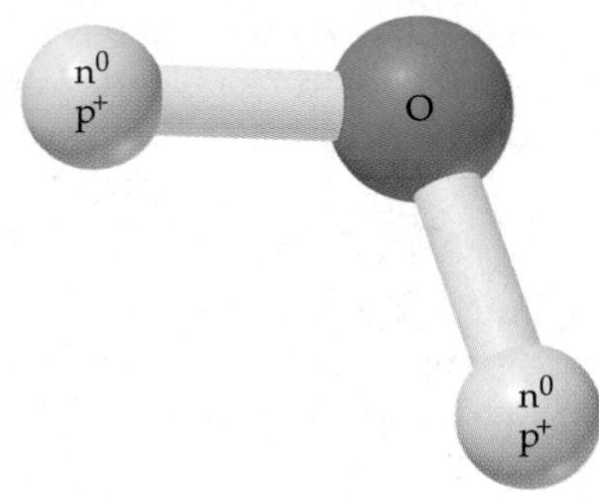

▲ Heavy water, D_2O

Heavy water is named deuterium oxide, and its formula is D_2O.

13.10 Hydrates

Experiment #13, Prentice Hall Laboratory Manual

Objectives
- To calculate the percentage of water in a hydrate.
- To calculate the water of hydration for a hydrate.

As mentioned in Section 13.9, a **hydrate** is a crystalline ionic compound containing water. Each formula unit in the hydrate has a specific number of water molecules attached to it. Common examples of hydrates include borax, $Na_2B_4O_7 \cdot 10H_2O$ and epsom salts, $MgSO_4 \cdot 7H_2O$. The dot (·) in the hydrate formula indicates that water molecules are bonded directly to each unit of hydrate. In epsom salts, $MgSO_4 \cdot 7H_2O$, for example, seven molecules of water are attached to each formula unit of $MgSO_4$.

Heating a hydrate produces an **anhydrous** compound and water. For example, when we heat copper(II) sulfate, it decomposes to give anhydrous copper(II) sulfate and water. The equation for the reaction is

$$CuSO_4 \cdot 5H_2O(s) \xrightarrow{\Delta} CuSO_4(s) + 5\,H_2O(l)$$

◀ **Figure 13.15 Hydrate and Anhydrous Copper Sulfate** Copper(II) sulfate pentahydrate, $CuSO_4 \cdot 5H_2O$, crystals are deep-blue, whereas anhydrous copper(II) sulfate, $CuSO_4$, is a white powder. A few drops of water produce the blue color of the hydrate (right).

The water molecules in the hydrate are referred to as the **water of hydration** (or water of crystallization). Thus, the number of waters of hydration for copper(II) sulfate is 5. Figure 13.15 shows the hydrate and anhydrous forms of copper(II) sulfate.

IUPAC prescribes rules for the nomenclature of hydrate compounds. According to these rules, the name of the anhydrous compound appears first, followed by the number of waters of hydration (as indicated by a Greek prefix) and the word "hydrate." For example, gypsum, $CaSO_4 \cdot 2H_2O$, is systematically named calcium sulfate dihydrate; the *di-* indicates two waters of hydration. The following example exercise further illustrates how to name hydrate compounds.

▲ **Turquoise** The natural mineral turqoise is an example of a tetrahydrate compound that contains copper, $CuAl_6(PO_4)_4(OH)_8 \cdot 4H_2O$.

Example Exercise 13.6 • Naming Hydrates

Supply a systematic name for each of the following hydrate compounds.

(a) $CaCl_2 \cdot 6H_2O$ (b) $FeSO_4 \cdot H_2O$

Solution

First, name the anhydrous compound and then indicate the water of hydration by a Greek prefix. (Refer to Table 7.4 if necessary.)

(a) $CaCl_2$ is a binary ionic compound. It is named calcium chloride. The Greek prefix for 6 is *hexa-*. Thus, the name of the hydrate is *calcium chloride hexahydrate*.

(b) $FeSO_4$ is a ternary ionic compound. Since iron has a variable ionic charge, it can be named using either the Stock system or the Latin system. Thus, $FeSO_4$ is named iron(II) sulfate or ferrous sulfate. There is 1 water, and so the hydrate is named *iron(II) sulfate monohydrate*, or *ferrous sulfate monohydrate*.

Self-Test Exercise

Supply a systematic name for each of the following hydrate compounds.

(a) zinc sulfate heptahydrate (b) sodium chromate tetrahydrate

Answers: (a) $ZnSO_4 \cdot 7H_2O$; (b) $Na_2CrO_4 \cdot 4H_2O$

Note According to IUPAC nomenclature rules, hydrates can also be named by indicating the water of hydration by number. For example, $MgSO_4 \cdot 7H_2O$ can be named either magnesium sulfate-7-water or magnesium sulfate heptahydrate.

Percent Composition of a Hydrate

In Section 9.7 we studied the percent composition of compounds. Here, we see that the percent composition of a hydrate is the ratio of the mass of water to the mass of the hydrate, all multiplied by 100.

$$\frac{\text{mass of water}}{\text{mass of hydrate}} \times 100 = \%\ H_2O$$

As an example, we'll find the percentage of water in gypsum, $CaSO_4 \cdot 2H_2O$. Using the periodic table, we find the molar mass of H_2O to be 18.02 g, and that of $CaSO_4$, 136.15 g. We find the percentage of water from the following ratio.

$$\frac{2(18.02)\ \cancel{g}}{136.15\ \cancel{g} + 2(18.02)\ \cancel{g}} \times 100 = 20.93\%$$

Notice that in the numerator we multiplied the molar mass of water by 2 because the compound is a dihydrate. In the denominator, we added twice the molar mass of water to that of the anhydrous compound. The following example exercise provides additional practice in calculating the percent composition of a hydrate.

Example Exercise 13.7 • Percentage of Water in a Hydrate

Calculate the percentage of water in each of the following hydrates.

(a) $CuSO_4 \cdot 5H_2O$ (b) $Na_2B_4O_7 \cdot 10H_2O$

Solution

In each example, first obtain the molar mass of the anhydrous compound from the periodic table. The molar mass of water is 18.02 g.

(a) The molar mass of $CuSO_4$ is 159.62 g (63.55 g + 32.07 g + 64.00 g). Since the hydrate has 5 waters of hydration, we have

$$\frac{5\ (18.02)\ \cancel{g}}{159.62\ \cancel{g} + 5\ (18.02)\ \cancel{g}} \times 100 = \%\ H_2O$$

$$= 36.08\%$$

(b) The molar mass of $Na_2B_4O_7$ is 201.22 g (45.98 g + 43.24 g + 112.00g). The hydrate has 10 waters of crystallization. Therefore, the ratio is

$$\frac{10\ (18.02)\ \cancel{g}}{201.22\ \cancel{g} + 10\ (18.02)\ \cancel{g}} \times 100 = \%\ H_2O$$

$$= 47.24\%$$

Self-Test Exercise

Calculate the percentage of water in each of the following hydrates.

(a) $NaC_2H_3O_2 \cdot 3H_2O$ (b) $Na_2S_2O_3 \cdot 5H_2O$

Answers: (a) 39.72%; (b) 36.30%

Determining the Formula of a Hydrate

In Section 9.8 we calculated the empirical formula for a compound from its percent composition. To determine the water of hydration for a hydrate we will proceed in a similar fashion.

The empirical formula of a compound is the simplest whole-number ratio of its elements. Similarly, the empirical formula of a hydrate is the simplest whole-number ratio of water molecules to the anhydrous compound. In the formula for washing soda, $Na_3PO_4 \cdot XH_2O$, X represents the water of hydration. To determine the value of X we must know the percent composition. Let's assume that $Na_3PO_4 \cdot XH_2O$ is found by experiment to contain 52.3% water. The equation for the decomposition reaction is as follows:

$$Na_3PO_4 \cdot XH_2O(s) \xrightarrow{\Delta} Na_3PO_4(s) + XH_2O(l)$$

Since the hydrate contains 52.3% water, the percentage of Na_3PO_4 is 47.7% (100.0 − 52.3 = 47.7%). As in an empirical formula, let's assume we have a 100.0-g sample of hydrate. Therefore, our sample has 52.3 g of water and 47.7 g of anhydrous compound. The next step is to calculate the moles of water.

$$52.3 \cancel{\text{g } H_2O} \times \frac{1 \text{ mol } H_2O}{18.02 \cancel{\text{g } H_2O}} = 2.90 \text{ mol } H_2O$$

From the periodic table, we find the molar mass of Na_3PO_4 to be 163.94 g. The moles of anhydrous compound are

$$47.7 \cancel{\text{g } Na_3PO_4} \times \frac{1 \text{ mol } Na_3PO_4}{163.94 \cancel{\text{g } Na_3PO_4}} = 0.291 \text{ mol } Na_3PO_4$$

We can write the mole ratio of the hydrate as $Na_3PO_4 \cdot (2.90/0.291)H_2O$, where the water of hydration is 2.90/0.291. This ratio simplifies to 10/1, meaning that the water of hydration is 10 and the formula of the hydrate is $Na_3PO_4 \cdot 10H_2O$.

Example Exercise 13.8 • Water of Crystallization for a Hydrate

Determine the water of crystallization for the hydrate of magnesium iodide. In an experiment, $MgI_2 \cdot XH_2O$ was found to contain 34.0% water.

Solution

We begin by writing an equation for the decomposition.

$$MgI_2 \cdot XH_2O(s) \xrightarrow{\Delta} MgI_2(s) + XH_2O(l)$$

Since the hydrate contains 34.0% water, the percentage of MgI_2 is 66.0% (100.0 − 34.0 = 66.0%). Assume that we have a 100.0-g sample of hydrate. Therefore, we have 34.0 g of water and 66.0 g of anhydrous compound. The moles of water are

$$34.0 \cancel{\text{g } H_2O} \times \frac{1 \text{ mol } H_2O}{18.02 \cancel{\text{g } H_2O}} = 1.89 \text{ mol } H_2O$$

The molar mass of MgI_2 is 278.11 g (24.31 g + 253.80 g). We find the moles of anhydrous compound as follows.

$$66.0 \cancel{\text{g } MgI_2} \times \frac{1 \text{ mol } MgI_2}{278.11 \cancel{\text{g } MgI_2}} = 0.237 \text{ mol } MgI_2$$

We can write the mole ratio of the hydrate as $MgI_2 \cdot (1.89/0.237)H_2O$. The ratio 1.89/0.237 reduces to 7.97 and rounds off to the whole number 8. The water of crystallization is 8, and the formula is $MgI_2 \cdot 8H_2O$. The name of the hydrate is magnesium iodide *octahydrate.*

Self-Test Exercise

Determine the water of crystallization for the hydrate of copper(II) fluoride. In an experiment, $CuF_2 \cdot XH_2O$ was found to contain 26.2% water.

Answer: $CuF_2 \cdot 2H_2O$

Chemistry Connection · Water Purification

Which ions are removed from hard water to produce soft water?

In some areas of the country, the minerals dissolved in water give it a high concentration of various ions that make the water suitable neither for drinking nor for agriculture. Such water is called **hard water**. Sometimes the mineral content of hard water is so great that it causes plumbing and corrosion problems. Hard water typically contains a high concentration of the following ions: Ca^{2+}, Mg^{2+}, Fe^{3+}, Cl^-, CO_3^{2-}, SO_4^{2-}, and PO_4^{3-}.

In homes having hard water, soap rings are sometimes found in bathtubs. A soap ring is formed by the cations in hard water reacting with soap to create a compound that deposits as an insoluble film. This problem can be eliminated by removing Ca^{2+}, Mg^{2+}, and Fe^{3+} from hard water by passing the water through a water softener. A water softener replaces Ca^{2+}, Mg^{2+}, and Fe^{3+} with Na^+; compounds containing Na^+ are generally soluble. Note, however, that although a water softener removes ions, **soft water** still has a high concentration of other ions. In addition to Na^+, soft water contains Cl^-, CO_3^{2-}, SO_4^{2-}, and PO_4^{3-}.

Even soft water can interfere with the results of a chemical analysis. Therefore, chemists routinely use water that contains no ions. They purify water by removing the minerals using an ion exchange system. Water purified by this method is called **deionized water**, or demineralized water. Water is deionized by passing it through a resin that has both cation and anion exchange components. First, cations such as Na^+ in the water are exchanged for hydrogen ions on the resin. Second, anions such as Cl^- in the water are exchanged for hydroxide ions.

$$Na^+(aq) + H(resin) \longrightarrow Na(resin) + H^+(aq)$$
$$Cl^-(aq) + (resin)OH \longrightarrow (resin)Cl + OH^-(aq)$$

Notice that the ion exchange resin produces both hydrogen ions and hydroxide ions which can readily combine to give water.

$$H^+(aq) + OH^-(aq) \longrightarrow H_2O(l)$$

The net result is that the resin removes *all ions* from water passing through the deionizing system.

Ion Exchange for Deionizing Water

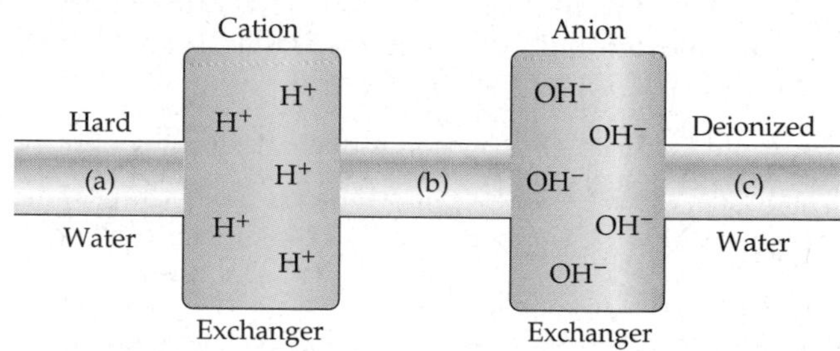

▲ Hard water is softened by exchanging Na^+ for Ca^{2+}, Mg^{2+}, and Fe^{3+}.

Hard water is softened by exchanging Na^+ for Ca^{2+}, Mg^{2+}, and Fe^{3+}.

Summary

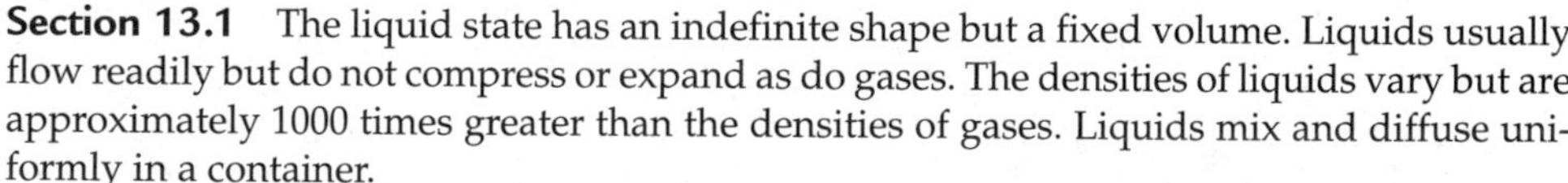

Section 13.1 The liquid state has an indefinite shape but a fixed volume. Liquids usually flow readily but do not compress or expand as do gases. The densities of liquids vary but are approximately 1000 times greater than the densities of gases. Liquids mix and diffuse uniformly in a container.

Section 13.2 There are three basic types of intermolecular attraction: dispersion forces, dipole forces, and hydrogen bonds. A **dispersion force** has the weakest attraction and results from temporary dipoles in molecules. A **dipole force** has a stronger attraction and results from permanent dipoles in molecules. A **hydrogen bond** exerts the strongest attraction and occurs between molecules having H—O or H—N bonds.

Section 13.3 When the **vapor pressure** of a liquid equals the atmospheric pressure, the liquid is at its **boiling point**. The normal boiling point of a liquid is the temperature at which the vapor pressure is 760 mm Hg. Liquids have a resistance to flow, and this property is called **viscosity**. The attraction between molecules at the surface of a liquid is called **surface tension**. Each of these properties is affected by the degree of attraction between molecules in the liquid. If the attraction is low, the vapor pressure is high. Conversely, if the attraction is high, the values for the boiling point, viscosity, and surface tension are also high.

Section 13.4 The solid state has a definite shape and a fixed volume. A solid that has a highly defined structure is called a **crystalline solid**. Solids do not compress or expand to any large degree. The density of a substance in the solid state is usually higher than that in the liquid state. Water and ammonia are two interesting exceptions to this rule. Solid ice and solid ammonia float on their respective liquids. Since the solid state contains particles that are fixed, solids do not mix or diffuse.

Section 13.5 There are three basic types of crystalline solids. **Ionic solids** are made up of ions. The ions are attracted to each other and form repeating geometric patterns. **Molecular solids** form crystals made up of molecules. **Metallic solids** are made up of metal atoms arranged in a definite pattern. The properties of ionic, molecular, and metallic solids differ. Metallic solids are good conductors of electricity. Ionic solids conduct electricity only when they are melted or in an aqueous solution. As a rule, molecular solids do not conduct electricity.

Section 13.6 Heat is necessary to change the physical state of a substance. The heat required to melt a substance is called the **heat of fusion**. Heat released when a substance freezes to a solid is called the *heat of solidification*. The heat of solidification of a substance has the same value as its heat of fusion. The heat required to convert a liquid to a gas at its boiling point is the **heat of vaporization**. When a vapor condenses to a liquid, it releases the same amount of heat energy that was necessary to vaporize the liquid. This energy is termed the *heat of condensation.*

Section 13.7 Water has unusual properties because of the two polar O—H bonds separated by a **bond angle** of 104.5°. The oxygen atom is the focus of a partial negative charge, and each hydrogen atom has a partial positive charge. The two polar bonds give rise to an overall **net dipole** for the water molecule.

Section 13.8 Hydrogen bonds between water molecules explain why water has extraordinarily high values for boiling point, viscosity, and surface tension. Hydrogen bonding also explains why the density of ice is less than that of water. For a small molecule, water has an unusually high melting point, boiling point, heat of fusion, and heat of vaporization.

Section 13.9 Water undergoes the following chemical reactions. (1) An electric current decomposes water into hydrogen and oxygen gases by **electrolysis**. (2) The active metals of Groups IA/1 and IIA/2 react with water to give a metal hydroxide and hydrogen gas. (3) A **metal oxide** reacts with water to give a metal hydroxide. (4) A **nonmetal oxide** combines with water to yield an acidic solution.

Table 13.7 Summary of the Reactions of Water

water	$\xrightarrow{\text{electricity}}$	hydrogen + oxygen
active metal + water	$\longrightarrow$	metal hydroxide + hydrogen
metal oxide + water	$\longrightarrow$	basic solution
nonmetal oxide + water	$\longrightarrow$	acidic solution

Water is produced from (1) the reaction of hydrogen and oxygen gases, (2) burning hydrocarbons, (3) neutralizing an acid with a base, and (4) heating a hydrate.

Table 13.8 Summary of the Reactions Producing Water

hydrogen + oxygen	$\xrightarrow{\text{spark}}$	water
hydrocarbon + oxygen	$\xrightarrow{\text{spark}}$	carbon dioxide + water
acid + base	$\longrightarrow$	salt + water
hydrate	$\xrightarrow{\Delta}$	anhydrous compound + water

Section 13.10 The name of a **hydrate** consists of the name of the **anhydrous** compound followed by a Greek prefix and the word "hydrate"; for example, $BaCl_2 \cdot 2H_2O$ is named barium chloride dihydrate. Determining the formula of a hydrate is similar to determining an empirical formula. From the percent composition of a hydrate, we can calculate the moles of water and anhydrous compound. The ratio of moles of water to moles of anhydrous compound gives the **water of hydration**.

Key Concepts*

1. Consider the following visual analogy. A beaker of marbles is covered with honey; the marbles move about slowly and randomly as they are shaken and overcome their stickiness. Which physical state is described by the analogy?
2. Explain why hard boiling an egg takes 10 minutes at 10,000 feet elevation and only 5 minutes at sea level.
3. Explain how the needle in the following photograph can float on water when the density of steel is many times greater than the density of water.

▲ Needle Floating on Water

4. Notice that in the following photograph water (left) has a concave lens at the top of the liquid, whereas mercury (right) has a convex lens. Explain this observation in terms of the attractive forces operating in a liquid.

▲ Water, H_2O ▲ Mercury, Hg

5. Which type of crystalline solid is hard and brittle, has a high melting point, and conducts electricity only when melted? Which type of crystalline solid has a low melting point, is insoluble in water, and is a nonconductor of electricity?
6. Calculate the number of calories necessary to convert 20.0 g of solid ice at its melting point to steam at 100°C.
7. A hydrate of sodium acetate is heated to give 39.7% water. What is the chemical formula for the hydrate?

Key Terms† 

Select the key term below that corresponds to each of the following definitions.

______ 1. an intermolecular attraction based on temporary dipoles
______ 2. an intermolecular attraction based on permanent dipoles
______ 3. an attraction between two molecules that each have a hydrogen atom bonded to an oxygen or a nitrogen atom
______ 4. the pressure exerted by vapor molecules above a liquid in a closed container when the rates of evaporation and condensation are equal
______ 5. the temperature at which the vapor pressure of a liquid is equal to the atmospheric pressure
______ 6. the resistance of a liquid to flow
______ 7. the tendency of a liquid to form spherical drops
______ 8. a solid substance composed of ions or molecules that repeat in a regular geometric pattern
______ 9. a crystalline solid composed of ions that repeat in a regular pattern
______ 10. a crystalline solid composed of molecules that repeat in a regular pattern
______ 11. a crystalline solid composed of metal atoms that repeat in a regular pattern
______ 12. the heat required to convert a solid to a liquid at its melting point

(a) anhydrous (*Sec. 13.10*)
(b) boiling point (*Sec. 13.3*)
(c) bond angle (*Sec. 13.7*)
(d) crystalline solid (*Sec. 13.4*)
(e) deionized water (*Sec. 13.10*)
(f) dipole force (*Sec. 13.2*)
(g) dispersion force (*Sec. 13.2*)
(h) electrolysis (*Sec. 13.9*)
(i) hard water (*Sec. 13.10*)
(j) heat of fusion (*Sec. 13.6*)
(k) heat of vaporization (*Sec. 13.6*)
(l) heavy water (*Sec. 13.9*)
(m) hydrate (*Sec. 13.10*)
(n) hydrogen bond (*Sec. 13.2*)
(o) ionic solid (*Sec. 13.5*)

* Answers to Key Concepts are in Appendix H.
† Answers to Key Terms are in Appendix I.

_____ 13. the heat required to vaporize a liquid to a gas at its boiling point

_____ 14. the angle formed by two atoms bonded to a central atom in a molecule

_____ 15. the overall direction of partial negative charge in a molecule having two or more dipoles

_____ 16. the chemical reaction produced by the passage of electric current through an aqueous solution

_____ 17. a compound that reacts with water to form a basic solution; also termed a basic oxide

_____ 18. a compound that reacts with water to form an acidic solution; also termed an acidic oxide

_____ 19. a molecule of water in which ordinary hydrogen atoms have been replaced by hydrogen atoms having a neutron

_____ 20. a substance that contains a specific number of water molecules attached to a formula unit in a crystalline compound

_____ 21. refers to a compound that does not contain water

_____ 22. the number of water molecules bound to a formula unit in a hydrate; also termed the water of crystallization

_____ 23. water containing a variety of cations and anions such as Ca^{2+}, Mg^{2+}, Fe^{3+}, CO_3^{2-}, SO_4^{2-}, and PO_4^{3-}

_____ 24. water containing sodium ions and a variety of anions

_____ 25. water purified by removing ions using an ion exchange method; also termed demineralized water

(p) metal oxide (*Sec. 13.9*)
(q) metallic solid (*Sec. 13.5*)
(r) molecular solid (*Sec. 13.5*)
(s) net dipole (*Sec. 13.7*)
(t) nonmetal oxide (*Sec. 13.9*)
(u) soft water (*Sec. 13.10*)
(v) surface tension (*Sec. 13.3*)
(w) vapor pressure (*Sec. 13.3*)
(x) viscosity (*Sec. 13.3*)
(y) water of hydration (*Sec. 13.10*)

Exercises‡

Properties of Liquids (Sec. 13.1)

1. List five general properties of the liquid state.

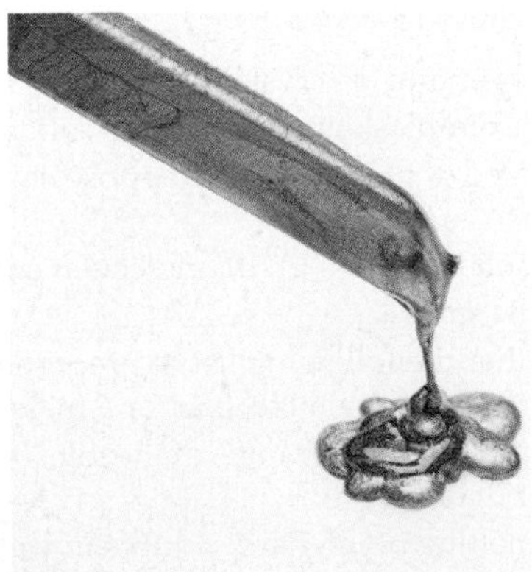

▲ **The Liquid State** As a substance changes from a solid to a liquid, it begins to flow.

2. Distinguish between a liquid and a gas at the molecular level.

3. Indicate the physical state (solid, liquid, or gas) for each of the following at the designated temperature.
   (a) H_2O at −20.0°C (b) H_2O at 120.0°C
   (c) NH_3 at −195.0°C (d) NH_3 at 0.0°C
   (e) $CHCl_3$ at −55.5°C (f) $CHCl_3$ at 100.0°C
   The melting points and boiling points for water, ammonia, and chloroform are as follows.

Element	Melting Point	Boiling Point
water, H_2O	0.0°C	100.0°C
ammonia, NH_3	−77.7°C	−33.4°C
chloroform, $CHCl_3$	−63.5°C	61.7°C

4. Indicate the physical state (solid, liquid, or gas) for each of the following noble gases at the designated temperature.
   (a) Ne at −248°C (b) Ne at −225°C
   (c) Ar at −187°C (d) Ar at −212°C
   (e) Kr at −100°C (f) Kr at −195°C
   The melting points and boiling points for neon, argon, and krypton are as follows.

Element	Melting Point	Boiling Point
neon, Ne	−248.7°C	−245.9°C
argon, Ar	−189.2°C	−185.7°C
krypton, Kr	−156.6°C	−152.3°C

‡Answers to odd-numbered Exercises are in Appendix J.

The Intermolecular Bond Concept (Sec. 13.2)

5. Which type of intermolecular attraction (dispersion force, dipole force, or hydrogen bond) exists in each of the following liquids?
   (a) C_8H_{18} (b) $CH_3—OH$
   (c) $CH_3—Cl$ (d) $CH_3—O—CH_3$
6. Which type of intermolecular attraction (dispersion force, dipole force, or hydrogen bond) exists in each of the following liquids?
   (a) $C_2H_5—O—CH_3$ (b) $C_2H_5—F$
   (c) C_4H_{10} (d) $HCOOH$
7. Predict which liquid in each pair has the higher vapor pressure.
   (a) CH_3COOH or C_2H_5Cl (b) C_2H_5OH or CH_3OCH_3
8. Predict which liquid in each pair has the higher boiling point.
   (a) CH_3COOH or C_2H_5Cl (b) C_2H_5OH or CH_3OCH_3
9. Predict which liquid in each pair has the higher viscosity.
   (a) CH_3COOH or C_2H_5Cl (b) C_2H_5OH or CH_3OCH_3
10. Predict which liquid in each pair has the higher surface tension.
   (a) CH_3COOH or C_2H_5Cl (b) C_2H_5OH or CH_3OCH_3

Vapor Pressure, Boiling Point, Viscosity, Surface Tension (Sec. 13.3)

11. Define and illustrate the concept of vapor pressure using water as an example.
12. Define and illustrate the concept of boiling point using water as an example.
13. Define and illustrate the concept of viscosity using water as an example.
14. Define and illustrate the concept of surface tension using water as an example.
15. If the molecules in a liquid are strongly attracted, which of the following properties has a high value?
   (a) vapor pressure (b) boiling point
   (c) viscosity (d) surface tension
16. If the molecules in a liquid are weakly attracted, which of the following properties has a high value?
   (a) vapor pressure (b) boiling point
   (c) viscosity (d) surface tension
17. What is the general relationship between the vapor pressure of a liquid and its temperature?
18. What is the general relationship between the boiling point of a liquid and its vapor pressure?
19. Refer to Figure 13.5 and estimate the vapor pressure of ether at each of the following temperatures.
   (a) 15°C (b) 30°C
20. Refer to Figure 13.5 and estimate the vapor pressure of ethanol at each of the following temperatures.
   (a) 30°C (b) 75°C
21. The vapor pressure of acetone is 1 torr at −59°C, 10 torr at −31°C, 100 torr at 8°C, 400 torr at 40°C, and 760 torr at 56°C. What is the normal boiling point of acetone?
22. The vapor pressure of methanol is 1 atm at 65°C, 2 atm at 84°C, 5 atm at 112°C, 10 atm at 138°C, and 20 atm at 168°C. What is the normal boiling point of methanol?

Properties of Solids (Sec. 13.4)

23. List five general properties of the solid state.
24. Distinguish between a solid and a liquid at the molecular level.
25. Indicate the physical state (solid, liquid, or gas) for each of the following metals after being placed in ice water or boiling water.
   (a) Ga in ice water (b) Ga in boiling water
   (c) Sn in ice water (d) Sn in boiling water
   (e) Hg in ice water (f) Hg in boiling water

   The melting points and boiling points for the three metals are as follows.

Element	Melting Point	Boiling Point
gallium, Ga	29.8°C	2403°C
tin, Sn	232.0°C	2270°C
mercury, Hg	−38.9°C	357°C

26. Indicate the physical state (solid, liquid, or gas) at the designated temperature for each of the following Group VIA/16 hydrogen compounds. Refer to Table 13.6 for melting point and boiling point data.
   (a) H_2S at −75.0°C (b) H_2S at −50.0°C
   (c) H_2Se at −50.0°C (d) H_2Se at −25.0°C
   (e) H_2Te at −51.5°C (f) H_2Te at 0.0°C

Crystalline Solids (Sec. 13.5)

27. List three examples of crystalline solids.
28. List three examples of noncrystalline solids.
29. State the type of particles that compose each of the following.
   (a) ionic solid (b) molecular solid
   (c) metallic solid
30. State whether the following list of properties is most descriptive of an ionic, a molecular, or a metallic solid.
   (a) wide melting point range, malleable, ductile, conductor of electricity
   (b) high melting point, hard, soluble in water, conductor of electricity when melted
   (c) low melting point, generally insoluble in water, nonconductor of electricity
31. Classify each of the following crystalline solids as ionic, molecular, or metallic.
   (a) zinc, Zn (b) zinc oxide, ZnO
   (c) phosphorus, P_4 (d) iodine monobromide, IBr
32. Classify each of the following crystalline solids as ionic, molecular, or metallic.
   (a) sulfur, S_8 (b) sulfur dioxide, SO_2
   (c) silver, Ag (d) silver nitrate, $AgNO_3$

▲ **Silver** The silver crystals forming on the copper wire are an example of a crystalline metallic solid.

Changes of Physical State (Sec. 13.6)

33. Draw the general shape of the temperature–energy graph for the heating of ethanol from −120° to 120°C. (*Given:* Mp = −117.3°C; Bp = 78.5°C)
34. Draw the general shape of the temperature–energy graph for the cooling of acetone from 100° to −100°C. (*Given:* Mp = −95.4°C; Bp = 56.2°C)
35. Calculate the amount of heat required to melt 125 g of ice at 0°C.
36. Calculate the amount of heat released when 75.5 g of steam condenses to a liquid at 100°C.
37. Calculate the amount of heat required to convert 25.0 g of water at 25.0°C to steam at 100.0°C.
38. Calculate the amount of heat released when 65.5 g of water at 55.5°C cools to ice at 0.0°C.
39. Calculate the amount of heat required to convert 115 g of ice at 0.0°C to steam at 100.0°C.
40. Calculate the amount of heat released when 155 g of steam at 100.0°C cools to ice at 0.0°C.
41. Calculate the amount of heat required to convert 38.5 g of ice at −20.0°C to steam at 100.0°C.
42. Calculate the amount of heat released when 90.5 g of steam at 110.0°C cools to ice at 0.0°C.
43. Calculate the amount of heat required to convert 100.0 g of ice at −40.0°C to steam at 125.0°C.
44. Calculate the amount of heat released when 0.500 kg of steam at 150.0°C cools to ice at −50.0°C.

Structure of Water (Sec. 13.7)

45. How many bonding and nonbonding electron pairs are in a water molecule?
46. Draw the electron dot and structural formulas for a molecule of water.
47. What is the observed bond angle in a water molecule?
48. The center of a tetrahedron forms an angle of 109° with its corners. If an oxygen atom is at the center of the tetrahedron and a hydrogen atom is at two of the corners, explain why the observed bond angle is less that 109°.
49. Indicate the two dipoles in a water molecule using the delta convention.
50. Draw the net dipole in a molecule of water.
51. Draw two molecules of hydrogen fluoride, HF, and diagram a hydrogen bond.
52. Draw two molecules of ammonia, NH_3, and diagram a hydrogen bond.

Physical Properties of Water (Sec. 13.8)

53. The density of solid ammonia is less than that of liquid ammonia. Does an "ammonia ice cube" float or sink?
54. A soft drink bottle is accidentally filled completely and then capped. What will happen if the soft drink is frozen solid?
55. Without referring to Table 13.6, predict which compound in each of the following pairs has the higher melting point.
    (a) H_2O or H_2S **(b)** H_2S or H_2Se
56. Without referring to Table 13.6, predict which compound in each of the following pairs has the higher boiling point.
    (a) H_2O or H_2Se **(b)** H_2S or H_2Te
57. Without referring to Table 13.6, predict which compound in each of the following pairs has the higher heat of fusion (cal/mol).
    (a) H_2O or H_2S **(b)** H_2S or H_2Se
58. Without referring to Table 13.6, predict which compound in each of the following pairs has the higher heat of vaporization (cal/mol).
    (a) H_2O or H_2Se **(b)** H_2S or H_2Te
59. In general, as the molar mass of Group VIA/16 hydrogen compounds increases, does each of the following increase or decrease?
    (a) melting point **(b)** boiling point
    (c) heat of fusion **(d)** heat of vaporization
60. Refer to the trends in Table 13.6 and estimate a value for the physical properties of radioactive H_2Po, that is, predict a value for the H_{fusion}, and H_{vapor}.

Chemical Properties of Water (Sec. 13.9)

61. Write a balanced chemical equation for the electrolysis of water.
62. Write a balanced equation for the reaction of hydrogen and oxygen gases.
63. Complete and balance the following equations.
    (a) $Li(s) + H_2O(l) \longrightarrow$
    (b) $Na_2O(s) + H_2O(l) \longrightarrow$
    (c) $CO_2(g) + H_2O(l) \longrightarrow$
64. Complete and balance the following equations.
    (a) $Cs_2O(s) + H_2O(l) \longrightarrow$
    (b) $Rb(s) + H_2O(l) \longrightarrow$
    (c) $P_2O_5(s) + H_2O(l) \longrightarrow$

65. Complete and balance the following equations.
(a) $Ba(s) + H_2O(l) \longrightarrow$
(b) $N_2O_3(g) + H_2O(l) \longrightarrow$
(c) $CaO(s) + H_2O(l) \longrightarrow$

66. Complete and balance the following equations.
(a) $Mg(s) + H_2O(l) \longrightarrow$
(b) $SrO(s) + H_2O(l) \longrightarrow$
(c) $N_2O_5(g) + H_2O(l) \longrightarrow$

67. Complete and balance the following equations.
(a) $C_3H_6(g) + O_2(g) \xrightarrow{\text{spark}}$
(b) $Na_2Cr_2O_7 \cdot 2H_2O(s) \xrightarrow{\Delta}$
(c) $HF(aq) + Ca(OH)_2(aq) \longrightarrow$

68. Complete and balance the following equations.
(a) $C_3H_6O(g) + O_2(g) \xrightarrow{\text{spark}}$
(b) $Ca(NO_3)_2 \cdot 4H_2O(s) \xrightarrow{\Delta}$
(c) $H_2CO_3(aq) + KOH(aq) \longrightarrow$

69. Complete and balance the following equations.
(a) $C_4H_{10}(g) + O_2(g) \xrightarrow{\text{spark}}$
(b) $Co(C_2H_3O_2)_2 \cdot 4H_2O(s) \xrightarrow{\Delta}$
(c) $HNO_3(aq) + Ba(OH)_2(aq) \longrightarrow$

70. Complete and balance the following equations.
(a) $C_4H_{10}O(g) + O_2(g) \xrightarrow{\text{spark}}$
(b) $KAl(SO_4)_2 \cdot 12H_2O(s) \xrightarrow{\Delta}$
(c) $H_3PO_4(aq) + NaOH(aq) \longrightarrow$

Hydrates (Sec. 13.10)

71. Supply a systematic name for each of the following hydrate compounds.
(a) $MgSO_4 \cdot 7H_2O$ (b) $Co(CN)_3 \cdot 3H_2O$
(c) $MnSO_4 \cdot H_2O$ (d) $Na_2Cr_2O_7 \cdot 2H_2O$

72. Supply a systematic name for each of the following hydrate compounds.
(a) $Sr(NO_3)_2 \cdot 6H_2O$ (b) $Co(C_2H_3O_2)_2 \cdot 4H_2O$
(c) $CuSO_4 \cdot 5H_2O$ (d) $Cr(NO_3)_3 \cdot 9H_2O$

73. Provide the formula for each of the following hydrate compounds.
(a) sodium acetate trihydrate
(b) calcium sulfate dihydrate
(c) potassium chromate tetrahydrate
(d) zinc sulfate heptahydrate

74. Provide the formula for each of the following hydrate compounds.
(a) sodium carbonate decahydrate
(b) nickel(II) nitrate hexahydrate
(c) cobalt(III) iodide octahydrate
(d) chromium(III) acetate monohydrate

75. Calculate the percentage of water in each of the following hydrates.
(a) $SrCl_2 \cdot 6H_2O$ (b) $K_2Cr_2O_7 \cdot 2H_2O$
(c) $Co(CN)_3 \cdot 3H_2O$ (d) $Na_2CrO_4 \cdot 4H_2O$

76. Calculate the percentage of water in each of the following hydrates.
(a) $MnSO_4 \cdot H_2O$ (b) $Sr(NO_3)_2 \cdot 6H_2O$
(c) $Co(C_2H_3O_2)_2 \cdot 4H_2O$ (d) $Cr(NO_3)_3 \cdot 9H_2O$

77. Determine the water of hydration for the following hydrates and write the chemical formula.
(a) $NiCl_2 \cdot XH_2O$ is found to contain 21.7% water.
(b) $Sr(NO_3)_2 \cdot XH_2O$ is found to contain 33.8% water.
(c) $CrI_3 \cdot XH_2O$ is found to contain 27.2% water.
(d) $Ca(NO_3)_2 \cdot XH_2O$ is found to contain 30.5% water.

78. Determine the water of hydration for the following hydrates and write the chemical formula.
(a) $SrCl_2 \cdot XH_2O$ is found to contain 18.5% water.
(b) $Ni(NO_3)_2 \cdot XH_2O$ is found to contain 37.2% water.
(c) $CoSO_4 \cdot XH_2O$ is found to contain 10.4% water.
(d) $Na_2B_4O_7 \cdot XH_2O$ is found to contain 30.9% water.

General Exercises

79. State the approximate percent of Earth's surface that is covered by water.

80. State the approximate percent of the human body that is water.

81. Refer to Figure 13.5 and determine the boiling point of water at an elevation where the atmospheric pressure is 650 torr.

82. Refer to Figure 13.5 and determine the boiling point of ethanol at an elevation where the atmospheric pressure is 0.5 atm.

83. The atmosphere on Venus has clouds of sulfuric acid. Explain why it is reasonable to predict that the "raindrops" on Venus are small spheres.

84. Explain why metallic solids are good conductors of electricity.

85. Ethylene glycol is a permanent antifreeze. Calculate the amount of heat released when 1250 g of liquid at 25.0°C cools to a solid at its melting point. (For ethylene glycol: specific heat = 0.561 cal/g × °C; Mp = −11.5°C; Bp = 197.6°C; H_{fusion} = 43.3 cal/g; H_{vapor} = 293 cal/g.)

86. Methanol is considered a temporary antifreeze because of its low boiling point. Calculate the amount of heat necessary to convert 1250 g of liquid at 25.0°C to a vapor at its boiling point. (For methanol: specific heat = 0.610 cal/g × °C; Mp = −97.9°C; Bp = 65.2°C; H_{fusion} = 23.7 cal/g; H_{vapor} = 293 cal/g.)

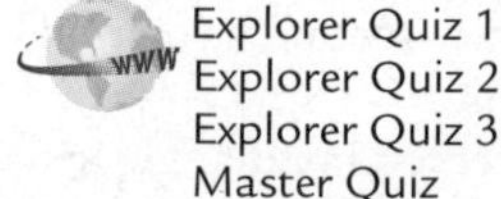

Explorer Quiz 1
Explorer Quiz 2
Explorer Quiz 3
Master Quiz

CHAPTER 14

Solutions

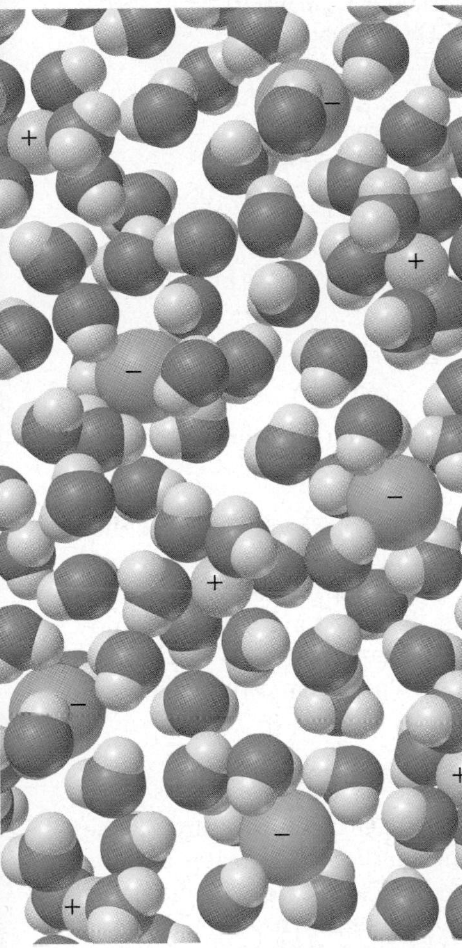

▲ In the above molecular art, one panel represents NaCl in solution and the other panel represents AgCl in solution. Which panel represents insoluble AgCl? (Na^+ shown in light gray, Ag^+ in dark gray, and Cl^- in green.)

This chapter lays an important foundation for the solution calculations found in the following chapter on Acids and Bases. The molecular art provides for the conceptual discussion of ionic and molecular substances undergoing the process of dissolution.

A solution is a homogeneous mixture. This means that a solution is the same throughout and that every sample of it has the same properties. A solution consists of a solute dissolved in a solvent. For instance, sugar dissolved in water is an aqueous solution. Other examples of solutions are carbon dioxide dissolved in a soft drink, and salt dissolved in the ocean. The relative proportions of the solute and the solvent can vary, but the solvent is always the greater quantity.

We usually think of solutions as liquids. However, solutes scattered throughout solvents exist in all three physical states; there are gaseous, liquid, and solid

Table 14.1 Types of Solutions

Solute	Solvent	Solution
Gaseous solutions		
gas	gas	humid atmosphere (water vapor in air)
liquid	gas	rainy atmosphere (water droplets in air)
solid	gas	snowy atmosphere (ice crystals in air)
Liquid solutions		
gas	liquid	carbonated drinks (CO_2 in water)
liquid	liquid	vinegar ($HC_2H_3O_2$ in water)
solid	liquid	salt water (NaCl in water)
Solid solutions		
gas	solid	sponge (air in sponge)
liquid	solid	dental fillings (Hg in Ag)
solid	solid	sterling silver (Cu in Ag)

solutions. Table 14.1 lists some common examples of solutions whose physical state corresponds to that of the solvent.

14.1 Gases in Solution

Objective · To state the effect of temperature and pressure on the solubility of a gas in a liquid.

Muriatic acid is available from the supermarket and is used to increase the acidity of swimming pools. The systematic name for muriatic acid is hydrochloric acid. It is produced by dissolving hydrogen chloride gas in water. Like muriatic acid, household ammonia is another common example of a gas dissolved in a liquid. Ammonia solutions have gaseous NH_3 dissolved in water. Champagne and soft drinks are also liquids containing dissolved gases. The dissolved gas in carbonated beverages is carbon dioxide, CO_2.

▲ **Deep-sea Divers** As divers descend, each 34 ft of depth increases the atmospheric pressure by 1 atm. Thus, at 34 ft the solubility of the air mixture in blood is two times that at sea level; at 68 ft the solubility of the air mixture is three times that at sea level.

Temperature Effects

Solutions of gases in liquids are greatly affected by changes in temperature. As the temperature increases, the kinetic energy of the solute gas becomes greater, and the gas molecules acquire more of a tendency to escape from the solvent. Thus, as the temperature goes up, the solubility of a gas in a liquid goes down.

A practical example of this principle is illustrated by a carbonated beverage. You have probably taken a beverage from the refrigerator and let it warm to room temperature. When you opened the beverage it foamed as gas escaped from the solution. Compare the foaming to what happens when you open a beverage just after taking it from the refrigerator. At colder temperatures, carbon dioxide is more soluble and foaming is minimal. Table 14.2 indicates the effect of temperature on the solubility of carbon dioxide in water.

Pressure Effects

The solubility of a gas in a liquid is strongly influenced by pressure. In 1803 the English chemist William Henry conducted experiments on the solubility of gases in liquids. He found that the solubility of a gas was proportional to the partial pressure of the gas above the liquid. This statement is known as **Henry's law**. As an example, consider carbonated beverages. When we double the partial pressure of

Table 14.2 Solubility of Carbon Dioxide in Water

Temperature	Pressure	Solubility of CO_2
Temperature effect		
0°C	1.00 atm	0.348 g/100 mL water
20°C	1.00 atm	0.176 g/100 mL water
40°C	1.00 atm	0.097 g/100 mL water
60°C	1.00 atm	0.058 g/100 mL water
Pressure effect		
0°C	1.00 atm	0.348 g/100 mL water
0°C	2.00 atm	0.696 g/100 mL water
0°C	3.00 atm	1.044 g/100 mL water

*Notice that higher temperatures decrease the solubility and that higher pressures increase the solubility.

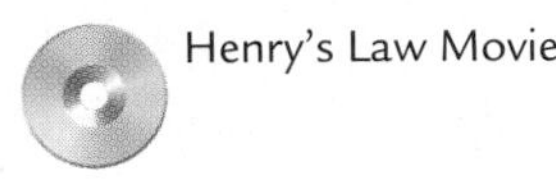

CO_2, we double the solubility. When we halve the partial pressure of CO_2, we cut the carbonation in half. Table 14.2 indicates the effect of pressure on the solubility of carbon dioxide in water.

If we examine the effect of pressure on the solubility of carbon dioxide as shown in Table 14.2, we see that the solubility doubles when the pressure increases from 1.00 atm to 2.00 atm and that the solubility triples when the pressure increases from 1.00 atm to 3.00 atm. Let's apply Henry's law and calculate the solubility of carbon dioxide when the pressure is 2.50 atm. The effect of increasing the pressure is to proportionally increase the solubility, which is given in the table as 0.348 g/100 mL water at 0°C and 1.00 atm.

$$0.348 \text{ g/100mL} \times \frac{2.50 \cancel{\text{atm}}}{1.00 \cancel{\text{atm}}} = 0.870 \text{ g/100mL}$$

Notice that the solubility of the carbon dioxide in water is given in g/100 mL. The solubility of a gas is often expressed as the volume of gas dissolved in solution. For example, the solubility of oxygen in water is listed in the *Handbook of Chemistry and Physics* as 4.78 cm^3/100 mL at 25°C and 1.00 atm. The following exercise further illustrates the application of Henry's law.

Example Exercise 14.1 • Henry's Law

Calculate the solubility of oxygen gas in water at 25°C and a partial pressure of 1150 torr. A reference book lists the solubility of oxygen in water at 25°C and 760 torr as 0.00414 g/100 mL.

Solution

Henry's law states that the solubility of oxygen gas is proportional to the partial pressure of the gas above the liquid. Since 1150 torr is greater than 760 torr, the solubility increases. We can write

$$\text{solubility} \times \text{pressure factor} = \text{new solubility}$$

$$0.00414 \text{ g/100 mL} \times \frac{1150 \cancel{\text{torr}}}{760 \cancel{\text{torr}}} = 0.00626 \text{ g/100 mL}$$

(continued)

Example Exercise 14.1 *(continued)*

Self-Test Exercise

Apply the solubility principles for a gas in a liquid to each of the following examples.

(a) One lake is at sea level, and another is at an elevation of 7500 feet. Assuming the two lakes are at the same temperature, which has the lesser concentration of oxygen dissolved from the air?

(b) A nuclear energy plant is located next to a lake in order to dissipate the heat that is produced. A neighboring lake at the same elevation is not heated by the nuclear power plant. Which lake has the lesser oxygen concentration?

Answers: (a) The lake at 7500 feet (atmospheric pressure is less at higher elevations); and (b) the lake heated by the nuclear power plant have less dissolved oxygen.

14.2 Liquids in Solution

Objective · To predict whether a liquid is soluble or insoluble in another liquid by applying the *like dissolves like* rule.

A **solution** is composed of a solute dissolved in a solvent. The **solute** is the lesser quantity, and the **solvent** the greater quantity. Let's consider the factors that affect the solubility of a liquid solute in a liquid solvent. As we will see, water is the most common solvent on Earth and every living system carries out reactions in aqueous solutions.

Dipoles

Polarity Activity

In Section 12.9, we learned that a polar molecule results from one or more polar bonds. In a water molecule, for example, the more electronegative oxygen atom has a partial negative charge (δ^-). The two hydrogen atoms are less electronegative than the oxygen atom and have a partial positive charge (δ^+). A region in a molecule having a positive and a negative charge is called a **dipole**. There are two dipoles in a water molecule. If we resolve the two dipoles into one force operating in a single direction, we have a **net dipole** for the molecule. The symbol for a net dipole is an arrow pointing toward the negative end of the molecule, as shown in Figure 14.1.

Polar and Nonpolar Solvents

A liquid composed of polar molecules is called a **polar solvent**. Water is the most common polar solvent. Although there are many exceptions, solvent molecules containing oxygen atoms, such as water, are usually polar. Methanol, CH_3OH, and ethanol, C_2H_5OH, are also examples of polar solvents.

A liquid composed of nonpolar molecules is a **nonpolar solvent**. Nonpolar solvents include hexane, C_6H_{14}, carbon tetrachloride, CCl_4, and ethyl ether, $C_4H_{10}O$. Table 14.3 lists some common polar and nonpolar solvents.

The Polar Water Molecule

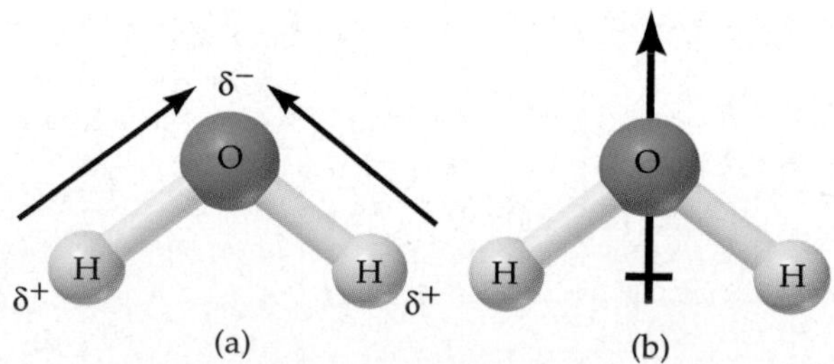

▶ Figure 14.1 The Polar Water Molecule (a) The more electronegative oxygen atom polarizes the O—H bond, which in turn creates two dipoles in a water molecule. (b) The two dipoles produce a net dipole for the entire water molecule.

Table 14.3 Selected Polar and Nonpolar Solvents

Polar Solvents	Nonpolar Solvents
water, H_2O	hexane, C_6H_{14}
methanol, CH_3OH	heptane, C_7H_{16}
ethanol, C_2H_5OH	toluene, C_7H_8
acetone, C_3H_6O	carbon tetrachloride, CCl_4
methyl ethyl ketone, C_4H_8O	chloroform, $CHCl_3$
formic acid, $HCHO_2$	methylene chloride, CH_2Cl_2
acetic acid, $HC_2H_3O_2$	ethyl ether, $C_4H_{10}O$*

*The general rule that oxygen-containing solvents are polar has some exceptions. For example, ethyl ether contains oxygen and yet is a nonpolar solvent. This is explained in the note at the end of this section.

Oil layer

Water layer

▲ **Immiscible Liquids** Oil and water are immiscible because oil molecules are nonpolar and water molecules are polar. A nonpolar solvent (oil) and a polar solvent (water) obey the *like dissolves like* rule.

Like Dissolves Like Rule

Polar solvents such as water, H_2O, and ethanol, C_2H_5OH, dissolve in one another. Nonpolar solvents such as hexane, C_6H_{14}, and chloroform, CCl_4, also dissolve in one another. We conclude that if two solvents are both polar, or both nonpolar, they dissolve in one another. This principle is the ***like dissolves like* rule**. This rule states that two liquids dissolve in one another if their molecules are similar in polarity.

Two solvents that are infinitely soluble in one another are said to be **miscible**. For example, water and ethanol are miscible. From the *like dissolves like* rule, it also follows that a polar solvent and a nonpolar solvent are not miscible. These liquids are said to be **immiscible**. Experimentally, mixtures of immiscible liquids separate into layers. For example, oil and water are immiscible and if mixed will separate into two layers with the oil floating on the water. The *like dissolves like* rule is summarized in Table 14.4.

▲ **Immiscible Liquids** Oil and water are immiscible liquids. The rainbow of color is produced by light reflecting off the oil film.

Table 14.4 Summary of the *Like Dissolves Like* Rule for Two Liquids

Solute	Polar Solvent	Nonpolar Solvent
polar	miscible	immiscible
nonpolar	immiscible	miscible

The following example exercise illustrates use of the *like dissolves like* rule to predict the miscibility of two liquids.

Example Exercise 14.2 • Miscibility Predictions

Predict whether each of the following solvents is miscible or immiscible with water, which is polar.

(a) methanol, CH_3OH (b) toluene, C_7H_8

Solution

Let's use the simplifying assumption that most solvents containing oxygen are polar. Thus, methanol is polar and toluene is nonpolar. Applying the *like dissolves like* rule gives the following.

(a) CH_3OH is polar and therefore is *miscible* with H_2O.
(b) C_7H_8 is nonpolar and therefore is *immiscible* with H_2O.

(continued)

Example Exercise 14.2 *(continued)*

Self-Test Exercise

Predict whether each of the following solvents is miscible or immiscible with water.

(a) trichloroethane, $C_2H_3Cl_3$ (b) glycerin, $C_3H_5(OH)_3$

Answers: (a) immiscible with H_2O; (b) miscible with H_2O

Note In our discussion we have classified two solvents as either miscible or immiscible; that is, two solvents are either soluble or insoluble in each other. In fact, two solvents can be partially soluble in one another. For example, a nonpolar solvent can partially dissolve in a polar solvent. Although ethyl ether is generally considered a nonpolar solvent, it partially dissolves in water (~7% soluble).

14.3 Solids in Solution

Objective · To predict whether a solid is soluble or insoluble in a liquid by applying the *like dissolves like* rule.

A solid substance, such as sugar, dissolves in a liquid, such as water, because the sugar and the water are attracted to each other. That is, the solute sugar particles are more strongly attracted to the solvent water molecules than they are to each other. The solute–solvent interaction is strongest when the polarities of the solute and the solvent are similar.

In general, a polar compound is likely to dissolve in a polar solvent such as water (Table 14.5). Table sugar, $C_{12}H_{22}O_{11}$, contains several oxygen atoms and is therefore a polar compound. Applying the *like dissolves like* rule, we can predict that polar table sugar, $C_{12}H_{22}O_{11}$, is soluble in water.

Similarly, molecules in a nonpolar compound are attracted by molecules in a nonpolar solvent. Thus, nonpolar compounds dissolve in nonpolar solvents. For example, grease dissolves in turpentine. This is because grease is a nonpolar compound and turpentine is a nonpolar solvent. Dissolving grease in turpentine is an illustration of the *like dissolves like* rule. Conversely, we cannot remove grease from our hands using water. Grease is a nonpolar compound, and water is a polar solvent.

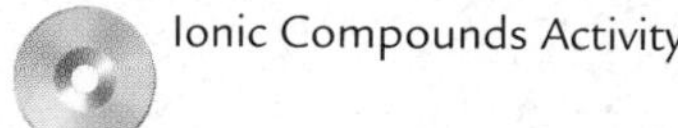

Since ionic compounds are composed of charged ions, they are similar to polar compounds in that they are both attracted to opposite charges. That is, ionic compounds are more soluble in a polar solvent than in a nonpolar solvent. Consider ordinary table salt, which is the ionic compound sodium chloride, NaCl. Table salt dissolves readily in water, which is a polar solvent. It does not dissolve readily in gasoline, which is a nonpolar solvent. This is because the solute (salt) is ionic and the solvent (gasoline) is nonpolar. It is important to note, however, that

Table 14.5 Summary of the *Like Dissolves Like* Rule for a Solid in a Liquid

Solid Solute	Polar Solvent	Nonpolar Solvent
polar	soluble	insoluble
nonpolar	insoluble	soluble
ionic	soluble	insoluble

many ionic compounds are only slightly soluble in water (refer to the solubility rules in Section 8.9).

The following example exercise illustrates the *like dissolves like* rule for a solid compound in water.

Example Exercise 14.3 • Solubility Predictions

Predict whether each of the following solid compounds is soluble or insoluble in water.

(a) fructose, $C_6H_{12}O_6$ (b) lithium carbonate, Li_2CO_3 (c) paradichlorobenzene, $C_6H_4Cl_2$

Solution

Generally, we can apply the *like dissolves like* rule to determine if a compound is soluble. Since water is a polar solvent, we can predict that water dissolves polar compounds and many ionic compounds.

(a) Fructose has six oxygen atoms and is a polar compound. We can predict that $C_6H_{12}O_6$ is *soluble* in water.
(b) Lithium carbonate contains lithium and carbonate ions; it is therefore an ionic compound. We can predict that Li_2CO_3 is *soluble* in water.
(c) Paradichlorobenzene does not contain oxygen and is a nonpolar compound. Thus, $C_6H_4Cl_2$ is *insoluble* in water.

Self-Test Exercise

Predict whether each of the following solid compounds is soluble or insoluble in water.

(a) naphthalene, $C_{10}H_8$ (b) cupric sulfate, $CuSO_4$ (c) lactic acid, $HC_3H_5O_3$

Answers: (a) insoluble; (b) soluble; (c) soluble

Some compounds contain both polar and nonpolar components. Although cholesterol, $C_{27}H_{46}O$, contains oxygen, it is considered a nonpolar compound. It is nonpolar because it has such a large number of carbon and hydrogen atoms. Using the *like dissolves like* rule, we can correctly predict that cholesterol is insoluble in water and soluble in a nonpolar substance. In the human body, cholesterol deposits in fatty tissue, which is nonpolar. The following example exercise further illustrates this principle.

Example Exercise 14.4 • Solubility Predictions

Predict whether each of the following vitamins is water-soluble or fat-soluble.

(a) vitamin A, $C_{20}H_{30}O$ (b) vitamin B_2, $C_{17}H_{20}N_4O_6$

Solution

Applying the *like dissolves like* rule, we predict that polar compounds are water-soluble and nonpolar compounds are fat-soluble.

(a) Vitamin A has one oxygen atom but is mostly nonpolar. We can predict that $C_{20}H_{30}O$ is nonpolar; thus, it is a *fat-soluble* vitamin.
(b) Vitamin B_2, riboflavin, has six oxygen atoms and is therefore polar. We can predict that $C_{17}H_{20}N_4O_6$ is polar; thus, it is a *water/soluble* vitamin.

Self-Test Exercise

Predict whether each of the following vitamins is soluble or insoluble in water.

(a) vitamin C, $C_6H_8O_6$ (b) vitamin D, $C_{27}H_{44}O$

Answers: (a) soluble; (b) insoluble

14.4 The Dissolving Process

Objective · To illustrate how an ionic or a molecular compound dissolves in water.

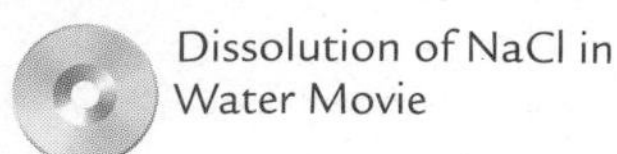

Let's try to visualize the dissolving process. When a solute crystal is dropped into a solution, the crystal begins to dissolve. This is because water molecules attack the crystal and begin pulling away part of it. Specifically, water molecules attack the edges and corners of the crystal.

As an example, suppose we drop a crystal of table sugar into water. Water molecules are attracted to the polar sugar molecule, $C_{12}H_{22}O_{11}$, and pull molecules of sugar into the solution. Several water molecules surround each sugar molecule in the solution. The sugar molecules, which are held within a cluster of water molecules, are said to be in a **solvent cage**. The number of water molecules in the solvent cage varies, depending on the concentration of the solute (Figure 14.2).

As another example, consider what happens when a crystal of table salt dissolves in water. According to the *like dissolves like* rule, polar water molecules are attracted to the ionic salt crystal. Once again, water molecules attack the edges of the crystal and begin pulling away part of it. For an ionic compound such as table salt, NaCl, the water molecules pull away positive and negative ions during the dissolving process. The negatively charged oxygen atom in a water molecule is attracted to the positively charged sodium ion, Na^+. The negatively charged chloride ion, Cl^-, is pulled into solution by the more positively charged hydrogen atoms in the water molecule. Figure 14.3 illustrates this process.

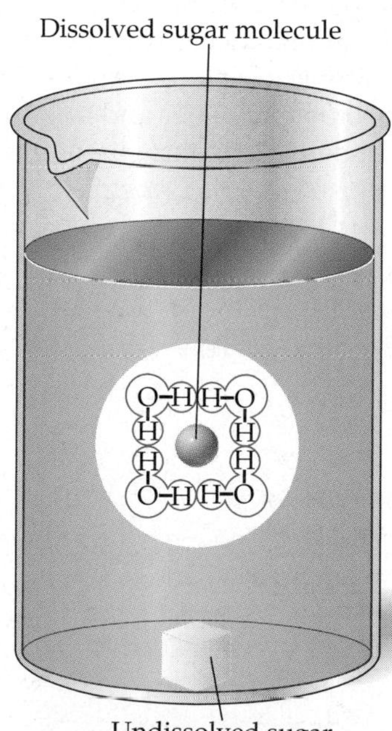

▲ **Figure 14.2 Table Sugar Dissolving in Water** Water molecules attack a sugar cube along the edges and at the corners. Each sugar molecule is pulled into solution and surrounded by several water molecules.

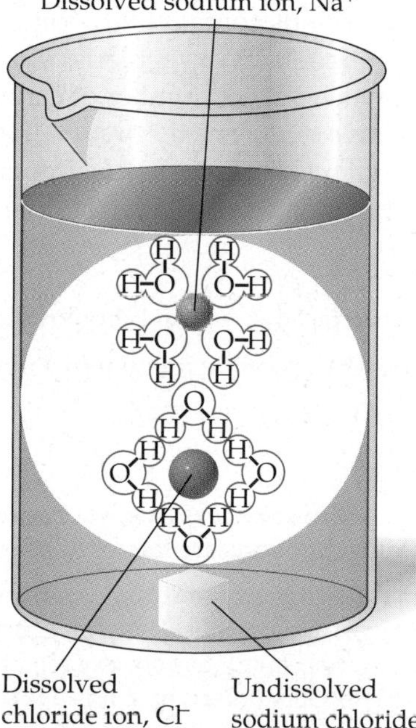

▲ **Figure 14.3 Table Salt Dissolving in Water** Note the orientation of the water molecules as they surround each ion. The Na^+ is attracted to the oxygen atom in a water molecule, whereas the Cl^- is attracted to the hydrogen atoms.

Table Sugar Dissolving in Water

Table Salt Dissolving in Water

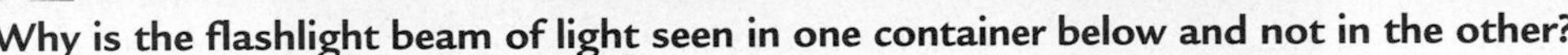

Chemistry Connection · Colloids

Why is the flashlight beam of light seen in one container below and not in the other?

A colloid is similar to a solution in that both contain dispersed particles. The size of the dispersed particle is what distinguishes a colloid from a solution. A **colloid** contains solute particles ranging in size from 1 to 100 nm, whereas a solution contains solute particles that are smaller than 1 nm in diameter. Colloids and solutions are stable, and their solute particles do not separate spontaneously from the solvent. However, a colloid contains particles that are large enough to scatter a beam of light; this phenomenon is called the **Tyndall effect**.

In theory, particles remain in solution because of a strong solute–solvent interaction. Because of their larger size, colloidal particles are held less strongly by the solvent. Colloidal particles are small enough to pass through ordinary filter paper but are too large to pass through a natural cell membrane, or a synthetic membrane like cellophane.

A practical application of this theory is kidney dialysis, which is based on the separation of colloidal particles from a solution. During dialysis, blood is circulated through an artificial kidney containing a semipermeable membrane. Protein molecules, nucleic acids, and cells are retained by the blood because of their large size. Small toxin molecules pass through the membrane and out of the blood. After 2-3 hours of dialysis, a kidney patient has blood that is free of small toxic molecules. We can summarize the distinctions between a colloid and a solution in the following ways.

Observation	Colloids	Solutions
particle type	large molecules	ions and molecules
particle size	1–100 nm	1 nm or less
particles and filter paper	do not separate	do not separate
semipermeable membrane	separate	do not separate
Tyndall effect	scatter light	do not scatter light

The beam of light is not visible as it passes through a true solution (left) but is readily visible as it passes through a colloidal dispersion of Fe_2O_3 particles (right).

14.5 Rate of Dissolving

Objective · To state the effect of temperature, stirring, and particle size on the rate at which a solid compound dissolves in water.

The rate at which a solid compound such as table sugar dissolves in a solution depends on three factors. We can increase the rate of dissolving by any one of the following.

1. heating the solution
2. stirring the solution
3. grinding the solid solute

By heating and stirring the solution or grinding the solute, we increase the rate at which solvent molecules attack the solute. *Heating the solution* increases the kinetic energy of the solution, and the solvent molecules move faster. In an aqueous

◀ **Figure 14.4 Rate of Dissolving Table Sugar** A cube of sugar suspended in water at 20°C dissolves slowly. An equal amount of powdered sugar stirred in water at 100°C dissolves immediately.

solution, water molecules attack the solute more frequently. Solute molecules are pulled into the solution faster, thus increasing the rate of dissolving. *Stirring the solution* increases the interaction between water molecules and the solute. Because the solute and solvent interact more often, the rate of dissolution is faster.

Grinding the solute into smaller crystals creates more surface area. As the solute crystals become smaller, the total surface area increases. We know that solvent molecules attack the surface of crystals along the edges. When smaller crystals and a greater surface area are created, more solute is exposed to attack by water molecules. The water molecules attack the solute more frequently, thus increasing the rate of dissolving (Figure 14.4).

14.6 Solubility and Temperature

Objective · To interpret a graph that shows the effect of temperature on the solubility of a solid compound in water.

If we heat a solution, the interaction between the solute and the solvent increases. In general, the solubility of a solid compound becomes greater as the temperature increases. A few compounds are exceptions, however, and become less soluble as the temperature increases. We will define the **solubility** of a compound as the maximum amount of solute that can be dissolved in 100 g of water at a given temperature. Figure 14.5 graphs the solubility of various compounds at different temperatures.

If you enjoy sugar in a hot beverage, you are probably familiar with an interesting illustration of the effect of temperature on solubility. Let's suppose you add sugar to hot coffee and then allow the coffee to cool. The solubility of the sugar decreases as the coffee cools. If you stir the cold coffee, you can detect crystals of solid sugar that have crystallized from solution at the cooler temperature.

Figure 14.5 illustrates that the effect of temperature on solubility varies with the compound. The solubility of table salt, NaCl, in water is only slightly affected by temperature. Its solubility is about 35 g of NaCl per 100 g of water at 20°C and

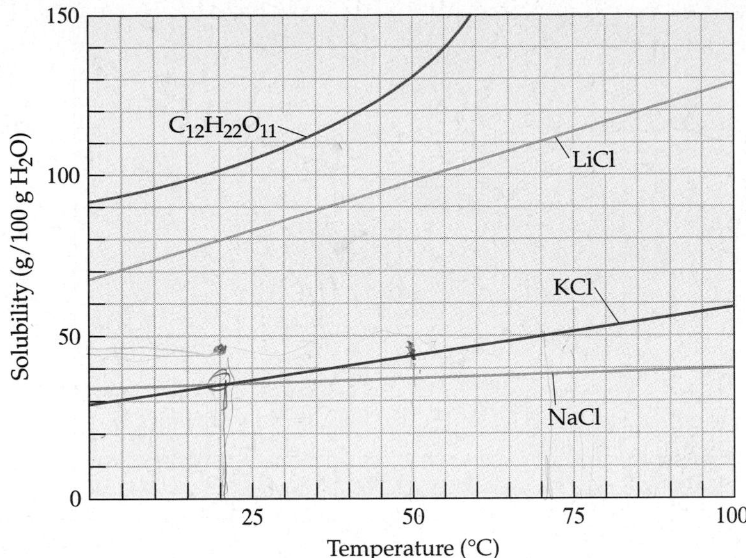

◀ **Figure 14.5 Solubility of Various Solid Compounds**
Although there are a few exceptions, solid compounds usually become more soluble as the temperature increases.

increases to 40 g/100 g water at 100°C. Interestingly, the solubility of table sugar, $C_{12}H_{22}O_{11}$, is greatly affected by an increase in temperature. At 20°C the solubility of table sugar is about 100 g/100 g water. At 55°C the solubility is 140 g/100 g water and increasing rapidly.

Example Exercise 14.5 • Determining Solubility from a Graph

Determine the solubility of each of the following solid compounds at 50°C as shown in Figure 14.5.

(a) NaCl (b) KCl (c) LiCl (d) $C_{12}H_{22}O_{11}$

Solution

From Figure 14.5, let's find the point at which the solubility of the compound intersects 50°C.

(a) The solubility of NaCl at 50°C is about 38 g/100 g water.
(b) The solubility of KCl at 50°C is about 45 g/100 g water.
(c) The solubility of LiCl at 50°C is about 98 g/100 g water.
(d) The solubility of $C_{12}H_{22}O_{11}$ at 50°C is about 130 g/100 g water.

Self-Test Exercise

Refer to the solubility behavior shown in Figure 14.5 and determine the minimum temperature required to obtain the following solutions.

(a) 35 g NaCl/100 g water (b) 45 g KCl/100 g water
(c) 100 g LiCl/100 g water (d) 140 g $C_{12}H_{22}O_{11}$/100 g water

Answers: (a) 20°C; (b) 55°C; (c) 55°C; (d) 55°C

14.7 Unsaturated, Saturated, Supersaturated

Objective • **To interpret a graph of temperature versus solubility and determine whether a solution is saturated, unsaturated, or supersaturated.**

In Section 14.6 we noted that the solubility of a compound usually increases as the temperature increases. As a further illustration, consider the solubility of sodium acetate, $NaC_2H_3O_2$. The solubility of sodium acetate at 55°C is 100 g of $NaC_2H_3O_2$ per 100 g of water. This is the maximum amount of solute that dissolves at 55°C.

Solubility of Sodium Acetate

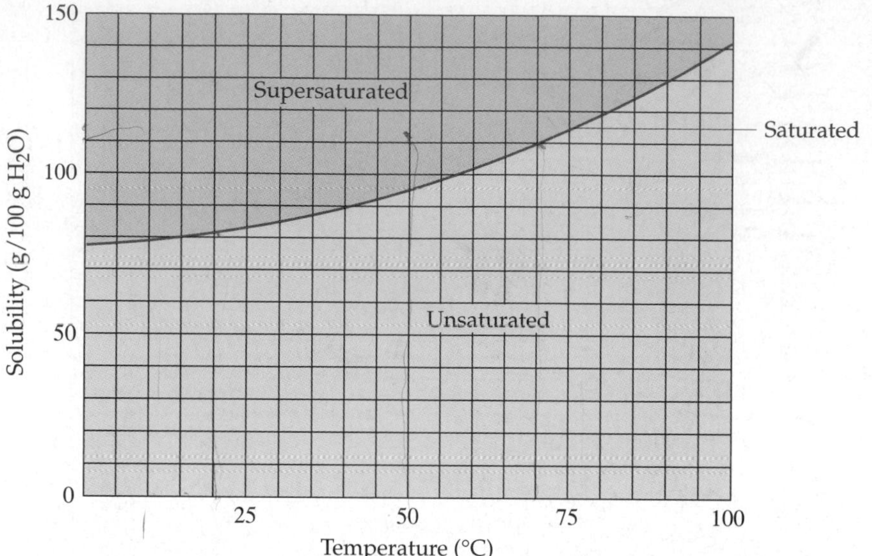

▶ **Figure 14.6 Solubility of Sodium Acetate** The curve represents a saturated solution of sodium acetate, $NaC_2H_3O_2$, at various temperatures. The region below the curve represents unsaturated solutions, whereas that above represents supersaturated solutions.

A solution containing the maximum amount of solute at a given temperature is said to be **saturated**. At 75°C the solubility increases. At that temperature, a solution containing 100 g of $NaC_2H_3O_2$ per 100 g of water is no longer saturated. Since more solute can be dissolved, the solution is said to be **unsaturated**. Figure 14.6 graphs the solubility of $NaC_2H_3O_2$ as a function of temperature.

Under special circumstances it is possible to exceed the maximum solubility of a compound. Solutions that contain more solute than ordinarily dissolves at a given temperature are said to be **supersaturated**. For example, recall that 100 g of $NaC_2H_3O_2$ can be dissolved in 100 g of water at 55°C. This solution is saturated. If we allow the solution to cool, being careful not to disturb the apparatus, the excess solute will remain in solution. It is possible to cool the solution to 20°C and still have 100 g of $NaC_2H_3O_2$ remain in solution. At 20°C the maximum solubility is only about 82 g/100 g water. However, since the excess solute remains in the solution, at 20°C the solution is supersaturated.

Supersaturated solutions are unstable. In our example, one tiny crystal can cause massive crystallization of $NaC_2H_3O_2$ from solution. As another example, consider rock candy, which is made by suspending a small sugar crystal in a supersaturated sugar solution. The small crystal turns into a large rock of sugar as excess solute deposits from the supersaturated solution. The following example exercise provides practice in obtaining information from a solubility graph.

Example Exercise 14.6 • Determining Saturation from a Graph

A sodium acetate solution contains 110 g of $NaC_2H_3O_2$ per 100 g of water. Refer to Figure 14.6 and determine whether the solution is unsaturated, saturated, or supersaturated at each of the following temperatures.

(a) 50°C (b) 70°C (c) 90°C

Solution

(a) At 50°C the solubility of $NaC_2H_3O_2$ is about 97 g/100 g water. Since the solution contains more solute, 110 g/100 g water, the solution is *supersaturated*.

(b) At 70°C the solubility is about 110 g/100 g water. Since the solution contains the same amount of solute, 110 g/100 g water, the solution is *saturated*.

(c) At 90°C the solubility is about 130 g/100 g water. Since the solution has only 110 g/100 g water, the solution is *unsaturated*.

(a)

(b)

(c)

▲ **Illustration of Supersaturation** (a) A single crystal of sodium acetate, $NaC_2H_3O_2$, is dropped into a supersaturated solution. (b) The small crystal causes extensive crystallization, and eventually, (c) the solute forms a solid mass of $NaC_2H_3O_2$.

Self-Test Exercise

A sodium acetate solution contains 80 g of $NaC_2H_3O_2$ per 100 g of water. Refer to Figure 14.6 and determine whether the solution is unsaturated, saturated, or supersaturated at each of the following temperatures.

(a) 0°C (b) 15°C (c) 45°C

Answers: (a) supersaturated; (b) saturated; (c) unsaturated

14.8 Mass Percent Concentration

Experiment #20, Prentice Hall Laboratory Manual

Objectives

- To write three pairs of unit factors given the mass percent concentration of a solution.
- To perform calculations that involve the mass of solute, mass of solvent, and mass percent concentration of a solution.

The concentration of a solution tells us how much solute is dissolved in a given volume of solution. We use the imprecise terms "dilute solution" and "concentrated solution" to describe the approximate concentration of a solution. We can also describe the precise concentration of a solution by comparing the mass of solute to the mass of solution. The **mass/mass percent (m/m %)** concentration is the mass of solute dissolved in 100 g of solution. This concentration expression can be written as follows.

$$\frac{\text{mass of solute}}{\text{mass of solution}} \times 100 = \text{m/m \%}$$

Alternatively, we can write the mass/mass percent concentration expression showing the two components that compose the denominator of the ratio.

$$\frac{\text{g solute}}{\text{g solute} + \text{g solvent}} \times 100 = \text{m/m\%}$$

If a student prepares a solution from 5.00 g of NaF dissolved in 95.0 g of water, what is the mass/mass percent concentration? Since NaF is the solute and water is the solvent, we can calculate the mass percent concentration as follows.

$$\frac{5.00 \text{ g NaF}}{5.00 \text{ g NaF} + 95.0 \text{ g H}_2\text{O}} \times 100 = \text{m/m}\,\%$$

$$\frac{5.00 \text{ g NaF}}{100.0 \text{ g solution}} \times 100 = 5.00\%$$

Although NaF is used to fluoridate public water, the concentration of fluoride in drinking water is about a million times less than the 5.00% in this example.

Writing Mass Percent Concentration Unit Factors

The unit analysis method of problem solving involves three steps: unknown units, relevant given value, and application of a unit factor. We can solve mass/mass percent calculations by using the solution concentration as a unit factor. For example, a 5.00% solution of NaF contains 5.00 g of solute in each 100.0 g of solution. We write this pair of unit factors as

$$\frac{5.00 \text{ g NaF}}{100.0 \text{ g solution}} \quad \text{and} \quad \frac{100.0 \text{ g solution}}{5.00 \text{ g NaF}}$$

Since the solution is composed of solute and solvent, we can write a second pair of unit factors. The solution contains 5.00 g of solute in 95.0 g of solvent.

$$\frac{5.00 \text{ g NaF}}{95.0 \text{ g water}} \quad \text{and} \quad \frac{95.0 \text{ g water}}{5.00 \text{ g NaF}}$$

Furthermore, we can express the ratio of the mass of solvent to the mass of solution in a third pair of unit factors. The solution contains 95.0 g of solvent in 100.0 g of solution.

$$\frac{95.0 \text{ g water}}{100.0 \text{ g solution}} \quad \text{and} \quad \frac{100.0 \text{ g solution}}{95.0 \text{ g water}}$$

Solving Mass Percent Concentration Problems

We can now apply unit factors derived from the mass/mass percent concentration to solution calculations. For example, intravenous injections of glucose are sometimes administered to patients with low blood sugar. If a normal IV glucose solution is 5.00%, what is the mass of solution that contains 25.0 g of glucose sugar?

Let's use the unit analysis method of problem solving. Step 1: The unknown quantity is g solution. Step 2: The relevant given value is 25.0 g sugar.

$$25.0 \text{ g sugar} \times \frac{\text{unit}}{\text{factor}} = \text{g solution}$$

The percent concentration provides the unit factor. Since the solution concentration is 5.00%, there is 5.00 g of solute in 100.0 g of solution. We can write the unit factors as

$$\frac{5.00 \text{ g sugar}}{100.0 \text{ g solution}} \quad \text{and} \quad \frac{100.0 \text{ g solution}}{5.00 \text{ g sugar}}$$

We select the second unit factor so as to cancel units properly.

$$25.0\ \cancel{\text{g sugar}} \times \frac{100.0\ \text{g solution}}{5.00\ \cancel{\text{g sugar}}} = 5.00 \times 10^2\ \text{g solution}$$

The following example exercises further illustrate calculations based on the mass/mass percent concentration of a solution.

Example Exercise 14.7 • Mass Percent Concentration

A glucose tolerance test is used to diagnose diabetes and hypoglycemia. If a patient is given a 30.0% glucose solution, what is the mass of glucose in 250.0 g of water?

Solution

Let's recall the unit analysis format of problem solving.

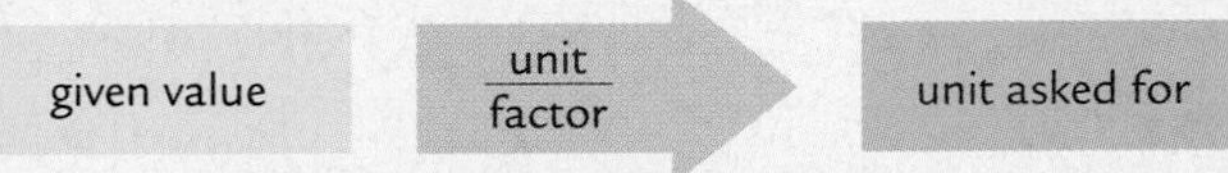

Step 1, The unit asked for in the answer is **g glucose**.
Step 2, The given value is **250.0 g water**.
Step 3, We apply a conversion factor to cancel units. Since the solution is 30.0%, there is 30.0 g of solute in 100.0 g of solution. Therefore, there is 30.0 g of glucose in 70.0 g of water. We can write the unit factors

$$\frac{30.0\ \text{g glucose}}{70.0\ \text{g water}} \quad \text{and} \quad \frac{70.0\ \text{g water}}{30.0\ \text{g glucose}}$$

In this example, we select the first unit factor to cancel the given units.

$$250.0\ \cancel{\text{g water}} \times \frac{30.0\ \text{g glucose}}{70.0\ \cancel{\text{g water}}} = 107\ \text{g glucose}$$

In a glucose tolerance test, a patient drinks a flavored glucose solution after fasting for 12 hours. A blood sample is taken at the start of the test and for several hours afterward to measure glucose levels. We can summarize as follows.

250.0 g water → mass/mass % concentration → 107 g glucose

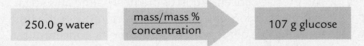

Self-Test Exercise

Given 125 g of a 15.0% sucrose solution, what is the mass of sucrose solute?

Answer: 18.8 g sucrose

Example Exercise 14.8 • Mass Percent Concentration

Intravenous saline injections are given to restore the electrolyte balance in trauma patients. What is the mass of water required to dissolve 2.00 g of NaCl for a 0.90% normal saline solution?

Solution

Step 1, The unit asked for in the answer is **g water**.
Step 2, The given value is **2.00 g NaCl**.
Step 3, We apply a conversion factor in order to cancel units. Since the solution concentration is 0.90%, there is 0.90 g of solute in 100.0 g of solution. Therefore, there is 0.90 g of NaCl in 99.1 g of water. We can write the unit factors as

(continued)

Example Exercise 14.8 *(continued)*

$$\frac{0.90 \text{ g NaCl}}{99.1 \text{ g water}} \quad \text{and} \quad \frac{99.1 \text{ g water}}{0.90 \text{ g NaCl}}$$

In this example, we select the second unit factor to cancel the given units

$$2.00 \cancel{\text{g NaCl}} \times \frac{99.1 \text{ g water}}{0.90 \cancel{\text{g NaCl}}} = 220 \text{ g glucose}$$

We can visualize the conversion as follows.

2.00 g NaCl → mass/mass % concentration → 220 g water

Self-Test Exercise

A 7.50% potassium chloride solution is prepared by dissolving enough of the salt to give 100.0 g of solution. What is the mass of water required?

Answer: 92.5 g water

Note The term "percent concentration" is ambiguous. It is used to express the mass of solute in 100 g of solution, the mass of solute in 100 mL of solution, and the volume of solute in 100 mL of solution. Thus, there are three interpretations of percent concentration expressions: mass–mass, mass–volume, and volume–volume. Unless specifically told otherwise, we usually assume that percent concentration refers to the mass–mass percent.

Experiment #20, Prentice Hall Laboratory Manual

14.9 Molar Concentration

Objectives
- To write a pair of unit factors given the molar concentration of a solution.
- To perform calculations that involve a mass of solute, volume of solution, and the molar concentration of a solution.

Solution Formation from a Solid Movie

We can describe the precise concentration of medical saline and sugar solutions in terms of mass–mass percent concentration. However, most of the time we express the precise concentration of solutions in terms of molar concentration. The molar concentration, or **molarity** (symbol M), relates the amount of solute to a given volume of solution. Specifically, molarity is the number of moles of a solute dissolved in one liter of a solution. We can express the molarity as follows.

$$\frac{\text{moles solute}}{\text{liter solution}} = M$$

As an example, consider a household drain cleaner that is a solution of caustic sodium hydroxide, NaOH. If a manufacturer prepares a solution from 24.0 g of NaOH dissolved in 0.100 L of solution, what is the molarity? We can calculate the molarity of NaOH (40.00 g/mol) as follows.

$$\frac{24.0 \cancel{\text{g NaOH}}}{0.100 \text{ L solution}} \times \frac{1 \text{ mol NaOH}}{40.00 \cancel{\text{g NaOH}}} = \frac{\text{mol NaOH}}{\text{L solution}}$$

$$= 6.00\, M \text{ NaOH}$$

Notice that we started the calculation with a ratio of two units (g/L) to obtain an answer that is a ratio of two units (mol/L).

Writing Molar Concentration Unit Factors

The unit analysis method of problem solving involves the application of a unit factor. We can solve molarity calculations by using the solution concentration as a unit factor. For example, a 6.00 *M* solution of NaOH contains 6.00 mol of solute in each liter of solution. We can write this pair of unit factors as

$$\frac{6.00 \text{ mol NaOH}}{1 \text{ L solution}} \quad \text{and} \quad \frac{1 \text{ L solution}}{6.00 \text{ mol NaOH}}$$

In the laboratory, the volume of a solution is usually measured in milliliters. Since there are 1000 mL in a liter, we can substitute for the 6.00 M solution as follows.

$$\frac{6.00 \text{ mol NaOH}}{1000 \text{ mL solution}} \quad \text{and} \quad \frac{1000 \text{ mL solution}}{6.00 \text{ mol NaOH}}$$

Solving Molar Concentration Problems

We can now apply unit factors derived from the molar concentration to solution calculations. Consider a solution prepared by dissolving potassium dichromate in water. Let's calculate the mass of solute dissolved in 250.0 mL of 0.100 *M* $K_2Cr_2O_7$ solution.

We begin with the unit analysis method of problem solving. Step 1: The unknown quantity is g $K_2Cr_2O_7$. Step 2: The relevant given value is 250.0 mL solution.

$$250.0 \text{ mL solution} \times \frac{\text{unit}}{\text{factor}} = \text{g } K_2Cr_2O_7$$

Since the solution concentration is 0.100 *M*, there is 0.100 mol of $K_2Cr_2O_7$ solute in each liter of solution. We can write two unit factors.

$$\frac{0.100 \text{ mol } K_2Cr_2O_7}{1000 \text{ mL solution}} \quad \text{and} \quad \frac{1000 \text{ mL solution}}{0.100 \text{ mol } K_2Cr_2O_7}$$

We choose the first unit factor so as to cancel units properly. In addition, we must use the molar mass of $K_2Cr_2O_7$ (294.20 g/mol) to convert from moles to grams.

$$250.0 \cancel{\text{ mL solution}} \times \frac{0.100 \cancel{\text{ mol } K_2Cr_2O_7}}{1000 \cancel{\text{ mL solution}}} \times \frac{294.20 \text{ g } K_2Cr_2O_7}{1 \cancel{\text{ mol } K_2Cr_2O_7}} = 7.36 \text{ g } K_2Cr_2O_7$$

The following example exercises further illustrate calculations based on the molar concentration of a solution.

(a)

(b)

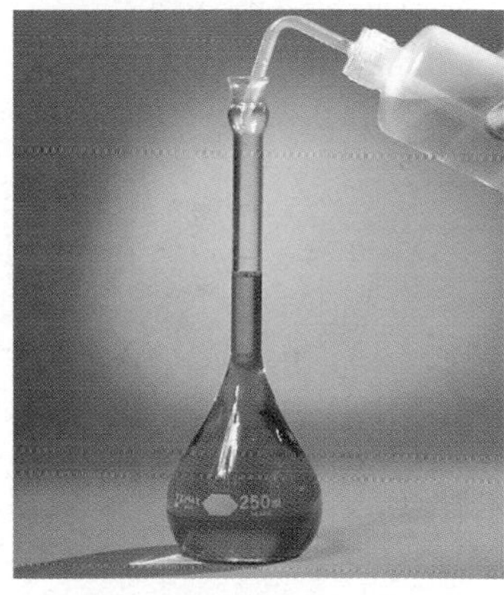

(c)

▲ **Preparing 0.100 *M* Potassium Dichromate** (a) A 250-mL volumetric flask contains 7.36 g of potassium dichromate, $K_2Cr_2O_7$. (b) Distilled water is added to the flask to dissolve the orange compound. (c) The volume of solution is adjusted to the calibration line on the flask, giving a 0.100 *M* $K_2Cr_2O_7$ solution.

Example Exercise 14.9 • Molar Concentration

How many milliliters of 12.0 *M* hydrochloric acid contain 7.30 g of HCl solute (36.46 g/mol)?

Solution

In molarity calculations, we relate the given value to moles and then convert to the unit asked for in the answer.

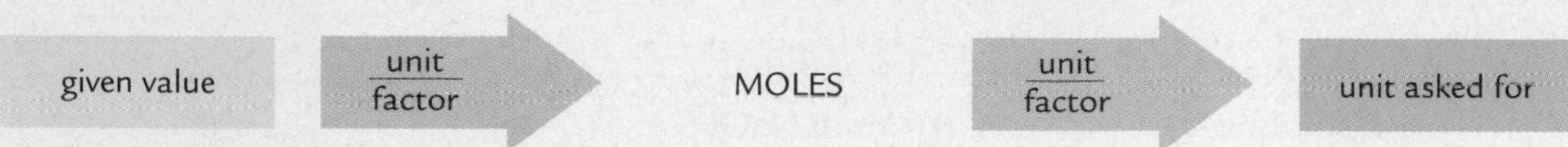

Step 1, The unit asked for in the answer is **mL acid**.

Step 2, The relevant given value is **7.30 g HCl**.

Step 3, The molar mass of HCl (36.46 g/mol) and the molarity of the solution (12.0 mol/1000 mL) provide the unit factors. We select unit factors so as to cancel units.

$$7.30 \text{ g HCl} \times \frac{1 \text{ mol HCl}}{36.46 \text{ g HCl}} \times \frac{1000 \text{ mL acid}}{12.0 \text{ mol HCl}} = 16.7 \text{ mL acid}$$

We can visualize the two conversions as follows.

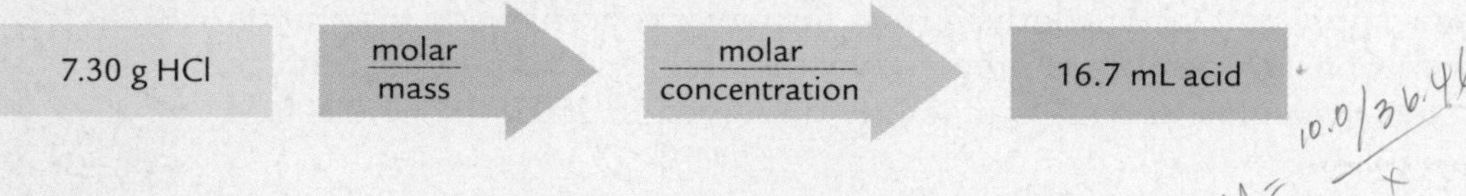

Self-Test Exercise

What volume of 6.00 M hydrochloric acid contains 10.0 g of HCl solute (36.46 g/mol)?

Answer: 45.7 mL acid

Example Exercise 14.10 • Molar Concentration

What is the mass of H_2SO_4 (98.09 g/mol) in 50.0 mL of 6.00 *M* sulfuric acid?

Solution

Step 1, The unit asked for in the answer is **g H_2SO_4**.

Step 2, The relevant given value is **50.0 mL acid**.

Step 3, The molarity (6.00 mol/1000 mL) and the molar mass of HCl (98.09 g/mol) provide the unit factors. We select unit factors to cancel units.

$$50.0 \text{ mL acid} \times \frac{6.00 \text{ mol } H_2SO_4}{1000 \text{ mL acid}} \times \frac{98.09 \text{ g } H_2SO_4}{1 \text{ mol } H_2SO_4} = 29.4 \text{ g } H_2SO_4$$

We can visualize the two conversions as follows.

Self-Test Exercise

What is the mass of H_2SO_4 (98.09 g/mol) in 25.0 mL of 18.0 *M* sulfuric acid?

Answer: 44.1 g H_2SO_4

Chemistry Connection · Water Fluoridation

What percentage of the people in the United States drink fluoridated water?

As early as the 1930s, studies by the U.S. Public Health Service indicated that fluoride can help prevent dental cavities. The action of fluoride is to make the enamel surface of teeth more resistant to decay.

Tooth enamel is made mainly of hydroxyapatite, $Ca_{10}(PO_4)_6(OH)_2$, the hardest substance in our bodies. However, bacteria in the mouth react with food to produce acids that attack tooth enamel. These acids can react with hydroxyapatite, causing enamel to dissolve and pits to form. Fluoride ions can prevent cavities by converting $Ca_{10}(PO_4)_6(OH)_2$ to $Ca_{10}(PO_4)_6F_2$, which is more resistant to acid attack.

In 1950 the Public Health Service officially endorsed the practice of adding sodium fluoride to public drinking water. Since then, there has been widespread fluoridation of public water supplies. Typically, fluoridated water has a concentration of fluoride ion that is less than 1 ppm (1 mg/L). A U.S. Department of Health census reveals that every state has drinking water with natural or controlled fluoridation. However, the number of people that drink fluoridated water varies widely: 2% in Utah, 22% in California, 36% in Florida, 40% in Washington, 59% in Texas, 66% in New York, and 86% in Illinois. In Canada, over half of the people drink water that has fluoride.

The benefits of fluoride in preventing tooth decay have been proven repeatedly. In one study, children who grew up drinking fluoridated water had an average of three teeth with cavities. Conversely, children who grew up without fluoridation had an average of ten teeth with cavities. In other studies, it has been shown that fluoride is most effective when it is available while teeth are still developing.

Despite proven benefits, not everyone agrees that fluoride should be added to the public water supply. One reason is that fluoride may have adverse side effects. For example, it has been reported that fluoridation causes brown mottling in the teeth of some children. It is recommended that the fluoride concentration in water be only 0.1 mg/L. Adding fluoride to the public water supply also raises the issue of free choice, thus sparking a political debate.

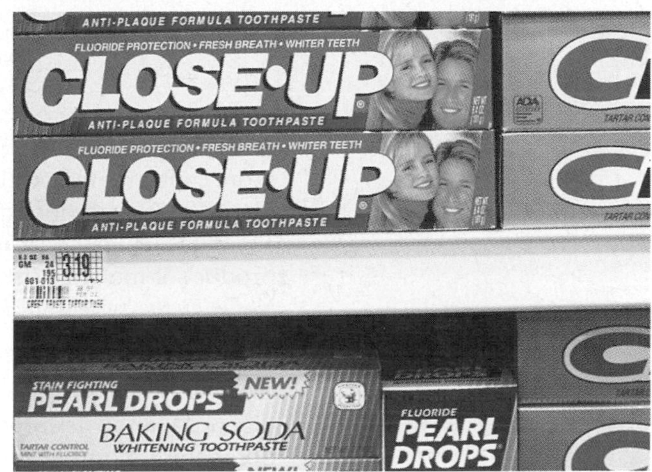

▲ **Toothpastes with Fluoride** Many toothpastes, including each of the brands shown in the photograph, contain a fluoride compound that helps strengthen the enamel surface of teeth.

As an alternative to fluoridation, people can choose to take inexpensive fluoride tablets. In addition, many toothpastes and mouthwashes are readily available sources of fluoride. Stannous fluoride, SnF_2, and sodium fluoride, NaF, are commonly found in toothpaste. Another alternative is in-office fluoride treatments by a dentist.

Currently, about 50% of the U.S. population drinks water that is naturally or artificially fluoridated.

14.10 Dilution of a Solution

Objective · To perform calculations that involve a solution undergoing a dilution in concentration.

In your chemistry laboratory, you may have noticed reagent bottles labeled 0.1 *M* HCl and 0.1 *M* NaOH. These solutions, whose concentrations are only approximate, are often prepared in the stockroom by diluting a more concentrated stock solution. For example, a stock solution of 6 *M* NaOH may be available to prepare a 0.1 M NaOH solution for an experiment.

When calculating the concentration of a diluted solution, you should remember the following principle: *The amount of solute does not change during dilution.* Stated mathematically, the moles of solute in the concentrated solution is equal to the moles of solute in the diluted solution. Suppose we want to prepare 5.00 L of 0.1 *M*

NaOH using a 6 *M* NaOH stock solution. First, we find the moles of diluted solution using 0.1 *M* NaOH as a unit factor.

$$5.00 \, \cancel{\text{L solution}} \times \frac{0.1 \text{ mol NaOH}}{1 \, \cancel{\text{L solution}}} = 0.5 \text{ mol NaOH}$$

Since we know the moles of NaOH remain constant during dilution, the moles of NaOH in the concentrated solution must be equal to the 0.5 mol of NaOH in the diluted solution. Using 6 *M* NaOH as a unit factor, we can calculate the volume of 6 *M* stock solution as follows.

$$0.5 \, \cancel{\text{mol NaOH}} \times \frac{1 \text{ L solution}}{6 \, \cancel{\text{mol NaOH}}} = 0.08 \text{ L NaOH}$$

In this example, we demonstrated how to prepare 5.00 L of 0.1 *M* NaOH from a more concentrated stock solution. That is, we simply dilute 0.08 L (~ 80 mL) of 6 *M* NaOH stock solution to a total volume of 5.00 L with distilled water.

Alternatively, we can solve this problem using a simple algebraic equation. The product of molarity (mol/L) and volume (L) gives units of mole. Since *the moles of NaOH in the concentrated and the dilute solutions are equal,* their products are also equal.

$$M_1 \times V_1 = M_2 \times V_2$$

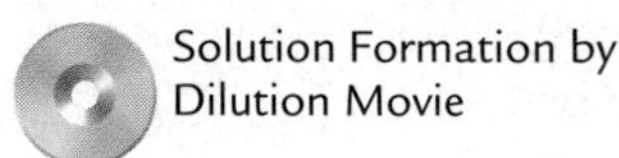
Solution Formation by Dilution Movie

where M_1 and V_1 represent the molarity and the volume of the concentrated solution before dilution, and M_2 and V_2 are the molarity and volume after dilution. In the preceding problem, we found the volume of a 6 *M* NaOH stock solution that must be diluted to give 5.00 L of 0.1 *M* NaOH. We can also find the volume using the above equation. After substituting the given values into the equation, we have

$$6 \, M \times V_1 = 0.1 \, M \times 5.00 \text{ L}$$

Solving for V_1,

$$\frac{0.1 \, M \times 5.00 \text{ L}}{6 \, M} = V_1$$

$$= 0.08 \text{ L NaOH}$$

The following example exercise provides additional practice in solving dilution problems involving the concentration of a solution.

Example Exercise 14.11 • Dilution of a Solution

Concentrated hydrochloric acid is available commercially as a 12 *M* solution. What is the molarity of an HCl solution prepared by diluting 50.0 mL of concentrated acid with distilled water to give a total volume of 2.50 L?

Solution

Let's find the moles of concentrated HCl before dilution.

$$50.0 \, \cancel{\text{mL solution}} \times \frac{12 \text{ mol HCl}}{1000 \, \cancel{\text{mL solution}}} = 0.60 \text{ mol HCl}$$

We know that the moles of HCl do not change during dilution; the moles of diluted HCl solution must be the same as the moles of concentrated HCl. Since the volume of diluted HCl is 2.50 L, we can calculate the diluted concentration.

$$\frac{0.60 \text{ mol HCl}}{2.50 \text{ L solution}} = \frac{0.24 \text{ mol HCl}}{1 \text{ L solution}} = 0.24\ M \text{ HCl}$$

Algebraic Solution
Alternatively, we can solve this problem using the equation

$$M_1 \times V_1 = M_2 \times V_2$$

Substituting, we have

$$12\ M \times 50.0 \text{ mL} = M_2 \times 2.50 \text{ L}$$

We must use the same units for volume; for example, we can convert 2.50 L to mL, which equals 2500 mL. Solving for M_2, we have

$$\frac{12\ M \times 50.0 \cancel{\text{mL}}}{2500 \cancel{\text{mL}}} = 0.24\ M \text{ HCl}$$

Self-Test Exercise
Battery acid is 18 *M* sulfuric acid. What volume of battery acid must be diluted with distilled water in order to prepare 1.00 L of 0.50 *M* H_2SO_4?

Answer: 28 mL H_2SO_4 (diluted to 1.00 L with distilled H_2O)

14.11 Solution Stoichiometry

Objective · To perform calculations that involve the concentration of a solution and a balanced chemical equation.

In Chapter 10, we performed mole calculations involving chemical equations; this topic is termed stoichiometry. Now, let's apply stoichiometry to chemical reactions involving the molar concentration of a solution. Suppose we are given 4.24 g of sodium carbonate, Na_2CO_3, which we react with 0.150 *M* hydrochloric acid, HCl. The balanced equation for the reaction is

$$Na_2CO_3(s) + 2\ HCl(aq) \longrightarrow 2\ NaCl(aq) + H_2O(l) + CO_2(g)$$

Let's calculate the volume of 0.150 *M* HCl that reacts with 4.24 g of Na_2CO_3. We can solve this problem in three steps. In the first step, we calculate the moles of Na_2CO_3. The molar mass of Na_2CO_3 is 105.99 g/mol. Therefore,

$$4.24 \cancel{\text{g } Na_2CO_3} \times \frac{1 \text{ mol } Na_2CO_3}{105.99 \cancel{\text{g } Na_2CO_3}} = 0.0400 \text{ mol } Na_2CO_3$$

In the second step, we convert moles of Na_2CO_3 to moles of HCl by applying the coefficients from the balanced equation. The relationship between the two substances is 1 mol Na_2CO_3 = 2 mol HCl. The unit factor conversion is

$$0.0400 \cancel{\text{mol } Na_2CO_3} \times \frac{2 \text{ mol HCl}}{1 \cancel{\text{mol } Na_2CO_3}} = 0.0800 \text{ mol HCl}$$

In the third step, we use the molar concentration of the hydrochloric acid solution as a unit factor to obtain the volume of HCl that reacts with Na_2CO_3. In this example, 0.150 *M* HCl gives two unit factors: 0.150 mol HCl/1 L solution, and the reciprocal, 1 L solution/0.150 mol HCl.

$$0.0800\ \cancel{\text{mol HCl}} \times \frac{1\ \text{L solution}}{0.150\ \cancel{\text{mol HCl}}} = 0.533\ \text{L solution}$$

It is also possible to solve this stoichiometry problem in one continuous calculation. We show the complete solution to the problem as

$$4.24\ \cancel{\text{g Na}_2\text{CO}_3} \times \frac{1\ \text{mol Na}_2\text{CO}_3}{105.99\ \cancel{\text{g Na}_2\text{CO}_3}} \times \frac{2\ \text{mol HCl}}{1\ \cancel{\text{mol Na}_2\text{CO}_3}} \times \frac{1\ \text{L solution}}{0.150\ \cancel{\text{mol HCl}}}$$
$$= 0.533\ \text{L solution}$$

The following example exercise provides additional practice in solving chemical calculations involving the concentration of a solution.

Example Exercise 14.12 • Solution Stoichiometry

Given that 37.5 mL of 0.100 *M* aluminum bromide solution reacts with silver nitrate solution, what is the mass of AgBr (187.77 g/mol) precipitate?

$$\text{AlBr}_3(aq) + 3\ \text{AgNO}_3(aq) \longrightarrow 3\ \text{AgBr}(s) + \text{Al(NO}_3)_3(aq)$$

Solution

Step 1, The unit asked for in the answer is **g AgBr**.

Step 2, The given value is **37.5 mL solution**.

Step 3, We apply unit conversion factors in order to cancel units. In this example, we can use molar concentration (0.100 mol $AlBr_3$/1000 mL) and molar mass (187.77 g/mol) as unit factors. From the balanced equation, we see that 1 mol of $AlBr_3$ produces 3 mol AgBr; therefore,

$$37.5\ \cancel{\text{mL solution}} \times \frac{0.100\ \cancel{\text{mol AlBr}_3}}{1000\ \cancel{\text{mL solution}}} \times \frac{3\ \cancel{\text{mol AgBr}}}{1\ \cancel{\text{mol AlBr}_3}} \times \frac{187.77\ \text{g}}{1\ \cancel{\text{mol AgBr}}} = 2.11\ \text{g AgBr}$$

We can visualize the overall solution as follows.

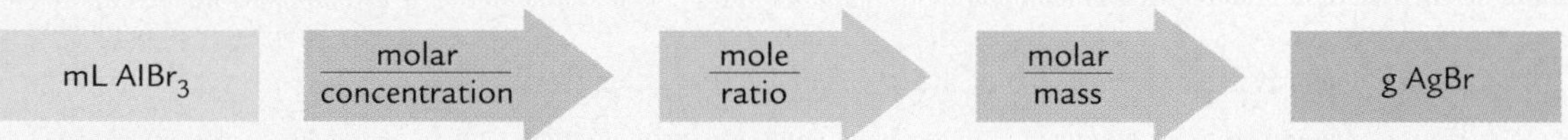

Self-Test Exercise

Given that 27.5 mL of 0.210 *M* lithium iodide reacts with 0.133 *M* lead(II) nitrate solution, what volume of $Pb(NO_3)_2$ is required for complete precipitation?

$$\text{Pb(NO}_3)_2(aq) + 2\ \text{LiI}(aq) \longrightarrow \text{PbI}_2(s) + 2\ \text{LiNO}_3(aq)$$

Answer: 21.7 mL $Pb(NO_3)_2$

Summary

Section 14.1 The most common solutions are formed from a solid, a liquid, or a gas dissolved in water. If the solute is a gas, its solubility is affected by temperature and pressure. Raising the temperature of the solution decreases the solubility of the dissolved gas. On the other hand, raising the partial pressure of the gas above the solution increases the amount of dissolved gas in the solution. In fact, **Henry's law** states that the solubility of a gas in a liquid is proportional to the partial pressure of the gas above the liquid.

Section 14.2 A **solution** consists of a **solute** dissolved in a **solvent**. A solution results from the interaction of solute and solvent molecules. A region in a polar molecule having a partial positive and partial negative charge is a **dipole**. In water, there are two dipoles that combine to produce a single **net dipole** for the molecule. Water is composed of polar H_2O molecules, hence it is considered a **polar solvent**. Hexane is composed of nonpolar C_6H_{14} molecules, and so it is considered a **nonpolar solvent**. Although it is a simplification, polar solvents are often small molecules that contain one or more oxygen atoms, for example, H_2O, CH_3OH, and C_2H_5OH.

We can predict the solubility of two liquids by applying the ***like dissolves like*** **rule**. This rule states that two liquids are **miscible** if they are both polar solvents, or both nonpolar solvents. Furthermore, two liquids are **immiscible** if one is a polar solvent and the other a nonpolar solvent. Since CH_3OH is polar, it is miscible with water. Since C_6H_{14} is nonpolar, it is immiscible with water.

Section 14.3 We can also usually predict whether or not a solid compound is soluble in a given solvent. According to the *like dissolves like* rule, a polar solvent dissolves a polar compound, and a nonpolar solvent dissolves a nonpolar compound. Ionic compounds are extreme examples of polar compounds and are therefore not soluble in nonpolar solvents. As we learned earlier in Section 8.9, some ionic compounds are water-soluble, but many are only slightly soluble.

Section 14.4 In the process of dissolving a solid compound, solvent molecules attack solute particles and pull them into solution. The solute particles are then surrounded by water molecules which form a **solvent cage**. Molecular compounds release molecules in solution, whereas ionic compounds release ions in solution. A **colloid** is similar to a solution except that the dissolved particles are larger. These larger colloid particles demonstrate an ability to scatter a beam of light; this phenomenon is called the **Tyndall effect**.

Section 14.5 The rate of dissolving a solid solute compound in solution is increased by three factors: *heating the solution, stirring the solution,* and *grinding the solute.*

Section 14.6 Generally, raising the temperature of a solution increases the **solubility** of a solid compound. By referring to a graph of solubility, we can determine the amount of solute that can dissolve in 100 g of water at a given temperature.

Section 14.7 A solution containing the maximum amount of dissolved solute possible at a given temperature is said to be **saturated**. If the concentration of the solution is less than its maximum solubility, it is said to be **unsaturated**. Under special circumstances, it is possible to exceed the maximum solubility of a solution. Such a solution is unstable and is said to be **supersaturated**.

Section 14.8 Often, a solution concentration is expressed in terms of **mass/mass percent (m/m %)**. The mass/mass percent concentration is the mass in grams of dissolved solute in 100 g of solution.

Section 14.9 A solution concentration can be expressed as **molarity** (*M*). The molar concentration, or molarity, is the number of moles of dissolved solute in one liter of solution (mol solute/L solution).

Section 14.10 Many dilute solutions in the chemistry laboratory are prepared from concentrated solutions. For example, a 6 *M* HCl solution can be diluted to 0.1 *M* HCl. When performing dilution calculations, it is important to realize that the amount of solute does not change when a solution is diluted.

Section 14.11 Solution stoichiometry calculations are similar to other stoichiometry calculations and require a balanced chemical equation. However, solution stoichiometry problems are different in that they involve the molar concentration of a solution.

Problem-Solving Organizer

Topic	Procedure	Example
Gases in Solution Sec. 14.1	Using Henry's law, the solubility of a gas is directly proportional to the partial pressure of the gas above the liquid.	If the solubility of a gas is 0.100 g/100 mL at 1.00 atm, what is it as 3.50 atm? $0.100 \text{ g/100 mL} \times \frac{3.50 \cancel{\text{atm}}}{1.00 \cancel{\text{atm}}} = 0.350 \text{ g/100 mL}$
Mass Percent Concentration Sec. 14.8	**1.** Write down the units asked for in the answer. **2.** Write down the related given value. **3.** Apply a unit factor to convert the given units to the units in the answer.	What mass of 10.0% sucrose solution contains 15.5 g of sugar? $15.5 \cancel{\text{g sugar}} \times \frac{100.0 \text{ g solution}}{10.0 \cancel{\text{g sugar}}} = 155 \text{ g solution}$
Molar Concentration Sec. 14.9	**1.** Write down the units asked for in the answer. **2.** Write down the related given value. **3.** Apply unit factors to convert the given units to the units in the answer.	What is the mass of sucrose, $C_{12}H_{22}O_{11}$ (342.34 g/mol), in 0.500 L of a 0.100 *M* solution? $0.500 \cancel{\text{L solution}} \times \frac{0.100 \cancel{\text{mol sucrose}}}{1 \cancel{\text{L solution}}}$ $\times \frac{342.34 \text{ g sucrose}}{1 \cancel{\text{mol sucrose}}} = 17.1 \text{ g sucrose}$
Dilution of a Solution Sec. 14.10	Solve dilution problems by unit analysis, or by the algebraic method: $M_1 \times V_1 = M_2 \times V_2$	If 50.0 mL of 6.0 *M* NaOH is diluted to a total volume of 250 mL, what is the molarity? $6.0\ M \times 50.0 \text{ mL} = M_2 \times 250 \text{ mL}$ Solving for M_2, $\frac{6.0\ M \times 50.0 \cancel{\text{mL}}}{250 \cancel{\text{mL}}} = 1.2\ M$
Solution Stoichiometry Sec. 14.11	**1.** Write down the units asked for in the answer. **2.** Write down the related given value. **3.** Apply unit factors to convert the given units to the units in the answer.	What is the mass of AgCl (143.32 g/mol) produced from 50.0 mL of 0.100 *M* KCl and excess $AgNO_3$? Given: $KCl(aq) + AgNO_3(aq) \longrightarrow AgCl(s) + KNO_3(aq)$ $50.0 \cancel{\text{mL KCl}} \times \frac{0.100 \cancel{\text{mol KCl}}}{1000 \cancel{\text{mL KCl}}} \times \frac{1 \cancel{\text{mol AgCl}}}{1 \cancel{\text{mol KCl}}}$ $\times \frac{143.32 \text{ g AgCl}}{1 \cancel{\text{mol AgCl}}} = 0.717 \text{ g AgCl}$

Key Concepts*

1. Which of the following explains why bubbles form on the inside of a pan of water when the pan of water is heated?
   (a) As the temperature increases, the vapor pressure increases.
   (b) As the temperature increases, the solubility of air in water decreases.
   (c) As the temperature increases, the atmospheric pressure decreases.
   (d) As the temperature increases, the kinetic energy of water decreases.
2. Which of the following explains why peanut butter can dissolve grease?
   (a) Peanut butter and grease are both polar.
   (b) Peanut butter and grease are both nonpolar.
   (c) Peanut butter is polar and grease is nonpolar.
   (d) Peanut butter is nonpolar and grease is polar.
3. Predict which of the following vitamins is water-soluble: vitamin A ($C_{20}H_{30}O$), vitamin C ($C_6H_8O_6$), vitamin D ($C_{27}H_{44}O$), vitamin E ($C_{29}H_{50}O_2$).
4. What is the solubility of $C_{12}H_{22}O_{11}$ at 20°C? (Refer to Figure 14.5.)
5. If a solution contains 100 g of $NaC_2H_3O_2$ in 100 g of water at 50°C, is the solution unsaturated, saturated, or supersaturated? (Refer to Figure 14.6.)
6. A 50.0-mL solution of sodium chloride, NaCl, weighing 50.320 g is evaporated to dryness leaving a white solute residue with a mass of 0.453 g. What is (a) the mass percent concentration, and (b) the molar concentration of the solution?
7. Examine the following illustration of table salt dissolving in water. Explain the solubility of NaCl in water at the level of individual ions.

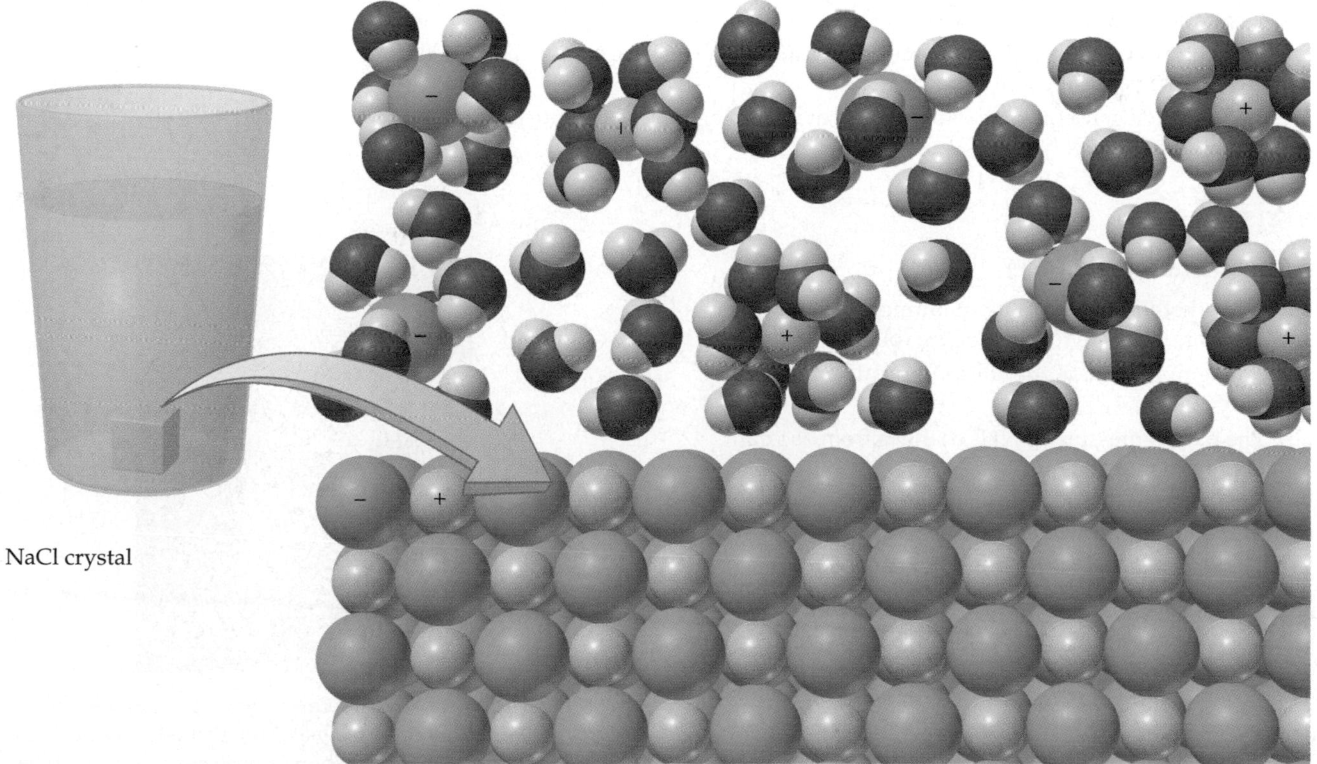

*Answers to Key Concepts are in Appendix H.

Key Terms†

Select the key term below that corresponds to each of the following definitions.

_____ 1. the solubility of a gas in a liquid is proportional to the partial pressure of the gas
_____ 2. the component of a solution that is present in the lesser quantity
_____ 3. the component of a solution that is present in the greater quantity
_____ 4. the general term for a solute dissolved in a solvent
_____ 5. a region in a polar molecule with a partial negative and a partial positive charge
_____ 6. the overall direction of partial negative charge in a molecule
_____ 7. a dissolving liquid composed of polar molecules
_____ 8. a dissolving liquid composed of nonpolar molecules
_____ 9. the principle that solubility is greatest when the polarity of the solute is similar to that of the solvent
_____ 10. refers to liquids that are soluble in one another
_____ 11. refers to liquids that are not soluble in one another and separate into two layers
_____ 12. a cluster of water molecules surrounding a solute particle in solution
_____ 13. a mixture in which the dispersed particles ranges from 1 to 100 nm
_____ 14. the scattering of a beam of light by colloid particles
_____ 15. the maximum amount of solute that can dissolve in a solvent at a given temperature
_____ 16. a solution containing the maximum amount of solute that can dissolve at a given temperature
_____ 17. a solution containing less than the maximum amount of solute that can dissolve at a given temperature
_____ 18. a solution containing more solute than can ordinarily dissolve at a given temperature
_____ 19. a solution concentration expression that relates the mass of solute in grams dissolved in each 100 g of solution
_____ 20. a solution concentration expression that relates the moles of solute dissolved in each liter of solution

(a) colloid (*Sec. 14.4*)
(b) dipole (*Sec. 14.2*)
(c) Henry's law (*Sec. 14.1*)
(d) immiscible (*Sec. 14.2*)
(e) *like dissolves like* rule (*Sec. 14.2*)
(f) mass/mass percent (m/m %) (*Sec. 14.8*)
(g) miscible (*Sec. 14.2*)
(h) molarity (*M*) (*Sec. 14.9*)
(i) net dipole (*Sec. 14.2*)
(j) nonpolar solvent (*Sec. 14.2*)
(k) polar solvent (*Sec. 14.2*)
(l) saturated (*Sec. 14.7*)
(m) solubility (*Sec. 14.6*)
(n) solute (*Sec. 14.2*)
(o) solution (*Sec. 14.2*)
(p) solvent (*Sec. 14.2*)
(q) solvent cage (*Sec. 14.4*)
(r) supersaturated (*Sec. 14.7*)
(s) Tyndall effect (*Sec. 14.4*)
(t) unsaturated (*Sec. 14.7*)

Exercises‡

Gases in Solution (Sec. 14.1)

1. Indicate whether the solubility of ammonia gas in water increases or decreases for each of the following.
   - **(a)** The temperature of the solution changes from 20° to 0°C.
   - **(b)** The partial pressure of NH_3 changes from 754 mm Hg to 775 mm Hg.
2. Indicate whether the solubility of hydrogen chloride gas in water increases or decreases for each of the following.
   - **(a)** The temperature of the solution changes from 20° to 50°C.
   - **(b)** The partial pressure of HCl changes from 29.0 in. Hg to 27.5 in. Hg.
3. If 1.45 g of carbon dioxide dissolves in a liter of champagne at 1.00 atm, what is the solubility of carbon dioxide at 10.0 atm?

▲ **Illustration of Henry's Law** After a can of soda is opened and poured into a glass, the soda fizzes as the carbonation escapes. The carbon dioxide gas is less soluble in the glass than when it was under pressure in the can.

† Answers to Key Terms are in Appendix I.
‡ Answers to odd-numbered Exercises are in Appendix J.

4. If the solubility of nitrogen is 1.90 cm^3/100 cm^3 blood at 1.00 atm, what is the solubility of nitrogen in a scuba diver's blood at a depth of 155 feet where the pressure is 5.50 atm?
5. The solubility of chlorine gas is 0.63 g Cl_2/100 g water at 25°C and 760 mm Hg. What is the solubility of chlorine gas in water at 25°C and 1200 mm Hg?
6. The solubility of nitrous oxide gas is 0.121 g N_2O/100 g water at 20°C and 1.00 atm. What is the partial pressure of nitrous oxide required to dissolve 1.18 g of the gas in 100 g of water at 20°C?

Liquids in Solution (Sec. 14.2)

7. State whether the solutes and solvents in each of the following combinations are miscible or immiscible.
   (a) polar solute + polar solvent
   (b) polar solute + nonpolar solvent
8. State whether the solutes and solvents in each of the following combinations are miscible or immiscible.
   (a) nonpolar solute + polar solvent
   (b) nonpolar solute + nonpolar solvent

▲ **Oil and Water** The oil floating on the water is a result of a crude oil spill off the Texas coast in the Gulf of Mexico.

9. Predict whether each of the following solvents is polar or nonpolar.
   (a) water, H_2O (b) hexane, C_6H_{14}
   (c) acetone, C_3H_6O (d) chloroform, $CHCl_3$
10. Predict whether each of the following solvents is polar or nonpolar.
    (a) isopropyl alcohol, C_3H_7OH
    (b) pentane, C_5H_{12}
    (c) xylene, $C_6H_4(CH_3)_2$
    (d) trichloroethane, $C_2H_3Cl_3$
11. Predict whether each of the following solvents is miscible or immiscible with water.
    (a) heptane, C_7H_{16}
    (b) methanol, CH_3OH
    (c) methyl ethyl ketone, C_4H_8O
    (d) toluene, C_7H_8
12. Predict whether each of the following solvents is miscible or immiscible with hexane, C_6H_{14}.
    (a) ethanol, C_2H_5OH
    (b) chloroform, $CHCl_3$
    (c) trichloroethylene, C_2HCl_3
    (d) acetic acid, $HC_2H_3O_2$
13. In the laboratory how could you quickly determine whether an unknown liquid is polar or nonpolar?
14. An oil-and-vinegar salad dressing separates into two layers. Explain why the two liquids are immiscible using the *like dissolves like* rule.

Solids in Solution (Sec. 14.3)

15. State whether the following combinations of solute and solvent are generally soluble or insoluble.
    (a) polar solute + polar solvent
    (b) nonpolar solute + polar solvent
    (c) ionic solute + polar solvent
16. State whether the following combinations of solute and solvent are generally soluble or insoluble.
    (a) polar solute + nonpolar solvent
    (b) nonpolar solute + nonpolar solvent
    (c) ionic solute + nonpolar solvent
17. Predict whether each of the following compounds is soluble or insoluble in water.
    (a) naphthalene, $C_{10}H_8$
    (b) potassium hydroxide, KOH
    (c) calcium acetate, $Ca(C_2H_3O_2)_2$
    (d) trichlorotoluene, $C_7H_5Cl_3$
    (e) glycine, $C_2H_5NO_2$
    (f) lactic acid, $HC_3H_5O_3$
18. Predict whether each of the following compounds is soluble or insoluble in hexane, C_6H_{14}.
    (a) trichloroethylene, C_2HCl_3
    (b) iron(III) nitrate, $Fe(NO_3)_3$
    (c) sulfuric acid, H_2SO_4
    (d) dodecane, $C_{12}H_{26}$
    (e) mothballs, $C_6H_4Cl_2$
    (f) tartaric acid, $H_2C_4H_4O_6$
19. Predict whether each of the following vitamins is water-soluble or fat-soluble.
    (a) vitamin B_1, $C_{12}H_{18}Cl_2N_4OS$
    (b) vitamin B_3, $C_6H_6N_2O$
    (c) vitamin B_6, $C_8H_{11}NO_3$
    (d) vitamin C, $C_6H_8O_6$
    (e) vitamin D, $C_{27}H_{44}O$
    (f) vitamin K, $C_{31}H_{46}O_2$

20. Predict whether each of the following compounds is water-soluble or fat-soluble.
    (a) cholesterol, $C_{27}H_{46}O$
    (b) citric acid, $C_6H_8O_7$
    (c) fructose, $C_6H_{12}O_6$
    (d) glycine, $CH_2(NH_2)COOH$
    (e) lactic acid, $CH_3CH(OH)COOH$
    (f) alanine, $CH_3CH(NH_2)COOH$

The Dissolving Process (Sec. 14.4)

21. Diagram a molecule of fructose, $C_6H_{12}O_6$, dissolved in water.
22. Diagram a crystal of sucrose, $C_{12}H_{22}O_{11}$, dissolving in aqueous solution.
23. Diagram a formula unit of each of the following substances dissolved in water.
    (a) lithium bromide, LiBr
    (b) calcium chloride, $CaCl_2$
24. Diagram a formula unit of each of the following substances dissolved in water.
    (a) cobalt(II) sulfate, $CoSO_4$
    (b) nickel(II) nitrate, $Ni(NO_3)_2$

Rate of Dissolving (Sec. 14.5)

25. What three factors increase the rate of dissolving of a solid substance in solution?
26. Indicate whether each of the following increases, decreases, or has no effect on the rate at which 10.0 g of sugar dissolves in a liter of water.
    (a) using water from the refrigerator
    (b) shaking the sugar and water
    (c) using powdered sugar rather than crystals
    (d) using tap water rather than distilled water

Solubility and Temperature (Sec. 14.6)

27. How many grams of each of the following solutes can dissolve in 100 g of water at 20°C? (Refer to Figure 14.5.)
    (a) NaCl (b) KCl
28. How many grams of each of the following solutes can dissolve in 100 g of water at 30°C? (Refer to Figure 14.5.)
    (a) LiCl (b) $C_{12}H_{22}O_{11}$
29. Determine the maximum solubility of each of the following solid compounds at 40°C. (Refer to Figure 14.5.)
    (a) NaCl (b) KCl
30. Determine the maximum solubility of each of the following solid compounds at 50°C. (Refer to Figure 14.5.)
    (a) LiCl (b) $C_{12}H_{22}O_{11}$
31. What is the minimum temperature required to dissolve each of the following? (Refer to Figure 14.5.)
    (a) 34 g NaCl in 100 g of water
    (b) 50 g KCl in 100 g of water
32. What is the minimum temperature required to dissolve each of the following? (Refer to Figure 14.5.)
    (a) 90 g LiCl in 100 g of water
    (b) 120 g $C_{12}H_{22}O_{11}$ in 100 g of water
33. At what temperature is each of the following solutions saturated? (Refer to Figure 14.5.)
    (a) 40 g NaCl/100 g water
    (b) 40 g KCl/100 g water
34. At what temperature is each of the following solutions saturated? (Refer to Figure 14.5.)
    (a) 105 g LiCl/100 g water
    (b) 140 g $C_{12}H_{22}O_{11}$/100 g water

Unsaturated, Saturated, Supersaturated (Sec. 14.7)

35. State whether each of the following solutions is saturated, unsaturated, or supersaturated. (Refer to Figure 14.6.)
    (a) 110 g $NaC_2H_3O_2$ in 100 g of water at 50°C
    (b) 110 g $NaC_2H_3O_2$ in 100 g of water at 70°C
    (c) 110 g $NaC_2H_3O_2$ in 100 g of water at 90°C
36. State whether each of the following solutions is saturated, unsaturated, or supersaturated. (Refer to Figure 14.5.)
    (a) 105 g $C_{12}H_{22}O_{11}$/100 g H_2O at 25°C
    (b) 120 g $C_{12}H_{22}O_{11}$/100 g H_2O at 50°C
    (c) 130 g $C_{12}H_{22}O_{11}$/100 g H_2O at 45°C
37. State whether each of the following solutions is saturated, unsaturated, or supersaturated. (Refer to Figure 14.5.)
    (a) 45 g KCl/100 g H_2O at 20°C
    (b) 45 g KCl/100 g H_2O at 50°C
    (c) 45 g KCl/100 g H_2O at 70°C
38. State whether each of the following solutions is saturated, unsaturated, or supersaturated. (Refer to Figure 14.5.)
    (a) 35 g NaCl/100 g H_2O at 0°C
    (b) 35 g NaCl/100 g H_2O at 25°C
    (c) 35 g NaCl/100 g H_2O at 100°C
39. The solubility of rock salt at 30°C is 40.0 g/100 g water. If a solution contains 10.0 g of rock salt in 25.0 g of water at 30°C, is the solution saturated, unsaturated, or supersaturated?
40. The solubility of sugar at 50°C is 100.0 g/100 g water. If a solution contains 95.0 g of sugar in 250 g of water at 50°C, is the solution saturated, unsaturated, or supersaturated?
41. Assume that 100 g of LiCl is dissolved in 100 g of water at 100°C and the solution is allowed to cool to 20°C. (Refer to Figure 14.5.)
    (a) How much solute remains in solution?
    (b) How much solute crystallizes from the solution?
42. Assume that 120 g of $C_{12}H_{22}O_{11}$ is dissolved in 100 g of water at 100°C and the solution is allowed to cool to 20°C. (Refer to Figure 14.5.)
    (a) How much solute remains in solution?
    (b) How much solute crystallizes from the solution?

Mass Percent Concentration (Sec. 14.8)

43. Calculate the mass/mass percent concentration for each of the following solutions.
    (a) 1.25 g NaCl in 100.0 g of solution
    (b) 2.50 g $K_2Cr_2O_7$ in 95.0 g of solution
    (c) 10.0 g $CaCl_2$ in 250.0 g of solution
    (d) 65.0 g sugar in 125.0 g of solution
44. Calculate the mass/mass percent concentration for each of the following solutions.
    (a) 20.0 g KI in 100.0 g of water
    (b) 2.50 g $AgC_2H_3O_2$ in 95.0 g of water
    (c) 5.57 g $SrCl_2$ in 225 g of water
    (d) 50.0 g sugar in 250.0 g of water
45. Write three pairs of unit factors for each of the following aqueous solutions given the mass/mass percent concentration.
    (a) 1.50% KBr (b) 2.50% $AlCl_3$
    (c) 3.75% $AgNO_3$ (d) 4.25% Li_2SO_4
46. Write three pairs of unit factors for each of the following aqueous solutions given the mass/mass percent concentration.
    (a) 3.35% $MgCl_2$ (b) 5.25% $Cd(NO_3)_2$
    (c) 6.50% Na_2CrO_4 (d) 7.25% $ZnSO_4$
47. What mass of solution contains each of the following amounts of dissolved solute?
    (a) 5.36 g of glucose in a 10.0% solution
    (b) 25.0 g of sucrose in a 12.5% solution
48. What mass of solution contains each of the following amounts of dissolved solute?
    (a) 35.0 g of sulfuric acid in a 5.00% solution
    (b) 10.5 g of acetic acid in a 4.50% solution
49. How many grams of solute are dissolved in the following solutions?
    (a) 85.0 g of 2.00% $FeBr_2$ solution
    (b) 105.0 g of 5.00% Na_2CO_3 solution
50. How many grams of solute are dissolved in the following solutions?
    (a) 10.0 g of 6.00% KOH solution
    (b) 50.0 g of 5.00% nitric acid, HNO_3
51. What mass of water is necessary to prepare each of the following solutions?
    (a) 250.0 g of 0.90% saline solution
    (b) 100.0 g of 5.00% sugar solution
52. What mass of water is necessary to prepare each of the following solutions?
    (a) 250.0 g of 10.0% NaOH solution
    (b) 100.0 g of 5.00% hydrochloric acid, HCl

Molar Concentration (Sec. 14.9)

53. Calculate the molar concentration for each of the following solutions.
    (a) 1.50 g NaCl in 100.0 mL of solution
    (b) 1.50 g $K_2Cr_2O_7$ in 100.0 mL of solution
    (c) 5.55 g $CaCl_2$ in 125 mL of solution
    (d) 5.55 g Na_2SO_4 in 125 mL of solution
54. Calculate the molar concentration for each of the following solutions.
    (a) 1.00 g KCl in 75.0 mL of solution
    (b) 1.00 g Na_2CrO_4 in 75.0 mL of solution
    (c) 20.0 g $MgBr_2$ in 250.0 mL of solution
    (d) 20.0 g Li_2CO_3 in 250.0 mL of solution
55. Write two pairs of unit factors for each of the following aqueous solutions given the molar concentration.
    (a) 0.100 *M* LiI (b) 0.100 *M* $NaNO_3$
    (c) 0.500 *M* K_2CrO_4 (d) 0.500 *M* $ZnSO_4$
56. Write two pairs of unit factors for each of the following aqueous solutions given the molar concentration.
    (a) 0.150 *M* KBr (b) 0.150 *M* $Ca(NO_3)_2$
    (c) 0.333 *M* $Sr(C_2H_3O_2)_2$ (d) 0.333 *M* NH_4Cl
57. What volume of each of the following solutions contains the indicated amount of dissolved solute?
    (a) 10.0 g solute in 0.275 *M* NaF
    (b) 10.0 g solute in 0.275 *M* $CdCl_2$
    (c) 10.0 g solute in 0.408 *M* K_2CO_3
    (d) 10.0 g solute in 0.408 *M* $Fe(ClO_3)_3$
58. What volume of each of the following solutions contains the indicated amount of dissolved solute?
    (a) 2.50 g solute in 0.325 *M* KNO_3
    (b) 2.50 g solute in 0.325 *M* $AlBr_3$
    (c) 2.50 g solute in 1.00 *M* $Co(C_2H_3O_2)_2$
    (d) 2.50 g solute in 1.00 *M* $(NH_4)_3PO_4$
59. What is the mass of solute dissolved in the indicated volume of each of the following solutions?
    (a) 1.00 L of 0.100 *M* NaOH
    (b) 1.00 L of 0.100 *M* $LiHCO_3$
    (c) 25.0 mL of 0.500 *M* $CuCl_2$
    (d) 25.0 mL of 0.500 *M* $KMnO_4$
60. What is the mass of solute dissolved in the indicated volume of each of the following solutions?
    (a) 2.25 L of 0.200 *M* $FeCl_3$
    (b) 2.25 L of 0.200 *M* KIO_4
    (c) 50.0 mL of 0.295 *M* $ZnSO_4$
    (d) 50.0 mL of 0.295 *M* $Ni(NO_3)_2$
61. What is the molar concentration of a saturated solution of calcium sulfate that contains 0.209 g of solute in 100 mL of solution?
62. What is the molar concentration of a saturated solution of calcium hydroxide that contains 0.185 g of solute in 100 mL of solution?
63. A hospital glucose solution is analyzed to verify its concentration. A 10.0-mL sample with a mass of 10.483 g is evaporated to dryness. If the solid glucose residue has a mass of 0.524 g, what is (a) the mass/mass percent

concentration, and (b) the molar concentration of the glucose, $C_6H_{12}O_6$, solution?

64. A hospital saline solution is analyzed to confirm its concentration. A 50.0-mL sample with a mass of 50.320 g is evaporated to dryness. If the solid sodium chloride residue has a mass of 0.453 g, find (a) the mass/mass percent concentration, and (b) the molar concentration of the NaCl solution.

Dilution of a Solution (Sec. 14.10)

65. Concentrated acetic acid is a 17 *M* solution. What is the molarity of an $HC_2H_3O_2$ solution prepared by diluting 10.0 mL of concentrated acid with distilled water to a total volume of 500.0 mL?

66. Concentrated nitric acid is available as a 16 *M* solution. What volume of nitric acid must be diluted with distilled water to prepare 2.25 L of 0.10 *M* HNO_3?

Solution Stoichiometry (Sec. 14.11)

67. Given that 50.0 mL of 0.100 *M* magnesium bromide reacts completely with 13.9 mL of silver nitrate solution according to the unbalanced equation

$$MgBr_2(aq) + AgNO_3(aq) \longrightarrow AgBr(s) + Mg(NO_3)_2(aq)$$

(a) What is the molarity of the $AgNO_3$ solution?
(b) What is the mass of AgBr precipitate?

68. Given that 24.0 mL of 0.170 *M* sodium iodide reacts with 0.209 *M* mercury(II) nitrate solution according to the unbalanced equation

$$Hg(NO_3)_2(aq) + NaI(aq) \longrightarrow HgI_2(s) + NaNO_3(aq)$$

(a) What volume of $Hg(NO_3)_2$ is required for complete precipitation?
(b) What is the mass of HgI_2 precipitate?

General Exercises

69. In foggy weather automobile headlights demonstrate the Tyndall effect. What is the approximate size of the water droplets in the fog?

70. At a concert in an auditorium the stage lights demonstrate the Tyndall effect. What can you conclude about the air in the concert auditorium?

71. Indicate whether a solution or a colloid produces each of the following observations.
(a) Dispersed particles separate in a centrifuge.
(b) Dispersed particles demonstrate the Tyndall effect.
(c) Dispersed particles pass through a semipermeable membrane.

72. Indicate whether a suspension or a colloid produces each of the following observations.
(a) Dispersed particles settle from solution.
(b) Dispersed particles scatter light.
(c) Dispersed particles separate with filter paper.

73. Scuba divers can experience "the bends" if they surface too quickly. The nitrogen in the air they breathe is more soluble in blood at depths and less soluble as the divers return to the surface. The pressure at 125 feet is 4.68 atm. If the solubility is 0.0019 g N_2 per 100 g of blood at normal pressure, what is the solubility of nitrogen at 125 feet?

74. Explain why scuba divers who dive deeper than 125 feet use a special gas mixture of oxygen and helium rather than oxygen and nitrogen.

75. Calculate the mass of sulfur dioxide gas in 1.00 L of saturated solution at 20°C. The solubility of SO_2 at 20°C is 22.8 g/100 mL.

76. Calculate the mass of chlorine gas in 500.0 mL of saturated solution at 20°C. The solubility of Cl_2 at 20°C is 0.63 g/100 mL.

77. Explain why bubbles form on the bottom and sides of a pan when water is heated.

78. Ethyl ether, $C_4H_{10}O$, is partially miscible with water. Explain why ether is only partially soluble even though the molecule contains a polar oxygen atom.

79. Listed below is the solubility of different alcohols in water. Expain the decrease in solubility.

Alcohol	Solubility
ethanol, C_2H_5OH	miscible
propanol, C_3H_7OH	97.2 g/100 mL H_2O
butanol, C_4H_9OH	7.9 g/100 mL H_2O
hexanol, $C_6H_{13}OH$	0.59 g/100 mL H_2O
decanol, $C_{10}H_{21}OH$	immiscible

80. Predict whether water or carbon tetrachloride, CCl_4, is a better solvent for each of the following household substances.
(a) grease **(b)** maple syrup
(c) food coloring **(d)** gasoline

81. Identify the solutes and solvents in the following solutions.
(a) 80 proof alcohol (40% ethanol in water)
(b) laboratory alcohol (95% ethanol in water)

82. Explain why grinding the solute increases the rate of dissolving of a solid substance in water.

83. If a household bleach solution contains 5.25% NaClO, what is the molarity of the sodium hypochlorite? (Assume the density is 1.04 g/mL.)

84. If a vinegar solution contains 5.25% $HC_2H_3O_2$, what is the molarity of the acetic acid? (Assume the density is 1.01 g/mL.)

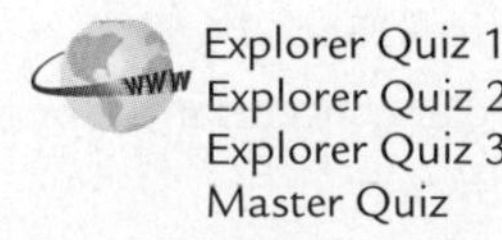

Explorer Quiz 1
Explorer Quiz 2
Explorer Quiz 3
Master Quiz

CHAPTERS 12–14

Cumulative Review

Key Concepts

1. What type of chemical bond results from the attraction between a metal cation and a nonmetal anion?
2. Which noble gas is isoelectronic with a calcium ion? with a chloride ion?
3. Which of the following is held together by covalent bonds: NaCl, MgO, CuS, AgBr, HI?
4. According to the general trends in the periodic table, predict which of the following molecules is nonpolar: H_2O, CO, NO, O_2, ICl?
5. What is the electron pair geometry for a water, H_2O, molecule? What is the molecular shape of a H_2O molecule?
6. A beaker of marbles is covered with molasses that roll about slowly as the beaker is shaken. Which physical state is described by this analogy?
7. Explain why aluminum foil can float on water even though the density of aluminum is 2.70 times greater than the density of water.
8. Which type of crystalline solid is malleable, ductile, electrical conductor, and insoluble in most solvents?
9. What is the amount of heat required to melt 125 g of ice at 0°C?
10. What is the percentage of water in nickel(II) chloride hexahydrate, $NiCl_2 \cdot 6H_2O$?
11. Why does gasoline float on water?
12. Predict which of the following vitamins is fat-soluble: vitamin A ($C_{20}H_{30}O$), vitamin C ($C_6H_8O_6$), vitamin D ($C_{27}H_{44}O$), vitamin E ($C_{29}H_{50}O_2$).
13. What is the solubility of NaCl at 50°C? (Refer to Figure 14.5.)
14. If a solution contains 100 g of $NaC_2H_3O_2$ in 100 g of water at 55°C, is the solution unsaturated, saturated, or supersaturated? (Refer to Figure 14.6.)
15. A 10.0-mL solution of glucose, $C_6H_{12}O_6$, with a mass of 10.483 g, is evaporated to dryness and has a mass of 0.524 g. What is (a) the mass percent concentration and (b) the molar concentration of the solution?

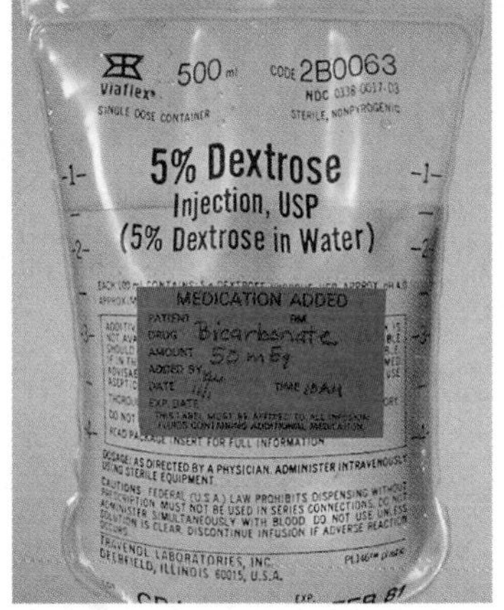

▲ Glucose (Dextrose) Solution Intravenous glucose, $C_6H_{12}O_6$, solutions are often administered to patients suffering from trauma and having low blood sugar.

Key Terms

State the key term that corresponds to each of the following descriptions.

_____ 1. a chemical bond characterized by sharing a pair of valence electrons

_____ 2. an atom must be surrounded by eight valence electrons to be stable

_____ 3. a bond composed of two electron pairs shared by two atoms

_____ 4. the ability of an atom to attract a shared pair of electrons

_____ 5. a bond in which a pair of electrons is shared unequally

_____ 6. a model that explains the shapes of molecules as a result of electron pairs about the central atom repelling each other

_____ 7. the angle formed by two atoms attached to the central atom in a molecule

_____ 8. the temperature at which the vapor pressure of a liquid is equal to the atmospheric pressure

_____ 9. an intermolecular attraction based on temporary dipoles

_____ 10. an intermolecular attraction based on permanent dipoles

_____ 11. an intermolecular attraction between a hydrogen atom and an oxygen atom

_____ 12. the heat required to vaporize a liquid to a gas at its boiling point

_____ 13. the overall direction of partial negative charge in a molecule

_____ 14. water containing sodium ions and a variety of anions
_____ 15. water purified by removing ions using an ion exchange method
_____ 16. a dissolving liquid composed of polar molecules
_____ 17. refers to liquids that are soluble in one another
_____ 18. a solution that contains the maximum amount of solute that will dissolve at a given temperature
_____ 19. a solution concentration expression that relates the mass of solute in grams dissolved in each 100 grams of solution
_____ 20. a solution concentration expression that relates the moles of solute dissolved in each liter of solution

Review Exercises

1. Which of the following statements is *not* true regarding the formation of an ionic bond between a metal and a nonmetal?
   (a) An ionic bond is formed by the attraction of metal and nonmetal ions.
   (b) The ionic radius of a metal atom is less than its atomic radius.
   (c) The ionic radius of a nonmetal atom is less than its atomic radius.
   (d) The simplest representative particle is a formula unit.
2. Which of the following statements is *not* true regarding the formation of a covalent bond between two nonmetal atoms?
   (a) Valence electrons are shared by two nonmetal atoms.
   (b) The bonding electrons are found delocalized about the molecule.
   (c) The bond length is equal to the sum of the two atomic radii.
   (d) Energy is released when a covalent bond is formed.
3. Write the electron dot formula and draw the structural formula for a molecule of sulfur trioxide, SO_3.
4. Write the electron dot formula and draw the structural formula for a polyatomic sulfite ion, SO_3^{2-}.
5. Predict the electron pair geometry, molecular shape, and bond angle for a molecule of NH_3 using VSEPR theory.
6. Which type of intermolecular bond attraction (dispersion force, dipole force, or hydrogen bond) exists in each of the following liquids?
   (a) C_6H_{14} **(b)** $CH_3—OH$ **(c)** $CH_3—O—CH_3$
7. If a liquid exhibits a strong intermolecular attraction, which of the following has a low value: vapor pressure, boiling point, viscosity, surface tension?
8. State the type of crystalline solid demonstrated by each of the following.
   (a) copper, Cu **(b)** sulfur, S_8 **(c)** copper sulfide, CuS
9. Calculate the number of calories necessary to convert 20.0 g of solid ice at its melting point to steam at 100°C.
10. An unknown hydrate of zinc sulfate, $ZnSO_4$, is heated to give 43.8% water. What is the water of hydration for the hydrate?
11. Apply the *like dissolves like* rule to predict which of the following is insoluble in water: NaI, $NaIO_2$, $NaIO_3$, $NaIO_4$, I_2.
12. What is the mass of solution that contains 10.0 g of dissolved solute in a 12.5% seawater sample?
13. What is the mass of sodium hydroxide dissolved in 0.500 L of 0.100 *M* NaOH solution?
14. What volume of 0.100 *M* hydrochloric acid contains 1.00 g of HCl?
15. If 25.0 mL of 0.100 *M* calcium chloride reacts completely with 45.9 mL of silver nitrate solution, what is the molarity of the $AgNO_3$ solution?

$$CaCl_2(aq) + AgNO_3(aq) \longrightarrow AgCl(s) + Ca(NO_3)_2(aq)$$

CHAPTER 15

Acids and Bases

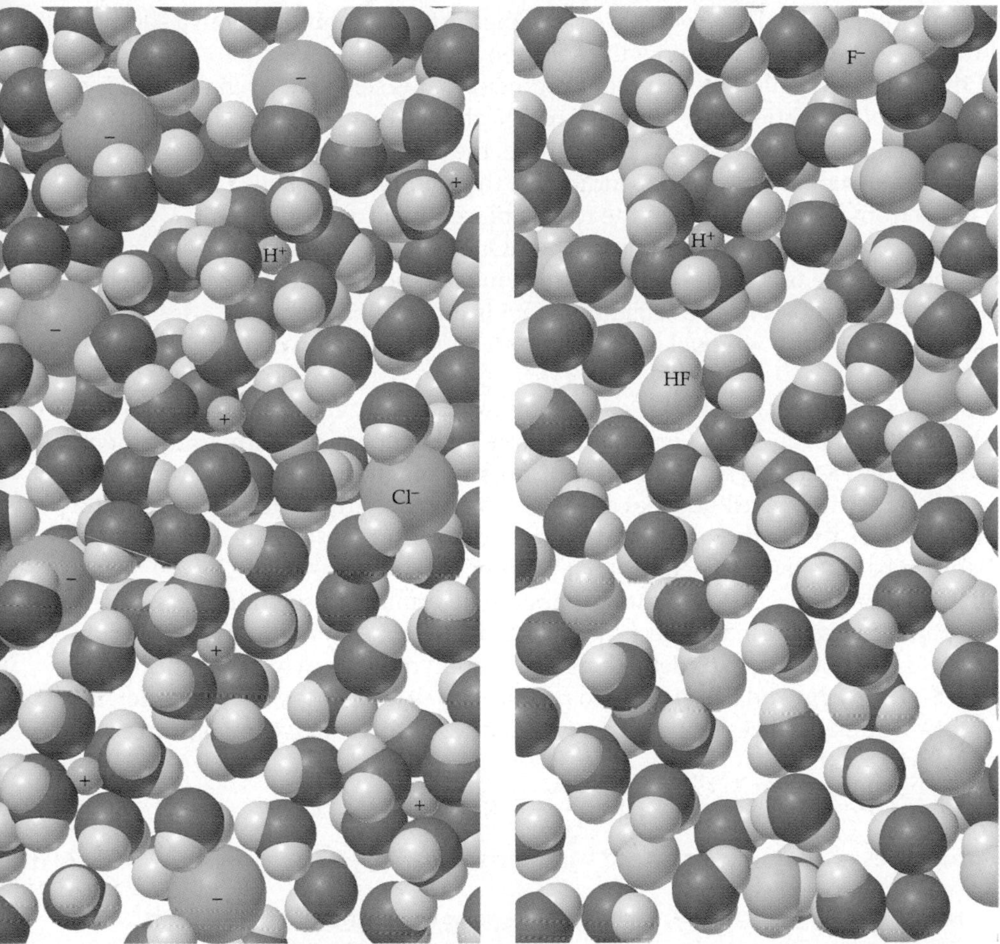

▲ One of these panels represents aqueous HCl, and the other panel represents aqueous HF. Which acid is highly ionized and is a strong acid? (H^+ is shown in light gray, Cl^- in green, and F^- in yellow.)

This chapter lays an important foundation for the equilibrium concept found in the following chapter on *Chemical Equilibrium*. The molecular art provides a visual introduction to the degree of ionization of strong and weak acids and bases in aqueous solution.

Acids and bases play an important role in our lives. The proper acidity of our blood and other body fluids is vital to our well-being and is carefully controlled by an elaborate buffering system. Many of the foods we eat are acidic. For instance, oranges and lemons contain citric acid, and vinegar is a solution of acetic acid. Vitamin C is ascorbic acid, and aspirin is acetylsalicylic acid. We adjust the acidity of our swimming pools with muriatic acid and fill our car batteries with sulfuric acid.

Baking soda and milk of magnesia are basic. Antacid tablets contain basic substances such as carbonates, bicarbonates, and hydroxides for neutralizing an acid, upset stomach. We use a dilute solution of basic ammonia to clean floors and a concentrated solution of caustic sodium hydroxide to clean drains and ovens.

15.1 Properties of Acids and Bases

Objectives
- To list the general properties of acids and bases.
- To classify a solution of given pH as strongly acidic, weakly acidic, neutral, weakly basic, or strongly basic.

Introduction to Acids and Bases Movie

An *acid* is any substance that produces hydrogen ions, H^+, in water. This property of acids allows us to test for acidity. In the laboratory we can determine if a solution is acidic using a strip of litmus paper. Blue litmus paper turns red in the presence of hydrogen ions. Therefore, if we put a piece of blue litmus paper in a solution and the paper turns red, we know the solution is acidic.

Litmus paper is made from a plant pigment that is sensitive to changes in hydrogen ion concentration. The colors of many plants are affected by the acidity of the soil in which they are grown. For example, the color of an orchid can vary from pale lavender to deep purple, depending on the acidity of the soil. The color of a rose fades from red to pink when the rose is placed in an acid solution.

Acids have their own special properties. One property is that they have a sour taste. The tart taste of a lemon, an apple, or vinegar shows us that these are acidic foods. The taste buds that are sensitive to acids are located along the edge of our tongue, which is why we pucker and roll our tongue when we taste something that is sour.

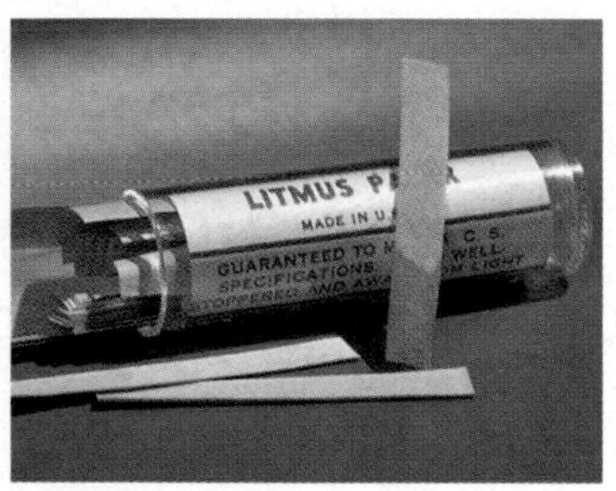

▲ **Litmus Paper** If a strip of red litmus paper turns blue when placed in a solution, it indicates that the solution is basic.

A *base* is any substance that produces hydroxide ions, OH^-, in solution. Bases also have special properties. For instance, they feel slippery or soapy to the touch and have a bitter taste. The bitter taste of milk of magnesia shows us that it is a basic substance. The taste buds that are sensitive to bases are located on the tongue toward the back of our throat, which is why we stick out our tongue and may feel nauseous when we taste something that is bitter.

In the laboratory, we use red litmus paper to determine if a solution is basic. Red litmus paper turns blue in the presence of hydroxide ion. Another property of acids and bases is their ability to undergo a neutralization reaction, which we studied in Section 8.11. An acid and a base react to produce a salt and water. For example, hydrochloric acid neutralizes a potassium hydroxide solution to give potassium chloride and water.

$$HCl(aq) + KOH(aq) \longrightarrow KCl(aq) + H_2O(l)$$

The properties of acids and bases are summarized in Table 15.1.

Table 15.1 Properties of Acids and Bases

Property	Acidic Solutions	Basic Solutions
taste	sour	bitter
feel	—	slippery, soapy
litmus paper	blue litmus turns red	red litmus turns blue
pH value	less than 7	greater than 7
neutralization reaction	react with bases to give a salt and water	react with acids to give a salt and water

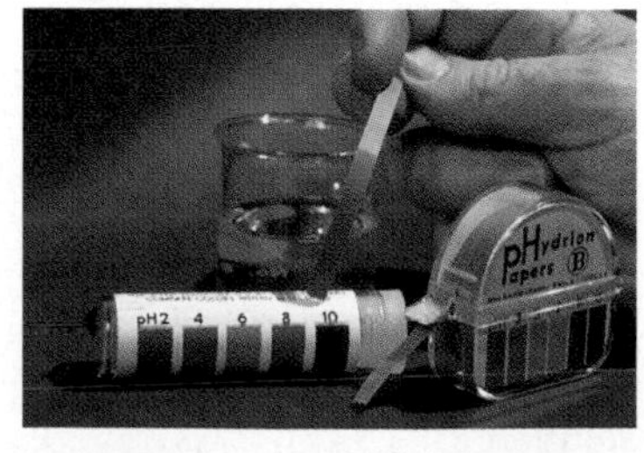

▲ **pHydrion Paper** After a strip of broad-range indicator paper is placed in a solution, the test strip is compared to a color chart in order to estimate the pH of the solution.

pH Scale

In Section 15.8 we will discuss the pH scale in detail. For now, all we need to know is that the pH value expresses the acidity of a solution. Most solutions have a pH between 0 and 14. For example, a 1 *M* HCl solution has a pH of 0, and a 1 *M* NaOH solution has a pH of 14. A pH of 7 is considered neutral. Pure distilled water has a pH of 7.

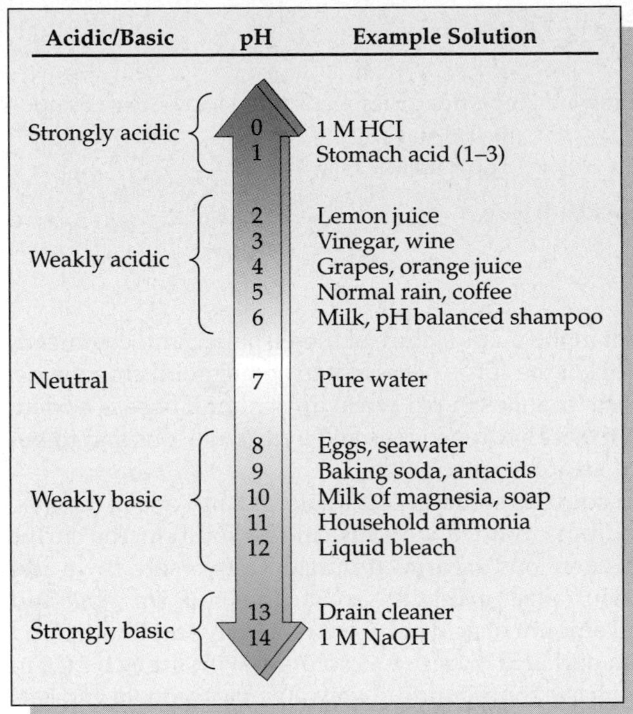

Acidic/Basic
pH
Example Solution
Strongly acidic
0 1 M HCl
1 Stomach acid (1–3)
Weakly acidic
2 Lemon juice
3 Vinegar, wine
4 Grapes, orange juice
5 Normal rain, coffee
6 Milk, pH balanced shampoo
Neutral
7 Pure water
Weakly basic
8 Eggs, seawater
9 Baking soda, antacids
10 Milk of magnesia, soap
11 Household ammonia
12 Liquid bleach
Strongly basic
13 Drain cleaner
14 1 M NaOH

◀ **Figure 15.1 The pH Scale** Note that the solutions may be strongly acidic (pH ~ 1), weakly acidic (pH ~ 4), neutral (pH ~ 7), weakly basic (pH ~ 10), or strongly basic (pH ~ 14).

On the pH scale, acidic solutions have a pH less than 7. As the acidity increases, the pH value *decreases*. Thus, a solution having a pH of 3 is more acidic than a solution with a pH of 4. Basic solutions have a pH greater than 7. As the basicity increases, the pH value *increases*. Thus, a solution having a pH of 11 is more basic than a solution with a pH of 10.

A solution can be classified as strongly acidic, weakly acidic, neutral, weakly basic, or strongly basic on the basis of pH. A strongly acidic solution has a pH between 0 and 2. A weakly acidic solution has a pH between 2 and 7. A weakly basic solution has a pH between 7 and 12. A strongly basic solution has a pH between 12 and 14. Figure 15.1 illustrates some common substances and their approximate pH.

The following example exercise illustrates the acid–base strength of a few common solutions.

Example Exercise 15.1 • The pH Scale

Indicate whether each of the following solutions is considered strongly acidic, weakly acidic, neutral, weakly basic, or strongly basic.

(a) gastric juice, pH 1.5
(b) oven cleaner, pH 13.5
(c) orange juice, pH 4.5
(d) deionized water, pH 7.0
(e) eggs, pH 7.5
(f) carbonated soda, pH 4.0

Solution

Refer to the guidelines in Figure 15.1 to classify each solution.

(a) Gastric juice has a pH between 0 and 2. It is *strongly acidic.*
(b) Oven cleaner has a pH of 13.5. It is *strongly basic.*
(c) Orange juice has a pH between 2 and 7. It is *weakly acidic.*
(d) Deionized water has a pH of 7.0 and is *neutral.*
(e) Eggs have a pH of 7.5 and is *weakly basic.*
(f) Carbonated soda has a pH of 4.0 and is *weakly acidic.*

(continued)

Example Exercise 15.1 *(continued)*

Self-Test Exercise

Indicate whether each of the following properties corresponds to an acid or to a base.

(a) sour taste
(b) slippery feel
(c) turns blue litmus paper red
(d) pH greater than 7

Answers: (a) acid; (b) base; (c) acid; (d) base

Buffers

Controlling acidity is important in the manufacture of foods, paper, and chemicals. In agriculture, controlling the pH of the soil is necessary to obtain good crop yields. A **buffer** is a solution that resists changes in pH when an acid or a base is added. A buffer solution can consume excess hydrogen ions and hydroxide ions and thereby helps to maintain the pH of an aqueous solution.

A buffer is composed of an aqueous solution of a weak acid and one of its salts, for example, citric acid and sodium citrate. If we add acid to a solution, the citrate ion combines with excess hydrogen ions to form citric acid. Conversely, if we add base to a solution, citric acid neutralizes excess hydroxide ions to form a salt and water. As long as the additional amount of acid or base is relatively small compared to the amount of buffer, we can add acid or base to a solution with little change in the pH. Of the many acidic and basic compounds in seawater, bicarbonate and carbonate ions are present in the highest concentrations and maintain the pH near 8. A similar buffer system controls the pH of blood, which must be kept near 7.4.

▲ **Hydrofluoric Acid, HF**
Although aqueous HF is a weak acid, it attacks glass. This design was created by coating the glass object with wax and then removing wax from the pattern areas. The object was dipped into hydrofluoric acid which etched the design into the glass.

15.2 Arrhenius Acids and Bases

Objectives
- To identify strong and weak Arrhenius acids and bases given the degree of ionization in aqueous solution.
- To identify the Arrhenius acid and base that react to produce a given salt.

In 1884 the Swedish chemist Svante Arrhenius suggested the first definitions for an acid and a base. He proposed that an **Arrhenius acid** is a substance that ionizes in water to produce hydrogen ions. Conversely, he proposed that an **Arrhenius base** is a substance that dissociates in water to release hydroxide ions.

Acids and bases are of varying strengths. The strength of an Arrhenius acid is measured by the degree of ionization in solution. **Ionization** is the process by which the molecules in a polar compound form cations and anions. The strength of an Arrhenius base is measured by the degree of dissociation in solution. **Dissociation** is the process whereby the already existing ions in an ionic compound simply separate. Thus, a molecule of HCl ionizes into H^+ and Cl^-, whereas a formula unit of NaOH dissociates into Na^+ and OH^-.

Arrhenius Acids

An acid is considered either strong or weak depending on how much it ionizes. According to the Arrhenius definition, a strong acid ionizes extensively to release hydrogen ions in solution. Hydrochloric acid is considered a strong acid because it ionizes nearly 100%. Acetic acid is a weak acid because it is only slightly ionized, about 1%. Table 15.2 lists some common Arrhenius acids.

All Arrhenius acids have a hydrogen atom attached to an acid molecule by a polar bond. Most acids are weak ternary oxyacids in which the H atom is attached to an O atom, for example, HOCl. When the acid molecule ionizes, it breaks this bond. In an aqueous solution, polar water molecules help the acid ionize by pulling

Table 15.2 Common Arrhenius Acids

Aqueous Acids	Percent Ionization	Acid Strength
hydrochloric acid, $HCl(aq)$	~100%	strong
nitric acid, $HNO_3(aq)$	~100%	strong
sulfuric acid, $H_2SO_4(aq)$	~100%	strong
acetic acid, $HC_2H_3O_2(aq)$	~1%	weak
carbonic acid, $H_2CO_3(aq)$	~1%	weak
hydrofluoric acid, $HF(aq)$	~1%	weak
phosphoric acid, $H_3PO_4(aq)$	~1%	weak

the hydrogen ion away from the acid molecule. We can show the ionization for hydrochloric acid, HCl, and hypochlorous acid, HOCl, as follows.

$$HCl(aq) + H_2O(l) \longrightarrow H_3O^+(aq) + Cl^-(aq) \qquad (\sim 100\%)$$
$$HOCl(aq) + H_2O(l) \longrightarrow H_3O^+(aq) + OCl^-(aq) \qquad (\sim 1\%)$$

Although chemists often refer to acids as solutions with hydrogen ions, the solutions actually contain H_3O^+. The **hydronium ion, H_3O^+**, is formed when an aqueous hydrogen ion attaches to a water molecule. For simplicity, we will usually designate a hydrogen ion in aqueous solution as $H^+(aq)$.

Arrhenius Bases

Recall that a base is considered either strong or weak depending on how much it dissociates. According to the Arrhenius definition, a strong base dissociates extensively to release hydroxide ions in solution. Sodium hydroxide is a strong base and dissociates nearly 100% in aqueous solution. Ammonium hydroxide is a weak base because it provides relatively few ions in solution. Table 15.3 lists some common Arrhenius bases.

When we dissolve sodium hydroxide in water, we obtain aqueous NaOH. In the dissolving process, NaOH dissociates into aqueous sodium ions and aqueous hydroxide ions. Aqueous NH_4OH is a weak base and provides few hydroxide ions in solution. We can show this process for the two Arrhenius bases is as follows.

$$NaOH(aq) \longrightarrow Na^+(aq) + OH^-(aq) \qquad (\sim 100\%)$$
$$NH_4OH(aq) \longrightarrow NH_4^+(aq) + OH^-(aq) \qquad (\sim 1\%)$$

The following example exercise illustrates the classification of acid–base strength based on ability to donate ions in solution.

Table 15.3 Common Arrhenius Bases

Aqueous Bases	Percent Dissociation	Base Strength
barium hydroxide, $Ba(OH)_2(aq)$	~100%	strong
calcium hydroxide, $Ca(OH)_2(aq)$	~100%	strong
lithium hydroxide, $LiOH(aq)$	~100%	strong
potassium hydroxide, $KOH(aq)$	~100%	strong
sodium hydroxide, $NaOH(aq)$	~100%	strong
ammonium hydroxide, $NH_4OH(aq)$*	~1%	weak

*Ammonium hydroxide is prepared by dissolving ammonia gas in water. It is referred to as an aqueous ammonia solution, $NH_3(aq)$, or as ammonia water, $NH_3 \cdot H_2O$. The name "ammonium hydroxide" is somewhat misleading since there is no evidence for ammonium hydroxide molecules in solution. For simplicity, however, we will refer to an ammonia solution as ammonium hydroxide, $NH_4OH(aq)$.

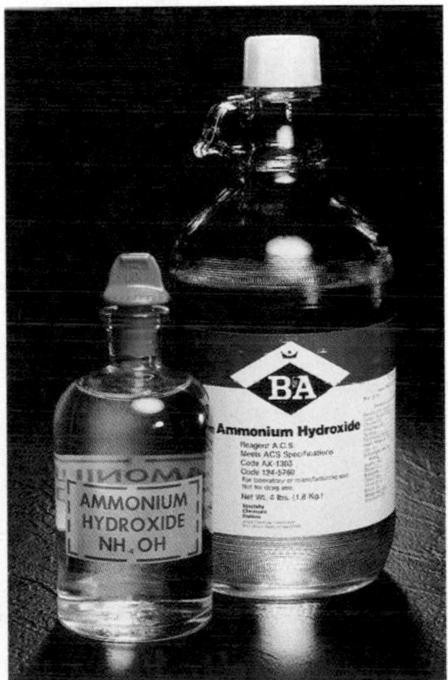

▲ **Ammonium Hydroxide, NH_4OH** Ammonia gas dissolved in water is commonly referred to as an ammonium hydroxide solution.

Example Exercise 15.2 • Strong and Weak Acids

Classify each of the following solutions as a strong or a weak Arrhenius acid given the degree of ionization.

(a) perchloric acid, $HClO_4(aq)$ ~100%
(b) hypochlorous acid, $HClO(aq)$ ~1%

Solution

To classify an aqueous solution as a strong or a weak acid we must know the amount of ionization.

(a) Perchloric acid is extensively ionized and is considered a *strong acid*. An aqueous solution is primarily $H^+(aq)$ and $ClO_4^-(aq)$.
(b) Hypochlorous acid is slightly ionized and is considered *weak acid*. An aqueous solution contains primarily molecules of HClO.

Self-Test Exercise

Classify each of the following solutions as a strong or a weak Arrhenius base given the degree of dissociation.

(a) magnesium hydroxide, $Mg(OH)_2(aq)$ ~1%
(b) strontium hydroxide, $Sr(OH)_2(aq)$ ~100%

Answers: (a) weak base; (b) strong base

Neutralization Reactions

We learned in Section 8.11 that an acid neutralizes a base to give a **salt** and water. For example, hydrochloric acid reacts with sodium hydroxide to give the salt, sodium chloride, and water. The equation for the reaction is

$$HCl(aq) + NaOH(aq) \longrightarrow NaCl(aq) + H_2O(l)$$

This reaction produces the aqueous salt NaCl. Different acids give other salts. When we completely neutralize sulfuric acid using aqueous sodium hydroxide, for example, we obtain sodium sulfate. The equation is

$$H_2SO_4(aq) + 2\,NaOH(aq) \longrightarrow Na_2SO_4(aq) + 2\,H_2O(l)$$

The hydrogen ions from an acid are neutralized by the hydroxide ions from a base. The hydrogen ions and hydroxide ions combine to form water. Note from the above neutralization reactions that Na_2SO_4 is the salt produced.

We can identify the Arrhenius acid and base that produce any given salt because each salt is composed of the cation from the reacting base and the anion from the acid. As an example, we can predict the neutralization reaction that produces the salt potassium acetate, $KC_2H_3O_2$. Since the salt contains potassium, the neutralized base must be potassium hydroxide, KOH. The acetate ion came from acetic acid, $HC_2H_3O_2$. The equation for the reaction is

$$HC_2H_3O_2(aq) + KOH(aq) \longrightarrow KC_2H_3O_2(aq) + H_2O(l)$$

The following example exercise further illustrates the neutralization of acids and bases to give a salt and water.

Example Exercise 15.3 • Neutralization Reactions

Determine the acid and base that produce each of the following salts. Write a balanced equation for the neutralization reaction.

(a) lithium fluoride, $LiF(aq)$ (b) calcium sulfate, $CaSO_4(aq)$

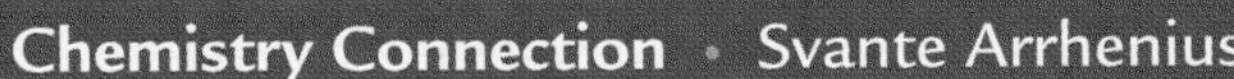

Chemistry Connection • Svante Arrhenius

What grade was Arrhenius awarded for his brilliant Ph.D. dissertation on the theory of ionic solutions?

Svante Arrhenius (1859–1927) was a child genius who taught himself to read by age 3. After graduating from high school at the top of his class, he enrolled at the University of Uppsala in his native Sweden. He majored in chemistry and did research on the passage of electricity through aqueous solutions.

▲ Svante August Arrhenius (1859–1927)

At 22, Arrhenius began his doctoral work at the University of Stockholm. For his graduate thesis, he continued to pursue his interest in the behavior of solutions. In particular, he found it puzzling that sodium chloride solutions conducted electricity, whereas sugar solutions did not. After carefully considering his observations, Arrhenius boldly proposed that a solution of sodium chloride conducts an electric current because it separates into charged particles in solution. Sugar does not conduct electricity because it does not form charged particles in solution.

Arrhenius also noted that the freezing point of water was lowered twice as much for a salt solution as it was for a sugar solution. He explained that when salt dissolves in solution, NaCl separates into sodium ions and chloride ions. When sugar dissolves in solution, it remains as molecules.

Arrhenius was aware, however, that there were difficulties with the concept of charged ions in solution. At the time, the scientific community considered atoms to be indivisible, electrically neutral particles.

In 1884 Arrhenius defended his Ph.D. dissertation on the theory of ionic solutions. After a grueling four-hour defense, he was awarded the lowest possible passing grade. The dissertation committee simply was not ready to accept the idea of ions in solution.

Arrhenius' career stagnated for almost a decade, and his ideas found little support. The ionic theory was eventually championed by Wilhelm Ostwald, a noted chemist, who invited Arrhenius to work with him. Gradually, evidence accumulated that supported the concept of ions. Most notably, the discovery of the electron in 1897 proved the existence of charged subatomic particles.

After the ionic theory gained credibility, Arrhenius was offered a professorship at the University of Stockholm. In 1899, he published the classic paper, "On the Dissociation of Substances in Aqueous Solutions." In 1903, after struggling for 20 years to establish his ideas, Arrhenius' was awarded the Nobel prize in chemistry for his ionic theory of solutions.

Arrhenius received a barely passing grade; the examining committee simply did not believe in the ionic theory of solutions.

Solution

(a) The salt LiF is produced from the neutralization of LiOH and HF acid. The equation for the reaction is

$$HF(aq) + LiOH(aq) \longrightarrow LiF(aq) + H_2O(l)$$

(b) Calcium sulfate is produced by the neutralization of $Ca(OH)_2$ and H_2SO_4. The equation for the reaction is

$$H_2SO_4(aq) + Ca(OH)_2(aq) \longrightarrow CaSO_4(aq) + 2\,H_2O(l)$$

Self-Test Exercise

Determine the acid and base that produce each of the following salts. Write a balanced equation for the neutralization reaction.

(a) potassium iodide, KI(aq) (b) barium nitrate, $Ba(NO_3)_2(aq)$

Answers: (a) HI is the acid, and KOH is the base:

$$HI(aq) + KOH(aq) \longrightarrow KI(aq) + H_2O(l)$$

(b) HNO_3 is the acid, and $Ba(OH)_2$ is the base:

$$2\,HNO_3(aq) + Ba(OH)_2(aq) \longrightarrow Ba(NO_3)_2(aq) + 2\,H_2O(l)$$

15.3 Brønsted–Lowry Acids and Bases

Objective · To identify the Brønsted–Lowry acid and base in a given neutralization reaction.

In 1923 the Danish chemist Johannes Brønsted and the English chemist Thomas Lowry independently proposed broader definitions than those of Arrhenius for an acid and a base. Brønsted and Lowry both defined an acid as a substance that is a hydrogen ion donor. Whereas Arrhenius defined an acid as a substance that donates a hydrogen ion in water, Brønsted and Lowry defined an acid as a substance that donates a hydrogen ion to any other substance. Since a hydrogen ion is simply a proton, a **Brønsted–Lowry acid** is also referred to as a **proton donor**.

Recall the Arrhenius definition of a base as a hydroxide ion donor in water. We know that hydroxide ions neutralize hydrogen ions to form water. There are, however, many substances that neutralize hydrogen ions in addition to hydroxide ions. Brønsted and Lowry proposed that a base is any substance that accepts a hydrogen ion. Thus, a **Brønsted–Lowry base** is a **proton acceptor**.

Although the Brønsted–Lowry definitions of an acid and a base are a bit different from the Arrhenius definitions, an acid and a base still neutralize each other. The following reactions illustrate the neutralization of a Brønsted–Lowry base by hydrochloric acid.

$$HCl(aq) + NaOH(aq) \longrightarrow NaCl(aq) + H_2O(l)$$

$$HCl(aq) + NH_3(aq) \longrightarrow NH_4Cl(aq)$$

$$HCl(aq) + H_2O(l) \longrightarrow H_3O^+(aq) + Cl^-(aq)$$

In the first equation, aqueous NaOH accepts a proton. According to the Brønsted–Lowry definition, it is therefore a base. In the second equation, aqueous NH_3 accepts a proton. It is also a base. As the third equation shows, even water can act as a Brønsted–Lowry base because it can accept a hydrogen ion.

Unlike Arrhenius acids and bases that depend on hydrogen ions and hydroxide ions dissolved in water, Brønsted–Lowry acids and bases depend on a particular reaction. For example, aqueous $NaHCO_3$ can act as a *base* by accepting a proton. In a different reaction, aqueous $NaHCO_3$ can act as an *acid* by donating a proton. The following examples illustrate.

$$HCl(aq) + NaHCO_3(aq) \longrightarrow NaCl(aq) + H_2CO_3(aq)$$

$$NaOH(aq) + NaHCO_3(aq) \longrightarrow Na_2CO_3(aq) + H_2O(l)$$

In the first reaction, $NaHCO_3$ accepts a proton from HCl and therefore acts as a Brønsted–Lowry base. In the second reaction, $NaHCO_3$ donates a proton to NaOH and therefore acts as a Brønsted–Lowry acid. A substance that is capable of both accepting and donating a proton is said to be **amphiprotic**. According to the Brønsted–Lowry theory, water is amphiprotic. That is, H_2O can act as an acid and donate a proton; H_2O can also act as a base and accept a proton to become the hydronium ion, H_3O^+. We can show the amphiprotic nature of water as follows.

$$\underset{\substack{\text{proton donor} \\ \text{acid}}}{H_2O(l)} + \underset{\substack{\text{proton acceptor} \\ \text{base}}}{H_2O(l)} \longrightarrow H_3O^+(aq) + OH^-(aq)$$

15.4 Acid–Base Indicators

Objective · To state the color of the following indicators in a solution of a given pH: phenolphthalein, methyl red, and bromthymol blue.

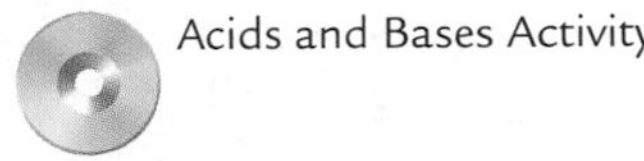
Acids and Bases Activity

We mentioned in Section 15.1 that litmus paper can be used to indicate whether a solution is acidic or basic. In addition, other paper test strips are commercially available that indicate the pH of a solution. These test strips are permeated with substances that change color depending on the pH of the test solution. Examples of such substances are grape juice, red cabbage, and flower petals such as those from roses, violets, and orchids.

Natural Indicators Movie

A solution that is pH-sensitive and changes color is referred to as an **acid–base indicator**. Methyl red, bromthymol blue, and phenolphthalein are common indicators. In solutions having a pH above 5, the indicator methyl red is yellow. In solutions with a pH below 5, this indicator is red. At a pH of 5, methyl red appears orange because only a portion of the indicator has been converted from yellow to red. In solutions having a pH of 9 or above, phenolphthalein is pink. In solutions having a pH below 9, this indicator is colorless. Dozens of acid–base indicators are available in the laboratory. A chemist selects an indicator on the basis of the pH at which it changes color (Figure 15.2). Table 15.4 lists three common indicators.

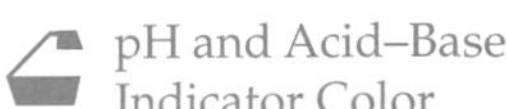
pH and Acid–Base Indicator Color

Table 15.4 Acid–Base Indicators

Indicator	Color Change	Color
methyl red	pH~5	below pH 5: red above pH 5: yellow
bromthymol blue	pH~7	below pH 7: yellow above pH 7: blue
phenolphthalein	pH~9	below pH 9: colorless above pH 9: pink

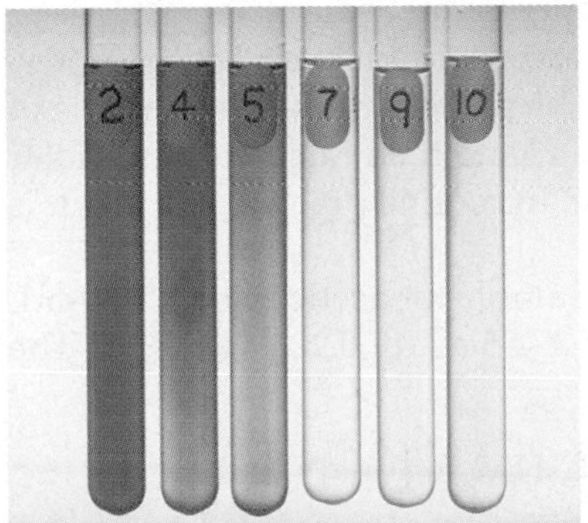

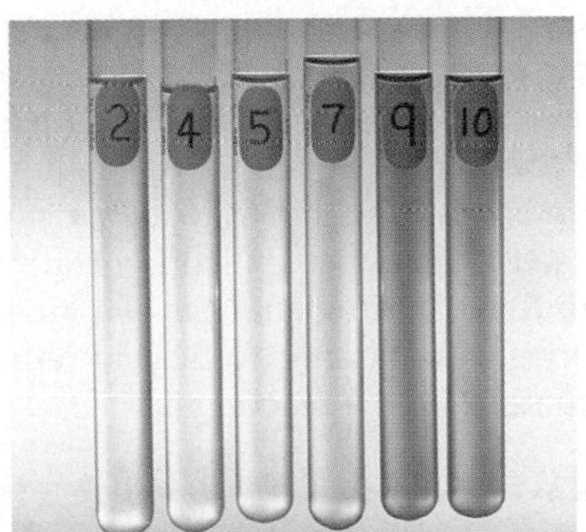

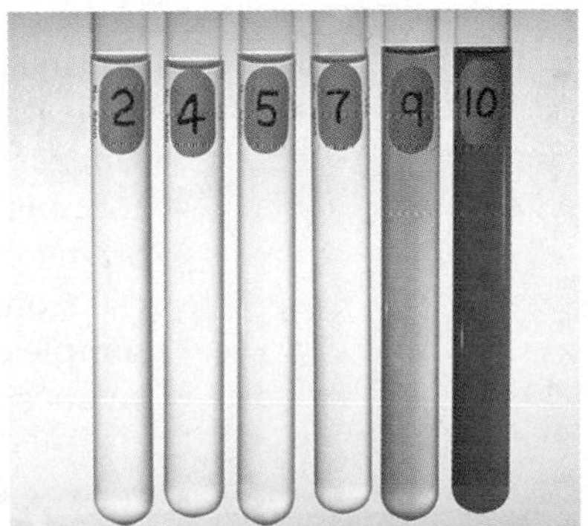

▲ **Figure 15.2 pH and Acid–Base Indicator Color** The color of three acid–base indicators at different pH values. Methyl red (left) changes from red to yellow at pH 5. Bromthymol blue (middle) changes from yellow to blue at pH 7. Phenolphthalein (right) changes from colorless to pink at pH 9.

The following example exercise illustrates the relationship of the color of an acid–base indicator and the pH of a given solution.

Example Exercise 15.4 • Acid–Base Indicators

State the color of the acid–base indicator in each of the following solutions.

(a) A solution at pH 3 contains a drop of methyl red.
(b) A solution at pH 8 contains a drop of bromthymol blue.
(c) A solution at pH 11 contains a drop of phenolphthalein.

Solution

Refer to Table 15.4 to determine the color of each solution.

(a) A pH 3 solution containing methyl red indicator is *red*.
(b) A pH 8 solution containing bromthymol blue indicator is *blue*.
(c) A pH 11 solution containing phenolphthalein indicator is *pink*.

Self-Test Exercise

State the pH at which each of the following acid–base indicators changes color.

(a) methyl red (b) bromthymol blue (c) phenolphthalein

Answers: (a) 5; (b) 7; (c) 9

Experiment #21, Prentice Hall Laboratory Manual

15.5 Acid–Base Titrations

Objectives
- To perform stoichiometry calculations that involve acid–base titrations.
- To convert the molarity of an acid or base to mass percent concentration.

Acid–Base Titration Movie; Acid–Base Titration Activity

Vinegar tastes sour because it contains acetic acid. To find the molar concentration of acetic acid in vinegar, we analyze the vinegar sample using a base solution of known concentration. In the laboratory, we can analyze acetic acid by a titration method. We use this **titration** process to deliver a measured volume of solution. To analyze a vinegar sample, we titrate the acetic acid by delivering a measured volume of sodium hydroxide solution from a buret. We use phenolphthalein as an indicator to signal when we have neutralized all the acetic acid.

After the acetic acid is neutralized, an extra drop of NaOH increases the pH dramatically and the phenolphthalein indicator changes from colorless to pink. When the indicator changes color, the titration is stopped. This point in the titration is the **endpoint**. Figure 15.3 illustrates the experimental procedure.

In laboratory analysis, a student uses a solution of sodium hydroxide to determine the concentration of acetic acid in vinegar. The student titrates the sample of vinegar to a phenolphthalein endpoint using NaOH and then calculates the molar concentration of acetic acid in the vinegar sample. In routine practice, the result is often expressed as mass percent concentration.

Consider the following analysis of a vinegar sample for acetic acid. A 10.0-mL sample of vinegar containing acetic acid requires 37.55 mL of 0.223 *M* NaOH. The balanced equation for the reaction is

$$HC_2H_3O_2(aq) + NaOH(aq) \longrightarrow NaC_2H_3O_2(aq) + H_2O(l)$$

This is a solution stoichiometry problem (Section 14.11). To find the molarity of the acetic acid, we must first find the moles of NaOH. The volume of NaOH is 37.55 mL, and the molar concentration is 0.223 *M*. The molarity of the NaOH can be written as the unit factor 0.223 mol NaOH/1000 mL solution.

$$37.55 \text{ \sout{mL solution}} \times \frac{0.223 \text{ mol NaOH}}{1000 \text{ \sout{mL solution}}} = 0.00837 \text{ mol NaOH}$$

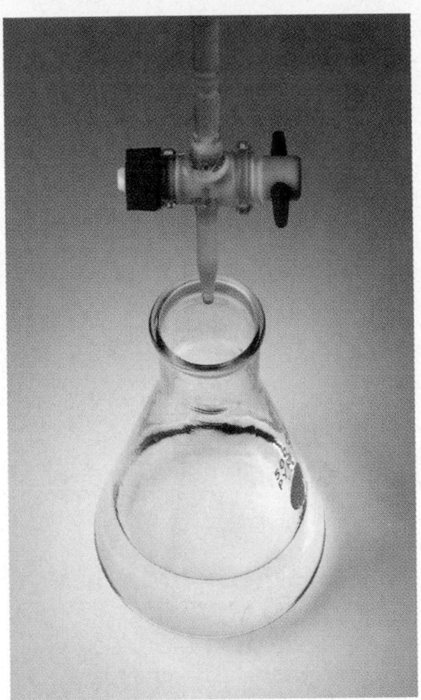

▲ **Figure 15.3 Titration of Acetic Acid with Sodium Hydroxide** (a) The flask contains a sample of acetic acid and a drop of phenolphthalein indicator. (b) The buret delivers a measured volume of sodium hydroxide solution into the flask. (c) The titration is complete when the solution attains a permanent pink color.

Titration of Acetic Acid with Sodium Hydroxide

Next, we find the moles of acetic acid titrated. From the balanced equation we see that 1 mol of NaOH base neutralizes 1 mol of $HC_2H_3O_2$.

$$0.00837 \cancel{\text{mol NaOH}} \times \frac{1 \text{ mol } HC_2H_3O_2}{1 \cancel{\text{mol NaOH}}} = 0.00837 \text{ mol } HC_2H_3O_2$$

Finally, we can calculate the molarity of the acetic acid. The volume of the $HC_2H_3O_2$ solution is 10.0 mL.

$$\frac{0.00837 \text{ mol } HC_2H_3O_2}{10.0 \cancel{\text{mL solution}}} \times \frac{1000 \cancel{\text{mL solution}}}{1 \text{ L solution}} = \frac{0.837 \text{ mol } HC_2H_3O_2}{1 \text{ L solution}}$$

$$= 0.837\, M \; HC_2H_3O_2$$

Now let's express the molar concentration of the acetic acid as a mass percent concentration. Given that the density of the vinegar is 1.01 g/mL and that the molar mass of acetic acid is 60.06 g/mol, we can proceed as follows.

$$\frac{0.837 \cancel{\text{mol } HC_2H_3O_2}}{1000 \cancel{\text{mL solution}}} \times \frac{60.06 \text{ g } HC_2H_3O_2}{1 \cancel{\text{mol } HC_2H_3O_2}} \times \frac{1 \cancel{\text{mL solution}}}{1.01 \text{ g solution}} \times 100$$

$$= \frac{0.0498 \text{ g } HC_2H_3O_2}{1 \text{ g solution}} \times 100 = 4.98\% \; HC_2H_3O_2$$

Therefore, the concentration of acetic acid in the vinegar solution is 0.837 *M*, which is equivalent to 4.98% $HC_2H_3O_2$.

In a different experiment, a student analyzes a household cleaning solution for its ammonia content. After titrating a sample of ammonia solution with an acid solution, the student calculates the concentration of the ammonia. The following example exercise illustrates the analysis of the ammonia solution.

Example Exercise 15.5 • Acid–Base Titration

If 25.30 mL of 0.277 *M* HCl is used to titrate 10.0 mL of aqueous ammonia to a methyl red endpoint, what is the molarity of the ammonia? When we write aqueous ammonia as NH_4OH, the balanced equation is

$$HCl(aq) + NH_4OH(aq) \longrightarrow NH_4Cl(aq) + H_2O(l)$$

Solution

To calculate the molarity of the NH_4OH, we must find the moles of NH_4OH. From the balanced equation, we see that 1 mol of HCl neutralizes 1 mol of NH_4OH. Since the acid solution is 0.277 *M* HCl, we have the unit factor 0.277 mol HCl/1000 mL solution.

$$25.30 \cancel{\text{ mL solution}} \times \frac{0.277 \cancel{\text{ mol HCl}}}{1000 \cancel{\text{ mL solution}}} \times \frac{1 \text{ mol } NH_4OH}{1 \cancel{\text{ mol HCl}}} = 0.00701 \text{ mol } NH_4OH$$

We obtain the molar concentration by dividing the moles NH_4OH by the volume titrated, 10.0 mL.

$$\frac{0.00701 \text{ mol } NH_4OH}{10.0 \cancel{\text{ mL solution}}} \times \frac{1000 \cancel{\text{ mL solution}}}{1 \text{ L solution}} = \frac{0.701 \text{ mol } NH_4OH}{1 \text{ L solution}} = 0.701\ M\ NH_4OH$$

Self-Test Exercise

If 38.30 mL of 0.250 *M* NaOH is used to titrate 25.0 mL of phosphoric acid, what is the molarity of the acid? The balanced equation is

$$H_3PO_4(aq) + 3\ NaOH(aq) \longrightarrow Na_3PO_4(aq) + 3\ H_2O(l)$$

Answer: 0.128 *M*

Let's try another type of problem and find the volume of base required to neutralize an acid. For example, we can find the volume of base needed to neutralize sulfuric acid in a battery acid sample. The following example exercise illustrates the analysis of battery acid by a titration method.

Example Exercise 15.6 • Acid–Base Titration

A 10.0-mL sample of 0.555 *M* battery acid, H_2SO_4, is titrated with 0.223 *M* NaOH. What volume of sodium hydroxide is required for the titration?

$$H_2SO_4(aq) + 2\ NaOH(aq) \longrightarrow Na_2SO_4(aq) + 2\ H_2O(l)$$

Solution

Step 1, The unit asked for in the answer is **mL NaOH.**
Step 2, The given value is **10.0 mL H_2SO_4.**
Step 3, We apply unit conversion factors to cancel units. In this example, we use the molar concentration of acid (0.555 mol H_2SO_4 per 1000 mL battery acid) and the molar concentration of base (0.223 mol NaOH per 1000 mL NaOH) as unit factors. From the balanced equation, we see that 1 mol of H_2SO_4 neutralizes 2 mol of NaOH.

$$10.0 \cancel{\text{ mL } H_2SO_4} \times \frac{0.555 \cancel{\text{ mol } H_2SO_4}}{1000 \cancel{\text{ mL } H_2SO_4}} \times \frac{2 \cancel{\text{ mol NaOH}}}{1 \cancel{\text{ mol } H_2SO_4}} \times \frac{1000 \text{ mL NaOH}}{0.223 \cancel{\text{ mol NaOH}}} = 49.8 \text{ mL NaOH}$$

We can visualize the overall solution as follows.

mL H_2SO_4 → molar concentration → mole ratio → molar concentration → mL NaOH

Self-Test Exercise

A 25.0-mL sample of hydrochloric acid is titrated with 0.125 *M* $Ba(OH)_2$. If 50.0 mL of barium hydroxide is required for the titration, what is the molar concentration of the acid? The balanced equation for the reaction is

$$2\ HCl(aq) + Ba(OH)_2(aq) \longrightarrow BaCl_2(aq) + 2\ H_2O(l)$$

Answer: 0.500 *M* HCl

15.6 Acid–Base Standardization

Objective · To perform stoichiometry calculations that involve a standard acid or base solution.

A **standard solution** is a solution in which the concentration is known precisely, for example, the concentration is known to three significant digits. Chemists routinely use standard solutions to analyze substances. For instance, a chemist may use a standard solution to analyze the neutralizing capacity of an antacid tablet or the tartness of a soft drink. Standard solutions are also used in manufacturing processes to assure quality. Here, chemical samples are selected randomly and analyzed using a standard solution. This is an important procedure in industry, and it is referred to as quality control (QC) or quality assurance (QA).

To standardize a solution of acid, we use a weighed quantity of a solid base. To standardize hydrochloric acid, for example, we can use solid sodium carbonate, Na_2CO_3. Let's find the molarity of hydrochloric acid if 25.50 mL of solution is required to neutralize 0.375 g of Na_2CO_3. The balanced equation for the reaction is

$$2\ HCl(aq) + Na_2CO_3(s) \longrightarrow 2\ NaCl(aq) + H_2O(l) + CO_2(g)$$

To calculate the molarity of the hydrochloric acid, we first find the number of moles of HCl. From the balanced equation we note that 2 mol of HCl reacts with 1 mol of Na_2CO_3 (105.99 g/mol).

$$0.375\ \cancel{g\ Na_2CO_3} \times \frac{1\ \cancel{mol\ Na_2CO_3}}{105.99\ \cancel{g\ Na_2CO_3}} \times \frac{2\ mol\ HCl}{1\ \cancel{mol\ Na_2CO_3}} = 0.00708\ mol\ HCl$$

To obtain the molarity of the acid, we divide the moles of HCl by the 25.50 mL of HCl required to neutralize the sodium carbonate.

$$\frac{0.00708\ mol\ HCl}{25.50\ \cancel{mL\ solution}} \times \frac{1000\ \cancel{mL\ solution}}{1\ L\ solution} = \frac{0.277\ mol\ HCl}{1\ L\ solution} = 0.277\ M\ HCl$$

To standardize a solution of base, we can use a weighed quantity of a solid acid. For example, we can use crystals of oxalic acid, $H_2C_2O_4$ to standardize aqueous sodium hydroxide. We first dissolve a weighed sample of $H_2C_2O_4$ in water and then neutralize it with a measured volume of a base. The following example exercise illustrates how to determine the concentration of a standard NaOH solution.

Example Exercise 15.7 • Acid–Base Standardization

What is the molarity of a sodium hydroxide solution if 32.15 mL of NaOH is required to neutralize 0.424 g of oxalic acid, $H_2C_2O_4$ (90.04 g/mol)?

$$H_2C_2O_4(s) + 2\ NaOH(aq) \longrightarrow Na_2C_2O_4(aq) + 2\ H_2O(l)$$

Solution

To calculate the molarity of the NaOH, we find the number of moles of $H_2C_2O_4$. The mass of $H_2C_2O_4$ is 0.424 g, and the molar mass is 90.04 g/mol. From the balanced equation we see that 2 mol of NaOH reacts with 1 mol of $H_2C_2O_4$.

$$0.424\ \cancel{g\ H_2C_2O_4} \times \frac{1\ \cancel{mol\ H_2C_2O_4}}{90.04\ \cancel{g\ H_2C_2O_4}} \times \frac{2\ mol\ NaOH}{1\ \cancel{mol\ H_2C_2O_4}} = 0.00942\ mol\ NaOH$$

We can obtain the molar concentration of the base from the moles of NaOH divided by the volume of base required to neutralize the oxalic acid.

(continued)

Example Exercise 15.7 *(continued)*

$$\frac{0.00942 \text{ mol NaOH}}{32.15 \text{ mL solution}} \times \frac{1000 \text{ mL solution}}{1 \text{ L solution}} = \frac{0.293 \text{ mol NaOH}}{1 \text{ L solution}}$$

$$= 0.293\ M \text{ NaOH}$$

Self-Test Exercise

If an unknown sample of oxalic acid, $H_2C_2O_4$, is neutralized by 33.50 mL of 0.293 *M* NaOH, what is the mass of the sample? Refer to the above balanced chemical equation for the neutralization of oxalic acid with sodium hydroxide.

Answer: 0.442 g $H_2C_2O_4$

Molar Mass of a Solid Acid or Solid Base

One application of a standard solution is to determine the molar mass of a solid acid or base. For example, the QA department of a soft drink company may want to verify that the proper amount of citric acid has been added to a new batch of soft drinks. We begin by dissolving crystals of citric acid in water and neutralize the solution using a standard base. Given the mass of citric acid and the volume of standard base, we can calculate the molar mass of citric acid.

The following example exercise illustrates calculation of the molar mass of a solid acid.

Example Exercise 15.8 • Molar Mass of an Acid

Citric acid, abbreviated H_3Cit, gives Kool-Aid its tart taste. If 36.10 mL of 0.293 *M* NaOH neutralize a 0.677-g sample of citric acid, what is the molar mass of citric acid? The balanced equation for the reaction is

$$H_3Cit(s) + 3\ NaOH(aq) \longrightarrow Na_3Cit(aq) + 3\ H_2O(l)$$

Solution

We are given that the volume of NaOH is 36.10 mL. Since the molar concentration is 0.293 *M*, we write the unit factor as 0.293 mol NaOH/1000 mL solution.

$$36.10 \text{ mL solution} \times \frac{0.293 \text{ mol NaOH}}{1000 \text{ mL solution}} = 0.0106 \text{ mol NaOH}$$

From the balanced chemical equation, we see that 3 mol of NaOH reacts with 1 mol of H_3Cit. Thus, the number of moles of citric acid is

$$0.0106 \text{ mol NaOH} \times \frac{1 \text{ mol } H_3Cit}{3 \text{ mol NaOH}} = 0.00353 \text{ mol } H_3Cit$$

We can visualize the conversion as follows.

mL base → molar / concentration → mole / ratio → mol acid

Since the molar mass is defined as grams per mole, we have

$$\frac{0.677 \text{ g } H_3Cit}{0.00353 \text{ mol } H_3Cit} = 192 \text{ g/mol}$$

The calculated molar mass of citric acid is 192 g/mol, and the chemical formula for citric acid is $H_3C_6H_5O_7$. When we determine the molar mass of citric acid by adding up the atomic masses from the periodic table, we obtain 192.14 g/mol.

Self-Test Exercise
Benzoic acid is a common food preservative, and its molecular formula is $HC_7H_5O_2$. If 22.55 mL of 0.293 *M* NaOH neutralize a 0.806-g sample of benzoic acid, what is the molar mass of the acid? The balanced equation for the reaction is

$$HC_7H_5O_2(s) + NaOH(aq) \longrightarrow NaC_7H_5O_2(aq) + H_2O(l)$$

Answer: 122 g/mol $HC_7H_5O_2$

15.7 Ionization of Water

Objectives
- To relate the ionization constant of water to $[H^+]$ and $[OH^-]$.
- To calculate the molar hydroxide ion concentration given $[H^+]$.

In Chapter 4, we learned that metals are good conductors of electricity. What would happen if we tested the electrical conductivity of an automobile key? The key should conduct an electric current because it is made of metal. In Figure 15.4a, we see that the key completes the circuit in the conductivity apparatus and that the bulb lights up. The explanation is as follows. Electricity is defined as the flow of electrons. Metal atoms, such as those in the key, hold their electrons loosely. Therefore, an electric current can push electrons from one atom to another. Thus, electricity flows from one end of the metal key to the other as electrons move from atom to atom.

What happens when we test the electrical conductivity of pure water? In Figure 15.4b, we see that pure water is a very poor conductor of electricity. Evidence suggests that it is difficult to push electrons from one water molecule to another. How then do we explain the observation that some electricity flows in pure water? We explain this observation by suggesting that there are ions present in pure water.

Arrhenius found that salt solutions are good conductors of electricity. He correctly concluded that *ions* in the solution are responsible for electrical conductivity. Since water is a very poor conductor, we can conclude that only a few ions are present in pure water (Figure 15.5).

We know that water molecules in the liquid state are free to move about and collide with one another. Although it happens only rarely, on occasion two water mol-

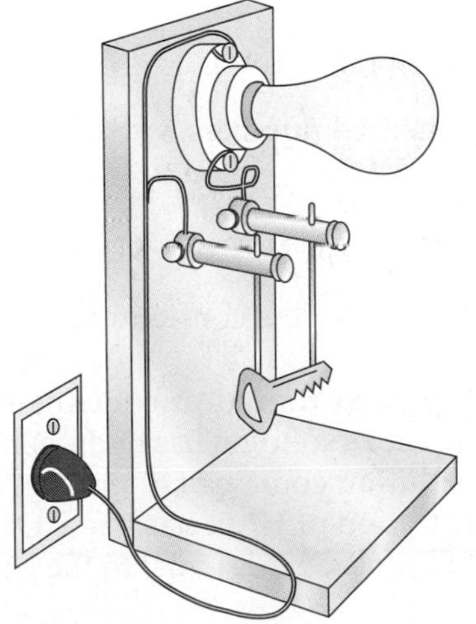

(a) Strong conductor

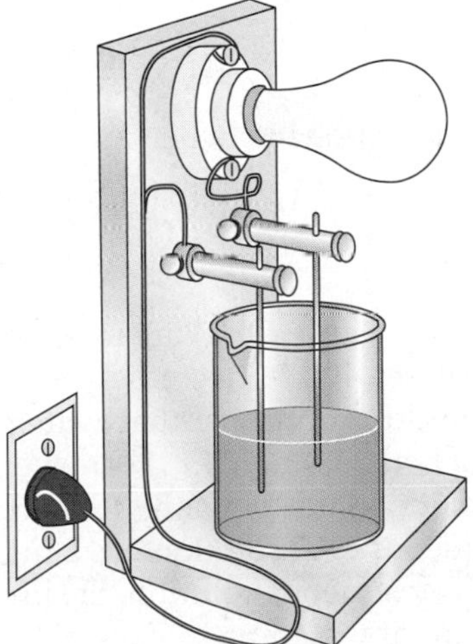

(b) Weak conductor

▲ **Figure 15.4 Conductivity Apparatus** (a) Metals are good conductors of electricity. A metal key completes the electric circuit and lights up the apparatus. (b) Pure water is a very poor conductor. It does, however, very weakly conduct an electric current.

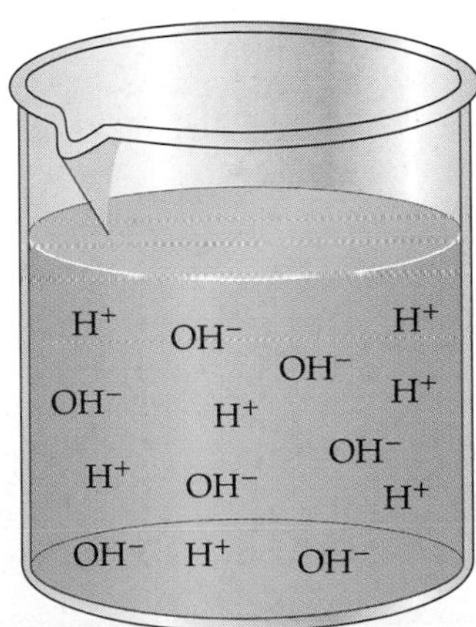

▲ **Figure 15.5 Ionization of Water** Pure water ionizes to give a few hydrogen ions and hydroxide ions. Water is a very weak conductor of electricity because only 1 molecule in about 500,000,000 forms ions.

ecules collide with sufficient energy for their bonds to break apart. This bond breaking produces a hydronium ion, H_3O^+, and a hydroxide ion, OH^-. Moreover, this autoionization is a dynamic process. While some molecules are breaking apart, other ions are combining to form water molecules. At any given moment, only about 1 water molecule in 500 million is present as ions. Apparently, however, these few ions in water are sufficient to conduct a very weak electric current. The chemical equation for the collision reaction is

$$H_2O(l) + H_2O(l) \longrightarrow \underset{\text{hydronium ion}}{H_3O^+(aq)} + \underset{\text{hydroxide ion}}{OH^-(aq)}$$

Alternatively, we can simplify the reaction by writing the ionization of water as

$$H_2O(l) \longrightarrow \underset{\text{hydrogen ion}}{H^+(aq)} + \underset{\text{hydroxide ion}}{OH^-(aq)}$$

The concentration of hydrogen ions in pure water is 1.0×10^{-7} mol/L at 25°C. If we know the concentration of hydrogen ions, we also know the concentration of hydroxide ions. Since the ionization of water gives one H^+ and one OH^- for each molecule of water that ionizes, the two concentrations are equal. Therefore, the concentration of OH^- must also be 1.0×10^{-7} mol/L at 25°C. Moreover, the molar concentration of H^+ multiplied by the molar concentration of OH^- equals a constant. This product is called the **ionization constant of water** (symbol K_w).

Let's calculate the value of K_w at 25°C. For convenience, we use brackets to symbolize molar concentration. Hence, $[H^+]$ is the symbol for the molar concentration of hydrogen ion. We calculate a value for K_w as follows.

If

$$[H^+] = 1.0 \times 10^{-7}$$

then

$$[OH^-] = 1.0 \times 10^{-7}$$

and

$$[H^+][OH^-] = (1.0 \times 10^{-7})(1.0 \times 10^{-7})$$

Since we add exponents when multiplying exponential numbers, we have

$$[H^+][OH^-] = 1.0 \times 10^{-14}$$

$$K_w = 1.0 \times 10^{-14} \qquad \text{(at 25°C)}$$

Notice that $[H^+]$ and $[OH^-]$ are inversely proportional to the constant, K_w. That is, if $[H^+]$ increases, $[OH^-]$ decreases.

We should emphasize that water is neutral even though it contains small amounts of both H^+ and OH^-. Moreover, every aqueous solution has hydrogen ions and hydroxide ions. Even a hydrochloric acid solution contains H^+ and OH^-. Although aqueous HCl solution has a high concentration of H^+, it has a few OH^- as well. In every aqueous solution, the product of H^+ and OH^- is equal to the ionization constant of water.

$$[H^+][OH^-] = 1.0 \times 10^{-14}$$

If an aqueous HCl solution is 0.1 *M*, we can calculate its hydroxide ion concentration.

$$[0.1][OH^-] = 1.0 \times 10^{-14}$$

and,

$$[OH^-] = 1.0 \times 10^{-13}$$

Although $[OH^-]$ is only 0.000 000 000 0001 *M*, the hydrochloric acid solution contains a trace of hydroxide ion!

To summarize, in an acidic aqueous solution $[H^+] > [OH^-]$. In a basic aqueous solution $[H^+] < [OH^-]$. In a neutral aqueous solution $[H^+] = [OH^-]$. The following example exercise further illustrates the presence of both hydrogen and hydroxide ions in aqueous solutions.

Example Exercise 15.9 • The Ionization Constant, K_w

Given the following hydrogen ion concentrations in aqueous solution, what is the molar concentration of hydroxide ion?

(a) $[H^+] = 1.4 \times 10^{-6}$　　(b) $[H^+] = 5.2 \times 10^{-11}$

Solution

From the ionization constant of water, we know that $[H^+][OH^-] = 1.0 \times 10^{-14}$.

(a) If $[H^+]$ is 1.4×10^{-6}, we can calculate $[OH^-]$ as follows.

$$K_w = [H^+][OH^-] = 1.0 \times 10^{-14}$$
$$1.4 \times 10^{-6}[OH^-] = 1.0 \times 10^{-14}$$

Dividing both sides of the equation by 1.4×10^{-6}, we have

$$[OH^-] = 7.1 \times 10^{-9}$$

(b) If $[H^+]$ is 5.2×10^{-11}, we can calculate $[OH^-]$ as follows.

$$K_w = [H^+][OH^-] = 1.0 \times 10^{-14}$$
$$5.2 \times 10^{-11}[OH^-] = 1.0 \times 10^{-14}$$

Dividing both sides of the equation by 5.2×10^{-11}, we have

$$[OH^-] = 1.9 \times 10^{-4}$$

Self-Test Exercise

Given the following hydroxide ion concentrations in aqueous solution, what is the molar concentration of hydrogen ion?

(a) $[OH^-] = 7.5 \times 10^{-4}$　　(b) $[OH^-] = 2.1 \times 10^{-10}$

Answers: (a) $[H^+] = 1.3 \times 10^{-11}$; (b) $[H^+] = 4.8 \times 10^{-5}$

15.8 The pH Concept

Objectives
- To explain the concept of pH.
- To relate integer pH values and the $[H^+]$ of a solution.

In Section 15.1, we introduced the term pH. Recall that pure water is neutral and has a pH of 7. We also know that an acid has a pH of less than 7. As the pH value *decreases*, the solution becomes more acidic. As the pH value *increases*, the solution becomes more basic.

Converting from $[H^+]$ to pH

In aqueous solution, the hydrogen ion concentration can vary from more than 1 *M* to less than 0.000 000 000 000 01 *M*. A pH scale is a convenient way to express this broad range of hydrogen ion concentrations. The pH scale expresses the molar hydrogen ion concentration, $[H^+]$, as a power of 10. In other words, the **pH** is the negative logarithm of the molar hydrogen ion concentration.

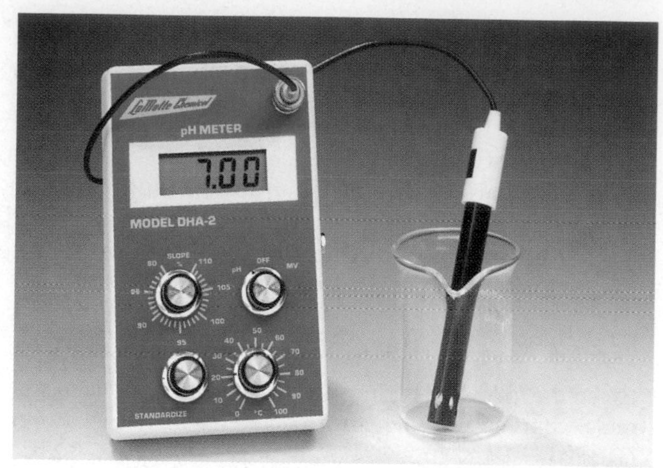

▶ **Digital pH Meter** The digital pH meter measures the hydrogen ion concentration in the beaker with the glass electrodes (shown). Pure boiled water has a hydrogen ion concentration of $1.0 \times 10^{-7}\,M$, which corresponds to a pH of 7.00.

$$\text{pH} = -\log\,[\text{H}^+]$$

For example, if the molar hydrogen ion concentration, $[\text{H}^+]$, is 0.1 *M*, then

$$\text{pH} = -\log 0.1$$

We do not need to understand logarithms to solve pH problems. All we need to know is that if we express the hydrogen ion concentration as a power of 10, the logarithm is the exponent. We can write 0.1 *M* hydrogen ion concentration as a power of 10 (that is, 10^{-1}) and find the pH as follows.

$$\begin{aligned}\text{pH} &= -\log 10^{-1}\\ &= -(-1) = 1\end{aligned}$$

Thus, the pH of a solution with a hydrogen ion concentration of 0.1 *M* is 1.

We've learned that the pH scale uses powers of 10 to express the acidity or basicity of a solution. Therefore, a solution with a pH of 3 is ten times more acidic than a solution with a pH of 4. Conversely, a solution with a pH of 12 is ten times more basic than a solution with a pH of 11.

The following example exercises provide practice in finding the pH of an aqueous solution given the molar hydrogen ion concentration.

Example Exercise 15.10 • Converting $[\text{H}^+]$ to pH

Calculate the pH of the following solutions given the molar hydrogen ion concentration.

(a) vinegar, $[\text{H}^+] = 0.001\,M$ (b) antacid, $[\text{H}^+] = 0.000\,000\,001\,M$

Solution

(a) The pH of vinegar is equal to $-\log\,[\text{H}^+]$.

$$\begin{aligned}\text{pH} &= -\log 0.001 &&= -\log 10^{-3}\\ &= -(-3) &&= 3\end{aligned}$$

The sour taste of vinegar indicates that an acid is present. The acid in vinegar is acetic acid, $HC_2H_3O_2$, and the pH is approximately 3.

(b) The pH of the antacid is equal to $-\log\,[\text{H}^+]$.

$$\begin{aligned}\text{pH} &= -\log 0.000\,000\,001 &&= -\log 10^{-9}\\ &= -(-9) &&= 9\end{aligned}$$

Self-Test Exercise
Calculate the pH for apple juice if the molar hydrogen ion concentration is 0.0001 *M*.

Answer: pH = 4

Converting from pH to $[H^+]$

We can calculate the hydrogen ion concentration given the pH of a solution. Let's rearrange the definition of pH to express the molar hydrogen ion concentration. $[H^+]$ is equal to ten raised to a negative pH value.

$$[H^+] = 10^{-pH}$$

If milk has a pH of 6, then $[H^+]$ can be expressed as

$$[H^+] = 10^{-6} = 0.000\,001\ M$$

The following example exercise provides further practice in converting a given pH to a molar hydrogen ion concentration.

Example Exercise 15.11 • Converting pH to $[H^+]$

Calculate the molar hydrogen ion concentration given the pH of the following solutions.
(a) lemon juice, pH 2 (b) tomato juice, pH 5

Solution

(a) The hydrogen concentration of lemon juice is equal to 10 raised to the negative value of the pH. Since the pH is 2,

$$[H^+] = 10^{-pH}$$
$$= 10^{-2} = 0.01\ M$$

(b) If the pH of tomato juice is 5, $[H^+]$ is expressed as

$$[H^+] = 10^{-pH}$$
$$= 10^{-5} = 0.000\,01\ M$$

Notice that lemon juice and tomato juice differ by three pH units. Thus, lemon juice is 1000 times more acidic than tomato juice.

Self-Test Exercise
Calculate the hydrogen ion concentration for milk of magnesia if the pH is 10.

Answer: $[H^+] = 0.000\,000\,000\,1\ M\ (1 \times 10^{-10}\ M)$

15.9 Advanced pH Calculations

Objective · To relate fractional pH values to the $[H^+]$ of a solution.

In the previous section, we defined pH as an exponential way of expressing the hydrogen ion concentration, but we considered only whole-number pH values. Many chemical reactions, however, must, be carefully controlled to a fraction of a pH unit. Biochemical reactions, for example, are extremely sensitive to small pH changes. In fact, the pH of our blood must be maintained within the narrow range of 7.3–7.5.

Converting from $[H^+]$ to pH

Recall the mathematical definition of pH. That is, pH is the negative logarithm of the molar hydrogen ion concentration.

$$pH = -\log [H^+]$$

Suppose we wish to express the pH of a solution having a hydrogen ion concentration of 0.00015 *M*. The pH expression is

$$\begin{aligned} pH &= -\log 0.00015 \\ &= -\log 1.5 \times 10^{-4} \end{aligned}$$

We can easily obtain the logarithm for a number using a scientific calculator (Appendix A). If we enter 1.5×10^{-4} and touch the log key, the display shows −3.82. After a change in sign, the answer is 3.82.

$$\begin{aligned} pH &= -(-3.82) \\ &= 3.82 \end{aligned}$$

For a solution having a hydrogen ion concentration of 0.00015 *M*, the pH is 3.82. Note that the number of significant digits in the given value must equal the number of decimal places in the log expression; in this example there are two significant digits and two decimal places in the log value. The following example exercise provides practice in finding the pH of an aqueous solution given the molar hydrogen ion concentration.

Example Exercise 15.12 • Converting $[H^+]$ to pH

Calculate the pH of the following solutions given the molar hydrogen ion concentration.

(a) stomach acid, $[H^+] = 0.020\ M$ (b) blood, $[H^+] = 0.000\,000\,048\ M$

Solution

(a) We can calculate the pH of stomach acid as follows.

$$\begin{aligned} pH &= -\log 0.020 \\ &= -(-1.70) \\ &= 1.70 \end{aligned}$$

The pH of stomach acid ranges from 1 to 3; hence, this sample is normal.

(b) We can calculate the pH of blood as follows.

$$\begin{aligned} pH &= -\log 0.000\,000\,048 \\ &= -\log 4.8 \times 10^{-8} \\ &= -(-7.32) \\ &= 7.32 \end{aligned}$$

The pH of blood ranges from 7.3 to 7.5; hence, this sample is normal.

Self-Test Exercise

Calculate the pH of grape juice if the hydrogen ion concentration is 0.000 089 *M*.

Answer: pH = 4.05

Converting from pH to $[H^+]$

Previously, we expressed $[H^+]$ by raising 10 to the negative pH value.

$$[H^+] = 10^{-pH}$$

If orange juice has a pH of 2.75, then $[H^+]$ can be expressed as

$$[H^+] = 10^{-2.75}$$

To find the number that corresponds to the fractional exponent, we obtain the inverse logarithm using a scientific calculator. If we enter −2.75 and touch the inverse log key, the display shows 0.0018, or 1.8×10^{-3}.

$$[H^+] = 10^{-2.75} = 0.0018\ M$$

The following example exercise provides further practice in converting a given pH to a molar hydrogen ion concentration.

Example Exercise 15.13 • Converting pH to $[H^+]$

Calculate the molar hydrogen ion concentration given the pH of the following solutions.

(a) acid rain, pH 3.68 (b) seawater, pH 7.85

Solution

(a) If acid rain has a pH of 3.68, we can find the $[H^+]$ as follows.

$$[H^+] = 10^{-3.68}$$

Using the inverse log key on a scientific calculator, we have

$$[H^+] = 2.1 \times 10^{-4}$$
$$= 0.000\,21\ M$$

(b) If seawater has a pH of 7.85, we can find the $[H^+]$ as follows.

$$[H^+] = 10^{-7.85}$$
$$= 1.4 \times 10^{-8}$$
$$= 0.000\,000\,014\ M$$

Self-Test Exercise

Calculate the hydrogen ion concentration in wine if the pH is 3.25.

Answer: $[H^+] = 0.000\,56\ M\ (5.6 \times 10^{-4}\ M)$

15.10 Strong and Weak Electrolytes

Objective • To illustrate a strong and a weak electrolyte in aqueous solution.

Water is a weak conductor of electricity because it is very slightly ionized. In the laboratory, we can use a conductivity apparatus to determine whether a substance in aqueous solution is a strong or a weak conductor of electricity. If an aqueous solution is a good conductor, it is called a **strong electrolyte**. If an aqueous solution is a poor conductor, it is called a **weak electrolyte**.

As a demonstration, we can test the electrical conductivity of hydrochloric acid and acetic acid using the apparatus in Figure 15.6. From the observations, we can conclude that hydrochloric acid is highly ionized because it is a strong electrolyte. Acetic acid is only slightly ionized because it is a weak electrolyte. Similar experiments using the conductivity apparatus demonstrate the electrolyte behavior of aqueous solutions.

Degree of Ionization in Aqueous Solution

We can measure the degree of **ionization**, by testing the conductivity of aqueous solutions. Strong electrolytes are highly ionized. Strong electrolytes include strong acids, strong bases, and soluble ionic compounds. Soluble ionic compounds dissolve by dissociating into ions.

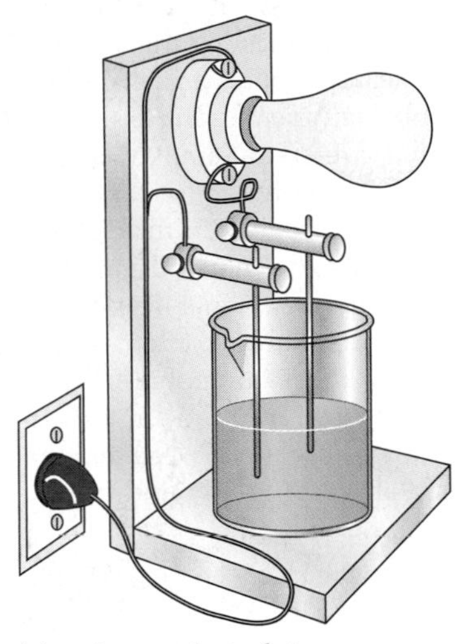

(a) Strong electrolyte

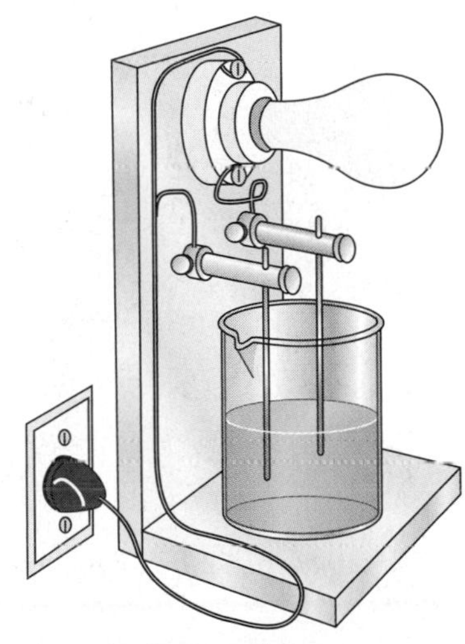

(b) Weak electrolyte

▲ **Figure 15.6 Conductivity of Aqueous Solutions** (a) An aqueous solution of hydrochloric acid is a strong electrolyte, and the bulb glows brightly. (b) An aqueous solution of acetic acid is a weak electrolyte, and the bulb glows dimly.

Chemistry Connection • Acid Rain

What is the primary source of the gases CO_2, NO_2, and SO_2 that dissolve in the atmosphere and contribute to acid rain?

Compare a 1935 photograph of a statue of George Washington with a current photograph. The striking difference is attributed to acid rain, which attacks marble statues by dissolving away the calcium carbonate in the stone.

The English chemist Robert Smith coined the term "acid rain" after studying the rainfall in London. He found that the air, heavily polluted from coal-burning, produced rain that was abnormally acidic. The term "acid rain" has persisted and today usually refers to rain having a pH 5 or below.

The gases that most contribute to acid rain are oxides of sulfur and nitrogen. Sulfur dioxide and sulfur trioxide in the atmosphere are released mainly by industrial steel plants and electric power plants. These plants burn low-grade coal, which has a high sulfur content. Most of the oxides of nitrogen are emitted from automobiles. When oxides of sulfur and nitrogen are released into the air, they dissolve in atmospheric water vapor. Subsequently, the rain that falls is composed of sulfuric acid and nitric acid raindrops.

Normal rain and acid rain are both acidic, but their pH values are different. Normal rain has a pH about 5.5. It is slightly acidic because carbon dioxide in the atmosphere dissolves in raindrops to form a dilute solution of carbonic acid. Severe acid rain can be a hundred times more acidic than normal rain, and an unhealthy pH of 2.8 has been recorded. In the northeastern United States, rainfall with a pH of 4 has been blamed on sulfur oxides released from the burning of coal.

Acid rain is a global problem, and countries all over the world are attempting to reduce it. One way is to reduce the emission of environmental pollutants. Canada has begun a program to reduce its sulfur dioxide emission by 50% within 10 years. The United States, Japan, and Germany are using high-grade coal in order to reduce their emissions. Ultimately, the problem of acid rain may be minimized by alternative energy sources. For instance, hydroelectric power and solar power are potential sources of energy that do not pollute the atmosphere.

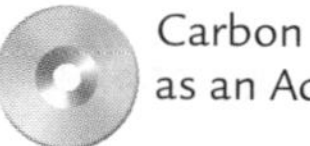

Carbon Dioxide Behaves as an Acid in Water Movie

◀ **Acid Rain** The photograph on the left shows a marble statue of George Washington in 1935. The recent photograph on the right shows the effect of acid rain, which attacks the marble statue by dissolving away calcium carbonate in the stone.

The combustion of fossil fuels produces CO_2, automobile emissions contain NO_2, and burning low-grade coal releases SO_2.

Electrolytes and Nonelectrolytes Movie; Strong and Weak Electrolytes Movie

Weak electrolytes are slightly ionized. They include weak acids, weak bases, and slightly soluble ionic compounds. Although an ionic compound may be classified as insoluble, it is actually very slightly soluble. A sufficient amount of the insoluble ionic compound dissolves in solution to act as a weak electrolyte. Table 15.5 lists examples of strong and weak electrolytes.

We can determine the degree of ionization in a solution from conductivity testing experiments. That is, we can distinguish between a highly ionized strong electrolyte and a slightly ionized weak electrolyte. Strong electrolytes ionize nearly completely in aqueous solution. For example, sodium chloride and calcium chloride dissociate into ions as follows.

Table 15.5 Strong and Weak Electrolytes

Strong Electrolytes	Weak Electrolytes
Strong Acids	*Weak Acids*
hydrochloric acid, $HCl(aq)$	hydrofluoric acid, $HF(aq)$
nitric acid, $HNO_3(aq)$	nitrous acid, $HNO_2(aq)$
sulfuric acid, $H_2SO_4(aq)$	sulfurous acid, $H_2SO_3(aq)$
perchloric acid, $HClO_4(aq)$	acetic acid, $HC_2H_3O_2(aq)$
	carbonic acid, $H_2CO_3(aq)$
	phosphoric acid, $H_3PO_4(aq)$
	most other acids
Strong Bases	*Weak Bases*
sodium hydroxide, $NaOH(aq)$	ammonium hydroxide, $NH_4OH(aq)$
potassium hydroxide, $KOH(aq)$	insoluble hydroxides such as
lithium hydroxide, $LiOH(aq)$	$Mg(OH)_2(s)$, $Al(OH)_3(s)$, $Fe(OH)_3(s)$
calcium hydroxide, $Ca(OH)_2(aq)$	most other bases
strontium hydroxide, $Sr(OH)_2(aq)$	
barium hydroxide, $Ba(OH)_2(aq)$	
Soluble Salts	*Insoluble Salts**
sodium chloride, $NaCl(aq)$	silver chloride, $AgCl(s)$
potassium carbonate, $K_2CO_3(aq)$	calcium carbonate, $CaCO_3(s)$
zinc sulfate, $ZnSO_4(aq)$	barium sulfate, $BaSO_4(s)$

*Insoluble salts are very slightly soluble in water and dissociate into a sufficient number of ions to give a very weak electrolyte solution.

$$NaCl(aq) \longrightarrow Na^+(aq) + Cl^-(aq)$$
$$CaCl_2(aq) \longrightarrow Ca^{2+}(aq) + 2\,Cl^-(aq)$$

We can estimate the amount of ionization in an aqueous solution given the electrolyte strength. That is, the number of ions in solution is proportional to the conductivity. Strong electrolytes are highly ionized, and so we write their formulas in the ionized form. Weak electrolytes are only slightly ionized, and so we write their formulas in the nonionized form. The nonionized formula is sometimes referred to as the molecular form. This is not quite accurate since ionic compounds exist as formula units, not molecules. The following example exercise illustrates how to write formulas for electrolytes in aqueous solution.

Example Exercise 15.14 • Strong and Weak Electrolytes

Write the ionized or nonionized formula for each of the following aqueous solutions given the electrolyte strength.

(a) $HNO_3(aq)$, strong (b) $NH_4OH(aq)$, weak (c) $K_2CO_3(aq)$, strong

Solution

(a) A nitric acid solution is a strong electrolyte and therefore is highly ionized. It is written in the ionized form as $H^+(aq)$ and $NO_3^-(aq)$.

(b) An ammonium hydroxide solution is a weak electrolyte and therefore is slightly ionized. It is written in the nonionized form as $NH_4OH(aq)$.

(c) A potassium carbonate solution is a strong electrolyte and therefore is highly ionized. It is written as $2\,K^+(aq)$ and $CO_3^{2-}(aq)$.

Self-Test Exercise

Write the ionized or nonionized formula for each of the following aqueous solutions given the electrolyte strength.

(a) $HF(aq)$, weak (b) $Ba(OH)_2(aq)$, strong (c) $CaCO_3(s)$, weak

Answers: (a) $HF(aq)$; (b) $Ba^{2+}(aq)$ and $2\,OH^-(aq)$; (c) $CaCO_3(s)$

Note In the examples, we considered aqueous solutions to be either strong or weak electrolytes. Some substances, however, are nonelectrolytes. That is, they do not conduct electricity at all. Examples include organic liquids such as alcohols. Although deionized water may not give an observable conductivity test with the apparatus, deionized water is considered a very weak electrolyte.

Experiment #22, Prentice Hall Laboratory Manual

15.11 Net Ionic Equations

Objective · To write a total ionic and a net ionic equation for a given chemical reaction.

In Section 15.10, we learned how to write ionized formulas for strong electrolytes, and nonionized formulas for weak electrolytes. The concept of ionization allows us to portray ionic solutions more accurately. That is, we can now write chemical equations by showing strong electrolytes in the ionized form.

Let's consider the neutralization reaction of hydrochloric acid and sodium hydroxide. The nonionized equation for the reaction is

$$HCl(aq) + NaOH(aq) \longrightarrow NaCl(aq) + H_2O(l)$$

By writing the ionized formula for the strong acid, strong base, and soluble salt, we can describe the reaction more accurately. Each substance in the **total ionic equation** is shown as it predominantly exists in solution. Table 15.5 lists highly ionized substances. Strong electrolytes include HCl, NaOH, and NaCl. Thus, the equation is

$$H^+(aq) + Cl^-(aq) + Na^+(aq) + OH^-(aq) \longrightarrow Na^+(aq) + Cl^-(aq) + H_2O(l)$$

Note that $Na^+(aq)$ and $Cl^-(aq)$ appear on both sides of the equation. These ions are called **spectator ions**. They are in the solution but do not participate in the reaction. We can simplify the total ionic equation by eliminating spectator ions. The resulting **net ionic equation** shows only the substances undergoing reaction. The net ionic equation for the above reaction is

$$H^+(aq) + OH^-(aq) \longrightarrow H_2O(l)$$

Note that we do not show the actual acid and base that were neutralized. This net ionic equation indicates that a strong acid and a strong base reacted to give water. In fact, the net ionic equation is identical for all strong acid and strong base reactions that yield a soluble salt.

Let's state a general procedure to be used when we write net ionic equations. Keep in mind that the net ionic equation gives us a good picture of substances undergoing reaction. That is, the net ionic equation helps us focus on only those substances that are undergoing reaction.

General Guidelines for Writing Net Ionic Equations

1. Complete and balance the nonionized chemical equation.
2. Convert the nonionized equation to a total ionic equation. Write strong electrolytes in the ionized form, and weak electrolytes in the nonionized form. Write water and dissolved gases in the nonionized form. Refer to Table 8.2 to determine if a compound is soluble or insoluble.

3. Cancel spectator ions to obtain the net ionic equation.
   (a) If canceling spectator ions eliminates all species, there is no reaction.
   (b) If the coefficients can be simplified, do so in order to have the simplest whole-number relationship.
4. Check (✓) each ion or atom on both sides of the equation. The total charge (positive or negative) on the reactants side of the equation must equal the total charge on the products side of the equation.

The following example exercises illustrate the procedure for writing balanced net ionic equations.

Example Exercise 15.15 • Net Ionic Equations

Write a net ionic equation for the reaction between nitric acid and aqueous potassium hydrogen carbonate.

$$HNO_3(aq) + KHCO_3(aq) \longrightarrow KNO_3(aq) + H_2O(l) + CO_2(g)$$

Solution

Step 1, We verify that the chemical equation is balanced. In this case, the coefficients are all 1 and the equation is balanced.

Step 2, We determine which species are highly ionized. From Table 15.5, we find that HNO_3 is a strong electrolyte. From Table 8.2, we find that $KHCO_3$ and KNO_3 are soluble strong electrolytes; H_2O and CO_2 are weak electrolytes. The total ionic equation is

$$H^+(aq) + NO_3^-(aq) + K^+(aq) + HCO_3^-(aq) \longrightarrow K^+(aq) + NO_3^-(aq) + H_2O(l) + CO_2(g)$$

Step 3, We obtain the net ionic equation by eliminating the spectator ions. In this example, K^+ and NO_3^- are in the aqueous solution but do not participate in the reaction. The net ionic equation is

$$H^+(aq) + HCO_3^-(aq) \longrightarrow H_2O(l) + CO_2(g)$$

Step 4, We check to verify that the net ionic equation is balanced. The equation is balanced because (a) the number of atoms of each element is the same on each side, and (b) the total charge on both sides is identical. In this case the net charge is zero.

Self-Test Exercise

Write a net ionic equation for the reaction between aqueous solutions of sulfuric acid and barium hydroxide.

$$H_2SO_4(aq) + Ba(OH)_2(aq) \longrightarrow BaSO_4(s) + 2\,H_2O(l)$$

Answer: In this example, there are no spectator ions to cancel.

$$2\,H^+(aq) + SO_4^{2-}(aq) + Ba^{2+}(aq) + 2\,OH^-(aq) \longrightarrow BaSO_4(s) + 2\,H_2O(l)$$

Example Exercise 15.16 • Net Ionic Equations

Write a net ionic equation for the reaction between aqueous solutions of silver nitrate and aluminum chloride.

$$AgNO_3(aq) + AlCl_3(aq) \longrightarrow AgCl(s) + Al(NO_3)_3(aq)$$

Solution

Step 1, We balance the chemical equation as follows.

$$3\,AgNO_3(aq) + AlCl_3(aq) \longrightarrow 3\,AgCl(s) + Al(NO_3)_3(aq)$$

(continued)

Example Exercise 15.16 *(continued)*

Step 2, We determine which of the species are strong electrolytes. From Table 8.2, we find that all the compounds are soluble except AgCl. Thus, AgCl is weakly ionized, and the total ionic equation is written

$$3\,Ag^+(aq) + 3\,NO_3^-(aq) + Al^{3+}(aq) + 3\,Cl^-(aq) \longrightarrow 3\,AgCl(s) + Al^{3+}(aq) + 3\,NO_3^-(aq)$$

Step 3, We cancel spectator ions. In this example, Al^{3+} and NO_3^- do not participate in the reaction.

$$3\,Ag^+(aq) + 3\,Cl^-(aq) \longrightarrow 3\,AgCl(s)$$

Step 4, We verify that the equation is balanced. Note that each species has a coefficient of 3 in this equation. Thus, after simplifying, the net ionic equation is written

$$Ag^+(aq) + Cl^-(aq) \longrightarrow AgCl(s)$$

Self-Test Exercise

Write a net ionic equation for the reaction between aqueous solutions of lithium sulfate and magnesium nitrate.

$$Li_2SO_4(aq) + Mg(NO_3)_2(aq) \longrightarrow MgSO_4(aq) + 2\,LiNO_3(aq)$$

Answer: $Li_2SO_4(aq) + Mg(NO_3)_2(aq) \longrightarrow$ no reaction

Summary

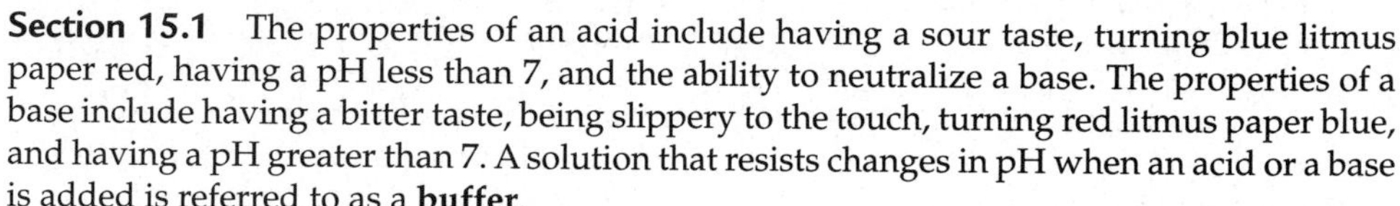

Section 15.1 The properties of an acid include having a sour taste, turning blue litmus paper red, having a pH less than 7, and the ability to neutralize a base. The properties of a base include having a bitter taste, being slippery to the touch, turning red litmus paper blue, and having a pH greater than 7. A solution that resists changes in pH when an acid or a base is added is referred to as a **buffer**.

Section 15.2 An **Arrhenius acid** is a substance that undergoes **ionization** in water to give hydrogen ions. Hydrochloric acid is a strong acid because it ionizes extensively. Acetic acid is a weak acid because it provides only a few hydrogen ions. A hydrogen ion attaches to a water molecule in solution to form a **hydronium ion**. An **Arrhenius base** is a substance that undergoes **dissociation** in water to give hydroxide ions. A **salt** is obtained from the reaction of an acid and a base. Sodium chloride, for example, is obtained from the neutralization reaction of sodium hydroxide and hydrochloric acid.

Section 15.3 A **Brønsted–Lowry acid** is a **proton donor**, and a **Brønsted–Lowry base** is a **proton acceptor**. A substance capable of either donating or accepting a proton is said to be **amphiprotic**.

Section 15.4 We can use litmus paper to indicate if a solution is acidic or basic. In the laboratory, there are dozens of solutions, each of which can act as an **acid–base indicator**. Three of the most common indicators are methyl red, bromthymol blue, and phenolphthalein. Methyl red changes color at a pH of 5, bromthymol blue changes at a pH of 7, and phenolphthalein at a pH of 9.

Section 15.5 A **titration** is a laboratory procedure for analyzing the amount of acid or base present in a solution. A measured sample of acid or base solution is delivered with a buret. An indicator is used to signal the neutralization point in the titration. When the **endpoint** is reached, the indicator changes color and the titration is stopped. The titration data allows the chemist to calculate the amount of acid or base.

Section 15.6 The concentration of a **standard solution** is known accurately to at least three significant digits. Standard solutions are used to analyze the amount of acid or base in a given sample. For example, a chemist can determine the concentration of acetic acid in vinegar or the amount of baking soda in an antacid tablet.

Section 15.7 On the basis of conductivity experiments, pure water is found to be a very weak electrolyte. Water ionizes to give only a few hydrogen and hydroxide ions. $[H^+]$ and $[OH^-]$ are equal, and both have a value of 1×10^{-7}. The **ionization constant of water,** K_w, is 1×10^{-14} at 25°C. In every aqueous solution, there is always a small number of both H^+ and OH^-.

Section 15.8 The **pH** of a solution expresses $[H^+]$ on an exponential scale. That is, a solution of pH 1 is 10 times more acidic than a solution of pH 2. To calculate the pH of a solution, we express $[H^+]$ as a power of 10 and change the sign. For example, if $[H^+]$ is 10^{-2} *M*, the pH is 2. Conversely, to find $[H^+]$ given the pH, we raise 10 to the negative pH. For example, if the pH is 3, $[H^+]$ is 10^{-3} *M*, that is, 0.001 *M*.

Section 15.9 To calculate the pH of solutions whose hydrogen ion concentrations are not exact powers of 10, we use the logarithm values found in a scientific calculator. That is, the pH is equal to $-\log[H^+]$. Conversely, to calculate the hydrogen ion concentration that corresponds to a given pH, we find the inverse logarithm on the calculator.

Section 15.10 A strong acid, a strong base, and a soluble ionic compound are each examples of a **strong electrolyte**. Therefore, these solutions are highly ionized. Hydrochloric acid is a strong acid and is better represented in aqueous solution as $H^+(aq)$ and $Cl^-(aq)$. Sodium chloride is soluble in water and is best represented as $Na^+(aq)$ and $Cl^-(aq)$. A weak acid, a weak base, and a slightly soluble ionic compound are each examples of a **weak electrolyte**. Therefore, these solutions are slightly ionized. Acetic acid is a weak acid and in aqueous solution should be written as $HC_2H_3O_2(aq)$. The slightly soluble ionic compound silver chloride is written as AgCl(s).

Section 15.11 A **net ionic equation** shows a solution reaction more accurately. The first step is to balance the chemical equation. Second, write a **total ionic equation** for the reaction. Strong electrolytes are written in the ionic form, and weak electrolytes in the nonionized form. Third, cancel **spectator ions** that are identical on both sides of the equation. Finally, verify that the net ionic equation is balanced and simplify the coefficients when possible.

Problem-Solving Organizer

Topic	Procedure	Example
Acid–Base Titrations Sec. 15.5	**1.** Write down the units asked for in the answer. **2.** Write down the related given value. **3.** Apply unit factors to convert the given units to the units in the answer.	If 10.00 mL of 0.100 *M* sulfuric acid neutralizes 15.50 mL of KOH, what is the molarity of the base? *Given:* $H_2SO_4(aq) + 2\ KOH(aq) \longrightarrow K_2SO_4(aq) + 2\ H_2O(l)$ $10.00\ \cancel{\text{mL solution}} \times \frac{0.100\ \cancel{\text{mol } H_2SO_4}}{1000\ \cancel{\text{mL solution}}} \times \frac{2\ \text{mol KOH}}{1\ \cancel{\text{mol } H_2SO_4}} = 0.00200\ \text{mol KOH}$ $\frac{0.00200\ \text{mol KOH}}{15.50\ \cancel{\text{mL solution}}} \times \frac{1000\ \cancel{\text{mL solution}}}{1\ \text{L solution}} = 0.129\ M\ \text{KOH}$
Acid–Base Standardization Sec. 15.6	**1.** Write down the units asked for in the answer. **2.** Write down the related given value. **3.** Apply unit factors to convert the given units to the units in the answer.	If 15.50 mL of KOH neutralizes 0.995 g of KHP (204.23 g/mol), what is the molarity of the base? *Given:* $KHP(s) + KOH(aq) \longrightarrow K_2P(aq) + H_2O(l)$ $0.995\ \cancel{\text{g KHP}} \times \frac{1\ \cancel{\text{mol KHP}}}{204.23\ \cancel{\text{g KHP}}} \times \frac{1\ \text{mol KOH}}{1\ \cancel{\text{mol KHP}}} = 0.00487\ \text{mol KOH}$ $\frac{0.00487\ \text{mol KOH}}{15.50\ \cancel{\text{mL solution}}} \times \frac{1000\ \cancel{\text{mL solution}}}{1\ \text{L solution}} = 0.314\ M\ \text{KOH}$

(continued)

Problem-Solving Organizer *(continued)*

Topic	Procedure	Example
Ionization of Water Sec. 15.7	Solve for $[H^+]$ or $[OH^-]$ in aqueous solution using the ionization of water equation: $K_w = [H^+][OH^-]$	If $[H^+]$ is 0.0015 *M* in an aqueous solution, what is the hydroxide ion concentration? $K_w = [H^+][OH^-] = 1.0 \times 10^{-14}$ $[0.0015][OH^-] = 1.0 \times 10^{-14}$ $[OH^-] = 6.7 \times 10^{-12}$
The pH Concept Sec. 15.8	Convert from $[H^+]$ to pH using the equation: $pH = -\log[H^+]$	If $[H^+]$ in an acid is 0.001 *M*, what is the pH? $pH = -\log[10^{-3}] = -(-3) = 3$
Advanced pH Calculations* Sec. 15.9	Convert from pH to $[H^+]$ using the equation: $[H^+] = 10^{-pH}$	If the pH of an acid is 3.20, what is $[H^+]$? $[H^+] = 10^{-3.20} = 0.00063\ M$

*Refer to Appendix A for help in using the log key on a scientific calculator.

Key Concepts*

1. Which of the following is a strong Arrhenius acid: $HC_2H_3O_2(aq)$, $HNO_3(aq)$, $H_2CO_3(aq)$, $H_3PO_4(aq)$?
2. Identify the Arrhenius acid and base that react to give lithium chloride, LiCl.
3. Identify the Brønsted–Lowry acid and base in the following reaction.

$$HNO_3(aq) + NH_3(aq) \longrightarrow NH_4NO_3(aq)$$

4. If 25.0 mL of 0.100 *M* H_2SO_4 is titrated with 0.200 *M* NaOH, what volume of sodium hydroxide is required to neutralize the acid?

$$H_2SO_4(aq) + NaOH(aq) \longrightarrow Na_2SO_4(aq) + H_2O(l)$$

5. If 0.424 g of sodium carbonate is titrated with 20.00 mL of HCl, what is the molar concentration of the hydrochloric acid?

$$HCl(aq) + Na_2CO_3(s) \longrightarrow NaCl(aq) + H_2O(l) + CO_2(g)$$

6. How can pure water be neutral since it contains both H^+ and OH^-?
7. If the hydrogen ion concentration in an aqueous acetic acid solution is 0.0015 *M*, what is the molar hydroxide ion concentration?
8. If the hydrogen ion concentration in an aqueous acetic acid solution is 0.0015 *M*, what is the pH?
9. Beakers X and Y (at the top of page 445) contain either nitric or nitrous acid. Based on the conductivity test shown, which beaker contains HNO_3? Which contains HNO_2?
10. Nitric acid reacts with sodium hydroxide to give sodium nitrate and water. Write the net ionic equation for the reaction.

$$HNO_3(aq) + NaOH(aq) \longrightarrow NaNO_3(s) + H_2O(l)$$

*Answers to Key Concepts are in Appendix H.

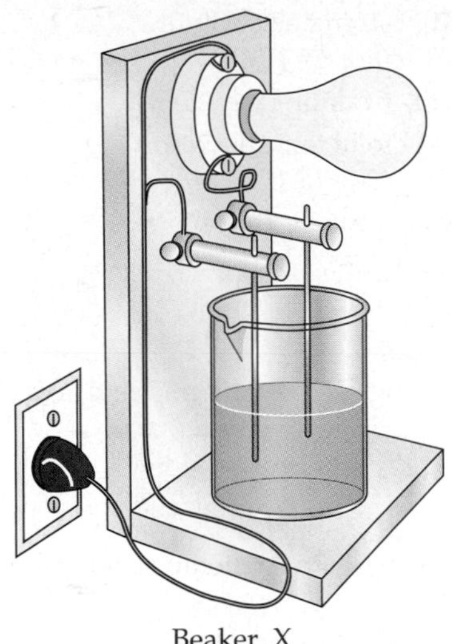
Beaker X

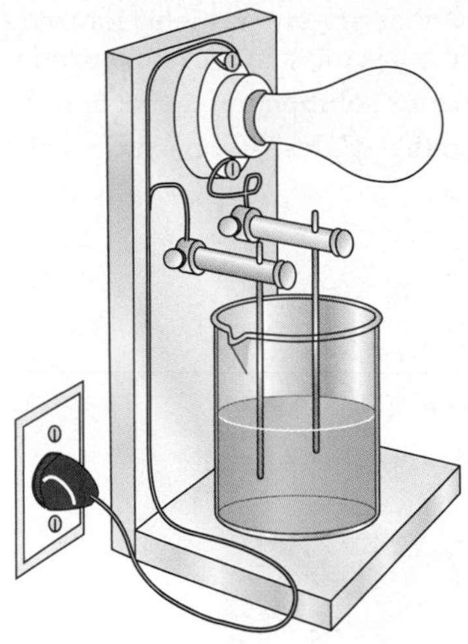
Beaker Y

Key Terms†

Select the key term below that corresponds to each of the following definitions.

_____ 1. a solution that resists changes in pH when an acid or a base is added
_____ 2. a substance that releases hydrogen ions when dissolved in water
_____ 3. a substance that releases hydroxide ions when dissolved in water
_____ 4. the process of a polar compound dissolving in water and forming positive and negative ions
_____ 5. the process of an ionic compound dissolving in water and separating into positive and negative ions
_____ 6. the ion that best represents the hydrogen ion in aqueous solution
_____ 7. a product obtained from a neutralization reaction in addition to water
_____ 8. a substance that donates a proton in an acid–base reaction
_____ 9. a substance that accepts a proton in an acid–base reaction
_____ 10. a term used interchangeably with hydrogen ion donor
_____ 11. a term used interchangeably with hydrogen ion acceptor
_____ 12. a substance capable of either accepting or donating a proton in an acid–base reaction
_____ 13. a chemical substance that changes color according to the pH of the solution
_____ 14. a procedure for delivering a measured volume of solution through a buret
_____ 15. the stage in a titration when the indicator changes color
_____ 16. a solution whose concentration has been established precisely
_____ 17. a constant that equals the product of the molar hydrogen ion concentration and the molar hydroxide ion concentration
_____ 18. the molar hydrogen ion concentration expressed on an exponential scale
_____ 19. an aqueous solution that is a good conductor of electricity
_____ 20. an aqueous solution that is a poor conductor of electricity

(a) acid–base indicator (*Sec. 15.4*)
(b) amphiprotic (*Sec. 15.3*)
(c) Arrhenius acid (*Sec. 15.2*)
(d) Arrhenius base (*Sec. 15.2*)
(e) Brønsted–Lowry acid (*Sec. 15.3*)
(f) Brønsted–Lowry base (*Sec. 15.3*)
(g) buffer (*Sec. 15.1*)
(h) dissociation (*Sec. 15.2*)
(i) endpoint (*Sec. 15.5*)
(j) hydronium ion (H_3O^+) (*Sec. 15.2*)
(k) ionization (*Sec. 15.2*)
(l) ionization constant of water, K_w (*Sec. 15.7*)
(m) net ionic equation (*Sec. 15.11*)
(n) pH (*Sec. 15.8*)
(o) proton acceptor (*Sec. 15.3*)
(p) proton donor (*Sec. 15.3*)
(q) salt (*Sec. 15.2*)
(r) spectator ions (*Sec. 15.11*)
(s) standard solution (*Sec. 15.6*)

† Answers to Key Terms are in Appendix I.

_____	21. a chemical equation that represents highly ionized substances in the ionic form and slightly ionized substances in the nonionized form	(t)	strong electrolyte (*Sec. 15.10*)
_____	22. ions that are in aqueous solution but do not appear in the net ionic equation	(u)	titration (*Sec. 15.5*)
_____	23. a chemical equation that represents an ionic reaction after spectator ions have been canceled	(v)	total ionic equation (*Sec. 15.11*)
		(w)	weak electrolyte (*Sec. 15.10*)

Exercises‡

Properties of Acids and Bases (Sec. 15.1)

1. State at least three general properties of acids.

▲ **Common Acids** An acid–base indicator is added to beakers of lime juice, vinegar, and soda water. The color of the indicator shows varying acidity in each solution.

2. State at least three general properties of bases.
3. Classify the following foods as acidic, basic, or neutral.
   (a) egg white, pH 7.9 **(b)** sour milk, pH 6.2
   (c) maple syrup, pH 7.0 **(d)** lime juice, pH 1.8
   (e) champagne, pH 3.8 **(f)** tomato juice, pH 4.1
4. Classify the following 0.1 *M* solutions as strongly acidic, weakly acidic, neutral, weakly basic, or strongly basic.
   (a) NaOH, pH 13.0 **(b)** NaCl, pH 7.0
   (c) Na_2CO_3, pH 11.7 **(d)** $NaHCO_3$, pH 8.3
   (e) H_2CO_3, pH 3.7 **(f)** HNO_3, pH 1.0

Arrhenius Acids and Bases (Sec. 15.2)

5. Classify each of the following Arrhenius acids as strong or weak given the degree of ionization.
   (a) chloric acid, $HClO_3(aq)$, ~100%
   (b) hypoiodous acid, HIO(aq), ~1%
   (c) hydrobromic acid, HBr(aq), ~100%
   (d) benzoic acid, $HC_7H_5O_2(aq)$, ~1%
6. Classify each of the following Arrhenius bases as strong or weak given the degree of ionization.
   (a) zinc hydroxide, $Zn(OH)_2(s)$, ~1%
   (b) lithium hydroxide, LiOH(aq), ~100%
   (c) iron(II) hydroxide, $Fe(OH)_2(s)$, ~1%
   (d) barium hydroxide, $Ba(OH)_2(aq)$, ~100%
7. Classify each of the following as an Arrhenius acid, an Arrhenius base, or a salt.
   (a) HClO(aq) **(b)** KOH(aq)
   (c) $K_2SO_4(aq)$ **(d)** $Sr(OH)_2(aq)$
8. Classify each of the following as an Arrhenius acid, an Arrhenius base, or a salt.
   (a) $HNO_3(aq)$ **(b)** $Mg(NO_3)_2(aq)$
   (c) $Ca(OH)_2(aq)$ **(d)** $H_2SO_3(aq)$
9. Identify the Arrhenius acid and Arrhenius base in each of the following neutralization reactions.
   (a) $HI(aq) + NaOH(aq) \longrightarrow NaI(aq) + H_2O(l)$
   (b) $HC_2H_3O_2(aq) + LiOH(aq) \longrightarrow LiC_2H_3O_2(aq) + H_2O(l)$
10. Identify the Arrhenius acid and Arrhenius base in each of the following neutralization reactions.
   (a) $2\,HClO_3(aq) + Ba(OH)_2(aq) \longrightarrow Ba(ClO_3)_2(aq) + 2\,H_2O(l)$
   (b) $H_2SO_4(aq) + 2\,KOH(aq) \longrightarrow K_2SO_4(aq) + 2\,H_2O(l)$
11. Determine the acid and base that were neutralized to produce each of the following salts.
   (a) sodium fluoride, NaF(aq)
   (b) magnesium iodide, $MgI_2(aq)$
   (c) calcium nitrate, $Ca(NO_3)_2(aq)$
   (d) lithium carbonate, $Li_2CO_3(aq)$
12. Determine the acid and base that were neutralized to produce each of the following salts.
   (a) potassium bromide, KBr(aq)
   (b) barium chloride, $BaCl_2(aq)$
   (c) cobalt(II) sulfate, $CoSO_4(aq)$
   (d) sodium phosphate, $Na_3PO_4(aq)$
13. Complete and balance the following neutralization reactions.
   (a) $HNO_3(aq) + Ca(OH)_2(aq) \longrightarrow$
   (b) $H_2CO_3(aq) + Ba(OH)_2(aq) \longrightarrow$
14. Complete and balance the following neutralization reactions.
   (a) $HC_2H_3O_2(aq) + Ca(OH)_2(aq) \longrightarrow$
   (b) $H_2SO_4(aq) + NH_4OH(aq) \longrightarrow$

‡Answers to odd-numbered Exercises are in Appendix J.

Brønsted–Lowry Acids and Bases (Sec. 15.3)

15. Identify the Brønsted–Lowry acid and base in each of the following neutralization reactions.
    (a) $HC_2H_3O_2(aq) + LiOH(aq) \longrightarrow LiC_2H_3O_2(aq) + H_2O(l)$
    (b) $NaCN(aq) + HBr(aq) \longrightarrow NaBr(aq) + HCN(aq)$

16. Identify the Brønsted–Lowry acid and base in each of the following neutralization reactions.
    (a) $2\ HClO_4(aq) + K_2CO_3(aq) \longrightarrow 2\ KClO_4(aq) + H_2O(l) + CO_2(g)$
    (b) $2\ NH_3(aq) + H_2SO_4(aq) \longrightarrow (NH_4)_2SO_4(aq)$

17. Identify the Brønsted–Lowry acid and base in each of the following neutralization reactions.
    (a) $HI(aq) + H_2O(l) \longrightarrow H_3O^+(aq) + I^-(aq)$
    (b) $HC_2H_3O_2(aq) + HS^-(aq) \longrightarrow H_2S(aq) + C_2H_3O_2^-(aq)$

18. Identify the Brønsted–Lowry acid and base in each of the following neutralization reactions.
    (a) $HCO_3^-(aq) + OH^-(aq) \longrightarrow CO_3^{2-}(aq) + H_2O(l)$
    (b) $NO_2^-(aq) + HClO_4(aq) \longrightarrow HNO_2(aq) + ClO_4^-(aq)$

19. Complete and balance the following Brønsted–Lowry neutralization reactions.
    (a) $HF(aq) + NaHS(aq) \longrightarrow$
    (b) $HNO_2(aq) + NaC_2H_3O_2(aq) \longrightarrow$

20. Complete and balance the following Brønsted–Lowry neutralization reactions.
    (a) $H_3O^+(aq) + SO_4^-(aq) \longrightarrow$
    (b) $H_2PO_4^-(aq) + NH_3(aq) \longrightarrow$

Acid–Base Indicators (Sec. 15.4)

21. Given the following pH values for solutions containing a drop of methyl red indicator, state the color of each solution.
    (a) pH 3 (b) pH 7

22. Given the following pH values for solutions containing a drop of bromthymol blue indicator, state the color of each solution.
    (a) pH 5 (b) pH 9

◀ **A Natural Indicator** A rose contains a natural acid–base indicator. After the indicator is extracted from the dark-red rose (right), the rose petals are a pale pink.

23. Given the following pH values for solutions containing a drop of phenolphthalein indicator, state the color of each solution.
    (a) pH 7 (b) pH 11

24. What is the color of phenolphthalein indicator in pure water?

25. What is the color of methyl red indicator in a solution of pH 5?

26. What is the color of bromthymol blue indicator in a solution of pH 7?

Acid–Base Titrations (Sec. 15.5)

27. If the titration of a 25.0-mL sample of hydrochloric acid requires 22.15 mL of 0.155 *M* sodium hydroxide, what is the molarity of the acid?
$$HCl(aq) + NaOH(aq) \longrightarrow NaCl(aq) + H_2O(l)$$

28. If the titration of a 25.0-mL sample of calcium hydroxide requires 34.45 mL of 0.100 *M* perchloric acid, what is the molarity of the base?
$$2\ HClO_4(aq) + Ca(OH)_2(aq) \longrightarrow Ca(ClO_4)_2(aq) + 2\ H_2O(l)$$

29. If 34.45 mL of 0.210 *M* NaOH neutralizes 50.0 mL of phosphoric acid, what is the molarity of the acid?
$$H_3PO_4(aq) + 3\ NaOH(aq) \longrightarrow Na_3PO_4(aq) + 3\ H_2O(l)$$

30. If 29.50 mL of 0.175 *M* nitric acid neutralizes 50.0 mL of ammonium hydroxide, what is the molarity of the base?
$$HNO_3(aq) + NH_4OH(aq) \longrightarrow NH_4NO_3(aq) + H_2O(l)$$

31. How many milliliters of 0.122 *M* sulfuric acid are required to completely neutralize 41.05 mL of 0.165 *M* KOH?
$$H_2SO_4(aq) + 2\ KOH(aq) \longrightarrow K_2SO_4(aq) + 2\ H_2O(l)$$

32. How many milliliters of 0.100 *M* barium hydroxide are required to completely neutralize 10.0 mL of 0.225 *M* nitrous acid?
$$2\ HNO_2(aq) + Ba(OH)_2(aq) \longrightarrow Ba(NO_2)_2(aq) + 2\ H_2O(l)$$

33. Given the molarity and density for each of the following acidic solutions, calculate the mass/mass percent concentration.
    (a) 6.00 *M* HCl (d = 1.10 g/mL)
    (b) 1.00 *M* $HC_2H_3O_2$ (d = 1.01 g/mL)
    (c) 0.500 *M* HNO_3 (d = 1.01 g/mL)
    (d) 3.00 *M* H_2SO_4 (d = 1.18 g/mL)

34. Given the molarity and density for each of the following basic solutions, calculate the mass/mass percent concentration.
    (a) 3.00 *M* NaOH (d = 1.12 g/mL)
    (b) 0.500 *M* KOH (d = 1.02 g/mL)
    (c) 6.00 *M* NH_3 (d = 0.954 g/mL)
    (d) 1.00 *M* Na_2CO_3 (d = 1.10 g/mL)

Acid–Base Standardization (Sec. 15.6)

35. What is the molarity of nitric acid if 41.25 mL of HNO_3 is required to neutralize 0.689 g of sodium carbonate?

$$2\ HNO_3(aq) + Na_2CO_3(s) \longrightarrow NaNO_3(aq) + H_2O(l) + CO_2(g)$$

36. What is the molarity of sulfuric acid if 32.35 mL of H_2SO_4 is required to neutralize 0.750 g of sodium hydrogen carbonate?

$$H_2SO_4(aq) + 2\ NaHCO_3(s) \longrightarrow Na_2SO_4(aq) + 2\ H_2O(l) + 2\ CO_2(g)$$

37. What is the molarity of hydrochloric acid if 20.95 mL of HCl is required to neutralize 1.550 g of sodium oxalate, $Na_2C_2O_4$?

$$2\ HCl(aq) + Na_2C_2O_4(s) \longrightarrow H_2C_2O_4(aq) + 2\ NaCl(aq)$$

38. What is the molarity of sodium hydroxide if 28.85 mL of NaOH is required to neutralize 0.506 g of oxalic acid, $H_2C_2O_4$?

$$H_2C_2O_4(s) + NaOH(aq) \longrightarrow NaHC_2O_4(aq) + H_2O(l)$$

39. What volume of 0.479 *M* lithium hydroxide is required to neutralize 0.627 g of oxalic acid, $H_2C_2O_4$?

$$H_2C_2O_4(s) + 2\ LiOH(aq) \longrightarrow Li_2C_2O_4(aq) + 2\ H_2O(l)$$

40. What volume of 0.167 *M* barium hydroxide is required to neutralize 1.655 g of potassium hydrogen phthalate, $KHC_8H_4O_4$ (204.23 g/mol)?

$$2\ KHC_8H_4O_4(s) + Ba(OH)_2(aq) \longrightarrow BaK_2(C_8H_4O_4)_2(aq) + 2\ H_2O(l)$$

41. Ascorbic acid, HAsc, is the chemical name for vitamin C. If 30.95 mL of 0.176 *M* NaOH neutralizes 0.959 g of acid, what is the molar mass of vitamin C?

$$HAsc(s) + NaOH(aq) \longrightarrow NaAsc(aq) + H_2O(l)$$

42. Tartaric acid, H_2Tart, gives grapes a sour taste. If 28.15 mL of 0.295 *M* NaOH neutralizes 0.623 g of acid, what is the molar mass of tartaric acid?

$$H_2Tart(s) + 2\ NaOH(aq) \longrightarrow Na_2Tart(aq) + 2\ H_2O(l)$$

43. Alanine is an amino acid. If 21.05 mL of 0.145 *M* NaOH is required to neutralize 0.272 g of alanine, what is the molar mass of the amino acid? (*Note*: One mole of alanine neutralizes one mole of acid.)

44. THAM is a base used to standardize acid solutions. If 21.35 mL of 0.115 *M* HCl is required to neutralize 0.297 g of THAM, what is the molar mass of the base? (*Note*: One mole of THAM neutralizes one mole of acid.)

Ionization of Water (Sec. 15.7)

45. Indicate each of the following for the ionization of pure water.
   (a) the simplified ionization equation
   (b) the ionization constant equation, K_w
   (c) the ionization constant at 25°C

46. Indicate each of the following for the ionization of pure water.
   (a) the molecular collision equation
   (b) the molar hydrogen ion concentration at 25°C
   (c) the molar hydroxide ion concentration at 25°C

47. Given the molar concentration of hydrogen ion, calculate the concentration of hydroxide ion.
   (a) $[H^+] = 0.025$ (b) $[H^+] = 0.000\ 017$

48. Given the molar concentration of hydrogen ion, calculate the concentration of hydroxide ion.
   (a) $[H^+] = 6.2 \times 10^{-7}$ (b) $[H^+] = 4.6 \times 10^{-12}$

49. Given the molar concentration of hydroxide ion, calculate the concentration of hydrogen ion.
   (a) $[OH^-] = 0.0016$ (b) $[OH^-] = 0.000\ 29$

50. Given the molar concentration of hydroxide ion, calculate the concentration of hydrogen ion.
   (a) $[OH^-] = 8.8 \times 10^{-8}$ (b) $[OH^-] = 4.6 \times 10^{-13}$

The pH Concept (Sec. 15.8)

51. Calculate the pH of each of the following given the molar hydrogen ion concentration.
   (a) soft drink, $[H^+] = 0.001\ M$
   (b) coffee, $[H^+] = 0.000\ 01\ M$

52. Calculate the pH of each of the following given the molar hydrogen ion concentration.
   (a) egg white, $[H^+] = 0.000\ 000\ 01\ M$
   (b) sour milk, $[H^+] = 0.000\ 001\ M$

53. Calculate the molar hydrogen ion concentration of each of the following given the pH.
   (a) shampoo, pH 6
   (b) pH balanced shampoo, pH 8

54. Calculate the molar hydrogen ion concentration of each of the following given the pH.
   (a) phosphate detergent, pH 9
   (b) nonphosphate detergent, pH 11

Advanced pH Calculations (Sec. 15.9)

55. Calculate the pH of each of the following given the molar hydrogen ion concentration.
   (a) carrots, $[H^+] = 0.000\ 007\ 9\ M$
   (b) peas, $[H^+] = 0.000\ 000\ 39\ M$

56. Calculate the pH of each of the following given the molar hydrogen ion concentration.
   (a) milk, $[H^+] = 0.000\ 000\ 30\ M$
   (b) eggs, $[H^+] = 0.000\ 000\ 016\ M$

57. Calculate the molar hydrogen ion concentration of each of the following biological solutions given the pH.
   (a) gastric juice, pH 1.80 (b) urine, pH 4.75

58. Calculate the molar hydrogen ion concentration of each of the following biological solutions given the pH.
   (a) saliva, pH 6.55 (b) blood, pH 7.50

59. Calculate the pH of each of the following given the molar hydroxide ion concentration.
(a) $[OH^-] = 0.11\ M$ (b) $[OH^-] = 0.000\ 55\ M$

60. Calculate the pH of each of the following given the molar hydroxide ion concentration.
(a) $[OH^-] = 0.000\ 031\ M$
(b) $[OH^-] = 0.000\ 000\ 000\ 66\ M$

61. Calculate the molar hydroxide ion concentration of each of the following solutions given the pH.
(a) pH 0.90 (b) pH 1.62

62. Calculate the molar hydroxide ion concentration of each of the following solutions given the pH.
(a) pH 4.55 (b) pH 5.20

Strong and Weak Electrolytes (Sec. 15.10)

63. State whether each of the following substances in aqueous solution is highly ionized or slightly ionized.
(a) strong acids (b) strong bases
(c) soluble ionic compounds

64. State whether each of the following substances in aqueous solution is highly ionized or slightly ionized.
(a) weak acids (b) weak bases
(c) slightly soluble ionic compounds

65. Classify each of the following acids as a strong or a weak electrolyte.
(a) $H_2CO_3(aq)$ (b) $H_2SO_4(aq)$
(c) $HI(aq)$ (d) $HNO_2(aq)$

66. Classify each of the following bases as a strong or a weak electrolyte.
(a) $KOH(aq)$ (b) $NH_4OH(aq)$
(c) $Ca(OH)_2(aq)$ (d) $Mg(OH)_2(s)$

67. Classify each of the following aqueous solutions as a strong or a weak electrolyte.
(a) $ZnCO_3(s)$ (b) $Sr(NO_3)_2(aq)$
(c) $K_2SO_4(aq)$ (d) $PbI_2(s)$

68. Classify each of the following aqueous solutions as a strong or a weak electrolyte.
(a) $Fe(C_2H_3O_2)_3(aq)$ (b) $AlPO_4(s)$
(c) $Ag_2CrO_4(s)$ (d) $CdSO_4(aq)$

69. Write the following acids in either the ionized or the nonionized form to best represent an aqueous solution.
(a) $HF(aq)$ (b) $HBr(aq)$
(c) $HNO_3(aq)$ (d) $HNO_2(aq)$

70. Write the following bases in either the ionized or the nonionized form to best represent an aqueous solution.
(a) $NaOH(aq)$ (b) $NH_4OH(aq)$
(c) $Ba(OH)_2(aq)$ (d) $Al(OH)_3(s)$

71. Write the following salts in either the ionized or the nonionized form to best represent an aqueous solution.
(a) $AgF(aq)$ (b) $AgI(s)$
(c) $Hg_2Cl_2(s)$ (d) $NiCl_2(aq)$

72. Write the following salts in either the ionized or the nonionized form to best represent an aqueous solution.
(a) $AlPO_4(s)$ (b) $Co(C_2H_3O_2)_3(aq)$
(c) $MnSO_4(aq)$ (d) $PbSO_4(s)$

Net Ionic Equations (Sec. 15.11)

73. List the four steps for writing balanced net ionic equations.

74. What is the term for ions that appear in the total ionic equation but are not present in the net ionic equation?

75. Write a balanced net ionic equation for each of the following acid–base reactions. Refer to Table 15.5 for electrolyte information.
(a) $HCl(aq) + KOH(aq) \longrightarrow KCl(aq) + H_2O(l)$
(b) $HC_2H_3O_2(aq) + Ca(OH)_2(aq) \longrightarrow Ca(C_2H_3O_2)_2(aq) + H_2O(l)$

76. Write a balanced net ionic equation for each of the following acid–base reactions. Refer to Table 15.5 for electrolyte information.
(a) $HF(aq) + Li_2CO_3(aq) \longrightarrow LiF(aq) + H_2O(l) + CO_2(g)$
(b) $H_2SO_4(aq) + Ba(OH)_2(aq) \longrightarrow BaSO_4(s) + H_2O(l)$

77. Write a balanced net ionic equation for each of the following solution reactions. Refer to Table 15.5 for electrolyte information.
(a) $AgNO_3(aq) + KI(aq) \longrightarrow AgI(s) + AgNO_3(aq)$
(b) $BaCl_2(aq) + K_2CrO_4(aq) \longrightarrow BaCrO_4(s) + KCl(aq)$

78. Write a balanced net ionic equation for each of the following solution reactions. Refer to Table 15.5 for electrolyte information.
(a) $Zn(NO_3)_2(aq) + NaOH(aq) \longrightarrow Zn(OH)_2(s) + NaNO_3(aq)$
(b) $MgSO_4(aq) + NH_4OH(aq) \longrightarrow Mg(OH)_2(s) + (NH_4)_2SO_4(aq)$

General Exercises

79. A drop of methyl red indicator and a drop of phenolphthalein are both added to a beaker of water. What is the resulting color of the water?

80. A drop of bromthymol blue indicator and a drop of phenolphthalein are both added to a beaker of water. What is the resulting color of the water?

81. The acid–base indicator methyl orange has a color range from pH 3.2 to pH 4.4. It appears red in strongly acidic solutions and yellow in basic solutions. Predict the color of the indicator in a solution having a pH of 3.8.

82. The acid–base indicator bromcresol green changes color from pH 3.8 to pH 5.4. It appears yellow in strongly acidic solutions and blue in basic solutions. Predict the color of the indicator in a solution having a pH of 4.6.

83. Identify an amphiprotic substance given the following reactions.

$$HCl(aq) + NaH_2PO_4(aq) \longrightarrow H_3PO_4(aq) + NaCl(aq)$$
$$NaOH(aq) + NaH_2PO_4(aq) \longrightarrow Na_2HPO_4(aq) + H_2O(l)$$

84. Identify an amphiprotic substance given the following reactions.

$$HCl(aq) + H_2O(l) \longrightarrow H_3O^+(aq) + Cl^-(aq)$$
$$NH_2^-(aq) + H_2O(l) \longrightarrow NH_3(aq) + OH^-(aq)$$

85. Cream of tartar is found in baking powder. If 42.10 mL of 0.100 *M* $Ba(OH)_2$ is required to neutralize 0.791 g of cream of tartar, what is the molar mass? (*Note*: One mole of cream of tartar neutralizes one mole of base.)

86. In 1 mL of water there are 3×10^{22} molecules of H_2O. How many hydrogen ions are in 1 mL of water? (*Hint*: In 1 billion water molecules, only 2 are ionized.)

87. What is the pH of a 0.50 *M* HCl solution?

88. What is the pH of a 0.25 *M* NaOH solution?

89. Write the net ionic equation for the reaction of the weak electrolytes acetic acid and ammonium hydroxide.

90. Write the net ionic equation for the reaction of the strong electrolytes nitric acid and potassium hydroxide.

Explorer Quiz 1
Explorer Quiz 2
Explorer Quiz 3
Master Quiz

CHAPTER 16

Chemical Equilibrium

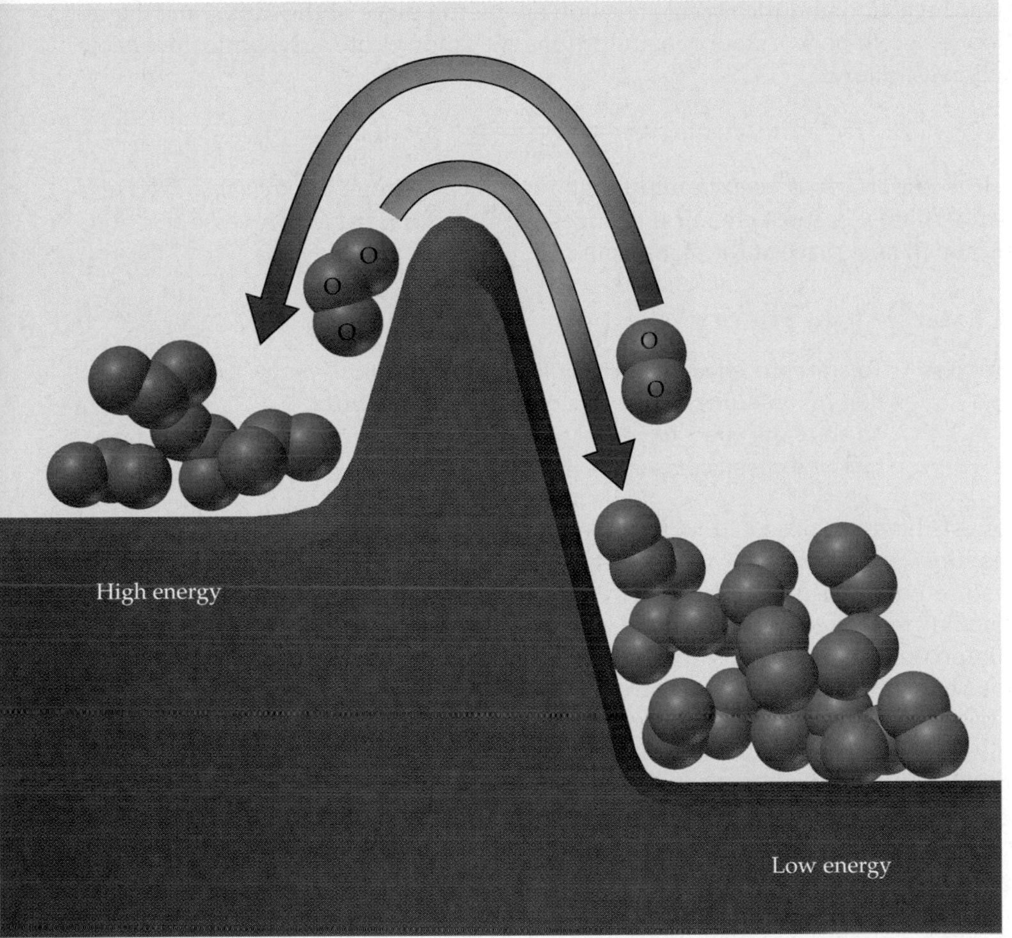

▲ In the above molecular art, ozone, O_3, is shown forming oxygen, O_2; and conversely, O_2 is changing to O_3. If the system is at equilibrium, which reaction is proceeding faster: the forward reaction to form O_2, or the reverse reaction?

 Oxygen–Ozone Equilibrium

This highly conceptual chapter is intentionally designed to follow the discussion in Chapter 15. The abundant molecular art helps the student to visualize the concept of equilibrium at the molecular level.

Previously, we have assumed that a chemical reaction continues until the reactants are used up and the reaction stops. In reality, most reactions continue to occur as an ongoing, reversible process. That is, a reaction occurs in the forward direction to give products, and simultaneously a reverse reaction occurs in the opposite direction to give the original reactants.

As an example of a reversible reaction, let's consider a chemical change that gives a precipitate (Figure 16.1). Initially, insoluble particles form in the aqueous solution. Then, some of the particles dissolve and go back into solution. Since some particles are precipitating while others are dissolving, the ongoing process is said to be dynamic. We can indicate this reversible chemical reaction as follows.

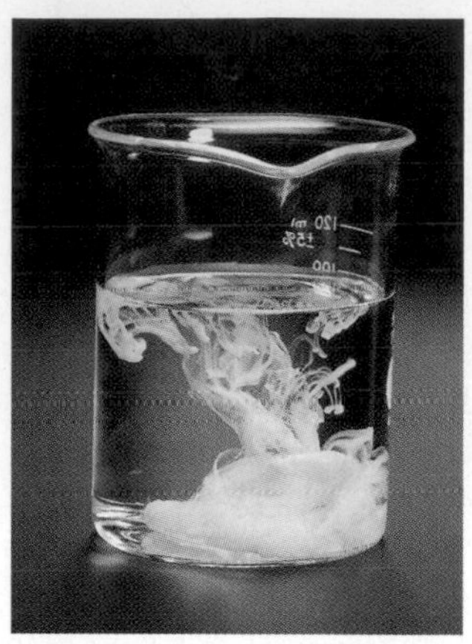

▲ **Figure 16.1 A Reversible Reaction** The reaction of lead nitrate and potassium iodide produces insoluble yellow, particles of lead iodide precipitate. Simultaneously, insoluble particles of lead iodide dissolve in the solution while other yellow particles precipitate from the solution.

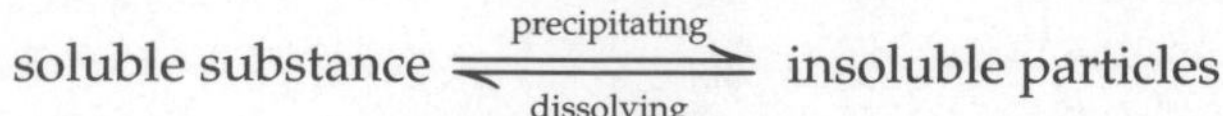

$$\text{soluble substance} \underset{\text{dissolving}}{\overset{\text{precipitating}}{\rightleftharpoons}} \text{insoluble particles}$$

Most chemical reactions involve a dynamic, reversible change. As reactants undergo a reaction to form products, the products react with each other to give the original reactants. Initially, the rate of the forward reaction is rapid. However, as the reaction proceeds, the concentrations of the reactants decrease and the concentrations of the products increase. With this change in concentrations, the rate of the forward reaction slows down, and the rate of the reverse reaction speeds up. When the forward reaction and the reverse reaction are taking place at the same rate, the reaction is said to be in a state of equilibrium. We can indicate a dynamic, reversible reaction as follows.

$$\text{reactants} \underset{\text{reverse reaction}}{\overset{\text{forward reaction}}{\rightleftharpoons}} \text{products}$$

It is important to keep in mind that a chemical change is a *dynamic, reversible process*. That is, in most chemical changes both the forward and reverse reactions continue to take place at the same time.

16.1 Collision Theory

Objectives
- **To state the effect of collision frequency, collision energy, and orientation of molecules on the rate of a chemical reaction.**
- **To state the effect of concentration, temperature, and a catalyst on the rate of a chemical reaction.**

Chemists have proposed a theoretical model for reactions which states that molecules must collide in order to react. If a collision is successful, molecules of reactants are changed into products. In a successful collision, existing bonds in a molecule are broken and new bonds are formed. This model is referred to as the **collision theory** of reactions. In a successful collision, molecules go through a transition state in which bonds are rearranged. Figure 16.2 illustrates the collision theory by showing molecules of nitrogen and oxygen colliding to form two molecules of nitrogen monoxide.

Collision theory proposes that three factors affect the rate of a chemical reaction. That is, three factors affect the rate of effective collisions: (1) collision frequency, (2) collision energy, and (3) orientation of the molecules. Let's discuss each factor in more detail.

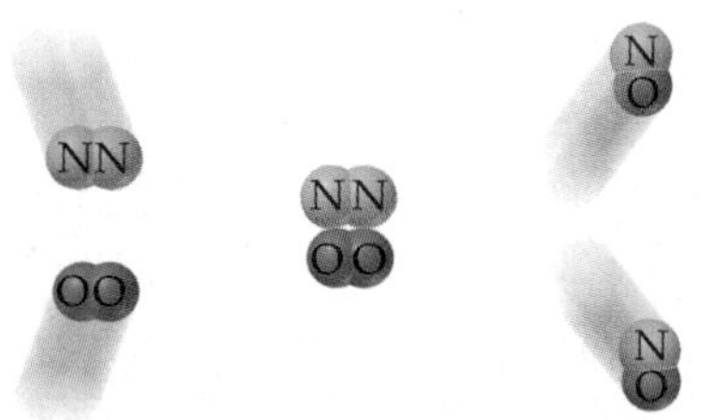

▲ **Figure 16.2 Collision Reaction** Nitrogen and oxygen gas molecules collide to form a temporary transition state. If the collision is successful, a molecule of N_2 and a molecule of O_2 produce two molecules of NO.

1. *Collision Frequency.* When we increase the frequency with which molecules collide, we increase the rate of the reaction. One way to increase the collision frequency is to increase the concentrations of molecules. A second way to increase collision frequency is to increase the temperature. As the temperature increases, the molecules move at a faster velocity and therefore collide more often.
2. *Collision Energy.* Even though molecules collide, they may not react. For a reaction to occur, molecules must collide with sufficient energy to form new bonds. When we increase the temperature, the molecules move at a faster velocity, and the collisions between molecules are more energetic. The rate of a reaction increases as the energy of collision increases.
3. *Collision Geometry.* Even though molecules may have an energetic collision, this does not guarantee that a reaction will occur. For a reaction to take place, the reacting molecules must be oriented in a favorable position. That is, the colliding molecules must strike each other in a particular way in order to undergo an effective collision. Otherwise, the molecules bounce off one another without reacting. The effect of collision geometry on the reaction of NO and O_3 molecules is shown in Figure 16.3.

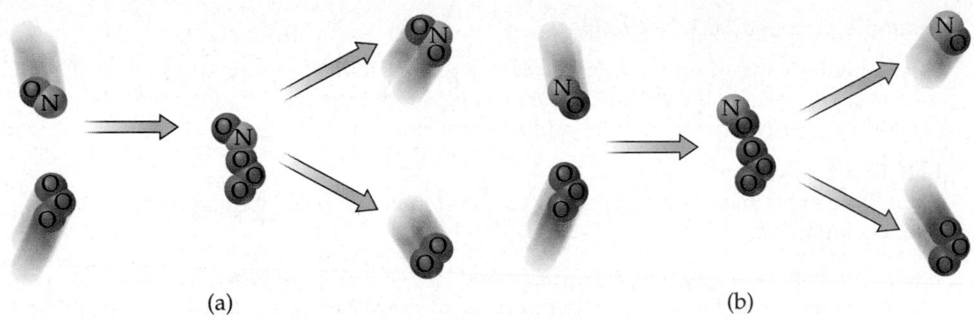

▲ **Figure 16.3 Effective versus Ineffective Collisions** (a) An effective collision of gaseous NO and O_3 produces NO_2 and O_2. (b) An ineffective collision of NO and O_3 molecules does not give products. If molecules do not have the correct orientation, they simply bounce off each other without forming new molecules.

Effective versus Ineffective Collisions

Effects of Concentration, Temperature, and Catalyst

So far, we have described the factors that influence the rate of reaction on the conceptual level. Now, we will discuss how to increase the rate of a chemical reaction on the practical level. Experimentally, three factors have been identified that affect the speed at which a chemical reaction proceeds. These factors are the concentration of reactants, the temperature of the reaction, and the presence of a catalyst. Let's consider each of these factors.

NO_2-N_2O_4 Equilibrium Movie

1. *Reactant Concentration.* When we increase the concentration of a reactant, the molecules are closer together and collide more frequently. It therefore follows that a chemical reaction proceeds faster as the concentrations of the reactants increase.

Temperature Dependence of Equilibrium Movie

2. *Reaction Temperature.* When we increase the temperature of the reaction, we increase the energy of the molecules. Increasing the temperature affects the rate of reaction in two ways. First, the collision frequency increases, and so the reaction speeds up. Second, the collision energy is greater, and so there are more molecules with enough energy to break bonds and form new molecules. Thus, as the temperature increases, the rate of reaction increases because of increased *collision frequency* and *collision energy.*

Catalysis, Surface Reaction Movies

3. *Catalyst.* If we add a catalyst, the rate of a chemical reaction increases. One way a catalyst speeds up a reaction is by increasing the number of effective collisions. In reactions involving hydrogen gas, Pt, Ni, and Zn metals can be used as catalysts. The rate of reaction increases because H_2 molecules are attracted to the surface of the metal catalyst, which in turn creates a more stable and favorable *collision geometry.*

In the following example exercise, the variables that affect the rate of a chemical reaction are summarized.

Rates of Reaction Activity

Example Exercise 16.1 • Rate of Reaction

When the temperature is increased, the rate of effective collision increases for two reasons. Explain these two reasons.

Solution

As discussed above, the two reasons are as follows.

(1) At higher temperatures molecules have a greater velocity, and so the collision frequency increases. As the number of collisions increases, so does the rate of effective collisions.

(continued)

Example Exercise 16.1 *(continued)*

(2) At higher temperatures molecules have a greater kinetic energy, and so the collision energy increases. If the collisions are more energetic, more molecules have the necessary energy to break bonds and form new ones.

Self-Test Exercise

State three experimental variables that can be changed in order to increase the rate of a chemical reaction.

Answer: The rate of reaction can be increased by (1) increasing the concentration of a reactant, (2) raising the reaction temperature, or (3) introducing a catalyst.

16.2 Energy Profiles of Chemical Reactions

Objectives
- To sketch the general energy profile for an endothermic and an exothermic reaction.
- To label the transition state, energy of activation, and heat of reaction on a given energy profile.

In air, nitrogen and oxygen gases are essentially unreactive. In a hot automobile engine, however, the two gases can react to produce nitrogen monoxide. To react, molecules of N_2 and O_2 must collide with sufficient energy to achieve the transition state before forming products. Since the transition state is a high-energy point, there is an energy barrier between the reactants and products. Figure 16.4 shows an analogy illustrating the energy changes that occur during a chemical reaction.

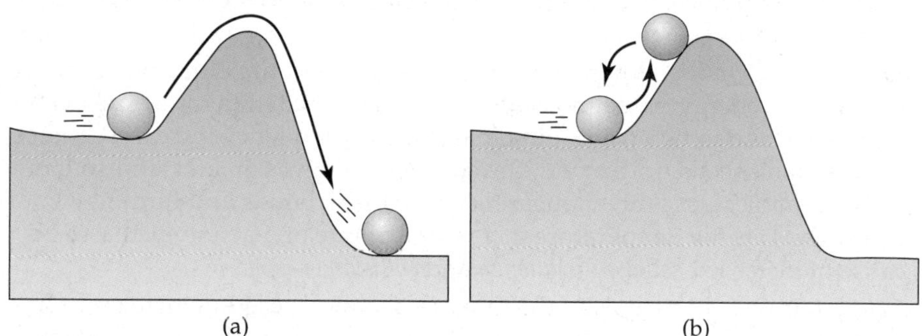

▲ **Figure 16.4 Energy Barrier Analogy for a Chemical Reaction** (a) The ball symbolizes molecules in a reaction that have sufficient energy to give products. If the ball has enough energy, it reaches the transition point at the top of the hill and rolls down the other side. (b) The ball symbolizes a reaction in which the molecules do not have sufficient energy to give products. If the ball does not have enough energy, it cannot reach the transition point and roll down the other side of the hill.

Endothermic Reaction Profiles

An **endothermic reaction** proceeds by consuming heat energy. As an example, consider the formation of nitrogen monoxide from nitrogen and oxygen gases. The reaction is

$$N_2(g) + O_2(g) + \text{heat} \rightleftharpoons 2\,NO(g)$$

Such a reaction, which absorbs heat energy in the process of going to completion, is called an endothermic reaction.

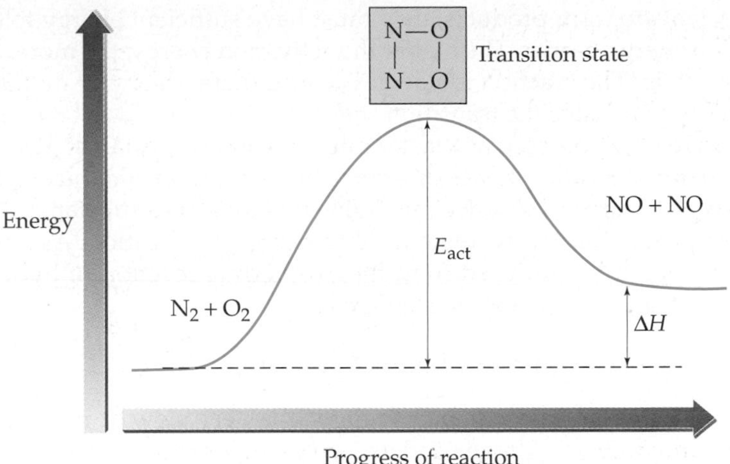

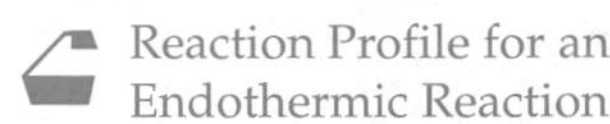

◀ **Figure 16.5 Reaction Profile for an Endothermic Reaction** Notice that the product of the reaction, NO, is at a higher energy level than the reactants, N_2 and O_2.

We can follow the progress of a chemical reaction by drawing an energy profile of reactants and products. A **reaction profile** shows the energy of reactants and products during the course of a reaction. The **transition state** is the highest point on the reaction profile and is the point where reactants and products have the same potential energy. A reaction profile for the reaction of nitrogen and oxygen is shown in Figure 16.5. The energy required for the reactants to achieve the transition state is called the **activation energy** (symbol $\boldsymbol{E_{act}}$). The energy difference between reactants and products is termed the **heat of reaction** (symbol $\boldsymbol{\Delta H}$).

Exothermic Reaction Profiles

An **exothermic reaction** proceeds by releasing heat energy. As an example of a reaction that proceeds by liberating heat energy, let's again consider that automobiles produce nitrogen monoxide, NO. Also consider that air contains ozone, O_3. The ozone in air converts the NO to NO_2, a brown gaseous component of smog.

$$NO(g) + O_3(g) \rightleftharpoons NO_2(g) + O_2(g) + \text{heat}$$

As you can see, the formation of nitrogen dioxide releases heat. Chemical processes that release heat energy are called exothermic reactions. As before, we can construct a reaction profile to follow the changes in energy for this reaction. Figure 16.6 illustrates the energy difference between reactants and products for an exothermic reaction.

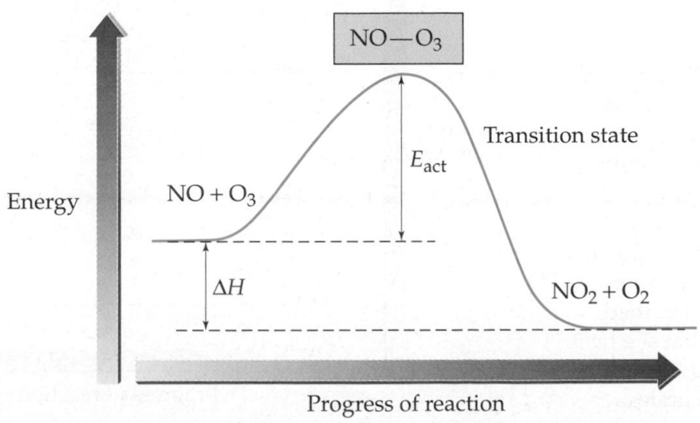

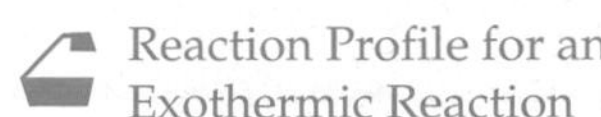

◀ **Figure 16.6 Reaction Profile for an Exothermic Reaction** Notice that the products of the reaction, NO_2 and O_2 are at a lower energy level than the reactants, NO and O_3.

For reactants to form products, they must have sufficient energy to overcome the activation energy barrier. The higher the activation energy, the more slowly the reaction proceeds. The reaction is slower because there are fewer molecules with enough energy to achieve the transition state.

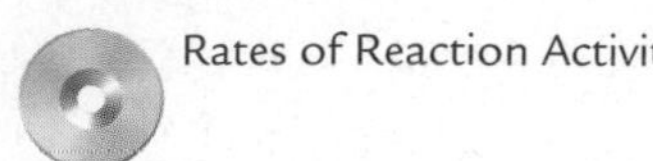

To help a reaction proceed at a faster rate, we can use a catalyst. Recall that we previously defined a catalyst as a substance that speeds up the rate of a reaction without being consumed. Now, we can define a **catalyst** as a substance that allows a reaction to proceed faster by lowering the energy of activation. As an example, consider how water is produced from the exothermic reaction of hydrogen and oxygen gases. The equation for the reaction is

$$2\,H_2(g) + O_2(g) \rightleftharpoons 2\,H_2O(g) + \text{heat}$$

Although this reaction produces heat energy, it is very slow at normal temperatures. When a mixture of hydrogen and oxygen gases is exposed to a flame or spark, the reaction is instantaneous. In fact, the instantaneous release of a large amount of energy creates an explosion. At room temperature, a powdered metal such as zinc dust acts as a catalyst and causes the reaction to occur rapidly without explosion. Figure 16.7 illustrates the reaction profile for hydrogen and oxygen gases with and without a catalyst.

Note that the heat of reaction is the same with and without a catalyst and that ΔH does not depend on the catalyst. Thus, the rate of a reaction is not related to the heat of reaction. Table 16.1 lists the rate of reaction and the heat of reaction for the formation of water from hydrogen and oxygen gases in the presence of various catalysts.

Table 16.1 Reaction of Hydrogen and Oxygen Gases

Catalyst	Rate of Reaction	Heat of Reaction*
none	very slow	54.6 kcal (228 kJ)
spark	explosive	54.6 kcal (228 kJ)
zinc dust	rapid	54.6 kcal (228 kJ)

*Energy released per mole of water produced at 25°C.

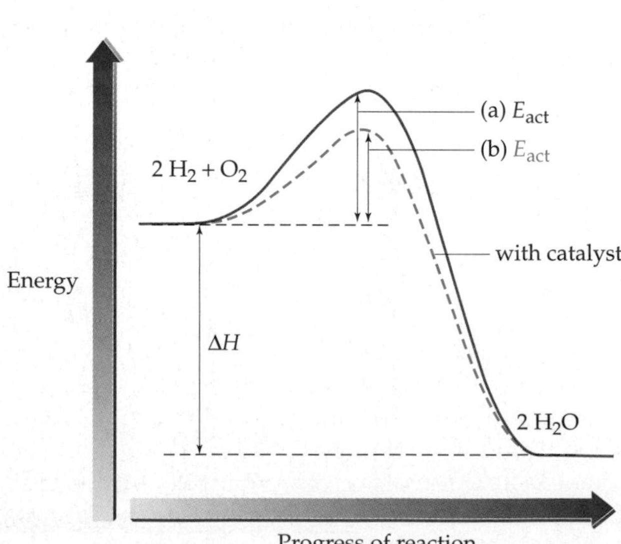

▶ **Figure 16.7 Effect of a Catalyst on Activation Energy** (a) Reaction profile for the reaction of H_2 and O_2 without a catalyst (solid line). (b) Reaction profile for the reaction of H_2 and O_2 with a zinc metal catalyst (dashed line). Notice that the heat of reaction, ΔH, is not affected by the catalyst.

The following example exercise further illustrates the relationship between reactants and products and the reaction profile.

Example Exercise 16.2 • Energy Profiles of Chemical Reactions

Ultraviolet light from the Sun converts oxygen molecules in the upper atmosphere to ozone molecules. Draw the reaction profile for the reaction.

$$3\,O_2(g) + \text{heat} \overset{\text{UV}}{\rightleftharpoons} 2\,O_3(g)$$

Solution

Since the reaction is endothermic, the reaction profile is

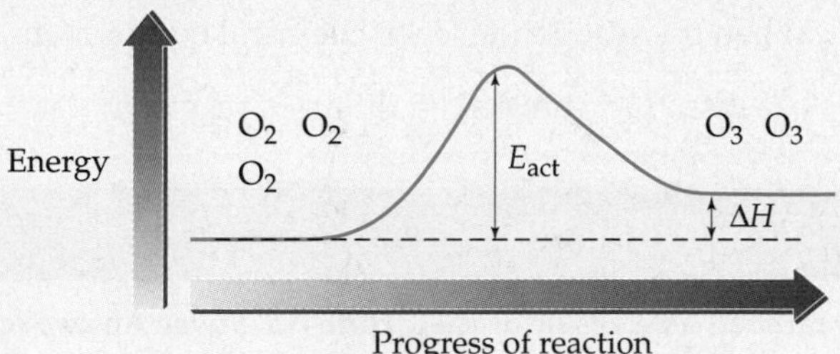

Self-Test Exercise

Nitrosyl bromide, NOBr, decomposes to give nitric oxide and bromine gas. Draw the reaction profile for the reaction.

$$2\,NOBr(g) \rightleftharpoons 2\,NO(g) + Br_2(g) + \text{heat}$$

Answer: Since the reaction is exothermic, the reaction profile is

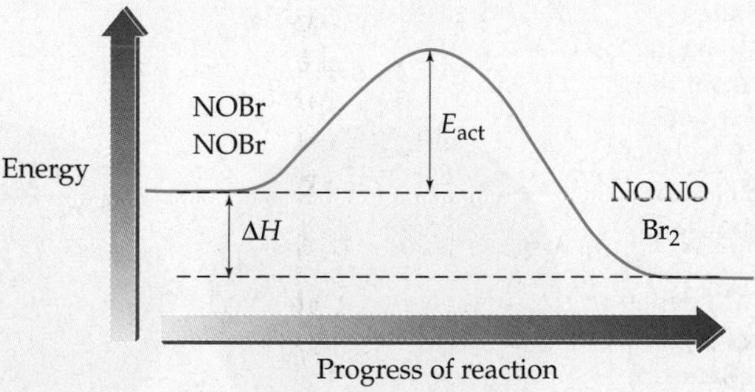

16.3 The Chemical Equilibrium Concept

Objectives
- To describe the equilibrium concept for a reversible reaction.
- To express the law of chemical equilibrium as an equation.

We have said that most chemical reactions are reversible processes. That is, a chemical change is a **reversible reaction** that can proceed simultaneously both in the forward direction toward products and in the reverse direction toward reactants. When the rate of the forward reaction is equal to the rate of the reverse reaction, a reaction is said to be in a state of **chemical equilibrium**.

It is important to understand that when a reaction is at equilibrium, it does not mean that the reaction has stopped. It also does not mean that the amounts of reactants and products are equal. Rather, a chemical reaction at equilibrium implies a *dynamic, reversible process.* That is, an ongoing forward reaction ($\longrightarrow$) and an ongoing reverse reaction ($\longleftarrow$) are taking place at the same speed.

As an example, consider the conversion of oxygen molecules, O_2, to ozone molecules, O_3, in the upper atmosphere. Ultraviolet light initiates the reaction. The equation for the reversible reaction is

$$\underset{\text{oxygen}}{3\,O_2(g)} \rightleftharpoons \underset{\text{ozone}}{2\,O_3(g)}$$

We can experimentally simulate this reversible reaction. We begin by pumping oxygen gas into a container that has been evacuated and contains no other gas. Next, we focus ultraviolet light on the oxygen gas in the container. Initially, the rate of the forward reaction ($rate_f$) is quite rapid and the rate of the reverse reaction ($rate_r$) is slow. As the reaction proceeds, the forward reaction slows down and the reverse reaction speeds up. When the reaction reaches chemical equilibrium, we have

Catalytic Destruction of Ozone Movie

$$rate_f\ (O_2\ \text{reaction}) = rate_r\ (O_3\ \text{reaction})$$

Chemistry Connection · The Ozone Hole

What animal population has been threatened as a result of the ozone hole over Antarctica?

All living species are protected from the Sun's harmful radiation by a layer of ozone molecules in the upper atmosphere. However, this ozone layer that shields the Earth's fragile life forms is being depleted. For some time, we have known about the ozone depletion over Antarctica referred to as an "ozone hole." Recently, scientists have found other holes in the ozone over heavily populated areas. Ozone depletion has been detected over Russia, Europe, Canada, and the United States.

In the stratosphere, about 50 miles above the Earth, ozone molecules absorb high-energy ultraviolet radiation from the Sun. A hole in the ozone layer is of great concern because the ozone layer protects us from the Sun's powerful rays. It is well known that ultraviolet radiation can cause skin cancer and cataracts. The Sun's harmful rays can also affect agriculture and cause lower crop yields.

In 1974 scientists at the University of California at Irvine suggested that chlorofluorocarbons (CFCs) could cause depletion of the ozone layer. CFCs are used mostly as propellants in aerosol cans and as coolants in air-conditioning units. These CFC compounds, such as Freon-12 (CF_2Cl_2), are quite inert. When released into the environment, they eventually diffuse up to the ozone layer.

CFCs destroy ozone molecules in a series of steps. First, ultraviolet radiation from the Sun strips a chlorine atom away from a CFC molecule. The resulting chlorine atom has an unpaired electron ($Cl\cdot$) and is referred to as a *free radical.*

Step 1: $\quad CFC \xrightarrow{UV} Cl\cdot$

A chlorine free radical is very reactive and can act as a catalyst for the further decomposition of ozone molecules, O_3, into oxygen.

Step 2: $\quad 2\,O_3 \xrightarrow{Cl\cdot} 3\,O_2$

Moreover, this process can occur repeatedly, and a single chlorine free radical can destroy hundreds of ozone molecules.

Ozone depletion is a worldwide problem, and several nations are working to solve it. Germany, Denmark, and the Netherlands have recently announced that they will halt production of CFCs and ban their use. The United States has resolved to phase out CFCs, but it is expensive to retrofit existing refrigeration units. However, the air-conditioning units in new homes and automobiles are being equipped with non-CFC gases.

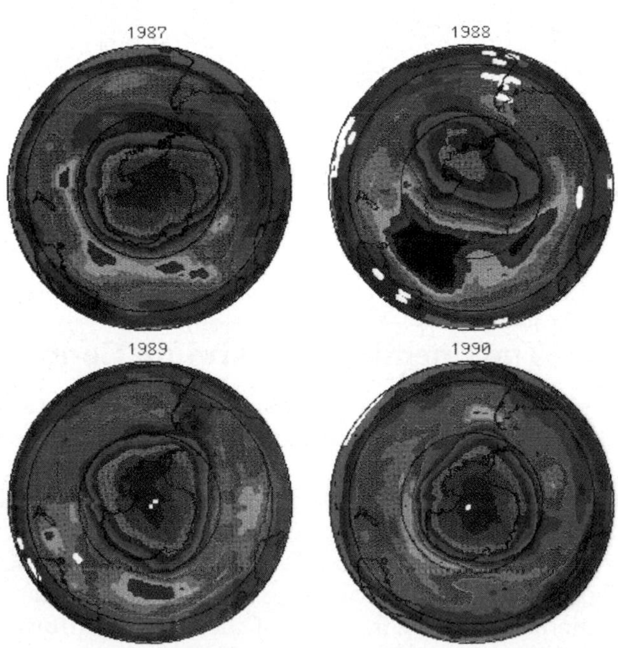

▲ Ozone levels in the stratosphere over Antarctica and the South Pole. The ozone concentration is lowest in the white and maroon regions and highest in the black areas.

The penguin population on Antarctica is threatened and may become extinct if ozone depletion persists.

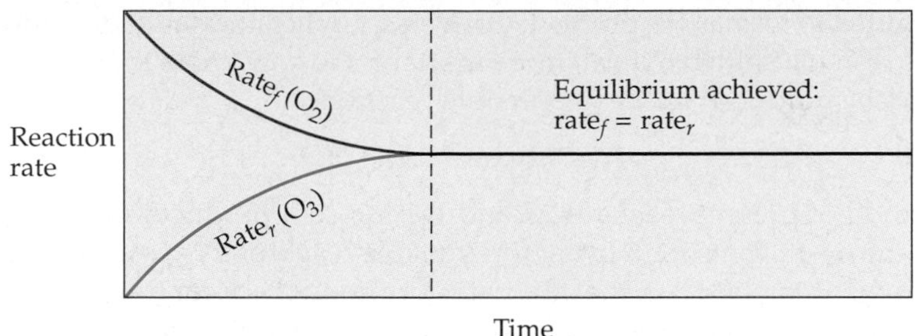

◀ **Figure 16.8 Chemical Equilibrium and Rates of Reaction** The rate of reaction for O_2 decreases with time. As the reaction proceeds, the rate of reaction for O_3 increases. At equilibrium, the rates of the forward and reverse reactions are the same.

We can now formally define the **rate of reaction** as the rate at which the concentrations of reactants decrease per unit time. Alternatively, we can define the rate of reaction as the rate at which the concentrations of products increase per unit time. The reaction that produces ozone in the upper atmosphere is extremely important; see ***Chemistry Connection • The Ozone Hole***. Figure 16.8 shows the changes in the rates of the forward and reverse reactions as the reversible oxygen–ozone reaction progresses.

Figure 16.9 provides a model to help us understand the equilibrium process. Oxygen molecules, O_2, react in a closed container to give ozone molecules, O_3. Simultaneously, some of the O_3 molecules decompose to give O_2 molecules. Initially, the O_2 concentration is high and the forward reaction is rapid. The O_3 concentration is low and the reverse reaction is slow. As equilibrium is approached, the forward reaction slows down and the reverse reaction speeds up. When the rates of the forward and reverse reactions are equal, the reaction is at equilibrium.

Law of Chemical Equilibrium

In 1864 the Norwegian chemists Cato Guldberg and Peter Waage observed that a change in the amount of substance participating in a reversible reaction produced a shift in the equilibrium. When they added more of a reactant substance, the

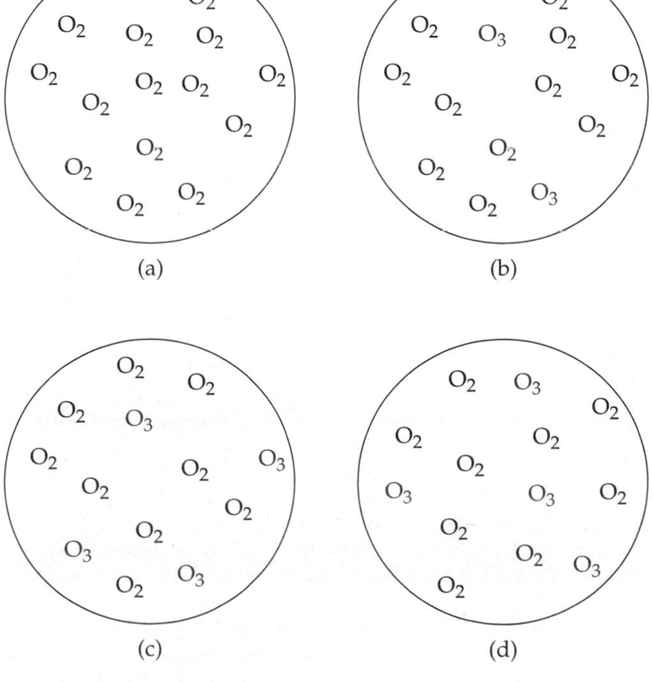

◀ **Figure 16.9 A Dynamic Equilibrium Model** (a) The container has only O_2 gas molecules. (b) As the reaction proceeds, O_3 is formed. (c) At equilibrium, the forward and reverse reactions occur at the same rate. (d) Later, the amounts of O_2 and O_3 are constant even though molecules of O_2 and O_3 continue to react.

reaction shifted to form more product. Conversely, when they added a product substance, the reaction shifted to form more reactant. We now know that a shift in equilibrium can be produced for any reversible reaction.

$$a\text{A} + b\text{B} \rightleftharpoons c\text{C} + d\text{D}$$

where A and B represent reactants, C and D represent products, and *a*, *b*, *c*, and *d* represent the coefficients for a general reversible reaction.

Moreover, it is observed that the ratio of product concentrations to reactant concentrations is constant for a given temperature. That is, the molar concentrations of the products (raised to the powers *c* and *d*) divided by the molar concentrations of the reactants (raised to the powers *a* and *b*), always gives the same value. Regardless of the initial amounts of reactants, at equilibrium the molar concentration ratio is a constant. This relationship, which applies to every reversible reaction, is known as the **law of chemical equilibrium**. Mathematically, we can express the law of chemical equilibrium as follows.

$$K_{eq} = \frac{[\text{C}]^c[\text{D}]^d}{[\text{A}]^a[\text{B}]^b}$$

where [A], [B], [C], and [D] represent the molar concentrations of reactants and products, and *a*, *b*, *c*, and *d* correspond to the coefficients in the balanced chemical equation. The ratio is equal to the **general equilibrium constant** (symbol K_{eq}), which expresses the molar concentration of each substance participating in the reaction at a given temperature.

Note An interesting model for understanding equilibrium is the flow of traffic in and out of a large city (Figure 16.10). Visualize commuters entering and leaving a large city at the same rate. This process is dynamic even though the number of people in the city remains constant. By analogy, a reversible reaction at equilibrium is dynamic even though the amounts of reactants and products are constant.

▶ **Figure 16.10 Equilibrium Analogy** Although commuters continue to enter and leave the city, the number of people in the city remains constant. In an analogous way, reactants and products continue to react even though the amounts are constant.

16.4 General Equilibrium Constant, K_{eq}

Objectives
- To write the equilibrium constant expression for a reversible reaction.
- To calculate an equilibrium constant, K_{eq}, from experimental data.

Let's construct the equilibrium constant expression for the following reversible reaction.

$$2\,A \rightleftharpoons B$$

To write the equilibrium expression, K_{eq}, we place the concentration of the product in the numerator and the concentration of the reactant in the denominator.

$$K_{eq} = \frac{[B]}{[A]}$$

The coefficients in the balanced equation are written as exponents in the K_{eq} expression. The coefficient of A is 2, and the coefficient of B is understood to be 1. These coefficients are written as the power of the corresponding concentration.

$$K_{eq} = \frac{[B]}{[A]^2}$$

The following example exercise provides additional practice in writing K_{eq} expressions for general cases of reversible reactions.

Example Exercise 16.3 • General Equilibrium Constant, K_{eq}

Write the equilibrium constant expression for the following reversible reaction.

$$A + 2\,B \rightleftharpoons 3\,C + D$$

Solution

Let's proceed in two steps. First, let's substitute into the K_{eq} expression. The ratio of product concentration to reactant concentration is

$$K_{eq} = \frac{[C]^c[D]^d}{[A]^a[B]^b}$$

Second, let's account for the coefficients of the balanced equation. The coefficients are 1, 2, 3, and 1, respectively.

$$K_{eq} = \frac{[C]^3[D]}{[A][B]^2}$$

Self-Test Exercise

Write the equilibrium constant expression for the following reversible reaction.

$$2\,A + 3\,B \rightleftharpoons C + 4\,D$$

Answer: $K_{eq} = \dfrac{[C][D]^4}{[A]^2[B]^3}$

If K_{eq} is much greater than 1 ($K_{eq} \gg 1$), the reaction occurs extensively and favors the product. Conversely, if K_{eq} is much less than 1 ($K_{eq} \ll 1$), the reaction does not occur extensively and favors the reactant.

Homogeneous Equilibria

In most reversible reactions, all the reactants and products are in the same physical state. A reversible reaction at equilibrium in which all the substances are in the same state is referred to as a **homogeneous equilibrium**. For example, low-grade coal has a high sulfur content, and burning it produces SO_2. In the atmosphere, oxygen converts SO_2 to SO_3 according to the following reversible reaction.

$$2\,SO_2(g) + O_2(g) \rightleftharpoons 2\,SO_3(g)$$

If we substitute the concentrations of reactants and products into the general equilibrium expression, we have

$$K_{eq} = \frac{[SO_3]}{[SO_2][O_2]}$$

The respective coefficients in the balanced equation are 2, 1, and 2. Thus, the equilibrium expression for the reaction is

$$K_{eq} = \frac{[SO_3]^2}{[SO_2]^2[O_2]}$$

Heterogeneous Equilibria

In the previous examples of gaseous equilibria, liquids and solids were not involved and did not appear in the K_{eq} expression. On occasion, a reactant or product may be in the solid or the liquid state. A reaction at equilibrium in which one of the substances is in a different physical state is referred to as a **heterogeneous equilibrium**.

As an example, consider hydrogen gas that can be manufactured by passing steam over hot charcoal. The equation for the reaction is as follows.

$$C(s) + H_2O(g) \rightleftharpoons CO(g) + H_2(g)$$

We can write the equilibrium constant expression as

$$K_{eq} = \frac{[CO][H_2]}{[C][H_2O]}$$

Note that charcoal is a solid but that all the other substances in the reaction are gases. When the reaction is studied experimentally, it is found that the amount of charcoal has no effect on the equilibrium. Therefore, the concentration of charcoal, [C], can be omitted from the equilibrium expression, giving

$$K_{eq} = \frac{[CO][H_2]}{[H_2O]}$$

Moreover, it is found that liquids and solids have no effect on any gaseous equilibrium. Therefore, liquids and solids do not appear in the equilibrium expression. The following example exercise provides additional practice in writing K_{eq} expressions for homogeneous and heterogeneous equilibria.

Example Exercise 16.4 • Writing K_{eq} Expressions

Write the equilibrium constant expression for each of the following reversible reactions.

(a) $2\,NO(g) + 2\,H_2(g) \rightleftharpoons N_2(g) + 2\,H_2O(g)$

(b) $NH_4NO_3(s) \rightleftharpoons N_2O(g) + 2\,H_2O(g)$

Solution

We can substitute into the K_{eq} expression as follows.

(a) Since the equation is balanced, we raise the concentration of each species to the power corresponding to the coefficient in the balanced equation.

$$K_{eq} = \frac{[N_2][H_2O]^2}{[NO]^2[H_2]^2}$$

(b) In this reaction NH_4NO_3 is a solid. Since NH_4NO_3 is not a gas, it does not appear in the equilibrium expression; thus, the equilibrium expression is

$$K_{eq} = [N_2O][H_2O]^2$$

Self-Test Exercise

Write the equilibrium constant expression for each of the following reversible reactions.

(a) $4\ NH_3(g) + 5\ O_2(g) \rightleftharpoons 4\ NO(g) + 6\ H_2O(g)$

(b) $CaCO_3(s) \rightleftharpoons CaO(s) + CO_2(g)$

Answers: (a) $K_{eq} = \frac{[NO]^4[H_2O]^6}{[NH_3]^4[O_2]^5}$; (b) $K_{eq} = [CO_2]$

Experimental Determination of K_{eq}

One of the most thoroughly investigated equilibrium reactions is involved in the formation of hydrogen iodide from hydrogen gas and iodine vapor.

$$\underset{\text{colorless gas}}{H_2(g)} + \underset{\text{purple vapor}}{I_2(g)} \rightleftharpoons \underset{\text{colorless gas}}{2\ HI(g)}$$

Experimentally, we can measure the molar concentration of each gas at equilibrium. If we start with 1.000 mol of H_2 and I_2 in a liter container, we find at equilibrium that $[H_2] = 0.212$, $[I_2] = 0.212$, and $[HI] = 1.576$. To calculate an experimental value for the equilibrium constant, K_{eq}, let's write the equilibrium expression.

$$K_{eq} = \frac{[HI]^2}{[H_2][I_2]}$$

Substituting, we have

$$K_{eq} = \frac{[1.576]^2}{[0.212][0.212]} = 55.3$$

Interestingly, if we start with 1.000 mol of HI and approach equilibrium from the opposite direction, we obtain the same value for K_{eq}. In fact, the K_{eq} value for the reaction is 55.2 regardless of the original concentrations of each gas. Table 16.2 presents data for four experiments involving the HI equilibrium.

Table 16.2 Experimental Determination of K_{eq}

Experiment	Initial Concentration			Equilibrium Concentration			Calculated K_{eq} at 425°C*
	$[H_2]$	$[I_2]$	$[HI]$	$[H_2]$	$[I_2]$	$[HI]$	$\frac{[HI]^2}{[H_2][I_2]}$
1	1.000	1.000	0	0.212	0.212	1.576	55.3
2	0.500	0.500	0	0.106	0.106	0.788	55.3
3	0	0	1.000	0.106	0.106	0.788	55.3
4	0	0	2.000	0.212	0.212	1.576	55.3

*The units for an equilibrium constant are usually omitted. In this example, the units in the numerator and denominator of the K_{eq} expression cancel.

Conceptually, we can visualize a closed vessel containing the three gases H_2, I_2, and HI in equilibrium. When we change the amount of any gas, the reaction shifts forward or backward in such a way that the ratio between the concentrations of reactants and products remains constant. The following example exercise further illustrates the calculation of an equilibrium constant from experimental data.

Example Exercise 16.5 • Experimental Equilibrium Constant, K_{eq}

In the upper atmosphere, nitric oxide and nitrogen dioxide participate in the following equilibrium.

$$2\,NO(g) + O_2(g) \rightleftharpoons 2\,NO_2(g)$$

Given the equilibrium concentrations for each gas at 25°C, what is K_{eq}?

$$[NO] = 1.5 \times 10^{-11}$$
$$[O_2] = 8.9 \times 10^{-3}$$
$$[NO_2] = 2.2 \times 10^{-6}$$

Solution

We first write the equilibrium expression for the reaction.

$$K_{eq} = \frac{[NO_2]^2}{[NO]^2[O_2]}$$

Second, we substitute the concentration values into the expression.

$$K_{eq} = \frac{[2.2 \times 10^{-6}]^2}{[1.5 \times 10^{-11}]^2[8.9 \times 10^{-3}]} = 2.4 \times 10^{12}$$

The large K_{eq} for this reaction indicates that the equilibrium overwhelmingly favors the formation of NO_2. However, K_{eq} does not indicate the rate of reaction. In the atmosphere, the conversion of NO to NO_2 takes place slowly.

Self-Test Exercise

Given the equilibrium concentrations for the gas mixture at 100°C, calculate the value of K_{eq} for the following reaction.

$N_2O_4(g)$	$\rightleftharpoons$	$2\,NO_2(g)$
0.00140 *M*		0.0172 *M*

Answer: $K_{eq} = 0.211$

Note In Section 11.3, we learned that gas pressure is proportional to the concentration of molecules. Therefore, we can write an equilibrium expression for a reversible gaseous reaction on the basis of partial gas pressures. The equilibrium expression for a reversible gaseous reaction can be written as

$$K_p = \frac{[P_C]^c[P_D]^d}{[P_A]^a[P_B]^b}$$

where K_p is the equilibrium pressure constant, and P_A, P_B, P_C, and P_D are the partial pressures of the gases participating in the equilibrium reaction.

16.5 Gaseous State Equilibria Shifts

Objective · To apply Le Chatelier's principle to reversible reactions in the gaseous state.

Henri Louis Le Chatelier (1850–1936) was trained in France as a mining engineer. When he graduated from college, he became a professor of chemistry at a mining school. His first interest was in the study of flames with the intention of preventing mine explosions. Eventually, he studied the effect of heat on reversible reactions and in 1888 stated the principle for which he is famous: "Every change of one of the factors of an equilibrium brings about a rearrangement of the system in such a direction as to minimize the original change."

The factors affecting a chemical equilibrium system are concentration, temperature, and pressure. A change in any of these factors shifts the equilibrium. For chemical reactions, **Le Chatelier's principle** can be stated as follows: When a reversible reaction at equilibrium is stressed by a change in concentration, temperature, or pressure, the equilibrium shifts to relieve the stress.

Le Chatelier's Principle Movie

Effect of Concentration

For any equilibrium system, we can experimentally shift the reaction forward or backward by changing the concentration of a reactant or product. Let's consider the following equilibrium.

$$\underset{\text{colorless}}{N_2O_4(g)} \rightleftharpoons \underset{\text{brown}}{2\ NO_2(g)}$$

If we increase the N_2O_4 reactant concentration, the equilibrium shifts to the right and the NO_2 product concentration increases. If we decrease the N_2O_4 concentration, the equilibrium shifts to the left and the NO_2 concentration decreases.

Conversely, if we increase the NO_2 product concentration, the equilibrium shifts to the left and the N_2O_4 reactant concentration increases. If we decrease the NO_2 concentration, the equilibrium shifts to the right and the N_2O_4 concentration decreases.

Effect of Temperature

Now, let's consider the effect of temperature on the following equilibrium.

Temperature Dependence of Equilibrium Movie

$$\underset{\text{colorless}}{N_2O_4(g)} + \text{heat} \rightleftharpoons \underset{\text{brown}}{2\ NO_2(g)}$$

When we change the temperature for a reversible reaction, we stress the equilibrium system and the equilibrium shifts to relieve the stress. In this example, if we heat this endothermic reaction, the equilibrium shifts to the *right*; the N_2O_4 concentration decreases and the NO_2 concentration increases. If we cool the reaction, the equilibrium shifts to the *left*. As the equilibrium shifts to the left, the NO_2 concentration decreases and the N_2O_4 concentration increases.

For simplicity, we can consider heat a reactant in an endothermic reaction. In an endothermic reaction, heat causes the equilibrium to shift to the right toward the products to relieve the stress. Likewise, we can consider heat a product in an exothermic reaction. In an exothermic reaction, heat causes the equilibrium to shift to the left toward the reactants to relieve the stress. Figure 16.11 illustrates the shift in equilibrium that results from heating the N_2O_4 reaction.

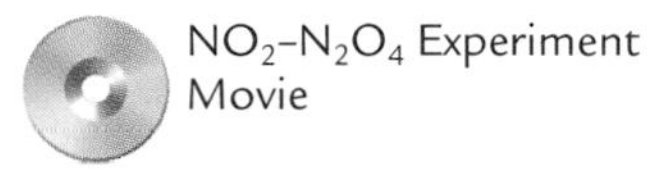
NO_2–N_2O_4 Experiment Movie

Note that N_2O_4 is a colorless gas and that NO_2 is a dark-brown gas. Since each gas has a distinctive color, we can observe an equilibrium shift by noting the color

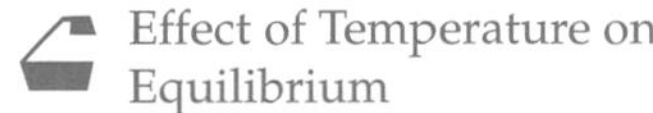
Effect of Temperature on Equilibrium

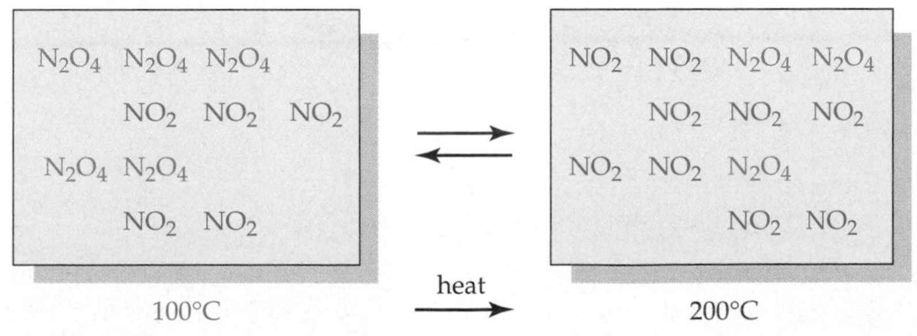

◀ **Figure 16.11 Effect of Temperature on Equilibrium**
(a) Initially, there are five molecules of N_2O_4 and five molecules of NO_2 in the equilibrium mixture. (b) If the temperature increases, the equilibrium shifts to the right to relieve the stress. As a result, the number of N_2O_4 molecules decreases and the number of NO_2 molecules increases.

(a)

(b)

(c)

▲ **Figure 16.12 Heating the N_2O_4–NO_2 Equilibrium** (a) At low temperatures, the N_2O_4–NO_2 equilibrium favors N_2O_4, which is a colorless gas. (b) If we heat the reaction, the equilibrium shifts to produce more NO_2, which is a brown gas. (c) Further heating produces more NO_2, and the gas mixture becomes a darker brown.

of the gaseous mixture. Figure 16.12 illustrates the shift in equilibrium resulting from heating the colorless N_2O_4 gas to give dark-brown NO_2 gas.

Effect of Pressure

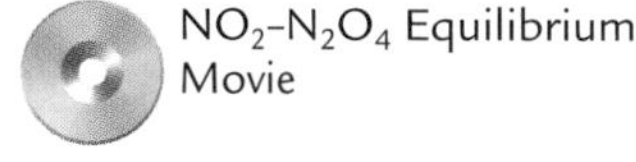

NO_2-N_2O_4 Equilibrium Movie

For many gaseous equilibrium systems, we can experimentally shift the reaction forward or backward by changing the volume, and therefore the pressures of the gases in the system. If we decrease the volume (increase the pressure), the equilibrium will shift to reduce the pressure by producing a smaller total number of molecules. Once again, let's consider the following equilibrium.

$$\underset{\text{colorless}}{N_2O_4(g)} \rightleftharpoons \underset{\text{brown}}{2\ NO_2(g)}$$

Notice that 1 molecule of N_2O_4 produces 2 molecules of NO_2. If we decrease the volume of the container, the gas pressure increases. To relieve the stress, the equilibrium shifts toward fewer molecules. That is, the equilibrium shifts to the left because there is 1 molecule of N_2O_4 for every 2 molecules of NO_2. Figure 16.13 illustrates the equilibrium shift for a decrease in volume.

Effect of Pressure on Equilibrium

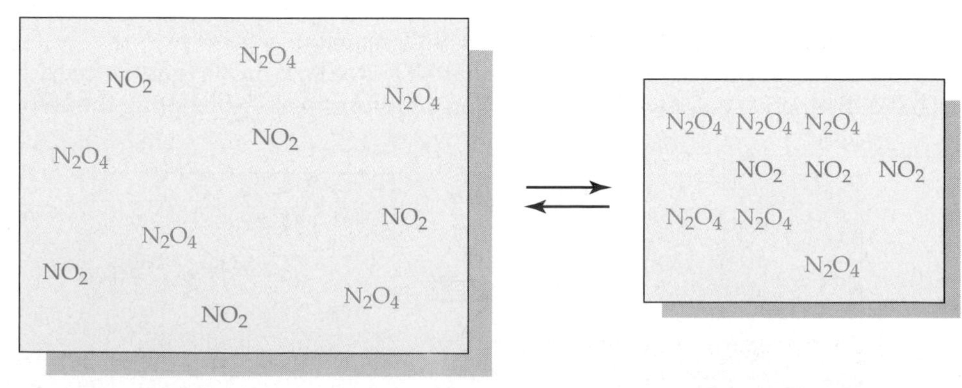

▶ **Figure 16.13 Effect of Pressure on Equilibrium** (a) Initially, there are 10 molecules in the N_2O_4–NO_2 equilibrium. (b) When the volume decreases and the pressures increases, the equilibrium shifts to relieve the stress. As a result, NO_2 is converted to N_2O_4 and the total number of molecules decreases.

If we increase the volume of the container, the gas pressure decreases. To relieve the stress, the equilibrium shifts toward more molecules. That is, the equilibrium shifts to the right because there are 2 molecules of NO_2 for 1 molecule of N_2O_4.

Effect of an Inert Gas

Let's consider what happens to the N_2O_4–NO_2 equilibrium if we add an inert gas such as helium. Since helium does not participate in the N_2O_4–NO_2 equilibrium, the partial gas pressures of N_2O_4 and NO_2 do not change. Since the pressures of N_2O_4 and NO_2 remain constant, there is no shift in equilibrium. The following example exercise further illustrates Le Chatelier's principle as applied to gaseous systems.

Example Exercise 16.6 • Gaseous State Equilibria Shifts

Charcoal reacts with steam according to the equation

$$C(s) + H_2O(g) + \text{heat} \rightleftharpoons CO(g) + H_2(g)$$

Predict the direction of equilibrium shift for each of the following stresses.

(a) $[H_2O]$ increases	(b) $[H_2O]$ decreases	(c) [CO] increases
(d) $[H_2]$ decreases	(e) temperature increases	(f) pressure increases
(g) adding a catalyst	(h) adding carbon	

Solution

Let's apply Le Chatelier's principle to each of the stresses.

(a) If $[H_2O]$ increases, the equilibrium shifts to the *right*.
(b) If $[H_2O]$ decreases, the equilibrium shifts to the *left*.
(c) If [CO] increases, the equilibrium shifts to the *left*.
(d) If $[H_2]$ decreases, the equilibrium shifts to the *right*.
(e) If the temperature increases, the equilibrium shifts to the *right*.
(f) If the pressure increases, the equilibrium shifts toward fewer gas molecules. The equilibrium shifts to the *left*.
(g) Adding a catalyst increases the rate of reaction but has no effect on the amount of reactant or product. Thus, there is *no shift* in equilibrium.
(h) Adding carbon has no effect on the gaseous equilibrium because carbon is a solid. Thus, there is *no shift* in equilibrium.

Self-Test Exercise

Methane, CH_4, reacts with limited oxygen according to the equation

$$2\,CH_4(g) + O_2(g) \rightleftharpoons 2\,CO(g) + 4\,H_2(g) + \text{heat}$$

Predict the direction of equilibrium shift for each of the following stresses.

(a) $[CH_4]$ increases	(b) [CO] increases	(c) $[O_2]$ decreases
(d) $[H_2]$ decreases	(e) temperature decreases	(f) volume decreases
(g) adding a catalyst	(h) adding neon gas	

Answers: (a) shifts right; (b) shifts left; (c) shifts left; (d) shifts right; (e) shifts right; (f) shifts left; (g) no shift; (h) no shift

16.6 Ionization Equilibrium Constant, K_i

Experiment #22, Prentice Hall Laboratory Manual

Objectives
- To write the equilibrium constant expression for a weak acid or a weak base.
- To calculate an ionization constant, K_i, from experimental data.

In Section 16.3, we developed the concept of a general equilibrium constant expression, K_{eq}. Here we will develop a special equilibrium constant expression for a weak acid or weak base. This type of equilibrium is characterized by a weak acid or base in equilibrium with its ions.

Ionization of a Weak Acid

Let's begin with an aqueous solution of a weak acid. For example, acetic acid in aqueous solution yields the following.

$$HC_2H_3O_2(aq) + H_2O(l) \rightleftharpoons H_3O^+(aq) + C_2H_3O_2^-(aq)$$

In aqueous solutions the concentration of water is constant. We can therefore simplify the reversible reaction as follows.

$$HC_2H_3O_2(aq) \rightleftharpoons H^+(aq) + C_2H_3O_2^-(aq)$$

In general, the equilibrium for weak acids lies overwhelmingly to the left. Typically, less than 1% of the parent molecules ionize. In the above reaction, very little ionization of acetic acid occurs. Although about 1% of the molecules form ions, 99% do not. Thus, the equilibrium strongly favors the reverse reaction.

When we substitute the acetic acid equilibrium into the general equilibrium constant expression, we obtain

$$K_i = \frac{[H^+][C_2H_3O_2^-]}{[HC_2H_3O_2]}$$

where K_i is the **ionization equilibrium constant** for the weak acid. Since a weak acid ionizes only slightly, K_i is a very small number. The following example exercise provides practice in writing equilibrium constant expressions.

Example Exercise 16.7 • Writing K_i Expressions

Write the equilibrium constant expression for each of the following weak acids.

(a) $HNO_2(aq) \rightleftharpoons H^+(aq) + NO_2^-(aq)$

(b) $H_2CO_3(aq) \rightleftharpoons H^+(aq) + HCO_3^-(aq)$

Solution

We substitute into the K_i expression as follows.

(a) We substitute the product concentrations into the numerator, and the reactant concentration into the denominator.

$$K_i = \frac{[H^+][NO_2^-]}{[HNO_2]}$$

(b) Carbonic acid, H_2CO_3, ionizes to give H^+ and HCO_3^-. Substituting into the ionization equilibrium expression, we have

$$K_i = \frac{[H^+][HCO_3^-]}{[H_2CO_3]}$$

Self-Test Exercise

Write the equilibrium constant expression for each of the following weak acids.

(a) $HClO_2(aq) \rightleftharpoons H^+(aq) + ClO_2^-(aq)$

(b) $H_3PO_4(aq) \rightleftharpoons H^+(aq) + H_2PO_4^-(aq)$

Answers: (a) $K_i = \frac{[H^+][ClO_2^-]}{[HClO_2]}$; (b) $K_i = \frac{[H^+][H_2PO_4^-]}{[H_3PO_4]}$

Ionization of a Weak Base

An aqueous solution of a weak base behaves similarly to an aqueous solution of a weak acid. That is, a weak base ionizes slightly and is in equilibrium with its ions. For example, aqueous ammonium hydroxide gives the following equilibrium.

$$NH_4OH(aq) \rightleftharpoons NH_4^+(aq) + OH^-(aq)$$

Typically, the equilibrium for weak bases lies overwhelmingly to the left. In ammonium hydroxide, less than 1% ionization occurs. Thus, the equilibrium strongly favors the reverse reaction. Substituting into the general equilibrium constant expression, we have

$$K_i = \frac{[NH_4^+][OH^-]}{[NH_4OH]}$$

Experimental Determination of K_i

In the laboratory, we can calculate the experimental equilibrium constant for a weak acid or a weak base from pH measurements. For example, we can use a pH meter to measure $[H^+]$ in a solution. K_i is then calculated from the hydrogen ion concentration.

Let's calculate the ionization constant, K_i, for acetic acid. In an experiment, a 0.100 *M* solution of acetic acid is found to have a hydrogen ion concentration of 0.00134 *M*. We begin by analyzing the equilibrium concentrations. For each acetic acid molecule that ionizes, we obtain one H^+ and one $C_2H_3O_2^-$. Therefore, $[H^+]$ must equal $[C_2H_3O_2^-]$, that is, 0.00134 *M*. The equilibrium reaction is

$$\begin{array}{ccc} HC_2H_3O_2(aq) & \rightleftharpoons & H^+(aq) + C_2H_3O_2^-(aq) \\ 0.100\,M & & 0.00134\,M \quad 0.00134\,M \end{array}$$

Next, we write the equilibrium constant expression by placing the ion concentrations in the numerator and the molar concentration of the weak acid in the denominator. That gives

$$K_i = \frac{[H^+][C_2H_3O_2^-]}{[HC_2H_3O_2]}$$

Substituting into the equilibrium expression for each concentration, we have

$$K_i = \frac{[0.00134][0.00134]}{[0.100]}$$

$$= 1.80 \times 10^{-5}$$

The ionization constant for acetic acid is a small value, 1.80×10^{-5}, because acetic acid is a weak acid and ionizes only slightly in aqueous solution.

16.7 Weak Acid–Base Equilibria Shifts

Objective · To apply Le Chatelier's principle to solutions of weak acids and weak bases.

In Section 16.5, we discussed Le Chatelier's principle in relation to equilibria of reversible reactions in the gaseous state. This principle also applies to equilibria in aqueous solutions. That is, a change in the concentration of any species in an aqueous solution causes the equilibrium to shift in order to relieve the stress.

Let's reconsider acetic acid, which ionizes very little in aqueous solution. We can indicate the weak acid equilibrium as follows.

$$HC_2H_3O_2(aq) \rightleftharpoons H^+(aq) + C_2H_3O_2^-(aq)$$

If we increase the molar concentration of $HC_2H_3O_2$, we stress the left side of the equilibrium. The reversible reaction responds to the stress by shifting to the right. The final result is that more $HC_2H_3O_2$ molecules ionize and the molar concentrations of H^+ and $C_2H_3O_2^-$ increase.

If we increase the molar concentration of H^+ by adding a strong acid, we stress the right side of the equilibrium. The equilibrium responds to the stress by shifting to the left. The final result is that the molar concentration of $HC_2H_3O_2$ increases, whereas the molar concentration of $C_2H_3O_2^-$ decreases. Since additional acid was added, the molar concentration of H^+ also increases.

We can also indicate a change in $[H^+]$ that corresponds to a change in pH. Since $[H^+]$ and pH are inversely related, $[H^+]$ decreases as pH increases. Therefore, if the pH of an acetic acid solution increases, $[H^+]$ decreases. A decrease in $[H^+]$ causes the equilibrium to shift to the right. Thus, $[HC_2H_3O_2]$ decreases, whereas $[C_2H_3O_2^-]$ increases.

If we increase the molar concentration of $C_2H_3O_2^-$ by adding $NaC_2H_3O_2$, we stress the right side of the equilibrium. As the solid $NaC_2H_3O_2$ dissolves, the equilibrium responds to the stress by shifting to the left. The final result is that $[HC_2H_3O_2]$ increases and $[H^+]$ decreases. Since $NaC_2H_3O_2$ was added, $[C_2H_3O_2^-]$ also increases.

If we add NaOH, we neutralize H^+ and stress the right side of the equilibrium. The NaOH lowers the H^+ concentration, which shifts the equilibrium to the right. After the equilibrium shift, the $HC_2H_3O_2$ concentration has decreased, whereas the $C_2H_3O_2^-$ concentration has increased.

Finally, consider what happens if we add $NaNO_3$ to aqueous acetic acid. Since sodium nitrate is a soluble salt, it dissolves, giving Na^+ and NO_3^- in solution. Neither of these ions participates in the acetic acid equilibrium. Thus, adding $NaNO_3$ has no effect, and there is no shift in equilibrium. The following example exercise further illustrates Le Chatelier's principle.

Example Exercise 16.8 • Weak Acid Equilibria Shifts

Hydrofluoric acid, HF, ionizes according to the equation

$$HF(aq) \rightleftharpoons H^+(aq) + F^-(aq)$$

Predict the direction of equilibrium shift for each of the following stresses.

(a) [HF] increases
(b) $[H^+]$ increases
(c) $[F^-]$ decreases
(d) adding aqueous HCl
(e) adding solid NaF
(f) adding solid NaCl
(g) adding solid NaOH
(h) pH decreases

Solution

Let's apply Le Chatelier's principle to each of the stresses.

(a) If [HF] increases, the equilibrium shifts to the *right*.
(b) If $[H^+]$ increases, the equilibrium shifts to the *left*.
(c) If $[F^-]$ decreases, the equilibrium shifts to the *right*.
(d) If HCl is added, $[H^+]$ increases and the equilibrium shifts to the *left*.
(e) If NaF is added, $[F^-]$ increases and the equilibrium shifts to the *left*.
(f) If NaCl is added, $[Cl^-]$ increases, which has no effect on the equilibrium. Thus, there is *no shift*.

(g) If NaOH is added, the OH^- neutralizes H^+ and lowers its concentration. Thus, the equilibrium shifts to the *right*.
(h) If the pH decreases, $[H^+]$ increases, and the equilibrium shifts to the *left*.

Self-Test Exercise
Hydrocyanic acid, HCN, ionizes according to the equation

$$HCN(aq) \rightleftharpoons H^+(aq) + CN^-(aq)$$

Predict the direction of equilibrium shift for each of the following stresses.
(a) [HCN] increases (b) $[H^+]$ decreases (c) $[CN^-]$ increases
(d) adding aqueous HNO_3 (e) adding solid KCN (f) adding solid KCl
(g) adding solid KOH (h) pH increases

Answers: (a) shifts right; (b) shifts right; (c) shifts left; (d) shifts left; (e) shifts left; (f) no shift; (g) shifts right; (h) shifts right

16.8 Solubility Product Equilibrium Constant, K_{sp}

Objectives
- To write the equilibrium constant expression for a slightly soluble ionic compound.
- To calculate a solubility product constant, K_{sp}, from experimental data.

Guldberg and Waage laid the foundation for the law of chemical equilibrium based on their study of insoluble barium carbonate in aqueous solution. After adding soluble potassium sulfate, they found that the insoluble precipitate contained both barium carbonate and barium sulfate. The Norwegian chemists proposed the following reversible reaction in aqueous solution.

$$BaCO_3(s) + K_2SO_4(aq) \rightleftharpoons BaSO_4(s) + K_2CO_3(aq)$$

Notice that $BaCO_3$ and $BaSO_4$ are both insoluble precipitates; however, each solid substance is very slightly soluble. In aqueous solution, $BaCO_3$ dissociates to give

$$BaCO_3(s) \rightleftharpoons Ba^{2+}(aq) + CO_3^{2-}(aq)$$

When $BaCO_3$ dissociates into ions, Ba^{2+} can recombine with SO_4^{2-} to give an insoluble $BaSO_4$ precipitate. Therefore, they concluded that seemingly insoluble precipitates in aqueous solution are actually *very slightly soluble*. Moreover, these insoluble ionic compounds are in dynamic equilibrium with their constituent ions.

Now, let's write the equilibrium reaction for very slightly soluble Ag_2SO_4 in aqueous solution. Ag_2SO_4 precipitate dissociates to give

$$Ag_2SO_4(s) \rightleftharpoons 2\,Ag^+(aq) + SO_4^{2-}(aq)$$

Notice that Ag_2SO_4 dissociates into 2 Ag^+ and 1 SO_4^{2-}. We can substitute this equilibrium into the general K_{eq} expression.

$$K_{eq} = \frac{[Ag^+]^2[SO_4^{2-}]}{[Ag_2SO_4]}$$

If an aqueous solution is saturated with insoluble Ag_2SO_4, the amount of precipitate has no effect on the equilibrium. As long as there is some precipitate, the concentrations of Ag^+ and SO_4^{2-} do not change. The amount of solid has no effect on the equilibrium, and we can write a special equilibrium expression as

$$K_{sp} = [Ag^+]^2[SO_4^{2-}]$$

where K_{sp} is the **solubility product equilibrium constant**. For an insoluble compound in aqueous solution, only the ions appear in the equilibrium expression; the precipitate never appears in the K_{sp} expression. The following example exercise provides further practice in writing solubility product expressions.

Example Exercise 16.9 • Writing K_{sp} Expressions

Write the solubility product expression for each of the following slightly soluble ionic compounds in aqueous solution.

(a) $CaCO_3(s) \rightleftharpoons Ca^{2+}(aq) + CO_3^{2-}(aq)$
(b) $Ca_3(PO_4)_2(s) \rightleftharpoons 3\,Ca^{2+}(aq) + 2\,PO_4^{3-}(aq)$

Solution

We can substitute into the K_{sp} expression as follows.

(a) Calcium carbonate dissociates into a calcium ion and a carbonate ion. Thus, $[Ca^{2+}]$ and $[CO_3^{2-}]$ are both raised to the first power.

$$K_{sp} = [Ca^{2+}][CO_3^{2-}]$$

(b) Calcium phosphate dissociates into 3 Ca^{2+} and 2 PO_4^{3-}. Thus, the Ca^{2+} concentration is raised to the third power, and the PO_4^{3-} concentration to the second power.

$$K_{sp} = [Ca^{2+}]^3[PO_4^{3-}]^2$$

Self-Test Exercise

Write the solubility product expression for each of the following slightly soluble ionic compounds in aqueous solution.

(a) $Al(OH)_3(s) \rightleftharpoons Al^{3+}(aq) + 3\,OH^-(aq)$
(b) $Al_2(CO_3)_3(s) \rightleftharpoons 2\,Al^{3+}(aq) + 3\,CO_3^{2-}(aq)$

Answers: (a) $K_{sp} = [Al^{3+}][OH^-]^3$; (b) $K_{sp} = [Al^{3+}]^2[CO_3^{2-}]^3$

Experimental Determination of K_{sp}

We can use several methods to find K_{sp} for a slightly soluble ionic compound. One method is to experimentally determine the ion concentrations in aqueous solution and calculate a value for the solubility product constant. For example, in a saturated solution of milk of magnesia, $Mg(OH)_2$, the hydroxide ion concentration is found to be 0.00032 *M*. To calculate the solubility product constant, let's first analyze the equilibrium.

$$Mg(OH)_2(s) \rightleftharpoons Mg^{2+}(aq) + 2\,OH^-(aq)$$

Notice that $Mg(OH)_2$ dissociates into 1 Mg^{2+} and 2 OH^-. The equilibrium constant expression for the dissociation is

$$K_{sp} = [Mg^{2+}][OH^-]^2$$

Since there is 1 Mg^{2+} for every 2 OH^-, $[Mg^{2+}]$ equals half of $[OH^-]$. The value of $[OH^-]$ is given as 0.00032 *M*, and so $[Mg^{2+}]$ is 0.00016 *M*. Substituting into the K_{sp} expression, we have

$$K_{sp} = [1.6 \times 10^{-4}][3.2 \times 10^{-4}]^2$$

Simplifying,

$$K_{sp} = 1.6 \times 10^{-11}$$

The solubility product constant for magnesium hydroxide is 1.6×10^{-11}.

16.9 Solubility Equilibria Shifts

Objective · To apply Le Chatelier's principle to a saturated solution of a slightly soluble ionic compound.

In Section 16.7, we discussed Le Chatelier's principle as it applies to ionization equilibria. This principle also applies to dissociation equilibria. That is, if we change the concentration of any ions participating in a dissociation equilibrium, there will be a shift to relieve the stress.

Let's consider the dissociation of an antacid tablet in aqueous solution. The tablet contains aluminum hydroxide, $Al(OH)_3$, which is only slightly soluble. The dissociation equilibrium is

$$Al(OH)_3(s) \rightleftharpoons Al^{3+}(aq) + 3\,OH^-(aq)$$

If the solution is saturated, Al^{3+} and OH^- are at maximum concentration. Therefore, if we add more aluminum hydroxide, $Al(OH)_3$, to the solution, it has no effect. Hence, there is no shift in equilibrium.

Suppose we add solid $AlCl_3$ to the solution to increase the Al^{3+} concentration. The solid dissolves to give more Al^{3+}. The aluminum ion stresses the right side of the reaction, and the equilibrium shifts to the left. Furthermore, more hydroxide ions precipitate, and the concentration of OH^- decreases.

Suppose we add solid NaCl to the aqueous solution. Sodium chloride dissolves to give sodium ions and chloride ions. Since neither Na^+ nor Cl^- participate in the equilibrium, NaCl has no effect. Thus, there is no shift in equilibrium, and the concentrations of Al^{3+} and OH^- remain constant.

Let's consider a less obvious stress on the equilibrium. Let's increase the H^+ concentration by adding aqueous HCl. Hydrochloric acid is a strong acid and ionizes to give H^+ and Cl^-. Although neither ion is shown in the equilibrium, H^+ neutralizes OH^-. Thus, the H^+ decreases the concentration of OH^- in the solution.

As the concentration of OH^- decreases, we stress the right side of the equation and shift the equilibrium to the right. As the equilibrium shifts, more $Al(OH)_3$ dissolves and the Al^{3+} concentration increases. The following example exercise further illustrates Le Chatelier's principle.

▲ **Saturated Solution of AgCl**
By adding various substances into the solution, we stress the solubility equilibrium. A stress may cause a Le Chatelier shift to the left (more precipitate) or to the right (less precipitate).

Example Exercise 16.10 • Solubility Equilibria Shifts

A saturated solution of silver chloride dissociates according to the equation

$$AgCl(s) \rightleftharpoons Ag^+(aq) + Cl^-(aq)$$

Predict the direction of equilibrium shift for each of the following stresses.

(a) $[Ag^+]$ increases
(b) $[Ag^+]$ decreases
(c) $[Cl^-]$ increases
(d) $[Cl^-]$ decreases
(e) adding solid $AgNO_3$
(f) adding solid $NaNO_3$
(g) adding solid AgCl
(h) pH decreases

Solution

Let's apply Le Chatelier's principle to each of the stresses.

(a) If $[Ag^+]$ increases, the equilibrium shifts to the *left*.
(b) If $[Ag^+]$ decreases, the equilibrium shifts to the *right*.
(c) If $[Cl^-]$ increases, the equilibrium shifts to the *left*.
(d) If $[Cl^-]$ decreases, the equilibrium shifts to the *right*.
(e) If $AgNO_3$ is added, $[Ag^+]$ increases and the equilibrium shifts to the *right*.
(f) If $NaNO_3$ is added, $[Na^+]$ and $[NO_3^-]$ increase, which has no effect on the equilibrium. Thus, there is *no shift*.
(g) If AgCl is added, there is no effect on the equilibrium because the solution is saturated with AgCl. Thus, there is *no shift*.

(continued)

Example Exercise 16.10 *(continued)*

(h) If the pH decreases, $[H^+]$ increases. Since the hydrogen ion has no effect on the equilibrium, there is *no shift.*

Self-Test Exercise

A saturated solution of magnesium hydroxide dissociates according to the equation

$$Mg(OH)_2(s) \rightleftharpoons Mg^{2+}(aq) + 2\,OH^-(aq)$$

Predict the direction of equilibrium shift for each of the following stresses.

(a) $[Mg^{2+}]$ increases (b) $[Mg^{2+}]$ decreases (c) $[OH^-]$ increases
(d) $[OH^-]$ decreases (e) adding solid KOH (f) adding solid KNO_3
(g) adding solid $Mg(OH)_2$ (h) pH decreases

Answers: (a) shifts left; (b) shifts right; (c) shifts left; (d) shifts right; (e) shifts left; (f) no shift; (g) no shift; (h) shifts right (neutralizes OH^-)

Summary

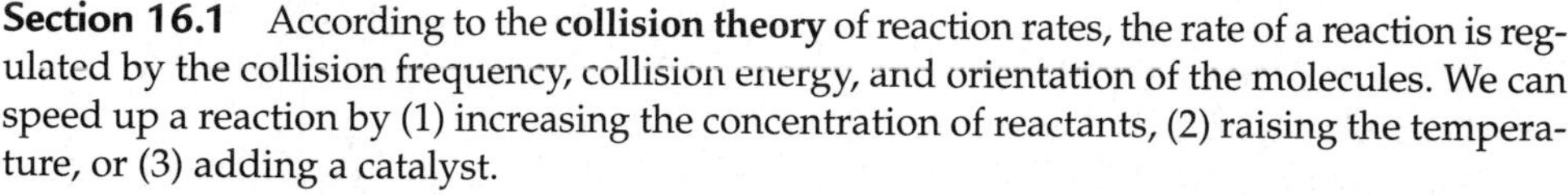

Section 16.1 According to the **collision theory** of reaction rates, the rate of a reaction is regulated by the collision frequency, collision energy, and orientation of the molecules. We can speed up a reaction by (1) increasing the concentration of reactants, (2) raising the temperature, or (3) adding a catalyst.

Section 16.2 An **endothermic reaction** absorbs heat energy, and an **exothermic reaction** releases heat energy. A **reaction profile** graphs the change in energy as reactants are converted to products. For molecules to react, they must have enough energy to achieve the **transition state.** The energy necessary to reach the transition state is called the **activation energy, E_{act}**. A **catalyst** can be defined as any substance that speeds up a reaction by lowering the activation energy. The difference between the energy of the reactants and the products is called the **heat of reaction, ΔH.**

Section 16.3 A **reversible reaction** takes place simultaneously in both the forward and reverse directions. The **rate of reaction** is the rate at which the concentration of the reactants decreases, or the concentration of the products increases, per unit time. When the rates of the forward and reverse reactions are equal, the reaction is at **chemical equilibrium**. The **law of chemical equilibrium** states that the molar concentrations of the products divided by reactants (each raised to a power corresponding to a coefficient in the balanced chemical equation) equals a **general equilibrium constant, K_{eq}**. The law of chemical equilibrium applies to every reversible reaction and is illustrated in Table 16.3.

Table 16.3 Selected Types of Chemical Equilibria

Type of Equilibrium	Equilibrium Expression
General Equilibrium	
$aA + bB \rightleftharpoons cC + dD$	$K_{eq} = \dfrac{[C]^c[D]^d}{[A]^a[B]^b}$
Ionization Equilibrium	
$HC_2H_3O_2(aq) \rightleftharpoons H^+(aq) + C_2H_3O_2^-(aq)$	$K_i = \dfrac{[H^+][C_2H_3O_2^-]}{[HC_2H_3O_2]}$
Dissociation Equilibrium	
$Al(OH)_3(s) \rightleftharpoons Al^{3+}(aq) + 3\,OH^-(aq)$	$K_{sp} = [Al^{3+}][OH^-]^3$

Section 16.4 In a **homogeneous equilibrium**, all the participating substances are in the same physical state, for example, an equilibrium in which all the reactants and products are in the gaseous state. In a **heterogeneous equilibrium**, one of the participating substances is in a different physical state, for example, an equilibrium in which a solid substance decomposes to give a mixture of gases.

Section 16.5 A change in concentration, temperature, or pressure can cause a stress to a reversible reaction. According to **Le Chatelier's principle**, a chemical reaction at equilibrium shifts in order to relieve a stress. The shifts in equilibrium for various stresses are summarized in Table 16.4.

Table 16.4 Summary of Chemical Equilibria Shifts

Stress on Equilibrium	Effect on Equilibrium
Concentration Increases (all systems)	
for a reactant	shifts to the right
for a product	shifts to the left
Temperature Increases (gaseous systems)	
for an endothermic reaction	shifts to the right
for an exothermic reaction	shifts to the left
Pressure Increases (gaseous systems)	
more reactant molecules	shifts to the right
more product molecules	shifts to the left
molecules of reactants and products are equal	no shift

Section 16.6 The law of chemical equilibrium applies to aqueous solutions of weak acids and weak bases. The **ionization equilibrium constant,** K_i, is equal to the molar concentrations of the ions divided by the molar concentration of the weak acid or base.

Section 16.7 Le Chatelier's principle applies to aqueous solutions of weak acids and weak bases. If we change the concentration of one of the ions in an aqueous solution, the equilibrium will shift to relieve the stress. If we add a substance that does not participate in the equilibrium, there will be no shift.

Section 16.8 The law of chemical equilibrium applies to saturated solutions of slightly soluble ionic compounds. The **solubility product equilibrium constant,** K_{sp}, is equal to the product of the molar concentrations of the ions in solution. The insoluble compound is not written in the K_{sp} expression because it has no effect on the equilibrium in a saturated solution.

Section 16.9 Le Chatelier's principle applies to saturated solutions of slightly soluble ionic compounds. If we change the concentration of one of the ions in aqueous solution, the equilibrium will shift to relieve the stress. If we add a substance that does not participate in the equilibrium, there will be no shift.

Key Concepts*

1. According to collision theory, which of the following factors influences the rate of a chemical reaction?
   (a) frequency of molecular collisions
   (b) energy of molecular collisions
   (c) geometry of molecular collisions
2. Which of the following factors increases the rate of a chemical reaction?
   (a) increasing concentration
   (b) increasing temperature
   (c) adding a catalyst
3. Which of the following reaction profiles illustrates an exothermic reaction.

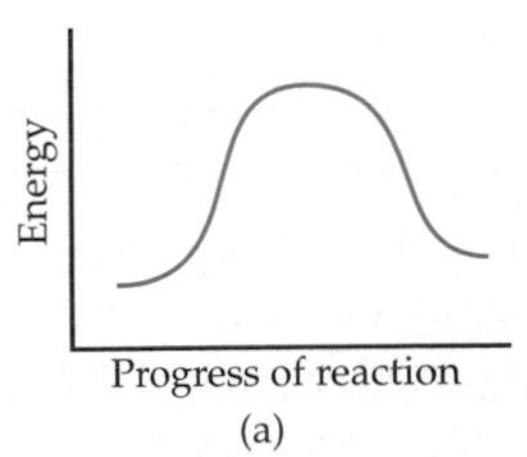

(a)

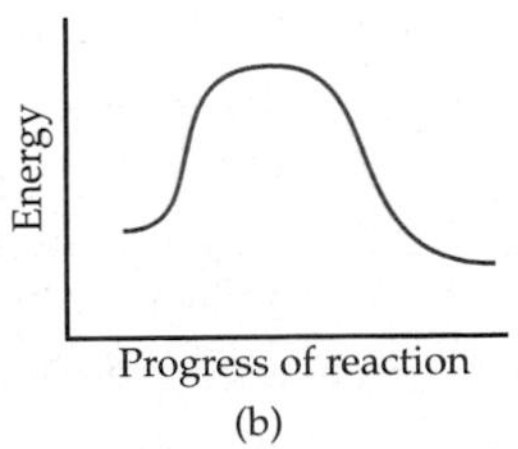

(b)

*Answers to Key Concepts are in Appendix H.

4. Which of the following is true *before* a reaction reaches chemical equilibrium?
   (a) The amount of reactants is decreasing.
   (b) The amount of products is decreasing.
   (c) The amounts of reactants and products are equal.
5. What is the general equilibrium constant expression, K_{eq}, for the following reversible reaction: $A + 3\,B \rightleftharpoons 2\,C$?
6. Which factor (concentration, temperature, or pressure) has no effect on the following equilibrium?

$$SO_3(g) + NO(g) + heat \rightleftharpoons SO_2(g) + NO_2(g)$$

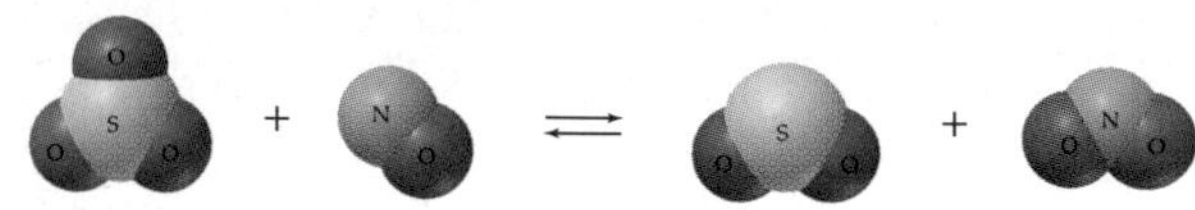

7. What is the ionization constant expression, K_i, for the following weak acid?

$$HX(aq) \rightleftharpoons H^+(aq) + X^-(aq)$$

8. If the hydrogen ion concentration in 0.100 *M* HX is 0.0038 M, what is the ionization constant for the weak acid?
9. What is the equilibrium constant expression for a saturated solution of slightly soluble silver chloride, AgCl?
10. If we add $Ca(NO_3)_2$, Na_2CO_3, and $CaCO_3$ to a saturated solution of calcium carbonate, which compound has no effect on the equilibrium?

$$CaCO_3(s) \rightleftharpoons Ca^{2+}(aq) + CO_3^{2-}(aq)$$

Key Terms† 

Select the key term below that corresponds to each of the following definitions.

_____ 1. the principle that the rate of a chemical reaction is controlled by the frequency and energy of molecules striking each other
_____ 2. a chemical reaction that consumes heat energy
_____ 3. a chemical reaction that liberates heat energy
_____ 4. a graph of the energy of reactants and products as a reaction occurs
_____ 5. the highest point on the reaction profile
_____ 6. the energy required for reactants to reach the transition state
_____ 7. the difference in energy between the reactants and the products
_____ 8. a substance that speeds up a reaction by lowering the energy of activation
_____ 9. a reaction that proceeds toward reactants and products simultaneously
_____ 10. the rate at which the concentrations of reactants decrease per unit time
_____ 11. a dynamic state for a reversible reaction in which the rates of the forward and reverse reactions are the same
_____ 12. the principle that the molar concentrations of the products in a reversible reaction divided by the molar concentrations of the reactants (each raised to a power corresponding to a coefficient in the balanced equation) is equal to a constant
_____ 13. the constant that expresses the molar equilibrium concentration of each substance participating in a reversible reaction at a given temperature
_____ 14. a type of equilibrium in which all participating species are in the same state
_____ 15. a type of equilibrium in which a participating species is in a different state
_____ 16. the statement that a chemical equilibrium stressed by a change in concentration, temperature, or pressure shifts to relieve the stress
_____ 17. the constant that expresses the molar equilibrium concentrations of ions in aqueous solution for a slightly ionized acid or base
_____ 18. the constant that expresses the molar equilibrium concentrations of ions in aqueous solution for a slightly dissociated ionic compound

(a) activation energy (E_{act}) (*Sec. 16.2*)
(b) catalyst (*Sec. 16.2*)
(c) chemical equilibrium (*Sec. 16.3*)
(d) collision theory (*Sec. 16.1*)
(e) endothermic reaction (*Sec. 16.2*)
(f) exothermic reaction (*Sec. 16.2*)
(g) general equilibrium constant (K_{eq}) (*Sec. 16.3*)
(h) heat of reaction (ΔH) (*Sec. 16.2*)
(i) heterogeneous equilibrium (*Sec. 16.4*)
(j) homogeneous equilibrium (*Sec. 16.4*)
(k) ionization equilibrium constant (K_i) (*Sec. 16.6*)
(l) law of chemical equilibrium (*Sec. 16.3*)
(m) Le Chatelier's principle (*Sec. 16.5*)
(n) rate of reaction (*Sec. 16.3*)
(o) reaction profile (*Sec. 16.2*)
(p) reversible reaction (*Sec. 16.3*)
(q) solubility product equilibrium constant (K_{sp}) (*Sec. 16.8*)
(r) transition state (*Sec. 16.2*)

† Answers to Key Terms are in Appendix I.

Exercises‡

Collision Theory (Sec. 16.1)

1. State the three factors that influence the rate of effective molecular collisions.
2. When the temperature is increased, the rate of effective collisions increases for two reasons. State the two reasons.
3. Draw a diagram showing an effective collision geometry between a hydrogen molecule and a bromine molecule.
4. Draw a diagram showing an ineffective collision geometry between a hydrogen molecule and a bromine molecule.
5. State the effect on the rate of reaction of each of the following.
   (a) increasing the concentration of a reactant
   (b) decreasing the temperature of the reaction
   (c) adding a catalyst
6. The Haber process uses a metal oxide catalyst to produce ammonia gas. Does the catalyst increase the amount of ammonia? Explain.
7. Why does a spark ignite an explosion in a coal mine but not in a charcoal barbecue?
8. Why do methane gas and chlorine gas react rapidly in sunlight and very slowly in the laboratory?

Energy Profiles of Chemical Reactions (Sec. 16.2)

9. Phosphorus pentachloride is used in the electronics industry to manufacture computer chips. Draw the energy profile for the following reaction.

$$PCl_5(g) + \text{heat} \rightleftharpoons PCl_3(g) + Cl_2(g)$$

10. Ozone slowly decomposes in the atmosphere to give oxygen gas. Draw the energy profile for the reaction.

$$2\,O_3(g) \rightleftharpoons 3\,O_2(g) + \text{heat}$$

11. Draw the energy profile for the following endothermic reaction.

$$H_2(g) + I_2(g) \rightleftharpoons 2\,HI(g)$$

Label the axes and indicate the reactants, products, transition state, activation energy, and energy of reaction.

12. Draw the energy profile for the following exothermic reaction.

$$H_2(g) + Cl_2(g) \rightleftharpoons 2\,HCl(g)$$

Label the axes and indicate the reactants, products, transition state, activation energy, and energy of reaction.

13. State the effect of a catalyst on the energy of activation, E_{act}.
14. State the effect of a catalyst on the heat of reaction, ΔH.
15. Consider the energy profile for a reversible endothermic reaction. Is E_{act} greater for the forward or the reverse reaction?
16. Consider the energy profile for a reversible exothermic reaction. Is E_{act} greater for the forward or the reverse reaction?

The Chemical Equilibrium Concept (Sec. 16.3)

17. Define the rate of a forward reaction in terms of (a) the reactant concentration and (b) the product concentration.
18. Define chemical equilibrium using the symbols $rate_f$ and $rate_r$.
19. Which of the following statements is true regarding the general equilibrium expression?
   (a) K_{eq} can be determined experimentally.
   (b) K_{eq} can be derived theoretically.
20. Which of the following statements is true regarding the general equilibrium expression?
   (a) K_{eq} is independent of temperature.
   (b) K_{eq} has all substances in the same physical state.

General Equilibrium Constant, K_{eq} (Sec. 16.4)

21. Write the general equilibrium constant expression for each of the following.
   (a) $2\,A \rightleftharpoons C$
   (b) $A + 2\,B \rightleftharpoons 3\,C$
   (c) $2\,A + 3\,B \rightleftharpoons 4\,C + D$
22. Write the general equilibrium constant expression for each of the following.
   (a) $3\,A \rightleftharpoons 2\,C$
   (b) $A + B \rightleftharpoons 2\,C$
   (c) $3\,A + 5\,B \rightleftharpoons C + 4\,D$
23. Does a substance in the solid state appear in the equilibrium expression for a gaseous state reaction?
24. Does a substance in the liquid state appear in the equilibrium expression for a gaseous state reaction?
25. Write the equilibrium constant expression for each of the following reversible reactions.
   (a) $H_2(g) + F_2(g) \rightleftharpoons 2\,HF(g)$
   (b) $4\,NH_3(g) + 7\,O_2(g) \rightleftharpoons 4\,NO_2(g) + 6\,H_2O(g)$
   (c) $ZnCO_3(s) \rightleftharpoons ZnO(s) + CO_2(g)$
26. Write the equilibrium constant expression for each of the following reversible reactions.
   (a) $H_2(g) + Br_2(g) \rightleftharpoons 2\,HBr(g)$
   (b) $4\,HCl(g) + O_2(g) \rightleftharpoons 2\,Cl_2(g) + 2\,H_2O(g)$
   (c) $CO(g) + 2\,H_2(g) \rightleftharpoons CH_3OH(l)$

‡Answers to Exercises are in Appendix J.

27. Given the equilibrium concentrations for each gas at 850°C, calculate the value of K_{eq} for the manufacture of sulfur trioxide.

$2\ SO_2(g)$	+	$O_2(g)$	$\rightleftharpoons$	$2\ SO_3(g)$
1.75 *M*		1.50 *M*		2.25 *M*

28. Given the equilibrium concentrations for each gas at 500°C, calculate the value of K_{eq} for the manufacture of ammonia.

$N_2(g)$	+	$3\ H_2(g)$	$\rightleftharpoons$	$2\ NH_3(g)$
0.400 *M*		1.20 *M*		0.195 *M*

Gaseous State Equilibria Shifts (Sec. 16.5)

29. Weather conditions affect the smog equilibrium in the atmosphere. What happens to the nitrogen dioxide concentration on (a) hot, sunny days and (b) cool, overcast days.

$$N_2O_4(g) + \text{heat} \rightleftharpoons 2\ NO_2(g)$$

30. The conditions for producing ammonia industrially are 500°C and 300 atm. What happens to the ammonia concentration if (a) the temperature increases and (b) the pressure increases?

$$N_2(g) + 3\ H_2(g) \rightleftharpoons 2\ NH_3(g) + \text{heat}$$

31. The industrial process for producing carbon monoxide gas involves passing carbon dioxide over hot charcoal.

$$C(s) + CO_2(g) + \text{heat} \rightleftharpoons 2\ CO(g)$$

Predict the direction of equilibrium shift for each of the following stresses.

(a) $[CO_2]$ decreases **(b)** $[CO]$ decreases
(c) solid charcoal is added **(d)** CO_2 gas is added
(e) temperature increases **(f)** temperature decreases
(g) pressure increases **(h)** pressure decreases

32. The industrial process for producing hydrogen gas involves reacting methane and steam at a high temperature.

$$CH_4(g) + H_2O(g) + \text{heat} \rightleftharpoons CO(g) + 3\ H_2(g)$$

Predict the direction of equilibrium shift for each of the following stresses.

(a) $[CH_4]$ increases **(b)** $[CO]$ increases
(c) $[H_2O]$ decreases **(d)** $[H_2]$ decreases
(e) xenon gas is added **(f)** temperature decreases
(g) pressure decreases **(h)** pressure increases

33. Coal-burning power plants release sulfur dioxide into the atmosphere. The SO_2 is converted to SO_3 by the nitrogen dioxide as follows.

$$SO_2(g) + NO_2(g) \rightleftharpoons SO_3(g) + NO(g) + \text{heat}$$

Predict the direction of equilibrium shift for each of the following stresses.

(a) $[SO_2]$ decreases **(b)** $[SO_3]$ decreases
(c) $[NO_2]$ increases **(d)** $[NO]$ increases
(e) temperature decreases **(f)** pressure increases
(g) pressure decreases **(h)** ultraviolet light

34. Smog contains formaldehyde that is responsible for an eye-burning sensation. Formaldehyde, CH_2O, is produced from the reaction of ozone and atmospheric ethylene, C_2H_4, as follows.

$$2\ C_2H_4(g) + 2\ O_3(g) \rightleftharpoons 4\ CH_2O(g) + O_2(g) + \text{heat}$$

Predict the direction of equilibrium shift for each of the following stresses.

(a) $[C_2H_4]$ increases **(b)** $[CH_2O]$ increases
(c) $[O_3]$ decreases **(d)** $[O_2]$ decreases
(e) temperature increases **(f)** pressure decreases
(g) volume decreases **(h)** metal catalyst

Ionization Equilibrium Constant, K_i (Sec. 16.6)

35. Write the equilibrium constant expression for each of the following weak acids.

(a) $HCHO_2(aq) \rightleftharpoons H^+(aq) + CHO_2^-(aq)$
(b) $H_2C_2O_4(aq) \rightleftharpoons H^+(aq) + HC_2O_4^-(aq)$
(c) $H_3C_6H_5O_7(aq) \rightleftharpoons H^+(aq) + H_2C_6H_5O_7^-(aq)$

36. Write the equilibrium constant expression for each of the following weak bases.

(a) $NH_2OH(aq) + H_2O(l) \rightleftharpoons NH_3OH^+(aq) + OH^-(aq)$
(b) $C_6H_5NH_2(aq) + H_2O(l) \rightleftharpoons C_6H_5NH_3^+(aq) + OH^-(aq)$
(c) $(CH_3)_2NH(aq) + H_2O(l) \rightleftharpoons (CH_3)_2NH_2^+(aq) + OH^-(aq)$

37. Nitrous acid, HNO_2, is used in the synthesis of selected organic compounds. If the hydrogen ion concentration of a 0.125 *M* solution is 7.5×10^{-3} *M*, what is the ionization constant for the acid?

38. Aqueous ammonium hydroxide, NH_4OH, is used as a household cleaning solution. If the hydroxide ion concentration of a 0.245 *M* solution is 2.1×10^{-3} *M*, what is the ionization constant for the base?

39. Hydrofluoric acid, HF, is used to etch silicon in the manufacture of computer chips. If the pH of a 0.139 *M* solution is 2.00, what is the ionization constant of the acid?

40. Hydrazine, N_2H_4, is a weak base and is used as a fuel in the space shuttle. If the pH of a 0.139 *M* solution is 11.00, what is the ionization constant of the base?

$$N_2H_4(aq) + H_2O(l) \rightleftharpoons N_2H_5^+(aq) + OH^-(aq)$$

Weak Acid–Base Equilibria Shifts (Sec. 16.7)

41. Given the chemical equation for the ionization of hydrofluoric acid:

$$HF(aq) \rightleftharpoons H^+(aq) + F^-(aq)$$

Predict the direction of equilibrium shift for each of the following stresses.

(a) increase [HF] **(b)** increase $[H^+]$
(c) decrease [HF] **(d)** decrease $[F^-]$

(e) add NaF solid (f) add HCl gas
(g) add NaOH solid (h) increase pH

42. Given the chemical equation for the ionization of nitrous acid:

$$HNO_2(aq) \rightleftharpoons H^+(aq) + NO_2^-(aq)$$

Predict the direction of equilibrium shift for each of the following stresses.

(a) decrease $[HNO_2]$ (b) decrease $[H^+]$
(c) increase $[HNO_2]$ (d) increase $[NO_2^-]$
(e) add KNO_2 solid (f) add KCl solid
(g) add KOH solid (h) increase pH

43. Given the chemical equation for the ionization of acetic acid:

$$HC_2H_3O_2(aq) \rightleftharpoons H^+(aq) + C_2H_3O_2^-(aq)$$

Predict the direction of equilibrium shift for each of the following stresses.

(a) increase $[HC_2H_3O_2]$ (b) increase $[H^+]$
(c) decrease $[HC_2H_3O_2]$ (d) decrease $[C_2H_3O_2^-]$
(e) add $NaC_2H_3O_2$ solid (f) add NaCl solid
(g) add NaOH solid (h) increase pH

44. Given the chemical equation for the ionization of ammonium hydroxide:

$$NH_4OH(aq) \rightleftharpoons NH_4^+(aq) + OH^-(aq)$$

Predict the direction of equilibrium shift for each of the following stresses.

(a) increase $[NH_4^+]$ (b) decrease $[OH^-]$
(c) increase $[NH_4OH]$ (d) decrease pH
(e) add NH_3 gas (f) add KCl solid
(g) add KOH solid (h) add NH_4Cl solid

Solubility Product Equilibrium Constant, K_{sp} (Sec. 16.8)

45. Write the solubility product expression for each of the following slightly soluble ionic compounds in a saturated aqueous solution.

(a) $AgI(s) \rightleftharpoons Ag^+(aq) + I^-(aq)$
(b) $Ag_2CrO_4(s) \rightleftharpoons 2\,Ag^+(aq) + CrO_4^{2-}(aq)$
(c) $Ag_3PO_4(s) \rightleftharpoons 3\,Ag^+(aq) + PO_4^{3-}(aq)$

46. Write the solubility product expression for each of the following slightly soluble ionic compounds in a saturated aqueous solution.

(a) $Cu_2CO_3(s) \rightleftharpoons 2\,Cu^+(aq) + CO_3^{2-}(aq)$
(b) $ZnCO_3(s) \rightleftharpoons Zn^{2+}(aq) + CO_3^{2-}(aq)$
(c) $Al_2(CO_3)_3(s) \rightleftharpoons 2\,Al^{3+}(aq) + 3\,CO_3^{2-}(aq)$

47. The cobalt ion concentration in a saturated solution of cobalt(II) sulfide, CoS, is 7.7×10^{-11} *M*. Calculate the value for the equilibrium constant.

48. The fluoride ion concentration in a saturated solution of magnesium fluoride, MgF_2, is 2.3×10^{-3} *M*. Calculate the value for the equilibrium constant.

49. The zinc ion concentration in a saturated solution of zinc phosphate, $Zn_3(PO_4)_2$, is 1.5×10^{-7} *M*. Calculate the equilibrium constant.

50. The hydroxide ion concentration in a saturated solution of iron(III) hydroxide, $Fe(OH)_3$, is 2.2×10^{-10} *M*. Calculate the equilibrium constant.

51. The K_{sp} values for $CaCO_3$ and CaC_2O_4 are 3.8×10^{-9} and 2.3×10^{-9}, respectively. In saturated solutions of $CaCO_3$ and CaC_2O_4, which has the higher calcium ion concentration?

52. The K_{sp} values for $MnCO_3$ and $Mn(OH)_2$ are 1.8×10^{-11} and 4.6×10^{-14}, respectively. In saturated solutions of $MnCO_3$ and $Mn(OH)_2$, which has the higher manganese ion concentration?

Solubility Equilibria Shifts (Sec. 16.9)

53. Teeth and bones are composed mainly of calcium phosphate, which dissociates slightly in aqueous solution as follows.

$$Ca_3(PO_4)_2(s) \rightleftharpoons 3\,Ca^{2+}(aq) + 2\,PO_4^{3-}(aq)$$

Predict the direction of equilibrium shift for each of the following stresses.

(a) increase $[Ca^{2+}]$ (b) increase $[PO_4^{3-}]$
(c) decrease $[Ca^{2+}]$ (d) decrease $[PO_4^{3-}]$
(e) add solid $Ca(NO_3)_2$ (f) add solid KNO_3
(g) add solid $Ca_3(PO_4)_2$ (h) add H^+

54. Cadmium sulfide dissociates slightly in aqueous solution as follows.

$$CdS(s) \rightleftharpoons Cd^{2+}(aq) + S^{2-}(aq)$$

Predict the direction of equilibrium shift for each of the following stresses.

(a) increase $[Cd^{2+}]$ (b) increase $[S^{2-}]$
(c) decrease $[Cd^{2+}]$ (d) decrease $[S^{2-}]$
(e) add solid $Cd(NO_3)_2$ (f) add solid $NaNO_3$
(g) add solid CdS (h) add H^+

55. Cupric hydroxide dissociates slightly in aqueous solution as follows.

$$Cu(OH)_2(s) \rightleftharpoons Cu^{2+}(aq) + 2\,OH^-(aq)$$

Predict the direction of equilibrium shift for each of the following stresses.

(a) increase $[Cu^{2+}]$ (b) increase $[OH^-]$
(c) decrease $[Cu^{2+}]$ (d) decrease $[OH^-]$
(e) add solid $Cu(OH)_2$ (f) add solid NaOH
(g) add solid NaCl (h) decrease pH

56. Strontium carbonate dissociates slightly in aqueous solution as follows.

$$SrCO_3(s) \rightleftharpoons Sr^{2+}(aq) + CO_3^{2-}(aq)$$

Predict the direction of equilibrium shift for each of the following stresses.

(a) increase $[Sr^{2+}]$ (b) increase $[CO_3^{2-}]$
(c) decrease $[Sr^{2+}]$ (d) decrease $[CO_3^{2-}]$

(e) add solid $SrCO_3$ (f) add solid $Sr(NO_3)_2$
(g) add solid KNO_3 (h) decrease pH

General Exercises

57. Although a clock with a swinging pendulum is not a chemical equilibrium system, explain how it represents a dynamic, reversible process.

58. Although a calculator charging and discharging is not a chemical equilibrium system, explain how it represents a dynamic, reversible process.

59. With regard to reaction rate, what is characteristic of a reversible reaction at equilibrium?

60. With regard to concentration, what is characteristic of a reversible reaction at equilibrium?

61. The N_2O_4–NO_2 reversible reaction is found to have the following equilibrium concentrations at 100°C. Calculate K_{eq} for the reaction.

$$\underset{4.5 \times 10^{-5}\,M}{N_2O_4(g)} \rightleftharpoons \underset{3.0 \times 10^{-3}\,M}{2\,NO_2(g)}$$

62. The N_2O_4–NO_2 reversible reaction is found to have the following equilibrium concentrations at 100°C. Calculate K_p for the reaction.

$$\underset{0.0014\text{ atm}}{N_2O_4(g)} \rightleftharpoons \underset{0.092\text{ atm}}{2\,NO_2(g)}$$

63. Given the chemical equation for the ionization of water:

$$H_2O(l) \rightleftharpoons H^+(aq) + OH^-(aq)$$

Predict the direction of equilibrium shift for each of the following stresses.
(a) increase $[H^+]$ (b) decrease $[OH^-]$
(c) increase pH (d) decrease pH

64. Given the chemical equation for the ionization of water:

$$H_2O(l) \rightleftharpoons H^+(aq) + OH^-(aq)$$

Predict the direction of equilibrium shift for each of the following stresses.
(a) add gaseous HCl (b) add solid NaOH
(c) add liquid H_2SO_4 (d) add solid NaF

65. Explain why ferric hydroxide, $Fe(OH)_3$, is more soluble in 0.1 *M* HCl than in water.

66. Explain why silver iodide, AgI, is more soluble in 0.1 *M* NaCl than in water.

67. A saturated solution of calcium hydroxide, $Ca(OH)_2$, has a pH of 12.35. Find the hydroxide ion concentration and calculate the equilibrium constant.

68. A saturated solution of zinc hydroxide, $Zn(OH)_2$, has a pH of 8.44. Find the hydroxide ion concentration and calculate the equilibrium constant.

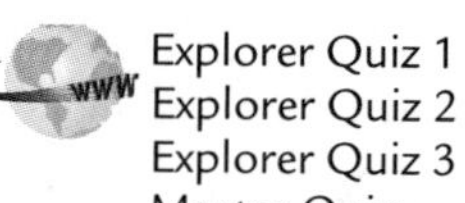

Explorer Quiz 1
Explorer Quiz 2
Explorer Quiz 3
Master Quiz

CHAPTERS 15–16

Cumulative Review

Key Concepts

1. Identify the Arrhenius acid and base that react to give sodium nitrate, $NaNO_3$.
2. Which of the following is an example of a strong Arrhenius acid?
   (a) $H_2S(aq)$ **(b)** $H_2SO_3(aq)$ **(c)** $H_2SO_4(aq)$ **(d)** $H_2O(l)$
3. Which of the following is an example of a weak Arrhenius base?
   (a) $NH_4OH(aq)$ **(b)** $LiOH(aq)$ **(c)** $Ca(OH)_2(aq)$ **(d)** $H_2O(l)$
4. Identify the Brønsted–Lowry acid and base in the following reaction.

$$NaHCO_3(aq) + NaHSO_4(aq) \longrightarrow Na_2SO_4(aq) + H_2O(l) + CO_2(g)$$

5. If the hydrogen ion concentration in acid rain is 0.00010 *M*, what is the molar hydroxide ion concentration?
6. If the hydrogen ion concentration of acid rain is 0.00010 *M*, what is the pH?
7. Which of the following are strong conductors of electricity in aqueous solution: H_2SO_4, $Ba(OH)_2$, $BaSO_4$, H_2O?
8. Sulfuric acid reacts neutralizes barium hydroxide to give barium sulfate and water. Write the net ionic equation for the reaction.

$$H_2SO_4(aq) + Ba(OH)_2(aq) \longrightarrow BaSO_4(s) + 2\,H_2O(l)$$

9. Which of the following is true *after* a reaction reaches chemical equilibrium?
   (a) The amount of reactants is decreasing.
   (b) The amount of products is increasing.
   (c) The amounts of reactants and products are constant.
10. In the following illustration, a liquid in a sealed beaker begins to evaporate (a). When the volume above the liquid becomes saturated with vapor, the vapor begins to condense to a liquid (b). What can we conclude about the system when it achieves a state of equilibrium (c)?

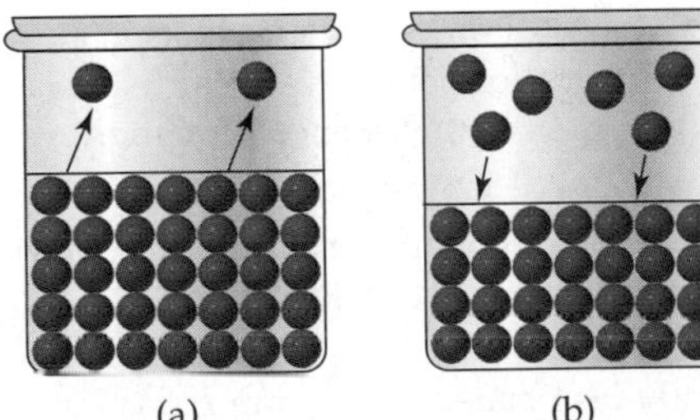

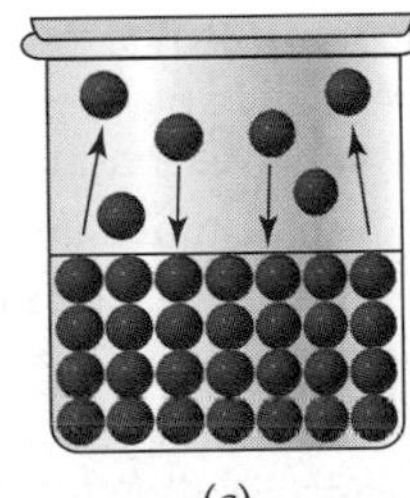

(a) (b) (c)

11. Which of the following factors increases the rate of the following reaction?

$$2\,NO(g) + O_2(g) \rightleftharpoons 2\,NO_2(g) + \text{heat}$$

   (a) increasing [NO] **(b)** decreasing temperature
   (c) adding argon gas **(d)** adding a catalyst
12. What is the general equilibrium constant expression, K_{eq}, for the following reversible reaction?

$$A + 2\,B \rightleftharpoons 3\,C$$

13. Which of the following stresses shifts the equilibrium toward the products?

$$2\,NO(g) + O_2(g) \rightleftharpoons 2\,NO_2(g) + \text{heat}$$

   (a) increasing [NO] **(b)** increasing temperature
   (c) increasing pressure **(d)** adding a catalyst

14. Which of the following stresses shifts the equilibrium toward the products?

$$NH_4OH(aq) \rightleftharpoons NH_4^+(aq) + OH^-(aq)$$

(a) increasing $[NH_4OH]$
(b) adding solid NH_4Cl
(c) adding aqueous HCl
(d) adding solid NaOH

15. Which of the following stresses shifts the equilibrium toward the products?

$$PbI_2(aq) \rightleftharpoons Pb^{2+}(aq) + 2\,I^-(aq)$$

(a) increasing $[Pb^{2+}]$
(b) decreasing $[I^-]$
(c) adding solid PbI_2
(d) decreasing pH

Key Terms

State the key term that corresponds to each of the following descriptions.

_____ 1. a solution that resists changes in pH when an acid or a base is added
_____ 2. a substance that releases hydrogen ions when dissolved in water
_____ 3. the ion that best represents the hydrogen ion in aqueous solution
_____ 4. a product from a neutralization reaction in addition to water
_____ 5. a substance that donates a proton in an acid–base reaction
_____ 6. a procedure for delivering a measured volume of solution through a buret
_____ 7. a constant that equals the product of the molar hydrogen ion concentration and the molar hydroxide ion concentration
_____ 8. the molar hydrogen ion concentration expressed on an exponential scale
_____ 9. an aqueous solution that is a good conductor of electricity
_____ 10. an ionic equation after spectator ions have been canceled
_____ 11. a chemical reaction that consumes heat energy
_____ 12. the highest point on the reaction profile
_____ 13. the energy required for reactants to reach the transition state
_____ 14. the difference in energy between the reactants and the products
_____ 15. a substance that speeds up a reaction by lowering the energy of activation
_____ 16. a reaction that proceeds toward reactants and products simultaneously
_____ 17. the principle that the molar concentrations of the products in a reversible reaction divided by the molar concentrations of the reactants (each raised to a power corresponding to a coefficient in the balanced equation) is a constant
_____ 18. a type of equilibrium in which all species are in the same state
_____ 19. the constant that expresses the molar equilibrium concentrations of ions in aqueous solution for a slightly ionized weak acid or base
_____ 20. the constant that expresses the molar equilibrium concentrations of ions in aqueous solution for a slightly dissociated ionic compound

Review Exercises

1. Which of the following is a general property of an acidic solution?

(a) tastes bitter
(b) feels slippery
(c) turns red litmus blue
(d) pH less than 7

2. Which of the following is a general property of a basic solution?

(a) tastes sour
(b) feels slippery
(c) neutralizes bases
(d) pH less than 7

3. Which of the following aqueous solutions is strongly basic?

(a) 0.1 *M* H_2SO_4
(b) 0.1 *M* H_2CO_3
(c) 0.1 *M* NH_4OH
(d) 0.1 *M* NaOH

4. What is the color of phenolphthalein indicator at pH 2? pH 7? pH 12?
5. If 25.0 mL of 0.100 *M* H_2SO_4 is titrated with 25.0 mL of NaOH, what is the molar concentration of the sodium hydroxide?

$$H_2SO_4(aq) + NaOH(aq) \longrightarrow Na_2SO_4(aq) + H_2O(l)$$

6. If 0.225 g of oxalic acid, $H_2C_2O_4$, is titrated with 25.0 mL of NaOH, what is the molar concentration of the sodium hydroxide?

$$H_2C_2O_4(s) + NaOH(aq) \longrightarrow Na_2C_2O_4(aq) + H_2O(l)$$

7. Vitamin C is ascorbic acid, HAsc. If 25.50 mL of 0.100 *M* sodium hydroxide neutralizes 0.446 g of HAsc, what is the molar mass of the acid?

$$HAsc(aq) + NaOH(aq) \longrightarrow NaAsc(aq) + H_2O(l)$$

8. What is the pH of an aqueous solution if the molar hydrogen ion concentration is 0.000 042 *M*?
9. What is the molar hydrogen ion concentration in a bleach sample that registers 9.55 on a pH meter?
10. Explain why the following energy profile illustrates an endothermic reaction.

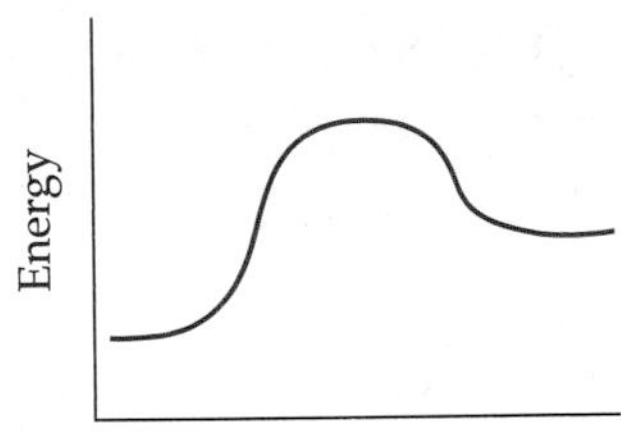

11. Which of the following is altered by a catalyst?
    (a) reactant concentration
    (b) product concentration
    (c) activation energy
    (d) heat of reaction
12. Write the equilibrium constant expression for the following reaction.

$$N_2(g) + 3\,H_2(g) \rightleftharpoons 2\,NH_3(g)$$

13. A 0.500 M sample of carbon monoxide, CO, combines with oxygen gas to give carbon dioxide. If the equilibrium concentration of CO is 1.4×10^{-16} *M* and that of O_2 is 8.6×10^{-15} *M*, what is the equilibrium constant for the reaction?

$$2\,CO(g) + O_2(g) \rightleftharpoons 2\,CO_2(g)$$

14. If the hydrogen ion concentration of a 0.100 *M* HClO solution is 0.000 055 *M*, what is the ionization constant for hypochlorous acid?
15. If the silver ion concentration in a saturated solution of silver carbonate is 0.00026 *M*, what is K_{sp} for Ag_2CO_3?

CHAPTER 17

Oxidation and Reduction

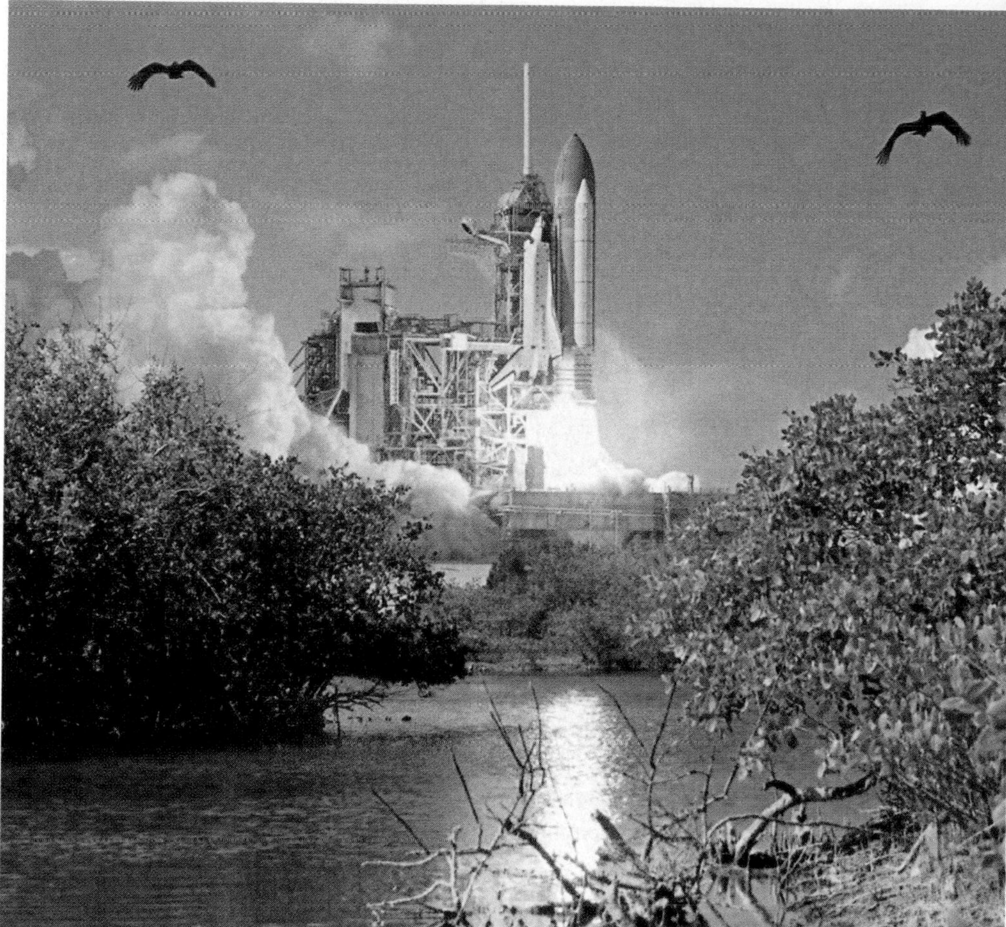

▲ A space shuttle at launch, being propelled by two booster rockets. What is the gas in the brown fuel tank, attached to the shuttle, which supplies fuel once the booster rockets have been released?

A discussion of the term "oxidation number" has been intentionally delayed until this chapter to simplify the concept of ions and their charges. In this chapter, it is helpful to point out that electron transfer in oxidation–reduction reactions is analogous to the familiar proton transfer (Brønsted–Lowry definition) in neutralization reactions.

Previously, we studied reactions involving the transfer of protons from an acid to a base, that is, *neutralization* reactions. Now we will consider reactions involving the transfer of electrons from one substance to another, that is, *oxidation–reduction* reactions.

Although we have not yet used the term "oxidation–reduction," many of the reactions we have already studied involve the transfer of electrons. In fact, whenever a metal reacts with a nonmetal, electrons are transferred. For instance, iron metal rusts when exposed to air and moisture (Figure 17.1). That is, iron reacts with oxygen in air to give iron oxide. The equation for the reaction is

(a)

(b)

▲ **Figure 17.1 Oxidation–Reduction Reactions** Examples of redox reactions include (a) iron rusting in a steel pipe and (b) silver tarnishing in a silver teakettle.

$$4\,Fe(s) + 3\,O_2(g) \longrightarrow 2\,Fe_2O_3(s)$$

Since iron oxide is an ionic compound, we can write Fe_2O_3 as Fe^{3+} and O^{2-}. In the process of becoming an ion, an iron atom loses three electrons. We say that the iron metal is *oxidized* as it loses electrons to the oxygen gas.

$$Fe \longrightarrow Fe^{3+} + 3\,e^-$$

As the reaction proceeds, we say that the oxygen is *reduced* as it gains electrons from the iron metal. In this example, each oxygen molecule reacts by gaining four electrons and becoming two oxide ions.

$$O_2 + 4\,e^- \longrightarrow 2\,O^{2-}$$

Let's consider another example. When silver metal is exposed to traces of hydrogen sulfide in the air, the silver tarnishes (Figure 17.1). The equation for the reaction is

$$2\,Ag(s) + H_2S(g) \longrightarrow Ag_2S(s) + H_2(g)$$

Since silver sulfide is an ionic compound, we can write the ions as Ag^+ and S^{2-}. In the process of becoming an ion, a silver atom loses an electron. We say that the silver metal is *oxidized* by hydrogen sulfide gas.

$$Ag \longrightarrow Ag^+ + e^-$$

Simultaneously, hydrogen sulfide is *reduced* as silver is oxidized. Hydrogen sulfide reacts by gaining two electrons and yielding a hydrogen molecule.

$$H_2S + 2\,e^- \longrightarrow H_2 + S^{2-}$$

In Chapter 8, we classified this type of reaction as a single-replacement reaction. Now, we can classify the reaction as a redox reaction. In addition, combination and decomposition reactions often involve an oxidation–reduction process.

17.1 Oxidation Numbers

Objective · To assign an oxidation number for an element in each of the following:
(a) metals and nonmetals
(b) monoatomic and polyatomic ions
(c) ionic and molecular compounds

In order to describe the number of electrons lost or gained by an atom, an **oxidation number** is assigned according to a set of rules (Table 17.1). All elements in the free state are electrically neutral. Therefore, their oxidation numbers have a value of zero. For example, Al, Cu, Fe, S, and P are each assigned an oxidation number of zero. Even nonmetals that exist as molecules, such as O_2 and Cl_2, have an oxidation number of zero.

The oxidation number of a monoatomic ion is equal to its ionic charge. For example, the oxidation number for Ca^{2+} is positive two (+2), and for Cl^-, negative one (−1). Note the distinction between the way ionic charges and oxidation numbers are written. Ionic charges are indicated by the number followed by the sign, and oxidation numbers are indicated by the sign followed by the number. For example, the oxidation numbers for Fe^{2+} and Fe^{3+} are +2 and +3, respectively.

Next, let's see how to assign oxidation numbers to elements in a compound. Recall that compounds are electrically neutral, and so the sum of the oxidation numbers of the individual atoms is 0. In binary ionic compounds, the oxidation numbers for the metal and the nonmetal correspond to their ionic charges. For example, in NaCl the oxidation number of Na^+ is +1, and that of Cl^- is −1. In AlF_3, the oxidation number of Al^{3+} is +3, and that of F^- is −1. In ternary ionic compounds, the situation is more complex, but the sum of the oxidation numbers is equal to 0.

In binary molecular compounds, the more electronegative element is assigned an oxidation number equal to the ionic charge of the free ion. Oxygen usually has an oxidation number of −2, and hydrogen usually has an oxidation number of +1. After assigning oxidation numbers to oxygen and hydrogen, we can determine the value for the other nonmetal in a binary molecular compound. For example, in NO the oxidation number of nitrogen is +2; in NO_2 the oxidation number of nitrogen is +4. We can state general rules for assigning oxidation numbers as listed in Table 17.1.

The rules for assigning oxidation numbers can be illustrated by listing some examples. Table 17.2 provides oxidation numbers for elements in the free state, in monoatomic ions, in ionic compounds, in molecular compounds, and in polyatomic ions. Note that each example is referenced to the appropriate rule in Table 17.1.

▶ **Metals in the Free State**
The oxidation number for any metal in the elemental state is 0. The metals shown are aluminum, copper, vanadium, zirconium, tin, and nickel (clockwise).

Table 17.1 Rules for Assigning Oxidation Numbers

1. A metal or a nonmetal in the free state has an oxidation number of 0.
2. A monoatomic ion has an oxidation number equal to its ionic charge.
3. A hydrogen atom is usually assigned an oxidation number of +1.
4. An oxygen atom is usually assigned an oxidation number of −2.
5. For a molecular compound, the more electronegative element is assigned a negative oxidation number equal to its charge as an anion.
6. For an ionic compound, the sum of the oxidation numbers for each atom is equal to 0.
7. For a polyatomic ion, the sum of the oxidation numbers for each atom is equal to the ionic charge on the polyatomic ion.

Table 17.2 Assigning Oxidation Numbers

Example	Oxidation Number
magnesium metal, Mg	Mg = 0 (rule 1)
bromine liquid, Br_2	Br = 0 (rule 1)
potassium ion, K^+	K = +1 (rule 2)
sulfide ion, S^{2-}	S = −2 (rule 2)
water, H_2O	H = +1 (rule 3)
	O = −2 (rule 4)
carbon tetrachloride, CCl_4	Cl = −1 (rule 5)
	C = +4 (rule 5)
barium chloride, $BaCl_2$	Ba = +2 (rule 6)
	Cl = −1 (rule 6)
nitrate ion, NO_3^-	O = −2 (rule 4)
	N = +5 (rule 7)

Oxidation Numbers in Compounds

Now, let's determine the oxidation number for an element in a compound. Oxalic acid, $H_2C_2O_4$, is used to clean jewelry. To find the oxidation number of carbon in $H_2C_2O_4$, we proceed as follows. First, we assign hydrogen an oxidation number of +1. Second, we assign oxygen an oxidation number of −2. Since compounds are electrically neutral, the sum of the oxidation numbers (ox no) is equal to 0.

$$2\,(\text{ox no H}) + 2\,(\text{ox no C}) + 4\,(\text{ox no O}) = 0$$

◀ **Nonmetals in the Free State** The oxidation number for any nonmetal in the elemental state is 0. The nonmetals shown are sulfur, white phosphorus stored under water, liquid bromine, and carbon (left to right).

Substituting the oxidation numbers of H and O, we have

$$2\,(+1) + 2\,(\text{ox no C}) + 4\,(-2) = 0$$

Simplifying and solving for the oxidation number of carbon, we have

$$+2 + 2\,(\text{ox no C}) + -8 = 0$$
$$2\,(\text{ox no C}) = +6$$
$$\text{ox no C} = +3$$

We find that the oxidation number of carbon in $H_2C_2O_4$ is +3. The following example exercise provides additional illustrations of how to determine oxidation numbers for an element in a compound.

Example Exercise 17.1 • Calculating Oxidation Numbers

Calculate the oxidation number for carbon in each of the following compounds.

(a) diamond, C
(b) dry ice, CO_2
(c) calcium carbonate, $CaCO_3$
(d) calcium bicarbonate, $Ca(HCO_3)_2$

Solution

We can begin by recalling that elements in the free state and compounds are electrically neutral and have no charge.

(a) In the free state, the oxidation number of C is 0.

(b) In CO_2, carbon is in a neutral molecular compound. Since we assign oxygen an oxidation number of −2, we can determine the oxidation number of carbon as follows.

$$\text{ox no C} + 2\,(\text{ox no O}) = 0$$
$$\text{ox no C} + 2\,(-2) = 0$$
$$\text{ox no C} = +4$$

(c) In $CaCO_3$, carbon is in an ionic compound. The oxidation number of Ca^{2+} is +2, and we assign oxygen an oxidation number of −2.

$$\text{ox no Ca} + \text{ox no C} + 3\,(\text{ox no O}) = 0$$
$$+2 + \text{ox no C} + 3\,(-2) = 0$$
$$\text{ox no C} = +4$$

(d) In $Ca(HCO_3)_2$, carbon is in an ionic compound with a polyatomic ion. The oxidation number of Ca^{2+} is +2. We assign hydrogen an oxidation number of +1, and oxygen an oxidation number of −2. Although this problem is more complex, the method is the same as the one used above.

$$\text{ox no Ca} + 2\,[\text{ox no H} + \text{ox no C} + 3\,(\text{ox no O})] = 0$$
$$+2 + 2\,[+1 + \text{ox no C} + 3(-2)] = 0$$
$$+2 + +2 + 2\,(\text{ox no C}) + -12 = 0$$
$$2\,(\text{ox no C}) = +8$$
$$\text{ox no C} = +4$$

Self-Test Exercise

Calculate the oxidation number for iodine in each of the following compounds.

(a) iodine, I_2
(b) potassium iodide, KI
(c) silver periodate, $AgIO_4$
(d) zinc iodate, $Zn(IO_3)_2$

Answers: (a) 0; (b) −1; (c) +7; (d) +5

◀ **Chromium Compounds** The oxidation number for Cr is +6 in yellow K_2CrO_4 and in orange $K_2Cr_2O_7$.

Oxidation Numbers in Polyatomic Ions

Next, let's determine the oxidation number of an element in a polyatomic ion. Consider the dichromate ion, $Cr_2O_7^{2-}$, which can be used as a fungicide in a fish aquarium. To find the oxidation number of chromium in $Cr_2O_7^{2-}$, we assign oxygen an oxidation number of −2. Since $Cr_2O_7^{2-}$ has an overall charge of −2, we write the equation

$$2\,(\text{ox no Cr}) + 7\,(\text{ox no O}) = -2$$

Substituting −2 for the oxidation number of O, we have

$$2\,(\text{ox no Cr}) + 7\,(-2) = -2$$

Simplifying and solving for the oxidation number of Cr, we have

$$2\,(\text{ox no Cr}) = +12$$
$$\text{ox no Cr} = +6$$

The following example exercise provides additional practice in calculating the oxidation number for an element in a polyatomic ion.

Example Exercise 17.2 • Calculating Oxidation Numbers

Calculate the oxidation number for sulfur in each of the following ions.

(a) sulfide ion, S^{2-} (b) sulfite ion, SO_3^{2-}
(c) sulfate ion, SO_4^{2-} (d) thiosulfate ion, $S_2O_3^{2-}$

Solution

We begin by recalling that the charge on an ion corresponds to the sum of the oxidation numbers.

(a) In S^{2-}, the oxidation number of sulfur is −2.
(b) In SO_3^{2-}, the polyatomic anion has a charge of 2−. We assign oxygen an oxidation number of −2 and write the equation

$$\text{ox no S} + 3\,(\text{ox no O}) = -2$$
$$\text{ox no S} + 3\,(-2) = -2$$
$$\text{ox no S} = +4$$

(continued)

Example Exercise 17.2 *(continued)*

(c) In SO_4^{2-}, the polyatomic anion has a charge of 2−. We assign oxygen an oxidation number of −2 and write the equation

$$\begin{aligned} \text{ox no S} + 4\,(\text{ox no O}) &= -2 \\ \text{ox no S} + 4\,(-2) &= -2 \\ \text{ox no S} &= +6 \end{aligned}$$

(d) In $S_2O_3^{2-}$, the polyatomic anion has a charge of 2−. We assign oxygen an oxidation number of −2 and write the equation

$$\begin{aligned} 2\,(\text{ox no S}) + 3\,(\text{ox no O}) &= -2 \\ 2\,(\text{ox no S}) + 3\,(-2) &= -2 \\ 2\,(\text{ox no S}) &= +4 \\ \text{ox no S} &= +2 \end{aligned}$$

Self-Test Exercise

Calculate the oxidation number for chlorine in each of the following ions.

(a) hypochlorite ion, ClO^-
(b) chlorite ion, ClO_2^-
(c) chlorate ion, ClO_3^-
(d) perchlorate ion, ClO_4^-

Answers: (a) +1; (b) +3; (c) +5; (d) +7

Experiment #23, Prentice Hall Laboratory Manual

17.2 Oxidation–Reduction Reactions

Objectives
- To identify the oxidized and reduced substances in a given redox reaction.
- To identify the oxidizing and reducing agents in a given redox reaction.

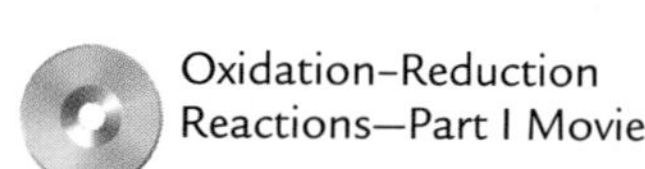

What do burning charcoal and igniting magnesium have in common? The answer is that both reactions involve oxygen gas and the transfer of electrons. Charcoal is oxidized to carbon dioxide, and magnesium metal is oxidized to magnesium oxide. We refer to a chemical change that involves the transfer of electrons as an oxidation–reduction reaction, or simply a **redox reaction**.

As an example, we can heat iron metal with sulfur powder to give the following redox reaction.

$$Fe(s) + S(s) \xrightarrow{\Delta} FeS(s)$$

Let's analyze the reaction as follows. The oxidation number of Fe changes from 0 to +2. Therefore, each iron atom loses 2 electrons. Simultaneously, the oxidation number of sulfur changes from 0 to −2. Thus, each sulfur atom gains 2 electrons. We can diagram the process as follows.

$$Fe \xrightarrow{\text{loses } 2\,e^-} Fe^{2+}$$

$$S \xrightarrow{\text{gains } 2\,e^-} S^{2-}$$

Since electrons are transferred during the chemical change, this is an example of a redox reaction. Notice that both iron and sulfur undergo a change in oxidation number.

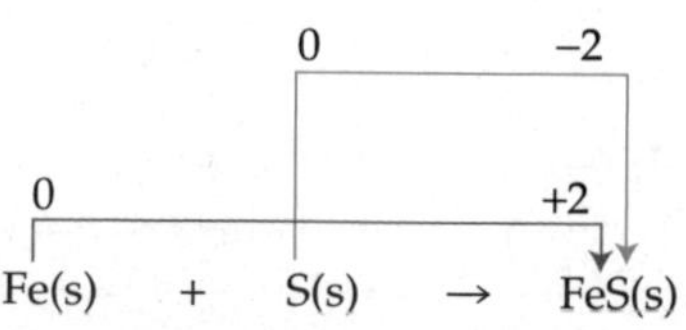

Oxidizing and Reducing Agents

By definition, the process of **oxidation** is characterized by the loss of electrons. Conversely, the process of **reduction** is characterized by the gain of electrons. In the above example, iron undergoes oxidation because it loses 2 electrons; sulfur undergoes reduction because it gains 2 electrons.

An **oxidizing agent** is a substance that causes the oxidation in a redox reaction by accepting the transferred electrons. When sulfur reacts with iron, S causes Fe to be oxidized from 0 to +2. Hence, S is the oxidizing agent. A **reducing agent** is any substance that causes reduction by donating the transferred electrons. When iron reacts with sulfur, Fe causes S to be reduced from 0 to −2. Hence, Fe is the reducing agent. We can illustrate the redox process as follows.

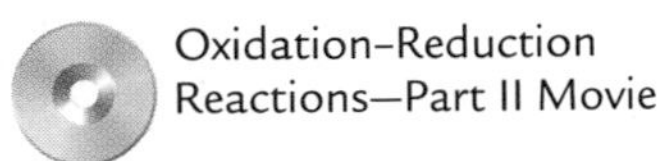

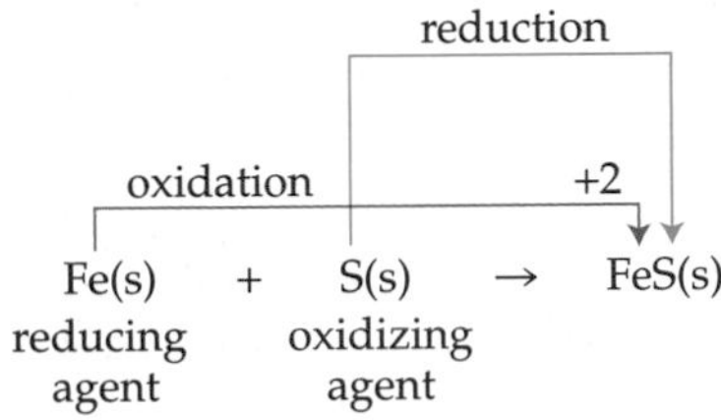

In a redox reaction one substance increases its oxidation number while another substance decreases its oxidation number. In most instances, the maximum positive oxidation number is +7; the minimum negative oxidation number is −4. Figure 17.2 illustrates the relationship of oxidation numbers in a redox reaction.

The following example exercise further illustrates redox reactions and the designation of oxidizing agents and reducing agents.

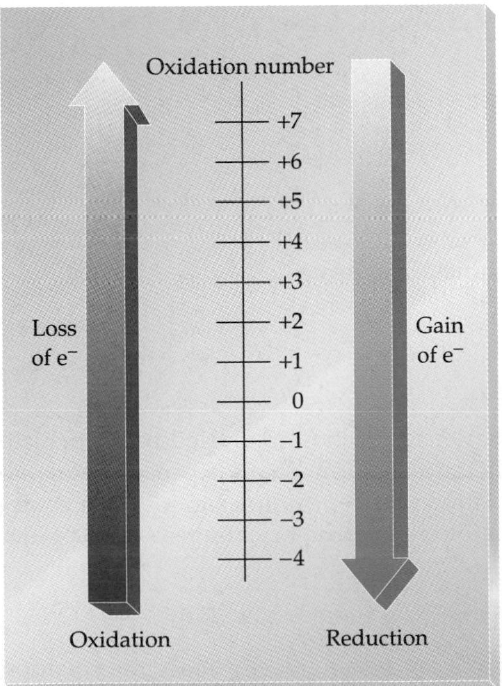

◀ **Figure 17.2 Oxidation and Reduction** The process of oxidation involves the loss of electrons (the oxidation number increases). The process of reduction involves the gain of electrons (the oxidation number decreases).

Example Exercise 17.3 • Oxidizing and Reducing Agents

An oxidation–reduction reaction occurs when a stream of hydrogen gas is passed over hot copper(II) sulfide.

$$CuS(s) + H_2(g) \xrightarrow{\Delta} Cu(s) + H_2S(g)$$

Indicate each of the following for the above redox reaction.

(a) substance oxidized (b) substance reduced
(c) oxidizing agent (d) reducing agent

Solution

By definition, the substance oxidized loses electrons, and its oxidation number increases. The substance reduced gains electrons, and its oxidation number decreases. After assigning oxidation numbers to each atom, we have the following.

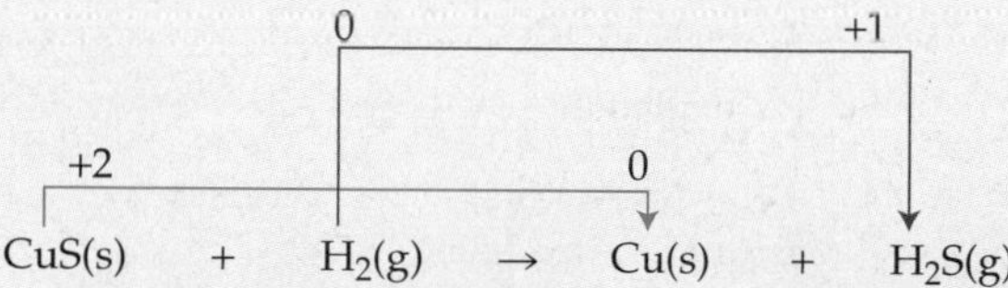

The oxidation number of hydrogen increased from 0 to +1. Thus, H_2 was *oxidized*. The oxidation number of copper decreased from +2 to 0. Thus, CuS was *reduced*. Note that the oxidation number of sulfur remained constant (−2).

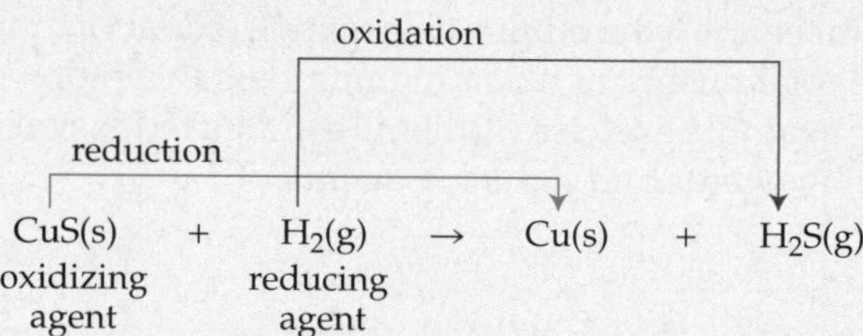

The *oxidizing agent* is CuS because it causes hydrogen to be oxidized from 0 to +1. The *reducing agent* is H_2 because it causes copper to be reduced from +2 to 0.

Self-Test Exercise

A redox reaction occurs when molten aluminum reacts with iron(III) oxide.

$$Fe_2O_3(l) + 2\,Al(l) \xrightarrow{\Delta} 2\,Fe(l) + Al_2O_3(l)$$

Indicate each of the following for the above redox reaction.

(a) substance oxidized (b) substance reduced
(c) oxidizing agent (d) reducing agent

Answers: (a) Al; (b) Fe_2O_3; (c) Fe_2O_3; (d) Al

Ionic Equations

Recall that we learned to write ionic equations in Section 15.11. Redox reactions in aqueous solution are most often shown in the ionic form. That's because ionic equations readily show us the change in oxidation number. As an example, iron analyses are routinely performed by a redox method. A typical reaction uses a potassium permanganate solution as follows.

$$5\,Fe^{2+}(aq) + MnO_4^-(aq) + 8\,H^+(aq) \longrightarrow 5\,Fe^{3+}(aq) + Mn^{2+}(aq) + 4\,H_2O(l)$$

Although the reaction appears complex, we can systematically study the equation for the oxidized and reduced substances. First, notice that the oxidation number of iron changes from +2 to +3. Since Fe^{2+} loses an electron, it is oxidized. Because Fe^{2+} is oxidized, it is the reducing agent in this reaction.

Next, let's identify the substance being reduced. When we calculate the oxidation number of manganese in permanganate ion, MnO_4^-, we find that the value is +7. On the product side of the equation, we find that the oxidation number of Mn^{2+} is +2. Since permanganate ion gains electrons, MnO_4^- is the substance reduced. Since MnO_4^- is reduced, it is the oxidizing agent in this reaction.

We can illustrate the oxidation and reduction processes for the redox reaction as follows.

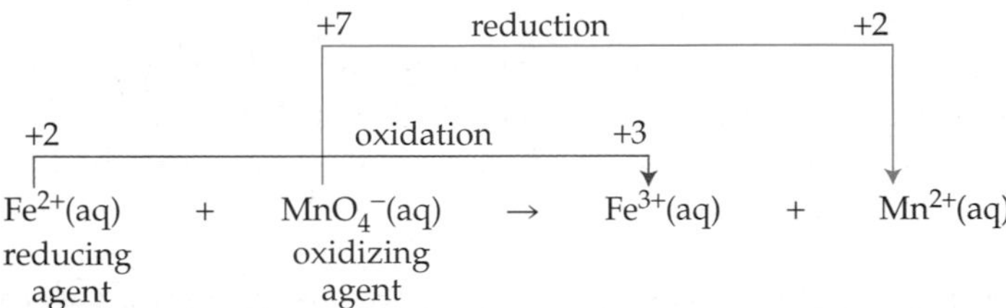

The following example exercise further illustrates determination of an oxidizing agent and a reducing agent in a redox reaction.

Example Exercise 17.4 • Oxidizing and Reducing Agents

The amount of iodine in a solution can be determined by a redox method using a sulfite solution.

$$I_2(s) + SO_3^{2-}(aq) + H_2O(l) \longrightarrow 2\,I^-(aq) + SO_4^{2-}(aq) + 2\,H^+(aq)$$

Indicate each of the following for the above reaction.

(a) substance oxidized
(b) substance reduced
(c) oxidizing agent
(d) reducing agent

Solution

We first notice that iodine is converted to iodide ion. Since iodine changes from the free state to a negative ion, it gains electrons. Therefore, I_2 is *reduced*, and it is the *oxidizing agent*.

The reducing agent is not as obvious. If we calculate the oxidation number for sulfur in SO_3^{2-} and SO_4^{2-}, we find a change from +4 to +6. Since sulfur changes from +4 to +6, it loses electrons. Therefore, SO_3^{2-} is *oxidized*, and it is the *reducing agent*. We can illustrate the oxidation and reduction processes for the redox reaction as follows.

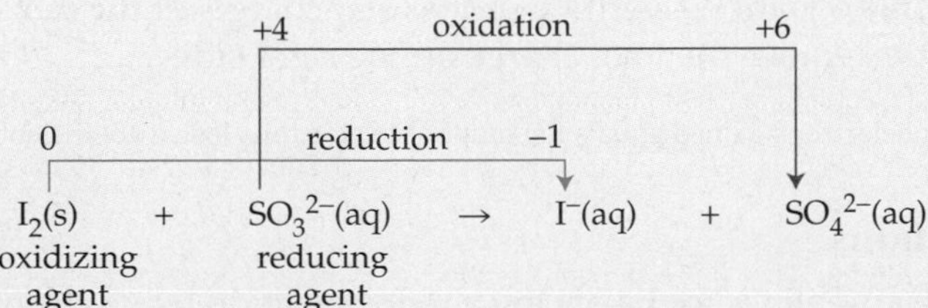

Self-Test Exercise

A redox reaction occurs when the tin(II) ion reacts with the iodate ion as follows.

$$6\,H^+(aq) + 3\,Sn^{2+}(aq) + IO_3^-(aq) \longrightarrow 3\,Sn^{4+}(aq) + I^-(aq) + 3\,H_2O(l)$$

Indicate each of the following for the above redox reaction.

(a) substance oxidized
(b) substance reduced
(c) oxidizing agent
(d) reducing agent

Answers: (a) Sn^{2+}; (b) IO_3^-; (c) IO_3^-; (d) Sn^{2+}

Note It is helpful to know the oxidation number of an element in order to determine the substances oxidized and reduced. In some instances, the oxidation number is obvious. In others, it must be calculated. With practice, you can readily determine the oxidation number even for complex substances.

17.3 Balancing Redox Equations: Oxidation Number Method

Objective · To write a balanced chemical equation for a redox reaction using the oxidation number method.

In Section 8.3, we learned to balance chemical equations by inspection. We can also balance simple redox reactions by inspection. It is difficult, however, to balance complex redox reactions by this method. Therefore, we will use the following general guidelines for balancing redox equations with an *oxidation number method.*

Guidelines for Balancing Redox Equations: Oxidation Number Method

1. Inspect the reactants and products to determine the substances undergoing a change in oxidation number. In some cases, it is not obvious and you must calculate the oxidation number.
   (a) Write the oxidation number above the element that is oxidized and above the element that is reduced.
   (b) Diagram the number of electrons lost by the oxidized substance and gained by the reduced substance.
2. Balance each element in the equation using a coefficient. Keep in mind that the total electron loss by oxidation must equal the total electron gain by reduction.
   (a) In front of the oxidized substance, place a coefficient equal to the number of electrons gained by the reduced substance.*
   (b) In front of the reduced substance, place a coefficient equal to the number of electrons lost by the oxidized substance.*
   (c) Balance the remaining elements.
3. After balancing the equation, verify that the coefficients are correct.
   (a) Place a check mark (✓) above the symbol for each element to indicate that the number of atoms is the same for reactants and products.
   (b) For ionic equations, verify that the total charge on the left side of the equation equals the total charge on the right side.

*If the number of electrons gained equals the number of electrons lost, a coefficient is not required.

Redox Equations

Let's apply the guidelines for balancing a redox reaction using the oxidation number method. Specifically, consider the reaction of a metal in an aqueous solution. For example, copper metal reacts with aqueous silver nitrate as follows.

$$Cu(s) + AgNO_3(aq) \longrightarrow Cu(NO_3)_2(aq) + Ag(s)$$

In this reaction, the oxidation number of copper increases from 0 to +2. Simultaneously, the oxidation number of silver decreases from +1 to 0. Thus, copper loses 2 electrons and silver gains 1 electron. We can diagram the loss and gain of electrons as follows.

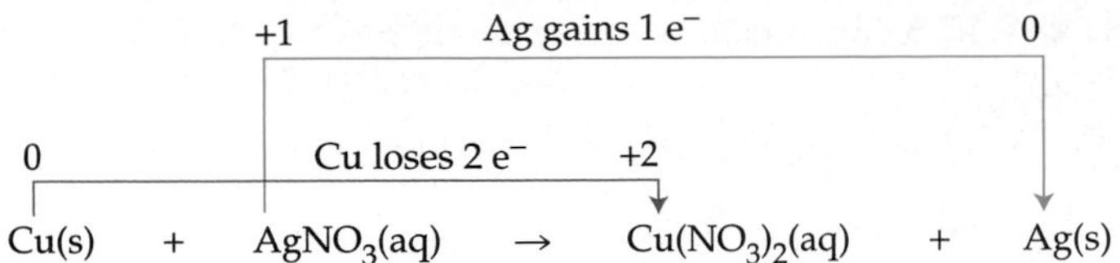

Since copper loses 2 electrons and silver gains only 1 electron, two silver atoms are required to balance the transfer of electrons. Therefore, we place the coefficient 2 in front of each silver substance. The balanced equation for the reaction is

$$Cu(s) + 2\,AgNO_3(aq) \longrightarrow Cu(NO_3)_2(aq) + 2\,Ag(s)$$

After placing the coefficients, we verify that the equation is balanced by checking (✓) each element in the reaction. Since the nitrate ion did not change, we can check off the polyatomic ion as a single unit.

$$\overset{\checkmark}{Cu}(s) + 2\,\overset{\checkmark}{Ag}\overset{\checkmark}{NO_3}(aq) \longrightarrow \overset{\checkmark}{Cu}(\overset{\checkmark}{NO_3})_2(aq) + 2\,\overset{\checkmark}{Ag}(s)$$

Since all the elements are balanced, we have verified that this is a balanced equation. The following example exercise provides additional practice in balancing redox equations.

Example Exercise 17.5 • Balancing Redox Reactions

An industrial blast furnace reduces iron ore, Fe_2O_3, to molten iron. Balance the following redox equation using the oxidation number method.

$$Fe_2O_3(l) + CO(g) \longrightarrow Fe(l) + CO_2(g)$$

Solution

In this reaction, the oxidation number of iron decreases from +3 in Fe_2O_3 to 0 in Fe. Simultaneously, the oxidation number of carbon increases from +2 to +4. Thus, each Fe gains 3 electrons while each C loses 2 electrons. We can diagram the redox process as follows.

+2 C loses 2 e⁻ +4

+3 Fe gains 3 e⁻ 0

$$Fe_2O_3(l) \quad + \quad CO(g) \quad \rightarrow \quad Fe(l) \quad + \quad CO_2(g)$$

Since the number of electrons gained and lost must be equal, we find the lowest common multiple. In this case, it is 6. Each Fe gains 3 electrons, and so we place the coefficient 3 in front of CO and CO_2.

$$Fe_2O_3(l) + 3\,CO(g) \longrightarrow Fe(l) + 3\,CO_2(g)$$

Each carbon atom loses 2 electrons, and so we place the coefficient 2 in front of each iron atom. Since Fe_2O_3 has two iron atoms, it does not require a coefficient.

$$Fe_2O_3(l) + 3\,CO(g) \longrightarrow 2\,Fe(l) + 3\,CO_2(g)$$

Finally, we verify that the equation is balanced. We check (✓) each element in the equation.

$$\overset{\checkmark}{Fe_2}\overset{\checkmark}{O_3}(l) + 3\,\overset{\checkmark\checkmark}{CO}(g) \longrightarrow 2\,\overset{\checkmark}{Fe}(l) + 3\,\overset{\checkmark\checkmark}{CO_2}(g)$$

Since all the elements are balanced, we have a balanced redox equation.

(continued)

▲ **A Penny in Nitric Acid** A copper penny reacts with concentrated nitric acid, HNO_3, to give a green Cu^{2+} solution and brown NO_2 gas.

Example Exercise 17.5 *(continued)*

Self-Test Exercise

Balance the following redox equation by the oxidation number method.

$$Cl_2O_5(g) + CO(g) \longrightarrow Cl_2(g) + CO_2(g)$$

Answer: $Cl_2O_5(g) + 5\,CO(g) \longrightarrow Cl_2(g) + 5\,CO_2(g)$

Ionic Equations

Now, let's apply the general guidelines for balancing a redox reaction to an ionic equation. As an example, we know that a copper penny reacts with nitric acid according to the following equation.

$$Cu(s) + H^+(aq) + NO_3^-(aq) \longrightarrow Cu^{2+}(aq) + NO_2(g) + H_2O(l)$$

In this ionic equation, we see that the oxidation number of copper increases from 0 to +2. Although it is not as evident, the oxidation number of nitrogen decreases from +5 to +4. Thus, Cu loses 2 electrons while each N gains 1 electron. We can diagram the loss and gain of electrons as follows.

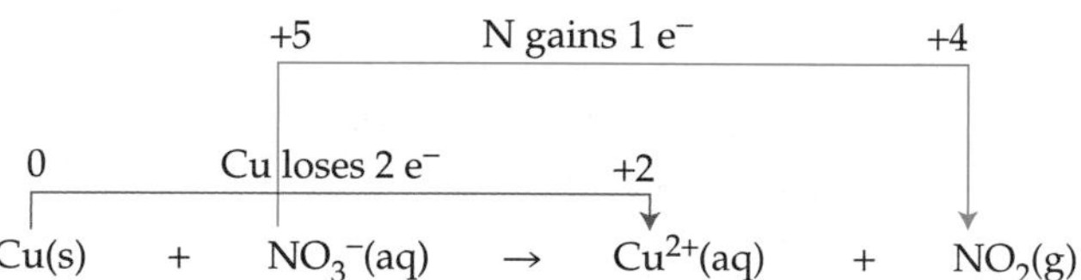

We can balance the loss and gain of electrons by placing the coefficient 2 in front of each nitrogen substance. It is understood that the coefficient of each copper is 1.

$$Cu(s) + H^+(aq) + 2\,NO_3^-(aq) \longrightarrow Cu^{2+}(aq) + 2\,NO_2(g) + H_2O(l)$$

In this example, H^+ and H_2O take part in the reaction. Thus, they also must be balanced. We complete balancing the equation by inspection. Since there is a total of six oxygen atoms as reactants, there must be six oxygen atoms in the products. Therefore, we need the coefficient 2 in front of H_2O. After balancing the remaining hydrogen atoms in the reaction, we have

$$Cu(s) + 4\,H^+(aq) + 2\,NO_3^-(aq) \longrightarrow Cu^{2+}(aq) + 2\,NO_2(g) + 2\,H_2O(l)$$

Now, we verify that the equation is balanced by checking (✓) each element in the redox reaction.

$$\overset{\checkmark}{Cu(s)} + \overset{\checkmark}{4\,H^+(aq)} + \overset{\checkmark\checkmark}{2\,NO_3^-(aq)} \longrightarrow \overset{\checkmark}{Cu^{2+}(aq)} + \overset{\checkmark\checkmark}{2\,NO_2(g)} + \overset{\checkmark\,\checkmark}{2\,H_2O(l)}$$

Finally, we check to verify that the total charge on the reactants equals the total charge on the products. On the left side of the equation, we have $+4 - 2 = +2$. On the right side of the equation, we have +2. Since the ionic charge on each side is +2, the equation is balanced. The following example exercise provides additional practice in balancing ionic redox equations.

For redox reactions, it is necessary to *verify both the mass and the charge*. That is, there must be the same number of atoms of each element in the reactants and products, and the total charge on the reactants must equal the total charge on the products.

Example Exercise 17.6 • Balancing Ionic Redox Reactions

Aqueous sodium iodide reacts with a potassium dichromate solution. Write a balanced equation for the following redox reaction.

$$H^+(aq) + I^-(aq) + Cr_2O_7^{2-}(aq) \longrightarrow I_2(s) + Cr^{3+}(aq) + H_2O(l)$$

Solution

In this special example, $Cr_2O_7^{2-}$ and I_2 each contain a subscript that affects electron transfer. Note that there are 2 atoms of chromium in the reactant and 2 atoms of iodine in the product. Let's balance the chromium and iodine atoms first.

$$H^+(aq) + 2\,I^-(aq) + Cr_2O_7^{2-}(aq) \longrightarrow I_2(s) + 2\,Cr^{3+}(aq) + H_2O(l)$$

In this reaction, the oxidation number of each iodine atom increases from −1 to 0, and the oxidation number of each chromium atom decreases from +6 to +3. We can show the loss and gain of electrons as

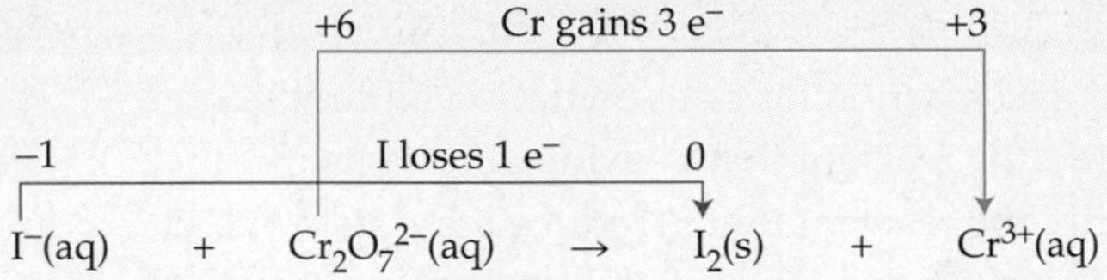

+6 Cr gains 3 e^- +3

−1 I loses 1 e^- 0

$$I^-(aq) + Cr_2O_7^{2-}(aq) \rightarrow I_2(s) + Cr^{3+}(aq)$$

There are 2 chromium atoms, and so the total electron gain is 6 electrons. Thus, the total electron loss must also be 6 electrons. Since there are 2 iodine atoms, we place the coefficients as follows.

$$H^+(aq) + 6\,I^-(aq) + Cr_2O_7^{2-}(aq) \longrightarrow 3\,I_2(s) + 2\,Cr^{3+}(aq) + H_2O(l)$$

Next, we balance the oxygen and hydrogen atoms. Since there are 7 oxygen atoms as reactants, we place the coefficient 7 in front of H_2O. This gives 14 hydrogen atoms, and so we place the coefficient 14 in front of H^+.

$$14\,H^+(aq) + 6\,I^-(aq) + Cr_2O_7^{2-}(aq) \longrightarrow 3\,I_2(s) + 2\,Cr^{3+}(aq) + 7\,H_2O(l)$$

We can verify that the equation is balanced by checking (✓) each element.

$$\overset{\checkmark}{14\,H^+(aq)} + \overset{\checkmark}{6\,I^-(aq)} + \overset{\checkmark\ \checkmark}{Cr_2O_7^{2-}(aq)} \longrightarrow \overset{\checkmark}{3\,I_2(s)} + \overset{\checkmark}{2\,Cr^{3+}(aq)} + \overset{\checkmark\ \checkmark}{7\,H_2O(l)}$$

Last, we verify that the ionic charges are balanced. On the left side of the equation, we have +14 − 6 − 2 = +6. On the right side of the equation, we have +6. Since the ionic charge on each side is +6, the equation is balanced.

Self-Test Exercise

Write a balanced equation for the following redox reaction.

$$H^+(aq) + MnO_4^-(aq) + NO_2^-(aq) \longrightarrow Mn^{2+}(aq) + NO_3^-(aq) + H_2O(l)$$

Answer:

$$6\,H^+(aq) + 2\,MnO_4^-(aq) + 5\,NO_2^-(aq) \longrightarrow 2\,Mn^{2+}(aq) + 5\,NO_3^-(aq) + 3\,H_2O(l)$$

Note Since the products of a redox reaction are often difficult to predict, you are not expected to complete the equation before balancing. Most often, you are given the reactants and products for a redox reaction.

17.4 Balancing Redox Equations: Half-Reaction Method

Objective · To write a balanced chemical equation for a redox reaction using the half-reaction method:
(a) in acidic solution
(b) in basic solution

In the preceding section, we learned how to balance redox equations using oxidation numbers. In this section, we are going to balance redox equations with a different procedure. This alternative method is referred to as the half-reaction method.

A **half-reaction** breaks down the redox reaction to show the oxidation process and the reduction process separately. The half-reaction method systematically balances the oxidation half-reaction and then balances the reduction half-reaction. We will use the following general guidelines for balancing redox equations with the half-reaction method.

Guidelines for Balancing Redox Equations: Half-Reaction Method

1. Write the half-reaction for both oxidation and reduction.
   (a) Identify the reactant that is oxidized and its product.
   (b) Identify the reactant that is reduced and its product.
2. Balance the atoms in each half-reaction using coefficients.
   (a) Balance all elements except oxygen and hydrogen.
   (b) Balance oxygen using H_2O.
   (c) Balance hydrogen using H^+.
   (*Note:* For reactions in a basic solution, add OH^- to neutralize H^+. For example, 2 OH^- neutralizes 2 H^+ to give 2 H_2O.)
   (d) Balance the ionic charges using e^-.
3. Multiply each half-reaction by a whole number so that the number of electrons lost by oxidation equals the number of electrons gained by reduction.
4. Add the two half-reactions together and cancel identical species, including electrons, on each side of the equation.
5. After balancing the equation, verify that the coefficients are correct.
   (a) Place a check mark (✓) above the symbol of each element to indicate that the number of atoms is the same for reactants and products.
   (b) For ionic equations, verify that the total charge on the left side of the equation equals the total charge on the right side.

Balancing Redox Equations in Acidic Solution

Let's apply the general guidelines for balancing a redox reaction in an acidic aqueous solution. As an example, zinc reacts in nitric acid to give nitrogen monoxide gas. The ionic equation for the redox reaction is

$$Zn(s) + NO_3^-(aq) \longrightarrow Zn^{2+}(aq) + NO(g)$$

Step 1: Write the oxidation and reduction half-reactions. Since the oxidation number of zinc increases, Zn is oxidized. Since the oxidation number of nitrogen decreases, NO_3^- is reduced.

$$\text{Oxidation:} \quad Zn \longrightarrow Zn^{2+}$$
$$\text{Reduction:} \quad NO_3^- \longrightarrow NO$$

Step 2: Balance the atoms and the charges for each half-reaction. The balanced half-reactions are

$$Zn \longrightarrow Zn^{2+} + 2\,e^-$$
$$3\,e^- + 4\,H^+ + NO_3^- \longrightarrow NO + 2\,H_2O$$

Note that electrons are added to the *product side* of the oxidation half-reaction and that electrons are added to the *reactant side* of the reduction half-reaction.

Step 3: Note that Zn loses 2 electrons and NO_3^- gains 3 electrons. So that the electron loss and gain will be equal, we multiply the Zn half-reaction by 3 and the NO_3^- by 2.

$$3\,Zn \longrightarrow 3\,Zn^{2+} + 6\,e^-$$
$$6\,e^- + 8\,H^+ + 2\,NO_3^- \longrightarrow 2\,NO + 4\,H_2O$$

Step 4: Now, we add the two half-reactions together and cancel the 6 electrons.

$$3\,Zn \longrightarrow 3\,Zn^{2+} + \cancel{6\,e^-}$$
$$\cancel{6\,e^-} + 8\,H^+ + 2\,NO_3^- \longrightarrow 2\,NO + 4\,H_2O$$

$$3\,Zn + 8\,H^+ + 2\,NO_3^- \longrightarrow 3\,Zn^{2+} + 2\,NO + 4\,H_2O$$

Step 5: Finally, let's check the atoms and ionic charges to verify that the equation is balanced. We have

Atoms:	3 Zn, 8 H, 2 N, 6 O	= 3 Zn, 2 N, 6 O, 8 H
Charges:	8 (+1) + 2(−1)	= 3(+2)
	+6	= +6

Since the atoms and ionic charges are equal, the redox equation is balanced. The following example exercise further illustrates how redox equations are balanced using the half-reaction method.

Example Exercise 17.7 • Balancing Ionic Redox Reactions

Write a balanced ionic equation for the reaction of iron(II) sulfate and potassium permanganate in acidic solution. The ionic equation is

$$Fe^{2+}(aq) + MnO_4^-(aq) \longrightarrow Fe^{3+}(aq) + Mn^{2+}(aq)$$

Solution

We can balance the redox reaction by the half-reaction method as follows.

Step 1, Since Fe^{2+} is oxidized from +2 to +3, MnO_4^- must be reduced. The two half-reactions are as follows.

Oxidation: $Fe^{2+} \longrightarrow Fe^{3+}$

Reduction: $MnO_4^- \longrightarrow Mn^{2+}$

Step 2, We can balance each half-reaction as follows.

$$Fe^{2+} \longrightarrow Fe^{3+} + e^-$$
$$5\,e^- + 8\,H^+ + MnO_4^- \longrightarrow Mn^{2+} + 4\,H_2O$$

Step 3, Since Fe^{2+} loses 1 e^- and MnO_4^- gains 5 e^-, we multiply the Fe^{2+} half-reaction by 5.

$$5\,Fe^{2+} \longrightarrow 5\,Fe^{3+} + 5\,e^-$$
$$5\,e^- + 8\,H^+ + MnO_4^- \longrightarrow Mn^{2+} + 4\,H_2O$$

(continued)

Example Exercise 17.7 *(continued)*

Step 4, Then we add the two half-reactions together and cancel the 5 e^-.

$$5\ Fe^{2+} \longrightarrow 5\ Fe^{3+} + \cancel{5\ e^-}$$
$$\cancel{5\ e^-} + 8\ H^+ + MnO_4^- \longrightarrow Mn^{2+} + 4\ H_2O$$

$$5\ Fe^{2+} + 8\ H^+ + MnO_4^- \longrightarrow 5\ Fe^{3+} + Mn^{2+} + 4\ H_2O$$

Step 5, Finally, let's check the atoms and ionic charges to verify that the equation is balanced. We have

Atoms:	5 Fe, 8 H, 1 Mn, 4 O	=	5 Fe, 1 Mn, 8 H, 4 O
Charges:	5(+2) + 8(+1) + (−1)	=	5(+3) + (+2)
	+17	=	+17

Since the atoms and ionic charges are equal for reactants and products, the redox equation is balanced.

Self-Test Exercise
Write a balanced ionic equation for the reaction of sodium nitrite and potassium permanganate in acidic solution. The ionic equation is

$$MnO_4^-(aq) + NO_2^-(aq) \longrightarrow Mn^{2+}(aq) + NO_3^-(aq)$$

Answer: $6\ H^+ + 2\ MnO_4^- + 5\ NO_2^- \longrightarrow 2\ Mn^{2+} + 5\ NO_3^- + 3\ H_2O$

Balancing Redox Equations in Basic Solution

Now, let's balance a redox reaction that takes place in a basic aqueous solution. For example, sodium iodide and potassium permanganate react in basic solution to produce iodine and manganese(IV) oxide. The ionic equation is

$$I^-(aq) + MnO_4^-(aq) \longrightarrow I_2(s) + MnO_2(s)$$

The procedure for balancing a redox equation in a base is similar to the procedure given in the general guidelines for balancing a redox reaction in an acid.

Step 1: First, we write the oxidation and reduction half-reactions. We note that iodine is oxidized from −1 to 0. Simultaneously, manganese is reduced from +7 to +4.

Oxidation: $I^- \longrightarrow I_2$

Reduction: $MnO_4^- \longrightarrow MnO_2$

Step 2: Second, we balance each half-reaction for both atoms and charges. In the oxidation half-reaction, we place the coefficient 2 in front of I^- and add 2 electrons to the product side. We balance the reduction half–reaction as follows: Add H_2O to balance O atoms, add H^+ to balance H atoms, and add e^- to balance charge. The balanced half-reactions are

$$2\ I^- \longrightarrow I_2 + 2\ e^-$$
$$3\ e^- + 4\ H^+ + MnO_4^- \longrightarrow MnO_2 + 2\ H_2O$$

Note that this redox reaction takes place in a basic solution, not an acidic solution. Therefore, we must follow guideline 2(c), which states that the H^+ should be neutralized with OH^-. In this example, we add 4 OH^- to each side and simplify the equation.

$$3\,e^- + 4\,H^+ + 4\,OH^- + MnO_4^- \longrightarrow MnO_2 + 2\,H_2O + 4\,OH^-$$
$$3\,e^- + 4\,H_2O + MnO_4^- \longrightarrow MnO_2 + 2\,H_2O + 4\,OH^-$$
$$3\,e^- + 2\,H_2O + MnO_4^- \longrightarrow MnO_2 + 4\,OH^-$$

Step 3: Note that 2 I^- lose 2 electrons and MnO_4^- gains 3 electrons. To equalize electron loss and gain, we multiply the I^- oxidation half-reaction by 3, and the MnO_4^- reduction half-reaction by 2.

$$6\,I^- \longrightarrow 3\,I_2 + 6\,e^-$$
$$6\,e^- + 4\,H_2O + 2\,MnO_4^- \longrightarrow 2\,MnO_2 + 8\,OH^-$$

Step 4: We add the two half-reactions together and cancel the 6 electrons.

$$6\,I^- \longrightarrow 3\,I_2 + \cancel{6\,e^-}$$
$$\cancel{6\,e^-} + 4\,H_2O + 2\,MnO_4^- \longrightarrow 2\,MnO_2 + 8\,OH^-$$

$$6\,I^- + 4\,H_2O + 2\,MnO_4^- \longrightarrow 3\,I_2 + 2\,MnO_2 + 8\,OH^-$$

Step 5: Finally, we check to verify that the equation is balanced.

Atoms:	6 I, 8 H, 12 O, 2 Mn	=	6 I, 2 Mn, 12 O, 8 H
Charges:	$6\,(-1) + 2(-1)$	=	$8\,(-1)$
	-8	=	-8

Since the atoms and ionic charges are equal for the reactants and products, this redox equation is balanced.

17.5 Predicting Spontaneous Redox Reactions

Objectives
- To predict the stronger oxidizing agent and reducing agent given a list of reduction potentials.
- To predict whether a redox reaction is spontaneous or nonspontaneous given a list of reduction potentials.

Let's consider a metal reacting in an aqueous solution. For example, consider the reaction of zinc metal in aqueous copper sulfate. The net ionic equation for the reaction is

$$Zn(s) + Cu^{2+}(aq) \longrightarrow Zn^{2+}(aq) + Cu(s)$$

It is experimentally observed that the reaction of zinc metal in aqueous copper sulfate proceeds spontaneously. Let's compare the two reduction half-reactions.

$$Cu^{2+}(aq) + 2\,e^- \longrightarrow Cu(s)$$
$$Zn^{2+}(aq) + 2\,e^- \longrightarrow Zn(s)$$

According to the overall redox reaction shown above, Zn(s) is oxidized and $Cu^{2+}(aq)$ is reduced. Therefore, we can conclude that the reduction of $Cu^{2+}(aq)$ has a greater tendency to occur than the reduction of $Zn^{2+}(aq)$.

Stated differently, Cu^{2+} has a greater tendency to gain electrons than Zn^{2+}. If we compare other combinations of metals and aqueous solutions, we can arrange a series of elements based on their ability to gain electrons. The strongest oxidizing agent is the substance that is most easily reduced. Conversely, the strongest reducing agent is the substance that is most easily oxidized.

The tendency for a substance to gain electrons and undergo reduction is called its **reduction potential**. The reduction potential for a substance is its relative strength as an oxidizing agent. Table 17.3 lists the reduction potentials for several substances in order of their ability to undergo reduction. A substance that is high in the table has a strong reduction potential and is easily reduced. A substance low in the table has a weak reduction potential and is not easily reduced.

Table 17.3 Reduction Potential of Selected Oxidizing and Reducing Agents

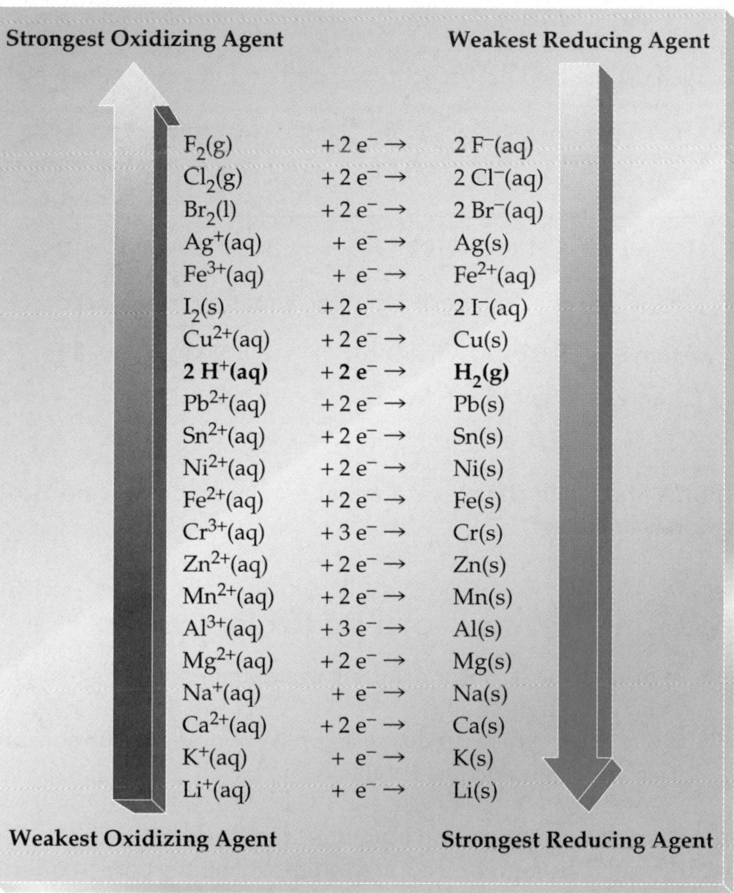

Strongest Oxidizing Agent		Weakest Reducing Agent
$F_2(g)$	$+ 2\,e^- \rightarrow$	$2\,F^-(aq)$
$Cl_2(g)$	$+ 2\,e^- \rightarrow$	$2\,Cl^-(aq)$
$Br_2(l)$	$+ 2\,e^- \rightarrow$	$2\,Br^-(aq)$
$Ag^+(aq)$	$+ e^- \rightarrow$	$Ag(s)$
$Fe^{3+}(aq)$	$+ e^- \rightarrow$	$Fe^{2+}(aq)$
$I_2(s)$	$+ 2\,e^- \rightarrow$	$2\,I^-(aq)$
$Cu^{2+}(aq)$	$+ 2\,e^- \rightarrow$	$Cu(s)$
$2\,H^+(aq)$	**$+ 2\,e^- \rightarrow$**	**$H_2(g)$**
$Pb^{2+}(aq)$	$+ 2\,e^- \rightarrow$	$Pb(s)$
$Sn^{2+}(aq)$	$+ 2\,e^- \rightarrow$	$Sn(s)$
$Ni^{2+}(aq)$	$+ 2\,e^- \rightarrow$	$Ni(s)$
$Fe^{2+}(aq)$	$+ 2\,e^- \rightarrow$	$Fe(s)$
$Cr^{3+}(aq)$	$+ 3\,e^- \rightarrow$	$Cr(s)$
$Zn^{2+}(aq)$	$+ 2\,e^- \rightarrow$	$Zn(s)$
$Mn^{2+}(aq)$	$+ 2\,e^- \rightarrow$	$Mn(s)$
$Al^{3+}(aq)$	$+ 3\,e^- \rightarrow$	$Al(s)$
$Mg^{2+}(aq)$	$+ 2\,e^- \rightarrow$	$Mg(s)$
$Na^+(aq)$	$+ e^- \rightarrow$	$Na(s)$
$Ca^{2+}(aq)$	$+ 2\,e^- \rightarrow$	$Ca(s)$
$K^+(aq)$	$+ e^- \rightarrow$	$K(s)$
$Li^+(aq)$	$+ e^- \rightarrow$	$Li(s)$
Weakest Oxidizing Agent		Strongest Reducing Agent

Spontaneous Reactions

We can interpret the information given in Table 17.3 as follows. The ability of a substance to be reduced decreases from top to bottom in the table. The strongest oxidizing agents are in the upper left portion of the table. Thus, F_2 has the highest potential for reduction and is the strongest oxidizing agent. Conversely, the ability of a substance to be oxidized decreases from top to bottom. The strongest reducing agents are in the lower right portion of the table. Thus, Li has the highest potential for oxidation and is the strongest reducing agent.

A metal ion can be reduced by any reducing agent lower in the table. For example, Sn^{2+} can be reduced by Ni or Fe, but not by Pb. Any metal below H_2 in the table can react with an acid and be oxidized. A metal above H^+ in the table cannot react with acid. For example, Pb, Sn, and Ni react with acid and undergo oxidation, whereas Cu and Ag do not react with acid.

A nonmetal can oxidize any reducing agent lower in the table. For example, Cl_2 can oxidize Br^- or Ag, but not F^-. In Table 17.3, we see that F_2 is the strongest oxidizing agent and that Li is the strongest reducing agent. However, H_2 is the strongest *nonmetal* reducing agent listed in the table.

In Table 17.3 we note that $Cu^{2+}(aq)$ is a stronger oxidizing agent than $Zn^{2+}(aq)$. We also note that Zn metal is a stronger reducing agent than Cu metal. Therefore, we can explain the above reaction as follows.

$Cu^{2+}(aq)$	+	$Zn(s)$	$\longrightarrow$	$Cu(s)$	+	$Zn^{2+}(aq)$
stronger oxidizing agent		stronger reducing agent		weaker reducing agent		weaker oxidizing agent

Moreover, we see that the reverse process has very little tendency to occur and is nonspontaneous. We can therefore conclude that there is no reaction between aqueous zinc ion and copper metal.

$Zn^{2+}(aq)$	+	$Cu(s)$	$\not\longrightarrow$	$Zn(s)$	+	$Cu^{2+}(aq)$
weaker oxidizing agent		weaker reducing agent		stronger reducing agent		stronger oxidizing agent

Now, let's consider a nonmetal reaction in aqueous solution. For example, consider what happens if we bubble chlorine gas through an aqueous sodium bromide solution. The net ionic equation for the possible reaction is

$$Cl_2(g) + 2\ Br^-(aq) \longrightarrow 2\ Cl^-(aq) + Br_2(l)$$

In Table 17.3 we see that $Cl_2(g)$ is a stronger oxidizing agent than $Br_2(l)$. We also see that $Br^-(aq)$ is a stronger reducing agent than $Cl^-(aq)$. Therefore, we can explain the above reaction as follows.

$Cl_2(g)$	+	$2\ Br^-(aq)$	$\longrightarrow$	$2\ Cl^-(aq)$	+	$Br_2(l)$
stronger oxidizing agent		stronger reducing agent		weaker reducing agent		weaker oxidizing agent

Since the reactants contain the stronger pair of oxidizing and reducing agents, the reaction takes place spontaneously. Is the reverse reaction spontaneous? Since the chloride ion and bromine liquid are the weaker pair of oxidizing and reducing agents, the reverse reaction is nonspontaneous. If a redox reaction is not spontaneous, we usually say that there is no reaction.

$2\ Cl^-(aq)$	+	$Br_2(l)$	$\not\longrightarrow$	$Cl_2(g)$	+	$2\ Br^-(aq)$
weaker reducing agent		weaker oxidizing agent		stronger oxidizing agent		stronger reducing agent

The following example exercise provides additional practice in predicting whether or not a redox reaction occurs spontaneously.

Example Exercise 17.8 • Predicting Spontaneous Redox Reactions

Predict whether the following reaction is spontaneous or nonspontaneous.

$$Ni^{2+}(aq) + Sn(s) \longrightarrow Ni(s) + Sn^{2+}(aq)$$

(continued)

Example Exercise 17.8 *(continued)*

Solution

Let's refer to the table of reduction potentials to predict whether or not the reaction is spontaneous. Table 17.3 lists $Ni^{2+}(aq)$ as a weaker oxidizing agent than $Sn^{2+}(aq)$. Moreover, Sn(s) is a weaker reducing agent than Ni(s).

$Ni^{2+}(aq)$	+	Sn(s)	$\longrightarrow$	Ni(s)	+	$Sn^{2+}(aq)$
weaker oxidizing agent		weaker reducing agent		stronger reducing agent		stronger oxidizing agent

Since the reactants are the weaker pair of oxidizing and reducing agents, the reaction is *nonspontaneous*. Conversely, the reverse reaction is spontaneous because the products are the stronger oxidizing and reducing agents.

Self-Test Exercise

Predict whether the following reaction is spontaneous or nonspontaneous.

$$Ni^{2+}(aq) + Al(s) \longrightarrow Ni(s) + Al^{3+}(aq)$$

Answer: spontaneous

17.6 Voltaic Cells

Objectives
- To indicate the anode and cathode in a given voltaic cell.
- To indicate the oxidation and reduction half-reactions in a given spontaneous electrochemical cell.

Batteries play an important role in our daily lives. They enable us to start an automobile simply by turning a switch. They allow us to use cordless electrical appliances. They extend the lives of heart patients who receive rhythmic pulses from pacemakers. All these devices use electrical energy supplied by batteries. Although different devices may use different types of batteries, all batteries involve a spontaneous redox reaction that produces electrical energy.

Redox Reactions in Voltaic Cells

Voltaic Cells Movie

The conversion of chemical energy to electrical energy in a redox reaction is called **electrochemistry**. As an example, consider the reaction of zinc metal in a copper(II) sulfate solution.

$$Zn(s) + CuSO_4(aq) \longrightarrow Cu(s) + ZnSO_4(aq)$$

Experimentally, we can physically separate the oxidation half-reaction from the reduction half-reaction. We do this by placing aqueous solutions of zinc sulfate and copper(II) sulfate in separate containers or compartments. Then, we place a Zn metal electrode in the first compartment containing $ZnSO_4$, and a Cu metal electrode in the second compartment containing $CuSO_4$. The two electrodes are connected by a wire, and so electrons are free to travel between the two compartments. Each compartment is called a **half-cell**, and together the apparatus is called an **electrochemical cell** (Figure 17.3).

It is important to note that the two compartments must be separated or there will be no flow of electrons through the wire. That is, the redox reaction occurs directly in solution, and reddish-brown copper deposits on the Zn metal.

Let's assume the zinc half-cell contains 1.00 *M* $ZnSO_4$ and the copper compartment has 1.00 *M* $CuSO_4$. Initially, the ionic concentrations are as follows.

$$\text{Zn half-cell:} \quad [Zn^{2+}] = 1.00\,M,\ [SO_4^{\ 2-}] = 1.00\,M$$
$$\text{Cu half-cell:} \quad [Cu^{2+}] = 1.00\,M,\ [SO_4^{\ 2-}] = 1.00\,M$$

As the reaction proceeds, the concentration of Zn^{2+} in the left compartment becomes greater than 1.00 *M*. The concentration of Cu^{2+} in the right compartment

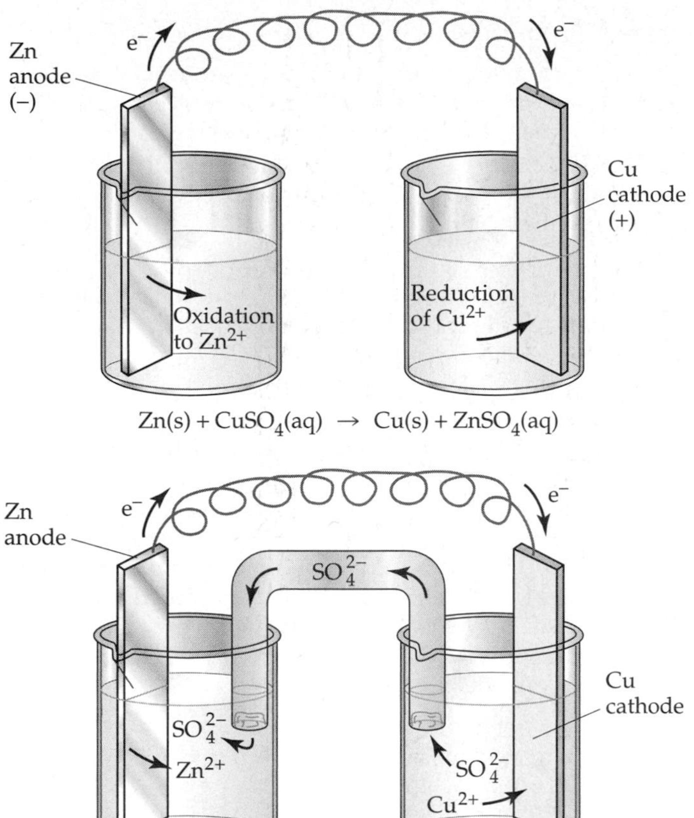

◀ **Figure 17.3 Redox Reaction in Half-Cells** In the compartment on the left, zinc metal is being oxidized. In the compartment on the right, copper ion is being reduced. This redox process allows electrons to flow from the zinc electrode to the copper electrode. As the reaction proceeds, the number of zinc ions in the left half-cell increases, and the number of copper(II) ions in the right half-cell decreases.

◀ **Figure 17.4 Voltaic Cell** If we connect the two half-cells by a salt bridge, negative sulfate ions travel from the right half-cell to the left half-cell. A salt bridge reduces the positive charge buildup in the left half-cell and the negative charge buildup in the right half-cell. Then the reaction can continue to operate spontaneously.

becomes less than 1.00 *M*. The concentration of SO_4^{2-} remains constant at 1.00 *M* in both compartments. As a result, the left half-cell develops a net positive charge as the right half-cell develops a net negative charge.

As charge develops in each compartment, the redox process comes to a halt. However, we can eliminate the charge buildup by introducing a **salt bridge**. A salt bridge placed between the compartments allows ions to travel between the two half-cells. The excess negative sulfate ions in the copper solution can move to the zinc solution, which has excess Zn^{2+}. Since a salt bridge eliminates charge buildup, the reaction can continue spontaneously. For spontaneous reactions that produce electrical energy, the cell is termed a **voltaic cell** (or galvanic cell). Figure 17.4 shows a voltaic cell based on the reaction of Zn in aqueous $CuSO_4$.

In an electrochemical cell, the electrode at which oxidation occurs is called the **anode**. The electrode at which reduction occurs is called the **cathode**. For the voltaic cell in Figure 17.4, the Zn electrode is the anode and the Cu electrode is the cathode. The following example exercise further illustrates the processes of oxidation and reduction that take place in a spontaneous electrochemical cell.

Example Exercise 17.9 • Voltaic Cells

Nickel can react with aqueous silver nitrate solution according to the following ionic equation.

$$Ni(s) + 2\,Ag^{+}(aq) \longrightarrow 2\,Ag(s) + Ni^{2+}(aq)$$

(continued)

Example Exercise 17.9 *(continued)*

Assume the half-reactions are separated into two compartments. A Ni electrode is placed in a compartment with 1.00 *M* $Ni(NO_3)_2$, and a Ag electrode is placed in a compartment with 1.00 *M* $AgNO_3$. Indicate each of the following.

(a) oxidation half-cell reaction
(b) reduction half-cell reaction
(c) anode and cathode
(d) direction of electron flow
(e) direction of NO_3^- in the salt bridge

Solution

Referring to Table 17.3, we see that Ag^+ has a higher reduction potential than Ni^{2+}. Therefore, the process is spontaneous and Ni is oxidized as Ag^+ is reduced. The two half-cell processes are

(a) Oxidation: $Ni \longrightarrow Ni^{2+} + 2\,e^-$
(b) Reduction: $Ag^+ + e^- \longrightarrow Ag$
(c) The anode is where oxidation occurs; thus, Ni is the anode.
The cathode is where reduction occurs; thus, Ag is the cathode.
(d) Since Ni loses two electrons while Ag^+ gains an electron, an electric current flows from the Ni anode to the Ag cathode. (Electrons flow toward the cathode.)
(e) As Ni is oxidized, $[Ni^{2+}]$ increases and the Ni compartment acquires a net positive charge. Simultaneously, $[Ag^+]$ decreases and the Ag compartment acquires a net negative charge. For the cell to operate continuously, we must use a salt bridge that allows NO_3^- ions to travel from the Ag compartment to the Ni compartment. (Anions flow toward the cathode.)

Self-Test Exercise

Iron can react with aqueous tin(II) sulfate solution according to the following ionic equation.

$$Fe(s) + Sn^{2+}(aq) \longrightarrow Sn(s) + Fe^{2+}(aq)$$

An Fe electrode is placed in a compartment with 1.00 *M* $FeSO_4$, and a Sn electrode is placed in another compartment with 1.00 *M* $SnSO_4$. Indicate each of the following.

(a) oxidation half-cell reaction
(b) reduction half-cell reaction
(c) anode and cathode
(d) direction of electron flow
(e) direction of NO_3^- in the salt bridge

Answers: (a) Oxidation: $Fe \longrightarrow Fe^{2+} + 2\,e^-$; (b) Reduction: $Sn^{2+} + 2\,e^- \longrightarrow Sn$; (c) Fe is the anode; Sn is the cathode; (d) Electrons flow from the Fe anode to the Sn cathode; (e) A salt bridge allows SO_4^{2-} ions to travel from the Sn compartment to the Fe anode compartment.

Note By convention, the anode in an electrochemical cell is usually shown as the left half-cell. If the anode is found on the left, electrons flow from the left half-cell to the right half-cell. To avoid a buildup of negative charge in the right half-cell, negative anions exit the right half-cell, travel through the salt bridge, and enter the left half-cell.

We should note that cations are also present in the salt bridge. Conversely, cations exit the left half-cell, travel through the salt bridge, and enter the right half-cell. In summary, anions travel toward the half-cell compartment where electrons are exiting (anode), and cations travel toward the half-cell compartment where electrons are entering (cathode).

Batteries

A **battery** is a general term for any electrochemical cell that spontaneously produces electrical energy. A battery has stored chemical energy, which can produce electrical energy measured in volts. A battery is made up of one or more voltaic cells. For example, the ordinary lead storage battery in an automobile is a series of six cells, each of which produces 2 volts. Operating together, the six cells generate 12 volts.

◀ **Figure 17.5 Common Batteries** Batteries spontaneously supply electrical energy. In the foreground, we see rechargeable nicad batteries, alkaline dry cells, and mercury dry cells. In the background is a lead storage battery.

The batteries we use in flashlights, electronic toys, and portable radios do not have electrodes immersed in an electrolyte solution. Rather, their electrodes are connected by a solid paste that can conduct electricity. A battery that does not use an electrolyte solution is called a **dry cell**. The common alkaline dry cell battery shown in Figure 17.5 was invented more than 100 years ago by the French chemist George Leclanché.

A more compact version of the dry cell is the mercury dry cell. In this battery, a small zinc cup is filled with a conducting paste of HgO and KOH. Oxidation takes place at the zinc anode, and reduction at the steel cathode. Mercury dry cells are used to power small devices such as electronic wristwatches and hearing aids.

An ordinary dry cell battery is not rechargeable, and the reactants stored in the battery are eventually depleted. However, a nicad battery, which is a special type of dry cell, can be recharged almost indefinitely. That is, electricity can be used to reverse the spontaneous reaction, and the original reactants and products are reformed. A nicad battery is so named because the redox reactions involve nickel and cadmium. Nicad batteries are used in rechargeable battery packs for calculators and portable electronic appliances. A collection of common batteries is shown in Figure 17.5.

17.7 Electrolytic Cells

Objectives
- To indicate the anode and cathode in a given electrolytic cell.
- To indicate the oxidation and reduction half-reactions in a given nonspontaneous electrochemical cell.

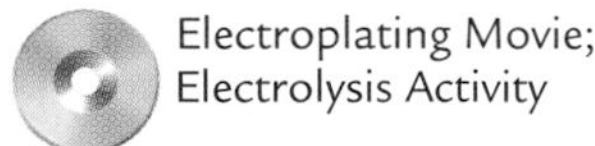

Electroplating Movie; Electrolysis Activity

So far we have discussed electrochemical cells that operate spontaneously. Equally as important are electrochemical cells that do not operate spontaneously. Although a redox reaction may be nonspontaneous, we can supply electricity and force the reaction to occur. This process is referred to as electrolysis. Moreover, if an electric current is necessary for a reaction to occur in an electrochemical cell, the cell is called an **electrolytic cell**. Thus, an electrolytic cell represents a nonspontaneous process.

Although the reaction of iron in an aqueous chromium solution is not spontaneous (Table 17.3), we can force the reaction to occur. In fact, chromium metal is commonly plated onto steel by an electrolytic method. In this process, an iron or steel object is placed in an aqueous chromium solution, and a direct electric current is applied to overcome the nonspontaneous reaction.

Many important elements, including sodium and chlorine, are produced from their compounds by electrolysis. In the preparation of sodium metal and chlorine gas, crystalline salt is first obtained by the evaporation of seawater. Next, the crystalline salt, NaCl, is placed in a crucible and heated until it melts. Then, two inert platinum metal electrodes are dipped into the molten sodium chloride and electricity is applied. The electric current forces the nonspontaneous reaction, and ionic NaCl decomposes into its elements.

Cathode reaction (reduction): $2\ Na^+ + 2\ e^- \longrightarrow 2\ Na$

Anode reaction (oxidation): $2\ Cl^- \longrightarrow Cl_2 + 2\ e^-$

We can write the overall equation for the electrolytic cell by adding the two half-reactions together. After canceling 2 electrons from each side of the equation, we have the equation for the nonspontaneous redox reaction.

$$2\ Na^+ + 2\ Cl^- \longrightarrow 2\ Na + Cl_2$$

The following example exercise illustrates nonspontaneous redox reactions that may occur in an electrolytic cell.

Example Exercise 17.10 • Electrolytic Cells

Aluminum metal is produced by passing an electric current through bauxite, Al_2O_3, dissolved in the molten mineral cryolite. Graphite rods serve as electrodes, and we can write the redox equation as follows.

$$3\ C(s) + 2\ Al_2O_3(l) \xrightarrow{\text{electricity}} 4\ Al(l) + 3\ CO_2(g)$$

Indicate each of the following for this nonspontaneous process.

(a) oxidation half-cell reaction
(b) reduction half-cell reaction
(c) anode and cathode
(d) direction of electron flow

Solution

In the equation, we see that C is oxidized to CO_2 and that Al_2O_3 is reduced to Al metal. The two half-cell processes are

(a) Oxidation: $C + 2\ O^{2-} \longrightarrow CO_2 + 4\ e^-$
(b) Reduction: $Al^{3+} + 3\ e^- \longrightarrow Al$
(c) By definition, oxidation always occurs at the anode. Thus, the anode is the C electrode where CO_2 gas is released. By definition, reduction occurs at the cathode. Thus, the cathode is the C electrode where Al metal is produced.
(d) The electrons flow from the anode, where CO_2 gas is released, to the cathode, where Al metal is produced.

Self-Test Exercise

Magnesium metal is produced by passing an electric current through molten $MgCl_2$ obtained from evaporated seawater. Carbon and platinum rods serve as electrodes, and we can write the redox equation as follows.

$$C(s) + 2\ MgCl_2(l) \xrightarrow{\text{electricity}} 2\ Mg(l) + CCl_4(g)$$

Chemistry Connection · Electric Vehicles

What percent of the General Motors EV1 weight is accounted for by the lead–acid battery pack?

In an effort to manufacture zero-emission vehicles, automobile companies are exploring a number of technologies. A German manufacturer is developing a fuel cell that combines hydrogen and oxygen gases directly and produces only electrical energy and pure water. A Japanese manufacturer is working on a process that utilizes lightweight voltaic cells to produce electrical energy. Recently, an American manufacturer offered the first commercial electric vehicle, the EV1.

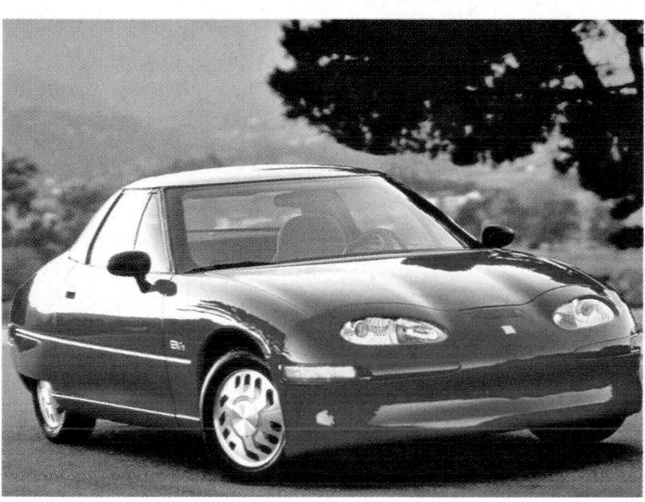

▲ The General Motors Electric Vehicle, EV1

The EV1 does not have an ordinary gasoline combustion engine. In fact, it does not have an exhaust system because it produces no emissions. The EV1 is powered by a lead–acid battery pack that contains 26 voltaic cells. The lead–acid batteries produce electrical energy by the following redox reaction.

$$Pb(s) + PbO_2(s) + 2\,H^+(aq) + 2\,HSO_4^-(aq) \longrightarrow 2\,PbSO_4(s) + 2\,H_2O(l)$$

Since a car uses most of its energy to move through the wind, the EV1 was designed in the shape of a raindrop. To make the car as aerodynamic as possible, the rear of the car is more narrow, the underside is enclosed, and the car is lowered to 5 inches. In addition, the body panels are made from a slippery composite material that reduces drag.

The EV1 uses few moving parts and is 100% recyclable. It has a rigid aluminum frame that weighs a mere 290 pounds. The total weight is 2970 pounds, but the battery pack accounts for 1175 pounds. To minimize energy loss due to friction, the EV1 is equipped with special low-rolling resistance tires. The tires self-seal punctures, thus saving weight by eliminating a jack and a spare tire.

The EV1 holds the electric car speed record at 183 mph, but production vehicles are electronically limited to 80 mph. The EV1 accelerates from 0 to 60 mph in less than 9 seconds. The estimated range between charges is 79 miles although the range varies greatly depending on driving conditions. Using the air conditioning and other accessories limits the range severely. The battery pack can be fully recharged in 3 hours at a cost of about a dollar.

The battery pack comprises approximately 40% of the total weight of the EV1.

Indicate each of the following for this nonspontaneous process:

(a) oxidation half-cell reaction
(b) reduction half-cell reaction
(c) anode and cathode
(d) direction of electron flow

Answers: (a) Oxidation: $C + 4\,Cl^- \longrightarrow CCl_4 + 4\,e^-$; (b) Reduction: $Mg^{2+} + 2\,e^- \longrightarrow Mg$; (c) The C electrode is the anode; the Pt electrode is the cathode; (d) The electrons flow from the C anode to the Pt cathode.

Summary 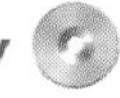

Section 17.1 An oxidation–reduction reaction is characterized by the transfer of electrons. When a substance undergoes oxidation, its **oxidation number** increases. Conversely, when a substance undergoes reduction, its oxidation number decreases. When Cu is oxidized to Cu^{2+}, its oxidation number changes from 0 to +2; when Cu^{2+} is reduced to Cu, its oxidation number changes from +2 to 0. Transition metals, as well as nonmetals, demonstrate a variety of oxidation numbers. For example, Mn has oxidation numbers of +7, +6, +4, +3, +2, and 0. The oxidation numbers of Cl include +7, +5, +3, +1, 0, and −1.

▲ **Reduction of a Rose** A red rose (top) undergoes a redox reaction with colorless SO_2 gas. After a few minutes, the color of the rose changes from red to pink (bottom).

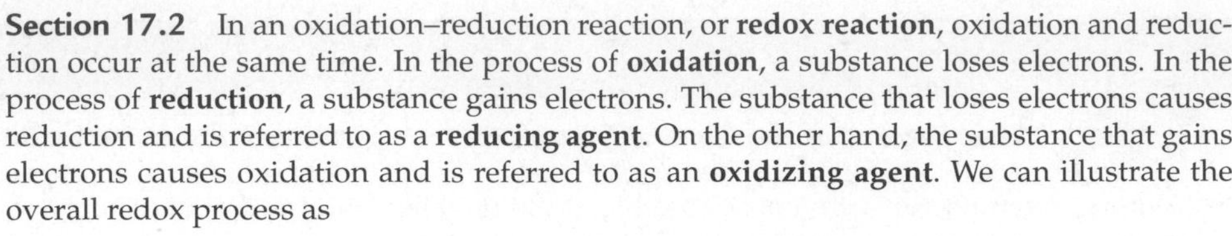

Section 17.2 In an oxidation–reduction reaction, or **redox reaction**, oxidation and reduction occur at the same time. In the process of **oxidation**, a substance loses electrons. In the process of **reduction**, a substance gains electrons. The substance that loses electrons causes reduction and is referred to as a **reducing agent**. On the other hand, the substance that gains electrons causes oxidation and is referred to as an **oxidizing agent**. We can illustrate the overall redox process as

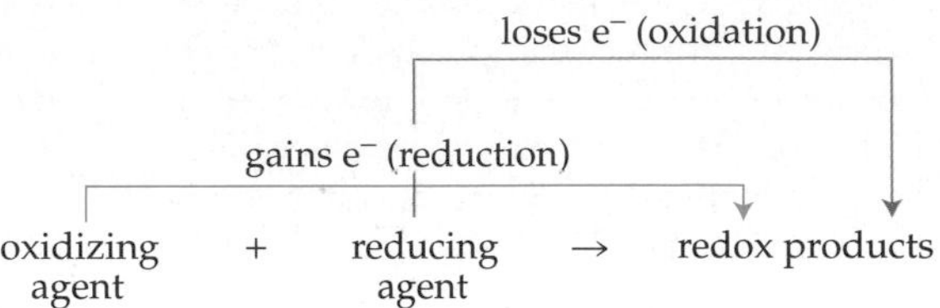

Section 17.3 To balance a redox equation, we keep track of the changes in oxidation number. This method allows us to balance a redox equation by determining changes in oxidation number for the oxidizing agent and reducing agents. If the oxidizing agent gains 2 electrons and the reducing agent loses 1 electron, the reducing agent must be multiplied by 2 to balance the electrons. The remaining elements in the equation are then balanced by inspection.

Section 17.4 Redox reactions can also be balanced by the half-reaction method. First, a balanced **half-reaction** is written for the oxidation process. Second, a balanced half-reaction is written for the reduction process. Each half-reaction is then multiplied by the number of electrons lost or gained in the opposite half-reaction. Thus, the number of electrons lost in oxidation equals the number of electrons gained in reduction. Finally, the two half-reactions are added together to give a balanced equation for the overall redox reaction.

Section 17.5 Half-reactions can be arranged according to their ability to undergo reduction. This tendency, called the **reduction potential**, is listed in Table 17.3 for selected oxidizing and reducing agents. A substance that is easily reduced is a strong oxidizing agent, and a substance that is easily oxidized is a strong reducing agent. For a redox reaction to be spontaneous, the oxidizing agent must be listed higher in the table than the reducing agent.

Section 17.6 **Electrochemistry** is the study of the interconversion of chemical and electrical energy. In an **electrochemical cell**, a chemical reaction produces electrical energy. An electrochemical cell is composed of two **half-cell** compartments. Oxidation occurs in one half-cell, and reduction in the other. If the redox reaction occurs spontaneously, the cell is referred to as a **voltaic cell**. The **anode** is the electrode where oxidation occurs, and the **cathode** is the electrode where reduction occurs. In the operation of an electrochemical cell, the electrodes are connected by a wire and the two half-cells are joined by a **salt bridge** that allows ions to travel from one compartment to the other. A **battery** is any device that supplies electricity and is composed of one or more voltaic cells. A **dry cell** is a special type of battery in which there is no conducting electrolyte solution.

Section 17.7 Whereas the redox reaction is spontaneous in a voltaic cell, the redox reaction is not spontaneous in an **electrolytic cell**. In an electrolytic cell, it is necessary to supply electrical energy to produce a chemical reaction. Active metals, such as Na, K, and Ca, are obtained from electrolytic cells. In addition, chlorine gas and bromine liquid are produced commercially by passing a direct electric current through an electrolytic cell.

Key Concepts*

1. What is the oxidation number of bromine in each of the following?
   (a) Br_2 (b) HBr
   (c) $CaBr_2$ (d) $HBrO_2$
2. Identify the oxidizing and reducing agents in the following redox reaction.

$$Co(s) + 2\,HCl(aq) \longrightarrow CoCl_2(aq) + H_2(g)$$

3. Identify the oxidizing and reducing agents in the following redox reaction.

$$F_2(g) + 2\,Cl^-(aq) \longrightarrow 2\,F^-(aq) + Cl_2(g)$$

4. Balance the following redox reaction using the oxidation number method.

$$Al_2O_3(s) + Cl_2(g) \longrightarrow AlCl_3(aq) + O_2(g)$$

5. Balance the following redox reaction in an acidic solution using the half-reaction method.

$$MnO_4^-(aq) + SO_3^{2-}(aq) \longrightarrow Mn^{2+}(aq) + SO_4^{2-}(aq)$$

6. Balance the following redox reaction in a basic solution using the half-reaction method.

$$MnO_4^-(aq) + SO_3^{2-}(aq) \longrightarrow MnO_2(s) + SO_4^{2-}(aq)$$

7. Refer to Table 17.3 and predict which of the following metals spontaneously reacts in an aqueous $FeSO_4$ solution.
   (a) Ag (b) Ni
   (c) Zn (d) Al
8. Refer to the following illustration of a voltaic cell and identify the following.
   (a) oxidation half-cell reaction
   (b) reduction half-cell reaction
   (c) anode and cathode
   (d) direction of electron flow
   (e) direction of SO_4^{2-} in the salt bridge

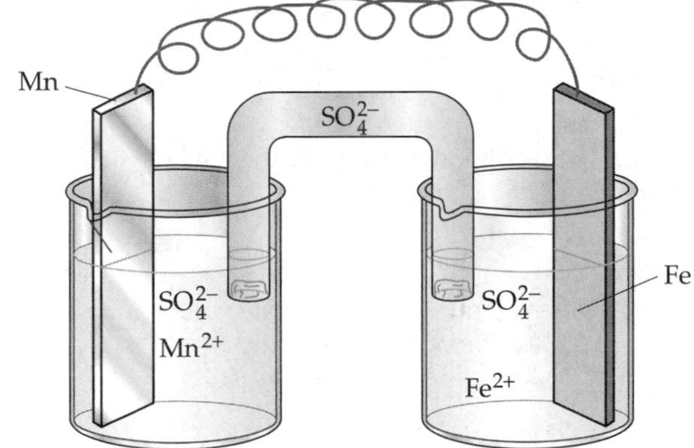

Key Terms†

Select the key term below that corresponds to each of the following definitions.

_____ 1. a value assigned to an atom in a substance that indicates whether the atom is electron poor or electron-rich compared to a free atom
_____ 2. a chemical reaction in which an electron transfer takes place
_____ 3. a chemical process in which a substance loses electrons
_____ 4. a chemical process in which a substance gains electrons
_____ 5. the substance undergoing reduction in a redox reaction
_____ 6. the substance undergoing oxidation in a redox reaction
_____ 7. a reaction that represents either an oxidation or a reduction process
_____ 8. the relative ability of a substance to undergo reduction
_____ 9. the study of the interconversion of chemical and electrical energy
_____ 10. a general term for an apparatus containing two electrodes in separate compartments connected by a wire
_____ 11. a portion of an electrochemical cell having a single electrode where either oxidation or reduction occurs.
_____ 12. a porous device that allows ions to travel between two half-cells to maintain an ionic charge balance in each compartment
_____ 13. an electrochemical cell in which a spontaneous redox reaction occurs and generates electrical energy
_____ 14. an electrochemical cell in which a nonspontaneous redox reaction occurs as a result of the input of direct electric current
_____ 15. the electrode in an electrochemical cell at which oxidation occurs

(a) anode (*Sec. 17.6*)
(b) battery (*Sec. 17.6*)
(c) cathode (*Sec. 17.6*)
(d) dry cell (*Sec. 17.6*)
(e) electrochemical cell (*Sec. 17.6*)
(f) electrochemistry (*Sec. 17.6*)
(g) electrolytic cell (*Sec. 17.7*)
(h) half-cell (*Sec. 17.6*)
(i) half-reaction (*Sec. 17.4*)
(j) oxidation (*Sec. 17.2*)
(k) oxidation number (*Sec. 17.1*)
(l) oxidizing agent (*Sec. 17.2*)
(m) redox reaction (*Sec. 17.2*)
(n) reducing agent (*Sec. 17.2*)
(o) reduction (*Sec. 17.2*)
(p) reduction potential (*Sec. 17.5*)
(q) salt bridge (*Sec. 17.6*)
(r) voltaic cell (*Sec. 17.6*)

*Answers to Key Concepts are in Appendix H.
†Answers to Key Terms are in Appendix I.

_______ 16. the electrode in an electrochemical cell at which reduction occurs

_______ 17. a general term for any electrochemical cell that produces electrical energy

_______ 18. an electrochemical cell in which the anode and cathode reactions do not take place in an aqueous solution

Exercises‡ 

Oxidation Numbers (Sec. 17.1)

1. State the oxidation number for each of the following metals in the free state.
   (a) Mg (b) Mn
   (c) K (d) Zn
2. State the oxidation number for each of the following nonmetals.
   (a) H_2 (b) He
   (c) P_4 (d) I_2
3. State the oxidation number for the metal in each of the following monoatomic cations.
   (a) Sr^{2+} (b) Sc^{3+}
   (c) Ti^{4+} (d) Ag^+
4. State the oxidation number for the nonmetal in each of the following monoatomic anions.
   (a) F^- (b) H^-
   (c) P^{3-} (d) Te^{2-}
5. Calculate the oxidation number for silicon in the following compounds.
   (a) SiO_2 (b) Si_2H_6
   (c) Si_3N_4 (d) $CaSiO_3$
6. Calculate the oxidation number for nitrogen in the following compounds.
   (a) NH_3 (b) N_2O_4
   (c) Li_3N (d) KNO_3
7. Calculate the oxidation number for carbon in the following polyatomic ions.
   (a) CO_3^{2-} (b) HCO_3^-
   (c) CN^- (d) CNO^-
8. Calculate the oxidation number for sulfur in the following polyatomic ions.
   (a) SO_3^{2-} (b) HSO_4^-
   (c) HS^- (d) $S_2O_8^{2-}$

Oxidation–Reduction Reactions (Sec. 17.2)

9. Supply the term that corresponds to each of the following.
   (a) a redox process characterized by electron loss
   (b) a redox process characterized by electron gain
10. Supply the term that corresponds to each of the following.
   (a) a substance that increases its oxidation number in a redox reaction
   (b) a substance that decreases its oxidation number in a redox reaction
11. Indicate the substances undergoing oxidation and reduction in the following redox reactions.
   (a) $Mn(s) + O_2(g) \longrightarrow MnO_2(s)$
   (b) $S(s) + O_2(g) \longrightarrow SO_2(g)$
12. Indicate the substances undergoing oxidation and reduction in the following redox reactions.
   (a) $Cd(s) + F_2(g) \longrightarrow CdF_2(s)$
   (b) $Sr(s) + Cl_2(g) \longrightarrow SrCl_2(s)$
13. Indicate the oxidizing and reducing agents in the following redox reactions.
   (a) $CuO(s) + H_2(g) \longrightarrow Cu(s) + H_2O(l)$
   (b) $PbO(s) + CO(g) \longrightarrow Pb(s) + CO_2(g)$
14. Indicate the oxidizing and reducing agents in the following redox reactions.
   (a) $Ca(s) + 2\,H_2O(l) \longrightarrow Ca(OH)_2(aq) + H_2(g)$
   (b) $Mg(s) + 2\,HCl(aq) \longrightarrow MgCl_2(aq) + H_2(g)$
15. Indicate the substances undergoing oxidation and reduction in the following ionic redox reactions.
   (a) $Al(s) + Cr^{3+}(aq) \longrightarrow Al^{3+}(aq) + Cr(s)$
   (b) $F_2(g) + 2\,Cl^-(aq) \longrightarrow 2\,F^-(aq) + Cl_2(g)$
16. Indicate the substances undergoing oxidation and reduction in the following ionic redox reactions.
   (a) $2\,Fe^{3+}(aq) + SO_3^{2-}(aq) \longrightarrow 2\,Fe^{2+}(aq) + SO_4^{2-}(aq)$
   (b) $Cl_2(g) + 2\,I^-(aq) \longrightarrow 2\,Cl^-(aq) + I_2(s)$
17. Indicate the oxidizing and reducing agents in the following redox reactions.
   (a) $Cr^{2+}(aq) + AgI(s) \longrightarrow Cr^{3+}(aq) + Ag(s) + I^-(aq)$
   (b) $Sn^{2+}(aq) + Hg^{2+}(aq) \longrightarrow Sn^{4+}(aq) + Hg_2^{2+}(aq)$
18. Indicate the oxidizing and reducing agents in the following redox reactions.
   (a) $H_2O_2(aq) + SO_3^{2-}(aq) \longrightarrow H_2O(l) + SO_4^{2-}(aq)$
   (b) $IO_3^-(aq) + Cu^+(aq) \longrightarrow Cu^{2+}(aq) + I_2(s)$

Balancing Redox Equations: Oxidation Number Method (Sec. 17.3)

19. Is it always true that the electron loss by oxidation equals the electron gain by reduction in a balanced redox equation?
20. Is it always true that the total ionic charge on the reactants equals the total ionic charge on the products in a balanced redox equation?
21. Write a balanced equation for each of the following redox reactions using the oxidation number method.

‡Answers to odd-numbered Exercises are in Appendix J.

(a) $Br_2(g) + NaI(aq) \longrightarrow I_2(s) + NaBr(aq)$
(b) $PbS(s) + O_2(g) \longrightarrow PbO(aq) + SO_2(g)$

22. Write a balanced equation for each of the following redox reactions using the oxidation number method.
(a) $Cl_2(g) + KI(aq) \longrightarrow I_2(s) + KCl(aq)$
(b) $Fe_2O_3(s) + CO(g) \longrightarrow Fe(s) + CO_2(g)$

23. Write a balanced equation for each of the following redox reactions using the oxidation number method.
(a) $MnO_4^-(aq) + I^-(aq) + H^+(aq) \longrightarrow Mn^{2+}(aq) + I_2(s) + H_2O(l)$
(b) $Cu(s) + H^+(aq) + SO_4^{2-}(aq) \longrightarrow Cu^{2+}(aq) + SO_2(g) + H_2O(l)$

24. Write a balanced equation for each of the following redox reactions using the oxidation number method.
(a) $Fe^{2+}(aq) + H_2O_2(aq) + H^+(aq) \longrightarrow Fe^{3+}(aq) + H_2O(l)$
(b) $Cr_2O_7^{2-}(aq) + Br^-(aq) + H^+(aq) \longrightarrow Cr^{3+}(aq) + Br_2(l) + H_2O(l)$

Balancing Redox Equations: Half-Reaction Method (Sec. 17.4)

25. Write a balanced half-reaction for each of the following in an acidic solution.
(a) $SO_2(g) \longrightarrow SO_4^{2-}(aq)$
(b) $AsO_3^{3-}(aq) \longrightarrow AsO_3^-(aq)$

26. Write a balanced half-reaction for each of the following in an acidic solution.
(a) $BrO_3^-(aq) \longrightarrow Br_2(l)$
(b) $H_2O_2(aq) \longrightarrow H_2O(l)$

27. Write a balanced half-reaction for each of the following in a basic solution.
(a) $ClO^-(aq) \longrightarrow Cl^-(aq)$
(b) $MnO_4^-(aq) \longrightarrow MnO_2(s)$

28. Write a balanced half-reaction for each of the following in a basic solution.
(a) $Ni(OH)_2(s) \longrightarrow NiO_2(s)$
(b) $NO_2^-(aq) \longrightarrow N_2O(g)$

29. Write a balanced equation for each of the following redox reactions in an acidic solution using the half-reaction method.
(a) $Zn(s) + NO_3^-(aq) \longrightarrow Zn^{2+}(aq) + NO(g)$
(b) $Mn^{2+}(aq) + BiO_3^-(aq) \longrightarrow MnO_4^-(aq) + Bi^{3+}(aq)$

30. Write a balanced equation for each of the following redox reactions in an acidic solution using the half-reaction method.
(a) $Sn^{2+}(aq) + IO_3^-(aq) \longrightarrow Sn^{4+}(aq) + I_2(s)$
(b) $AsO_3^{3-}(aq) + Br_2(l) \longrightarrow AsO_4^{3-}(aq) + Br^-(aq)$

31. Write a balanced equation for each of the following redox reactions in a basic solution using the half-reaction method.
(a) $MnO_4^-(aq) + S^{2-}(aq) \longrightarrow MnO_2(s) + S(s)$
(b) $Cu(s) + ClO^-(aq) \longrightarrow Cu^{2+}(aq) + Cl^-(aq)$

32. Write a balanced equation for each of the following redox reactions in a basic solution using the half-reaction method.
(a) $Cl_2(g) + BrO_2^-(aq) \longrightarrow Cl^-(aq) + BrO_3^-(aq)$
(b) $MnO_2(s) + O_2(g) \longrightarrow MnO_4^-(aq) + H_2O(l)$

33. Chlorine can undergo a redox reaction in which it is simultaneously oxidized and reduced. Write a balanced equation for the following in an acidic solution.

$$Cl_2(aq) \longrightarrow Cl^-(aq) + HOCl(aq)$$

34. Nitrous acid, HNO_2, can undergo a redox reaction in which it is simultaneously oxidized and reduced. Write a balanced equation for the following in an acidic solution.

$$HNO_2(aq) \longrightarrow NO_3^-(aq) + NO(g)$$

35. Chlorine can undergo a redox reaction in which it is simultaneously oxidized and reduced. Write a balanced equation for the following in a basic solution.

$$Cl_2(aq) \longrightarrow ClO_2^-(aq) + Cl^-(aq)$$

36. Sulfur can undergo a redox reaction in which it is simultaneously oxidized and reduced. Write a balanced equation for the following in a basic solution.

$$S(s) \longrightarrow SO_3^{2-}(aq) + S^{2-}(aq)$$

Predicting Spontaneous Redox Reactions (Sec. 17.5)

37. Refer to Table 17.3 and indicate which substance in each of the following pairs has the greater tendency to be reduced.
(a) $Pb^{2+}(aq)$ or $Zn^{2+}(aq)$ (b) $Fe^{3+}(aq)$ or $Al^{3+}(aq)$
(c) $Ag^+(aq)$ or $I_2(s)$ (d) $Cu^{2+}(aq)$ or $Br_2(l)$

38. Refer to Table 17.3 and indicate which substance in each of the following pairs has the greater tendency to be oxidized.
(a) Li(s) or K(s) (b) Al(s) or Mg(s)
(c) $Fe^{2+}(aq)$ or $I^-(aq)$ (d) $Br^-(aq)$ or $Cl^-(aq)$

39. Refer to Table 17.3 and indicate which substance in each of the following pairs is the stronger oxidizing agent.
(a) $F_2(g)$ or $Cl_2(g)$ (b) $Ag^+(aq)$ or $Br_2(l)$
(c) $Cu^{2+}(aq)$ or $H^+(aq)$ (d) $Mg^{2+}(aq)$ or $Mn^{2+}(aq)$

40. Refer to Table 17.3 and indicate which substance in each of the following pairs is the stronger reducing agent.
(a) Cu(s) or Cr(s) (b) $H_2(g)$ or Cu(s)
(c) Cu(s) or $I^-(aq)$ (d) $Cl^-(aq)$ or $H_2(g)$

41. Refer to Table 17.3 and state whether each of the following reactions is spontaneous or nonspontaneous.
(a) $Br_2(g) + LiF(aq) \longrightarrow F_2(s) + LiBr(aq)$
(b) $Al(NO_3)_3(aq) + Mn(s) \longrightarrow Mn(NO_3)_2(aq) + Al(s)$

42. Refer to Table 17.3 and state whether each of the following reactions is spontaneous or nonspontaneous.
(a) $Cr(s) + HCl(aq) \longrightarrow CrCl_3(aq) + H_2(g)$
(b) $FeCl_3(aq) + NaI(aq) \longrightarrow FeCl_2(aq) + NaCl(aq) + I_2(s)$

43. Refer to Table 17.3 and state whether each of the following ionic redox reactions is spontaneous or nonspontaneous.
(a) $Mg(s) + Sn^{2+}(aq) \longrightarrow Mg^{2+}(aq) + Sn(s)$
(b) $H^+(aq) + Ni(s) \longrightarrow H_2(g) + Ni^{2+}(aq)$
44. Refer to Table 17.3 and state whether each of the following ionic redox reactions is spontaneous or nonspontaneous.
(a) $H^+(aq) + I^-(aq) \longrightarrow H_2(g) + I_2(s)$
(b) $Fe^{3+}(aq) + I^-(aq) \longrightarrow Fe^{2+}(aq) + I_2(s)$
45. Given the following redox reactions, list A, B, and C according to which is higher in the table of reduction potentials.

$$A(s) + B^+(aq) \longrightarrow B(s) + A^+(aq)$$
$$B(s) + C^+(aq) \longrightarrow C(s) + B^+(aq)$$

46. Given the following redox reactions, list X, Y, and Z according to which is higher in the table of reduction potentials.

$$X_2(aq) + 2\,Y^-(aq) \longrightarrow Y_2(aq) + 2\,X^-(aq)$$
$$Y_2(aq) + 2\,Z^-(aq) \longrightarrow Z_2(aq) + 2\,Y^-(aq)$$

Voltaic Cells (Sec. 17.6)

47. Sketch the following voltaic cell showing the two compartments, the anode and cathode, the wire connecting the two electrodes, and a salt bridge.

$$Ni(s) + 2\,AgNO_3(aq) \longrightarrow 2\,Ag(s) + Ni(NO_3)_2(aq)$$

48. Sketch the following voltaic cell showing the two compartments, the Pt anode and Pt cathode, the wire connecting the two electrodes, and a salt bridge.

$$Cl_2(g) + 2\,KBr(aq) \longrightarrow 2\,KCl(aq) + Br_2(l)$$

49. Diagram the direction of electron flow and the movement of nitrate ions in the salt bridge that connects the two half-cells in Exercise 47.
50. Diagram the direction of electron flow and the movement of potassium ions in the salt bridge that connects the two half-cells in Exercise 48.
51. The spontaneous redox reaction of tin and copper(II) sulfate solution occurs according to the following ionic equation.

$$Sn(s) + Cu^{2+}(aq) \longrightarrow Cu(s) + Sn^{2+}(aq)$$

Assume the half-reactions are separated into two compartments. A tin electrode is immersed in 1.00 *M* $SnSO_4$, and a copper electrode is placed in 1.00 *M* $CuSO_4$. Indicate each of the following in this voltaic cell.
(a) oxidation half-cell reaction
(b) reduction half-cell reaction
(c) anode and cathode
(d) direction of electron flow
(e) direction of SO_4^{2-} in the salt bridge
52. The spontaneous redox reaction of manganese and iron(III) chloride solution occurs according to the following ionic equation.

$$Mn(s) + 2\,Fe^{3+}(aq) \longrightarrow Mn^{2+}(aq) + 2\,Fe^{2+}(aq)$$

Assume the half-reactions are separated into two compartments. A manganese electrode is immersed in 1.00 *M* $MnCl_2$, and a platinum electrode is placed in 1.00 *M* $FeCl_3$. Indicate each of the following in this voltaic cell.
(a) oxidation half-cell reaction
(b) reduction half-cell reaction
(c) anode and cathode
(d) direction of electron flow
(e) direction of Cl^- in the salt bridge
53. The spontaneous redox reaction of magnesium and manganese (II) nitrate solution occurs according to the following equation.

$$Mg(s) + Mn(NO_3)_2(aq) \longrightarrow Mn(s) + Mg(NO_3)_2(aq)$$

Assume the half-reactions are separated into two compartments. A magnesium electrode is immersed in 1.00 *M* $Mg(NO_3)_2$, and a manganese electrode is placed in 1.00 *M* $Mn(NO_3)_2$. Indicate each of the following in this voltaic cell.
(a) oxidation half-cell reaction
(b) reduction half-cell reaction
(c) anode and cathode
(d) direction of electron flow
(e) direction of NO_3^- in the salt bridge
54. The spontaneous redox reaction of hydrogen gas and iodine occurs according to the following equation.

$$H_2(g) + I_2(s) \longrightarrow 2\,H^+(aq) + 2\,I^-(aq)$$

Assume the half-reactions are separated into two compartments. A hydrogen gas Pt electrode is immersed in 1.00 *M* HI, and a second Pt electrode is placed in 1.00 *M* $KI–I_2$ solution. Indicate each of the following in this voltaic cell.
(a) oxidation half-cell reaction
(b) reduction half-cell reaction
(c) anode and cathode
(d) direction of electron flow
(e) direction of I^- in the salt bridge

Electrolytic Cells (Sec. 17.7)

55. Sketch the following electrolytic cell showing the two compartments, the anode and cathode, the wire connecting the two electrodes, and a salt bridge.

$$Ni(s) + Cd(NO_3)_2(aq) \longrightarrow Cd(s) + Ni(NO_3)_2(aq)$$

56. Sketch the following electrolytic cell showing the two compartments, the anode and cathode, the wire connecting the two electrodes, and a salt bridge.

$$Cu(s) + 2\,HBr(aq) \longrightarrow CuBr_2(aq) + H_2(g)$$

57. Diagram the direction of electron flow and the movement of nitrate ions in the salt bridge that connects the two half-cells in Exercise 55.

58. Diagram the direction of electron flow and the movement of bromide ions in the salt bridge that connects the two half-cells in Exercise 56.

59. The nonspontaneous redox reaction of nickel and iron(II) sulfate solution occurs according to the following ionic equation.

$$Ni(s) + Fe^{2+}(aq) \longrightarrow Fe(s) + Ni^{2+}(aq)$$

Assume the half-reactions are separated into two compartments. A nickel electrode is immersed in 1.00 *M* $NiSO_4$, and an iron electrode in 1.00 *M* $FeSO_4$. Indicate each of the following in this electrolytic cell.

(a) oxidation half-cell reaction
(b) reduction half-cell reaction
(c) anode and cathode
(d) direction of electron flow
(e) direction of SO_4^{2-} in the salt bridge

60. The nonspontaneous redox reaction of lead and zinc nitrate solution occurs according to the following ionic equation.

$$Pb(s) + Zn^{2+}(aq) \longrightarrow Zn(s) + Pb^{2+}(aq)$$

Assume the half-reactions are separated into two compartments. A zinc electrode is immersed in 1.00 *M* $Zn(NO_3)_2$, and a lead electrode in 1.00 *M* $Pb(NO_3)_2$. Indicate each of the following in this electrolytic cell.

(a) oxidation half-cell reaction
(b) reduction half-cell reaction
(c) anode and cathode
(d) direction of electron flow
(e) direction of NO_3^- in the salt bridge

61. The nonspontaneous redox reaction of chromium and aluminum acetate solution occurs according to the following equation.

$$Cr(s) + Al(C_2H_3O_2)_3(aq) \longrightarrow Al(s) + Cr(C_2H_3O_2)_3(aq)$$

Assume the half-reactions are separated into two compartments. An aluminum electrode is immersed in 1.00 *M* $Al(C_2H_3O_2)_3$, and a chromium electrode in 1.00 *M* $Cr(C_2H_3O_2)_3$. Indicate each of the following in this electrolytic cell.

(a) oxidation half-cell reaction
(b) reduction half-cell reaction
(c) anode and cathode
(d) direction of electron flow
(e) direction of $C_2H_3O_2^-$ in the salt bridge

62. The nonspontaneous redox reaction of chlorine gas and sodium fluoride occurs according to the following equation.

$$Cl_2(g) + 2\,NaF(s) \longrightarrow 2\,NaCl(aq) + F_2(g)$$

Assume the half-reactions are separated into two compartments. A chlorine gas Pt electrode is immersed in 1.00 *M* NaCl, and a second Pt electrode in 1.00 *M* NaF. Indicate each of the following in this electrolytic cell.

(a) oxidation half-cell reaction
(b) reduction half-cell reaction
(c) anode and cathode
(d) direction of electron flow
(e) direction of Na^+ in the salt bridge

General Exercises

63. Calculate the oxidation number for sulfur in sodium thiosulfate, $Na_2S_2O_3$.

64. Calculate the oxidation number for sulfur in the tetrathionate ion, $S_4O_6^{2-}$.

65. The redox reaction between cobalt metal and aqueous mercuric nitrate produces cobalt(II) nitrate and droplets of liquid mercury metal. Write a net ionic equation for this reaction.

66. The redox reaction between liquid mercury and aqueous gold(III) acetate produces mercury(I) acetate and gold metal. Write a net ionic equation for this reaction.

67. The redox reaction between zinc metal and sulfuric acid produces zinc sulfate and hydrogen gas. Write a net ionic equation for this reaction.

68. The redox reaction between potassium metal and water produces potassium hydroxide and hydrogen gas. Write a net ionic equation for this reaction.

69. Write a balanced equation for the reaction of a copper penny in a dilute nitric acid solution.

$$Cu(s) + NO_3^-(aq) \longrightarrow Cu^{2+}(aq) + NO(g)$$

70. Write a balanced equation for the reaction of ethyl alcohol, C_2H_5OH, and acidic potassium dichromate solution.

$$C_2H_5OH(aq) + Cr_2O_7^{2-}(aq) \longrightarrow HC_2H_3O_2(aq) + Cr^{3+}(aq) + H_2O(l)$$

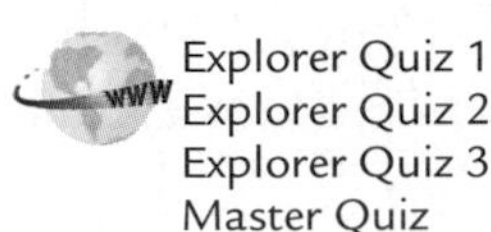

Explorer Quiz 1
Explorer Quiz 2
Explorer Quiz 3
Master Quiz

CHAPTER 18

Nuclear Chemistry

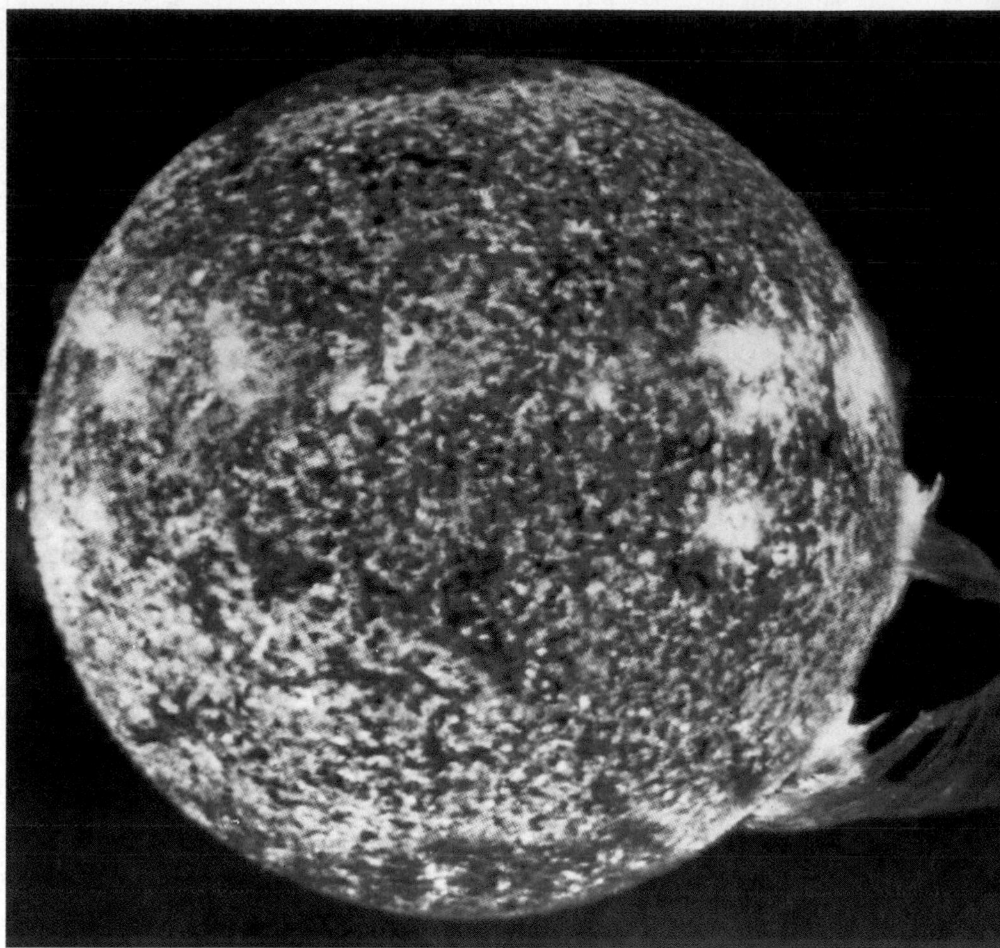

▲ The interior of the Sun acts as a nuclear reactor to fuse hydrogen nuclei at temperatures above 10 million K. What element is produced as a result of this fusion reaction?

Although nuclear chemistry is a topic that is closely associated with the material on atomic theory, this topic can be taken up at any point in an introductory chemistry course following Chapter 5.

The tremendous energy requirements of the industrial nations of the world have created a need to develop additional energy sources. To compensate for declining oil and natural gas resources, there is great interest in developing alternative energy sources. Nuclear energy is one possibility. The energy released from the splitting of atomic nuclei, most notably uranium nuclei, is called nuclear fission energy, or simply nuclear energy. Although the United States has enough uranium to meet its nuclear energy needs through the middle of the twenty-first century, the threat of nuclear accident and the disposal of radioactive waste are problems of public concern.

Interestingly, the United States uses less nuclear energy than France or Sweden. In the United States, about 20% of all energy comes from nuclear fission reactors. And in Canada, England, Germany, and Japan, the percentage is comparable to that in the United States. In France and Sweden, however, 50–70% of all energy comes from nuclear fission.

18.1 Natural Radioactivity

Objective · **To state the properties of alpha, beta, and gamma radiation.**

In 1896 the French physicist Henri Becquerel (1852–1908) noticed a fogged image on a photographic plate he had just removed from its heavy, black protective paper. Moreover, the image was that of a uranium crystal that had been resting on the paper wrapped around the photographic plate. Becquerel concluded that the uranium crystal had given off high-energy radiation to produce the fogged image. After further experiments, he discovered that the uranium crystal emitted small, charged particles in addition to high-energy radiation.

The Becquerel rays were of immediate interest to the French physicist Pierre Curie (1859–1906) and his soon-to-be wife, Marie Sklodowska (1867–1934). In fact, it was Marie who coined the name **radioactivity** for Becquerel rays. In 1898 the Curies discovered two other radioactive elements, polonium and radium. However, it took 4 years of meticulous labor to isolate a single gram of radium from 8 tons of uranium ore. In 1903 Pierre and Marie Curie shared the Nobel prize in physics with Becquerel for their work with radioactivity.

▲ **Marie Curie** In 1903, Marie Curie shared the Nobel prize in physics with her husband and Henri Becquerel. In 1911 she won an unshared Nobel prize in chemistry.

By 1903 the British physicist Ernest Rutherford had identified and named three different types of radioactivity. One type, an alpha ray, consists of a stream of positive particles. A second type, a beta ray, consists of negative particles. The third type, a gamma ray, does not consist of particles but rather is high-energy light similar to X-rays. Rutherford was able to separate alpha, beta, and gamma rays experimentally by passing the radiation from a radioactive sample through an electric field. The apparatus he used is illustrated in Figure 18.1.

Separation of Alpha, Beta, and Gamma Rays Movie

After studying the three types of radiation, Rutherford identified alpha particles as helium nuclei. That is, an **alpha particle** (symbol α) is identical to a helium nucleus containing two protons and two neutrons. He found that a **beta particle** (symbol β) is identical to an electron. A **gamma ray** (symbol γ) is a stream of very-high-energy photons. Table 18.1 lists the properties of alpha, beta, and gamma radiation.

Separation of Alpha, Beta, and Gamma Rays

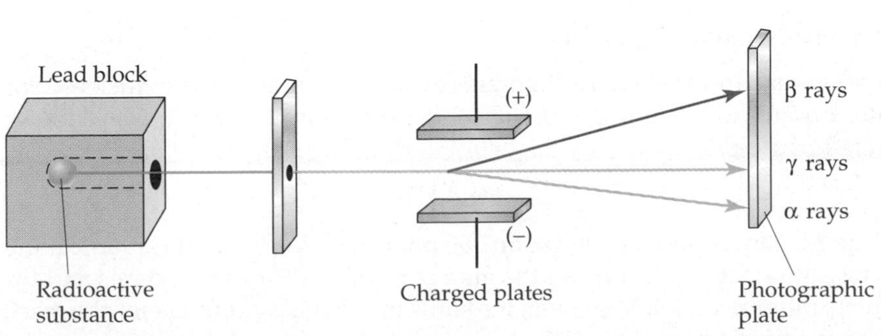

◀ **Figure 18.1 Separation of Alpha, Beta, and Gamma Rays** Notice that the alpha ray is deflected toward the negative electrode, and the beta ray toward the positive electrode. This indicates that the alpha ray is positively charged and the beta ray is negatively charged. The gamma ray is not affected because it has no charge.

Table 18.1 Properties of Nuclear Radiation

Radiation	Identity	Approximate Velocity*	Shielding Required	Penetrating Power
alpha, α	helium nucleus	$\leq 10\%\ c$	paper, clothing	low, stopped by the skin
beta, β	electron	$\leq 90\%\ c$	30 cm wood, aluminum foil	medium, ~1 cm of flesh
gamma, γ	high-energy radiation	$100\%\ c$	10 cm lead, 30 cm concrete	high, passes through body

*The letter c is the symbol for the velocity of light, 3×10^8 m/s.

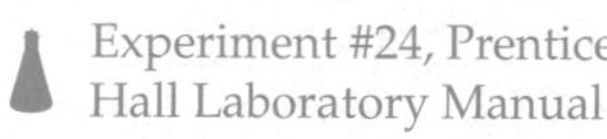

18.2 Nuclear Equations

Objectives
- To illustrate the following types of radiation using atomic notation: alpha, beta, gamma, positron, neutron, and proton.
- To write balanced nuclear equations involving natural radioactivity.

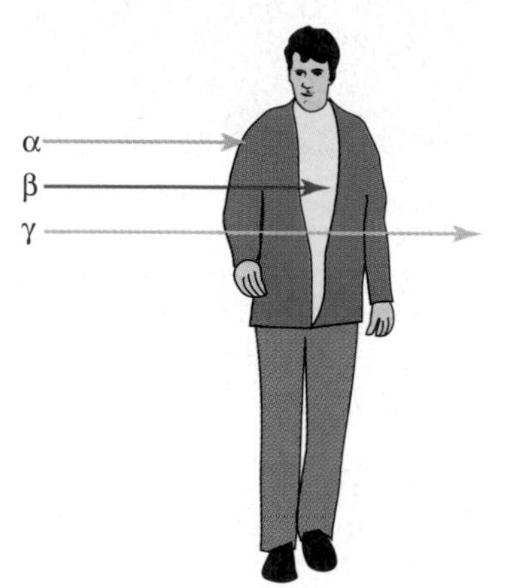

▲ **Radiation's Penetrating Power** Alpha particles are stopped by the skin, while beta particles can penetrate about a centimeter. Gamma rays easily pass through the human body.

In Section 5.4, we introduced the atomic notation used to indicate the composition of an atomic nucleus. In atomic notation, the symbol for the element is preceded by a subscript and superscript. The subscript indicates the number of protons (atomic number). The superscript indicates the number of protons and neutrons (mass number). In the following example, the **atomic number** (symbol Z) and **mass number** (symbol A) are indicated by the subscript and superscript, respectively.

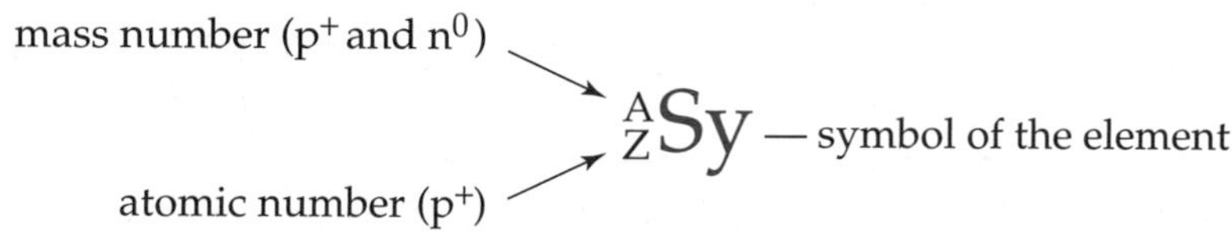

In this chapter, we will use the term **nuclide** when referring to the nucleus of a specific atom. For example, strontium-90 is a radioactive nuclide that contains a total of 90 protons and neutrons. From the periodic table, we find that strontium has an atomic number of 38. Thus, the atomic notation for strontium-90 is

If we subtract the atomic number from the mass number $(90 - 38 = 52)$, we find that strontium-90 has 52 neutrons in addition to 38 protons. Although strontium-90 is radioactive, other strontium nuclides are not. For example, Sr-86, Sr-87, and Sr-88 are stable. Thus, we can conclude that the number of neutrons affects the stability of a nucleus.

Balancing Nuclear Equations

A **nuclear reaction** involves a high-energy change in an atomic nucleus. For example, a uranium-238 nucleus changes into a thorium-231 nucleus by releasing a helium-4 particle.

$$^{235}_{92}U \longrightarrow {}^{231}_{90}Th + {}^{4}_{2}He$$

To write a balanced **nuclear equation**, we must account for all the atomic numbers and mass numbers. The sums of the atomic numbers for the reactants and for the products must be equal. Moreover, the sums of the mass numbers for the reactants and for the products must be equal. The guidelines for balancing a nuclear equation are as follows.

General Guidelines for Balancing a Nuclear Equation

1. The total of the atomic numbers (subscripts) on the left side of the equation must equal the sum of the atomic numbers on the right side.

$$\text{sum of atomic numbers for reactants} = \text{sum of atomic numbers for products}$$

2. The sum of the mass numbers (superscripts) on the left side of the equation must equal the sum of the mass numbers on the right side.

$$\text{sum of mass numbers for reactants} = \text{sum of mass numbers for products}$$

3. After completing the equation by writing all the nuclear particles in atomic notation, a coefficient may be necessary to balance:
(a) atomic numbers of reactants and products
(b) mass numbers of reactants and products

To balance the atomic numbers and mass numbers in a nuclear equation, we must be able to write the atomic notation for each particle in the reaction. Table 18.2 lists some common particles involved in nuclear reactions.

Table 18.2 Particles in Nuclear Reactions

Particle	Notation	Mass	Charge
alpha, α	${}^{4}_{2}\text{He}$	~4 amu	2+
beta, β^-	${}^{0}_{-1}\text{e}$	~0 amu	1−
gamma*, γ	${}^{0}_{0}\gamma$	0 amu	0
positron, β^+	${}^{0}_{+1}\text{e}$	~0 amu	1+
neutron, n^0	${}^{1}_{0}\text{n}$	~1 amu	0
proton, p^+	${}^{1}_{1}\text{H}$	~1 amu	1+

*Gamma rays are high-energy light waves, not particles, which have no mass.

Alpha Emission

Let's write a balanced nuclear equation for the radioactive decay of radium-226 by alpha emission. We can write the decay process using symbols as follows.

$${}^{226}_{88}\text{Ra} \longrightarrow {}^{A}_{Z}\text{X} + {}^{4}_{2}\text{He}\ (\alpha \text{ particle})$$

To balance the number of protons, the atomic number of X (represented by Z) must be 86 ($88 = 86 + 2$). Therefore, we can write

$${}^{226}_{88}\text{Ra} \longrightarrow {}^{A}_{86}\text{X} + {}^{4}_{2}\text{He}$$

When we refer to the periodic table, we find that element 86 is radon, Rn. Thus,

$${}^{226}_{88}\text{Ra} \longrightarrow {}^{A}_{86}\text{Rn} + {}^{4}_{2}\text{He}$$

To balance the mass numbers for reactants and products, the A value for the Rn nuclide must be 222 ($226 = 222 + 4$). The balanced equation is

$${}^{226}_{88}\text{Ra} \longrightarrow {}^{222}_{86}\text{Rn} + {}^{4}_{2}\text{He}$$

Note that alpha emission decreases the atomic number by 2 and therefore produces the element that is two places lower in the periodic table.

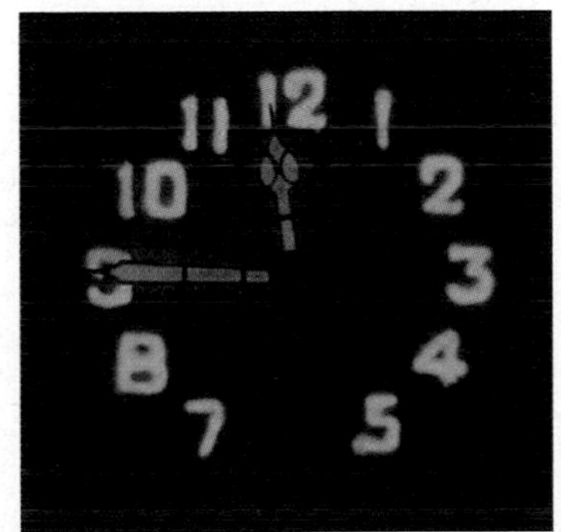

▲ **Natural Radioactivity** Before the hazards of radiation exposure were fully understood, radioactive radium was used on the face of a watch. Radiation from the radium causes the watch to glow in the dark.

Beta Emission

Although radium-226 decays by alpha emission, radium-228 decays by beta emission. We can write this decay process in symbols as follows.

$$^{228}_{88}\text{Ra} \longrightarrow {}^{A}_{Z}\text{X} + {}^{0}_{-1}\text{e}\ (\beta\ \text{particle})$$

To balance the number of protons, the atomic number of X must be 89 $(88 = 89 - 1)$. Since nuclide X has an atomic number of 89, the periodic table shows us that the element is actinium, Ac. The equation is

$$^{228}_{88}\text{Ra} \longrightarrow {}^{A}_{89}\text{Ac} + {}^{0}_{-1}\text{e}$$

To balance the mass numbers for reactants and products, the A value for the Ac nuclide must be 228 since a beta particle has no appreciable mass $(228 = 228 + 0)$. The balanced equation for the reaction is

$$^{228}_{88}\text{Ra} \longrightarrow {}^{228}_{89}\text{Ac} + {}^{0}_{-1}\text{e}$$

Note that beta emission increases the atomic number by 1, thus producing the next higher element in the periodic table. The net result is that one neutron in the radium-228 nucleus decays into a proton as it releases an electron.

Gamma Emission

Now let's try a more complex example. Uranium-233 decays by releasing both alpha particles and gamma rays. We can write the decay process as follows.

$$^{233}_{92}\text{U} \longrightarrow {}^{A}_{Z}\text{X} + {}^{4}_{2}\text{He} + {}^{0}_{0}\gamma\ (\gamma\ \text{ray})$$

To balance subscripts, the atomic number of X must be 90 $(92 = 90 + 2 + 0)$. Since nuclide X has an atomic number of 90, we see from the periodic table that the element is thorium, Th. The equation is

$$^{233}_{92}\text{U} \longrightarrow {}^{A}_{90}\text{Th} + {}^{4}_{2}\text{He} + {}^{0}_{0}\gamma$$

To balance the mass numbers for reactants and products, the A value for the Th nuclide must be 229 $(233 = 229 + 4 + 0)$. The balanced equation for the reaction is

$$^{233}_{92}\text{U} \longrightarrow {}^{229}_{90}\text{Th} + {}^{4}_{2}\text{He} + {}^{0}_{0}\gamma$$

Note that the atomic number and mass number of a gamma ray are zero and have no effect on the equation. The gamma ray is included, however, to indicate that high-energy radiation is emitted during the decay process.

Positron Emission

In 1932 a new particle was discovered. Its mass is identical to that of an electron, but its charge is $+1$ rather than -1. The newly discovered particle was given the name **positron** (symbol β^+), for positive electron. A positron is said to be an example of antimatter. When positrons strike electrons, they immediately annihilate each other. A collision between a positron and an electron causes mutual destruction and the release of two gamma rays.

Sodium-22 is a positron emitter and is used in medicine to diagnose small lesions in the brain. The equation for the radioactive decay is as follows.

$$^{22}_{11}\text{Na} \longrightarrow {}^{A}_{Z}\text{X} + {}^{0}_{+1}\text{e}\ (\text{positron})$$

To balance the atomic numbers of reactants and products, the atomic number of X must be 10 (11 = 10 + 1). Referring to the periodic table, we see that the atomic number 10 corresponds to neon, Ne. The equation is

$$^{22}_{11}\text{Na} \longrightarrow {}^{A}_{10}\text{Ne} + {}^{0}_{+1}\text{e}$$

To balance the mass numbers of reactants and products, the mass number of Ne must be 22 since a positron has approximately zero mass (22 = 22 + 0). The balanced equation for the reaction is

$$^{22}_{11}\text{Na} \longrightarrow {}^{22}_{10}\text{Ne} + {}^{0}_{+1}\text{e}$$

Note that positron emission decreases the atomic number by 1, thus producing the next lower element in the periodic table. The net result is that one proton in the sodium-22 nucleus decays into a neutron while releasing a positron.

Electron Capture

A few large, unstable nuclides decay by a process called **electron capture** (symbol **EC**). In this process, a heavy, positively charged nucleus strongly attracts an electron close to the nucleus. If the attraction is sufficient, the negatively charged electron is captured by the nucleus and a positive proton is converted to a neutral neutron.

As an example, consider an unstable lead-205 nuclide that captures an electron as follows.

$$^{205}_{82}\text{Pb} + {}^{0}_{-1}\text{e} \longrightarrow {}^{A}_{Z}\text{X}$$

To balance the atomic numbers, the atomic number of X must be 81 (82 − 1 = 81). From the periodic table we see that element 81 is thallium, Tl. Since the mass of an electron is 0, the mass number does not change (205 + 0 = 205). The balanced equation for the reaction is

$$^{205}_{82}\text{Pb} + {}^{0}_{-1}\text{e} \longrightarrow {}^{205}_{81}\text{Tl}$$

The following example exercise further illustrates how to write balanced nuclear equations for naturally decaying nuclides.

Example Exercise 18.1 • Writing Nuclear Equations

Write a balanced nuclear equation for each of the following radioactive decay reactions.

(a) Radon-222 decays by alpha and gamma emission.
(b) Barium-133 decays by electron capture.

Solution

To write a balanced nuclear equation we must have

(1) equal sums of atomic numbers for reactants and products
(2) equal sums of mass numbers for reactants and products

(a) We can write the equation for the decay of radon-222 by alpha and gamma emission as follows.

$$^{222}_{86}\text{Rn} \longrightarrow {}^{218}_{84}\text{Po} + {}^{4}_{2}\text{He} + {}^{0}_{0}\gamma$$

Since the sums of the atomic numbers (86) and mass numbers (222) are the same for reactants and products, the equation is balanced.

(b) We can write the balanced nuclear equation for the decay of barium-133 by electron capture as follows.

$$^{133}_{56}\text{Ba} + {}^{0}_{-1}\text{e} \longrightarrow {}^{133}_{55}\text{Cs}$$

(continued)

Example Exercise 18.1 *(continued)*

Since the sums of the atomic numbers (55) and mass numbers (133) are the same for reactants and products, the equation is balanced.

Self-Test Exercise

Identify nuclide X in the following nuclear reactions that represents the radioactive decay product from (a) beta emission and (b) positron emission.

(a) ${}^{99}_{43}\text{Tc} \longrightarrow {}^{Z}_{A}\text{X} + {}^{0}_{-1}\text{e}$ (b) ${}^{30}_{15}\text{P} \longrightarrow {}^{Z}_{A}\text{X} + {}^{0}_{+1}\text{e}$

Answers: (a) ${}^{99}_{44}\text{Ru}$; (b) ${}^{30}_{14}\text{Si}$

18.3 Radioactive Decay Series

Objective · To illustrate a radioactive decay series given the partial scheme for uranium-238, uranium-235, or thorium-232.

In Section 18.2, we discussed the radioactive decay of unstable nuclides. During this process, nuclides become stable by emitting radiation. Many nuclides become stable by emitting radiation in a single step. Some heavy nuclides, however, must go through a series of decay steps to reach a nuclide that is stable. This stepwise disintegration of a radioactive nuclide until a stable nucleus is reached is called a **radioactive decay series**.

Natural Decay Series

The decay series for three naturally occurring radioactive nuclides have been studied extensively. They are uranium-238, uranium-235, and thorium-232. The following diagram suggests that these radioactive nuclides decay in a series of steps.

$$\text{U-238} \longrightarrow \longrightarrow \longrightarrow \text{Pb-206}$$
$$\text{U-235} \longrightarrow \longrightarrow \longrightarrow \text{Pb-207}$$
$$\text{Th-232} \longrightarrow \longrightarrow \longrightarrow \text{Pb-208}$$

We use the term **parent–daughter nuclides** to describe a parent decaying nucleus and the resulting daughter nucleus. Although the diagram does not show the number of decay steps necessary to reach a stable nuclide, the actual sequence has been determined for each nuclide. In the case of uranium-238, 14 decay steps are required to become stable. Uranium-235 requires 11 steps, and thorium-232 requires 10 steps.

The first step in the radioactive decay of uranium-238 is the emission of an alpha particle to give thorium-234.

$${}^{238}_{92}\text{U} \longrightarrow {}^{234}_{90}\text{Th} + {}^{4}_{2}\text{He}$$

parent nuclide — daughter product — alpha particle

The second step is the radioactive decay of Th-234 into Pa-234 by the emission of a beta particle.

$${}^{234}_{90}\text{Th} \longrightarrow {}^{234}_{91}\text{Pa} + {}^{0}_{-1}\text{e}$$

The third step in the series is the decay of Pa-234 by beta emission. Several more steps in the decay series are required before a stable nuclide, Pb-206, is reached. Figure 18.2 illustrates the entire decay series for uranium-238. Notice that the atomic number *decreases* by 2 after each α emission (${}^{4}_{-1}\text{He}$), whereas the atomic number *increases* by 1 for each β emission (${}^{0}_{-1}\text{e}$).

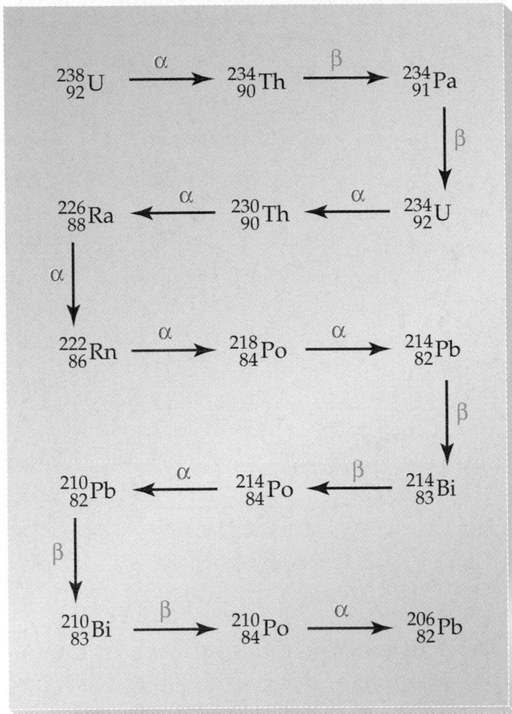

◀ **Figure 18.2 Radioactive Decay Series for Uranium-238** Uranium-238 requires 14 decay steps before reaching a stable nuclide. Notice that several radioactive nuclides of Po, Pb, and Bi are unstable and decay. Only Pb-206, which ends the series, is stable.

Note When we examine the periodic table, we see that a mass number, rather than an atomic mass, is listed for all the elements following bismuth. Thus, there are no stable nuclides of any elements higher than bismuth. All the elements beyond atomic number 83 are unstable and therefore undergo spontaneous radioactive decay.

18.4 Radioactive Half-Life

Objective · To relate the amount of radioactive sample, or its radioactivity, to a given half-life.

In 1908 Hans Geiger, while working for Rutherford, developed an instrument for measuring radiation. Geiger filled a tube with a gas that ionizes when struck by radiation. The ionized gas particles allow passage of a pulse of electricity that can be converted to an audio signal heard as a click. Each click of a *Geiger counter* indicates an emission from the radioactive sample (Figure 18.3). A digital counter registers the number of clicks, providing a measure of the number of decaying nuclei. The number of nuclei that disintegrate in a given period of time is called the **activity** of the sample.

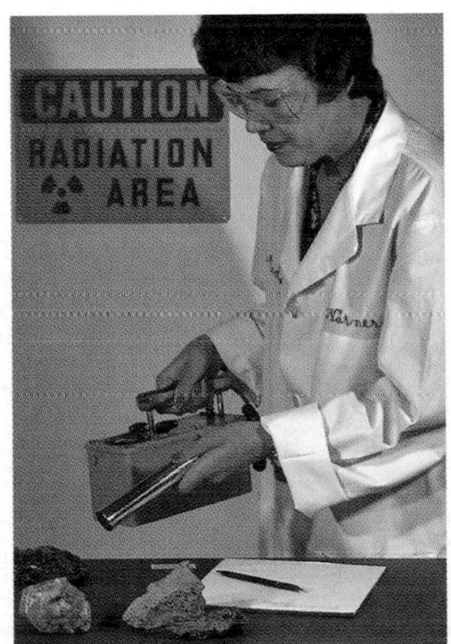

▲ **Figure 18.3 A Geiger Counter** A geologist measures the radioactivity of a rock sample. The Geiger counter gives a click each time a nucleus decays radioactively.

Half-Life

Scientists have observed that the radioactivity of all samples decreases with time. In fact, radioactive decay slows down in a systematic progression. If we begin, for example, with a sample that has an activity of 1200 disintegrations per minute (dpm), after a given period of time the radioactivity will drop to 600 dpm. Moreover, after the same period of time, the activity will drop from 600 dpm to 300 dpm.

The amount of time required for the activity of a sample to decrease by half is called the **half-life** (symbol $t_{1/2}$). If the half-life of a sample is 4 hours, it requires 4

Decay Curve and Half-Life

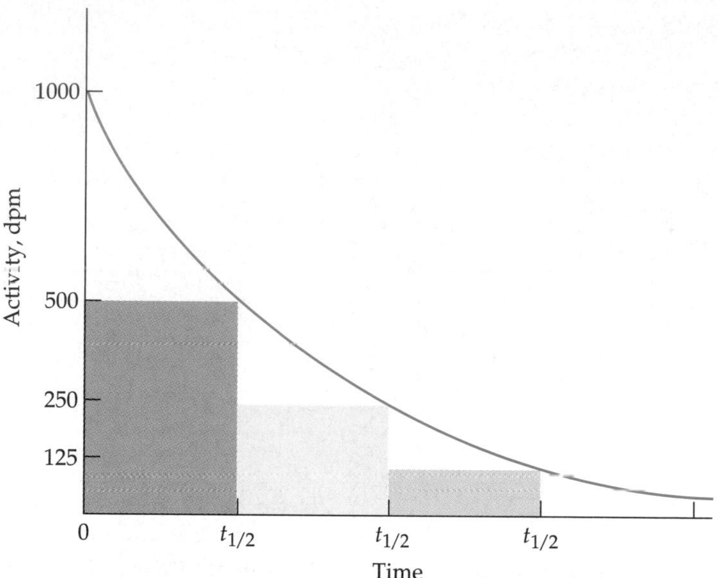

▶ **Figure 18.4 Decay Curve and Half-Life** The time required for the activity of a radioactive sample to decrease from 1000 to 500 dpm is one half-life. Notice that it requires the same amount of time, one half-life, for the activity to drop from 500 to 250 dpm. One half-life elapses each time the activity (dpm) of a sample drops by 50%.

Radioactive Decay Activity

hours for the activity to drop by half. That is, it takes 4 hours for each of the following half-life changes: 1200 to 600 dpm, 600 to 300 dpm, and 300 to 150 dpm. Since three half-lives are required for the activity to drop from 1200 to 150 dpm, the total time required is

$$3\cancel{t}_{1/2} \times \frac{4 \text{ hours}}{1\cancel{t}_{1/2}} = 12 \text{ hours}$$

The decrease in activity can be illustrated by the decay curve in Figure 18.4. The curve shows that each time the activity drops by 50%, one half-life has elapsed.

Radioactive Waste

Disposing of radioactive waste material is a concern because dangerous levels of radioactivity can persist for thousands of years. For example, the radioactive fuel from a plutonium-239 nuclear reactor emits high levels of radioactivity long after it is too weak to run a nuclear reactor. Let's find out how long it takes for the radioactivity of plutonium waste to decrease from 20,000 to 625 dpm.

The half-life of Pu-239 is 24,000 years. By definition, we know that the activity of a radioactive sample drops by one-half after each half-life. We can list the activity of the sample corresponding to the elapsed time.

Activity (dpm)	Elapsed Time, $t_{1/2}$	Elapsed Time (years)
20,000	—	0
10,000	1	24,000
5,000	2	48,000
2,500	3	72,000
1,250	4	96,000
625	5	120,000

We see that it takes five half-lives for the radioactivity of plutonium to drop from 20,000 to 625 dpm. Given that the half-life is 24,000 years, we can calculate the amount of time required for the radiation to drop to 625 dpm. The equation is

$$5\ \cancel{t_{1/2}} \times \frac{24{,}000 \text{ years}}{1\ \cancel{t_{1/2}}} = 120{,}000 \text{ years}$$

The activity of a sample is proportional to the number of radioactive nuclei, and thus to the mass of the sample. The time required for radioactive substances to decay to a safe level varies according to the nuclide. When debating the use of nuclear energy, we must consider the problem of radioactive waste disposal and its potential effect on the environment for thousands of years.

Example Exercise 18.2 • Radioactive Half-LIfe

Iodine-131, in a sodium iodide tablet, is used to measure the activity of the thyroid gland. If a tablet initially contains 88 mg of I-131, what is the mass of the nuclide that remains after 24 days. The half-life of I-131 is 8 days.

Solution

This problem asks us to find the mass of I-131 remaining after 24 days. Let's begin by determining the number of half-lives that elapse. One half-life equals 8 days. Thus,

$$24\ \cancel{\text{days}} \times \frac{1\ t_{1/2}}{8\ \cancel{\text{days}}} = 3\ t_{1/2}$$

To calculate the mass of I-131 remaining after three half-lives, we decrease the mass of I-131 by one-half for each half-life. We can proceed as follows.

$$88 \text{ mg I-131} \times \tfrac{1}{2} \times \tfrac{1}{2} \times \tfrac{1}{2} = 11 \text{ mg I-131}$$

Self-Test Exercise

Iodine-128, in a potassium iodide solution, is used to treat a thyroid condition. If the initial activity is 10,000 dpm and the activity drops to 1250 dpm after 75 minutes, what is the half-life of I-128?

Answer: $t_{1/2}$ = 25 minutes

Note Examine the decay curve for the radioactive material in Figure 18.4. Notice that the number of disintegrations does not decrease in a straight line. The decay curve is said to decrease logarithmically. To avoid calculations requiring logarithms, we will consider only decay periods corresponding to whole-number half-lives, for example, 1, 2, or 3 half-lives.

18.5 Radionuclide Applications

Objective · To describe radiocarbon dating, uranium–lead dating, and radionuclide applications in agriculture and medicine.

A nuclide that is unstable is called a **radionuclide**. Different radionuclides decay by emitting different types of radiation, for example, alpha, beta, gamma, and positron emission. These different types of radioactive emission have many applications in agriculture, industry, medicine, and research.

A nucleus of a specific atom is called a nuclide or radionuclide, however, the terms isotope and radioisotope are commonly used.

Radiocarbon Dating

Carbon occurs naturally as two stable nuclides, carbon-12 and carbon-13. It also appears as an unstable nuclide, carbon-14, that decays by beta emission. The fact that carbon-14 is unstable can be used to estimate the age of any substance containing carbon. Scientists do this by measuring the carbon-14 radioactivity. This process is called *carbon-14 dating* or *radiocarbon dating*. Radiocarbon dating has been applied to fossils such as wood, bone, and any substance that was once part of a living system. It is considered reliable up to 50,000 years. Older objects usually emit too little radiation for accurate measurement.

Radiocarbon dating depends on a fundamental assumption: *the amount of carbon-14 in the atmosphere has remained constant for the last 50,000 years.* If this assumption is not true, then radiocarbon dating is not valid. We know carbon-14 is continuously produced in the Earth's upper atmosphere. The process begins when cosmic rays strike molecules and scatter high-energy neutrons. These neutrons in turn collide with nitrogen in the atmosphere. This collision produces carbon-14 and a proton. The nuclear equation is as follows.

$$^{14}_{7}\text{N} + ^{1}_{0}\text{n} \longrightarrow ^{14}_{6}\text{C} + ^{1}_{1}\text{H}$$

Carbon-14 atoms in the atmosphere combine with oxygen molecules to give radioactive CO_2. This carbon dioxide is then incorporated into plant life through the process of photosynthesis. The food chain begins with plants and continues through animals and eventually humans. A substance that contains carbon-14 emits beta radiation according to the following nuclear equation.

$$^{14}_{6}\text{C} \longrightarrow ^{14}_{7}\text{N} + ^{0}_{-1}\text{e}$$

In a living plant or animal, the radiation from radiocarbon is 15.3 dpm per gram of carbon. When a plant or animal dies, it ceases to intake carbon-14 from the atmosphere, and the radiation level begins to decrease. After one half-life has elapsed, the activity declines to about 7.7 dpm. The half-life of carbon-14 is 5730 years. Therefore, if a radioactive sample has an activity of 7.7 dpm, its estimated age is 5730 years.

Recently, a piece of wood was found on Mt. Ararat in eastern Turkey. It was suspected that the wood might be from Noah's Ark. The wood sample underwent carbon-14 dating and gave an activity of 12.5 dpm per gram of carbon. This activity corresponds to that of a sample about 1000 years old. Thus, using radiocarbon dating, scientists concluded that the wood was not part of Noah's Ark.

When the Dead Sea Scrolls were found, radiocarbon dating was used to verify their authenticity. Actually, it was the linen wrapping around the Old Testament writings that gave an age-dating value of 2000 years. This value corresponds to 0 A.D. and is consistent with biblical history. Thus, the Dead Sea Scrolls are believed to be authentic.

In a similar fashion, Stonehenge in England was age-dated using charcoal from various campsites. The charcoal samples gave an activity of 9.5 dpm per gram of carbon, which corresponds to a sample about 3800 years old. Thus, the Stonehenge site is believed to have been active in about 1800 B.C. Figure 18.5 illustrates radiocarbon dating.

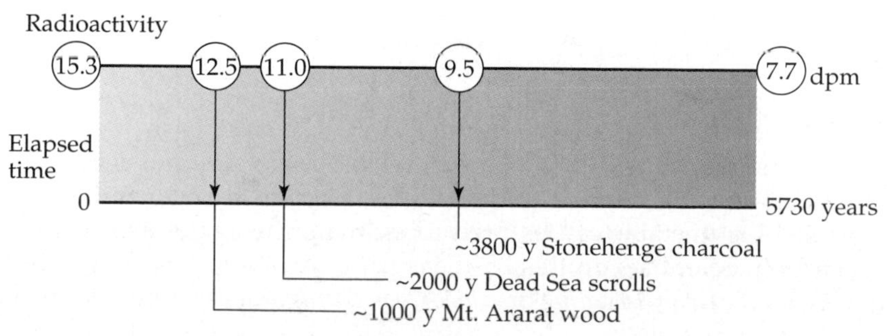

▶ **Figure 18.5 Radiocarbon Dating** The activity from carbon-14 in living tissue is about 15.3 dpm. When tissue expires, it continues to decay radioactively but no longer incorporates carbon-14 from the atmosphere. After 5730 years, half the carbon-14 nuclides in the sample have decayed, and the activity is about 7.7 dpm.

Uranium–Lead Dating

Naturally occurring uranium-238 radioactively decays to lead-206 in a series of 14 steps. The half-life for the overall process is 4.5 billion years. We can indicate the net reaction according to the following equation.

$$^{238}_{92}\text{U} \longrightarrow {}^{206}_{82}\text{Pb} + 8\,{}^{4}_{2}\text{He} + 6\,{}^{0}_{-1}\text{e}$$

The ratio of uranium-238 to lead-206 in uranium samples provides an estimate of the age of very old geological events. For example, a meteorite that fell in Mexico is found to have a uranium-238/lead-206 ratio of 1:1. A ratio of 1:1 corresponds to a single half-life. Thus, the meteorite is about 4.5 billion years old.

The uranium–lead dating technique has also been used to estimate the age of the Earth and of lunar rock samples collected during the *Apollo* missions to the Moon. Uranium–lead dating indicates that the Earth and the Moon are about the same age. Thus, scientists have evidence for the theory that the Moon may have originally been part of the Earth and broke away while the Earth was in an early molten stage.

Agricultural Applications

Pesticides like DDT can effectively control insects, but they often persist in the environment for a long time. Some of these pesticides are toxic to animals and humans. As an alternative, insect populations can be controlled using radiation. Gamma rays from a radionuclide such as cobalt-60 have been employed to sterilize male insects. When large numbers of sterilized males are released in an infested area, they mate with the females, but fertilization does not occur. The insect population is reduced and thus effectively controlled. A decrease in the insect population results in an increase in agricultural harvests, and the consumer benefits.

▲ **Irradiation of Food** Gamma irradiation destroys microorganisms in food and extends freshness. All these mushrooms were picked at the same time, but the ones on the right were irradiated with gamma rays.

Gamma irradiation of processed food destroys microorganisms. For example, cobalt-60 irradiation of pork destroys the parasite that causes trichinosis; irradiation of chicken destroys the parasite that causes salmonella. Irradiation of other foods can extend shelf life without the use of preservatives.

Medical Applications

Today, more than 100 radionuclides are available for medical diagnoses and treatments. An interesting example is plutonium-238, which decays by gamma emission. The gamma rays are used to power pacemakers in patients with irregular heartbeats. The radionuclide is sealed in a stainless steel case and implanted in the chest of the patient. The radiation from a plutonium-238 source can power a pacemaker for up to 10 years before it needs to be replaced. Several other radionuclide applications are found in **Chemistry Connection** • ***Nuclear Medicine.***

18.6 Artificial Radioactivity

Objective • To write balanced nuclear equations involving artificial radioactivity.

When a nuclide is bombarded with an atomic particle, the nuclide is converted to a different element. That is, a nuclear reaction occurs. Causing a nuclear reaction by particle bombardment creates artificial radioactivity. The conversion of one element to another by a nuclear reaction is called **transmutation**.

In 1919 Ernest Rutherford discovered the first transmutation reaction. When he bombarded nitrogen-14 with alpha particles, he detected the release of a proton, $^{1}_{1}\text{H}$. After careful analysis, he concluded that oxygen had also been produced. The equation for the transmutation is

$$^{14}_{7}\text{N} + {}^{4}_{2}\text{He} \longrightarrow {}^{17}_{8}\text{O} + {}^{1}_{1}\text{H}$$

Chemistry Connection · Nuclear Medicine

Which radionuclide emits gamma rays that are used for radiation therapy?

The term *nuclear medicine* refers to the use of radionuclides for medical purposes. For example, iodine-131 is used to measure the activity of the thyroid gland, which requires iodine to regulate metabolism. Patients diagnosed as having a thyroid condition are given an iodine-131 tablet, and after 24 hours, a radiation scan is performed. The amount of radiation detected at the base of the throat, where the thyroid is located, is an indication of thyroid activity.

Xenon-133 is used to diagnose respiratory problems. A patient inhales air containing the radionuclide. Both oxygen and xenon-133 are taken up by healthy lung tissue. As the chest is scanned for radiation, active areas of the lungs are revealed. An area where respiration is impaired is indicated by less radiation.

Iron-59 is used to diagnose anemia, a condition marked by a low red blood cell count. Red blood cells contain hemoglobin, which in turn contains iron atoms. A patient is given iron-59, and the radionuclide bonds to the hemoglobin. When the patient is checked for radiation, a scan reveals radioactive hemoglobin, providing an indirect count of red blood cells.

Brain tumors can be diagnosed using technetium-99, which concentrates in brain tissue. When a patient is given technetium-99, rapidly dividing cancer cells incorporate more of the radionuclide than normal cells. A scan shows radiation hot spots that can help locate the tumor.

Breast cancer can be treated with iridium-192. A hypodermic needle is first injected into the tumor, followed by insertion of an iridium-192 bead through the needle. The needle is then withdrawn, and gamma rays from the radionuclide destroy the tumor and a small amount of surrounding healthy tissue.

Although there is public concern about the use of radioactive substances, nuclear medicine is considered safe. In fact, diagnoses and treatments involving medical radionuclides save or prolong thousands of lives each year.

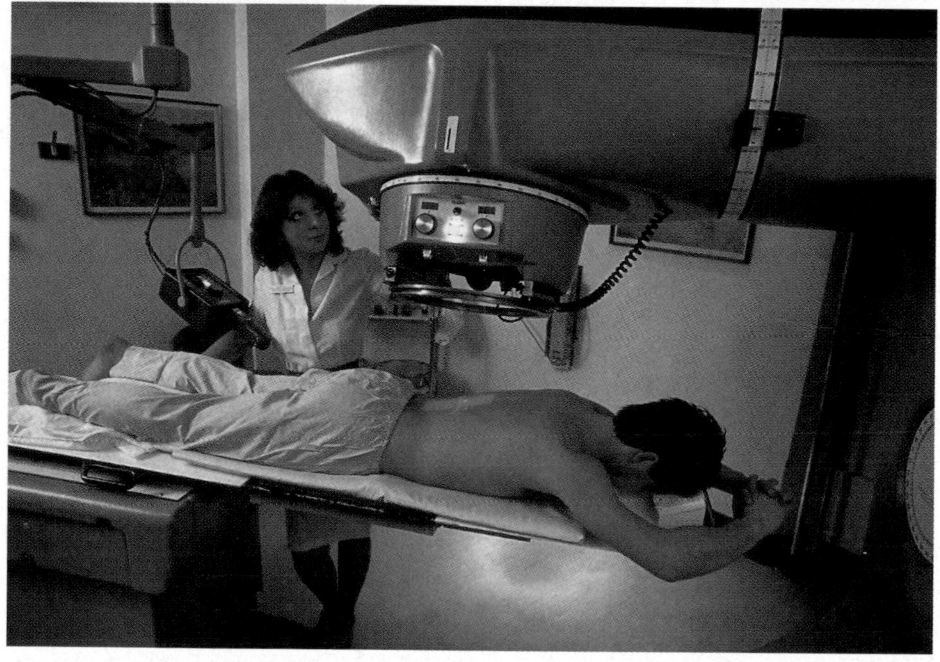

◀ A cancer patient receiving radiation therapy using gamma rays from a cobalt-60 source.

Cobalt-60 is the gamma-ray source used for radiation therapy treatments.

In 1932 James Chadwick, a student of Rutherford's, fired alpha particles at a sheet of beryllium metal. He found that some of the beryllium was changed to carbon and that a neutral particle was released. Chadwick called the neutral particle a neutron. Interestingly, Rutherford had predicted a neutral particle two decades earlier; however, at that time the neutron was too elusive to be confirmed experimentally. The equation for the transmutation is

$$^{9}_{4}\text{Be} + ^{4}_{2}\text{He} \longrightarrow ^{12}_{6}\text{C} + ^{1}_{0}\text{n}$$

Synthesis of New Elements

After the discovery of the neutron, it was proposed that a neutral particle could easily invade another nucleus. In 1940 this proposal proved correct, as uranium was transmuted into a new element. The new element was named neptunium for the planet Neptune. (Neptune lies just beyond Uranus, the planet for which uranium is named.) The equation for the synthesis of neptunium is

$$^{238}_{92}U + ^{1}_{0}n \longrightarrow ^{239}_{93}Np + ^{0}_{-1}e$$

The elements beyond uranium in the periodic table do not occur naturally. Rather, they have been created in the laboratory. These transuranium elements are synthesized by accelerating a light projectile particle to a high velocity and firing it at a heavy nuclide that serves as a target nucleus. One instrument used to accelerate light particles to sufficient energy to overcome the repulsion of the target nucleus is a *cyclotron.* Most of the elements beyond uranium have been synthesized using the cyclotron at the University of California at Berkeley. In fact, element 103 is named lawrencium (symbol Lr) for the American physicist who invented the cyclotron, Ernest O. Lawrence.

Another instrument for accelerating particles is the *linear accelerator.* For example, the linear accelerator at Stanford University was completed in 1966 and is about two miles long. It is capable of accelerating positive or negative projectile particles to a velocity approaching the speed of light. Linear accelerators and cyclotrons are sometimes referred to as "atom smashers" because they fire high-velocity particles that smash into target nuclides (Figure 18.6).

In 1964 Soviet physicists claimed a successful synthesis of element 104. They named the element kurchatovium (symbol Ku) in honor of their team leader, Igor Kurchatov. They used a linear accelerator to send a neon-22 projectile smashing into a plutonium target nucleus. Although the half-life of the new element was only 0.3 s, the Russians proposed the following reaction.

$$^{242}_{94}Pu + ^{22}_{10}Ne \longrightarrow ^{260}_{104}Ku + ^{1}_{0}n$$

Notice that the mass numbers in the equation are not balanced. The total of the reactant superscripts is 264 (242 + 22), but the total of the product superscripts is only 261 (260 + 1). To balance the equation, we must use 4 neutrons. The balanced nuclear equation is

$$^{242}_{94}Pu + ^{22}_{10}Ne \longrightarrow ^{260}_{104}Ku + 4\,^{1}_{0}n$$

◀ **Figure 18.6 Fermi National Accelerator** The Fermi accelerator, outside Chicago, is one of the most powerful in the world. It accelerates particles to a high velocity around an oval that is about 4 miles in circumference.

When the Americans tried to confirm the Russian experiment, they were unsuccessful, which cast doubt on the credibility of element 104. In 1969, however, scientists at the University of California at Berkeley reported a different nuclide of element 104 which has a half-life of 5 s. The Berkeley group named the element rutherfordium (symbol Rf) in honor of Ernest Rutherford. They induced the reaction by firing a carbon-12 projectile at a californium target nucleus.

$$^{249}_{98}\text{Cf} + ^{12}_{6}\text{C} \longrightarrow ^{257}_{104}\text{Rf} + ^{1}_{0}\text{n}$$

When we check the equation, we find that the atomic numbers are balanced but the mass numbers are not. The superscript total is 261 on the left side, but only 258 on the right side. To balance the equation, we place the coefficient 4 before the neutrons. The balanced nuclear equation is

$$^{249}_{98}\text{Cf} + ^{12}_{6}\text{C} \longrightarrow ^{257}_{104}\text{Rf} + 4\,^{1}_{0}\text{n}$$

The following example exercise provides additional practice in writing balanced nuclear equations for the synthesis of new elements.

Example Exercise 18.3 • Writing Nuclear Equations

Write a balanced nuclear equation for the following nuclear synthesis reaction.

$$\text{Bi-209} + \text{Fe-?} \longrightarrow \text{Mt-266} + 1 \text{ neutron}$$

Solution

To write a nuclear equation, we must balance atomic numbers and mass numbers. Let's begin by writing the equation using atomic notation. We obtain the atomic number of each nuclide from the periodic table.

$$^{209}_{83}\text{Bi} + ^{?}_{26}\text{Fe} \longrightarrow ^{266}_{109}\text{Mt} + ^{1}_{0}\text{n}$$

We see that the sum of the atomic numbers is 109 on both sides of the equation. Since the sum of the mass numbers on the right side is 267, the sum on the left side must be 267. Thus, the mass number for Fe must be 58 (267 − 209). The balanced nuclear equation is

$$^{209}_{83}\text{Bi} + ^{58}_{26}\text{Fe} \longrightarrow ^{266}_{109}\text{Mt} + ^{1}_{0}\text{n}$$

Self-Test Exercise

Write a balanced nuclear equation for the following nuclear synthesis reaction.

$$\text{Cf-249} + \text{O-18} \longrightarrow \text{Sg-?} + 4 \text{ neutrons}$$

Answer: $^{249}_{98}\text{Cf} + ^{18}_{8}\text{O} \longrightarrow ^{263}_{106}\text{Sg} + 4\,^{1}_{0}\text{n}$

18.7 Nuclear Fission

Objectives
- To describe the process in a nuclear fission reaction.
- To write a nuclear equation for a given fission reaction.

Spontaneous **nuclear fission** is the process whereby a heavy nucleus splits into lighter nuclei and releases a large amount of energy. Many heavy nuclei are unstable and decay by spontaneous nuclear fission. For example, californium-252 can split apart spontaneously into lighter nuclei as follows.

$$^{252}_{98}\text{Cf} \longrightarrow ^{142}_{56}\text{Ba} + ^{106}_{42}\text{Mo} + 4\,^{1}_{0}\text{n} + \text{energy}$$

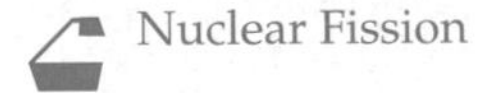

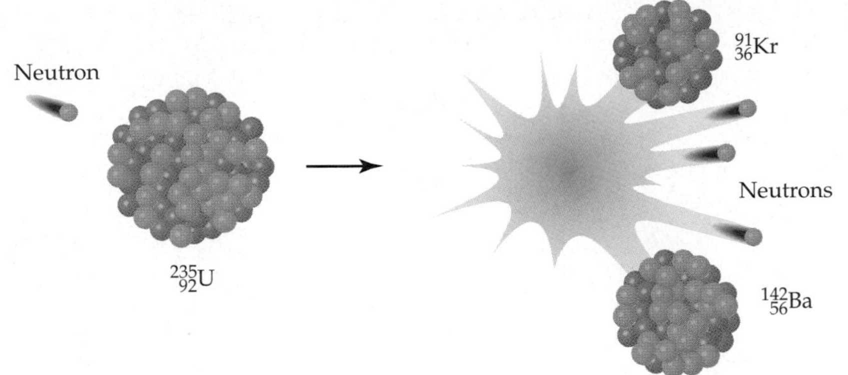

◀ **Figure 18.7 Nuclear Fission** A slow-moving neutron collides with a uranium-235 nucleus to form uranium-236. The resulting nucleus is unstable and splits into two smaller nuclei, releasing a few neutrons. In addition to particles, the fission reaction releases a huge burst of energy.

There are only a few nuclides that can be made to undergo fission, for example, uranium-235, uranium-233, and plutonium-239. The fission process is induced by a slow-moving neutron. When the neutron strikes a heavy nucleus and is captured, the resulting nucleus becomes unstable. The unstable nucleus then decays by splitting into lighter nuclei and releasing a few neutrons. The fission of uranium-235 can occur as follows.

$$^{1}_{0}n + ^{235}_{92}U \longrightarrow ^{137}_{52}Te + ^{97}_{40}Zr + 2\,^{1}_{0}n + \text{energy}$$

After uranium-235 is struck by a neutron, the resulting unstable nucleus can fission in a number of ways, one of which is depicted in Figure 18.7. Analogously, a glass jar can split apart in various ways after being shot by a pellet. For example, uranium-235 can also give the following nuclei as products.

$$^{1}_{0}n + ^{235}_{92}U \longrightarrow ^{141}_{56}Ba + ^{92}_{36}Kr + 3\,^{1}_{0}n + \text{energy}$$

Nuclear Chain Reaction

The process of nuclear fission is analogous to a single domino knocking over three dominos, which in turn can ultimately topple an entire row of dominos. For each nucleus that undergoes fission, two or three neutrons are usually released. The emitted neutrons can cause a second fission reaction. In turn, the second fission releases neutrons that can initiate a third fission. If the fission process continues to repeat itself and is self-sustaining, it is called a **chain reaction**. We can indicate the growth in the number of neutrons in a chain reaction of uranium-235 as follows.

$$1\,^{1}_{0}n \longrightarrow 3\,^{1}_{0}n \longrightarrow 9\,^{1}_{0}n \longrightarrow 27\,^{1}_{0}n$$

We can illustrate a chain reaction as shown in Figure 18.8. Starting with a single uranium-235 nucleus and 1 neutron, the chain reaction quickly escalates to give fission products and 9 neutrons.

For a chain reaction to occur, there must be a fairly large mass of fissionable nuclide. Recall that atoms are mostly empty space. If the mass is too small, fission neutrons simply pass out of the fissionable substance without striking a second nucleus and causing it to fission. The minimum mass of fissionable nuclide required to sustain a chain reaction is called the **critical mass**. For uranium-235, the critical mass is about a kilogram. This amount corresponds to a sphere about the size of a softball.

Nuclear Chain Reaction

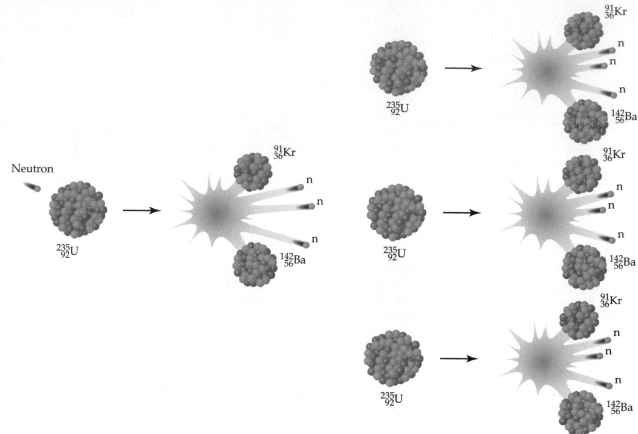

▲ **Figure 18.8 Nuclear Chain Reaction** A single neutron can cause the fission of a uranium-235 nucleus that releases three neutrons. In turn, the three emitted neutrons can cause 3 additional nuclei to fission with the release of 3 neutrons (9 total), and so on. It is important to recognize that this is only one possible scenario and that many other fission nuclides are produced.

In 1939 nuclear fission was discovered in Germany, and the possibility of a sustained chain reaction was experimentally confirmed 3 years later. In 1942 the Italian physicist Enrico Fermi directed an international group of scientists who carried out the first self-sustaining nuclear chain reaction. This landmark event took place on a converted squash court beneath the University of Chicago and is commemorated by a plaque at the university (Figure 18.9).

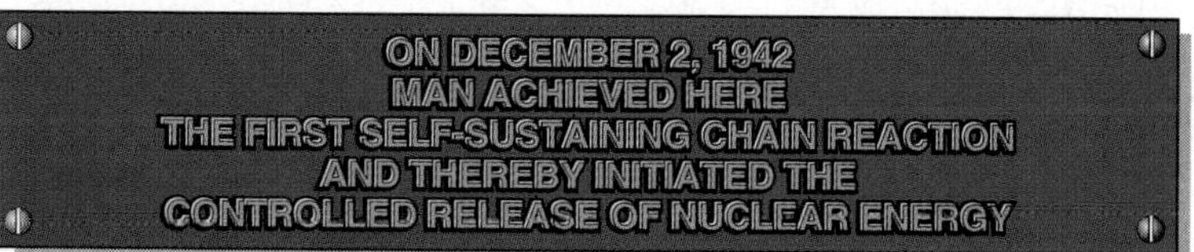

▶ **Figure 18.9 Commemoration Plaque** This large plaque at the University of Chicago signifies the ushering in of the nuclear age.

Uranium-235 Enrichment

Naturally occurring uranium is only 0.7% fissionable uranium-235. Most of the uranium, 99.3%, is nonfissionable uranium-238. To obtain a chain reaction, it is necessary to enrich the natural mixture with the fissionable nuclide. That is, the mixture must be enriched from 0.7% to about 3% uranium-235. The enrichment process is started by reacting uranium with fluorine gas to give gaseous uranium hexafluoride.

$$^{235}U(s) + 3\ F_2(g) \longrightarrow {}^{235}UF_6(g)$$

$$^{238}U(s) + 3\ F_2(g) \longrightarrow {}^{238}UF_6(g)$$

The nuclides $^{235}UF_6$ and $^{238}UF_6$ are too similar to separate chemically, however, the two compounds can be separated physically by gaseous diffusion. This is possible because the lighter nuclide, $^{235}UF_6$, diffuses slightly faster than the heavier nuclide, $^{238}UF_6$. When the first gaseous fraction that diffuses through the tubing is collected, the sample is richer in the faster-diffusing nuclide, $^{235}UF_6$. The collected fraction of uranium compounds is then reduced to elemental uranium enriched in ^{235}U.

18.8 Nuclear Fusion

Objectives
- To describe the process of nuclear fusion using deuterium and tritium.
- To write a nuclear equation for a given fusion reaction.

Nuclear fusion is the process of combining two light nuclei into a heavier nucleus. It is more difficult to initiate a fusion reaction than a fission reaction, but a fusion reaction releases more energy than a comparable fission reaction. In addition, nuclear fusion is a cleaner process than fission because it produces very little radioactive waste.

In essence, the Sun is a huge nuclear fusion reactor operating at millions of degrees. The Sun is made up of 73% hydrogen, 26% helium, and only 1% of all other elements. Three of the fusion reactions thought to occur on the Sun are

$${}^{1}_{1}H + {}^{1}_{1}H \longrightarrow {}^{2}_{1}H + {}^{0}_{+1}e + \text{energy}$$

$${}^{2}_{1}H + {}^{1}_{1}H \longrightarrow {}^{3}_{2}He + \text{energy}$$

$${}^{3}_{2}He + {}^{1}_{1}H \longrightarrow {}^{4}_{2}He + {}^{0}_{+1}e + \text{energy}$$

Extreme temperatures are needed to overcome the strong repulsive forces of light, positively charged nuclei. Experimentally, a powerful laser beam is used to generate the necessary heat to initiate nuclear fusion. A major practical problem is how to confine a reaction that takes place at millions of degrees. Since ordinary materials vaporize at this temperature, a strong magnetic field is used to confine the fusion reaction. Thus, nuclear fusion is said to occur in a "magnetic bottle."

Nuclear fusion has been achieved in research fusion reactors on a limited scale (Figure 18.10). To date, continuous nuclear fusion has not been achieved, and no commercial fusion reactors are in operation. If the technology obstacles can be overcome, the fusion of two deuterium nuclei could be used to supply the world with electrical energy.

Deuterium, ${}^{2}_{1}H$, is the nuclide of hydrogen that contains one proton and one neutron. Deuterium is available from the sea in the form of heavy water molecules, D_2O (Section 13.9). It is estimated that more energy is potentially available from the deuterium in one cubic mile of seawater than from all the current petroleum reserves in the world. A typical fusion reaction of deuterium is as follows.

$${}^{2}_{1}H + {}^{2}_{1}H \longrightarrow {}^{4}_{2}He + \text{energy}$$

Chemistry Connection · Nuclear Power Plant

What fissionable nuclides are used as fuel in a nuclear reactor?

Nuclear energy is an attractive source of energy because of its enormous potential compared to that of energy obtained from ordinary reactions. For example, the fission of 1 g of uranium-235 produces about 12 million times more energy than the combustion of 1 g of gasoline!

◀ **Nuclear Power Plant** A nuclear power plant in southern California that generates electrical energy.

A nuclear power plant generates electrical energy by a nuclear chain reaction. Nuclear fission occurs in the *reactor core,* which contains cylindrical metal *fuel rods.* Each fuel rod is filled with a fissionable nuclide, such as uranium-235, plutonium-239, or uranium-233. The fuel rods are separated by *control rods* made of boron or cadmium. These rods absorb neutrons and thus regulate the rate of fission. When the control rods are removed, more of the fuel rods are exposed, and the fission rate increases. When the control rods are inserted, more neutrons are absorbed and the rate slows, thus preventing the reactor core from overheating.

A liquid *coolant* such as water circulates through the reactor core to absorb heat. If the coolant is D_2O, the reactor is called a heavy water reactor. If the coolant is ordinary H_2O, the reactor is called a light water reactor. The coolant also serves to slow neutrons released from nuclear fission. Slower neutrons are more easily captured by the fissionable nuclides in the fuel rods. Since the coolant affects the rate of capture, it acts as a *moderator* for the fission reaction.

In a nuclear power plant, the nuclear reactor is housed in a concrete building to lessen the chance of an accidental release of radiation into the environment. When the fuel rods are exhausted, they are still highly radioactive and must be stored safely. Many nuclear power plants have on-site underwater vaults for temporary storage. Later, the fuel rods are recycled and the nuclear fuel recovered.

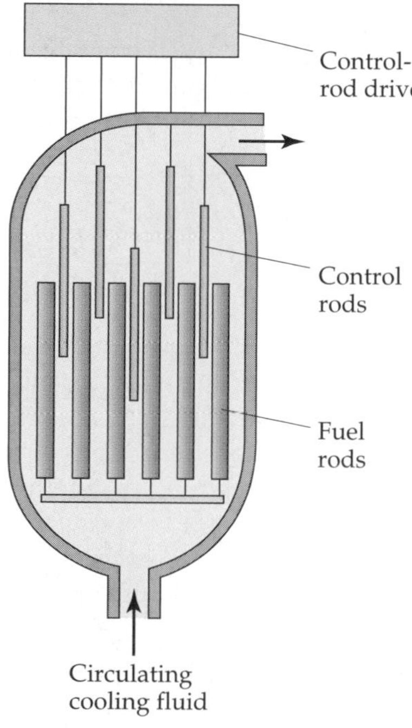

◀ **Nuclear Reactor Core** Notice the control rods inserted between the fuel rods. Lowering the control rods slows the chain reaction by absorbing neutrons released by the fuel rods.

 Nuclear Reactor Core

The fissionable nuclides used in nuclear reactors are U-235, Pu-239, and U-233.

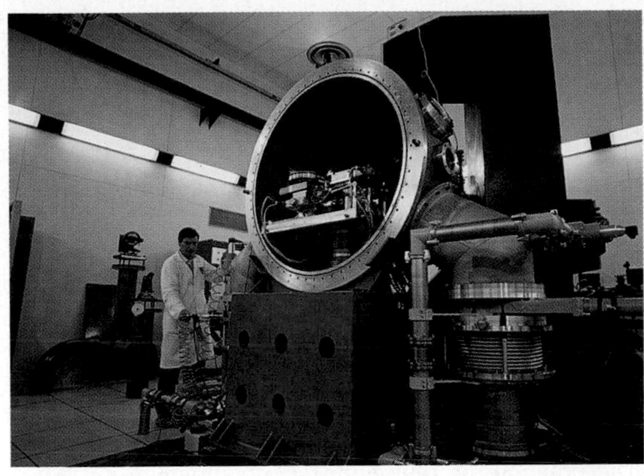

◀ **Figure 18.10 Nuclear Fusion Reactor** A nuclear fusion reactor is essentially a large "magnetic bottle" that contains small nuclei heated to millions of degrees by a powerful pulse laser.

Another promising nuclear fusion reaction involves deuterium and tritium. **Tritium,** ${}^{3}_{1}H$, is the nuclide of hydrogen that contains one proton and two neutrons. Here again, the fusion of nuclei requires temperatures above 10,000,000 K and is initiated by a powerful laser. A typical fusion reaction of deuterium and tritium produces an alpha particle and a neutron, while releasing a huge amount of energy.

$${}^{3}_{1}H + {}^{2}_{1}H \longrightarrow {}^{4}_{2}He + {}^{1}_{0}n + \text{energy}$$

Summary

Section 18.1 An unstable atomic nucleus can disintegrate by emitting **radioactivity** in the form of alpha, beta, or gamma radiation. An **alpha particle** has a positive charge and is identical to a helium-4 nucleus. A **beta particle** has a negative charge and is identical to an electron. A **gamma ray** is a form of high-energy light that has neither mass nor charge.

Section 18.2 The composition of a nucleus is indicated using atomic notation, where the **atomic number** of the element is indicated by a subscript and the **mass number** by a superscript. In this chapter, we have used the term **nuclide** to refer to an atom with a specific nuclear composition.

A **nuclear reaction** involves a change in a nucleus, for example, **electron capture** or **positron** emission. A **nuclear equation** represents the nuclear change using atomic notation for each particle. To have a balanced nuclear equation, two criteria are necessary: (1) the sums of the atomic numbers must be the same on each side of the equation, and (2) the sums of the mass numbers must be equal for reactants and products.

Section 18.3 If a nuclide disintegrates through the emission of radiation in more than one step, the overall process is called a **radioactive decay series**. The decaying nucleus is referred to as the parent nuclide, and the resulting nucleus is the daughter nuclide. Together, they are referred to as **parent–daughter nuclides**. In nature, there are three important radioactive decay series: uranium-235, uranium-238, and thorium-232. Each of these nuclides continues to decay in a series of steps until a stable nuclide of lead is produced.

Section 18.4 We can keep track of the number of decaying nuclei in a sample using a Geiger counter. The **activity** of the sample refers to the number of unstable nuclei that decay in a given period of time. The activity of a sample decreases with time because there are fewer nuclei to decay. However, the time required for 50% of the radioactive nuclei to decay is constant and is called the **half-life**. After each half-life, only 50% of the radioactive nuclei remain.

Section 18.5 Scientists have found important applications for a broad selection of radioactive nuclides. **Radionuclides** are available for age dating, for improving agricultural harvests, and for providing energy. In addition, there is a long and growing list of radionuclides that are used for medical diagnosis and treatment. For example, cobalt-60 is used in radiation therapy for cancer patients.

Section 18.6 Scientists can induce radioactivity by firing light, accelerated nuclides at heavy nuclei. In the event that the atomic number changes and the new nucleus corresponds to a different element, the change is called a **transmutation** reaction. With particle accelerators, it is possible to cause hundreds of nuclear reactions. A particle accelerator, such as a *cyclotron* or a *linear accelerator*, accelerates a projectile nuclide and sends it crashing into a target nucleus. This technique is called "atom smashing" and has created hundreds of synthetic nuclides.

Section 18.7 A neutron striking an unstable nucleus can induce **nuclear fission**; that is, the nucleus can split into two or more particles while releasing energy. The three most important nuclides that undergo fission are uranium-235, plutonium-239, and uranium-233. When a nucleus fissions, two or three neutrons are usually released. These neutrons can in turn cause a second and a third nuclear fission. If a single neutron initiates continuous fission, the process is called a **chain reaction**. For a chain reaction to occur, there must be a sufficient amount of fissionable nuclide. The minimum size of sample that can support a chain reaction is called the **critical mass**.

Section 18.8 The process of combining two light nuclei into a single nucleus is called **nuclear fusion**. In the interior of the Sun at temperatures above 10,000,000°C, there is continuous nuclear fusion. In the laboratory, scientists have had limited success and to date have not been able to create continuous nuclear fusion. One promising fusion reaction is that involving two **deuterium**, ^{2}H, nuclei. Another possibility is the fusion of deuterium and **tritium**, ^{3}H, nuclei. The technological problems are considerable because the fusion process requires temperatures of millions of degrees. However, nuclear fusion does not produce radioactive waste and is therefore an attractive possibility as a future energy source.

Key Concepts*

1. Which type of natural radiation corresponds to each of the following descriptions?
   (a) passes through the body and requires lead shielding
   (b) penetrates skin and requires aluminum shielding
   (c) stopped by skin and requires heavy cloth shielding
2. Write a nuclear equation for each of the following natural decay reactions.
   (a) Pu-238 decays by alpha emission.
   (b) Fe-59 decays by beta emission.
   (c) Mg-23 decays by positron emission.
   (d) Au-183 nucleus decays by electron capture.
3. In the uranium-238 decay series, a parent nuclide disintegrates into a Pb-210 daughter product by emitting an alpha particle. What is the parent nuclide?
4. When 100 mg of technetium-99 is administered for medical diagnosis, how much of the nuclide remains after 12 hours? ($t_{1/2}$ = 6 hours)
5. What event is recorded each time a "click" is registered by a Geiger counter?
6. In 1996 a team of physicists in Germany synthesized element 112. If nuclide-277 and a neutron were produced by firing a projectile into Pb-208, what was the projectile nuclide?
7. A single neutron causes uranium-235 to fission and release three neutrons. Assuming each of the neutrons causes a fission that releases two neutrons, how many neutrons are released in the second fission step?
8. Which of the following is the approximate size of a uranium sample that has critical mass and can sustain a chain reaction: a marble, a softball, a basketball?
9. The nuclear fusion of two protons produces a deuterium nucleus and nuclide X. What is nuclide X?
10. Write a nuclear equation corresponding to the following illustration.

82 p^+ 128 n^0 → 83 p^+ 127 n^0 + ○

* Answers to Key Concepts are in Appendix H.

Key Terms†

Select the key term below that corresponds to each of the following definitions.

_____ 1. the emission of particles or energy from an unstable nucleus
_____ 2. a nuclear radiation identical to a helium-4 nucleus
_____ 3. a nuclear radiation identical to an electron
_____ 4. a nuclear radiation identical to high-energy light
_____ 5. the value that indicates the number of protons in a nucleus
_____ 6. the value that indicates the number of protons and neutrons in a nucleus
_____ 7. an atom with a specific number of protons and neutrons
_____ 8. a high-energy change involving the atomic nucleus
_____ 9. a shorthand representation using atomic notation to describe a nuclear reaction
_____ 10. a nuclear radiation identical to an electron but opposite in charge
_____ 11. nuclear decay whereby a heavy nuclide draws an electron into its nucleus
_____ 12. the stepwise disintegration of a radionuclide until a stable nucleus is reached
_____ 13. a decaying nucleus and the resulting nuclide
_____ 14. the number of radionuclides in a sample that disintegrate in a given period of time, for example, 500 dpm
_____ 15. the time required for 50% of the radionuclides in a given sample to decay
_____ 16. an atom whose nucleus is unstable and decays radioactively
_____ 17. the conversion of one element to another by a nuclear reaction
_____ 18. a nuclear reaction in which a nucleus splits into two or more lighter nuclei
_____ 19. a fission reaction in which the neutrons released initiate a second reaction, which in turn initiates a third reaction, and so on
_____ 20. the minimum mass of a fissionable nuclide that can sustain a chain reaction
_____ 21. a nuclear reaction in which two light nuclei combine into a single nucleus
_____ 22. the nuclide of hydrogen having one neutron
_____ 23. the nuclide of hydrogen having two neutrons

(a) activity (*Sec. 18.4*)
(b) alpha particle (α) (*Sec. 18.1*)
(c) atomic number (Z) (*Sec. 18.2*)
(d) beta particle (β) (*Sec. 18.1*)
(e) chain reaction (*Sec. 18.7*)
(f) critical mass (*Sec. 18.7*)
(g) deuterium (*Sec. 18.8*)
(h) electron capture (EC) (*Sec. 18.2*)
(i) gamma ray (γ) (*Sec. 18.1*)
(j) half-life ($t_{1/2}$) (*Sec. 18.4*)
(k) mass number (A) (*Sec. 18.2*)
(l) nuclear equation (*Sec. 18.2*)
(m) nuclear fission (*Sec. 18.7*)
(n) nuclear fusion (*Sec. 18.8*)
(o) nuclear reaction (*Sec. 18.2*)
(p) nuclide (*Sec. 18.2*)
(q) parent–daughter nuclides (*Sec. 18.3*)
(r) positron (*Sec. 18.2*)
(s) radioactive decay series (*Sec. 18.3*)
(t) radioactivity (*Sec. 18.1*)
(u) radionuclide (*Sec. 18.5*)
(v) transmutation (*Sec. 18.6*)
(w) tritium (*Sec. 18.8*)

Exercises‡

Natural Radioactivity (Sec. 18.1)

1. State the three principal types of natural radiation emitted from a radioactive nucleus.

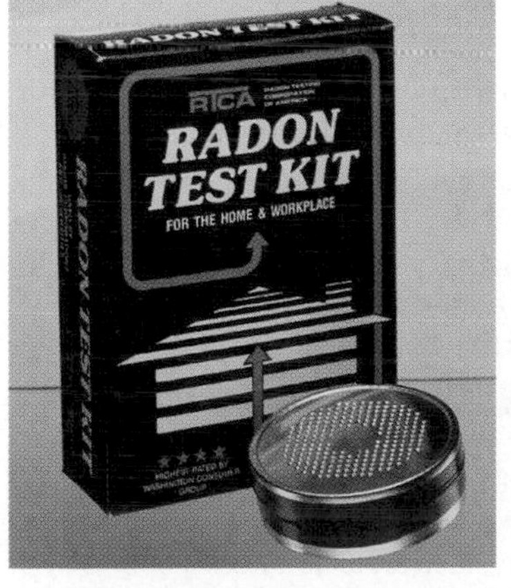

◀ **Radon Test Kit** The Earth releases radioactive radon gas that may collect in the foundation beneath a home. Although the gas is present only in trace amounts, it can accumulate in basements and areas with minimal ventilation and pose a health hazard.

2. State the type of radiation that is deflected toward the following electrodes in an electric field.
   (a) positive electrode **(b)** negative electrode
3. Which nuclear emission is identical to a helium-4 nucleus?
4. Which nuclear emission is identical to an electron?
5. Which nuclear emission is a form of radiant light energy?
6. Which nuclear emission is not affected by an electric field?

Nuclear Equations (Sec. 18.2)

7. Represent each of the following using atomic notation.
   (a) alpha particle **(b)** beta particle
   (c) gamma ray **(d)** positron
   (e) neutron **(f)** proton
8. What is the approximate mass and relative charge of each of the following?
   (a) γ **(b)** n^0
   (c) $\beta-$ **(d)** β^+
   (e) α **(f)** p^+

† Answers to Key Terms are in Appendix I.
‡ Answers to odd-numbered Exercises are in Appendix J.

9. Write an equation for each of the following natural radioactive decay reactions.
   (a) Pt-175 decays by alpha emission.
   (b) Al-28 decays by beta emission.
   (c) Co-55 decays by positron emission.
   (d) Ti-44 decays by electron capture.

10. Write an equation for each of the following natural radioactive decay reactions.
   (a) W-160 decays by alpha emission.
   (b) P-32 decays by beta emission.
   (c) O-15 decays by positron emission.
   (d) Fe-55 decays by electron capture.

11. Identify X, an unknown radioactive nuclide, given the following information.
   (a) Nuclide X decays by alpha emission to give ^{217}Rn.
   (b) Nuclide X decays by beta emission to give ^{43}Ca.
   (c) Nuclide X decays by positron emission to give ^{19}F.
   (d) Nuclide X decays by electron capture to give ^{37}Cl.

12. Identify X, an unknown radioactive nuclide, given the following information.
   (a) Nuclide X decays by alpha emission to give ^{218}Ra.
   (b) Nuclide X decays by beta emission to give ^{56}Fe.
   (c) Nuclide X decays by positron emission to give ^{73}Br.
   (d) Nuclide X decays by electron capture to give ^{133}Cs.

Radioactive Decay Series (Sec. 18.3)

13. In the final step of the uranium-238 disintegration series, the parent nuclide decays into lead-206 and an alpha particle. Identify the parent nuclide.

14. The uranium-238 decay series begins with the emission of an alpha particle. The daughter product emits a beta particle to give which nuclide?

15. In the final step of the uranium-235 disintegration series, the parent nuclide decays into lead-207 and a beta particle. Identify the parent nuclide.

16. The uranium-235 decay series begins with the emission of an alpha particle. The daughter product emits a beta particle to give which nuclide?

17. In the final step of the thorium-232 disintegration series, the parent nuclide decays into lead-208 and an alpha particle. Identify the parent nuclide.

18. The thorium-232 decay series begins with the emission of an alpha particle. The daughter product emits a beta particle to give which nuclide?

19. Supply each of the following emission particles in the 10-step decay series for radioactive thorium-232.

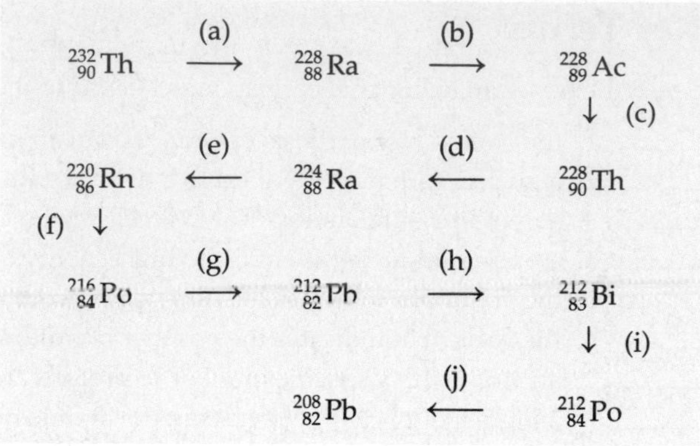

$$
\begin{array}{ccccc}
{}^{232}_{90}\text{Th} & \xrightarrow{(a)} & {}^{228}_{88}\text{Ra} & \xrightarrow{(b)} & {}^{228}_{89}\text{Ac} \\
 & & & & \downarrow (c) \\
{}^{220}_{86}\text{Rn} & \xleftarrow{(e)} & {}^{224}_{88}\text{Ra} & \xleftarrow{(d)} & {}^{228}_{90}\text{Th} \\
(f)\downarrow & & & & \\
{}^{216}_{84}\text{Po} & \xrightarrow{(g)} & {}^{212}_{82}\text{Pb} & \xrightarrow{(h)} & {}^{212}_{83}\text{Bi} \\
 & & & & \downarrow (i) \\
 & & {}^{208}_{82}\text{Pb} & \xleftarrow{(j)} & {}^{212}_{84}\text{Po}
\end{array}
$$

20. Supply each of the following decaying nuclides in the 11-step decay series for radioactive neptunium-237.

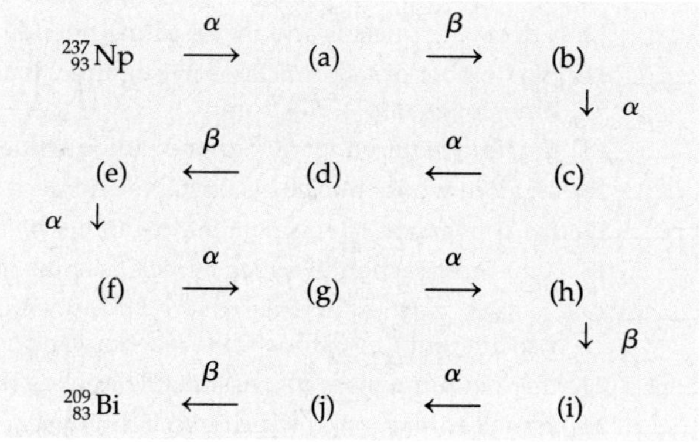

$$
\begin{array}{ccccc}
{}^{237}_{93}\text{Np} & \xrightarrow{\alpha} & (a) & \xrightarrow{\beta} & (b) \\
 & & & & \downarrow \alpha \\
(e) & \xleftarrow{\beta} & (d) & \xleftarrow{\alpha} & (c) \\
\alpha\downarrow & & & & \\
(f) & \xrightarrow{\alpha} & (g) & \xrightarrow{\alpha} & (h) \\
 & & & & \downarrow \beta \\
{}^{209}_{83}\text{Bi} & \xleftarrow{\beta} & (j) & \xleftarrow{\alpha} & (i)
\end{array}
$$

Radioactive Half-life (Sec. 18.4)

21. How many half-lives have elapsed if only 25% of the original radionuclides remains?

22. What percentage of a given radioactive nuclide remains after three half-lives?

23. Spent Pu-239 fuel rods from a nuclear reactor are quite radioactive. If 15 half-lives are required for the nuclide to reach a safe level, how long must the rods be stored? ($t_{1/2}$ = 24,400 years)

24. A nuclear fission reactor produces Sr-90 as radioactive waste. If 20 half-lives are required for the nuclide to reach a safe level, how long must the nuclear waste be stored? ($t_{1/2}$ = 28.8 years)

25. If the carbon-14 radioactivity of an ancient wooden artifact is 6.25% of that of a reference sample, what is the estimated age of the artifact? ($t_{1/2}$ = 5730 years)

26. An archaeologist finds a fossil bone with a carbon-14 reading of 60 dpm. If a recent similar bone gives 240 dpm, what is the age of the fossil? ($t_{1/2}$ = 5730 years)

27. Sodium-24 in the form of NaCl is given as an injection to measure the sodium electrolyte balance. If 80 mg of the medical radionuclide is injected, how much Na-24 remains after 60 hours? ($t_{1/2}$ = 15 hours)

28. If 160 mg of technetium-99 is administered for a medical diagnosis, how much of the nuclide remains after 24 hours? ($t_{1/2}$ = 6 hours)
29. Small beads of iridium-192 are sealed in a plastic tube and inserted through a needle into breast tumors. If an Ir-192 sample has an initial activity of 560 dpm, how much time is required for the activity to drop to 35 dpm? ($t_{1/2}$ = 74 days)
30. If an iron-59 sample has an initial activity of 200 dpm, how much time is required for the activity to drop to 25 dpm? ($t_{1/2}$ = 45 days)
31. The initial radioactivity of a cobalt-60 sample was 1200 dpm, and after 21.2 years the activity dropped to 75 dpm. What is the half-life of the radionuclide?
32. If 2400 μg of hydrogen-3 decay to 600 μg after 24.8 years, what is the half-life of this radionuclide that is used as a chemical tracer?

Radionuclide Applications (Sec. 18.5)

33. What radionuclide technique can be used to estimate the age of fossils up to 50,000 years old?
34. What fundamental assumption must be made when using radiocarbon dating?
35. What radionuclide technique can be used to estimate the age of geological events up to a few billion years?
36. What radionuclide can be used for insect sterilization and pest control?
37. What γ-emitting radionuclide is used to power pacemakers for heart patients?
38. What γ-emitting radionuclide can be used to diagnose and locate inactive lung tissue in the respiratory system?
39. What β-emitting radionuclide can be used to measure the activity of the thyroid gland?
40. What β-emitting radionuclide can be used to diagnose anemia by attaching it to a hemoglobin molecule?
41. What γ-emitting radionuclide can be used to diagnose and locate brain tumors?
42. What γ-emitting radionuclide can be used to treat breast cancer?

Artificial Radioactivity (Sec. 18.6)

43. Bombarding Na-23 with a proton produces a radioactive nuclide and a neutron. What is the radionuclide?
44. Bombarding Br-81 with gamma rays gives a radioactive nuclide and a neutron. What is the radionuclide?
45. Bombarding Li-6 with a neutron produces a radioactive nuclide and an alpha particle. What is the radionuclide?
46. Bombarding U-238 with a hydrogen-2 nucleus gives a radioactive nuclide, a beta particle, and two neutrons. What is the radionuclide?
47. Firing a neutron at a target nucleus gives Mn-56 and an alpha particle. What is the target nuclide?
48. Firing a hydrogen-2 nucleus at a target nucleus gives I-131 and a neutron. What is the target nuclide?
49. Firing an accelerated particle at a boron-10 target nucleus produces N-14 and a gamma ray. What is the projectile particle?
50. Firing an accelerated particle at a Mg-26 target nucleus produces Mg-27 and a proton. What is the projectile particle?
51. In 1967 a Russian team smashed a neon nucleus into an americium target and claimed they had synthesized a new element, X. Given the equation for the reaction, what nuclide did they create?

$$^{243}_{95}\text{Am} + ^{22}_{10}\text{Ne} \longrightarrow ^{A}_{Z}\text{X} + 5\,^{1}_{0}\text{n}$$

52. In 1970 a group at the University of California at Berkeley fired a nitrogen nucleus into a californium target and claimed they had synthesized a new element, X. Given the equation for the reaction, what nuclide did they create?

$$^{249}_{98}\text{Cf} + ^{15}_{7}\text{N} \longrightarrow ^{A}_{Z}\text{X} + 4\,^{1}_{0}\text{n}$$

53. In 1974 a team of Russian physicists smashed a chromium-54 nucleus into a lead-207 target. If a new element, X, and one neutron were produced from the collision, what nuclide did they create?
54. In 1976 a team of German physicists smashed a chromium-54 nucleus into a bismuth-209 target. If a new element, X, and one neutron were produced from the collision, what nuclide did they create?

Nuclear Fission (Sec. 18.7)

55. If a neutron causes a fission reaction that releases two neutrons, how many neutrons have been produced after the third fission step? (Assume each step in the fission process releases two neutrons.)
56. If a neutron causes a fission reaction that releases three neutrons, how many neutrons have been produced after the third fission step? (Assume each step in the fission process releases three neutrons.)
57. The fission of uranium-235 produces 2.4 neutrons per nucleus. Why is there a fractional value for the number of neutrons released?
58. Why must uranium ore be enriched in U-235 before it can be used in the fuel rods of a nuclear reactor?
59. How many neutrons are produced from the following fission reaction?

$$^{235}_{92}\text{U} + ^{1}_{0}\text{n} \longrightarrow ^{144}_{54}\text{Xe} + ^{90}_{38}\text{Sr} + ^{1}_{0}\text{n}$$

60. How many neutrons are produced from the following fission reaction?

$$^{235}_{92}\text{U} + ^{1}_{0}\text{n} \longrightarrow ^{142}_{56}\text{Ba} + ^{91}_{36}\text{Kr} + ^{1}_{0}\text{n}$$

61. How many neutrons are produced from the following fission reaction?

$$^{239}_{94}\text{Pu} + ^{1}_{0}\text{n} \longrightarrow ^{137}_{55}\text{Cs} + ^{100}_{39}\text{Y} + ^{1}_{0}\text{n}$$

62. How many neutrons are produced from the following fission reaction?

$$^{239}_{94}Pu + ^{1}_{0}n \longrightarrow ^{142}_{57}La + ^{94}_{37}Rb + ^{1}_{0}n$$

63. What nuclide undergoes nuclear fission to give barium-143, krypton-88, and three neutrons?

64. What nuclide undergoes nuclear fission to give xenon-142, strontium-90, and two neutrons?

Nuclear Fusion (Sec. 18.8)

65. The nuclear fusion of two deuterium nuclei gives a tritium nucleus, a positron, and particle X. Identify particle X.

66. The nuclear fusion of two helium-3 nuclei gives two protons and particle X. Identify particle X.

67. The nuclear fusion of a helium-3 nucleus and particle X releases an alpha particle and a positron. Identify particle X.

68. The nuclear fusion of a lithium-7 nucleus and particle X releases two alpha particles and a neutron. Identify particle X.

69. The nuclear fusion of two identical nuclides produces a deuterium nucleus and a positron. What nuclide underwent fusion?

70. The nuclear fusion of two identical nuclides produces an alpha particle and a gamma ray. What nuclide underwent fusion?

General Exercises

71. Plutonium-241 undergoes a 13-step decay series. If an alpha particle is emitted in the first step, what is the daughter product?

72. Plutonium-241 undergoes a 13-step decay series. If an alpha particle is emitted in the first step followed by a beta particle, what is the granddaughter product?

73. Nuclide X has a half-life of 23 s and decays by beta emission to produce Xe-136 and a neutron. Write a nuclear equation for the reaction.

74. Nuclide X has a half-life of 2.6 years and decays by electron capture to produce Mn-55 and a gamma ray. Write a nuclear equation for the reaction.

75. Crater Lake in Oregon was formed by volcanic explosion. If a tree charred by the explosion gave an activity of ~7.7 dpm and the half-life of C-14 is 5730 years, what is the approximate age of Crater Lake?

76. Uranium–lead dating of lunar rock samples gave a U-238 to Pb-206 ratio of ~1.0. If the half-life of U-238 is 4.5 billion years, what is the approximate age of the lunar rocks?

77. A radionuclide ionizes the air in a smoke detector, which in turn allows the flow of electricity. When smoke particles enter the detector, they disrupt the electric current and set off an alarm. What α-emitting radionuclide is used to ionize the air in a smoke detector? (*Hint:* The radionuclide decays into Np-237.)

78. A PET scan is a medical imaging technique for the circulatory system. First, a radionuclide is injected that decays by positron emission. When the positron is emitted, it strikes an electron, causing mutual annihilation and the release of two gamma rays. Write an equation for the annihilation reaction.

79. Element 99 was first synthesized by bombarding uranium-238 with neutrons. The reaction yielded Es-253 and seven beta particles. How many neutrons are necessary to balance the equation?

80. Element 100 was first synthesized by bombarding uranium-238 with neutrons. The reaction yielded Fm-255 and eight beta particles. How many neutrons are necessary to balance the equation?

81. In 1955 a single atom of a new element was synthesized using the cyclotron at the University of California at Berkeley. If Es-253 reacted with an alpha particle to release a neutron, what new nuclide was produced?

82. In the Middle Ages, alchemists attempted to transmute base metals into gold. Recently, platinum has been transmuted into gold by bombarding Pt-198 with a deuterium nucleus to give a platinum nuclide and a proton. If the platinum nuclide decays by beta emission, what nuclide results?

83. A neutron induces fission of lithium-6 in a hydrogen bomb and the release of energy, a tritium nucleus, and nuclide X. Write the nuclear equation and identify nuclide X.

84. The nuclear fusion of deuterium and tritium in a hydrogen bomb releases an alpha particle, energy, and nuclide X. Write the nuclear equation and identify nuclide X.

WWW Explorer Quiz 1
Explorer Quiz 2
Explorer Quiz 3
Master Quiz

CHAPTER 19

Organic Chemistry

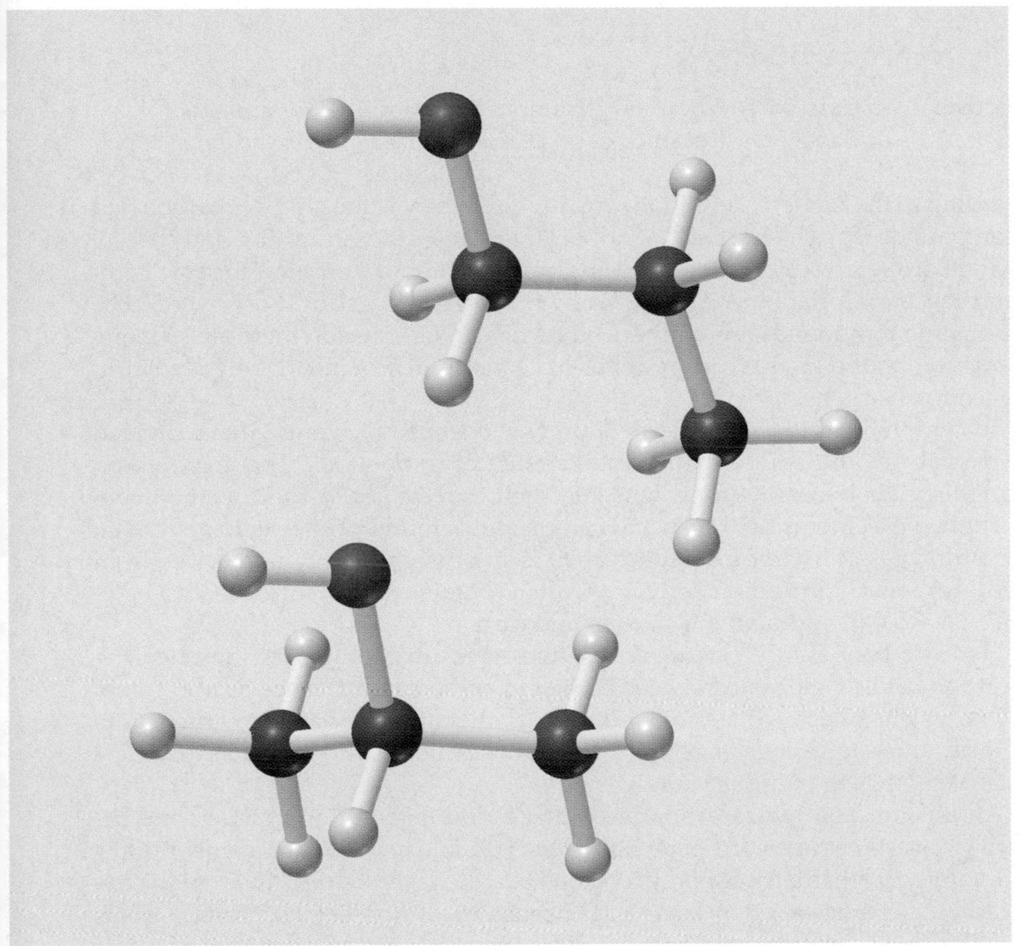

▲ Organic compounds demonstrate the ability to from different molecules having the same formula; that is, they can form isomers. Which of the above isomers of C_3H_7OH is ordinary rubbing alcohol?

In the early 1800s, chemists believed that inorganic compounds were different from organic compounds. Inorganic compounds originate from rocks and minerals obtained from the earth. Organic compounds, on the other hand, originate from plants and animals. Since organic compounds are derived from living organisms, they were thought to contain a life force. This popular idea was referred to as the vital force theory.

In 1828 the German chemist Friedrich Wöhler heated ammonium cyanate in the absence of oxygen and obtained a most unexpected product—urea, a compound present in urine. The result was surprising because ammonium cyanate was considered an *inorganic* compound, whereas urea is an *organic* compound. The reaction is

This chapter on organic chemistry is an invaluable introduction and overview for students who are preparing for a second semester of organic chemistry and biochemistry.

$$\underset{\text{ammonium cyanate}}{NH_4OCN} \xrightarrow{\Delta} \underset{\text{urea}}{NH_2{-}\overset{\overset{\displaystyle O}{\|}}{C}{-}NH_2}$$

Previously, urea had been isolated only from animals. Wöhler's experiment helped disprove the vital force theory, although reluctant skeptics insisted that the vital force from Wöhler's hands had contaminated the result.

19.1 Hydrocarbons

Experiment # 25, Prentice Hall Laboratory Manual

Objectives
- To classify a hydrocarbon as saturated, unsaturated, or aromatic.
- To classify a hydrocarbon as an alkane, an alkene, an alkyne, or an arene.

According to the modern definition, **organic chemistry** is the study of carbon and its compounds. The chief sources of carbon are petroleum, natural gas, and coal, all of which are referred to as fossil fuels. Currently, about 7 million organic compounds account for about 90% of all known substances. In addition, over 50,000 new organic compounds are synthesized each year. One reason there are so many organic compounds is that carbon atoms have the ability to link together and form long, complex chains.

To comprehend the vast subject of organic chemistry, chemists have divided the massive amount of information into families of compounds. The first division distinguishes between hydrocarbons and their derivatives. A **hydrocarbon** contains only the elements hydrogen and carbon and is found primarily in petroleum and natural gas. A **hydrocarbon derivative** is an organic compound derived from a hydrocarbon. Hydrocarbon derivatives often contain oxygen, nitrogen, or a halogen (F, Cl, Br, I) in addition to hydrogen and carbon.

Hydrocarbons can be classified as saturated or unsaturated. A **saturated hydrocarbon** has a single bond between each of its carbon atoms. Since a carbon atom has four valence electrons, it can bond to four other atoms. That is, a carbon atom can form four single covalent bonds. If a hydrocarbon compound has all single bonds, it belongs to the *alkane* family.

An **unsaturated hydrocarbon** has either a double bond or a triple bond between two carbon atoms. If a compound has a double bond, it is a member of the *alkene* family. If it has a triple bond, it belongs to the *alkyne* family. An **aromatic hydrocarbon** is characterized by having a benzene ring composed of six carbon atoms bonded to form a circular compound. If a compound contains a benzene ring, it is a member of the *arene* family. Figure 19.1 shows the classification of hydrocarbons.

We can construct molecular models of hydrocarbons to gain an appreciation of their three-dimensional structure. Ball-and-stick models are commonly used to show the arrangement of atoms in a molecule. Figure 19.2 illustrates the four classes of compounds with selected molecular models.

19.2 Alkanes

Objective
- To write names, formulas, and reactions for simple alkanes.

Alkanes are a family of compounds whose names end in the suffix *-ane*. Alkanes are saturated hydrocarbons, and each member of the family shares the same general molecular formula: C_nH_{2n+2}, where n is the number of carbon atoms. Thus, the number of hydrogen atoms is twice the number of carbon atoms plus the two end

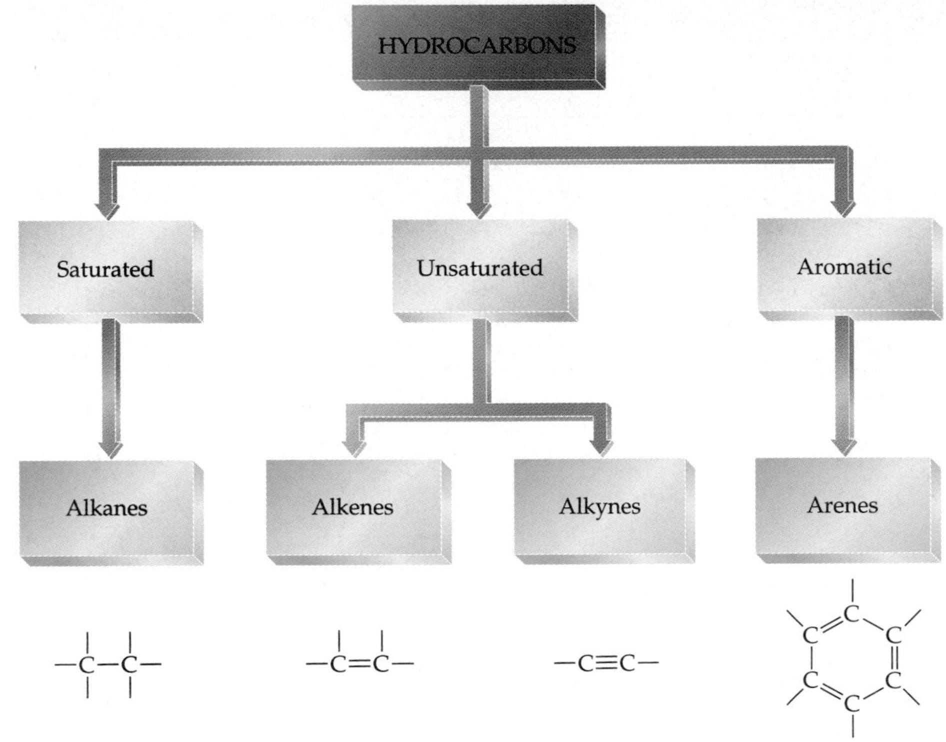

HYDROCARBONS

Saturated | Unsaturated | Aromatic

Alkanes | Alkenes | Alkynes | Arenes

Classification of Hydrocarbons

◀ **Figure 19.1 Classification of Hydrocarbons** A hydrocarbon can be classified as (a) an alkane, (b) an alkene, (c) an alkyne, or (d) an arene.

▲ **Figure 19.2 Molecular Models of Hydrocarbons** The photographs show molecular models of (a) an alkane, (b) an alkene, (c) an alkyne, and (d) an arene.

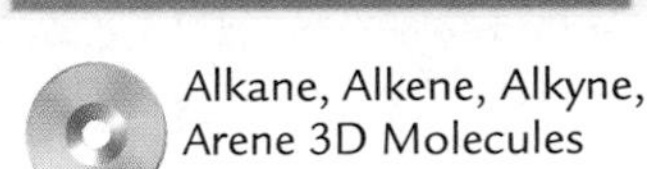
Alkane, Alkene, Alkyne, Arene 3D Molecules

carbon atoms. For example, the fifth member of the alkane family has 12 hydrogen atoms ($2 \times 5 + 2 = 12$) and the molecular formula is C_5H_{12}.

Alkane Family

The lower members (C_1–C_{10}) of the alkane family are used as gaseous and liquid fuels. The higher members (C_{20}–C_{40}) are waxy solids. These solid alkanes are referred to as paraffins and are used to make waxes and candles. Table 19.1 lists the first ten members of the alkane family.

The molecular and structural formulas of alkanes are shown in Table 19.1. For example, the molecular formula of propane is C_3H_8. The condensed **structural formula** is $CH_3—CH_2—CH_3$. To see the structure even more clearly, we can write an expanded structural formula. For simplicity, we can omit the hydrogen atoms and write a skeletal structural formula. Each of the following formulas can be used to represent a molecule of propane, C_3H_8.

Table 19.1 Alkanes

IUPAC Name*	Molecular Formula	Structural Formula	Boiling Point
methane	CH_4	CH_4	−161°C
ethane	C_2H_6	$CH_3{-}CH_3$	−89°C
propane	C_3H_8	$CH_3{-}CH_2{-}CH_3$	−44°C
butane	C_4H_{10}	$CH_3{-}CH_2{-}CH_2{-}CH_3$	−1°C
pentane	C_5H_{12}	$CH_3{-}CH_2{-}CH_2{-}CH_2{-}CH_3$	36°C
hexane	C_6H_{14}	$CH_3{-}CH_2{-}CH_2{-}CH_2{-}CH_2{-}CH_3$	68°C
heptane	C_7H_{16}	$CH_3{-}CH_2{-}CH_2{-}CH_2{-}CH_2{-}CH_2{-}CH_3$	98°C
octane	C_8H_{18}	$CH_3{-}CH_2{-}CH_2{-}CH_2{-}CH_2{-}CH_2{-}CH_2{-}CH_3$	125°C
nonane	C_9H_{20}	$CH_3{-}CH_2{-}CH_2{-}CH_2{-}CH_2{-}CH_2{-}CH_2{-}CH_2{-}CH_3$	151°C
decane	$C_{10}H_{22}$	$CH_3{-}CH_2{-}CH_2{-}CH_2{-}CH_2{-}CH_2{-}CH_2{-}CH_2{-}CH_2{-}CH_3$	174°C

*Note that the Greek prefixes we learned previously (*penta-*, *hexa-*, *hepta-*, *octa-*, *nona-*, *deca-*) are used to form the root names of alkanes C_5H_{12} through $C_{10}H_{22}$.

Boiling Points Activity

$CH_3{-}CH_2{-}CH_3$

condensed structural formula

```
    H   H   H
    |   |   |
H — C — C — C — H
    |   |   |
    H   H   H
```

expanded structural formula

```
  |   |   |
— C — C — C —
  |   |   |
```

skeletal structural formula

Structural Isomers

Each structural formula in Table 19.1 assumes that the molecular formula is a straight chain and does not have branches. However, many carbon chains do have branches. When branching occurs, a different compound results. Two compounds that have the same molecular formula but a different structural formula are called **isomers**.

Isomerism Movie

Butane has the molecular formula C_4H_{10}. However, we can draw the structural formula in two ways: $CH_3CH_2CH_2CH_3$ and $CH_3CH(CH_3)CH_3$. These formulas represent the two isomers—butane and "isobutane."[1] In addition to different structural formulas, isomers have different physical properties such as melting points and boiling points.

$CH_3{-}CH_2{-}CH_2{-}CH_3$

butane
Mp = −138°C, Bp = −0.5°C

```
        CH3
        |
CH3 — CH — CH3
```

"isobutane"
Mp = −159°C, Bp = −12°C

As the number of carbons in the molecule increases, the number of possible isomers increases. For example, C_5H_{12} has 3 isomers, C_6H_{14} has 5 isomers, and $C_{10}H_{22}$ has 75 isomers! To draw the structural isomers corresponding to a molecular formula, we use the following guidelines.

Guidelines for Drawing Hydrocarbon Isomers

1. Draw a continuous chain of carbon atoms. For C_5H_{12}, the chain is five carbon atoms long. To simplify, we omit hydrogen atoms and show the skeletal formula.

   C—C—C—C—C

[1] This text places a chemical name in quotes to indicate a common name rather than a systematic name.

2. To construct a different isomer, draw a shorter continuous chain; in this case, the chain is now four carbon atoms long. Attach the fifth carbon to either carbon atom in the middle of the chain.

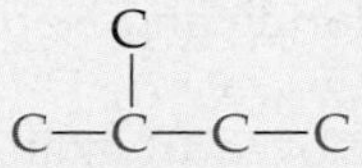

```
   C
   |
C—C—C—C
```

It makes no difference which middle carbon atom we attach the fifth carbon to since we can flip the molecule over without changing the structure. The following structure is identical to the one above.

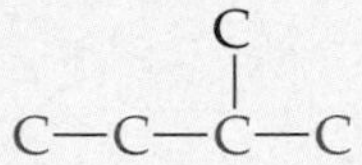

```
     C
     |
C—C—C—C
```

3. To construct another isomer, shorten the chain to three carbon atoms. Attach the fourth and fifth carbon atoms to the central carbon atom.

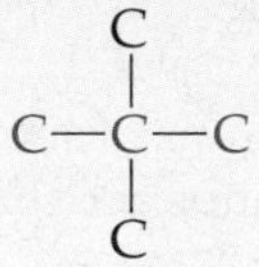

```
   C
   |
C—C—C
   |
   C
```

4. For higher members of the alkane series that contain more than five carbons, apply a similar procedure. Although the process can become complex, this systematic procedure is very effective.

Alkyl and Aryl Groups

When a hydrogen atom is removed from an alkane, an **alkyl group** (symbol **R**—) results. The name of the alkyl group is formed by changing the suffix *-ane* of the alkane to the suffix *-yl*. For example, methane becomes a methyl group.

$$CH_4 \quad - \mathbf{H} \quad = \quad CH_3—$$

meth**ane** → meth**yl**

Similarly, removing a hydrogen atom from ethane gives an ethyl group.

$$CH_3—CH_3 \quad - \mathbf{H} \quad = \quad CH_3—CH_2—$$

eth**ane** → eth**yl**

Removal of a hydrogen atom from propane can occur in two ways. The structural formula of propane, $CH_3—CH_2—CH_3$, shows that there are six hydrogen atoms on the end carbons and two hydrogen atoms on the central carbon. Since the end carbon atoms are identical, we can remove any one of the six hydrogen atoms to produce a propyl group.

$$CH_3—CH_2—CH_3 \quad - \mathbf{H} \quad = \quad CH_3—CH_2—CH_2—$$

prop**ane** → prop**yl**

Since the middle two hydrogen atoms are equivalent, we can remove either of these atoms to form an isopropyl group.

```
                               |
CH3—CH2—CH3  − H  =  CH3—CH—CH3
```

prop**ane** → isoprop**yl**

It is easy to see that we can substitute a chlorine atom (substituent group) onto a propane molecule at two different positions. If the chlorine is attached to an end carbon, the result is propyl chloride, $CH_3—CH_2—CH_2—Cl$. If the chlorine atom is attached to the central carbon, the result is isopropyl chloride, $CH_3—CH(Cl)—CH_3$. Propyl chloride and isopropyl chloride are structural isomers and have unique properties.

When a hydrogen atom is removed from an aromatic hydrocarbon, the result is an **aryl group** (symbol **Ar—**). The most common aryl group results after removal of a hydrogen atom from a benzene ring. The resulting aryl group is called a **phenyl group**.

benzene, C_6H_6 phenyl, C_6H_5—

Now, let's summarize the structures of substituent groups. The names and formulas of selected alkyl and aryl groups are shown in Table 19.2.

Table 19.2 Alkyl and Aryl Groups

Name	Molecular Formula	Structural Formula	Example
methyl	CH_3—	CH_3—	$CH_3—Cl$ "methyl chloride"
ethyl	C_2H_5—	$CH_3—CH_2$—	$CH_3—CH_2—Cl$ "ethyl chloride"
propyl	C_3H_7—	$CH_3—CH_2—CH_2$—	$CH_3—CH_2—CH_2—Cl$ "propyl chloride"
isopropyl	C_3H_7—	$(CH_3)_2CH$—	$CH_3—CH—Cl$ CH_3 "isopropyl chloride"
phenyl	C_6H_5—	H H C=C H—C C— C—C H H	H H C=C H—C C—Cl C—C H H "phenyl chloride"

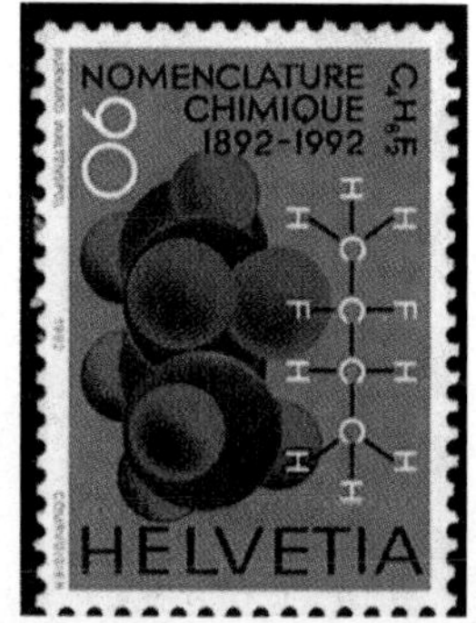

▲ **Swiss Postage Stamp** This stamp commemorates the 100th anniversary of the proposal of the systematic nomenclature rules for organic compounds in Geneva, Switzerland.

Nomenclature of Alkanes

In 1892 the International Union of Chemistry met in Geneva, Switzerland, and recommended a systematic set of rules for naming organic compounds. Although these rules are revised periodically by the International Union of Pure and Applied Chemistry (IUPAC), the basic guidelines for naming organic compounds are still referred to as the Geneva convention. In naming an alkane, we will apply the following nomenclature guidelines.

Nomenclature of Alkenes

An alkene is named after the corresponding alkane. That is, the name of an alkene is formed by changing the suffix *-ane* to the suffix *-ene*. For example, ethane is changed to ethene. The rules for naming alkenes are similar to the rules for naming alkanes. We will use the following nomenclature guidelines.

Guidelines for Naming Alkenes

1. Name an alkene for the longest continuous carbon chain that contains the double bond. Consider the following alkene.

$$\begin{array}{l} \quad\quad\;\; CH_2-CH_3 \\ \quad\quad\;\; | \\ CH_2{=}C-CH_2-CH_2-CH_3 \end{array}$$

   Although the longest chain is six carbon atoms, only five carbon atoms are in the chain with the double bond. Thus, the compound is a *pentene*.

2. Number the longest continuous chain starting from the end closest to the double bond. Thus, we number from left to right to give *1-pentene*.

$$\begin{array}{l} \quad\quad\;\; CH_2-CH_3 \\ \quad\quad\;\; | \\ CH_2{=}C-CH_2-CH_2-CH_3 \\ \;\;1 \quad\;\; 2 \quad\;\; 3 \quad\;\;\; 4 \quad\;\;\; 5 \end{array}$$

3. Indicate the position of each attached alkyl group(s) by name and number. Since there is an ethyl group, CH_3-CH_2-, on the second carbon, the compound is named *2-ethyl-1-pentene*.
4. If there are two or more of the same group on the chain, use the prefix *di-* for two, *tri-* for three, or *tetra-* for four. In this case, there is one substituent, and so the name remains *2-ethyl-1-pentene*.

Example Exercise 19.2 • Nomenclature of Alkenes

Give the IUPAC name for the following structure.

$$\begin{array}{l} \quad\quad\;\; CH_3 \\ \quad\quad\;\; | \\ CH_3-C-CH_2-C{=}CH-CH_3 \\ \quad\quad\;\; | \quad\quad\quad\;\; | \\ \quad\quad\;\; CH_3 \quad\quad CH_3 \end{array}$$

Solution

Let's apply the systematic rules for naming alkenes.

1. The longest continuous carbon chain containing the double bond has six carbons, and so the compound is a *hexene*.
2. To assign the double bond the lowest value, we start numbering from the right. This compound is a *2-hexene*.

$$\begin{array}{l} \quad\quad\;\; CH_3 \\ \quad\quad\;\; | \\ CH_3-C-CH_2-C{=}CH-CH_3 \\ \quad\quad\;\; | \quad\quad\quad\;\; | \\ \quad\quad\;\; CH_3 \quad\quad CH_3 \\ \;\;6 \quad\;\; 5 \quad\;\;\; 4 \quad\;\;\; 3 \quad\; 2 \quad\;\;\; 1 \end{array}$$

(continued)

Example Exercise 19.2 *(continued)*

3. Each of the three branch substituents is a methyl group. Thus, the compound is a *methyl-2-hexene.*
4. Since there are three methyl groups, we use the prefix *tri-* and indicate the position of each substituent. The IUPAC name for this alkene is *3,5,5-trimethyl-2-hexene.*

Self-Test Exercise

Draw the condensed structural formula for 2-methyl-2-pentene.

Answer:

$$\begin{array}{l} CH_3{-}\underset{\displaystyle |\atop \displaystyle CH_3}{C}{=}CH{-}CH_2{-}CH_3 \end{array}$$

Alkyne Family

The first member of the alkyne family is $CH{\equiv}CH$. Its IUPAC name is ethyne, but it is commonly called "acetylene." Oxygen and acetylene gases are used for oxyacetylene welding. When oxygen and acetylene gases are ignited, the temperature reaches 3000°C. The second member of the alkyne family is $CH{\equiv}C{-}CH_3$. Its IUPAC name is propyne, but it is commonly referred to as "methyl acetylene."

The next member of the alkyne family, C_4H_6, is butyne. Butyne has two structural isomers because there are two possible positions for the triple bond. The triple bond can be between the first and second carbon atoms, $CH{\equiv}C{-}CH_2{-}CH_3$, or between the second and third, $CH_3{-}C{\equiv}C{-}CH_3$. The name of the first isomer is 1-butyne, and the name of the second is 2-butyne. Table 19.4 lists the first few members of the alkyne family.

Table 19.4 Alkynes

IUPAC Name	Common Name	Molecular Formula	Structural Formula	Boiling Point
ethyne	"acetylene"	C_2H_2	$CH{\equiv}CH$	−84°C
propyne	"methyl acetylene"	C_3H_4	$CH_3{-}C{\equiv}CH$	−23°C
1-butyne	"ethyl acetylene"	C_4H_6	$CH_3{-}CH_2{-}C{\equiv}CH$	8°C
2-butyne	"dimethyl acetylene"	C_4H_6	$CH_3{-}C{\equiv}C{-}CH_3$	27°C

Nomenclature of Alkynes

Like alkenes, alkynes are named after the corresponding alkane. That is, the name of an alkyne is formed by changing the suffix *-ane* to the suffix *-yne.* For example, ethane is changed to ethyne. The rules for naming complex alkynes are similar to the general rules for naming alkenes. The following example exercise illustrates.

Example Exercise 19.3 • Nomenclature of Alkynes

Write the IUPAC name for the following structure.

$$CH_3{-}\underset{\displaystyle |\atop \displaystyle CH_3}{CH}{-}C{\equiv}CH$$

Solution

We can follow the systematic rules for naming alkenes, but we substitute the suffix *-yne* for alkynes.

1. The longest continuous carbon chain containing the triple bond has four carbons, and so the compound is a *butyne*.
2. To assign the triple bond the lowest value, we start numbering from the right. This compound is *1-butyne*.

$$\underset{4}{CH_3}-\underset{\underset{3}{CH_3}}{\overset{}{CH}}-\underset{2}{C}\equiv\underset{1}{CH}$$

3. Since there is a methyl group on the third carbon, the complete name of the compound is *3-methyl-1-butyne*.

Self-Test Exercise
Draw the condensed structural formula for 1-pentyne.

Answer: $CH\equiv C-CH_2-CH_2-CH_3$

Reactions of Alkenes and Alkynes

Like alkanes, alkenes and alkynes also burn with oxygen to provide energy. For example, acetylene, C_2H_2, and oxygen gases give the following combustion reaction in oxyacetylene welding.

$$2\,C_2H_2(g) + 5\,O_2(g) \xrightarrow{\text{spark}} 4\,CO_2(g) + 2\,H_2O(g)$$

Alkenes and alkynes are more reactive than alkanes because of the double and triple bonds. Unlike an alkane, an alkene or alkyne can readily undergo an **addition reaction**. That is, the double and triple bonds in alkenes and alkynes can add hydrogen molecules to the unsaturated bonds in an alkene or an alkyne. Such a reaction is called a *hydrogenation reaction*. Hydrogenation reactions occur readily at room temperature and atmospheric pressure, although a metal catalyst such as Ni or Pt is required.

Hydrogenation Movie

Consider the following stepwise addition of hydrogen gas to acetylene. First, ethyne is converted to ethene; then, ethene is converted to ethane.

$$\underset{\text{ethyne}}{CH\equiv CH(g)} + H_2(g) \xrightarrow{\text{Ni}} \underset{\text{ethene}}{CH_2=CH_2(g)}$$

$$\underset{\text{ethene}}{CH_2=CH_2(g)} + H_2(g) \xrightarrow{\text{Ni}} \underset{\text{ethane}}{CH_3=CH_3(g)}$$

In addition to adding hydrogen, alkenes and alkynes can react by adding a halogen. For example, alkenes and alkynes can add bromine. In general, a catalyst is not necessary. Consider the following stepwise addition of bromine to ethyne.

Addition of Bromine Movie

$$\underset{\text{ethyne}}{CH\equiv CH(g)} + Br_2(l) \longrightarrow \underset{\text{1,2-dibromoethene}}{CHBr=CHBr(l)}$$

$$\underset{\text{1,2-dibromoethene}}{CHBr=CHBr(l)} + Br_2(l) \longrightarrow \underset{\text{1,1,2,2-tetrabromoethane}}{CHBr_2-CHBr_2(l)}$$

Note Although there are hundreds of thousands of hydrocarbons, it is possible to have a general understanding of their chemical reactions. All hydrocarbons undergo combustion reactions. Unsaturated hydrocarbons undergo additional reactions as well as combustion reactions.

▲ **August Kekulé** This stamp commemorates Kekulé's discovery of the cyclic nature of benzene in 1865.

19.4 Aromatic Hydrocarbons

Objective · To write names and formulas for simple aromatic hydrocarbons.

In 1825 the English scientist Michael Faraday isolated a substance from the fuel oil used in gaslights. After analysis, the formula of the compound was determined to be C_6H_6. A few years later, the same compound was isolated from the natural product benzoin, a topical antiseptic. This compound, C_6H_6, was originally called benzin, but the name evolved into benzine, and eventually benzene.

As other organic compounds were isolated and identified, chemists observed that those containing a benzene structure often had a fragrant odor. Thus, these compounds were termed aromatic because of their pleasant aroma. Further analysis revealed that the benzene structure and the fragrant odor were not related. The term persisted, however, and today an aromatic hydrocarbon is one that contains the benzene structure. **Arenes** are a family of compounds that are aromatic hydrocarbons containing a benzene ring.

In 1865 the German chemist August Kekulé proposed that benzene, C_6H_6, had a cyclic structure with three double bonds. He suggested that six carbon atoms formed a ring with alternating single and double bonds. The structure is called the Kekulé structure of benzene. For simplicity, the benzene ring is usually shown by a hexagon enclosing a circle to represent the delocalized double bonds.

Isomers of Disubstituted Benzene

Using the delocalized electron model of benzene, we can represent the structure of dichlorobenzene as follows.

Notice that the two chlorine atoms are next to each other on the benzene ring. When two substituents are adjacent on a benzene ring, their positions are indicated by the prefix *ortho*-. If the substituents are separated by a carbon, their positions are indicated by the prefix *meta*-. If the substituents are opposite each other on the benzene ring, their positions are indicated by the prefix *para*-.

Since *ortho*, *meta*, and *para* compounds have the same molecular formula but are different, they are structural isomers. Recall that isomers have different physical properties. We can illustrate these isomers as follows.

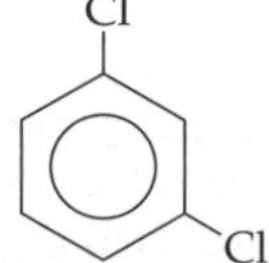

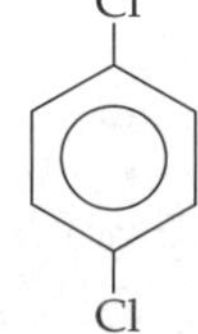

ortho-dichlorobenzene
Mp = –17°C, Bp = 181°C

meta-dichlorobenzene
Mp = –25°C, Bp = 174°C

para-dichlorobenzene
Mp = 53°C, Bp = 175°C

Example Exercise 19.4 • Benzene Isomers

Draw the structural formula for *meta*-dimethylbenzene.

Solution

The compound has an aromatic ring because it contains benzene. It also has two $—CH_3$ groups on the benzene ring in the *meta* position.

CH_3

CH_3

Self-Test Exercise

Give the systematic IUPAC name for F— —I

Answer: *para*-fluoroiodobenzene

19.5 Hydrocarbon Derivatives

Objective · To identify the functional group in each of the following hydrocarbon derivatives: organic halides, alcohols, phenols, ethers, amines, aldehydes, ketones, carboxylic acids, esters, and amides.

In the 1800s chemists began to realize that there was a huge number of organic compounds. As more compounds were identified and their properties recorded, the subject became overwhelming. According to the German chemist Friedrich Wöhler, organic chemistry seemed to be a "dreadful endless jungle." Gradually, chemists began to realize that only a few structural features were shared by all organic compounds. Therefore, millions of organic compounds could be classified into a few categories. Each of these categories is referred to as a **class of compounds**. Moreover, all the compounds in a given class have similar chemical properties.

Classes of compounds are represented by a general formula, which indicates the unique structural feature that characterizes a given class. For instance, the general formula for the alcohol class of compounds is R—OH. The symbol R— represents any alkyl group, such as methyl, CH_3—, or ethyl, C_2H_5—. Each member of the alcohol class is made up of a hydrocarbon (R—) bonded to an —OH group. All alcohol compounds have similar chemical properties because they have similar structural features.

Hydrocarbon Derivatives

A hydrocarbon derivative is an organic compound derived from a hydrocarbon. Recall that a hydrocarbon derivative contains hydrogen, carbon, and additional elements such as oxygen, nitrogen, or a halogen. As a result, there are many classes of hydrogen derivatives. Within each class, members have similar names and properties. The 10 classes of hydrocarbon derivatives we will study are organic halides (R—X), alcohols (R—OH), phenols (Ar—OH), ethers (R—O—R′), amines (R—NH_2), aldehydes (R—CHO), ketones (R—COR′), carboxylic acids (R—COOH), esters (R—COOR′), and amides (R—$CONH_2$).

The symbol R represents any alkyl group, and Ar represents any aryl group. The designations R′ and R″ indicate different alkyl groups, and Ar′ indicates a different aryl group.

The structural feature in a molecule that characterizes a class of compounds is called the **functional group**. For alkenes, the functional group is the double bond

A functional group in a molecule determines its function, that is, its chemical reactivity.

(C=C). For an alcohol (R—OH), the functional group is —OH. For an ether (R—O—R′), the functional group is the oxygen atom.

Of these 10 hydrocarbon derivatives, 5 have a carbon atom double-bonded to an oxygen atom (C=O) and 5 do not (Figures 19.3 and 19.4). A carbon atom joined to an oxygen atom by a double bond is called a **carbonyl group**. The presence of a carbonyl group gives distinct properties to a given class of compounds, such as aldehydes or ketones.

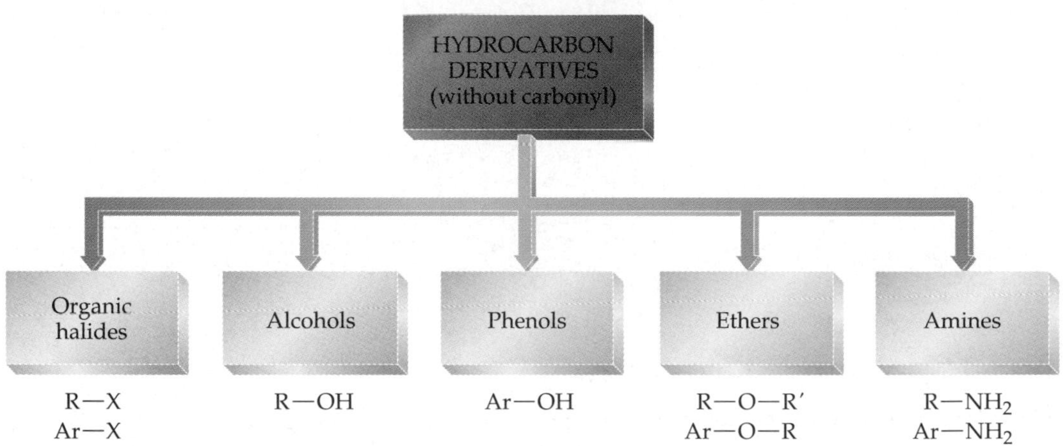

Hydrocarbon Derivatives—without a Carbonyl Group

▲ Figure 19.3 Hydrocarbon Derivatives (without a Carbonyl Group) For organic halides, X can be F, Cl, Br, or I. For alcohols, R can be any alkyl group that is not aryl (Ar); for phenols, the —OH group must be attached directly to a benzene ring.

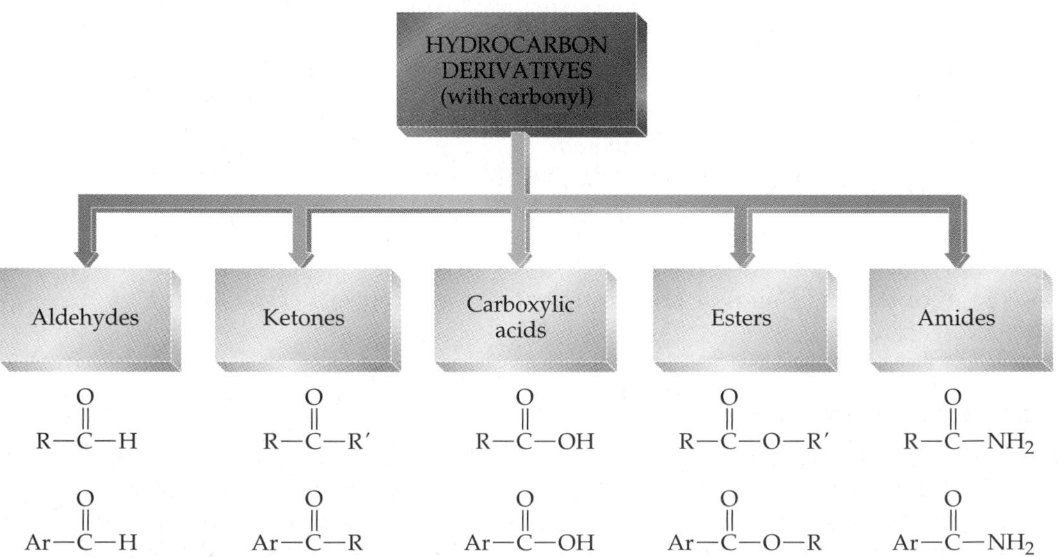

Hydrocarbon Derivatives—with a Carbonyl Group

▲ Figure 19.4 Hydrocarbons Derivative (with a Carbonyl Group) Aldehydes, ketones, carboxylic acids, esters, and amides all contain a carbonyl group with different attached groups. The symbols R and Ar can be any alkyl or aryl substituent. Except in ketones, R can also be a hydrogen atom.

Example Exercise 19.5 • Classifying Hydrocarbon Derivatives

Indicate the class of hydrocarbon derivative for each of the following compounds.

(a) $CH_3—Br$ (b) $CH_3—CH_2—NH_2$

(c) $CH_3—\overset{\overset{\displaystyle O}{\|}}{C}—H$ (d) 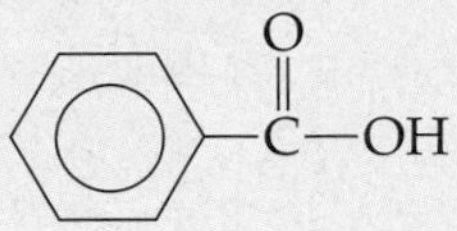 $C_6H_5—\overset{\overset{\displaystyle O}{\|}}{C}—OH$

Solution

Let's first identify the functional group in each compound. We can then refer to the general formulas to determine the class of compound.

(a) This hydrocarbon derivative does not have a carbonyl group. Since the functional group is a halogen (—Br),

$CH_3—Br$ is an *organic halide.*

(b) This hydrocarbon derivative does not have a carbonyl group. Since the functional group is an amine ($—NH_2$),

$CH_3—CH_2—NH_2$ is an *amine.*

(c) This hydrocarbon derivative has a carbonyl group. Since the functional group has a carbonyl attached to a H atom,

$CH_3—\overset{\overset{\displaystyle O}{\|}}{C}—H$ is an *aldehyde.*

(d) This hydrocarbon derivative has a carbonyl group attached to a benzene ring. Since the functional group has a carbonyl attached to an OH,

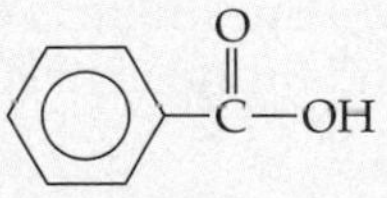

$C_6H_5—\overset{\overset{\displaystyle O}{\|}}{C}—OH$ is a *carboxylic acid.*

Self-Test Exercise

Indicate the class of hydrocarbon derivative for each of the following compounds.

(a) $(CH_3)_2CH—OH$ (b) 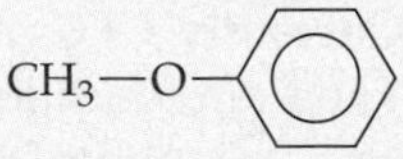 $CH_3—O—C_6H_5$

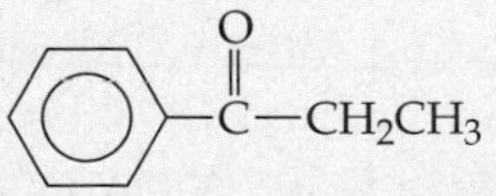

(c) $C_6H_5—\overset{\overset{\displaystyle O}{\|}}{C}—CH_2CH_3$ (d) $H—\overset{\overset{\displaystyle O}{\|}}{C}—NH_2$

Answers: (a) alcohol; (b) ether; (c) ketone; (d) amide

19.6 Organic Halides

Objective · To write names and formulas for simple organic halides.

If a halogen (F, Cl, Br, I) atom replaces a hydrogen atom in a hydrocarbon, it gives rise to a class of compounds called *organic halides* (Figure 19.5). In this class of compounds, the hydrocarbon group can be alkyl, R—X, or aryl, Ar—X. For example,

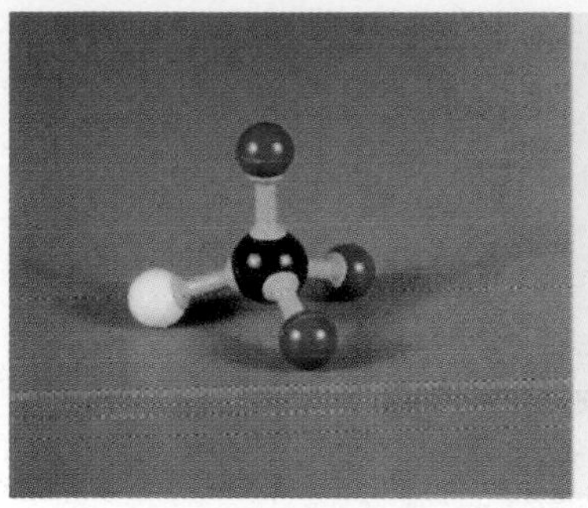

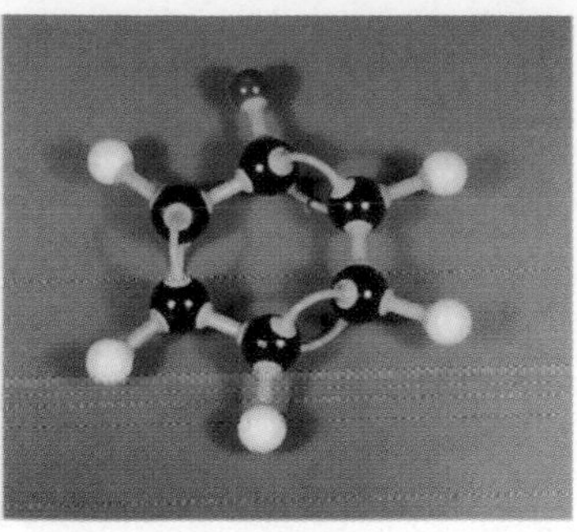

▲ **Figure 19.5 Molecular Models of Organic Halides** Model structures of (a) chloroform, $CHCl_3$; (b) ethyl iodide, C_2H_5I; and (c) phenyl bromide, C_6H_5Br.

```
      Cl              H   H            Br
      |               |   |            |
  H—C—Cl          H—C—C—I        (benzene ring)
      |               |   |
      Cl              H   H
```

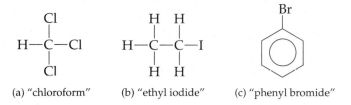

(a) "chloroform" (b) "ethyl iodide" (c) "phenyl bromide"

Organic halides are used primarily as industrial and household solvents. Although carbon tetrachloride, CCl_4, was once widely used for dry cleaning and spot removing, it has been replaced because of its toxicity. Now less toxic solvents, such as trichloroethane, $C_2H_3Cl_3$, are used in dry cleaning.

Chloroform, $CHCl_3$, is a common solvent originally used for general anesthesia. When its vapor was found to be harmful to the respiratory system, it was replaced by safer anesthetics, chiefly halothane, $CF_3CHClBr$. Ethyl chloride, CH_3CH_2Cl, is used as a mild topical anesthetic. After it is sprayed on the skin, ethyl chloride rapidly evaporates and produces a cold sensation that numbs the skin.

Chlorinated fluorocarbons (CFCs) are organic halides composed of chlorine, fluorine, and carbon. The best known example is Freon-12, CF_2Cl_2, which has been used as an aerosol propellant and a refrigerant gas. Carbon dioxide has replaced Freon-12 in most aerosol cans. Currently, CFCs are being replaced by hydrofluorocarbons (HFCs) in refrigeration and air-conditioning units. This change was mandated after CFCs in the upper atmosphere were found to be responsible for depletion of the ozone layer.

Organic halides are found in many pesticides, including Aldrin, Chlordane, and DDT. In 1972 the Environmental Protection Agency banned DDT because of its threat to wildlife. Before the ban, DDT was responsible for boosting agricultural harvests and reducing insect populations that carried malaria and yellow fever.

According to IUPAC nomenclature, organic halides are named for the halogen substituent on the parent alkane. For example, CH_3CH_2Br is named bromoethane. Alternatively, we can designate an acceptable common name for the compound by stating the alkyl group attached to the bromine atom; that is, we can also name CH_3CH_2Br "ethyl bromide."

Most organic halides are essentially nonpolar, although some are slightly polar. In general, organic halides have physical properties similar to those of alkanes. That is, they have low boiling points, are soluble in hydrocarbon solvents, and are insoluble in water. Table 19.5 briefly describes a few simple organic halides.

Table 19.5 Organic Halides*

IUPAC Name	Common Name	Molecular Formula	Boiling Point	Solubility in Water
chloromethane	"methyl chloride"	$CH_3—Cl$	−24°C	insoluble
chloroethane	"ethyl chloride"	$CH_3—CH_2—Cl$	12°C	insoluble
1-chloropropane	"propyl chloride"	$CH_3—CH_2—CH_2—Cl$	47°C	insoluble
2-chloropropane	"isopropyl chloride"	$CH_3—CH(Cl)—CH_3$	36°C	insoluble
chlorobenzene	"phenyl chloride"	$C_6H_5—Cl$	132°C	insoluble

*Most low-molecular-weight organic halides are liquids at room temperature.

19.7 Alcohols, Phenols, and Ethers

Objective · To write names and formulas for simple alcohols, phenols, and ethers.

Alcohols, phenols, and ethers all contain oxygen in addition to hydrogen and carbon. Alcohols and phenols contain an —OH group and, in general, are polar. Ethers, however, contain a C—O—C functional group and are mostly nonpolar.

Alcohols and Phenol

When an —OH group replaces a hydrogen atom on an alkane, the result is an *alcohol*, R—OH. When an —OH group replaces a hydrogen atom on a benzene ring, the result is a *phenol*, Ar—OH. The —OH group on an alcohol or a phenol is called a **hydroxyl group** (Figure 19.6). For example,

$$\begin{array}{c} OH \\ | \\ H—C—H \\ | \\ H \end{array} \qquad \begin{array}{c} H\ \ OH \\ |\quad\ | \\ H—C—C—H \\ |\quad\ | \\ H\ \ \ H \end{array}$$

OH

(a) "methyl alcohol" (b) "ethyl alcohol" (c) phenol

Alcohols are everyday substances with many practical uses. The simplest family member, methyl alcohol, CH_3OH, is used as a solvent and fuel. Ethyl alcohol, CH_3CH_2OH, is blended with gasoline to give gasohol. It is sometimes called grain alcohol because it is produced by the fermentation of carbohydrates found in various grains such as corn, rye, and barley. Isopropyl alcohol, $CH_3CH(OH)CH_3$, is used as rubbing alcohol and for sterilizing medical instruments.

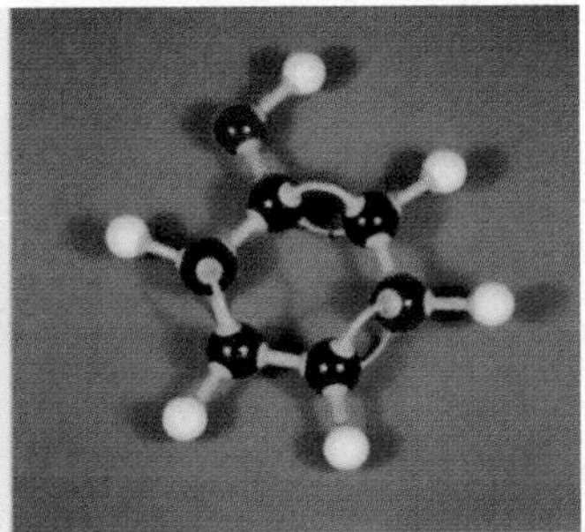

▲ **Figure 19.6 Molecular Models of Alcohols and Phenol** Model structures of (a) methyl alcohol, CH_3OH; (b) ethyl alcohol, C_2H_5OH; and (c) phenol, C_6H_5OH.

Polyhydroxy alcohols have two or more hydroxyl groups on the same molecule. Ethylene glycol, $HOCH_2CH_2OH$, and glycerol, $HOCH_2CH(OH)CH_2OH$, are important examples. Ethylene glycol is the main ingredient in permanent antifreeze. Glycerol, also called glycerin, is a sweet, syrupy liquid used as a moisturizer in cosmetics and candy.

According to IUPAC nomenclature, the name of an alcohol is derived from the name of the parent alkane. The name is formed by dropping the final *e* and adding the suffix *-ol*. For example, CH_3OH has an —OH group on a methane molecule; the IUPAC name is methanol, and the common name is "methyl alcohol." In CH_3CH_2OH, the —OH is on an ethane molecule; the systematic name is ethanol, and the common name is "ethyl alcohol."

If an —OH group is attached to an aromatic ring, the resulting alcohol belongs to a special class of compounds called *phenols*, Ar—OH. The simplest member of this class, phenol, C_6H_5OH, is an important industrial chemical. Phenol is used in the manufacture of plastics and for the preparation of dyes. It is somewhat acidic and is referred to as "carbolic acid." Phenol has disinfectant properties and is used as an antiseptic in medical offices and waste treatment plants.

Most alcohols are quite polar because of the —OH group. Alcohols, like water, can form hydrogen bonds between an oxygen atom and a hydrogen atom in separate molecules. Figure 19.7 depicts hydrogen bonding between alcohol molecules.

As a result of hydrogen bonding, the physical properties of alcohols are quite different from those of the parent alkane; alcohols have high boiling points and are soluble in polar solvents. Table 19.6 lists a few simple alcohols as well as phenol.

Table 19.6 Alcohols and Phenol

IUPAC Name	Common Name	Molecular Formula	Boiling Point	Solubility in Water
methanol	"methyl alcohol"	$CH_3—OH$	65°C	soluble
ethanol	"ethyl alcohol"	$CH_3—CH_2—OH$	78°C	soluble
1-propanol	"propyl alcohol"	$CH_3—CH_2—CH_2—OH$	97°C	soluble
2-propanol	"isopropyl alcohol"	$CH_3—CH(OH)—CH_3$	82°C	soluble
phenol	"carbolic acid"	$C_6H_5—OH$	182°C	~7%

Ethers

If two alkyl groups are attached to an oxygen atom, we have a class of compounds called *ethers*, R—O—R′. In the ether class of compounds, the attached groups may be alkyl or aryl. That is, an ether may be R—O—R, R—O—Ar, or Ar—O—Ar. Commercially, the most significant member of this class is diethyl ether, $CH_3CH_2—O—CH_2CH_3$. It is usually referred to as "ethyl ether," or simply "ether."

"Ethyl ether" is a common laboratory solvent for organic reactions. It was once used as a general surgical anesthetic but has been replaced because it is highly flammable and causes patients to become nauseous. "Ethyl ether" is synthesized by

Hydrogen Bonding in Alcohols

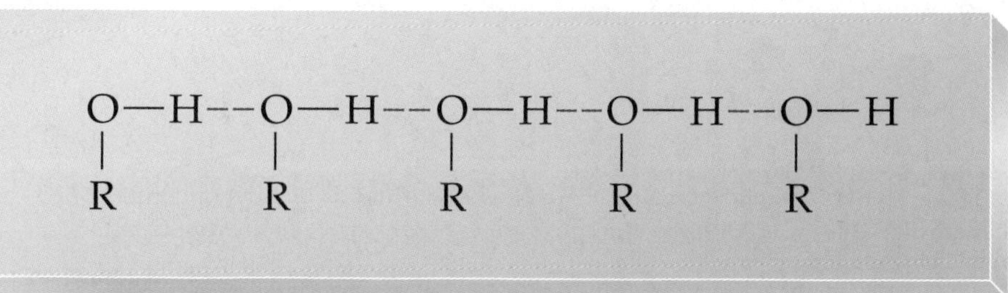

▶ Figure 19.7 Hydrogen Bonding in Alcohols Notice that hydrogen bonds connect alcohol molecules in a long chain. In effect, this increases the molecular weight of the alcohol.

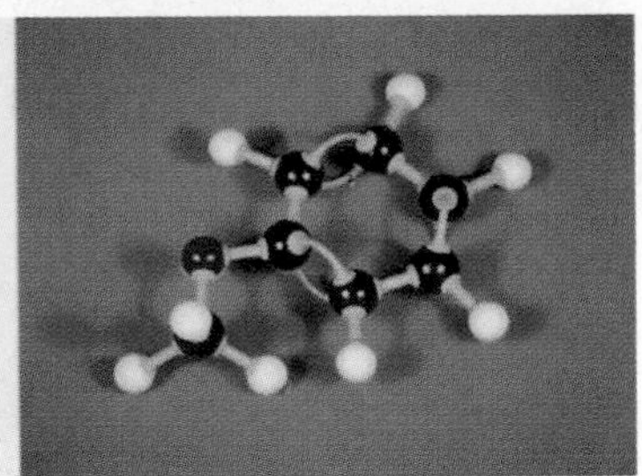

▲ **Figure 19.8 Molecular Models of Ethers** Model structures of (a) diethyl ether, $C_2H_5—O—C_2H_5$; (b) methyl ethyl ether, $CH_3—O—C_2H_5$; and (c) methyl phenyl ether, $CH_3—O—C_6H_5$.

dehydrating "ethyl alcohol." In the reaction, sulfuric acid is used to split out a molecule of water from two molecules of alcohol as follows.

$$\underset{\text{"ethyl alcohol"}}{CH_3CH_2—\mathbf{OH}} + \underset{\text{"ethyl alcohol"}}{\mathbf{H}—OCH_2CH_3} \xrightarrow{H_2SO_4} \underset{\text{"diethyl ether"}}{CH_3CH_2—O—CH_2CH_3} + \mathbf{H_2O}$$

In the ether class of compounds, the attached groups may be alkyl or aryl. That is, an ether may be R—O—Ar or Ar—O—Ar′, as well as R—O—R′ (Figure 19.8). The IUPAC names of ethers are not often used, and we will refer to an ether by indicating the two attached groups. For example, the name of $CH_3—O—CH_2CH_3$ is "methyl ethyl ether." Similarly, $CH_3—O—C_6H_5$ is named "methyl phenyl ether." Figure 19.8 illustrates the model structure of selected ethers.

In general, the physical properties of ethers lie between those of nonpolar hydrocarbons and those of polar alcohols of comparable molecular weight. Ethers are relatively nonpolar and do not hydrogen-bond. Table 19.7 briefly describes a few simple ethers.

Table 19.7 Ethers*

IUPAC Name	Common Name	Molecular Formula	Boiling Point	Solubility in Water
—	"dimethyl ether"	$CH_3—O—CH_3$	−25°C	soluble
—	"methyl ethyl ether"	$CH_3—O—CH_2CH_3$	8°C	soluble
—	"diethyl ether"	$CH_3CH_2—O—CH_2CH_3$	35°C	~7%
—	"diphenyl ether"	$C_6H_5—O—C_6H_5$	259°C	insoluble

*Most low molecular-weight ethers are liquids at room temperature; diphenyl ether is a solid.

19.8 Amines

Objective · **To write names and formulas for simple amines.**

If an alkyl replaces a hydrogen in ammonia, NH_3, the derivative is called an *amine*, $R—NH_2$ (Figure 19.9). Further substitution may occur to give amines with two or three alkyl substituents; $R_2—NH$ or $R_3—N$. In this class of compounds, the hydrocarbon group can be alkyl or aryl. For example,

$CH_3—CH_2—NH_2$ (a) "ethylamine"

$CH_3—NH$ with CH_3 on N (b) "dimethylamine"

$C_6H_5—NH_2$ (benzene ring with NH_2) (c) "phenylamine"

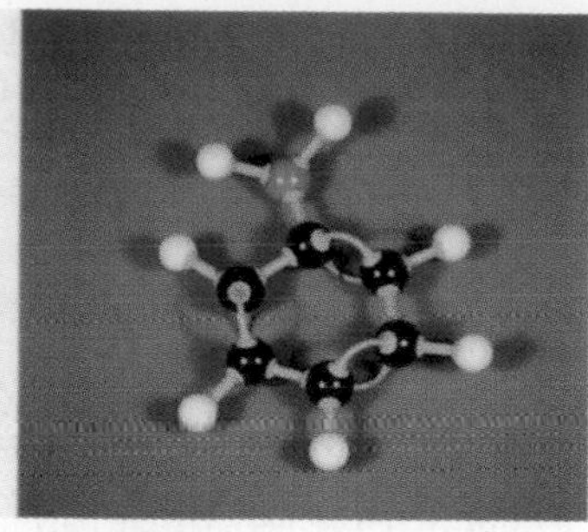

▲ **Figure 19.9 Molecular Models of Amines** Model structures of (a) ethylamine, $C_2H_5NH_2$; (b) dimethylamine, $(CH_3)_2NH$; and (c) phenylamine, $C_6H_5NH_2$.

The amine group is present in many compounds that have biological activity. Amines are found in B-complex vitamins, the hormone epinephrine, and the dental anesthetic novocaine, as well as in nicotine, amphetamine, and cocaine. Phenylamine, $C_6H_5NH_2$, is an important chemical used in the manufacture of dyes.

The IUPAC names of amines are not often used. Thus, we will refer to an amine by its common name and indicate the attached alkyl or aryl groups. The common name of CH_3NH_2 is "methylamine," and the name of $CH_3CH_2NH_2$ is "ethylamine." Similarly, the common name of $C_6H_5NH_2$ is "phenylamine."

Most amines are polar because of the $—NH_2$ group. In fact, amines are similar to alcohols in that they can form intermolecular hydrogen bonds. As we might expect, the physical properties of amines and alcohols are comparable. That is, they both have high boiling points and are soluble in water. Table 19.8 briefly describes a few simple amines.

Table 19.8 Amines*

IUPAC Name	Common Name	Molecular Formula	Boiling Point	Solubility in Water
—	"methylamine"	$CH_3—NH_2$	−7°C	soluble
—	"ethylamine"	$CH_3—CH_2—NH_2$	17°C	soluble
—	"propylamine"	$CH_3—CH_2—CH_2—NH_2$	48°C	soluble
—	"isopropylamine"	$CH_3—CH(NH_2)—CH_3$	33°C	soluble
—	"phenylamine"	$C_6H_5—NH_2$	185°C	~5%

*Most low-molecular-weight amines are liquids at room temperature.

19.9 Aldehydes and Ketones

Objective · To write names and formulas for simple aldehydes and ketones.

Many aldehydes and ketones have an appealing taste and a fragrant odor. For this reason they are frequently used as flavorings in food and candy and as fragrances in inhalants and perfumes. They are also used as solvents and as starting chemicals for organic synthesis.

Aldehydes

Aldehydes contain a carbonyl group (C═O) and, in general, are polar. In *aldehydes*, RCHO, the carbonyl group is attached to a hydrogen atom and either an alkyl or an aryl group (Figure 19.10). For example,

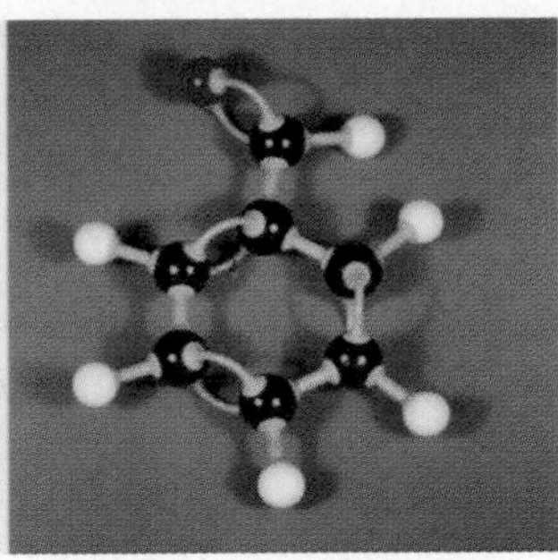

▲ **Figure 19.10 Molecular Models of Aldehydes** Model structures of (a) formaldehyde, HCHO; (b) acetaldehyde, CH_3CHO; and (c) benzaldehyde, C_6H_5CHO.

$$\underset{\text{(a) "formaldehyde"}}{H-\overset{\overset{\displaystyle O}{\|}}{C}-H} \qquad \underset{\text{(b) "acetaldehyde"}}{CH_3-\overset{\overset{\displaystyle O}{\|}}{C}-H} \qquad \underset{\text{(c) benzaldehyde}}{C_6H_5-\overset{\overset{\displaystyle O}{\|}}{C}-H}$$

The simplest aldehyde, formaldehyde, HCHO, is one of the top 50 industrial chemicals. Formaldehyde is perhaps best known as the solution used to preserve specimens in biology classes. The preservative is actually Formalin, an aqueous solution of formaldehyde. Formaldehyde is manufactured by the oxidation of methanol using a variety of metal catalysts.

According to IUPAC nomenclature, the name of an aldehyde is derived from the name of the parent alkane. The name is formed by dropping the final *e* and adding the suffix *-al*. For example, the IUPAC name for HCHO is methanal, and the name of CH_3CHO is ethanal. We can also refer to CH_3CHO by its common name "acetaldehyde."

Low-molecular-weight aldehydes are slightly polar because of the carbonyl group. They are not as polar as alcohols and do not hydrogen bond. Accordingly, the physical properties of aldehydes fall between those of polar alcohols and nonpolar hydrocarbons. In general, aldehydes are less soluble in water and have lower boiling points than similar alcohols. Table 19.9 lists a few simple aldehydes.

Table 19.9 Aldehydes*

IUPAC Name	Common Name	Molecular Formula	Boiling Point	Solubility in Water
methanal	"formaldehyde"	$H-CHO$	−21°C	soluble
ethanal	"acetaldehyde"	CH_3-CHO	21°C	soluble
propanal	"propionaldehyde"	CH_3-CH_2-CHO	49°C	soluble
benzaldehyde	"benzoic aldehyde"	C_6H_5-CHO	179°C	~1%

*Most low-molecular-weight aldehydes are liquids at room temperature.

Ketones

Ketones contain a carbonyl group (C═O) and, in general, are polar. In *ketones*, RCOR′, either an alkyl group or an aryl group is attached on each side of the carbonyl group (Figure 19.11). For example,

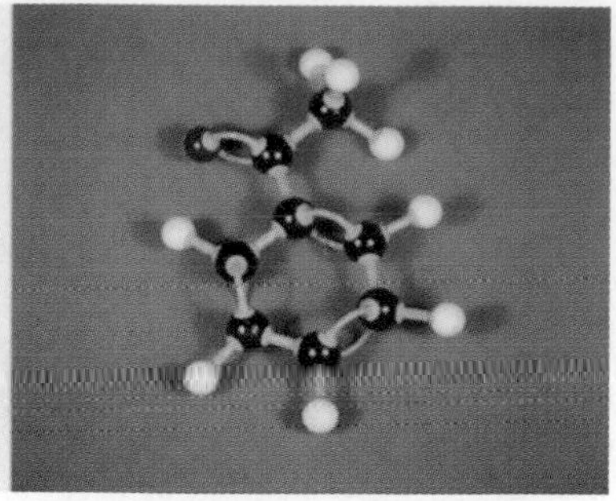

▲ **Figure 19.11 Molecular Models of Ketones** Model structures of (a) acetone, CH_3COCH_3; (b) butanone, $CH_3COC_2H_5$; and (c) acetophenone, $CH_3COC_6H_5$.

$$CH_3-\overset{\overset{\displaystyle O}{\|}}{C}-CH_3 \qquad CH_3-\overset{\overset{\displaystyle O}{\|}}{C}-CH_2-CH_3 \qquad CH_3-\overset{\overset{\displaystyle O}{\|}}{C}-C_6H_5$$

(a) "dimethyl ketone" or "acetone" (b) "methyl ethyl ketone" (c) "methyl phenyl ketone"

Acetone, CH_3COCH_3, is the simplest ketone and is one of the top 50 industrial chemicals. It evaporates rapidly, and its vapor is quite flammable. Although acetone is soluble in water, it can dissolve a wide range of polar and nonpolar compounds. Methyl ethyl ketone, $CH_3COCH_2CH_3$, is a solvent in the paint and varnish industry. Ketones are produced by the oxidation of alcohols. For example, large quantities of acetone are obtained from the oxidation of isopropyl alcohol.

According to IUPAC nomenclature, the name of a ketone is derived from the name of the parent alkane. The name is formed by dropping the final *e* and adding the suffix *-one*. For example, the IUPAC name for CH_3COCH_3 is propanone, and for $CH_3COCH_2CH_3$, butanone. We can also refer to CH_3COCH_3 by either of its common names, "dimethyl ketone" or "acetone."

Low-molecular-weight ketones are slightly polar because of the carbonyl group. They are not as polar as alcohols and do not hydrogen bond. In general, ketones are less soluble in water and have lower boiling points than similar alcohols. Table 19.10 lists three of the most common ketones.

Table 19.10 Ketones*

IUPAC Name	Common Name	Molecular Formula	Boiling Point	Solubility in Water
propanone	"acetone" or "dimethyl ketone"	$CH_3-CO-CH_3$	56°C	soluble
butanone	"methyl ethyl ketone"	$CH_3-CO-CH_2CH_3$	80°C	soluble
acetophenone	"methyl phenyl ketone"	$CH_3-CO-C_6H_5$	202°C	~1%

*Most low-molecular-weight ketones are liquids at room temperature.

19.10 Carboxylic Acids, Esters, and Amides

Objective · To write names and formulas for simple carboxylic acids, esters, and amides.

Carboxylic acids, esters, and amides are similar in that members of each class of compound contain a carbonyl group (C=O) and, in general, are polar. Carboxylic acids undergo many reactions, some of which produce esters and amides. Thus, esters and amides are derived from carboxylic acids.

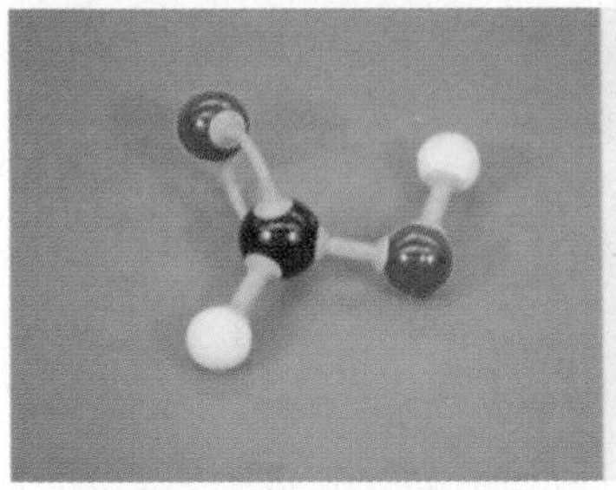
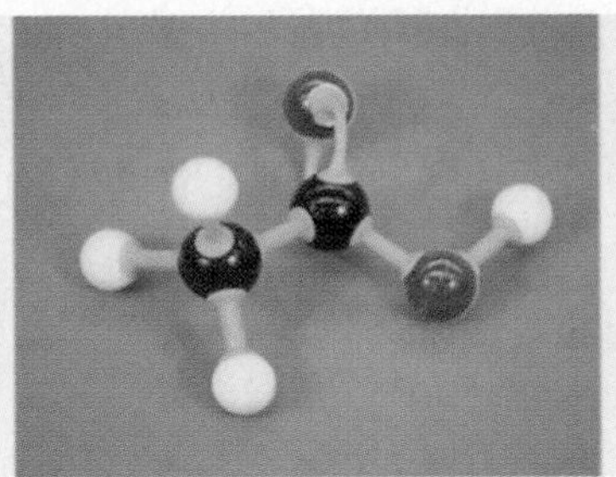
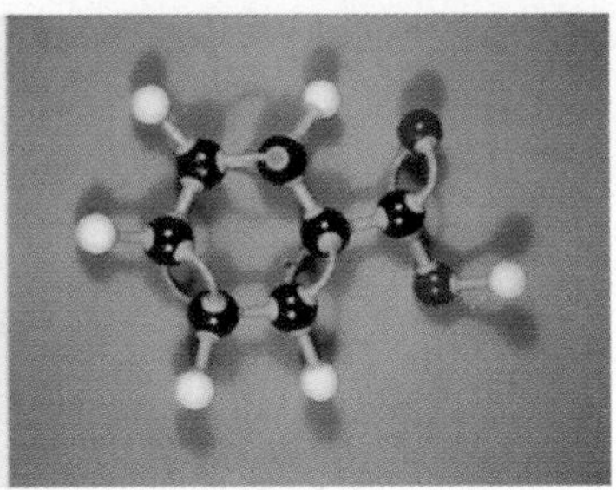

▲ **Figure 19.12 Molecular Models of Carboxylic Acids** Model structures of (a) formic acid, HCOOH; (b) acetic acid, CH_3COOH; and (c) benzoic acid, C_6H_5COOH.

Carboxylic Acids

In a *carboxylic acid*, RCOOH, we find a carbonyl group and a hydroxyl group bonded together to form a **carboxyl group**, —COOH (Figure 19.12). In this class of compounds, the hydrocarbon group can be alkyl or aryl. For example,

$$\mathrm{H{-}\overset{\overset{\displaystyle O}{\|}}{C}{-}OH} \qquad \mathrm{CH_3{-}\overset{\overset{\displaystyle O}{\|}}{C}{-}OH} \qquad \mathrm{C_6H_5{-}\overset{\overset{\displaystyle O}{\|}}{C}{-}OH}$$

(a) "formic acid" (b) "acetic acid" (c) benzoic acid

The simplest carboxylic acid is formic acid, HCOOH. Formic acid was first extracted from ants. Its name, in fact, is derived from the Latin *formica*, meaning "ant." Formic acid is responsible for the stinging sensation of a red ant bite. It is used industrially as a leather tanning agent and for coagulating rubber latex.

Acetic acid was first isolated from vinegar, and its name is derived from the Latin *acetum*, meaning "sour." Acetic acid gives vinegar its sour taste and is the most common carboxylic acid. It is used for manufacturing other chemicals and for preparing foods such as pickles, mayonnaise, and salad dressing.

According to IUPAC nomenclature, the name of an carboxylic acid is derived from the name of the parent alkane. The name is formed by dropping the final *e* and adding the suffix *-oic acid*. For example, the IUPAC name for HCOOH is methanoic acid, and for CH_3COOH, ethanoic acid. We also refer to CH_3COOH by its common name, "acetic acid."

Inasmuch as carboxylic acids have a polar — COOH group, they are capable of forming intermolecular hydrogen bonds. The properties of carboxylic acids are similar to those of alcohols, and they have high boiling points. Since the carboxyl group is quite polar, we would predict that carboxylic acids are usually soluble in water. When we refer to Table 19.11, we see that our predictions are correct.

Table 19.11 Carboxylic Acids*

IUPAC Name	Common Name	Molecular Formula	Boiling Point	Solubility in Water
methanoic acid	"formic acid"	H—COOH	101°C	soluble
ethanoic acid	"acetic acid"	CH_3—COOH	118°C	soluble
propanoic acid	"propionic acid"	CH_3—CH_2—COOH	141°C	soluble
benzoic acid	—	C_6H_5—COOH	249°C	~1%

*Most low-molecular-weight carboxylic acids are liquids at room temperature; benzoic acid is a solid which melts at 122°C.

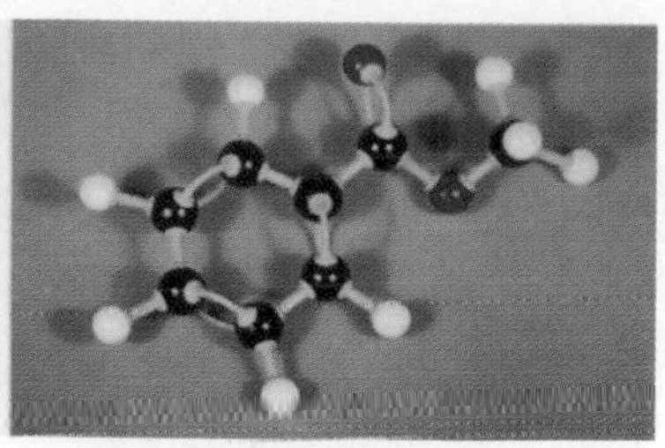

▲ **Figure 19.13 Molecular Models of Esters** Model structures of (a) ethyl formate, $HCOOC_2H_5$; (b) phenyl acetate, $CH_3COOC_6H_5$; and (c) methyl benzoate, $C_6H_5COOCH_3$.

Esters

In *esters*, RCOOR′, the carbonyl is attached to an —R and to an —OR′ group (Figure 19.13). In this class of compounds, the hydrocarbon group can be alkyl or aryl.

$$\mathrm{H{-}\overset{\overset{\displaystyle O}{\|}}{C}{-}O{-}CH_2CH_3} \qquad \mathrm{CH_3{-}\overset{\overset{\displaystyle O}{\|}}{C}{-}O{-}C_6H_5} \qquad \mathrm{C_6H_5{-}\overset{\overset{\displaystyle O}{\|}}{C}{-}O{-}CH_3}$$

(a) "ethyl formate" (b) "phenyl acetate" (c) methyl benzoate

Esters typically have a pleasant fruity odor. For example, butyl acetate has the aroma of banana, ethyl butyrate has the smell of pineapple, and ethyl formate has the taste of rum.

An ester is produced by the reaction of a carboxylic acid and an alcohol. In the presence of sulfuric acid, the reaction takes place by splitting out a molecule of water. For example, acetic acid and ethyl alcohol react to give ethyl acetate and water.

$$\mathrm{CH_3{-}\overset{\overset{\displaystyle O}{\|}}{C}{-}\mathbf{OH}} + \mathbf{HO}\mathrm{{-}CH_2CH_3} \xrightarrow{H_2SO_4} \mathrm{CH_3{-}\overset{\overset{\displaystyle O}{\|}}{C}{-}O{-}CH_2CH_3} + \mathbf{H_2O}$$

ethanoic acid "acetic acid" ethanol "ethyl alcohol" ethyl ethanoate "ethyl acetate"

▲ **The Fragrance of Esters** The aroma of banana is butyl acetate, that of apple is methyl butyrate, and that of orange is octyl acetate.

According to IUPAC nomenclature, the name of an ester is formed from the names of the parent alcohol and carboxylic acid and the suffix *-oate.* For example, the IUPAC name for $HCOOC_2H_5$ is formed from the name of the parent alcohol (ethyl alcohol) and the parent acid (methanoic acid). After adding the suffix *-oate,* the resulting name is ethyl methanoate. The common name for the ester $HCOOC_2H_5$ is "ethyl formate."

Esters are slightly polar because of the carbonyl group. They are not as polar as alcohols and do not hydrogen bond. Consequently, their physical properties fall between those of polar alcohols and those of nonpolar hydrocarbons. In general, esters are not soluble in water and have lower boiling points than similar alcohols. Table 19.12 describes a few common esters.

Table 19.12 Esters

IUPAC Name	Common Name	Molecular Formula	Boiling Point	Solubility in Water
methyl methanoate	"methyl formate"	$H{-}CO{-}OCH_3$	32°C	soluble
methyl ethanoate	"methyl acetate"	$CH_3{-}CO{-}OCH_3$	57°C	soluble
ethyl ethanoate	"ethyl acetate"	$CH_3{-}CO{-}OCH_2CH_3$	77°C	soluble
methyl benzoate	—	$C_6H_5{-}CO{-}OCH_3$	200°C	~1%

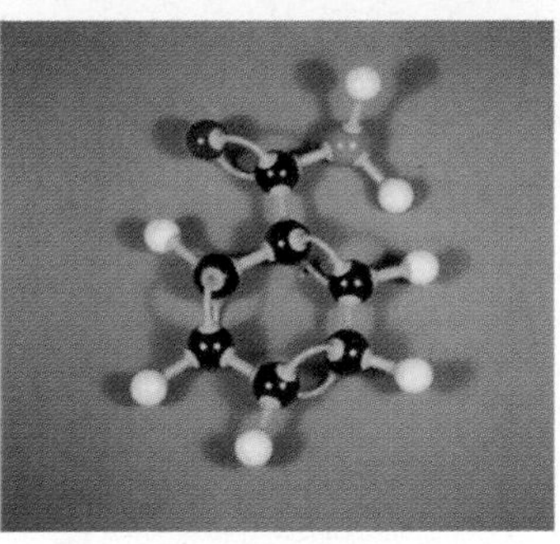

▲ **Figure 19.14 Molecular Models of Amides** Model structures of (a) formamide, $HCONH_2$; (b) acetamide, CH_3CONH_2; and (c) benzamide, $C_6H_5CONH_2$.

Amides

In *amides*, $RCONH_2$, the carbonyl is attached to an —R group and to an $—NH_2$ group (Figure 19.14). In this class of compounds, the hydrocarbon group can be alkyl or aryl.

$$H—\overset{\overset{\displaystyle O}{\|}}{C}—NH_2 \qquad CH_3—\overset{\overset{\displaystyle O}{\|}}{C}—NH_2 \qquad C_6H_5—\overset{\overset{\displaystyle O}{\|}}{C}—NH_2$$

(a) "formamide" (b) "acetamide" (c) benzamide

An amide is synthesized from a carboxylic acid and ammonia, NH_3. In the presence of heat, a molecule of water is lost as the amide is formed. For example, acetic acid and ammonia react to split out water and give acetamide.

$$CH_3—\overset{\overset{\displaystyle O}{\|}}{C}—\mathbf{OH} + \mathbf{H}—NH_2 \xrightarrow{\Delta} CH_3—\overset{\overset{\displaystyle O}{\|}}{C}—NH_2 + \mathbf{H_2O}$$

ethanoic acid "acetic acid" — ammonia — ethanamide "acetamide"

According to IUPAC nomenclature, the name of an amide compound ends in the suffix *-amide*. The name of an amide is formed by dropping *-oic acid* from the name of the parent acid and adding the suffix *-amide*. For example, the amide derivative of ethanoic acid is CH_3CONH_2; it is named ethanamide. The amide derivative of propanoic acid is $CH_3CH_2CONH_2$; it is named propanamide.

Inasmuch as amides have a polar $—CONH_2$ group, they are capable of forming intermolecular hydrogen bonds. The properties of amides are similar to those of carboxylic acids, and amides are usually soluble in water. If we refer to Table 19.13, we see that amides are solids at room temperature.

Table 19.13 Amides*

IUPAC Name	Common Name	Molecular Formula	Melting Point	Solubility in Water
methanamide	"formamide"	$H—CONH_2$	3°C	soluble
ethanamide	"acetamide"	$CH_3—CONH_2$	82°C	soluble
propanamide	"propionamide"	$CH_3CH_2—CONH_2$	79°C	soluble
benzamide	"benzoic acid amide"	$C_6H_5—CONH_2$	132°C	~1%

*Amides are generally solids at room temperature.

Chemistry Connection • Giant Molecules

What is the difference between a polyamide and polyester thread?

A *polymer* is a giant molecule made up of many small molecules (monomers) joined in a long chain. Polymers are an important product of the petrochemical industry. There are two basic types of polymers: addition and condensation.

An addition polymer is synthesized from an alkene. In an addition polymerization reaction, an alkene molecule adds to a second alkene molecule, which in turn adds a third molecule, and so on. For instance, an ethylene molecule, $CH_2{=}CH_2$, reacts to give a long-chain molecule called *polyethylene*. During the reaction, the double bond "springs open" and joins to another ethylene molecule as follows.

$$\underset{\text{ethylene}}{\overset{\text{monomer}}{CH_2{=}CH_2}} + \underset{\text{ethylene}}{\overset{\text{monomer}}{CH_2{=}CH_2}} + \cdots \longrightarrow \underset{\text{polyethylene}}{\overset{\text{polymer}}{[-CH_2-CH_2-]_n}}$$

The number of ethylene units (n) in the chain is typically large (~1000–100,000). Although the actual number of ethylene molecules can vary and affects the properties of the polymer, chemists can regulate the length of the chain by the choice of catalyst and reaction conditions. Polyethylene is used to make soft bottles and plastic bags.

▶ **Plastic Wrap** Plastic cling wrap and plastic bags are often made from polyethylene.

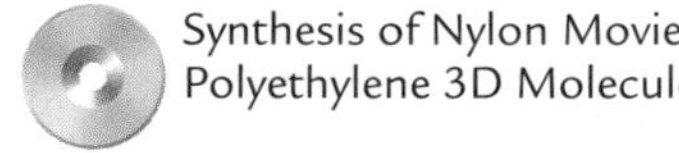

Synthesis of Nylon Movie; Polyethylene 3D Molecule

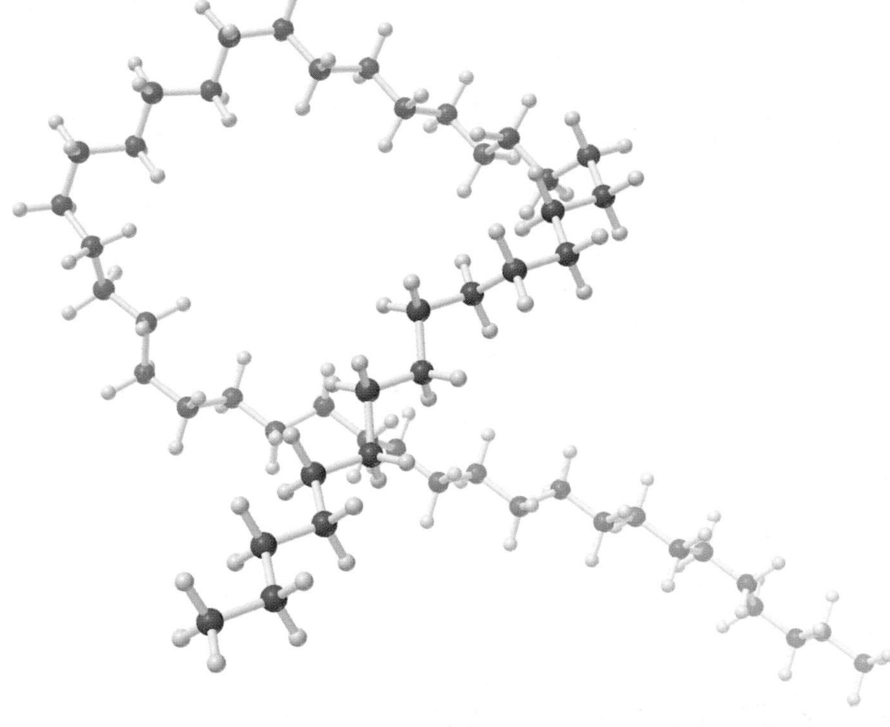

Polyethylene

◀ **Polyethylene** Ethylene, $CH_2{=}CH_2$, polymerizes to produce a long continuous polyethylene chain.

Similarly, *polypropylene* is synthesized from propylene, $CH_2{=}CH{-}CH_3$. Polypropylene is a rugged plastic used to make luggage and plastic toys. Another familiar polymer, *Teflon*, is synthesized from $CF_2{=}CF_2$. Teflon has a slippery feel and is used as a lubricant and to coat nonstick cookware. Other addition polymers are used to manufacture a variety of consumer products including plastic plumbing, fiberoptic cables, simulated leather, and Krazy Glue.

A condensation polymer is synthesized from hydrocarbon derivatives. Polyamides and polyesters are two of the most important types of condensation polymers. A *polyamide* is produced by the reaction of a carboxylic acid and an amine. For example, nylon is a polyamide that can be synthesized as follows.

$$\underset{\textit{adipic acid}}{HO{-}\overset{\overset{\displaystyle O}{\|}}{C}{-}(CH_2)_4{-}\overset{\overset{\displaystyle O}{\|}}{C}{-}OH} + \underset{\textit{hexamethylene diamine}}{H_2N{-}(CH_2)_6{-}NH_2} + \cdots \longrightarrow$$

$$\underset{\textit{nylon 6,6}}{[{-}\overset{\overset{\displaystyle O}{\|}}{C}{-}(CH_2)_4{-}\overset{\overset{\displaystyle O}{\|}}{C}{-}NH{-}(CH_2)_6{-}NH{-}]_n} + H_2O$$

Nylon polymer is extruded as fibrous threads used to make ropes, sheer fabrics, and nylon stockings. The term "nylon" has become a generic term as the reactants can vary slightly for similar nylon polymers. Different polyamides are used to manufacture bulletproof vests and fire-retardant fabric.

A *polyester* is produced by the reaction of a carboxylic acid and an alcohol. For example, the reaction of *para*-phthalic acid and ethylene glycol splits out a water molecule and gives the polyester we know as Dacron.

$$\underset{\textit{para-phthalic acid}}{HO{-}\overset{\overset{\displaystyle O}{\|}}{C}{-}C_6H_4{-}\overset{\overset{\displaystyle O}{\|}}{C}{-}OH} + \underset{\textit{ethylene glycol}}{HO{-}CH_2{-}CH_2{-}OH} + \cdots \longrightarrow$$

$$\underset{\textit{Dacron}}{[{-}\overset{\overset{\displaystyle O}{\|}}{C}{-}C_6H_4{-}\overset{\overset{\displaystyle O}{\|}}{C}{-}O{-}CH_2{-}CH_2{-}O{-}]_n} + H_2O$$

Dacron polymer fibers are extruded and can be woven with cotton to give a blended clothing fabric. When the fiber is extruded as a thin film, it is known as Mylar. Mylar is used to make magnetic audio- and videotape. The polymer is also used for party balloons that can stay inflated for weeks. Different polyesters are used to manufacture crash helmets and bulletproof windows.

A synthetic polyamide thread is produced from a carboxylic acid and an amine, whereas a polyester thread is made from a carboxylic acid and an alcohol.

Summary 

Section 19.1 **Organic chemistry** is the study of compounds that contain carbon. There are millions of organic compounds, but fortunately, they can be divided into a few families. **Hydrocarbons** contain only hydrogen and carbon, whereas **hydrocarbon derivatives** contain additional elements such as O, N, or a halogen. **Saturated hydrocarbons** include alkanes; **unsaturated hydrocarbons** include alkenes and alkynes. **Aromatic hydrocarbons** contain a benzene ring.

Section 19.2 All **alkanes** have single bonds and the general molecular formula C_nH_{2n+2}. The **structural formula** shows the arrangement of atoms in a molecule. Molecules having the same molecular formula but a different structural formula are called **isomers**. The name of each member of the alkane family ends in the suffix *-ane*. If a H atom is removed from an alkane, an **alkyl group** results. If a H atom is removed from an aromatic hydrocarbon, an **aryl group** results. If a H atom is removed from a benzene molecule, a **phenyl group** results. When an alkane burns with oxygen in air, it is said to undergo a **combustion reaction**.

Section 19.3 All **alkenes** have at least one double bond and the general molecular formula C_nH_{2n}. All **alkynes** have at least one triple bond and the general molecular formula C_nH_{2n-2}. The name of each member of the alkene family ends in the suffix *-ene*. The names of alkynes end in the suffix *-yne*. Alkenes and alkynes can undergo an **addition reaction**. That is, hydrogen and halogens react with alkenes and alkynes by adding a molecule of H_2, Cl_2, Br_2, or I_2 to an unsaturated bond.

Section 19.4 All **arenes** contain a benzene ring. If there are two substituents attached to the ring, there are three possible isomers: *ortho*, *meta*, and *para*. For example, if two bromine atoms are in adjacent positions on the ring, the isomer is called *ortho*-dibromobenzene. If the bromine atoms are separated by a carbon on the ring, the isomer is called *meta*-dibromobenzene. If the bromine atoms are on opposite sides of the ring, the isomer is called *para*-dibromobenzene.

Section 19.5 A family of compounds in which all the members have the same structural feature and similar chemical properties is called a **class of compounds**. The atom or group of atoms that characterizes a class of compounds is referred to as a **functional group**. There are 10 classes of hydrocarbon derivatives, 5 of which have a **carbonyl group** (C=O) and 5 of which do not.

Section 19.6 *Organic halides* are compounds containing hydrogen, carbon, and a halogen. They are generally nonpolar and insoluble in water. The IUPAC systematic name for an organic halide is derived from the names of the halogen and the parent alkane. For example, CH_3Br is named bromomethane, and its common name is "methyl bromide."

Section 19.7 *Alcohols* and *phenols* are generally polar and are soluble in water. They have unusually high boiling points because the **hydroxyl group** (—OH) can hydrogen bond. The systematic name for an alcohol is formed by applying the suffix *-ol* to the name of the parent alkane. For example, CH_3OH is named methanol, and CH_3CH_2OH is named ethanol. The most important member of the phenol class is C_6H_5OH; its systematic name is phenol. *Ethers* are generally nonpolar and only slightly soluble in water. Ethers are usually referred to by their common names; for example, $CH_3CH_2—O—CH_2CH_3$ is referred to as "diethyl ether."

Section 19.8 *Amines* are generally polar and are soluble in water. They have unusually high melting points and boiling points because they can hydrogen bond. The common name for an amine is formed by adding "amine" to the name of the parent alkyl group. For example, the name of CH_3NH_2 is "methylamine"; $CH_3CH_2NH_2$ is named "ethyl amine." The name of the aromatic amine $C_6H_5NH_2$ is "phenyl amine."

Section 19.9 *Aldehydes* and *ketones* are slightly polar, and those with low molecular weights are soluble in water. The systematic name for an aldehyde is formed by applying the suffix *-al* to the name of the parent alkane. For example, HCHO is named methanal but is usually referred to by its common name "formaldehyde." The systematic name for a ketone is formed by applying the suffix *-one* to the name of the parent alkane. For example, CH_3COCH_3 is named propanone, but is referred to as "acetone" or "dimethyl ketone."

Section 19.10 *Carboxylic acids* are soluble in water and have high boiling points because the **carboxyl group** (—COOH) can hydrogen bond. The systematic name for an acid is formed by applying the suffix *-oic acid* to the name of the parent alkane. For example, CH_3COOH is named ethanoic acid, but it is usually referred to as "acetic acid." *Esters* are slightly polar, and only the smallest members are soluble in water. These compounds are made by the reaction of a carboxylic acid and an alcohol. The systematic name for an ester is formed by applying the suffix *-oate* to the name of the parent carboxylic acid. For example, CH_3COOCH_3 is named methyl ethanoate. *Amides* are polar and can hydrogen bond. Amides are made from the reaction of a carboxylic acid and ammonia. The systematic name for an amide is formed by applying the suffix *-amide* to the name of the parent carboxylic acid. For example, CH_3COONH_2 is synthesized from ethanoic acid and is named ethanamide.

Key Concepts*

1. Which of the following are unsaturated hydrocarbons?
(a) hexane (b) hexene
(c) hexyne (d) benzene

2. Give a systematic IUPAC name for each of the following.
(a) $CH_3—CH_2—CH_2—CH_2—CH_2—CH_3$
(b) $CH_3—CH{=}CH—CH_2—CH_2—CH_3$
(c) $CH_3—CH_2—C{\equiv}C—CH_2—CH_3$
(d) C_6H_6

3. Draw the structural formula for each of the following.
(a) methyl alcohol (b) propylamine
(c) isopropyl iodide (d) diethyl ether

4. Draw the structural formula for each of the following.
(a) acetaldehyde (b) acetic acid
(c) phenylacetate (d) acetamide

5. Identify the five functional groups present in thyroxine (thyroid hormone).

HO—(diiodophenyl)—O—(diiodophenyl)—$CH_2—CH(NH_2)—C(=O)—OH$

6. Draw the alcohol and ether isomers with the molecular formula C_2H_6O.

7. Draw the aldehyde and ketone isomers with the molecular formula C_3H_6O.

8. Draw the carboxylic acid and ester isomers with the molecular formula $C_2H_4O_2$.

9. Note the nomenclature suffix and identify the probable class of compound for each of the following.
(a) limonene (b) iodoform
(c) cholesterol (d) codeine
(e) cinnamal (f) progesterone
(g) ethyl octanoate (h) thioacetamide

10. Draw the condensed structural formula for each of the following molecules.

(a)

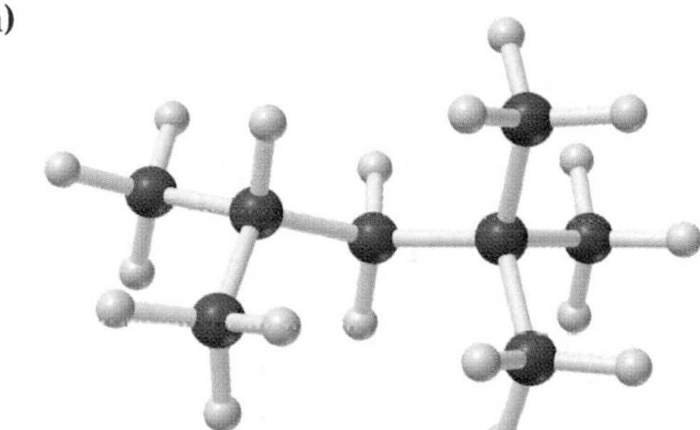

(b)

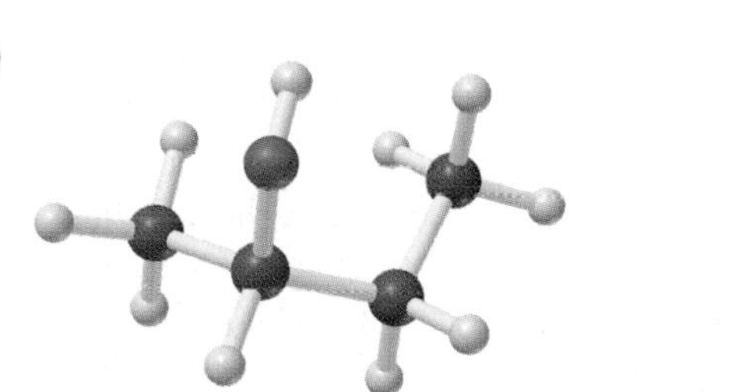

Key Terms†

Select the key term below that corresponds to each of the following definitions.

_____ 1. the study of carbon and its compounds
_____ 2. an organic compound containing only hydrogen and carbon
_____ 3. an organic compound containing carbon, hydrogen, oxygen,...

(a) addition reaction (*Sec. 19.3*)
(b) alkanes (*Sec. 19.2*)
(c) alkenes (*Sec. 19.3*)
(d) alkyl group (R—) (*Sec. 19.2*)
(e) alkynes (*Sec. 19.3*)

*Answers to Key Concepts are in Appendix H.
†Answers to Key Terms are in Appendix I.

_____ **4.** a hydrocarbon containing all single bonds

_____ **5.** a hydrocarbon containing a double bond or triple bond

_____ **6.** a hydrocarbon containing a benzene ring

_____ **7.** a family of compounds that are saturated hydrocarbons

_____ **8.** a family of compounds that are unsaturated hydrocarbons and have a double bond

_____ **9.** a family of compounds that are unsaturated hydrocarbons and have a triple bond

_____ **10.** a family of compounds that are aromatic hydrocarbons

_____ **11.** a formula that shows the arrangement of atoms in a molecule

_____ **12.** compounds with the same molecular formula but different structures

_____ **13.** a hydrocarbon fragment that results after the removal of a H atom from an alkane

_____ **14.** a hydrocarbon fragment that results after the removal of a H atom from an arene

_____ **15.** a hydrocarbon fragment that results after the removal of a H atom from benzene

_____ **16.** a chemical reaction in which a hydrocarbon reacts rapidly with O_2

_____ **17.** a chemical reaction in which an unsaturated hydrocarbon reacts with H_2 or Br_2

_____ **18.** a family of compounds in which all members have the same structural feature

_____ **19.** a structural feature that characterizes a class of compounds

_____ **20.** a structural feature composed of a carbon–oxygen double bond

_____ **21.** the functional group in an alcohol or phenol, —OH

_____ **22.** the functional group in a carboxylic acid, —COOH

(f) arenes (*Sec. 19.4*)
(g) aromatic hydrocarbon (*Sec. 19.1*)
(h) aryl group (Ar—) (*Sec. 19.2*)
(i) carbonyl group (*Sec. 19.5*)
(j) carboxyl group (*Sec. 19.10*)
(k) class of compounds (*Sec. 19.5*)
(l) combustion reaction (*Sec. 19.2*)
(m) functional group (*Sec. 19.5*)
(n) hydrocarbon (*Sec. 19.1*)
(o) hydrocarbon derivative (*Sec. 19.1*)
(p) hydroxyl group (*Sec. 19.7*)
(q) isomers (*Sec. 19.2*)
(r) organic chemistry (*Sec. 19.1*)
(s) phenyl group (*Sec. 19.2*)
(t) saturated hydrocarbon (*Sec. 19.1*)
(u) structural formula (*Sec. 19.2*)
(v) unsaturated hydrocarbon (*Sec. 19.1*)

Exercises‡

Hydrocarbons (Sec. 19.1)

1. What is the approximate percentage of organic compounds?
2. What is the approximate number of identified organic compounds?
3. What is the primary source of hydrocarbons?
4. What are three types of fossil fuels?
5. Classify each of the following hydrocarbons as saturated or unsaturated.
   (a) alkanes **(b)** alkenes
   (c) alkynes
6. What type of bonds characterize the following types of hydrocarbons?
   (a) alkanes **(b)** alkenes
   (c) alkynes

Alkanes (Sec. 19.2)

7. Which of the following molecular formulas represents an alkane?
   (a) $C_{10}H_{22}$ **(b)** $C_{14}H_{28}$
8. What is the molecular formula for each of the following alkanes?
   (a) $C_{12}H_?$ **(b)** $C_{20}H_?$
9. State the name of each of the following alkanes.
   (a) $CH_3—CH_2—CH_2—CH_3$
   (b) $CH_3—CH_2—CH_2—CH_2—CH_2—CH_3$
   (c) $CH_3—CH_2—CH_2—CH_2—CH_2—CH_2—CH_2—CH_3$
   (d) $CH_3—CH_2—CH_2—CH_2—CH_2—CH_2—CH_2—CH_2—CH_2—CH_3$
10. Draw the condensed structural formula for each of the following alkanes.
    (a) propane **(b)** pentane
    (c) heptane **(d)** nonane
11. Draw the condensed structural formula for the five isomers of hexane, C_6H_{14}.
12. Draw the condensed structural formula for the nine isomers of heptane, C_7H_{16}.
13. Draw the two isomers of bromopropane, C_3H_7Br.
14. Draw the four isomers of dibromopropane, $C_3H_6Br_2$.
15. Name each of the following alkyl groups.
    (a) $CH_3—$ **(b)** $CH_3CH_2—$
16. Draw the structure of each of the following alkyl groups.
    (a) propyl **(b)** isopropyl
17. Give the IUPAC names for each of the following structural formulas.

    (a)
$$\begin{array}{l} \qquad\qquad\qquad\qquad\qquad\quad CH_3 \\ \qquad\qquad\qquad\qquad\qquad\quad | \\ CH_3—CH_2—CH_2—CH—CH_3 \end{array}$$

‡Answers to odd-numbered Exercises are in Appendix J.

(b) $$\begin{array}{ccccccccccc} & & & & & & \mathrm{CH_2{-}CH_3} & & & & \\ & & & & & & | & & & & \\ \mathrm{CH_3} & - & \mathrm{CH} & - & \mathrm{CH_2} & - & \mathrm{CH} & - & \mathrm{CH_2} & - & \mathrm{CH_3} \\ & & | & & & & & & & & \\ & & \mathrm{CH_3} & & & & & & & & \end{array}$$

(c) $$\begin{array}{ccccccccccccc} & & & & & & \mathrm{CH_3} & & & & & & \\ & & & & & & | & & & & & & \\ \mathrm{CH_3} & - & \mathrm{CH} & - & \mathrm{CH_2} & - & \mathrm{C} & - & \mathrm{CH_2} & - & \mathrm{CH_2} & - & \mathrm{CH_3} \\ & & | & & & & | & & & & & & \\ & & \mathrm{CH_3} & & & & \mathrm{CH_3} & & & & & & \end{array}$$

(d) $$\begin{array}{ccccccccccccccc} & & & & & & & & \mathrm{CH_3} & & & & & & \\ & & & & & & & & | & & & & & & \\ \mathrm{CH_3} & - & \mathrm{CH_2} & - & \mathrm{CH_2} & - & \mathrm{CH} & - & \mathrm{CH} & - & \mathrm{CH} & - & \mathrm{CH_2} & - & \mathrm{CH_3} \\ & & & & & & | & & & & | & & & & \\ & & & & & & \mathrm{CH_3} & & & & \mathrm{CH_3} & & & & \end{array}$$

18. Draw the condensed structural formula for each of the following compounds.
    (a) 2-methylpropane (b) 2,2-dimethylbutane
    (c) 3-ethylheptane (d) 3,3-diethylpentane
19. Write a balanced equation for the complete combustion of each of the following.
    (a) CH_4 (b) C_5H_{12}
    (c) C_3H_8 (d) C_7H_{16}

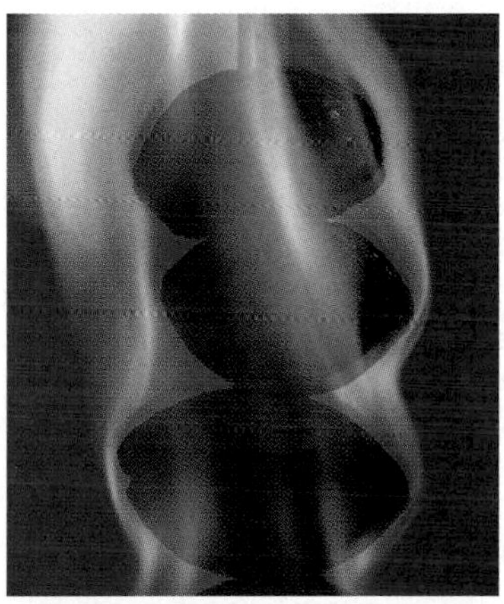

▲ **Combustion Reaction** Charcoal reacts with oxygen in air to give carbon dioxide, water, and heat.

20. Write a balanced equation for the complete combustion of each of the following.
    (a) ethane (b) butane
    (c) hexane (d) octane

Alkenes and Alkynes (Sec. 19.3)

21. Which of the following molecular formulas is an alkene?
    (a) $C_{10}H_{22}$ (b) $C_{14}H_{28}$
22. What is the molecular formula for each of the following alkenes?
    (a) $C_{12}H_?$ (b) $C_{20}H_?$
23. State the name of each of the following alkenes.
    (a) $CH_2{=}CH{-}CH_2{-}CH_3$
    (b) $CH_2{=}CH{-}CH_2{-}CH_2{-}CH_3$
    (c) $CH_3{-}CH{=}CH{-}CH_2{-}CH_2{-}CH_3$
    (d) $CH_3{-}CH_2{-}CH_2{-}CH{=}CH{-}CH_2{-}CH_2{-}CH_3$
24. Draw the condensed structural formula for each of the following alkenes.
    (a) 2-pentene (b) 3-heptene
    (c) 1-propene (d) 4-nonene
25. Which of the following molecular formulas is an alkyne?
    (a) $C_{10}H_{18}$ (b) $C_{14}H_{24}$
26. What is the molecular formula for each of the following alkynes?
    (a) $C_{12}H_?$ (b) $C_{20}H_?$
27. State the name of each of the following alkynes.
    (a) $CH_3{-}C{\equiv}C{-}CH_2{-}CH_3$
    (b) $CH{\equiv}C{-}CH_2{-}CH_2{-}CH_3$
    (c) $CH_3{-}CH_2{-}C{\equiv}C{-}CH_2{-}CH_3$
    (d) $CH_3{-}CH_2{-}CH_2{-}CH_2{-}C{\equiv}C{-}CH_3$
28. Draw the condensed structural formula for each of the following alkynes.
    (a) 1-propyne (b) 4-octyne
    (c) 2-butyne (d) 3-decyne
29. Draw the two structural isomers of straight-chain pentene, C_5H_{10}.
30. Draw the two structural isomers of straight-chain pentyne, C_5H_8.
31. Give the IUPAC name for each of the following structural formulas.

(a) $$\begin{array}{ccccccccc} & & & & & & \mathrm{CH_3} & & \\ & & & & & & | & & \\ \mathrm{CH_3} & - & \mathrm{CH} & = & \mathrm{CH} & - & \mathrm{CH} & - & \mathrm{CH_3} \end{array}$$

(b) $$\begin{array}{ccccccccccc} & & \mathrm{CH_3} & & & & \mathrm{CH_3} & & & & \\ & & | & & & & | & & & & \\ \mathrm{CH_3} & - & \mathrm{CH} & - & \mathrm{CH_2} & - & \mathrm{C} & = & \mathrm{CH} & - & \mathrm{CH_3} \end{array}$$

32. Draw the condensed structural formula for each of the following compounds.
    (a) 2-methyl-1-propene (b) 3,3-dimethyl-1-butyne
33. Give the IUPAC name for each of the following structural formulas.

(a) $$\begin{array}{ccccccccccccc} & & & & & & \mathrm{CH_3} & & & & & & \\ & & & & & & | & & & & & & \\ \mathrm{CH_3} & - & \mathrm{C} & \equiv & \mathrm{C} & - & \mathrm{C} & - & \mathrm{CH_2} & - & \mathrm{CH_2} & - & \mathrm{CH_3} \\ & & & & & & | & & & & & & \\ & & & & & & \mathrm{CH_3} & & & & & & \end{array}$$

(b) $$\begin{array}{ccccccccccccccc} & & & & & & & & \mathrm{CH_3} & & & & & & \\ & & & & & & & & | & & & & & & \\ \mathrm{CH_3} & - & \mathrm{CH_2} & - & \mathrm{CH_2} & - & \mathrm{CH} & - & \mathrm{CH} & - & \mathrm{CH} & - & \mathrm{C} & \equiv & \mathrm{CH} \\ & & & & & & | & & & & | & & & & \\ & & & & & & \mathrm{CH_3} & & & & \mathrm{CH_3} & & & & \end{array}$$

34. Draw the condensed structural formula for each of the following compounds.
    (a) 3-ethyl-2-heptene (b) 4,4-dimethyl-2-pentyne
35. Write a balanced equation for each of the following reactions.

(a) $CH_2{=}CH_2 \quad + \quad O_2 \longrightarrow$
(b) $CH_3-CH{=}CH_2 \quad + \quad H_2 \longrightarrow$
(c) $CH_3-CH{=}CH-CH_3 \quad + \quad Br_2 \longrightarrow$

36. Write a balanced equation for each of the following reactions.
(a) $CH{\equiv}C-CH_3 \quad + \quad O_2 \longrightarrow$
(b) $CH_3-C{\equiv}CH \quad + \quad 2\,H_2 \longrightarrow$
(c) $CH_3-C{\equiv}C \quad CH_3 \quad + \quad 2\,Br_2 \longrightarrow$

Aromatic Hydrocarbons (Sec. 19.4)

37. Draw the two Kekulé structures of benzene.

38. Draw the delocalized electron structure of benzene.

39. Draw the three isomers of the solvent xylene, $C_6H_4(CH_3)_2$.

40. Draw the three isomers of dinitrobenzene, $C_6H_4(NO_2)_2$.

Hydrocarbon Derivatives (Sec. 19.5)

41. Identify the general formula for each of the following hydrocarbon derivatives.
(a) $R-O-R'$ (b) $R-X$
(c) $Ar-OH$ (d) $R-NH_2$

42. Identify the general formula for each of the following hydrocarbon derivatives.
(a) $R-\overset{\overset{O}{\|}}{C}-OH$ (b) $R-\overset{\overset{O}{\|}}{C}-H$
(c) $R-\overset{\overset{O}{\|}}{C}-O-R'$ (d) $R-\overset{\overset{O}{\|}}{C}-NH_2$

43. Indicate the class of compound for each of the following hydrocarbon derivatives.
(a) CH_3-NH_2 (b) CH_3-CH_2-F
(c) $CH_3-\overset{\overset{O}{\|}}{C}-O-CH_3$ (d) $H-\overset{\overset{O}{\|}}{C}-OH$
(e) $C_6H_5-\overset{\overset{O}{\|}}{C}-H$ (f) $CH_3-CH_2-O-C_6H_5$
(g) $C_6H_5-\overset{\overset{O}{\|}}{C}-CH_3$ (h) $C_6H_5-\overset{\overset{O}{\|}}{C}-NH_2$

44. Indicate the class of compound for each of the following hydrocarbon derivatives.
(a) CH_3-OH
(b) $CH_3-CH_2-O-CH_2-CH_3$
(c) $CH_3-\overset{\overset{O}{\|}}{C}-NH_2$
(d) $CH_3-\overset{\overset{O}{\|}}{C}-CH_2-CH_3$
(e) $CH_3-\overset{\overset{O}{\|}}{C}-O-C_6H_5$
(f) $CH_3-C_6H_4-OH$
(g) C_6H_5-Cl
(h) $C_6H_5-CH_2-OH$

Organic Halides (Sec. 19.6)

45. Give the common name for each of the following organic halides.
(a) CH_3-I (b) CH_3-CH_2-Br
(c) $CH_3-CH_2-CH_2-F$ (d) $(CH_3)_2-CH-Cl$

46. Draw the structure of each of the following organic halides.
(a) dichloromethane (b) iodoethane
(c) 2-fluoropropane (d) 1,2-dibromopropane

47. Draw the structure of the common dry-cleaning solvent 1,1,1-trichloroethane.

48. Draw the structure of trichloroethylene (TCE), a common solvent used for degreasing electronic components.

49. Indicate whether organic halides are generally soluble or insoluble in each of the following solvents.
(a) water (b) hydrocarbons

50. Indicate whether organic halides have boiling points that are generally higher, lower, or similar to those of the following. (Assume the molecular mass is comparable.)
(a) alcohols (b) hydrocarbons

Alcohols, Phenols, and Ethers (Sec. 19.7)

51. Give the systematic IUPAC name for each of the following alcohols.
(a) $CH_3-CH_2-CH_2-CH_2-OH$
(b) $CH_3-CH_2-CH(OH)-CH_3$
(c) $CH_2(OH)-CH_2-CH_2-CH_3$
(d) $CH_3-CH(OH)-CH_2-CH_3$

52. Draw the structure of each of the following alcohols.
(a) methyl alcohol (b) ethyl alcohol
(c) propyl alcohol (d) isopropyl alcohol

53. Give an acceptable name for each of the following phenols.
(a) C_6H_5-OH (b) $CH_3-C_6H_4-OH$

54. Draw the structure for each of the following phenols.
(a) *ortho*-chlorophenol (b) *meta*-ethylphenol

55. Give the common name for each of the following ethers.
(a) CH_3-O-CH_3 (b) $CH_3-O-CH_2CH_3$

(c) $CH_3CH_2CH_2{-}O{-}CH_2CH_2CH_3$

(d) 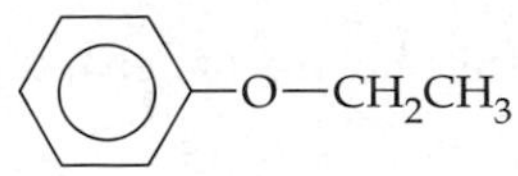 $C_6H_5{-}O{-}CH_2CH_3$

56. Draw the structure of each of the following ethers.
  (a) diisopropyl ether **(b)** diphenyl ether
  (c) ethyl propyl ether **(d)** methyl phenyl ether

57. Ethyl alcohol and dimethyl ether have the same molecular mass. Explain why the boiling point of ethyl alcohol (78°C) is much higher than the boiling point of dimethyl ether (–25°C).

58. Methanol, ethanol, and propanol are soluble in water. Explain why butanol is only slightly soluble.

59. There are two compounds that have the molecular formula C_2H_6O. Draw the structure of each isomer and classify the hydrocarbon derivative.

60. There are three compounds that have the molecular formula C_3H_8O. Draw the structure of each isomer and classify the hydrocarbon derivative.

Amines (Sec. 19.8)

61. Give the common name for each of the following amines.
  (a) $CH_3CH_2NH_2$ **(b)** $CH_3CH_2CH_2NH_2$

62. Draw the structure of each of the following amines.
  (a) methylamine **(b)** isopropylamine

63. Ethylamine and propane have about the same molecular mass. Explain why the boiling point of ethylamine (17°C) is much higher than that of propane (–44°C).

64. Ethylamine and diethylamine are very soluble in water. Explain why triethylamine is only slightly soluble.

Aldehydes and Ketones (Sec. 19.9)

65. Give the systematic IUPAC name for each of the following aldehydes.

(a) $H{-}\overset{\displaystyle O}{\overset{\|}{C}}{-}H$

(b) $CH_3{-}\overset{\displaystyle O}{\overset{\|}{C}}{-}H$

(c) $CH_3{-}CH_2{-}\overset{\displaystyle O}{\overset{\|}{C}}{-}H$

(d) $CH_3{-}CH_2{-}CH_2{-}\overset{\displaystyle O}{\overset{\|}{C}}{-}H$

66. Draw the structure of each of the following aldehydes.
  (a) formaldehyde **(b)** acetaldehyde
  (c) propionaldehyde **(d)** benzaldehyde

67. Give the systematic IUPAC name for each of the following ketones.

(a) $CH_3{-}\overset{\displaystyle O}{\overset{\|}{C}}{-}CH_3$

(b) $CH_3{-}\overset{\displaystyle O}{\overset{\|}{C}}{-}CH_2{-}CH_3$

(c) $CH_3{-}CH_2{-}CH_2{-}\overset{\displaystyle O}{\overset{\|}{C}}{-}CH_3$

(d) $CH_3{-}CH_2{-}\overset{\displaystyle O}{\overset{\|}{C}}{-}CH_2{-}CH_3$

68. Draw the structure of each of the following ketones.
  (a) methyl ethyl ketone **(b)** diethyl ketone
  (c) methyl phenyl ketone **(d)** diphenyl ketone

69. Indicate whether the following compounds have boiling points that are generally higher or lower than that of a hydrocarbon of comparable molecular mass.
  (a) aldehydes **(b)** ketones

70. Indicate whether the following compounds are generally soluble or insoluble in water.
  (a) aldehydes **(b)** ketones

71. An aldehyde and a ketone each have the molecular formula C_3H_6O. Draw the structure of each isomer.

72. An aromatic aldehyde and an aromatic ketone each have the molecular formula C_8H_8O. Draw the structure of each isomer.

Carboxylic Acids, Esters, and Amides (Sec. 19.10)

73. Give the systematic IUPAC name for each of the following carboxylic acids.

(a) $H{-}\overset{\displaystyle O}{\overset{\|}{C}}{-}OH$

(b) $CH_3{-}\overset{\displaystyle O}{\overset{\|}{C}}{-}OH$

(c) $CH_3{-}CH_2{-}\overset{\displaystyle O}{\overset{\|}{C}}{-}OH$

(d) $CH_3{-}CH_2{-}CH_2{-}\overset{\displaystyle O}{\overset{\|}{C}}{-}OH$

74. Draw the structure of each of the following carboxylic acids.
  (a) formic acid **(b)** acetic acid
  (c) propionic acid **(d)** benzoic acid

75. Give the systematic IUPAC name for each of the following esters.

(a) $H{-}\overset{\displaystyle O}{\overset{\|}{C}}{-}O{-}CH_2{-}CH_3$

(b) $CH_3{-}\overset{\displaystyle O}{\overset{\|}{C}}{-}O{-}CH_3$

(c) $CH_3-CH_2-\overset{O}{\overset{\|}{C}}-O-CH_2-CH_3$

(d) $H-\overset{O}{\overset{\|}{C}}-O-C_6H_5$

76. Draw the structure of each of the following esters.
(a) propyl formate (b) ethyl acetate
(c) phenyl propionate (d) ethyl benzoate

77. Give an acceptable name for each of the following amides.

(a) $H-\overset{O}{\overset{\|}{C}}-NH_2$ (b) $C_6H_5-\overset{O}{\overset{\|}{C}}-NH_2$

78. Draw the structure of each of the following amides.
(a) acetamide (b) propionamide

79. Which of the following classes of compounds can hydrogen bond?
(a) carboxylic acids (b) esters
(c) amides

80. Which of the following classes of compounds is generally soluble in water?
(a) carboxylic acids (b) esters
(c) amides

81. Propionic acid and methyl acetate have the same molecular formula, $C_3H_6O_2$. Predict which isomer has the higher boiling point.

82. Butanoic acid and ethyl acetate have the same molecular formula, $C_4H_8O_2$. Predict which isomer has the higher boiling point.

83. Write a balanced equation for each of the following reactions.

(a) $CH_3-\overset{O}{\overset{\|}{C}}-OH + CH_3-OH \xrightarrow{H_2SO_4}$

(b) $C_6H_5-\overset{O}{\overset{\|}{C}}-OH + NH_3 \xrightarrow{\Delta}$

84. Write a balanced equation for each of the following reactions.

(a) $C_6H_5-\overset{O}{\overset{\|}{C}}-OH + CH_3OH \xrightarrow{H_2SO_4}$

(b) $H-\overset{O}{\overset{\|}{C}}-OH + NH_3 \xrightarrow{\Delta}$

85. What ester is produced from the reaction of phenol and acetic acid?

86. What carboxylic acid and alcohol react to give ethyl formate?

87. What amide is produced from the reaction of ammonia and formic acid?

88. What carboxylic acid reacts with ammonia to give propionamide?

General Exercises

89. Give the IUPAC name for each of the following incorrectly named hydrocarbons.
(a) ethylmethane (b) propylethane

90. Give the IUPAC name for each of the following incorrectly named hydrocarbons.
(a) methylethane (b) isopropylmethane

91. Predict whether each of the following compounds is saturated or unsaturated as indicated by the name.
(a) eicosane (b) dodecene

92. Predict whether each of the following compounds is saturated or unsaturated as indicated by the name.
(a) cyclohexene (b) cyclopropane

93. The following pairs of compounds have about the same molecular mass. Predict which compound in each pair has the higher boiling point.
(a) CH_3-O-CH_3 or CH_3-CH_2-OH
(b) $CH_3-CH_2-CH_2-NH_2$ or $CH_3-CH_2-CH_2-F$

94. The following pairs of compounds have about the same molecular mass. Predict which compound in each pair has the higher boiling point.
(a) CH_3COOH or $HCOOCH_3$
(b) $CH_3CH_2CONH_2$ or CH_3COOCH_3

95. Note the nomenclature suffix and predict the probable class of hydrocarbon derivative for each of the following.
(a) chlordane (insecticide)
(b) cortisone (anti-inflammatory drug)
(c) acetylsalicylic acid (aspirin)
(d) sulfanilamide (antibacterial preparation)
(e) nicotine (in tobacco leaves)

96. Note the nomenclature suffix and predict the probable class of hydrocarbon derivative for each of the following.
(a) cresol (aromatic disinfectant)
(b) benzyl acetate (jasmine aroma)
(c) vinyl ether (general anesthetic)
(d) chloral hydrate (veterinary anesthetic)
(e) ethynyl estradiol (oral contraceptive)

Explorer Quiz 1
Explorer Quiz 2
Explorer Quiz 3
Master Quiz

CHAPTER 20
Biochemistry

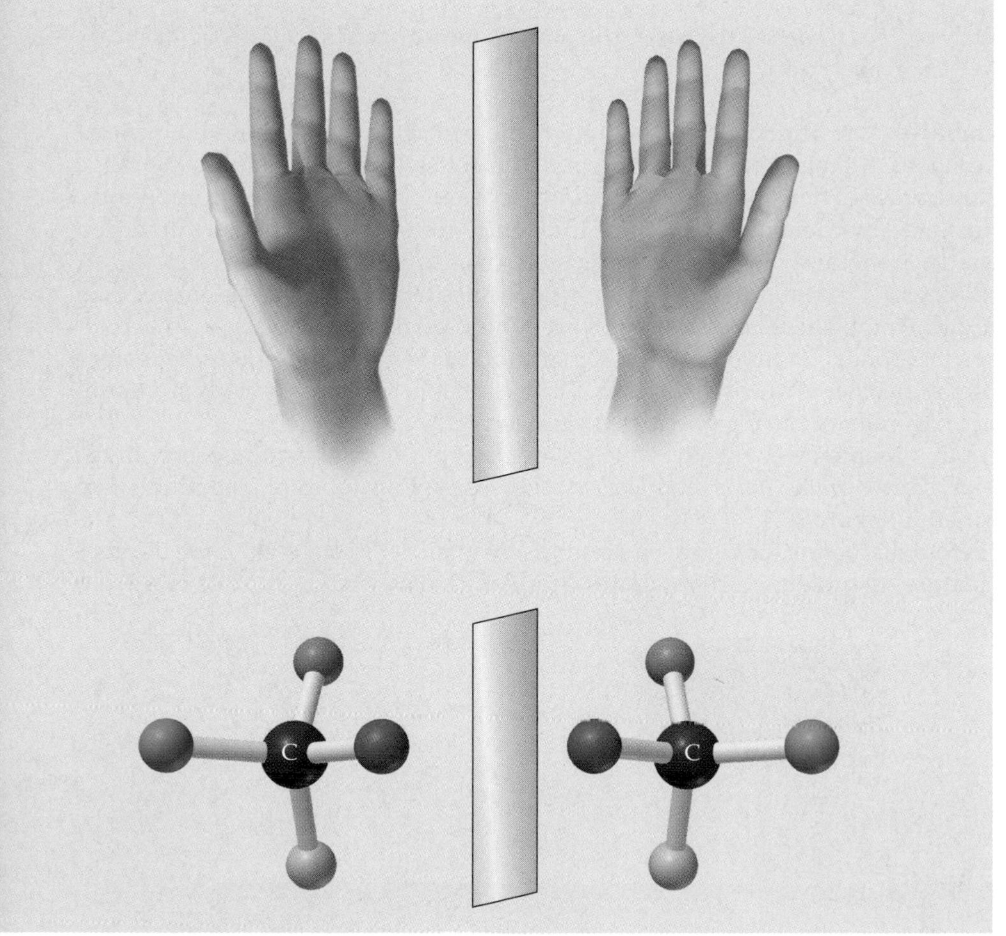

▲ Our hands illustrate the concept of nonidentical mirror images. Examine the two molecular models carefully. Are the two molecules an example of identical or nonidentical mirror images?

This chapter on biochemistry is an invaluable introduction and overview for students who are preparing for a second semester of organic chemistry and biochemistry.

Early in our discussion, we learned that about 30 chemical elements are essential for life. Figure 4.7 shows these *essential elements* that are necessary to support life for one or more biological species. The elements found in significant amounts in humans are oxygen, carbon, hydrogen, nitrogen, phosphorus, chlorine, sulfur, calcium, magnesium, sodium, and potassium. In addition, there are several *trace elements* that are present in only small amounts but nonetheless are necessary to sustain life. These trace elements include iron, zinc, cobalt, iodine, and a few others.

The study of substances derived from plants and animals is the branch of chemistry referred to as *biochemistry*. Many biochemists are employed in the pharmaceutical industry, which manufactures drugs that regulate chemical reactions in living systems. Some biochemists are referred to as molecular biologists because they study the relationship between an individual molecule and its biological role in determining the characteristics of an entire organism. As we proceed into the twenty-first century, applications of biotechnology and genetic engineering offer the potential for transforming our lives in ways that previously were unimaginable.

20.1 Biological Compounds

Objective · To recognize the basic structure of a protein, carbohydrate, lipid, and nucleic acid.

Biochemistry can be defined as the study of biological compounds and their chemical reactions. Although biological compounds are often large and complex, they have the same types of functional groups we encountered in simple organic compounds. For instance, biochemical compounds frequently contain an alcohol, an amine, an aldehyde, a ketone, a carboxylic acid, an ester, or an amide functional group.

Biological compounds are often large molecules with molar masses greater than 1,000,000 g/mol. These large molecules are composed of small molecules linked together in a long, continuous chain. A giant molecule containing many repeating smaller molecules is called a **polymer**. Protein, starch, and nucleic acids are examples of naturally occurring polymers found in plants and animals.

In this chapter, we will study the following types of biochemical compounds: *proteins*, *carbohydrates*, *lipids*, and *nucleic acids.* These four types of compounds are classified in Figure 20.1.

Before discussing biological compounds in detail, let's briefly describe the general features of proteins, carbohydrates, lipids, and nucleic acids. A **protein** is a nat-

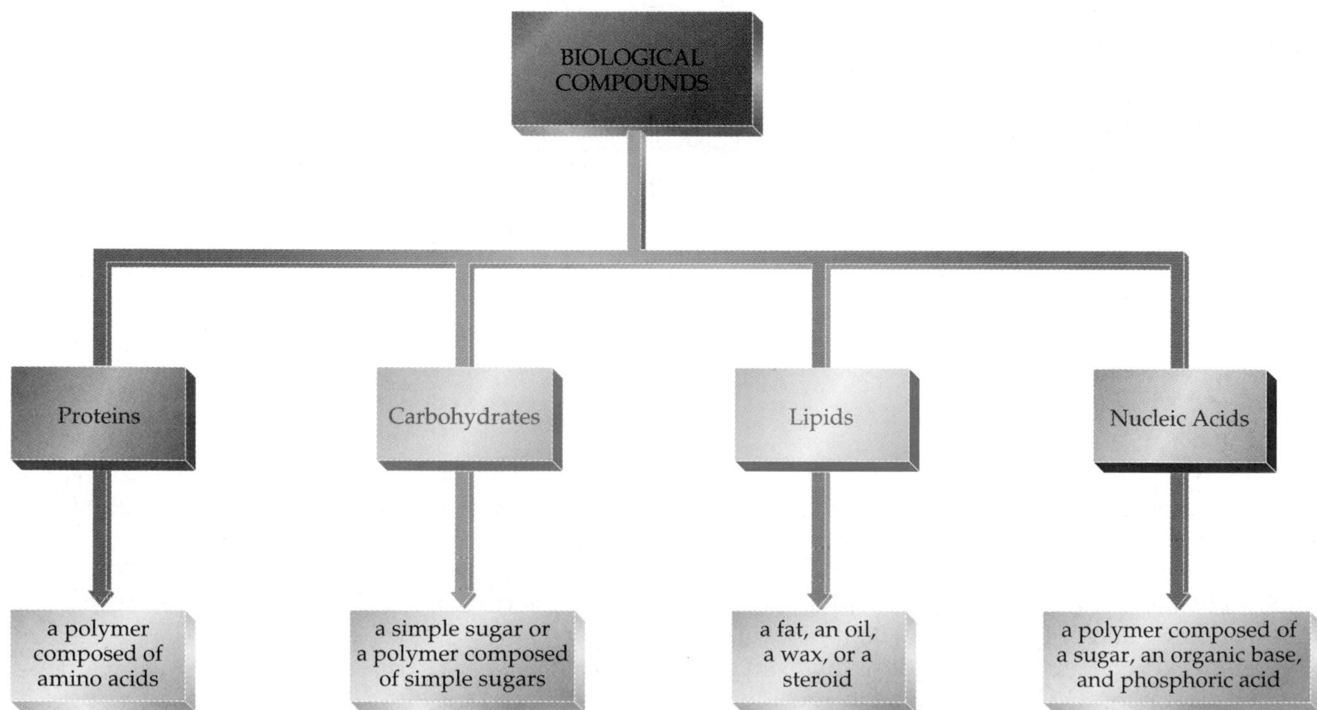

Classification of Biological Compounds

▲ **Figure 20.1 Classification of Biological Compounds** Biochemical compounds can be classified as one of the following: protein, carbohydrate, lipid, or nucleic acid.

urally occurring polymer composed of many amino acids. An amino acid has both an amine and a carboxylic acid functional group. In a protein, small amino acid molecules are linked together by amide bonds. Each of these amide bonds is called a *peptide linkage*.

Amino Acids and Peptide Linkage

--- Amino acid —C(=O)—NH— Amino acid ---

Peptide linkage

A **carbohydrate** is a compound that can be either a simple sugar or a polymer composed of simple sugar molecules. A carbohydrate usually contains either an aldehyde or a ketone functional group. Furthermore, a typical carbohydrate is characterized by having several alcohol groups: Glucose is a simple sugar that contains an aldehyde and five alcohol groups; fructose is a simple sugar that contains a ketone and five alcohol groups.

Starch is a carbohydrate polymer composed of glucose units linked together in a large molecule. In carbohydrates containing more than one sugar unit, the sugar molecules are joined together by an —O— linkage. The —O— bond between sugar molecules in a carbohydrate is called a *glycoside linkage,* which we can show as follows.

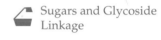

Sugars and Glycoside Linkage

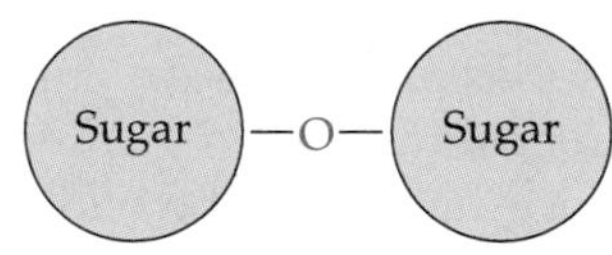

Glycoside linkage

A **lipid** is a water-insoluble compound usually composed of an alcohol and one or more carboxylic acid molecules. The lipids include animal fats and vegetable oils. Fats and oils are esters of glycerol, an alcohol that has three —OH groups. As a result, each molecule of a fat or an oil contains three ester groups resulting from three carboxylic acid molecules joined to one glycerol molecule. We can show the *ester linkages* in a fat or oil molecule as follows.

Lipids and Ester Linkages

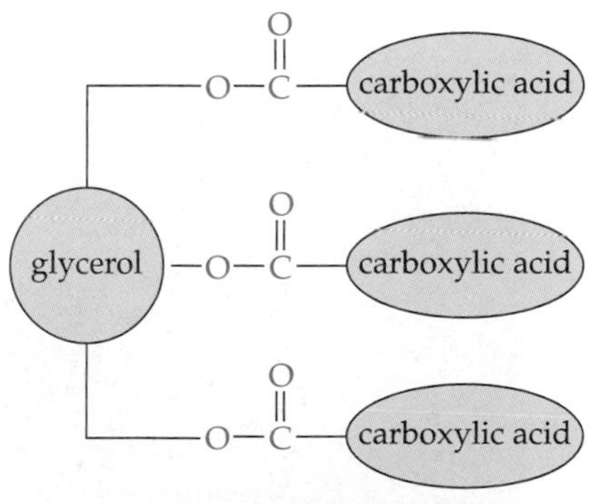

Ester linkages

A **nucleic acid** is a biochemical compound that contains a sugar molecule attached to a nitrogen-containing molecule. The nitrogen-containing molecule has an amine functional group and is termed an *organic base.* In turn, the sugar molecules are joined together in a long polymer chain. In a nucleic acid molecule, there may be millions of sugar molecules (each bearing an organic base) linked by phosphate ester bonds. We can show the *phosphate linkage* in a typical nucleic acid as follows.

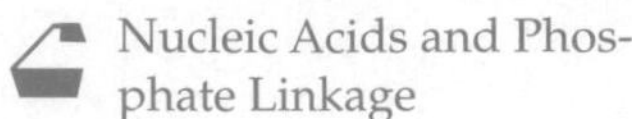
Nucleic Acids and Phosphate Linkage

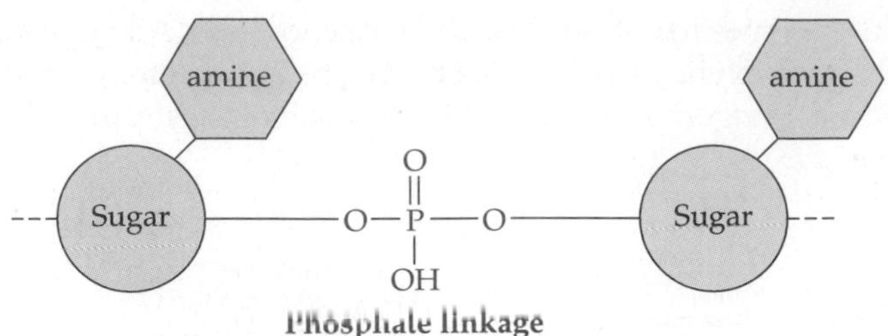

Phosphate linkage

We will now examine the composition of these four types of biological compounds in more detail. That is, we will discuss the structure and function of proteins, carbohydrates, lipids, and nucleic acids.

20.2 Proteins

Objectives
- To identify a peptide linkage between amino acids.
- To recognize the primary, secondary, and tertiary structure of proteins.

Proteins and Amino Acids Movie

Proteins play a crucial role in virtually all biological processes. In fact, the word "protein" is derived from the Greek word *proteios,* meaning "of first importance." The shapes of protein molecules have been observed to be either long fibers or compact globules. Moreover, biochemists have found that the molecular shape of a protein is related to its biological function.

Long protein fibers are responsible for the structure, support, and motion of an organism. Accordingly, protein fibers are found in muscle tissue and cartilage. In addition, protein fibers compose the hair, skin, and nails. On the other hand, compact protein globules are responsible for metabolic processes. For example, many hormones are globular proteins such as those secreted by the pituitary gland and the pancreas. Globular proteins are involved in controlling growth, transporting small molecules and ions, and providing immune protection from foreign substances.

Human protein is composed of 20 different amino acids. Of these 20 amino acids, only 8 are essential and must be included in the diet. The remaining 12 amino acids are nonessential, and the human body can synthesize them from one of the 8 essential amino acids. An **amino acid** has an amine group, $—NH_2$, attached to a carbon that is also bonded to a carboxyl group, —COOH. Since the amine group is

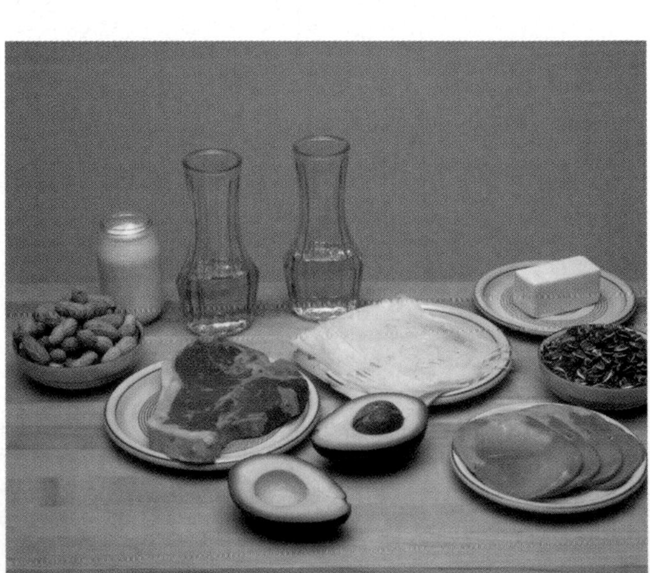
▶ **Protein Foods** Meat, fish, poultry, cheese, and milk products are all food sources rich in protein.

always attached to the carbon next to the carboxyl group (alpha carbon, α), an amino acid is referred to as an α-amino acid. The α-amino acids have the following general structural formula.

$$\begin{array}{c} \overset{\alpha}{C}H-\overset{O}{\overset{\|}{C}}-OH \\ NH_2-\overset{|}{\underset{R}{}}\end{array}$$

$$NH_2-\underset{\displaystyle R}{\underset{|}{\overset{\alpha}{C}H}}-\overset{\displaystyle O}{\overset{\|}{C}}-OH$$

Alpha amino acid

Amino acids differ from one another according to the R— group attached to the α-carbon. This R— group is referred to as the side chain. In some amino acids, the R— group is neutral, and in other amino acids the side chain has either acidic or basic properties. The structures of the 20 α-amino acids found in human protein are shown in Figure 20.2. For convenience, chemists abbreviate the names of amino acids by a three-letter symbol; for example, the name of the amino acid glycine is abbreviated Gly.

We mentioned previously that proteins consist of amino acids linked together by amide bonds. In a protein, an amide bond is called a **peptide linkage** or a peptide bond. A peptide linkage has the following general structural features.

$$\text{(amine end)}\quad NH_2-\underset{\displaystyle R}{\underset{|}{CH}}-\overset{\displaystyle O}{\overset{\|}{C}}-NH-\underset{\displaystyle R}{\underset{|}{CH}}-\overset{\displaystyle O}{\overset{\|}{C}}-OH\quad \text{(carboxyl end)}$$

Peptide linkage

In a **dipeptide**, 2 amino acid units are joined by a peptide bond. In a tripeptide, there are 3 units, and in a **polypeptide**, there are up to 50 amino acid units. Human insulin, for example, is a polypeptide composed of 48 amino acid units. We can illustrate a specific dipeptide by joining the 2 amino acids glycine and alanine.

$$\underset{\text{glycine (Gly)}}{H_2N-\underset{\displaystyle H}{\underset{|}{CH}}-\overset{\displaystyle O}{\overset{\|}{C}}-\boxed{OH\ +\ H_2N}}\underset{\text{alanine (Ala)}}{-\underset{\displaystyle CH_3}{\underset{|}{CH}}-\overset{\displaystyle O}{\overset{\|}{C}}-OH} \longrightarrow$$

$$H_2N-\underset{\displaystyle H}{\underset{|}{CH}}-\overset{\displaystyle O}{\overset{\|}{C}}-NH-\underset{\displaystyle CH_3}{\underset{|}{CH}}-\overset{\displaystyle O}{\overset{\|}{C}}-OH + H_2O$$

glycylalanine (Gly-Ala)

The dipeptide of glycine and alanine is named glycylalanine and is abbreviated Gly-Ala. The sequence of amino acids in a peptide or protein is always listed beginning with the amine end of the chain on the left. The other amino acids in the chain are listed in order, and the amino acid with the free carboxyl group is mentioned last. In the aforementioned dipeptide example, the amine end of the peptide chain is glycine (Gly) and the carboxyl end is alanine (Ala).

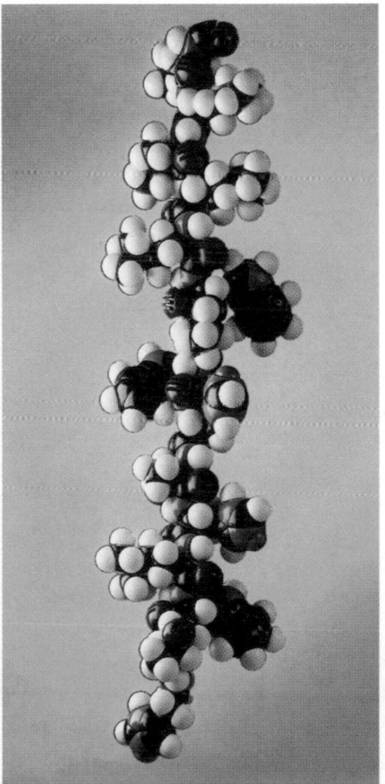

▲ **α-Helix** Molecular model showing a portion of a protein chain whose secondary structure is arranged in an α-helix.

Primary Structure of Proteins

Proteins are polypeptides that may contain hundreds or thousands of amino acid units linked together in a long-chain molecule. Egg whites, for example, contain a protein chain with nearly 400 amino acid molecules linked by peptide bonds.

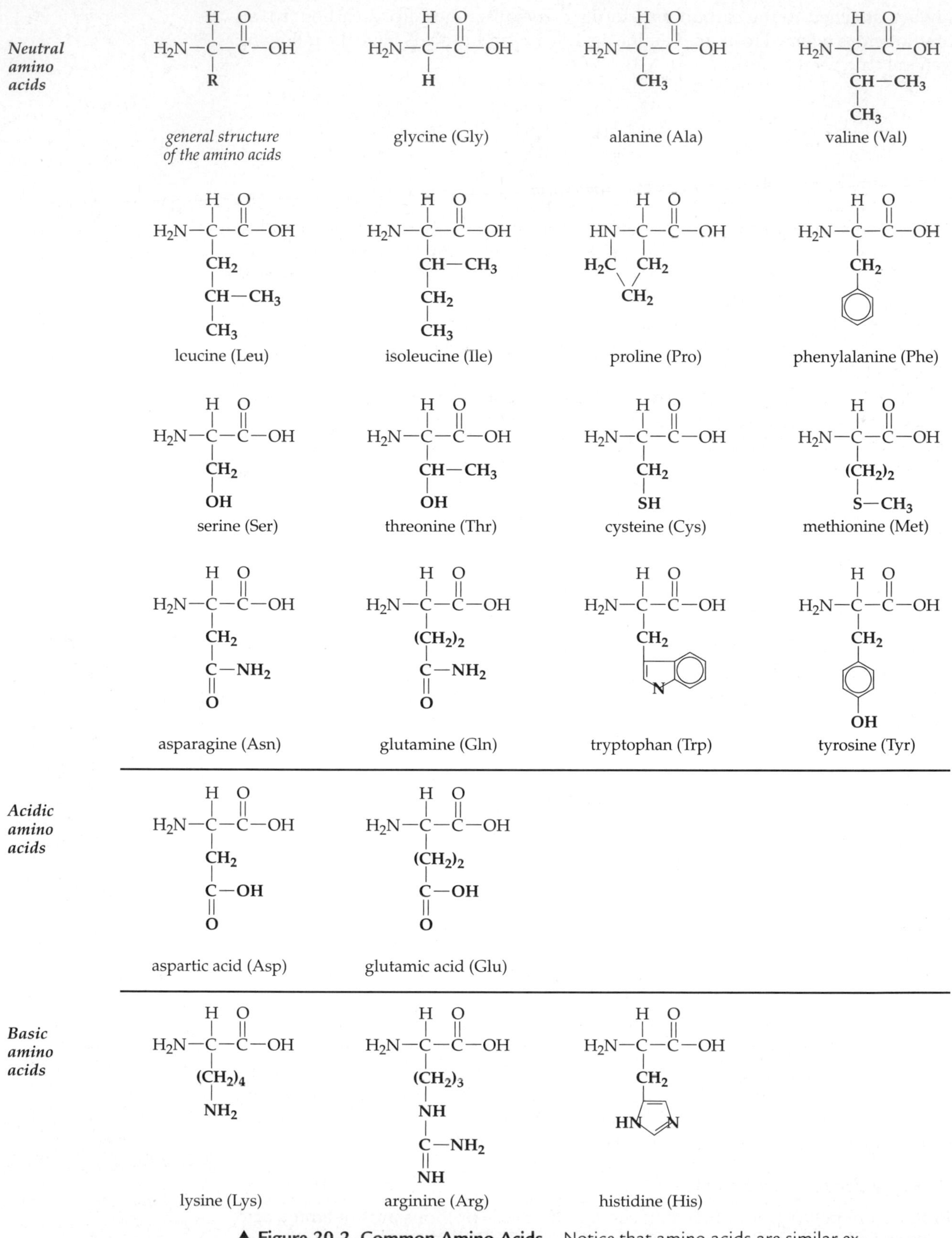

▲ **Figure 20.2 Common Amino Acids** Notice that amino acids are similar except for the R— group. The side chain is responsible for the amino acid having neutral, acidic, or basic properties.

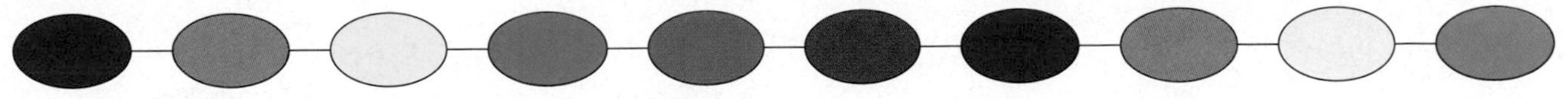

▲ **Figure 20.3 Primary Structure of a Protein** The primary structure of a protein refers to the sequence of amino acids present in the polymer chain. (Each ellipse represents an amino acid.)

Primary Structure of a Protein

One of the most amazing scientific achievements of the twentieth century was the unraveling of the mystery of protein synthesis in living organisms. In the 1960s biochemists discovered how living organisms synthesize huge protein molecules in which thousands of amino acids are always in the exact same sequence. When an organism synthesizes a protein, the sequence of amino acids is determined by a genetic code embodied in the structure of a nucleic acid.

Using currently available instruments, biochemists can determine the exact sequence of amino acids in small proteins. The sequence of amino acids in a peptide chain is referred to as the *primary structure* of the peptide. Figure 20.3 shows the primary structure of a polypeptide with several amino acids.

The replacement of a single amino acid by another amino acid in the primary structure of a protein can alter its biological activity. Even a small change in the primary amino acid sequence can drastically alter the biochemistry of a protein. For example, the disease sickle-cell anemia results from an amino acid substitution in hemoglobin. Hemoglobin is the protein responsible for oxygen transport through the circulatory system. Sickle-cell anemia results when the amino acid glutamic acid is replaced by the amino acid valine.

Secondary Structure of Proteins

Heme 3D Molecule

When a peptide chain twists and bends, the protein acquires a *secondary structure.* One type of secondary structure exhibited by proteins is referred to as an α-helix. This is the structure of protein molecules found in hair and wool fibers. An α-helix structure is analogous to that of a coiled Slinky toy. That is, an α-helix is a coiled chain of amino acids. Figure 20.4 illustrates the α-helix secondary structure of a hypothetical polypeptide composed of several amino acids.

Another type of secondary structure exhibited by proteins is referred to as a pleated sheet. In this arrangement, protein chains are antiparallel; that is, the carboxyl ends of adjacent protein chains run in opposite directions. This is the structure of protein molecules found in muscle and silk fibers. A pleated sheet structure is analogous to a sheet of paper folded into an accordion shape. Figure 20.5 illustrates the pleated sheet secondary structure for a pair of hypothetical polypeptides, such as silk and collagen.

What maintains the secondary structure of a protein? The answer is hydrogen bonding (Section 13.3). Hydrogen bonds between amino acids are responsible for the secondary structure of a protein. In an α-helix, the hydrogen bonds are between

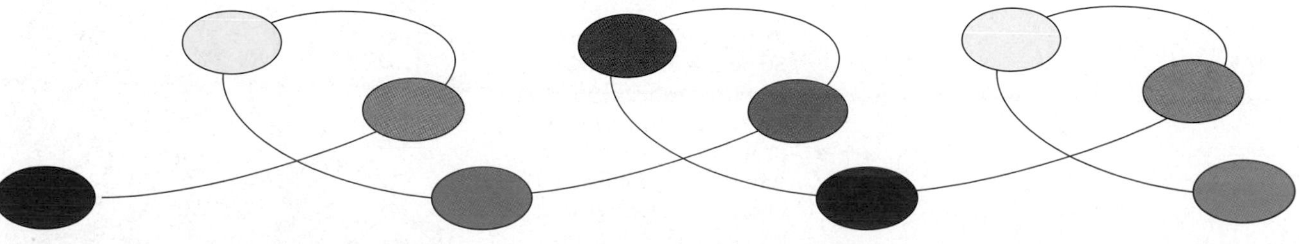

▲ **Figure 20.4 Secondary Structure of a Protein: α-Helix** The secondary structure of a protein may be in the shape of a coiled chain of amino acids referred to as an α-helix. (Each ellipse represents an amino acid.)

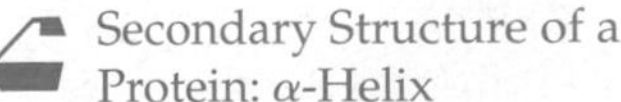

Secondary Structure of a Protein: α-Helix

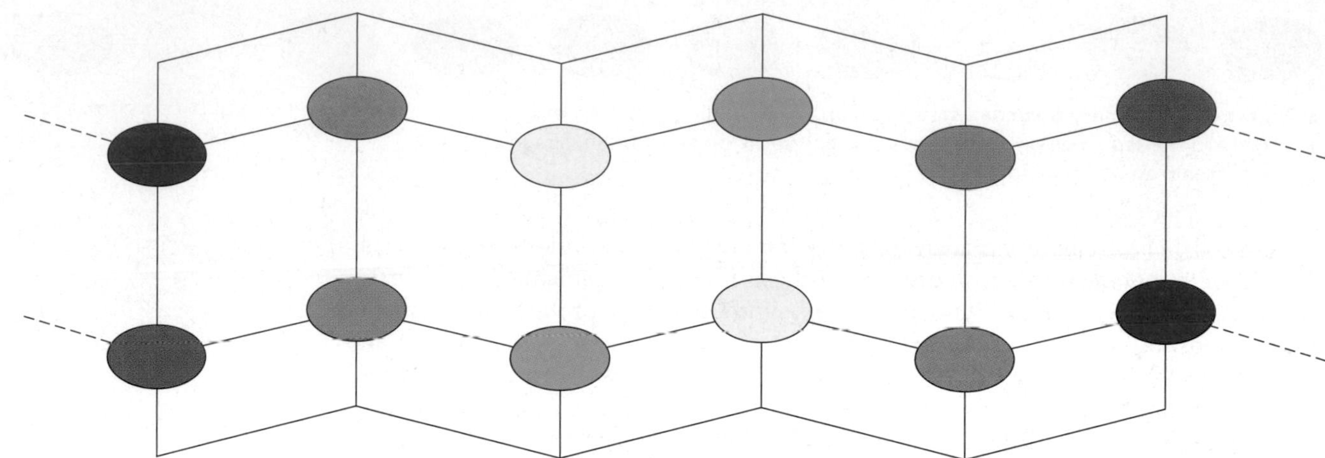

Secondary Structure of a Protein: Pleated Sheet

▲ Figure 20.5 Secondary Structure of a Protein: Pleated Sheet The secondary structure of a protein may be in the shape of folded chains of amino acids referred to as a pleated sheet. Notice that the protein chains are parallel but aligned in opposite directions. (Each ellipse represents an amino acid.)

nearby amino acid units in the same coiled protein chain. In a pleated sheet, the hydrogen bonds are between amino acids in neighboring protein chains.

Hydrogen bonds are weaker than ordinary covalent bonds, and the secondary structure of a protein can collapse with a change in pH or temperature. The process of breaking hydrogen bonds in a protein molecule is called *denaturation*. A practical example of denaturing a protein is boiling an egg in hot water.

Tertiary Structure of Proteins

The overall structure of a protein chain is referred to as its *tertiary structure*. The tertiary structure of a protein may be long and extended or compact and folded. As an analogy, the tertiary structure of a protein may be similar to a Slinky that is straight and extended or a Slinky that is folded into a knot. The twisting of an α-helix or a pleated sheet illustrates a folded tertiary structure. The formation of a twisted helix results in a protein molecule with the appearance of a small globule.

The tertiary structure of a protein is held together by intermolecular forces, primarily hydrogen bonds. Figure 20.6 illustrates the tertiary structure of myoglobin. Myoglobin is a globular protein that stores oxygen in animal tissue until it is required for a metabolic activity.

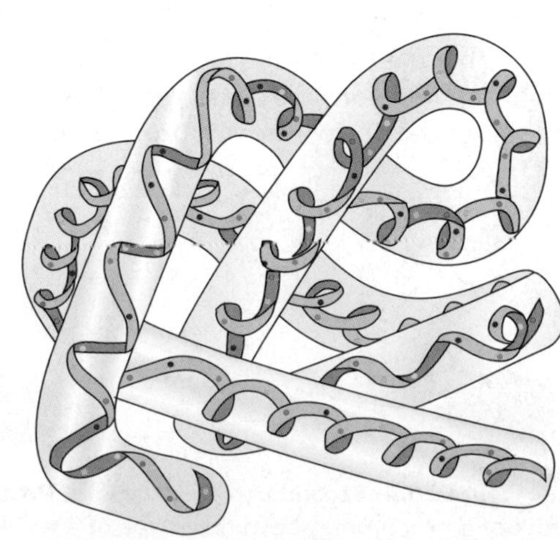

▶ Figure 20.6 Tertiary Structure of a Protein A twisted α-helix can create the tertiary structure of a protein. This model structure shows the coiled secondary structure twisted into a tertiary structure. (Each dot represents an amino acid.)

20.3 Enzymes

Objective · To explain the role of an enzyme using the lock-and-key model.

A single human cell may contain many enzymes that are crucial for life processes. An **enzyme** is a protein that acts as a biological catalyst. Enzymes are remarkably selective for specific molecules. An enzyme can speed up a biochemical reaction so that the rate is a million times faster than it would be in the absence of the enzyme. In fact, without enzymes, these reactions would take place so slowly that life would be impossible. Furthermore, only a very small amount of enzyme catalyst is necessary because a biochemical reaction can occur in microseconds, and the enzyme is then free to repeat its role as catalyst.

Biochemical reactions require sensitive conditions and cannot withstand the conditions of extreme pH or high temperature often associated with inorganic and organic chemical reactions. If the pH or temperature is too high or too low, an enzyme becomes inactive and a biochemical reaction cannot take place. A chemical that renders an enzyme inactive is called an *inhibitor*. Nerve gas is an inhibitor that operates by blocking an enzyme responsible for breaking down lethal toxins that accumulate naturally in the body.

The function of an enzyme is complex, but we can use a **lock-and-key model** to explain the mechanism. The lock-and-key theory proposes that an enzyme molecule has a given structural feature similar in shape to the shape of the reacting molecule. This location on the enzyme where a reaction can occur is called the *active site,* and the reacting molecule is referred to as the *substrate.* Figure 20.7 illustrates the lock-and-key model using a lock and key as an analogy.[1]

An enzyme-catalyzed reaction takes place in two steps. First, the substrate (S) binds to the active site on the enzyme (E). Second, the substrate molecule undergoes a reaction and the enzyme releases two or more substrate products (P_1 and P_2).

Step 1: $E + S \longrightarrow ES$
Step 2: $ES \longrightarrow E + P_1 + P_2$

Figure 20.8 illustrates this two-step process for an enzyme-catalyzed reaction of a substrate molecule to give two products.

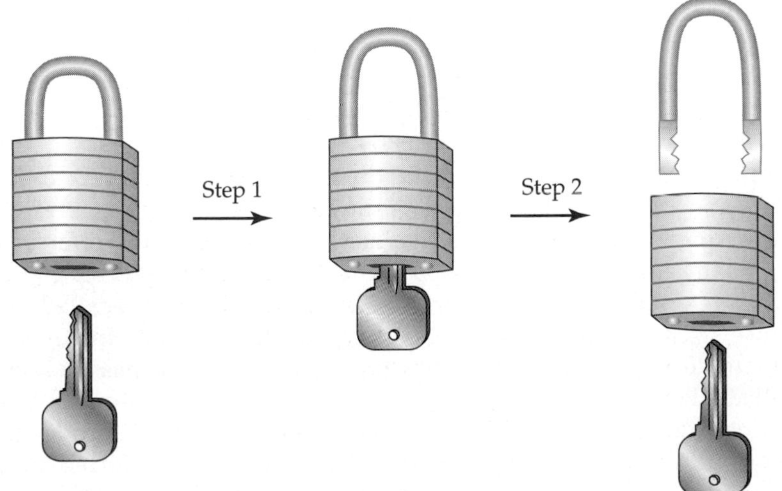

◀ Figure 20.7 Analogy for Enzyme Catalysis In Step 1, the key (enzyme) is inserted into the lock (substrate). In Step 2, the key opens the lock and releases the two parts (products).

[1] Although this illustrates the current model, it can be noted that in the original lock-and-key model proposed by Emil Fischer in 1894, the key corresponded to the substrate and the lock to the enzyme.

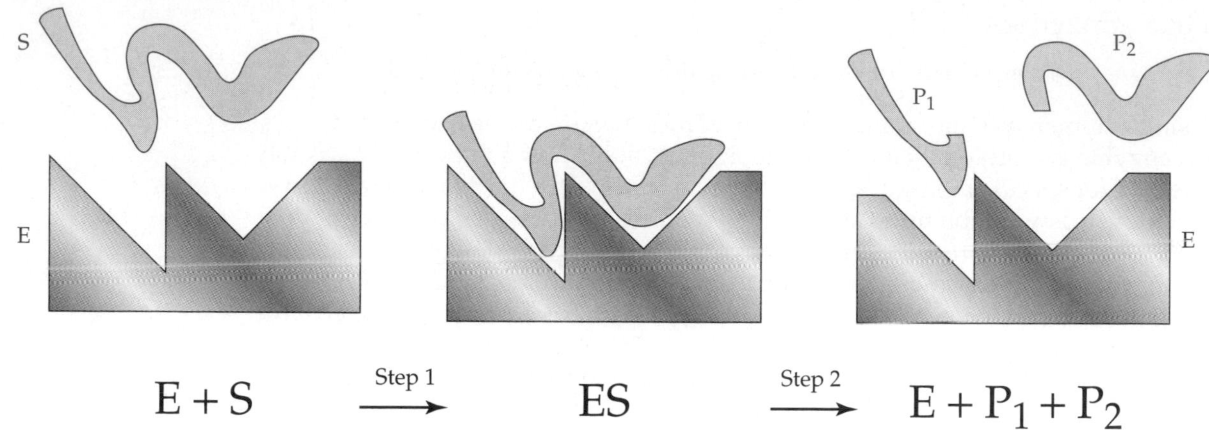

▲ **Figure 20.8 Enzyme Catalysis** In Step 1, the substrate molecule (S) binds to the active site on the enzyme (E). In Step 2, the enzyme cleaves the substrate and releases the two products, P_1 and P_2.

After the enzyme releases the products, the enzyme is then free to bind to another substrate molecule and repeat the process all over again. Enzyme catalysis is quite rapid, and a trace amount of enzyme can yield a large amount of product in a short period of time. You can demonstrate enzyme catalysis with an ordinary saltine cracker. Although the cracker is not sweet, it contains starch, which is a polymer of glucose sugar. If you place the cracker in your mouth, in a short time you will notice a sweet taste. The explanation is that saliva contains the enzyme ptyalin, which begins the digestion process by breaking down starch molecules into sugar units.

20.4 Carbohydrates

Objectives
- To recognize the basic structure of mono-, di-, and polysaccharides.
- To identify a glycoside linkage between monosaccharides.

Carbohydrates are an important food source for animals and provide the structure of plants. The word "carbohydrate" literally means "hydrates of carbon" ($C \cdot H_2O$), and indeed many carbohydrates have the empirical formula CH_2O. The names of carbohydrates usually end in the suffix *-ose,* for example, glucose, fructose, and sucrose. A carbohydrate is characterized by having an aldehyde or ketone functional group and, in addition, several hydroxyl (—OH) groups.

A simple sugar molecule typically has three to six carbons, an aldehyde or a ketone group, and a few hydroxyl groups. A simple sugar molecule is referred to as a **monosaccharide**. If a monosaccharide has an aldehyde group, it is referred to as an *aldose;* if it has a ketone group, it is called a *ketose*. If the sugar molecule contains five carbons, it is referred to as a *pentose;* if it contains six carbons, it is called a *hexose*.

▲ **Carbohydrate Foods** Bread, fruits, and vegetables are all food sources rich in carbohydrates.

Glucose, $C_6H_{12}O_6$, and fructose, $C_6H_{12}O_6$, are both hexoses. However, glucose has an aldehyde group and is referred to as an aldose; fructose has a ketone group and is called a ketose. Ribose, $C_5H_{10}O_5$, is an important pentose found in selected nucleic acids. Figure 20.9 illustrates the structures of these three simple sugars.

Glucose occurs in grapes and blood and, accordingly, is referred to as grape sugar or blood sugar. Fructose occurs naturally in fruit and honey and is commonly referred to as fruit sugar. Although they have the same molecular formula, fructose is twice as sweet as glucose. Fructose and glucose are monosaccharides and do not hydrolyze; that is, they do not break down in aqueous acid.

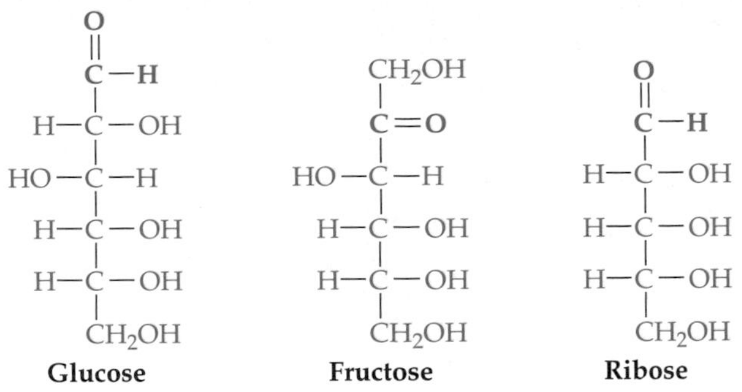

Common Monosaccharides

◀ **Figure 20.9 Common Monosaccharides** Glucose and fructose are both hexoses, but notice that glucose is an aldose and fructose is a ketose. Glucose and ribose are both aldoses, but notice that glucose is a hexose and ribose is a pentose.

A **disaccharide** is composed of two simple sugars. For instance, maltose is malt sugar, which is composed of two glucose molecules. Lactose is milk sugar, which is comprised of glucose and galactose. Sucrose is ordinary table sugar derived from sugarcane and sugar beets. Sucrose is comprised of glucose and fructose. In the presence of an enzyme or aqueous acid, a disaccharide breaks down into its two component monosaccharides. The equation for the acid *hydrolysis* of maltose, lactose, and sucrose is

$$C_{12}H_{22}O_{11} + H_2O \xrightarrow{H^+} C_6H_{12}O_6 + C_6H_{12}O_6$$

maltose	+	water	⟶	glucose	+	glucose
lactose	+	water	⟶	glucose	+	galactose
sucrose	+	water	⟶	glucose	+	fructose

Structures of Sugars in Aqueous Solution

In aqueous solution, sugar molecules usually exist as ringlike structures. In water, glucose forms a cyclic structure by joining a hydroxyl group to the aldehyde group. The resulting ring structure has five carbon atoms and one oxygen atom. Figure 20.10 illustrates the formation of a glucose ring in solution.

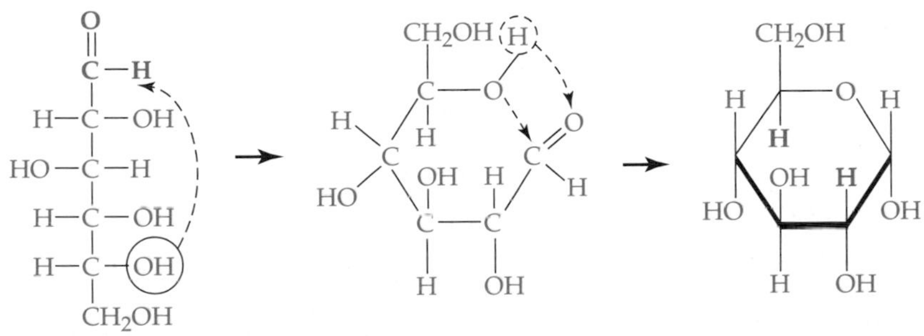

Glucose in Aqueous Solutions

◀ **Figure 20.10 Glucose in Aqueous Solution** In aqueous solution, a glucose molecule forms a ring structure containing five carbon atoms and one oxygen atom.

In the formation of a disaccharide, two simple monosaccharides split out water and join together by a special —O— bond called a **glycoside linkage**. For instance, a molecule of galactose and glucose can split out water and form a molecule of lactose.

In the process of digestion, the enzyme lactase breaks down lactose sugar to give galactose and glucose. It is interesting to note that adults have a less active form of this enzyme than infants. As a result, many adults cannot break down lactose and consequently may suffer indigestion. This condition, known as lactose intolerance, is treated by avoiding milk products or by taking a lactase supplement in the form of a liquid or tablet. Figure 20.11 illustrates the formation and breakage of a glycoside linkage in lactose.

Galactose

Glucose

$+H_2O$ $-H_2O$

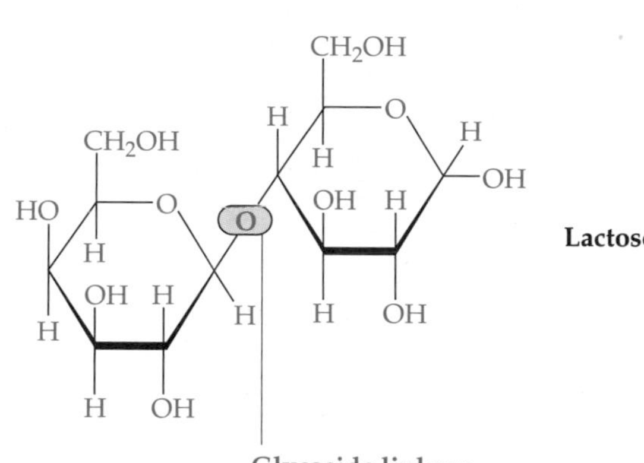

Formation and Breakage of a Glycoside Linkage

▶ **Figure 20.11 Formation and Breakage of a Glycoside Linkage** Lactose is composed of two simple sugars, glucose and galactose. In the formation of a glycoside linkage, a molecule of galactose and glucose split out a water molecule. In the reverse reaction, the glycoside linkage is cleaved in the presence of an enzyme or acid.

Polysaccharides

A **polysaccharide** is a polymer composed of hundreds or thousands of monosaccharide units joined by glycoside linkages. Starch and cellulose are important examples of polysaccharides. Starch is a polysaccharide composed of a single repeating unit. When starch undergoes hydrolysis and is broken down with dilute acid, the only product obtained is glucose. Figure 20.12 illustrates a segment of a starch polymer showing a simplified representation of glucose molecules.

▲ **Figure 20.12 A Segment of Starch** Starch is a polymer consisting of repeating glucose units. In addition to the straight-chain molecule shown, a starch molecule has numerous branches that are also composed of glucose units.

In the human body, glucose can form *glycogen*, a small polysaccharide stored in the muscles and liver as an energy resource. Excess glucose, however, is converted to lipids and stored in the body as fat. When energy is expended for physical activity, glycogen is released. The glycogen in turn is broken down to release glucose molecules. At the cellular level, glucose molecules are further metabolized to produce carbon dioxide, water, and energy for the body.

Cellulose is a polysaccharide found in plants and trees. When cellulose is broken down with dilute acid, the only product obtained is glucose. Thus, starch and cellulose are both polysaccharides composed exclusively of glucose units. Although the distinction is subtle, the glucose units in starch and cellulose are not linked together in exactly the same way. In cellulose every other glucose unit is flipped upside down. Figure 20.13 illustrates a segment of a cellulose polymer showing a simplified version of glucose molecules.

--- O – [glucose] – O – O – [glucose] – O – O – [glucose] – O – O – [glucose] – O ---

▲ **Figure 20.13 A Segment of Cellulose** Cellulose is a polymer consisting of repeating glucose units. In addition to the straight-chain molecule shown, a cellulose molecule has numerous branches composed of alternating inverted glucose units.

The subtle structural difference between starch and cellulose reveals that nature is highly selective. That is, animals possess an enzyme that digests starch but not cellulose. However, bacteria in the digestive tract of grazing animals and termites secrete enzymes that can cleave the inverted glycoside linkages in cellulose. Thus, cows can digest grass and termites can digest wood, but humans cannot digest either because we lack an enzyme to break the —O— bonds in cellulose.

20.5 Lipids

Objectives · To recognize the structure of a triglyceride, a phospholipid, and a wax.
· To recognize the basic ring structure of a steroid.

Unlike most proteins, enzymes, and carbohydrates, lipids are biochemical compounds that are not water soluble. Lipids include the familiar compounds we refer to as fats, oils, and waxes. They also include steroids such as sex hormones and the water-insoluble vitamins, that is, vitamins A, D, E, and K.

Triglycerides (Triacylglycerols)

A **triglyceride**[2] is a lipid formed from glycerol, a trihydroxy alcohol, and three long-chain carboxylic acids. A carboxylic acid with a long hydrocarbon chain is referred to as a **fatty acid**. When three fatty acids (RCOOH, R′COOH, R″COOH) react with glycerol, $C_3H_5(OH)_3$, a triglyceride is produced that has three ester linkages. We can represent the general reaction for the formation of a triglyceride as follows.

$$\begin{matrix} CH_2-OH & + & HO-\overset{O}{\overset{\|}{C}}-R & & CH_2-O-\overset{O}{\overset{\|}{C}}-R & \\ | & & & & | & \\ CH-OH & + & HO-\overset{O}{\overset{\|}{C}}-R' & \longrightarrow & CH-O-\overset{O}{\overset{\|}{C}}-R' & + \; 3\,H_2O \\ | & & & & | & \\ CH_2-OH & + & HO-\overset{O}{\overset{\|}{C}}-R'' & & CH_2-O-\overset{O}{\overset{\|}{C}}-R'' & \\ \text{glycerol} & & \text{fatty acids} & & \text{triglyceride (triacylglycerol)} & \end{matrix}$$

The notation R, R′, and R″ represents long hydrocarbon chains on the three fatty acids. If the triglyceride is obtained from an animal source, the fatty acid chains are mostly saturated and have few double bonds. A semisolid lipid obtained from an animal source is called a **fat**. Table 20.1 lists some common saturated fatty acids present in animal fat.

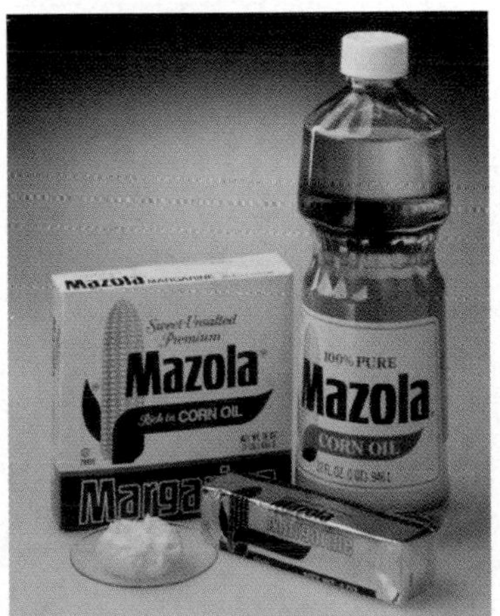

▲ **Lipid Foods** Vegetable oils, margarine, and "rich, creamy" foods are all sources of lipids.

[2]Although the term "triglyceride" is common, the term "triacylglycerol" is the more acceptable term.

Table 20.1 Common Saturated Fatty Acids

Name	Structure	Source
lauric acid	$CH_3—(CH_2)_{10}—COOH$	coconut oil
myristic acid	$CH_3—(CH_2)_{12}—COOH$	whale oil
palmitic acid	$CH_3—(CH_2)_{14}—COOH$	palm oil
stearic acid	$CH_3—(CH_2)_{16}—COOH$	animal fat

The following structure illustrates an example of a saturated triglyceride found in animal fat.

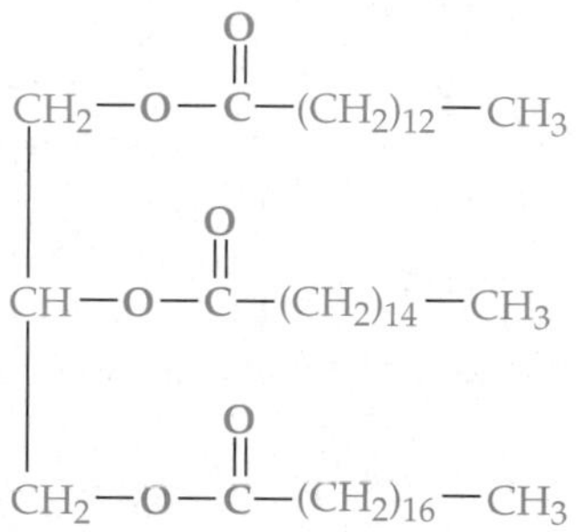

$$\begin{array}{ll} CH_2—O—\overset{\overset{\displaystyle O}{\|}}{C}—(CH_2)_{12}—CH_3 & \text{myristic acid ester} \\ | & \\ CH—O—\overset{\overset{\displaystyle O}{\|}}{C}—(CH_2)_{14}—CH_3 & \text{palmitic acid ester} \\ | & \\ CH_2—O—\overset{\overset{\displaystyle O}{\|}}{C}—(CH_2)_{16}—CH_3 & \text{stearic acid ester} \end{array}$$

A liquid lipid obtained from a plant source is called an **oil**. If a triglyceride is obtained from a plant source, the fatty acid chains are mostly unsaturated and have one or more double bonds. Table 20.2 lists some common unsaturated fatty acids that occur in vegetable oils.

Table 20.2 Common Unsaturated Fatty Acids

Name	Structure	Source
oleic acid	$CH_3—(CH_2)_7—CH{=}CH—(CH_2)_7—COOH$	olive oil, peanut oil
linoleic acid	$CH_3—(CH_2)_4—CH{=}CH—CH_2—CH{=}CH—(CH_2)_7—COOH$	soybean oil, safflower oil
linolenic acid	$CH_3—CH_2—CH{=}CH—CH_2—CH{=}CH—CH_2—CH{=}CH—(CH_2)_7—COOH$	linseed oil

The following structure illustrates an example of an unsaturated triglyceride found in vegetable oils. Note that the fatty acids have one, two, and three double bonds, respectively.

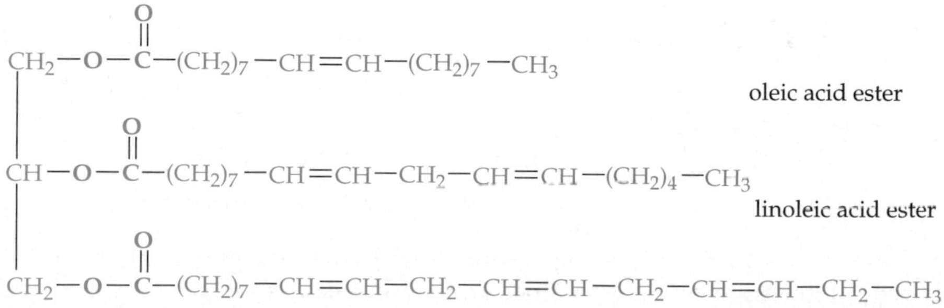

$$\begin{array}{ll} CH_2—O—\overset{\overset{\displaystyle O}{\|}}{C}—(CH_2)_7—CH{=}CH—(CH_2)_7—CH_3 & \text{oleic acid ester} \\ | & \\ CH—O—\overset{\overset{\displaystyle O}{\|}}{C}—(CH_2)_7—CH{=}CH—CH_2—CH{=}CH—(CH_2)_4—CH_3 & \text{linoleic acid ester} \\ | & \\ CH_2—O—\overset{\overset{\displaystyle O}{\|}}{C}—(CH_2)_7—CH{=}CH—CH_2—CH{=}CH—CH_2—CH{=}CH—CH_2—CH_3 & \text{linolenic acid ester} \end{array}$$

The distinction between a fat and an oil, based on the saturation of the fatty acids, is a generalization with numerous exceptions. That is, even fats have a small percentage of unsaturated fatty acids, and oils contain a few saturated fatty acids. Moreover, a natural source of a fat or oil is a complex mixture of triglycerides.

A precise analysis of a fat or oil indicates the average percentage of each fatty acid. Corn oil, for example, averages about 2% myristic acid, 8% palmitic acid, 5% stearic acid, 35% oleic acid, and 50% linoleic acid. Thus, corn oil averages about 15% saturated fatty acids and 85% unsaturated fatty acids.

Note The medical community has linked a diet high in saturated animal fat to coronary disease. Heart patients, in particular, are advised to maintain a low-fat diet and to replace butter (a saturated fat) with margarine (a polyunsaturated oil). Although the structures of butter and margarine are similar, the unsaturated fatty acids in margarine favor lower levels of cholesterol and triglycerides in the blood.

Saponification of Fats and Oils

The ester linkage in a triglyceride can be broken by treatment with aqueous sodium hydroxide. The products of the reaction are glycerol and three sodium salts of fatty acids. Sodium salts of fatty acids constitute what we refer to as soap. This reaction is called **saponication**, which literally means "soap making." We can represent the general reaction for the saponication of a triglyceride as follows.

$$\begin{array}{l} CH_2-O-\overset{O}{\overset{\|}{C}}-R \\ | \\ CH-O-\overset{O}{\overset{\|}{C}}-R' \\ | \\ CH_2-O-\overset{O}{\overset{\|}{C}}-R'' \end{array} + 3\,NaOH \longrightarrow \begin{array}{l} CH_2-OH \\ | \\ CH-OH \\ | \\ CH_2-OH \end{array} \begin{array}{l} + \; \mathbf{RCOO^-Na^+} \\ \\ + \; \mathbf{R'COO^-Na^+} \\ \\ + \; \mathbf{R''COO^-Na^+} \end{array}$$

triglyceride glycerol + soap

Soap has a polar ionic "head" (COO^-Na^+) and a long nonpolar "tail" (R, R', or R"). Although oil and water do not mix, we can use soap to remove an oily residue from our hands. Soap orients its fatty acid nonpolar "tail" toward the oil drop on our hands, and its polar "head" toward the water. Soap thereby provides a link between nonpolar oil molecules and polar water molecules (Figure 20.14). Thus, washing our hands with soap and water can remove an oily residue even though oil and water are immiscible.

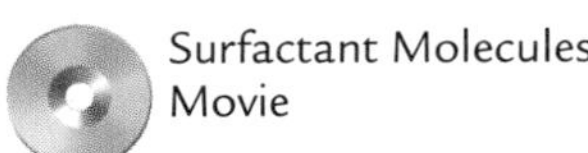

$$CH_3-CH_2-CH_2-CH_2-CH_2-CH_2-CH_2-CH_2-CH_2-CH_2-CH_2-CH_2-CH_2-COO^-Na^+$$

(nonpolar "tail") (polar "head")

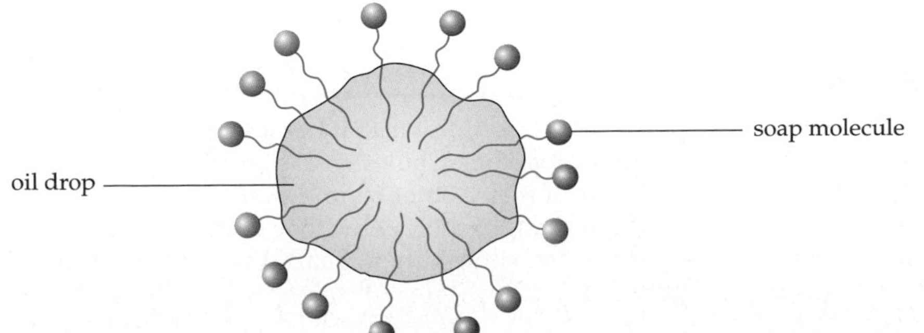

◀ Figure 20.14 The Action of Soap A soap molecule ($R-COO^-Na^+$) has a long nonpolar "tail" (R, R', or R") and a polar "head" (COO^-Na^+). Soap is effective in removing oil because of its dual nature; soap has a "tail" that is soluble in oil and a "head" that is soluble in water.

Phospholipids

A **phospholipid** is a glyceride found in living cells and cell membranes. Unlike fats and oils, which are triglycerides, phospholipids have *two* fatty acids esterified to glycerol. The third —OH in glycerol is linked to phosphoric acid. Thus, the properties of phospholipids are different from those of fats and oils. With the addition of a phosphate group, a phospholipid has an ionic "head" that is soluble in water, as well as a nonpolar "tail" that is insoluble. In fact, the phospholipid lecithin is used to emulsify margarine and chocolate, which are otherwise insoluble in water. The following example structure illustrates a typical composition of lecithin.

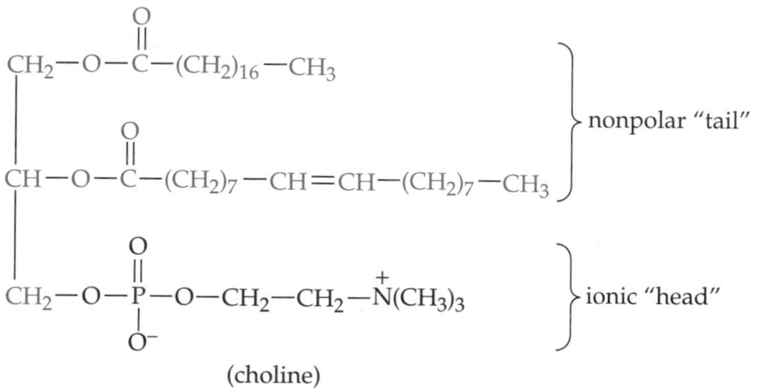

$$\begin{array}{l} CH_2-O-\overset{\displaystyle O}{\overset{\|}{C}}-(CH_2)_{16}-CH_3 \\ | \\ CH-O-\overset{\displaystyle O}{\overset{\|}{C}}-(CH_2)_7-CH{=}CH-(CH_2)_7-CH_3 \\ | \\ CH_2-O-\overset{\displaystyle O}{\overset{\|}{P}}-O-CH_2-CH_2-\overset{+}{N}(CH_3)_3 \\ \qquad\qquad\ \ | \\ \qquad\qquad\ O^- \end{array}$$

nonpolar "tail" (fatty acid chains); ionic "head" (phosphate–choline group)

(choline)

The phosphate ester in lecithin is bonded to choline, $HO-CH_2CH_2N(CH_3)_3$, as well as glycerol. Choline is a B-complex vitamin found in egg yolks and is essential to liver function.

Waxes

A paraffin wax is obtained from heavy crude oil and is a mixture of long-chain alkanes. A lipid **wax**, however, is a mixture of naturally occurring esters of fatty acids and long-chain alcohols. An example of a lipid wax is beeswax, which is found in honeycombs and is used to make candles and shoe polish. Another example is carnauba wax, an ingredient in automobile and furniture polish. A lipid wax is water-insoluble and occurs naturally as the slippery coating on fruits, vegetables, and plant leaves. Since waxes are insoluble in water, they can be used to waterproof fabrics and wood surfaces. The following example structure illustrates one of the typical esters found in beeswax.

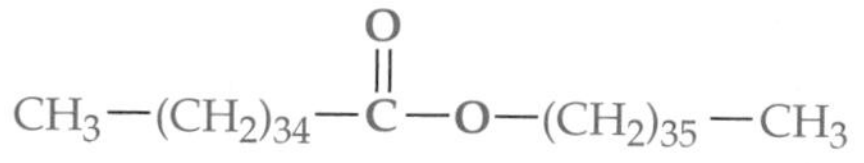

$$CH_3-(CH_2)_{34}-\overset{\displaystyle O}{\overset{\|}{C}}-O-(CH_2)_{35}-CH_3 \qquad \text{Beeswax}$$

Waxes are esters found in nature and, like triglycerides, demonstrate variety in their fatty acid composition. A typical analysis of carnauba wax reveals a 12-carbon unsaturated fatty acid and a broad mixture of long-chain alcohols.

Steroids

A **steroid** belongs to a special class of lipids that has a structure composed of four rings of carbon atoms fused together. Cholesterol is an example of an important steroid and is the precursor of many other steroid molecules essential to human life. On the other hand, high levels of cholesterol have been linked to the formation of plaque in arteries, which can contribute to heart disease. We can diagram the ring structure of cholesterol as follows.

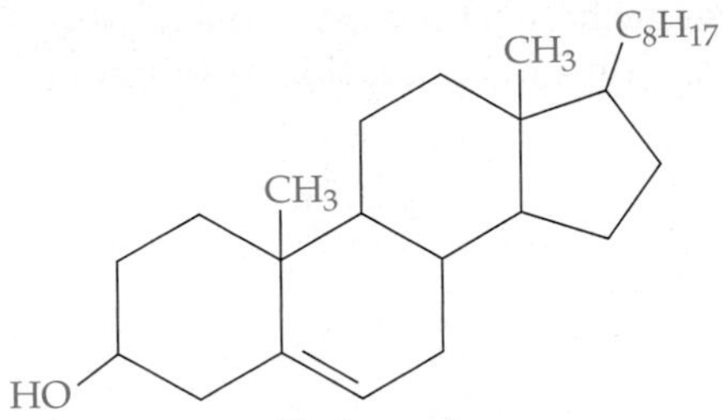

Cholesterol

As mentioned previously, sex hormones are also steroids. This includes the hormones testosterone and progesterone secreted by the testes and ovaries, respectively. Testosterone is responsible for the sex characteristics of males, while progesterone helps regulate ovulation in females. It is interesting to note that the two hormones are quite similar and differ only slightly in molecular structure.

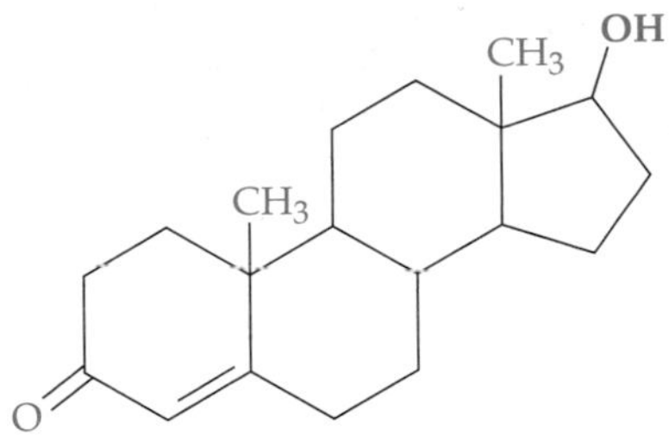

Testosterone

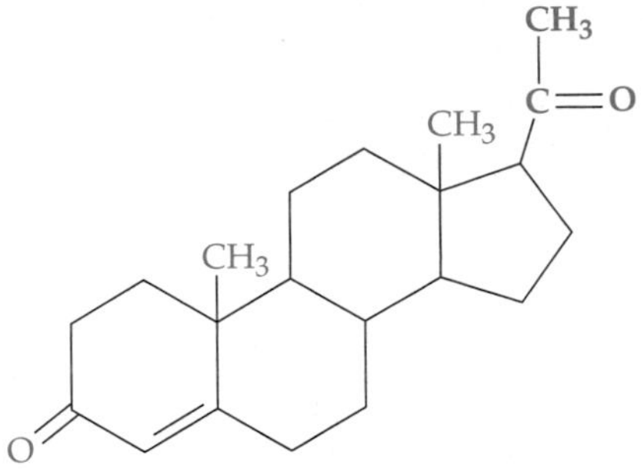

Progesterone

20.6 Nucleic Acids

Objectives
- To state the three components of a nucleotide.
- To describe the double-helix structure of DNA.

Nucleic acids are biochemical compounds found in every living cell. Nucleic acids carry the genetic information responsible for the reproduction of a species. The nucleic acid deoxyribonucleic acid (DNA) is found in the chromosomes inside the nuclei of cells. DNA molecules are responsible for storing the genetic code that controls the biosynthesis of proteins and other biochemical compounds.

The nucleic acid ribonucleic acid (RNA) is responsible for transmitting genetic information such as the instructions for protein synthesis. In protein synthesis, a molecule of messenger RNA (mRNA) is made in the cell nucleus using a section of DNA as a template. That is, a portion of the DNA molecule transcribes a genetic code into a mRNA molecule for an amino acid sequence.

Then the mRNA molecule moves out of the nucleus into the surrounding cytoplasm. It is here that a smaller, different type of RNA called transfer RNA (tRNA) builds a protein chain according to the coded mRNA instructions. Each tRNA molecule has a distinctive structure that is selective for a specific amino acid. Thus, one by one, tRNA molecules build a chain of amino acids that eventually becomes a protein.

Structurally, a nucleic acid is a polymer molecule composed of many repeating units, each of which is called a **nucleotide**. All nucleotides are not identical, but each one consists of three similar component molecules: a five-carbon sugar, a nitrogen-containing organic base, and a molecule of phosphoric acid. We can diagram the general components of a nucleotide as follows.

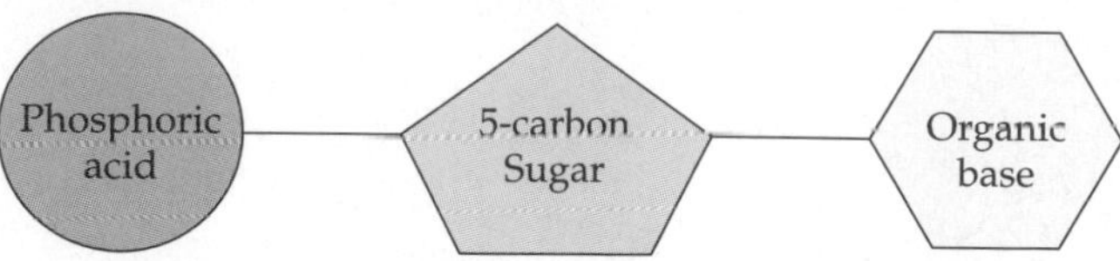

Chemistry Connection • Vitamins

Which vitamins can be toxic if ingested in excessive amounts?

Vitamins are biochemical compounds required for normal metabolism. With the exception of vitamin D, they are not synthesized in the body and therefore must be included in the diet. If a vitamin is not present in the diet, its absence can lead to a vitamin-deficiency disease.

The U.S. Food and Nutrition Board has established minimal vitamin requirements. The board periodically publishes a new listing of the Recommended Dietary Allowance (RDA) for vitamins based on current research. These RDA values are intended as a guideline for estimating the needs of adults and children and may vary depending on individual circumstances. Ingesting polluted air or water can increase vitamin requirements. Some infectious diseases, accompanied by a high fever, may also result in increased vitamin requirements.

In most instances, a diet containing fresh fruits, vegetables, dairy products, and meat ensures an adequate intake of necessary vitamins. Vitamin supplements, however, may be recommended by a physician during periods of stress, growth, or pregnancy. There is a growing body of evidence that vitamins A, C, and E may lessen the risk of cancer. These vitamins are antioxidants, and the so-called "free-radical theory" proposes that oxidizing molecular radicals build up in tissues and can alter the structure of DNA. Vitamins C and E, in conjunction with the mineral selenium, help to clear the tissues of toxic free radicals.

◀ A sampling of various vitamins that are available as pills, tablets, and gelatin capsules.

The consumption of megadoses of vitamin supplements has recently raised some controversy. Excess amounts of water-soluble vitamins (B_1, B_2, B_3, B_5, B_6, B_{12}, and C) are rapidly excreted in the urine and are rarely toxic. Excess amounts of fat-soluble vitamins (A, D, E, and K), however, can accumulate in body fat tissue and lead to toxic symptoms.

Taking supplemental vitamin C has been controversial since 1970 when Linus Pauling (see Chapter 12 Chemistry Connection) and others began advocating large daily doses of vitamin C to strengthen the immune system and prevent the common cold. Although vitamin C is relatively nontoxic, daily doses in excess of a gram may lead to skin rashes and diarrhea in some individuals. A deficiency of vitamin C, ascorbic acid, can lead to the disease scurvy which is characterized by loose teeth and bleeding gums. Citrus fruits prevent and cure this disease. In the early 1800s the British navy began issuing rations of lemons and limes to sailors who acquired the familiar nickname "limeys."

Excess amounts of fat-soluble vitamins (A, D, E, and K) can accumulate in body fat tissue and lead to toxic symptoms. For instance, an excess of vitamin A can cause irritability, vomiting, loss of appetite, headache, and dry skin. On the other hand, a deficiency of vitamin A, retinol, has long been known to cause night blindness.

Vitamin D is synthesized by the body when the skin is exposed to sunlight. An excess of vitamin D can cause weakness, fatigue, headache, nausea, and vomiting. Vitamin D deficiencies are rare in tropical climates and are more common in northern regions. A deficiency of vitamin D results in failure to absorb calcium and phosphorus, causing malformation of the bones. In children, the disease is known as rickets and is characterized by a deformed rib cage and bow legs.

A supplement of vitamin E, alpha-tocopherol, has been advocated for many disorders, especially coronary artery disease. In moderate excess, it is considered relatively nontoxic. Vitamin K is available in milk products and is essential for the clotting of blood. In addition, a wide variety of foods such as vegetables, eggs, and fish contain vitamin K, and so a deficiency of this vitamin is rare.

Fat-soluble vitamins (A, D, E, and K) can be toxic when taken in excessive amounts.

Deoxyribose (DNA)

Ribose (RNA)

◀ **Figure 20.15 Nucleic Acid Sugars** Notice that deoxyribose, the sugar in DNA, and ribose, the sugar in RNA, differ by a single hydroxyl group (—OH).

The nucleotides found in DNA and RNA are similar. However, in DNA the sugar is deoxyribose and in RNA the sugar is ribose. The five-carbon sugars are identical except that ribose has one more hydroxyl group. Figure 20.15 shows the structures of these two sugars in their cyclic form.

Four nitrogen-containing organic bases are found in DNA and RNA. Figure 20.16 illustrates the structures of these four nitrogen-containing bases. The four bases in DNA and RNA are identical with one exception. In DNA, the bases are adenine (A), cytosine (C), guanine (G), and thymine (T). In RNA, the organic base uracil (U) is substituted for thymine (T).

Now that we have introduced the structures of the three basic components of a nucleotide, we can show how the sugar molecule, the organic base, and the phosphoric acid group are related. Figure 20.17 illustrates typical structures for a DNA nucleotide and an RNA nucleotide.

Let's consider a small portion of the large DNA molecule and diagram a segment of a single DNA chain. That is, let's string together a few nucleotides to represent a portion of the nucleic acid polymer. For simplicity, we will not specify a particular base although we know that the base in DNA must be A, C, G, or T. Figure 20.18 illustrates a portion of a hypothetical DNA chain.

Nucleic Acid Organic Bases

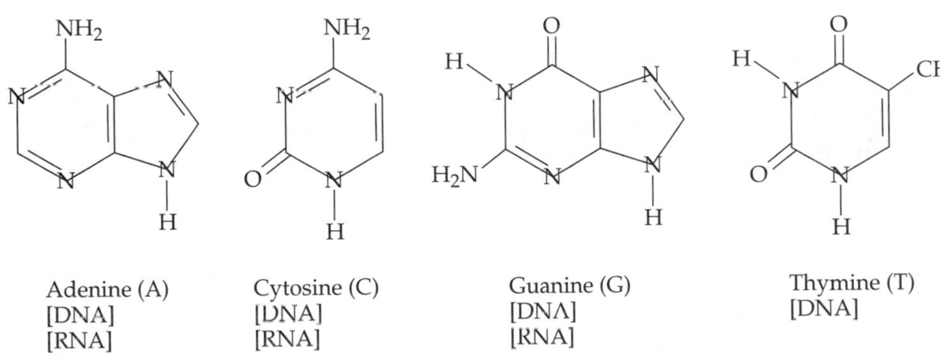

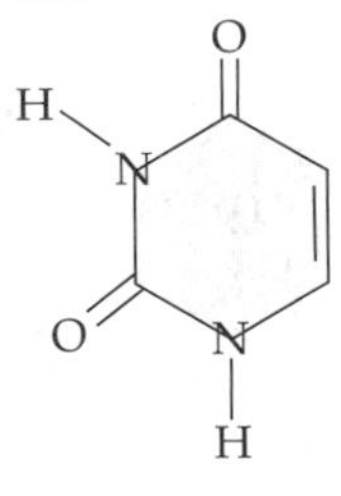

Uracil (U)

[RNA]

▲ **Figure 20.16 Nucleic Acid Organic Bases** Note that DNA contains the organic bases adenine (A), cytosine (C), guanine (G), and thymine (T). In RNA, the organic base uracil (U) is substituted for thymine.

DNA and RNA Nucleotides

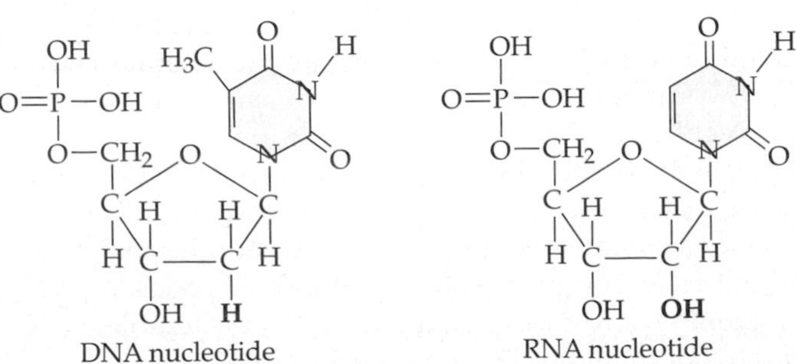

◀ **Figure 20.17 DNA and RNA Nucleotides** Note that the DNA nucleotide contains deoxyribose sugar, and that the RNA nucleotide contains ribose sugar. In this example, deoxyribose in the DNA nucleotide is attached to the organic base thymine, and ribose in the RNA nucleotide is attached to the base uracil.

DNA 3D Molecule

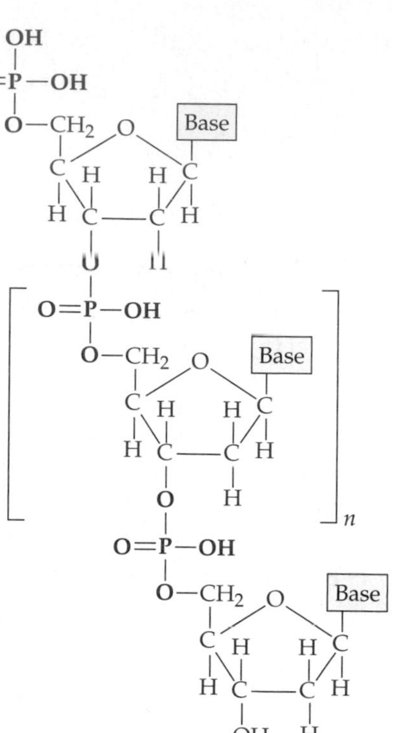

A Segment of DNA

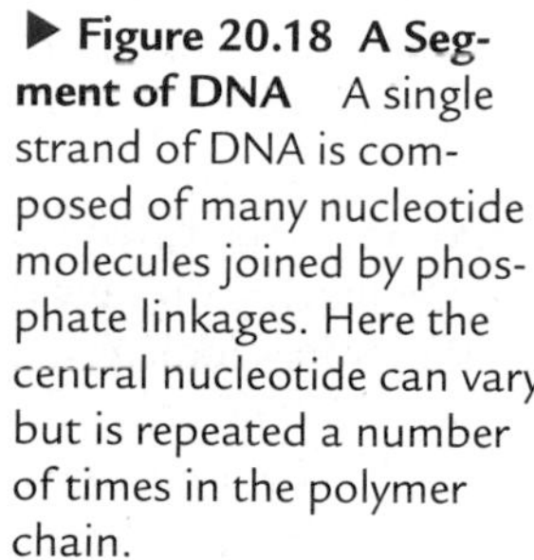

▶ **Figure 20.18 A Segment of DNA** A single strand of DNA is composed of many nucleotide molecules joined by phosphate linkages. Here the central nucleotide can vary but is repeated a number of times in the polymer chain.

The DNA Double Helix

The two strands of the DNA helix are held together by hydrogen bonds between the organic bases attached to deoxyribose sugar. Specifically, adenine (A) on one strand *always* hydrogen bonds to thymine (T) on the other strand. In Figure 20.19, note that there are two hydrogen bonds between each pair of these organic bases (A═T or T═A). Moreover, cytosine (C) on one strand *always* hydrogen bonds to guanine (G) on the other strand. Note that there are three hydrogen bonds between each pair of these organic bases (C≡G or G≡C).

Biochemists propose that a DNA molecule undergoes *replication* by first unwinding and breaking hydrogen bonds between organic bases in the double helix. Then, each single strand of DNA synthesizes a complementary strand of DNA one nucleotide at a time. This is accomplished as follows.

Since the organic bases always pair in the same way, A═T or C≡G, each new nucleotide in a growing single strand of DNA must complement the existing nucleotide in the template strand. That is, a new nucleotide in the growing strand must contain adenine (A) if it is to pair with thymine (T) on the template strand. Alternatively, a new nucleotide in the growing strand must contain cytosine (C) if it is to pair with guanine (G) on the template strand.

When the synthesis is complete, each of the original two strands of DNA has produced complementary single strands of DNA. In turn, each original single strand of DNA and its complement can hydrogen-bond and regenerate a new double helix. The final result is two strands of double-helix DNA, each of which is identical to the original DNA molecule.

Similarly, biochemists propose the process of *transcription* for a molecule of DNA synthesizing a molecule of RNA. That is, a string of nucleotides in a strand of DNA can code for a complementary string of nucleotides in RNA. As with replication, cytosine (C) codes for guanine (G), guanine (G) codes for cytosine (C), and thymine (T) for adenine (A); however, adenine (A) in the template DNA strand will code for uracil (U), not thymine (T), in the growing RNA strand.

▲ **DNA Double Helix** Molecular model showing a portion of DNA whose two strands of nucleotides are twisted into an α-helix.

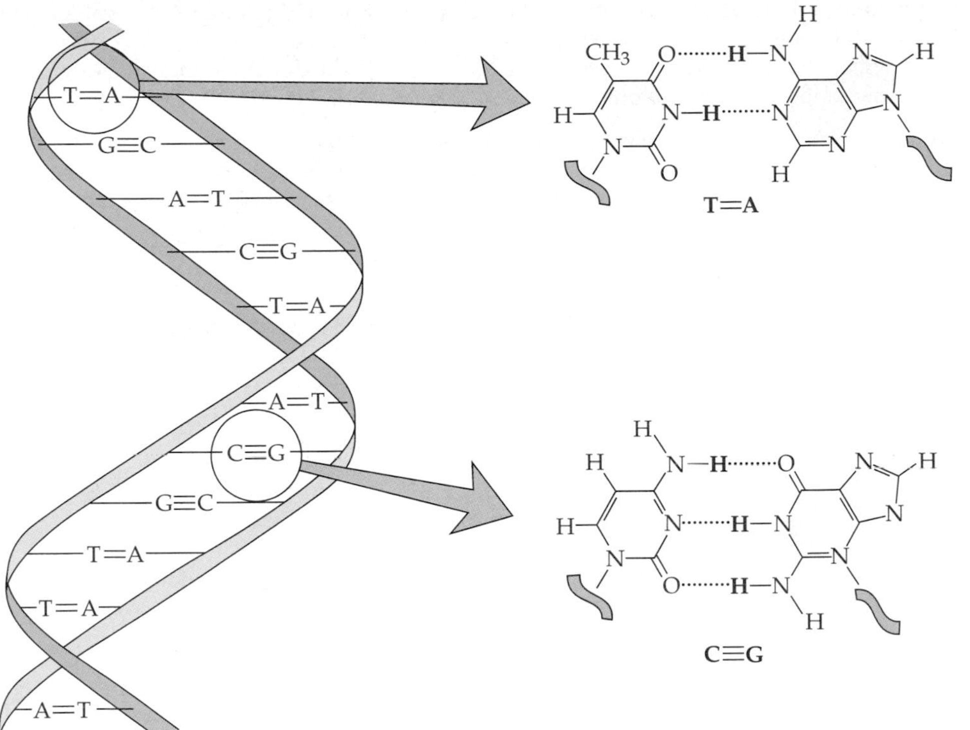

◀ **Figure 20.19 The DNA Double Helix** The two DNA strands of nucleotides are held together by hydrogen bonds. The hydrogen bonds are always between the same organic bases; that is, there are two hydrogen bonds between thymine (T) and adenine (A), or three hydrogen bonds between cytosine (C) and guanine (G).

Summary

Section 20.1 **Biochemistry** is the study of biological compounds and their chemical reactions. A biochemical compound is often composed of small molecules linked together in a long chain called a **polymer**. A **protein** is a polymer composed of amino acids. A **carbohydrate** is a compound consisting of one or more simple sugar molecules. A **lipid** is a water-insoluble biological compound such as a fat, oil, wax, or steroid. A **nucleic acid** is a polymer composed of a sugar molecule, an organic base, and phosphoric acid. Although biological compounds can be large and complex, they have familiar functional groups such as amines, alcohols, and esters.

Section 20.2 A carboxylic acid having an amine group is called an **amino acid**. A **dipeptide** is two amino acids joined by an amide bond called a **peptide linkage**. A **polypeptide** has several amino acids joined by peptide linkages. A protein is a large polypeptide that can contain thousands of amino acids. The structure of a protein is related to its biological function. Extended protein molecules are responsible for shape and structure, whereas globular proteins often regulate hormonal activity.

Section 20.3 An **enzyme** is a protein that acts as a catalyst for biochemical reactions. An enzyme is highly specific for a given biological molecule and is quite sensitive to changes in pH and temperature. Although the mechanism of enzyme activity is complex, the **lock-and-key model** explains the selectivity of enzymes. To speed up a biochemical reaction, the shape of the substrate molecule must fit the contour of the active site on an enzyme.

Section 20.4 A carbohydrate is a biological compound that contains one or more sugar molecules. Glucose is a **monosaccharide** corresponding to one simple sugar molecule. Sucrose is a **disaccharide** and contains molecules of glucose and fructose joined by a **glycoside linkage**. Starch is a **polysaccharide** and contains many molecules of glucose joined by glycoside linkages.

Section 20.5 Many lipids contain glycerol, an alcohol with three hydroxyl groups. A **fatty acid** is a carboxylic acid bearing a long hydrocarbon chain. A **triglyceride** is an ester of glycerol and three fatty acids. An animal **fat** is a triglyceride with mostly saturated fatty acids. A vegetable **oil** is a triglyceride with mainly unsaturated fatty acids. Triglycerides react with sodium hydroxide to produce soap and glycerol in a chemical reaction called **saponification**.

A **phospholipid** contains two fatty acids and phosphoric acid joined to glycerol by ester linkages. A lipid **wax** is a simple ester of a fatty acid and a long-chain alcohol. A **steroid** is a lipid with a characteristic structure of four rings of carbon atoms. Steroids include many hormones and fat-soluble vitamins.

Section 20.6 A nucleic acid is a biological compound that carries genetic information and controls protein synthesis. Nucleic acids are polymers that have repeating units called nucleotides. A **nucleotide** is composed of a sugar, an organic base, and phosphoric acid. DNA and RNA are examples of nucleic acids. DNA and RNA are based on slightly different nucleotides whose composition is summarized in Table 20.3.

Table 20.3 DNA versus RNA Nucleotides

Component	DNA	RNA
Pentose sugar	deoxyribose	ribose
Organic bases	adenine (A), cytosine (C), guanine (G), thymine (T)	adenine (A), cytosine (C), guanine (G), uracil (U)
Inorganic acid	phosphoric acid	phosphoric acid

Key Concepts*

1. What type of biological compound corresponds to each of the following descriptions?
   (a) a polymer of amino acids
   (b) a water-insoluble fat, oil, or wax
   (c) a simple sugar or a polymer of simple sugar molecules
   (d) a polymer composed of a sugar, an organic base, and H_3PO_4
2. Draw the structure for the dipeptide glycylserine (Gly-Ser).
3. State whether the primary, secondary, or tertiary structure of a protein molecule corresponds to each of the following descriptions.
   (a) a helix of amino acids in a spiral protein chain
   (b) the sequence of amino acids in a spiral protein chain
   (c) a twisted helix of amino acids in a spiral protein chain
4. In the lock-and-key model of enzyme catalysis, what do the following represent?
   (a) the lock (b) the key
5. Complete the following reaction for the acid hydrolysis of sucrose and label the two products.

$$\underset{\text{sucrose}}{C_{12}H_{22}O_{11}} + \underset{\text{water}}{H_2O} \xrightarrow{H^+}$$

6. Complete the following reaction for the saponification of an animal fat. Label the two products and identify the fatty acids in the animal fat.

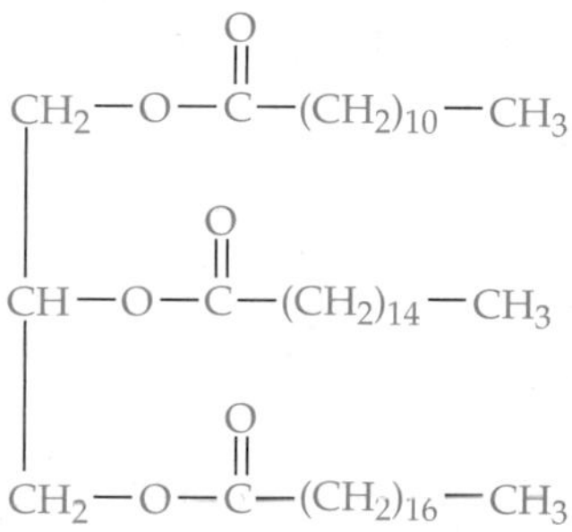

$+ \; 3\,NaOH \longrightarrow$

7. One of the components of beeswax is an ester composed of the following fatty acid and alcohol: fatty acid $CH_3—(CH_2)_{34}—COOH$; alcohol, $CH_3—(CH_2)_{35}—OH$. Draw the structure of the lipid ester.
8. Identify the sugar, organic base, and inorganic acid in each of the following.
   (a) a DNA nucleotide (b) an RNA nucleotide

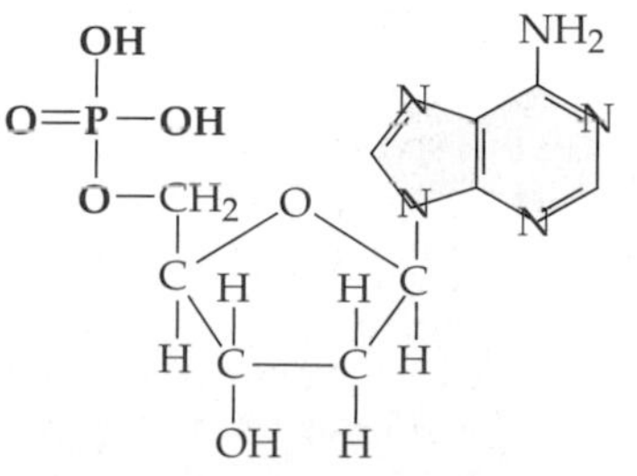

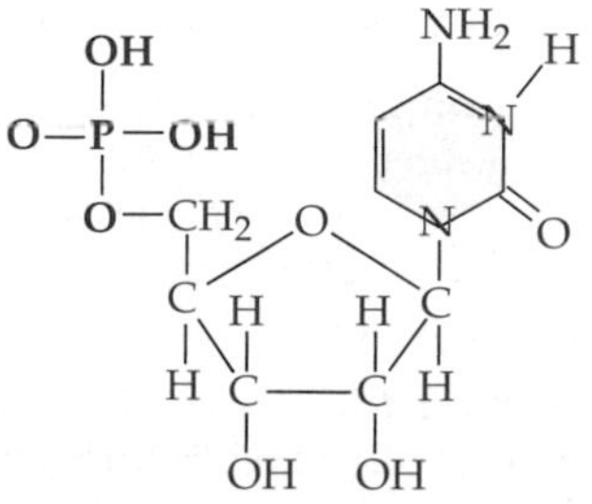

*Answers to Key Concepts are in Appendix H.

Key Terms[†] 

Select the key term below that corresponds to each of the following definitions.

_____ **1.** the study of biological compounds and their chemical reactions
_____ **2.** a giant molecule composed of many small repeating units
_____ **3.** a biological compound that is a polymer of amino acids
_____ **4.** a biological compound composed of one or more sugar molecules
_____ **5.** a biological compound that is insoluble in water, such as a fat, oil, wax, or steroid
_____ **6.** a biological compound that is a polymer and can transmit genetic information
_____ **7.** a carboxylic acid with an amine group on the alpha carbon
_____ **8.** an amide bond that joins two amino acids
_____ **9.** two amino acids joined by a peptide linkage
_____ **10.** 10–50 amino acids joined by peptide linkages
_____ **11.** a protein molecule that catalyzes a biochemical reaction
_____ **12.** a theory that explains enzyme catalysis
_____ **13.** a carbohydrate composed of a simple sugar molecule
_____ **14.** a carbohydrate composed of two simple sugar molecules
_____ **15.** a carbohydrate composed of many simple sugar molecules
_____ **16.** an —O— bond that joins two simple sugars
_____ **17.** a carboxylic acid with a long hydrocarbon chain
_____ **18.** a lipid composed of glycerol and three fatty acids
_____ **19.** a triglyceride from an animal source that has mostly saturated fatty acids
_____ **20.** a triglyceride from a plant source that has mostly unsaturated fatty acids
_____ **21.** a chemical reaction of a fat or oil with sodium hydroxide to produce soap
_____ **22.** a lipid composed of glycerol, two fatty acids, and phosphoric acid
_____ **23.** a lipid composed of a fatty acid and a long-chain alcohol
_____ **24.** a lipid hormone composed of four rings of carbon atoms
_____ **25.** a repeating unit in a nucleic acid composed of a sugar, a base, and phosphoric acid

(a) amino acid (*Sec. 20.2*)
(b) biochemistry (*Sec. 20.1*)
(c) carbohydrate (*Sec. 20.1*)
(d) dipeptide (*Sec. 20.2*)
(e) disaccharide (*Sec. 20.4*)
(f) enzyme (*Sec. 20.3*)
(g) fat (*Sec. 20.5*)
(h) fatty acid (*Sec. 20.5*)
(i) glycoside linkage (*Sec. 20.4*)
(j) lipid (*Sec. 20.1*)
(k) lock-and-key model (*Sec. 20.3*)
(l) monosaccharide (*Sec. 20.4*)
(m) nucleic acid (*Sec. 20.1*)
(n) nucleotide (*Sec. 20.6*)
(o) oil (*Sec. 20.5*)
(p) peptide linkage (*Sec. 20.2*)
(q) phospholipid (*Sec. 20.5*)
(r) polymer (*Sec. 20.1*)
(s) polypeptide (*Sec. 20.2*)
(t) polysaccharide (*Sec. 20.4*)
(u) protein (*Sec. 20.1*)
(v) saponification (*Sec. 20.5*)
(w) steroid (*Sec. 20.5*)
(x) triglyceride (*Sec. 20.5*)
(y) wax (*Sec. 20.5*)

Exercises[‡]

Biological Compounds (Sec. 20.1)

1. Identify the type of biological compound having each of the following structures.

--- Amino acid —C(=O)—NH— Amino acid ---

--- Sugar(amine) —O—P(=O)(OH)—O— Sugar(amine) ---

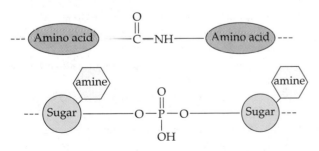

2. Identify the type of biological compound having each of the following structures.

glycerol —O—C(=O)— carboxylic acid (three times)

Sugar —O— Sugar

[†] Answers to Key Terms are in Appendix I.

[‡] Answers to odd-numbered Exercises are in Appendix J.

3. Identify the type of linkage that joins repeating units in each of the following.
   (a) protein (b) nucleic acid
4. Identify the type of linkage that joins each of the following.
   (a) two sugars in a carbohydrate
   (b) glycerol and a carboxylic acid in a lipid
5. Identify one or more organic functional groups present in each of the following biological compounds.
   (a) amino acid (b) sugar
   (c) fatty acid (d) organic base
6. Identify one or more organic functional groups that characterize each of the following biological compounds.
   (a) protein (b) carbohydrate
   (c) lipid (d) nucleic acid

Proteins (Sec. 20.2)

7. What type of bonds are responsible for the primary structure of a protein?
8. What type of bonds are responsible for the secondary structure of a protein?
9. Explain the difference between the primary and secondary structure of a protein.
10. Explain the difference between the secondary and tertiary structure of a protein.
11. Draw the structure of the dipeptide formed by two molecules of the amino acid alanine. Circle the peptide bond.
12. Draw the structure of the dipeptide formed by two molecules of the amino acid cysteine. Circle the peptide bond.
13. Draw the structure of the dipeptide seryltyrosine (Ser-Tyr).
14. Draw the structure of the dipeptide tyrosylserine (Tyr-Ser).
15. Indicate all the possible sequences for a tripeptide containing arginine, histidine, and methionine; use the standard abbreviation for each amino acid.
16. Indicate all the possible sequences for a tripeptide containing asparagine, glutamine, and tryptophan; use the standard abbreviation for each amino acid.
17. The artificial sweetener NutraSweet is the methyl ester of a dipeptide. Identify the two amino acids given the structure for NutraSweet.

```
          O              O
          ||             ||
H2N—CH—C—NH—CH—C—O—CH3
      |            |
     CH2          CH2
      |            |
     COOH      (benzene ring)
```

18. A segment of a protein chain contains the following tripeptide. Identify the three amino acids given the structure for the tripeptide.

```
           O              O              O
           ||             ||             ||
[—NH—CH—C—NH—CH—C—NH—CH—C—]
       |              |              |
     (CH2)2         (CH2)4          CH2
       |              |              |
     COOH            NH2             OH
```

19. The two hormones vasopressin and oxytocin are polypeptides with a similar amino acid composition. Given the primary sequence for each hormone, identify the amino acids that differ in each polypeptide.

   vasopressin: Cys-Tyr-Phe-Gln-Asn-Cys-Pro-Arg-Gly
   oxytocin: Cys-Tyr-Ile-Gln-Asn-Cys-Pro-Leu-Gly

20. Vasopressin raises blood pressure in humans, and oxytocin induces lactation in mammals. Suggest a reason why the biological activity of the polypeptides differs.

Enzymes (Sec. 20.3)

21. Describe step 1 in the enzyme catalysis for the reaction of a substrate.
22. Describe step 2 in the enzyme catalysis for the reaction of a substrate.
23. In the lock-and-key model, what do the "teeth" on the key represent?
24. In the lock-and-key model, what do the separate lock and clasp represent?
25. What is the term for a molecule that blocks an enzyme and prevents a substrate reaction?
26. What is the term for a location on an enzyme that conforms to the shape of the substrate?
27. State three characteristics of an enzyme.
28. Under what experimental conditions is an enzyme likely to be inactive?

Carbohydrates (Sec. 20.4)

29. What are the functional groups in an aldose sugar?
30. What are the functional groups in a ketose sugar?
31. How many carbon atoms are in a molecule of pentose sugar?
32. How many carbon atoms are in a molecule of hexose sugar?
33. What is the distinction between a monosaccharide and a disaccharide?
34. What is the distinction between a disaccharide and a polysaccharide?
35. Draw the open-chain structure for a glucose molecule.
36. Draw the open-chain structure for a fructose molecule.
37. Draw the cyclic structure for a glucose molecule in aqueous solution.
38. Draw the cyclic structure for a galactose molecule in aqueous solution.

39. Complete the following reaction for the acid hydrolysis of maltose and label the two products.

$$\underset{\text{maltose}}{C_{12}H_{22}O_{11}} + \underset{\text{water}}{H_2O} \xrightarrow{H^+}$$

40. Complete the following reaction for the acid hydrolysis of lactose and label the two products.

$$\underset{\text{lactose}}{C_{12}H_{22}O_{11}} + \underset{\text{water}}{H_2O} \xrightarrow{H^+}$$

41. What is glycogen? What is the repeating monosaccharide in glycogen?
42. What is the biological function of glycogen in the body? Where is glycogen stored in the body?

Lipids (Sec. 20.5)

43. State whether each of the following characteristics indicates a fat, an oil, or both.
    (a) animal source **(b)** plant source
    (c) semisolid **(d)** insoluble in water
44. State whether each of the following constituents typifies a fat, an oil, or both.
    (a) saturated fatty acids **(b)** unsaturated fatty acids
    (c) glycerol **(d)** ester linkages
45. Draw the structural formula for tristearin, the triglyceride of stearic acid.
46. Draw the structural formula for triolein, the triglyceride of oleic acid.
47. Identify the fatty acids in the following triglyceride.

$$\begin{array}{l} CH_2-O-\overset{O}{\overset{\|}{C}}-(CH_2)_{14}-CH_3 \\ | \\ CH-O-\overset{O}{\overset{\|}{C}}-(CH_2)_{10}-CH_3 \\ | \\ CH_2-O-\overset{O}{\overset{\|}{C}}-(CH_2)_{12}-CH_3 \end{array}$$

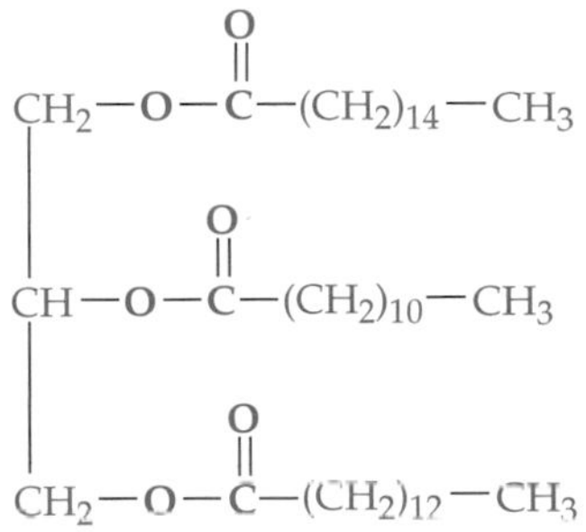

48. Identify the fatty acids in the following triglyceride.

$$\begin{array}{l} CH_2-O-\overset{O}{\overset{\|}{C}}-(CH_2)_{16}-CH_3 \\ | \\ CH-O-\overset{O}{\overset{\|}{C}}-(CH_2)_7-CH=CH-CH_2-CH=CH-(CH_2)_4-CH_3 \\ | \\ CH_2-O-\overset{O}{\overset{\|}{C}}-(CH_2)_7-CH=CH-CH_2-CH=CH-CH_2-CH=CH-CH_2-CH_3 \end{array}$$

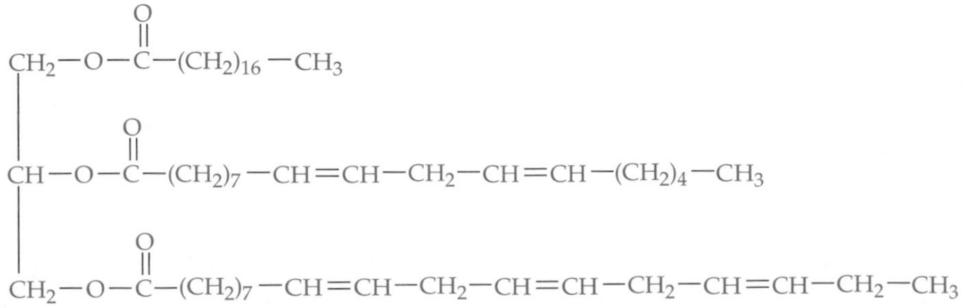

49. Carnauba wax is a lipid obtained from the leaves of palm trees and used in polishes for metal and wood surfaces. If the given structure is treated with aqueous NaOH, what are the formulas of the two products?

$$CH_3-(CH_2)_{24}-\overset{O}{\overset{\|}{C}}-O-(CH_2)_{29}-CH_3 \qquad \text{Carnauba wax}$$

50. Lanolin is a lipid wax obtained from sheep's wool and used in ointments and cosmetic lotions. If lanolin is treated with aqueous NaOH, what two general types of organic structures are obtained?
51. Which vitamins belong to the lipid class of biological compounds?
52. Draw the unique ring structure that is typical of steroid compounds.

Nucleic Acids (Sec. 20.6)

53. What are the three general components of a DNA nucleotide?
54. What are the three general components of an RNA nucleotide?
55. What are the four organic nitrogen bases in DNA?
56. What are the four organic nitrogen bases in RNA?
57. Structurally, what is the difference between a molecule of deoxyribose (in DNA) and a molecule of ribose (in RNA)?
58. Structurally, what is the difference between a molecule of thymine (in DNA) and a molecule of uracil (in RNA)?
59. How many strands of nucleotides are in a DNA molecule?
60. How many strands of nucleotides are in an RNA molecule?
61. What intermolecular force holds the DNA double helix together?
62. How many hydrogen bonds are there between adenine and thymine nucleotides in the DNA double helix? between cytosine and guanine nucleotides?
63. During DNA replication, an adenine base in the template strand codes for which base in the complementary strand?
64. During DNA replication, a guanine base in the template strand codes for which base in the complementary strand?

65. During RNA transcription, an adenine base in the template strand codes for which base in the growing strand?
66. During RNA transcription, a cytosine base in the template strand codes for which base in the growing strand?

General Exercises

67. What is the overall shape of a protein that provides strength, such as a protein in muscle tissue?
68. What is the overall shape of a protein that serves a metabolic role, such as a protein for oxygen transport?
69. What happens to a protein when it undergoes denaturation in aqueous acid?
70. What happens to a polysaccharide when in undergoes hydrolysis in aqueous acid?
71. Predict the type of biological compound that is the main component of animal horns, hooves, claws, and feathers.
72. Chitin is a polymer found in the skeletons of insects. If the repeating unit in chitin is a polyhydroxy aldehyde with an amine group, which type of biological compound is chitin?
73. How many nucleotide sequences are possible for an RNA trinucleotide having an adenine base and two cytosine bases?
74. How many nucleotide sequences are possible for an RNA trinucleotide having an adenine base, a guanine base, and a uracil base?
75. Explain how a single strand of DNA acts as a template to synthesize a complementary strand of DNA and regenerate a double helix.
76. Explain how tRNA builds a protein chain using mRNA as a template.
77. How many amino acid sequences are possible for a tetrapeptide containing glycine, alanine, serine, and tryptophan?

WWW
Explorer Quiz 1
Explorer Quiz 2
Explorer Quiz 3
Master Quiz

APPENDIX A

The Scientific Calculator

A handheld calculator is essential for many of the calculations in this textbook. Given a choice of calculators, choose a scientific calculator rather than a business calculator. Most scientific calculators have the following keys that you will find helpful when performing typical chemical calculations.

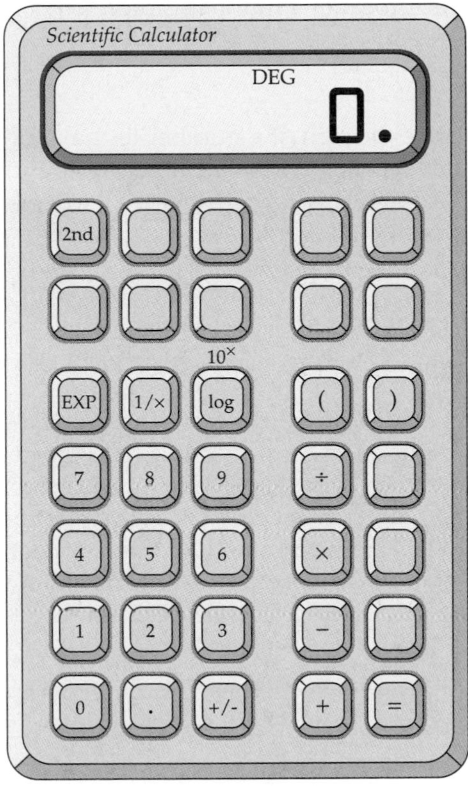

Arithmetic Operations

- *Basic Function Keys:* + − × ÷ =

Example		Key			Display
(a) 85.8 + 6.43	**85.8**	+	**6.43**	=	*92.23*
(b) 297 − 11.04	**297**	−	**11.04**	=	*285.96*
(c) 0.882 × 6.02	**.882**	×	**6.02**	=	*5.30964*
(d) 768 ÷ 0.16	**768**	÷	**.16**	=	*4800*

- *Second Function Key:* [2nd] or [SHIFT] or [INV]

Scientific calculators have many more preprogrammed functions than keys. Thus, a single key must serve two or more functions. The first function of a key is printed directly on the keypad; the second function of a key is usually printed above the keypad. To use the second function, first touch the **2nd** key and then touch the function key. The **2nd** function key may also be labeled a **SHIFT** key or an **INV** key.

- *Reciprocal Key:* [1/X] or [X^{-1}]

The reciprocal of a number is 1 divided by that number. For example, the reciprocal of 4 is 1/4, which appears in the display as 0.25. The reciprocal of 100 is 1/100 or 0.01. To obtain the reciprocal of a number, simply enter the number and touch the **1/X** key.

- *Change Sign Key:* [+ / −]

This key is used to change the sign of a number to the opposite sign. If a number is positive, the change-of-sign key will make it negative, and a minus sign will appear. If the number is negative, this key will change it to a positive number, and the minus sign will disappear.

Exponential Operations

- *Exponent Key:* [EXP] or [EEX] or [EE]

Using exponents is a shorthand way of expressing very large and very small values. A positive exponent indicates a value that is 10 or greater, while a negative exponent indicates a value that is less than 1 (Sections 2.6 and 2.7).

To enter an exponential number, first enter the numerical portion and then touch the **EXP** key followed by the exponent. If the exponent is negative, touch the change-of-sign key. The following examples illustrate how to enter exponential numbers and a possible display on a scientific calculator.

Example	Key				Display
(a) 1.29×10^{2}	**1.29**	[EXP]	**2**		*1.29* ^{02}
(b) 6.02×10^{23}	**6.02**	[EXP]	**23**		*6.02* ^{23}
(c) 5.87×10^{-7}	**5.87**	[EXP]	**7**	[+ / −]	*5.87* $^{-07}$
(d) 1.00×10^{-14}	**1.00**	[EXP]	**14**	[+ / −]	*1.00* $^{-14}$

Note Never enter the $\times$ **10** portion of an exponential number. Touching the **EXP** key automatically considers the following number a power of 10.

Note If an exponent is a small number, some calculators will automatically display the value in numerical form. For example, a calculator may display 1.29×10^{2} as *129* rather than *1.29* ^{02}.

If your calculator has a **SCI** or a **MODE** key, it is possible to program your calculator to give a display that is always in scientific notation. In fact, you may be able to fix the number of significant digits shown in the display.

If your calculator has a **FSE** key, you can choose fixed decimal notation (**F**), scientific notation (**S**), or engineering notation (**E**). For details about your model of calculator, refer to the instruction booklet or ask your instructor.

Chain Calculations

A chain calculation is a computation that requires more than one arithmetic operation. The following example exercise provides practice in using your calculator to solve problems involving chain calculations.

Example Exercise •

Use your calculator to solve the following problems and express the answer in scientific notation.

(a) $5.76 \times 10^2 \times 3.15 \times 10^{-4}$

(b) $1.02 \times 10^5 \div 3.13 \times 10^2$

(c) $9.53 \times 10^3 \times \dfrac{1.02 \times 10^2}{4.95 \times 10^9}$

(d) $1.98 \times 10^2 \times \dfrac{2.34 \times 10^5}{8.67 \times 10^{-1}}$

Solution

	Key	Display
(a)	**5.76** [EXP] **2** [×] **3.15** [EXP] **4** [+/−] [=]	*1.81* $^{-01}$
(b)	**1.02** [EXP] **5** [÷] **3.13** [EXP] **2** [=]	*3.26* ^{02}
(c)	**9.53** [EXP] **3** [×] **1.02** [EXP] **2** [÷] **4.95** [EXP] **9** [=]	*1.96* $^{-04}$
(d)	**1.98** [EXP] **2** [×] **2.34** [EXP] **5** [÷] **8.67** [EXP] **1** [+/−] [=]	*5.34* ^{07}

Thus, the answers to the calculations expressed in scientific notation are

(a) 1.81×10^{-1}

(b) 3.26×10^2

(c) 1.96×10^{-4}

(d) 5.34×10^7

Self-Test

Use your calculator to solve the following problems and express the answer in scientific notation.

(a) $9.41 \times 10^{-1} \times 6.98 \times 10^5$

(b) $1.67 \times 10^3 \div 2.32 \times 10^{-6}$

(c) $8.59 \times 10^{-7} \times \dfrac{7.36 \times 10^{27}}{6.32 \times 10^{21}}$

(d) $2.75 \times 10^5 \times \dfrac{5.92 \times 10^5}{1.45 \times 10^{20}}$

Answers: (a) 6.57×10^5; (b) 7.20×10^8; (c) 1.00; (d) 1.12×10^{-9}

Logarithmic Operations

• *Base 10 Logarithm Key:* [LOG] or [LOG_{10}]

The base 10 logarithm of a number is the power to which 10 must be raised in order to equal that number. In this text, we use logarithms to calculate the pH of solutions (Sections 15.8 and 15.9). The following examples show the conversion of $[H^+]$ to pH, that is, $-\log [H^+]$.

Example	Key	Display
(a) $[H^+] = 3.8 \times 10^{-5}$	**3.83** [EXP] 5 [+/−] [LOG] [+/−]	*4.42*
(b) $[H^+] = 1.6 \times 10^{-3}$	**1.59** [EXP] 3 [+/−] [LOG] [+/−]	*2.80*

Answers:
(a) pH = 4.42; (b) pH = 2.80

- *Base 10 Antilogarithm Key:* [2nd] [LOG] or [10^X]

Obtaining the inverse logarithm, or antilogarithm, is the reverse operation of finding the logarithm of a number. In this text, we use this operation when converting pH into the molar hydrogen ion concentration of an acid solution. The following examples show the conversion of pH to $[H^+]$.

Example	Key	Display
(a) pH = 8.15	**8.15** [+/−] [2nd] [LOG]	*7.1* $^{-09}$
(b) pH = 1.55	**1.55** [+/−] [2nd] [LOG]	*2.8* $^{-02}$

Answers:
(a) $[H^+] = 7.1 \times 10^{-9}$; (b) $[H^+] = 2.8 \times 10^{-2} = 0.028$

APPENDIX B

Weights and Measures

Metric System Exact Equivalents

Length:	1 meter (m)	≡	100 centimeters (cm)
	1 meter (m)	≡	1000 millimeters (mm)
	1 kilometer (km)	≡	1000 meters (m)
Mass:	1 gram (g)	≡	1000 milligrams (mg)
	1 kilogram (kg)	≡	1000 grams (g)
	1 metric ton (t)	≡	1000 kilograms (kg)
Volume:	1 liter (L)	≡	1000 milliliters (mL)
	1 liter (L)	≡	10 deciliters (dL)
	1 milliliter (mL)	≡	1 cubic centimeter (cm^3)

English System Exact Equivalents

Length:	1 foot (ft)	≡	12 inches (in.)
	1 mile (mi)	≡	1760 yards (yd)
	1 mile (mi)	≡	5280 feet (ft)
Mass:	1 troy pound (t lb)	≡	12 troy ounces (t oz)
	1 pound (lb)	≡	16 ounces (oz)
	1 ton	≡	2000 pounds (lb)
Volume:	1 quart (qt)	≡	32 fluid ounces (fl oz)
	1 quart (qt)	≡	2 pints (pt)
	1 gallon (gal)	≡	4 quarts (qt)

Metric–English Approximate Equivalents

Length:	1 inch (in.)	=	2.54 centimeters (cm)
Mass:	1 pound (lb)	=	454 grams (g)
Volume:	1 quart (qt)	=	946 milliliters (mL)
Time:	1 second (sec)	=	1.00 second (s)

Temperature Equivalents

$\frac{9}{5}$°C + 32	=	°F
$\frac{5}{9}$(°F − 32)	=	°C
°C + 273	=	K

Energy Equivalents

1 calorie (cal)	=	4.184 joules (J)
1 kilocalorie (kcal)	≡	1000 calories (cal)
1 kilocalorie (kcal)	=	4.184 kilojoules (kJ)

APPENDIX C

Physical Constants

Avogadro's number	=	6.02×10^{23}
Absolute zero		
Kelvin scale	=	0 K
Celsius scale	=	−273.15°C
Standard temperature and pressure (STP)		
Standard temperature	≡	0°C (273 K)
Standard atmospheric pressure	≡	1 atm
	≡	760 mm Hg (760 torr)
	≡	76 cm Hg
	=	29.9 in. Hg
	=	14.7 psi
	=	101 kPa
Molar volume of a gas (STP)	=	22.4 L/mol
Ideal gas constant, *R*	=	0.0821 L · atm/mol · K
Ionization constant of water, K_w	=	1.00×10^{-14}
Mass of proton	=	1.0073 amu
Mass of neutron	=	1.0087 amu
Mass of electron	=	0.00055 amu
Velocity of light	=	3.00×10^{8} m/s

APPENDIX D

Activity Series of Metals

Most Active:	Li
	K
	Ba
	Sr
	Ca
	Na
	Mg
	Al
	Mn
	Zn
	Fe
	Cd
	Co
	Ni
	Sn
	Pb
	(H)
	Cu
	Ag
	Hg
Least Active:	Au

APPENDIX E

Solubility Rules for Ionic Compounds

Compounds containing the following ions are generally *soluble* in water:

1. Alkali metal ions and ammonium ions; Li^+, Na^+, K^+, NH_4^+
2. Acetate ion: $C_2H_3O_2^-$
3. Nitrate ion: NO_3^-
4. Halide ions (X): Cl^-, Br^-, I^-. (AgX, Hg_2X_2, and PbX_2 are insoluble exceptions.)
5. Sulfate ion: SO_4^{2-}. ($SrSO_4$, $BaSO_4$, and $PbSO_4$ are insoluble exceptions.)

Compounds containing the following ions are generally *insoluble* in water:*

6. Carbonate ion, CO_3^{2-}. (See rule 1 exceptions, which are soluble.)
7. Chromate ion, CrO_4^{2-}. (See rule 1 exceptions, which are soluble.)
8. Phosphate ion, PO_4^{3-}. (See rule 1 exceptions, which are soluble.)
9. Sulfide ion, S^{2-}. (CaS, SrS, BaS, and rule 1 exceptions are soluble.)
10. Hydroxide ion, OH^-. [$Ca(OH)_2$, $Sr(OH)_2$, $Ba(OH)_2$, and rule 1 exceptions are soluble.]

*These compounds are actually slightly soluble, or very slightly soluble, in water.

APPENDIX F

Vapor Pressure of Water

Temperature (°C)	Vapor Pressure (mm Hg)	Temperature (°C)	Vapor Pressure (mm Hg)	Temperature (°C)	Vapor Pressure (mm Hg)
0	4.6	21	18.7	35	42.2
5	6.5	22	19.8	40	55.3
10	9.2	23	21.1	45	71.9
12	10.5	24	22.4	50	92.5
14	12.0	25	23.8	55	118.0
16	13.6	26	25.2	60	149.4
17	14.5	27	26.7	70	233.7
18	15.5	28	28.4	80	355.1
19	16.5	29	30.0	90	525.8
20	17.5	30	31.8	100	760.0

APPENDIX G

Properties of Water

Density of H_2O:	0.99987 g/mL at 0°C
	1.00000 g/mL at 4°C
	0.99707 g/mL at 25°C
Heat of fusion at 0°C:	80.0 cal/g (335 J/g)
Heat of solidification at 0°C:	80.0 cal/g (335 J/g)
Heat of vaporization at 100°C:	540 cal/g (2260 J/g)
Heat of condensation at 100°C:	540 cal/g (2260 J/g)
Specific heat of ice:	0.50 cal/g × °C (2.1 J/g × °C)
of water:	1.00 cal/g × °C (4.18 J/g × °C)
of steam:	0.48 cal/g × °C (2.0 J/g × °C)
Ionization constant, K_w:	1.00×10^{-14} at 25°C

APPENDIX H

Answers to Key Concept Exercises

Chapter 1

1. The principal difference between ancient and modern chemistry is the application of the scientific method. Ancient chemistry was based on speculation, while modern chemistry is based on planned experiments and tested explanations of the results.

2. We can distinguish a theory from a law by asking the question: Is the proposed statement measurable? If it is, the statement is a natural law; if it is not, the statement is a scientific theory.

3. Although the line segments appear to be of equal length, line *AB* is about 10 mm longer than *BC*.

4. Although Box A appears to be longer and more narrow, both Box A and B have the same dimensions; that is, each box measures 30 mm in length, 15 mm in width, and 10 mm in height.

Chapter 2

1. No instrument is capable of making an exact measurement. Since no measurement is exact, a length, mass, or volume measurement can never be stated with absolute certainty.

2. ruler A: 1.9 ± 0.1 cm; ruler B: 1.90 ± 0.05 cm

3. diameter of a 1¢ coin: ~2 cm; thickness of a 10¢ coin: ~0.1 cm

4. mass of a 1¢ coin: ~3 g; mass of a 25¢ coin: ~5 g

5. volume of 10 drops: ~0.5 mL: volume of a quart: ~1000 mL

6. Mass is independent of gravity, and so the mass of the astronaut is the *same* on Uranus and on Mars.

7. Weight is affected by gravity, and so the weight of the astronaut is *more* on the heavier planet Uranus where gravity is greater.

8. Metric–metric relationships can be exactly equivalent; metric–English relationships cannot because the two systems use different reference standards. Thus, **(a)** 1 meter ≡ 1000 millimeters and **(b)** 1 meter = 1.09 yards.

9. **(a)** Since 1 meter and 1000 millimeters are exactly equivalent, there is an infinite number of significant digits in the relationship. **(b)** Since 1 meter and 1.09 yards are approximately equal, there are only three significant digits in the relationship.

10. Step 1: Write down the units asked for in the answer. Step 2: Write down the given value in the problem that is related to the units in the answer. Step 3: Apply a unit factor to convert the units in the given value to the units in the answer.

Chapter 3

1. Statements **(a)**, **(b)**, **(c)**, **(d)**, and **(e)** are all true.

2. None; there is no basic unit of length in the English system.

3. The meter is the basic unit of length in the metric system.

4. Since 100 cm ≡ 1 m, there is an infinite number of significant digits in the unit factor 100 cm/1 m.

5. Since 39.4 in. = 1 m, there are three significant digits in the unit factor 39.4 in./1 m.

6. A cube 1 cm on a side (1 cm^3) has a volume exactly equal to 1 mL.

7. A cube 10 cm on a side (1000 cm^3) has a volume exactly equal to 1 L.

8. A solid object sinks in a liquid that is less dense and floats on a liquid that is more dense. Thus, a cork rests on ether (L_1), beeswax rests on water, a silver coin rests on mercury (L_2), and a gold coin sinks in mercury and rests on the bottom of the cylinder.

9. The Kelvin temperature scale is assigned a value of zero for the coldest possible temperature; thus, −100 K cannot exist.

10. Foods cook faster at higher temperatures. If a pie crust cooks faster than the fruit filling, the filling has a higher specific heat; that is, 1 g of fruit filling requires more calories to raise the temperature 1°C.

Chapter 4

1. **(a)** solid; **(b)** liquid; **(c)** gas

2. **(a)** compound; **(b)** element; **(c)** homogeneous mixture; **(d)** heterogeneous mixture

3. **(a)** element; **(b)** element; **(c)** compound; **(d)** homogeneous mixture

4. Lithium is a metallic solid under normal conditions.

5. Chlorine is a nonmetallic gas under normal conditions.

6. None; the properties of compounds are not similar to those of their constituent elements.

7. When an aspirin tablet dissolves in water without producing bubbles, it is an example of a physical change.

8. When an Alka-Seltzer tablet dissolves in water while producing bubbles, it is an example of a chemical change.

9. Applying the conservation of mass law, the mass of oxygen is found to be 0.658 g (1.658 g − 1.000 g).

10. As the temperature of a gas increases, the kinetic energy of the molecules increases. As the kinetic energy of a gas increases, the velocity of the molecules increases.

Chapter 5

1. If an atom (~10^{-8} cm) were magnified to the size of a golf ball (~5 cm), the golf ball would be about 5×10^8 times larger than the atom. If a golf ball were magnified 5×10^8 times, it would be about the size of Earth.

2. In the analogy, the marble represents an atomic nucleus, and the Astrodome represents the relative size of an entire atom.

3. In the analogy, the missile represents an alpha particle, and the planets in the solar system represent atomic nuclei.

4. Atoms of different elements cannot have the same atomic number. However, atoms of different elements can have the same mass number.

5. In the analogy, an ocean wave is to a drop of water as a light wave is to a particle of light.

6. The statements **(a)**, **(b)**, and **(d)** are true. Statement **(c)** is false, as an electron must occupy a quantum level having fixed energy.

7. The statements **(a)**, **(b)**, **(c)**, and **(d)** are all true.

8. The atomic theory has become increasingly more sophisticated since its inception in 1803.
(a) In the Dalton model, an atom is a simple sphere.
(b) In the Thomson model, the mass of an atom is homogeneous but the atom contains subatomic particles.
(c) In the Rutherford model, an atom contains a positively charged nucleus surrounded by negatively charged electrons.
(d) In the Bohr model, the electrons in an atom occupy a quantum level of fixed energy.
(e) In the Heisenberg model, the exact location of an electron is uncertain although it must occupy an energy boundary called an orbital.
(f) In the current model, the composition of the nucleus is complex and the nucleus contains numerous highly unstable particles.

Chapter 6

1. The modern periodic law states that properties repeat when the elements are arranged according to increasing atomic number.

2. The elements having the lowest atomic masses are as follows: alkali metal (Li), alkaline earth metal (Be), halogen (F), noble gas (He), and transuranium element (Np).

3. The elements having the lowest atomic masses are as follows: semimetal (B), lanthanide (Ce), actinide (Th), rare earth metal (Sc), and inner transition element (Ce).

4. The alkali metal that has the most metallic character (Fr) also has the largest atomic radius. The halogen that has the least metallic character (F) also has the smallest atomic radius.

5. The predicted atomic radius for Po is 0.170 nm (0.027 nm + 0.143 nm).

6. The predicted boiling point for Rn is −62°C (45°C −107°C).

7. Since boron is in the same group as Al, the predicted formula of the oxide is similar, that is, B_2O_3.

8. Since boron is in Group IIIA/13, the predicted number of valence electrons is 3.

9. The Se^{2-} ion has 36 electrons and is isoelectronic with Kr.

10. The predicted electron configuration for Co^{3+} is $1s^2\,2s^2\,2p^6\,3s^2\,3p^6\,3d^6$.

11. The predicted electron configuration for Br^- is $1s^2\,2s^2\,2p^6\,3s^2\,3p^6\,4s^2\,3d^{10}\,4p^6$.

12. The yellow substance appears to be common sulfur. The gray substance is antimony because a semimetal appears metallic. By elimination, the orange substance must be antimony sulfide, Sb_2S_3.

Chapter 7

1. NaCl is a binary ionic compound, HCl is binary molecular compound, HCl(aq) is a binary acid, NaClO is ternary ionic, and HClO(aq) is a ternary oxyacid.

2. Na^+ is a monoatomic cation, Cl^- is a monoatomic anion, ClO^- is a polyatomic anion, and ClO_3^- is a polyatomic anion.

3. mercuric chloride, $HgCl_2$

4. mercuric chlorate, $Hg(ClO_3)_2$

5. NaCl, sodium chloride; HCl, hydrogen chloride;

6. $NaClO_3$, sodium chlorate; $NaClO_4$, sodium perchlorate

7. NaClO, sodium hypochlorite; $NaClO_2$, sodium chlorite

8. HCl(aq), hydrochloric acid

9. $HClO_3$(aq), chloric acid; $HClO_4$(aq), perchloric acid

10. HClO(aq), hypochlorous acid; $HClO_2$(aq), chlorous acid

Chapter 8

1. Each statement—**(a)**, **(b)**, and **(c)**—indicates a reaction is producing a gas. **(a)** A glowing splint bursting into flames indicates O_2(g). **(b)** An ammonia smell indicates NH_3(g). **(c)** A fizzing solid indicates a gas.

2. Each statement—**(a)**, **(b)**, and **(c)**—indicates a reaction in aqueous solution. **(a)** A gas is released. **(b)** A precipitate is formed. **(c)** A permanent color change.

3. A color change is *not* evidence of an exothermic reaction. Heat, light, and an explosion are evidence that energy is released.

4. Carbon does *not* occur naturally as a diatomic molecule; the nonmetals that do are H_2, N_2, O_2, F_2, Cl_2, Br_2, and I_2.

5. $C_2H_5OH(g) + 3\,O_2(g) \longrightarrow 2\,CO_2(g) + 3\,H_2O(g)$

6. The metals that react with aqueous $CuSO_4$ include Ca, Mg, and Zn.

7. The metals that react with an aqueous acid include Ca, Mg, and Zn.

8. The metals that react with water at 25°C include Ca.

9. The compounds that are insoluble in water include AgCl and Ag_2CO_3.

10. $2\ KI(aq) + Pb(NO_3)_2(aq) \longrightarrow PbI_2(s) + 2\ KNO_3(aq)$

11. $HCl(aq) + NaOH(aq) \longrightarrow NaCl(s) + HOH(l)$

Chapter 9

1. No. At the rate of one number per nanosecond, the computer can only count to 3.15×10^{18} in 100 years.

2. No. A billion iron atoms weigh 9.28×10^{-14} g, which is much less than a microgram (1×10^{-6} g).

3. A single iron atom has a mass of 55.85 amu.

4. Avogadro's number of iron atoms has a mass of 55.85 g.

5. **(a)** 6.02×10^{23} Fe atoms, **(b)** 6.02×10^{23} O_2 molecules, and **(c)** 6.02×10^{23} FeO formula units.

6. The mass of 1 mol of $C_6H_{12}O_6$ is 180.18 g.

7. All gases—including He, H_2, and CH_4—occupy 22.4 L at STP.

8. The volume of 1 mol of CO_2 is 22.4 L at STP.

9. The N_2 balloon sinks while the He balloon floats because the density of nitrogen gas is about 7 times greater than that of helium gas. The molar mass of N_2 is 28 g/mol, while the molar mass of He is only 4 g/mol.

10. The empirical formula is SO_2 (0.500 mol S + 0.500 mol O_2).

11. The molecular formula of galactose is $C_6H_{12}O_6$.

Chapter 10

1. The coefficients in a balanced equation indicate the ratios of **(a)** moles of reactants and products and **(c)** volumes of gases. The coefficients do not indicate the ratios of **(b)** actual masses but rather the ratios of molar masses.

2. According to the balanced equation for the chemical reaction: **(a)** 5.00 mol O_2; **(b)** 4.00 mol NO; **(c)** 4.69 g O_2; **(d)** 2.63 L NO; **(e)** 6.25 L O_2; **(f)** 5.00 L NO

3. According to the balanced equation for the chemical reaction: **(a)** 2.00 mol NO_2; **(b)** 3.07 g NO_2; **(c)** 2.00 L of NO_2

4. 93.3% yield

5. **(a)** 4 ZnS, 2 S; Zn is limiting reactant. **(b)** 4 ZnS, 1 Zn; S is limiting reactant. **(c)** 5 ZnS; Zn and S are both limiting reactants.

6. after reaction: 1.00 mol Cu, 0.00 mol S, 2.00 mol CuS.

7. after reaction: 0.00 mol Cu, 0.50 mol S, 1.50 mol CuS.

Chapter 11

1. As the steam cools from a gas to a liquid, the internal pressure in the can decreases, and the external atmospheric pressure crushes the can.

2. Drinking through a straw reduces the gas pressure above the liquid. Thus, the atmospheric pressure forces the liquid higher in the straw.

3. number of molecules

4. The pressure decreases because the volume increases.

5. The volume increases because the temperature increases.

6. The temperature increases because the pressure increases.

7. You cry first because the ammonia gas, NH_3, has a lower molecular mass and thus a faster molecular velocity.

Chapter 12

1. An *ionic bond* results from the attraction between a metal cation and a nonmetal anion; a *covalent bond* is formed between two nonmetal atoms.

2. Neon is isoelectronic with a magnesium ion; krypton is isoelectronic with a bromide ion.

3. LiCl, MgO, CuS, and AlP

4. HCl, NO, CO, and IF

5. **(a)** atoms; **(b)** molecules; **(c)** formula units

6. There are 18 valence electrons and 6 nonbonding pairs of electrons in one sulfur dioxide molecule.

7. There are 26 valence electrons and 10 nonbonding pairs of electrons in one sulfite ion.

8. A molecule of O_3 is nonpolar, and CH_4 is only very slightly polar.

9. HClO, $HClO_2$, $HClO_3$, and $HClO_4$ each contain one or more coordinate covalent bonds.

10. The electron pair geometry for H_2S is tetrahedral; the molecular shape is angular (bent).

Chapter 13

1. A beaker containing marbles covered with honey is analogous to the liquid physical state in which individual particles have restricted movement owing to molecular attraction.

2. At sea level, water boils at 100°C. At 10,000 ft, water boils at about 90°C because of the lower atmospheric pressure. Food cooks faster at a higher temperature, and you can boil an egg faster at 100°C than at 90°C.

3. The needle floats on water because of surface tension. If you disturb the water, the needle will sink because it is more dense.

4. Water forms a concave lens because there is intermolecular attraction between polar water molecules and glass. Mercury forms a convex lens because there is repulsion between mercury atoms and glass.

5. An *ionic* crystalline solid has a high melting point and conducts electricity only when melted. A *molecular* crystalline solid has a low melting point, is insoluble in water, and is a nonconductor of electricity.

6. 1600 cal + 2000 cal + 10,800 cal = 14,400 cal

7. In 100.0 g of hydrate, there are 39.7 g of H_2O and 60.3 g of $NaC_2H_3O_2$. The mole ratio of H_2O to $NaC_2H_3O_2$ is 2.20/0.735. The chemical formula for the hydrate is $NaC_2H_3O_2 \cdot 3\ H_2O$.

Chapter 14

1. Bubbles form on the inside of a pan of water when the water is heated because the solubility of air in water is less as the temperature increases.

2. Peanut butter can dissolve grease because peanut butter and grease are both nonpolar.

3. Vitamin C ($C_6H_8O_6$), is water-soluble because it is polar. Vitamins A, D, and E are fat-soluble because they are nonpolar.

4. The solubility of $C_{12}H_{22}O_{11}$ at 20°C is 100 g/100 g water.

5. A solution with 100 g of $NaC_2H_3O_2$ in 100 g of water at 50°C is supersaturated.

6. 0.900%, 0.155 *M*

7. Table salt is an ionic solid composed of sodium and chloride ions. When salt dissolves in water, the polar water molecules are attracted to the ions in the salt crystal. Specifically, the negative end of the H_2O molecule is attracted to Na^+ and the positive end of the H_2O is attracted to a Cl^-.

Chapter 15

1. Aqueous HNO_3 is a strong Arrhenius acid and ionizes ~100% in solution.

2. Aqueous HCl and LiOH react to give the salt LiCl.

3. Brønsted–Lowry acid, HNO_3 (aq); Brønsted–Lowry base, NH_3 (aq)

4. 25.0 mL of 0.200 *M* NaOH neutralizes 25.0 mL of 0.100 *M* H_2SO_4

5. 0.400 *M* HCl

6. Water is neutral because $[H^+]$ and $[OH^-]$ are equal.

7. The molar hydroxide ion concentration is 6.7×10^{-12}.

8. The pH is 2.82.

9. The bright conductivity light indicates Beaker X contains a strong electrolyte such as HNO_3, which is highly ionized. The dim conductivity light indicates Beaker Y contains a weak electrolyte such as HNO_2, which is slightly ionized.

10. $H^+(aq) + OH^-(aq) \longrightarrow H_2O(l)$

Chapter 16

1. **(a)**, **(b)**, and **(c)**. The frequency, energy, and geometry of molecular collisions all influence the rate of a chemical reaction.

2. **(a)**, **(b)**, and **(c)**. Increasing concentration or temperature, or adding a catalyst, all increase the rate of a chemical reaction.

3. **(b)** Notice that the progress of the reaction goes from higher to lower energy, which characterizes an exothermic reaction.

4. **(a)** Before a reaction reaches chemical equilibrium, the amount of reactants is decreasing (and the amount of products is increasing).

5. The general equilibrium constant expression is $K_{eq} = [C]^2/[A]^1\,[B]^3$.

6. Although a change in concentration or temperature affects the given equilibrium, a change in *pressure has no effect* because there are the same number of molecules (two) on each side of the equation.

7. The ionization constant expression is $K_i = [H^+][X^-]/[HX]$.

8. The ionization constant for the weak acid is $K_i = 1.4 \times 10^{-4}$.

9. The solubility constant expression is $K_{sp} = [Ag^+][Cl^-]$.

10. $Ca(NO_3)_2$ shifts the equilibrium to the right, and Na_2CO_3 shifts the equilibrium to the left; however, $CaCO_3$ has no effect on the equilibrium because the solution is already saturated.

Chapter 17

1. **(a)** 0; **(b)** −1; **(c)** −1; **(d)** +3

2. oxidizing agent, HCl; reducing agent, Co

3. oxidizing agent, F_2; reducing agent, Cl^-

4. $2\,Al_2O_3(s) + 6\,Cl_2(g) \longrightarrow 4\,AlCl_3(aq) + 3\,O_2(g)$

5. $2\,MnO_4^-(aq) + 5\,SO_3^{2-}(aq) + 6\,H^+(aq) \longrightarrow 2\,Mn^{2+}(aq) + 5\,SO_4^{2-}(aq) + 3\,H_2O(l)$

6. $2\,MnO_4^-(aq) + 3\,SO_3^{2-}(aq) + 2\,H^+(aq) \longrightarrow 2\,MnO_2(s) + 3\,SO_4^{2-}(aq) + H_2O(l)$

7. Table 17.3 lists Zn and Al as stronger reducing agents than Fe^{2+}(aq). Thus, Zn and Al react spontaneously with $FeSO_4$(aq); Ag and Ni do not.

8. In the voltaic electrochemical cell:
(a) $Mn(s) \longrightarrow Mn^{2+} + 2\,e^-$
(b) $Fe^{2+} + 2\,e^- \longrightarrow Fe(s)$
(c) Anode, Mn; cathode, Fe
(d) Electrons flow from the Mn anode to the Fe cathode.
(e) SO_4^{2-} ions flow from the Fe compartment to the Mn compartment.

Chapter 18

1. **(a)** gamma ray; **(b)** beta ray; **(c)** alpha ray

2. ${}^{238}_{94}Pu \longrightarrow {}^{234}_{92}U + {}^{4}_{2}He$
${}^{59}_{26}Fe \longrightarrow {}^{59}_{27}Co + {}^{0}_{-1}e$
${}^{23}_{12}Mg \longrightarrow {}^{23}_{11}Na + {}^{0}_{+1}e$
${}^{183}_{79}Au + {}^{0}_{-1}e \longrightarrow {}^{183}_{78}Pt$

3. ${}^{214}_{84}Po$

4. 25 mg

5. Each click indicates a decaying radionuclide.

6. ${}^{70}_{30}Zn$

7. $6\,n^0$

8. softball

9. ${}^{0}_{+1}e$

10. ${}^{210}_{82}Pb \longrightarrow {}^{210}_{83}Bi + {}^{0}_{-1}e$

Chapter 19

1. **(b)** and **(c)**. Hexene and hexyne are unsaturated hydrocarbons.

2. **(a)** hexane; **(b)** 2-hexene; **(c)** 3-hexyne; **(d)** benzene

3. **(a)** CH_3-OH **(b)** $CH_3-CH_2-CH_2-NH_2$
(c) $CH_3-CH(CH_3)-I$ **(d)** $CH_3-CH_2-O-CH_2-CH_3$

4. **(a)** $CH_3-\overset{\overset{\large O}{\|}}{C}-H$ **(b)** $CH_3-\overset{\overset{\large O}{\|}}{C}-OH$

(c) $CH_3-\overset{\overset{\large O}{\|}}{C}-O-C_6H_5$ (phenyl ring) **(d)** $CH_3-\overset{\overset{\large O}{\|}}{C}-NH_2$

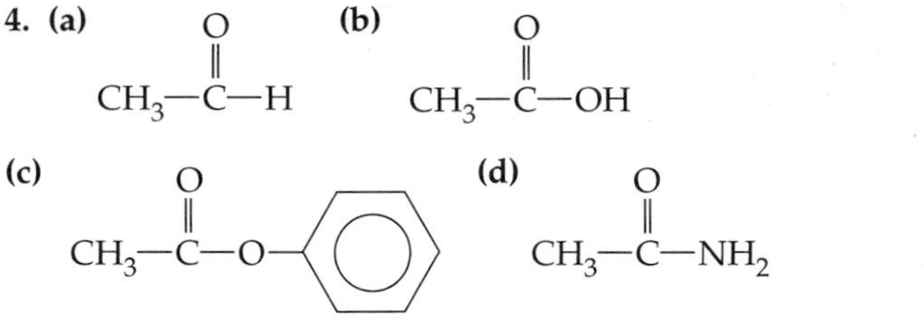

5. Thyroxine contains the following functional groups (left to right): phenol, organic halide, ether, amine, carboxylic acid.

6. alcohol, CH_3-CH_2-OH; ether, CH_3-O-CH_3

7. aldehyde, CH_3-CH_2-CHO; ketone, $CH_3-CO-CH_3$

8. carboxylic acid, $CH_3-CO-OH$; ester, $H-CO-OCH_3$

9. **(a)** alkene; **(b)** organic halide; **(c)** alcohol (phenol); **(d)** amine; **(e)** aldehyde; **(f)** ketone; **(g)** ester; **(h)** amide

10. **(a)** $CH_3-CH(CH_3)-CH_2-C(CH_3)_2-CH_3$
(b) $CH_3-CH(OH)-CH_2-CH_3$

Chapter 20

1. **(a)** protein; **(b)** lipid; **(c)** carbohydrate; **(d)** nucleic acid

2.

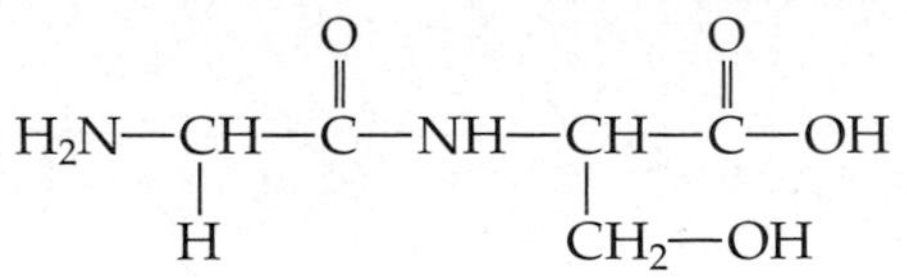

$$H_2N-\underset{\underset{\large H}{|}}{CH}-\overset{\overset{\large O}{\|}}{C}-NH-\underset{\underset{\large CH_2-OH}{|}}{CH}-\overset{\overset{\large O}{\|}}{C}-OH$$

3. **(a)** secondary structure; **(b)** primary structure; **(c)** tertiary structure

4. **(a)** the substrate molecule; **(b)** the enzyme

5. $$\underset{\text{sucrose}}{C_{12}H_{22}O_{11}} + \underset{\text{water}}{H_2O} \xrightarrow{H^+} \underset{\text{glucose}}{C_6H_{12}O_6} + \underset{\text{fructose}}{C_6H_{12}O_6}$$

6.

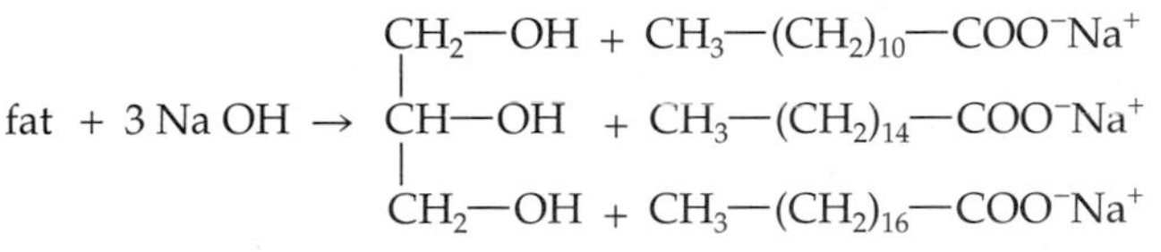

$$\text{fat} + 3\,\text{Na OH} \rightarrow \begin{array}{l} CH_2-OH \\ | \\ CH-OH \\ | \\ CH_2-OH \end{array} \begin{array}{l} + CH_3-(CH_2)_{10}-COO^-Na^+ \\ \\ + CH_3-(CH_2)_{14}-COO^-Na^+ \\ \\ + CH_3-(CH_2)_{16}-COO^-Na^+ \end{array}$$

The two products are glycerol and soap. The fat contains fatty acid esters of lauric acid, palmitic acid, and stearic acid.

7.

$$CH_3-(CH_2)_{34}-\overset{\overset{\large O}{\|}}{C}-O-(CH_2)_{35}-CH_3$$

8. **(a)** The DNA nucleotide contains the sugar deoxyribose, the organic base adenine, and phosphoric acid. **(b)** The RNA nucleotide contains the sugar ribose, the organic base cytosine, and phosphoric acid.

APPENDIX

Answers to Key Term Exercises

Chapter 1

1. i; **2.** d; **3.** j; **4.** e; **5.** k; **6.** g; **7.** a; **8.** c; **9.** f; **10.** h; **11.** b

Chapter 2

1. g; **2.** a; **3.** d; **4.** h; **5.** e; **6.** p; **7.** f; **8.** t; **9.** o; **10.** i; **11.** m; **12.** c; **13.** k; **14.** n; **15.** r; **16.** s; **17.** l; **18.** b; **19.** q; **20.** j

Chapter 3

1. e; **2.** o; **3.** n; **4.** h; **5.** m; **6.** q; **7.** j; **8.** v; **9.** f; **10.** w; **11.** p; **12.** u; **13.** c; **14.** x; **15.** d; **16.** r; **17.** i; **18.** t; **19.** g; **20.** b; **21.** l; **22.** s; **23.** a; **24.** k

Chapter 4

1. z; **2.** cc; **3.** l; **4.** o; **5.** p; **6.** dd; **7.** a; **8.** h; **9.** n; **10.** b; **11.** g; **12.** t; **13.** s; **14.** m; **15.** v; **16.** bb; **17.** c; **18.** w; **19.** r; **20.** u; **21.** e; **22.** y; **23.** f; **24.** d; **25.** x; **26.** aa; **27.** q; **28.** j; **29.** i; **30.** k

Chapter 5

1. g; **2.** i; **3.** u; **4.** r; **5.** d; **6.** e; **7.** q; **8.** c; **9.** o; **10.** b; **11.** a; **12.** z; **13.** n; **14.** p; **15.** y; **16.** w; **17.** h; **18.** t; **19.** f; **20.** l; **21.** k; **22.** m; **23.** j; **24.** x; **25.** v; **26.** s

Chapter 6

1. r; **2.** h; **3.** q; **4.** b; **5.** c; **6.** i; **7.** p; **8.** t; **9.** u; **10.** o; **11.** a; **12.** j; **13.** s; **14.** v; **15.** f; **16.** d; **17.** w; **18.** g; **19.** k; **20.** m; **21.** l; **22.** n; **23.** e

Chapter 7

1. h; **2.** i; **3.** d; **4.** p; **5.** e; **6.** b; **7.** c; **8.** q; **9.** f; **10.** a; **11.** l; **12.** n; **13.** o; **14.** j; **15.** m; **16.** g; **17.** k

Chapter 8

1. h; **2.** o; **3.** n; **4.** d; **5.** q; **6.** g; **7.** s; **8.** r; **9.** f; **10.** l; **11.** i; **12.** v; **13.** j; **14.** k; **15.** u; **16.** m; **17.** p; **18.** c; **19.** b; **20.** a; **21.** e; **22.** t

Chapter 9

1. a; **2.** g; **3.** e; **4.** f; **5.** j; **6.** b; **7.** d; **8.** i; **9.** c; **10.** h

Chapter 10

1. i; **2.** b; **3.** d; **4.** h; **5.** m; **6.** f; **7.** g; **8.** o; **9.** j; **10.** l; **11.** c; **12.** e; **13.** a; **14.** n; **15.** k

Chapter 11

1. j; **2.** u; **3.** b; **4.** c; **5.** t; **6.** h; **7.** o; **8.** d; **9.** e; **10.** k; **11.** f; **12.** s; **13.** v; **14.** g; **15.** q; **16.** w; **17.** p; **18.** l; **19.** r; **20.** i; **21.** a; **22.** n; **23.** m

Chapter 12

1. e; **2.** aa; **3.** u; **4.** o; **5.** g; **6.** n; **7.** q; **8.** d; **9.** b; **10.** c; **11.** s; **12.** k; **13.** y; **14.** x; **15.** j; **16.** z; **17.** r; **18.** w; **19.** m; **20.** h; **21.** v; **22.** t; **23.** i; **24.** f; **25.** bb; **26.** l; **27.** p; **28.** a

Chapter 13

1. g; **2.** f; **3.** n; **4.** w; **5.** b; **6.** x; **7.** v; **8.** d; **9.** o; **10.** r; **11.** q; **12.** j; **13.** k; **14.** c; **15.** s; **16.** h; **17.** p; **18.** t; **19.** l; **20.** m; **21.** a; **22.** y; **23.** i; **24.** u; **25.** e

Chapter 14

1. c; **2.** n; **3.** p; **4.** o; **5.** b; **6.** i; **7.** k; **8.** j; **9.** e; **10.** g; **11.** d; **12.** q; **13.** a; **14.** s; **15.** m; **16.** l; **17.** t; **18.** r; **19.** f; **20.** h

Chapter 15

1. g; **2.** c; **3.** d; **4.** k; **5.** h; **6.** j; **7.** q; **8.** e; **9.** f; **10.** p; **11.** o; **12.** b; **13.** a; **14.** u; **15.** i; **16.** s; **17.** l; **18.** n; **19.** t; **20.** w; **21.** v; **22.** r; **23.** m

Chapter 16

1. d; **2.** e; **3.** f; **4.** o; **5.** r; **6.** a; **7.** h; **8.** b; **9.** p; **10.** n; **11.** c; **12.** l; **13.** g; **14.** j; **15.** i; **16.** m; **17.** k; **18.** q

Chapter 17

1. k; **2.** m; **3.** j; **4.** o; **5.** l; **6.** n; **7.** i; **8.** p; **9.** f; **10.** e; **11.** h; **12.** q; **13.** r; **14.** g; **15.** a; **16.** c; **17.** b; **18.** d

Chapter 18

1. t; **2.** b; **3.** d; **4.** i; **5.** c; **6.** k; **7.** p; **8.** o; **9.** l; **10.** r; **11.** h; **12.** s; **13.** q; **14.** a; **15.** j; **16.** u; **17.** v; **18.** m; **19.** e; **20.** f; **21.** n; **22.** g; **23.** w

Chapter 19

1. r; **2.** n; **3.** o; **4.** t; **5.** v; **6.** g; **7.** b; **8.** c; **9.** e; **10.** f; **11.** u; **12.** q; **13.** d; **14.** h; **15.** s; **16.** l; **17.** a; **18.** k; **19.** m; **20.** i; **21.** p; **22.** j

Chapter 20

1. b; **2.** r; **3.** u; **4.** c; **5.** j; **6.** m; **7.** a; **8.** p; **9.** d; **10.** s; **11.** f; **12.** k; **13.** l; **14.** e; **15.** t; **16.** i; **17.** h; **18.** x; **19.** g; **20.** o; **21.** v; **22.** q; **23.** y; **24.** w; **25.** n

APPENDIX J

Answers to Odd-Numbered Exercises

Chapter 1

1. yin and yang
3. Robert Boyle
5. A *hypothesis* is an initial proposal that is tentative, whereas a *theory* is a proposal that has been extensively tested.
7. Statement **(a)** is a *theory* because it explains the composition of an atom. Statement **(b)** is a *theory* because it explains a change in the neutron. Statement **(c)** is a *natural law* because gas pressures can be measured. Statement **(d)** is a *natural law* because gas volumes can be measured.
9. Antoine Lavoisier
11. Doctors and nurses receive training in chemistry, as well as dentists, physical therapists, chiropractors, dietitians, veterinarians, scientists, engineers, and many others.
13. A solution to the nine-dot problem using three straight lines is shown below; the assumption regards the angle of the lines and the size of the dots.

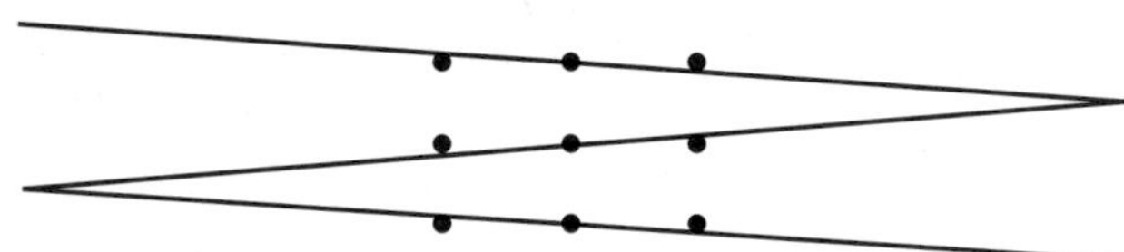

15. By flipping the image, we can view the blocks as stacked upward or as hanging downward.

Chapter 2

1. **(a)** length; **(b)** volume; **(c)** mass; **(d)** volume; **(e)** time; **(f)** volume
3. **(a)** 6.4 cm to 6.6 cm; **(b)** 0.50 g to 0.52 g; **(c)** 9.9 mL to 10.1 mL; **(d)** 35.4 s to 35.6 s
5. **(a)** 1; **(b)** 3; **(c)** 3; **(d)** 4
7. **(a)** 2; **(b)** 2; **(c)** 1; **(d)** 1
9. **(a)** 3; **(b)** 2; **(c)** 4; **(d)** 1
11. **(a)** 0; **(b)** 0; **(c)** 0; **(d)** 0
13. **(a)** 31.5; **(b)** 214,000; **(c)** 5.16; **(d)** 77.5
15. **(a)** 61.2; **(b)** 362; **(c)** 2160; **(d)** 0.367
17. **(a)** 72.15 cm; **(b)** 49.7 cm
19. **(a)** 1.3 g; **(b)** 23.1 g
21. **(a)** 67 g; **(b)** 13.01 g
23. **(a)** 7.67 cm^2; **(b)** 10.1 cm^2; **(c)** 28 cm^2; **(d)** 400 cm^2
25. **(a)** 8.8 g/mL; **(b)** 3.0 g/mL; **(c)** 4.26 g/mL; **(d)** 9.124 g/mL
27. **(a)** 2^6; **(b)** $(1/2)^3$ or 2^{-3}
29. **(a)** 105; **(b)** $(1/10)^3$ or 10^{-3}
31. **(a)** 1000; **(b)** 0.000 000 1
33. **(a)** 1×10^9; **(b)** 1×10^{-8}
35. **(a)** 10; **(b)** 0.1
37. **(a)** 8.0916×10^7; **(b)** 1.5×10^{-8}; **(c)** 3.356×10^{14}; **(d)** 9.27×10^{-13}
39. 2.69×10^{22} neon atoms
41. **(a)** 2 nickels = 1 dime; **(b)** 1 nickel = 5 pennies
43. **(a)** 1 mile = 5280 feet; **(b)** 2000 pounds = 1 ton
45. **(a)** $\frac{2 \text{ nickels}}{1 \text{ dime}}$ and $\frac{1 \text{ dime}}{2 \text{ nickels}}$; **(b)** $\frac{1 \text{ nickel}}{5 \text{ pennies}}$ and $\frac{5 \text{ pennies}}{1 \text{ nickel}}$
47. **(a)** $\frac{1 \text{ mile}}{5280 \text{ feet}}$ and $\frac{5280 \text{ feet}}{1 \text{ mile}}$; **(b)** $\frac{2000 \text{ pounds}}{1 \text{ ton}}$ and $\frac{1 \text{ ton}}{2000 \text{ pounds}}$
49. **(a)** 1 mile ≡ 5280 feet; **(d)** 1 week ≡ 7 days
51. **(a)** 3; **(b)** infinite; **(c)** infinite; **(d)** 2
53. 120 furlongs
55. 0.150 grams
57. 5.0 cups
59. 39.0 months
61. 2.19×10^{10} troy pounds
63. 5.87×10^{12} miles/year
65. 1.73%
67. 0.236 g
69. 40.5%
71. 16.9 g water
73. 1.17×10^{25} g oxygen; 6.09×10^{24} g silicon; 1.78×10^{24} g aluminum
75. 94.99% copper; 5.01% zinc
77. 10.0 mL
79. 23.0 amu
81. 258.1 g
83. **(a)** 3.52×10^6; **(b)** 1.91×10^{-6}
85. 1.67356×10^{-24} g
87. 200 pounds
89. 3.5×10^{13} hours
91. 454 grams of feathers
93. 1 pound of feathers

Chapter 3

1. All the statements are true.

3. **(a)** meter (m); **(b)** gram (g); **(c)** liter (L); **(d)** second (s)

5. (a) *kilo-* (k); **(b)** *mega-* (M); **(c)** *deci-* (d); **(d)** *micro-* (μ)

7. **(a)** km; **(b)** Gg; **(c)** μL; **(d)** ms

9. **(a)** millimeter; **(b)** kilogram; **(c)** milliliter; **(d)** microsecond

11. **(a)** 1000 mm = 1 m; **(b)** 1 g = 1×10^6 μg; **(c)** 1 L = 100 cL; **(d)** 1×10^9 s = 1 Gs

13. **(a)** $\frac{1 \times 10^9\text{m}}{1\text{ Gm}}$ and $\frac{1\text{ Gm}}{1 \times 10^9\text{m}}$; **(b)** $\frac{1\text{ g}}{1000\text{ mg}}$ and $\frac{1000\text{ mg}}{1\text{g}}$; **(c)** $\frac{1 \times 10^6\mu\text{L}}{1\text{ L}}$ and $\frac{1\text{ L}}{1 \times 10^6\mu\text{L}}$; **(d)** $\frac{100\text{ cs}}{1\text{ s}}$ and $\frac{1\text{ s}}{100\text{ cs}}$

15. **(a)** 1550 m; **(b)** 48.6 cg; **(c)** 0.125 L; **(d)** 1×10^{-7} s

17. **(a)** 1.25×10^5 Mm; **(b)** 2.55 dg; **(c)** 1.45×10^{-3} cL; **(d)** 1.56×10^9 ns

19. 3.5×10^4 μs

21. **(a)** 2.54 cm = 1 in.; **(b)** 454 g = 1 lb; **(c)** 946 mL = 1 qt; **(d)** 1 s = 1 sec

23. **(a)** 170 cm; **(b)** 459 g; **(c)** 473 mL; **(d)** 8.00×10^2 s

25. **(a)** 1.8 m; **(b)** 79.5 kg; **(c)** 0.330 gal; **(d)** 2.53 min

27. 22 km/L

29. 1.58×10^4 cm³

31. 1.26 cm

33. **(a)** 1000 mL; **(b)** 1000 cm³

35. 6.80 L

37. 0.5 mL

39. 130 mL

41. **(a)** sink; **(b)** float

43. **(a)** rise; **(b)** fall

45. **(a)** 170 g; **(b)** 1.6 g

47. **(a)** 0.160 mL; **(b)** 1.12 mL

49. **(a)** 0.788 g/mL; **(b)** 2.8 g/mL

51. **(a)** 32°F; **(b)** 273 K; **(c)** 0°C

53. **(a)** 38°C; **(b)** −137°C

55. **(a)** 768 K; **(b)** 88 K

57. Temperature is a measure of the average kinetic energy in a system, while heat is the total energy in a system.

59. 1.9×10^4 cal

61. 67.5 cal

63. 0.0306 cal/g × °C

65. 56.5 g

67. 59.2°C

69. 380 disks

71. **(a)** infinite; **(b)** 3; **(c)** infinite; **(d)** 3

73. 5.87×10^{12} miles/year

75. The 1500-meter race (0.93 mile) is shorter and faster.

77. 7720 tablets

79. 5333 yd²

81. 5.0 g/mL

83. 0.801

85. 62.4 lb/ft³

87. −252°C; 21 K

89. 1.0×10^2 kcal

91. 27.8 cal; 52.9°C

93. 1.36×10^4 kg/m³

95. 3.90 cm

Chapter 4

1. **(a)** definite; **(b)** indefinite; **(c)** indefinite

3. **(a)** negligible; **(b)** negligible; **(c)** significant

5. **(a)** melting; **(b)** vaporizing; **(c)** deposition

7. **(a)** absorbed; **(b)** absorbed; **(c)** released

9. A pure substance has definite and constant properties. A heterogeneous mixture has properties that vary within a given sample; a homogeneous mixture has properties that vary from sample to sample.

11. **(a)** pure substance; **(b)** mixture; **(c)** mixture; **(d)** pure substance

13. **(a)** element; **(b)** heterogeneous mixture; **(c)** compound; **(d)** homogeneous mixture

15. oxygen, silicon, and aluminum

17. **(a)** Li; **(b)** Ar; **(c)** Mg; **(d)** Mn; **(e)** F; **(f)** Na; **(g)** Cu; **(h)** Ni

19. **(a)** chlorine; **(b)** neon; **(c)** cadmium; **(d)** germanium; **(e)** cobalt; **(f)** radium; **(g)** chromium; **(h)** tellurium

21. **(a)** 1; **(b)** 56; **(c)** 79; **(d)** 53; **(e)** 35; **(f)** 13; **(g)** 19; **(h)** 50

23. **(a)** metal; **(b)** nonmetal; **(c)** metal; **(d)** metal or nonmetal

25. **(a)** metal; **(b)** nonmetal; **(c)** metal; **(d)** nonmetal

27. **(a)** metal; **(b)** nonmetal; **(c)** nonmetal; **(d)** semimetal

29. **(a)** metal; **(b)** semimetal; **(c)** nonmetal; **(d)** metal

31. **(a)** solid; **(b)** gas; **(c)** gas; **(d)** solid

33. **(a)** solid; **(b)** liquid; **(c)** gas; **(d)** solid

35. The mass ratio of the elements in the mineral is 5:1:4.

37. **(a)** 9 carbon atoms; 8 hydrogen atoms; 4 oxygen atoms **(b)** 17 carbon atoms; 20 hydrogen atoms; 4 nitrogen atoms; 6 oxygen atoms

39. **(a)** $C_{20}H_{30}O$; **(b)** $C_{12}H_{18}Cl_2N_4OS$

41. **(a)** 21 atoms; **(b)** 12 atoms

43. **(a)** physical; **(b)** physical; **(c)** physical; **(d)** chemical

45. **(a)** physical; **(b)** chemical; **(c)** chemical; **(d)** physical

47. **(a)** chemical; **(b)** physical; **(c)** physical; **(d)** physical

49. **(a)** physical; **(b)** chemical; **(c)** physical; **(d)** chemical

51. **(a)** physical; **(b)** chemical; **(c)** chemical; **(d)** physical

53. **(a)** physical; **(b)** chemical; **(c)** physical; **(d)** chemical

55. **(a)** physical; **(b)** chemical; **(c)** physical; **(d)** chemical

57. **(a)** chemical; **(b)** physical; **(c)** chemical; **(d)** physical

59. 3.94 g

61. 0.055 g

63. When the roller coaster is at its highest point, *potential energy is maximum* and *kinetic energy is minimum.*

65. gaseous state
67. increases
69. increases
71. 1500 kcal
73. 807 cal
75. 48.0 kcal
77. heat, light, chemical, electrical, mechanical, and nuclear
79. **(a)** nuclear $\longrightarrow$ heat;
(b) heat $\longrightarrow$ mechanical;
(c) mechanical $\longrightarrow$ mechanical;
(d) mechanical $\longrightarrow$ electrical
81. **(a)** chemical $\longrightarrow$ electrical;
(b) electrical $\longrightarrow$ mechanical;
(c) mechanical $\longrightarrow$ mechanical;
(d) chemical $\longrightarrow$ heat
83. E = energy; m = mass; c = speed of light
85. homogeneous mixture
87. germanium
89. **(a)** Fe; **(b)** Pb; **(c)** Sn; **(d)** Au
91. physical change
93. 3250 cal
95. slightly less

Chapter 5

1. Refer to Section 5.1.
3. **(1)** Atoms are divisible. **(2)** All atoms of an element are not identical.
5. electron (e^-)
7. electron (1−); proton (1+)
9. electrons
11. Atoms have a dense nucleus with a positive charge.
13. Protons and neutrons are located in the center of the atom, while electrons surround the nucleus.
15. electron, 1−; proton, 1+; neutron, 0
17. **(a)** 2 n°; **(b)** 16 n°; **(c)** 5 n°; **(d)** 24 n°
19. **(a)** 4 n°; **(b)** 21 n°; **(c)** 50 n°; **(d)** 117 n°
21.

Atomic Notation	Atomic Number	Mass Number	Number of Protons	Number of Neutrons	Number of Electrons
$^{4}_{2}He$	2	4	2	2	2
$^{21}_{10}Ne$	10	21	10	11	10
$^{50}_{22}Ti$	22	50	22	28	22
$^{197}_{79}Au$	79	197	79	118	79

23.

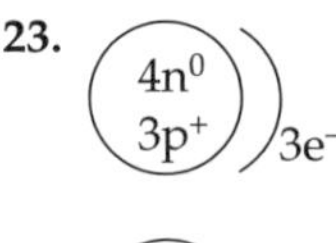

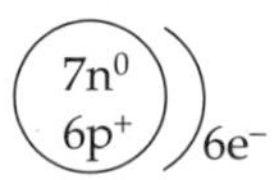

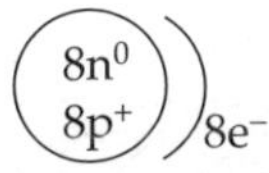

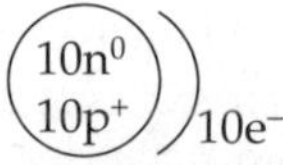

25. carbon-12
27. The masses of atoms are determined relative to that of carbon-12.
29. 26.98 amu
31. **(a)** 3.5 g; **(b)** 3.0 g
33. 6.941 amu
35. 55.85 amu
37. ^{35}Cl
39. violet light
41. orange light
43. 450 nm
45. 450 nm
47. photon
49. **(a)** continuous; **(b)** quantized
51. **(a)** continuous; **(b)** quantized
53.

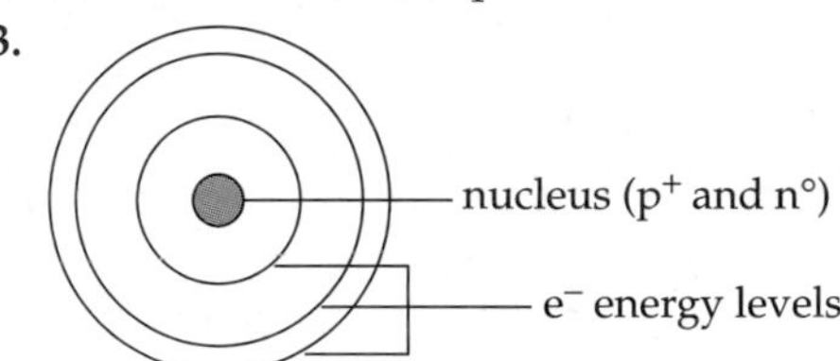

55. 5 $\longrightarrow$ 2
57. blue-green
59. 5 $\longrightarrow$ 1
61. violet
63. **(a)** 1; **(b)** 1
65. **(a)** ultraviolet; **(b)** red; **(c)** infrared
67. spectral emission lines
69. **(a)** 1*s*; **(b)** 2*s* 2*p*; **(c)** 3*s* 3*p* 3*d*; **(d)** 4*s* 4*p* 4*d* 4*f*
71. **(a)** 2 e^-; **(b)** 6 e^-; **(c)** 10 e^-; **(d)** 14 e^-
73. 8 e^-
75. 1*s* 2*s* 2*p* 3*s* 3*p* 4*s* 3*d* 4*p* 5*s* 4*d* 5*p*
77. (a) $1s^2$; **(b)** $1s^2\ 2s^2$; **(c)** $1s^2\ 2s^2\ 2p^6\ 3s^2\ 3p^6\ 4s^2\ 3d^7$; **(d)** $1s^2\ 2s^2\ 2p^6\ 3s^2\ 3p^6\ 4s^2\ 3d^{10}\ 4p^6\ 5s^2\ 4d^{10}$
79. **(a)** Li; **(b)** Si; **(c)** Ti; **(d)** Sr
81. An orbit is the path traveled by an electron of given energy around the nucleus of an atom; an orbital is a region around the nucleus in which there is a high probability of finding an electron of given energy.
83.

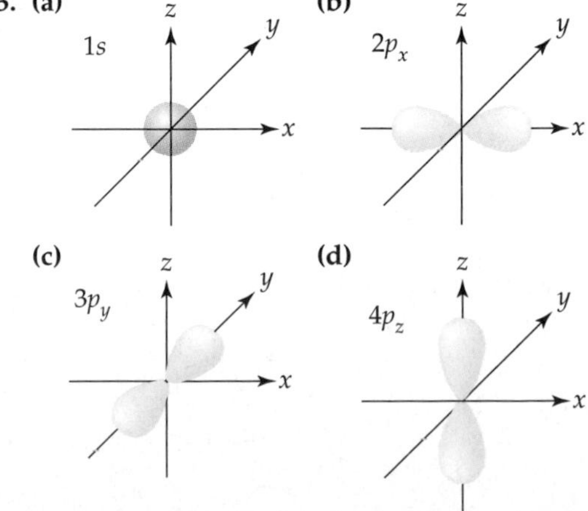

85. **(a)** 3*s*; **(b)** $3p_x$; **(c)** equal; **(d)** equal
87. **(a)** 5*s*; **(b)** 4*p*
89. **(a)** 2 e^-; **(b)** 2 e^-; **(c)** 2 e^-; **(d)** 2 e^-
91. 9.09×10^{-28} g
93. 70.93 amu
95. no stable isotope
97. **(a)** ultraviolet; **(b)** visible; **(c)** infrared
99. The 3*d* sublevel is filled and is therefore more stable.

Chapter 6

1. copper (Cu)
3. Germanium had not been discovered.
5. increasing atomic mass
7. increasing atomic number
9. groups and families
11. representative elements
13. inner transition elements
15. nonmetals
17. **(a)** IA/1; **(b)** IIA/2; **(c)** VIIA/17; **(d)** VIIIA/18
19. lanthanides
21. rare earth elements
23. **(a)** 1; **(b)** 11; **(c)** 13; **(d)** 3; **(e)** 15; **(f)** 5; **(g)** 17; **(h)** 7
25. **(a)** Ge; **(b)** Na; **(c)** I; **(d)** Pm
27. **(a)** Sb; **(b)** Mg; **(c)** Br; **(d)** Th
29. increases
31. increases
33. **(a)** Na; **(b)** P; **(c)** Ca; **(d)** Kr
35. **(a)** Al; **(b)** K; **(c)** Ba; **(d)** Fe
37. ~0.230 nm; ~1.38 g/mL; ~14.5°C
39. ~0.097 nm; ~3.27 g/mL; ~152.2°C
41. **(a)** Li_2O; **(b)** CaO; **(c)** Ga_2O_3; **(d)** SnO_2
43. **(a)** CdO; **(b)** ZnS; **(c)** HgS; **(d)** CdSe
45. **(a)**SO_3; **(b)** TeO_3; **(c)** SeS_3; **(d)** TeS_3
47. *s* sublevels
49. *d* sublevels
51. 4*f* sublevel
53. **(a)** 1*s*; **(b)** 3*s*; **(c)** 4*f*; **(d)** 4*p*; **(e)** 5*s*; **(f)** 2*p*; **(g)** 5*p*; **(h)** 6*s*
55. **(a)** $1s^2\ 2s^1$
(b) $1s^2\ 2s^2\ 2p^5$
(c) $1s^2\ 2s^2\ 2p^6\ 3s^2$
(d) $1s^2\ 2s^2\ 2p^6\ 3s^2\ 3p^3$
(e) $1s^2\ 2s^2\ 2p^6\ 3s^2\ 3p^6\ 4s^2$
(f) $1s^2\ 2s^2\ 2p^6\ 3s^2\ 3p^6\ 4s^2\ 3d^5$
(g) $1s^2\ 2s^2\ 2p^6\ 3s^2\ 3p^6\ 4s^2\ 3d^{10}\ 4p^1$
(h) $1s^2\ 2s^2\ 2p^6\ 3s^2\ 3p^6\ 4s^2\ 3d^{10}\ 4p^6\ 5s^1$
57. **(a)** 1; **(b)** 3; **(c)** 5; **(d)** 7
59. **(a)** 1; **(b)** 3; **(c)** 5; **(d)** 7; **(e)** 2; **(f)** 4; **(g)** 6; **(h)** 8
61. **(a)** H· ; **(b)** ·Ḃ· ;
(c) ·Ṅ: ; **(d)** :F̤: ;
(e) Ca· ; **(f)** ·Ṡi· ;
(g) ·Ö: ; **(h)** :Är:
63. decreases
65. Group VIIIA/18
67. **(a)** Mg; **(b)** S; **(c)** Sn; **(d)** N
69. **(a)** Cs; **(b)** Ar; **(c)** Al; **(d)** I
71. **(a)** 1+; **(b)** 2+; **(c)** 3+; **(d)** 4+
73. **(a)** 1+; **(b)** 3+; **(c)** 2−; **(d)** 1−
75. **(b)** Ca^{2+}; **(c)** S^{2-}
77. **(a)** $1s^2\ 2s^2\ 2p^6$
(b) $1s^2\ 2s^2\ 2p^6\ 3s^2\ 3p^6$
(c) $1s^2\ 2s^2\ 2p^6\ 3s^2\ 3p^6\ 3d^6$
(d) $1s^2\ 2s^2\ 2p^6\ 3s^2\ 3p^6\ 4s^2\ 3d^{10}\ 4p^6\ 5s^2\ 4d^{10}\ 5p^6$
79. **(a)** $1s^2\ 2s^2\ 2p^6$
(b) $1s^2\ 2s^2\ 2p^6\ 3s^2\ 3p^6$
(c) $1s^2\ 2s^2\ 2p^6$
(d) $1s^2\ 2s^2\ 2p^6\ 3s^2\ 3p^6\ 4s^2\ 3d^{10}\ 4p^6\ 5s^2\ 4d^{10}\ 5p^6$
81. scandium (Sc)
83. **(a)** Group IA; **(b)** Group IB; **(c)** Group IIIB; **(d)** Group IIIA
85. ~0.284 nm; ~2.21 g/mL; ~17.9°C
87. **(a)** [Kr] $5s^2$; **(b)** [Kr] $5s^2\ 4d^6$; **(c)** [Kr] $5s^2\ 4d^{10}\ 5p^3$; **(d)** [Xe] $6s^1$
89. When an alkali metal atom loses one electron, it assumes a noble gas electron configuration, and an alkaline earth metal does not.
91. Although hydrogen and Group IA/1 metals each have one valence electron in an *s* sublevel, in hydrogen the negatively charged electron occupies the 1*s* sublevel, which is closest to the positively charged nucleus.

Chapter 7

1. **(a)** binary ionic; **(b)** binary molecular; **(c)** ternary ionic; **(d)** ternary oxyacid; **(e)** binary acid
3. **(a)** monoatomic anion; **(b)** polyatomic cation; **(c)** polyatomic anion; **(d)** monoatomic cation
5. **(a)** potassium ion; **(b)** barium ion; **(c)** silver ion; **(d)** cadmium ion
7. **(a)** mercury(II) ion; **(b)** copper(II) ion; **(c)** iron(II) ion; **(d)** cobalt(III) ion
9. **(a)** cuprous ion; **(b)** ferric ion; **(c)** stannous ion; **(d)** plumbic ion
11. **(a)** fluoride ion; **(b)** iodide ion; **(c)** oxide ion; **(d)** phosphide ion
13. **(a)** hypochlorite ion; **(b)** sulfite ion; **(c)** acetate ion; **(d)** carbonate ion
15. **(a)** OH^-; **(b)** NO_2^-; **(c)** $Cr_2O_7^{2-}$; **(d)** HCO_3^-
17. **(a)** LiCl; **(b)** Ag_2O; **(c)** Cr_2O_3; **(d)** SnI_4
19. **(a)** KNO_3; **(b)** $(NH_4)_2Cr_2O_7$; **(c)** $Al_2(SO_3)_3$; **(d)** $Cr(ClO)_3$
21. **(a)** $Sr(NO_2)_2$; **(b)** $Zn(MnO_4)_2$; **(c)** $CaCrO_4$; **(d)** $Cr(ClO_4)_3$

23. **(a)** magnesium oxide; **(b)** silver bromide; **(c)** cadmium chloride; **(d)** aluminum sulfide

25. **(a)** copper(II) oxide; **(b)** iron(II) oxide; **(c)** mercury(II) oxide; **(d)** tin(II) oxide

27. **(a)** cuprous oxide; **(b)** ferric oxide; **(c)** mercurous oxide; **(d)** stannic oxide

29. **(a)** RbCl; **(b)** NaBr

31. **(a)** GaN; **(b)** AlAs

33. **(a)** lithium permanganate; **(b)** strontium perchlorate; **(c)** calcium chromate; **(d)** cadmium cyanide

35. **(a)** copper(II) sulfate; **(b)** iron(II) chromate; **(c)** mercury(II) nitrite; **(d)** lead(II) acetate

37. **(a)** cuprous sulfate; **(b)** ferric chromate; **(c)** mercurous nitrite; **(d)** plumbic acetate

39. **(a)** Fr_2SO_4; **(b)** Na_2SO_3

41. **(a)** $Ra(ClO_3)_2$; **(b)** $Ba(BrO_3)_2$

43. **(a)** sulfur trioxide; **(b)** diphosphorus trioxide; **(c)** dinitrogen oxide; **(d)** tricarbon dioxide

45. **(a)** N_2O_5; **(b)** CCl_4; **(c)** IBr; **(d)** H_2S

47. **(a)** hydrobromic acid; **(b)** hydroiodic acid

49. **(a)** chlorous acid; **(b)** phosphoric acid

51. **(a)** $HC_2H_3O_2(aq)$; **(b)** $H_3PO_3(aq)$

53. **(a)** $HClO_2$ (aq); **(b)** HBrO(aq)

55. **(a)** 0; **(b)** 2+; **(c)** 3+; **(d)** 0

57. SiO_3^{2-}

59.

Ions	F^-	O^{2-}	N^{3-}
Ag^+	AgF silver fluoride	Ag_2O silver oxide	Ag_3N silver nitride
Hg_2^{2+}	Hg_2F_2 mercury(I) fluoride	Hg_2O mercury(I) oxide	$(Hg_2)_3N_2$ mercury(I) nitride
Al^{3+}	AlF_3 aluminum fluoride	Al_2O_3 aluminum oxide	AlN aluminum nitride

61.

Ions	MnO_4^-	SO_3^{2-}	PO_4^{3-}
Cu^+	$CuMnO_4$ copper(I) permanganate	Cu_2SO_3 copper(I) sulfite	Cu_3PO_4 copper(I) phosphate
Cd^{2+}	$Cd(MnO_4)_2$ cadmium permanganate	$CdSO_3$ cadmium sulfite	$Cd_3(PO_4)_2$ cadmium phosphate
Cr^{3+}	$Cr(MnO_4)_3$ chromium(III) permanganate	$Cr_2(SO_3)_3$ chromium(III) sulfite	$CrPO_4$ chromium(III) phosphate

63. **(a)** *-ide*; **(b)** *-ic* acid

65. **(a)** *-ite*; **(b)** *-ous* acid

67. **(a)** *-ate*; **(b)** *-ic* acid

69. **(a)** H_2O; **(b)** NaClO; **(c)** NaOH; **(d)** $NaHCO_3$

71. **(a)** boron trifluoride; **(b)** silicon tetrachloride; **(c)** diarsenic pentaoxide; **(d)** diantimony trioxide

73. $Ca(C_2H_3O_2)_2$

75. $LrCl_3$

Chapter 8

1. **(a)**, **(b)**, and **(c)**

3. **(a)**, **(b)**, **(c)**, and **(d)**

5. H_2, N_2, O_2, F_2, C_{12}, Br_2, and I_2

7. $2\,Fe(s) + 3\,Cl_2(g) \longrightarrow 2\,FeCl_3(s)$

9. $ZnCO_3(s) \longrightarrow ZnO(s) + CO_2(g)$

11. $Mg(s) + Co(NO_3)_2(aq) \longrightarrow Mg(NO_3)_2(aq) + Co(s)$

13. $LiBr(aq) + AgNO_3(aq) \longrightarrow AgBr(s) + LiNO_3(aq)$

15. $HC_2H_3O_2(aq) + KOH(aq) \longrightarrow KC_2H_3O_2(aq) + HOH(l)$

17. **(e)**

19. **(a)** $4\,Co(s) + 3\,O_2(g) \longrightarrow 2\,Co_2O_3(s)$
(b) $2\,LiClO_3(s) \longrightarrow 2\,LiCl(s) + 3\,O_2(g)$
(c) $Cu(s) + 2\,AgC_2H_3O_2(aq) \longrightarrow Cu(C_2H_3O_2)_2(aq) + 2\,Ag(s)$
(d) $Pb(NO_3)_2(aq) + 2\,LiCl(aq) \longrightarrow PbCl_2(s) + 2\,LiNO_3(aq)$
(e) $3\,H_2SO_4(aq) + 2\,Al(OH)_3(aq) \longrightarrow Al_2(SO_4)_3(aq) + 6\,HOH(l)$

21. **(a)** $H_2CO_3(aq) + 2\,NH_4OH(aq) \longrightarrow (NH_4)_2CO_3(aq) + 2\,HOH(l)$
(b) $Hg_2(NO_3)_2(aq) + 2\,NaBr(aq) \longrightarrow Hg_2Br_2(s) + 2\,NaNO_3(aq)$
(c) $Mg(s) + 2\,HC_2H_3O_2(aq) \longrightarrow Mg(C_2H_3O_2)_2(aq) + H_2(g)$
(d) $2\,LiNO_3(s) \longrightarrow 2\,LiNO_2(s) + O_2(g)$
(e) $2\,Pb(s) + O_2(g) \longrightarrow 2\,PbO(s)$

23. **(a)** combination; **(b)** decomposition; **(c)** single replacement; **(d)** double replacement; **(e)** neutralization

25. **(a)** neutralization; **(b)** double replacement; **(c)** single replacement; **(d)** decomposition; **(e)** combination

27. **(a)** $4\,Fe(s) + 3\,O_2(g) \longrightarrow 2\,Fe_2O_3(s)$
(b) $2\,Sn(s) + O_2(g) \longrightarrow 2\,SnO(s)$

29. **(a)** $2\,C(s) + O_2(g) \longrightarrow 2\,CO(g)$
(b) $4\,P(s) + 5\,O_2(g) \longrightarrow 2\,P_2O_5(s)$

31. **(a)** $2\,Cu(s) + Cl_2(g) \longrightarrow 2\,CuCl(s)$
(b) $Co(s) + S(s) \longrightarrow CoS(s)$

33. **(a)** $4\,Cr(s) + 3\,O_2(g) \longrightarrow 2\,Cr_2O_3(s)$
(b) $2\,Cr(s) + N_2(g) \longrightarrow 2\,CrN(s)$

35. **(a)** $4\,Li + O_2 \longrightarrow 2\,Li_2O$
(b) $2\,Ca + O_2 \longrightarrow 2\,CaO$

37. **(a)** $2\,Na + I_2 \longrightarrow 2\,NaI$
(b) $3\,Ba + N_2 \longrightarrow Ba_3N_2$

39. **(a)** $2\,AgHCO_3(s) \longrightarrow Ag_2CO_3(s) + H_2O(g) + CO_2(g)$
(b) $Ba(HCO_3)_2(s) \longrightarrow BaCO_3(s) + H_2O(g) + CO_2(g)$

41. **(a)** $K_2CO_3(s) \longrightarrow K_2O(s) + CO_2(g)$
(b) $MnCO_3(s) \longrightarrow MnO(s) + CO_2(g)$

43. **(a)** $Ca(NO_3)_2(s) \longrightarrow Ca(NO_2)_2(s) + O_2(g)$
(b) $2\ Ag_2SO_4(s) \longrightarrow 2\ Ag_2SO_3(s) + O_2(g)$

45. **(a)** $2\ KHCO_3(s) \longrightarrow K_2CO_3(s) + H_2O(g) + CO_2(g)$
(b) $Zn(HCO_3)_2(s) \longrightarrow ZnCO_3(s) + H_2O(g) + CO_2(g)$

47. **(a)** $2\ NaClO_3(s) \longrightarrow 2\ NaCl(s) + 3\ O_2(g)$
(b) $Ca(NO_3)_2(s) \longrightarrow Ca(NO_2)_2(s) + O_2(g)$

49. **(b)** Zn; **(d)** Mg

51. **(a)** Ni; **(b)** Zn; **(d)** Al

53. **(a)** Li; **(c)** Ca

55. **(a)** $Cu(s) + Al(NO_3)_3(aq) \longrightarrow NR$
(b) $2\ Al(s) + 3\ Cu(NO_3)_2(aq) \longrightarrow 3\ Cu(s) + 2\ Al(NO_3)_3(aq)$

57. **(a)** $Ni(s) + Pb(C_2H_3O_2)_2(aq) \longrightarrow Pb(s) + Ni(C_2H_3O_2)_2(aq)$
(b) $Pb(s) + Ni(C_2H_3O_2)_2(aq) \longrightarrow NR$

59. **(a)** $Mg(s) + 2\ HCl(aq) \longrightarrow MgCl_2(s) + H_2(g)$
(b) $Mn(s) + 2\ HNO_3(aq) \longrightarrow Mn(NO_3)_2(aq) + H_2(g)$

61. **(a)** $2\ Li(s) + 2\ H_2O(l) \longrightarrow 2\ LiOH(aq) + H_2(g)$
(b) $Ba(s) + 2\ H_2O(l) \longrightarrow Ba(OH)_2(aq) + H_2(g)$

63. **(a)** $Zn(s) + Pb(NO_3)_2(aq) \longrightarrow Pb(s) + Zn(NO_3)_2(aq)$
(b) $Cd(s) + Fe(NO_3)_2(aq) \longrightarrow NR$

65. **(a)** $Zn(s) + 2\ HNO_3(aq) \longrightarrow Zn(NO_3)_2(aq) + H_2(g)$
(b) $Cd(s) + 2\ HNO_3(aq) \longrightarrow Cd(NO_3)_2(aq) + H_2(g)$

67. **(a)** $2\ K(s) + 2\ H_2O(l) \longrightarrow 2\ KOH(aq) + H_2(g)$
(b) $Ba(s) + 2\ H_2O(l) \longrightarrow Ba(OH)_2(aq) + H_2(g)$

69. **(b)** soluble; **(d)** soluble

71. **(a)** insoluble; **(c)** insoluble; **(d)** insoluble

73. **(a)** $ZnCl_2(aq) + 2\ NH_4OH(aq) \longrightarrow Zn(OH)_2(s) + 2\ NH_4Cl(aq)$
(b) $NiSO_4(aq) + Hg_2(NO_3)_2(aq) \longrightarrow Hg_2SO_4(s) + Ni(NO_3)_2(aq)$

75. **(a)** $MgSO_4(aq) + BaCl_2(aq) \longrightarrow BaSO_4(s) + MgCl_2(aq)$
(b) $2\ AlBr_3(aq) + 3\ Na_2CO_3(aq) \longrightarrow Al_2(CO_3)_3(s) + 6\ NaBr(aq)$

77. **(a)** $NaOH(aq) + HNO_3(aq) \longrightarrow NaNO_3(aq) + HOH(l)$
(b) $3\ Ba(OH)_2(aq) + 2\ H_3PO_4(aq) \longrightarrow Ba_3(PO_4)_2(s) + 6\ HOH(l)$

79. **(a)** $2\ HF(aq) + Ca(OH)_2(aq) \longrightarrow CaF_2(aq) + 2\ HOH(l)$
(b) $H_2SO_4(aq) + 2\ LiOH(aq) \longrightarrow Li_2SO_4(aq) + 2\ HOH(l)$

81. **(a)** $3\ Fe(s) + 4\ H_2O(g) \longrightarrow Fe_3O_4(s) + 4\ H_2(g)$
(b) $4\ FeS(s) + 7\ O_2(g) \longrightarrow 2\ Fe_2O_3(s) + 4\ SO_2(g)$

83. **(a)** $F_2(g) + 2\ NaBr(aq) \longrightarrow Br_2(l) + 2\ NaF(aq)$
(b) $Sb_2S_3(s) + 6\ HCl(aq) \longrightarrow 2\ SbCl_3(aq) + 3\ H_2S(g)$

85. **(a)** $CH_4(g) + 2\ O_2(g) \longrightarrow CO_2(g) + 2\ H_2O(g)$
(b) $C_3H_8(g) + 5\ O_2(g) \longrightarrow 3\ CO_2(g) + 4\ H_2O(g)$

87. $4\ HCl(g) + O_2(g) \longrightarrow 2\ Cl_2(g) + 2\ H_2O(g)$

89. **(1)** $S(s) + O_2(g) \longrightarrow SO_2(g)$
(2) $2\ SO_2(g) + O_2(g) \xrightarrow{Pt} 2\ SO_3(g)$
(3) $SO_3(g) + H_2O(l) \longrightarrow H_2SO_4(aq)$

Chapter 9

1. **(a)** 1.01 amu; **(b)** 6.94 amu; **(c)** 12.01 amu; **(d)** 30.97 amu

3. **(a)** 1.01 g; **(b)** 6.94 g; **(c)** 12.01 g; **(d)** 30.97 g

5. **(a)** 6.02×10^{23} atoms Mn; **(b)** 6.02×10^{23} formula units $Mn(NO_3)_2$

7. **(a)** 1 mol Cu atoms; **(b)** 1 mol $CuSO_4$ formula units

9. **(a)** 2.02×10^{23} atoms Ti; **(b)** 6.74×10^{22} molecules CO_2; **(c)** 1.16×10^{24} formula units $ZnCl_2$

11. **(a)** 0.0689 mol Fe; **(b)** 0.00550 mol Br_2; **(c)** 6.96×10^{-4} mol $Cd(NO_3)_2$

13. **(a)** 200.59 g/mol; **(b)** 28.09 g/mol; **(c)** 159.80 g/mol; **(d)** 123.88 g/mol

15. **(a)** 175.33 g/mol; **(b)** 110.27 g/mol; **(c)** 233.00 g/mol; **(d)** 452.80 g/mol

17. **(a)** 98.3 g Hg; **(b)** 0.540 g N_2; **(c)** 1.75 g $BaCl_2$

19. **(a)** 2.31×10^{22} atoms K; **(b)** 8.84×10^{21} molecules O_2; **(c)** 1.75×10^{21} formula units $AgClO_3$

21. **(a)** 1.50×10^{-23} g/atom; **(b)** 3.82×10^{-23} g/atom; **(c)** 9.79×10^{-23} g/atom; **(d)** 1.24×10^{-22} g/atom

23. 0°C and 1 atm

25. **(a)** 0.901 g/L; **(b)** 3.17 g/L; **(c)** 2.05 g/L; **(d)** 5.71 g/L

27. **(a)** 30.0 g/mol; **(b)** 27.6 g/mol; **(c)** 121 g/mol; **(d)** 45.9 g/mol

29.

Gas	Molecules	Mass	Volume at STP
Fluorine, F_2	6.02×10^{23}	38.00 g	22.4 L
Hydrogen fluoride, HF	6.02×10^{23}	20.01 g	22.4 L
Silicon tetrafluoride, SiF_4	6.02×10^{23}	104.09 g	22.4 L
Oxygen difluoride, OF_2	6.02×10^{23}	54.00 g	22.4 L

31. **(a)** 1.40 L He; **(b)** 4.04 L N_2

33. **(a)** 1.60 g H_2S; **(b)** 18.1 g N_2O_3

35. **(a)** 2.69×10^{21} molecules H_2; **(b)** 1.89×10^{21} molecules NH_3

37.

Gas	Molecules	Atoms	Mass	Volume at STP
N_2	1.35×10^{23}	2.70×10^{23}	6.28 g	5.02 L
NO_2	1.35×10^{23}	4.05×10^{23}	10.3 g	5.02 L
NO	1.35×10^{23}	2.70×10^{23}	6.75 g	5.02 L
N_2O_4	1.35×10^{23}	8.10×10^{23}	20.6 g	5.02 L

39. 60.86% C, 4.39% H, 34.75% O

41. 67.296% C, 6.991% H, 4.618% N, 21.09% O
43. 30.20% C, 5.08% H, 20.16% S, 44.57% Cl
45. 13.59% Na, 35.51% C, 4.78% H, 8.284% N, 37.84% O
47. SnO_2
49. HgO
51. Co_2S_3
53. **(a)** MnF_2; **(b)** CuCl; **(c)** $SnBr_2$
55. C_2HCl_3
57. $C_9H_8O_4$
59. $C_6H_{10}O_4$
61. CH_3O, $C_2H_6O_2$
63. CHCl, $C_6H_6Cl_6$
65. C_5H_7N, $C_{10}H_{14}N_2$
67. 6.02×10^{23} electrons
69. 5-g nickel coin
71. the Earth
73. Ga_2O_3
75. 2.99×10^{-23} cm^3 H_2O
77. 2.11×10^{22} atoms C
79. 11 atoms C
81. 6.04×10^{23} atoms Cu/mol

Chapter 10

1. **(a)** 1 mole C; **(b)** 2 liters D
3. **(a)** 25.0 g C; **(b)** 25.0 g B
5. **(a)** 202.22 g $\longrightarrow$ 202.22 g; **(b)** 533.36 g $\longrightarrow$ 533.36 g
7. 0.250 mol O_2 react; 0.500 mol H_2O produced
9. 0.500 mol Cl_2 react; 0.333 mol $FeCl_3$ produced
11. 0.350 mol C_3H_8 react; 1.05 mol CO_2 produced
13. mass–mass problem
15. mass–volume problem
17. mass–mass problem
19. 2.94 g ZnO
21. 5.21 g $BiCl_3$
23. 2.09 g Ag
25. 6.38 g Hg
27. 1.68 g $Ca_3(PO_4)_2$
29. 366 mL CO_2
31. 262 mL CO_2
33. 0.244 g Mg
35. 0.167 g H_2O_2
37. 1.00 L O_2
39. 125 mL I_2
41. 22.5 mL N_2
43. 1750 mL Cl_2
45. 50.0 L SO_3
47. 500.0 cm^3 N_2
49. The limiting reactant is N_2, which produces 2.00 mol of NO.
51. The limiting reactant is NO, which produces 1.00 mol of NO_2.
53. The limiting reactant is H_2, which produces 5.00 mol of H_2O.
55. The limiting reactant is C_2H_6, which produces 3.00 mol of H_2O.
57. **(1)** The limiting reactant is Co, which produces 1.50 mol of CoS. After the reaction: Co = 0.00 mol; S = 0.50 mol; CoS = 1.50 mol. **(2)** The limiting reactant is S, which produces 2.00 mol of CoS. After the reaction: Co = 1.00 mol; S = 0.00 mol; CoS = 2.00 mol.
59. 23.0 g Fe
61. 77.6 g Fe
63. 0.742 g $MgSO_4$
65. 0.693 g H_2O
67. 90.0 mL NO_2
69. 30.0 mL N_2O_3
71. 2.50 L SO_3
73. 20.0 L Cl_2
75. 89.7%
77. 106%
79. g/mol
81. 0.929 g Cr_2O_3
83. 655 g Al
85. 0.593 L H_2S
87. 124 L H_2
89. 16.3 g H_2O
91. The limiting reactant is *gasoline*; the excess reactant is *oxygen*.

Chapter 11

1. Refer to Section 11.1 in the text.
3. **(a)** 1 atm; **(b)** 760 mm Hg; **(c)** 760 torr; **(d)** 76 cm Hg
5. **(a)** 3990 mm Hg; **(b)** 3990 torr; **(c)** 399 cm Hg; **(d)** 157 in. Hg
7. **(a)** 0.963 atm; **(b)** 732 mm Hg; **(c)** 73.2 cm Hg; **(d)** 732 torr
9. volume, temperature, and number of molecules
11. **(a)** Molecules collide less frequently. **(b)** Molecules collide with higher frequency and more energy. **(c)** More molecules have more collisions.
13. **(a)** pressure decreases; **(b)** pressure increases; **(c)** pressure decreases
15.

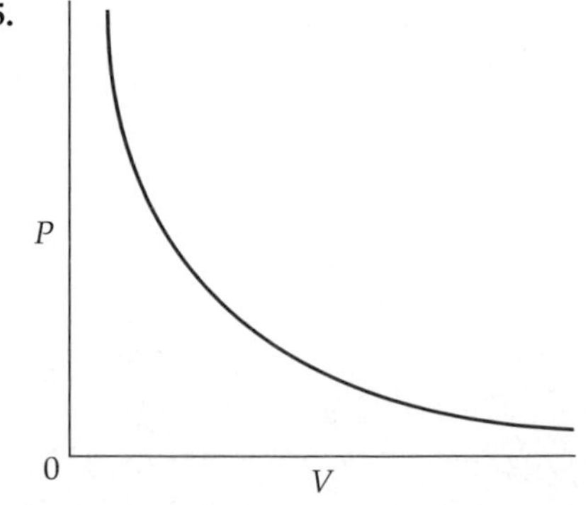

17. 0.286 atm

19. 17.0 psi

21.

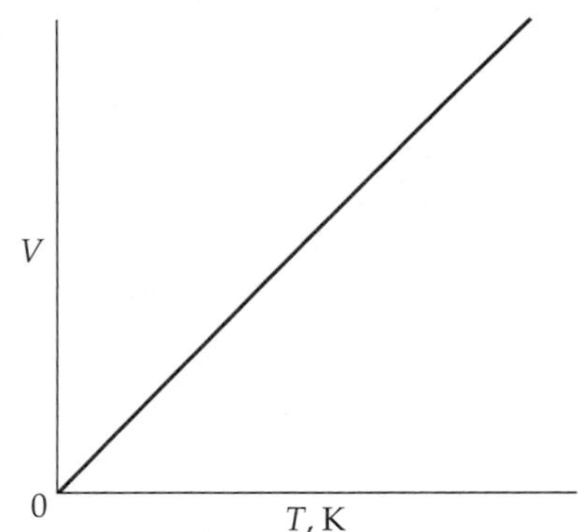

23. 363 mL O_2

25. 109 cm^3 F_2

27.

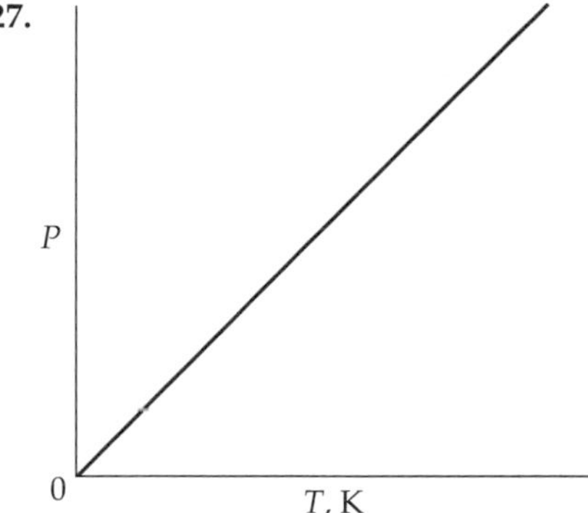

29. 1230 torr

31. 104 cm Hg

33. 94.3 mL H_2

35. 5.31 L air

37. 7820 torr

39. 1290 mm Hg

41. 880 K (607°C)

43. As the temperature of a liquid increases, so does its vapor pressure.

45. **(a)** 23.8 mm Hg; **(b)** 92.5 mm Hg

47. 752 mm Hg

49. 135 mm Hg

51. A gas can be collected in a flask containing water and will displace an equal volume of liquid.

53. 748.5 torr

55. Refer to Section 11.1 in the text.

57. high temperature, and low pressure

59. **(a)** all gases are equal; **(b)** all gases are equal; **(c)** He atoms; **(d)** Ar atoms

61. 0 mm Hg

63. 245 atm

65. 2.78 mol N_2O

67. 48.0 g/mol

69. 12.5 g/mol

71. 3.92 g Cl_2

73. 37,000 lb

75. 15 psi

77. 38.5 mL O_2

79. 12.3 atm

81. 5.11 L NO

83. 62.4 torr · L/mol · K

85. Helium atoms are lighter than nitrogen molecules and move faster over the vocal cords; this produces a higher-pitched voice.

Chapter 12

1. A metal loses valence electrons, and a nonmetal gains valence electrons. The attraction between the resulting cation and anion is an ionic bond.

3. Mg (2 valence e^-); Mg^{2+} (0 valence e^-); S (6 valence e^-); S^{2-} (8 valence e^-)

5. **(a)** covalent; **(b)** ionic; **(c)** covalent; **(d)** ionic

7. **(a)** molecule; **(b)** formula unit; **(c)** molecule; **(d)** formula unit

9. **(a)** atom; **(b)** molecule; **(c)** molecule; **(d)** formula unit

11. **(a)** 1+; **(b)** 2+; **(c)** 4+; **(d)** 3+

13. **(a)** 1−; **(b)** 1−; **(c)** 2−; **(d)** 3−

15. **(a)** $1s^2$
(b) $1s^2\,2s^2\,2p^6$
(c) $1s^2\,2s^2\,2p^6\,3s^2\,3p^6$
(d) $1s^2\,2s^2\,2p^6$

17. **(a)** $1s^2\,2s^2\,2p^6\,3s^2\,3p^6$
(b) $1s^2\,2s^2\,2p^6\,3s^2\,3p^6\,4s^2\,3d^{10}\,4p^6\,5s^2\,4d^{10}\,5p^6$
(c) $1s^2\,2s^2\,2p^6\,3s^2\,3p^6$
(d) $1s^2\,2s^2\,2p^6\,3s^2\,3p^6$

19. **(a)** He; **(b)** Ar; **(c)** Ar; **(d)** Rn

21. **(a)** He; **(b)** Ne; **(c)** Ar; **(d)** Ne

23. **(a)** Ar; **(b)** Xe; **(c)** Ar; **(d)** Ar

25. **(a)** Li atom; **(b)** Mg atom; **(c)** F ion; **(d)** O ion

27. **(a)** An ionic bond is formed by the attraction of metal and nonmetal ions.

29. **(a)** sum of the atomic radii; **(b)** sum of the atomic radii

31. None are true.

33. **(a)** H:H H—H

(b) $:\ddot{\underset{..}{F}}:\ddot{\underset{..}{F}}:$ F—F

(c) $H:\ddot{\underset{..}{Br}}:$ H—Br

(d) $\begin{matrix} & H & \\ H: & \underset{..}{N} & :H \end{matrix}$ $\begin{matrix} & H & \\ & | & \\ H— & N & —H \end{matrix}$

35. 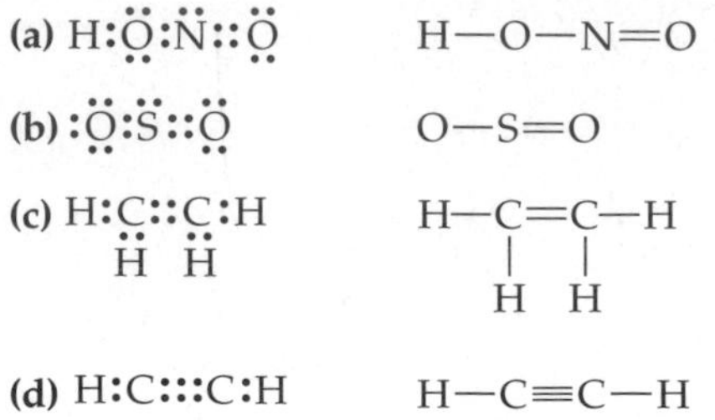 **(a)** H:O:N::O H—O—N=O

(b) :O:S::O O—S=O

(c) H:C::C:H (with H under each C) H—C=C—H (with H under each C)

(d) H:C:::C:H H—C≡C—H

37. 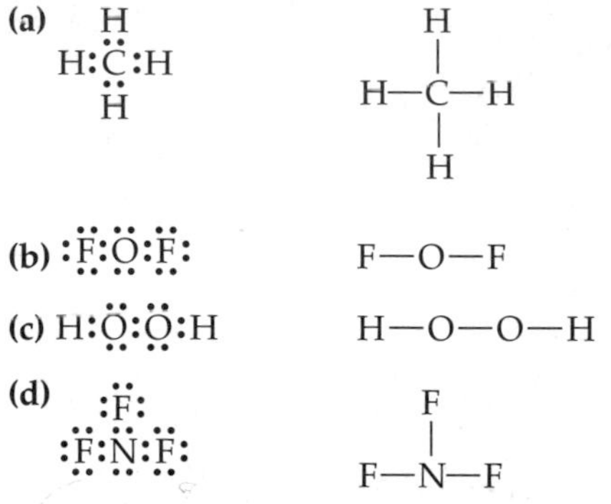 **(a)** H:C:H (with H above and below C) H—C—H (with H above and below C)

(b) :F:O:F: F—O—F

(c) H:O:O:H H—O—O—H

(d) :F:N:F: (with :F: above N) F—N—F (with F above N)

39. 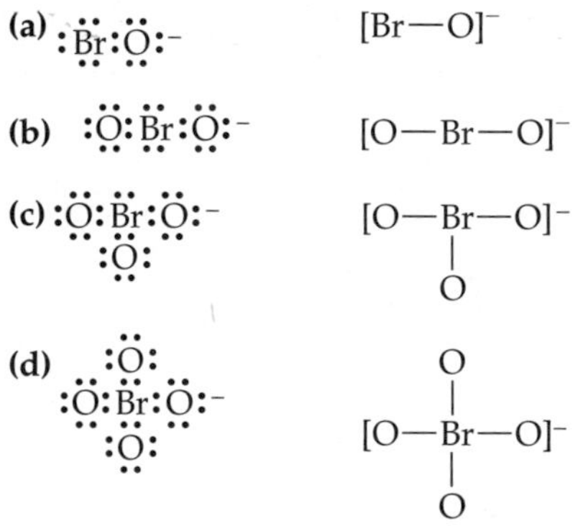 **(a)** :Br:O:$^-$ [Br—O]$^-$

(b) :O:Br:O:$^-$ [O—Br—O]$^-$

(c) :O:Br:O:$^-$ (with :O: below Br) [O—Br—O]$^-$ (with O below Br)

(d) :O:Br:O:$^-$ (with :O: above and below Br) [O—Br—O]$^-$ (with O above and below Br)

41. 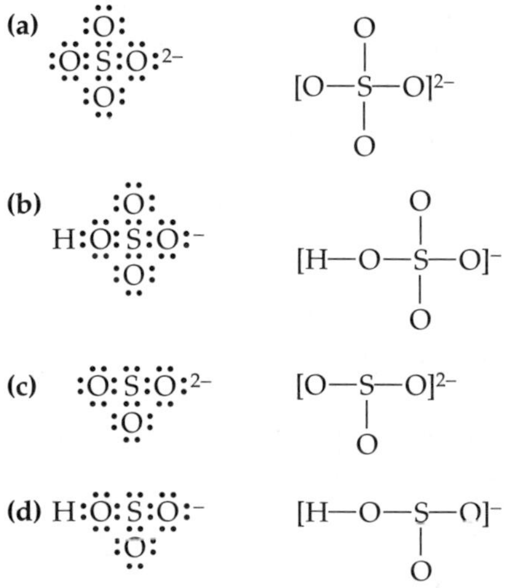 **(a)** :O:S:O:$^{2-}$ (with :O: above and below S) [O—S—O]$^{2-}$ (with O above and below S)

(b) H:O:S:O:$^-$ (with :O: above and below S) [H—O—S—O]$^-$ (with O above and below S)

(c) :O:S:O:$^{2-}$ (with :O: below S) [O—S—O]$^{2-}$ (with O below S)

(d) H:O:S:O:$^-$ (with :O: below S) [H—O—S—O]$^-$ (with O below S)

43. 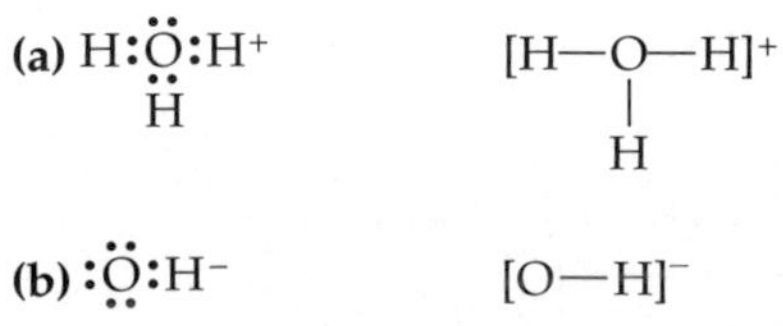 **(a)** H:O:H$^+$ (with H below O) [H—O—H]$^+$ (with H below O)

(b) :O:H$^-$ [O—H]$^-$

(c) H:S:$^-$ [H—S]$^-$

(d) :C:::N:$^-$ [C≡N]$^-$

45. decreases

47. nonmetals

49. **(a)** Cl; **(b)** O; **(c)** Se; **(d)** F

51. **(a)** 0.2; **(b)** 1.2; **(c)** 0.5; **(d)** 0.3

53. **(a)** δ^+ H—S δ^-; **(b)** δ^- O—S δ^+; **(c)** δ^+ N—F δ^-; **(d)** δ^+ S—Cl δ^-

55. **(a)**, **(b)**, and **(d)**

57. H_2, N_2, and O_2

59.  H:Br:O: H:Br—O:

61. 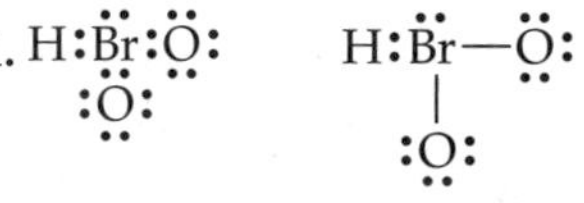 H:Br:O: (with :O: below Br) H:Br—O: (with :O: below Br)

63. 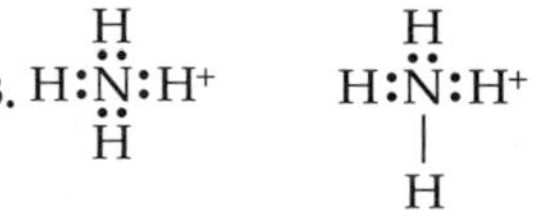 H:N:H$^+$ (with H above and below N) H:N:H$^+$ (with H above N and H bonded below)

65. 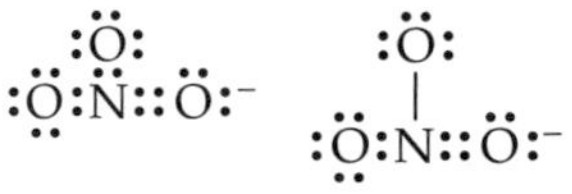 :O:N::O:$^-$ (with :O: above N) :O:N::O:$^-$ (with :O: bonded above N)

67. tetrahedral, tetrahedral, 109.5°

69. tetrahedral, trigonal pyramidal, 107°

71. tetrahedral, bent, 104.5°

73. Each C—F bond is polar, but the symmetric tetrahedral arrangement of the four bonds produces a nonpolar molecule.

75. **(a)** atom; **(b)** molecule; **(c)** formula unit

77. **(a)** Sr_3As_2; **(b)** RaO; **(c)** $Al_2(CO_3)_3$; **(d)** $Cd(OH)_2$

79. The sodium ion has one less energy level (3*s*).

81. 1.0 (polar)

83. 0 (nonpolar)

85. δ^+ Ge—Cl δ^-

87. 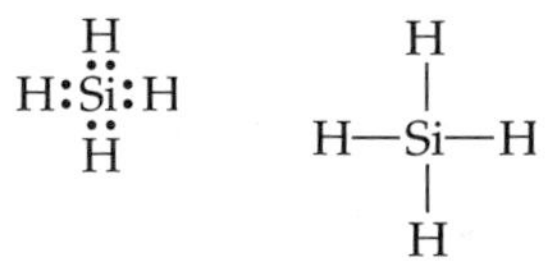 H:Si:H (with H above and below Si) H—Si—H (with H above and below Si)

89. 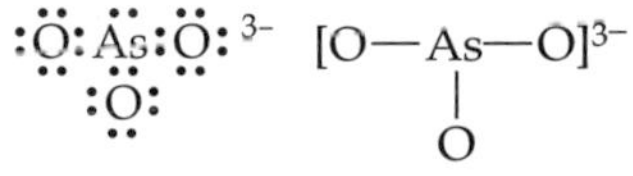 :O:As:O:$^{3-}$ (with :O: below As) [O—As—O]$^{3-}$ (with O below As)

91. All valence electrons move about the entire molecule.

93. O=S—O and O—S=O

95. 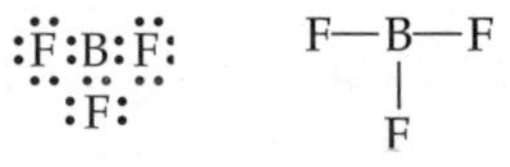 :F:B:F: (with :F: below B) F—B—F (with F below B)

97. :Ö:Ẍe:Ö: O—Xe—O

99. The molecular shape of H_3O^+ is trigonal pyramidal; H_2O is bent.

Chapter 13

1. Refer to Section 13.1 in the text.
3. **(a)** solid; **(b)** gas; **(c)** solid; **(d)** gas; **(e)** liquid; **(f)** gas
5. **(a)** dispersion forces; **(b)** dispersion forces, hydrogen bonds; **(c)** dispersion forces, dipole forces; **(d)** dispersion forces, dipole forces
7. **(a)** C_2H_5Cl; **(b)** CH_3OCH_3
9. **(a)** CH_3COOH; **(b)** C_2H_5OH
11. The vapor pressure of water is the pressure exerted by water vapor molecules when the rate of evaporation is equal to the rate of condensation.
13. The intermolecular attraction in water is strong; thus, the viscosity of water is higher than that of other liquids with molecules the same size.
15. **(b)**, **(c)**, and **(d)**
17. As the temperature of a liquid increases, the vapor pressure increases.
19. **(a)** ~70 mm Hg; **(b)** ~300 mm Hg
21. 56°C
23. Refer to Section 13.4 in the text.
25. **(a)** solid; **(b)** liquid; **(c)** solid; **(d)** solid; **(e)** liquid; **(f)** liquid
27. NaCl, Cl_2, and Na
29. **(a)** ions; **(b)** molecules; **(c)** atoms
31. **(a)** metallic; **(b)** ionic; **(c)** molecular; **(d)** molecular
33.

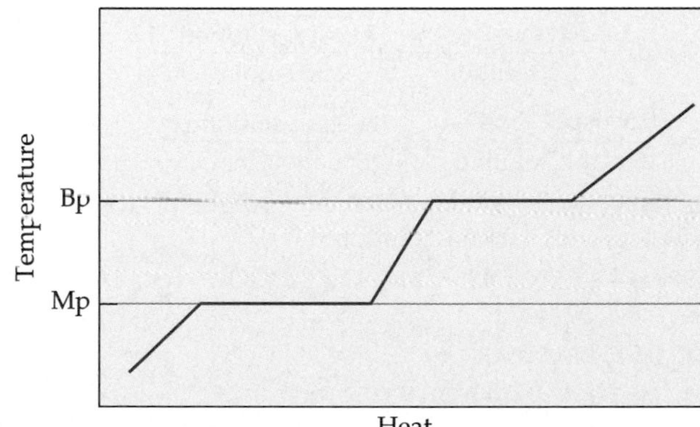

35. 1.00×10^4 cal (10.0 kcal)
37. 15,400 cal (15.4 kcal)
39. 82,800 cal (82.8 kcal)
41. 28,100 cal (28.1 kcal)
43. 75,200 cal (75.2 kcal)
45. two bonding pairs and two nonbonding pairs
47. 104.5°
49. H(δ+)—O(δ−)—H(δ+)
51. H—F··········H—F (↑ H bond)
53. floats
55. **(a)** H_2O; **(b)** H_2Se
57. **(a)** H_2O; **(b)** H_2Se
59. **(a)** increases; **(b)** increases; **(c)** increases; **(d)** increases
61. $2\,H_2O(l) \longrightarrow 2\,H_2(g) + O_2(g)$
63. **(a)** $2\,Li(s) + 2\,H_2O(l) \longrightarrow 2\,LiOH(aq) + H_2(g)$
(b) $Na_2O(s) + H_2O(l) \longrightarrow 2\,NaOH(aq)$
(c) $CO_2(g) + H_2O(l) \longrightarrow H_2CO_3(aq)$
65. **(a)** $Ba(s) + 2\,H_2O(l) \longrightarrow Ba(OH)_2(aq) + H_2(g)$
(b) $N_2O_3(g) + H_2O(l) \longrightarrow 2\,HNO_2(aq)$
(c) $CaO(s) + H_2O(l) \longrightarrow Ca(OH)_2(aq)$
67. **(a)** $2\,C_3H_6(g) + 9\,O_2(g) \longrightarrow 6\,CO_2(g) + 6\,H_2O(g)$
(b) $Na_2Cr_2O_7 \cdot 2\,H_2O(s) \longrightarrow Na_2Cr_2O_7(s) + 2\,H_2O(g)$
(c) $2\,HF(aq) + Ca(OH)_2(aq) \longrightarrow CaF_2(aq) + 2\,HOH(l)$
69. **(a)** $2\,C_4H_{10}(g) + 13\,O_2(g) \longrightarrow 8\,CO_2(g) + 10\,H_2O(g)$
(b) $Co(C_2H_3O_2)_2 \cdot 4\,H_2O(s) \longrightarrow Co(C_2H_3O_2)_2(s) + 4\,H_2O(g)$
(c) $2\,HNO_3(aq) + Ba(OH)_2(aq) \longrightarrow Ba(NO_3)_2(aq) + 2\,HOH(l)$
71. **(a)** magnesium sulfate heptahydrate; **(b)** cobalt(III) cyanide trihydrate; **(c)** manganese(II) sulfate monohydrate; **(d)** sodium dichromate dihydrate
73. **(a)** $NaC_2H_3O_2 \cdot 3\,H_2O$; **(b)** $CaSO_4 \cdot 2\,H_2O$; **(c)** $K_2CrO_4 \cdot 4\,H_2O$; **(d)** $ZnSO_4 \cdot 7\,H_2O$
75. **(a)** 40.55% H_2O; **(b)** 10.91% H_2O; **(c)** 28.28% H_2O; **(d)** 30.80% H_2O
77. **(a)** $NiCl_2 \cdot 2\,H_2O$; **(b)** $Sr(NO_3)_2 \cdot 6\,H_2O$; **(c)** $CrI_3 \cdot 9\,H_2O$; **(d)** $Ca(NO_3)_2 \cdot 4\,H_2O$
79. ~75%
81. ~95°C
83. spherical "raindrops"
85. 79,700 cal (79.7 kcal)

Chapter 14

1. **(a)** increases; **(b)** increases
3. 14.5 g CO_2/1 L
5. 0.99 g Cl_2/100 g H_2O
7. **(a)** miscible; **(b)** immiscible
9. **(a)** polar; **(b)** nonpolar; **(c)** polar; **(d)** nonpolar
11. **(a)** immiscible; **(b)** miscible; **(c)** miscible; **(d)** immiscible
13. Add several drops of the unknown to water. If the unknown liquid is polar, it will be miscible; if it is nonpolar, it will be immiscible.

15. **(a)** soluble; **(b)** insoluble; **(c)** soluble

17. **(a)** insoluble; **(b)** soluble; **(c)** soluble; **(d)** insoluble; **(e)** soluble; **(f)** soluble

19. **(a)** water-soluble; **(b)** water-soluble; **(c)** water-soluble; **(d)** water-soluble; **(e)** fat-soluble; **(f)** fat-soluble

21.

```
         H2O
          |
H2O---C6H12O6---H2O
          |
         H2O
```

23. **(a)**

```
H           H                       H
 \         /                         \
  O---Li+---O          H---Br−---H    O
 /         \          /          \   /
H           H        O            ---
                      \
                       H
```

(b)

```
H           H                       H
 \         /                         \
  O---Ca2+---O         H---Cl−---H    O
 /         \          /          \   /
H           H        O            ---
                      \
                       H
```

25. **(1)** heating the solution; **(2)** stirring the solution; **(3)** grinding the solute

27. **(a)** ~35 g NaCl; **(b)** ~35 g KCl

29. **(a)** ~36 g NaCl/100 g H_2O; **(b)** ~41 g KCl/100 g H_2O

31. **(a)** ~0°C; **(b)** ~70°C

33. **(a)** ~100°C; **(b)** ~35°C

35. **(a)** supersaturated; **(b)** saturated; **(c)** unsaturated

37. **(a)** supersaturated; **(b)** saturated; **(c)** unsaturated

39. saturated

41. **(a)** ~80 g solute; **(b)** ~20 g solute

43. **(a)** 1.25%; **(b)** 2.63%; **(c)** 4.00%; **(d)** 52.0%

45. **(a)** $\dfrac{1.50\text{ g KBr}}{100.00\text{ g solution}}$ and $\dfrac{100.00\text{ g solution}}{1.50\text{ g KBr}}$

$\dfrac{98.50\text{ g H}_2\text{O}}{100.00\text{ g solution}}$ and $\dfrac{100.00\text{ g solution}}{98.50\text{ g H}_2\text{O}}$

$\dfrac{1.50\text{ g KBr}}{98.50\text{ g H}_2\text{O}}$ and $\dfrac{98.50\text{ g H}_2\text{O}}{1.50\text{ g KBr}}$

(b) $\dfrac{2.50\text{ g AlCl}_3}{100.00\text{ g solution}}$ and $\dfrac{100.00\text{ g solution}}{2.50\text{ g AlCl}_3}$

$\dfrac{97.50\text{ g H}_2\text{O}}{100.00\text{ g solution}}$ and $\dfrac{100.00\text{ g solution}}{97.50\text{ g H}_2\text{O}}$

$\dfrac{2.50\text{ g AlCl}_3}{97.50\text{ g H}_2\text{O}}$ and $\dfrac{97.50\text{ g H}_2\text{O}}{2.50\text{ g AlCl}_3}$

(c) $\dfrac{3.75\text{ g AgNO}_3}{100.00\text{ g solution}}$ and $\dfrac{100.00\text{ g solution}}{3.75\text{ g AgNO}_3}$

$\dfrac{96.25\text{ g H}_2\text{O}}{100.00\text{ g solution}}$ and $\dfrac{100.00\text{ g solution}}{96.25\text{ g H}_2\text{O}}$

$\dfrac{3.75\text{ g AgNO}_3}{96.25\text{ g H}_2\text{O}}$ and $\dfrac{96.25\text{ g H}_2\text{O}}{3.75\text{ g AgNO}_3}$

(d) $\dfrac{4.25\text{ g Li}_2\text{SO}_4}{100.00\text{ g solution}}$ and $\dfrac{100.00\text{ g solution}}{4.25\text{ g Li}_2\text{SO}_4}$

$\dfrac{95.75\text{ g H}_2\text{O}}{100.00\text{ g solution}}$ and $\dfrac{100.00\text{ g solution}}{95.75\text{ g H}_2\text{O}}$

$\dfrac{4.25\text{ g Li}_2\text{SO}_4}{95.75\text{ g H}_2\text{O}}$ and $\dfrac{95.75\text{ g H}_2\text{O}}{4.25\text{ g Li}_2\text{SO}_4}$

47. **(a)** 53.6 g solution; **(b)** 2.00×10^2 g solution

49. **(a)** 1.70 g $FeBr_2$; **(b)** 5.25 g Na_2CO_3

51. **(a)** 247.8 g H_2O; **(b)** 95.00 g H_2O

53. **(a)** 0.257 *M* NaCl; **(b)** 0.0510 *M* $K_2Cr_2O_7$; **(c)** 0.400 *M* $CaCl_2$; **(d)** 0.313 *M* Na_2SO_4

55. **(a)** $\dfrac{0.100\text{ mol LiI}}{1\text{ L solution}}$ and $\dfrac{1\text{ L solution}}{0.100\text{ mol LiI}}$

$\dfrac{0.100\text{ mol LiI}}{1000\text{ mL solution}}$ and $\dfrac{1000\text{ mL solution}}{0.100\text{ mol LiI}}$

(b) $\dfrac{0.100\text{ mol NaNO}_3}{1\text{ L solution}}$ and $\dfrac{1\text{ L solution}}{0.100\text{ mol NaNO}_3}$

$\dfrac{0.100\text{ mol NaNO}_3}{1000\text{ mL solution}}$ and $\dfrac{1000\text{ mL solution}}{0.100\text{ mol NaNO}_3}$

(c) $\dfrac{0.500\text{ mol K}_2\text{CrO}_4}{1\text{ L solution}}$ and $\dfrac{1\text{ L solution}}{0.500\text{ mol K}_2\text{CrO}_4}$

$\dfrac{0.500\text{ mol K}_2\text{CrO}_4}{1000\text{ mL solution}}$ and $\dfrac{1000\text{ mL solution}}{0.500\text{ mol K}_2\text{CrO}_4}$

(d) $\dfrac{0.500\text{ mol ZnSO}_4}{1\text{ L solution}}$ and $\dfrac{1\text{ L solution}}{0.500\text{ mol ZnSO}_4}$

$\dfrac{0.500\text{ mol ZnSO}_4}{1000\text{ mL solution}}$ and $\dfrac{1000\text{ mL solution}}{0.500\text{ mol ZnSO}_4}$

57. **(a)** 0.866 L solution; **(b)** 0.198 L solution; **(c)** 0.177 L solution; **(d)** 0.0800 L solution

59. **(a)** 4.00 g NaOH; **(b)** 6.80 g $LiHCO_3$; **(c)** 1.68 g $CuCl_2$; **(d)** 1.98 g $KMnO_4$

61. 0.0154 *M* $CaSO_4$

63. **(a)** 5.00%; **(b)** 0.291 *M* $C_6H_{12}O_6$

65. 0.34 *M* $HC_2H_3O_2$

67. 0.719 *M* $AgNO_3$; 1.88 g AgBr

69. 1–100 nm

71. **(a)** colloid; **(b)** colloid; **(c)** solution

73. 0.0089 g N_2/100 g blood

75. 228 g SO_2

77. Air is less soluble in hot water and leaves the solution to form bubbles on the inside of the pan.

79. The polar —OH on an alcohol can hydrogen-bond with water, causing it to be soluble. If the nonpolar C_xH_y— portion of the molecule is large, the molecule is less polar and is immiscible with water.

81. **(a)** ethyl alcohol (solute), water (solvent); **(b)** ethyl alcohol (solvent), water (solute)

83. 0.733 *M* NaClO

Chapter 15

1. sour taste, pH < 7, turn blue litmus red

3. **(a)** basic; **(b)** acidic; **(c)** neutral; **(d)** acidic; **(e)** acidic; **(f)** acidic

5. **(a)** strong; **(b)** weak; **(c)** strong; **(d)** weak

7. **(a)** acid; **(b)** base; **(c)** salt; **(d)** base

9. **(a)** HI(aq), NaOH(aq); **(b)** $HC_2H_3O_2$(aq), LiOH(aq)

11. **(a)** HF(aq), NaOH(aq); **(b)** HI(aq), $Mg(OH)_2$ (aq); **(c)** HNO_3(aq), $Ca(OH)_2$(aq); **(d)** H_2CO_3(aq), LiOH(aq)

13. **(a)** $2\ HNO_3(aq) + Ca(OH)_2(aq) \longrightarrow Ca(NO_3)_2(aq) + 2\ H_2O(l)$

(b) $H_2CO_3(aq) + Ba(OH)_2(aq) \longrightarrow BaCO_3(aq) + 2\ H_2O(l)$

15. **(a)** $HC_2H_3O_2$(aq), LiOH(aq); **(b)** HBr(aq), NaCN(aq)

17. **(a)** HI(aq), H_2O(l); **(b)** $HC_2H_3O_2$(aq), HS^-(aq)

19. **(a)** $HF(aq) + NaHS(aq) \longrightarrow H_2S(aq) + NaF(aq)$

(b) $HNO_2(aq) + NaC_2H_3O_2(aq) \longrightarrow NaNO_2(aq) + HC_2H_3O_2(aq)$

21. **(a)** red; **(b)** yellow

23. **(a)** colorless; **(b)** pink

25. orange

27. 0.137 *M* HCl

29. 0.0482 *M* H_3PO_4

31. 27.8 mL H_2SO_4

33. **(a)** 19.9% HCl; **(b)** 5.95% $HC_2H_3O_2$; **(c)** 3.12% HNO_3; **(d)** 24.9% H_2SO_4

35. 0.315 *M* HNO_3

37. 1.104 *M* HCl

39. 29.0 mL LiOH

41. 176 g/mol

43. 89.2 g/mol

45. **(a)** $H_2O(l) \longrightarrow H^+(aq) + OH^-(aq)$;

(b) $K_w = [H^+][OH^-]$; **(c)** $K_w = 1.0 \times 10^{-14}$

47. **(a)** $[OH^-] = 4.0 \times 10^{-13}$; **(b)** $[OH^-] = 5.9 \times 10^{-10}$

49. **(a)** $[H^+] = 6.3 \times 10^{-12}$; **(b)** $[H^+] = 3.4 \times 10^{-11}$

51. **(a)** pH = 3; **(b)** pH = 5

53. **(a)** $[H^+] = 0.000\ 001\ M$; **(b)** $[H^+] = 0.000\ 000\ 01\ M$

55. **(a)** pH = 5.10; **(b)** pH = 6.41

57. **(a)** $[H^+] = 0.016\ M$; **(b)** $[H^+] = 0.000\ 018\ M$

59. **(a)** pH = 13.04; **(b)** pH = 10.74

61. **(a)** $[OH^-] = 7.7 \times 10^{-14}$; **(b)** $[OH^-] = 4.2 \times 10^{-13}$

63. **(a)** highly ionized; **(b)** highly ionized; **(c)** highly ionized

65. **(a)** weak; **(b)** strong; **(c)** strong; **(d)** weak

67. **(a)** weak; **(b)** strong; **(c)** strong; **(d)** weak

69. **(a)** HF(aq); **(b)** H^+(aq) and Br^-(aq); **(c)** H^+(aq) and NO_3^-(aq); **(d)** HNO_2(aq)

71. **(a)** Ag^+(aq) and F^-(aq); **(b)** AgI(s); **(c)** Hg_2Cl_2(s); **(d)** Ni^{2+}(aq) and 2 Cl^-(aq)

73. **(1)** Complete and balance the nonionized equation. **(2)** Convert the nonionized equation to the total ionic equation. **(3)** Cancel spectator ions to obtain the net ionic equation. **(4)** Check (✓) each ion or atom on both sides of the equation.

75. **(a)** $H^+aq) + OH^-(aq) \longrightarrow H_2O(l)$

(b) $C_2H_3O_2(aq) + OH^-(aq) \longrightarrow C_2H_3O_2^-(aq) + H_2O(l)$

77. **(a)** $Ag^+(aq) + I^-(aq) \longrightarrow AgI(s)$

(b) $Ba^{2+}(aq) + CrO_4^{2-}(aq) \longrightarrow BaCrO_4(s)$

79. yellow

81. orange (red + yellow)

83. NaH_2PO_4

85. 188 g/mol

87. pH = 0.30

89. $NH_4OH(aq) + HC_2H_3O_2(aq) \longrightarrow NH_4^+(aq) + C_2H_3O_2^-(aq) + H_2O(l)$

Chapter 16

1. collision frequency, collision energy, and collision geometry

3.

H Br
H Br

5. **(a)** increases; **(b)** decreases; **(c)** increases

7. more surface area

9.

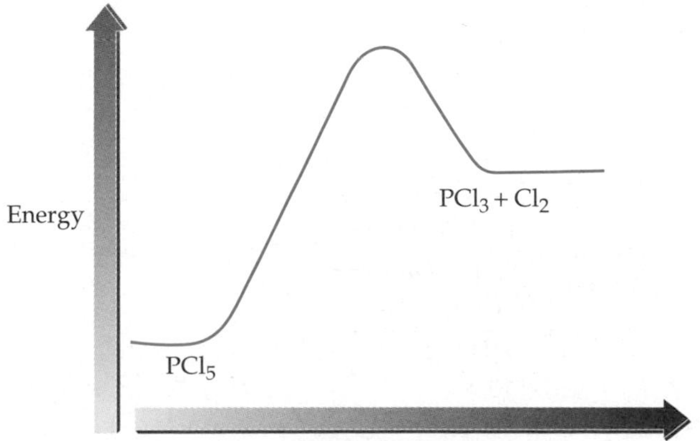

11.

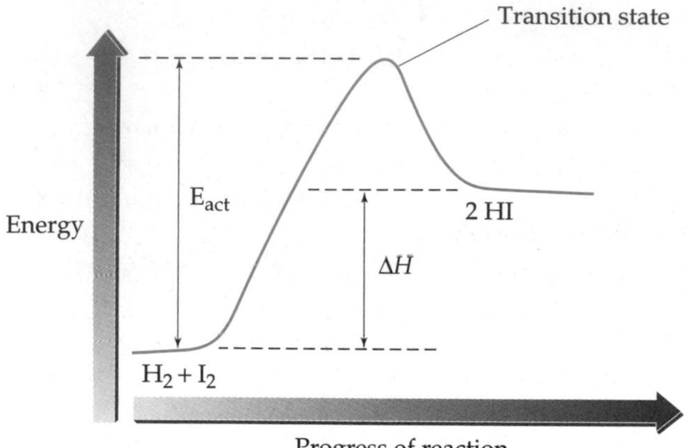

13. lowers E_{act}

15. forward direction

17. **(a)** The rate of the forward reaction corresponds to the rate of *decrease* in the concentration of reactant(s). **(b)** The rate of the forward reaction corresponds to the rate of *increase* in the concentration of product(s).

19. **(a)** and **(b)**

21. **(a)** $K_{eq} = \dfrac{[C]}{[A]^2}$

(b) $K_{eq} = \dfrac{[C]^3}{[A][B]^2}$

(c) $K_{eq} = \dfrac{[C]^4[D]}{[A]^2[B]^3}$

23. no

25. **(a)** $K_{eq} = \dfrac{[HF]^2}{[H_2][F_2]}$

(b) $K_{eq} = \dfrac{[NO_2]^4[H_2O]^6}{[NH_3]^4[O_2]^7}$

(c) $K_{eq} = [CO_2]$

27. $K_{eq} = 1.10$

29. **(a)** increases; **(b)** decreases

31. **(a)** left; **(b)** right; **(c)** no shift; **(d)** right; **(e)** right; **(f)** left; **(g)** left; **(h)** right

33. **(a)** left; **(b)** right; **(c)** right; **(d)** left; **(e)** right; **(f)** no shift; **(g)** no shift; **(h)** no shift

35. **(a)** $K_i = \dfrac{[H^+][CHO_2^-]}{[HCHO_2]}$

(b) $K_i = \dfrac{[H^+][HC_2O_4^-]}{[H_2C_2O_4]}$

(c) $K_i = \dfrac{[H^+][H_2C_6H_5O_7^-]}{[H_3C_6H_5O_7]}$

37. $K_i = 4.5 \times 10^{-4}$

39. $K_i = 7.2 \times 10^{-4}$

41. **(a)** right; **(b)** left; **(c)** left; **(d)** right; **(e)** left; **(f)** left; **(g)** right; **(h)** right

43. **(a)** right; **(b)** left; **(c)** left; **(d)** right; **(e)** left; **(f)** no shift; **(g)** right; **(h)** right

45. **(a)** $K_{sp} = [Ag^+][I^-]$

(b) $K_{sp} = [Ag^+]^2[CrO_4^{2-}]$

(c) $K_{sp} = [Ag^+]^3[PO_4^{3-}]$

47. $K_{sp} = 5.9 \times 10^{-21}$

49. $K_{sp} = 3.4 \times 10^{-35}$

51. The calcium ion concentration in a saturated solution of $CaCO_3$ is slightly greater than in a saturated solution of CaC_2O_4.

53. **(a)** left; **(b)** left; **(c)** right; **(d)** right; **(e)** left; **(f)** no shift; **(g)** no shift; **(h)** right

55. **(a)** left; **(b)** left; **(c)** right; **(d)** right; **(e)** no shift; **(f)** left; **(g)** no shift; **(h)** right

57. A swinging pendulum represents a *dynamic reversible process* as it is in constant motion and swings back and forth in opposite directions.

59. rate of forward reaction = rate of reverse reaction

61. $K_{eq} = 0.20$

63. **(a)** left; **(b)** right; **(c)** right; **(d)** left

65. $Fe(OH)_3$ is more soluble in aqueous HCl than in water because of the following reaction:
$Fe(OH)_3(s) + 3\,H^+(aq) \rightleftharpoons Fe^{3+}(aq) + 3\,H_2O(l)$.

67. $K_{sp} = 5.3 \times 10^{-6}$

Chapter 17

1. **(a)** 0; **(b)** 0; **(c)** 0; **(d)** 0

3. **(a)** +2; **(b)** +3; **(c)** +4; **(d)** +1

5. **(a)** +4; **(b)** +3; **(c)** +4; **(d)** +4

7. **(a)** +4; **(b)** +4; **(c)** +2; **(d)** +4

9. **(a)** oxidation; **(b)** reduction

11. **(a)** oxidized, Mn; reduced, O_2;
(b) oxidized, S; reduced, O_2

13. **(a)** oxidizing agent, CuO; reducing agent, H_2;
(b) oxidizing agent, PbO; reducing agent, CO

15. **(a)** oxidized, Al; reduced, Cr^{3+};
(b) oxidized, Cl^-; reduced, F_2

17. **(a)** oxidizing agent, AgI; reducing agent, Cr^{2+};
(b) oxidizing agent, Hg^{2+}; reducing agent, Sn^{2+}

19. yes

21. **(a)** $Br_2(l) + 2\,NaI(aq) \longrightarrow 2\,NaBr(aq) + I_2(s)$
(b) $2\,PbS(s) + 3\,O_2(g) \longrightarrow 2\,PbO(s) + 2\,SO_2(g)$

23. **(a)** $2\,MnO_4^-(aq) + 10\,I^-(aq) + 16\,H^+(aq) \longrightarrow 2\,Mn^{2+}(aq) + 5\,I_2(s) + 8\,H_2O(l)$
(b) $Cu(s) + SO_4^{2-}(aq) + 4\,H^+(aq) \longrightarrow Cu^{2+}(aq) + SO_2(g) + 2\,H_2O(l)$

25. **(a)** $SO_2(g) + 2\,H_2O(l) \longrightarrow SO_4^{2-}(aq) + 4\,H^+(aq) + 2\,e^-$
(b) $AsO_3^{3-}(aq) \longrightarrow AsO_3^-(aq) + 2\,e^-$

27. **(a)** $ClO^-(aq) + H_2O(l) + 2\,e^- \longrightarrow Cl^-(aq) + 2\,OH^-(aq)$
(b) $MnO_4^-(aq) + 2\,H_2O(l) + 3\,e^- \longrightarrow MnO_2(s) + 4\,OH^-(aq)$

29. **(a)** $3\,Zn(s) + 2\,NO_3^-(aq) + 8\,H^+(aq) \longrightarrow 3\,Zn^{2+}(aq) + 2\,NO(g) + 4\,H_2O(l)$
(b) $2\,Mn^{2+}(aq) + 5\,BiO_3^-(aq) + 14\,H^+(aq) \longrightarrow 2\,MnO_4^-(aq) + 5\,Bi^{3+}(aq) + 7\,H_2O(l)$

31. **(a)** $3\,S^{2-}(aq) + 2\,MnO_4^-(aq) + 4\,H_2O(l) \longrightarrow 3\,S(s) + 2\,MnO_2(s) + 8\,OH^-(aq)$
(b) $Cu(s) + ClO^-(aq) + H_2O(l) \longrightarrow Cu^{2+}(aq) + Cl^-(aq) + 2\,OH^-(aq)$

33. $Cl_2(g) + H_2O(l) \longrightarrow Cl^-(aq) + HOCl(aq) + H^+(aq)$

35. $4\,Cl_2(g) + 8\,OH^-(aq) \longrightarrow 2\,ClO_2^-(aq) + 6\,Cl^-(aq) + 4\,H_2O(l)$

37. **(a)** $Pb^{2+}(aq)$; **(b)** $Fe^{3+}(aq)$; **(c)** $Ag^+(aq)$; **(d)** $Br_2(l)$

39. **(a)** $F_2(g)$; **(b)** $Br_2(l)$; **(c)** $Cu^{2+}(aq)$; **(d)** $Mn^{2+}(aq)$

41. **(a)** nonspontaneous; **(b)** spontaneous

43. **(a)** spontaneous; **(b)** spontaneous

45. C > B > A

47.

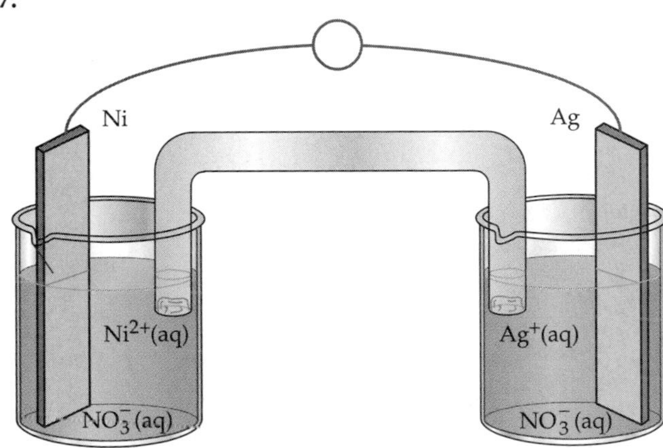

49.

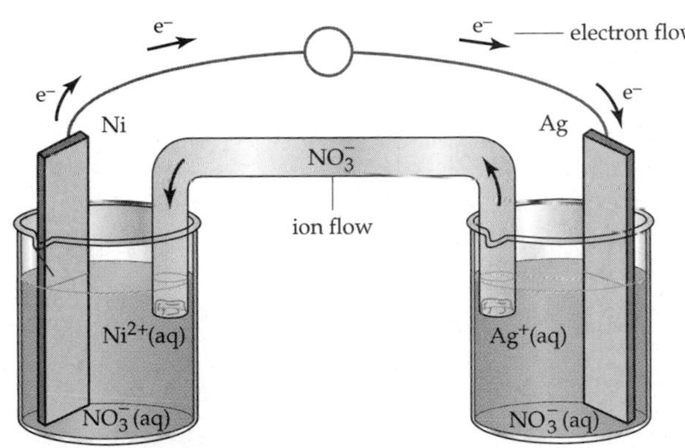

51. **(a)** $Sn(s) \longrightarrow Sn^{2+}(aq) + 2\,e^-$
(b) $Cu^{2+}(aq) + 2\,e^- \longrightarrow Cu(s)$
(c) Sn anode, Cu cathode
(d) Sn anode $\longrightarrow$ Cu cathode
(e) Cu half-cell $\longrightarrow$ Sn half-cell

53. **(a)** $Mg(s) \longrightarrow Mg^{2+}(aq) + 2\,e^-$
(b) $Mn^{2+}(aq) + 2\,e^- \longrightarrow Mn(s)$
(c) Mg anode, Mn cathode
(d) Mg anode $\longrightarrow$ Mn cathode
(e) Mn half-cell $\longrightarrow$ Mg half-cell

55.

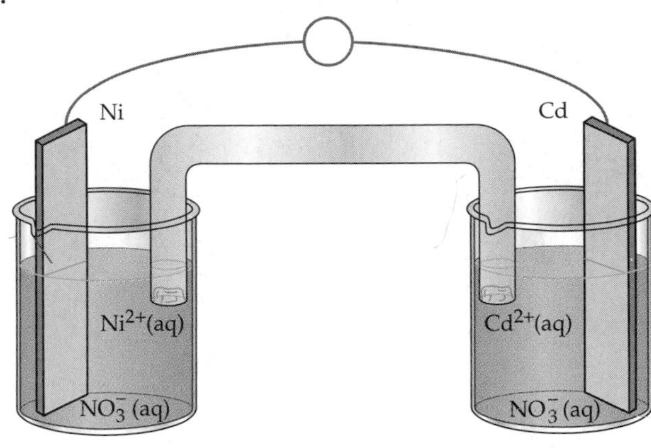

57.

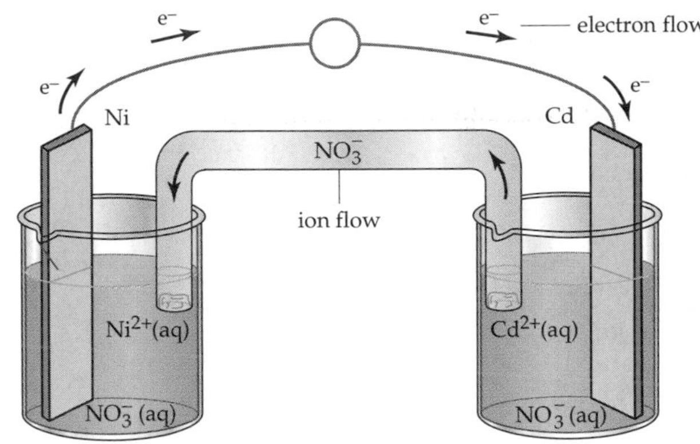

59. **(a)** $Ni(s) \longrightarrow Ni^{2+}(aq) + 2\,e^-$
(b) $Fe^{2+}(aq) + 2\,e^- \longrightarrow Fe(s)$
(c) Ni anode, Fe cathode
(d) Ni anode $\longrightarrow$ Fe cathode
(e) Fe half-cell $\longrightarrow$ Ni half-cell

61. **(a)** $Cr(s) \longrightarrow Cr^{3+}(aq) + 3\,e^-$
(b) $Al^{3+}(aq) + 3\,e^- \longrightarrow Al(s)$
(c) Cr anode, Al cathode
(d) Cr anode $\longrightarrow$ Al cathode
(e) Al half-cell $\longrightarrow$ Cr half-cell

63. +2

65. $Co(s) + Hg^{2+}(aq) \longrightarrow Co^{2+}(aq) + Hg(l)$

67. $Zn(s) + 2\,H^+(aq) \longrightarrow Zn^{2+}(aq) + H_2(g)$

69. $Cu(s) + 4\,H^+(aq) + 2\,NO_3^-(aq) \longrightarrow Cu^{2+}(aq) + 2\,NO_2(g) + 2\,H_2O(l)$

Chapter 18

1. alpha, beta, and gamma

3. alpha particle (α)

5. gamma ray (γ)

7. **(a)** ${}^{4}_{2}He$; **(b)** ${}^{0}_{-1}e$; **(c)** ${}^{0}_{0}\gamma$; **(d)** ${}^{0}_{+1}e$; **(e)** ${}^{1}_{0}n$; **(f)** ${}^{1}_{1}H$;

9. **(a)** ${}^{175}_{78}Pt \longrightarrow {}^{171}_{76}Os + {}^{4}_{2}He$; **(b)** ${}^{28}_{13}Al \longrightarrow {}^{28}_{14}Si + {}^{0}_{-1}e$; **(c)** ${}^{55}_{27}Co \longrightarrow {}^{55}_{26}Fe + {}^{0}_{+1}e$; **(d)** ${}^{44}_{22}Ti + {}^{0}_{-1}e \longrightarrow {}^{44}_{21}Sc$;

11. **(a)** ${}^{221}_{88}Ra$; **(b)** ${}^{43}_{19}K$; **(c)** ${}^{19}_{10}Ne$; **(d)** ${}^{37}_{18}Ar$

13. ${}^{210}_{84}Po$

15. ${}^{207}_{81}Tl$

17. ${}^{212}_{84}Po$

19. **(a)** α; **(b)** β; **(c)** β **(d)** α; **(e)** α; **(f)** α; **(g)** α **(h)** β; **(i)** β; **(j)** α

21. 2 half-lives

23. 366,000 years

25. 22,900 years old

27. 5 mg

29. 296 days

31. 5.30 years/half-life

33. radiocarbon dating

35. uranium–lead dating

37. plutonium-238

39. iodine-131

41. technetium-99

43. ${}^{23}_{12}Mg$

45. ${}^{3}_{1}M$

47. ${}^{59}_{27}Co$

49. ${}^{4}_{2}He$

51. ${}^{260}_{105}X$

53. ${}^{260}_{106}X$

55. 8 neutrons

57. The number of neutrons varies, but the *average* number is 2.4.

59. 2 neutrons

61. 3 neutrons

63. ${}^{233}_{92}U$

65. ${}^{1}_{0}n$

67. ${}^{1}_{1}H$

69. ${}^{1}_{1}H$

71. ${}^{237}_{92}U$

73. ${}^{137}_{53}I$

75. ~6000 years

77. ${}^{241}_{95}Am$

79. 15 neutrons

81. ${}^{256}_{101}Md$

83. ${}^{4}_{2}He$

Chapter 19

1. ~90%

3. petroleum

5. **(a)** saturated; **(b)** unsaturated; **(c)** unsaturated

7. **(a)** $C_{10}H_{22}$

9. **(a)** butane; **(b)** hexane; **(c)** octane; **(d)** decane

11. $CH_3-CH_2-CH_2-CH_2-CH_2-CH_3$

$$\begin{array}{l} \qquad\ \ CH_3 \\ \qquad\quad | \\ CH_3-CH-CH_2-CH_2-CH_3 \end{array}$$

$$\begin{array}{l} \qquad\qquad\quad\ CH_3 \\ \qquad\qquad\qquad | \\ CH_3-CH_2-CH-CH_2-CH_3 \end{array}$$

$$\begin{array}{l} \qquad\ \ CH_3\ CH_3 \\ \qquad\quad |\qquad | \\ CH_3-CH-CH-CH_3 \end{array}$$

$$\begin{array}{l} \qquad\ \ CH_3 \\ \qquad\quad | \\ CH_3-C-CH_2-CH_3 \\ \qquad\quad | \\ \qquad\ \ CH_3 \end{array}$$

13. $CH_3-CH_2-CH_2-Br$ and $CH_3-CH(Br)-CH_3$

15. **(a)** methyl; **(b)** ethyl

17. **(a)** 2-methylpentane; **(b)** 2-methyl-4-ethylhexane; **(c)** 2,4,4-trimethylheptane; **(d)** 3,4,5-trimethyloctane

19. **(a)** $CH_4 + 2\,O_2 \longrightarrow CO_2 + 2\,H_2O$
(b) $C_5H_{12} + 8\,O_2 \longrightarrow 5\,CO_2 + 6\,H_2O$
(c) $C_3H_8 + 5\,O_2 \longrightarrow 3\,CO_2 + 4\,H_2O$
(d) $C_7H_{16} + 11\,O_2 \longrightarrow 7\,CO_2 + 8\,H_2O$

21. **(b)** $C_{14}H_{28}$

23. **(a)** 1-butene; **(b)** 1-pentene; **(c)** 2-hexene; **(d)** 4-octene

25. **(a)** $C_{10}H_{18}$

27. **(a)** 2-pentyne; **(b)** 1-pentyne; **(c)** 3-hexyne; **(d)** 2-heptyne

29. $CH_2=CH-CH_2-CH_2-CH_3$ and $CH_3-CH=CH-CH_2-CH_3$

31. **(a)** 4-methyl-2-pentene; **(b)** 3,5-dimethyl-2-hexene

33. **(a)** 4,4-dimethyl-2-heptyne; **(b)** 3,4,5-trimethyl-1-octyne

35. **(a)** $CH_2=CH_2 + 3\,O_2 \longrightarrow 2\,CO_2 + 2\,H_2O$
(b) $CH_3-CH=CH_2 + H_2 \longrightarrow CH_3-CH_2-CH_3$
(c) $CH_3-CH=CH-CH_3 + Br_2 \longrightarrow CH_3-CHBr-CHBr-CH_3$

37. 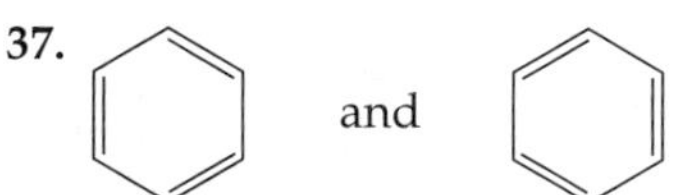

39. 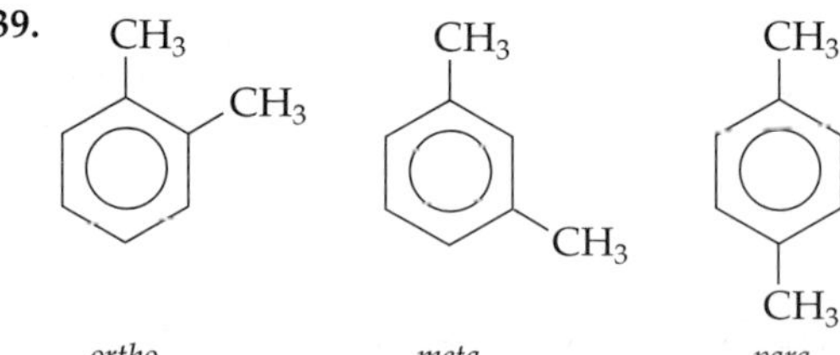

41. **(a)** ether; **(b)** organic halide; **(c)** phenol; **(d)** amine

43. **(a)** amine; **(b)** organic halide; **(c)** ester; **(d)** carboxylic acid; **(e)** aldehyde; **(f)** ether; **(g)** ketone; **(h)** amide

45. **(a)** "methyl iodide"; **(b)** "ethyl bromide"; **(c)** "propyl fluoride"; **(d)** "isopropyl chloride"

47. $CH_3—CCl_3$

49. **(a)** insoluble; **(b)** soluble

51. **(a)** 1-butanol; **(b)** 2-butanol; **(c)** 1-butanol; **(d)** 2-butanol

53. **(a)** phenol; **(b)** *para*-methylphenol

55. **(a)** "dimethyl ether"; **(b)** "methyl ethyl ether"; **(c)** "dipropyl ether"; **(d)** "phenyl ethyl ether"

57. Ethyl alcohol molecules can hydrogen-bond.

59. $CH_3—CH_2—OH$ and $CH_3—O—CH_3$

61. **(a)** "ethylamine"; **(b)** "propylamine"

63. Ethylamine molecules can hydrogen-bond.

65. **(a)** methanal; **(b)** ethanal; **(c)** propanal; **(d)** butanal

67. **(a)** 2-propanone; **(b)** 2-butanone; **(c)** 2-pentanone; **(d)** 3-pentanone

69. **(a)** higher boiling point; **(b)** higher boiling point

71.

$$CH_3CH_2—\overset{\overset{\displaystyle O}{\|}}{C}—H \text{ and } CH_3—\overset{\overset{\displaystyle O}{\|}}{C}—CH_3$$

73. **(a)** methanoic acid; **(b)** ethanoic acid; **(c)** propanoic acid; **(d)** butanoic acid

75. **(a)** ethyl methanoate; **(b)** methyl ethanoate; **(c)** ethyl propanoate; **(d)** phenyl methanoate

77. **(a)** methanamide ("formamide"); **(b)** benzamide

79. **(a)** and **(c)**

81. propionic acid

83. **(a)**

$$CH_3—\overset{\overset{\displaystyle O}{\|}}{C}—OH + CH_3—OH \xrightarrow{H_2SO_4} CH_3—\overset{\overset{\displaystyle O}{\|}}{C}—O—CH_3 + H_2O$$

(b)

$$C_6H_5—\overset{\overset{\displaystyle O}{\|}}{C}—OH + NH_3 \xrightarrow{\Delta} C_6H_5—\overset{\overset{\displaystyle O}{\|}}{C}—NH_2 + H_2O$$

85. phenyl ethanoate ("phenyl acetate")

87. methanamide ("formamide")

89. **(a)** propane; **(b)** pentane

91. **(a)** saturated; **(b)** unsaturated

93. **(a)** $CH_3CH_2—OH$; **(b)** $CH_3CH_2CH_2—NH_2$

95. **(a)** organic halide; **(b)** ketone; **(c)** carboxylic acid; **(d)** amide; **(e)** amine

Chapter 20

1. **(a)** protein; **(b)** nucleic acid

3. **(a)** peptide; **(b)** phosphate

5. **(a)** carboxylic acid, amine; **(b)** aldehyde, ketone, alcohol; **(c)** carboxylic acid; **(d)** amine

7. covalent amide bond

9. The primary structure of a protein refers to the amino acid sequence. The secondary structure of a protein refers to the shape of the polymer chain.

11.

$$H_2N—\underset{\underset{\displaystyle CH_3}{|}}{CH}—\overset{\overset{\displaystyle O}{\|}}{C}—NH—\underset{\underset{\displaystyle CH_3}{|}}{CH}—\overset{\overset{\displaystyle O}{\|}}{C}—OH$$

13.

$$H_2N—\underset{\underset{\displaystyle CH_2—OH}{|}}{CH}—\overset{\overset{\displaystyle O}{\|}}{C}—NH—\underset{\underset{\displaystyle CH_2—C_6H_4—OH}{|}}{CH}—\overset{\overset{\displaystyle O}{\|}}{C}—OH$$

15. **(1)** Arg-His-Met; **(2)** Arg-Met-His; **(3)** His-Arg-Met; **(4)** His-Met-Arg; **(5)** Met-Arg-His; **(6)** Met-His-Arg

17. aspartic acid and phenylalanine

19. In oxytocin, *isoleucine* replaces phenylalanine, and *leucine* replaces arginine.

21. In step 1, the substrate molecule binds to the active site on the enzyme.

23. In the model, the "teeth" represent an active site on the enzyme molecule.

25. inhibitor

27. **(1)** An enzyme is selective for a given molecule and reacts to give specific products. **(2)** An enzyme can accelerate a biochemical reaction by a factor of a million or more. **(3)** Only a trace amount of enzyme is necessary to catalyze a biochemical reaction.

29. aldehyde and alcohol groups

31. five

33. A disaccharide has two monosaccharide molecules joined by a glycoside linkage.

35. See textbook Figure 20.9.

37. See textbook Figure 20.10.

39.

$$\underset{\text{maltose}}{C_{12}H_{22}O_{11}} + \underset{\text{water}}{H_2O} \xrightarrow{H^+} \underset{\text{glucose}}{C_6H_{12}O_6} + \underset{\text{glucose}}{C_6H_{12}O_6}$$

41. Glycogen is a small polysaccharide composed of glucose molecules.

43. **(a)** fat; **(b)** oil; **(c)** fat; **(d)** fat and oil

45.

$$\begin{array}{l} CH_2—O—\overset{\overset{\displaystyle O}{\|}}{C}—(CH_2)_{16}—CH_3 \\ | \\ CH_2—O—\overset{\overset{\displaystyle O}{\|}}{C}—(CH_2)_{16}—CH_3 \\ | \\ CH_2—O—\overset{\overset{\displaystyle O}{\|}}{C}—(CH_2)_{16}—CH_3 \end{array}$$

47. **(a)** palmitic acid, $CH_3—(CH_2)_{14}—COOH$; **(b)** lauric acid, $CH_3—(CH_2)_{10}—COOH$; **(c)** myristic acid, $CH_3—(CH_2)_{12}—COOH$

49. $CH_3—(CH_2)_{24}—COO^-Na^+$ and $CH_3—(CH_2)_{29}—OH$

51. A, D, E, and K

53. deoxyribose sugar, an organic base, and phosphoric acid

55. adenine (A), cytosine (C), guanine (G), and thymine (T)

57. The structures are identical except that deoxyribose is missing an —OH group.

59. two

61. hydrogen bonds

63. thymine (T)

65. uracil (U)

67. long and extended

69. The protein loses its secondary and tertiary structure and acquires a random shape.

71. protein

73. three

75. A DNA molecule replicates by first unwinding and breaking hydrogen bonds between the two strands in the double helix. Second, each single strand of DNA acts as a template to synthesize a complementary strand of DNA. For example, a nucleotide in the template strand that contains adenine (A) will code for thymine (T) on the growing strand; cytosine (C) will code for guanine (G); and so on. After the synthesis is complete, the template strand and complementary strand are joined by hydrogen bonds and the resulting structure is an exact replication of the original DNA double helix.

77. 24

Glossary

A

absolute zero The theoretical temperature at which the kinetic energy of a gas is zero. (*Sec. 11.10*)

acid A substance that releases hydrogen ions (H^+) when dissolved in water. (*Sec. 8.11*)

acid–base indicator chemical substance that changes color according to the pH of the solution. (*Sec. 15.4*)

acidic oxide See *nonmetal oxide.*

actinide series Elements with atomic numbers 90 through 103. (*Sec. 6.3*)

activation energy (E_{act}) The energy necessary for reactants to achieve the transition state in order to form products. (*Sec. 16.2*)

active metal A metal that is sufficiently active to react with water at room temperature; active metals include Li, Na, K, Ca, Sr, and Ba. (*Sec. 8.8*)

activity The number of radioactive nuclei in a sample that disintegrate in a given period of time, for example, 500 disintegrations per minute (dpm). (*Sec. 18.4*)

activity series A relative order of metals arranged according to their ability to undergo reaction; also called an electromotive series. (*Sec. 8.7*)

actual yield The amount of product experimentally obtained from given amounts of reactants. (*Sec. 10.9*)

addition reaction A chemical reaction in which an unsaturated hydrocarbon adds a molecule, such as H—H or Br—Br, across a double or triple bond. (*Sec. 19.3*)

alchemy A pseudoscience that attempted to convert a base metal such as lead into gold; a medieval science that sought to discover a universal cure for disease and a magic potion for immortality. (*Sec. 1.2*)

alkali metals Group IA/1 elements, excluding hydrogen. (*Sec. 6.3*)

alkaline earth metals Group IIA/2 elements. (*Sec. 6.3*)

alkanes A family of compounds that are saturated hydrocarbons. (*Sec. 19.2*)

alkenes A family of compounds that are unsaturated hydrocarbons and have a double bond. (*Sec. 19.3*)

alkyl group (R—) A saturated hydrocarbon fragment that results after the removal of a hydrogen atom from an alkane. (*Sec. 19.2*)

alkynes A family of compounds that are unsaturated hydrocarbons and have a triple bond. (*Sec. 19.3*)

alloy A homogeneous mixture of two or more metals. (*Sec. 4.2*)

alpha particle (α or ^4_2He) A particle in an alpha ray that is identical to a helium-4 nucleus. (*Sec. 18.1*)

amino acid A carboxylic acid with an amine group on the alpha carbon; also referred to as an α-amino acid. (*Sec. 20.2*)

amphiprotic A substance capable of either accepting a proton or donating a proton in acid–base reactions. (*Sec. 15.3*)

anhydrous A term for a compound that does not contain water. (*Sec. 13.10*)

anion A negatively charged ion. (*Sec. 7.1*)

anode The electrode in an electrochemical cell at which oxidation occurs. (*Sec. 17.6*)

aqueous solution A homogeneous mixture of a substance dissolved in water. (*Secs. 7.1 and 8.1*)

arenes A family of compounds that are aromatic hydrocarbons. (*Sec. 19.4*)

aromatic hydrocarbon A hydrocarbon containing a benzene ring. (*Sec. 19.1*)

Arrhenius acid A substance that releases hydrogen ions when dissolved in water. (*Sec. 15.2*)

Arrhenius base A substance that releases hydroxide ions when dissolved in water. (*Sec. 15.2*)

artificial radioactivity See *transmutation.*

aryl group (Ar—) An aromatic hydrocarbon fragment that results after the removal of a hydrogen atom from an arene. (*Sec. 19.2*)

atmospheric pressure The pressure exerted by the air molecules in Earth's atmosphere. The atmospheric pressure at sea level is 1 atm or 760 mm Hg. (*Sec. 11.2*)

atom The smallest particle that represents an element. (*Sec. 4.3*)

atomic mass The weighted average mass of all the naturally occurring isotopes of an element. (*Sec. 5.6*)

atomic mass unit (amu) A unit of mass exactly equal to 1/12 the mass of a carbon-12 atom. (*Sec. 5.5*)

atomic notation A symbolic method for expressing the composition of an atomic nucleus; the mass number and atomic number are indicated to the left of the chemical symbol for the element. (*Sec. 5.4*)

atomic nucleus A region of very high density in the center of the atom that contains protons and neutrons. (*Sec. 5.3*)

atomic number (*Z*) A number that identifies a specific element. (*Sec. 4.4*) A value that indicates the number of protons in the nucleus of an atom. (*Secs. 5.4 and 18.2*)

Avogadro's number (*N*) The value that corresponds to the number of carbon atoms in 12.01 g of carbon; 6.02×10^{23} particles. (*Sec. 9.1*)

Avogadro's theory The principle that equal volumes of gases, at the same temperature and pressure, contain equal numbers of molecules. (*Secs. 9.5 and 10.1*)

B

barometer An instrument for measuring atmospheric pressure. (*Sec. 11.2*)

base A substance that releases hydroxide ions (OH^-) when dissolved in water. (*Sec. 8.11*)

basic oxide See *metal oxide.*

battery A general term for any electrochemical cell that produces electric energy spontaneously. (*Sec. 17.6*)

beta particle (β or ${}_{-1}^{0}e$) A particle in a beta ray that is identical to an electron. (*Sec. 18.1*)

binary acid A compound that contains hydrogen and a nonmetal dissolved in water. (*Sec. 7.1*)

binary ionic A compound that contains one metal and one nonmetal. (*Sec. 7.1*)

binary molecular A compound that contains two nonmetals. (*Sec. 7.1*)

biochemistry The study of chemical substances derived from plants and animals. (*Sec. 1.2*) The study of biological compounds and their reactions. (*Sec. 20.1*)

Bohr atom A model of the atom that describes an electron circling the atomic nucleus in an orbit of specific energy. (*Sec. 5.8*)

boiling point (Bp) The temperature at which the vapor pressure of a liquid is equal to the atmospheric pressure. (*Sec. 13.3*)

bond angle The angle formed by two atoms bonded to the central atom in a molecule. (*Secs. 12.9 and 13.7*)

bond energy The amount of energy required to break a given bond in a mole of substance in the gaseous state. (*Sec. 12.3*)

bond length The distance between the nuclei of two atoms that are joined by a covalent bond. (*Sec. 12.3*)

bonding electrons The valence electrons in a molecule that are shared between two atoms. (*Sec. 12.4*)

Boyle's law The statement that the pressure and the volume of a gas are inversely proportional at constant temperature; that is, $P_1V_1 = P_2V_2$. (*Sec. 11.4*)

Brønsted–Lowry acid A substance that donates a proton in an acid–base reaction. (*Sec. 15.3*)

Brønsted–Lowry base A substance that accepts a proton in an acid–base reaction. (*Sec. 15.3*)

buffer A solution that resists changes in pH when an acid or a base is added. (*Sec. 15.1*)

C

calorie (cal) The amount of heat required to raise the temperature of 1 g of water 1°C. (*Sec. 3.9*)

canal ray A stream of positive particles produced in a cathode-ray tube. (*Sec. 5.2*)

carbohydrate A biological compound that contains one or more sugar molecules; a sugar is a polyhydroxy aldehyde or ketone. (*Sec. 20.4*)

carbonyl group (—C=O) The structural unit composed of a carbon–oxygen double bond found in aldehydes, ketones, and other organic compounds. (*Sec. 19.5*)

carboxyl group (—COOH) The functional group in a carboxylic acid. (*Sec. 19.10*)

catalyst A substance that increases the rate of reaction but can be recovered without being permanently changed. (*Sec. 8.2*) A substance that allows a reaction to proceed faster by lowering the energy of activation. (*Sec. 16.2*)

cathode The electrode in an electrochemical cell at which reduction occurs. (*Sec. 17.6*)

cathode ray A stream of negative particles produced in a cathode-ray tube. (*Sec. 5.2*)

cation A positively charged ion. (*Sec. 7.1*)

Celsius degree (°C) The basic unit of temperature in the metric system. (*Sec. 3.8*)

centimeter (cm) A common unit of length in the metric system of measurement that is equal to 1/100 of a meter. (*Sec. 2.1*)

chain reaction A fission reaction in which the neutrons released initiate a second reaction, which in turn initiates a third reaction, and so on. (*Sec. 18.7*)

Charles' law The statement that the volume and the Kelvin temperature of a gas are directly proportional at constant pressure; that is, $V_1/T_1 = V_2/T_2$. (*Sec. 11.5*)

chemical bond The general term for the attraction between two ions or the affinity between two atoms. (*Sec. 12.1*)

chemical change A modification of a substance that alters the chemical composition. (*Sec. 4.7*)

chemical equation A shorthand representation using formulas and symbols to describe a chemical reaction. (*Sec. 8.2*)

chemical equilibrium A dynamic state of a reversible reaction in which the rates of the forward and reverse reactions are equal and the amounts of the reactants and products are constant. (*Sec. 16.3*)

chemical formula An abbreviation for the name of a chemical compound that indicates the number of atoms of each element; for example, H_2O is the formula for water. (*Sec. 4.5*)

chemical property A characteristic of a substance that cannot be observed without changing the chemical formula of the substance. (*Sec. 4.6*)

chemical reaction The process of undergoing a chemical change. (*Sec. 8.1*)

chemical symbol An abbreviation for the name of a chemical element; for example, Cu is the symbol for copper. (*Sec. 4.3*)

chemistry The branch of science that studies the composition and properties of matter. (*Sec. 1.2*)

class of compounds A family of compounds in which all the members have the same structural feature (that is, an atom or group of atoms) and similar chemical properties. (*Sec. 19.5*)

coefficient A digit in front of a chemical formula in a chemical equation; a digit that serves to balance the number of atoms of each element on the left side with those on the right side of a chemical equation. (*Sec. 8.3*)

collision theory The principle that the rate of a chemical reaction is regulated by the collision frequency, collision energy, and orientation of molecules striking each other. (*Sec. 16.1*)

colloid A homogeneous mixture in which the diameter of the dispersed particles ranges from 1 to 100 nm. (*Sec. 14.4*)

combination reaction A type of reaction in which two substances produce a single compound. (*Sec. 8.4*)

combined gas law The statement that the pressure exerted by a gas is inversely proportional to its volume and directly proportional to its Kelvin temperature; that is, $P_1V_1/T_1 = P_2V_2/T_2$. (*Sec. 11.7*)

combustion reaction A chemical reaction in which a hydrocarbon reacts rapidly with oxygen (burns) to produce carbon dioxide, water, and heat. (*Sec. 19.2*)

compound A pure substance that can be broken down into two or more simpler substances by chemical reaction. (*Sec. 4.2*)

continuous spectrum A single, broad band of radiant energy. (*Sec. 5.6*)

coordinate covalent bond A bond in which an electron pair is shared but both electrons have been donated by a single atom. (*Sec. 12.8*)

core The portion of the atom that includes the nucleus and inner electrons that are not available for bonding; also termed the kernel of the atom. (*Sec. 6.8*)

core notation A method of writing the electron configuration in which the inner core electrons are represented by a noble gas symbol in brackets followed by the valence electrons; for example, [Ne] $3s^2$. (*Sec. 6.10*)

covalent bond A chemical bond characterized by the sharing of one or more pairs of valence electrons. (*Sec. 12.1*)

critical mass The minimum size mass of a fissionable nuclide that is necessary to sustain a continuous chain reaction. (*Sec. 18.7*)

crystalline solid A solid substance composed of ions or molecules that repeat in a regular geometric pattern. (*Sec. 13.4*)

cubic centimeter (cm^3) A unit of volume occupied by a cube 1 cm on a side; 1 cm^3 is exactly equal to 1 mL. (*Sec. 3.5*)

D

Dalton's law The statement that the pressure exerted by a mixture of gases is equal to the sum of the individual pressures exerted by each gas; that is, $P_1 + P_1 + P_1 + \cdots = P_{total}$. (*Sec. 11.9*)

decay series See *radioactive decay series.*

decomposition reaction A type of reaction in which a single compound produces two or more substances. (*Sec. 8.4*)

degree Fahrenheit See *Fahrenheit degree.*

degree Celsius See *Celsius degree.*

deionized water Water purified by removing ions using an ion exchange method; also termed demineralized water. (*Sec. 13.10*)

delta (δ) notation A method of indicating the partial positive charge (δ^+) and the partial negative charge (δ^-) in a polar covalent bond. (*Sec. 12.6*)

density (*d*) The amount of mass in one unit volume of matter. (*Sec. 3.7*)

deposition A direct change in state from a gas to a solid without forming a liquid. (*Sec. 4.1*)

deuterium (^2_1H) The nuclide of hydrogen with one neutron in the nucleus. (*Sec. 18.8*)

diatomic molecule A particle composed of two nonmetal atoms. (*Sec. 8.2*) A molecule composed of two nonmetal atoms held together by a covalent bond. (*Sec. 12.7*)

dipeptide A molecule containing two amino acids joined by a peptide linkage. (*Sec. 20.2*)

dipole A region in a molecule having partial negative charge and partial positive charge resulting from a polar bond. (*Sec. 14.2*)

dipole force A type of attraction between two molecules that have permanent dipoles. (*Sec. 13.2*)

directly proportional The association between two variables that have a similar relationship; for example, if one variable doubles, the other variable doubles. (*Sec. 11.5*)

disaccharide A carbohydrate composed of two simple sugar molecules joined by a glycoside linkage. (*Sec. 20.4*)

dispersion force A type of attraction between two molecules that have temporary dipoles. A temporary dipole arises from the unequal distribution of charge as electrons move around a molecule; also called a London dispersion force. (*Sec. 13.2*)

dissociation The process of an ionic compound dissolving in water and separating into positive and negative ions; for example, NaOH dissolves in water to give sodium ions and hydroxide ions. (*Sec. 15.2*)

double bond A bond between two atoms composed of two shared electron pairs. A double bond is represented as two dashes between the symbols of two atoms. (*Sec. 12.4*)

double-replacement reaction A type of reaction in which two cations in different compounds exchange anions. (*Sec. 8.4*)

dry cell An electrochemical cell in which the anode and cathode reactions do not take place in an aqueous solution. (*Sec. 17.6*)

ductile The property of a metal that allows it to be drawn into a wire. (*Sec. 4.4*)

E

elastic collision An impact between gas molecules in which the total energy remains constant. (*Sec. 11.10*)

electrochemical cell A general term for an apparatus containing two solutions with electrodes in separate compartments that are connected by a conducting wire and salt bridge. (*Sec. 17.6*)

electrochemistry The study of the interconversion of chemical and electric energy from redox reactions. (*Sec. 17.6*)

electrolysis The chemical reaction produced by passing electricity through an aqueous solution. (*Sec. 13.9*)

electrolytic cell An electrochemical cell in which a nonspontaneous redox reaction occurs as a result of the input of direct electric current. (*Sec. 17.7*)

electromotive series See *activity series.*

electron (e^-) A negatively charged subatomic particle having a negligible mass. (*Sec. 5.2*)

electron capture (EC) A nuclear decay reaction in which a heavy nuclide attracts one of its inner core electrons into the nucleus. (*Sec. 18.2*)

electron configuration A shorthand description of the arrangement of electrons by sublevels according to increasing energy. (*Sec. 5.10*)

electron dot formula A representation of an atom and its valence electrons that shows the chemical symbol surrounded by a dot for each valence electron. (*Sec. 6.8*) A diagram of a molecule or polyatomic ion that shows the chemical symbol of each atom surrounded by two dots on a side for each pair of bonding or nonbonding electrons. An electron dot formula is also called a Lewis structure. (*Sec. 12.4*)

electron pair geometry The geometric shape formed by bonding and nonbonding electron pairs surrounding the central atom in a molecule. (*Sec. 12.9*)

electronegativity The ability of an atom to attract a pair of electrons in a chemical bond. (*Sec. 12.6*)

element A pure substance that cannot be broken down by ordinary chemical reaction. (*Sec. 4.2*)

emission line spectrum Several narrow bands of radiant energy that result from excited atoms releasing energy. (*Sec. 5.8*)

empirical formula A chemical formula that expresses the simplest whole-number ratio of the atoms in a molecule, or ions in a formula unit. (*Sec. 9.8*)

endothermic reaction A chemical reaction that absorbs heat energy. (*Secs. 8.1 and 16.2*)

endpoint The stage in a titration when the indicator changes color permanently. (*Sec. 15.4*)

energy level An orbit of specific energy that electrons occupy at a fixed distance from the nucleus; also referred to as a main energy level; designated 1, 2, 3, 4, ... (*Sec. 5.8*)

energy sublevel An electron energy level resulting from the splitting of a main energy level; designated *s*, *p*, *d*, *f*, ... (*Sec. 5.9*)

English system A nondecimal system of measurement without any basic unit for length, mass, or volume. (*Sec. 3.1*)

enzyme A protein molecule that catalyzes a specific biochemical reaction. (*Sec. 20.3*)

equilibrium constant See *general equilibrium constant* (K_{eq}), *ionization equilibrium constant* (K_i), or, *solubility equilibrium product constant* (K_{sp}).

evaporation See *vaporization*.

exact equivalent (≡) A statement that relates two values that are exactly equivalent; for example, 1 yd ≡ 36 in. and 1 m ≡ 100 cm. (*Secs. 2.8 and 3.2*)

exothermic reaction A chemical reaction that releases heat energy. (*Secs. 8.1 and 16.2*)

experiment A scientific procedure for collecting data and recording observations under controlled conditions. (*Sec. 1.1*)

exponent A number written as a superscript indicating that a value is multiplied by itself; for example, $10^4 = 10 \times 10 \times 10 \times 10$ and $cm^3 = cm \times cm \times cm$. (*Sec. 2.6*)

F

Fahrenheit degree (°F) A basic unit of temperature in the English system. (*Sec. 3.8*)

fat A triglyceride from an animal source that contains mostly saturated fatty acids. (*Sec. 20.5*)

fatty acid A carboxylic acid with a long hydrocarbon chain. (*Sec. 20.5*)

formula unit The simplest representative particle in a compound composed of ions. (*Sec. 7.4*) The simplest representative particle in a substance held together by ionic bonds. (*Sec. 12.1*)

frequency The number of times a light wave travels a complete cycle in 1 s. (*Sec. 5.6*)

functional group An atom or group of atoms that characterizes a class of compounds and contributes to their similar chemical properties. (*Sec. 19.5*)

G

galvanic cell See *voltaic cell*.

gamma ray ($^{0}_{0}\gamma$) A powerful type of radiant energy with a very short wavelength. (*Sec. 18.1*)

gas density The ratio of mass per unit volume of a gas, usually expressed in grams per liter. (*Sec. 9.5*)

gas pressure A measure of the frequency and energy of gas molecules colliding with the walls of their container. (*Sec. 11.2*)

Gay-Lussac's law The statement that the pressure and the Kelvin temperature of a gas are directly proportional at constant volume; that is, $P_1/T_1 = P_2/T_2$. (*Sec. 11.6*)

Gay-Lussac's law of combining volumes See *law of combining volumes*.

general equilibrium constant (K_{eq}) A constant that expresses the molar equilibrium concentration of each substance participating in a reversible reaction at a given temperature. (*Sec. 16.3*)

glycoside linkage An —O— bond that joins two simple sugar molecules. (*Sec. 20.4*)

gram (g) The basic unit of mass in the metric system. (*Secs. 2.1 and 3.1*)

group A vertical column in the periodic table; a family of elements with similar properties. (*Sec. 6.3*)

H

[H^+] The symbol for the molar concentration of the hydrogen ion. (*Sec. 15.7*)

half-cell A portion of an electrochemical cell having a single electrode where either oxidation or reduction occurs. (*Sec. 17.6*)

half-life ($t_{1/2}$) The amount of time required for 50% of the radioactive nuclei in a given sample to decay. (*Sec. 18.4*)

half-reaction A reaction that represents either an oxidation or a reduction process separately and indicates the number of electrons lost or gained. (*Sec. 17.4*)

halogens Group VIIA/17 elements. (*Sec. 6.3*)

hard water Water containing a variety of cations and anions such as Ca^{2+}, Mg^{2+}, Fe^{3+}, CO_3^{2-}, SO_4^{2-}, and PO_4^{3-}. (*Sec. 13.10*)

heat The flow of energy from a system at a higher temperature to a system at a lower temperature. Heat is a measure of the total energy in a system. (*Sec. 3.9*)

heat of fusion (H_{fusion}) The heat required to convert a solid to a liquid at its melting point. Conversely, the heat released when a liquid changes to a solid is called the heat of solidification (H_{solid}). (*Sec. 13.6*)

heat of reaction (ΔH) The difference in heat energy between the reactants and the products for a given chemical reaction. (*Sec. 16.2*)

heat of vaporization (H_{vapor}) The heat required to convert a liquid to a gas at its boiling point. Conversely, the heat released when a gas changes to a liquid is called the heat of condensation (H_{cond}). (*Sec. 13.6*)

heavy water (D_2O) A molecule of water in which ordinary hydrogen atoms (^1_1H) are replaced by hydrogen atoms with a neutron (^2_1H). (*Sec. 13.9*)

Henry's law The principle that the solubility of a gas in a liquid is proportional to the partial pressure of the gas above the liquid. (*Sec. 14.1*)

heterogeneous equilibrium A type of equilibrium in which one or more of the participating species is in a different physical state, for example, an equilibrium between an insoluble precipitate and its ions in aqueous solution. (*Sec. 16.4*)

heterogeneous mixture Matter having indefinite composition and properties that vary within the sample. (*Sec. 4.2*)

homogeneous equilibrium A type of equilibrium in which all the participating species are in the same physical state; for example, an equilibrium between reactants and products in the gaseous state. (*Sec. 16.4*)

homogeneous mixture Matter having indefinite composition and properties that vary from sample to sample; examples include alloys, solutions, and mixtures of gases. (*Sec. 4.2*)

hydrate A substance that contains a specific number of water molecules attached to a formula unit in a crystalline compound. (*Sec. 13.10*)

hydrocarbon An organic compound containing only hydrogen and carbon. (*Sec. 19.1*)

hydrocarbon derivative An organic compound containing carbon, hydrogen, and another element such as oxygen, nitrogen, or a halogen. (*Sec. 19.1*)

hydrogen bond A type of attraction between molecules that have a hydrogen atom bonded to a highly electronegative atom such as oxygen or nitrogen. (*Sec. 13.2*)

hydronium ion (H_3O^+) The ion that results when a hydrogen ion attaches to a water molecule by a coordinate covalent bond. The hydronium ion is the predominant form of the hydrogen ion in aqueous acid solution. (*Sec. 15.2*)

hydroxyl group (—OH) The functional group in an alcohol or phenol. (*Sec. 19.7*)

hypothesis An initial, tentative proposal of a scientific principle that attempts to explain the meaning of a set of data collected in an experiment. (*Sec. 1.1*)

I

IUPAC nomenclature The system of rules set forth by the International Union of Pure and Applied Chemistry for naming chemical compounds. (*Sec. 7.1*)

ideal gas A theoretical gas that obeys the kinetic theory under all conditions of temperature and pressure. (*Sec. 11.10*)

ideal gas constant The proportionality constant R in the equation $PV = nRT$. (*Sec. 11.11*)

ideal gas law The principle stated by the relationship $PV = nRT$; also called the ideal gas equation. (*Sec. 11.11*)

immiscible A term that refers to liquids that are not soluble in one another and separate into two layers. (*Sec. 14.2*)

inner transition elements The elements in the lanthanide and actinide series. (*Sec. 6.3*)

inorganic chemistry The study of chemical substances that do not contain carbon. (*Sec. 1.2*)

inorganic compound A compound that does not contain carbon. (*Sec. 7.1*)

instrument A device for recording a measurement such as length, mass, volume, time, or temperature. (*Sec. 2.1*)

International System (SI) A sophisticated system of measurement that is more comprehensive than the metric system and has seven base units. (*Sec. 3.1*)

inversely proportional The association between two variables that have an opposite relationship; for example, if one variable doubles, the other variable is halved. (*Sec. 11.4*)

ion An atom (or group of atoms) that has a positive or negative charge resulting from the gain or loss of valence electrons. (*Sec. 6.9*)

ionic bond A chemical bond characterized by the attraction between a cation and an anion. (*Sec. 12.1*)

ionic charge A term for the positive charge on a metal atom that has lost electrons, or the negative charge on a nonmetal atom that has gained electrons. (*Sec. 6.10*)

ionic solid A crystalline solid composed of ions that repeat in a regular pattern. (*Sec. 13.5*)

ionization The process of a polar compound dissolving in water and forming positive and negative ions; for example, HCl dissolves in water to give hydrogen ions and chloride ions. (*Sec. 15.2*)

ionization constant of water (K_w) A constant that equals the product of the molar hydrogen ion concentration and the molar hydroxide ion concentration in water; $K_w = 1.0 \times 10^{-14}$ at 25°C. (*Sec. 15.7*)

ionization equilibrium constant (K_i) A constant that expresses the molar equilibrium concentrations of ions in aqueous solution for a slightly ionized acid or base. (*Sec. 16.6*)

ionization energy The amount of energy necessary to remove an electron from a neutral atom in the gaseous state. (*Sec. 6.9*)

isoelectronic A term for two or more ions (or ions and an atom) having the same electron configuration; for example, Na^+ and F^- each have 10 electrons and their electron configurations are identical to that of the noble gas neon. (*Sec. 6.10*)

isomers Compounds with the same molecular formula but different structural formulas. Structural isomers have different physical and chemical properties. (*Sec. 19.2*)

isotopes Atoms having the same atomic number but a different mass number. Atoms of the same element that differ by the number of neutrons in the nucleus. (*Sec. 5.4*)

J

joule (J) A unit of energy in the SI system; 1 cal = 4.184 J. (*Sec. 3.9*)

K

Kelvin unit (K) The basic unit of temperature in the SI system. (*Sec. 3.8*)

kernel See *core.*

kinetic energy The energy associated with the motion of particles; the energy associated with the mass and velocity of a particle. (*Sec. 4.2*)

kinetic theory A theoretical description of gas molecules demonstrating ideal behavior. (*Sec. 11.10*)

L

lanthanide series The elements with atomic numbers 58 through 71. (*Sec. 6.3*)

Latin system A naming system that designates the variable charge on a metal cation with the suffix *-ic* or *-ous* attached to the stem of the Latin name. (*Sec. 7.2*)

law See *scientific law.*

law of chemical equilibrium The principle that the product of the molar concentrations of the products in a reversible reaction divided by the product of the molar concentrations of the reactants (each raised to a power corresponding to a coefficient in the balanced equation) is equal to a constant. The law of chemical equilibrium can be written mathematically as $K_{eq} = [C]^a [D]^d/[A]^a[B]^b$. (*Sec. 16.3*)

law of combining volumes The principle that volumes of gases that combine in a chemical reaction, at the same temperature and pressure, are in the ratio of small whole numbers; also called Gay-Lussac's law of combining volumes. (*Sec. 10.6*)

law of conservation of energy The principle that states that energy can neither be created nor destroyed. (*Sec. 4.10*)

law of conservation of mass The principle that states that mass can neither be created nor destroyed. (*Sec. 4.8*) The statement that the total mass of the reactants in a chemical reaction is equal to the total mass of the products. (*Sec. 10.1*)

law of conservation of mass and energy The principle that states that the total mass and energy in the universe are constant. The statement that the total mass and energy, before and after a chemical change, are constant. (*Sec. 4.10*)

law of definite composition The principle that states that a compound always contains the same elements in the same proportion by mass. (*Sec. 4.5*)

Le Châtelier's principle The statement that any reversible reaction at equilibrium, when stressed by a change in concentration, temperature, or pressure, shifts to relieve the stress. (*Sec. 16.5*)

Lewis structure See *electron dot formula.*

light A specific term that refers to visible radiant energy, that is, violet through red. A general term that refers to both visible and invisible radiant energy. (*Sec. 5.6*)

***like dissolves like* rule** The general principle that solute–solvent interaction is greatest when the polarity of the solute is similar to that of the solvent. (*Sec. 14.2*)

limiting reactant The substance in a chemical reaction that controls or limits the maximum amount of product formed. (*Sec. 10.7*)

line spectrum See *emission line spectrum.*

lipid A biological compound such as a fat, oil, wax, or steroid that is insoluble in water. (*Sec. 20.1*)

liter (L) The basic unit of volume in the metric system equal to the volume of a cube 10 cm on a side. (*Sec. 3.1*)

lock-and-key model The theory that explains enzyme catalysis; a theory that states that the shape of the substrate molecule must fit the contour of the active site on an enzyme. (*Sec. 20.3*)

London force See *dispersion force.*

M

main-group elements See *representative elements.*

malleable The property of a metal that allows it to be hammered or machined into a foil. (*Sec. 4.4*)

mass The quantity of matter in an object that is measured by a balance. (*Sec. 2.1*)

mass/mass percent (m/m%) A solution concentration expression involving the mass of solute in grams dissolved in each 100 g of solution. (*Sec. 14.8*)

mass–mass problem A type of stoichiometry calculation that relates the masses of two substances according to a balanced equation. (*Sec. 10.3*)

mass number (*A*) A value that indicates the number of protons and neutrons in the nucleus of a given atom. (*Secs. 5.4 and 18.2*)

mass–volume problem A type of stoichiometry calculation that relates the mass of a substance to the volume of a gas according to a balanced equation. (*Sec. 10.3*)

measurement A numerical value with attached units that expresses a physical quantity such as length, mass, volume, time, or temperature. (*Sec. 2.1*)

metal An element that is generally shiny in appearance, has a high density, has a high melting point, and is a good conductor of heat and electricity. (*Sec. 4.4*)

metal oxide A compound that reacts with water to form an alkaline solution; also termed a basic oxide. (*Sec. 13.9*)

metallic solid A crystalline solid composed of metal atoms that repeat in a regular pattern. (*Sec. 13.5*)

metalloid See *semimetal.*

meter (m) The basic unit of length in the metric system of measurement. (*Sec. 3.1*)

metric system A decimal system of measurement using prefixes and a basic unit to express physical quantities such as length, mass, and volume. (*Sec. 3.1*)

milliliter (mL) A common unit of volume in the metric system of measurement that is equal to 1/1000 of a liter. (*Sec. 2.1*)

miscible A term that refers to liquids that are soluble in one another. (*Sec. 14.2*)

molar mass (MM) The mass of 1 mol of substance expressed in grams; the mass of Avogadro's number of atoms, molecules, or formula units. (*Sec. 9.3*) The mass of 1 mol of any substance expressed in grams. (*Sec. 10.1*)

molar volume The volume occupied by 1 mol of gas at STP; at 0°C and 1.00 atm the volume of 1 mol of gas is 22.4 L. (*Secs. 9.5 and 10.5*)

molarity (M) A solution concentration expression involving the number of moles of solute dissolved in each liter of solution. (*Sec. 14.9*)

mole (mol) The amount of substance that contains Avogadro's number of particles, that is, an amount of substance that contains 6.02×10^{23} particles. (*Sec. 9.2*)

mole ratio The ratio of moles of reactants and products according to the coefficients in the balanced chemical equation. (*Sec. 10.1*)

molecular formula A chemical formula that expresses the actual number of atoms of each element in a molecule. (*Sec. 9.9*)

molecular shape The geometric shape formed by the atoms bonded to the central atom in a molecule; also called molecular geometry. (*Sec. 12.9*)

molecular solid A crystalline solid composed of molecules that repeat in a regular pattern. (*Sec. 13.5*)

molecule A single particle composed of nonmetal atoms. (*Sec. 4.5*) The simplest representative particle in a compound composed of nonmetals. (*Sec. 7.7*) The simplest representative particle in a substance held together by covalent bonds. (*Sec. 12.1*)

monoatomic ion A single atom that bears a positive or a negative charge as the result of gaining or losing valence electrons. (*Secs. 7.1 and 12.5*)

monosaccharide A carbohydrate composed of a simple sugar molecule; a simple sugar is characterized by an aldehyde or ketone functional group and usually several hydroxyl groups. (*Sec. 20.4*)

N

natural law An extensively tested proposal of a scientific principle that states a measurable relationship under different experimental conditions. A natural law is often expressed as an equation, for example, $P_1V_1 = P_2V_2$. (*Sec. 1.1*)

net dipole The overall direction of partial negative charge in a molecule having two or more dipoles. (*Secs. 13.7 and 14.2*)

net ionic equation A chemical equation that portrays an ionic reaction after spectator ions have been canceled from the total ionic equation. (*Sec. 15.11*)

neutralization reaction A type of reaction in which an acid and a base produce a salt and water. (*Sec. 8.4*)

neutron (n^0) A neutral subatomic particle having a mass of 1 amu, which is approximately equal to the mass of a proton. (*Sec. 5.3*)

noble gases The relatively unreactive Group VIIIA/18 elements. (*Sec. 6.3*)

nonbonding electrons The valence electrons in a molecule that are not shared. (*Sec. 12.4*)

nonmetal An element that is generally dull in appearance, has a low density, has a low melting point, and is not a good conductor of heat and electricity. (*Sec. 4.4*)

nonmetal oxide A compound that reacts with water to form an acidic solution; also termed an acidic oxide. (*Sec. 13.9*)

nonpolar bond A covalent bond in which one or more pairs of electrons are shared equally. (*Sec. 12.7*)

nonpolar solvent A dissolving liquid composed of nonpolar molecules. (*Sec. 14.2*)

nonsignificant digits The digits in a measurement that exceed the certainty of the instrument. (*Sec. 2.3*)

nuclear equation A shorthand representation using atomic notation to describe a nuclear reaction. (*Sec. 18.2*)

nuclear fission A nuclear reaction in which a heavy nucleus splits into two or more lighter nuclei and release energy. (*Sec. 18.7*)

nuclear fusion A nuclear reaction in which two light nuclei combine into a single heavy nucleus and release energy. (*Sec. 18.8*)

nuclear reaction A high-energy change involving the atomic nucleus. (*Sec. 18.2*)

nucleic acid A biological compound that is a polymer and carries genetic information. (*Sec. 20.1*)

nucleotide A repeating unit in a nucleic acid composed of a sugar, an organic base, and phosphoric acid. (*Sec. 20.6*)

nucleus See *atomic nucleus*.

nuclide A specific atom with a designated number of protons and neutrons in its nucleus. (*Sec. 18.2*)

O

octet rule The statement that an atom must be surrounded by eight valence electrons in order to be stable. A hydrogen atom is an exception to this rule and shares only two electrons. (*Sec. 12.1*)

oil A triglyceride from a plant source that contains mostly unsaturated fatty acids. (*Sec. 20.5*)

orbital A region in space surrounding the nucleus of an atom in which there is a high probability (~95%) of finding an electron with a given energy. (*Sec. 5.10*)

organic chemistry The study of chemical substances that contain carbon. (*Sec. 1.2*) The study of carbon and its compounds. (*Sec. 19.1*)

oxidation A process in which a substance undergoes an increase in oxidation number. A process characterized by losing electrons. (*Sec. 17.2*)

oxidation number A positive or negative value assigned to an atom in a substance according to a set of rules. A value that indicates whether an atom is electron-poor or electron-rich compared to a free atom. (*Sec. 17.1*)

oxidation–reduction reaction See *redox reaction*.

oxidizing agent A substance that causes the oxidation of another substance in a redox reaction. The substance that is reduced in a redox reaction. (*Sec. 17.2*)

oxyanion A polyatomic anion that contains one or more elements combined with oxygen, for example, NO_3^- and SO_4^{2-}. (*Sec. 7.3*)

P

parent–daughter nuclides A term for a decaying nuclide and the resulting nuclide that is produced. (*Sec. 18.3*)

partial pressure The pressure exerted by an individual gas in a mixture of two or more gases. (*Sec. 11.9*)

peptide linkage An amide bond that joins two amino acid molecules in a peptide or protein. (*Sec. 20.2*)

percent (%) The ratio of a single quantity compared to an entire sample, all multiplied by 100; an expression of parts per 100 parts. (*Sec. 2.10*)

percent composition A list of the elements present in a compound and the percent by mass of each element. (*Sec. 9.7*)

percent yield The ratio between the actual yield and the theoretical yield, all times 100. (*Sec. 10.9*)

period A horizontal row in the periodic table; a series of elements with properties that vary from metallic to nonmetallic. (*Sec. 6.3*)

periodic law The properties of elements recur in a repeating pattern when the arranged according to increasing atomic number. (*Sec. 6.2*)

periodic table A chart that arranges elements according to their properties that includes metals, nonmetals, and semimetals. (*Sec. 4.4*)

pH The molar concentration of hydrogen ion expressed on an exponential scale. The negative logarithm of the molar hydrogen ion concentration. (*Sec. 15.8*)

phenyl group An aromatic hydrocarbon fragment that results after the removal of a hydrogen atom from a benzene molecule. (*Sec. 19.2*)

phospholipid A lipid that has two fatty acids and phosphoric acid joined to glycerol by ester bonds. (*Sec. 20.5*)

photon A particle of radiant energy. (*Sec. 5.7*)

physical change A modification of a substance that does not alter the chemical composition, for example, a change in physical state. (*Sec. 4.7*)

physical property A characteristic that can be observed without changing the chemical formula of a substance. (*Sec. 4.6*)

physical state The condition of matter existing as a solid, liquid, or gas. (*Sec. 4.1*)

polar bond A covalent bond in which one or more pairs of electrons are shared unequally. (*Sec. 12.6*)

polar solvent A dissolving liquid composed of polar molecules. (*Sec. 14.2*)

polyatomic ion A group of atoms that has a positive or negative charge resulting from the gain or loss of valence electrons. (*Sec. 7.1*) A group of atoms that has a positive or negative charge and is held together by covalent bonds. (*Sec. 12.9*)

polymer A giant molecule composed of small molecules joined together in a long, continuous chain. (*Sec. 20.1*)

polypeptide A large molecule composed of 10–50 amino acids joined by peptide linkages. (*Sec. 20.2*)

polysaccharide A carbohydrate polymer composed of simple sugar molecules joined by glycoside linkages. (*Sec. 20.4*)

positron ($^{0}_{+1}e$) A nuclear radiation identical in mass, but opposite in charge, to that of an electron. The term positron is derived from "positive electron." (*Sec. 18.2*)

potential energy The stored energy that matter possesses owing to its position or chemical composition. (*Sec. 4.9*)

power of 10 A positive or negative exponent of 10. (*Sec. 2.6*)

precipitate An insoluble solid substance produced by a chemical reaction in aqueous solution. (*Sec. 8.1*)

product A substance resulting from a chemical reaction. (*Sec. 8.2*)

protein A biological compound that is a polymer composed of amino acids linked by peptide bonds. (*Sec. 20.1*)

proton (p^+) A positively charged subatomic particle having a mass of 1 amu, which is approximately 1836 times the mass of an electron. (*Sec. 5.2*)

proton acceptor A term for a Brønsted–Lowry base that is used interchangeably with hydrogen ion acceptor. (*Sec. 15.3*)

proton donor A term for a Brønsted–Lowry acid that is used interchangeably with hydrogen ion donor. (*Sec. 15.3*)

pure substance See *substance*.

Q

quantum level See *energy level*.

quantum mechanical atom A model of the atom that describes an electron in terms of its probability of being found in a particular location around the nucleus. (*Sec. 5.10*)

R

radiant energy spectrum A range of light energy extending from short-wavelength gamma rays through long-wavelength microwaves. (*Sec. 5.6*)

radioactive decay series The stepwise disintegration of a radioactive nucleus until a stable nucleus is reached. (*Sec. 18.3*)

radioactivity The emission of particles or energy from an unstable atomic nucleus. (*Sec. 18.1*)

radionuclide An atom whose nucleus is unstable and decays radioactively. (*Sec. 18.5*)

rare earth elements The elements with atomic numbers 21, 39, 57, and 58 through 71. (*Sec. 6.3*)

rate of reaction The rate at which the concentrations of reactants decrease, or the concentrations of products increase, per unit time. (*Sec. 16.3*)

reactant A substance undergoing a chemical reaction. (*Sec. 8.2*)

reaction profile A graph of the energy of reactants and products as a reaction progresses. (*Sec. 16.2*)

real gas An actual gas that deviates from ideal behavior at low temperature and high pressure. (*Sec. 11.10*)

reciprocal The relationship between a fraction and its inverse, for example, 1 yd/3 ft and 3 ft/1 yd. (*Sec. 2.8 and 3.2*)

redox reaction A chemical reaction that involves electron transfer between two reacting substances. (*Sec. 17.2*)

reducing agent A substance that causes the reduction of another substance in a redox reaction. The substance that is oxidized in a redox reaction. (*Sec. 17.2*)

reduction A process in which a substance undergoes a decrease in oxidation number. A process characterized by gaining electrons. (*Sec. 17.2*)

reduction potential The relative ability of a substance to undergo reduction; the relative strength of an oxidizing agent. (*Sec. 17.5*)

representative elements The Group A (1, 2, and 13–18) elements in the periodic table; also termed main-group elements. (*Sec. 6.3*)

reversible reaction A reaction that proceeds simultaneously in both the forward direction toward products and the opposite direction toward reactants. (*Sec. 16.3*)

rounding off The process of eliminating digits that are not significant. (*Sec. 2.3*)

S

salt An ionic compound produced by an acid–base reaction. The product of a neutralization reaction in addition to water. (*Secs. 8.11 and 15.2*)

salt bridge A porous device that allows ions to travel between two half-cells to maintain an ionic charge balance in each compartment. (*Sec. 17.6*)

saponification A chemical reaction of a fat or oil with sodium hydroxide to produce soap and glycerol. (*Sec. 20.5*)

saturated hydrocarbon A hydrocarbon containing a single bond between each carbon atom. (*Sec. 19.1*)

saturated solution A solution that contains the maximum amount of solute that can dissolve at a given temperature. (*Sec. 14.7*)

science The methodical exploration of nature and the logical explanation of the observations. (*Sec. 1.1*)

scientific method A systematic investigation that involves performing an experiment, proposing a hypothesis, testing the hypothesis, and finally stating a theory or law that explains a scientific principle. (*Sec. 1.1*)

scientific notation A method for expressing numbers by moving the decimal place after the first significant digit and indicating the number of decimal moves by a power of 10. (*Sec. 2.7*)

second (s) The basic unit of time in the metric system. (*Sec. 3.1*)

semimetal An element that is generally metal-like in appearance and has properties midway between those of a metal and of a nonmetal; also called a metalloid. (*Sec. 4.4*)

SI See *International System.*

significant digits The digits in a measurement known with certainty plus one digit that is estimated; also referred to as significant figures. (*Sec. 2.2*)

single bond A bond between two atoms composed of one shared electron pair. A single bond is represented as a dash between the symbols of two atoms. (*Sec. 12.4*)

single-replacement reaction A type of reaction in which a more active element displaces a less active element from a solution or compound. (*Sec. 8.4*)

soft water Water containing sodium cations, Na^+, and a variety of anions such as CO_3^{2-}, SO_4^{2-}, and PO_4^{3-}. (*Sec. 13.10*)

solubility The maximum amount of solute that can dissolve in a solvent at a given temperature; usually expressed in grams of solute per 100 g of solvent. (*Sec. 14.6*)

solubility product equilibrium constant (K_{sp}) A constant that expresses the molar equilibrium concentrations of ions in aqueous solution for a slightly dissociated ionic compound. (*Sec. 16.8*)

solute The component of a solution that is present in the lesser quantity. (*Sec. 14.2*)

solution The general term for a solute dissolved in a solvent. A solution is an example of a homogeneous mixture. (*Sec. 14.2*)

solvent The component of a solution that is present in the greater quantity. (*Sec. 14.2*)

solvent cage A cluster of solvent molecules surrounding a solute molecule or ion in solution. (*Sec. 14.4*)

specific gravity (sp gr) The ratio of the density of a liquid compared to the density of water at 4°C; a unitless expression. (*Sec. 3.7*)

specific heat The amount of heat required to raise the temperature of 1 g of any substance 1°C; the specific heat of water is 1.00 cal/g × °C. (*Sec. 3.9*)

spectator ions Those ions that are in aqueous solution but neither participate in a reaction nor appear as reactants or products in the net ionic equation. (*Sec. 15.11*)

standard conditions See *standard temperature and pressure (STP).*

standard solution A solution whose concentration has been established accurately (usually by titration to three or four significant digits). (*Sec. 15.6*)

standard temperature and pressure (STP) A temperature of 0°C and a pressure of 1 atm. (*Secs. 9.5 and 10.5*) A temperature of 273 K and a pressure of 760 mm Hg for a gas. (*Sec. 11.7*)

steroid A lipid hormone composed of four rings of carbon atoms fused into a single molecular structure. (*Sec. 20.5*)

Stock system A naming system that designates the variable charge on a metal cation with Roman numerals in parentheses. (*Sec. 7.2*)

stoichiometry The relationship between quantities (mass of substance or volume of gas) in a chemical reaction according to the balanced chemical equation. (*Sec. 10.3*)

strong electrolyte An aqueous solution that is a good conductor of electricity, for example, strong acids, strong bases, and soluble salts. (*Sec. 15.10*)

structural formula A diagram of a molecule or polyatomic ion in which each atom is represented by its chemical symbol and a dash for each pair of bonding electrons. (*Sec. 12.4*) A chemical formula that shows the arrangement of atoms in a molecule. (*Sec. 19.2*)

sublimation A direct change in state from a solid to a gas without forming a liquid. (*Sec. 4.1*)

subscript A digit in a chemical formula that represents the number of atoms or ions occurring in the substance. (*Sec. 8.3*)

substance Matter having definite composition and constant properties. (*Sec. 4.2*)

supersaturated solution A solution that contains more than the maximum amount of solute that can ordinarily dissolve at a given temperature. (*Sec. 14.7*)

surface tension The resistance of a liquid to spread out and its tendency to form spherical drops with minimum surface area. (*Sec. 13.3*)

T

temperature A measure of the average energy of individual particles in a system. (*Sec. 3.8*)

ternary ionic A compound that contains three elements including at least one metal. (*Sec. 7.1*)

ternary oxyacid A compound that contains hydrogen, a nonmetal, and oxygen dissolved in water. (*Sec. 7.1*)

theoretical yield The amount of product that is calculated to be obtained from given amounts of reactants. (*Sec. 10.9*)

theory An extensively tested proposal of a scientific principle that explains the behavior of nature. A theory offers a model, for example the atomic theory, to describe nature. (*Sec. 1.1*)

titration A laboratory procedure for delivering a measured volume of solution through a buret. (*Sec. 15.5*)

torr A unit of gas pressure equal to 1 mm Hg. (*Sec. 11.2*)

total ionic equation A chemical equation that portrays highly ionized substances in the ionic form and slightly ionized substances in the nonionized form. (*Sec. 15.11*)

transition elements The Group B (3–12) elements in the periodic table. (*Sec. 6.3*)

transition state The highest point on a reaction profile at which there is the greatest potential energy. (*Sec. 16.2*)

transmutation The conversion of one element to another by a nuclear reaction. (*Sec. 18.6*)

transuranium elements The elements with atomic numbers beyond 92. All the elements following uranium are synthetic and do not occur naturally. (*Sec. 6.3*)

triglyceride A lipid that has three fatty acids joined to glycerol by ester bonds; also called a triacylglycerol. (*Sec. 20.5*)

triple bond A bond between two atoms composed of three shared electron pairs. A triple bond is represented as three dashes between the symbols of two atoms. (*Sec. 12.4*)

tritium (^3_1H) The nuclide of hydrogen with two neutrons in the nucleus. (*Sec. 18.8*)

Tyndall effect The phenomenon of scattering of a beam of light by colloid particles. (*Sec. 14.4*)

U

uncertainty The degree of inexactness in a measurement obtained from an instrument. (*Sec. 2.1*)

uncertainty principle The statement that it is impossible to precisely measure both the location and energy of a particle at the same time. (*Sec. 5.10*)

unit analysis method A systematic procedure for solving problems that converts the units in a given value to the units in the answer; also referred to as dimensional analysis. (*Secs. 2.9 and 3.3*)

unit equation A statement that relates two values that are equivalent, for example, 1 ft = 12 in. and 1 in. = 2.54 cm. (*Secs. 2.8 and 3.2*)

unit factor A ratio of two quantities that are equivalent and can be applied to convert from one unit to another, for example, 1 m/100 cm. (*Secs. 2.8 and 3.2*)

unsaturated hydrocarbon A hydrocarbon containing a double bond or a triple bond between two carbon atoms. (*Sec. 19.1*)

unsaturated solution A solution that contains less than the maximum amount of solute that can dissolve at a given temperature. (*Sec. 14.7*)

V

vacuum A gaseous volume that does not contain molecules. (*Sec. 11.2*)

valence electrons The electrons that occupy the outermost *s* and *p* sublevels of an atom. (*Sec. 6.7*) The electrons in the highest *s* and *p* sublevels in an atom that undergo reaction and form chemical bonds. (*Sec. 12.1*)

valence shell electron pair repulsion See *VSEPR theory*.

vapor pressure The pressure exerted by vapor molecules above a liquid in a closed container when the rates of evaporation and condensation are equal. (*Secs. 11.8 and 13.3*)

viscosity The resistance of a liquid to flow. (*Sec. 13.3*)

visible spectrum A range of light energy observed as violet, blue, green, yellow, orange, or red; the wavelengths of light from 400 to 700 nm. (*Sec. 5.6*)

voltaic cell An electrochemical cell in which a spontaneous redox reaction occurs and generates electrical energy; also called a galvanic cell. (*Sec. 17.6*)

volume by displacement A technique for determining the volume of a solid or a gas by measuring the volume of water it displaces. (*Sec. 3.6*) A technique for determining the volume of a gas by measuring the amount of water it displaces. (*Sec. 11.9*)

volume–volume problem A type of stoichiometry calculation that relates the volumes of two gases (at the same temperature and pressure) according to a balanced equation. (*Sec. 10.3*)

VSEPR theory A model that explains the shapes of molecules as a result of bonding and nonbonding electron pairs around the central atom repelling each other. (*Sec. 12.9*)

W

water of hydration Water molecules bound to a formula unit in a hydrate; also called water of crystallization. (*Sec. 13.10*)

wavelength The distance a light wave travels to complete 1 cycle. (*Sec. 5.6*)

wax A lipid that contains a fatty acid joined to an alcohol by an ester linkage. (*Sec. 20.5*)

weak electrolyte An aqueous solution that is a poor conductor of electricity, for example, weak acids, weak bases, and slightly soluble salts. (*Sec. 15.10*)

weight The force exerted by gravity on an object. The *mass* of an object is fixed, whereas the *weight* of an object varies with elevation and the distance from the center of Earth. (*Sec. 2.1*)

Photo Credits

(UP = Unnumbered Photograph); (T = Top); (B = Bottom); (R = Right); (L = Left); (M = Middle)

Chapter 1: **Page 1** AP/Wide World Photos (CO1) **Page 2** Corbis (1.1) **Page 3** Stamp from the private collection of Professor C. M. Lang, photography by Gary J. Shulfer, University of Wisconsin—Stevens Point. "Ireland;" Scott Standard Postage Stamp Catalogue, Scott Pub. Co., Sidney, Ohio. (UP) **Page 4** Stamp from the private collection of Professor C. M. Lang, photography by Gary J. Shulfer, University of Wisconsin—Stevens Point. "1993, Republic of Mali;" Scott Standard Postage Stamp Catalogue, Scott Pub. Co., Sidney, Ohio. (UP) **Page 5** Cornelis Bega (Dutch, 1620-1664), The Alchemist. From the collection of Roy Eddleman. (UP) **Page 6** (T) Photo Researchers, Inc./Will & Deni McIntyre/Science Source (1.5a) **Page 6** (M) Photo Researchers, Inc./Simon Fraser/Searle Pharmaceuticals/Science Photo Library (1.5b) **Page 6** (B) The Stock Market/Charles West (1.5c) **Page 7** (T) Photo Lennart Nilsson/Albert Bonniers Forlag (1.6a) **Page 7** (M) Stock Boston/Mark Burnett (1.6b) **Page 7** (B) Pearson Education/PH College/Donald Clegg and Roxy Wilson (1.6c) **Page 8** (T) Farmland Industries, Inc. (1.7a) **Page 8** (B/R) Photo Researchers, Inc./Michael P. Gadomski (1.7b) **Page 8** (B/L) The Stock Market/Craig Tuttle (1.7c) **Page 9** (T) Shell Oil Company (1.8a) **Page 9** (R) Photo Researchers, Inc./Martin Bond/Science Photo Library (1.8b) **Page 9** (B/L) Fundamental Photographs/Peticolas/Megna (1.8c)

Chapter 2: **Page 15** Photo Researchers, Inc./Blair Seitz (CO2) **Page 18** (T) Ohaus Corporation (2.3a) **Page 18** (L) Ohaus Corporation (2.3b) **Page 18** (R) Sartorius Corporation (2.3c) **Page 21** Texas Instruments Incorporated (UP) **Page 26** (T) NASA Headquarters (UP) **Page 26** (B) Custom Medical Stock Photo, Inc./Dr. Gopal Murti/Science Photo Library (UP) **Page 27** The Image Bank/Al Hamdan (UP) **Page 28** Alcoa (UP) **Page 29** Pearson Education/PH College/McCracken Photographers (UP) **Page 31** (T) Fundamental Photographs/Richard Megna (2.11a) **Page 31** (B) Fundamental Photographs/Richard Megna (2.11b) **Page 33** Photo Researchers, Inc./Ted Clutter (UP) **Page 34** Tom Pantages (UP) **Page 40** Photo Researchers, Inc./Tom McHugh (UP) **Page 41** Phototake NYC/NASA (UP)

Chapter 3: **Page 42** Carey Van Loon (CO3) **Page 44** International Bureau of Weights and Measures, Sevres, France (3.1) **Page 46** International Bureau of Weights and Measures, Sevres, France (UP) **Page 48** Liaison Agency, Inc./Lien/Nibauer Photography (UP) **Page 51** Fundamental Photographs/Richard Megna (UP) **Page 52** Mary Teresa Giancoli (UP) **Page 54** Corbis/AFP (UP) **Page 56** AP/Wide World Photos/IIHS (UP) **Page 65** Fundamental Photographs/Richard Megna (UP) **Page 74** Mary Teresa Giancoli (UP) **Page 75** Pearson Education/PH College/Tom Bochsler Photography Limited (UP) **Page 76** Photo Researchers, Inc./Rich Treptow (UP) **Page 77** Specimen from North Museum, Franklin and Marshall College. Photo by Runk/Schoenberger/Grant Heilman Photography, Inc. (UP)

Chapter 4: **Page 78** Corbis (CO4) **Page 79** (T) Pearson Education/PH College/Tom Bochsler Photography Limited (UP) **Page 79** (B) Pearson Education/PH College/Tom Bochsler Photography Limited (UP) **Page 81** Fundamental Photographs/Richard Megna (UP) **Page 82** Pearson Education/PH College/Tom Bochsler Photography Limited (UP) **Page 95** (T/L) Fundamental Photographs/Richard Megna (4.11a) **Page 95** (T/R) Stock Boston/Stephen Frisch (4.11b) **Page 95** (M) Fundamental Photographs/Richard Megna (4.11c) **Page 95** (B) Visuals Unlimited/Albert Copley (4.11d) **Page 96** (L) Fundamental Photographs/Richard Megna (4.12a) **Page 96** (M) Pearson Education/PH College/McCracken Photographers (4.12b) **Page 96** (R) Fundamental Photographs/Richard Megna (4.12c) **Page 97** Pearson Education/PH College/Donald Clegg and Roxy Wilson (UP) **Page 102** Stamp from the private collection of Professor C. M. Lang, photography by Gary J. Shulfer, University of Wisconsin, Stevens Point. "1979, China—People's Republic (Scott #1468);" Scott Standard Postage Stamp Catalogue, Scott Pub. Co., Sidney, Ohio. (UP) **Page 103** Pearson Education/PH College/Tom Bochsler Photography Limited (UP) **Page 105** Rainbow/T. J. Florian (UP) **Page 107** Tom Stack & Associates/John Cancalosi (UP) **Page 108** Fundamental Photographs/Richard Megna (UP)

Chapter 5: **Page 113** Corbis (UP) **Page 121** Archive Photos/Reuters/Lampen (UP) **Page 126** Stamp from the private collection of Professor C. M. Lang, photography by Gary J. Shulfer, University of Wisconsin, Stevens Point. "1963, Denmark (Scott #409);" Scott Standard Postage Stamp Catalogue, Scott Pub. Co., Sidney, Ohio. (UP) **Page 130** Fundamental Photographs/Michael Dalton (UP) **Page 134** American Institute of Physics/Emilio Segre Visual Archives/Paul Ehrenfest (UP) **Page 138** Stock Boston/Bob Daemmrich (UP) **Page 141** Fundamental Photographs/Richard Megna (UP) **Page 142** Fundamental Photographs/Richard Megna (UP)

Chapter 6: **Page 144** Fundamental Photographs/Richard Megna (CO6) **Page 146** Stamp from the private collection of Professor C. M. Lang, photography by Gary J. Shulfer, University of Wisconsin, Stevens Point. "1957, Russia (Scott #1906) and 1969, Russia (Scott #3607);" Scott Standard Postage Stamp Catalogue, Scott Pub. Co., Sidney, Ohio. (UP) **Page 166** Pearson Education/PH College/Tom Bochsler Photography Limited (UP) **Page 172** Peter Arnold, Inc./Manfred Kage (UP)

Chapter 7: **Page 175** Fundamental Photographs/Richard Megna (CO7) **Page 177** Phototake NYC/Karl Hartmann/Traudel Sachs (UP) **Page 178** Fundamental Photographs/Richard Megna (UP) **Page 179** Jacques Louis David (1748-1825). Antoine Laurent Lavoisier (1743-1794) and his wife (Marie Anne Pierrette Paulze, 1758-1836). Oil on canvas. H. 102-1/4 in. W. 76-5/8 in. The Metropolitan Museum of Art, Purchase, Mr. and Mrs. Charles Wrightsman Gift, 1977. (UP) **Page 183** Corbis/Kevin Schafer (UP) **Page 184** Stock Boston/Margaret Ross (UP) **Page 185** Charles H. Corwin (UP) **Page 186** (T) Grant Heilman Photography, Inc./Barry L. Runk (UP) **Page 186** (B) Grant Heilman Photography, Inc./Runk/Schoenberger (UP) **Page 187** Chip Clark (UP) **Page 189** Photo Researchers, Inc./Jim W. Grace (UP) **Page 191** The Image Works (7.4) **Page 192** Fundamental Photographs/Richard Megna (UP) **Page 193** Fundamental Photographs/Richard Megna (UP) **Page 195** Fundamental Photographs/Richard Megna (UP) **Page 197** Pearson Education/PH College/Tom Bochsler Photography Limited (UP) **Page 198** Department of Mineral Sciences, National Museum of Natural History, Smithsonian Institution (UP)

Chapter 8: **Page 201** (L/R) Fundamental Photographs/Richard Megna (CO8) **Page 203** (L) Pearson Education/PH College/McCracken Photographers (8.1a) **Page 203** (M) Grant Heilman Photography, Inc./Runk/Schoenberger (8.1b) **Page 203** (R) Pearson Education/PH College/McCracken Photographers (8.1c) **Page 210** Fundamental Photographs/Robert Mathena (UP) **Page 211** Pearson Education/PH College/Tom Bochsler Photography Limited (UP) **Page 212** Pearson Education/PH College/Tom Bochsler Photography Limited (UP) **Page 213** Pearson Education/PH College/Tom Bochsler Photography Limited (UP) **Page 216** Fundamental Photographs/Richard Megna (UP) **Page 217** Pearson Education/PH College/Tom Bochsler Photography Limited (UP) **Page 218** (T) Pearson Education/PH College/McCracken Photographers (UP) **Page 218** (B) Pearson Education/PH College/Tom Bochsler Photography Limited (UP) **Page 219** (T) Fundamental Photographs/Richard Megna (UP) **Page 219** (B) Color-Pic, Inc./Ed Degginger (UP) **Page 221** Pearson Education/PH College/McCracken Photographers (UP) **Page 222** Pearson Education/PH College/McCracken Photographers (UP) **Page 223** Fundamental Photographs/Richard Megna (UP) **Page 224** (L) Photo Researchers, Inc./Peter Sorula (UP)

Page 224 (R) Dr. H. Eugene LeMay, Jr. (UP) **Page 228** Fundamental Photographs (UP) **Page 229** Pearson Education/PH College/Tom Bochsler Photography, Ltd. (UP) **Page 231** Pearson Education/PH College/Tom Bochsler Photography Limited (UP) **Page 233** Fundamental Photographs/ Richard Megna (UP) **Page 235** Photo Researchers, Inc./Lawrence Migdale/Science Source (UP)

Chapter 9: **Page 236** NASA/Johnson Space Center (UP) **Page 237** Carey Van Loon (UP) **Page 239** Fundamental Photographs/Richard Megna (UP) **Page 240** Pearson Education/PH College/Tom Bochsler Photography, Ltd. (UP) **Page 244** Carey Van Loon (UP) **Page 253** Fundamental Photographs/Richard Megna (UP) **Page 255** Fundamental Photographs/Richard Megna (UP)

Chapter 10: **Page 260** Carey Van Loon (CO10) **Page 261** Stamp from the private collection of Professor C. M. Lang, photography by Gary J. Shulfer, University of Wisconsin, Stevens Point. "Italy #714 (1956);" Scott Standard Postage Stamp Catalogue, Scott Pub. Co., Sidney, Ohio. (UP) **Page 266** Stone/Robin Smith (UP) **Page 267** (T) Fundamental Photographs/Richard Megna (UP) **Page 267** (B) Pearson Education/PH College/ McCracken Photographers (UP) **Page 269** Pearson Education/PH College/McCracken Photographers (UP) **Page 271** Stone/Donald Johnston (UP) **Page 272** Fundamental Photographs/ Robert Mathena (UP) **Page 274** (B) Stamp from the private collection of Professor C. M. Lang, photography by Gary J. Shulfer, University of Wisconsin, Stevens Point. "1978, Sweden (Scott #1271);" Scott Standard Postage Stamp Catalogue, Scott Pub. Co., Sidney, Ohio. (UP) **Page 274** (T) Fundamental Photographs/Richard Megna (UP) **Page 275** (T) Grant Heilman Photography, Inc./Grant Heilman (UP) **Page 275** (B) Pearson Education/PH College/Tom Bochsler Photography Limited (UP) **Page 277** Fundamental Photographs/Richard Megna (UP) **Page 280** Photo Researchers, Inc./Rosenfeld Images Ltd./Science Photo Library (UP) **Page 281** Pearson Education/PH College/McCracken Photographers (UP) **Page 282** Pearson Education/PH College/McCracken Photographers (UP) **Page 286** Fundamental Photographs/Richard Megna (UP) **Page 287** Color-Pic, Inc./Ed Degginger (UP)

Chapter 11: **Page 291** Corbis/James Marshall (CO11) **Page 292** FPG International LLC/Richard Price (UP) **Page 294** Stock Boston/Michael A. Dwyer (UP) **Page 299** Johann Kerseboom, Robert Boyle (1627-1691). © 1689. Oil on canvas. The Granger Collection. (UP) **Page 301** Fundamental Photographs/Richard Megna (UP) **Page 308** Corbis/Jim Sugar Photography (UP) **Page 318** Stock Boston/Keren Su (UP) **Page 319** Color-Pic, Inc./Phil Degginger (UP) **Page 321** Peter Arnold, Inc./Norbert Wu (UP)

Chapter 12: **Page 343** AP/Wide World Photos (UP) **Page 348** General Electric Corporate Research & Development Center (UP) **Page 355** (L) Carey Van Loon (UP) **Page 355** (R) Carey Van Loon (UP)

Chapter 13: **Page 356** Stone/Delphine Star (CO13) **Page 357** Pearson Education Corporate Digital Archive (UP) **Page 362** (T) Color-Pic, Inc./Phil Degginger (UP) **Page 362** (B) Photo Researchers, Inc./Hermann Eisenbeiss (UP) **Page 365** (T/L) Fundamental Photographs/Paul Silverman (UP) **Page 365** (T/R) Fundamental Photographs/Paul Silverman (UP) **Page 365** (B) Tom Stack & Associates/Allen B. Smith (UP) **Page 366** (T) Grant Heilman Photography, Inc./Runk/ Schoenberger (UP) **Page 366** (B) Fundamental Photographs/Paul Silverman (UP) **Page 371** Charles H. Corwin (UP) **Page 373** Color-Pic, Inc./Ed Degginger (UP) **Page 375** (L) Fundamental Photographs/Richard Megna (UP) **Page 375** (R) Phototake NYC/Karl Hartmann/Sachs (UP) **Page 380** (L) Carey Van Loon (UP) **Page 380** (R) Color-Pic, Inc./Ed Degginger (UP) **Page 381** Fundamental Photographs/Peticolas/Megna (UP) **Page 383** Fundamental Photographs/Richard Megna (UP)

Chapter 14: **Page 386** Photo Researchers, Inc./Douglas Faulkner (UP) **Page 389** Fundamental Photographs/Paul Silverman (UP) **Page 394** Fundamental Photographs/Richard Megna (UP) **Page 397** Fundamental Photographs/ Richard Megna (UP) **Page 401** Pearson Education/PH College/Tom Bochsler Photography Limited (UP) **Page 403** PhotoEdit/Michael Newman (UP) **Page 410** Phototake NYC/Yoav Levy (UP) **Page 411** Photo Researchers, Inc./Sam Pierson, Jr. (UP) **Page 415** Pearson Education/PH College (UP)

Chapter 15: **Page 418** (T) Pearson Education/PH College/Tom Bochsler Photography Limited (UP) **Page 418** (B) Pearson Education/PH College/ Tom Bochsler Photography Limited (UP) **Page 420** Fundamental Photographs/Richard Megna (UP) **Page 421** Carey Van Loon (UP) **Page 423** Corbis (UP) **Page 425** Pearson Education/PH College/Tom Bochsler Photography Limited (15.2) **Page 427** Fundamental Photographs/ Richard Megna (15.3) **Page 434** Fundamental Photographs/Richard Megna (UP) **Page 438** (L) Fundamental Photographs/NYC Parks Photo Archive (UP) **Page 438** (R) Fundamental Photographs/Kristen Brochmann (UP) **Page 446** Fundamental Photographs/Richard Megna (UP) **Page 447** Pearson Education/PH College/Tom Bochsler Photography Limited (UP)

Chapter 16: **Page 452** Carey Van Loon (16.1) **Page 458** NASA/Goddard Space Flight Center (UP) **Page 460** AP/Wide World Photos/Paul Sakuma (16.10) **Page 466** Fundamental Photographs/Richard Megna (UP) **Page 473** Fundamental Photographs/Richard Megna (UP)

Chapter 17: **Page 484** AP/Wide World Photos/Chris O'Meara (CO17) **Page 485** (L) Photo Researchers, Inc./Jack Dermid (17.1) **Page 485** (R) Photo Researchers, Inc./James L. Amos (17.1) **Page 486** Color-Pic, Inc./Ed Degginger (UP) **Page 487** Color-Pic, Inc./Ed Degginger (UP) **Page 489** Fundamental Photographs/Richard Megna (UP) **Page 496** Fundamental Photographs/Richard Megna (UP) **Page 507** Fundamental Photographs/Richard Megna (17.5) **Page 509** General Motors Media Archives (UP) **Page 510** Pearson Education/PH College/Tom Bochsler Photography Limited (UP)

Chapter 18: **Page 516** NASA Headquarters (CO18) **Page 517** The Burndy Library, Dibner Institute for the History of Science and Technology (UP) **Page 519** Fundamental Photographs/ Richard Megna (UP) **Page 523** Color-Pic, Inc./ Ed Degginger (18.3) **Page 527** Fundamental Photo- graphs/Peticolas/ Megna (UP) **Page 528** Photo Researchers, Inc./ Martin Dohrn/Science Photo Library (UP) **Page 529** Fermilab Visual Media Services (18.6) **Page 534** Comstock (UP) **Page 535** Phototake NYC (18.10) **Page 537** Fundamental Photographs/ Richard Megna (UP)

Chapter 19: **Page 543** Charles H. Corwin (19.2) **Page 546** Government of Switzerland (UP) **Page 552** Stamp from the private collection of Professor C. M. Lang, photography by Gary J. Shulfer, University of Wisconsin, Stevens Point. "Belgium," Scott Standard Postage Stamp Catalogue, Scott Pub. Co., Sidney, Ohio. (UP) **Page 556** Charles H. Corwin (19.5) **Page 557** Charles H. Corwin (19.6) **Page 559** Charles H. Corwin (19.8) **Page 560** Charles H. Corwin (19.9) **Page 561** Charles H. Corwin (19.10) **Page 562** Charles H. Corwin (19.11) **Page 563** Charles H. Corwin (19.12) **Page 564** Charles H. Corwin (19.13) **Page 564** PhotoDisc, Inc./F. Schussler (UP) **Page 565** Charles H. Corwin (19.14) **Page 566** Fundamental Photo-graphs/Richard Megna (UP) **Page 571** Photo Researchers, Inc./Will and Deni McIntyre (UP)

Chapter 20: **Page 578** Stock Boston/ Stephen Frisch (UP) **Page 579** Fundamental Photographs/Richard Megna (UP) **Page 584** Stock Boston/Tim Barnwell (UP) **Page 587** Fundamental Photographs/Richard Megna (UP) **Page 592** Photo Researchers, Inc./Gary Retherford (UP) **Page 594** Fundamental Photographs/Richard Megna (UP)

Index

D

N

Q

R

S

T

X

Y

Z